About Pearson

Pearson is the world's learning company, with presence across 70 countries worldwide. Our unique insights and world-class expertise comes from a long history of working closely with renowned teachers, authors and thought leaders, as a result of which, we have emerged as the preferred choice for millions of teachers and learners across the world.

We believe learning opens up opportunities, creates fulfilling careers and hence better lives. We hence collaborate with the best of minds to deliver you class-leading products, spread across the Higher Education and K12 spectrum.

Superior learning experience and improved outcomes are at the heart of everything we do. This product is the result of one such effort.

Your feedback plays a critical role in the evolution of our products and you can contact us – reachus@pearson.com. We look forward to it.

OBJECTIVE CHEMISTRY
for
NEET
(National Eligibility Cum Entrance Test)
and Other Medical Entrance Examinations

Volume I

Dr A.K. Singhal
U.K. Singhal

Editor—Acquisitions: Abhilash Ayyappan
Senior Editor—Development: Nandini Basu
Senior Editor—Production: G. Sharmilee

ISBN 978-93-325-8620-8

First Impression

Published by Pearson India Education Services Pvt. Ltd, CIN: U72200TN2005PTC057128, formerly known as TutorVista Global Pvt. Ltd, licensee of Pearson Education in South Asia.

Head Office: 15th Floor, Tower-B, World Trade Tower, Plot No. 1, Block-C, Sector 16, Noida 201 301, Uttar Pradesh, India.
Registered Office: 4th Floor, Software Block, Elnet Software City, TS 140, Block 2 & 9, Rajiv Gandhi Salai, Taramani, Chennai 600 113, Tamil Nadu, India.
Fax: 080-30461003, Phone: 080-30461060
Website: in.pearson.com, Email id: companysecretary.india@pearson.com

Compositor: Data Tech Solutions
Printed in India at Thomson Press India Ltd.

Dedicated to
My Grand Parents, Parents and Teachers

Contents

* *Sample Papers and Mock Test Papers are included in Volume II*

Preface

Objective Chemistry for NEET and other Medical Entrance Examinations, Volumes I and II, are the recommended preparatory books for students aspiring to crack the prestigious medical entrance examinations. It provides class-tested material and practice problems which help the aspirants to understand important theories, concepts and in addition to develop problem solving skills for the examination.

The author has strictly adhered to the latest pattern of NEET and developed the content in line with previous years' papers from AIPMT/PMT/NEET and other competitive medical entrance examinations from last 15 years. All chapters in this series conform to the pattern laid out in NCERT textbooks.

The objective of this series is to provide the course material in a structured and useful manner that could help students prepare in an organized manner. The unique feature of this series is its huge repository of multiple-choice questions that cover all important topics from respective chapters. Each practice exercise is further divided into three stages: Practice Exercise I (to revise the concepts) and Practice Exercise II (to sharpen the concepts) and Practice Exercise III (concept crackers). In addition, there is a separate section on 'AIIMS Essentials' which is designed to prepare students for "Assertion and Reason" type of questions. Witten in lucid language, the book aims to assist students in understanding the concepts without the help of any instructor.

Features of the book

- Structured as per class XI and XII syllabus of NCERT
- Chapter-wise NCERT exemplar questions
- Assertion and Reason questions to aid in preparing for AIIMS and other similar examinations
- Previous years' questions embedded in every chapter
- Mock tests and sample papers for practice
- Solutions for the latest NEET question paper

This book has been written to pass on my fascination with descriptive physical chemistry to the next generation of learners. Thus comments of readers, both students and instructors, will be sincerely appreciated. Any suggestions for added or updated additional readings would also be welcome. Students can share their feedback at reachus@ pearson.com

Dr A.K. Singhal
U.K. Singhal

Acknowledgements

The contentment that accompanies the successful completion of my work would remain essentially incomplete if I fail to mention the people whose constant guidance and support has encouraged me.

I am grateful to all my revered teachers, especially, the late J.K. Mishra, Dr D.K. Rastogi, the late A.K. Rastogi and my honourable guide, Dr S.K. Agarwala. Their knowledge and wisdom assisted me in no small measure in presenting this work.

I am thankful to my colleagues and friends, Deepak Bhatia, Er Vikas Kaushik, Er A.R. Khan, Vipul Agarwal, Er Ankit Arora, Er Wasim, Manoj Singhal, Vijay Arora (Director, Dronacharya), Anupam Shrivastav, Ajay Verma, Ashutosh Tripathi, Vivek Shukla, Satish Gupta (ALLEN, Jaipur), Chandan Kumar (MENTORS, Patna), Sandeep Singhal (PACE, Delhi).

I am indebted to my father, B.K. Singhal, mother Usha Singhal, brothers, Amit Singhal and Katar Singh, who have been my motivation at every step. Their never-ending affection has provided me with moral support and encouragement while writing this book.

Last but not the least, I express my deepest gratitude to my wife Urmila and my little, but witty-beyond-years daughters Khushi and Shanvi, who always supported me during my work.

Dr A.K. Singhal
U.K. Singhal
Singhal.atul1974@gmail.com

About the Examination

National Eligibility Cum Entrance Test (NEET) is conducted by the Central Board of Secondary Education (CBSE) for the 15% 'All India Quota' seats in the Medical and Dental Colleges in India. The merit list of this entrance test is also used by state governments/universities/institutions for admission in the medical/dental colleges against seats under their control.

Time Management

In order to successfully crack NEET, one must prepare profoundly. Consistent hard work, practice and perseverance are needed to crack this examination. Statistically, NEET is one of the toughest examinations in India. During this examination, one must exercise clinical precision with speed since the average time available to respond to a question is hardly a minute. The aspirants should be conceptually clear in the subject owing to the negative marking in the examination. A better strategy to solve the paper would be to go for the easiest questions first and then gradually progress to the more complicated ones.

Regular practice of MCQs will assist the aspirants in preparing for the examination. In a nutshell, hard work, conceptual clarity and self-assessment are the essential ingredients to achieve success in NEET. NCERT texts play an important role in understanding the concepts. Student need to read up on all concepts/theories in a regular and systematic manner.

NEET is a 3-hour examination, with a total of 180 multiple-choice questions (MCQ) to be answered. The exam paper will be divided into four sections—Physics, Chemistry, Zoology and Botany. Each question carries 4 marks.

Time Management tips

Step 1, first 30 minutes: In order to solve maximum questions in the limited time frame, attempt the easy questions first. You may start from Physics/Chemistry Section, where some problems may be answered by comparing the units of the options.

Step 2, next 60 minutes: Attempt Section II or III. For a few questions where dimensional analysis fails, recognize the proper limits of the answer and mark the correct answers.

Step 3, next 60 minutes: In some problems, solution may not be obtained from a simple formula and hence we need to calculate the formula. At that time try cause-and-result method to solve the problem and use the estimation rule.

Step 4, last 30 minutes: Try the difficult questions of Physics and Chemistry and revise the complete answer script.

Trend Analysis from 2007–2017

Ch. No.	Chapter Name	Number of Question(s) in											
		2007	2008	2009	2010	2011	2012	2013	2014	2015		2016	2017
1.	Basic Concepts of Chemistry	2	4	1	1	-	-	3	2	-	2	1	1
2.	Structure of Atom	1	2	3	-	3	2	1	3	2	1	2	1
3.	Chemical Bonding	2	3	4	6	3	4	8	2	3	2	3	4
4.	Solution	2	-	1	3	3	1	1	1	2	1	2	-
5.	States of Matter	2	5	2	1	3	3	1	2	2	3	1	2
6.	Nuclear Chemistry												
7.	Chemical Equilibrium	1	4	-	1	2	-	1	1	1	1	1	2
8.	Ionic Equilibrium	3	1	3	4	2	2	3	1	1	2	4	2
9.	Thermochemistry & Thermodynamics	2	3	2	-	4	3	1	2	2	2	2	2
10.	Chemical Kinetics	3	2	4	3	1	3	1	-	2	1	1	1
11.	Electrochemistry	2	3	3	4	4	1	3	3	1	1	1	2
12.	Surface Chemistry: Colloids & Caralysts	1	-	-	-	1	2	-	2	1	-	1	1
13.	Principle of Metallurgical Operations	2	-	-	-	2	3	-	-	1	1	-	1
14.	Chemical Periodicity	2	1	1	2	-	1	-	1	1	2	-	1
15.	Chemical Study of Elements	5	4	6	2	2	8	1	2	3	4	-	1
16.	Tramsition Metals & f-Block Elements	2	-	2	2	2	1	1	5	6	3	-	2
17.	Coordination Compounds	2	1	2	4	1	1	-	3	2	3	1	2
18.	Chemical Analysis	-	-	4	2	-	1						
19.	General Organic Chemistry	5	7	4	3	6	3	6	1	4	5	-	5
20.	Hydrocarbons	2	1	2	4	1	1	-	3	2	3	1	2
21.	Halogen Compounds & Alcohols	-	-	4	2	-	1	-	1	1	3	6	3
22.	Ether, Aldehydes, Ketones, Acids & their Derivatives & Nitrogen Compounds	6	4	2	5	3	4	4	5	5	4	9	3
23.	Polymers	1	1	1	1	1	2	3	2	1	1	1	1
24.	Biomolecules and Chemistry in Everyday Life	2	2	3	2	2	2	1	3	1	-	3	1
25.	Environmental Chemistry	-	1	-	-	1	1	-	1	-	-	-	
Total Questions		**50**	**50**	**50**	**50**	**50**	**50**	**45**	**45**	**45**	**45**	**45**	**40**

IMPORTANT FACTS TO MEMORIZE

Most abundant element in human body	Oxygen
Most abundant gas in sun	H_2
Most abundant gases in universe	H_2, He
Element having maximum tendency for catenation	Li
Most abundant metal of earth	Al
Rarest element of earth	Astatine (At)
Most abundant element of earth	Oxygen (O)
Element containing no neutron	$_1H^1$
Amphoteric metals	Zn, Al, Sn, Pb
Noble metals	Au, Pt
Elements having highest tensile strength	Boron
Metals showing highest oxidation number	Ru, Os
Non-metal having highest m.p., b.p.	Diamond
Lightest element	H
Heaviest naturally occurring element	^{238}U
Best electricity conductor among metals	Ag
Best conductor among non metals	Graphite
Most reactive solid element	Li
Most reactive liquid element	Cs
Most reactive gaseous element	F
Amphoteric non metal	Si
Elements showing diagonal relationship	Li-Mg, Be-Al, B-Si
Highest electronegativity	F
Highest ionization potential	He
Lowest ionization potential	Cs
Lowest electron affinity	Noble gases (zero)
Highest electron affinity	Cl
Non-metals having metallic lusture	iodine, graphite
Element sublime on heating	I
Coolant in nuclear reactors	D_2O
Most poisonous element	Pu
Liquid non-metal	Br_2
Total number of radioactive elements in periodic table	25
Volatile d-block elements	Zn, Cd, Hg
Liquid metals s	Hg, Ga, Cs, Fr, Eka
Element kept in water	P
Elements kept in kerosene oil	Na, K, I_2, Cs
Metal with highest m.p.	W
Metal with lowest m.p.	Hg
Non metal with highest refractive index	Diamond
Lowest refractive index	Air
Lowest b.p.	H_2
Heaviest solid metal	Os
Lightest solid metal	Li
Lightest solid non metal	B
Hardest naturally occurring non metal	Diamond
Hardest artificial substance good conductor of heat	B_4C (norbide)
Strongest basic hydroxide	CsOH
Strongest basic oxide	CsO_2 (caesium peroxide)
Most stable metal carbonate	Cs_2CO_3
Element with highest radioactivity	Ra
Strongest reducing agent	Azide (N_3^-)
Strongest oxidising agent	OF_2
Smallest anion	H^- (hydrides)
Smallest atomic size	H
Largest atomic size	Cs
Element with maximum number of isotopes	Ag (46)
Element with minimum number of isotopes	H (3)
Element with maximum number of allotropes	Sn
Liquid element of radioactive nature	Francium (Fr)
Poorest conductor of current	Pb (metal), S (non-metal)
Metalloids elements	B, Si, Ge, As, Sb, Te
Dry ice	CO_2
Most recently elements name by IUPAC	Ds (atomic number = 110)
'All purpose' grease	Lithium stearate
Old name of astatine	Albamine
Most abundant gas in atmosphere	N_2
Rarest gas in atmosphere	Rn
Lightest gas in atmosphere	H_2

Groups containing higher no. of gaseous elements	Group 18 (Noble gases)	Compound with maximum covalent nature	H_2, N_2, O_2, Cl_2 (diatomic gases)	Strongest acid	HSO_3F + 90 % SbF_5 called magic acid
Most electrovalent compound	CsF	First noble prize in chemistry	Van't Hoff	Most reactive form of P	White
Bad conductor of electricity	Mica	Softest form of carbon	Lamp black	Least reactive form of P	Red
Lightest radio isotope	Tritium ($_1H^3$)	Latest allotrope of carbon	fullerene or bucky ball	Purest form of silica	Quartz
				Most ductile metal	Gold

IMPORTANT CHEMICAL COMPOUNDS

IMPORTANT ORES

Fe

Magnetite ⟶ Fe_3O_4
Limonite ⟶ 3 $Fe_2O_3.3H_2O$
Iron Pyrite ⟶ FeS_2
Haematite ⟶ Fe_2O_3
Copper Pyrite ⟶ $CuFeS_2$
Spathic Iron ⟶ $FeCO_3$

Cu

Cuprite or Ruby copper ⟶ Cu_2O
Copper Glance ⟶ Cu_2S
Malachite ⟶ $Cu(OH)_2.CuCO_3$
Azurite ⟶ $Cu(OH)_2.2CuCO_3$

Pb

Galena ⟶ PbS
Anglesite ⟶ $PbSO_4$
Stolzite ⟶ $PbWO_4$
Cerrusite ⟶ $PbCO_3$
Wulfenite ⟶ $PbMnO_4$

Ag

Argentite or Silver Glance ⟶ Ag_2S
Pyragurite ⟶ 3 Ag_2S. Sb_2S_3
Proustite ⟶ $3Ag_2S$. As_2O_3
Horn Silver ⟶ AgCl

Mg

Magnesite ⟶ $MgCO_3$
Carnalite ⟶ $MgCl_2$. KCl. $6H_2O$
Kiesserite ⟶ $MgSO_4$. H_2O
Schonite ⟶ $MgSO_4$. $K_2SO_4.6H_2O$
Dolomite ⟶ $MgCO_3.CaCO_3$
Epsomite ⟶ $MgCO_3.7H_2O$
Kainite ⟶ $MgSO_4$. KCl. $3H_2O$
Olivine ⟶ Mg_2SiO_4
Spinel ⟶ $Mg_2Al_2O_4$

Al

Corundum ⟶ Al_2O_3
Diaspore ⟶ $Al_2O_3.H_2O$
Bauxite ⟶ Al_2O_3. $2H_2O$
Cryolite ⟶ Na_3AlF_6
Felspar ⟶ $KAlSi_3O_8$
Talc ⟶ Mg_3 H_2 $(SiO_3)_4$
Asbestos ⟶ $CaMg_3$ $(SiO_3)_4$
Mica ⟶ K Al Si_3 O_{10} $(OH)_2$
Kaolinite ⟶ $Al(OH)_4$ Si_2O_5
Turquoise ⟶ $AlPO_4.Al(OH)_3$. H_2O
Alum Stone or Alunite ⟶ K_2SO_4. $Al_2(SO_4)_3$. $4Al(OH)_3$

Be

Beryl ⟶ 3BeO $Al_2O_3.6SiO_2$ or $Be_3Al_2Si_6O_8$
Phenacite ⟶ 2BeO SiO_2 or Be_2SiO_4
Cryso beryl ⟶ BeO Al_2O_3

Ca

Anhydrite ⟶ $CaSO_4$
Dolomite ⟶ $CaCO_3$.$MgCO_3$
Fluorapatite ⟶ $3Ca_3$ $(PO_4)_2$. CaF_2
Gypsum ⟶ $CaSO_4.2H_2O$

Sr

Strontionite ⟶ $SrCO_3$
Celestite ⟶ $SrSO_4$

Ores of Ba

Barytes ⟶ $BaSO_4$
Whitherite ⟶ $BaCO_3$

Ores of Ra

Carnotite ⟶ K_2O, U_2O_3, $(VO_4)_2$.$3H_2O$
Pitch blende ⟶ Uranium oxide

Ores of Zn

Zinc blends or Sphalerite ⟶ ZnS
Zincite or Red zinc ⟶ ZnO
Calamine or Zinc spar ⟶ $ZnCO_3$
Franklinite ⟶ $ZnO.Fe_2O_3$
Willemite ⟶ Zn_2SiO_4
Hg ⟶ HgS (Cinabar)
Cassetirite or Tin Stone ⟶ SnO_2
Rutile ⟶ TiO_2
Ilmenite ⟶ $FeOTiO_2$
Beryl ⟶ $3BeO.Al_2O_3$. $6SiO_2$
Florapatite ⟶ $3Ca_3$ $(PO_4)_2$ .CaF_2

SOME IMPORTANT ALLOYS

1. *Alclad* It is an alloy of aluminium and used in making sea planes.
2. *Alnico* It is an alloy of steel (77%), nickel (2%), aluminium (20%) and cobalt (1%). It is used in making permanent magnets.
3. *Aluminium bronze* It is an alloy of 90% copper and 10% aluminium. It is used in making coins, trays and picture frames.
4. *Aluminium bronze* contains 90% Cu and 10% Al phosphor bronze consisting of 94% Cu, 5% Sn and 1% P is hard, elastic and used for pump rods, valves, axle bearing and certain other equipments. These are malleable, corrosion resistant and suitable for cold working.
5. *Babbit metal* It is an alloy of Sn (88-90%), Sb (7-4%) and Cu (3-7%). Hard babbit composition is (a) Sn = 91%, Pb = 4.5% and Sb = 4.5% and (b) Sn = 83%, Cu = 8.5% and Sb = 8.5%. It is used as bearing metal.
6. *Bell metal* It is an alloy of Cu and Sn having 80% Cu and 20% Sn. It is hard, brittle and sonorous. It is used for fabricating machine parts and bells, gongs etc.
7. *Brass* It is an alloy of 70% copper and 30% zinc. It is used in making utensils etc.
8. *Britania metal or pewter* It is an alloy of Sn (85-95%), Sb (6-10%), and Cu (1-3%). It is used for making cups, mugs and other utensils.
9. *Bronzes* These are mostly the alloys of copper and tin and contain 75–90% Cu and 25–10% Sn. These bronzes are mainly used for making coins, statues an special type of utensils.
10. *Constantan* It is an alloy of nickel (40%) and copper (60%). It is used in electrical work such as for making resistance boxes and thermo couples etc.
11. *Delta metal* It is an alloy of Cu (55%), Zinc (41%) and Fe(4%). It is used in making ships, bearing, and properllers.
12. *Duralumin* It is an alloy of Al(95.5%), Copper(4%), Mg(0.5%) and Mn(0.5%). It is used in making aeroplanes parts.
13. *Dutch metal* It is an alloy of copper and Zinc and is used in gold coverings.
14. *Electron* It is an alloy of Mg and Zn with small amounts of Al, Cu and Mn. It is a hard metal alloy used for making propellers of engines and air-crafts.
15. *Ferro alloys – Ferro molubdenum* is an alloy of Mo. Usually the percentage of Mo is less than 1%, but about 1.5-2% Mo has been used for high speed steel and 5% Mo in resisting steels. 6-10% Mo is used for preparing special steels. *Ferro silicon* has a composition of Si = 90-95%, C = 0.15%, S = 0.01% and P = 0.05%.
16. *Ferro manganese or Spiegeleisen* has average composition Mn = 78-82%, C = 7.5%, P = 0.35%, S = 0.5% and Si = 1.25%.
17. *Ferro nikel* contains Ni = 2.5-5%. It is hard, tough and rustless. It is used in the manufacture of cables. Propeller shaft, armor plates etc.
18. *Ferro titanium* has a composition of Ti = 38-45%, C = 0.1-6%, Si = 15-25% and Al = 9-10%, *Ferro tungsten* is hard and strong and contains W = 14-20%. It is used in the manufacture of high speed tools.
19. *Ferro vanadium* has composition V = 30-40%, C = 3.5%, P = 0.25%, S = 0.4%, Si = 13% and Al = 1.5%. It has high tensils strength and is used for making springs, axles, shafts etc. Both tungsten and vanadium make steel hard. Such hard steel is used for making high speed tools.
20. *German silver or Nickel silver* These are Cu–Zn – Ni alloys containing about 50% Cu, 25% Zn and 25% Ni. Nickel is used for fancy articles, forks, spoons, cigarette cases etc.
21. *Gun metal* It is an alloy of Cu, Sn and Zn. It contains 88% Cu, 10%Sn and 2%Zn. It is used in making guns. Gears and bearings.
22. *Invar* It has 64% Fe, 35% Ni and some traces of Mn, C. It is used in making pendulum rods.
23. *Magnalium* It is an alloy of 90% Al and 10% Mg. It is used in making balance beams.
24. *Monel metal* It is an alloy of 30% Cu, 67%Ni and 3%Mn or Fe. It is used for the construction of household sinks and containers and alkali-resisting equipments.
25. *Nichrome* It is an alloy of Ni(60%), Cr(15%) and Fe(25%) and is used in making electrical resistance.
26. *Nickel coinage alloy* for coinage purpose, an alloy of Ni(25%) and Cu (75%) is used. Another coinage alloy is silver coinage which contains 5% nickel.
27. *Pewter* It is an alloy of 24% Pb and 76% Sn and used for making utensils.
28. *Rose metal* It is an alloy of Bi(50%), Pb(25%) and Sn(25%). It is used in making stereo metal in printing and safety Plugs in boilers.
29. *Silicon Bronze* contains upto 4% Si and upto 1% Fe, Mn. Zn and Al, but does not contain Sn. They have strength like mild steel, excellent corrosion resistance and also have welding prop.
30. *Solder* It is an alloy of 67% Sn and 33% Pb. It is used in soldering. Soft solder contains 3-80% Pb and 97-20% Sn. This tin-lead alloy is used for joining metal parts because of its low melting point.
31. *Stainless steel* It contains Cr (about 11%) and Ni (about 7%). It is used in making utensils and surgical instruments.
32. *Sterite* It is an alloy of chromium, tungsten and nickel and is used for the manufacture of high speed tools and cutlery. This alloy is also used for making surgical instruments.
33. *Tiscor* It contains maximum 0.1% carbon, Mn = 0.1 - 0.4% (maximum), Cr = 0.7 – 1.1%, Cu = 0.3 - 0.5%, Si = 0.5 – 1.0% , P = 0.1 – 0.2 % and S = 0.05% .These two alloys (Tiscorn and Tiscor) are high strength engineering steels prepared by *Tata Iron and Steel company* in India.
34. *Tiscorn* It contains maximum 0.3% carbon, Mn = 0.5-1.3%, Cr = 1.00%, Ca = 0.25-0.6%, Si = 0.3% (maximum), P = 0.05% (maximum) and S = 0.05% (maximum).
35. *Type metal* It is an alloy of Pb(82%), Sb(15%) and Sn(3%) and is used for making type for printing.
36. *Wood metal* It is alloy of Pb(25%), Sn (12.5%), Cd (12.5%) and Bi (50%). It is used as automatic sprinkles. It melts in hot water as its melting point is 68°C.

IMPORTANT COMPOUNDS

1. Agate is silicon dioxide, SiO_2.
2. Ammonal is a mixture of ammonium nitrate and Al powder (NH_4NO_3 + Al). It is used as an explosive.
3. Alum is $(NH_4)_2SO_4.Al_2(SO_4)_3.24H_2O$. It is used as mordant by dyers of clothes. Potash alum is $K_2SO_4.Al_2(SO_4)_3.24H_2O$.
4. Aqua fortis is nitric acid, HNO_3.
5. Antichlor is sodium thiosulphate, $Na_2S_2O_3$. $5H_2O$. It is also called Hypo.
6. Aqua-Regia is a mixture of conc. HNO_3 and conc. HCl in the ratio of 1 : 3. It is also known as kingly water.
7. Baking soda is sodium bicarbonate, $NaHCO_3$.
8. Barytes is barium sulphate, $BaSO_4$.
9. Brine is sodium chloride (NaCl) solution.
10. Blue vitriol is copper sulphate, $CuSO_4.5H_2O$.
11. Bone ash is mainly calcium phosphate, $Ca_3(PO_4)_2$.
12. Borax is the name of sodium tetraborate hydrate $Na_2B_4O_7.10H_2O$. Borax ($Na_2B_4O_7$) is also called tincal.
13. $B(OH)_3$ is an acid.
14. Brown Ring is of $FeSO_4.NO$
15. Butter of tin is $SnCl_4$ $.5H_2O$.
16. B.O.D. is biological oxygen demand.
17. Cuprite is Cu_2O.
18. Calomel is Hg_2Cl_2.
19. Caustic potash is KOH.
20. Caustic soda is sodium hydroxide, NaOH.
21. Chile saltpeter is sodium nitrate, $NaNO_3$.
22. Cinnabar is HgS.
23. Carbonic acid is hydrogen carbonate, H_2CO_3.
24. Carbolic acid is phenol, C_6H_5OH.
25. Carborundum is silicon carbide, SiC.
26. Copper glance is Cu_2S.
27. Carbogen is a mixture of 1% CO_2 and O_2. It is used as antidote for for CO poisoning.
28. Corrosive sublimate is mercuric chloride, $HgCl_2$.
29. Corundum is aluminium oxide, Al_2O_3.
30. Chromyl chloride is CrO_2Cl_2.
31. Cream of tartar is $KHC_4H_4O_6$.
32. Cyanogen is C_2N_2.
33. Dead burnt plaster is anhydrous $CaSO_4$.
34. Dry ice is solid carbon dioxide (CO_2).
35. Epsom salt is the name of magnesium sulphate, $MgSO_4$ $.7H_2O$.
36. Energy is a mixture of Al_2O_3 and Fe_2O_3.
37. Eka aluminium is gallium.
38. Fluorspar is CaF_2.
39. Freon is CCl_2F_2.
40. Formalin is 40% formaldehyde (HCHO).
41. Fremy's salt is potassium hydrogen fluoride KHF_2.
42. Foul air is nitrogen, N_2. It is also called azote.
43. Fischer's salt is potassium cobalt nitric $K_3[Co(NO_2)_6]$.
44. Fowler's solution is $NaAsO_2$ solution.
45. Fusion mixture is Na_2CO_3 + K_2CO_3.
46. Fluorine is called super halogen.
47. Fulminating gold is $Au(NH_2)$ = NH.
48. Grain alcohol is ethyl alcohol, C_2H_5OH.
49. Grape sugar is dextrose, $C_6H_{12}O_6$.
50. Glauber's salt is the name of sodium sulphate, $Na_2SO_4.10H_2O$.
51. Gypsum is calcium sulphate, $CaSO_4.2H_2O$
52. Green vitriol is ferrous sulphate, $FeSO_4.7H_2O$
53. Gammexane is benzene hexachloride (BHC), $C_6H_6Cl_6$.
54. Gun powder is a mixture of sulphur, charcoal and nitre.
55. Graham's salt is $(NaPO_3)_3$.
56. Hydrolith is calcium hydride, CaH_2.
57. Halite is common rock salt (NaCl).
58. Horn silver is AgCl.
59. Hair salt is $Al_2(SO_4)_3.18H_2O$.
60. Hypo is $Na_2S_2O_35H_2O$.
61. King of chemicals is H_2SO_4.
62. Kali : Germans used the word 'Kali' for Potash.
63. Limestone is calcium carbonate, $CaCO_3$.
64. Lunar caustic is silver nitrate, $AgNO_3$.
65. Laughing gas is nitrous oxide, N_2O. It is also known as laughing grites.
66. Lithia water is aqueous solution of lithium bicarbonate ($LiHCO_3$).
67. Lapis Lazuli is blue coloured mineral used as semi-precious stone. It is sodium alumino silicate.
68. Milk of magnesia is magnesium hydroxide, $Mg(OH)_2$.
69. Marshall's acid is persulphuric acid, $H_2S_2O_8$.
70. Milk of lime is calcium hydroxides, $Ca(OH)_2$. It is also called slaked lime.
71. Magnesite is $MgCO_3$.
72. Microscopic salt is $Na(NH_4)HPO_4.4H_2O$.
73. Mica is $KH_2Al_2(SiO_4)_3$
74. Magnesia is MgO.
75. Mosaic gold is SnS_2.
76. Marsh gas or fire damp is CH_4.
77. Mohr's salt is $FeSO_4.(NH_4)_2SO_4.6H_2O$.
78. Muriatic acid is hydrochloric acid, HCl
79. Nessler's reagent is K_2HgI_4. It contains $HgCl_2$, KI and NaOH. The ions present in it is HgI_4^{2-}.
80. Nitre Cake is $NaHSO_4$
81. Nitrolim is $CaCN_2$ + C (graphite)
82. Norweigian saltpetre is basic calcium nitrate, $Ca(NO_3)_2$.
83. Oxone is sodium peroxide, Na_2O_2.
84. Oil of vitriol is sulphuric acid, H_2SO_4.
85. Oleum is fuming sulphuric acid, concentrated H_2SO_4 + SO_3.
86. Oil of mirbane is $C_6H_5NO_2$.
87. Oil of winter green is methyl salicylate.
88. Plaster of paris is calcium sulphate hemihydrate, $CaSO_4$. $^1/_2$ H_2O.
89. Philosopher's wool is zinc oxide, ZnO.
90. Phosgene is carbonyl chloride, $COCl_2$.
91. Picric acid is 2,4,6-trinitrophenol.
92. Paris green is double salt of copper acetate and copper arsenate.
93. Pearl white is BiOCl and is used as a pungent.
94. Prussian blue is Fe_4 $[Fe(CN)_6]_3$.
95. Pearl ash is K_2CO_3.
96. Perhydrol is 30% H_2O_2.
97. Prussic acid HCN.
98. Quick lime is calcium oxide, CaO.
99. Quartz is silicon dioxide, SiO_2.
100. Quick silver is mercury, Hg.
101. Realgar is As_2S_3.
102. Rectified spirit is 95% ethyl alcohol, C_2H_5OH.
103. Reagar is As_4S_4.
104. Red lead is lead tetroxide, Pb_3O_4. It is also called Minium.
105. Red liquor is aluminium acetate, $(CH_3COO)_3$ Al.
106. Rochelle salt is sodium potassium tartrate, $NaKC_4H_4O_6$.
107. Rock salt is NaCl.
108. Ruby or sapphire or Emery is Al_2O_3.
109. Salt cake is sodium sulphate, Na_2SO_4.
110. Sand is silicon dioxide, SiO_2.
111. Scheeb's green is $CuHAsO_3$.
112. Selidlitz powder is $NaHCO_3$.
113. Smelting salt is $(NH_4)_2CO_3$.
114. Soda lime is a mixture of NaOH and CaO.
115. Soda ash is sodium carbonate, Na_2CO_3.
116. Sodamide is $NaNH_2$.
117. Spirit of wine is C_2H_5OH.
118. Spirit of salt is HCl.

119. Sterling silver is a solution of Cu and Ag.
120. Stranger gas is xenon (Xe).
121. Sugar of lead is lead acetate, $(CH_3COO)_2Pb$.
122. Super phosphate of lime contains Ca $(H_2PO_4).H_2O$ and $2CaSO_4.2H_2O$
123. Syvine is KCl.
124. Tartar emetic is potassium antimony tartrate, $K(SbO)C_4H_4O_6$.
125. TNT is trinitrotoluene, an explosive.
126. Tear gas is chloropicrin, CCl_3NO_2.
127. TEL is tetra ethyl lead, $Pb(C_2H_5)_4$.
128. Thomas slag is calcium phosphate, $Ca_3(PO_4)_2$.
129. Tincture of iodine is I_2 and KI solution in alcohol.
130. Thermite is a mixture of iron oxide (Fe_3O_4) and Al powder.
131. Tinstone or Cassiterite is SnO_2.
132. Vinegar is dilute acetic acid, CH_3COOH.
133. Washing soda is Na_2CO_3.
134. Water glass is sodium silicate, Na_2SiO_3.
135. White lead is $Pb(OH)_2.2PbCO_3$.
136. Wackenroder's liquid is $H_2SO_4 + H_2S$.
137. White vitriol is Zinc sulphate, $ZnSO_4.7H_2O$
138. Wood spirit is CH_3OH.
139. Verdigris is the name of basic copper acetate, $(CH_3COO)_2$ Cu $Cu(OH)_2$.
140. Yellow ammonium sulphide is $(NH_4)_2Sx$.
141. Zincite is ZnO.

WATER SOLUBILITY OF SOME COMMON INORGANIC COMPOUNDS

1. All Na^+, K^+ and NH_4^+ compounds are soluble.
2. All nitrates, nitrites and acetates are soluble.
3. All chlorides, except AgCl, $PbCl_2$ and Hg_2Cl_2 are soluble. $PbCl_2$ is soluble in hot water.
4. All bromides, except AgBr, $PbBr_2$, Hg_2Br_2 and $HgBr_2$ are soluble.
5. All iodides, except AgI, PbI_2, Hg_2I_2 and HgI_2 are soluble.
6. All sulphates except $BaSO_4$, $CaSO_4$, $SrSO_4$, $PbSO_4$, Hg_2SO_4 and Ag_2SO_4 are soluble.
7. All carbonates except those of group 1 elements and $(NH_4)_2CO_3$ are insoluble.
8. All hydroxides except those of group 1 elements, $Ba(OH)_2$, $Ca(OH)_2$ and $Sr(OH)_2$ are insoluble.
9. All sulphides except those of group 1 & 2 elements and $(NH_4)_2S$ are insoluble.
10. All phosphates except those of group 1 elements and $(NH_4)_3PO_4$ are insoluble.
11. All sulphites, except those of group 1 elements and $(NH_4)_2SO_3$ are insoluble.

ACTION OF HEAT ON SOME SALTS

$ZnCO_3$ (white) $\rightarrow$ $ZnO + CO_2$ (yellow-hot; white-cold)
$CuCO_3$ (green) $\longrightarrow$ $CuO + CO_2$ (black)
$CaCO_3 \longrightarrow CaO + CO_2$
$2Ag_2CO_3 \longrightarrow 4Ag + 2CO_2 + O_2$
$2NaHCO_3 \longrightarrow Na_2CO_3 + CO_2 + H_2O$
$NH_4HCO_3 \longrightarrow NH_3 + CO_2 + H_2O$
$2FeSO_4 \longrightarrow Fe_2O_3 + SO_2 + SO_3$
$2CaSO_4.2H_2O$ 393K $2CaSO_4.\frac{1}{2}H_2O + 3H_2O$ (Plaster of paris)
$CuSO_4.5H_2O$ (blue) $\longrightarrow$ $CuSO_4 + 5H_2O$ (white)
$CuSO_4 \longrightarrow CuO + SO_3$
$(NH_4)_2Cr_2O_7$ (orange) $\longrightarrow$ $N_2 + Cr_2O_3 + 4H_2O$ (green)
$2KMnO_4 \longrightarrow K_2MnO_4 + MnO_2 + O_2$
$2KClO_3 \longrightarrow 2KCl + 3O_2$
$4K_2Cr_2O_7 \longrightarrow 4K_2CrO_4 + 2Cr_2O_3 + 3O_2$
$(COO)_2Fe \longrightarrow FeO + CO + CO_2$ (black)
$2Ag_2O \longrightarrow 4Ag + O_2$
$2HgO$ (red) $\longrightarrow$ $2Hg + O_2$ (silver deposit)
$2Pb_3O_4$ (red) $\longrightarrow$ $6PbO + O_2$ (yellow)
$2PbO_2$ (brown) $\longrightarrow$ $2PbO + O_2$ (black)
$2NaNO_3 \longrightarrow 2NaNO_2 + O_2$
$NH_4NO_2 \longrightarrow N_2 + 2H_2O$
$2AgNO_3 \longrightarrow 2Ag + 2NO_2 + O_2$
$2Cu(NO_3)_2$ (brown) $\longrightarrow$ $2CuO + 4NO_2 + O_2$
$2Zn(NO_3)_2$ (white) $\rightarrow$ $2ZnO + 4NO_2 + O_2$ yellow-hot white-cold (brown)
$2Ca(NO_3)_2 \longrightarrow 2CaO + 4NO_2 + O_2$
$2Pb(NO_3)_2$ (white) $\longrightarrow$ $2PbO + 4NO_2 + O_2$ (yellow) (brown)
$2Mg(NO_3)_2 \longrightarrow 2MgO + 4NO_2 + O_2$
$NH_4NO_3 \longrightarrow N_2O + 2H_2O$

IMPORTANT PROCESSES

1. Bosch process $\longrightarrow$ H_2
2. Down, Castner $\longrightarrow$ Na
3. Nelson, Castner – Kellner, Solvey Droney, Lowing $\longrightarrow$ NaOH
4. Ammonia soda process (Solvay process) $\longrightarrow$ Na_2CO_3, $NaHCO_3$
5. Leblanc, Pretch process $\longrightarrow$ K_2CO_3
6. MacArthur forest or Cyanidation $\longrightarrow$ Ag, Au
7. Perk, Pattinson $\longrightarrow$ Ag
8. Cupellation $\longrightarrow$ Ag (Purification)
9. Mund's process $\longrightarrow$ Ni (Purification)
10. Baeyer's or Serpeck's process $\longrightarrow$ Al
11. Hoope's process $\longrightarrow$ Al (Purification)
12. Hall, Heroult Process $\longrightarrow$ Al
13. Gold schmidt crocess $\longrightarrow$ Thermite Welding
14. Carter's process $\longrightarrow$ White lead
15. Haeber's process $\longrightarrow$ NH_3
16. Deacon's process $\longrightarrow$ Cl_2
17. Contact, Lead Chamber process $\longrightarrow$ H_2SO_4
18. Berkland-Eyde, Ostwald $\longrightarrow$ HNO_3, NO
19. Kaldo, L.D. $\longrightarrow$ steel
20. Corey – House $\longrightarrow$ Alkane
21. Oxo $\longrightarrow$ R – OH
22. Dow's process $\longrightarrow$ Mg
23. Pidgeon $\longrightarrow$ Mg
24. Cyanamide $\longrightarrow$ NH_3
25. IMI $\longrightarrow$ Ti
26. frash $\longrightarrow$ 'S'
27. Siemene, Basemer Thomass $\longrightarrow$ Steel.
28. Lane's process $\longrightarrow$ H_2
29. Gossage process $\longrightarrow$ NaOH

IMPORTANT ABBREVIATIONS

ADP	Adenosine diphosphate
AMP	Adenosine monophosphate
amu (u)	Atomic mass unit
ANTU	Alpha naphthyl thiurea
ATP	Adenosine triphosphate
b.p.	Boiling point
BCC	Body centred cubic
BHC	Benzene hexachloride
BO	Bond order
BOD	Biological oxygen process
BOP	Basic oxygen process
CAN	Calcium ammonium nitrate
CCP	Cubic close packing
DDT	Dichloro diphenyl trichloroethane
DNA	Deoxyribonucleic acid
DNP	Dihydroxy phenylhydrazine
DOPA	Dihydroxy phenylalanine
EDTA	Ethylene diammine tetra acetic acid
EMF	Electro motive force
EPM	Electron probe microanalysis
EPR	Ethylene-propylene rubber
ESR	Electron spin resonance
FAD	Flavin adenine dinucleotide
FCC	Face centred cubic
FP	Freezing point
GAW	Gram atomic weight
GMV	Gram molecular volume
GSC	Gas-solid chromatography
GTP	Guanosine triphosphate
HCP	Hexagonal close packing
HEED	High energy electron diffraction
Ie	Ionisation energy
IP	Ionization potential
IR	Infrared
IUPAC	International union of pure and applied chemistry
Kcal	Kilocalories
LAH	Lithium aluminium hydride
LD	Linnz-Donnewitz process Process
LEED	Low energy electron diffraction
LNG	Liquid natural gas
LPF	Liquefied petroleum gas
LR	Laboratory reagent
LSD	Lysergic acid diethylanide
LTA	Lead tetra acetate
m	Molality
M	Molarity
m.p.	Melting point
MEK	Methyl ethyl ketone
MeV	Mega electron volt
MMH	Monomethyl hydrazine
MO	Molecular orbital
NAA	Neutron activation analysis
NAD	Nicotinamide adenine dinucleotide
NADP	Nicotinamide adenine dinucleotide phosphate
NHE	Normal hydrogen electrode
NMR	Nuclear magnetic resonance
NPK	Nitrogen, phosphorous and potassium
NTP	Normal temperature and pressure
OR	Optical rotation
ORD	Optical rotatory dispersion
PCTFE	Polymono chloro tri fluoro ethylene
ppb	Parts per billion
ppm	Parts per million
PTEE	Polytetra fluoroethane
PVC	Polyvinyl chloride
rms	Root mean square
RNA	Ribonucleic acid
STP	Standard temperature and pressure
TCA	Tricarboxylic acid
TEL	Tetraethyl lead
TLC	Thin layer chromatography
TNM	Tetranitro methane
TNT	Trinitro toluene
UDMH	unsymmetrical dimethyl hydrazine
UV	Ultraviolet
VBT	Valence bond theory
VSEPR	Valence shell electron pair repulsion
WC	Tungsten carbide

MINERALS WITH METALLIC LUSTURE

Mineral (Formula)	Colour	Crystal System	Uses and Their Properties
Augite [(Ca, Na) (Mg, Fe, Al) (Al, Si)$_2$ O$_6$]	Black	Monoclinic	Square or 8-sided cross section
Bornite [Cu_5FeS]	Bronze, tarnishes to dark blue, purple	Tetragonal	Source of copper; called "Peakcock ore" due to its purple shine when it tarnishes
Chalcopyrite ($CuFeS_2$)	Brassy to golden yellow	Tetragonal	Main ore of copper
Chromite ($FeCr_2O_4$)	Black or brown	Cubic	Ore of chromium, stainless steel, metallurgical bricks
Copper (Cu)	Copper red	Cubic	Pipes, coins, gutters, wire, cooking utensils, jewellery, decorative plaques; malleable and ductile
Corundum (Al_2O_3)	Colourless, blue, brown, green, white, pink, red	Hexagonal	Gemstones; ruby is red, sapphire is blue, industrial abrasive
Feldspar (orthoclase) ($KAlSi_3O_8$)	Colourless, white to gray, green and yellow	Monoclinic	Insoluble in acids, used in the manufacture of porcelain

Mineral (Formula)	Colour	Crystal System	Uses and Their Properties
Feldspar (plagioclase) ($NaAlSi_3O_8$) ($CaAl_2Si_2O_8$)	Gray, green white	Triclinic	Used in ceramics; striations present on some faces
Fluorite (CaF_2)	Colourless, white, blue, green, red, yellow, purple	Cubic	Used in the manufacture of optical equipment, glows under ultraviolet light
Galena (PbS)	Gray	Cubic	Source of lead, shields for X-rays, fishing equipment sinkers, used in pipes
Garnet (Mg, Fe, Ca)$_3$ ($Al_2Si_3O_{12}$)	Deep yellow-red, green, black	Cubic	Used in jewellery, also used as an abrasive
Gold (Au)	Pale to golden	Cubic	Medicines, jewellery, money, gold leaf, fillings for teeth, does not tarnish
Graphite (C)	Black to grey	Hexagonal	Pencil lead, rods to control some small nuclear reactions, lubricants for locks, battery poles
Haematite (Fe_2O_3)	Black or reddish brown	Hexagonal	Source of iron; roasted in a blast furnace, converted to "pig" iron, made into steel
Hornblende [Ca Na (Mg, Al, Fe)$_5$ (Al, Si)$_2$ Si_6 O_{22} $(OH)_2$]	Green to black	Monoclinic	Will transmit light on thin edges; 6-sided cross section
Limonite (hydrous iron oxides)	Yellow, brown, black	-	Source of iron, weathers easily colouring matter of soils
Magnetite (Fe_3O_4)	Black	Cubic	Source of iron, naturally magnetic, called lodestone
Olivine [(Mg, Fe)$_2$ SiO_4]	Olive green	Ortho-rhombic	Gemstones, refractory sand
Pyrite (FeS_2)	Light, brassy, yellow	Cubic	Source of iron, "fools's gold" alters to limonite
Pyrrhotite (FeS)	Bronze	Hexagonal	Often found with pentlandite, an ore of nickel; may be magnetic
Quartz (SiO_2)	Colourless, various colorus	Hexagonal	Used in glass manufacture, electronic equipment, radios, computers, watches, gemstones
Silver (Ag)	Silvery white, tarnishes to black	Cubic	Coins, jewellery, silverplate, fillings for teeth, wires, malleable and ductile
Topaz [(Al_2SiO_4) (F, OH)$_2$]	White, pink yellow, pale blue, colourless	Ortho-rhombic	Valuable gemstone
Bauxite (hydrous aluminium compound)	Gray, red, white, brown	-	Source of aluminum; used in paints, aluminium foil, and airplane parts
Biotite [K(Mg,Fe)$_3$ $AlSi_3O_{10}$ $(OH)_2$]	Black to dark brown	Monoclinic	Occurs in large flexible plates
Calcite ($CaCO_3$)	Colourless, white pale, blue	Hexagonal	Fizzes when HCl is added; used in cement and other building materials
Dolomite [$CaMg(CO_3)_2$]	Colourless, white, pink, green, gray, black	Hexagonal	Concrete and cement, used as an ornamental building stone
Gypsum ($CaSO_4.2H_2O$)	Colourless, gray, white, brown	Monoclinic	Used extensively in the preparation of plaster of paris, alabaster, and dry wall for building construction
Halite (NaCl)	Colourless, red, white, blue	Cubic	Salt; very soluble in water; a preservative
Kaolinite [$Al_4Si_2O_5(OH)_4$]	White, red, reddish brown, black	Triclinic	Clays; used in ceramics and in china dishes; common in most soils; often microscopic sized particles
Muscovite [$KAl_3Si_3O_{10}(OH)_2$]	White, light gray, yellow, rose, green	Monoclinic	Occurs in large flexible plates, used as an insulator in electrical equipment, lubricant
Sphalerite (ZnS)	Brown	Cubic	Main ore of zinc; used in paints, dyes, and medicine
Sulphur (S)	Yellow	Ortho-rhombic	Used in medicine, fungicides for plants, vulcanisatoin of rubber, production of sulphuric acid
Talc [$Mg_3(OH)_2$ Si_4O_{10}]	White, greenish	Monoclinic	Easily cut with fingernail; used for talcum powder; soapstone is used in paper and for table tops

Name	Starting Material	Reaction Condition	Products
Aldol condensation	Aldehyde having α-hydrogen atom	conc. NaOH	aldol & ketol
Benzoin condensation	aromatic aldehyde	KCN	benzoin
Beckmann rearrangement	aromatic ketoxime	180-200°C, PCl_5 or P_4O_{10} or SO_2Cl_2 or H_2SO_4	N-substituted acetamide
Coupling reaction	C_6H_5NNCl + phenol (or aniline)	NaOH (in case of phenol), HCl (in case of aniline) NaOH	Azo dye
Claisen condensation	(aldehyde, ketone or ester Containing α-hydrogen atom) + (aromatic aldehyde or ketone)		unsaturated aro-matic aldehyde or ketone
Clemmensen reduction	aldehyde or ketone	Zn/Hg + conc HCl	hydrocarbon
Carbylamine reaction	primary amine	$CHCl_3$ + NaOH	carbylamine
Cannizzaro reaction n	aldehyde not having α-hydrogen atom	aq. KOH or alc. KOH	alcohol & acid
Diels-Alder reaction	α, β-unsaturated carbonyl compound + conjugated diene	100°C	cyclic addition product
Diazotization	aromatic primary amine	$NaNO_2$/conc. HCl, 0 – 5°C	C_6H_5NNCl
Etard reaction	toluene	CrO_2Cl_2/Cs_2	benzaldehyde
Friedel-Craft reaction	C_6H_6 + RX (or RCOCl)	anhy. $AlCl_3$	alkyl & acyl benzene
Fitting reaction	aryl halide	Na/dry ether	diphenyl
Gattermann-Koch reaction	C_6H_6 + CO + HCl	anhy. $AlCl_3$	aromatic aldehyde
Gattermann reaction	C_6H_5NNCl	Cu powder + HCl	halobenzene
Hunsdiecker reaction	Ag salt of carboxylic acid + Br_2	CCl_4, 80°C	alkyl or aryl bromide
Hofmann mustard oil reaction	primary aliphatic amine + CS_2	$HgCl_2$	$CH_3CH_2-N=C=S$ (mustard oil)
Hoffmann bromamide reaction	acid amide	Br_2/NaOH	primary amine
Hell-Volhard-Zelinsky reaction	carboxylic acid having α-hydrogen atom	Br_2/red P	α-halogenated carboxylic acid
Haloform reaction	CH_3CHO, methyl ketone	NaOH/X_2 or NaOX	haloform
Kolbe-Schmidt reaction	C_6H_5OH	CO_2, 125°C, 4-7 atm, H^+	salicylic acid
Kolbe electrolytic reaction	alkali metal salt of carboxylic acid	electrolysis	alkane, alkene & alkyne
Libermann nitroso reaction	secondary amine	$NaNO_2$/HCl	$R_2N-N=O$
Mendius reaction	alkyl or aryl cyanide	Na/C_2H_5OH	primary amine
Meerwein-Ponndorf verley reduction	ketone	$[(CH_3)_2CHO]_3Al$ + $(CH_3)_2CHOH$	sec. alcohol
Perkin condensation	aromatic aldehyde + $(CH_3CO)_2O$	CH_3COONa, H^+/H_2O	α, β-unsaturated aromatic acid
Rosenmund reduction	acid chloride	H_2, Pd/$BaSO_4$, S, Boiling xylene	aldehyde
Reamer-Tiemann reaction	phenol	NaOH + $CHCl_3$ or CCl_4, dil. HCl	salicylaldehyde or salicylic acid
Stephen reaction	alkyl cyanide	$SnCl_2$/HCl	aldehyde
Schotten-Baumann reaction	(phenol or aniline or alcohol) + C_6H_5COCl	NaOH	benzoylated product
Sandmeyer reaction	C_6H_5NNCl	CuCl/HCl or CuBr/HBr or CuCN/KCN, heat	halo or cyano-benzene
Sabatier-Senderens reaction	unsaturated hydrocarbon	raney Ni/H_2, 200-300°C	alkane
Ullmann reaction	iodobenzene + Cu	heat	diphenyl
Wurtz –Fitting reaction	alkyl halide + aryl halide	Na/dry ether	alkyl benzene
Williamson synthesis	alkyl halide + sodium alkoxide or sodium phenoxide	heat	ether

INCREASING ORDERS

1. Decreasing ionic size $Mg^{2+} < O^{2-} < Na^{+} < F^{-}$
2. Increasing acidic property $ZnO < Na_2O_2 < P_2O_5 < MgO$
3. Increasing first ionization potential Mg< Al< Si< Na
4. Increasing bond length $F_2 < N_2 < Cl_2 < O_2$
5. Increasing size $Cl^{-} < S^{2-} < Ca^{2+} <$ Ar
6. Increasing acid strength $HClO_3 < HClO_4 < HClO_2 <$ HClO
7. Increasing bond strength HCl< HBr< HF< HI
8. Increasing oxidation number of iodine $I_2 <$ HI< $HIO_4 <$ ICI
9. Increasing thermal stability HOCl< $HOClO_2 <$ $HOClO_3 <$ HOClO
10. Increasing bond enthalpy $N_2 < O_2 < F_2 < Cl_2$
11. Increasing acidic character $CO_2 < N_2O_5 < SiO_2 < SO_3$
12. Increasing ionic size $N^{3-} < Na^{+} < F^{-} < O^{2-} < Mg^{2+}$
13. Increasing basic character MgO< SrO< $K_2O <$ NiO< Cs_2O
14. Increasing extent of hydrolysis $CCl_4 < MgCl_2 < AlCl_3 < PCl_5 < SiCl_4$
15. Increasing strength of hydrogen bonding (X….H–X) O< S< F< Cl< N
16. Increasing ionic radii in water $Li^{+} < Na^{+} < K^{+} < Rb^{+} < Cs^{+}$
17. Increasing molar conductivity in water $Li^{+} < Na^{+} < K^{+} < Rb^{+} < Cs^{+}$
18. Increasing reactivity with water Li< Na< K< Rb< Cs
19. Increasing reactivity with hydrogen Li< Na< K< Rb< Cs
20. Increasing melting point Li< Na< K< Rb< Cs
21. Increasing basic nature of hydroxides LiOH< NaOH< KOH< RbOH< CsOH
22. Increasing thermal stability of hydroxides LiOH< NaOH< KOH< RbOH< CsOH
23. Increasing covalent character LiCl< LiBr< LiI
24. Increasing ionic character $CaCl_2 < BeCl_2 < MgCl_2 < BaCl_2 < SrCl_2$
25. Increasing solubility $BeCO_3 < MgCO_3 < CaCO_3 < BaCO_3$
26. Increasing solubility $BeF_2 < MgF_2 < CaF_2 < BaF_2$
27. Increasing solubility $Be(OH)_2 < Mg(OH)_2 < Ca(OH)_2 < Ba(OH)_2$
28. Increasing basicity $Be(OH)_2 < Mg(OH)_2 < Ca(OH)_2 < Ba(OH)_2$
29. Increasing hydration of ions $Be^{2+} < Mg^{2+} < Ca^{2+} < Sr^{2+} < Ba^{2+}$
30. Increasing reactivity with water Be< Mg< Ca< Sr< Ba
31. Increasing reactivity towards air Be< Mg< Ca< Sr< Ba
32. Increasing solubility $BeSO_4 < MgSO_4 < CaSO_4 < SrSO_4 < BaSO_4$
33. Increasing ionic character $BCl_3 < AlCl_3 < GaCl_3$
34. Increasing strength of Lewis acid $wBF_3 < BCl_3 < BBr_3$
35. Increasing strength of Lewis acid $AlCl_3 < GaCl_3 < InCl_3$
36. Increasing reducing power $GeCl_2 < SnCl_2 < PbCl_2$
37. Increasing oxidizing power $GeCl_4 < SnCl_4 < PbCl_4$
38. Increasing number of hybrid orbitals of C< Si< Sn
39. Increasing basic character $NH_3 < AsH_3 < SbH_3 < PH_3$
40. Increasing thermal stability $NH_3 < AsH_3 < SbH_3 < PH_3$
41. Increasing acidic strength $HNO_3 < H_3PO_4 < H_3AsO_4 < H_3SbO_4$
42. Increasing solubility in water $HNO_3 < H_3PO_4 < H_3AsO_4 < H_3SbO_4$
43. Increasing order of +5 oxidation state N< P< As< Sb and Bi
44. Increasing extent of hydrolysis $NCl_3 < PCl_3 < AsCl_3$ $SbCl_3 < BiCl_3$
45. Increasing stability of hydrides $H_2O < H_2S < H_2Se < H_2Te$
46. Increasing poisonous nature $H_2S < H_2Se < H_2Te < H_2Po$
47. Increasing acidic strength $H_2O < H_2S < H_2Se < H_2Te$
48. Increasing strength of oxoacids $H_2SO_3 < H_2SeO_3 < H_2TeO_3$
49. Increasing stability of oxoacids $H_2SO_3 < H_2SeO_3 < H_2TeO_3$
50. Increasing strength of oxoacids $H_2SO_4 < H_2SeO_4 < H_2TeO_4$
51. Increasing stability of oxoacids $H_2SO_4 < H_2SeO_4 < H_2TeO_4$
52. Increasing electron affinity F< Cl< Br< I
53. Increasing reducing power HF< HCl< HBr< HI
54. Increasing affinity for hydrogen $F_2 < Cl_2 < Br_2 < I_2$
55. Increasing acidity HF< HCl< HBr< HI
56. Increasing melting point HF< HCl< HBr< HI
57. Increasing boiling point HF< HCl< HBr< HI
58. Increasing stability $HFO_3 < HClO_3 < HBrO_3 < HIO_3$
59. Increasing covalent character $TiCl_2 < TiCl_3 < TiCl_4$
60. Increasing magnetic moment $Ti^{3+} < Ni^{2+} < Cr^{2+} < Co^{2+} < Zn^{2+}$
61. Increasing ionic character $VCl_2 < VCl_3 < VCl_4$
62. Increasing electropositivity Hg< Na< Cu< Li
63. Increasing density Fe< Pb< Al< Au
64. Increasing basic characteristics $Li_2O <$ BeO $< B_2O_3 < CO_2$
65. Increasing electronegativity As< P< S< Cl
66. Increasing ionization energy N< O< F
67. Increasing atomic size S< O< Se< C
68. Increasing acidity HOCl< HOBr< HOI
69. Increasing density $H_2S < O_2 < CO_2 < NH_3 < H_2$
70. Increasing thermal stability HF< HCl< HBr< HI
71. Increasing bond enthalpy $N_2 < O_2 < F_2 < Cl_2$
72. Increasing melting point $CaF_2 < CaCl_2 < CaBr_2 < CaI_2$
73. Increasing oxidizing power O< S< Se< Te
74. Increasing oxidizing power F< Cl< Br< I
75. Increasing size B< Be< Li< Na
76. Increasing single bond length N – N< O – O< F - F
77. Increasing stability of hydrides LiH< NaH< KH< CsH
78. Decreasing pH of aqueous solution of LiCL< $BeCl_2 < MgCl_2 < AlCl_4$
79. Increasing acidic oxide $Al_2O_3 <$ MgO< $SiO_2 < P_4O_{10}$

80. Increasing basicity $F^- < Cl^- < Br^- < I^-$

81. Increasing basic strength $F^- < OH^- < NH_2^- < CH_3^-$

82. Increasing boiling point $NH_3 < PH_3 < AsH_3 < SbH_3$

83. Increasing ionization energy B< C< N< O

84. Increasing thermal stability $BeCO_3 < MgCO_3 < CaCO_3 < BaCO_3$

85. Increasing paramagnetism Ca< Al< N< O

86. Increasing ionic character LiBr< NaBr< KBr< RbBr< CsBr

87. Increasing hydration energy $Be^{2+} < Mg^{2+} < Ca^{2+} < Ba^{2+} < Sr^{2+}$

88. Increasing bond angle $NH_3 < PH_3 < AsH_3$

89. Increasing bond angle $NF_3 < PH_3 < AsF_3$

90. Increasing bond angle $H_2O < H_2S < H_2Se$

91. Increasing bond angle $NF_3 < NCl_3$

92. Increasing bond angle $NO_2^+ < NO_2 < NO_2$

93. Increasing bond angle $NH_3 < NF_3$

94. Increasing bond angle $PH_3 < PF_3$

95. Increasing electronegativity $O^+ < O < O^-$

96. Decreasing dipole moment: $CH_3Cl < CH_2Cl_2 < CHCl_3$ and CCl_4

97. Decreasing ionic nature: $MCl < MCl_2 < MCl_3$

98. Increasing strength of H-bonding: O< F< Cl< S & N

99. Increasing N – O bond length: $NO_2^+ < NO_2^-$ and NO_3^-

100. Increasing bond order: $N_2 < F_2 < O_2 < O_2^- < O_2^+$

IMPORTANT FORMULAE OF REPLACEMENT OF GROUPS

- $>C{=}C< \xrightarrow{\text{reduction}} >CH-CH<$ (alkanes → alkanes)
- $-C{\equiv}C- \xrightarrow{\text{reduction}} -CH-CH_2-$ (alkanes → alkanes)
- $-X \xrightarrow{\text{Aq. KOH/NaOH}} -OH$
- $-X \xrightarrow{\text{reduction}} -H$
- $-X \xrightarrow{\text{KCN}} -CH$
- $-CH_2OH \xrightarrow[\text{mild oxidizing agent}]{\text{controlled oxidation}} -CHO$ (1° alcohols → aldehydes)
- $-CH_2OH \xrightarrow{\text{oxidation}} -COOH$ (1° alcohols → acids)
- $>CHOH \xrightarrow{\text{oxidation}} >C=O$ (2° alcohols → ketones)
- $>COH \xrightarrow{\text{oxidation}} >C=O$ (3° alcohols → ketones)
- $-OH \xrightarrow{PCl_5/\ PCl_3/\ SOCl_2/\ HCl} -Cl$
- $-H \xrightarrow{X_2} -X$
- $-CHO \xrightarrow{H_2} -CH_2OH$ (aldehydes → 1° alcohols)
- $>C=O \xrightarrow{H_2} >CHOH$ (ketones → 2° alcohols)
- $>CHO$ or $\xrightarrow[\text{reduction}]{\text{clemmenson}} -CH_2- >C=O$ (alkanes; carbonyl compounds)
- $-CHO \xrightarrow{\text{oxidation}} -COOH$ (aldehydes → acids)
- $>C=O \xrightarrow{\text{oxidation}} -COOH$ (ketones → acids)
- $-CO- \xrightarrow{NH_2OH} =NOH$
- $-COOH \xrightarrow{Ca/Ca(OH)_2} (-COO)_2Ca$
- $-COOH \xrightarrow{PCl_5/PCl_3/SOCl_2} COCl$
- $-COOH \xrightarrow{NH_3} -CONH_2$
- $-COONH_4 \xrightarrow{\text{heat}} -CONH_2$
- $(-COO)_2Ca \xrightarrow{\text{heat}} -CO-$ or $-CHO$
- $-C\equiv N \xrightarrow{\text{reduction}} -CH_2-NH_2$ (cyanides → 1° amine)
- $R-NC \xrightarrow{\text{reduction}} R-NH-CH_3$ (Isocyanides → 2° amine)
- $-NO_2 \xrightarrow{\text{reduction}} -NH_2$ (nitro → 1° amine)
- $=NOH \xrightarrow{\text{reduction}} -NH_2$ (oximes → 1° amine)
- $-NH_2 \xrightarrow{NaNO_2 + HCl/HNO_2} -OH$
- $>NH \xrightarrow{NaNO_2 + HCl/HNO_2} -N-N=O$
- $-CN \xrightarrow[\text{hydrolysis}]{\text{acid or alkaline}} -COOH$ (cyanides → acids)

IMPORTANT REAGENTS AND MIXTURES

Ammonal	Al powder + NH_4NO_3 [used as an explosive]
Amatol	NH_4NO_3 (80%) + trinitrotoluene (20%) [used as an explosive]
Aqua regia	Conc. HNO_3 + conc. HCl (1 : 3) [used as a laboratory reagent]
Bordeaux mixture	A solution of $CuSO_4$ + lime [used to kill moulds and fungi on plants]
Black ash	Na_2CO_3 + CaS [impure Na_2CO_3 obtained in Leblanc process]
Benedict solution	A solution of $CuSO_4.5H_2O$ 1 NaOH + sodium citrate [used for detecting aldehydes]
Baeyer reagent	Alkaline $KMnO_4$ solution [used for detecting ethylene and acetylene linkages]
Baking powder	$NaHCO_3$ + sodium potassium tartarate
Carbon oil	Vegetable oil + lime water [used for treatment of burns]
Carbogen	O_2 (90-95%) + CO_2 (5-10%) [used for artificial respiration]
Coal gas	H_2 (47%) + CH_4 (32%) + CO (7%) + N_2 (4%) + C_2H_4 (3%) + C_2H_2 (2%) + CO_2 (1%) + other gases (4%) [used to produce reducing atmosphere in metallurgical operations]
Euchlorine	Cl_2 + ClO_2
Fusion mixture	Na_2CO_3 + K_2CO_3 [used as a laboratory reagent]

Name	Composition / Use
Freezing mixture	NaCl + ice [used for lowering temperature]
Fehling solution	$CuSO_4.5H_2O$ + NaOH + sodium potassium tartarate [used for detecting aldehydes]
Fenton reagent	H_2O_2 + few drops of $FeCl_3$
Gun powder	KNO_3 (75%) + S (12%) + charcoal (13%) [used as an explosive]
Granite	Mica + rock + clay + sand
Gobar gas	CH_4 + CO + H_2 [used as a domestic fuel]
Ignition mixture	BaO_2 + Mg + Al [used in aluminothermit process]
Lucas reagent	Conc. HCl + anhy. $ZnCl_2$ [used for distinguishing three types of alcohols]
Lithopone	ZnS + $BaSO_4$ [used as a white paint]
Lindlar catalyst	Pd/$BaSO_4$, S [used for hydrogenation of alkyne to alkene]
Mortar	(slaked lime + sand) (1 : 3 in water)
Molish reagent	An ethanolic solution of a-naphthol [used for detecting carbohydrates]
Milk of magnesia	An aqueous suspension of $Mg(OH)_2$ [used as a antacid]
Methylated spirit	Rectified spirit (85-90%) + CH_3OH (10-15%) + pyridine + acetone [used as a solvent]
Matte	Cu_2S + FeS
Nitrophos	$Ca(H_2PO_4)_2$ + $Ca(NO_3)_2$ [used as a fertilizer]
Nitrolim	$CaCN_2$ + graphite [used as a fertilizer]
Nitro chalk	NH_4NO_3 + $CaCO_3$ [used as a fertilizer]
Nessler reagent	A solution of K_2HgI_4 + KOH [used for detecting $NH_4^!$ ion]
Oil gas	H_2 (50-55%) + CH_4 (25-30%) + CO (10-12%) + CO_2 (3%) [used in laboratory]
Purple of cassius	Colloidal sol of Au + $Sn(OH)_4$ [used for colouring of glass and pottery red]
Producer gas	N_2 (52-55%) + CO (22-30%) + H_2 (8-12%) + CO_2 (3%) [used as a fuel]
Power alcohol	(petrol + C_2H_5OH) (4 : 1) + little benzene [used as a motor fuel]
Rectified spirit	C_2H_5OH (95.87%) + H_2O (4.13%) [used as a solvent]
Super phosphate of lime	$Ca(H_2PO_4)_2$ + $CaSO_4$ [used as a fertilizer]
Sublimated white lead	PbO + $PbSO_4$ + ZnO [used as a white paint]
Sorel's cement	MgO + $MgCl_2$ [used as a substitute for tiles]
Sodalime	NaOH + $Ca(OH)_2$ [used in decarboxylation of carboxylic acids]
Soda bleach	Na_2O_2 + HCl [used for bleaching of fabrics]
Schweitzer reagent	$[Cu(NH_3)_4]SO_4$ [used in the manufacture of artificial silk]
Schiff's reagent	An aqueous solution of rosaniline hydrochloride whose red colour has been discharged by SO_2 [used for detecting aldehydes]
Thomas slag	$Ca_3(PO_4)_2$ + $CaSiO_3$ [used as a fertilizer]
Tincture of iodine	I_2 + KI + C_2H_5OH + water [used as an antiseptic]
Thermite mixture	Al powder + metal oxide [used in metallurgy]
Water gas	H_2 (51%) + CO (41%) + N_2 (4%) + CO_2 (4%) [Wused for manufacture of CH_3OH]

IMPORTANT INDUSTRIAL PROCESSES

Bosch process	Manufacture of H_2
Birkeland-Eyde process	Manufacture of HNO_3
Bessemer process	Manufacture of steel
Baeyer process	Manufacture of Al
Cupellation process	Purification of Ag
Contact process	Manufacture of H_2SO_4
Castner-Kellner process	Manufacture of NaOH
Castner process	Manufacture of Na
Carter process	Manufacture of white lead (basic lead carbonate)
Down process	Manufacture of Na
Dow process	Manufacture of phenol
Deacon process	Manufacture of Cl_2
Goldschmidt process	Thermite welding
Hoop process	Purification of Al
Hasenclever process	Manufacture of bleaching powder
L. D. process	Manufacture of steel
Leblanc process	Manufacture of K_2CO_3
Lead chamber process	Manufacture of H_2SO_4
Merck process	Manufacture of H_2O_2
MacArthur-Forest process	Manufacture of Ag
Nelson process	Manufacture of NaOH
Ostwald process	Manufacture of HNO_3
Open hearth process	Manufacture of steel
Parke process	Manufacture of Ag
Pattinson process	Manufacture of Ag
Solvay ammonia soda process	Manufacture of Na_2CO_3 an $NaHCO_3$
Serpeck process	Manufacture of Al

ELEMENTS, THEIR ATOMIC NUMBER AND MOLAR MASS

Element	Symbol	Atomic Number	Molar Mass/ (g mol^{-1})
Actinium	Ac	89	227.03
Aluminium	Al	13	26.98
Americium	Am	95	(243)
Antimony	Sb	51	121.75
Argon	Ar	18	39.95
Arsenic	As	33	74.92
Astatine	At	85	210
Barium	Ba	56	137.34
Berkelium	Bk	97	(247)
Beryllium	Be	4	9.01
Bismuth	Bi	83	208.98
Bohrium	Bh	107	(264)
Boron	B	5	10.81
Bromine	Br	35	79.91
Cadmium	Cd	48	112.40
Caesium	Cs	55	132.91
Calcium	Ca	20	40.08
Californium	Cf	98	251.08
Carbon	C	6	12.01
Cerium	Ce	58	140.12
Chlorine	Cl	17	35.45
Chromium	Cr	24	52.00
Cobalt	Co	27	58.93
Copper	Cu	29	63.54
Curium	Cm	96	247.07
Dubnium	Db	105	(263)
Dysprosium	Dy	66	162.50
Einsteinium	Es	99	(252)
Erbium	Er	68	167.26
Europium	Eu	63	151.96
Fermium	Fm	100	(257.10)
Fluorine	F	9	19.00
Francium	Fr	87	(223)
Gadolinium	Gd	64	157.25
Gallium	Ga	31	69.72
Germanium	Ge	32	72.61
Gold	Au	79	196.97
Hafnium	Hf	72	178.49
Hassium	Hs	108	(269)
Helium	He	2	4.00
Holmium	Ho	67	164.93

Element	Symbol	Atomic Number	Molar Mass/ (g mol^{-1})
Hydrogen	H	1	1.0079
Indium	In	49	114.82
Iodine	I	53	126.90
Iridium	Ir	77	192.2
Iron	Fe	26	55.85
Krypton	Kr	36	83.80
Lanthanum	La	57	138.91
Lawrencium	Lr	103	(262.1)
Lead	Pb	82	207.19
Lithium	Li	3	6.94
Lutetium	Lu	71	174.96
Magnesium	Mg	12	24.31
Manganese	Mn	25	54.94
Meitneium	Mt	109	(268)
Mendelevium	Md	101	258.10
Mercury	Hg	80	200.59
Molybdenum	Mo	42	95.94
Neodymium	Nd	60	144.24
Neon	Ne	10	20.18
Neptunium	Np	93	(237.05)
Nickel	Ni	28	58.71
Niobium	Nb	41	92.91
Nitrogen	N	7	14.0067
Nobelium	No	102	(259)
Osmium	Os	76	190.2
Oxygen	O	8	16.00
Palladium	Pd	46	106.4
Phosphorus	P	15	30.97
Platinum	Pt	78	195.09
Plutonium	Pu	94	(244)
Poonium	Po	84	210
Potassium	K	19	39.10
Praseodymium	Pr	59	140.91
Promethium	Pm	61	(145)
Protactinium	Pa	91	231.04
Radium	Ra	88	(226)
Radon	Rn	86	(222)
Rhenium	Re	75	186.2
Rhodium	Rh	45	102.91
Rubidium	Rb	37	85.47
Ruthernium	Ru	44	101.07

(Continued)

Element	Symbol	Atomic Number	Molar Mass/ (g mol^{-1})
Rutherfordium	Rf	104	(261)
Samarium	Sm	62	150.35
Scandium	Sc	21	44.96
Seaborgium	Sg	106	(266)
Selenium	Se	34	78.96
Silicon	Si	14	28.08
Silver	Ag	47	107.87
Sodium	Na	11	22.99
Strontium	Sr	38	87.62
Sulphur	S	16	32.06
Tantalum	Ta	73	180.95
Technetium	Tc	43	(98.91)
Tellurium	Te	52	127.60
Terbium	Tb	65	158.92
Thallium	Tl	81	204.37

Element	Symbol	Atomic Number	Molar Mass/ (g mol^{-1})
Thorium	Th	90	232.04
Thulium	Tm	69	168.93
Tin	Sn	50	118.69
Titanium	Ti	22	47.88
Tungsten	W	74	183.85
Ununbium	Uub	112	(277)
Ununnilium	Uun	110	(269)
Unununium	Uuu	111	(272)
Uranium	U	92	238.03
Vanadium	V	23	50.94
Xenon	Xe	54	131.30
Ytterbium	Yb	70	173.04
Yttrium	Y	39	88.91
Zinc	Zn	30	65.37
Zirconium	Zr	40	91.22

* *The value given in paranthesis is the molar mass of the isotope of largest known half life.*

LOGARITHMS TABLE

N	0	1	2	3	4	5	6	7	8	9	1	2	3	4	5	6	7	8	9
10	0000	0043	0086	0128	0170						5	9	13	17	21	26	30	34	38
						0212	0253	0294	0334	0374	4	8	12	16	20	24	28	32	36
11	0414	0453	0492	0531	0569						4	8	12	16	20	23	27	31	35
						0607	0645	0682	0719	0755	4	7	11	15	18	22	26	29	33
12	0792	0828	0864	0899	0934						3	7	11	14	18	21	25	28	32
						0969	1004	1038	1072	1106	3	7	10	14	17	20	24	27	31
13	1139	1173	1206	1239	1271						3	6	10	13	16	19	23	26	29
						1303	1335	1367	1399	1430	3	7	10	13	16	19	22	25	29
14	1461	1492	1523	1553	1584						3	6	9	12	15	19	22	25	28
						1614	1644	1673	1703	1732	3	6	9	12	14	17	20	23	26
15	1761	1790	1818	1847	1875						3	6	9	11	14	17	20	23	26
						1903	1931	1959	1987	2014	3	6	8	11	14	17	19	22	25
16	2041	2068	2095	2122	2148						3	6	8	11	14	16	19	22	24
						2175	2201	2227	2253	2279	3	5	8	10	13	16	18	21	23
17	2304	2330	2355	2380	2405						3	5	8	10	13	15	18	20	23
						2430	2455	2480	2504	2529	3	5	8	10	12	15	17	20	22
18	2553	2577	2601	2625	2648						2	5	7	9	12	14	17	19	21
						2672	2695	2718	2742	2765	2	4	7	9	11	14	16	18	21
19	2788	2810	2833	2856	2878						2	4	7	9	11	13	16	18	20
						2900	2923	2945	2967	2989	2	4	6	8	11	13	15	17	19
20	3010	3032	3054	3075	3096	3118	3139	3160	3181	3201	2	4	6	8	11	13	15	17	19
21	3222	3243	3263	3284	3304	3324	3345	3365	3385	3404	2	4	6	8	10	12	14	16	18
22	3424	3444	3464	3483	3502	3522	3541	3560	3579	3598	2	4	6	8	10	12	14	15	17
23	3617	3636	3655	3674	3692	3711	3729	3747	3766	3784	2	4	6	7	9	11	13	15	17
24	3802	3820	3838	3856	3874	3892	3909	3927	3945	3962	2	4	5	7	9	11	12	14	16
25	3979	3997	4014	4031	4048	4065	4082	4099	4116	4133	2	3	5	7	9	10	12	14	15
26	4150	4166	4183	4200	4216	4232	4249	4265	4281	4298	2	3	5	7	8	10	11	13	15
27	4314	4330	4346	4362	4378	4393	4409	4425	4440	4456	2	3	5	6	8	9	11	13	14
28	4472	4487	4502	4518	4533	4548	4564	4579	4594	4609	2	3	5	6	8	9	11	12	14
29	4624	4639	4654	4669	4683	4698	4713	4728	4742	4757	1	3	4	6	7	9	10	12	13
30	4771	4786	4800	4814	4829	4843	4857	4871	4886	4900	1	3	4	6	7	9	10	11	13
31	4914	4928	4942	4955	4969	4983	4997	5011	5024	5038	1	3	4	6	7	8	10	11	12
32	5051	5065	5079	5092	5105	5119	5132	5145	5159	5172	1	3	4	5	7	8	9	11	12
33	5185	5198	5211	5224	5237	5250	5263	5276	5289	5302	1	3	4	5	6	8	9	10	12
34	5315	5328	5340	5353	5366	5378	5391	5403	5416	5428	1	3	4	5	6	8	9	10	11
35	5441	5453	5465	5478	5490	5502	5514	5527	5539	5551	1	2	4	5	6	7	9	10	11
36	5563	5575	5587	5599	5611	5623	5635	5647	5658	5670	1	2	4	5	6	7	8	10	11
37	5682	5694	5705	5717	5729	5740	5752	5763	5775	5786	1	2	3	5	6	7	8	9	10
38	5798	5809	5821	5832	5843	5855	5866	5877	5888	5899	1	2	3	5	6	7	8	9	10
39	5911	5922	5933	5944	5955	5966	5977	5988	5999	6010	1	2	3	4	5	7	8	9	10
40	6021	6031	6042	6053	6064	6075	6085	6096	6107	6117	1	2	3	4	5	6	8	9	10
41	6128	6138	6149	6160	6170	6180	6191	6201	6212	6222	1	2	3	4	5	6	7	8	9
42	6232	6243	6253	6263	6274	6284	6294	6304	6314	6325	1	2	3	4	5	6	7	8	9
43	6335	6345	6355	6365	6375	6385	6395	6405	6415	6425	1	2	3	4	5	6	7	8	9
44	6435	6444	6454	6464	6474	6484	6493	6503	6513	6522	1	2	3	4	5	6	7	8	9
45	6532	6542	6551	6561	6471	6580	6590	6599	6609	6618	1	2	3	4	5	6	7	8	9
46	6628	6637	6646	6656	6665	6675	6684	6693	6702	6712	1	2	3	4	5	6	7	7	8
47	6721	6730	6739	6749	6758	6767	6776	6785	6794	6803	1	2	3	4	5	5	6	7	8
48	6812	6821	6830	6839	6848	6857	6866	6875	6884	6893	1	2	3	4	4	5	6	7	8
49	6902	6911	6920	6928	6937	6946	6955	6964	6972	6981	1	2	3	4	4	5	6	7	8

(Continued)

LOGARITHMS TABLE

N	0	1	2	3	4	5	6	7	8	9	1	2	3	4	5	6	7	8	9
50	6990	6998	7007	7016	7024	7033	7042	7050	7059	7067	1	2	3	3	4	5	6	7	8
51	7076	7084	7093	7101	7110	7118	7126	7135	7143	7152	1	2	3	3	4	5	6	7	8
52	7160	7168	7177	7185	7193	7202	7210	7218	7226	7235	1	2	2	3	4	5	6	7	7
53	7243	7251	7259	7267	7275	7284	7292	7300	7308	7316	1	2	2	3	4	5	6	6	7
54	7324	7332	7340	7348	7356	7364	7372	7380	7388	7396	1	2	2	3	4	5	6	6	7
55	7404	7412	7419	7427	7435	7443	7451	7459	7466	7474	1	2	2	3	4	5	5	6	7
56	7482	7490	7497	7505	7513	7520	7528	7536	7543	7551	1	2	2	3	4	5	5	6	7
57	7559	7566	7574	7582	7589	7597	7604	7612	7619	7627	1	2	2	3	4	5	5	6	7
58	7634	7642	7649	7657	7664	7672	7679	7686	7694	7701	1	1	2	3	4	4	5	6	7
59	7709	7716	7723	7731	7738	7745	7752	7760	7767	7774	1	1	2	3	4	4	5	6	7
60	7782	7789	7796	7803	7810	7818	7825	7832	7839	7846	1	1	2	3	4	4	5	6	6
61	7853	7860	7768	7875	7882	7889	7896	7903	7910	7917	1	1	2	3	4	4	5	6	6
62	7924	7931	7938	7945	7952	7959	7966	7973	7980	7987	1	1	2	3	3	4	5	6	6
63	7993	8000	8007	8014	8021	8028	8035	8041	8048	8055	1	1	2	3	3	4	5	5	6
64	8062	8069	8075	8082	8089	8096	8102	8109	8116	8122	1	1	2	3	3	4	5	5	6
65	8129	8136	8142	8149	8156	8162	8169	8176	8182	8189	1	1	2	3	3	4	5	5	6
66	8195	8202	8209	8215	8222	8228	8235	8241	8248	8254	1	1	2	3	3	4	5	5	6
67	8261	8267	8274	8280	8287	8293	8299	8306	8312	8319	1	1	2	3	3	4	5	5	6
68	8325	8331	8338	8344	8351	8357	8363	8370	8376	8382	1	1	2	3	3	4	4	5	6
69	8388	8395	8401	8407	8414	8420	8426	8432	8439	8445	1	1	2	2	3	4	4	5	6
70	8451	8457	8463	8470	8476	8482	8488	8494	8500	8506	1	1	2	2	3	4	4	5	6
71	8513	8519	8525	8531	8537	8543	8549	8555	8561	8567	1	1	2	2	3	4	4	5	5
72	8573	8579	8585	8591	8597	8603	8609	8615	8621	8627	1	1	2	2	3	4	4	5	5
73	8633	8639	8645	8651	8657	8663	8669	8675	8681	8686	1	1	2	2	3	4	4	5	5
74	8692	8698	8704	8710	8716	8722	8727	8733	8739	8745	1	1	2	2	3	4	4	5	5
75	8751	8756	8762	8768	8774	8779	8785	8791	8797	8802	1	1	2	2	3	3	4	5	5
76	8808	8814	8820	8825	8831	8837	8842	8848	8854	8859	1	1	2	2	3	3	4	5	5
77	8865	8871	8876	8882	8887	8893	8899	8904	8910	8915	1	1	2	2	3	3	4	4	5
78	8921	8927	8932	8938	8943	8949	8954	8960	8965	8971	1	1	2	2	3	3	4	4	5
79	8976	8982	8987	8993	8998	9004	9009	9015	9020	9025	1	1	2	2	3	3	4	4	5
80	9031	9036	9042	9047	9053	9058	9063	9069	9074	9079	1	1	2	2	3	3	4	4	5
81	9085	9090	9096	9101	9106	9112	9117	9122	9128	9133	1	1	2	2	3	3	4	4	5
82	9138	9143	9149	9154	9159	9165	9170	9175	9180	9186	1	1	2	2	3	3	4	4	5
83	9191	9196	9201	9206	9212	9217	9222	9227	9232	9238	1	1	2	2	3	3	4	4	5
84	9243	9248	9253	9258	9263	9269	9274	9279	9284	9289	1	1	2	2	3	3	4	4	5
85	9294	9299	9304	9309	9315	9320	9325	9330	9335	9340	1	1	2	2	3	3	4	4	5
86	9345	9350	9355	9360	9365	9370	9375	9380	9385	9390	1	1	2	2	3	3	4	4	5
87	9395	9400	9405	9410	9415	9420	9425	9430	9435	9440	0	1	1	2	2	3	3	4	4
88	9445	9450	9455	9460	9465	9469	9474	9479	9484	9489	0	1	1	2	2	3	3	4	4
89	9494	9499	9504	9509	9513	9518	9523	9528	9533	9538	0	1	1	2	2	3	3	4	4
90	9542	9547	9552	9557	9562	9566	9571	9576	9581	9586	0	1	1	2	2	3	3	4	4
91	9590	9595	9600	9605	9609	9614	9619	9624	9628	9633	0	1	1	2	2	3	3	4	4
92	9638	9643	9647	9652	9657	9661	9666	9671	9675	9680	0	1	1	2	2	3	3	4	4
93	9685	9689	9694	9699	9703	9708	9713	9717	9722	9727	0	1	1	2	2	3	3	4	4
94	9731	9736	9741	9745	9750	9754	9759	9763	9768	9773	0	1	1	2	2	3	3	4	4
95	9777	9782	9786	9791	9795	9800	9805	9809	9814	9818	0	1	1	2	2	3	3	4	4
96	9823	9827	9832	9836	9841	9845	9850	9854	9859	9863	0	1	1	2	2	3	3	4	4
97	9868	9872	9877	9881	9886	9890	9894	9899	9903	9908	0	1	1	2	2	3	3	4	4
98	9912	9917	9921	9926	9930	9934	9939	9943	9948	9952	0	1	1	2	2	3	3	4	4
99	9956	9961	9965	9969	9974	9978	9983	9987	9997	9996	0	1	1	2	2	3	3	3	4

CHAPTER 1

Some Basic Concepts of Chemistry

Chapter Outline

■ Chemistry and its Importance ■ Physical Quantities and their Measurements ■ Matter ■ Laws of Chemical Combinations ■ Avogadro's Law ■ Mass ■ Mole Concept ■ Equivalent Weight ■ Mole Fraction ■ Chemical Equation and Stoichiometry of Chemical Reactions

Chemistry and its Importance

Chemistry is the science of molecules and their transformations that deals with the study of matter, its composition, the changes that matter undergoes and the relation between changes in composition and energy. Chemistry plays a vital and centralized role in Science. It has a vital role in fulfiling human needs for food, health care products, life saving drugs, etc. Cancer treatment may become curable by using cis-platin taxol, etc. AZT (Aziodothymidine) is a boon for AIDS victims. There is no substitute of antiseptics like Detol, insecticides like DDT and BHC and antipyretics like Paracetamol in our life even today.

Physical Quantities and their Measurements

In order to describe and interpret the behaviour of chemical species, we not only require chemical properties but also few physical properties. Physical properties are mass, length and temperature time electric current.

Further to express the measurement of any physical quantity we require its numerical value as well as its unit. Hence, the magnitude of a physical quantity can be given as Magnitude of physical quantity = Its numerical value × Unit.

Table 1.1 SI Units

Measure	Unit
Length (l)	Metre (m)
Mass (m)	Kilogram (kg)
Time (t)	Second (s)
Temperature (T)	Kelvin (K)
Current (i)	Ampere (A)
Intensity (I)	Candela (Cd)
Amount of Substance (n)	Mole (mol)

Table 1.2 Derived Units

Measure		Derivation
Concentration (C or S)	=	$\frac{\text{Mass of solute}}{\text{Mass of solution}}$ $\frac{\text{mol}}{\text{m}^3} = \text{mol m}^{-3}$
Volume (V)	= =	Length × Height × Breadth $\text{m} \times \text{m} \times \text{m} = \text{m}^3$
Density (d)	=	$\frac{\text{Mass}}{\text{Volume}} = \frac{\text{kg}}{\text{m}^3} = \text{kg m}^{-3}$
Velocity (v)	=	$\frac{\text{Distance}}{\text{Time}} = \frac{\text{m}}{\text{sec}} = \text{m sec}^{-1}$
Acceleration (a)	= =	$\frac{\text{Change in velocity}}{\text{Time}}$ $\frac{\text{m sec}^{-1}}{\text{sec}} = \text{m sec}^{-2}$
Force (F)	= =	Mass × Acceleration = m × a kg m sec^{-1} = Newton (N)
Pressure (P)	= =	$\frac{\text{Force}}{\text{Area}} = \frac{\text{kg m sec}^{-2}}{\text{m}^2}$ $\text{kg m}^{-1}\text{ sec}^{-2}$ = Pascal (Pa)
Work (W)	= =	Force × Displacement = F × d $\text{kg m}^2\text{ sec}^{-2}$ = Joule

FACTS TO REMEMBER

- Plane angle (Radian, that is, 'rad')
- Solid angle (Steradian, that is, 'str')

Few Prefixes Used for Subsidiary Units

Sub multiples

1 micro (μ) = 10^{-6}

1 nano (n) = 10^{-9}

1 femto (f) = 10^{-15}

1 atto (a) = 10^{-18}

1 zepto (z) = 10^{-21}

I yocta (y) = 10^{-24}

1 Giga (G) = 10^{9}

1 Tetra (T) = 10^{12}

1 Exa (E) = 10^{18}

1 Zetta (Z) = 10^{21}

1 Yotta (Y) = 10^{24}

1 litre = 10^{-3} m^3 = 1 dm^3

1 atmosphere= 760 mm or torr

= 101.325 Pa or Nm^{-2}

1 bar = 10^5 Nm^{-2} = 10^5 Pa

1 calorie = 4.184 joule

1 eV (electron volt) = 1.602×10^{-19} joule

1 joule = 10^{-7} erg

So, 1 eV = 1.602×10^{-12} erg

1 cal > 1 J > 1 erg > 1 eV

'Barn' is a unit of area to measure the cross section of nucleus.

1 Barn = 10^{-28} m^2 ≈ 10^{-24} cm^2

Precision and Accuracy

- The measurements are considered accurate when the average value of different measurements is closer to the actual value. An individual measurement is considered more accurate when it differs slightly from the actual value.
- When the values of different measurements are close to each other as well as to the average value, such measurements are called **precise.**
- In fact, precision is simply the measurement of reproductability of an experiment.

Uncertainty in measurement and significant figures

There are some uncertainties in values during measurement of matter. In order to make accurate measurements, we use significant figures.

The total number of digits in a number including the last digit with uncertain value is known as the number of significant figures, for example, 14.3256 ± 0.0001 has six significant figures.

Rules to determine significant numbers

- All non-zero digits as well as the zeros present between the non-zero digits are significant, for example, 6003 has four significant figures.
- Zeros to the LHS of the first non-zero digit in a given number are not significant figures, for example, 0.00336 has only three significant figures.
- In a number ending with zeros, if the zeros are present at right of the decimal point then these zeros are also significant figures, for example, 33.600 has five significant figures.
- Zeros at the end of a number without a decimal are not counted as significant figures, for example, 12600 has just three significant figures.
- The result of division or multiplication must be reported to the same number of significant figures as possessed by the least precise term, for example, $3.331 \times 0.011 = 0.036641 \approx 0.037$.
- The result of subtraction or addition must be reported to the same number of significant figures as possessed by the least precise term, for example, 5.1 + 7.21 + 8.008 > 20.318 ≈ 20.32.

Rounding-off non-significant figures

Rounding-off non-significant figures means dropping of the uncertain or non-significant digits in a number. It is possible as follows:

- If the rightmost digit to be rounded-off is > 5, then the preceding number is increased by one, for example, 3.17 is rounded off to 3.2.
- If the rightmost digit to be rounded-off is < 5, then the preceding number is kept unchanged, for example, 5.12 is rounded off to 5.1.
- If the rightmost digit to be rounded-off is equal to 5, the preceding number is kept as such in case of an even value. However, in case of an odd value it is increased by one, for example, 4.45 is rounded-off to 4.4; 5.35 is rounded off to 5.4.

Exponential notation or scientific notation

In case a number ends in zeros that are not to the right of decimal point it is not essential that zeros are significant. For example, 290 has 2 or 3 significant figures and 19500 has 3, 4 or 5 significant figures.

This confusion can be removed when the values are expressed in terms of scientific notations, for example, 19500 can be written as 1.95×10^4 (3 significant figures), 1.950×10^4 (4 significant figures), 1.9500×10^4 (5 significant figures). In this kind of notation, every number can be written as

$N \times 10^n$
Here,
n = Integer,
N = Number with non-zero digit to the left of the decimal point.

For example, 0.00069 can be expressed as 6.9×10^{-4} (2 significant figures).

MATTER

Any species having mass and occupying space is known as matter. It can exist in the three physical states, namely, solid, liquid and gas.

Pencil, air and water, justify the physical states and are all composed of matter.

- At the bulk level or macroscopic level, we can further classify matter as mixtures or pure substances.

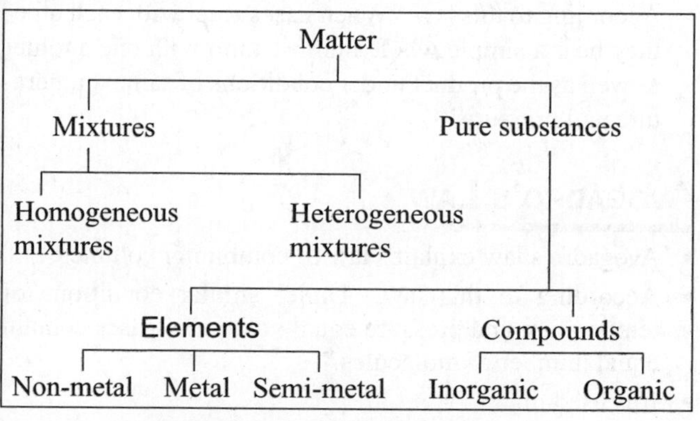

FIGURE 1.1 Classification of Matter

Mixture

A mixture is composed of two or more substances which are known as its components or constituents (in any ratio). The components of the mixture can be separated with the help of physical separation methods like filtration, crystallization, distillation.

- A mixture is further classified into two categories—homogeneous and heterogeneous.
- In a homogeneous mixture, all the components undergo complete mixing forming a uniform composition as, air or sugar solution.
- In a heterogeneous mixture, the composition formed due to the mixing of components is not entirely uniform like in the case of grains mixed with dust, etc.

Pure Substance

Pure substances have fixed compositions and their constituents cannot be separated by using simple physical methods of separation.

- A pure substance can be further classified into an element or a compound.
- An **element** is composed of one type of particle which could either be atoms or molecules. Na, Cu and Ag have only one type of atoms.
- A **compound** is formed by the combination of two or more atoms or different elements. For example, H_2O, CO_2.

Dalton's Atomic Theory

An atom is the smallest particle of an element which is neutral in nature, retains all the properties of the element and takes part in a chemical reaction. The word atom was introduced by Dalton (*alamos* means undivided).

The Dalton's atomic theory was proposed by Dalton on the basis of laws of chemical combination.

Main assumptions

- Matter (of any type) is composed of atoms.
- An atom is the smallest, fundamental, undivided particle. (Building block of any species).
- An atom can neither be created nor destroyed and cannot be transformed into atom of other element.
- Atoms of an element have similar size, energy and properties while atoms of different element differ in these aspects.
- Atoms combine in whole number ratios to form a molecule, therefore, a molecule is the smallest identity that exists individually.

Modern view about atom

According to modern view:

- An atom is divisible into other smaller particles which are known as subatomic particles. It can also combine in non-whole number ratio as in the case of non-stoichiometric compounds (Berthollide compounds) like $Fe_{0.93}O$.

- Atoms of same element also differ in mass and mass related properties as in the case of isotopes.
- A chemical reaction involves rearrangement of atoms.

Molecule

The term molecule was introduced by Avogadro. It is the smallest particle (identity) of matter that can exist independently and retains all the properties of the substance. Normally the diameter of the molecules is in the range of 4–20 Å and the molecular mass is between 2–1000.

- In case of macromolecules, the diameter is in the range of 50–250 Å and the molecular weight may be in lakhs.

Berzelius Hypothesis

According to the Berzelius hypothesis, "Equal volumes of all the gases contain same number of atoms under the similar conditions of temperature and pressure."

This hypothesis on application to law of combining volume confirms that atoms are divisible which is in contrary to Dalton's theory.

Laws of Chemical Combinations

Law of Conservation of Mass

- Law of conservation of mass was proposed by Lavoisier in 1774.
- It was verified by Landolt.
- According to this law, "In a chemical change, the total mass of the products is equal to the total mass of the reactants, that is, mass is neither created nor destroyed." For example, when a solution with calculated weight of $AgNO_3$ and NaCl is mixed, white precipitates of AgCl are formed while $NaNO_3$ remains in solution. The weight of the solution remains the same before and after this experiment.
- It is not applicable to nuclear reactions.

Law of Constant Composition or Law of Definite Proportion

- Law of constant composition was proposed by Proust in 1779.
- It was verified by Star and Richards.
- According to this law, "A chemical compound always contains same elements combined together in same proportion by mass." For example, NaCl extracted from sea water or achieved from deposits will have 23 g Na and 35.5 g of chlorine in its one mole.
- It is not applicable to non-stoichiometric compounds like $Fe_{0.93}O$.

Law of Multiple Proportion

- Law of multiple proportion was proposed by Dalton in 1804.
- It was verified by Berzilius.
- According to this law, "Different weights of an element that combine with a fixed weight of another element bear a simple whole number ratio." For example, in case of CO, and CO_2 weight of oxygen which combines with 12 g of carbon is in 1:2 ratio.
- It is applicable when same compound is prepared from different isotopes of an element. For example, H_2O, D_2O.

Law of Reciprocal Proportion

- Law of reciprocal proportion was proposed by Richter in 1792.
- It was verified by Star.
- According to this law, "When two different elements undergo combination with same weight of a third element, the ratio in which they combine will either be same or some simple multiple of the ratio in which they combine with each other."
- It is also known as **Law of equivalent proportion** which states "Elements always combine in terms of their equivalent weight."

Law of Combining Volume

- Law of combining volume was proposed by Gay Lussac.
- It applies to gases.
- According to this law, "When gases react with each other they bear a simple whole number ratio with one another as well as the product under conditions of same temperature and pressure."

Avogadro's Law

- Avogadro's law explains law of combining volumes.
- According to this law, "Under similar conditions of temperature and pressure equal volume of gases contain equal number of molecules."
- It is used in:
  1. Deriving molecular formula of a gas
  2. Determining atomicity of a gas
  3. Deriving a relation
     Molecular mass = 2 × Vapour Density
     (M = 2 × V.D.)
  4. Deriving the gram molecular volume
- Avogadro number (N_0 or N_A) = 6.023×10^{23}.

- Avogadro number of gas molecules occupies 22.4 litre or 22400 mL or cm^3 volume at STP.
- The number of molecules in 1 cm^3 of a gas at STP is equal to Loschmidt number, that is, 2.68×10^{19}.
- Reciprocal of Avogadro number is known as **avogram.**

MASS

Mass can be expressed in terms of atoms or molecules as follows:

Atomic Mass

Atomic mass is the relative mass of an atom which shows the number of times an atom is heavier than $\frac{1}{12}$ mass of C-12.

$$\text{Atomic Mass} = \frac{\text{Mass of one atom of an element}}{\frac{1}{12}\text{ mass of one C-atom}}$$

- The atomic mass of any element expressed in grams is called g.a.m. (gram atomic mass) or gram atom.
- A gram atom has number of atoms of the element.

  Atomic mass = E × V

  Here, E = Equivalent weight

  V = Valency
- $\text{Atomic mass} = \frac{6.4}{\text{Specific heat in calories}}$

  This is known as **Dulong Petit's Law.**

Atomic Mass Unit

The quantity of 1/12 mass of an atom of C-12 represents it and it is abbreviated as a.m.u

$$1\text{ A.m.u} = \frac{1.99\times10^{-23}}{12} = 1.66\times10^{-24}$$

$$\text{Atomic Mass} = \frac{\text{Mass of one atom of an element}}{1\text{ A.m.u}}$$

Here 1.99×10^{-23}g is wt. of one C-12-atom.

Average Atomic Mass:

$$(\text{At. Mass})\text{ Av.} = \frac{M_1\times a + M_2\times b + M_3\times C}{a+b+c}$$

Here M_1, M_2 and M_3 are masses of isotopes and a, b, c are their percentage ratio.

Note: Rest methods are discussed in chapter on Solution.

Molecular Mass

Molecular mass represents the total mass of a molecule, that is, number of times a molecule is heavier than $\frac{1}{12}$ weight of C-12 atom or $\frac{1}{16}$ weight of one O-atom.

- It is non-variable.

Determination of molecular mass

Vapour density method

Mol. mass = 2 × V.D.

$$\text{V.D.} = \frac{W\times 22400}{\text{Volume at STP (in mL)}}$$

Here, W = Weight of substance in g

V.D. = Vapour density

Graham's diffusion method

$$\frac{r_1}{r_2} = \sqrt{\left(\frac{M_2}{M_1}\right)}$$

Here r_1 and r_2 are rates of diffusion/effusion for two species while M_1 and M_2 are their molecular masses, respectively.

Colligative properties method

$$\pi V = \frac{W}{m}RT$$

Here, π = Osmotic pressure in atm

V = Volume in litre

W = Weight in gram

R = Universal gas constant

T = Given temperature

m = Molar mass

MOLE CONCEPT

- Mole is a unit which represents 6.023×10^{23} particles, atoms, molecules or ions, etc., irrespective of their nature.
- Mole is related to the mass of substance, the volume of gaseous substance and the number of particles.
- Volume of one mole of any gas is equal to 22.4 litres or 22.4 dm^3 at STP. It is known as **molar volume.**
- $\text{Mole} = \frac{W}{M}$

$$= \frac{\text{Wt. of substance in g}}{\text{Molar mass of substance}}\text{ (g.m.m.)}$$

Here, g.m.m. = Gram molecular mass

$$\text{Mole} = \frac{\text{Vol. of substance in litre}}{22.4\text{ L}}$$

Facts to Remember

A mole of any substance (like N_2) stands for:

- 6.023×10^{23} molecules of N_2
- $2 \times 6.023 \times 10^{23}$ atoms of nitrogen
- 28 g of nitrogen
- 22.4 litre of N_2 at STP

To Find Total Number of Identities

Total Number of Molecules = mole (n) × N_A

Total Number of Atoms in one molecule = mole (n) × N_A × Number of atoms present in one molecule

Total Number of Electrons in one molecule = mole (n) × N_A × Number of electron present in one molecule

Total charge on any ion = mole (n) × N_A × charge on one ion × 1.6×10^{-19} C

Equivalent Weight

- Equivalent weight is the weight of an element or a compound which combines with or displaces 1.008 part by weight of H_2 or 8 part by weight of O_2 or 35.5 part by weight of Cl_2.
- Equivalent weight is a number and when it is denoted in grams, it is called **gram equivalent.**
- It depends upon the nature of chemical reaction in which substances take part.

Methods to Find Equivalent Weight

For acids $E = \dfrac{\text{Molecular weight}}{\text{Protocity or basicity of acid}}$

For example, for H_3PO_4, $E = \dfrac{M}{3}$

For H_3PO_3, $E = \dfrac{M}{2}$,

For H_2SO_4, $E = \dfrac{M}{2}$, for $\begin{array}{c}COOH\\ |\\ COOH\end{array} \times 2H_2O$, $E = \dfrac{M}{2}$

For H_3PO_2, $E = \dfrac{M}{1}$,

For bases $E = \dfrac{\text{Molecular weight}}{\text{Acidity or number of } OH^- \text{ ions}}$

For example, for $Ca(OH)_2$, $E = \dfrac{M}{2}$

For $Al(OH)_3$, $E = \dfrac{M}{3}$

For ions $E = \dfrac{\text{Molecular weight}}{\text{Charge on ion}}$

For example, for SO_4^{2-}, $E = \dfrac{M}{2}$

For PO_4^{3-}, $E = \dfrac{M}{3}$

For compounds $E = \dfrac{\text{Molecular weight}}{\text{Valency of cation or anion}}$

For example, for $CaCO_3$, $E = \dfrac{M}{2}$; Na_2CO_3, $E = \dfrac{M}{2}$

For $AlCl_3$, $E = \dfrac{M}{3}$

For redox reactions

$$E = \frac{\text{Molecular weight}}{\text{Total change in oxidation number}}$$

Let's take $KMnO_4$ as an example.

(a) In acidic medium, $E = \dfrac{M}{5}$

$$2K\overset{+7}{Mn}O_4 + 3H_2SO_4 \longrightarrow K_2SO_4 + 2\overset{+2}{Mn}SO_4 + 3H_2O + 5\,[O]$$

5 units change in oxidation number

(b) In basic medium, $E = \dfrac{M}{1}$

$$2K\overset{+7}{Mn}O_4 + 2KOH \longrightarrow 2K_2\overset{+6}{Mn}O_4 + H_2O + [O]$$

one unit change in oxidation number

(c) In neutral medium, $E = \dfrac{M}{3}$

$$2K\overset{+7}{Mn}O_4 + H_2O \longrightarrow 2KOH + 2\overset{+4}{Mn}O_2 + 3[O]$$

3 units change in oxidation number

For acidic salts $E = \dfrac{\text{Molecular weight}}{\text{Number of replaceable H-atoms}}$

For H_3PO_4, for example,

$$Ca(OH)_2 + H_3PO_4 \longrightarrow CaHPO_4 + 2H_2O$$

$E = \dfrac{M}{2}$ (M = Molar Mass of Acid)

Some Other methods

(a) Hydrogen displacement method

$$E = \frac{W \times 11200}{\text{Volume of } H_2 \text{ at NTP}}$$

(b) Oxide formation method

$$E = \frac{\text{Wt. of metal}}{\text{Wt. of oxygen}} \times 8$$

Weight of oxygen = Weight of metal oxide – Weight of metal

(c) Chloride formation method

$$E = \frac{\text{Wt. of metal}}{\text{Wt. of chloride}} \times 35.5$$

Weight of chloride = Weight of metal chloride − Weight of metal

(d) Double decomposition method

$$\frac{\text{Eq. wt. of salt taken}}{\text{Eq. wt. of salt ppt.}} = \frac{\text{Wt. of salt taken}}{\text{Wt. of salt ppt.}}$$

(e) Metal displacement method

$$\frac{E_1}{E_2} = \frac{W_1}{W_2}$$

Mole Fraction

- Mole fraction is the ratio of moles of one component to the total number of moles present in the solution. It is expressed by X, for example, for a binary solution of two components A and B.

$$X_A = \frac{n_A}{n_A + n_B}$$

$$X_B = \frac{n_B}{n_A + n_B}$$

$$X_A + X_B = 1$$

- Mole fraction of solute $(X_2) = \frac{n_2}{n_1 + n_2}$
- Here, n_1 and n_2 represent number of moles of solvent and solute respectively.
- Mole fraction does not depend upon temperature as both the solute and the solvent are expressed by weight.

Chemical Equation and Stoichiometry of Chemical Reactions

- A balanced chemical reaction represents a stoichiometric equation.
- In a stoichiometric equation, the coefficient of reactants and products represents their stoichiometric amounts.
- The reactant which is completely used up during an irreversible reaction is called the **limiting reagent** while the reactant left is called the **excess reagent,** for example, 20 g of calcium is burnt in 32 g of O_2, then Ca is the limiting reagent while O_2 is the excess reagent.
- Stoichiometric calculations help in finding whether the production of a particular substance is economically feasible or not.
- These stoichiometric calculations are of following four types:
  1. Calculations based on weight–weight relationship
  2. Calculations based on weight–volume relationships
  3. Calculations based on volume–volume relationships
  4. Calculations based on weight–volume–energy relationships
- If the amount of the reactant in a particular reaction is known, then the amount of the other substance needed in the reaction or the amount of the product formed in the reaction can be calculated.
- For stoichiometric calculations the following steps must be considered:
  1. A balanced chemical equation using chemical formulas of reactants and products must be written.
  2. Here, the coefficients of balanced chemical equation provide the mole ratio of the reactants and products.
  3. This mole ratio is convertible into weight–weight (w/w) ratio, weight–volume (w/v) ratio or volume–volume (v/v) ratio. These are called percentage by weight, percentage by volume and percentage by strength respectively.

Empirical and Molecular Formulas

Empirical formula It gives the simplest ratio of the number of atoms of different elements present in one molecule of a compound. It does not represent the actual number of atoms of different elements present in one molecule of the compound.

Calculation of the empirical formula It involves these steps one by one:

(i) First determine the percentage composition by weight of each element present in the compound.

$$C\% = \frac{12}{44} \times \frac{\text{Wt. of } CO_2}{\text{Wt. of organic comp.}} \times 100$$

$$H\% = \frac{2}{18} \times \frac{\text{Wt. of } H_2O}{\text{Wt. of organic comp.}} \times 100$$

Duma's method:

$$N\% = \frac{28}{22400} \times \frac{\text{Volume of } N_2 \text{ at S.T.P.}}{\text{Wt. of organic comp.}} \times 100$$

OR

$$N\% = \frac{\text{Volume of } N_2 \text{ at S.T.P.}}{8 \times \text{Wt. of organic comp.}}$$

Kjeldahl's Method

$$N\,\% = \frac{1.4 \times N \times V}{\text{Wt. of org. compound}}$$

Here N = normality of acid the used to neutralize ammonia

V = volume of acid the used to neutralize ammonia

(ii) Now the percentage of each element is divided by its atomic weight to get the relative number of atoms of each element.

(iii) These relative numbers obtained are divided by smallest number to get the simplest ratio numbers.

(iv) If the simplest ratio number is not a whole number it should be multiplied by a suitable integer to get a whole number.

(v) The ratio of these simple whole numbers gives the empirical formula of the compound.

Molecular formula

It shows the actual number of atoms of different elements present in one molecule of the compound.

- $n = \frac{\text{Molecular weight}}{\text{Empirial formula weight}}$
- Molecular formula = empirical formula × n.
- Molecular weight of a substance can be determined by following methods like Vapour density method, Elevation in boiling point method, Depression in freezing point methods, etc.
- The molecular weight of volatile compounds is determined by Victor Mayer's method.

  Molecular Weight = 2 × Vapour density
- Molecular weight = empirical formula wt. × 'n'.
- For some compounds the molecular formula and empirical formula may be same also.

POINTS TO REMEMBER

1. Giorgi introduced MKS system.
2. π has infinite number of significant numbers.
3. 1 mole of $H_2O \neq$ 22400 mL or cc of H_2O (since it is liquid)

   1 mole of H_2O = 18 cc of H_2O (as density of H_2O = 1g/cc)
4. Mass of one mole of e^- = Mass of one $e^- \times N_A$
   $= 9.1 \times 10^{-31} \times 6.02 \times 10^{23}$
   = 0.55 mg
5. 20 carat gold is a mixture having 20 parts by weight of gold and 4 parts by weight of copper.
6. Some substances like $CuSO_4.5H_2O, Na_2CO_3.10H_2O$ have a tendency to lose water in air. These are called **efflorescent substances** and this tendency is called **efflorescence**.
7. Some solid substances like NaOH, KOH, which have a tendency to absorb moisture greatly from air and to get wet are called **deliquescent** and this tendency is called **deliquescence**.
8. Hygroscopic substances like quicklime (CaO) anhydrous P_2O_5, etc., absorb moisture from air.
9. Compounds having similar chemical composition in the same crystalline form are called isomorphs. For example, all alums

   $[M_2SO_4\ M_2\ (SO_4)_3\ 24H_2O]$

   Here, M = Monovalent (K)

   M = Trivalent (Al)

   $FeSO_4\ 7H_2O$ (Green vitriol) and

   $ZnSO_4\ 7H_2O$
10. Different crystalline forms of a substance are called **polymorphs** and this phenomenon is called polymorphism.

    For example, ZnS → Zinc blende

    Wurtzite
11. To find equivalent weight of an acid, Silver Salt Formation Method is used.

    $$\frac{\text{Eq. wt. of RCOOAg}}{108} = \frac{\text{Wt. of RCOOAg}}{\text{Wt. of Ag}}$$

    Eq. wt. of R–COOH
    = Eq. wt. of RCOOAg – 107
12. Equivalent weight (E) = Weight deposited by 96500 coulombs or 1 Faraday.

Important Formula

1. $\text{Mole (n)} = \frac{W}{M} = \frac{V}{22.4\ L} = \frac{N}{N_0}$

   Here, W = Weight

   M = Molecular weight

   N = Number of atoms/molecules

   V = Volume in litre

2. Molar mass (M) = 2 × V.D. (Vapour density)

$$V.D. = \frac{W \times 22400}{\text{Volume at STP (in mL)}}$$

Here, W = Weight of substance in g

3. $\frac{r_1}{r_2} = \sqrt{\left(\frac{m_2}{M_1}\right)}$

Here r_1, r_2 are rates of diffusion for two species while M_1, M_2 are their molecular masses respectively.

4. $\pi V = \frac{W}{M} RT$

Here, π = Osmotic pressure in atm
V = Volume in litre
W = Weight in gram
R = Universal gas constant
T = Given temperature

5. Equivalent weight (E)

$$E = \frac{M}{H^+ \text{ or } OH^- \text{ or Charge}}$$

$$E = \frac{M}{\text{Change in oxidation number}}$$

$$E = \frac{M}{\text{Number of replaceable hydrogen atoms}}$$

$$\frac{E_1}{E_2} = \frac{W_1}{W_2}$$

6. Specific gravity $= \frac{\text{Mass of Liquid}}{\text{Volume of Liquid}}$

7. $M = \frac{\text{\% by mass} \times d \times 10}{\text{M. wt. of solute}}$; $N = \frac{\text{\% by mass} \times d \times 10}{\text{Eq. wt. of solute}}$

$$m = \frac{M \times 100}{(1000 \times d - M \times \text{M. Wt.})}$$

(Here 'd' is density of solution in g cm^{-3}, M is molarity, N is normality and 'm' is molality)

8. Vapour Density $= \frac{\text{Mass of } V_{(L)} \text{ of Gas}}{\text{Mass of } V_{(L)} \text{ of } H_2}$

Solved Numericals

Mole Concept

1. If 1 Faraday was to be 60230 coulombs instead of 96500 coulombs, what will be the charge on an electron?

Solution:

As One mole electron carries 1 Faraday charge.

6.023×10^{23} electrons carry = 60230 C

So 1 electron carries $= \frac{60230 \text{ C}}{6.023 \times 10^{23}}$

$= 1 \times 10^{-19}$ C.

2. If a piece of copper weighs 0.635 g, how many atoms does it contain?

Solution:

Number of moles of Cu in 0.635 g

$= \frac{0.635 \text{ g}}{63.5 \text{ g mol}^{-1}} = 10^{-2}$ mol

1 mole Cu contains 6.023×10^{23} atoms of Cu
10^{-2} mole Cu contains $6.023 \times 10^{23} \times 10^{-2}$

$= 6.023 \times 10^{21}$ atoms of Cu.

3. Calculate the number of atoms of oxygen present in 88 g of CO_2. What would be the mass of CO having the same number of oxygen atoms?

Solution:

Number of moles of $CO_2 = \frac{88 \text{ g}}{44 \text{ g mol}^{-1}}$

= 2 moles

1 mole of CO_2 contains 2 moles of oxygen atoms,
2 moles of CO_2 will contain 4 moles of oxygen atoms.
Number of oxygen atoms $= 4 \times 6.023 \times 10^{23}$
$= 2.5092 \times 10^{24}$

1 mole oxygen atom is present in 1 mole of CO,
4 moles oxygen atoms are present in 4 moles of CO
Its mass is 4 (12 + 16) = 112 g.

4. Calculate the total number of electrons present in 1.6 g of methane.

Solution:

Molecular mass of methane = 16 g mol^{-1}
16 g CH_4 contains 6.02×10^{23} molecules of CH_4 1.6 g CH_4 contains 6.02×10^{22} molecules of CH_4

As one molecule of CH_4 contains (6 + 4) = 10 electrons, 6.02×10^{22} molecules of CH_4 will have $10 \times 6.02 \times 10^{22} = 6.02 \times 10^{23}$ electrons.

5. How many atoms of carbon has a young man given to his bride-to-be if the engagement ring contains 0.5 carat diamond? (1 carat = 200 mg)

Solution:

Mass of diamond (C) $= 0.5 \times 200$ mg
$= 100 \text{ mg} = 100 \times 10^{-3}$ g
$= 0.1$ g

Number of mole of C $= \dfrac{0.1 \text{ g}}{12 \text{ g mol}^{-1}}$
$= 1/120$ mole

Number of C atoms $= \dfrac{1}{120} \times 6.023 \times 10^{23}$
$= 5.02 \times 10^{21}$

6. A mixture of aluminium and zinc weighing 1.67 g was completely dissolved in acid and the evolved 1.69 litres of hydrogen gas was measured at 273 K and one atmosphere pressure. What was the mass of aluminium in the original mixture?

Solution:

Let the mass of aluminium in the sample be 'A' g. The mass of Zn = (1.67 – A) g

The volume of H_2 at NTP given by Al (1)

$= \dfrac{3 \times 22.4 \times A}{2 \times 27}$ L

The volume of H_2 at NTP given by Zn

$= \dfrac{(1.67 - A)\ 22.4}{65.4}$ L (2)

From equation (1) and (2),

$$\frac{3 \times 22.4 \times A}{54} + \frac{(1.67 - A)\ 22.4}{65.4} = 1.69$$

$142.2 \times A = 176.26$

$A = 1.248$ g

7. Find the equivalent mass of H_3PO_4 in the reaction:

$$Ca(OH)_2 + H_3PO_4 \longrightarrow CaHPO_4 + 2H_2O$$

Solution:

As in this reaction only two hydrogen atoms are replaced so its equivalent mass will be given by the following expression:

Equivalent mass of H_3PO_4

$= \dfrac{\text{Molecular mass of } H_3PO_4}{2}$

$= \dfrac{98}{2} = 49$

8. How many years would it take to spend Avogadro number of rupees at the rate of 10 lac rupees per second?

Solution:

Avogadro number $= 6.023 \times 10^{23}$
Total rupees $= 6.023 \times 10^{23}$
Rate of spending = 10 lac rupees/s $= 10^6$ Rs/s
Number of years to spend all the rupees

$= \dfrac{6.023 \times 10^{23} \text{ Rs}}{10^6 \times 60 \times 60 \times 24 \times 365 \text{ Rs/year}}$

$= 1.90988 \times 10^{10}$ years

9. Oxygen is present in a one litre flask at a pressure of 7.6×10^{-10} mm of Hg. Calculate the number of oxygen molecules in the flask at 0°C.

Solution:

Since, PV = nRT

$\dfrac{7.6 \times 10^{-10} \text{ atm} \times 1 \text{ L}}{760}$

$= n \times 0.0821 \text{ L atm K}^{-1} \text{ mol}^{-1} \times 273 \text{ K}$

$n = \dfrac{7.6 \times 10^{-10} \text{ L atm}}{760 \times 0.0821 \text{ L atm K}^{-1} \text{ mol}^{-1} \times 273 \text{ K}}$

$= \dfrac{10^{-12} \text{ mol}}{22.41}$

Number of molecules $= (6.02 \times 10^{23} \text{ mol}^{-1}) \times \dfrac{10^{-12} \text{ mol}}{22.41}$

$= 2.68 \times 10^{10}$

10. 2.68×10^{-3} moles of a solution containing an ion A^{n+} required 1.61×10^{-3} moles of MnO_4^- for the oxidation of A^{n+} to AO_3^- in an acidic medium. What is the value of n?

Solution:

1.61×10^{-3} M $KMnO_4 \equiv 2.68 \times 10^{-3}$ M solution of A^{n+}

M/5 $KMnO_4 = \dfrac{2.68 \times 10^{-3} \text{ M}}{1.61 \times 10^{-3} \text{ M}} \times \dfrac{M}{5}$

$= 0.33$ M solution of A^{n+}

$0.33 \text{ M} = \dfrac{M}{5 - n}$

$5 - n = \dfrac{1}{0.33} = 3$

$n = 2$

Concentration Terms

11. 50 mL of 10 N H_2SO_4, 25 mL of 12 N HCl and 40 mL of 5 N HNO_3 are mixed and the volume of the mixture is made 1000 mL by adding water. Find the normality of the resulting solution.

Solution:

$N_1V_1 + N_2V_2 + N_3V_3 = N_RV_R$

$10 \times 50 + 12 \times 25 + 5 \times 40 = N_R \times 1000$

$500 + 300 + 200 = N_R \times 1000$

$N_R = 1$

Hence, the normality of resulting solution is 1.

12. How many millilitres of 0.5 M H_2SO_4 are needed to dissolve 0.5 g of copper (II) carbonate?

Solution:

$N_1 V_1 = N_2 V_2$

N_1 = Normality of $H_2SO_4 = 0.5 \times 2 = 1$ N

V_1 = Volume of H_2SO_4

N_2 = Normality of copper (II) carbonate

$= \dfrac{0.5 \times 2}{123.5}$ N

V_2 = Volume of copper (II) carbonate

= 1000 ml

So, $1.0 \times V_1 = \dfrac{0.5 \times 2}{123.5} \times 1000$

Therefore, $V_1 = 8.09$ mL.

13. 50 litres of water containing $Ca(HCO_3)_2$ when converted into soft water required 22.2 g $Ca(OH)_2$. Calculate the amount of $Ca(HCO_3)_2$ present per litre of hard water.

Solution:

Reaction

$$\underset{162\text{ g}}{Ca(HCO_3)_2} + \underset{74\text{ g}}{Ca(OH)_2} \longrightarrow 2CaCO_3 + 2H_2O$$

As 74 g $Ca(OH)_2$ reacts with 162 g of $Ca(HCO_3)_2$

So 22.2 g $Ca(OH)_2$ will react with

$\dfrac{162 \times 22.2}{74}$

$= 48.6$ g $Ca(HCO_3)_2$

50 L water contains = 48.6 g $Ca(HCO_3)_2$

1 L water contains $= \dfrac{48.6}{50}$

$= 0.972$ g L^{-1}

14. The formula weight of an acid is 82. In a titration, 100 cm^3 of a solution of this acid containing 39.0 g of the acid per litre were completely neutralized by 95.0 cm^3 of aqueous NaOH containing 40.0 g of NaOH per litre. What is the basicity of the acid?

Solution:

Normality of NaOH = 1

Normality of acid $= \dfrac{1 \times 95}{100} = 0.95$

Suppose the equivalent mass of the acid is E.

$\dfrac{39}{E} = 0.95$

$E = 41$

Therefore, basicity = 82/41 = 2.

15. One g of impure Na_2CO_3 is dissolved in water and the solution is made upto 250 mL. To 50 mL of this made up solution, 50 mL of 0.1 N HCl is added and the mixture after shaking well required 10 mL of 0.16 N sodium hydroxide solution for complete neutralization. Calculate the per cent purity of the sample of Na_2CO_3.

Solution:

Strength of the Na_2CO_3 solution = 4 g L^{-1}

Suppose the normality of Na_2CO_3 solution = N_x

As after mixing Na_2CO_3 and HCl solution, NaOH solution is added, so, according to the normality equation

$50 \times N_x + 0.16 \times 10 = 50 \times 0.1$

$N_x = 0.068$ N

Strength (g L^{-1}) = Normality × Equivalent mass

$= 0.068 \times 53 = 3.6$ g L^{-1}

So, purity of $Na_2CO_3 = \dfrac{3.6 \times 100}{4}$

= 90%.

16. Calculate the volume of water to be added to a 100 mL of 5N solution to make it 0.01 N.

Solution:

According to normality equation,

$N_1 V_1 = N_2 V_2$

$0.01 \times V_1 = 5 \times 100$

$V_1 = \dfrac{5 \times 100}{0.01} = 50000$ mL

So, volume of water to be added = 50000 – 100

= 49900 mL

= 49.9 L.

17. A small amount of $CaCO_3$ completely neutralizes 525 mL of 0.1 N HCl and no acid is left in the end. After converting all calcium chloride to $CaSO_4$, how much plaster of paris can be obtained?

Solution:

525 mL of 0.1 N HCl = 525 mL of 0.1 N $CaCl_2$
= 525 mL of 0.1 N plaster of paris

Molecular mass of plaster of paris = 145

Equivalent mass of plaster of paris $= \frac{145}{2}$
= 72.5

Mass of plaster of paris in 525 mL of 0.1 N solution

$$= \frac{N \times E \times V}{1000}$$

$$= \frac{0.1 \times 72.5 \times 525}{1000}$$

= 3.806 g

Calculations Based on Reactions

18. Metallic tin in the presence of HCl is oxidized by $K_2Cr_2O_7$ solution to stannic chloride. What volume of decinorrnal dichromate solution would be reduced by 1 g of Sn?

Solution:

$$\underset{3 \times 118.7\ g}{3Sn} + \underset{2 \times 294\ g}{2K_2Cr_2O_7} + 28\ HCl \longrightarrow 3SnCl_4 + 4KCl + 4CrCl_3 + 14H_2O$$

$K_2Cr_2O_7$ required for 1 g of Sn $= \frac{2 \times 294}{3 \times 118.7}$
= 1.65 g

19. How many grams of CaO are required to neutralize 852 g of P_4O_{10}?

Solution:

The reaction is as follows:

$$6CaO + P_4O_{10} \longrightarrow 2Ca_3(PO_4)_2$$

852 g P_4O_{10} ≡ 3 mol P_4O_{10}

As 1 mole of P_4O_{10} neutralizes 6 moles of CaO

3 moles of P_4O_{10} will neutralize 18 moles of CaO

So, mass of CaO = 18 × 56 = 1008 g.

20. Find the weight of iron which will be converted into its oxide by the action of 18 g of steam.

Solution:

The reaction is

$$3Fe + 4H_2O \longrightarrow Fe_3O_4 + 4H_2$$

As 4 moles steam reacts with 3 moles Fe

So 1 mole (18 g) steam reacts with 3/4 moles Fe

$= \frac{3}{4}$ mole × 56 g mol^{-1}

= 42 g Fe.

21. The mineral haematite is Fe_2O_3. Haematite ore contains unwanted material called gangue in addition of Fe_2O_3. If 5.0 kg of ore contains 2.78 kg of Fe, what per cent of ore is Fe_2O_3?

Solution:

2Fe	≡	Fe_2O_3
2 × 55.85 g		59.7 g

2 × 55.85 g Fe is present in 159.7 g Fe_2O_3

2.78 kg Fe is present in

$$= \frac{159.7\ g \times 2.78\ kg}{2 \times 55.85\ g}$$

= 3.97 kg Fe_2O_3

As 5 kg ore contains = 3.97 kg Fe_2O_3

So, 100 kg ore contains $= \frac{3.97 \times 100}{5}$
= 79.4 kg Fe_2O_3

Thus, percentage of Fe_2O_3 in ore = 79.4%.

22. What should be the weight of $NaNO_3$ to make 50 mL of an aqueous solution so that it contains 70 mg Na mL^{-1}?

Solution:

Molecular mass of $NaNO_3$ = 23 + 14 + 3 × 16
= 85 g mol^{-1}

23 mg Na is present in 85 mg of $NaNO_3$

70 mg Na is present in $= \frac{85 \times 70}{23}$
= 258.7 mg $NaNO_3$

1 mL solution contains 258.7 mg $NaNO_3$

50 mL solution contains

$$\frac{258.7\ mg \times 50\ mL}{1\ mL}$$

= 13935 mg
= 13.935 g.

29. Suppose the two carbonates are MCO_3 and M^1CO_3

As M = 13.6% by wt. so W_{Sample} = 2.5 g.

$$W_m = \frac{13.6}{100} \times 2.5 = 0.34\ g$$

$$n_{CO_2} = \frac{W_{CO_2}}{44} = \frac{1.32}{44} = 0.03$$

$$W_{CO_3^{2-}} = 0.03 \times 60 = 1.8\ g$$

$$\text{Hence \% of M} = \frac{2.5 - 0.34 - 1.8}{2.5} \times 100 = 14.4\%$$

CHAPTER-END EXERCISES

PRACTICE QUESTIONS – I

1. The number of significant figures for the three numbers 193 cm, 0.193 cm, 0.0193 cm are:
   (a) 2, 3 and 4 respectively
   (b) 3, 3 and 3 respectively
   (c) 3, 5 and 4 respectively
   (d) 3, 4 and 4 respectively

2. The equivalent weight of an acid is equal to:
   (a) Molecular weight/acidity
   (b) Molecular weight/basicity
   (c) Molecular weight × basicity
   (d) Molecular weight × acidity

3. Avogadro's number is the number of molecules present in:
   (a) 1 litre of molecule
   (b) 1 g of molecule
   (c) Gram molecule mass
   (d) 1 g atom

4. A mole of any substance is related to:
   (a) Number of particles
   (b) Volume of gaseous substances
   (c) Mass of a substance
   (d) All of these

5. 25 mL of a solution of barium hydroxide on titration with 0.1 molar solution of hydrochloric acid gave a titre value of 35 mL. The molarity of barium hydroxide solution was:
   (a) 0.07 (b) 0.14
   (c) 0.28 (d) 0.35

6. Number of atoms in 560 g of Fe (atomic mass 56 g mol^{-1}) is:
   (a) Twice that of 70 g N
   (b) Half that of 20 g H
   (c) Both are correct
   (d) None is correct

7. The number of oxygen atoms in 6.4 g of SO_2 is:
   (a) 6×10^{23} (b) 11×10^{23}
   (c) 12×10^{23} (d) 1.2×10^{23}

8. What volume of hydrogen gas at 273 K and 1 atm pressure will be consumed in obtaining 21.6 g of elemental boron (atomic mass = 10.8) from the reduction of boron trichloride by hydrogen?
   (a) 89.6 L (b) 67.2 L
   (c) 44.8 L (d) 22.4 L

9. The number of moles of KCl in 1000 mL of 3 molar solution is:
   (a) 2 (b) 3
   (c) 4 (d) 6

10. The specific heat of metal is 0.16. Its atomic weight is:
    (a) 1.6 (b) 16
    (c) 32 (d) 48

11. The correct relationship between molecular mass and vapour density is:
    (a) V.D. = 2M (b) $V.D. = \frac{M}{2}$
    (c) $M = (V.D.)^{\frac{1}{2}}$ (d) $V.D. = M^2$

12. How many moles of potassium chlorate should be decomposed completely to obtain 67.2 litres of oxygen at STP?
    (a) 1 (b) 2
    (c) 3 (d) 4

13. How many grams of phosphoric acid is required to complete neutralize 120 g of sodium hydroxide?
    (a) 0.98 (b) 98
    (c) 89 (d) 49

14. The hydrated salt $Na_2CO_3\ n\ H_2O$ undergoes 63% loss in mass on heating and becomes anhydrous. The value of *n* is:
    (a) 4 (b) 6
    (c) 8 (d) 10

15. The vapour density of a mixture having NO_2 and N_2O_4 is 27.6. The mole fraction NO_2 in the mixture is:
    (a) 1.6 (b) 0.8
    (c) 2.4 (d) 0.6

16. Among the following pairs of compounds, the one that illustrates the law of multiple proportions is:
    (a) Cu and $CuSO_4$ (b) CuO and Cu_2O
    (c) H_2S and SO_2 (d) NH_3 and NCl_3

17. The value of 1 amu is equal to:
(a) 1.66×10^{-24} g (b) 12.00×10^{-24} g
(c) 1.992×10^{-24} g (d) 1.0 g

18. Normality of 0.04 M H_2SO_4 is:
(a) 0.02 N (b) 0.01 N
(c) 0.04 N (d) 0.08 N

19. Which among the following is the heaviest?
(a) One mole of oxygen
(b) One molecule of sulphur trioxide
(c) 100 amu of uranium
(d) 44 g of carbon dioxide

20. The number of e^- present in 3.6 ml drop of water with a density of 1 g/ml as?
(a) $2N_A$ (b) N_A (c) $N_A/2$ (d) $4N_A$

21. A boy drinks 500 mL of 9% glucose solution. The number of glucose molecules he has consumed are [mol. wt. of glucose = 180].
(a) 0.5×10^{23} (b) 1.0×10^{23}
(c) 1.5×10^{23} (d) 2.0×10^{23}

22. The pollution of SO_2 in air is 10 ppm by volume. The volume of SO_2 per litre of air is:
(a) 10^{-2} mL (b) 10^{-3} mL
(c) 10^{-4} mL (d) 10^{-6} mL

23. The molarity of pure water is:
(a) 55.56 M (b) 5.56 M
(c) 1.0 M (d) 0.01 M

24. The number of grams of a dibasic acid (molecular weight 200) present is 100 mL of its aqueous solution to give decinormal strength is:
(a) 1 g (b) 2 g
(c) 3 g (d) 4 g

25. Normality of 0.3 M H_3PO_4 solution is:
(a) 0.1 N (b) 0.45 N
(c) 0.6 N (d) 0.9 N

26. 2 g of O_2 at NTP occupies the volume:
(a) 1.4 L (b) 2.8 L
(c) 11.4 L (d) 3.2 L

27. Which has maximum number of oxygen atoms?
(a) 1 g of O
(b) 1 g of O_2
(c) 1 g of O_3
(d) All have same number of O-atoms

28. 7.5 g of a gas occupies 5.6 litres as STP. The gas is:
(a) CO (b) NO
(c) CO_2 (d) N_2O

29. 50 g of calcium carbonate was completely burnt in air. What is the weight (in grams) of the residue?
(a) 28 (b) 2.8
(c) 46 (d) 4.8

30. At STP the density of a gas (mol. wt. = 45) in g/L is:
(a) 11.2 (b) 1000
(c) 2 (d) 22.4

31. One mole of a substance present in 1 kg of solvent. The correct statement regarding above solution is:
(a) Strength by weight
(b) Molar concentration
(c) Molal concentration
(d) Normality

32. How many moles of acidified $FeSO_4$ can be completely oxidized by one mole of $KMnO_4$?
(a) 20 (b) 10
(c) 5 (d) 0.5

33. A compound possess 8% sulphur by mass. The least molecular mass is:
(a) 200 (b) 400
(c) 155 (d) 355

34. The vapour density of ozone is:
(a) 24 (b) 16
(c) 48 (d) 72

35. The weight of one molecule of a compound $C_{60}H_{122}$ is:
(a) 1.3×10^{-20} g
(b) 5.01×10^{-21} g
(c) 3.72×10^{23} g
(d) 1.4×10^{-21} g

36. 1000 g calcium carbonate solution contains 10 g carbonate. The concentration of solution is:
(a) 10 ppm (b) 100 ppm
(c) 1000 ppm (d) 10,000 ppm

37. One mole of CH_4 contains:
(a) 4.0 g atoms of hydrogen
(b) 3.0 g atom of carbon
(c) 6.02×10^{23} atoms of hydrogen
(d) 1.81×10^{23} molecules of CH_4

38. Number of moles of a solute per kilogram of a solvent is called:
(a) Normality
(b) Formality
(c) Molality
(d) Molarity

39. The maximum number of molecules is present in:
(a) 15 L of O_2 gas at STP
(b) 10 L of H_2 gas at STP
(c) 1.5 g of H_2 gas
(d) 5 g of CO_2 gas

40. If we consider that $\frac{1}{6}$, in place of $\frac{1}{12}$, mass of carbon atom is taken to be the relative atomic mass unit, the mass of one mole of a substance will:
(a) Decrease twice
(b) Increase two fold
(c) Remain unchanged
(d) Be a function of the molecular mass of the substance

41. The incorrect statement for 14 g of CO is:
(a) It occupies 2.24 litre at NTP
(b) It corresponds to 0.5 mol of CO
(c) It corresponds to same mol of CO and N_2
(d) It corresponds to 3.01×10^{23} molecules of CO

42. Area of nuclear cross section is measured in 'Barn'. It is equal to:
(a) 10^{-28} m^2 (b) 10^{-18} m^2
(c) 10^{-8} m^2 (d) 10^{-34} m^2

43. Which of the following statement is correct?
(a) 1 mole of electrons weighs 5.4 mg
(b) 1 mole of electrons weighs 5.4 kg
(c) 1 mole of electrons weighs 0.54 mg
(d) 1 mole of electrons has 1.6×10^{-19} C of charge

44. Which of the following pairs of gases contain equal number of molecules?
(a) CO_2 and NO_2
(b) CO and $(CN)_2$
(c) NO and CO
(d) N_2O and CO_2

45. The samples of NaCl are produced when Na combines separately with two isotopes of chlorine Cl^{35} and Cl^{37}. Which law is illustrated?
(a) Law of constant volume
(b) Law of multiple proportions
(c) Law of reciprocal proportions
(d) None of these

46. Which of the following is the odd one with regard to mass?
(a) 1 g atom of sulphur
(b) 0.5 moles of CO_2
(c) 1 mole of O_2
(d) 3×10^{23} molecules of SO_2

47. A breakfast cereal in advertised to contain 110 mg of sodium per 100 g of the cereal. The per cent of sodium in the cereal is:
(a) 0.110%
(b) 0.0110%
(c) 11.0%
(d) 0.22%

48. Express 145.6 L of chlorine in terms of gram moles.
(a) 6.5 g moles
(b) 4.5 g moles
(c) 0.65 g moles
(d) 9.5 g moles

49. The number of significant figures in 306.45 and 40440 are respectively:
(a) 4, 5 (b) 5, 5
(c) 5, 4 (d) 4, 6

50. The quantity of $\frac{PV}{K_BT}$ represents the:
(a) Molar mass of a gas
(b) Number of molecules in a gas
(c) Mass of gas
(d) Number of moles of a gas

51. Which is the correct order of micro, nano, femto and pico here?
(a) Micro ≤ Nano ≤ Pico ≤ Femto
(b) Pico ≤ Femto ≤ Nano ≤ Micro
(c) Femto ≤ Pico ≤ Nano ≤ Micro
(d) Femto ≤ Nano ≤ Micro ≤ Pico

52. Find the number of atoms present in 0.016 g of methane.
(a) 0.5 N_0 (b) 0.005 N_0
(c) N_0 (d) 1.6 N_0

53. 15 litre atmosphere is equal to:
(a) 1.515×10^8 erg
(b) 15.15×10^9 erg
(c) 1.515×10^{10} erg
(d) 15.15×10^{12} erg

54. If equal moles of water and urea are taken in a vessel what will be the mass percentage of urea in the solution?
(a) 22.086 (b) 11. 536
(c) 46.146 (d) 23.076

PRACTICE QUESTIONS – II

55. Mixture X = 0.02 mol of $[Co(NH_3)_5SO_4]Br$ and $[Co(NH_3)_5Br]SO_4$ was prepared in 2 litre of solution.
1 litre of mixture X + excess $AgNO_3 \longrightarrow Y$
1 litre of mixture X + excess $BaCl_2 \longrightarrow Z$
Number of moles of Y and Z are:
(a) 0.02, 0.01 (b) 0.01, 0.01
(c) 0.01, 0.02 (d) 0.02, 0.02

56. To neutralize completely 20 mL of 0.1 M aqueous solution of phosphorus acid, the volume of 0.1 M aqueous KOH solution required is:
(a) 10 mL (b) 40 mL
(c) 60 mL (d) 80 mL

57. The amount of zinc required to produce 224 mL of H_2 at STP on treatment with dilute H_2SO_4 will be (Zn = 65):
(a) 65.0 g (b) 0.65 g
(c) 6.35 g (d) 0.065 g

58. 6.02×10^{20} molecules of urea are present in 100 mL of its solution. The concentration of urea solution is:
(a) 0.02 M (b) 0.001 M
(c) 0.01 M (d) 0.1 M

59. When 18 g of glucose is dissolved in 180 g of water then the mole fraction of glucose is:
(a) 0.0099 (b) 0.0999
(c) 0.9999 (d) 0.9111

60. The weight of 1×10^{22} molecules of $CuSO_4{\cdot}2H_2O$ is:
(a) 42.42 g (b) 41.42 g
(c) 44.44 g (d) 48.94 g

61. Density of a 2.05 M solution of acetic acid in water is 1.02 g/mL. The molality of the solution is:
(a) 1.14 mol kg^{-1} (b) 3.28 mol kg^{-1}
(c) 2.28 mol kg^{-1} (d) 0.44 mol kg^{-1}

62. The reaction,
$2Al(s) + 6HCl\,(aq) \longrightarrow 2Al^{3+}\,(aq) + 6Cl^-\,(aq) + 3H_2\,(g)$
(a) 33.6 L H_2 (g) is produced regardless of temperature and pressure for every mole of Al that reacts
(b) 67.2 L H_2 (g) at STP is produced for every mole of Al that reacts
(c) 11.2 L H_2 (g) at STP is produced for every mole of HCl (aq) consumed
(d) 6 L HCl (aq) is consumed for every 3L H_2 (g) produced

63. What is the molarity of H_2SO_4 solution that has a density 1.84 g/cc at 35°C and contains 98% H_2SO_4 by weight?
(a) 1.84 M (b) 81.4 M
(c) 18.4 M (d) 184 M

64. The number of moles of oxygen in one litre of air containing 21% oxygen by volume, in standard conditions, is:
(a) 0.176 mol (b) 0.32 mol
(c) 0.0093 mol (d) 2.20 mol

65. Number of atoms in 4.25 g of NH_3 is approximately:
(a) 6×10^{23} (b) 15×10^{23}
(c) 1.5×10^{23} (d) 2.5×10^{23}

66. The amount of NH_3 formed by the combustion of 2 L of N_2 and 2 L of H_2 is:
(a) 2 L (b) 1 L
(c) 0.66 L (d) 1.33 L

67. The amount of O_2 formed at N.T.P by the complete combustion of 1 kg coal is:
(a) 22.4 L (b) 2240 L
(c) 1866 L (d) 100 L

68. A gaseous mixture contains O_2 and N_2 in the ratio 1:4 by weight. Then the ratio of their number of molecules in the mixture is:
(a) 3:32 (b) 7:32
(c) 1:4 (d) 3:16

69. What is the Mass of 0.5 moles of O_3 molecules?
(a) 16 g (b) 20 g
(c) 48 g (d) 24 g

70. 0.30 g of a volatile liquid displaces 90.0 cm^3 of air at STP in the Victor Meyer's method. The molecular mass of the liquid is:
(a) 54.44 g (b) 34.64 g
(c) 64.76 g (d) 74.66 g

71. A metal oxide having 40% oxygen. The equivalent weight of metal is:
(a) 24 (b) 12
(c) 36 (d) 20

72. If 0.50 mole of $BaCl_2$ is mixed with 0.20 mole of Na_3PO_4, the maximum number of moles of $Ba_3(PO_4)_2$ that can be formed is:
(a) 0.10 (b) 0.20
(c) 0.30 (d) 0.40

73. The equivalent weight of $MnSO_4$ is half its molecular weight when it is converted to:
(a) MnO (b) MnO_4^{2-}
(c) MnO_2 (d) MnO_4^-

74. An aqueous solution of 6.3 g oxalic acid dehydrate is made up to 250 ml. The volume of 0.1 N NaOH required to completely neutralize 10 mL of this solution is:
(a) 4 mL (b) 20 mL
(c) 40 mL (d) 60 mL

75. In the standardization of $Na_2S_2O_3$ using $K_2Cr_2O_7$ by iodometry, the equivalent weight of $K_2Cr_2O_7$ is:
(a) Same as mol. wt. (b) $\frac{\text{Mol. wt.}}{2}$
(c) $\frac{\text{Mol. wt.}}{4}$ (d) $\frac{\text{Mol. wt.}}{6}$

76. The number of molecules in 4.25 g of ammonia is:
(a) 1.5×10^{23} (b) 2.5×10^{23}
(c) 3.5×10^{23} (d) 15×10^{23}

77. The volume in litres of CO_2 liberated at STP, when log of 90% pure limestone is heated completely is:
(a) 2.24 (b) 22.4
(c) 2.016 (d) 20.16

78. The weight of a single atom of oxygen is:
(a) 5.057×10^{23} g (b) 1.556×10^{23} g
(c) 2.656×10^{-23} g (d) 4.538×10^{-23} g

79. From the complete decomposition of 20 g $CaCO_3$ at STP the volume of CO_2 obtained is:
(a) 2.24 L (b) 4.48 L
(c) 44.8 L (d) 48.4 L

80. 5 g of CH_3COOH is dissolved in one litre of ethanol. Suppose there is no reaction between them. If the density of ethanol is 0.789 g/mL then the molality of resulting solution is:
(a) 0.0256 (b) 0.1056
(c) 1.1288 (d) 0.2076

81. 800 g of a 40% solution by weight was cooled. 100 g of solute precipitated. The percentage composition of remaining solution is:
(a) 31.4% (b) 57.6%
(c) 45.8% (d) 41.4%

82. 0.25 mol of P_4 molecules contains________atoms.
(a) 1.764×10^{23} (b) 6.02×10^{19}
(c) 6.023×10^{23} (d) 8.086×10^{23}

83. How many grams of CH_3OH would have to be added to water to prepare 150 mL of a solution that is 2.0 M CH_3OH?
(a) 9.6 g (b) 906 g
(c) 4.3×10^2 g (d) 9.6×10^3 g

84. The oxide of an element contains 67.67% of oxygen and the vapour density of its volatile chloride is 79. Equivalent weight of the element is:
(a) 2.46 (b) 3.82
(c) 4.36 (d) 4.96

85. The molar concentration of 20 g of NaOH present in 5 litre of solution is:
(a) 0.1 mol/L (b) 0.2 mol/L
(c) 1.0 mol/L (d) 2.0 mol/L

86. Volume of a gas at NTP is 1.12×10^{-7} cc. The number of molecules in it is:
(a) 3.01×10^{12} (b) 3.01×10^{18}
(c) 3.01×10^{24} (d) 3.01×10^{30}

87. Maximum number of molecules will be in:
(a) 1 g of H_2
(b) 10 g of H_2
(c) 22 g of O_2
(d) 44 g of CO_2

88. How many grams of KCl must be added to 75 g of water to produce a solution with a molality of 2.25?
(a) 1.257 g (b) 125.7 g
(c) 12.57 g (d) 25.14 g

89. The equivalent weight of phosphoric acid (H_3PO_4) in the reaction:

$NaOH + H_3PO_4 \longrightarrow NaH_2PO_4 + H_2O$ is
(a) 89 (b) 98
(c) 59 (d) 29

90. 2.76 g of silver carbonate (At. mass of Ag = 108) on being heated strongly yields a reduce weight:
(a) 2.32 g (b) 3.32 g
(c) 1.36 g (d) 0.32

91. 4 g caustic soda is dissolved in 100 cc of solution. The normality of solution is:
(a) 1 (b) 0.8
(c) 0.6 (d) 0.10

92. The volume of 1.0 g of hydrogen in litres at NTP is:
(a) 22.4 L (b) 1.12 L
(c) 11.2 L (d) 44.56 L

93. The normality of H_2SO_4 having 50 milliequivalent in 2 L solution is:

(a) 1.025 (b) 1.25
(c) 0.050 (d) 0.025

94. 120 g of urea is present in 5 litre of solution. The active mass of urea is:

(a) 0.06 (b) 0.2
(c) 0.4 (d) 1.4

95. The normality of orthophosphoric acid having purity of 70% be weight and specific gravity 1.54 is:

(a) 11 N (b) 22 N
(c) 33 N (d) 44 N

96. 10^{21} molecules are removed from 200 mg of CO_2. The moles of CO_2 left are:

(a) 2.88×10^{-3} (b) 28.8×10^{-3}
(c) 288×10^{-3} (d) 28.8×10^{3}

97. What is the volume (in litres) of oxygen at STP required for complete combustion of 32 g of CH_4? (mol. wt. of $CH_4 = 16$)

(a) 89.6 (b) 189.6
(c) 98.4 (d) 169.5

98. Two grams of sulphur is completely burnt in oxygen to form SO_2, In this reaction, what is the volume (in litres) of oxygen consumed at STP? (At. wt. of sulphur and oxygen are 32 and 16, respectively)

(a) $\frac{22.414}{16}$ (b) $\frac{16}{22.441}$
(c) $\frac{32.414}{18}$ (d) $\frac{42.414}{16}$

99. How many water molecules are there in one drop of water (volume = 0.0018 mL) at room temperature?

(a) 4.86×10^{17} (b) 6.023×10^{24}
(c) 2.584×10^{19} (d) 6.023×10^{19}

100. 'X' litres of carbon monoxide is present at STP. It is completely oxidized to CO_2. The volume of CO_2 formed is 11.207 litres at STP. What is the value of 'X' in litres?

(a) 32.2 (b) 21.2
(c) 10.2 (d) 11.2

101. Which has maximum number of molecules?

(a) 1.7 g NH_3 (b) 2 g He
(c) 4.6 g NO_2 (d) 3.2 g SO_2

102. One mole of fluorine is reacted with two mole of hot and concentrated KOH. The products formed are KF, H_2O and O_2. The molar ratio of KF, H_2O and O_2 respectively is:

(a) 1:2:1 (b) 1:2:2
(c) 0.5:1:2 (d) 2:1:0.5

103. The total number of protons in 10 g of calcium carbonate is ($N_0 = 6.023 \times 10^{23}$):

(a) 3.01×10^{24} (b) 4.06×10^{24}
(c) 30.1×10^{24} (d) 3.01×10^{23}

104. What is the volume (in litre) of oxygen required at STP to completely convert 1.5 moles of sulphur to sulphur dioxide?

(a) 33.6 (b) 43.6
(c) 11.2 (d) 23.6

105. In acidic medium, dichromate ion oxidize ferrous ion to ferric ion. If the gram molecular weight of potassium dichromate is 294 g, its equivalent weight is:

(a) 19 (b) 49
(c) 99 (d) 294

106. 10 g of $CaCO_3$ is completely decomposed to X and CaO. X is passed into an aqueous solution containing one mole of sodium carbonate. What is the number of moles of sodium bicarbonate formed?

(Mol. wt. of $CaCO_3$ = 100, Na_2CO_3 = 106, $NaHCO_3 = 84$)

(a) 0.010 (b) 0.2
(c) 0.4 (d) 10

107. When 3 g H_2 reacts with 29 g O_2, the amount of H_2O formed is?

(a) 9 g (b) 18 g
(c) 36 g (d) 27 g

108. How many moles are present in 2.5 litre of 0.2 M H_2SO_4?

(a) 0.25 (b) 0.5
(c) 0.75 (d) 0.2

109. How many gram of dibasic acid (mol. wt. 200) should be present in 100 mL of the aqueous solution to give 0.1 normality?

(a) 1 g (b) 1.5 g
(c) 0.5 g (d) 20 g

110. 'X' gram of calcium carbonate was completely burnt in air. The weight of solid residue formed is 28 g. What is the value of 'X' (in grams)?

(a) 50 (b) 100
(c) 150 (d) 200

111. One mole of acidified $K_2Cr_2O_7$ on reaction with excess KI will liberate ...... moles (s) of I_2.

(a) 2 (b) 3
(c) 6 (d) 7

112. 0.59 g of the silver salt of an organic acid (molar mass 210) on ignition gave 0.36 g of pure silver. The basicity of the acid is:

(a) 2 (b) 3
(c) 4 (d) 5

113. $KMnO_4$ (mol. wt. = 158) oxidizes oxalic acid in acidic medium to CO_2 and water as follows:

$$5C_2O_4^{2-} + 2MnO_4^- + 16H^+ \longrightarrow 10CO_2 + 2Mn^{2+} + 8H_2O$$

What is the equivalent weight of $KMnO_4$?

(a) 158 (b) 31.6
(c) 39.5 (d) 79

114. Sodium bicarbonate on heating decomposes to form sodium carbonate, CO_2 and water. If 0.2 moles of sodium bicarbonate is completely decomposed, how many moles of sodium carbonate is formed?

(a) 0.1 (b) 0.2
(c) 0.05 (d) 0.025

115. A purified pepsin was subjected to amino acid analysis. The amino acid present in the smallest amount was lysine, $C_6H_{14}N_2O_2$ and the amount of lysine was found to be 0.431 g per 100 g of protein. The minimum molecular mass of protein is:

(a) 34 u (b) 3400 u
(c) 34,000 u (d) 3400,000 u

116. Find the number of valence electrons present in 0.53 g of Na_2CO_3.

(a) 3.01×10^{23} (b) 1.2046×10^{23}
(c) 12.046×10^{23} (d) 6.023×10^{23}

117. Given below is the graphical representation of volumes occupied by several gases at STP find out which gas/gases is/are not placed at the correct position?

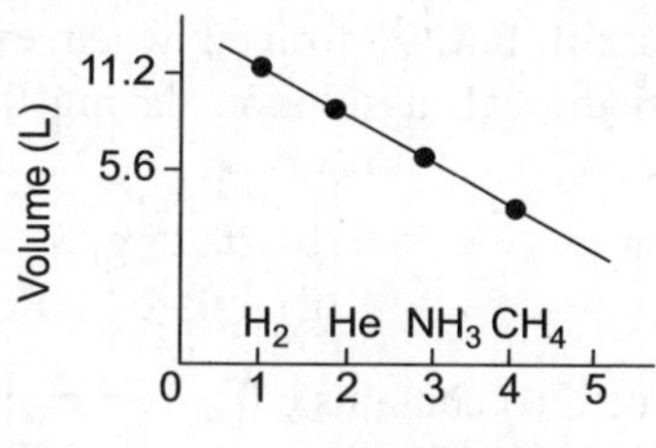

(a) He, NH_3 (b) CH_4, He
(c) NH_3, H_2 (d) NH_3, CH_4

Practice Questions – III

118. If 3.02×10^{19} molecules are removed from 98 mg of H_2SO_4, then the number of moles of H_2SO_4 left are:

(a) 0.1×10^{-3} (b) 5×10^{-4}
(c) 1.2×10^{-4} (d) 1.5×10^{-3}

119. A metal oxide has the formula M_2O_3. It can be reduced by hydrogen to give free metal and water. 0.1595 g of the metal oxide requires 6 mg of hydrogen for complete reduction. What is the atomic weight of metal?

(a) 54.4 (b) 46.56
(c) 55.8 (d) 58.5

120. A chloride of a metal (M) has 65.5% of chlorine. 100 ml of vapour of the chloride of metal at STP weighs 0.72 g. The molecular formula of this metal chloride is:

(a) MCl_3 (b) MCl_4
(c) M_2Cl_3 (d) MCl_5

121. The strength of 0.01 M Na_2CO_3 solution in terms of molality is nearly ______ when the density of the solution is 1.1 g/mL.

(a) 9×10^{-3} (b) 1.8×10^{-3}
(c) 4.5×10^{-3} (d) 1.1×10^{-3}

122. One mole of magnesium in the vapour state absorbed 1200 kJ mol^{-1} of energy. If the first and second ionization energies of Mg are 750 and 1450 kJ mol^{-1} respectively, the final composition of the mixture is:

(a) 86% Mg^+ + 14% Mg^{2+}
(b) 36% Mg^+ + 64% Mg^{2+}
(c) 69% Mg^+ + 31% Mg^{2+}
(d) 31% Mg^+ + 69% Mg^{2+}

123. For the formation of 3.65 g of HCl, what volume of H_2 and Cl_2 are needed at N.T.P?

(a) 1.12 L, 1.12 L (b) 1.12 L, 2.24 L
(c) 3.65 L, 1.83 L (d) 1 L, 1 L

124. The volume of carbon dioxide gas evolved at STP by heating 7.3 g of $Mg(HCO_3)_2$ will be:

(a) 1100 mL (b) 1120 mL
(c) 2230 mL (d) 3240 mL

125. The amount of Zinc (atomic weight = 65) necessary to produce 224 mL of H_2 by the reaction with an acid will be:

(a) 0.65 g (b) 7.6 g
(c) 6.5 g (d) 8.5 g

126. When a mixture of Na_2CO_3 and $NaHCO_3$ was heated at 423 K, 112 ml of CO_2 was formed only. What is the % of Na_2CO_3 here in the mixture:

(a) 84% (b) 16%
(c) 32% (d) 68%

127. The mass of $BaCO_3$ formed when excess of CO_2 is passed through a solution having 0.205 mole of $Ba(OH)_2$ is:

(a) 40.5 g (b) 20.25 g
(c) 81 g (d) 4.05 g

128. 500 mL of NH_3 contains 6.0×10^{23} molecules at STP. How many molecules are present in 100 mL of CO_2 at STP?

(a) 6×10^{23} (b) 1.5×10^{23}
(c) 1.2×10^{23} (d) None of these

129. How many moles of magnesium phosphate, $Mg_3(PO_4)_2$ will contain 0.25 mole of oxygen atoms?

(a) 0.02 (b) 3.125×10^{-2}
(c) 1.25×10^{-2} (d) 2.5×10^{-2}

130. The number of gram molecules of oxygen in 6.02×10^{24} CO molecule‘s’ is:

(a) 10 g molecules (b) 5 g molecules
(c) 1 g molecules (d) 0.5 g molecules

131. The number of oxalic acid molecules in 100 ml of 0.02 N oxalic acid solution is:

(a) 6.023×10^{22} (b) 10^{-3}
(c) 6.022×10^{20} (d) None of these

132. In the reaction,

$4NH_3 (g) + 5O_2 (g) \longrightarrow 4NO (g) + 6H_2O (l)$

When 1 mol of ammonia and 1 mol of O_2 are made to react to completion then:

(a) 1.0 mol of H_2O is produced
(b) 1.0 mol of NO will be produced
(c) All the ammonia will be consumed
(d) All the oxygen will be consumed

133. Pressure in a mixture of 4 g of O_2 and 2g of H_2 confined in a container of 1 litre capacity at 0°C is:

(a) 25.2 atm (b) 35.6 atm
(c) 15.4 atm (d) 48.2 atm

134. What is the volume (in litres) of CO_2 liberated at STP, when 2.12 g of sodium carbonate (mol. wt. = 106) is treated with excess dilute HCl?

(a) 11.2 (b) 2.12
(c) 0.448 (d) 4.26

135. The density (in g mL^{-1}) of a 3.60 M sulphuric acid solution, that is, 29% H_2SO_4 (molar mass = 98 g mol^{-1}) by mass will be:

(a) 1.88 (b) 1.22
(c) 1.45 (d) 1.64

136. The weight of one molecule of a compound $C_{60}H_{122}$ is:

(a) 1.3×10^{-20} g
(b) 5.01×10^{-21} g
(c) 3.72×10^{23} g
(d) 1.4×10^{-21} g

137. Study the following table:

TABLE 1.3

Compound (mol. wt.)	Wt. of compound (in g) taken
1. CO_2 (44)	4.4
2. NO_2 (46)	2.3
3. H_2O_2 (34)	6.8
4. SO_2 (64)	1.6

Which two compounds have least weight of oxygen?

(a) 1 and 2 (b) 1 and 3
(c) 4 and 4 (d) 3 and 4

138. The decomposition of a certain mass of $CaCO_3$ gave 11.2 dm^3 of CO_2 gas at STP. The mass of KOH required to completely neutralize the gas is:

(a) 56 g (b) 28 g
(c) 42 g (d) 20 g

139. The empirical formula of a commercial ion exchange resin is $C_8H_7SO_3Na$. The resin can be used to soften water according to the reaction $Ca^{+2} + 2C_8H_7SO_3Na \rightarrow (C_8H_7SO_3)_2 Ca + 2Na^+$. What would be the maximum uptake to Ca^{+2} by the resin expressed in mole/g resin?

(a) 0.0024 (b) 0.0246
(c) 0.246 (d) 24.6

Answer Keys

1. (b)	2. (b)	3. (c)	4. (d)	5. (a)	6. (c)	7. (d)	8. (b)	9. (b)	10. (b)
11. (b)	12. (b)	13. (b)	14. (d)	15. (b)	16. (b)	17. (a)	18. (d)	19. (d)	20. (a)
21. (c)	22. (a)	23. (a)	24. (a)	25. (d)	26. (a)	27. (d)	28. (b)	29. (a)	30. (c)
31. (c)	32. (c)	33. (b)	34. (a)	35. (d)	36. (d)	37. (a)	38. (c)	39. (a)	40. (a)
41. (d)	42. (a)	43. (c)	44. (d)	45. (d)	46. (b)	47. (a)	48. (a)	49. (b)	50. (b)
51. (c)	52. (b)	53. (b)	54. (d)	55. (b)	56. (b)	57. (b)	58. (c)	59. (a)	60. (b)
61. (c)	62. (c)	63. (c)	64. (c)	65. (c)	66. (d)	67. (c)	68. (b)	69. (d)	70. (d)
71. (b)	72. (a)	73. (c)	74. (c)	75. (d)	76. (a)	77. (c)	78. (c)	79. (b)	80. (b)
81. (a)	82. (c)	83. (a)	84. (b)	85. (a)	86. (a)	87. (d)	88. (c)	89. (b)	90. (a)
91. (a)	92. (c)	93. (d)	94. (c)	95. (c)	96. (a)	97. (a)	98. (a)	99. (d)	100. (d)
101. (b)	102. (d)	103. (a)	104. (a)	105. (b)	106. (b)	107. (d)	108. (b)	109. (a)	110. (a)
111. (b)	112. (b)	113. (b)	114. (a)	115. (c)	116. (b)	117. (a)	118. (b)	119. (c)	120. (a)
121. (a)	122. (c)	123. (a)	124. (b)	125. (a)	126. (b)	127. (a)	128. (c)	129. (b)	130. (b)
131. (c)	132. (d)	133. (a)	134. (c)	135. (b)	136. (d)	137. (c)	138. (b)	139. (a)	

Hints and Explanations for Selective Questions

2. Equivalent weight of acid

$= \frac{\text{Mol. wt. of acid}}{\text{Basicity}}$

5. Let molarity of $Ba(OH)_2 = M_1$

Normality $= 2M_1$

Molarity of HCl $= 0.1\text{ M} = 0.1\text{ N}$

$N_1V_1 = N_2V_2$

$2M_1 \times 25 = 0.1 \times 35$

$M_1 = 0.07\text{ M}$

6. As $n_{Fe} = \frac{560}{56} = 10,\ n_{N_2} = \frac{70}{14} = 5$

So, number of atoms of Fe are twice that of N-atoms.

7. 6.4 g $SO_2 = 0.1$ mol SO_2

$= 6 \times 10^{22}$ molecules

$= 2 \times 6 \times 10^{22}$ atoms of oxygen

$= 1.2 \times 10^{23}$ atoms.

8. $\underset{2\text{ mol}}{2BCl_3} + \underset{3\text{ mol}}{3H_2} \longrightarrow \underset{2\text{ mol}}{2B} + 6HCl$

21.6 g = 2 mol

21.6 g B = 2 mol B ≅ 3 mol H_2

As PV = nRT

So $V = \frac{nRT}{P} = \frac{3 \times 0.0821 \times 273}{1} = 67.2\text{ L}$

12. $2KClO_3 \longrightarrow 2KCl + 3O_2$

2 moles of $KClO_3$ gives here 3 × 22.4, that is, 67.2 L of O_2.

13. $H_3PO_4 + 3NaOH \longrightarrow Na_3PO_4 + 3H_2O$

That is 98 g react, with 120 g of NaOH.

14. $Na_2CO_3\,nH_2O$. Its weight = 106 + 18 n

Weight of anhydrous $Na_2CO_3 = 106$

% loss in weight $= \frac{18n \times 100}{106 + 18n} = 63$

Hence, n = 10.

15. $V.D._{max}$

$= X_{NO_2}\,(V.D.)_{NO_2} + X_{N_2O_4}\,(V.D.)_{N_2O_4}$

$27.6 = X \times 23 + (1 - X) \times 46$

$X_{NO_2} = 0.8$.

16. As in CuO and Cu_2O the O:Cu is 1:1 and 1:2 respectively. This is law of multiple proportion.

18. For H_2SO_4,

$N = 2 \times M = 2 \times 0.04 = 0.08\text{ N}$

20. Total $e^- = \frac{3.6}{18} \times N_A \times 10 = 2\,N_A$

27. As all have same value of n × N_A × number of atom in one species

28. 7.5 g of gas occupies 5.6 L at STP.
7.5 × 4 g of gas occupies 22 L at STP 30 g is the molecular weight of NO gas.

29. $CaCO_3 \longrightarrow CaO\ (\text{residue}) + CO_2 \uparrow$
100 g ⟶ 56 g
50 g gives 56/2 = 28 g.

32. $5Fe^{2+} + MnO_4^- + 8H^+ \longrightarrow 5Fe^{3+} + Mn^{2+} + 4H_2O$
One mole of $KMnO_4$ = 5 moles of $FeSO_4$

40. 1 atomic mass unit on the scale of 1/6 of C-12 = 2 amu on the scale of 1/12 of C-12.
Numerically, the mass of a substance will become half of the normal scale.

43. 1 mole of electrons weighs
$= 9.1 \times 10^{-31}$ kg $\times 6.023 \times 10^{23}$
$= 54 \times 10^{-8}$ kg
$= 54 \times 10^{-8} \times 1000 = 54 \times 10^{-5}$ g
$= 54 \times 10^{-5} \times 10^3$ mg
$= 54 \times 10^{-2}$ mg
= 0.54 mg.

46. As CO_2 has 22 g weight here while rest have 32 g weight.

47. $110\text{ mg} = 110\text{ mg} \times \dfrac{1\text{ g}}{1000\text{ mg}} = 0.110\text{ g}$

$\%\text{ of sodium} = \dfrac{0.110\text{ g}}{100\text{ mg}} \times 100 = 0.110\%$

52. $\text{Number of atoms} = \dfrac{0.016}{16} \times N_A \times 5A = 0.005\ N_A$

53. As 1 litre atm $= 1.01 \times 10^9$ erg
So 15 litre atm $= 15 \times 1.01 \times 10^9$ erg
$= 15.15 \times 10^9$ erg

54. Weight % of urea
$$= \frac{\text{Weight of urea}}{\text{Weight of urea and } H_2O} \times 100$$
$$= \frac{18}{78} \times 100 = 23.076.$$

55. $\underset{\text{Excess}}{Ag^+} + \underset{0.01\text{ mole}}{Br^-} \longrightarrow \underset{0.01\text{ mole}}{AgBr}$

$\underset{\text{Excess}}{Ba^{2+}} + \underset{0.01\text{ mole}}{SO_4^{2-}} \longrightarrow \underset{0.01\text{ mole}}{BaSO_4}$

56. H_3PO_3 is a dibasic acid (containing two ionisable protons attached to Oxygen directly)
$H_3PO_3 \rightleftharpoons 2H^+ + HPO_4^{2-}$
0.1 M H_3PO_3 = 0.2 N H_3PO_3
0.1 M KOH = 0.1 N KOH
$N_1V_1 = N_2V_2$
(KOH) (H_3PO_3)
$1.1\ V_1 = 0.2 \times 20$
$V_1 = 40$ mL.

58. $N_A = 6.02 \times 10^{23} = 1$ mol
6.02×10^{20} molecules = 0.001 mol in 100 mL (0.1 L) solution
$$\text{Molar concentration} = \frac{\text{Vol}}{\text{Volume in L}}$$
$$= \frac{0.001}{0.1} = 0.01\text{ M.}$$

60. 6.022×10^{23} molecules of $CuSO_4.5H_2O$ (1 mole) have weight = 249.5 g
1×10^{22} molecules of $CuSO_4.5H_2O$ have
$$\text{Weight} = \frac{249.5 \times 1 \times 10^{22}}{6.022 \times 10^{23}}$$
$= 41.42$ g

61. $$m = \frac{\text{Moles of } CH_3COOH}{\text{Wt. of solvent in kg}} = \frac{2.05 \times 1000}{897}$$
$= 2.285$
Wt. of solvent = Wt. of solution – Wt. of solute
$= [1000 \times 1.02 - 20.5 \times 60] = 897$ g.

62. $2Al\ (s) + 6HCl\ (aq) \longrightarrow 2Al^{3+}\ (aq) + 6Cl^-\ (aq) + 3H_2\ (g)$
For each mole of HCl reacted, 0.5 mole H_2 gas is formed at STP.
As 1 mole of an ideal gas occupies 22.4 L at STP.
So Volume of H_2 gas formed at STP per mole of HCl reacted is 22.4 × 0.5 litre.
= 11.2 L

65. $$\text{Number of moles of } NH_3 = \frac{\text{Massing}}{\text{Mol. mass}}$$
$$= \frac{4.25}{17} = 0.25\text{ mol}$$
Number of atoms $= 0.25 \times 6 \times 10^{23}$
$= 1.50 \times 10^{23}$ atoms.

66. $\underset{\substack{12\text{gm} \\ 1000\text{gm}}}{C} + O_2 \xrightarrow{\Delta} \underset{\substack{22.44\text{L} \\ ?}}{CO_2}$

As 12 g 'C' gives $= 22.4$ L CO_2

So 1000 g 'C' gives $= \dfrac{22.4 \times 1000}{12}$

$= 1866.66$ L CO_2

68. $\dfrac{\text{No. of moles of } O_2}{\text{No. of moles of } N_2}$

$= \dfrac{\text{Wt./mol. wt. of } O_2}{\text{Wt./mol. wt. of } N_2}$

$= \dfrac{\frac{1}{32}}{\frac{4}{28}} = \dfrac{1}{32} \times \dfrac{28}{4} = \dfrac{7}{32}$

69. $w = n \times M = 0.5 \times 48$

$= 24g.$

71. $E_M = \dfrac{W_M}{W_O} \times 8 = \dfrac{60}{40} \times 8$

$= 12$

72. $3BaCl_2 + 2Na_3PO_4 \longrightarrow Ba_3(PO_4)_2 + 6NaCl$

0.5 0.2

Limiting reagent is Na_3PO_4 (0.2 mol), which gives 0.1 mol of $Ba_3(PO_4)_2$.

73. $\text{Eq. mass} = \dfrac{\text{Molecular weight}}{\text{Change in oxidation no. of Mn}}$

$= \dfrac{\text{Mol. wt.}}{4-2} = \dfrac{\text{Mol. wt.}}{2}$

(oxidation number of Mn in $MnSO_4 = +2$ and Mn in $MnO_2 = +4$)

74. Equivalents of $H_2C_2O_4 .2H_2O$ in 10 mL = equivalents of NaOH

$\dfrac{\frac{6.3}{2}}{126} \times \dfrac{10}{250} = 0.1 \times V$ (in litre)

$V = 0.04L = 40$ mL.

75. In iodometry, $K_2Cr_2O_7$ liberates I_2 from iodides (NaI or KI) which is titrated with $Na_2S_2O_3$ solution.

$K_2Cr_2O_7 + I^- + H^+ \longrightarrow Cr^{3+} + I_2$

Here, one mole of $K_2Cr_2O_7$ accepts 6 mole of electrons.

$\text{Equivalent weight} = \dfrac{\text{Molecular weight}}{6}$

76. Mol. wt. of ammonia = 14 + (3 × 1) = 17

As wt. of ammonia = 4.25 g

So, 17 g of ammonia contains 6.02×10^{23} atoms

4.25 g of ammonia contain

$= \dfrac{6.02 \times 10^{23} \times 4.25}{17} = 1.5 \times 10^{23}$

77. $CaCO_3 \longrightarrow CaO + CO_2$

100 g $\longrightarrow$ 22.4

10 g 100% pure $\longrightarrow$ 2.24

Incase of 90% purity volume (x)

$X = \dfrac{90 \times 2.24}{100} = 2.016$ L

78. Atomic wt. of oxygen = 16 g

Wt. of 6.02×10^{23} atoms of oxygen is 16 g

So, weight of 1 atom $= \dfrac{16}{6.02 \times 10^{23}}$

$= 2.656 \times 10^{-23}$ g.

87. Number of moles of $CO_2 = \dfrac{44}{44} = 1$

Number of moles of $O_2 = \dfrac{24}{32} = 0.75$

Number of moles of $H_2 = \dfrac{1}{2} = 0.5$

$H_2 = \dfrac{10}{28} = 0.357$

Number of molecules × moles hence CO_2 has max. number of molecules.

88. Here W = 75 g

M = 74.5 m = 2.25 w = ?

As $m = \dfrac{w \times 1000}{M \times W}$

So $w = \dfrac{m \times M \times W}{1000} = \dfrac{2.25 \times 74.5 \times 75}{1000} = 12.57$ g

89. Mol. wt. of H_3PO_4 is 98 and change in its valence = 1.

Eq. wt. of $H_3PO_4 = \dfrac{\text{Mol. wt.}}{\text{Change in valency}}$

$= 98/1 = 98$

90. $Ag_2CO_3 \longrightarrow Ag_2O + CO_2$

276 g 232 g

2.76 g of Ag_2CO_3 shall give residue of $Ag_2O = 2.32$ g.

97. $CH_4 + 2O_2 \longrightarrow CO_2 + 2H_2O$

16 g 2 × 22.4

32 g $\dfrac{2 \times 22.4 \times 32}{16} = 4 \times 22.4$

$= 89.6$ litre of O_2.

98. $S + O_2 \longrightarrow SO_2$

32 g of sulphur burns in 22.4 litre of O_2.

2 g of sulphur burns in $\dfrac{2 \times 22.4}{32}$

$= \dfrac{22.4}{16}$ litre of O_2

99. Volume of one drop of water = 0.0018 mL.
As the density of water is 18 g/cc,
So, 18 mL of water = 1 mole of water
$= 6.02 \times 10^{23}$ molecules of water.
Thus 0.0018 mL of water $= 18 \times 10^{-4}$ g of water molecules of water
$= (6.023 \times 10^{23}) \times 10^{-4}$
$= 6.023 \times 10^{19}$ molecules.

100. $CO + 1/2\,O_2 \longrightarrow CO_2$
22.4 L   22.4 L
X   11.2

$$X = \frac{22.4 \times 11.2}{22.4} = 11.2 \text{ litre}$$

101. 2 g He has maximum molecules here as 'n' (mole) is maximum for it.

102. $F_2 + 2KOH \longrightarrow 2KF + H_2O + 1/2\,O_2$ is
The ratio is 2:1:0.5.

104. $S + O_2 \longrightarrow SO_2$
1 mol requires 22.4 litres of O_2 at STP.
1.5 mol requires $1.5 \times 22.4 = 33.6$ L of O_2 at STP.

105. In acidic medium $2Cr^{6+}$ changes to $2Cr^{3+}$

$$\text{Equivalent wt.} = \frac{\text{Molecular weight}}{\text{Change in odixation state}}$$

$$= \frac{294}{6} = 49$$

106. 10 g i.e., 0.1 mol $CaCO_3$ is decomposed to 0.1 mol CO_2 and 0.1 mol CaO.
$Na_2CO_3 + H_2O + CO_2 \longrightarrow 2NaHCO_3$
1 mol   0.1 mol   0.2
Because here, limiting reagent is CO_2. $NaHCO_3$ is always twice the number of moles of CO_2.

108. $n = M \times V = 0.2 \times 2.5 = 0.5$

110. $CaCO_3\,(s) \longrightarrow CaO\,(s) + CO_2\,(g)$
$100\text{ g} \longrightarrow 56\text{ g}$

$$X = \frac{100 \times 28}{56} = 50 \text{ g}$$

111. $Cr_2O_7^{2-} + 14H^+ + 6e^- \longrightarrow 2Cr^{3+} + 7H_2O$
$6I^- \longrightarrow 3I_2 + 6e^-$
$Cr_2O_7^{2-} + 14H^+ + 6I^- \longrightarrow 2Cr^{3+} + 3I_2 + 7H_2O$
One mole of acidified $K_2Cr_2O_7$ on reaction with excess KI will liberate 3 moles of I_2.

113. $MnO_4^- + 8H^+ + 5e^- \longrightarrow Mn^{2+} + 4H_2O$

$$\text{Eq. wt. of } KMnO_4 \text{ in acidic medium} = \frac{\text{Mol. wt.}}{5} = \frac{158}{5} = 31.6.$$

114. $2NaHCO_3 \longrightarrow Na_2CO_3 + H_2O + CO_2$
2 mol of $NaHCO_3$ on complete decomposition gives 1 mol of Na_2CO_3
So, 0.2 mol of $NaHCO_3$ on complete decomposition gives 0.1 mol of $Na_2CO_3^-$

115. Molecular mass of lysine $C_6H_{14}N_2O_2 = 146$ u
Mass of lysine per gram of macromolecule
$= 0.431/100 = 0.00431$ g
Minimum molecular mass

$$= \frac{\text{Molecular mass of lysine}}{\text{Mass of lysine per gram of macromolecule}}$$

$$= \frac{146 \text{ u}}{0.00431} = 34{,}000 \text{ u}$$

116. Total number of valence electron in Na_2CO_3 $= 2 \times 8 + 24 = 40$
$Na^+ = 8$, $CO_3^{2-} = 24$
Number of valence electron in 106 g of Na_2CO_3 $= 40 \times 6.023 \times 10^{23}$
Number of valence electron in 0.53 g of Na_2CO_3 $= 0.2 \times 6.023 \times 10^{23} = 1.2046 \times 10^{23}$.

118. As moles of H_2SO_4 in 98 mg of H_2SO_4

$$= \frac{1}{98} \times 0.098 = 0.001$$

$$\text{Moles of } H_2SO_4 \text{ removed} = \frac{30.2 \times 10^{20}}{6.023 \times 10^{23}} = 0.5 \times 10^{-3} = 0.0005$$

Hence, moles of H_2SO_4 left $= 0.001 - 0.005$
$= 0.5 \times 10^{-3} = 5 \times 10^{-4}$.

119. The equation
$M_2O_3 + 3H_2 \longrightarrow 2M + 3H_2O$
0.1596 g   0.006 g
0.006 g of H_2 reduces 0.1596 g of M_2O_3
6 g of H_2 reduces

$$= \frac{0.1596}{0.0006} \times 6 \text{ g of } M_2O_3$$

$= 159.6$ g of M_2O_3.

Hence, molecular weight of $M_2O_3 = 159.6$ g
So atomic weight of metal $= 159.6 - 48$
$= 55.8$ g.

120. Molecular weight of the metal chloride

$= \frac{0.72 \times 22400}{100} = 161.28$ g

Weight of chlorine in metal chloride

$= \frac{65.5 \times 161.28}{100} = 105.64$ g

So, Mole atoms of chlorine $= \frac{105.64}{35.5} = 3$

Hence, metal chloride is MCl_3

121. $m = \frac{M \times 1000}{(1000 \times d - M \times \text{molecular wt.})}$

$= \frac{10^{-2} \times 1000}{(1000 \times 1.1 - 10^{-2} \times 106)}$

$= \frac{10}{1100 - 1.06} = \frac{10}{1098.94}$

$= 9 \times 10^{-3}$

122. Number of moles of one g of Mg $= \frac{1}{24}$

$= 0.0417$

1 g of Mg (g) absorbs $= \frac{1200}{24} = 50$ kJ

Energy required to convert Mg (g) to Mg^+ (g)

$= 0.0417 \times 750 = 31.275$ kJ

Remaining energy $= 50 - 31.275$

$= 18.725$ kJ

Number of moles of Mg^{2+} formed $= \frac{18.725}{1450}$

$= 0.013$

Thus remaining Mg^+ will be

$= 0.0417 - 0.013 = 0.0287$

So, % $Mg^+ = \frac{0.0287}{0.0417} \times 100 = 68.82\%$

% $Mg^{2+} = 100 - 68.82 = 31.18\%$

123. $\underset{\substack{1M \\ 22.4L}}{H_2} + \underset{\substack{1M \\ 22.4L}}{Cl_2} \rightleftharpoons \underset{\substack{2M \\ 2\times36.5gm}}{2HCl}$

As 22.4L of H_2, Cl_2 can give = 73g HCl

So 3.65g HCl $= \frac{22.4 \times 3.65}{73} = 1.12$L

124. Wt. of $Mg(HCO_3)_2 = 7.3$ g

$Mg(HCO_3)_2 = MgCO_3 + H_2O + CO_2$

Mol. wt. of $Mg(HCO_3)_2$

$= 24 + (1 + 12 + 48) \times 2 = 146$ g

As 146 g of $Mg(HCO_3)_2$ evolves 22.4 L of CO_2 at STP.

So, 7.3 g of $Mg(HCO_3)_2$ evolves

$\frac{7.3 \times 22.4}{146} = 1.12$ L

125. Atomic weight of Zn = 95 and volume of H_2 = 224 mL

$Zn + H_2SO_4 \longrightarrow ZnSO_4 + H_2$

$\frac{\text{Volume of } H_2}{22400} = \frac{\text{Weight of Zn}}{\text{Atomic weight of Zn}}$

Wt. of Zn $= \frac{224 \times 65}{22400} = 0.65$ g.

126. Let Na_2CO_3 = w g

$NaHCO_3$ = (1–w) g

$\underset{2 \times 84gm}{2NaHCO_3} \xrightarrow{\Delta} Na_2CO_3 + \underset{22.4\,L\,CO_2}{CO_2} + H_2O$

As 168g $NaHCO_3$ gives = 22.4 L CO_2

So (1–w) g $= \frac{22.4\,(1 - w)}{168}$

As $\frac{22.4\,(1 - w)}{168} = 0.112$ L (Given)

on solving w = 0.16g = 16%

127. $\underset{197gm \text{ or } 1 \text{ mole}}{Ba(OH_2)} + CO_2 \longrightarrow \underset{1 \text{ mole}}{BaCO_3} + H_2O$

As 1 mole of $Ba(OH)_2$ = 1 mole $BaCO_3$

So 0.205 of $Ba(OH)_2 = \frac{197 \times 0.205}{1}$

= 40.5g $BaCO_3$

128. $\frac{PV_1}{n_1RT} = \frac{PV_2}{n_2RT}$

$\frac{V_1}{n_1} = \frac{V_2}{n_2}$

$\frac{500}{1} = \frac{100}{n_2}$

$n_2 = \frac{1}{5}$

As 1 mole of CO_2 contains 6.0×10^{23} molecules.

So $\frac{1}{5}$ mole of CO_2 contains 1.2×10^{23} molecules.

129. 8 mole of O ≡ 1 mole $Mg_3(PO_4)_2$

0.25 mole O $= \frac{1 \times 0.25}{8}$

$= 3.125 \times 10^{-2}$ mole $Mg_3(PO_4)_2$

130. 6.02×10^{23} CO molecules = 1 g molecule

6.02×10^{24} CO molecules $= \frac{1 \times 6.02 \times 10^{24}}{6.02 \times 10^{23}}$

= 10 g molecules of CO

Now, 10 g molecules of CO contains oxygen

= 10 g atoms $= \frac{10}{2}$ g molecules.

131. 1000 mL of 0.02 N oxalic acid has

$= 0.02$ g eq

So, 100 mL of 0.02 N oxalic acid has

$= 0.002$ g eq $= 0.001$ mol

$(M = N \times \text{basicity})$

Thus, number of molecules

$= 10^{-3} \times 6.022 \times 10^{23}$

$= 6.022 \times 10^{20}$.

132. According to stoichiometry, they should react as follows:

$$4NH_3 (g) + 5O_2 (g) \longrightarrow 4NO (g) + 6H_2O (l)$$

4 mol	5 mol	4 mol	6 mol
0.8 mol	1 mol	0.8 mol	1.2 mol

In this reaction 1 mole of O_2 and 0.8 mole of NH_3 are consumed. there by indicating complete consumption of O_2.

133. Number of moles of O_2 in mixture

$$= \frac{\text{Weight of } O_2}{\text{Mol. wt. of } O_2} = \frac{4}{32} = \frac{1}{8}$$

Number of moles of H_2 in mixture $= \frac{2}{2} = 1$

Thus total moles in mixture $= 1 + \frac{1}{8} = \frac{9}{8}$

From ideal gas equation PV = nRT

$$P = \frac{9}{8} \times 0.0821 \times 273 = 25.2 \text{ atm}$$

134. $Na_2CO_3 + 2HCl \longrightarrow 2NaCl + H_2O + CO_2$

106 g gives 22.4 L

2.12 g gives $\frac{2.12 \times 22.4}{106} = \frac{22.4}{50}$

$= 0.448$ L

135. Let the density of solution be 'd'

Molarity of solution given = 36

1 litre of solution contains 3.6 moles of H_2SO_4

1 litre of solution contains 3.6×98 g of H_2SO_4

As the solution is 29% by mass.

100 g solution contains 29 g H_2SO_4

100/d mL solution contains 29 g H_2SO_4

100 mL solution contains 3.6×98 g H_2SO_4

$$3.6 \times 98 = \frac{29 \times d}{100} \times 1000$$

$d = 1.22$

Alternatively

$$m = \frac{\% \times d \times 10}{\text{Mol. wt.}}$$

$$d = \frac{3.6 \times 98}{29 \times 10}$$

$= 1.22$

136. As Molecular weight of $C_{60}H_{122}$

$= (12 \times 60) + 122 = 842$.

So, weight of one molecule

$$= \frac{\text{Mol. wt. of } C_{60}H_{12}}{\text{Avogadro's number}} = \frac{842}{6.023 \times 10^{23}}$$

$= 1.4 \times 10^{-21}$ g.

138. Weight of 11.2 dm^3 of CO_2 gas at STP is

$$\frac{44}{2} = 22 \text{ g.}$$

$KOH + CO_2 \longrightarrow KHCO_3$

56 g   44 g

KOH required for complete neutralization of 22 g

$$CO_2 = \frac{56}{44} \times 22 = 28 \text{ g.}$$

Previous Years' Questions

1. Assuming fully decomposed, the volume of CO_2 released at STP on heating 9.85 g of $BaCO_3$ (atomic mass, Ba = 137) will be:

(a) 1.12 L (b) 4.84 L
(c) 2.12 L (d) 2.06 L

[2000]

2. Specific volume of cylindrical virus particle is 6.02×10^{-2} cc/g. Whose radius and length are 7 Å and 10 Å respectively. If $N_A = 6.02 \times 10^{23}$, find molecular weight of virus:

(a) 15.4 kg/mol (b) 1.54×10^4 kg/mol
(c) 4.68×10^4 kg/mol (d) 2.08×10^3 kg/mol

[2001]

3. Molarity of liquid HCl if density of solution is 1.17 g/cc, is:

(a) 32.05 (b) 12.15
(c) 3.05 (d) 22.10

[2001]

4. Percentage of Se in peroxidase anhydrous enzyme is 0.5% by weight (at. wt. = 78.4) then minimum molecular weight of peroxidase anhydrous enzymes is:
(a) 1.568×10^3 (b) 1.568×10^4
(c) 25.68 (d) 4.316×10^4
[2001]

5. Which has maximum number of molecules?
(a) 7 g N_2 (b) 2 g H_2
(c) 18 g NO_2 (d) 16 g O_2
[2002]

6. In Haber's process, 30 L of dihydrogen and 30 L of dinitrogen were taken for reaction which yielded only 50% of the expected product. What will be the composition of gaseous mixture under the aforesaid condition in the end?
(a) 10 L ammonia, 25 L nitrogen, 15 L hydrogen
(b) 20 L ammonia, 20 L nitrogen, 15 L hydrogen
(c) 20 L ammonia, 25 L nitrogen, 25 L hydrogen
(d) 20 L ammonia, 25 L nitrogen, 30 L hydrogen
[2003]

7. The oxidation states of sulphur in the anions SO_3^{2-}, $S_2O_4^{2-}$ and $S_2O_6^{2-}$ follow the order:
(a) $SO_3^{2-} < S_2O_4^{2-} < S_2O_6^{2-}$
(b) $S_2O_4^{2-} < SO_3^{2-} < S_2O_6^{2-}$
(c) $S_2O_4^{2-} < S_2O_6^{2-} < SO_3^{2-}$
(d) $S_2O_6^{2-} < S_2O_4^{2-} < SO_3^{2-}$
[2003]

8. The maximum number of molecules is present in:
(a) 15 L of H_2 gas at STP
(b) 5 L of N_2 gas at STP
(c) 1.5 g of H_2 gas
(d) 5 g of O_2 gas
[2004]

9. The number of moles of $KMnO_4$ reduced by one mole of KI in alkaline medium is:
(a) 2 (b) 1
(c) 5 (d) 6
[2005]

10. The mass of carbon anode consumed (giving only CO_2) in the production of 270 kg of aluminium metal from bauxite by the Hall process is (atomic mass of Al = 27):
(a) 180 kg (b) 270 kg
(c) 145 kg (d) 90 kg
[2005]

11. An element, x has the following isotopic composition;
^{200}X:90%
^{199}X:8.0%
^{202}X:2.0%
The weighted average atomic mass of the naturally occurring element x is closest to:
(a) 200 amu (b) 201 amu
(c) 202 amu (d) 199 amu
[2007]

12. The number of moles of $KMnO_4$ that will be needed to react with one mole of sulphite ion in acidic solution is:
(a) 3/5 (b) 4/5
(c) 2/5 (d) 1
[2007]

13. Concentrated aqueous sulphuric acid is 98% H_2SO_4 by mass and has a density of 1.80 g mL^{-1}. Volume of acid required to make one litre of 0.1 M H_2SO_4 solution is:
(a) 11.10 mL
(b) 16.65 mL
(c) 22.20 mL
(d) 5.55 mL
[2007]

14. How many moles of lead (II) chloride will be formed from a reaction between 6.5 g of PbO and 3.2 g of HCl?
(a) 0.011 (b) 0.029
(c) 0.044 (d) 0.333
[2008]

15. Volume occupied by one molecule of water (density = 1 g cm^{-3}) is:
(a) 3.0×10^{-23} cm^3 (b) 5.5×10^{-23} cm^3
(c) 9.0×10^{-23} cm^3 (d) 6.023×10^{-23} cm^3
[2008]

16. Number of moles of MnO_4^- required to oxidize one mole of ferrous oxalate completely in acidic medium will be:
(a) 7.5 moles (b) 0.2 moles
(c) 0.6 moles (d) 0.4 moles
[2008]

17. 10 g H_2 and 64 g O_2 were filled in a steel vessel and exploded. Amount of water produced in this reaction will be?
(a) 3 mole (b) 4 mole
(c) 1 mole (d) 2 mole
[2009]

18. An increase in equivalent conductance of a strong electrolyte with dilute dilution is mainly due to:
(a) Increase in number of ions
(b) Increase in ionic mobility of ions
(c) 100% ionization of electrolyte at normal dilution
(d) Increase in both i.e., number of ions and ionic mobility of ions
[2010]

19. The number of atoms in 0.1 mole of a triatomic gas is:
(a) 1.8×10^{22} (b) 6.02×10^{23}
(c) 1.806×10^{23} (d) 3.6×10^{23}
[2010]

20. Mole fraction of the solution in a 1 molal aqueous solution is:
(a) 0.0344 (b) 1.77
(c) 0.177 (d) 0.0177
[2011]

21. Find the molar concentration of Na^+ ions in a solution of Na_2CO_3 having 25.3 g Na_2CO_3 in 250 ml.
(a) 0.95 (b) 1.91
(c) 0.475 (d) 19.1
[2011]

22. How many grams of concentrated nitric acid solution should be used to prepare 250 mL of 2.0 M HNO_3? The concentrated acid is 70% HNO_3.
(a) 90 g conc. HNO_3 (b) 70 g conc. HNO_3
(c) 54 g conc. HNO_3 (d) 45.0 g conc. HNO_3
[2013]

23. 6.02×10^{20} molecules of urea are present in 100 mL of its solution. The concentration of solution is:
(a) 0.01 M (b) 0.001 M
(c) 0.1 M (d) 0.02 M
[2013]

24. When 22.4 L of H_2 is mixed with11.2 L of Cl_2 (g) at STP the moles of HCL formed is equal to:
(a) 1 mole of HCl (g)
(b) 2 moles of HCl (g)
(c) 0.5 moles of HCl (g)
(d) 1.5 moles of HCl (g)
[2014]

25. 1 g of Mg is burnt with 0.56 g of oxygen in a closed versel which reactant is left in excess and how much?
(a) Mg 0.16 g
(b) O_2 0.16 g
(c) Mg 0.44 g
(d) O_2 0.28 g
[2014]

26. If Avogadro number Na is changed from 6.022×10^{23} to 6.022×10^{20} mol^- this would change:
(a) The definition of mass in units of grams
(b) The mass of one mole of carbon
(c) The ratio of chemical species to each other in a balanced equation
(d) The ratio of elements to each other in a compound.
[2015]

27. What is the mole fraction of the solute in a 1.00 M aqueous solution?
(a) 0.177 (b) 0.0177
(c) 1.77 (b) 0.0354
[2016]

28. The Number of water molecules is maximum in:
(a) 1.8 g H_2O (b) 18 g H_2O
(c) 18 moles of water (d) 18 molecules of water
[2015]

29. What is the mass of participate formed when 50 ml of 16.9% Solution of $AgNO_3$ is mixed with 50 ml of 5.8% NaCl solution? (Ag = 107.8, N = 14, O = 16)
(a) 7 g (b) 14 g
(c) 3.5 g (d) 28 g
[2015]

30. Suppose the elements X and Y combine to form two compound XY_2 and X_3Y_2. When 0.1 mole of XY_2 weights 10 g and 0.05 mole of X_3Y_2 weight 9 g, the atomic weights of X and Y are?
(a) 20, 30 (b) 30, 20
(c) 40, 30 (d) 60, 40
[2016]

Answer Keys

1. (a)	2. (a)	3. (a)	4. (b)	5. (b)	6. (a)	7. (b)	8. (a)	9. (b)	10. (d)
11. (a)	12. (c)	13. (d)	14. (b)	15. (a)	16. (d)	17. (b)	18. (b)	19. (c)	20. (d)
21. (b)	22. (d)	23. (a)	24. (a)	25. (a)	26. (b)	27. (b)	28. (c)	29. (a)	30. (c)

Hints and Explanations

1. The decomposition of $BaCO_3$ occurs as follows:
$BaCO_3 \rightarrow BaO + CO_2$
197 g   22.4 lit
As 197 g of $BaCO_3$ released CO_2 = 22.4 lit

So 1 g of $BaCO_3$ released $CO_2 = \frac{22.4}{197}$ litre

i.e., 9.85 g of $BaCO_3$ released CO_2

$$= \frac{22.4}{197} \times 9.85 = 1.12 \text{ litre}$$

2. Specific volume (volume of 1 g) cylindrical virus particle = 6.02×10^{-2} cc/g
Radius of virus (r) = 7Å = 7×10^{-8} cm
Length of virus = 10×10^{-8} cm
Volume of virus = $\pi r^2 l$

$$= \frac{22}{7} \times (7 \times 10^{-8})^2 \times 10 \times 10^{-8}$$
$$= 154 \times 10^{-23} \text{ cc}$$

Wt. of one virus particle = $\frac{\text{Volume}}{\text{Specific volume}}$

So mol. wt. of virus = wt. of N_A particle

$$= \frac{154 \times 10^{-23}}{6.02 \times 10^{-2}} \times 6.02 \times 10^{23}$$

= 15400 g /mol = 15.4 kg/mol.

3. As 1 c.c. of the HCl solution contains
= 1.17 g of HCl
So 1000 c.c. of HCl solution will contain

$$= \frac{1.17 \times 1000}{1}$$
= 1170 g of HCl

Hence 36.5 g of HCl in 1000 c.c. = 1 M

So 1170 g of HCl in 1000 c.c. = $\frac{1170}{36.5}$ = 32.05 M

4. Suppose the mol. wt. of enzyme = a
So in 100 g of enzyme wt. of Se = 0.5 g

Hence in 'a' g of enzyme wt. of Se = $\frac{0.5}{100} \times$ 'a'

Hence $78.4 = \frac{0.5 \times a}{100}$ $a = 15680 = 1.568 \times 10^4$

5. As 2 g of H_2 means one mole of H_2, so it contains 6.023×10^{23} molecules.
As rest have less moles than it so they will have less number of molecules.

6.

N_2	+	$3H_2$	$\rightarrow$	$2NH_3$
1 mole		3 mole		2 mole
or 1/2 mole		3/2 mole		1 mole
or $10 \times 1/2 = 5$		$10 \times 3/2 = 15$		$10 \times 1 = 10$

As only 50% ammonia formation is expected so composition of gaseous mixture under the above mentioned condition is as follows:
$H_2 = 30 - 15 = 15$ L
$N_2 = 30 - 5 = 25$ L
$NH_3 = 10$ L

7. As oxidation state of 'S' in
$S_2O_4^{2-} = +3$
$SO_3^{2-} = +4$
$S_2O_6^{2-} = +5$
So increasing order of oxidation state of 'S' is
$S_2O_4^{2-} < SO_3^{2-} < S_2O_6^{2-}$

8. As 15 L of hydrogen will have maximum moles here so it will have maximum number of molecules
i.e., $\frac{15}{22.4} \times 6.02 \times 10^{23}$
$= 4.03 \times 10^{23}$

9. $\overset{+7}{KMnO_4} \xrightarrow{OH^-} K_2\overset{+6}{MnO_4}$
As change in oxidation number of Mn in basic medium is 1. So 1 mole of KI will reduce one mole of $KMnO_4$.

10. As Eq. of Al = Eq. of C
$$\frac{270 \times 10^3}{27/3} = \frac{w}{12/4}.$$
(As $C^0 \rightarrow C^{4+} + 4e$)
On solving, we get
w = 90 kg.

11. Average atomic mass
$$= \frac{\% \text{ atomic wt.} + \% \times \text{atomic wt.}}{100}$$
$$= \frac{200 \times 90 + 199 \times 8 + 202 \times 2}{90 + 8 + 2}$$

$$= \frac{18000+1592+404}{100} = 199.96 \text{ amu}$$

$= 200$ amu.

12. $2MnO_4^- + 6H^+ + 5SO_3^{2-} \longrightarrow 2Mn^{2+} + 5SO_4^{2-} + 3H_2O$

As 5 moles of SO_3^{2-} are oxidized by 2 moles of MnO_4^-.
So 1 mole of SO_3^{2-} is oxidized by 2/5 moles of MnO_4^-.

13. H_2SO_4 is 98% by weight.
Weight of $H_2SO_4 = 98$ g
Weight of solution $= 100$ g

$$\text{Volume of solution} = \frac{\text{Mass}}{\text{Density}} = \frac{100}{1.80} \text{ mL}$$

$= 55.55$ ml $= 0.0555$ L

$$\text{Molarity of solution} = \frac{98}{98 \times 0.0555} \text{ M}$$

$= 18.02$ M

Let V ml of this H_2SO_4 are used to prepare one litre of 0.1 M H_2SO_4.
mM of conc. H_2SO_4 = mM of dil. H_2SO_4

$V \times 18.02 = 1000 \times 0.1$

$$V = \frac{1000 \times 0.1}{18.02} = 5.55 \text{ mL}$$

14. $PbO \quad + \quad 2HCl \longrightarrow PbCl_2 + H_2O$

X mol   2X mol   X mol

$\frac{6.5}{224}$ mol $\frac{3.2}{36.5}$ mol

$= 0.029$ $= 0.087$ mol

Thus, 0.029 moles of lead (II) chloride will be formed from a reaction between 6.5 g of PbO and 3.2 g of HCl.

15. Weight of 6.023×10^{23} molecule of water = 18 g
As volume occupied by 6.023×10^{23} molecule of water (density = 1g cm^{-3}) will be

$$= \frac{18 \text{ g}}{1 \text{ g cm}^{-3}} = 18 \text{ cm}^3 \text{ or ml}$$

So volume occupied by one molecule of water

$$= \frac{18}{6.023 \times 10^{23}} = 2.988 \times 10^{-23}$$

$= 3.0 \times 10^{-23}$ cm^3

16. $[5e + MnO_4^- + 8H^+ \longrightarrow Mn^{2+} + 4H_2O \text{ ..(i)}] \times 2$

$[C_2O_4^{2-} \longrightarrow 2e + 2CO_2 \text{ ....(ii)}] \times 5$

$2MnO_4^- + 16H^+ + 5C_2O_4^{2-} \longrightarrow 2Mn^{2+} + 10CO_2$

As 2 moles of MnO_4^- required to oxidize 5 moles of oxalate

So number of moles of MnO_4^- required to oxidize 1 mole of oxalate = 2/5 = 0.4

17. $H_2 + 1/2\, O_2 \longrightarrow H_2O$
2 g 16 g   18 g
As 2 g H_2 needs = 16 g O_2
So 10 g H_2 needs = 80 g O_2
Now we can say that
64 g O_2 needs 8 g of H_2 So
H_2 left is 2 g
It means here O_2 is the limiting reagent and H_2 is excess reactant.
Now amount of H_2O produced from 64 g O_2 =

$$\frac{18}{16} \times 64 = 72 \text{ g } H_2O \quad n_{H_2O} = \frac{72}{18} = 4 \text{ mole}$$

18. Equivalent conductance increases with dilution due to increase in ionic mobility of ions.

19. Number of atoms $= n \times 3 \times N_A$

$= 0.1 \times 3 \times 6.02 \times 10^{23}$

$= 1.806 \times 10^{23}$

20. $n_{solute} = 1 \quad W = 1 \text{ kg} = 1000 \text{ g}$

$$N_{solvent} = \frac{1000}{18} = 55.56$$

$$X_{solute} = \frac{n}{n+N} = \frac{1}{1+55.56} = 0.0177$$

21. $$\text{As } M = \frac{W \times 1000}{\text{Mol.Wt.} \times V_{ml}} = \frac{25.3 \times 1000}{106 \times 250} = 0.955$$

$[CO_3^{2-}] = 0.955$ m

$(Na^+) = 2 \times 0.955 = 1.91$ m

22. $$W = \frac{M \times \text{Mol}\cdot\text{Wt.}\cdot XV}{1000} = \frac{2 \times 63 \times 250}{1000} = \frac{63}{2} \text{ g}$$

$$\text{As } HNO_3 \text{ is 70\% so } W = \frac{63}{2} \times \frac{70}{100} = 45 \text{ g}$$

23. $$n = \frac{N}{N_A} = \frac{6.02 \times 10^{20}}{6.02 \times 10^{23}} = 10^{-3}$$

$$M = \frac{n}{v} = \frac{10^{-3} \times 1000}{100} = 0.01$$

24. $H_2 + Cl_2 \longrightarrow 2HCl$

22.4   11.2 L   2 mole

1 mole   0.5 mole (L.R.)

As 1 mole Cl_2 = 2 moles of HCl

So, 0.5 mole of Cl_2 = 1 mole of HCl

i.e., 1 mole of HCl (g) is formed.

25. $Mg + 1/2\ O_2 \longrightarrow MgO$

24 g   16 g   40 g

1 g   0.56

As 24 g Mg = 16 g O_2

So 1 g Mg $= \frac{16}{24} = 0.66$ g O_2

Hence O_2 is limiting Reagent

As 16 g O_2 = 24 g Mg

$0.56\ O_2 = \frac{0.56 \times 24}{16} = 0.86$

Hence Mg lost = 1 – 0.86

= 0.16 g

26. As 6.022×10^{23} = 1 mole C = 12g C

So 6.022×10^{20} = 0.001 mole = 0.012 g

Hence mass of carbon changes = = 0.012 g

27. As $m = \frac{n \times 1000}{N \times 18}$

$1 = \frac{n}{N} \times \frac{1000}{18}$

$\frac{n}{N} = \frac{18}{1000}$

$\frac{n}{n+N} = \frac{18}{1000+18} = 0.0177$

28. Number of molecules of directly proportional to number of moles

As 18 mole means → 18 N_A molecules of water which is maximum here

Number of molecules = $n \times N_A$

= 18 N_A (18 mole water)

29. As 16.9% solution means 16.9 g $AgNO_3$ is per cent in 100 ml

So in 50 ml = 8.45 g $AgNO_3$

in 50 ml = 2.9 g NaCl

$AgNO_3 + NaCl \longrightarrow AgCl + NaNO_3$

$\frac{8.45}{169.8}$ $\frac{2.9}{58.5}$ 0   0

0.049   0.049   0   0

Mass of AgCl = 0.049 × 143.5 = 7 g

30. Suppose masses of X and Y are Ax and Ay

then for $XY_2 \rightarrow n_{XY2} = 0.1 = \frac{10}{Ax + 2Ay}$

Ax + 2Ay = 100   (1)

For $XY_2 \rightarrow n_{XY_2} = 0.05 = \frac{9}{3Ax + 2Ay}$

3Ax + 2Ay = 180   (2)

From equation (1) and (2)

On solving Ax = 40 g/mole

Ay = 30 g/mole

NCERT EXEMPLAR

1. Two students performed the same experiment separately and each one of them recorded two readings of mass which are given below. Correct reading of mass is 3.0 g. On the basis of given data, mark the correct option out of the following statements.

Student	Readings	
	(i)	(ii)
A	3.01	2.99
B	3.05	2.95

(a) Results of both the students are neither accurate nor precise.

(b) Results of student A are both precise and accurate.

(c) Results of student B are neither precise nor accurate.

(d) Results of student B are both precise and accurate.

2. A measured temperature on Fahrenheit scale is 200 °F. What will this reading be on Celsius scale?

(a) 40 °C   (b) 94 °C

(c) 93.3 °C   (d) 30 °C

3. What will be the molarity of a solution, which contains 5.85 g of NaCl(s) per 500 mL?

(a) 4 mol L^{-1} (b) 20 mol L^{-1}

(c) 0.2 mol L^{-1} (d) 2 mol L^{-1}

4. If 500 mL of a 5M solution is diluted to 1500 mL, what will be the molarity of the solution obtained?

(a) 1.5 M (b) 1.66 M
(c) 0.017 M (d) 1.59 M

5. The number of atoms present in one mole of an element is equal to Avogadro number. Which of the following element contains the greatest number of atoms?

(a) 4g He (b) 46g Na
(c) 0.40g Ca (d) 12g He

6. If the concentration of glucose ($C_6H_{12}O_6$) in blood is 0.9 g L^{-1}, what will be the molarity of glucose in blood?

(a) 5 M (b) 50 M
(c) 0.005 M (d) 0.5 M

7. What will be the molality of the solution containing 18.25 g of HCl gas in 500 g of water?

(a) 0.1 m (b) 1 M
(c) 0.5 m (d) 1 m

8. One mole of any substance contains 6.022×10^{23} atoms/molecules. Number of molecules of H_2SO_4 present in 100 mL of 0.02M H_2SO_4 solution is ______.

(a) 12.044×10^{20} molecules
(b) 6.022×10^{23} molecules
(c) 1×10^{23} molecules
(d) 12.044×10^{23} molecules

9. What is the mass percent of carbon in carbon dioxide?

(a) 0.034% (b) 27.27%
(c) 3.4% (d) 28.7%

10. The empirical formula and molecular mass of a compound are CH_2O and 180 g respectively. What will be the molecular formula of the compound?

(a) $C_9H_{18}O_9$ (b) CH_2O
(c) $C_6H_{12}O_6$ (d) $C_2H_4O_2$

11. If the density of a solution is 3.12 g mL^{-1}, the mass of 1.5 mL solution in significant figures is _______.

(a) 4.7g (b) 4680×10^{-3}g
(c) 4.680g (d) 46.80g

12. Which of the following statements about a compound is incorrect?

(a) A molecule of a compound has atoms of different elements.

(b) A compound cannot be separated into its constituent elements by physical methods of separation.

(c) A compound retains the physical properties of its constituent elements.

(d) The ratio of atoms of different elements in a compound is fixed.

13. Which of the following statements is correct about the reaction given below:

$$4Fe(s) + 3O_2(g) \rightarrow 2Fe_2O_3(g)$$

(a) Total mass of iron and oxygen in reactants = total mass of iron and oxygen in product therefore it follows law of conservation of mass.

(b) Total mass of reactants = total mass of product; therefore, law of multiple proportions is followed.

(c) Amount of Fe_2O_3 can be increased by taking any one of the reactants (iron or oxygen) in excess.

(d) Amount of Fe_2O_3 produced will decrease if the amount of any one of the reactants (iron or oxygen) is taken in excess.

14. Which of the following reactions is not correct according to the law of conservation of mass.

(a) $2Mg(s) + O_2(g) \rightarrow 2MgO(s)$
(b) $C_3H_8(g) + O_2(g) \rightarrow CO_2(g) + H_2O(g)$
(c) $P_4(s) + 5O_2(g) \rightarrow P_4O_{10}(s)$
(d) $CH_4(g) + 2O_2(g) \rightarrow CO_2(g) + 2H_2O$ (g)

15. Which of the following statements indicates that law of multiple proportion is being followed.

(a) Sample of carbon dioxide taken from any source will always have carbon and oxygen in the ratio 1:2.

(b) Carbon forms two oxides namely CO2 and CO, where masses of oxygen which combine with fixed mass of carbon are in the simple ratio 2:1.

(c) When magnesium burns in oxygen, the amount of magnesium taken for the reaction is equal to the amount of magnesium in magnesium oxide formed.

(d) At constant temperature and pressure 200 mL of hydrogen will combine with 100 mL oxygen to produce 200 mL of water vapour.

Answer Keys

1. (b)	2. (c)	3. (c)	4. (b)	5. (d)	6. (c)	7. (d)	8. (a)	9. (b)	10. (c)
11. (a)	12. (c)	13. (d)	14. (b)	15. (b)					

Hints and Explanations for Selective Questions

1. Average of both students is close to the correct reading. But readings of student A are close to each other and above to the average value.

2. °F = (9/5)° C + 32

3. M = number of moles of solute / volume of solution in L.

5. Number of atoms = Moles × N_A

6. Molar mass = 12 × 6 + 1 × 12+16 × 6 =180

Molarity = 0.90 g L^{-1} / 180 g mol^{-1} = 0.005 M

8. Number of molecules = n × 6.022 × 10^{23}

= 0.002 × 6.022 × 10^{23}

= 12.044 × 10^{20} molecules

12. A compound does not retain the physical or chemical properties of its constituent elements.

14. In non-balanced equation mass of reactant and products are different which is against the law of conservation of mass.

AIIMS ESSENTIALS

Assertion and Reason

In the following questions, two statements (Assertion) A and Reason (R) are given. Mark

(a) if A and R both are correct and R is the correct explanation of A;

(b) if A and R both are correct but R is not the correct explanation of A;

(c) A is true but R is false;

(d) A is false but R is true;

(e) A and R both are false.

1. (A) : One mole of NaCl contains 6.023 × 10^{23} molecules of sodium chloride.

(R) : 58.5 g of NaCl also contains 6.023 × 10^{23} molecules of NaCl.

2. (A) : 22.4 L of N_2 at NTP and 5.6 L O_2 at NTP contain equal number of molecules.

(R) : Under similar conditions of temperature and pressure all gases contain equal number of molecules.

3. (A) : Number of g-molecules of SO_2Cl_2 in 13.5 g of sulphuryl chloride is 0.2.

(R) : Gram molecules is equal to those molecules which are expressed in gram.

4. (A) : In CO molecule 12 parts by mass of carbon combine with 16 parts by mass of oxygen and in CO_2, 12 parts by mass of carbon combine with 32 parts by mass of oxygen.

(R) : When two elements combine separately with a fixed mass or a third element, then the ratio of their masses in which they do so is either the same or whole number multiple of the ratio in which they combine with each other.

5. In the reaction

(A) : $2\ KOH + H_3PO_4 \rightarrow K_2H\ PO_4 + 2H_2O$

Equivalent weight of the acid H_3PO_4 is equal to $\frac{M}{2}$ [M = mol. wt. of H_3PO_4]

(R) : As $E = \frac{M}{n - \text{factor}}$

6. (A) : Molarity of a solution and molality of a solution both change with density
(R) : Density of the solution changes when percentage by mass of solution changes.

7. (A) : The percentage of nitrogen in urea is 46.6%.
(R) : Urea is ionic compound.

8. (A) : 0.28 g of N_2 has equal volume as 0.44 g of another gas at same conditions of temperature and pressure.
(R) : molecular mass of another gas is 44 g mol^{-1}.

9. (A) : In $MnO_4^- + 5Fe^{2+} + 8H^+ \rightarrow Mn^{2+} + 5Fe^{3+} + 4H_2O$, MnO_4^- acts as oxidizing agent and Fe^{2+} acts as reducing agent.
(R) : The reactions involving simultaneous loss or gain of electron among the reacting species are called oxidation reduction reactions.

10. (A) : Equivalent mass of a base which contains one mole of replaceable OH^- ion in a molecule.
(R) : It is the mass of a base which completely reacts with one gram equivalent mass of an acid.

11. (A) : Decomposition of H_2O_2 is a disproportionation reaction.
(R) : H_2O_2 molecule simultaneously undergoes both oxidation and reduction.

12. (A) : On increasing temperature, normality decreases like molarity.
(R) : On increasing temperature, volume increases as a result, normality decreases.

13. (A) : The molality of the solution does not change with change in temperature.
(R) : The molality is expressed in units of moles per 1000 g of solvent.

14. (A) : In the balanced redox reaction x Cu_2O + y NO_3^- + 14 $H^+ \rightarrow 6Cu^{2+} + 2NO + 7H_2O$, the n-factor of Cu_2O and NO_3^- is 2 and 3 respectively.
(R) : Since reciprocal of n-factor's ratio is molar ratio and so, x:y = 3:2

15. (A) : For the reaction $Na_2CO_3 + HCl \rightarrow NaCl + NaHCO_3$ The equivalent weight of Na_2CO_3 is 106.
(R) : Because the n-factor Na_2CO_3 is 1 and equivalent weight = $\dfrac{\text{Molecular weight}}{\text{N-factor}}$

16. (A) : A bottle is labeled as "10 V" of H_2O_2. So its percentage strength is 5%.
(R) : % strength of H_2O_2 is nothing but it is the number of grams of H_2O_2 in 100 mL solution of H_2O_2 and it is related with volume strength as under 1 "vol" of H_2O_2 = 0.303%.

Answer Keys

1. (b)	2. (e)	3. (e)	4. (b)	5. (a)	6. (a)	7. (c)	8. (b)	9. (a)	10. (a)
11. (a)	12. (a)	13. (a)	14. (a)	15. (a)	16. (d)				

CHAPTER 2

Structure of Atoms

Chapter Outline

■ Atom and Atomic Theory ■ Thomson's Atomic Model ■ Rutherford's Atomic Model ■ Electronic Structure of Atoms ■ Spectrum ■ Bohr's Atomic Model ■ De Broglie Equation and Dual Nature Theory ■ Heisenberg's Uncertainty Principle ■ Schrodinger Wave Equation ■ Quantum Numbers ■ Aufbau Principle ■ Pauli Exclusion Principle ■ Hund's Rule of Maximum Multiplicity

Atom and Atomic Theory

- Atom (Greek: *atomos)* means indivisible or uncuttable.
- The words atom and atomic theory were introduced by John Dalton in 1808 while explaning the atomic model.

The following are the main postulates of atomic theory as proposed by Dalton:

- Atom is the smallest, indivisible fundamental particle.
- It can neither be created nor destroyed however it is interconvertible into another atom.
- Atoms undergo chemical combination to form molecules but in a whole number ratio, for example, molecule is the smallest identity that exists individually.
- Atoms of the same element have same mass, size, properties etc., but atoms of different element differ in these properties.

J. J. Thomson introduced significant modifications in the prevalent atomic model. His conclusions are the basis of the modern view of atomic theory, according to which:

- Atom is divisible into sub atomic particles.
- Atoms can also combine in non-whole number ratio in non stoichiometric compounds, for example, $Fe_{0.93}O$.
- Atoms of same element also differ in properties related to mass as in case of isotopes and atoms of different elements have also same mass related properties in case of Isobars.

Sub Atomic Particles

Discussed below are the sub atomic particles or the constituents of atoms.

Electrons ($_{-1}e^0$ or e^-): The electron was discovered by J.J. Thomson during the study of cathode rays (stream of electrons).

The term electron was coined by J. L. Stoney.

Mass: Its mass is $\frac{1}{1837}$ of H-atom 0.00055 amu, 9.1×10^{-31} kg.

Location: It is present in the extra nuclear region of the atom.

Charge: It is confirmed by Miliikan's oil drop experiment and its value is 1.6×10^{-19} Q or -4.8×10^{-10} esu.

Thomson gave the **e/m ratio** or **specific charge ratio** for an electron which equals to 1.76×10^8 C/g

- e/m ratio or specific charge ratio or Thomson ratio follows the order mentioned: $_{-1}e^- >, p^1 > {}_2\alpha^4 > {}_0n^1$ e/m ratio for a neutron is zero as it is a chargeless particle.
- e/m ratio for cathode rays is constant and independent of the nature of gas in discharge tube and nature of cathode or electrodes of tube, that is, electron is universal fundamental particle.
- Mass of one mole of electron is 0.55 mg.
- Mass of moving $e^- = \frac{m_{rest}}{\sqrt{1-(v/c)^2}}$

- Electron can leave its orbit for a maximum duration of 10^{-8} sec.
- Particle nature of electron is confirmed by scintillation effect on ZnS screen.
- Photoelectric effect, emission of β-particle, thermal emission or heating of metal filament confirm the fundamental particle nature of electron.

Density: Its density is 2.17×10^{17} g/cm^3.

Proton ($_1H^1$ or p) The proton was discovered by Goldstein during the study of anode rays or canal rays or +ve rays which originate in the region between cathode and anode in a discharge tube. The term proton was coined by Rutherford.

- For anode rays e/m ratio depends upon nature of gas taken in a discharge tube and it is maximum for hydrogen gas.

Charge: It carries a charge of $+ 1.6 \times 10^{-19}$ C or $+ 4.8 \times 10^{-10}$esu.

Mass: Its mass is same as that of H-atom, that is, 1.673×10^{-27} kg or 1.0072 amu.

- e/m ratio is 9.5×10^4C/g

  1 amu = 1/12 mass of C-atom

  1 amu = 1.66×10^{-27} kg = 931.5 MeV

Neutron ($_0n^1$): The neutron was discovered by Chadwick.

$$_4Be^9 + {_2He^4} \longrightarrow {_6C^{12}} + {_0n^1}$$

Neutron was discovered late because it is chargeless.

- It is heavier than an electron and a proton.
- An isolated neutron is unstable and disintegrates as follows:

$$_0n^1 \longrightarrow {_{-1}e^0} + {_1p^1} + \upsilon$$

- It is stable inside the nucleus but unstable outside it.
- It is the best projectile as it is chargeless.
- The H-atom has no neutron.

Mass Its mass is 1.674×10^{-27} kg or 1.0086 amu.

Density Its density is 1.5×10^{14} g/cc.

Meson or pion: This subatomic particle was discovered by Yukawa in cosmic rays (in 1935) (π^+, π^-).

- These are present in the intermediate region between $10^{-13} - 10^{-15}$cm of atom.
- Mesons hold neutrons and protons together and inter convert them. These are inter nuclear forces or exchange forces.

$$n \rightleftharpoons p + \pi^-$$
$$p \rightleftharpoons n + \pi^+$$
$$p \rightleftharpoons p + \pi^\circ$$
$$n \rightleftharpoons n + \pi^\circ$$

- Neutral pions were discovered by Keemer.
- Now kaons (K^+, K^-, K^0) and eta-meson have also been discovered.
- Mass of $\pi \approx 270 \times$ Mass of electron.
- π mesons are called pions.

Neutrino and antineutrino: They are chargeless, very light particles and their existence was postulated by Fermi and Pauli. Neutrino has an anti-clockwise spin while anti-neutrino has a clockwise spin.

Quarks: Electrons are made up of small particles called quarks. According to Murray Gellmann, quarks have fractional charges, colour and flavour. For example, upward and downward quark etc.

Positron or antielectron: (e^+, β^+ or $_{+1}e^\circ$) These are highly unstable and generate γ-rays on combining with electrons. Positron was discovered by Carl Anderson.

Antiproton (p^-): It was discovered by Segree and Wei-gland by p–p and p–n type of collisions.

v-Particles: These were discovered by Rochester and Butler. These may be v^+, v^- or v^0.

Atomic Number (Z)

Moseley postulated that the frequency of the X-rays was related to the charge present on the nucleus of the atom of the element used as anticathode and found that

$$\sqrt{\upsilon} = a(z - b)$$

where υ is the frequency, z is the nuclear charge and a and b are constants.

a = Probability constant (Depends upon nature of metal)

b = A constant having same values for all lines of X-ray spectrum.

The number of unit positive charges carried by the nucleus of an atom is called the **atomic number** of the element.

$Z = p = e^-$ (for atoms)

$Z = p$ (for ions)

Some facts about atomic number (Z)

- It is always a whole number, and a permanent value for each element.
- It is the serial number of elements in the periodic table, that is, it determines their exact position.
- Periodic properties are related to the atomic number of elements.

Note: Atomic weight (A) = Mass number = n + p

Isotopes

- The concept of isotopes was formulated by Soddy.
- These are atoms of same element with same atomic number (Z) but different mass numbers (A).
- They have same chemical properties.
- They have same number of protons.
- Rate of lighter isotopes > rate of heavy isotope (isotopic effect).
- Isotopes are separated by fractional distillation, thermal diffusion, gaseous diffusion methods.
- Isotopes can be detected by Thomson Parabola method, Aston's mass spectrograph method, Dempster method.
- Some examples of isotopes

$_1H^1$	$_1H^2$	$_1H^3$
	C_{12}	C_{14}
	O_{16}	O_{18}
	Cl_{35}	Cl_{37}

- Heaviest isotopes are radioactive in nature, for example, C-14, O-18 are radioactive isotopes of carbon and oxygen.

Isobars

Isobars are atoms of different elements with same mass number (A) but different atomic numbers (Z).

- They have different number of protons and neutrons but the sum of their protons and neutrons is same.
- They have different chemical properties but physical properties related to mass are same, for example, $_{18}Ar^{40}$, $_{19}K^{40}$, $_{20}Ca^{40}$

Isotones

Isotones are species having different atomic number (Z) and mass number (A) but same number of neutrons.
For example,

	$_6C^{14}$	$_7N^{15}$	$_8O^{16}$
Number of neutrous	14–6 = 8	15–7 = 8	16–8 = 8

Isoelectronics

Isoelectronics are species having same number of electrons, for example, Ne, Na^+ O^{-2}, CH_4, NH_4^+, all have 10 electrons.

Isoesters

Isoesters are species having same number of atoms as well as same number of electrons, that is, all isoesters are isoelectronic species but all isoelectronic species may not be isoesters.

	CN^-	N_2	CO	NO^+
Number of atoms	2	2	2	2
Number of e^-	14	14	14	14

Isomorphs

Isomorphs are compounds that have same crystalline structures. For example, all alums, green vitriol ($FeSO_4.7H_2O$), blue vitriol ($CuSO_4.5H_2O$).

Isodiaphers

Isodiaphers are atoms having same isotopic number or they have same (n–Z) or (A–2Z) value.
For example,

$_9F^{19}$	$_{19}K^{39}$
A = 19	A = 39
Z = 9	Z = 19
n = 10	n = 20
n – Z = 1	n – Z = 1
A – 2Z = 1	A – 2Z = 1

Thomson's Atomic Model

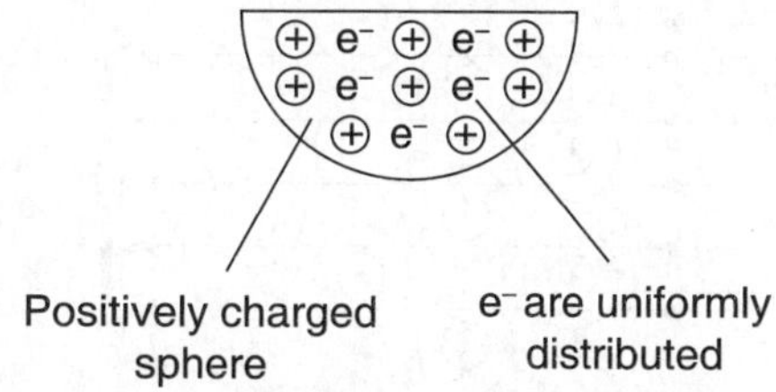

Thomson atomic model was introduced by Thomson.

- According to this model, an atom occupies a spherical shape with a radius of nearly 10^{-10} metre in which the positive charged is uniformly distributed and electrons are embedded uniformly in this positively charged sphere just like raisins are present in pudding.
- The important feature of this model is that the mass of the atom is assumed to be uniformly distribute about the atom.
- The model explains the neutrality of an atom but fails to explain the other aspects relating to the atomic structure.
- The Thomson atomic model was completely rejected by the Rutherford and Mordson experiment.

RUTHERFORD'S ATOMIC MODEL

During an experiment involving α-particle bombardment on a 10^{-4} mm thick gold foil, Rutherford observed that:

1. Most part of the atom is empty as most of the α-particles pass straight through the foil without any deflection.
2. Electrons occupy position in this empty space (extra nuclear region).
3. Centrally located, solid, compact, small part having all positive charge and nearly the whole mass. It is called the nucleus (as only few α-particles get deflected up to a maximum of 180°)
4. The value of radius can be find out as follows:

   $r = 1.3 \times 10^{-13} \times A^{\frac{1}{3}}$ cm.

   Radius of atom $= 10^{-10}$ m or 10^{-8} cm

   Radius of nucleus $= 10^{-15}$ m or 10^{-13} cm

   Atomic radius is greater than the nucleus radius by 10^5 times.

   Density of nucleus is 10^{17} kg/m^3 or 10^{14} g/cm^3 and volume of nucleus is 10^{-39} cm^3.
5. Centrifugal force develops between electrons and the nucleus so electrons revolve around the nucleus as planets move around the sun.

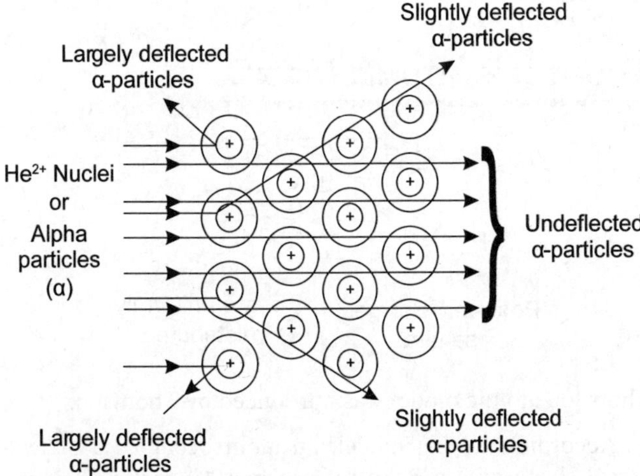

FIGURE 2.1 The Rutherford Scattering Experiment

FACTS TO REMEMBER

- Scattering of α-particles $\propto \dfrac{1}{\sin^4(\theta/2)}$

Strengths

- It explained the existence of a nucleus.
- It explains the movement of electrons as a circulatory rotation of electron around the nucleus.

Limitations

- It cannot explain the stability of an atom.
- It cannot explain the number and velocity of electrons.
- According to this theory, atomic spectrum is continuous and non linear but in fact it is linear and discontinuous.

ELECTRONIC STRUCTURE OF ATOMS

Wave Theory

Wave theory is described in brief as follows:

- All radiant energy propagates in terms of waves.
- Radiant energy is a form of electromagnetic waves.
- Radiations are associated with electric and magnetic fields, perpendicular to one another.
- In the propagation of an electromagnetic radiation, there is only propagation of wave but not of the medium.

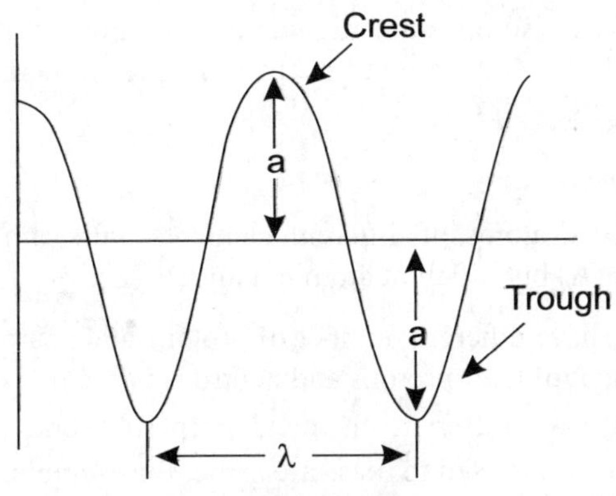

FIGURE 2.2

Wavelength It is the distance between any two successive crests or two successive troughs of a wave.

It is denoted by λ and it is measured in Å (Angstroms) or nm (nanometres).

Frequency It is the number of waves per second passing at a given point. It is denoted by υ or f. The units of frequency are Hertz or cycles sec^{-1}.

Velocity of light It is the distance travelled by one wave in one second.

- Velocity of light = frequency × wavelength

  $c = \upsilon\lambda$

  $\lambda \propto 1/\upsilon$
- Light and all the other electromagnetic radiations travel in vacuum or air with same velocity.

Wave number It is the number of waves spread in one centimetre and it is denoted by $\bar{v}$. It is increase of wavelength.

$\bar{\upsilon} = \frac{1}{\lambda}$, as $\bar{\upsilon} = \frac{c}{\lambda}$

So, $\upsilon = c\bar{\upsilon}$

- Its units are cm^{-1} or m^{-1}.

TABLE 2.1 Electronmagnetic Spectrum

Wavelength (in m)	Name	Frequency (cycles sec.$^{-1}$)
10^{-11}	gamma rays	10^{19}
10^{-9}	X-rays	10^{17}
10^{-8}	ultra violet	10^{16}
10^{-7}	visible	10^{15}
10^{-6}	infrared	10^{14}
10^{-2}	microwave	10^{10}
1	TV	10^{8}
10^{1}–10^{2}	radio	10^{7}–10^{6}

Corpuscular Theory of Light

- Corpuscular theory of light was first introduced by Newton.
- According to this theory, light is propagated in the form of small invisible particles.

Quantum Theory

Quantum theory was proposed by Max Planck and later extended by Einstein.

- It states that a hot **vibrating** body does not emit or absorb energy continuously but does so discontinuously in the form of small energy packets or bundles known as **quanta** or **photons** in case of light energy.
- The energy of radiation (E) is directly proportional to frequency of radiation (υ).

  $E \propto \upsilon$

  $E = h\upsilon = \frac{hc}{\lambda}$

  Here, h is Planck constant and its value is 6.6253×10^{-34} Js or kg m^2 s^{-1}.
- Absorption or emission of energy in the form of integral multiples of quantum is known as quantization of energy, $E = nh\upsilon$
- A hollow sphere coated inside with platinum black and having a small hole in its wall acts as a black body. It is a perfect absorber and perfect emitter of radiant energy.
- At a given temperature, the intensity of radiation increases with wavelength, approaches maximum and then starts decreasing.
- Planck's quantum theory could explain only black body radiations, so Einstein extended the quantum theory to all types of electromagnetic radiations.

Photoelectric Effect

Photoelectric effect was introduced by P. Lenard and explained by Einstein.

- It is the emission of electrons from a metal surface on exposing it to radiation of suitable frequency.
- It is readily shown by alkali metals like potassium and calcium, which have very low value of ionization energy.
- When a photon strikes a metal surface, its energy is absorbed by the metal, consequently emission of electrons takes place.

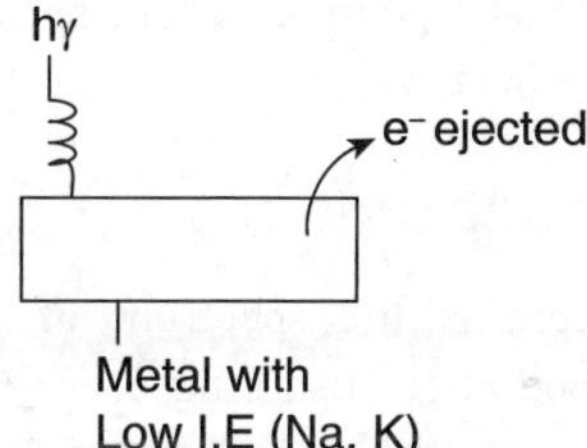

A part of the energy of a photon is used to overcome the attractive forces and the remaining energy is used in increasing the kinetic energy of electron.

K.E. $(1/2\ mv^2) = E - \Phi$

Here E = energy of radiation

Φ = work function (minimum energy needed for electron emission)

$$K.E. = h\upsilon - h\upsilon_0$$

$$= hc\left(\frac{1}{\lambda} - \frac{1}{\lambda_0}\right)$$

ν^0 or υ_0 = threshold frequency (minimum frequency of light radiation needed for electron emission)

- If $\nu > \nu_0$

  K.E. is positive, therefore, electrons are emitted.
- If $\upsilon = \upsilon_0$

  K.E. = 0, therefore no electron is emitted.
- If $\upsilon < \upsilon_0$

  K.E. is negative, therefore, there is no emission of electrons.

- Photoelectric effect ∝ Frequency of light radiation
- Kinetic energy of electrons does not depend upon intensity of light. However, number of ejected electrons depends upon intensity of light.

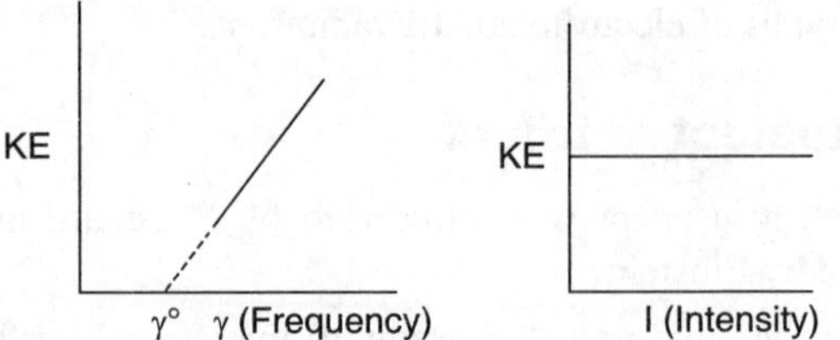

SPECTRUM

Spectrum is the arrangement of components of electromagnetic radiations in the increasing order of wavelength or decreasing order of frequency, when it is passed through a prism. The apparatus used to record the spectrum is called a **spectrometer** or **spectrograph**.

Emission Spectrum

Emission spectrum is the spectrum of radiation emitted by a substance after absorbing energy of a particular wavelength, or alternatively, the spectrum formed by the emission of energy in the form of light radiation is called emission spectrum.

- Emission spectrum consists of bright lines or bands on a dark background.
- Emission spectrum is formed due to the emission of light by the excited atoms or molecules.

Continuous Spectrum

- When no sharp boundaries are observed between adjacent radiations, the spectrum is continuous.
- The spectrum of incandescent white light obtained by heating a solid to very high temperature is a continuous spectrum.
- When white light passes through a prism, it gives a continuous spectrum of seven colours (VIBGYOR).
- In a continuous spectrum, each colour fades into the next colour as in a rainbow.

Absorption Spectrum

- The spectrum formed by the absorption of energy in the form of light radiation is called absorption spectrum.
- It can be understood as the photographic negative of an emission spectrum, that is, the missing wavelength leaving the spaces in continuous spectrum. In this sense, absorption spectrum is just the opposite of the emission spectrum.
- Absorption spectrum is formed due to the excitation of atoms or molecules by absorbing energy.
- Absorption spectrum consists of dark lines or bands on a bright background.

Discontinuous or Line Spectrum

- The emission spectrum of atoms in gaseous phase containing bright lines with sharp boundaries is called a line spectrum. It is a discontinuous spectrum having emitted radiations of a particular wavelength.
- The dark lines in the absorption spectrum and bright lines in the emission spectrum of a particular substance appear at the same places (same wavelengths).
- Each element has its own characteristic line spectrum like fingerprints by which it can be identified.
- The line spectrum of two element can never be identical as
    1. They do not have same number of electrons.
    2. They have different mass number.
    3. They have different energy level skill.
    4. They have different number of valence electron.
- Line spectrum is used in chemical analysis and identification of unknown atoms. Rb, Cs, In, Ga, He are identified by this method.
- Line spectra are given by atoms so they are known as atomic spectra whereas band spectra are given by molecules so they are known as molecular spectra.

BOHR'S ATOMIC MODEL

Bohr introduced circular orbit concept based on Planck's quantum theory. According to Bohr's model

- Around the nucleus there are circular regions called orbits or shells.

K	L	M	N	O.....
n = 1	2	3	4	5.....

→

Energy and distance from nucleus increase from K onwards

- Every orbit has a fixed amount of energy so it is also referrred to as an energy level.
- An electron revolves around the nucleus without any loss of energy in a particular orbit of definite energy that is why orbit is called stationary state also.
- Angular momentum (mvr) in each orbit is quantized

$mvr = n\frac{h}{2\pi} = n\hbar$

Here h is Planck constant.

$n \rightarrow$	1	2	3	4
$m\upsilon r \rightarrow$	$\frac{h}{2\pi}$	$\frac{h}{\pi}$	$1.5\frac{h}{\pi}$	$2\frac{h}{\pi}$

- When electron changes its orbit, energy change occurs in quanta.

$\Delta E = E_2 - E_1 = h\upsilon \text{ or } = \frac{hc}{\lambda}$

$\mathbf{E_2 - E_1 > E_3 - E_2 > E_4 - E_3 > E_5 - E_4 >}$

Few Important Terms

Quanta It is a small bundle of any type of energy.

Photon It is a small massless bundle of light energy and not a material body.

Excited state Here electron jumps from a lower to a higher orbit or energy level by absorbing energy in quanta.

De-excited state Here electron jumps from higher to lower energy level by releasing energy in quanta.

Merits of Bohr's Atomic Model

It can explain linear nature of spectrum of hydrogen like species (that is with one e"), for example, H, He^+, Li^{+2} etc.

Spectral series for hydrogen atom The atoms of hydrogen in a gas discharge tube, emit radiations whose spectrum shows line characteristics (line spectra) and lies in the infrared, visible and ultraviolet region of the electromagnetic spectrum.

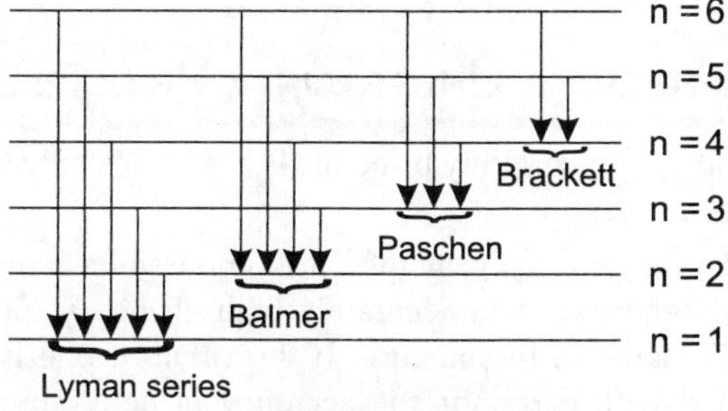

Figure 2.3

Table 2.2 Type of Lines Obtained in a Hydrogen Spectrum

	n_1	n_2	Spectrum
Lyman	1	2....∞	uv
Balmer	2	3....∞	Visible
Paschen	3	4....∞	Near IR
Brackett	4	5....∞	Far IR
Pfund	5	6....∞	Far IR
Humphries	6	7....∞	Far IR

- For radiations with the longest wavelength or least energy, n_2 must be nearest (next value) to n_1.
- For shortest wavelength and highest energy, n_2 must be infinite, that is, it must be close to the series limit. Balmer series contains four important lines whose wavelengths are 6563Å, 4862 Å, 4341 Å and 4102 Å. These are called H_α, H_β, H_γ, and H_δ lines respectively.
- When $n_2 = \infty$, the line produced is the limiting line of series

Calculation of energy and wavelength:

$$\Delta E \text{ or } \frac{hc}{\lambda} = \frac{2\pi^2 Me^4 Z^2}{n^2} = \left(\frac{1}{n_1^2} - \frac{1}{n_2^2}\right)$$

$$\bar{v} = \frac{1}{\lambda} = \frac{2\pi^2 Me^4 Z^2}{n^3 c} = \left(\frac{1}{n_1^2} - \frac{1}{n_2^2}\right)$$

$$\bar{v} = \frac{1}{\lambda} = R_H Z^2 \left(\frac{1}{n_1^2} - \frac{1}{n_2^2}\right)$$

$\bar{v}$ = Wave number, λ = Wavelength

n_1 = Lower orbit, n_2 = Higher orbit

R_H = Rydberg constant = 109678 cm^{-1} or $1.1 \times 10^7\ m^{-1}$

- In the calculation of radius of nth orbit

$$r_n = \frac{n^2h^2}{4\pi^2 mKZe^2}$$

$$r_n = 0.53 \times \frac{n^2}{Z} \text{ Å}$$

Here n = nth orbit

- In the calculation of velocity of electron in nth orbit

$$v_n = 2.165 \times 10^6 \times \frac{Z}{n} \text{ m/sec.}$$

- In the calculation of energy of electron in nth orbit

$$E_n = -\frac{2\pi^2 mZ^2e^2K^2}{n^2h^2}$$

$$E_n = -\frac{313.52}{n^2} Z^2 \text{ Kcal/mol}$$

$$E_n = -13.6\ \frac{Z^2}{n^2} \text{ eV/atom}$$

$$= -2.178 \times 10^{-18} \times \frac{Z^2}{n^2} \text{ J/atom}$$

Here $K = 1/4\pi\varepsilon_0$
e = Charge on electron
m = Mass of electron
h = Planck constant

Bohr value = value in first orbit of hydrogen atom
Bohr radius = 0.53 Å
Bohr velocity = 2.165×10^6 m/s
Bohr energy = –13.6e V/atom

- Number of revolutions = $\frac{V_n}{2\pi r_n}$
- To find Total Number of Spectral Lines:
  Case I: when an e^- jumps from n^{th} orbit to ground state

Zeeman Effect

It is the splitting of main spectrum line into several lines in a strong magnetic field.

Stark Effect

It is the splitting of main spectrum line into several lines in a strong electric field.

$$\text{Number of lines} = \frac{n(n-1)}{2}$$

$$\text{If } n = 4, \text{ Number of lines} = \frac{4(4-1)}{2} = 6$$

Case II: When e^- comes from n_2 to n_1

$$\text{Number of lines} = \frac{(n_2 - n_1)(n_2 - n_1 + 1)}{2}$$

If $n_2 = 5$, $n_1 = 2$

$$\text{Number of lines} = \frac{(5-2)(5-2+1)}{2} = 6$$

Shortcomings of Bohr's Theory

- It is not applicable for species having more than one electron like Li, He.
- It cannot explain fine spectrum of H, Li^{++} also.
- It does not explain Zeeman and Stark effects.
- It cannot explain chemical bonding.
- It can explain only particle nature, that is, no explanation of wave nature (against de Broglie's, Heisenberg's, Planck's theory).

De Broglie Equation and Dual Nature Theory

- De Broglie introduced dual nature theory for matter, that is for the subatomic particles (e^-, p, n).
- According to this theory, matter has both particle as well as wave nature.
- Wave nature of electron was confirmed by Davisson and Germer by diffraction pattern experiment on nickel surface. The patterns obtained were found to be same as those predicted by Bragg for X-rays.

$E = mc^2$ (1) for particle

$E = h\nu$

$E = \frac{hc}{\lambda}$ (2) for wave

Using (1) and (2)

$$\frac{hc}{\lambda} = mc^2$$

$$\frac{1}{\lambda} = \frac{mc^2}{hc}$$

$$\lambda = \frac{h}{mc}$$

if $c = v$

$$\lambda = \frac{h}{mv}$$

$$\lambda = \frac{h}{p} \quad (\text{As } p = mv)$$

$\lambda \propto \frac{1}{p}$ **(At constant temperature)**

- If a particle has more mass or momentum it has less wave nature.
  e > p > > n
- Heavy objects have less wavelength due to more mass.
- Relation between wavelength and kinetic energy:

$$\lambda = \frac{h}{\sqrt{2m\,K.E.}} = \frac{h}{\sqrt{2m\,e.V}}$$

As K.E = $\frac{1}{2}mv^2 = eV$ (charge into potential difference)

$$\frac{1}{2}mv^2 = eV$$

$$v^2 = \frac{eV}{m}$$

$$v = \sqrt{\frac{2eV}{m}}$$

Facts to Remember

- This equation can be used only when microscopic objects are taken and their energy is expressed in terms of speed only.

Heisenberg's Uncertainty Principle

Heisenberg's uncertainty principle is applicable only to subatomic particles.

- According to this principle, it is impossible to measure simultaneously and accurately both change in position and change in momentum. If the value of one is determined with certainty, the accuracy in determining the other value is compromised.

- It is also called **principle of indeterminancy**.

  $\Delta x.\ \Delta p \geq \frac{h}{4\pi}$

  $\Delta x.\ m\ \Delta v \geq \frac{h}{4\pi}$

  Here Δx = Uncertainty in position

  Δv = Uncertainty in velocity

  $\Delta x.\ \Delta v \geq \frac{h}{4\pi m}$

  $\Delta E.\ \Delta t \geq \frac{h}{4\pi}$

  Here ΔE = change in energy

  Δt = change in time

- If change in position is zero, change in momentum will be infinite and vice versa.
- It applies when both Δx and Δp are along the same axis.

  Concept of Probability: As a according to this priniciple Δx and ΔP of an e^- can not be find accurately at a particular time so the idea of definite paths i.e., orbits as suggested by Bohr has no more significance. This leads to the concept of probability. According to which it is possible to state or predict the probability of locating an electron of a specific energy in a given region of space around the nucleus at a given time. This leads to the concept of 'orbital'.

SCHRODINGER WAVE EQUATION

Schrodinger wave equation explains three dimensional wave nature of electron as follows:

$$\frac{\delta^2\Psi}{\delta x^2} + \frac{\delta^2\Psi}{\delta Y^2} + \frac{\delta^2\Psi}{\delta Z^2} + \frac{8\pi^2 m}{h^2}[E - V]\,\Psi = 0$$

$$\overline{\nabla}^2\ \Psi + \frac{8\pi^2 m}{h^2}[E - V]\,\Psi = 0$$

Here X, Y, Z are three coordinate axis.

E is total energy.

V is potential energy ($V = -Ze^2/r$)

m is mass of electron wave.

$\overline{\nabla}^2\ \psi$ is Laplacian operator.

Significance of ψ and ψ 2

ψ : It has no physical significance. It represents amplitude of electron-wave or boundary surface of an orbital.

ψ^2: It is the probable electron density or it is the probability of finding electrons in any region (3 dimensional space around the nucleus). If ψ^2 is positive, electrons are present and if ψ^2 is zero electrons are absent.

- An orbital is represen ed by $\psi\ \psi^*$ or ψ^2 for showing electron density.
- The probability of finding electron at a distance r from the nucleus in the region dr is given by $4\ \pi\ r^2\ dr\ \psi^*$.
- In case where the energy of an atom or a molecule does not change, another equation given by Schrodinger becomes applicable.

  $\hat{H}\ \psi = \hat{E}\ \psi$

  $\hat{H} = \hat{T} + \hat{V}$:

  $(\hat{T} + \hat{V})\psi = E\ \psi$

  Here, H = Hamiltonion factor

  V = Potential energy

  T = Kinetic energy

  E = Total energy

Orbital

An orbital is a variedly shaped, three dimensional region around the nucleus where the probability of finding electrons is maximum.

An orbital possesses the following features:

- It represents both the particle and wave nature of an electron.
- The probability of finding an electron in an orbital is more than 90%.
- Except the s orbital, the rest are directional in nature.
- An orbital can accommodate a maximum of 2 electrons with opposite/anti-parallel spins.

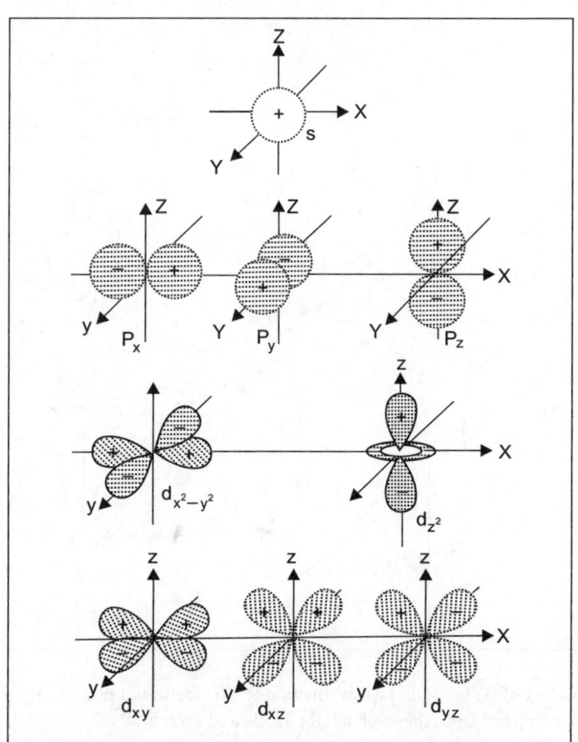

FIGURE 2.4 Shapes of Different Orbitals

The solution of Schrodinger Wave Equation, for the single electron hydrogen atom, yields the wave function of an atomic orbital, that is, ψ (r, θ, φ). It is a mathematical function of r, θ and φ which are the three coordinates of an electron. It can further be factorized into three separate parts, each of which is a function of only one coordinate. $\psi (r, \theta, \varphi) = R(r)\, \Theta(\theta)\, \phi(\varphi)$

Here, R(r) is the radial function which shows the dependence of an orbital on the distance V of the electron from the nucleus. The radial function depends upon the quantum numbers n and *l*.

Θ (θ) and ϕ (φ) depict the angular dependence of the orbital on θ, and φ respectively. It is dependent on the quantum numbers *l* and m.

The total wave functions y can now be represented as

$$\psi (r, \theta, \varphi) = R_{nl}(r)\, \Theta_{lm'}(\theta)\, \phi_{m}(\varphi)$$

Radial part   Angular part

Degenerate Orbital

Orbitals with same value of n and of same sub-shell are degenerate orbitals, for example,

I. $2p_x$, $2p_y$ $2p_z$

II. $3d_{xy}$, $3d_{yz}$, $3d_{xz}$.

Radial Probability Distribution Function

- It is the probability of finding electrons in the region between r and dr δr.
- It is given by $4 \pi r^2 \psi^2 \delta r$. It is direction independent. Here, r = radius or distance from nucleus

  δr = thickness

Note: It is zero electron density region.

$$\psi^2 = 0$$

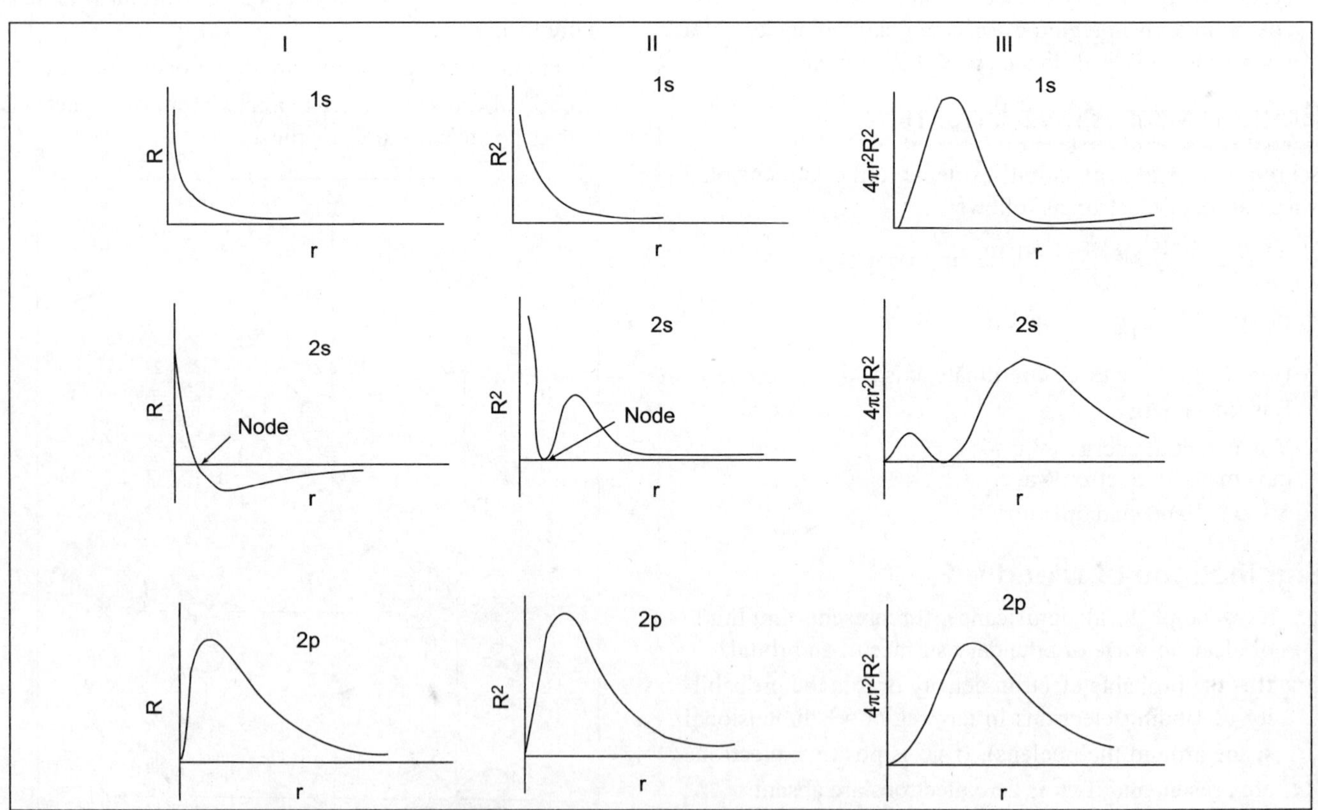

The plots of (I) the redial wave function R; (II) the radial probability density R^2 and (III) the radial density function $4\pi r^2 R^2$ as a function of distance r of the electron from the nucleus for 1s, 2s and 2p orbitals.

FIGURE 2.5 Various Probability Curves

Radial node It is the region between two successives orbitals.

Nodal point It is a point where two lobes join and the electron density here is zero.

Nodal plane It is the plane which separates two lobes. Electron density here is zero.

- No s orbital has a nodal plane or a point.
- p_x, p_y p_z, $d_z$orbitals have one nodal plane or point.
- Rest of d orbitals (d_{xy} d_{yz} d_{xz} d_{x2-y2}) have two nodal plane or points.

Calculation of Nodes

Node = n–1

Radial nodes or spherical nodes = n–*l*–1

Angular or non-spherical = *l*

For example, 1s: Node = 1–1 = 0

For example, 2s: Node = 2–1 = 1
Angular node = 0
Radial node = 2–0–1 = 1

For example, 3s: Node = 3–1 = 2
Angular node = 0
Radial node = 3–0–1 = 2

For example, 4d: Node = 4–1 = 3
Angular node = 0
Radial node = 4–0–1 = 3

Quantum Numbers

Quantum numbers are set of four numbers used to address an electron as follows:

1. To determine size, distance from nucleus and energy of electron in **orbit**.
2. To decide shape, energy of electron in a **sub-orbit**.
3. To find number of orbitals and their directional position or orientation in space.
4. To find number of electrons and their spin.

Principle Quantum Number

- Principle quantum number was introduced by Bohr.
- It is denoted by n.

It determines:

1. Size of **orbit (shell)**
2. Distance of **orbit** from nucleus
3. Energy of electron in an **orbit**
4. Maximum number of electron in any **orbit**
5. Angular momentum of electron in an **orbit**

$$mvr = \frac{nh}{2\pi} = n\hbar$$

mvr = angular momentum in **orbit**.

h = Planck constant

- Values of n = 1,2 ,3, 4 .... ∞
  n ≠ 0, –ve, fractional

Azimuthal or Secondary or Angular Momentum Quantum Number

- Angular momentum quantum number was introduced by Sommerfield. It is denoted by '*l*'.
- It determines shape of sub-orbit, energy in sub shell and angular momentum of an electron in any orbital.

$$mvr = \sqrt{\ell(\ell+1)}\frac{h}{2\pi} \text{ or } \hbar \text{ or } \hbar$$

For any s orbital e^- (mvr) = 0 (as $\ell = 0$)

For any p orbital e^- (mvr) = $\sqrt{\ell(\ell+1)}\,\hbar = \sqrt{2}\,\hbar$

- This explains the existence of many closely packed spectral lines in hydrogen spectrum.
- Values of *l* = 0 to n – 1

n: 0	1	2	3	4
l: 0	0	0, 1	0, 1, 2	0, 1, 2, 3

• ***l*: 0**	**1**	**2**	**3**	**4**
s	**p**	**d**	**f**	**g**

Magnetic Quantum Number

- Magnetic quantum number was introduced by Land and Zeeman. It is denoted by m.

It determines:

1. Number of orbitals
   m (total) value = n^2 (when n is given)
   = 2*l* + 1 when '*l*'is given.
   when n = 2
   m (total) = 2^2 = 4
   when *l* = 2
   m (total) = 2 × 2 + 1 = 5
2. Directional position of orbitais or orientation in shape and Zeemann effect
3. Angular magnetic momentum
   $$mvr = m\frac{h}{2\pi} = m\hbar$$
4. Values of m = –*l* to + *l* including zero

E.g., *l* ⟶	0	1	2
m ⟶	0	–1, 0, +1	–2, –1, 0, +1, +2

Spin Quantum Number

- Spin quantum number was introduced by Uhlenbeck and Goudsmith.
- It is the intrinsic angular momentum measurement which is used to differentiate between two electrons in the same orbital.
- Schrodinger could not explain the spin of an electron.
- For each value of m, s has two different values + 1/2 and –1/2.

  +1/2 –1/2

  α-spin β-spin
- +1/2 and –1/2 are just to mechanical spin states with no classical importance now as other spin states are also possible.
- Total values of s or total number of electrons $= 2m = 2n^2 = 2(l + 1)$

AUFBAU PRINCIPLE

Aufbau is a German word which means 'to build up' so it is also called building up rule.

$(n + l)$ Rule

According to Aufbau principle electrons are filled in different orbitais from lower (n + *l*) value orbital to higher $(n + l)$ value orbital as follows:

1 s < 2s < 2p < 3s < 3p < 4s < 3d < 4p < 5s < 4d < 5p < 6s <

[Increasing order of (n + *l*) value or energy]

In case (n + *l*) value is same, electron is filled first in orbital with lower value of n, for example,

3p	4s
n = 3	n =4

electrons first goes into 3p subshell.

FACTS TO REMEMBER

	1s	2s	2p	3s
(n + *l*)	1 + 0	2 + 0	2 + 1	3 + 0
	3p	4s	3d	4p
	3 + 1	4 + 0	3 + 2	4 + 1

Violation of Aufba's Rule

	ns¹	np⁴	
	[1]	[1l] [1] [1]	
As	ns^1	np^4	Configuration
	is not possible		

Electronic configuration It shows distribution of electrons invarious orbits sub-orbits and orbitals.

- First fill electrons in orbitals according to (n + *l*) rule (initial configuration).
- If atomic number is >20, arrange the orbital in increasing order of (n + *l*) value (for final configuration).

  For example, $_{26}Fe$

  $1s^2, 2s^2, 2p^6, 3s^2, 3p^6, 4s^2, 3d^6$

  $1s^2, 2s^2, 2p^6, 3s^2, 3p^6, 3d^6, 4s^2$

  (final configuration)

Some Exceptional Configurations

Fully filled and half filled orbitals are more stable due to greater number of exchanges, higher exchange energy and symmetry.

Case I

$(n – 1)d^4, ns^2$ incomplete d-orbital so d-orbital is less stable	:	$(n – 1)d^5, ns^1$ both orbitals are half filled so more stable.

Example:

$_{24}Cr$: $[_{18}Ar]\ 3d^4\ 4s^2$ [×]

$[_{18}Ar]\ 3d^5\ 4s^1$ [√]

The number of exchanges that can take place in d^4 configuration are as follows:

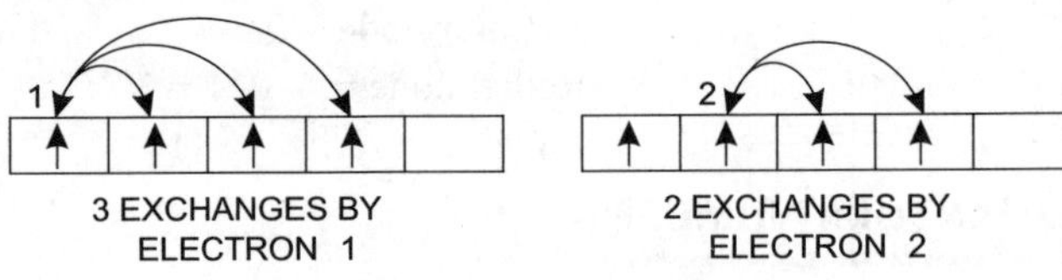

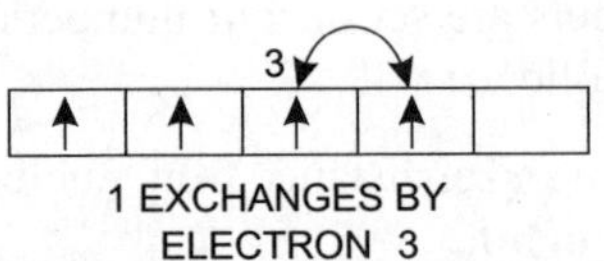

Total number of exchanges = 3 + 2 + 1 = 6

The number of exchanges that can take place in d^5 configuration are as follows:

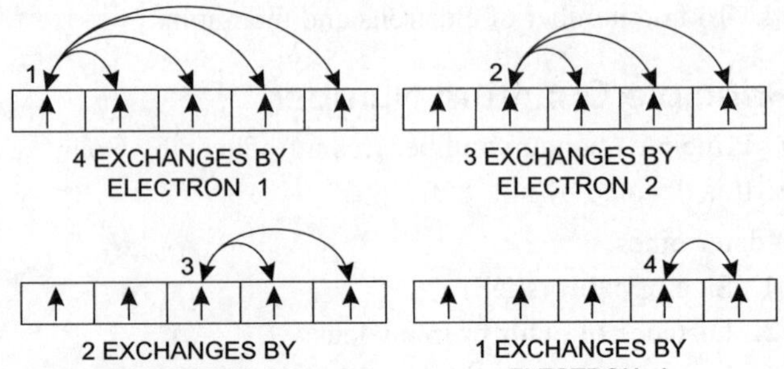

Total number of exchanges = 4 + 3 + 2 + 1 = 10

Case II

$(n-1)\,d^9\,ns^2$	:	$(n-1)\,d^{10}$	ns^{-1}
incompletely		fully	half
filled d orbital		filled	filled
[less stable]		[more stable]	

Example:

$_{29}Cu$: $[_{18}Ar]\,3d^9\,4s^2$ [×]

$_{29}Cu$: $[_{18}Ar]\,3d^{10}\,4s^1$ [√]

Some Other Exceptional Configurations

$_{41}Nb$: $1s^2\,2s^2\,2p^6\,3s^2\,3p^6\,3d^{10}\,4s^2\,4p^6\,4d^4\,5s^1$

$_{42}Mo$: $1s^2\,2s^2\,2p^6\,3s^2\,3p^6\,3d^{10}\,4s^2\,4p^6\,4d^5\,5s^1$

$_{44}Ru$: $1s^2\,2s^2\,2p^6\,3s^2\,3p^6\,3d^{10}\,4s^2\,4p^6\,4d^7\,5s^1$

or

$[Kr]\,4d^7\,5s^1$

$_{45}Rh$: $[Kr]\,4d^8\,5s^1$

$_{46}Pd$: $[Kr]\,4d^{10}\,5s^0$

$_{47}Ag$: $[Kr]\,4d^{10}\,5s^1$

$_{57}La$: $[Xe]\,4f^0\,5d^1\,6s^2$

$_{58}Cc$: $[Xc]\,4f^2\,5d^0\,6s^2$

$_{61}Pm$: $[Xe]\,4f^5\,5d^0\,6s^2$

$_{78}Pt$: $[Xe]\,4f^{14}\,5d^9\,6s^1$

$_{79}Au$: $[Xe]\,4f^{14}\,5d^{10}\,6s^1$

Pauli Exclusion Principle

1. Any orbital can have a maximum of two electrons with opposite or anti-parallel spin.

   [⇅] or [⇵]

2. No two electrons of an atom can have the same set of all the four quantum numbers, for example, if n, *l*, m are same, the value of s will be different.

Applications of Pauli Exclusion Principle

To find maximum number of electrons in:

1. Orbit: maximum number of electrons = $2n^2$

   $n = 1$ for K $e^- = 2 \times 1^2 = 2\,e^-$

   $n = 2$ for L $e^- = 2 \times 2^2 = 8e^-$

   $n = 3$ for $Me^- = 2 \times 3^2 = 18\,e^-$

2. Sub-orbit:

s	p	d	f	g
2	6	10	14	18 electrons

3. Orbital: maximum two electrons are possible in any orbital.

Violation of Pauli's Law:

When any orbital has two e^- with parallel spin.

[⇈ | 1 | 1]

When any orbit/sub-orbit/orbital has electrons more than maximum limit.

E.g., N ⇒ $1s^7$. It is not possible as first orbit, s-sub-orbit can have max. 2 e^-.

Hund's Rule of Maximum Multiplicity

According to Hund's rule, "In degenerate orbitals, electrons are filled in such a way that maximum stability and multiplicity can be obtained for the former. First, each degenerate orbital is singly occupied then pairing of electron takes place."

The following would be a violation of Hund's rule:

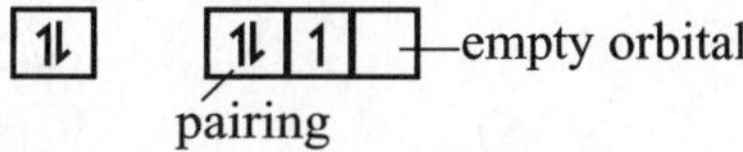

Applications of Hund's Rule

- To write correct and stable electronic configuration, for example,

  $_{24}Cr$: $[Ar]\,3d^4\,4s^2$ [×]

  $[Ar]\,3d^5\,4s^1$

- To find number of unpaired electrons (n), for example,

  $_{24}Cr$: $[Ar]\,3d^5\,4s^1$

  [1 | 1 | 1 | 1 | 1] [1]

  $3d^5$ $4s^1$

  n = 6

- To find multiplicity (m) and spin(s):

  $m = (n + l) = (2s+1)$

  Example:

  $_{24}Cr$: $[_{18}Ar]\,3d^5\,4s^1$

  $m = 6 + 1 = 7$

- Number of spins = n/2 = 6/2 = 3.
- To find magnetic nature and magnetic moment

  n = 0 diamagnetic

  n = 1, 2, 3 paramagnetic

  n = 4, 5, 6 ferromagnetic or highly paramagnetic

  magnetic moment $(\mu) = \sqrt{n(n+2)}$ B. M.

  B. M. = Bohr magneton

- To find colour

  1. If n = 0, ion is colourless
     For example, Cu^+ Zn^{+2}, Cd^{+2}
  2. If n = 1, 2, 3, ion is coloured
     For example, Cu^{+2}, Fe^{+2}, Fe^{+3}

Table 2.3

S.No.	Elements	Symbol	Atomic number	Electronic configuration
1	Hydrogen	H	1	$1s^1$
2	Helium	He	2	$1s^2$
3	Lithium	Li	3	$1s^2, 2s^1$
4	Beryllium	Be	4	$1s^2, 2s^2$
5	Boron	B	5	$1s^2, 2s^2\ 2p^1$
6	Carbon	C	6	$1s^2, 2s^2\ 2p^2$
7	Nitrogen	N	7	$1s^2, 2s^2\ 2p^3$
8	Oxygen	O	8	$1s^2, 2s^2\ 2p^4$
9	Fluorine	F	9	$1s^2, 2s^2\ 2p^5$
10	Neon	Ne	10	$1s^2, 2s^2\ 2p^6$
11	Sodium	Na	11	$1s^2, 2s^2\ 2p^6, 3s^1$
12	Magnesium	Mg	12	$1s^2, 2s^2\ 2p^6, 3s^2$
13	Aluminium	Al	13	$1s^2, 2s^2\ 2p^6, 3s^2\ 3p^1$
14	Silicon	Si	14	$1s^2, 2s^2\ 2p^6, 3s^2\ 3p^2$
15	Phosphorous	P	15	$1s^2, 2s^2\ 2p^6, 3s^2\ 3p^3$
16	Sulphur	S	16	$1s^2, 2s^2\ 2p^6, 3s^2\ 3p^4$
17	Chlorine	Cl	17	$1s^2, 2s^2\ 2p^6, 3s^2\ 3p^5$
18	Argon	Ar	18	$1s^2, 2s^2\ 2p^6, 3s^2\ 3p^6$
19	Potassium	K	19	$1s^2, 2s^2\ 2p^6, 3s^2\ 3p^6, 4s^1$
20	Calcium	Ca	20	$1s^2, 2s^2\ 2p^6, 3s^2\ 3p^6, 4s^2$
21	Scandium	Sc	21	$1s^2, 2s^2\ 2p^6, 3s^2\ 3p^6\ 3d^1, 4s^2$
22	Titanium	Ti	22	$1s^2, 2s^2\ 2p^6, 3s^2\ 3p^6\ 3d^2, 4s^2$
23	Vanadium	V	23	$1s^2, 2s^2\ 2p^6, 3s^2\ 3p^6\ 3d^3, 4s^2$
24	Chromium	Cr	24	$1s^2, 2s^2\ 2p^6, 3s^2\ 3p^6\ 3d^5, 4s^1$
25	Manganese	Mn	25	$1s^2, 2s^2\ 2p^6, 3s^2\ 3p^6\ 3d^5, 4s^2$
26	Iron	Fe	26	$1s^2, 2s^2\ 2p^6, 3s^2\ 3p^6\ 3d^6, 4s^2$
27	Cobalt	Co	27	$1s^2, 2s^2\ 2p^6, 3s^2\ 3p^6\ 3d^7, 4s^2$
28	Nickel	Ni	28	$1s^2, 2s^2\ 2p^6, 3s^2\ 3p^6\ 3d^8, 4s^2$
29	Copper	Cu	29	$1s^2, 2s^2\ 2p^6, 3s^2\ 3p^6\ 3d^{10}, 4s^1$
30	Zinc	Zn	30	$1s^2, 2s^2\ 2p^6, 3s^2\ 3p^6\ 3d^{10}, 4s^2$
31	Gallium	Ga	31	$1s^2, 2s^2\ 2p^6, 3s^2\ 3p^6\ 3d^{10}, 4s^2\ 4p^1$
32	Germanium	Ge	32	$1s^2, 2s^2\ 2p^6, 3s^2\ 3p^6\ 3d^{10}, 4s^2\ 4p^2$
33	Arsenic	As	33	$1s^2, 2s^2\ 2p^6, 3s^2\ 3p^6\ 3d^{10}, 4s^2\ 4p^3$
34	Selenium	Se	34	$1s^2, 2s^2\ 2p^6, 3s^2\ 3p^6\ 3d^{10}, 4s^2\ 4p^4$
35	Bromine	Br	35	$1s^2, 2s^2\ 2p^6, 3s^2\ 3p^6\ 3d^{10}, 4s^2\ 4p^5$
36	Krypton	Kr	36	$1s^2, 2s^2\ 2p^6, 3s^2\ 3p^6\ 3d^{10}, 4s^2\ 4p^6$

POINTS TO REMEMBER

1. The radius r, of the nucleus of an atom, is related to its mass number (A) according to the equation $r = R_0 A^{1/3}$
   where R_0 is a constant having value $= 1.4 \times 10^{-15}$ m.
2. The shapes of the seven f-orbtials are omplicated but their designations as follows:
   $f_{x^3}, f_{y^3}, f_{z^3}, f_{x(z^2-y^2)}, f_{y(z^2-x^2)}, f_{z(y^2-x^2)}, f_{xyz}$
3. The folllowing are not possible $n \neq 1, n < 1, m > 1$. If any of these is not obeyed arrangement of quantum numbers set is not permissible.
4. Since ℓ cannot be equal to n so orbitals 1 p, 2d, 3f, 4g, 5h do not exist.
5. In case of H-atom (Hydrogenic species) energy depends only upon n-value so
   $1s < 2s \simeq 2p < 3s \simeq 3p \simeq 3d <$
6. Total energy (E_T) for any e^- in any orbit:
   $$E_T = E_{K.E} + E_{P.E} = \frac{1}{2} MV^2 + E_{P.E}$$
   $$= \frac{1}{2}\frac{Ze^2}{r} + \left[\frac{Ze^2}{r}\right]$$
   $$E_T = -\frac{1}{2}\frac{Ze^2}{r}$$
   $$E_K = -E_T$$
7. Total number of electrons in any orbit is given as
   $$2n^2 \text{ or } \sum_{\ell=0}^{\ell=(n-1)} (2\ell+1)$$
8. Exchange Energy:
   The electrons with parallel spins present in the degenerate orbitals tend to exchange their position and during it energy released is known as exchange energy.
9. The two electrons in different orbitals are for apart so suffer less $e^- = e^-$ repulsion than they are present in same orbital with anti-parallel spins.
10. The limitation of de-Broglie concept is that it can be applied only on objects having force free environment it means not applicable for any electron present in the atom.

Important formula

- $E = Nh\upsilon = \frac{Nhc}{\lambda}$
- When electron changes its orbit energy change occurs in quanta.
  $\Delta E = E_2 - E_1 = h\upsilon$ or $\frac{hc}{\lambda}$ (quanta)
- Number of spectrum lines $= \frac{n(n-1)}{2}$
  Here, n is orbit from which electron jumps or higher energy level.

Rydberg Formula

$$\Delta E \text{ or } \frac{hc}{\lambda} = \frac{2\pi^2 Me^4 Z^2}{n^2}\left(\frac{1}{n_1^2} - \frac{1}{n_2^2}\right)$$

$$\bar{v} = 1/\lambda = \frac{2\pi^2 Me^4 Z^2}{n^3 c}\left(\frac{1}{n_1^2} - \frac{1}{n_2^2}\right)$$

$$\bar{v} = \frac{1}{\lambda} = R_H Z^2\left(\frac{1}{n_1^2} - \frac{1}{n_2^2}\right)$$

$\bar{v}$ = Wave number, λ = Wavelength
n, = Lower orbit, n_2 = Higher orbit
R_H = Rydberg constant
R_H = 109678 cm^{-1} or 1.1×10^7 m.

- Calculation of radius of nth orbit
  $r_n = 0.53 \times \frac{n^2}{Z}$ Å [for H = Z = 1]
  Here, n = nth orbit
- Calculation of velocity of electron in nth orbit:
  $v_n = 2.165 \times 10^6 \times \frac{Z}{n}$ m/sec.
- Calculation of energy of electron in nth orbit:
  $E_n = \frac{-313.52}{n^2} Z^2$ kcal/mol
  $E_n = -13.6 \frac{Z^2}{n^2}$ eV/atom
  $= -2.178 \times 10^{-8} \times \frac{Z^2}{n^2}$ J/atom
- Number of revolutions $= \frac{v_n}{2\pi r_n}$

De Broglie Equation

$$\lambda = \frac{h}{mc} = \frac{h}{mv}$$

$$\lambda = \frac{h}{\sqrt{2m\,K.E.}}$$

$$K.E. = \frac{1}{2} mv^2 = eV \text{ (Potential difference)}$$

$$\frac{1}{2} mv^2 = eV$$

$$v^2 = 2\frac{eV}{m}$$

$v = \sqrt{2eV/m}$

$\Delta x \,.\, m\,\Delta v \geq \frac{h}{4\pi}$.

- Here Δx = Uncertainty in position,
  Δv = Uncertainty in velocity.

$\Delta x \,.\, \Delta v \geq \frac{h}{4\pi m}$

Calculation of Nodes

Node = n–1

Radial nodes or spherical nodes = n–l–1

Angular or non-spherical = l

Azimuthal or Secondary or Angular Mometum

- Angular momentum of electron in any orbital.
  $\sqrt{l(l+1)}\,.\,\frac{h}{2\pi}$ or $\sqrt{l(l+1)}.\hbar$
- K.E. ($1/2\ mv^2$) = E – Φ
  Here, E = Energy of radiation
  Φ = Work function (minimum energy needed for electron emission)
  K.E. = $h\upsilon - h\upsilon_0$
  υ_0 = Threshold frequency (minimum frequency of light radiation needed for electron ejection)

SOLVED NUMERICALS

Wave Nature, Wavelength

1. 3×10^{18} photons of a certain light radiation produce 1.5 J of energy. Find out the wavelength of this radiation ($h = 6.626 \times 10^{-34}$ Js).

Solution:

$E = h\upsilon$ (for one photon)

$E = Nh\upsilon$ (for N photons)

$E = \frac{Nhc}{\lambda}$

$\lambda = \frac{N\,hc}{E} = \frac{3 \times 10^{18} \times 6.62 \times 10^{-34} \times 3 \times 10^{8}}{1.5}$

$= 39.75 \times 10 - 8$ m = 3975 Å.

2. Determine the wavelength of a photon of electromagnetic radiation having energy 2.99×10^{12} erg mol^{-1} ($h = 6.625 \times 10^{-34}$Js).

Solution:

$\lambda = \frac{N\,hc}{E}$

$= \frac{6.023 \times 10^{23} \times 6.625 \times 10^{-34} \times 3 \times 10^{8}}{2.99 \times 10^{12}}$

= 4000 Å.

Spectral Series and Rydberg Formula

3. What transition in the hydrogen spectrum would have the same wavelength as the Balmer transition n = 4 to n = 2 of He$^+$ spectrum?

Solution:

For He$^+$,

$\frac{1}{\lambda} = R_H Z^2 \left(\frac{1}{2^2} - \frac{1}{4^2}\right)$

For H,

$\frac{1}{\lambda} = R_H \left(\frac{1}{n_1^2} - \frac{1}{n_2^2}\right)$

Since λ is same

$Z^2\,[1/2^2 - 1/4^2] = [1/n_1^{\,2} - 1/n_2^{\,2}]$

Z = 2

$[1/1^2 - 1/2^2] = [1/n_1^{\,2} - 1/n_2^{\,2}]$

$n_1 = 1$ and $n_2 = 2$

4. Find out the wavelength of H_α line of hydrogen spectrum in the Balmer series. ($R_H = 109676$ cm^{-1}).

Solution:

$\frac{1}{\lambda} = R_H \left(\frac{1}{n_1^2} - \frac{1}{n_2^2}\right)$

For Balmer series $n_1 = 2$ and for H,

$n_2 = 3$ (first line of Balmer series)

$\frac{1}{\lambda} = 109676 \left(\frac{1}{2^2} - \frac{1}{3^2}\right)$

$= 109676\ (1/4 - 1/9)$

$= 109676 \times \frac{5}{36}$ cm^{-1}

$= \frac{36}{5 \times 109676}$ cm

$= 65.65 \times 10^{-6}$ cm

= 6564 Å

5. Find out the wavelength of radiations emitted produced in a line in Lyman series, when an electron falls from fourth stationary state in hydrogen atom. ($R_H = 1.1 \times 10^7 m^{-1}$)

Solution:

$$\frac{1}{\lambda} = R_H\left(\frac{1}{n_1^2} - \frac{1}{n_2^2}\right)$$

For Lyman series $n_1 = 1, n_2 = 4$

$$\frac{1}{\lambda} = 1.1 \times 10^7\left(\frac{1}{1^2} - \frac{1}{4^2}\right)$$

$$= 0.9696 \times 10^{-7}\ m$$

6. Calculate the wavelength for the shortest wavelength transition in the Balmer series of atomic hydrogen.

Solution:

$$\frac{1}{\lambda} = R_H\left(\frac{1}{n_1^2} - \frac{1}{n_2^2}\right)$$

$$= 109677\ (1/4 - 0) = \frac{109677}{4}$$

$$= 27419.25\ cm^{-1}$$

7. Wavelength of high energy transition of H-atoms is 91.2 nm. Calculate the corresponding wavelength of He^+.

Solution:

For H-atom

$$\frac{1}{\lambda} = R_H\left(\frac{1}{n_1^2} - \frac{1}{n_2^2}\right) \quad (1)$$

For He^+ ion:

$$\frac{1}{\lambda^2} = R_H Z^2\left(\frac{1}{n_1^2} - \frac{1}{n_2^2}\right) \quad (2)$$

By (1) and (2),

$$\frac{\lambda_2}{\lambda_1} = \frac{1}{z^2}$$

$$\text{or } \lambda_{He+} = \lambda_H \times \frac{1}{Z^2} = 91.2 \times \frac{1}{2^2} = 22.8\ nm$$

De Broglie Relation and Uncertainty Principle

8. Calculate the momentum of a moving particle which has a de Broglie wavelength of 2 Å.

Solution:

According to de Broglie equation,

$$\lambda = \frac{h}{p}$$

$$p = \frac{h}{\lambda}$$

$$= \frac{6.62 \times 10^{-34}}{2 \times 10^{-10}} = 3.31 \times 10^{-24}\ kg\ ms^{-1}$$

9. Show that the wavelength of a 150 g rubber ball moving with a velocity 50 m sec.$^{-1}$ is short enough to be observed ($h = 6.62 \times 10^{-34}$ Js).

Solution:

$$\lambda = \frac{h}{mv} = \frac{6.62 \times 10^{-34}}{0.150 \times 50} = 8.82 \times 10^{-35}\ m$$

$$= 8.82 \times 10^{-33}\ cm$$

The wavelength is short enough to be observed.

10. A cricket ball weighing 100 g is to be located within 0.1 Å. What is the uncertainty in its velocity?

Solution:

$$\Delta x \,.\, m \,.\, \Delta v = \frac{h}{4\pi}$$

Δx = Location of ball= 0.1 Å = 0.1×10^{-10} m

m = Mass of ball = 100 g = 0.1 kg

Δv = Uncertainty in velocity

$h = 6.626 \times 10^{-34}$ Js

$$0.1 \times 10^{-10} \times 0.1\ \Delta v = \frac{6.626 \times 10^{-34}}{4 \times 3.14}$$

$$\Delta v = 0.527 \times 10^{-22}\ ms^{-1}$$

11. Calculate the minimum uncertainty in velocity of a particle of mass 1.1×10^{-27} kg if uncertainty in its position is 3×10^{-10} cm ($h = 6.62 \times 10^{-34}$ kg m^2 s^{-1})

Solution:

$$\Delta x \,.\, m \,.\, \Delta v = \frac{h}{4\pi}$$

$$\Delta v = \frac{h}{4\pi}\frac{1}{m\,\Delta x}$$

$$= \frac{6.62 \times 10^{-34}}{4 \times 3.14 \times 1.1 \times 10^{-27} \times 3 \times 10^{-8}}$$

$$= 1.59\ ms^{-1}$$

12. A bulb emits light of λ = 4500 Å. The bulb is rated as 150 watt and 8% of the energy is emitted as light. How many photons are emitted by the bulb per second?

Solution:

$$\text{Energy of one photon} = \frac{hc}{\lambda}$$

$$= \frac{6.625 \times 10^{-34} \times 3 \times 10^8\ J}{4500 \times 10^{-10}}$$

$$= 4.42 \times 10^{-19}\ J$$

$$\text{Energy emitted by the bulb} = 150 \times \frac{8}{100}\ J$$

$$n \times 4.42 \times 10^{-19} = 150 \times \frac{8}{100}$$

$$n = 27.2 \times 10^{18}$$

Miscellaneous

13. State the number of protons, neutrons and electrons in C^{12} and C^{14}.

Solution:

The atomic number of C^{12} is 6. So in it number of electrons = 6

Number of protons = 6

Number of neutrons = 12 – 6 = 6

In $_6C^{14}$

Number of electrons = 6

Number of protons = 6

Number of neutrons = 14 – 6 = 8.

14. Find the ratio of the difference in energy between the first and the second Bohr orbit to that between the second and the third Bohr orbit.

Solution:

$\Delta E = 1312$ kj [(1/n² initial) – (1/n² final)]

$\Delta E_{2\to 1} = E_{n=2} - E_{n=1}$

$= 1312$ kJ $[1/2^2 - 1/3^2]$

$= 1312$ kJ (5/36)

$$\frac{\Delta E_{2\to1}}{\Delta E_{3\to2}} = \frac{3/4}{5/36} = \frac{3}{4} \times \frac{36}{5} = \frac{27}{5}$$

15. Find the ratio of the radii of the first three Bohr orbits.

Solution:

Radius of the electron orbit in a particular state,

$$r_n = \frac{n^2}{Z}(5.29 \times 10^{-11} \text{ m})$$

$= n^2 (5.29 \times 10^{-11}$ m) for H atom

So, $r_1:r_2:r_3 = 1^2:2^2:3^2 = 1:4:9$.

16. A compound of vanadium has a magnetic moment of 1.73 B.M. Work out the electronic configuration of the vanadium ion in the compound.

Solution:

Magnetic moment = 1.73 B.M. = $\sqrt{n(n+2)}$

Here n = number of unpaired electrons n = 1

It means that in the compound vanadium ion must have one unpaired electrons.

$_{23}V = 1s^2, 2s^2 2p^6, 3s^2 3p^6 4s^2 3d^3$

$V^{+4} = 1s^2, 2s^2 2p^6, 3s^2 3p^6 3d^1$

CHAPTER-END EXERCISES

Practice Questions – I

1. Millikan's oil drop method is used to determine:

(a) Velocity of electron
(b) Mass of electron
(c) Charge of electron
(d) e/m ratio of electron

2. Wavelength of spectral line emitted is inversely proportional to:

(a) Energy (b) Velocity
(c) Radius (d) Quantum number

3. To explain the structure of atom the first use of quantum theory was made by:

(a) Heisenberg (b) Bohr
(c) Planck (d) Einstein

4. What is the packet of energy called?

(a) Proton (b) Photon
(c) Positron (d) Electron

5. Which one of the following forms a colourless solution in aqueous medium?

(a) Fe^{+2} (b) Cr^{3+}
(c) Mn^{+2} (d) Zn^{2+}

6. The incorrect configuration is:

(a) K = [Ar] $4s^1$ (b) Cr = [Ar] $3d^5 4s^1$
(c) Cr = [Ar] $3d^4 4s^2$ (d) Cu = [Ar] $3d^{10} 4s^1$

7. Which of the following elements has least number of electrons in its M shell?

(a) Mn (b) Ni
(c) K (d) Sc

8. In which orbital the angular momentum of an electron is zero?

(a) 2p (b) 2s
(c) 3d (d) 4f

9. Chromium is represented by the electronic configuration:

(a) [Ne] $3s^2 3p^6 3d^1 4s^2$
(b) [Ne] $3s^2 3p^6 3d^2 4s^1$
(c) [Ne] $3s^2 3p^6 3d^5 4s^1$
(d) [Ne] $3s^2 3p^6 4s^2 3d^4$

10. The correct set of quantum numbers is:

(a) n = 2, l = 1, m = –2, s = 0
(b) n = 2, l = –2, m = 1, s = $+\frac{1}{2}$

(c) n = 2, l = 2, m = –1, s = – –

(d) n = 2, l = 1, m = 0, s = + $\frac{1}{2}$

11. The values of four quantum numbers of valence electron of an element are n = 4, l = 0, m = 0 and s = + $\frac{1}{2}$. The element is:

(a) Ti (b) K
(c) Na (d) Sc

12. Set of isoelectronic species is:

(a) H_2, CO_2, CN^-, O^- (b) N, H_2S, CO
(c) N_2, CO, CN^-, O_2^{+2} (d) Ca, Mg, Cl

13. Number of orbitals in L energy level:

(a) 1 (b) 2
(c) 3 (d) 4

14. Ratio of radii of second and first Bohr orbits of H atom is:

(a) 2 (b) 4
(c) 3 (d) 4

15. Dual nature of particle was given by:

(a) Bohr theory
(b) Thomson model
(c) Heisenberg principle
(d) De Broglie equation

16. The outermost configuration of most electronegative element is:

(a) ns^2 np^5 (b) ns^2 np^6
(c) ns^2 np^4 (d) ns^2 np^6

17. Which one of the following expressions represent the electron probability function D?

(a) $4\pi r$ dr ψ^2 (b) $4\pi r^2$ dr ψ
(c) $4\pi r^2$ dr ψ^2 (d) $4\pi r$ dr ψ

18. If the electron of a hydrogen atom is present in the first orbit, the total energy of the electron is

(a) $-e^2/2r$ (b) $-e^2/r$
(c) $-e^2/r^2$ (d) $-e^2/2r^2$

19. Quantum numbers of an atom can be defined on the basis of:

(a) Aufbau's principle
(b) Heisenberg's uncertainity principle
(c) Hund's rule
(d) Pauli's exclusion principle

20. The number of waves in 3rd orbit of H-atom is:

(a) 1 (b) 2
(c) 3 (d) 4

21. Which of the following does not travel with the speed of light?

(a) De Broglie waves (b) X-rays
(c) Gamma rays (d) All of these

22. How many electrons can be accommodated in a p-orbital?

(a) 6 electrons (b) 2 electrons
(c) 4 electrons (d) None of these

23. The quantum number 'm' of a free gaseous atom is associated with:

(a) The effective volume of the orbital
(b) The shape of the orbital
(c) The spatial orientation of the orbital
(d) The energy of the orbital in the absence of a magnetic field

24. In ground state, an element has 13 electrons in its M shell. The element is:

(a) Copper (b) Iron
(c) Nickel (d) Chromium

25. Which of the following pair of ions have same para magnetic moment?

(a) Cu^{2+}, Ti^{3+} (b) Ti^{3+}, Ni^{2+}
(c) Ti^{4+}, Cu^{2+} (d) Mn^{2+}, Cu^{2+}

26. The total spin resulting from a d^3 configuration is:

(a) 3 (b) $\frac{3}{2}$
(c) 1 (d) Zero

27. The number of nodal planes in a p_x orbital is

(a) 1 (b) 2
(c) 3 (d) 0

28. How many d-electrons are present in Cr^{2+} ion?

(a) 5 (b) 6
(c) 3 (d) 4

29. l = 3, then the values of magnetic quantum numbers are?

(a) ± 1, ± 2, ± 3 (b) 0, ± 1, ± 2, ± 3
(c) –1, –2, –3 (d) 0, + 1, + 2, + 3

30. Which one of the following pairs of ions have the same electronic configuration?

(a) Sr^{3+}, Cr^{3+} (b) Fe^{3+}, Mn^{2+}
(c) Fe^{3+}, Co^{3+} (d) Cr^{3+}, Fe^{3+}

31. The radius of hydrogen atom is 0.53 Å. The radius of Li^{2+} is of:

(a) 1.27 Å (b) 0.17 Å
(c) 0.57 Å (d) 0.99 Å

32. The atomic number of an element is 35. What is the total number of electrons present in all the p orbitals of the ground state atom of that element?

(a) 17 (b) 11
(c) 23 (d) 6

33. The atomic number of an element is 17. The number of orbitals electron pairs in its valence shell is:

(a) 3 (b) 4
(c) 6 (d) 8

34. The correct representation for d orbital is:

(a) $(n-1)\,d^{1-9}\,ns^{1}$ (b) $(n-1)\,d^{1-10}\,ns^{1-2}$
(c) $(n-1)\,d^{1-5}$ (d) $(n-1)\,d^{1-10}\,ns^{2}$

35. An element has 2 electrons in its K shell, 8 electrons in L shell, 13 electrons in M shell and one electron in N shell. The element is:

(a) Cr (b) Fe
(c) V (d) Ti

36. Effective magnetic moment of Sc^{3+} ion is:

(a) 0 (b) 1.73
(c) 2.83 (d) 3.87

37. Which of the following is not possible?

(a) $n = 2, l = 1, m = 0$
(b) $n = 2, l = 0, m = -1$
(c) $n = 3, l = 0, m = 0$
(d) $n = 3, l = 1, m = -1$

38. The most probable radius (in pm) for finding the electron in He^{+} is:

(a) 105.8 (b) 52.9
(c) 26.5 (d) 0.0

39. In which of the following pairs are both the ions coloured in aqueous solution?

(a) Sc^{3+}, Co^{2+} (b) Ni^{2+}, Cu^{+}
(c) Ni^{2+}, Ti^{3+} (d) Sc^{3+}, Ti^{3+}
(Atomic number Sc = 21, Ti = 22, Ni = 28, Cu = 29, Co = 27)

40. When potassium metal is exposed to violet light:

(a) There is no effect
(b) Ejection of electron takes place
(c) The absorption of electrons takes place
(d) Ejection of some potassium atoms occurs

41. The atomic number of Ni and Cu are 28 and 29 re spectively. The electronic configuration $1s^2\,2s^2\,2p^6\,3s^2\,3p^6\,3d^{10}$ represents:

(a) Cu^{+} (b) Cu^{2+}
(c) Ni^{2+} (d) Ni

42. Rutherford's experiment, which established the nuclear model of the atom, used a beam of:

(a) β-particles, which impinged on a metal foil and got absorbed.
(b) γ-rays, which impinged on a metal foil and ejected electrons.
(c) Helium atoms, which impinged on a metal foil and got scattered.
(d) Helium nuclei, which impinged on a metal foil and got scatterd.

43. The quantum number $+\frac{1}{2}$ and $-\frac{1}{2}$ for the electron spin represent:

(a) Rotation of the electron in clockwise and anti-clockwise direction respectively.
(b) Rotation of the electron in anti clockwise and clockwise direction respectively.
(c) Magnetic moment of the electron pointing up and down respectively.
(d) Two quantum mechanical spin states which have no classical analogues.

44. The electronic configuration of an element is $1s^2\,2s^2\,2p^6\,3s^2\,3p^6\,3d^5\,4s^1$. This represents:

(a) Excited state (b) Ground state
(c) Cationic form (d) Anionic form

45. The angular momentum (L) of an electron in a Bohr orbit is given as:

(a) $L = nh/2\pi$
(b) $L = \sqrt{[l(l+1)\,h/2\,\pi]}$
(c) $L = mg/2\,\pi$
(d) $L = h/4\,\pi$

46. Which one of the following is the standard for atomic mass?

(a) ${}_{1}H^{1}$ (b) ${}_{6}C^{12}$
(c) ${}_{6}C^{14}$ (d) ${}_{8}O^{16}$

47. Which of the following relates to photons both as wave motion and as a stream of particles?

(a) Interference (b) $E = mc^2$
(c) Diffraction (d) $E = h\upsilon$

48. Electromagnetic radiation with maximum wavelength is:

(a) Radiowave (b) X-ray
(c) Infraredt (d) Ultraviolet

49. Rutherford's scattering experiment is related to the size of the:

(a) Atom (b) Nucleus
(c) Neutron (d) Electron

50. The principal quantum number of an atom is related to the:
(a) Orientation of the orbital in space
(b) Spin angular momentum
(c) Orbital angular momentum
(d) Size of the orbital

51. The orbital diagram in which both the Pauli's exclusion principle and Hund's rule are violated is:
(a) [↑↓] [↑↑ | ↑ |]
(b) [↑↓] [↑↓ | ↑↓ | ↑↓]
(c) [↑↓] [↓ | ↓ |]
(d) [↑↓] [↑↓ | ↑↓ | ↑]

52. The maximum number of 3d electrons having $s = +\frac{1}{2}$ are:
(a) 10 (b) 5
(c) 14 (d) 7

53. The fourth electron of Be atom will have which of the following quantum numbers?
(a) $1, 0, 0, \frac{1}{2}$ (b) $2, 0, 0, -\frac{1}{2}$
(c) $2, 1, 0, +\frac{1}{2}$ (d) $1, 1, 1, +\frac{1}{2}$

54. The electron density between 1s and 2s orbital is:
(a) High (b) Low
(c) Zero (d) None of these

55. A p-orbital can accommodate upto:
(a) Six electrons
(b) Four electrons
(c) Two electrons with opposite spins
(d) Two electrons with parallel spins

Practice Questions – II

56. If the wavelength of an electromagnetic radiation is 2000 Å. What is the energy in ergs?
(a) 9.92×10^{-19} (b) 9.94×10^{-12}
(c) 4.97×10^{-12} (d) 4.97×10^{-19}

57. The de Broglie wavelength of the electron in the ground state of hydrogen atom is [K.E.= 13.6 eV]; $1 eV = 1.602 \times 10^{-19}$ J:
(a) 33.28 nm (b) 3.328 nm
(c) 0.3328 nm (d) 0.0332 nm

58. Uncertainty in position of a particle of 25 g in space is 10^{-5} m. Hence uncertainty in velocity (ms^{-1}) is (Planck constant $h = 6.6 \times 10^{-34}$ Js):
(a) 2.1×10^{-28} (b) 2.1×10^{-34}
(c) 0.5×10^{-34} (d) 5.0×10^{-24}

59. The total number of electrons present in all the s orbitals, all the p orbitals and all the d orbitals of cesium ion are respectively:
(a) 12, 20, 22 (b) 8, 22, 24
(c) 10, 24, 20 (d) 8, 26, 10

60. Energy of H-atom in the ground state is –13.6 eV, hence energy in the second excited state is:
(a) –6.8 eV (b) –3.4eV
(c) –1.51 eV (d) –4.53 eV

61. The de Broglie wavelength of a Tennis ball of mass 60 g moving with a velocity of 10 metres per second is approximately:
(Planck constant $h = 6.63 \times 10^{-31}$ Js):
(a) 10^{-33} m (b) 10^{-31}m
(c) 10^{-16} m (d) 10^{-25} m

62. The relationship between energy E, of the radiation with a wavelength 8000 Å and the energy of the radiation with a wavelength 16000 Å is:
(a) $E_1 = 2E_2$ (b) $E_1 = 4E_2$
(c) $E_1 = 6E_2$ (d) $E_1 = E_2$

63. The atomic numbers of elements X, Y, Z are 19. 21 and 25 respectively. The number of electrons present in the 'M' shells of these elements follow the order:
(a) $Z > Y > X$ (b) $X > Y > Z$
(c) $Z > X > Y$ (d) $Y > Z > X$

64. An electron is moving in Bohr's fourth orbit, its de-Broglie wavelength is *X*. What is the circumference of the fourth orbit?
(a) 2λ (b) $2/\lambda$
(c) 3λ (d) 4λ

65. The correct order of number of unpaired electrons in the ion Cu^{2+} Ni^{2+}, Fe^{3+} and Cr^{3+} is:
(a) $Cu^{2+} > Ni^{2+} > Cr^{3+} > Fe^{3+}$
(b) $Ni^{2+} > Cu^{2+} > Fe^{3+} > Cr^{3+}$
(c) $Fe^{3+} > Cr^{3+} > Ni^{2+} > Cu^{2+}$
(d) $Fe^{3+} > Cr^{3+} > Cu^{2+} > Ni^{2+}$

66. Find the magnetic moment of a divalent ion in aqueous solution if its atomic number is 25:

(a) 6.9 B.M. (b) 5.9 B.M.
(c) 4.9 B.M. (d) 3.0 B.M.

67. The magnetic moment of Cu^{2+} ion is:
(a) 2.6 (b) 2.76
(c) 1.73 (d) 0

68. Which of the following sets of quantum numbers is correct for an electron in 4f orbital?
(a) $n = 4, l = 3, m = +4, s = +\frac{1}{2}$
(b) $n = 4, l = 4, m = -4, s = -\frac{1}{2}$
(c) $n = 4, l = 3, m = +1, s = +\frac{1}{2}$
(d) $n = 3, l = 2, m = -2, s = +\frac{1}{2}$

69. The energy ratio of a photon of wavelength 3000Å and 6000Å is:
(a) 1:1 (b) 2:1
(c) 1:2 (d) 1:4

70. The orbital angular momentum for an electron re volving in an orbit is given by $\sqrt{l(l+1)}\ h/2\pi$. This momentum for an s electron will be given by:
(a) $+\frac{1}{2}.h/2\pi$ (b) Zero
(c) $h/2\pi$ (d) $\sqrt{2}.h/2\pi$

71. The de Broglie wavelength associated with a particle of mass 10^{-6} kg moving with a velocity of 10 ms^{-1} is:
(a) 6.63×10^{-7} m (b) 6.63×10^{-16} m
(c) 6.63×10^{-21} m (d) 6.63×10^{-29} m

72. Which one of the following grouping represents a collection of isoelectronic species? (At. numbers Cs–55, Br–35)
(a) Na^+, Ca^{2+}, Mg^{2+} (b) N^{3-}, F^-, Na^+
(c) Be, Al^{3+}, Cl^- (d) Ca^{2+}, Cs^+, Br

73. The velocity of an electron in the second shell of hydrogen atom is:
(a) $10.94 \times 10^6\ ms^{-1}$ (b) $18.88 \times 10^6\ ms^{-1}$
(c) $1.888 \times 10^6\ ms^{-1}$ (d) $1.094 \times 10^6\ ms^{-1}$

74. Consider the ground state of Cr atom (Z = 24). The numbers of electrons with the azimuthal quantum numbers, $l = 1$ and 2 are, respectively:
(a) 12 and 4 (b) 12 and 5
(c) 16 and 4 (d) 16 and 5

75. If the nitrogen atom has electronic configuration $1s^7$, it would have energy lower than that of the normal ground state configuration $1s^2\ 2s^2\ 2p^3$, because the electrons would be closer to the nucleus. Yet $1s^7$ is not observed because it violates:
(a) Heisenberg uncertainty principle
(b) Hund's rule
(c) Pauli's exclusion principle
(d) Bohr postulates of stationary orbits

76. Radial nodes present in 3s and 2p orbitals are respectively:
(a) 0, 2 (b) 2, 0
(c) 2, 1 (d) 1,2

77. The radius of which of the following orbits is same as that of the first Bohr's orbit of hydrogen atom?
(a) He^+ (n = 2)
(b) Li^{2+} (n = 2)
(c) Li^{2+}(n = 3)
(d) Be^{3+} (n = 2)

78. The wavelength associated with a golf ball weighing 200 g and moving at a speed of 5 m/h is of the order:
(a) 10^{-10} m (b) 10^{-20} m
(c) 10^{-30} m (d) 10^{-40} m

79. Ground state electronic configuration of nitrogen atom can be represented by:
(1) [↿⇂] [↿⇂] [↿|↿|↿]
(2) [↿⇂] [↿⇂] [↿|⇃|↿]
(3) [↿⇂] [↿⇂] [↿|⇃|⇃]
(4) [↿⇂] [↿⇂] [⇃|⇃|⇃]
(a) 1 only (b) 1, 2
(c) 1, 4 (d) 2, 3

80. Which of the following statement(s) are correct?
(1) The electronic configuration of Cr is [Ar] $3d^5\ 4s^1$ (atomic number of Cr = 24)
(2) The magnetic quantum number may have a negative value
(3) In silver atom, 23 electrons have a spin of one type and 24 of the opposite type (atomic number of Ag = 47)
(4) The oxidation state of nitrogen in HN_3, is –3
(a) 1, 2, 3 (b) 2, 3, 4
(c) 3, 4 (d) 1, 2, 4

81. For a d electron, the orbital angular momentum is:
(a) $\sqrt{6}h$ (b) $\sqrt{2}h$
(c) H (d) 2h

82. The orbital angular momentum of an electron in 2s orbital is:

(a) $+\frac{1}{2}\cdot\frac{h}{2\pi}$ (b) zero
(c) $\frac{h}{2\pi}$ (d) $\sqrt{2}\,\frac{h}{2\pi}$

83. The mass of an electron is m. Its charge is e and it is accelerated from rest through a potential difference V. The velocity acquired by the electron will be:

(a) $\sqrt{V/m}$ (b) $\sqrt{eV/m}$
(c) $\sqrt{2eV/m}$ (d) None

84. An atom A has the electronic configuration of $1s^2\ 2s^2\ 2p^1$. Atom B has the electronic configuration of $1s^2\ 2s^2\ 2p^1$. The empirical formula of the compound obtained from the reaction of A and B is:

(a) AB (b) AB_3
(c) A_3B_3 (d) A_2B_6

85. The radius of an atomic nucleus is of the order of:

(a) 10^{-8} cm (b) 10^{-13} cm
(c) 10^{-10} cm (d) 10^{-15} cm

86. Bohr model can explain:

(a) The solar spectrum.
(b) The spectrum of hydrogen molecule.
(c) The spectrum of hydrogen atom only.
(d) Spectrum of an atom or ion containing one electron only.

87. Which electronic level would allow the hydrogen atom to absorb a photon but not to emit a photon?

(a) 1s (b) 2s
(c) 2p (d) 3s

88. Correct set of four quantum numbers for the valence (outermost) electron of rubidium (Z = 37) is:

(a) 5, 0, 0, ±1/2 (b) 6, 0, 0, +1/2
(c) 5, 1, 1, ±1/2 (d) 5, 1, 0, ±1/2

89. The increasing order (lowest first) for the values of e/m (charge/mass) for electron (e), proton (p), neutron (n) and alpha particle (a) is:

(a) n, p, a , e (b) n, p, e, a
(c) n, a, p, e (d) e, p, n, a

90. The ionization energy of hydrogen atom is 13.6 eV. What will be the ionization energy of He^+?

(a) 13.6 eV (b) 54.4 eV
(c) 122.4eV (d) zero

91. If S_1 be the specific charge (e/m) of cathode rays and S_2 be that of positive rays then which is true?

(a) $S_1 = S_2$ (b) $S_1 < S_2$
(c) $S_1 > S_2$ (d) None of these

92. Predict the total spin in Ni^{2+} ion

(a) ±5/2 (b) ±3/2
(c) ±1/2 (d) ±1

93. The orbital diagram in which Aufbau principle is violated is:

(a) [↑↓] [↑↓|↑|] (b) [↑] [↑↓|↑|↑]
(c) [↑↓] [↑|↑|↑] (d) [↑↓] [↑↓|↑↓|↑]

94. Wave function vs distance from nucleus graph of an orbital is given below:

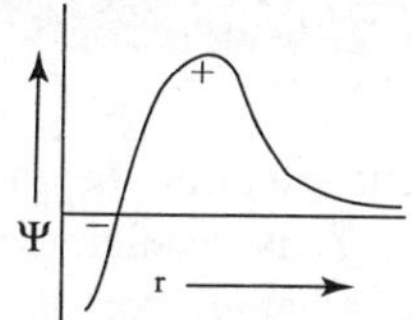

The number of nodal sphere of this orbital is:

(a) 1 (b) 2
(c) 3 (d) 4

95. For the electronic transition from n = 2 → n = 1, which of the following will produce shortest wave length?

(a) Li^{2+} ion (b) D atom
(c) He^+ ion (d) H atom

96. Which of the following curves may represent the speed of the electron in a hydrogen atom as a function of the principal quantum number n?

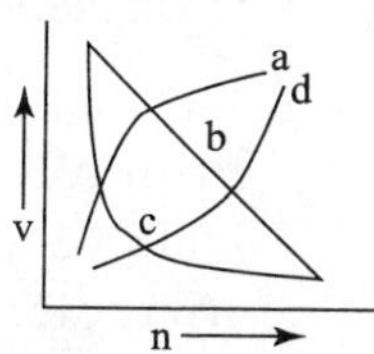

(a) D (b) C
(c) B (d) A

97. Spin of electron is:

(a) Rotation of electron about it's own axis.
(b) Clock wise and anticlock wise rotation of electron.
(c) An intrinsic characteristics of electron connect with its magnetic field.
(d) A quantum number which depend upon direction and speed of rotation of electron.

98. Probability of finding the electron ψ^2 of s orbital doesn't depend upon:

(a) Azimuthal quantum number.
(b) Energy of s orbital.

(c) Principal quantum number.
(d) Distance from nucleus (r).

99. The charge cloud of a single electron in a $2p_x$ atomic orbital has two lobes of electron density. This means:

(a) There is a high probability of locating the electron in a 2px atomic orbital at values of $x > 0$
(b) There is a great probability of finding a p electron right at the nucleus
(c) There is a high probability of locating it values of $x < 0$ but no probability at alloy locating if any where in the yz plane along which $x = 0$.
(d) Both (a) and (c)

100. The wavelength of the de Broglie wave of the electron revolving in the fifth orbit of the hydrogen atom is (r_o is the Bohr's radius = 0.529 Å).

(a) $20\ r_0$ (b) $(10\ \pi)\ r_0$
(c) $5\ \pi\ r_0$ (d) $15\ \pi\ r_0$

Practice Questions – III

101. A monoenergetic electron beam with a de Broglie wavelength of x Å is accelerated till its wavelength is halved. By what factor is its kinetic energy changed?

(a) 8 (b) 6
(c) 4 (d) 3

102. The de Broglie wavelength associated with a ball of mass 1 kg having a kinetic energy 0.5 J is:

(a) 6.626×10^{-34} m (b) 13.2×10^{-34} m
(c) 10.38×10^{-21} m (d) 6.626 Å

103. Match list I with list II and select the correct answer using the codes given below the lists:

List I (Metal ions)	List II (Magnetic moment)
1. Cr^{3+}	(i) $\sqrt{35}$
2. Fe^{2+}	(ii) $\sqrt{30}$
3. Ni^{2+}	(iii) $\sqrt{24}$
4. Mn^{2+}	(iv) $\sqrt{15}$
	(v) $\sqrt{8}$

The correct matching is:

	1	2	3	4
(a)	(i)	(iii)	(v)	(iv)
(b)	(ii)	(iii)	(v)	(i)
(c)	(iv)	(iii)	(v)	(i)
(d)	(iv)	(v)	(iii)	(i)

104. The size of a microscopic particle is one micron and its mass is 6×10^{-13} gm. If its position may be mea sured to within 0.1% of its size, the uncertainty in velocity, in cm s^{-1}, is approximately:

(a) $10^{-6}/3\ \pi$ (b) $10^{-7}/2\pi$
(c) $10^{-5}/4\ \pi$ (d) $10^{-7}/4\pi$

105. The electrons, identified by quantum numbers n and l (i) $n = 4$. $l = 1$ (ii) $n = 4$, $l = 0$ (iii) $n = 3$, $l = 2$ (iv) $n = 3$, $l = 1$ can be placed in order of increasing energy from the lowest to highest as:

(a) (iv) < (ii) < (iii) < (i)
(b) (ii) < (iv) < (i) < (iii)
(c) (i) < (iii) < (ii) < (iv)
(d) (iii) < (i) < (iv) < (ii)

106. What is the wavelength of the radiation emitted produced in a line in the Lyman series when an electron falls from fourth stationary state in hydrogen atom? ($R_H = 1.1 \times 10^7\ m^{-1}$)

(a) 96.97 nm (b) 969.7 nm
(c) 9.697 nm (d) None of these

107. Rearrange the following (I to IV) in the order of in creasing masses and choose the correct answer from (a), (b), (c), (d). (atomic masses: N =14, O = 16, Cu = 63).

I. 1 molecule of oxygen
II. 1 atom of nitrogen
III. 1×10^{10} g molecular weight of oxygen
IV. 1×10^{-18} g atomic weight of copper

(a) II < I < IV < III (b) IV < III < II < I
(c) II < III < I < IV (d) III < IV < I < II

108. The ionization energy of He^+ is 19.6×10^{-18} J $atom^{-1}$. Calculate the energy of the first stationary state of Li^{2+}.

(a) 19.6×10^{-18} J $atom^{-1}$
(b) 4.41×10^{-18} J $atom^{-1}$
(c) 19.6×10^{-19} J $atom^{-1}$
(d) 4.41×10^{-17} J $atom^{-1}$

109. The masses of photons corresponding to the first lines of the Lyman and the Balmer series of the atomic spectrum of hydrogen are in the ratio of

(a) 4:1 (b) 27:5
(c) 1:4 (d) 5:27

110. An X-ray tube is operated at 50,000 volts. The shortest wavelength limit of the X-rays produced is

(a) 0.1245 Å (b) 0.3485 Å
(c) 0.2485 Å (d) 0.03456 Å

111. The ratio of the difference between 2nd and 3rd Bohr's orbit energy to that between 3rd and 4th orbit energy is

(a) 7/20 (b) 20/7
(c) 27/9 (d) 9/27

112. An electron in a hydrogen atom in its ground state absorbs 1: 50 times as much energy as the minimum required for it to escape from the atom. What is the wavelength of the emitted electron?

(a) 4.7 Å (b) 4.70 pm
(c) 6.3 Å (d) 8.4 Å

113. If the radius of the first Bohr orbit is 'a', then de Broglie wavelength of electron in 3rd orbit is nearly:

(a) 2π a (b) 6π a
(c) 3a (d) a/3

114. If the shortest wavelength of H atom in Lyman series is 'a', then longest wavelength in Balmer series of He^+ is:

(a) a/4 (b) 5a/9
(c) 4a/9 (d) 9a/5

115. In hydrogen atom, an orbit has a diameter of about 16.92 A. What is the maximum number of electrons that can be accommodated?

(a) 32 (b) 16
(c) 48 (d) 72

116. Energy levels A, B, C of a certain atom corresponds to increasing values of energy, i.e., $E_A < E_B < E_C$. If X_1, X_2 and X_3 are the wavelengths of radiations cor responding to the transitions C to B, B to A and C to A respectively, which of the following statement is correct?

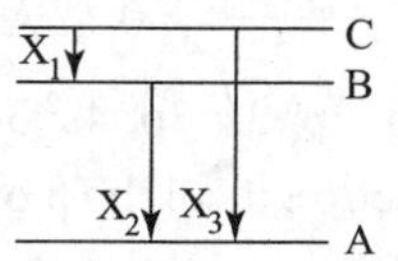

(a) $X_1 + X_2 + X_3 = 0$ (b) $X_3 = X_1 + X_2$
(c) $X_3^2 = X_1^2 + X_2^2$ (d) $X_3 = \dfrac{X_1 X_2}{X_1 + X_2}$

117. A 1000 watt radio transmitter operates at a frequen cy of 880 kc/sec. How many photons per sec. does it emit? [$h = 6.626 \times 10^{-34}$ Js]

(a) 2.51×10^{30} (b) 2.27×10^{28}
(c) 1.72×10^{30} (d) 1.77×10^{27}

118. How many moles of electrons weigh one kilogram? (mass of electron = 9.108×10^{-31} kg, Avogadro number = 6.023×10^{23})

(a) 6.023×10^{23} (b) $\dfrac{1}{9.108} \times 10^{31}$
(c) $\dfrac{6.023}{9.108} \times 10^{54}$ (d) $\dfrac{1}{9.108 \times 6.023} \times 10^{8}$

119. Match the following:

List I	List II
1. Number of values of '*l*' for an energy level ....(n – 1)	(1) 0, 1, 2
2. Actual values of '*l*' for a particular type of orbital	(2) + *l* ....+ 2, + 1, 0, –*l*
3. Number of 'm' values	(3) –2 .... –*l*
4. Actual values of 'n' for a particular type of orbital	(4) n

The correct matching is:

	1	2	3	4
(a)	(4)	(3)	(1)	(2)
(b)	(4)	(1)	(3)	(2)
(c)	(1)	(2)	(4)	(3)
(d)	(2)	(3)	(1)	(4)

120. Calculate the wavelength and energy of the radiation emitted for the electronic transition from infinity (∞) to stationary state first of the hydrogen atom. ($R_H = 1.09678 \times 10^7$ m^{-1}, $h = 6.6256 \times 10^{-34}$ Js)

(a) 2.18×10^{-21} kJ (b) 3.18×10^{-22} kJ
(c) 1.18×10^{-23} kJ (d) 2.18×10^{-31} kJ

121. In Bohr series of lines of hydrogen spectrum, the third line from the red end corresponds to which one of the following inter-orbit jumps of the electron for Bohr orbits in an atom of hydrogen?

(a) $3 \longrightarrow 2$ (b) $5 \longrightarrow 2$
(c) $4 \longrightarrow 1$ (d) $2 \longrightarrow 5$

122. The wavelength of the radiation emitted, when in a hydrogen atom electron falls from infinity to stationary state one, would be (Rydberg constant $= 1.097 \times 10^7\ m^{-1}$)
(a) 91 nm (b) 192 nm
(c) 406 nm (d) 9.1×10^{-8} nm

123. Which of the following sets of quantum numbers represents the highest energy of an atom?
(a) $n = 3, l = 2, m = 1, s = +1/2$
(b) $n = 4, l = 0, m = 0, s = +1/2$
(c) $n = 3, l = 0, m = 0, s = +1/2$
(d) $n = 3, l = 1, m = 1, s = +1/2$

124. In a multielectron atom, which of the following orbitals described by the three quantum numbers will have the same energy in the absence of magnetic field and electric fields?
(a) $n = 1, l = 0, m = 0$
(b) $n = 2, l = 0, m = 0$
(c) $n = 3, l = 1, m = 1$
(d) $n = 3, l = 2, m = 1$
(e) $n = 3, l = 2, m = 0$
(a) b and c (b) d and e
(c) c and d (d) a and b

125. Of the following sets which one does not contain iso electronic species?
(a) PO_4^{3-}, SO_4^{2-}, ClO_4^-
(b) CN^-, N_2, C_2^{2-}
(c) SO_3^{2-}, CO_3^{2-}, NO_3^-
(d) BO_3^{3-}, CO_3^{2-}, NO_3^-

Answer Keys

1. (c)	2. (a)	3. (b)	4. (b)	5. (d)	6. (c)	7. (c)	8. (b)	9. (c)	10. (d)
11. (b)	12. (c)	13. (d)	14. (b)	15. (d)	16. (a)	17. (c)	18. (a)	19. (d)	20. (c)
21. (a)	22. (b)	23. (c)	24. (d)	25. (a)	26. (b)	27. (a)	28. (d)	29. (b)	30. (b)
31. (b)	32. (a)	33. (a)	34. (b)	35. (a)	36. (a)	37. (b)	38. (c)	39. (c)	40. (b)
41. (a)	42. (d)	43. (d)	44. (b)	45. (a)	46. (b)	47. (d)	48. (a)	49. (b)	50. (d)
51. (a)	52. (b)	53. (b)	54. (c)	55. (c)	56. (b)	57. (c)	58. (a)	59. (c)	60. (c)
61. (a)	62. (a)	63. (a)	64. (d)	65. (c)	66. (b)	67. (c)	68. (c)	69. (b)	70. (b)
71. (d)	72. (b)	73. (d)	74. (b)	75. (c)	76. (b)	77. (d)	78. (c)	79. (c)	80. (a)
81. (a)	82. (b)	83. (c)	84. (a)	85. (b)	86. (d)	87. (a)	88. (a)	89. (c)	90. (b)
91. (c)	92. (d)	93. (b)	94. (a)	95. (a)	96. (b)	97. (c)	98. (a)	99. (d)	100. (b)
101. (c)	102. (a)	103. (c)	104. (d)	105. (a)	106. (a)	107. (a)	108. (d)	109. (b)	110. (c)
111. (b)	112. (a)	113. (b)	114. (d)	115. (a)	116. (d)	117. (c)	118. (d)	119. (b)	120. (a)
121. (b)	122. (a)	123. (a)	124. (b)	125. (c)					

Hints and Explanations for Selective Questions

5. As Zn^{2+} has no un-paired electron so it is colourless.

12. As all have 14 electrons so they are iso electronic.

14. As $r_n \propto n^2$. So $r_2 = 4 r_1$

16. As halogens are most electronegative so the configuration is $ns^2\ np^5$.

20. As number of wave = n = 3

24. As $_{24}Cr = 2, 8, 13, 1$ i.e., M shell has 13 electrons.

25. As n = 1 in them so they have same magnetic moment.

26. d^3 configuration means n = 3

$$\text{spin} = \frac{n}{2} = \frac{3}{2}$$

30. Both Fe^{3+} and Mn^{2+} have [Ar] $3d^5$ configuration.

31. $r_n = 0.53 \times \frac{n^2}{Z} = \frac{0.53 \times 1^2}{3} = 0.17$ Å

32. $(Z = 35): 1s^2\ 2s^2\ 2p^6\ 3s^2\ 3p^6\ 4s^2\ 3d^{10}\ 4p^5$

Number of electron in all the p orbitals

$= 6 \times 2 + 5 = 17$

36. As it has no unpaired electrons so it is dramagnetic and it is zero.

37. As values of m is from –1 to +1 including zero.

39. As Ni^{2+}, Ti^{3+} have 2, 1 unpaired electrons respectively so both are coloured.

41. It represent the configuration of Cu^+.

43. These represent two quantum mechanical spin states which have no classical analogues.

49. The central part consisting whole of the positive charge and most of the mass, called nucleus, is extremely small in size compared to the size of the atom.

50. The principal quantum number (n) is related to the size of the orbital (n = 1, 2, 3....)

55. One p orbital can accommodate up to two electrons with opposite spin while p-subshell can accommodate upto six electrons.

56. $E = hc/\lambda = \frac{6.626 \times 10^{-27} \times 3 \times 10^6}{2000 \times 10^{-8}}$

$= 9.94 \times 10^{-12}$ ergs

58. $\Delta x.\ \Delta v \geq \frac{h}{4\pi m}$ or $\Delta v = \frac{h}{4\pi m.\Delta x}$

59. Cs^+:$1s^2\ 2s^2\ 2p^6\ 3s^2\ 3p^6\ 4s^2\ 3d^{10}\ 4p^6\ 5s^2\ 4d^{10}\ 5p^6$
Number of electrons in s orbitals = 5 × 2 = 10
Number of electrons in p orbitals = 4 × 6 = 24
Number of electrons in d orbitals = 10 × 2 = 20

60. $E_n = \frac{-13.6}{n^2}$ eV

For second excited state n = 3,

$E_3 = -\frac{13.6}{9} = -1.51$ eV

61. $\lambda = h/mv$

$\lambda = \frac{6.626 \times 10^{-34}}{60 \times 10^{-3} \times 10} = 10^{-33}$ m (nearly)

63. As Z has 13 electrons, Y has 9 electrons and X has 8 electrons in M shell.

64. As circumference = $n\lambda = 4\lambda$

65. As they have 5, 3, 2, 1 number of unpaired electrons respectively.

66. It is Mn^{2+} having five unpaired electrons so its magnetic moment is:

$\mu = \sqrt{5\,(5 + 2)} = 5.9$ B.M.

67. As it has only one unpaired electron so magnetic moment is: li $= \sqrt{n(n+2)} =$

$\sqrt{1\,(1 + 2)} = 1.73$ B.M.

68. Any sub-orbit is represented as nl such that n is the principle quantum number (in the form of values) and l is the azimuthal quantum number (its name). Value of l < n, 1 0 1 2 3 4

s p d f g

Value of m: $-l$, –l, + 1 ....0, $+ l$
Value of s: + 1/2 or – 1/2
Thus for 4f: n = 4, l = 3, m = any value between –3 To + 3.

69. $E_1/E_2 = \lambda_2/\lambda_1$
= 6000/3000 = 2:1

70. As $\ell = 0$ so mvr is also zero.

71. Use the relation; $\lambda = \frac{h}{mv} = \frac{6.626 \times 10^{-34}}{10^{-6} \times 10} = 6.626 \times 10^{29}$

72. As N^{-3}, F^-, Na^+ all have 10 electrons.

73. $v_n = 2.188 \times 10^6 \times \frac{Z}{n} = 2.188 \times 10^6 \times \frac{1}{2}$

$= 1.094 \times 10^6\ ms^{-1}$

74. E.C of Cr (Z = 24) is $1s^2\ 2s^2\ 2p^6\ 3s^2\ 3p^6\ 3d^5\ 4s^1$ so electrons with l = 1 (p), are 12 with l = 2 (d), are 5

75. As maximum number of electrons in any orbit, sub-orbit or orbital is decided by Pauli's law.

78. $\lambda = h/mv = \frac{6.626 \times 10^{-34}}{200 \times 10^{-3} \times 5}$

$= 6.626 \times 10^{-32}$ m

81. As mvr $= \sqrt{2l(l + 1)}\ h$

Here l = 2 so mvr $= \sqrt{6}\ \hbar$

82. Here l = 0 so mvr = 0.

84. Both A and B have the valencies 3 So, the formula is AB.

85. The radius of nucleus is of the order of 1.5×10^{-13} to 6.5×10^{-13} cm or 1.5 to 6.5 Fermi (1 Fermi = 10^{-13} cm)

86. Bohr model can explain spectrum of atoms/ions containing one electron only.

87. It is the 1s level, the ground state, where the H-atom can only absorb a photon and go to higher excited states.

88. Rb has the configuration, $1s^2\ 2s^2\ 2p^2\ 3s^2\ 3p^6\ 3d^{10}\ 4s^2\ 4p^6\ 5s^1$; so n = 5, l = 0, m = 0 and s = $\frac{1}{2}$

90. I.E. of He^+ = 13.6 eV × Z^2
= 13.6 eV × 4 = 54.4 eV

91. Mass of positively charged ions in positive rays is more than mass of electrons.

92. Number of unpaired electrons in Ni^{2+} is two.

Total spin = $\pm \frac{1}{2} \times$ Number of unpaired electrons

93. s-subshell should be filled first as it possesses lower energy level than p-subshell.

95. Li^{2+} ion has the shortest wavelength.

100. Radius of the 5th orbit = $5^2 r_0 = 25\ r_0$

Circumference = $2\ \pi\ (25\ r_0) = 50\ \pi r_0 = 5\ \lambda$,

$n\lambda = 2\pi\ r_n \qquad \lambda = (10\ \pi)\ r_n$

101. $\lambda = \dfrac{h}{\sqrt{2mE}}$

$\lambda E^{1/2}$ = constant.

So $\lambda_1 \sqrt{E_1} = \lambda_2 \sqrt{E_2}$

$\lambda_1/\lambda_1 = \sqrt{(E_2/E_1)} = 2$

Therefore, $E_2/E_1 = 4$

102. $\lambda = h/\sqrt{2m\ E}$

$= \dfrac{6.626 \times 10^{-34}}{\sqrt{2 \times 1 \times 0.5}}$

$= 6.626 \times 10^{-34}$ m

103. As Cr^{3+} Fe^{2+} Ni^{2+}, Mn^{2+} have 3, 4. 2, 5 unpaired electron respectively.

104. $\Delta p.\ \Delta x = h/2\pi$

$m.\ \Delta v.\ \Delta x = h/2\pi$

$\Delta\ v = \dfrac{1}{4\ \pi} \times \dfrac{h}{m.\ \Delta x}$

$m = 6 \times 10^{-13}$ g.

$\Delta\ x = \dfrac{0.1}{100} \times 10^{-4}\ cm = 10^{-7}\ cm$

$\Delta v = \dfrac{1}{4\pi} \times \dfrac{6.626 \times 10^{-27}}{6 \times 10^{-13} \times 10^{-7}}\ cm\ s^{-1}$

$= \frac{1}{4}\ \pi \times 1.1 \times 10^{-7}\ cm\ s^{-1} = \dfrac{10^{-7}}{4\pi} cm\ s^{-1}$

106. $n_1 = 1$, $n_2 = 4$

Now $1/\lambda = R_H\ [l/n_1{}^2 - 1/n_2{}^2]$

$= 1.1 \times 10^{-7}\ [l/l^2 - 1/4^2]$

After solving

$\lambda = 96.9$ nm.

108. I.E. of $He^+ = E \times 2^2$ (as Z for He = 2)

I.E. of $Li^{2+} = E \times 3^2$ (as Z for Li = 3)

Hence, $\dfrac{\text{I.E.}\ (He^+)}{\text{I.E.}\ (Li^{2+})} = \dfrac{4}{9}$

Therefore, I.E. $(Li^{2+}) = \frac{9}{4} \times$ I.E. (He^+)

$= \frac{9}{4} \times 19.6 \times 10^{-18}$

$= 4.41 \times 10^{-17}$ J $atom^{-1}$

109. $\Delta E = mc^2$.

For Lymann series, $m_1c^2 \propto (l/l^2 - 1/2^2)$

For Balmer series, $m_2c^2 \propto (1/2^2 - 1/3^2)$

$m_1/m_2 = 3 \times 36/4 \times 5$

$m{:}m_2 = 27{:}5$

110. Energy of electrons striking the anti-cathode = $50000 \times 1.6 \times 10^{-19}$ J

$= 8.0 \times 10^{-15}$ J

$h\dfrac{c}{\lambda} = \dfrac{6.626 \times 10^{-34} \times 3 \times 10^{8}}{\lambda}$

$= \dfrac{19.88 \times 10^{-26}}{\lambda}$ J

$\lambda = \dfrac{19.88 \times 10^{-26}}{8 \times 10^{-15}}\ m = 2.485 \times 10^{-11}$ m

$= 0.2485$ Å.

111. As $E_n = -1312/n^2$ kJ/mole

$\Delta E_{3-2} = -1312[(1/n_2^2) - (1/n_1^2)]$

$= -1312\ (1/9 - 1/4)$

$= \dfrac{-1312 \times -5}{36}$

Similarly,

$\Delta E_{4-3} = -1312\ (1/16 - 1/9)$

$= \dfrac{-1312 \times -7}{16 \times 9}$

$\dfrac{\Delta E_{3-2}}{\Delta E_{4-3}} = \dfrac{20}{7}$

112. As 13.6 eV is needed for ionization, 20.4 eV (13.6×1.5) must have been absorbed. Of this, 6.8 eV is converted to kinetic energy.

$\lambda = \dfrac{h}{\sqrt{(2\ m\ K.E.)}}$

$= \dfrac{6.626 \times 10^{-34}\ Js}{\sqrt{(2 \times 9.1 \times 10^{-31} \times 6.8 \times 1.6 \times 10^{-19}\ J)}}$

$= 4.70 \times 10^{-10}$ m = 4.70 Å

113. $r_n = r_1 \times n^2$

$r_3 = 3^2 a = 9a$

$mvr_3 = \dfrac{3\ h}{2\pi}$

$mv = \dfrac{3h}{2\pi\ .\ 9a} = \dfrac{h}{6\pi a}$

$\lambda = \dfrac{h}{mv} = \dfrac{h.6\pi\ a}{h} = 6\pi\ a.$

114. For shortest λ of Lyman series of H $n_2 = \infty$

$\frac{1}{\lambda_L} = R_H \frac{1}{1^2} - \frac{1}{\infty^2}$

For longest λ of Balmer series of He^+ $n_2 = 3$

$\frac{1}{\lambda_B} = Z^2 R_H \frac{1}{2^2} - \frac{1}{3^2}$

So, $\frac{\lambda_B}{\lambda_A} = \frac{1 \times 36}{5 \times 4}$

Hence, $\lambda_B = 9\lambda_L/5 = 9a/5$.

115. $r = \frac{6.92}{2} = 8.46$ Å

$r_n = r^1 . n^2$

$n^2 = \frac{r_n}{r_1} = \frac{8.46 \text{ Å}}{0.529 \text{ Å}} = 16$

n = 4 (N – shell)

Number of electrons = $2n^2 = 2 \times 4^2 = 32$

117. Frequency = 880 kc/sec. = 880×10^3 cycles/sec. h = 6.626×10^{-34}Js

$h\upsilon = [6.626 \times 10^{-34}] \times 880 \times 10^3$ J

$= 583.1 \times 10^{-30}$ J

Power transmitted = 1000 watts = 1000 J/s

Number of photons emitted per sec.

$= \frac{1000}{583 \times 10^{-30}} = 1.72 \times 10^{30}$

118. As mass of one electron = 9.108×10^{-31} kg. So, 1 kg of electrons = $1/9.108 \times 10^{-31}$

$= \frac{10^{31}}{9.108 \times 6.022 \times 10^{23}}$ mol

$= \frac{10^{31-23}}{9.108 \times 6.022}$ mol $= \frac{10^8}{9.108 \times 6.022}$ mol

120. $n_1 = 1, n_2 = \infty$

$\frac{1}{\lambda} = R_H \left(\frac{1}{n_1^2} - \frac{1}{n_2^2}\right)$

$\frac{1}{\lambda} = 1.09678 \times 10^7 \left(\frac{1}{1^2} - \frac{1}{\infty^2}\right)$

$\lambda = 9.11 \times 10^{-8}$ m

$E = hv = h \times c/\lambda$ ($c = 3 \times 10^8$ m sec.$^{-1}$)

$= \frac{6.6256 \times 10^{-34} \times 10^8}{9.11 \times 10^{-8}} = 2.18 \times 10^{-18}$ J

$= 2.18 \times 10^{-21}$ kj

121. The lines at the red end suggest Balmer series. These are obtained for the jumps n = 2 from n = 3 and second line from n = 4 and third line from n = 5, that is, 5 ⟶ 2.

122. $\frac{1}{\lambda} = v_H = R_H \frac{1}{n_1^2} - \frac{1}{n_2^2}$

$= 1.097 \times 10^7 \frac{1}{1^2} - \frac{1}{\infty^2}$

now $\lambda = \frac{1}{1.097 \times 10^7}$ m $= 9.11 \times 10^{-8}$ m

$= 91.1 \times 10^{-9}$ m

= 91.1 nm. (1 nm = 10^{-9} m)

123. Using (n + ℓ) rule here,

n = 3, ℓ = 2, m = 1, s = $\frac{1}{2}$ has maximum energy

124. Orbitals having same (n + *l*) value in the absence of electric and magnetic field will have same energy.

125. Number of electrons in SO_3^{2-} = 16 + 8 × 3 + 2 = 42

Number of e⁻ in CO_3^{2-} = 6 + 8 × 3 + 2 = 32

Number of e⁻ in NO_3^- = 7 + 8 × 3 + 1 = 32

These are not isoelectronic species as number of electrons are not same.

Previous Years' Questions

1. Set of isoelectronic species is
   (a) H_2, CO_2, CN^-, O^- (b) N, H_2S, CO
   (c) N_2, CO, CN^-, O_2^{+2} (d) Ca, Mg, Cl

   [2000]

2. Electron energy of a photon is given as: ΔE/atom = 3.03×10^{-19} J atom^{-1} then, the wavelength of the photon is
   (a) 6.56 nm (b) 65.6 nm
   (c) 656 nm (d) 0.656 nm

   Given, h (Planck's constant) = 6.63×10^{-34} J/s
   c (velocity of light) = 3.00×10^8 ms^{-1}

   [2000]

3. Which one of the following forms a colourless solution in aqueous medium?
   (a) V^{3+} (b) Cr^{3+}
   (c) Ti^{3+} (d) Sc^{3+}
   (Atomic number Sc = 21, Ti = 22, V = 23, Cr = 24)

   [2000]

4. The following quantum numbers are possible for how many orbital n = 3, l = 2, m = +2?
(a) 1 (b) 2
(c) 3 (d) 4
[2001]

5. Which of the following is isoelectronic?
(a) CO_2, NO_2 (b) NO_2^-, CO_2
(c) CN^-, CO (d) SO_2, CO_2
[2002]

6. In hydrogen atom, energy of first excited state is –3.4 eV. Then kinetic energy of same orbit of hydrogen atom
(a) + 3.4 eV (b) + 6.8 eV
(c) –13.6 eV (d) + 13.6 eV
[2002]

7. The ions O^{2-}, F^-, Na^+, Mg^{2+} and Al^{3+} are isoelectronic. Their ionic radii show:
(a) An increase from O^{2-} to F^- and then decrease from Na^+ to Al^{3+}
(b) An decrease from O^{2-} to F^- and then increase from Na^+ to Al^{3+}
(c) A significant increase from O^{2-} to Al^{3+}
(d) A significant decrease from O^{2-} to Al^{3+}
[2003]

8. The value of planck's constant is 6.63×10^{-34} Js. The velocity of light is 3.0×10^8 ms^{-1}. which value is closest to the wavelength in nanometers of a quantum of light with frequency of 8×10^{15} s^{-1}?
(a) 4×10^1 (b) 3×10^7
(c) 2×10^{-25} (d) 5×10^{-18}
[2003]

9. Among the following series of transition metal ions, the one where all metal ion have $3d^2$ electronic configuration is
(a) Ti^{3+}, V^{2+}, Cr^{3+}, Mn^{4+}
(b) Ti^+, V^{4+}, Cr^{6+}, Mn^{7+}
(c) Ti^{4+}, V^{3+}, Cr^{2+}, Mn^{3+}
(d) Ti^{2+}, V^{3+}, Cr^{4+}, Mn^{5+}
(At. wt. Ti = 22; V = 23; Cr = 24; Mn=25)
[2004]

10. The frequency of radiation emitted when the electron falls from n = 4 to n = 1 in a hydrogen atom will be (Given ionization energy of H = 2.18×10^{-18} J $atom^{-1}$ and h = 6.625×10^{-34} Js)
(a) 1.54×10^{15} s^{-1} (b) 1.03×10^{15} s^{-1}
(c) 3.08×10^{15} s^{-1} (d) 2.00×10^{15} s^{-1}
[2004]

11. The energy of second Bohr orbit of the hydrogen atom is –328 kJ mol^{-1}, hence the energy of fourth bohr orbit would be
(a) –164 kJ mol^{-1} (b) –41 kJ mol^{-1}
(c) –82 kJ mol^{-1} (d) –1312 kJ mol^{-1}
[2005]

12. The orientation of an atomic orbital is governed by
(a) Magnetic quantum number
(b) Principal quantum number
(c) Azimuthal quantum number
(d) Spin quantum number
[2006]

13. Given: the mass of electron is 9.11×10^{-31} kg
Planck constant is 6.626×10^{-34} Js,
the uncertainty involved in the measurement of velocity within a distance of 0.1 Å is
(a) 5.79×10^8 ms^{-1} (b) 5.79×10^5 ms^{-1}
(c) 5.79×10^6 ms^{-1} (d) 5.79×10^7 ms^{-1}
[2006]

14. In which of the following pairs are both the ions coloured in aqueous solution?
(a) Sc^{3+}, Co^{2+} (b) Ni^{2+}, Cu^+
(c) Ni^{2+}, Ti^{3+} (d) Sc^{3+}, Ti^{3+}
(Atomic number Sc = 21, Ti = 22, Ni = 28,Cu = 29, Co = 27)
[2006]

15. Consider the following sets of quantum numbers:

	n	*l*	*m*	*s*
(1)	3	0	0	+ 1/2
(2)	2	2	1	+ 1/2
(3)	4	3	–2	– 1/2
(4)	1	0	–1	– 1/2
(5)	3	2	3	+ 1/2

Which of the following sets of quantum number is not possible?
(a) 2, 3 and 4 (b) 1, 2, 3 and 4
(c) 2, 4 and 5 (d) 1 and 3
[2007]

16. If uncertainty in position and momentum are equal, then uncertainty in velocity is:
(a) $1/m\sqrt{(h/\pi)}$ (b) $\sqrt{(h/\pi)}$
(c) $1/2m\sqrt{(h/\pi)}$ (d) $\sqrt{(h/2\pi)}$
[2008]

17. The measurement of the electron position is associated with an uncertainty in momentum, which is equal to 1×10^{-18} g cm s^{-1}. The uncertainty in electron velocity is, (mass of an electron is 9×10^{-28} g)
(a) 1×10^9 cm s^{-1} (b) 1×10^{11} cm s^{-1}
(c) 1×10^5 cm s^{-1} (d) 1×10^6 cm s^{-1}
[2008]

18. Maximum number of electrons in a sub-shell of an atom is determined by the following?

(a) $2\ell+1$ (b) $4\ell+2$
(c) $2n^2$ (d) 4ℓ

[2009]

19. Which of the following is not permissible arrangement of electrons in an atom?

(a) $n=5, \ell=3\ m=0, s=+1/2$
(b) $n=3\ \ell=2, m=-2, s=-1/2$
(c) $n=3, \ell=2, m=-3, s=-1/2$
(d) $n=4, \ell=0, m=0, s=-1/2$

[2009]

20. The energy absorbed by each molecule (A_2) of a substance is 4.4×10^{-19} J and bond energy per molecule is 4×10^{-19} $_J$. The kinetic energy of the molecule per atom will be?

(a) 2.2×10^{-19} $_J$ (b) 2×10^{-19} $_J$
(c) 2×10^{-20} $_J$ (d) 4×10^{-20} $_J$

[2009]

21. Which one of the following ions has electronic configuration $[Ar]3d^6$?

(a) Co^{3+} (b) Ni^{3+}
(c) Mn^{3+} (d) Fe^{3+}
(At numbers, Mn = 25, Fe = 26, Co = 27, Ni = 28)

[2010]

22. If n = 6, the correct sequence for filling of electrons will be

(a) $ns \rightarrow np \rightarrow (n-1)d \rightarrow (n-2)f$
(b) $ns \rightarrow (n-2)f \rightarrow (n-1)d \rightarrow np$
(c) $ns \rightarrow (n-1)d \rightarrow (n-2)f \rightarrow np$
(d) $ns \rightarrow (n-2)f \rightarrow np \rightarrow (n-1)d$

[2011]

23. The energies E_1 and E_2 of two radiations are 2.5 ev and 5.0 ev respectively. The relation between their wavelengths i.e., l_1 and l_2 will be

(a) $l_1 = l_2$ (b) $l_1 = 2l_2$
(c) $l_1 = 4l_2$ (d) $\lambda_1 = \frac{1}{2}\lambda_2$

[2011]

24. Maximum number of electrons in a sub-shell with $\ell = 3$ and n = 4 is

(a) 10 (b) 12
(c) 14 (d) 16

[2012]

25. The correct set of four quantum numbers for the valence electron of rubidium atom (z = 37) is

(a) 5,0,0, + 1/2 (b) 5,1,0, + 1/2
(c) 5,1,1, + 1/2 (d) 6,0,0, + 1/2

[2012]

26. The orbital angular momentum of a p-electron is given as

(a) $\sqrt{3}\frac{h}{2\pi}$ (b) $\sqrt{\frac{3}{2}\frac{h}{\pi}}$
(c) $\sqrt{6}\sqrt{\frac{h}{2\pi}}$ (d) $\frac{h}{\sqrt{2}\pi}$

[2012]

27. The value of Planck's constant is 6.63×10^{-34} Js. The speed of light is 3×10^{17} nm s^{-1}. Which value is closest the wavelength in nanometer of a quantum of light with frequency of $6\times10^{15}\ s^{-1}$?

(a) 25 (b) 50
(c) 75 (d) 10

[2013]

28. What is the maximum numbers of electrons that can be associated with the following set of quantum numbers?
$n = 3, l = 1$ and $m = -1$.

(a) 6 (b) 4
(c) 2 (d) 10

[2013]

29. What is the maximum number of orbitals that can be identified with the following quantum numbers?
$n = 3, l = 1$ and $m = 0$

(a) 1 (b) 2
(c) 3 (d) 4

[2014]

30. Calculate the energy in joule corresponding to light of wavelength 45 nm ($h = 6.63\times10^{-34}$ JS $V = 3\times10^8$ m/s)

(a) 4.42×10^{-15} (b) 4.42×10^{-18}
(c) 6.67×10^{11} (d) 6.67×10^{15}

[2014]

31. Magnetic moment 2.84 B.M is given by
(Atomic number of Ni = 28, Ti = 22, Cr = 24, Co = 27)

(a) Ni^{2+} (b) Ti^{3+}
(c) Cr^{3+} (d) Co^{2+}

[2015]

32. The angular momentum of electrons in d orbital is equal to

(a) Zero (b) $\sqrt{2}\,\hbar$

(c) $\sqrt{6}\,\hbar$ (d) $2\sqrt{3}\,\hbar$

[2015]

33. Two electrons occupying the same orbital are distinguished by

(a) Principal quantum number

(b) Azimuthal quantum number

(c) Magnetic quantum number

(d) Spin quantum number

[2016]

34. How many electrons can fit in the orbital for which $n = 3$ and $\ell = 1$?

(a) 2 (b) 6

(c) 10 (d) 14 **[2016]**

35. Which of the following pairs of orbitals will have electron density along the axes?

(a) $d_{z^2}, d_{x^2-y^2}$ (b) d_{Z^2}, d_{xz}

(c) d_{xz}, d_{yz} (d) $d_{xy}, d_{x^2-y^2}$

[2016]

Answer Keys

1. (c)	2. (c)	3. (d)	4. (b)	5. (c)	6. (a)	7. (d)	8. (a)	9. (d)	10. (c)
11. (c)	12. (a)	13. (c)	14. (c)	15. (c)	16. (c)	17. (a)	18. (b)	19. (c)	20. (c)
21. (a)	22. (b)	23. (b)	24. (c)	25. (a)	26. (d)	27. (b)	28. (c)	29. (a)	30. (b)
31. (c)	32. (c)	33. (d)	34. (a)	35. (a)					

Hints and Explanations

1. N_2, CO, CN^-, O_2^{+2} All these species have 14 electrons so they are isoelectronic.

2. Use the relation $\Delta E = hc/\lambda$

$$\lambda = \frac{h}{\Delta E} = \frac{6.63\times10^{-34}\times3\times10^{8}}{3.03\times10^{-19}} = 656 \text{ nm.}$$

3. As Sc^{3+} has no unpaired electron so no d-d electron is possible in it Hence, it is colourless.

4. It may be any 3d-orbital like $3d_{xy}$ or $3d_{yz}$.

5. CN^-, CO as both have 14 electrons so they are isoelectronic.

6. Kinetic energy in any orbital (n) = –En = –(–3.4) = + 3.4 eV

7. Radius α 1/Zeff

a 1/Magnitude of positive charge

Hence order of radii is $O^{2-} > F^- > Na^+ > Mg^{2+} > Al^{3+}$

8. $v = c/\lambda = \dfrac{3\times10^{8}}{8\times10^{15}} = 37.5 \times 10^{-9}$ m

= 37.5 nm ≈ 4×10^1 nm

9. Ti^{2+}: $1s^2, 2s^2\,2p^6, 3s^2\,3p^6\,3d^2, 4s^0$

V^{3+}: $1s^2, 2s^2\,2p^6, 3s^2\,3p^6\,3d^2, 4s^0$

Cr^{4+}: $1s^2, 2s^2\,2p^6, 3s^2\,3p^6\,3d^2, 4s^0$

Mn^{5+}: $1s^2, 2s^2\,2p^6, 3s^2\,3p^6\,3d^2, 4s^0$

So Ti^{2+}, V^{3+}, Cr^{4+}, Mn^{5+} have same electronic configuration here.

10. Ionization energy of H = $E_\alpha - E_1$

2.18×10^{-18} J atom^{-1} = $0 - E_1$

As E_1 (Energy of Ist orbit of H-atom)

$= -2.18 \times 10^{-18}$ J atom^{-1}

So $E_n = \dfrac{-2.18\times10^{-18}}{n^2}$ J-atom^{-1}

For H-atom, Z = 1

$\Delta E = E_4 - E_1$

$= \dfrac{-2.18\times10^{-18}}{4^2} - \dfrac{-2.18\times10^{-18}}{1^2}$

$= -2.18 \times 10^{-18} \times [1/4^2 - 1/1^2]$

$\Delta E = h\nu = -2.18 \times 10^{-18} \times -\dfrac{15}{16}$

$= +2.0437 \times 10^{-18}$ J atom^{-1}

$$\nu = \frac{\Delta E}{h} = \frac{2.0437\times10^{-18}\,\text{J-atom}^{-1}}{6.625\times10^{-34}\,\text{J}\frac{n!}{r!(n-r)!}\text{sec}}$$

$= 3.084 \times 10^{15}$ sec.$^{-1}$

11. As $E_2 = -\frac{K}{2^2} = -\frac{1312}{4} = -328$ kJ mol^{-1}

$E_4 = -\frac{K}{4^2} = -\frac{1312}{16} = -82$ kj mol^{-1}

12. The orientation of an atomic orbital is governed by magnetic quantum number.

13. $\Delta V = h/4\pi m \times \Delta X$

$$= \frac{6.626\times10^{-34}}{4\times3.14\times9.1\times10^{-31}\times0.1\times10^{-10}} = 5.8\times10^{6}\ \text{ms}^{-1}$$

14. As Ni^{2+}, Ti^{3+} has 2, 1 unpaired electrons respectively so both are coloured.

16. According to Heisenberg uncertainty principle.

$\Delta p\ \Delta x \geq \frac{h}{4\pi}$

$m\Delta v\ \Delta x \geq \frac{h}{4\pi}$

$(m\Delta v)^2 \geq \frac{h}{4\pi}$

$\Delta v \geq \frac{1}{2m}\sqrt{(h/\pi)}$

17. Uncertainty in momentum

$(m\Delta v) = 1 \times 10^{-18}$ g cm s^{-1}

Uncertainty in velocity (Δv)

$$= \frac{1\times10^{-18}}{9\times10^{-28}} = 1.1\times10^{9}\ \text{cm s}^{-1}$$

18. Maximum number of e^- = $2(2l + 1)$ in axy sub-shell $= 4l + 2$

19. As m = ± value of l(with zero)

Here $l = 2$ so m can not be –3.

20. K.E per atom can be given as

$$(\text{K.E})_{\text{Atom}} = \frac{E_{\text{Total}} - E_{\text{B.E}}}{2}$$

$$= \frac{(4.4\times10^{-19})-(4\times10^{-19})}{2}$$

$= 2 \times 10^{-20}$ Joule

23. As $E \propto \frac{1}{\lambda}$ so $\frac{E_1}{E_2} = \frac{\lambda_2}{\lambda_1}$

$\frac{2.5}{5} = \frac{\lambda_2}{\lambda_1}$ or $l_1 = 2l_2$

24. As n = 4 ℓ = 3 it means 4f sub-shell, which may have 14 electrons.

25. ${}_{37}Rb \rightarrow [Kr]\ 5s^1$

n = 0, ℓ = 0, m = 0, s = + 1/2

26. As orbital angular momentum

$= \sqrt{\ell(\ell+1)}\ \frac{h}{2\pi}$

$= \sqrt{1(1+1)}\ \frac{h}{2\pi} = \frac{h}{\sqrt{2}\pi}$

27. $\lambda = \frac{c}{\nu} = \frac{3\times10^{17}}{6\times10^{15}}$

= 50 nm

28. $n = 3$ $\quad \ell = 1$ $\quad$ m = –1

Hence it is a 3p-orbital so it can have maximum 2 electrons.

29. $n = 3$, $\ell = 1$, $m = 0$ means $3p_z$ orbital.

30. $E = \frac{hc}{l} = \frac{6.63\times10^{-34}\times3\times10^{8}}{45\times10^{-9}\ \text{M}}$

$= 4.42 \times 10^{-18}$ J

31. $\mu = \sqrt{n(n+2)}$

$2.84 = \sqrt{n(n+2)}$

n = 3

In $Cr^{3+} \rightarrow (Ar)\ 3d^3$

$n = 3$

32. $mvr = \sqrt{\ell(\ell+1)}\ \hbar$

For d-orbitale$^-$ $\ell = 2$

$mvr = \sqrt{2(2+1)}\ \hbar$

$= \sqrt{6}\ \hbar$

33. Using Pauli law no two e^- can have a same set of n, ℓ, m and s. If n, ℓ, m are same, s-values are different e.g. $_4Be \rightarrow 1s^2, 2s^2$

First	Second e^-
$n \rightarrow 2$	2
$\ell \rightarrow 0$	0
$m \rightarrow 0$	0
$s \rightarrow +1/2$	$-1/2$

34. Using Pauli's Exclusion Principle and orbital (here $3p_{x/y}$) can have maximum $2e^-$.

35. In case of d_{z^2}, $d_{x^2-y^2}$ e^- density is along with the axes while in d_{xy}, d_{yz}, d_{xz} it is in between two axis at 45°.

NCERT EXEMPLAR

1. Which of the following conclusions could not be derived from Rutherford's α -particle scattering experiement?
   (a) Most of the space in the atom is empty.
   (b) The radius of the atom is about 10^{-10} m while that of nucleus is 10^{-15} m.
   (c) Electrons move in a circular path of fixed energy called orbits.
   (d) Electrons and the nucleus are held together by electrostatic forces of attraction.
2. Which of the following options does not represent ground state electronic configuration of an atom?
   (a) $1s^2\ 2s^2\ 2p^6\ 3s^2\ 3p^6\ 3d^8\ 4s^2$
   (b) $1s^2\ 2s^2\ 2p^6\ 3s^2\ 3p^6\ 3d^9\ 4s^2$
   (c) $1s^2\ 2s^2\ 2p^6\ 3s^2\ 3p^6\ 3d^{10}\ 4s^1$
   (d) $1s^2\ 2s^2\ 2p^6\ 3s^2\ 3p^6\ 3d^5\ 4s^1$
3. The probability density plots of 1s and 2s orbitals are given in Fig. 2.6:

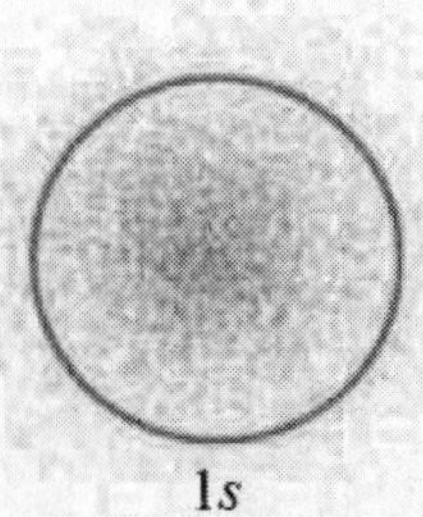

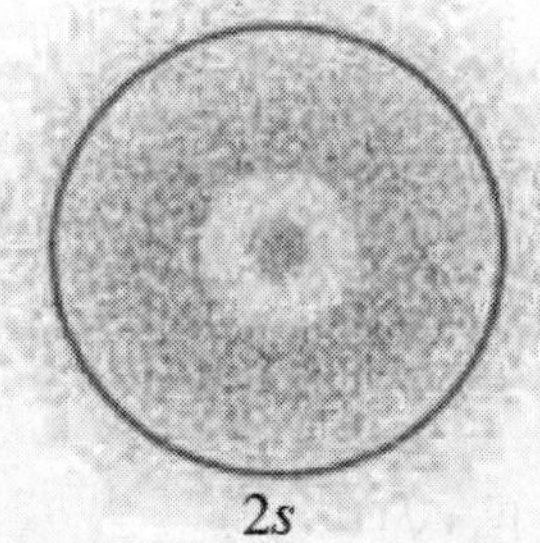

FIGURE 2.6

The density of dots in a region represents the probability density of finding electrons in the region.
On the basis of above diagram which of the following statements is incorrect?
   (a) 1*s* and 2*s* orbitals are spherical in shape.
   (b) The probability of finding the electron is maximum near the nucleus.
   (c) The probability of finding the electron at a given distance is equal in all directions.
   (d) The probability density of electrons for 2*s* orbital decreases uniformly as distance from the nucleus increases.
4. Which of the following statement is not correct about the characteristics of cathode rays?
   (a) They start from the cathode and move towards the anode.
   (b) They travel in straight line in the absence of an external electrical or magnetic field.
   (c) Characteristics of cathode rays do not depend upon the material of electrodes in cathode ray tube.
   (d) Characteristics of cathode rays depend upon the nature of gas present in the cathode ray tube.
5. Which of the following statements about the electron is incorrect?
   (a) It is a negatively charged particle.
   (b) The mass of electron is equal to the mass of neutron.
   (c) It is a basic constituent of all atoms.
   (d) It is a constituent of cathode rays.
6. Which of the following properties of atom could be explained correctly by Thomson Model of atom?
   (a) Overall neutrality of atom.
   (b) Spectra of hydrogen atom.
   (c) Position of electrons, protons and neutrons in atom.
   (d) Stability of atom.
7. Two atoms are said to be isobars if.
   (a) They have same atomic number but different mass number.
   (b) They have same number of electrons but different number of neutrons.
   (c) They have same number of neutrons but different number of electrons.
   (d) Sum of the number of protons and neutrons is same but the number of protons is different.

8. The number of radial nodes for $3p$ orbital is ________.

(a) 3 (c) 4
(b) 2 (d) 1

9. Number of angular nodes for $4d$ orbital is ________.

(a) 4 (b) 3
(c) 2 (d) 1

10. Which of the following is responsible to rule out the existence of definite paths or trajectories of electrons?
(1) Pauli's exclusion principle.
(2) Heisenberg's uncertainty principle.
(3) Hund's rule of maximum multiplicity.
(4) Aufbau principle.

11. Total number of orbitals associated with third shell will be ________.

(a) 2 (b) 4
(c) 9 (d) 3

12. Orbital angular momentum depends on ________.

(a) l (b) n and l
(c) n and m (d) m and s

13. Chlorine exists in two isotopic forms, Cl-37 and Cl-35 but its atomic mass is 35.5. This indicates the ratio of Cl-37 and Cl-35 is approximately

(a) 1:2 (b) 1:1
(c) 1:3 (d) 3:1

14. The pair of ions having same electronic configuration is ________.

(a) Cr^{3+}, Fe^{3+} (b) Fe^{3+}, Mn^{2+}
(c) Fe^{3+}, Co^{3+} (d) Sc^{3+}, Cr^{3+}

15. For the electrons of oxygen atom, which of the following statements is correct?

(a) Z_{eff} for an electron in a $2s$ orbital is the same as Z_{eff} for an electron in a $2p$ orbital.
(b) An electron in the $2s$ orbital has the same energy as an electron in the $2p$ orbital.
(c) Z_{eff} for an electron in $1s$ orbital is the same as Z_{eff} for an electron in a $2s$ orbital.
(d) The two electrons present in the $2s$ orbital have spin quantum numbers ms but of opposite sign.

16. If travelling at same speeds, which of the following matter waves have the shortest wavelength?

(a) Electron
(b) Alpha particle (He^{2+})
(c) Neutron
(d) Proton

Answer Keys

1. (c)	2. (b)	3. (d)	4. (d)	5. (b)	6. (a)	7. (d)	8. (d)	9. (c)	10. (b)
11. (c)	12. (a)	13. (c)	14. (b)	15. (d)	16. (b)				

Hints and Explanations for Selective Questions

1. Concept of circular path was proposed by Bohr but not derived from Rutherford's scattering experiment.

3. the density of electrons for $2s$ orbit first increases and then decreases and followed by increasing again.

5. Mass of electron (9.1×10^{-31} kg) is very small when compared it with mass of neutron (1.674×10^{-27} kg).

8. $n = 3-1-1 = 1$

11. $n^2 = 3^2 = 9$

14. Fe^{3+}, Mn^{2+} have 23 electrons. Since both have same number of electrons so same electronic configuration.

16. for same value of frequency, larger the value of mass m, wavelength gets shorter.

AIIMS ESSENTIALS

ASSERTION AND REASON

In the following questions, two statements (Assertion) A and Reason (R) are given. Mark

(a) If A and R both are correct and R is the correct explanation of A;

(b) If A and R both are correct but R is not the correct explanation of A;

(c) A is true but R is false;

(d) A is false but R is true,

(e) A and R both are false.

1. (A) : The shortest wavelength of transition of Lyman series is observed when electron jumps from orbit number, $n = \infty$ to $n = 1$.
   (R) : Since the wavelength of transition is given by $hc/\lambda = \Delta E$

2. (A) : The kinetic energy of the photo-electron ejected increases with increase in intensity of incident light.
   (R) : Increase in intensity of incident light increases the rate of emission.

3. (A) : Limiting line in the Balmer series of H-spectrum has a wavelength of 364.4 nm.
   (R) : Limiting line is obtained for a jump of electron from $n = \infty$ to $n = 2$.

4. (A) : $3d_{z^2}$ orbital is spherically symmetrical.
   (R) : $3d_{z^2}$ orbital is the only d-orbital which is spherical in shape.

5. (A) : Shapes of the orbitals are represented by boundary surface diagrams of contrast probability density.
   (R) : Boundary surface diagram helps in interpreting and visualizing an atomic orbital.

6. (A) : Half-filled and fully-filled degenerate orbitals are more stable.
   (R) : Extra stability is due to the symmetry of degenerate orbitals.

7. (A) : A spectral line will be observed for a 2px – 2py transition.
   (R) : The energy is released in the form of wave of light when electron drops from 2px to 2py orbital.

8. (A) : The angular momentum of d-orbital is $\sqrt{6}\, h/2\pi$.
   (R) : d-orbitals have double dumb-bell shaped except dz^2.

9. (A) : Electron can have wavelength equal to that of proton of Ve = 1836 Vp.
   (R) : Electron has mass 1/1836 of the mass of proton.

10. (A) : The kinetic energy of photoelectrons increases with increase in frequency of incident light.
    (R) : The number of photoelectron ejected increases with increase in intensity of light.

11. (A) : Fluorescence is the emission of light which has been absorbed by the molecule.
    (R) : It may continue for appreciable time after the exciting light is switched off.

12. (A) : Fe^{2+} has 24 electrons hence, its electronic configuration, is similar to that of Cr(24) [Ar] $3d^5\, 4s^1$.
    (R) : All the five unpaired electrons in 3d gives stability to the ion.

13. (A) : On heating a solid for a longer time, radiations become white and then blue as the temperature becomes very high.
    (R) : Radiations emitted go from a lower frequency to higher frequency as the temperature increases.

14. (A) : Cu^{2+} is a coloured ion.
    (R) : Every ion with unpaired electron is coloured.

15. (A) : The paramagnetism of Cu^+ ion is zero.
    (R) : Paramagnetism is given by the relation $\mu = \sqrt{S(S+1)}$ magnetons where S is the total spin.

16. (A) : Electrons are negatively charged.
    (R) : The application of electric and magnetic field deflected the rays in the discharge tube towards the cathode.

17. (A) : Electronic configuration of K (19) is $1s^2\, 2s^2\, 2p^6\, 3s^2\, 3p^6\, 4s^1$.

(R) : Energy of 4s < 3d hence, 4s is filled before 3d as decided by Aufbau rule.

18. (A) : Atoms can neither be created not destroyed.
(R) : Under similar conditions of temperature and pressure, equal volumes of gases do not contain equal number of atoms.

19. (A) : For n = 3, 1 may be 0, 1 and 2 and may be 0; 0±l; 0, ±1 and ±2
(R) : For each value of n, there are 0 to (n – 1) possible values of 1; and for each value of 1, there are 0 to ±l values of m.

20. (A) : The electronic configuration of nitrogen atom is represented as:

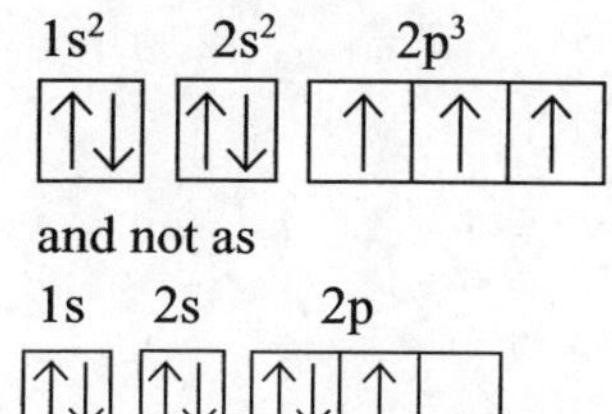

(R) : The electronic configuration of the group state of an atom is the one which has the greatest multiplicity.

Answer Keys

1. (a)	2. (d)	3. (a)	4. (e)	5. (a)	6. (b)	7. (e)	8. (b)	9. (a)	10. (b)
11. (c)	12. (d)	13. (a)	14. (c)	15. (c)	16. (a)	17. (a)	18. (c)	19. (a)	20. (a)

Classification of Elements and Periodicity Properties

Chapter Outline

■ Modern Periodic Law ■ Nomenclature of Elements with Atomic Number > 100 ■ Trends in Periodic Properties of Elements

Modern Periodic Law

According to the modern periodic law, "The physical and chemical properties of elements are the periodic function of their atomic numbers."

Modern Periodic Table

- Periodic table is an arrangement of elements with similar properties placed together.
- Elements have been arranged in order of increasing atomic number.
- It has seven horizontal rows known as **periods**.
- There are sixteen vertical columns which are called **groups** or **families**.
- The transition elements are placed in the middle of the long periods.
- Isotopes of an element are assigned a single place as they have the same atomic number.
- Dissimilar metals are placed in different groups. For example, noble metals and alkali metals are separated by placing them in group I and group II respectively.

Periodic table

1	H	1.008
Atomic number, Z	Element symbol	Relative atomic mass, A_r

1	2	3	4	5	6	7	8	9	10	11	12	13	14	15	16	17	18
1 H 1.008																	2 He 4.00
3 Li 6.94	4 Be 9.01											5 B 10.81	6 C 12.01	7 N 14.01	8 O 16.00	9 F 19.00	10 Ne 20.18
11 Na 22.99	12 Mg 24.31											13 Al 26.98	14 Si 28.09	15 P 30.97	16 S 32.06	17 Cl 35.45	18 Ar 39.95
19 K 39.10	20 Ca 40.08	21 Sc 44.96	22 Ti 47.90	23 V 50.94	24 Cr 52.01	25 Mn 54.94	26 Fe 55.85	27 Co 58.93	28 Ni 58.69	29 Cu 63.54	30 Zn 65.41	31 Ga 69.72	32 Ge 72.59	33 As 74.92	34 Se 78.96	35 Br 79.91	36 Kr 83.80
37 Rb 85.47	38 Sr 87.62	39 Y 88.91	40 Zr 91.22	41 Nb 92.91	42 Mo 95.94	43 Tc 98.91	44 Ru 101.07	45 Rh 102.91	46 Pd 106.42	47 Ag 107.87	48 Cd 112.40	49 In 114.82	50 Sn 118.71	51 Sb 121.75	52 Te 127.60	53 I 126.90	54 Xe 131.30
55 Cs 132.91	56 Ba 137.34	La–L u	72 Hf 178.49	73 Ta 180.95	74 W 183.85	75 Re 186.21	76 Os 190.23	77 Ir 192.22	78 Pt 195.08	79 Au 196.97	80 Hg 200.59	81 Tl 204.37	82 Pb 207.19	83 Bi 208.98	84 Po 210	85 At 210	86 Rn 222
87 Fr 223	88 Ra 226.03	Ac–L r	104 Rf [261]	105 Db [262]	106 Sg [266]	107 Bh [264]	108 Hs [277]	109 Mt [268]	110 Ds [271]	111 Rg [272]	112 Uub [285]						

Lanthanoid s	57 La 138.91	58 Ce 140.12	59 Pr 140.91	60 Nd 144.24	61 Pm 146.92	62 Sm 150.36	63 Eu 151.96	64 Gd 157.25	65 Tb 158.92	66 Dy 162.50	67 Ho 164.93	68 Er 167.26	69 Tm 168.93	70 Yb 173.04	71 Lu 174.97
Actinoid s	89 Ac 227.03	90 Th 232.04	91 Pa 231.04	92 U 238.03	93 Np 237.05	94 Pu 239.05	95 Am 241.06	96 Cm 244.07	97 Bk 249.08	98 Cf 252.08	99 Es 252.09	100 Fm 257.10	101 Md 258.10	102 No 259	103 Lr 262

Long Form of Periodic Table

- Long form of periodic table is also known as extended form of periodic table or Moseley's periodic table.
- It was developed by Range, Werner and Burey on the basis of electronic configuration of elements.

Features

- O is divided into four blocks, namely, s, p, d and of, on the basis of electronic configuration.
- It has seven horizontal rows known as periods.
- The first period is shortest, having only two elements (H, He).
- The second and the third periods are also short and have eight elements each.
- The fourth and the fifth periods are long and contain eighteen elements each.
- The sixth period is the longest and contains thirty two elements.
- The seventh period is incomplete and contains twenty six elements.
- Group III B (3rd vertical row) is the largest group having thirty two elements which include lanthanides and actinides.
- The first element of each period is an alkali metal while the last element is an inert gas.

Genesis of periodic Classification

- *Telluric helix (1862)* The idea of periodic arrangement of elements was introduced for the first time by Chancourtois.
- *Unitary hypothesis (1915)* According to Prout's unitary hypothesis, all elements are made up of hydrogen atoms.

	Main Groups		Transitional Groups										Main Groups					
	1 1A	2 2A	3 3B	4 4B	5 5B	6 6B	7 7B	8	9 8B	10	11 1B	12 2B	13 3A	14 4A	15 5A	16 6A	17 7A	18 8A
1	1 H																	2 He
2	3 Li	4 Be											5 B	6 C	7 N	8 O	9 F	10 Ne
3	11 Na	12 Mg											13 Al	14 Si	15 P	16 S	17 Cl	18 Ar
4	19 K	20 Ca	21 Sc	22 Ti	23 V	24 Cr	25 Mn	26 Fe	27 Co	28 Ni	29 Cu	30 Zn	31 Ga	32 Ge	33 As	34 Se	35 Br	36 Kr
5	37 Rb	38 Sr	39 Y	40 Zr	41 Nb	42 Mo	43 Tc	44 Ru	45 Rh	46 Pd	47 Ag	48 Cd	49 In	50 Sn	51 Sb	52 Te	53 I	56 Xe
6	55 Cs	56 Ba	57 La	72 Hf	73 Ta	74 W	75 Re	76 Os	77 Ir	78 Pt	79 Au	80 Hg	81 Tl	82 Pb	83 Bi	84 Po	85 At	86 Rn
7	87 Fr	88 Ra	89 Ac	104 Rf	105 Db	106 Sg	107 Bh	108 Hs	109 Mt	110 Uun	111 Uuu	112 Uub		114 Uuq				

Lanthanides	58 Ce	59 Pr	60 Nd	61 Pm	62 Sm	63 Eu	64 Gd	65 Tb	66 Dy	67 Ho	68 Er	69 Tm	70 Yb	71 Lu
Actinides	90 Th	91 Pa	92 U	93 Np	94 Pu	95 Am	96 Cm	97 Bk	98 Cf	99 Es	100 Fm	101 Md	102 No	103 Lr

- *Law of triads (1864)* According to this law given by Dobereiner, elements can be arranged in groups of three, in which the atomic weight of the second (middle) element is the arithmetic mean of the first and third element. The difference in the atomic weight of consecutive elements is atom constant.

  For example,

  $$_{3}Li^{9},\ _{11}Na^{23},\ _{19}K^{39}$$
  $$_{4}Be^{9},\ _{12}Mg^{24},\ _{20}Ca^{24}$$

- *Law of octaves (1864)* According to this law given by Newland, "If the elements are arranged in the order of increasing atomic weights the eighth element starting from a given one, is a kind of repetition of the first." This is similar to the octave in music in which the eighth note is similar to the first. For example, starting from lithium the eighth element is sodium so according so according to this law, the properties of lithium and sodium shiuld be similar.
- *Lothar Meyer (1869)* He plotted various physical properties of elements against their atomic weights and obtained a curve. He observed that similar elements occupied similar positions on that curve.
- *Mendeleev's periodic law* According to this law, "The physical and chemical properties of elements are the periodic functions of their atomic weights."

Type of Elements

There are four type of known elements:

- The s–block and p–block elements are called **representative elements.**
- The d–blocks elements are called **transition elements.**
- The f–blocks elements are called **inner transition elements.**

s-block elements

- The elements having ns^1 and ns^2 electronic configurations in their outermost shell are called s–block elements.
- Elements with ns^1 configuration are called **group 1** elements (alkali metals).

- Elements with ns^2 configuration are called **group 2** elements (alkaline earth metals).
- They are highly reactive and readily form univalent or bivalent positive ions respectively, by losing the valence electrons.
- Elements of this block are soft, malleable and good conductors of heat and electricity.
- Elements of this block have the largest atomic and ionic radii but lowest ionization energies in their respective periods.
- They show fixed valency and oxidation states.
- The loss of the outermost electron(s) occurs readily to form M^+ (in case of alkali metals) or M^{+2} ions (in case of alkaline earth metals).
- Except for compounds of beryllium, all other compounds of the elements of this block are predominantly ionic.
- They are soft metals having low melting and boiling points.
- These metals and their salts, impart characteristic colour to the flame. For example, sodium salts impart a golden yellow colour to the flame.
- The elements of this group have a large size, strong reducing nature, are highly electropositive in nature, have very low electronegativity, ionization energy and electron affinity.

p-block elements

- The elements whose atoms have incomplete p orbitals in their outermost shell or in which, the last electron enters any p orbital are known as p–block elements.
- The general outer electronic configuration for these elements varies from $ns^2\ np^{1\text{-}6}$.
- Elements of groups 13 (IIIA), 14 (1VA), 15 (VA), 16 (VIA), 17 (VIIA), 18 (VIIIA) are p–block elements.
- Group 13 (IIIA) elements have one electron in their p orbital, whereas group 18 (VIIIA) (inert gas) have 6 electrons in their outer p orbitals. The outer p orbitals in an inert gas are fully filled with electrons.

Boron	$1s^2\ 2s^2\ 2p^1$
Oxygen	$1s^2\ 2s^2\ 2p^4$
Neon	$1s^2\ 2s^2\ 2p^6$

- They include both metals and non metals but there is a regular gradation from metallic to non metallic character as we move from left to right across the period.
- Element of this block do not impart any specific colour to flame.
- Except fluorine and inert gases, all other elements of this block show variable oxidation states.

Group	IIIA	IVA	VA	VIA	VIIA
O.S.	+3	+4	+5	+6	+7
		to	to	to	to
		–4	–3	–2	–1

- They have high ionization energies and these values tend to increase as we move from left to right across the period.
- They form covalent compounds, mostly like oxides, halides, sulphides, carbonates.
- Except metals, other elements of this block are insulators.
- A number of elements of this block, like C, Si, Ge, S, O, show catenation property and allotropy.
- As we move from left to right, there is a gradation from reducing to oxidizing properties.

d-block elements

- The d–block lies between s–and p–blocks. These are the elements which have incompletely filled d orbitals.
- d–block elements are called transition elements as their properties lie in between those of and p-block elements.
- A transition element is the one whose atom or at least one of its ion has an incomplete dorbital or in which, the outermost electron is present in a d orbital.
- Their general outer electronic configuration is

$$(n-1)\ d^{1-10}\ ns^{1-2}$$

- These elements are present in between group 2 to group 13, in three series of ten elements each.
- They show variable valency and oxidation state be cause of the participation of ns and (n–l)d electrons in their chemical bond formation. This is because these electrons have nearly similar energies.
- These are metals with high values of melting points, boiling points, densities, thermal stabilities and hardness.
- They are ductile and malleable.
- They are good conductors of heat and electricity due to the presence of mobile or free electrons.
- They form coloured ions and complexes.

- These metals and their ions are generally paramagnetic in nature because of the presence of unpaired electrons.
- These metals form a number of alloys as they have similar sizes.
- These metals and their compounds are widely used as catalysts.
- These metals also form non–stoichiometric and interstitial compounds with small sized atoms like H, C, N, O which can easily fit in the vacant sites of the lattices of these metals. For example, $Fe_{0.93}$ O, ZrH_2, WC.

f-block elements

- The elements placed in two separate rows at the bottom of the periodic table are f–block elements.
- They have incompletely filled f orbitals in their electronic configurations.
- The elements from cerium to lutetium having incomplete 4f orbitals are **lanthanones** or **lanthanoids** or **lanthamides.**
- The elements from thorium to lawrencium having incomplete 5f orbitals are **actinoids** or **actinides.**
- In addition to incompletely filled 5f orbitals, they also have 1 to 2 incomplete d orbitals in their penultimate shells.
- Their general outer electronic configuration is $(n-2)\ f^{1-14}\ (n-1)\ d^{1-2}\ ns^2$.
- Many actinides have been synthesized only in nanogram or lesser quantities by nuclear reactions and their chemistry has not been fully studied.
- Many of them are synthetic elements.
- The elements that appear after uranium in the periodic table are called **transuranium elements.**
- They are metals having high melting and boiling points.
- They show variable oxidation states (variable valency). However, their most common and stable oxidation state is +3.
- They form coloured ions and complexes.
- Actinides are radioactive in nature.

NOMENCLATURE OF ELEMENTS WITH ATOMIC NUMBER > 100

The IUPAC proposed a system for naming element with Z > 100 by using these rules as follows:

1. The name are derived by using roots for the three digits in the atomic number of the element and adding the ending–ium. The roots for the numbers are:

0	1	2	3	4	5	6	7	8	9
nil	un	bi	tri	quad	pent	hex	sept	oct	enn

2. In certain cases the names are shortened; for example, bi ium and tri ium are shortened to bium and trium, and enn nil is shortened to ennil.
3. The symbol for the element is made up from the first letters from the roots which make up the name. The strange mixture of Latin and Greek roots has been chosen to ensure that the symbols are all different.

Atomic No.	Name	Symbol	Atomic No.	Name	Symbol
101	un–nil–unium	Unu	113	un–un–trium	Uut
102	un–nil–bium	Unb	114	un–un–quadium	Uuq
103	un–nil–trium	Unt	115	un–un–pentium	Uup
104	un–nil–quadium	Unq	116	un–un–hexium	Uuh
105	un–nil–pentium	Unp	117	un–un–septium	Uus
106	un–nil–hexium	Unh	118	un–un–octium	Uuo
107	un–nil–septium	Uns	119	un–un–ennium	Uue
108	un–nil–octium	Uno	120	un–bi–nilium	Ubn
109	un–nil–ennium	Une	130	un–tri–nilium	Utn
110	un–un–nilium	Uun	140	un–quad–nilium	Uqn
111	un–un–unium	Uuu	150	un–pent–nilium	Upn
112	un–un–bium	Uub			

Trends in Periodic Properties of Elements

The properties of the elements vary periodically with their atomic numbers when we move from left to right across the period or from top to bottom in any group.

Atomic Size or Radius

- Atomic size is the distance between the center of the nucleus of the atom to its outermost shell of electrons, that is , the penultimate shell electron.
- The absolute value of atomic radius cannot be determined because
  1. It is not possible to locate the exact position of electrons in an atom, as an orbital does not have a sharp boundary.
  2. It is not possible to isolate an individual atom for its size determination due to its extremely small size.
- In a group of atoms, the probability distribution of electrons is influenced by the presence of neighbouring atoms, hence, size of an atom may change while going from one environment to another.
- The values of atomic radii are derived from bond lengths, measured by various techniques such as X–rays diffraction, electron diffraction, infrared spectroscopy and nuclear magnetic resonance spectroscopy.
- As the absolute value of atomic size cannot be determined, it is expressed in terms of operational definitions such as ionic radius, covalent radius, van der Waals radius and metallic radius.

Covalent radius It is half of the distance between two successive nuclei of a covalent molecule made of two same atoms or similar atoms in a molecule. It is also called single bond covalent radius.

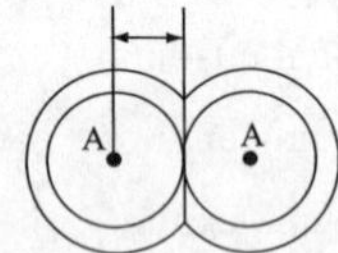

FIGURE 3.1 Covalent Radius

- If the bond length in between the two atoms say A – A is ‘a’, then

  Covalent radius (r_{cov})

  = ½ [internuclear distance between two covalently bonded similar atoms]

  = ½ a.

Van der waals radius It is one half of the distance between the nuclei of two unbonded isolated atoms lying adjacent to each other and, belonging to two neighbouring molecules of an element in the solid state.

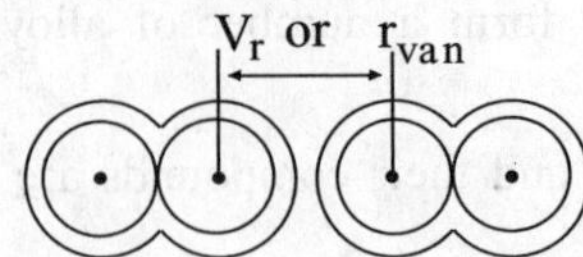

FIGURE 3.2 Van der Waals Radius

Metallic radius It is half of the distance between any two successive nuclei of two adjacent metal atoms in a closely packed crystal lattice.

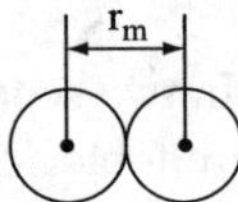

FIGURE 3.3 Metallic Radius

Ionic radius It is the effective distance from the centre of the nuclei of an ion up to which it has an influence over the electron cloud.

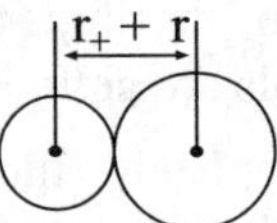

FIGURE 3.4 Ionic Radius

> **Facts To Remember**
>
> Van der Waals radius > metallic radius > anionic radius > covalent radius > cationic radius

Factors affecting size

Effective nuclear charge and screening effect The force of attraction present between the nucleus and the valence electrons is reduced by the force of repulsion exerted by inner shell electrons on these valence electrons. It is called **screening** or **shielding effect** and the net force is called **effective nuclear charge**. It is denoted to Z_{eff}

$$Z_{eff} = Z - \sigma$$

$\sigma \propto$ Number of inner shell electrons

Here, Z is atomic number and σ is screening constant.

$$\text{Size} \propto \frac{1}{\text{Effective nuclear charge}}$$

Size $\propto$ Number of orbits

Size $\propto$ Number of inner shell electrons (σ)

Size ∝ Electronic repulsion

$$\text{Size} \propto \frac{1}{\text{Bond order or multiplicity}}$$

Variation in the value of radii

In a period On moving from left to right, atomic size decreases as number of atomic orbitals remains same, while number of electron increases; as a result, effective nuclear charge increases. The order of atomic radius for various groups is given as

IA > IIA > IIIA > IVA > VA ≈ VI > VII < zero group.

Example:

Li > Be > B > C > N ≈ O > F

Na > Mg > Al > Si > P > S > Cl

- Only van der Waals radius is considered in case of zero group. It is always greater than covalent radius.

In a group Atomic radius increases from top to bottom as the number of shells or orbitals increase and screening effect increases so Z_{eff} decreases. For example,

Li < Na < K < Rb < Cs

F < Cl < Br < I

Size of a cation is always smaller than its atom During cation formation, the outermost orbital is destroyed and number of valence electron decreases so Z_{eff}, increases and size decreases.

- Size of cation ∝ $1/Z_{eff}$ or magnitude of positive charge
- $M^{+3} < M^{+2} < M^{+} < M$

  For example, $Fe^{+3} < Fe^{+2} < Fe$

Size of an anion is always greater than the size of its atom As during anion formation electrons are taken up so Z_{eff} decreases and size increases.

- Size of anion ∝ magnitude of negative charge

  $M^{-4} > M^{-3} > M^{-2} > M^{-} > M$

  For example, $O^{-2} > O^{-} > O$
- In general, anion > atom > cation

  For example, $X^{-} > X > X^{+}$
- In general, for isoelectronic species, ionic size decreases as follows:

  $M^{-4} > M^{-3} > M^{-2} > M^{-} > M^{+} > M^{+2} > M^{+3} > M^{+4}$

  For example,

  $C^{-4} > N^{-3} > O^{-2} > F^{-} > Na^{+} > Mg^{+2} > Al^{+3} > Si^{+4}$

  $P^{-3} > S^{-2} > Cl^{-} > K^{+} > Ca^{+2} > Sc^{+3} > Ti^{+4} > V^{+5}$

Period/Group	I	II	III	IV	V	VI	VII
1	H 0.32						
2	Li 1.23	Be 0.89	B 0.80	C 0.77	N 0.75	O 0.73	F 0.72
3	Na 1.54	Mg 1.36	Al 1.20	Si 1.17	P 1.10	S 1.04	Cl 0.99
4	K 2.03	Ca 1.74	Ga 1.26	Ge 1.22	As 1.20	Se 1.16	Br 1.14
5	Rb 2.16	Sr 1.91	In 1.44	Sn 1.41	Sb 1.40	Te 1.36	I 1.33
6	Cs 2.35	Ba 1.98	Tl 1.48	Pb 1.47	Bi 1.46	Po 1.46	

Increases (down the group) ↓ ; Decreases (across the period) →

Values of Atomic radius in p.m.

Facts To Remember

If the covalent bond present between atoms have different electronegativities, then atomic radius is determined by the formula given by Shoemaker and Stevenson.

$D_{(A-B)} = r_A + r_B - 0.09\,(X_A - X_B)$

Here X_A and X_B are electronegativities of A and B respectively.

Ionization Enthalpy

- Ionization enthalpy is also called or ionization energy or ionization potential.
- It is defined as the energy required to remove the most loosely bound electron from an isolated atom in the gaseous state resulting in the formation of a positive ion.

$$M - 1e^{-} \longrightarrow M^{+} - I_1$$
$$M^{+} - 1e^{-} \longrightarrow M^{+2} - I_2$$
$$M^{+2} - 1e^{-} \longrightarrow M^{+3} - I_3$$

- Here I_1, I_2 and I_3 are the first, second and third ionization enthalpies respectively.

$$I_1 < I_2 < I_3 < I_4 \longrightarrow$$

The value increases as removal of the electron becomes more and more difficult.

- Its units are eV/atom or kcal/mole or kJ/mole.

Factors Affecting Ionization Energy

- Ionization energy ∝ Nuclear charge (Z_{eff})
- $\text{Ionization energy} \propto \frac{1}{\text{Atomic size}}$
- $\text{Ionization energy} \propto \frac{1}{\text{Penetration effect of the electrons}}$

This means that the closer an orbital is to the nucleus, more will be the value of ionization energy, due to greater force of attraction between electrons and the nucleus.The magnitude of ionzation energy for orbitals increases in the following orders

$$s > p > d > f$$

- Ionization energy $\propto \frac{1}{\text{Screening effect of the inner electrons}}$

  Due to greater screening effect Z_{eff} decreases and removal of valence electron becomer easier.

- Ionization energy $\propto$ Stable electronic configuration.

Fully filled or half-filled orbital's electronic configurations are stable. Removal of electron is difficult from such configurations hence a higher value of ionization energy is observed in such cases.

$$N > O$$

$$2p^3 \; 2p^4$$

$$Be > B$$

$$2s^2 \; 2p^1$$

Variation in value of ionization energy

In a period On moving from left to right in a period, the ionization energy increases as, Z_{eff} increases and size decreases, so removal of the electron becomes more and more difficult. Order for first ionization energy in any period is

$s^1 \quad s^2 \quad p^1 \quad p^2 \quad p^3 \quad p^4 \quad p^5 \quad p^6$

IA < IIA > IIIA < IVA < VA > VIA < VIIA < VIIIA (maximum)

- In case of group IIA, the value of first ionization energy is higher than that of group IIIA, as in group IIA, ns^2 (fully filled state) is present but in group IIIA, the configuration is np1 (incomplete) which is less stable.
- In case of group VA, the value of first ionization energy is higher than that of group VIA, as in group VA, $ns^2\ np^3$ configuration (half filled more stable state) is present but the configuration in group VIA is $ns^2\ np^4$ (incomplete, less stable state).
- In any given period, an inert gas has the maximum value of first ionization energy, due to it stable octet state.
- The largest gap in between IE_1 and IE_2 is observed case of alkali metals, as ns^1 configurations changes into inert gas configuration ($ns^2\ np^6$).

For example,

IE_1 : Li < Be > B < C < N > O < F < Ne

Na < Mg > Al< Si < P > S < CI < Ar

IE_2 : Li > Ne > O > F > N > B > C > Be

Na > Ar > CI > S > P > Al > Si > Mg

In a group On moving from top to bottom in a group, ionization energy decreases as Z_{eff} decreases. Moreover, the size increases, so the removal of electron becomes more and more easy. For example,

He > Ne > Ar > Kr > Xe

F > CI > Br > I

Li > Na > K > Rb > Cs

B > Ga > Al > Tl > In

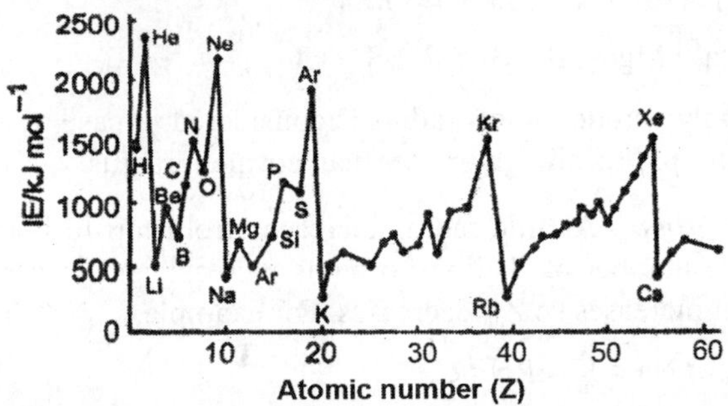

FIGURE 3.5 Variation of First Ionization Enthalpies with Atomic Number for Elements with Z = 1 to 60

FACTS TO REMEMBER

In the periodic table, helium has the maximum value of first ionization energy (1500 eV) while cesium has the lowest value for the same.

The largest gap between IE_2 and IE_3 is observed in alkaline earth metals as configuration changes from ns_1 to inert gas configuration (ns^2np^6).

$Be \longrightarrow Be+ \longrightarrow Be^2 + \longrightarrow Be^{3}+$

$1s^2\ 2s^2\ IE_1 \quad 2s^1 \quad IE_2 \quad 1s^2 \quad IE_3$

$\leftrightarrow$ for IA $\rightarrow I_1 <<< I_2$ IIIA $\rightarrow I_3 <<< I_4$

Importance of ionization energy The elements having low ionization energies are reducing agents, basic in nature, form cations, form ionic compounds and show maximum photo-electric effect. For example, K and Cs show maximum pho-toelectric effect and are therefore used in photoelectric cells.

Electron Gain Enthalpy or Electron Affinity (EA)

- Electron gain enthalpy is the amount of energy released, when a neutral isolated gaseous atom accepts an electron to form a gaseous anion.

$$M + 1e^- \longrightarrow M^- + E_1$$
$$M^- + 1e^- \longrightarrow M^{-2} + E_2$$
$$M^{-2} + 1e^- \longrightarrow M^{-3} + E_3$$

- Here E_1, E_2 and E_3 stand for the first, second and third electron gain enthalpies respectively.

FACTS TO REMEMBER

$$\xrightarrow{E_1 > E_2 > E_3}$$

The value of electron affinity decreases as the addition of electron becomes more and more difficult and possible only by absorbing some part of energy, that is, E_2 becomes endothermic in comparison to E_1. (E_2 is exoergonic and E_2 is endoergonic). For example,

$$O + e^- = \xrightarrow[E_1]{} O^- \xrightarrow[E_2]{+e^-} O^{2-}$$
$$E_2 << E_1$$

- Its units are eV/atom or Kcal/mole or kJ/mole.
- Electron affinity cannot be determined directly, however its value can be obtained indirectly from Born–Haber cycle.

Factors affecting electron affinity

- Electron affinity $\propto \dfrac{1}{\text{Atomic size}}$
- Electron affinity $\propto$ Effective nuclear charge
- Electron affinity $\propto \dfrac{1}{\text{Screening effect}}$
- Electron affinity $\propto \dfrac{1}{\text{Stable electronic configuration}}$
- The electron affinity of inert gases is zero, this is because of the presence of $ns^2\ np^6$ configuration in the valence shell, so there is no possibility of adding an extra electron.
- Electron affinity of Mg, Be is practically zero due to extra stability of completely filled s orbitals in them.
- If an atom has half filled orbitals, its electron affinity will be very low (in case of fifth group).

Variation in value of electron affinity

In a period On moving left to right in a period, electron affinity increases as Z_{eff} increases and size decreases.

In general, electron affinity follows the following trend:

Halogens > oxygen family > carbon family > nitrogen family > metals of group I and XIII > metals of group II > zero group.

- The sequence of electron affinity in second period is as follows:

$$Be < N < Li < B < C < O < F$$

- The sequence of electron affinity in third period is as follows:

$$Mg < Na < Al < P < Si < S < Cl.$$

In a group On moving down the group, the electron affinity decreases, as Z_{eff} decreases and size increases.

- Following are few exceptions to the rule:

E_a of F	<	E_a of Cl
E_a of C	<	E_a of Si
E_a of N	<	E_a of P
E_a of O	<	E_a of S

- In case of group II elements, value of electron affinity is less, Since due to their small size and greater electronic repulsion, addition of electron becomes difficult.

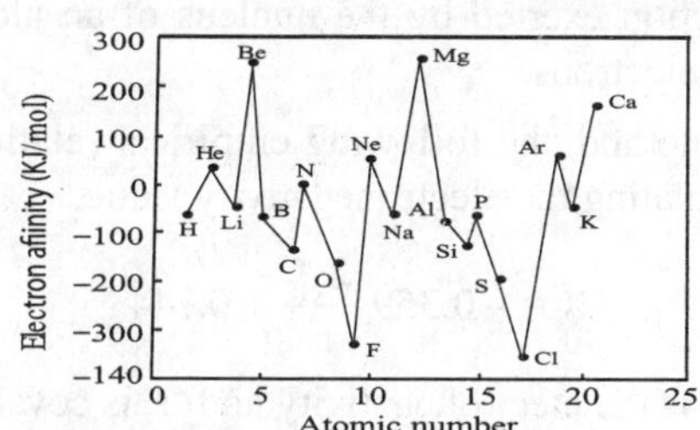

FIGURE 3.6 Electron Affinity Vs Atomic number

FACTS TO REMEMBER

- The sequence of electron affinity of group VII is as follows: I < Br < F < Cl
- Oxidizing power of an element $\propto$ Electron affinity
- Reactivity of non-metals $\propto$ Electron affinity

Electronegativity

- According to Pauling (1931), electonegativity is the power or tendency of an atom in a molecule to attract the shared pair of electrons towards itself.
- His observation was based on the reaction of the following type:

$\frac{1}{2}(A-A) + \frac{1}{2}(B-B)\ A-B$

The bond dissociation energy of (A – B) is higher than the mean of bond dissociation energies of (A – A) and (B – B). Their difference Δ is related to the difference in electronegativities of A and B respectively.

$$\Delta = E(A-B) - \frac{1}{2}[(E(A-A) + E(B-B)]^{\frac{1}{2}}$$

$$= 23(X_A - X_B)^2$$

$$\Delta = 23(X_A - X_B)^2$$

$$0.208\sqrt{\Delta} = (X_A - X_B)$$

Here E(A – B), E(A – A) and E(B – B) represent bond dissociation energies of A-B, A-A and B-B respectively. X_A and X_B are electronegativities of A and B respectively.

- Pauling took geometrical mean instead of arithmetic mean of E(A – A) and E(B – B) and introduced following empirical relation:

$$E(A-B) - (E(A-A) \times E(B-B))^{\frac{1}{2}} = \Delta$$

$$= 30(X_A - X_N)$$

$$0.182\sqrt{\Delta} = X_A - X_B$$

- **Allred and rochow's method** According to this method, electronegativity is the electrostatic force of attraction exerted by the nucleus of an atom on the valence electrons.
- In this method, the following empirical relation is used for calculating the electronegativity value.

$$X = -0.359\frac{Z_{eff}}{r^2} + 0.744$$

- Here, X is the electronegativity and 'r' is covalent radius of the atom.
- **Mulliken's method**

$$X = \frac{1E + EA}{2}$$

Here, IE = Ionization energy in eV

EA = Electron affinity in eV

When these are expressed in kJ/mol replace 2 by 540.

Factors affecting electronegativity

- Electronegativity $\propto Z_{eff}$

$$\propto \frac{1}{\text{Size}}$$

- Ioniztion energy and electron affinity
- Electrogenetivity $\propto$ Inozation energy

$\propto$ Electron affinity

- *Charge on an atom* A cation will be more electronegative than the parent atom, which in turn will be more electronegative than its anion. Higher the positive charge on an atom (oxidation state), greater will be its electronegativity. For example, electronegativity of Fe^{+3} is greater than that of Fe^{+2}
- *Effect of substitution* The electronegativity of an atom depends upon the nature of substituent attached to that atom. For example, carbon atom in CF_3I acquires greater positive charge than CH_3I. Therefore, C–atom in CF_3I is more electronegative than in CH_3I.
- The difference in electronegativity of an atom caused by substituents, results in different chemical behaviour of that atom.
- Electronegativity $\propto$ s percentage. Therefore, the electronegativity follows the order $sp > sp^2 > sp^3$. Consequently, the value of electronegativity follows the order

$$C \equiv C > C = C > C - C$$

Most

Electronegative

Variation in electronegativity value

In a period On moving left to right in a period, the electronegativity increases as Z_{eff} increases and size decreases.

- Order for first electronegativity in any period follows the order.

IA < IIA > IIIA < IVA < VA > VIA < VIIA

Minimum Maximum

For Example:

Li < Be < B < C < N < O < F

Na < Mg < Al < Si < P < S < Cl

- In any given period, halogens have the maximum, while alkali metals have the lowest value of electronegativity.
- Electronegativity of zero group elements is zero because of the stable octet state, they have no tendency to attract electrons.

In a group On moving from top to bottom in a group, electronegativity decreases as Z_{eff} decreases and size increases. For example, F > CI > Br > I

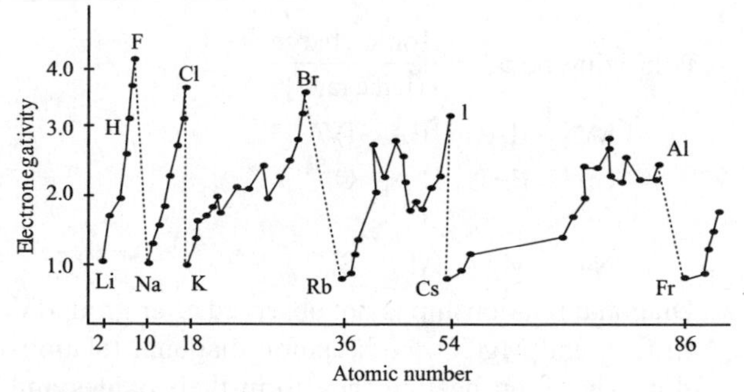

FIGURE 3.7 Electronegativity Values for Different Periods

FACTS TO REMEMBER

- **Decreasing order of electronegativity**
- **F > O > N > Cl > C > B**
- **4 3.5 3 2.97 2.5 2**
- **Metalloids have nearly 2 times the value of electronegativity as compared to the above elements.**

Applications of electronegativity

- Calculation of partial ionic character in a covalent bond. It depends upon two factors:
  1. The electronegativity difference between the two bonded atoms.
  2. Dipole moment of the compound.
- In a covalent bond, ionic character percentage is calculated through **Hannay and Smyth equation** as follows:

$$= 16(X_A - X_B) + 3.5\,(X_A - X_B)^2$$

Ionic character percentage

$$= 1 - e^{-1/4}\,(X_A - X_B)2$$

- When electronegativity difference is greater than 1.7 the compound will be ionic in nature.
- CsF is most ionic due to a very large electronegativity difference of 3.3 between its constituent atoms, (the highest amongst any other combination of elements.)

Per cent Ionic Character of a Bond

Per cent ionic character

$$= \frac{\text{Actual dipole moment of the bond}}{\text{Dipole moment of a pure ionic bond}} \times 100$$

Nature of Oxides and Hydroxides

On moving left to right in a period, basic nature decreases while acidic nature increases. For example,

Na_2O, MgO, Al_2O_3, SiO_2, P_2O_5, SO_2, Cl_2O_7

Most basic ... Most acidic

$NaOH > Mg(OH)_2 > Al(OH)_3 > Si(OH)_4$

Most basic ... Most acidic

On moving down the group, basic nature increases. For example in group IA, Cs_2O and CsOH will be most basic.

Nature of Oxy-acids

In a period, the strength of the oxy-acids formed by nonmetals increases from left to right.

Period II

H_3BO_3 H_2CO_3 HNO_3 → Strength increases

Period III

H_2SiO_3 H_3PO_4 H_2SO_4 $HClO_4$ → Strength increases

- On moving down the group, the strength of the oxyacids of non-metals decreases.

Group V

HNO_3 H_3PO_4 H_3AsO_4 → Strength decreases

Group VII

$HClO_4$ $HBrO_4$ HIO_4 → Strength decreases

Nature of Hydrides

The nature of the hydrides changes from basic to acidic on moving from left to right in a period.

NH_3	H_2O	HF
Weak base	Neutral	Weak acid
PH_3	H_2S	HCl
Very weak base	Weak acid	Strong acid

Atomic Volume

Atomic volume is the volume occupied by one mole of atoms of an element in solid state as its melting point. It is obtained by dividing the gram atomic mass with density of the element.

- In a period, on moving from left to right, first the atomic
- volume decreases to a minimum value and then start increasing.
- It normally increases down the group.

Boiling Point, Melting Point and Density

Boiling point, melting point and density increase to a maximum value and then decrease as we move along a period. On moving down the group, their values increase.

Hydration Energy (ΔH_y) and Lattice Energy (U)

$$\Delta H_Y \text{ or } U \propto \text{Charge on ion}$$

$$\propto \frac{1}{\text{Size of ion}}$$

These increase from left to right in a period and decrease down the group. For example,

$$Li^+ < Be^{2+} < B^{3+}$$

$$Li^+ > Na^+ > K^+$$

Ionic Mobility

$$\text{Ionic mobility} \propto \frac{1}{\text{Charge on ion}}$$

$$\propto \text{Size of Hydrated Ion.}$$

This means that it deceases from left to right in a period and increases down the group. For example,

$$Li^+ > Be^{2+} > B^{3+}$$

$$Li^+ < Na^+ < K^+ < Rb^+ < Cs^+$$

POINTS TO REMEMBER

Diagonal relationship Certain period II elements show some similarities with the period III elements which are diagonal to them. This phenomenon is referred to as **diagonal relationship**. It is occurs due to the similar ionic sizes, electronegativities and polarizing power.

$$\text{Polarizing power} = \frac{\text{Ionic charge}}{(\text{Ionic raius})^2}$$

IA	IIA	IIIA	IVA
Li	Be	B	C
↘	↘	↘	
Na	Mg	Al	Si

- Diagonal relationship is not observed after group IV. Li_2CO_3 and $MgCO_3$ both exhibit diagonal relationship where, on heating they form their oxides and CO_2.

$$Li_2CO_3 \xrightarrow{\Delta} Li_2O + CO_2$$

$$MgCO_3 \xrightarrow{\Delta} MgO + CO_2$$

BeO and Al_2O_3 are amphoteric oxides.

- Carbides of Be and Al, on hydrolysis, give methane gas.

$$Al_4C_3 + 12H_2O \rightarrow 4A1(OH)_3 + 3CH_4$$

$$Be_2C + 2H_2O \rightarrow 2BeO + CH_4$$

- When the difference between ionization energy and electron affinity is high, a covalent bond is formed.
- H, Be, Al, Ga, In, Tl, Sn, Pb, Sb, Bi forms amphoteric oxides.
- Fe, CO and Ni have nearly same Z_{eff} value, so they have almost same size and ionization energy.
- Properties for which, the value increases from left to right in a period and decreases from top to bottom in a group are: non-metallic nature, electronegative nature, oxidizing nature, electronegativity, ionization energy, electron affinity, lattice energy and hydration energy.
- Properties for which the value decreases from left to right and increases from top to bottom in a group are: metallic nature, electropositive nature, reducing nature, basic nature, radius or size and ionic mobility.

CHAPTER-END EXERCISES

PRACTICE QUESTIONS – I

1. The element with atomic number 12 belongs to .... group and .... period.

(a) IA, third
(b) IIIA third
(c) IIA, third
(d) IIA, second

2. An element having atomic number 56 belongs to:

(a) Lanthanides
(b) Actinides
(c) Alkaline earth metals
(d) None of these

3. The maximum valency of an element having atomic number seven is:
(a) 1 (b) 3
(c) 5 (d) 7

4. The number of elements present in the fifth period of periodic table is:
(a) 10 (b) 8
(c) 32 (d) 18

5. An example of amphoteric oxide is:
(a) Ti_2O_2 (b) MgO
(c) $C1_2O_7$ (d) Al_2O_3

6. The elements with atomic numbers 9, 17, 35, 53 and 85.
(a) Noble gases (b) Light metals
(c) Halogens (d) Heavy metals

7. Which one of the following is a metalloid?
(a) Phosphorus (b) Antimony
(c) Nitrogen (d) Bismuth

8. The correct order of atomic radii is:
(a) F < K < Na < Li
(b) Li < Na < K < F
(c) F < Li < Na < K
(d) Na < K < Li < F

9. Which one of the following is an amphoteric oxide?
(a) ZnO (b) Na_2O
(c) SO_2 (d) B_2O_3

10. Which of the following oxides is most basic?
(a) Bi_2O_3 (b) SeO_2
(c) $A1_2O_3$ (d) Sb_2O_3

11. The outermost configuration of the most electronegative elements is
(a) $ns^2\ np^5$ (b) $ns^2\ np^6$
(c) $ns^2\ np^4$ (d) $ns^2\ np^3$

12. Which of the following is the atomic number of a metal?
(a) 32 (b) 34
(c) 36 (d) 38

13. Which of the following has maximum ionization potential?
(a) Al (b) P
(c) Si (d) Mg

14. The first ionization potential will be maximum for:
(a) Uranium (b) Iron
(c) Hydrogen (d) Lithium

15. The most acidic oxide is:
(a) MgO (b) CaO
(c) Na_2O (d) Al_2O_3

16. The atomic radius decreases in a period due to
(a) Increase in nuclear attraction
(b) Decrease in nuclear attraction
(c) Increase in number of electrons
(d) Decrease in number of electrons

17. The correct order of decreasing first ionization potential is
(a) Ca > K > Rb > Cs
(b) Cs > Rb > K > Ca
(c) Ca > Cs > Rb > K
(d) K > Rb > Cs > Ca

18. The electron affinity of Be is similar to
(a) He (b) B
(c) Li (d) Na

19. Which one of the following represents the correct order of electronegativity?
(a) P > O > N
(b) N > P > O
(c) O > N > P
(d) N > O > P

20. The transition metal with least atomic number is:
(a) Os (b) Zr
(c) Pt (d) Ru

21. Correct order of radii is:
(a) Na < Li < K (b) N < Be < B
(c) $Cl^- < S^{2-} < P^{3-}$ (d) $Fe^{3+} < Fe^{2+} < Fe^{4+}$

22. The element having highest electron affinity is:
(a) Bromine (b) Iodine
(c) Fluorine (d) Chlorine

23. Which of the following is a favourable factor for cation formation?
(a) High electron affinity
(b) High electronegativity
(c) Small atomic size
(d) Low ionization potential

24. The order of first ionization energies of the elements Li, Be, B, Na is:
(a) Be > Li > B > Na
(b) B > Be > Li > Na
(c) Na > Li > B > Be
(d) Be > B > Li > Na

25. Which of the following is an inert gas?

(a) O_2 (b) Argon
(c) N_2 (d) H_2

26. The outermost electronic configuration of the most electronegative element is:

(a) $ns^2 np^3$ (b) $ns^2 np^4$
(c) $ns^2 np^5$ (d) $ns^2 np^6$

27. Which one of the following hydroxides is insoluble in water?

(a) $Ca(OH)_2$ (b) $Ba(OH)_2$
(c) $Mg(OH)_2$ (d) $Sr(OH)_2$

28. The electronic configuration of four different elements is given below. Identify the group IV element among these.

(a) [He] $2s^1$ (b) [Ne]$3s^2$
(c) [Ne] $3s^2 3p^2$ (d) [Ne] $3s^2 3p^5$

29. The lanthanide contraction is responsible for the fact that

(a) Zr, Hf have about the same radius
(b) Zr, Y have about the same radius
(c) Zr, Nb have same oxidation state
(d) Zr, Zn have same oxidation state

30. Sodium forms Na^+ ion but it does not form Na^{2+} because of:

(a) Very low value of $(IE)_1$ and $(IE)_2$
(b) Very high value of $(IE)_1$ and $(IE)_2$
(c) Low value of $(IE)_1$ and low value of $(IE)_2$
(d) Low value of $(IE)_1$ and high value of $(IE)_2$

31. Which of the electronic configuration represents a noble gas?

(a) $ls^2, 2s^2, 2p^6, 3s^2, 3p^6$
(b) $ls^2, 2s^2, 2p^6, 3s^2, 3p^3$
(c) $ls^2, 2s^2, 2p^6, 3s^2, 3p^1$
(d) $ls^2, 2s^2, 2p^6, 3s^2, 3p^4$

32. Ionic radius (in A) of As^{3+}, Sb^{3+} and Bi^{3+} follow the order:

(a) $As^{3+} > Sb^{3+} > Bi^{3+}$
(b) $Sb^{3+} > Bi^{3+} > As^{3+}$
(c) $Bi^{3+} > As^{3+} > Sb^{3+}$
(d) $Bi^{3+} > Sb^{3+} > As^{3+}$

33. Identify the least stable ion amongst the following:

(a) Li^- (b) Be^-
(c) B^- (d) C^-

34. The set representing the correct order of first ionization potential is:

(a) K > Na > Li (b) Br > Mg > Ca
(c) B > C > N (d) Ge > Si > C

35. Which of the following oxides is amphoteric in character?

(a) SnO_2 (b) CO_2
(c) CaO (d) SiO_2

36. Property of alkaline earth metals that increases with their atomic number is:

(a) Ionization energy
(b) Solubility of their hydroxides
(c) Solubility of their sulphates
(d) Electronegativity

37. With reference to the concept of ionization energy, which one of the following set is correct?

(a) Cs > U > B (b) U > K > Cs
(c) Cs < U < K (d) B > U > K

38. For electron affinity of halogens which of the following is correct?

(a) F > I (b) F > CI
(c) Br > Cl (d) Br > F

39. The electronic configuration of the most electronegative element is:

(a) $ls^2, 2s^2, 2p^5$
(b) $Is^2, 2s^2, 2p^4, 3s^1$
(c) $ls^2, 2s^2, 2p^6, 3s^1, 3p^5$
(d) $ls^2, 2s^2, 2p^6, 3s^2, 3p^5$

40. Let electronegativity, ionization energy and electronicaffinity be represented as EN, IP and EA respectively. Which one of the following equation is correct according to Mulliken?

(a) $EN = IP \times EA$
(b) $EN = \frac{IP}{EA}$
(c) $EN = \frac{IP + EA}{2}$
(d) $EN = IP - EA$

41. Which of the following order is correct for the size of Fe^{3+}, Fe and Fe^{2+}?

(a) $Fe < Fe^{2+} < Fe^{3+}$
(b) $Fe^{2+} <$ Fe3+ $<$ Fe
(c) $Fe < Fe^{3+} < Fe^{2+}$
(d) $Fe^{3+} < Fe^{2+} < Fe$

42. In the following, the element with the highest ionization energy is:
(a) [Ne] $3s^2$ $3p^1$ (b) [Ne] $3s^2$ $3p^3$
(c) [Ne] $3s^2$ $3p^2$ (d) [Ne] $3s^2$ $3p^4$

43. Why Sc (Z = 21) is not considered as a transition element?
(a) Properties of Sc are similar to alkali metals
(b) 3d orbitals are empty in its stable compound
(c) Stable oxidation number of Sc is +2
(d) Atomic volume of Sc is very large

44. Which is not correctly given here?
(a) F > O > N (Electro negativity)
(b) F > O > N (Ionisation energy)
(c) F > O > N (Electron offinity)
(d) F > O > N (Reactivity)

45. Which corresponds to the most electropositive character?
(a) [Xe] $6s^1$ (b) [Xe] $6s^2$
(c) [He] $2s^1$ (d) [He] $2s^1$

46. Arrange S, O and Se in the ascending order of electron affinity.
(a) S < Se < O (b) Se < O < S
(c) Se < S < O (d) S < O < Se

47. Let IP stand for ionization potential. Then element for which the value of IP_1, IP_2, IP_3, IP_4 are 50, 100, 250 and 2051 eV respectively, is
(a) Al (b) Na
(c) Mg (d) Si

48. Correct order of hydrolysis is:
(a) $PC1_5 < A1Cl_3 < CCl_4 < SiCl_4$
(b) $CCl_4 < AlCl_3 < SiCl_4 < PCl_5$
(c) $AICl_3 < PCl_3 < CCl_4 < SiCl_4$
(d) $CCl_4 < AICl_3 < PCl_5 < SiCl_4$

49. The correct order of electronegativity for O, O^+ and O^{--} is:
(a) $O^- > O > O^+$ (b) $O > O^+ > O^-$
(c) $O^+ > O^- > O$ (d) $O^+ > O > O^-$

50. Which of the following statements is true about effective nuclear charge?
(a) Z_{eff} decreases from top to bottom
(b) Z_{eff} increases from top to bottom
(c) Z_{eff} increases as we move from left to right in periodic table
(d) $Z_{eff} = Z \times \sigma$ (here σ is screening constant)

51. The ionization potential of K^+ is 2.4 eV. The electron affinity of K^+ will be
(a) 1.2eV (b) 2.4eV
(c) 4.2eV (d) 8.4eV

52. Let IP stand for ionization potential. The IP_1 and IP_2 of Mg are 178 and 348 kcal mol^{--1}. The energy required for the following reaction is:
$Mg \rightarrow Mg^{2+} + 2e^-$
(a) +178 kcal (b) +526 kcal
(c) –170 kcal (d) –526 kcal

53. The correct order of ionic radii is:
(a) $I^+ > I^- > I$ (b) $I > I^+ > I^-$
(c) $I^+ > I > I^-$ (d) $I^- > I > I^+$

54. The first ionization energy in electron volts of nitrogen and oxygen atoms are respectively given by:
(a) 14.6, 13.6 (b) 13.6, 14.6
(c) 13.6, 13.6 (d) 14.6, 14.6

55. The electronic configurations of some elements are given below. The element with highest electron affinity is:
(a) ls^2, $2s^2$, $2p^3$ (b) ls^2, $2s^2$, $2p^4$
(c) ls^2, $2s^2$, $2p^5$ (d) ls^2, $2s^2$, $2p^2$

56. First second and third ionization potential values are 100 eV, 150 eV and 1500 eV. Element can be:
(a) Be (b) B
(c) F (d) Na

57. Arrange P^{3-}, S^{2-}, H^-, I^- in the order of increasing ionic radii:
(a) P^{3-}, S^{2-}, 1, H^- (b) S^{2-}, P^{3-}, H^-, I^-
(c) S^{2-}, H^-, P^{3-}, I^- (d) H^-, S^{2-}, P^{3-}, I^-

58. The correct value of ionisation energies (in kJ mol^{-1}) of Si, P, CI and S respectively are:
(a) 786, 1012,999,1256
(b) 1012, 786, 999, 1256
(c) 786, 1012,1256,999
(d) 786, 999, 1012, 1256

59. In which of the following process is energy librated
(a) $Cl \rightarrow Cl^+ + e$ (b) $HCl \rightarrow H^+ + Cl^-$
(c) $Cl + e \rightarrow Cl^-$ (d) $O^- + e \rightarrow O^{2-}$

60. Which of the following is mismatched with reference to third period?
(a) Largest size sodium
(b) Strongest oxidantchlorine

(c) Ionization potential of greater than nitrogen phosphorus
(d) Ionization potential of greater than sulphur phosphorus

61. The electron affinities of N, O, S and Cl are:
(a) O ≈ Cl < N ≈ S (b) O < S < Cl < N
(c) N < O < S < Cl (d) O < N < Cl < S

PRACTICE QUESTIONS – II

62. While moving down a group in the periodic table, which of the following would be true?
(1) All the atoms have the same number of valence electrons
(2) Gram atomic volume increases
(3) Electronegativity decreases
(4) Metallic character decreases and the basic nature of their oxides decreases.
Select the correct answer by using the following codes:
(a) 1,2 and 3 (b) 2, 3 and 4
(c) 2 and 3 (d) 1 and 3

63. Enthalpy change in the following process is $M + e^- \rightarrow M^-$, $\Delta H = X$ kJ mole^{-1}.
Which of the following process have enthalpy change = X kJ mole^{-1}?
(a) $M^- \rightarrow M + e^-$ (b) $M^+ + e^- \rightarrow M$
(c) $M^{2+} + e^- \rightarrow M^+$ (d) $M + e^- \rightarrow M^-$

64. The formation of O^{2-} (g) starting from O (g) is endothermic by 603 kJ mol^{-1}. If electron affinity of O (g) is –141 kJ mol^{-1}, the second electron affinity of oxygen would be:
(a) +744 kJ mol^{-1} (b) –744 kJ mol^{-1}
(c) +462 kJ mol^{-1} (d) –462 kJ mol^{-1}

65. Sum of first three ionization energies of Al is 53.0 eV atom^{-1} and the sum of first two ionization energies of: Na is 52.2 eV atom^{-1}. Out of Al (III) and Na (II)
(a) Al (III) is more stable than Na (II)
(b) Na (II) is more stable than Al (III)
(c) Both are equally unstable
(d) Both are equally stable

66. The correct order of first ionization potential is:
(a) F > He > Mg > N > O
(b) He > F > N > O > Mg
(c) He > O > F > N > Mg
(d) N > F > He > O > Mg

67. The correct sequence of the ionic radii of the following is:
(a) $I^- > S^{2-} > Cl^- > O^{2-} > F^-$
(b) $S^{2-} > I^- > O^{2-} > Cl^- > F^-$
(c) $I^- > Cl^- > S^{2-} > O^{2-} > F^-$
(d) $I^- > S^{2-} > Cl^- > F^- > O^{2-}$

68. The second ionization energies of the C, N, O and F atoms are such that:
(a) O > N > F > C
(b) F > O > N > C
(c) O > O > N > F
(d) O > F > N > C

69. Which of the following arrangement shows the correct order of increasing stability?
(a) $N^{2+} < As^{2}+ < Sb^{2+} < Bi^{2+}$
(b) $Zn^{2+} < Cu^{2+} < As^{2+} < Bi^{2+}$
(c) $Cu_{2+} < CO^{2+} < P^{2+} < N^{2+}$
(d) $C^{2+} < Ge^{2+} < Sn^{2+} < Pb^{2+}$

70. The first, second, third and fourth ionization energies of a given element are 0.80, 2.43, 3.66 and 25.03 kJ mol^{-1} respectively. The element is:
(a) Boron (b) Carbon
(c) Aluminium (d) Nitrogen

71. Which of the following arrangements show the correct order of increasing lattice energy?
(a) $BaSO_4 < SrSO_4 < CaSO_4 < MgSO_4$
(b) $MgCO_3 < CaCO_3 < SrCO_3 < BaCO_3$
(c) LiF < LiCl < LiBr < LiI
(d) NaF < KF < RbF < CsF

72. The correct order of radii is
(a) N < Be < B (b) $F^- < O^{2-} < N^{3-}$
(c) Na < Li < K (d) $Fe^{+3} < Fe^{+2} < Fe^{+4}$

73. Which one of the following arrangements does not truly represent the property indicated against it?
(a) $Br_2 < Cl_2 < F_2$: electronegativity
(b) $Br_2 < F_2 < Cl_2$: electron affinity
(c) $Br_2 < Cl_2 < F_2$: bond energy
(d) $Br_2 < Cl_2 < F_2$: oxidizing power

74. Correct order of ionization energy among the elements Be, B, C, N, O is

(a) B < Be < C < O < N
(b) B < Be < C < N < O
(c) Be < B < C < N < O
(d) Be < B < O < N < C

75. Which of the following orders is incorrect?
(a) $NH_3 < PH_3 < AsH_3 \rightarrow$ acidic nature.
(b) Li < Be < B < C $\rightarrow$ first ionization energy
(c) $Al_2O_3 > MgO < Na_2O < K_2O \rightarrow$ basic nature.
(d) $Li^+ < Na^+ < K^+ < Cs^+ \rightarrow$ ionic radius

76. In which of the following arrangements, the order is not according to the property indicated against it?
(a) I < Br < F < Cl
increasing electron gain enthalpy
(b) Li < Na < K < Rb
increasing metallic radius
(c) B < C < N < O
increasing first ionization energy
(d) $Al^{3+} < Mg^{2+} < Na^+ < F^-$
increasing ionic size

77. Which of the following statements are correct?
(1) HF is a stronger acid than HCl.
(2) Among halide ions, iodide is the most powerful reducing agent.
(3) Fluorine is the only halogen that does not show a variable oxidation state.
(4) HOCl is a stronger acid than HOBr.
(a) 2 and 4 (b) 2 and 3
(c) 1,2 and 3 (d) 2, 3 and 4

78. Which one of the following sets of ions represents a collection of isoelectronic species?
(a) $K^+, Cl^-, Ca^{2+}, Sc^{3+}$
(b) $Ba^{2+}, Sr^{2+}, K^+, Ca^{2+}$
(c) $N^{3-}, O^{2-}, F^-, S^{2-}$
(d) $Li^+, Na^+, Mg^{2+}, Ca^{2+}$

79. Which one of the following groupings represents a collection of isoelectronic species? (At. number of Cs = 55, Br = 35)
(a) Na^+, Ca^{2+}, Mg^{2+} (b) N^{3-}, F^-, Na^+
(c) Be, Al^{3+}, F^- (d) Ca^{2+}, Cs^+, Br^-

80. The atomic numbers of V, Cr, Mn and Fe are respectively 23, 24, 25 and 26. Which one of these may be expected to have the highest second ionization enthalpy?
(a) V (b) Cr
(c) Mn (d) Fe

81. The electronic configuration of elements A, B and C are [He] $2s^1$, [Ne] $3s^1$ and [Ar] $4s^1$ respectively. Which one of the following order is correct for the first ionization potentials (in kJ mol^{-1}) of A, B and C?
(a) A > B > C (b) C > B > A
(c) B > C > A (d) C > A > B

82. The electronic affinity values (in kJ mol^{-1}) of three halogens X, Y and Z are respectively –349, –333 and –325. Then X, Y and Z respectively are
(a) F_2, Cl_2 and Br_2
(b) Cl_2, F_2 and Br_2
(c) Cl_2, Br_2 and F_2
(d) Br_2, Cl_2 and F_2

83. A sudden large difference between the values of second and third ionization energies of elements would be associated with which of the following electronic configuration?
(a) $1s^2\ 2s^2\ 2p^6\ 3s^4$
(b) $1s^2\ 2s^2\ 2p^6\ 3s^2$
(c) $1s^2\ 2s^2\ 2p^6\ 3s^2\ 3p^1$
(d) $1s^2\ 2s^2\ 2p^6\ 3s^2\ 3p^2$

84. Identify the correct order in which the ionic radius of the following ions increases:
(1) F^- (2) Na^+ (3) N^{3-}
(a) 3, 1, 2 (b) 1, 2, 3
(c) 2, 3, 1 (d) 2, 1, 3

85. Identify the correct order in which the covalent radius of the following elements increases:
(1) Ti (2) Ca (3) Sc
(a) (1), (2), (3) (b) (2), (1), (3)
(c) (1), (3), (2) (d) (3), (2), (1)

86. Amongst the following elements (whose electronic configurations are given below), the correct increasing order of ionization energy is
(1) [Ne] $3s^2\ 3p^1$ (2) [Ne] $3s^2\ 3p^3$
(3) [Ne] $3s^2\ 3p^2$ (4) [Ar] $3d^{10}\ 4s^2\ 4p^3$
(a) 3 < 1 < 2 < 4 (b) 4 < 2 < 3 < 1
(c) 1 < 3 < 4 < 2 (d) 2 < 4 < 1 < 3

87. Atomic radii of fluorine and neon in Angstorm units are respectively given by
(a) 1.60, 1.60 (b) 0.72, 0.72
(c) 0.72, 1.60 (d) 1.60, 0.72

88. Identify the correct order of acidic strengths of CO_2, CuO, CaO and H_2O.
(a) $CaO < CuO < H_2O < CO_2$

(b) $H_2O < CuO < CaO < CO_2$
(c) $CaO < H_2O < CuO < CO_2$
(d) $H_2O < CO_2 < CaO < CuO$

89. The correct statement among the following is:
(a) The first ionization potential of Al is less than the first ionization potential of Mg.
(b) The second ionization potential of Mg is greater than the second ionization potential of Na.
(c) The first ionization potential of Na is less than the first ioniation potential of Mg.
(d) The third ionization potential of Mg is greater than the third ionization potential of Al.

90. The statement that is not correct for periodic classification of elements is
(a) The properties of elements are a periodic function of their atomic numbers.
(b) Non-metallic elements are less in number than metallic elements.
(c) The first ionization energies of elements along a period do not vary in a regular manner with increase in atomic number.
(d) For transition elements, the d-subshells are filled with electrons monotonically with increase in atomic number.

91. The first ionization potential of Na, Mg, Al and Si are in the order:
(a) Na > Mg > Al < Si
(b) Na > Mg > Al > Si
(c) Na < Mg < AKSi
(d) Na < Mg >AKSi

92. The electronegativity of the following elements increases in the order:
(a) Si, P, C, N
(b) N, Si, C, P
(c) P, Si, N, C
(d) C, N, Si, P

93. Ionisation energies of fluorine and neon in KJ/Mole units are given respectively by:
(a) 1681, 1681 (b) 2081, 2081
(c) 1681, 2081 (d) 2081, 1681

94. The element with the highest first ionization potential is:
(a) Nitrogen (b) Oxygen
(c) Boron (d) Carbon

95. The correct order of second ionization potential of carbon, nitrogen, oxygen and fluorine is:
(a) O > N > F > C
(b) O > F > N > C
(c) F > O > N > C
(d) C > N > O > F

96. The correct increasing bond angle among BF_3, PF_3 and $C1F_3$ follows the order:
(a) $BF_3 < PF_3 < CIF_3$
(b) $PF_3 < BF_3 < CIF_3$
(c) $CIF_3 < PF_3 < BF_3$
(d) $BF_3 < PF_3 < CIF_3$

97. Which ionization potential in the following equations involves the greatest amount of energy?
(a) $Na \rightarrow Na^+ + e^-$
(b) $K \rightarrow K^+ e^-$
(c) $C^{2+} \rightarrow C^{3+} + e^-$
(d) $Ca^+ \rightarrow Ca^{2+} + e^-$

98. Among the elements W, X, Y and Z having atomic numbers 9, 10, 11 and 12 respectively, the correct order of ionization energies is:
(a) W > Y > X > Z
(b) X > W > Z > Y
(c) X > Z > Y > W
(d) Z > Y > X > W

Practice Questions – III

99. Which of the following orders are correct?
(1) Thermal stability
$BeCO_3 < MgCO_3 < CaCO_3 < SrCO_3 < BaCO_3$
(2) Basic nature
LiOH > NaOH > KOH > RbOH > CsOH
(3) Solubility in water
$BeSO_4 < MgSO_4 < CaSO_4 < SrSO_4 < BaSO_4$
(4) Melting point
NaCl > KC1 > RbCI >CsCl > LiCl
(a) (1), (4) (b) (1), (2), (4)
(c) (2), (3) (d) (1), (2) (3), (4)

100. Match the following:

List I	List II
1. $Li < Na < K < Rb$	(1) Increasing order energy
2. $Li < Be > B < C$	(2) Decreasing order of metallic nature
3. $F_2 > Cl_2 > Br_2 > I_2$	(3) Increasing order of size
4. $F_2 > O_2 > Cl_2 > s$	(4) Increasing order of electron affinity

The correct matching is:

	1	2	3	4
(a)	(1)	(4)	(2)	(3)
(b)	(3)	(1)	(4)	(2)
(c)	(3)	(1)	(2)	(4)
(d)	(1)	(2)	(3)	(4)

101. Match the following:

List I	List II
1. Element with highest	(1) I_2 electronegativity
2. Element with highest	(2) Br_2 electron affinity
3. Liquid non metal	(3) Cl_2
4. metallic solid	(4) F_2

The correct matching is:

	(1)	(2)	(3)	(4)
(a)	(3)	(2)	(1)	(4)
(b)	(4)	(3)	(1)	(2)
(c)	(2)	(3)	(4)	(1)
(d)	(1)	(2)	(3)	(4)

102. Calculate the electronegativity of fluorine from the following data.

$E_{H-H} = 104.2$ Kcal mol^{-1}

$E_{F-F} = 36.6$ Kcal mol^{-1}

$E_{H-F} = 134.6$ Kcal mol^{-1}

$X_H = 2.1$

(a) 2.86 (b) 3.76
(c) 1.86 (d) 3.26

103. For the gaseous reaction, $K + F \rightarrow K^+ + F^-$, ΔH was calculated to be 19 kcal under conditions where the cations and anions were prevented by electrostatic separation from combining with each other. The ionization potential of K is 4.3 eV. What is the electron affinity of F?

(a) 3.21 (b) 4.28
(c) 3.48 (d) 1.48

104. The electronic configurations of four elements are given below:

(1) $1s^2\ 2s^2\ 2p^5$
(2) $1s^2\ 2s^2\ 2p^4$
(3) $1s^2\ 2s^2\ 2p^3$
(4) $1s^2\ 2s^2\ 2p^6\ 3s^2\ 3p^4$

Which of the following arrangements gives the correct order in terms of increasing electronegativity of the elements?

(a) $3 < 2 < 4 < 1$
(b) $2 > 3 > 1 > 4$
(c) $4 < 3 < 2 < 1$
(d) $1 < 2 < 3 < 4$

105. The successive ionization energy values for an element X are given below:

(a) 1st ionization energy = 410 kJ mol^{-1}
(b) 2nd ionization energy = 820 kJ mol^{-1}
(c) 3rd ionization energy = 1100 kJ mol^{-1}
(d) 4th ionization energy = 1500 kJ mol^{-1}
(e) 5th ionization energy = 3200 kJ mol^{-1}

Find out the number of valence electron for the atom, X.

(a) 4 (b) 3
(c) 5 (d) 2

106. The electronic configuration of four elements are

(1) $[Xe]6s^1$ (2) $[Xe]4f^{14}\ 5d^1\ 6s^2$
(3) $[Ar]\ 4s^2\ 4p^5$ (4) $[Ar]\ 3d^7\ 4s^2$

Which one of the following statements about these elements is not correct?

(a) (1) is a strong reducing agent
(b) (2) is a d block element
(c) (3) has high electron affinity
(d) (4) shows variable oxidation state

107. Which of the following statements is/are not true about the diagonal relationship of Be and Al?

(1) Both react with NaOH to liberate hydrogen
(2) Their oxides are basic
(3) They are made passive by nitric acid
(4) Their carbides give acetylene on treatment with water

(a) Only (1) (b) (2) and (3)
(c) Only (4) (d) (2) and (4)

108. Pick out the statement(s) which is are not true about the diagonal relationship of Li and Mg.

(1) Polarizing powers of Li^+ and Mg^{2+} are almost same.
(2) Like Li, Mg decomposes water very fast.
(3) LiCl and $MgCl_2$ are deliquescent.
(4) Like Li, Mg readily reacts with liquid bromine at ordinary temperature.

(a) (1) and (4) (b) (2) and (4)
(c) Only (2) (d) Only (1)

109. The correct statement among the following is:
(a) The first ionization potential of Al is less than the first ionization potential of Mg
(b) The second ionization potential of Mg is greater than the second ionization potential of Na
(c) The first ionization potential of Na is less than the first ioniation potential of Mg
(d) The third ionization potential of Mg is greater than the third ionization potential of Al

110. The statement that is not correct for periodic classification of elements is:
(a) The properties of elements are a periodic function of their atomic numbers
(b) Non–metallic elements are less in number than metallic elements
(c) The first ionization energies of elements along a period do not vary in a regular manner with in crease in atomic number
(d) For transition elements, the d–subshells are filled with electrons monotonically with increase in atomic number

111. Ce^{3+}, La^{3+}, Pm^{3+} and Yb^{3+} have ionic radii in the in creasing order as:
(a) $La^{3+} < Ce^{3+} < Pm^{3+} < Yb^{3+}$
(b) $Yb^{3+} < Pm^{3+} < Ce^{3+} < La^{3+}$
(c) $La^{3+} = Ce^{3+} < Pm^{3+} < Yb^{3+}$
(d) $Yb^{3+} < Pm^{3+} < La^{3+} < Ce^{3+}$

112. The radius of La^{3+} (atomic number = 57) is 1.06 A. Which one of the following given values will be closest to the radius of Lu^{3+} (atomic number = 71)?
(a) 0.85 A (b) 1.06 A
(c) 1.40 A (d) 1.60 A

113. Among Al_2O_3, SiO_2, P_2O_3 and SO_2 the correct order of acidic strength is:
(a) $Al_2O_3 < SiO_2 < P_2O_3 < SO_2$
(b) $SO_2 < P_2O_3 < SiO_2 < Al_2O_3$
(c) $SiO_2 < SO_2 < Al_2O_3 < P_2O_3$
(d) $Al_2O_3 < SiO_2 < SO_2 < P_2O_3$

114. The formation of the oxide ion O^{2-} (g) require first an exothermic and then an endothermic step as shown below:
$O (g) + e^- \rightarrow O^- (g); \Delta H^0 = -142 \text{ kJ mol}^{-1}$
$O^- (g) + e^- \rightarrow O^{2-} (g); \Delta H^0 = 844 \text{ kJ mol}^{-1}$
This is because,
(a) Oxygen is more electronegative
(b) Oxygen has high electron affinity
(c) O^- ion has comparatively larger size than oxygen atom
(d) O^- ion will tend to resist the addition of another electron

115. The increasing order of the first ionization enthalpies of the element B, P, S and F (lowest first) is:
(a) F < S < P < B
(b) P < S < B < F
(c) B < P < S < F
(d) B < S < P < F

116. The ionization enthalpy of hydrogen atom is 1.312 × 106 J mol^{-1}. The energy required to excite the electron in the atom from n = 1 to n = 2 is:
(a) 8.51×10^5 J mol^{-1}
(b) 6.56×10^5 J mol^{-1}
(c) 7.56×10^5 J mol^{-1}
(d) 9.84×10^5 J mol^{-1}

117. The first four ionization energy values of an element are 190, 578, 872 and 5960 k.cal. The Number of valence electrons in the element is?
(a) 1 (b) 2
(c) 3 (d) 4

118. Which of the following are correct here?
(1) E.N → F > O > N > P
(2) I.E → F > N > O > P
(3) E.A → F > O > P > N
(4) Size → F > P > N> O
(a) 1, 2 (b) 1, 2, 3
(c) 1, 2, 4 (d) 1, 2, 3, 4

119. An element has successive ionization energies as 940, 2080, 3090, 4140, 7030, 7870, 16000 and 19500 kJ/mole. The element belongs to group number?
(a) 14 (b) 15
(c) 16 (d) 17

120. If you are given Avogadro's number of atoms of a gas 'P' and half of the atoms are converted into by energy ΔH. The ionization energy of 'P' can be given as
(a) $\frac{2\Delta H}{N_A}$ (b) $\frac{\Delta H}{N_A}$
(c) $\frac{2N_A}{\Delta H}$ (d) $\frac{N_A}{\Delta H}$

121. For Aluminium the values of I, I_2, I_3 are 578, 1817 and 2745 kJ/mole respectively. What will be the energy needed to convert all the atoms of Al to Al^{3+} present in 135 mg of Al vapours.

(a) 51.4 kJ (b) 25.7 kJ
(c) 77.1 kJ (d) 12.85 kJ

122. Consider the ground state configuration of these Elements:

P → [He] $2p^3$
Q → [He] $2p^4$
R → [Ne] $3p^5$
S → [Ne] $3p^6$

the correct order of Electron affinity is?

(a) P > Q > R > S
(b) Q > R > P > S
(c) R > Q > P > S
(d) Q > P > R > S

123. Ionization energy of sodium is same as that of

(a) Electronegativity of Na^+
(b) Electron affinity of Ne
(c) Ionisation energy of Mg
(d) Electron affinity of Na^+

Answer Keys

1. (c)	2. (c)	3. (d)	4. (d)	5. (d)	6. (c)	7. (b)	8. (c)	9. (a)	10. (a)
11. (a)	12. (d)	13. (c)	14. (c)	15. (d)	16. (a)	17. (a)	18. (a)	19. (c)	20. (b)
21. (c)	22. (d)	23. (d)	24. (d)	25. (b)	26. (c)	27. (b)	28. (c)	29. (a)	30. (d)
31. (a)	32. (d)	33. (b)	34. (b)	35. (a)	36. (b)	37. (c)	38. (a)	39. (a)	40. (c)
41. (d)	42. (b)	43. (b)	44. (b)	45. (a)	46. (c)	47. (a)	48. (b)	49. (d)	50. (c)
51. (b)	52. (b)	53. (d)	54. (a)	55. (c)	56. (a)	57. (d)	58. (d)	59. (c)	60. (c)
61. (c)	62. (a)	63. (a)	64. (a)	65. (a)	66. (b)	67. (a)	68. (d)	69. (d)	70. (a)
71. (b)	72. (b)	73. (c)	74. (a)	75. (b)	76. (c)	77. (d)	78. (a)	79. (b)	80. (b)
81. (a)	82. (b)	83. (b)	84. (d)	85. (c)	86. (c)	87. (c)	88. (a)	89. (b)	90. (d)
91. (d)	92. (a)	93. (c)	94. (a)	95. (b)	96. (c)	97. (c)	98. (b)	99. (a)	100. (c)
101. (b)	102. (b)	103. (c)	104. (c)	105. (a)	106. (b)	107. (d)	108. (b)	109. (b)	110. (d)
111. (b)	112. (a)	113. (a)	114. (d)	115. (d)	116. (d)	117. (c)	118. (b)	119. (c)	120. (a)
121. (b)	122. (c)	123. (d)							

Hints and Explanations for Selective Questions

1. It is magnesium, a third period and second group element.

3. As maximum valency equals to group number.

10. Here Bi_2O_3 is the most basic oxide while rest are amphoteric oxides.

15. Here acidic nature decreases in the order

$Al_2O_3 > MgO > CaO > Na_2O$

18. Both have zero electron affinity.

27. As solubility decreases down the group.

29. In each vertical column of transition elements, the elements of second and third transition series resemble each other more closely than the elements of first and second transition series on account of lanthanide contraction.

The pairs of elements such as Zr–Hf, Mo–W, Nb–Ta possess almost the same properties.

31. It is the configuration of Argon.

33. Be (Is^2 2s^2) due to its fully filled 2s–sub shell has least tendency to take up an electron. As such Be is least stable.

34. As down the group ionization energy decreases.

35. CaO–basic, CO_2 and SiO_2–acidic, SnO_2–amphoteric, as it reacts both with acids and bases.

$SnO_2 + 4HCl \rightarrow SnCl_4 + 2H_2O$

$SnO_2 + 2NaOH \rightarrow Na_2SnO_3 + H_2O$

36. Solubility of hydroxides of group II increases down the group.

37. Cs and K are the group one elements. Since ionization energy of an element decreases on moving down the group, therefore, the value of ionization energy for Cs is less than that for K.

38. As in case of halogens the electron affinity decreases as follows Cl > F > Br > I.

39. It is fluorine, the most electronegative element.

41. As size of cation $\propto$ charge on cation.

42. Half-filled orbitals and fully filled orbitals are more stable ad hence they have higher ionization energies. So [Ne] $3s^2$ $3p^3$ has the highest ionization energy.

44. As the correct ionisation energy order is F > N > O.

45. The configuration [Xe] $6s^1$ is of the first member of the sixth period and thus, the elements belongs to group one. Elements of group one have maximum metallic character, which increases on moving down the group.

46. The electron affinity decreases on moving down the group.

52. Removal of two electrons (one by one) from an atom requires energy = $IP_1 + IP_2$

= 178 + 348

= 526 kcal

53. As anions are always larger in size than their parent atom and cations are always smaller in size than their parent atom.

61. Chorine being a group 17 element has maximum electronegativity. Nitrogen has zero electron affinity as its extra stability is associated with its half filled orbitals. Sulphur has more electron affinity than oxygen because the effect of small size of oxygen atom is more dominated by the repulsion of electrons already present in 2p orbitals of O atom.

62. Fourth statement is not correct. The correct statement is metallic character increases and the basic nature of their oxides increases.

63. In $M^- \rightarrow M + e^-$, an electron is removed and for it energy is required. This makes it an endothermic process, therefore, enthalpy is postive (X).

64. $O (g) + 2e^- \rightarrow O^{2-} (g), \Delta H = 603$ kJ mol^1..........(1)

$O (g) + e^- \rightarrow O^-, \Delta H = -141$ kJ mol^{-1}...............(2)

Equations (1) and (2) give

$O^- + e^- \rightarrow O^{2-}, \Delta H = 603 - (-141)$

$= 744$ kJ mol^{-1}

65. As ionization energy is not the only criteria for the stability of an oxidation state.

67. Radii of anions carrying same charge decrease from left to right in a period and increase down the group.

68. Half filled $2p^3$ subshell of O^+ is more stable than $2p^4$ subshell of F^+ So second ionization energy of Oxygen will greater than that of F^+.

72. For the isoelectronic ions, the size decreases with increase in nuclear charge.

75. As the correct order of ionization energy is

Li < B < Be < C.

76. Nitrogen has stable configuration. So, ionization enthalpy of nitrogen is greater than that of oxygen.

So, the correct order of increasing first ionization enthalpy is B < C < O < N.

77. HF is not a stronger acid than HC1 because fluorine is more electronegative than chlorine, therefore it does not donate hydrogen easily as is done in the case of HCl.

80. Cr:[Ar]

$3d^5$ $4s^1$

Cr^+: [Ar]

$3d^5$ $4s^1$

(by first IP)

This is a stable electronic configuration. Hence, formation of Cr^{2+} by second ionization potential requires maximum enthalpy.

81. As ionization energy decreases with the increase in number of orbitals or down the group.

82. As in case of halogens the electron affinity decreases as follows Cl > F > Br > I.

86. As both (2) and (4) have exactly half filled configuration, but (2) has smaller size than (4), so it has higher ionization energy.

87. Atomic radius of neon being van der Waals radius, is bigger than that of fluorine which is infact its covalent radius.

88. CaO is basic while CO_2 is most acidic of these. The increasing acidic strength order is:

CaO < CuO < H_2O < CO_2.

89. Second ionization energy of Na is greater than that of Mg because second electron is to be removed from a stable, noble gas configuration in case of sodium.

91. First ionization energy of Mg is higher than that of Na because of increased nuclear charge and also that of Al because in Mg a 3s electron has to be removed while in Al it is a 3p electron. The first ionization energy of Si is, however, higher than those of Mg and Al because of its increased nuclear charge. So the order is Na < Mg > Al < Si

92. Si and P are in the 3rd period while C and N are in the 2nd period. Elements in 2nd period have higher electronegativities than those in the 3rd period. Since N has smaller size and a higher nuclear charge than C, its electronegativity is higher than that of C. Similarly the electronegativity of P is higher than that of Si. Thus the order is Si, P, C, N.

93. Atomic radius of neon being van der Waals radius higher than that of fluorine which is infact its covalent radius.

94. Amongst B, C, N, and O, N has the highest value of first ionization energy, due to its half filled 2p orbital which is more stable.

95. The correct order is O > F > N > C

96. BF_3 It has sp^2 hybridization and bond angle is 120° PF_3 It has sp^3 hybridization and bond angle is less than 109° 28' but greater than 100+ due to lone pairbond pair repulsions. CIF_3 has sp^3d hybridization and bond angle is 90°

97. As C^{2+} has $Is^2\ 2s^2$ for example, fully filled and stable configuration.

99. As basic nature increase down the group. Solubility of alkaline earth metal sulphates decrease down the group.

102. Let the electronegativity of fluorine be X_F. According to Pauling equation.

$$X_F - X_H = 0.208\,[E_{H\text{-}F} - \tfrac{1}{2}(E_{F\text{-}F} + E_{H\text{-}H})]^{1/2}$$

In this equation dissociation energies are taken in kcal mol^{-1}

$$X_F - 2.1 = 0.208\,[134.6 - \tfrac{1}{2}\,(104.2 + 36.6)]^{1/2}$$

$= 3.76$

103. $K \rightarrow K^+ + e \qquad \Delta E_1 = 4.3$ eV

$F + e \rightarrow F^- \qquad \Delta E_2 = -E$ eV

$$\frac{19.0}{23.06} = \Delta E_1 - \Delta E_2 = 4.3 - E$$

$0.82 = 4.3 - E$

$E = 3.48$

109. IE_2 of Na is greater than Mg because second electron is to be removed from stable noble gas configuration in case of sodium.

111. r_n (radius) $\propto \dfrac{1}{Z}$

113. As acidic nature increases left to right in a period with increase in electronegativity.

114. It is because of electronic repulsion.

116. $\Delta E = E_2 - E_1$

$$= \frac{1.312 \times 10^6}{22} - \frac{(-1.312 \times 10^6)}{1}$$

$= 9.84 \times 10^5$ J mol^{-1}

119. As the largest jump is in between I_6 and I_7 so, the element must have 6 valence electrons it means it must be a VI A or 16th group element.

121. In order to convert total Al into Al^{3+}, the required energy = 578 + 1817 + 2745 = 5140 kJ/mole

$$h_{Al} = \frac{135}{1000 \times 27} = 0.005$$

As for 1 mole Energy needed = 5140 kJ

So for 0.005 mole Energy needed = 5140 × 0.005

= 25.7 kJ

122. Here the correct order can be given as:

R > Q > P > S

(Cl) (O) (N) (Ar)

PREVIOUS YEARS' QUESTIONS

1. Which one of the following arrangements does not truly represents the property indicated against it?
(a) $Br_2 < Cl_2 < F_2$: eletronegativity
(b) $Br_2 < F_2 < Cl_2$: eletronegativity
(c) $Br_2 < Cl_2 < F_2$: bond energy
(d) $Br_2 < Cl_2 < F_2$: oxidizing power

[2000]

2. Correct order of (IE) among the elements Be, B, C, N, O is:
(a) B < Be < C < O < N
(b) B < Be < C < N < O
(c) Be < B < C < N < O
(d) Be < B < O < N < C

[2001]

3. Which of the following order is wrong?
(a) $NH_3 < PH_3 < AsH_3$ – Acidic
(b) Li < Be < B < C – $(IE)_1$
(c) $Al_2O_3 > MgO < Na_2O < K_2O$ – Basic
(d) $Li^+ < Na^+ < K^+ < Cs^+$ – Ionic radius

[2002]

4. General electronic confguration of lanthanides is:
(a) $(n-2)\,f^{1-14}\,(n-1)\,s^2\,p^6\,d^{0-1}\,ns^2$
(b) $(n-2)\,f^{0-14}\,(n-1)\,d^{0-1}\,ns^2$
(c) $(n-2)\,f^{0-14}\,(n-1)\,d^{10}\,ns^2$
(d) $(n-2)\,f^{0-1}\,(n-1)\,f^{1-14}\,ns^2$

[2002]

5. An atom has electronic configuration: $1s^2\,2s^2\,2p^6\,3s^2\,3p^6\,3d^3\,4s^2$, you will place it in
(a) V group (b) IV group
(c) II group (d) III group

[2002]

6. The ions O^{2-}, F^-, Na^+, Mg^{2+} and Al^{3+} are isoelectronic. Their ionic radii show:
(a) An increase from O^{2-} to F^- and then decrease from Na^+ to Al^{3+}
(b) A decrease from O^{2-} to F^- and then increase from Na^+ to Al^{3+}
(c) A significant increase from O^{2-} to Al^{3+}
(d) A significant decrease from O^{2-} to Al^{3+}

[2003]

7. Ionic radii are:
(a) Inversely proportional to effective nuclear charge
(b) Incersely proportional to square of effective nuclear charge
(c) Directly proportional to effective nuclear charge
(d) Directly proportional to square of effective nuclear charge

[2004]

8. Which one of the following arrangements represents the correct order of electron gain enthalpy (with negative sign) of the given atomic species?
(a) S < O < Cl < F (b) Cl < F < S < O
(c) F < Cl < O < S (d) O < S < F < Cl

[2005]

9. Four successive members of the first row transition elements are listed below with their atomic numbers. Which one of them is expected to have the highest third ionization energy?
(a) Iron (Z = 26)
(b) Vanadium (Z = 23)
(c) Manganese (Z = 25)
(d) Chromium (Z = 24)

[2005]

10. The correct order regarding the electronegativity of hybrid orbitals of carbon is:
(a) $sp < sp^2 > sp^3$
(b) $sp < sp^2 < sp^3$
(c) $sp > sp^2 < sp^3$
(d) $sp > sp^2 > sp^3$

[2006]

11. Which one of the following orders is not in accordance with the property stated against it?
(a) $F_2 > Cl_2 > Br_2 > I_2$; electronegativity
(b) $F_2 > Cl_2 > Br_2 > I_2$; bond dissociation energy
(c) $F_2 > Cl_2 > Br_2 > I_2$ oxidizing power
(d) HI > HBr > HCl > HF acidic property in water

[2006]

12. With which of the following electronic configuration an atom has the lowest ionization enthalpy?
(a) $1s^2\,2s^2\,2p^5$ (b) $1s^2\,2s^2\,2p^3$
(c) $1s^2\,2s^2\,2p^6\,3s^1$ (d) $1s^2\,2s^2\,2p^6$

[2007]

13. Identify the correct order of the size of the following:

(a) $Ca^{2+} < K^+ < Ar < S^{2-} < Cl^-$
(b) $Ca^{2+} < K^+ < Ar < Cl^- < S^{2-}$
(c) $Ar < Ca^{2+} < K^+ < Cl^- < S^{2-}$
(d) $Ca^{2+} < Ar < K^+ Cl^- < S^{2-}$

[2007]

14. The correct order of decreasing second ionization enthalpy of Ti (22) V (23), Cr(24) and Mn (25) is:

(a) Mn > Cr > Ti > V (b) Ti > V > Cr > Mn
(c) Cr > Mn > V > Ti (d) V > Mn > Cr > Ti

[2008]

15. The correct order of decreasing ionic radii among the following isoelectronic species is?

(a) $K^+ > Ca^{2+} > Cl^- > S^{-2}$
(b) $Ca^{2+} > K^+ > S^{-2} > Cl^-$
(c) $Cl^- > S^{2-} > K^+ > Ca^{2+}$
(d) $S^{-2} > Cl^- > K^+ > Ca^{2+}$

[2010]

16. Amoung the elements Ca, Mg, P, Cl the order of increasing atomic radii is?

(a) Mg < Ca < Cl < P
(b) Cl < P < Mg < Ca
(c) P < Cl < Ca < Mg
(d) Ca < Mg < P < Cl

[2010]

17. Which of the following represents the correct oredr of increasing electron gain enthalpy with negative sign for the elements O, S, F, and Cl?

(a) S < O < Cl < F
(b) Cl < F < O < S
(c) O < S < F < Cl
(d) F < S < O < Cl

[2011]

18. What is the value of electron gain enthalpy of Na^+ if IE_1 of Na is 5.1 e.v?

(a) 2.55 e.v (b) 10.2 e.v
(c) –5.1 e.v (d) –10.2 e.v

[2011]

19. Which of the following orders of ionic radie is correctly represented.

(a) $H^- > H > H^+$ (b) $Na^+ > F^- > O^{2-}$
(c) $Al^{3+} > Mg^{2+} > N^{3-}$ (d) $O^- > S^{2-} > Cl^-$

[2014]

20. The species Ar, K^+ and Ca^{2+} contains the same number of electrons in which order do their radie increase?

(a) $Ar < K^+ < Ca^{2+}$ (b) $Ca^{2+} < Ar < K^+$
(c) $Ca^{2+} < K^+ < Ar$ (d) $K^+ < Ar < Ca^{2+}$

[2015]

21. In which of the following options the order of arrangement does not agree with the variation of property indicated against it

(a) C < B < N < O (Increasing first ionisation energy)
(b) I < Br < Cl < F (Increasing electron gain enthalpy)
(c) Li < Na < K < Rb (Increasing metallic radius)
(d) $Al^{3+} < Mg^{2+} < Na^+ < F^-$ (Increasing Ionic size)

[2016]

Answer Keys

1. (c)	2. (a)	3. (b)	4. (b)	5. (a)	6. (d)	7. (a)	8. (b)	9. (c)	10. (d)
11. (b)	12. (c)	13. (b)	14. (c)	15. (d)	16. (b)	17. (c)	18. (c)	19. (a)	20. (c)
21. (a)									

HINTS AND EXPLANATIONS

1. As the bond energy of F_2 is less than Cl_2 and Br_2 due to smaller size and more electronic repulsion so the correct order must be as follows:

   $Cl_2 > Br_2 > F_2$

2. The ionization potential decreases as the size of atom increases and Z_{eff} decreases. So

   B < Be < C < O < N

3. As the correct order of ionization energy is

   Li < B < Be < C

4. General electronic configuration of lanthanides is given as

   $(n-2)\ f^{0-14}\ (n-1)\ d^{0-1}\ ns^2$

5. It is Vanadium present in fifth group.

6. Radius $\alpha\ 1/Z_{eff}$

   α 1/Magnitude of positive charge

   Hence order of radii is

   $O^{2-} > F^- > Na^+ > Mg^{2+} > Al^{3+}$

7. Ionic radii are inversely proportional to effective nuclear charge (Z_{eff})

   Ionic radii = $\alpha\ 1/Z_{eff}$

8. Cl has highest affinity but with negative sign its value is lowest.

Cl	< F	< S	< O
–349	–328	–200	–141kJ/mole

9. As it has half filled 3d–orbital (more stable configuration state) so it will have more ionization energy as more energy is required to remove electron here.

10. As the s–character increases in hybridized orbitals so its electronegativity increases.

    Electonegativity α s–percentage

    Hence $sp > sp^2 > sp^3$

11. Bond dissociation energy order is as follows:

Cl_2 >	Br_2 >	F_2 >	I_2
242.6	192.8	158.8	151.1 kJ mol^{-1}

12. $1s^2, 2s^2\ 2p^6, 3s^1$ (na) As it is largest in size with lowest effective nuclear charge so it will have lowest ionization energy.

13. Among isoelectronic ions, ionic radii of anions is more than that of cations. Further size of the anion increases with increase in negative charge and size of the cation decreases with increase in positive charge.

    Here the correct order of size is as follows:

    $Ca^{2+} < K^+ < Ar < Cl^- < S^{2-}$

14. As ionization enthalpy (both first and second) increases from left to right across the period. Only chromium is exceptional due to the stable half filled configuration ($3d^5$) so the correct order of ionization energy is:

    Cr > Mn > V > Ti

15. Size $\propto$ magnitude of – Charge

    $\propto \dfrac{1}{\text{magnitude of + Charge}}$

    Hence → $S^{2-} > Cl^- > K^+ > Ca^{2+}$.

16. As size decreases from left to right in any period as Ca > Mg > P > Cl.

17. Cl has highest affinity but with negative sign its value is lowest.

Cl	< F	< S	< O
–349	–328	–200	–141kJ/mole

18. E.G.E = – I.E

    = – 5.1 e.v

19. As anion > atom > cation

    $H^- > H > H^+$

20. The correct order of radie is as follows $Ca^{2+} < K^+ < Ar$

21. The correct order of first ionisation energy is as follows B < C < O < N

    Rest orders are correct.

NCERT EXEMPLAR

1. Consider the isoelectronic species, Na^+, Mg^{2+}, F^- and O^{2-}. The correct order of increasing length of their radii is _________.
   (a) $F^- < O^{2-} < Mg^{2+} < Na^+$
   (b) $Mg^{2+} < Na^+ < F^- < O^{2-}$
   (c) $O^{2-} < F^- < Na^+ < Mg^{2+}$
   (d) $O^{2-} < F^- < Mg^{2+} < Na^+$

2. Which of the following is not an actinoid?
   (a) Curium (Z = 96) (b) Californium (Z = 98)
   (c) Uranium (Z = 92) (d) Terbium (Z = 65)

3. The order of screening effect of electrons of *s*, *p*, *d* and *f* orbitals of a given shell of an atom on its outer shell electrons is:
   (a) $s > p > d > f$ (b) $f > d > p > s$
   (c) $p < d < s > f$ (d) $f > p > s > d$

4. The first ionisation enthalpies of Na, Mg, Al and Si are in the order:
   (a) Na < Mg > Al < Si
   (b) Na > Mg > Al > Si
   (c) Na < Mg < Al < Si
   (d) Na > Mg > Al < Si

5. The electronic configuration of gadolinium (Atomic number 64) is:
   (a) [Xe] $4f^3\ 5d^5\ 6s^2$ (b) [Xe] $4f^7\ 5d^2\ 6s^1$
   (c) [Xe] $4f^7\ 5d^1\ 6s^2$ (d) [Xe] $4f^8\ 5d^6\ 6s^2$

6. The statement that is not correct for periodic classification of elements is:
   (a) The properties of elements are periodic function of their atomic numbers.
   (b) Non metallic elements are less in number than metallic elements.
   (c) For transition elements, the 3*d*-orbitals are filled with electrons after 3*p*-orbitals and before 4*s*-orbitals.
   (d) The first ionisation enthalpies of elements generally increase with increase in atomic number as we go along a period.

7. Among halogens, the correct order of amount of energy released in electron gain (electron gain enthalpy) is:
   (a) F > Cl > Br > I (b) F < Cl < Br < I
   (c) F < Cl > Br > I (d) F < Cl < Br < I

8. The period number in the long form of the periodic table is equal to:
   (a) Magnetic quantum number of any element of the period.
   (b) Atomic number of any element of the period.
   (c) Maximum Principal quantum number of any element of the period.
   (d) Maximum Azimuthal quantum number of any element of the period.

9. The elements in which electrons are progressively filled in 4*f*-orbital are called:
   (a) Actinoids (b) Transition elements
   (c) Lanthanoids (d) Halogens

10. Which of the following is the correct order of size of the given species:
   (a) $I > I^- > I^+$ (b) $I^+ > I^- > I$
   (c) $I > I^+ > I^-$ (d) $I^- > I > I^+$

11. The formation of the oxide ion, O^{2-} (g), from oxygen atom requires first an exothermic and then an endothermic step as shown below:
   $O(g) + e^- \rightarrow O^-(g)$; $H^V = -141$ kJ mol^{-1}
   $O^-(g) + e^- \rightarrow O^{2-}(g)$; $H^V = +780$ kJ mol^{-1}
   Thus process of formation of O^{2-} in gas phase is unfavourable even though O^{2-} is isoelectronic with neon. It is due to the fact that,
   (a) Oxygen is more electronegative.
   (b) Addition of electron in oxygen results in larger size of the ion.
   (c) Electron repulsion outweighs the stability gained by achieving noble gas configuration.
   (d) O^- ion has comparatively smaller size than oxygen atom.

12. Comprehension given below is followed by some multiple choice questions. Each question has one correct option. Choose the correct option.
   In the modern periodic table, elements are arranged in order of increasing

   atomic numbers which is related to the electronic configuration. Depending upon the type of orbitals receiving the last electron, the elements in the periodic table have been divided into four blocks, viz, *s, p, d* and *f*. The modern periodic table consists of 7 periods and 18 groups. Each period begins with the filling of a new energy shell. In accordance with the Arfbau principle, the seven periods (1 to 7) have 2, 8, 8, 18, 18, 32 and 32 elements respectively. The seventh period is still incomplete. To avoid the periodic table being too long, the two series of *f*-block elements, called lanthanoids and actinoids are placed at the bottom of the main body of the periodic table.

(a) The element with atomic number 57 belongs to
(1) *s*-block (2) *p*-block
(3) *d*-block (4) *f*-block

(b) The last element of the *p*-block in 6th period is represented by the outermost electronic configuration.
(1) $7s^2\,7p^6$ (2) $5f^{14}\,6d^{10}\,7s^2\,7p^0$
(3) $4f^{14}\,5d^{10}\,6s^2\,6p^6$ (4) $4f^{14}\,5d^{10}\,6s^2\,6p^4$

(c) Which of the elements whose atomic numbers are given below, cannot be accommodated in the present set up of the long form of the periodic table?
(1) 107 (2) 118
(3) 126 (4) 102

(d) The electronic configuration of the element which is just above the element with atomic number 43 in the same group is ________.
(1) $1s^2\,2s^2\,2p^6\,3s^2\,3p^6\,3d^5\,4s^2$
(2) $1s^2\,2s^2\,2p^6\,3s^2\,3p^6\,3d^5\,4s^3\,4p^6$
(3) $1s^2\,2s^2\,2p^6\,3s^2\,3p^6\,3d^6\,4s^2$
(4) $1s^2\,2s^2\,2p^6\,3s^2\,3p^6\,3d^7\,4s^2$

(e) The elements with atomic numbers 35, 53 and 85 are all ________.
(1) Noble gases (2) Halogens
(3) Heavy metals (4) Light metals

13. Electronic configurations of four elements A, B, C and D are given below :
(A) $1s^2\,2s^2\,2p^6$ (B) $1s^2\,2s^2\,2p^4$
(C) $1s^2\,2s^2\,2p^6\,3s^1$ (D) $1s^2\,2s^2\,2p^5$
Which of the following is the correct order of increasing tendency to gain electron :
(a) A < C < B < D (b) A < B < C < D
(c) D < B < C < A (d) D < A < B < C

Answer Keys

1. (b) **2. (d)** **3. (a)** **4. (a)** **5. (c)** **6. (c)** **7. (c)** **8. (c)** **9. (c)** **10. (d)**
11. (c) **12. (a) (3), (b) (3), (c) (3), (d) (1), (e) (2)** **13. (a)**

Hints and Explanations for Selective Questions

1. Ionic radii decreases with increase in nuclear charge.

3. screening effect decreases proceeding from s-orbital to f-orbital.

6. 3d orbitals are filled with electrons after 3p and 4s orbitals but before 4p-orbitals.

7. electron gain enthalpy decreases as atomic size increases.

10. Anion > parent atom > cation

11. there is always repulsion when similar charges approach each other but force of repulsion is overcome by applying external energy.

12. a) Configuration is $5d^1\,6s^2$

b) Configuration is $4f^{14}\,5d^{10}\,6s^2\,6p^2$

c) In present form of periodic table elements till 118 are accumulated.

d) $1s^2\,2s^2\,2p^6\,3s^2\,3p^6\,3d^5\,4s^2$

AIIMS ESSENTIALS

Assertion and Reason

In the following questions, two statements (Assertion) A and Reason (R) are given. Mark

(a) If A and R both are correct and R is the correct explanation of A;

(b) If A and R both are correct but R is not the correct explanation of A;

(c) A is true but R is false;

(d) A is false but R is true

1. (A) : Bond dissociation energy of F_2 is less than Cl_2.

(R) : Due to smaller size of fluorine there is greater electron repulsion between the F atoms than Cl atoms.

2. (A) : Ions K^+, S^{2-}, Sc^{3+} are isoelectronic.

(R) : In each ion the total number of electrons are 18.

3. (A) : Ionisation energy of Magnesium is more than that of Aluminium.

(R) : In Aluminium 3p–orbital is full filled whereas in Magnesium it is not full filled.

4. (A) : Fluorine is more electronegative than chlorine.

(R) : Fluorine is smaller in size than chlorine.

5. (A) : Ionisation energy for s–electrons is more than the p–electrons for the same shell.

(R) : s–electrons are closer to the nucleus than p-electrons hence more tightly attached.

6. (A) : Li and Mg show diagonal relationship.

(R) : Li and Mg have same atomic radius.

7. (A) : He and Be both have the same outer electronic configuration like ns^2 type.

(R) : Both are chemically inert.

8. (A) : The first ionization energy of N is more stable than that of O.

(R) : Oxygen after losing one electron gets a stable electronic configuration.

9. (A) : For noble gases in the solid state the crystal radii are actually Van der Waal's radii.

(R) : In crystals of noble gas no chemical forces operate between the atom.

10. (A) : Electron gain enthalpy of oxygen is less than that of fluorine but greater than that of nitrogen.

(R) : Ionization enthalpy is as follows N > O > F.

11. (A) : Second ionization enthalpy will be higher than the first ionization enthalpy.

(R) : Ionization enthalpy is a quantitative measure of the tendency of an element to lose electron.

12. (A) : Noble gases have large positive electron gain enthalpy.

(R) : Electron has to enter the next higher principle quantum level.

Answer Keys

1. (a)	**2. (a)**	**3. (c)**	**4. (b)**	**5. (a)**	**6. (c)**	**7. (c)**	**8. (b)**	**9. (b)**	**10. (c)**
11. (b)	**12. (a)**								

CHAPTER 4

Chemical Bonding and Molecular Structure

CHAPTER OUTLINE

■ Valency ■ Chemical Bond ■ Ionic or Kernel Bond ■ Covalent Bond ■ Coordinate or Semi-polar Bond ■ Hydrogen Bond ■ Metallic Bonding ■ Resonance ■ Hybridization ■ VSEPR (Valence Shell Electron Pair Repulsion Theory) ■ Molecular Orbital Theory

VALENCY

Valency is a property of atoms whereby they form chemical bond among themselves. The term valency was introduced by Frankland and it means 'power to combine.' Hence, it is the power of an atom to combine with another atom. Atoms do so by either giving up or accepting electrons in their outermost shell. Modern or electronic concept of valency was given by Kossel and Lewis; it was completed by Langmuir.

Valency (V) = No. of valence electrons

For instance, the electronic configuration for the group IA element sodium (Na), is 2, 8, 1. Here, the number of valence electron is 1 and hence its valency is 1.

If the number of valence electrons is more than 4, then we use the following relationship to determine the valency:

$V = V\ e^{-} - 8$ (number of valence electrons minus 8)

For example, the configuration of nitrogen (N) is 2, 5. According to the above relationship, its valency will be given as

$V = 5 - 8 = -3$ (negative sign signifies the tendency to accept electrons)

CHEMICAL BOND

Chemical bond is the force of attraction that binds two atoms together. A chemical bond balances the force of attraction and force of repulsion at a particular distance.

A chemical bond is formed to:

- Attain the octet state
- Minimize energy
- Gain stability
- Decrease reactivity

When two atoms come close to each other, forces of attraction and repulsion operate between them. The distance at which the attractive forces overcome repulsive forces is called **bond distance**. Here, potential energy for the system is lowest, hence the bond is formed.

Types of Bonds

Following are the six types of chemical bonds. Here, they are listed in a decreasing order of their respective bond strengths.

1. Ionic bond
2. Covalent bond
3. Coordinate bond

4. Metallic bond
5. Hydrogen bond
6. Van der Waals bond

Metallic bond, hydrogen bond and van der Waals bond are interactions.

Octet Rule

It was introduced by Lewis and Kossel. According to this rule, each atom tries to obtain the octet state, that is, a state with eight valence electrons.

Exceptions to the octet rule

- Transition metal ions like Cr^{+3}, Mn^{+2}, Fe^{+2}.
- Pseudo inert gas configuration cations like Zn^{2+}, Cd^{2+}.

Contraction of octet state

- Here central atom is electron deficient or does not have an octet state. For example,

BeX_2	BX_3	AX_3	$Ge(CH_3)_3$
4	6	6	$6e^-$

Expansion of octet state

- Here central atom has more than 8 e^- due to empty d orbital. For example, PCl_5, SF_6, OsF_8, ICl_3
- Odd electronic species like NO, NO_2, ClO_2
- Inter halogens compounds like IF_7, BrF_3
- Compounds of xenon such as XeF_2, XeF_4, XeF_6

IONIC OR KERNEL BOND

Ionic bond is formed by the complete transfer of valence electrons from a metal to a non-metal. This was first studied by Kossel.

For example,

$$\underset{(2,8,1)}{Na} + \underset{(2,8,7)}{Cl} \longrightarrow \underset{(2,8)\,(2,8)}{Na^+\,Cl^-}$$

$$\underset{(2,8,2)}{Mg} + \underset{(2,6)}{O} \longrightarrow \underset{(2,8)\,(2,8)}{Mg^{+2}\,O^{-2}}$$

$$\underset{(2,8,3)}{Al} + \underset{(2,5)}{N} \longrightarrow \underset{(2,8)\,(2,8)}{A^{+3}\,N^{-3}}$$

- Number of electrons transferred is equal to electrovalency.
- Maximum number of electrons transferred by a metal to non-metal is three, as is the case of AlF_3, (Al metal transfers three electrons to F).
- During electron transfer, the outermost orbit of metal is destroyed and the remaining portion is called core or kernel, so this bond is also called kernel bond.
- Nature of ionic bond is electrostatic or coloumbic force of attraction.
- It is a non-directional bond.

Conditions for the Formation of an Ionic Bond

The process of bond formation must be exothermic (ΔH = –ve) and for it the essential conditions are

- Metal must have low ionization energy.
- Non-metal must have high electron affinity.
- Ions must have high lattice energy.
- Cation should be large with low electronegativity.
- Anion must be small with high electronegativity.

Born–Haber Cycle

The formation of an ionic compound in terms of energy can be shown by Born–Haber cycle. It is also used to find lattice energy, ionization energy and electron affinity.
For example,

$$M(s) \xrightarrow[+s]{\text{Sublimation}} M(g) \xrightarrow[+I]{\text{Ionization}} M^+(g) + e^-$$

$${}^1/_2\,X_2 \xrightarrow[+s]{\text{decomposition}} X(g) \xrightarrow[+E]{\text{Addition of e}} X^-(g)$$

$$M^1(g)\ 1\ X^-(g) \xrightarrow[-U]{\text{Crystal formation}} MX(g)$$

$$\Delta H_f = S + \frac{1}{2}D + I - E - U$$

Here,

S = Heat of sublimation

D = Heat of dissociation

I = Ionization enthalpy

E = Electron gain enthalpy or electron affinity

U = Lattice energy

- For the formation of an ionic solid, energy must be released during its formation, that is, ΔH must be negative for it.

  $-E - U > S + ½\,D + I$

Properties of Ionic Compounds

1. Ionic compounds have solid crystalline structures (flat surfaces), with definite geometry, due to strong electrostatic force of attraction as constituents are arranged in a definite pattern.
2. These compounds are hard in nature.

$$\text{Hardness} \propto \text{Electrostatic force of attraction} \propto \text{Charge on ion} \propto \frac{1}{\text{Ionic radius}}$$

3. Ionic compounds have high value of boiling point, melting point and density due to strong electrostatic force of attraction.

Boiling point, melting point $\propto$ Electrostatic force of attraction

$$\text{Volatile nature} \propto \frac{1}{\text{Electorstatic force of attraction}}$$

4. Ionic compounds show isomorphism, that is, they have same crystalline structure. For example, all alums, NaF and MgO.
5. These are conductors in fused, molten or aqueous state due to presence of free ions. In solid state, these are non-conductors as no free ions are present.
6. They show fast ionic reactions as activation energy is zero for ions.
7. They do not show space isomerism due to nondirectional nature of ionic bond.
8. Lattice energy (U) is released during the formation of an ionic solid molecule from its constituent ions.

Lattice Energy

Lattice energy is also the energy needed to break an ionic solid molecule into its constitutent ions. It is denoted by U.

$$U \propto \text{Charge on ion} \propto \frac{1}{\text{Size of ion}}$$

Hence, lattice energy for the following compounds increases in the order shown below:

$$NaCl < MgCl_2 < AlCl_3 < SiCl_4$$

As charge on a metal atom increases, its size decreases.

In case of univalent and bivalent ionic compounds, lattice energy decreases as follows:

Bi-bi > Uni-bi or Bi-uni > Uni-uni

For example,

$$MgO > MgCl_2 > NaCl.$$

9. Ionic compounds are soluble in polar solvents like water due to the high dielectric constant of these solvents, therefore, force of attraction between ions are destroyed and they dissolve in the solvent.

Facts Related to Solubility

- If ΔH (hydration) > Lattice energy then ionic compound is soluble.
- If ΔH (hydration) < Lattice energy then ionic compound is insoluble
- If ΔH (hydration) = Lattice energy then the compound is at equilibrium state

Some Solubility Orders

a. $LiX < NaX < KX < RbX < CsX$

b. $LiOH < NaOH < KOH < RbOH < CsOH$

c. $BeX_2 < MgX_2 < CaX_2 < BaX_2$

d. $Be(OH)_2 < Mg(OH)_2 < Ca(OH)_2 < Ba(OH)_2$

e. $BeSO_4 > MgSO_4 > CaSO_4 > SrSO_4 > BaSO_4$

f. $AlF_3 > AlCl_3 > AlBr_3 > AlI_3$

- Crystals of high ionic charges are less soluble. For example, compounds of CO_3^{-2}, SO_4^{-2}, PO_4^{-3} are less soluble.

  Compounds Ba^{+2}, Pb^{+2} are insoluble as lattice energy > ΔH_{hy}

  Compounds of Ag (salt) are insoluble as lattice energy > ΔH_{hy}
- Presence of common ions decrease solubility. For example, solubility of AgCl decreases in presence of $AgNO_3$ or KCl, due to presence of common ions that is, Ag^+ and Cl^- respectively.

Covalent Bond

A covalent bond is formed by equal sharing of electrons between two similar or different atoms.

- If atoms are same or their electronegativity is same, the covalent bond between them is non-polar. For example,

$$X\text{–}X, O = O, N \equiv N$$

- If atoms are different or have different value of electronegativity, the covalent bond formed between them is polar. For example,

$$\overset{+\delta}{H} - \overset{-\delta}{O} - \overset{+\delta}{H}, \quad \overset{+\delta}{H} - \overset{-\delta}{X}$$

- Here, number of electrons shared or covalent bonds represent covalency.
- One atom can share a maximum of three electrons with the other atom.
- The nature of covalent bond is explained on the basis of Heitler–London's valence bond theory, Pauling–Slater's overlapping theory and Hund–Mullikan's theory.
- Orbital concept of covalent bond was introduced by Heitler and London. According to this concept, "Covalent bond is formed due to half-filled atomic orbitals having electrons with opposite spin to each other."
- Due to overlapping, the potential energy of system decreases.
- The internuclear distance with maximum overlapping and greater decrease of potential energy is known as **bond length**.

Energy consideration of covalent bond When two hydrogen atoms H_A and H_B with respective electrons e_A and e_B approach each other, following attractive and repulsive force start operatings.

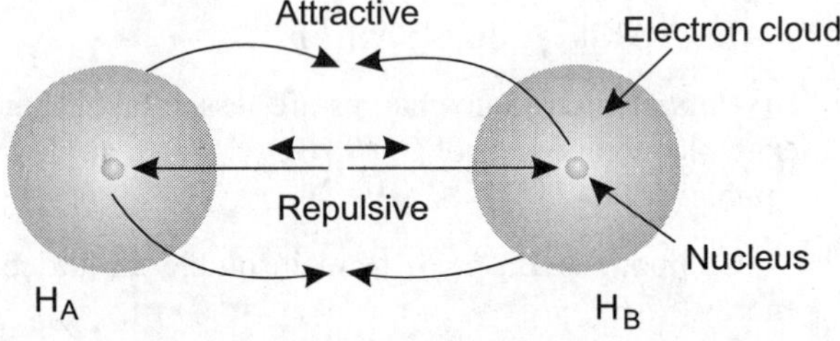

FIGURE 4.1 Formation of Covalent bond

Here, attractive forces between H_Ae_A and H_Be_B and H_Be_A and H_Ae_B.

Repulsive forces are between e_A and e_B and between nucleus of H_A and H_B.

It is observed that attractive force are more than a repulsive forces which results in decreased energy, so the potential energy of the system decreases.

The minimum energy point corresponds to critical distance between two nuclei, when maximum lowering of energy takes place. This distance is called **bond length** e.g., in H–H, bond length is 74 pm.

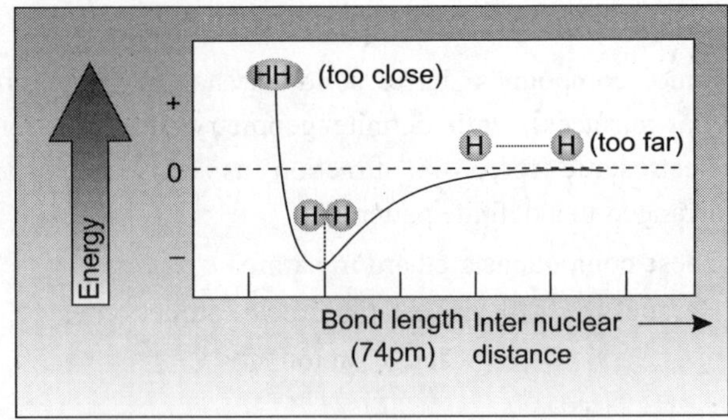

FIGURE 4.2 Variation of Energy and Bond length

Features of Covalent Compounds

1. Covalent compounds mostly occur in liquid and gaseous state, but if molecular weight of the compound is high, they may occur in solid state also. For example,

Molecular wt	Glucose	Sugar
	180	342
	(less solid)	(more solid)

2. 'Like dissolves like', that is, non-polar solute dissolves in non-polar solvent. For example, CCl_4 dissolves in organic solvents.
   Similarly, polar solutes dissolves in polar solvent. For example, alcohol and ammonia dissolve in water.
3. Covalent compounds have lower boiling point and melting point values than those of ionic compounds. This is because covalent bond is weak van der Waals force in nature.

Comparable to	KOH	>	HX
	Strong ionic force of attarction		Weak van der Waals forces

   Boiling point and melting point ∝ Hydrogen bonding
   ∝ Molecular weight

   For example,

HF	>	HI	>	HBr	>	HCl
Due to H-bonding				As molecular weight decreases		

4. Covalent compounds are non-conductors due to absence of free ions, but graphite is a conductor, as in graphite, free electrons are available in its hexagonal sheet like structure.
   In case of diamond, the structure is tetrahedral so free electrons are not available. It is therefore not a conductor.
5. Covalent bond is directional, so these compounds can show space isomerism.

6. When cation and anion are close to each other, the shape of anion is distorted by the cation, this is called **polarization**. Due to this, covalent nature develops in an ionic molecule.

$$\text{Polarization} \propto \text{Covalent nature} \propto \frac{1}{\text{Ionic nature}}$$

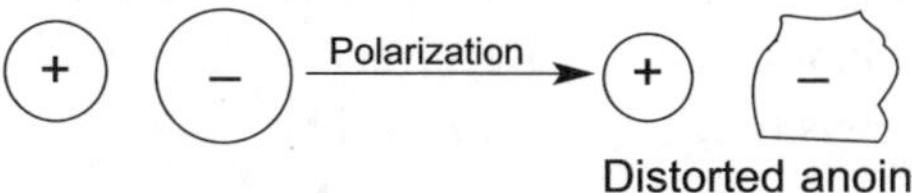

FIGURE 4.3 Effect of Polarization

Fajan's rule

Polarization or covalent nature is explained by the following rules:

Charge on cation polarization, covalent nature or polarizing power of cation ∝ charge on cation. That is, greater the charge on cation, greater will be its polarizing power and more will be covalent nature. For example, $SiCl_4 > AICl_3 > MgCl_2 > NaCl$

Size of Cation When charge is same and anion is common, consider it covalent nature $\propto \frac{1}{\text{Size of cation}}$

That is, smaller cation has more polarizing power. For example,

LiCl > NaCl > KCl > RbCl > CsCl

Max. covalent	Max. ionic
Least ionic	Least covalent

$Li^+ < Na^+ < K+ < Rb^+ < Cs^+$

Smallest in size	Largest in size

Size of anion This property is taken into account when the charges are same and the cation is common.

- Polarization or covalent nature ∝ size of anion. Hence, larger anions are more polarized.

  For example, LiF , LiCl , LiBr . LiI

  As F^- , Cl^- , Br^- , I^-

  Larger the size of anion easier, will be its polarization.
- A cation with 18 valence electrons has more polarizing power than a cation with 8 valence electrons.

Example:

group IB	>	group IA
Cu^+	>	Na^+
Ag^+	>	K^+
group IIB	>	group IIA
Zn^{+2}	>	Mg^{+2}

For example,

ZnO	>	MgO
Zn^{+2}		Mg^{+2}
2, 8, 18		2, 8

FACTS TO REMEMBER

As the covalent nature increases, the intensity of the colour increases. For example, $FeCl_3$ is reddish-brown while $FeCl_2$ is greenish-yellow.

Sigma (σ) Bond

Sigma bond is formed by axial or head to head or linear overlapping between two s–s or s–p or p–p orbitals.

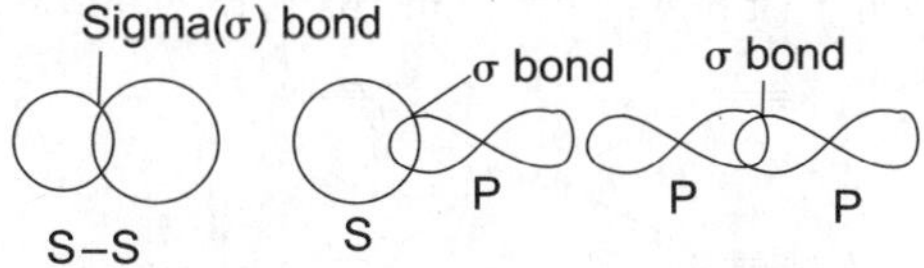

FIGURE 4.4 Formation of Sigma Bond

1. Sigma bond is stronger and therefore less reactive, due to more effective and stronger overlapping than pi bond.
2. The minimum and maximum number of sigma bond between two bonded atoms is one.
3. Stability ∝ Number of sigma bonds.
4. Reactivity $\propto \frac{1}{\sigma}$
5. In sigma bond, free rotation of the atoms is possible.
6. Sigma bond determines the shape of molecule.

Pi (π) Bond

Pi bond is formed by lateral or sidewise overlapping between two p orbitals.

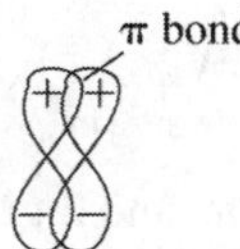

FIGURE 4.5 Formation of pi (p)Bond

1. It is a weak or less stable bond, and therefore more reactive, due to less effective overlapping.
2. Minimum and maximum number of pi bonds between two bonded atoms are 0 and 2 respectively.
3. Stability $\propto \dfrac{1}{\text{Number of pi bonds}}$
4. Reactivity $\propto$ Number of pi bonds.
5. In case of a pi bond, free rotation is not possible.
6. It does not determine the shape of a molecule but shortens bond length.

To Find the Number of Sigma and Pi Bonds in a Molecule

(Single bond) = 1 σ
(Double bond) = 1σ, 1π
(Triple bond) = 1σ, 2π
For example,

1. Enolic form of acetone

```
 H\                 H
H —C — C = C<
 H/     |           H
       :O—H
```

It has 9σ and 1π bond, and 2 lone pairs of electrons.

2. $C_2(CN)_4$

```
N≡C             C≡N
    \          /
      C=C
    /          \
N≡C             C≡N
```

It has 9σ and 9π bond, and 4 lone pairs of electrons.

3.

```
          H
          |
          C
       //    \
   H– C        C –H
      |        ||
   H– C        C –H
       \\    /
          C
          |
          H
```

It has 12σ and 3π bonds. It has 15 σ and 3π bonds

4. Buta-1, 3-diene

```
H\              /H
   C=C–C=C
H/   |  |  \H
     H  H
```

It has 9s and 2π bonds

5. CaC_2 Ca^{+2} $[C \equiv C]^{2-}$ 1s, 2π

Coordinate or Semi-polar Bond

Coordinate bond is a special type of bond which is formed by donation of electron pair from donor to receiver, that is, it involves partial transfer or unequal sharing of electrons. It is denoted as ($\rightarrow$) from donor to receiver.

A:	+ B $\longrightarrow$ (A $\rightarrow$ B)
Donor or	Receiver
Lewis base	Lewis acid

e.g., formation of $NH_3 . BF_3$ Molecule

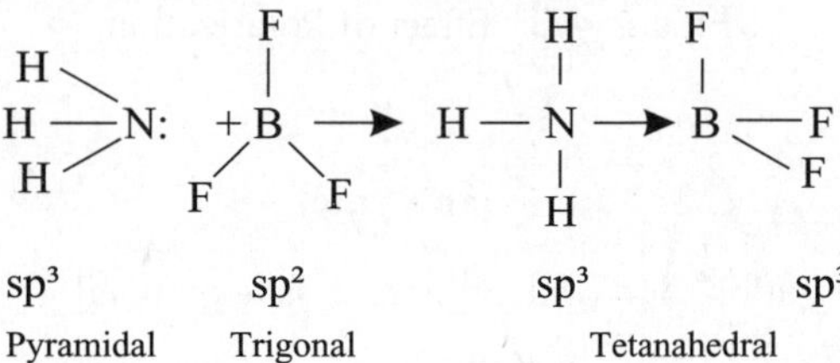

- Coordinate bond is intermediate between ionic and covalent bonds, but more closely resembles a covalent bond.
- The properties of coordinate compounds are more close to covalent compounds. For example, Properties of ligands like NH_4^+, H_3O^+, NH_3 resemble BF_3 (a covalent compound) and complexes like $K_3Fe(CN)_6$.
- Sugden or singlet linkage is formed by donation of one electron and denoted by ($\neg$). For example, PCl_5, SF_6, IF_7.

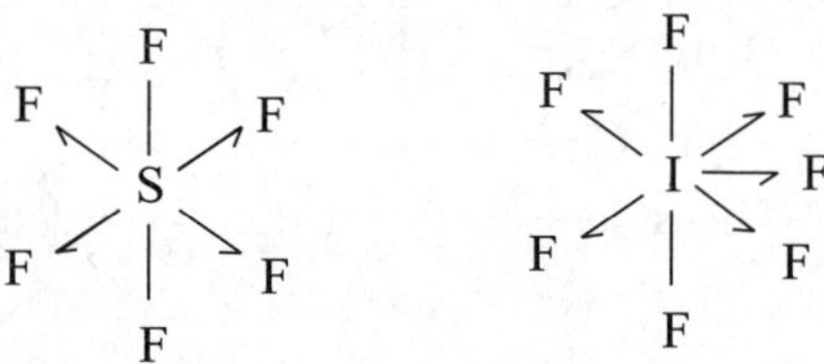

It was studied by Lorry and Sidwick.

Hydrogen Bond

- Hydrogen bond was introduced by Latimer and Rodebush.
- It is a weak interaction denoted by dotted (—) lines between hydrogen and a highly electronegative andsmall sized atom like F, O and N. Here, the hydrogen atom is covalently bonded to any of these.
- The nature of a hydrogen bond is either dipole–dipole type, ion–dipole type or dipole–induced dipole type
- HCl has no H-bonding as chlorine is large in size.

- H-bond strength for the following order is 10 kCal per mole, 7 kCal per mole and 2 kCal mole respectively. HF > H_2O > NH_3

Hydrogen bonding is of following two types:

Intermolecular H-bonding

Intermolecular H-bonding is formed between two or more different molecules of the same or different types.

For example, HF, H_2O, NH_3, R–OH, R–COOH, R–NH_2 R_2>NH, Saccharides like Glucose, Sucrose etc.

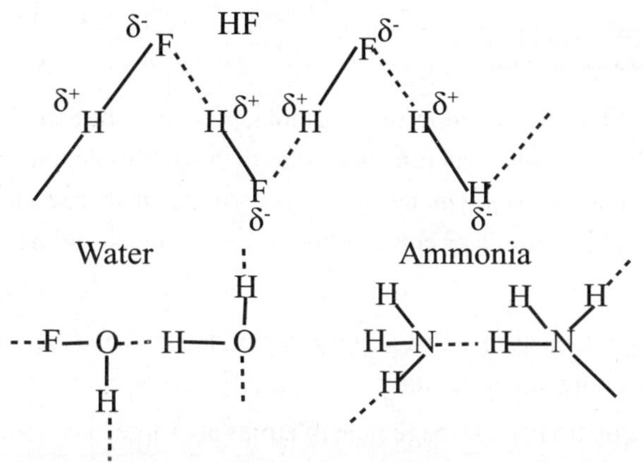

FIGURE 4.6 Intermolecular H-bonding in Water and Ammonia

Facts Related to Intermolecular Hydrogen Bonding

- One water molecule can form hydrogen bonding with four other water molecules.
- Due to hydrogen bonding in water, the water molecules are closely packed, so water has less volume but more density than ice where an open cage like structure is observed.

- Water has maximum density at 4°C as above 4°C some hydrogen bonds are broken leading to a decrease in the density.
- Two ice cubes when pressed against each other, form one block due to hydrogen bonding.

Effects of Intermolecular H-bonding

- Increase in boiling point, melting point, solubility, thermal stability, viscosity, surface tension and occurence liquid state is observed as molecules get associated more closely due to inter molecular H-bonding.
- HF is a liquid and has a higher boiling point than other HX molecules which are gases at room temperature (Here X = halogens)
- Alcohols are highly soluble in water in any proportion and have higher boiling points than others which are very less soluble in water.
- Glycerol is highly viscous with a high boiling point.
- Acids have higher boiling point and solubility than their corresponding acid derivatives.
- In DNA and RNA, the complementary strands are held together by intermolecular H-bonding between the nitrogenous bases of the two strands.
- Nucleic acid and proteins are held together by hydrogen bonds.
- KHF_2 or HF_2^- exists due to hydrogen bonding, but formation of other HX_2^- (for example HCl_2) is not possible, due to absence of hydrogen bonding because of large sizes of the halogen atoms.
- The extent of hydrogen bonding in water is higher than H_2O, so it has a higher boiling point than HF.

$$H_2O > HF > NH_3$$

- Acids can dimerize due to intermolecular hydrogen bonding. For example, acetic acid dimerizes in benzene.

FIGURE 4.7 Dimerization of Acetic Acid

Intramolecular H-bonding or Chelation

Intramolecular H-bonding or chelation is formed within a molecule. For example,

Other examples are pyridine-2-carbonaldoxime and o-hydroxybenzoic acid.

Effects of Intramolecular H-bonding

- Due to this bonding, the boiling point and acidic nature of the molecule decreases but its volatile nature increases.
- O-nitrophenol has a low boiling point and reduced acidic nature, but is more volatile than p-nitrophenol. A mixture of both these componds can be separated by steam distillation method.

METALLIC BONDING

- The concept of metallic bonding was introduced by Drude and Lorenz in the form of **electron-sea model.**
- Metallic bond is the force of interaction between the mobile electrons and positively charged kernels found in metal atoms which holds the atom together.
- Metallic bond strength ∝ Number of valence electrons or charge on nucleus

Conditions for Formation of Metallic Bond

1. The metal should have low ionization energy.
2. The number of vacant orbitals should be enough in the metal.

Properties Related to Metallic Bond

- A metal has a bright lusture because of to and fro oscillations of mobile electrons on the surface of metal.
- Metals are ductile (can be drawn into wires) and maleable (can be beaten into sheets) as the metallic bond is nondirectional and the atomic kernels of metals are slippery.
- Metals have high thermal and electrical conductivities due to the presence of mobile electrons. On increasing temperature, the condustivity decreases, as the increase of temperature causes vibration of kernels which in turn pushes the mobile electrons away from the kernels.
- Boiling point, melting point, hardness, density of metal α Metallic bond strength.

Therefore, alkali metals are soft and can be cut with a knife due to weak metallic bonding.

Hg is liquid possessing the lowest melting point –38.5° (among metals) due to very weak metallic bond.

- Iridium and osmium have very high densities due to strong metallic bonding.
- Tungsten has a very high melting point due to strong metallic bonding.

Demerits of Electron-Sea Model

- Electron-Sea model cannot explain heat of atomization, heat of fusion, hardness and melting point in a proper way.
- It cannot explain why Cu is 50 times more conductive than Bi, and why Na is soft and Os is hard.

  Melting point of mercury is 234 K and that of tungsten is 3275K.

RESONANCE

- When all the properties of a molecule cannot be explained by a single structural formula, then such molecules are represented by many structural formulas that are canonical structures or contributing or resonating structures.
- It is observed due to the delocalization of π electrons.
- Canonical structures for a given molecule, have the same arrangement of atoms.
- Position and arrangement of atoms are same in canonical structures, they only differ in the distribution of electrons.
- Canonical structures are depicted by the symbol (⟷) betweem them.
- Canonical structures should be planar or nearly planar.
- Total number of paired and unpaired electrons are also same in canonical structures. For example,

1. $CH_3—CH_2—Cl: \longleftrightarrow CH_3—\overset{+}{C}H_2— Cl^-:$
2. $CH_2=CH—CH=CH_2$
   $\updownarrow$
   $\overset{+}{C}H_2—CH=CH—\overset{-}{C}H_2$
   $\updownarrow$
   $\overset{-}{C}H_2—CH=CH—\overset{+}{C}H_2$
3. $:\ddot{C}=\ddot{O}: \longleftrightarrow :\overset{+}{C}=\overset{-}{\ddot{O}}: \longleftrightarrow :\overset{-}{C}\equiv\overset{+}{O}:$
4. $O=C=O \longleftrightarrow \bar{O}-\overset{+}{C}=O \longleftrightarrow O=\overset{+}{C}-\bar{O}$

   or

   $:\overset{+}{O}\equiv C-\overset{-}{\ddot{O}}: \longleftrightarrow :\overset{-}{\ddot{O}}-C\equiv\overset{+}{O}:$

5. 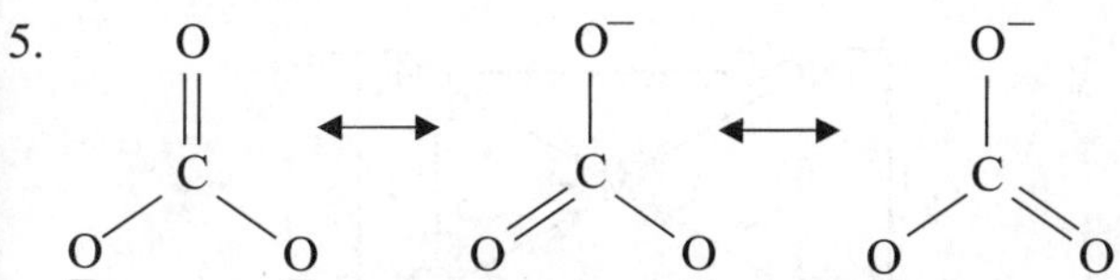

- Resonance changes bond length, for example, in benzene C–C = 1.39 Å, which is an intermediate value between (C–C) = 1.54 Å, (C=C) = 1.34 Å

Resonance Energy

- Resonance energy = Energy of most stable canonical structure – Resonance hybrid energy.
- Resonance energy ∝ Number of canonical structure

  Resonance energy ∝ Stability

  Resonance energy $\propto \frac{1}{\text{Reactivity}}$

  Resonance energy = Expected heat of hydrogenation – Actual heat of hydrogenation.
- Due to high resonance energy, benzene is quite stable and undergoes electrophilic substitution reactions. It does not undergo addition reactions, although it has double bonds (due to delocalization of π electrons or resonance).
- Benzene has 36 kcal/mole of resonance energy.
- Resonance energy of CO_2 is 154.9 kJ.
- In tautomerism, arrangement of atoms is different for its different arrangements but in resonance, the arrangement of atoms is same.

Stability of Different Canonical Structures

1. A non-polar structure is always more stable than a polar structure. In the following example, the structures are arranged in a decreasing order of stability.

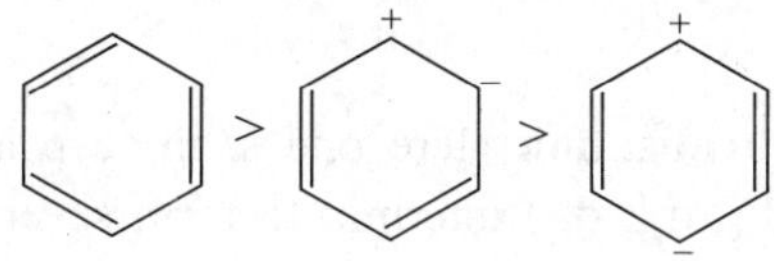

In the last two structures, the charges are apart so they are less stable.

2. Greater the number of covalent bonds, greater will be the stability. Therefore,

$$CH_3\text{–}C\equiv \overset{+}{O} > CH_3\text{–}C{=}\overset{+}{O}$$

3. The canonical structure in which positive charge on electro positive atom and negative charge on the electro negative atom is more stable. Therefore,

$$R_2\overset{+}{C}-\overset{-}{O} > R_2C{=}\overset{+}{O}$$

4. The canonical structure in which each atom has an octet state is more stable. Therefore,

$$\underset{8}{CH_3\text{–}C\equiv \overset{+}{O}} > \underset{6}{CH_3\text{–}\overset{+}{C}{=}O}$$

5. If like charges are closer then the structure will be unstable.

$$:\ddot{N}{=}\overset{+}{N}{=}O \longleftrightarrow \underset{\text{Most Stable}}{N{\equiv}\overset{+}{N}{-}\ddot{O}:} \longleftrightarrow \underset{\text{Unstable}}{:\overset{+}{N}{-}\overset{+}{N}{\equiv}O}$$

Types of Resonance

1. *Isovalent resonance* The canonical structures have same number of bonds and same type of charges. For example, SO_2, NO_2, CO_3^{-2}
2. *Heterovalent resonance* Here, the canonical structures have different number of bonds and charges. For example, buta- 1, 3-diene, vinyl cyanide.

HYBRIDIZATION

- Pauling and Slater introduced this concept to explain the shape of molecules which could not be explained by the valence bond theory.
- It is the intermixing or re-distribution of energy among two or more half-filled, fully filled, incompletely filled or empty orbitals of comparable energy, to form same number of hybrid orbitals. Hybrids have identical energies and similar shapes.

Facts About Hybridization

- Number of atomic orbitals taking part in hybridization is equal to number of hybrids formed.
- Electrons do not undergo hybridization.
- A hybrid bond is always a sigma bond.
- A hybrid bond is always stronger than a non-hybrid bond.
- Hybridization occurs at the time of bond formation.
- Hybridization ∝ Overlapping (for enough overlapping, orbitals must be at an approppriate distance from each other, that is, neither very close nor very far).

- Hybridization increases stability and decreases reactivity and energy of a molecules
- Hybridization occurs in the central atom in a molecule (NH_3, H_2O, CH_4). Here the central atoms are N, O, C respectively
- Hybridization does not occur in isolated atoms but in bonded atoms.

Types of Hybridization

1. **sp hybridization** Here, one s and one p orbitals form two sp hybrid orbitals after intermixing.
   - Shape of molecule is linear and bond angle is 180°. For example, X–M–X (M = Be, Zn, Hg)

$$\underset{sp}{H-C}\equiv\underset{sp}{C-H}$$

Some other examples are CO_2, and CS_2.

2. **sp² hybridization** Here, one s and two p orbitals intermix to form three new sp² hybrid orbitals.
   - Shape of these species is trigonal or coplanar and the bond angle is 120°. For example, BX_3, C_2H_4

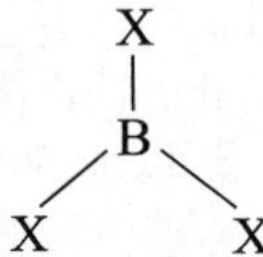

3. **sp³ hybridization** Here, one s and three p orbitals intermix to give four new sp3 hybrid orbitals.

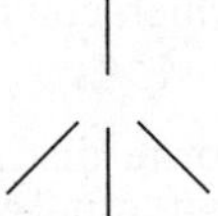

   - Shape of the species is tetrahedral and bond angle is 109° 28′. For example, C_2H_6, CX_4, SiX_4, NH_4^+, BX_4^-, NH_3, PH_3, H_2O, H_2S,

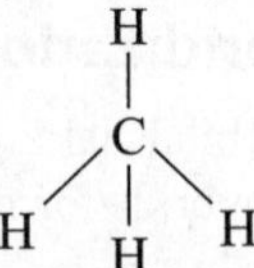

4. **dsp² hybridization** Here, one s, two p and one d orbitals ($d_{x^2-y^2}$) intermix to give four new dsp2 hybrid orbitals.

   - Shape of the species is square planar and bond angle is 90°. For example, $[Ni(CN)_4]^{-2}$, $[Cu(CN)_4]^{2-}$

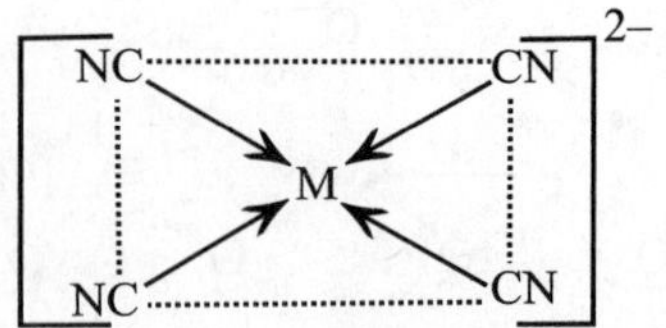

5. **sp³d hybridization** Here, one s, three p and one d-orbital (dz^2) intermix to form five new sp³d hybrid orbitals.

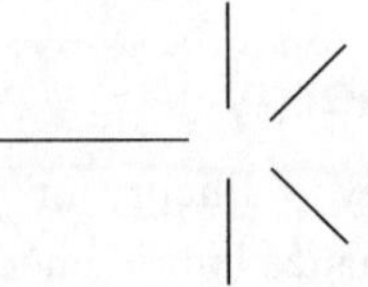

   - Shape of the species is trigonal bipyramidal and bond angles are 90°, 120°. For example, PX_5, XeF_2, I_3^-

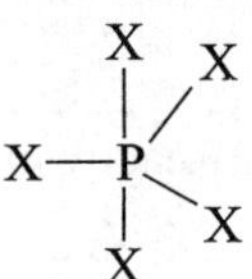

6. **sp³d² hybridization** Here, one s, three p and two dorbital (d_{z^2} and $d_{x^2-y^2}$) intermix to form six new sp³d² hybrid orbitals.

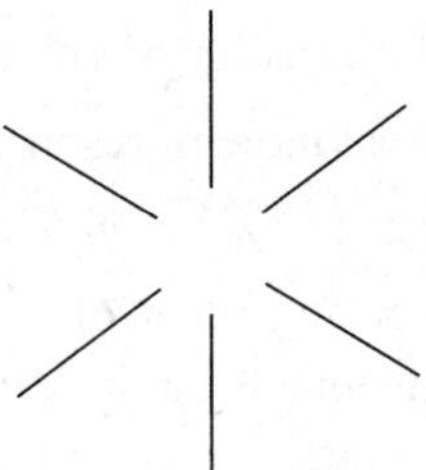

   - Shape of the species is octahedral and bond angle is 90°. For example, SF_6, XeF_4

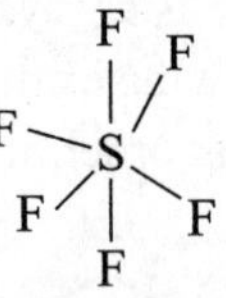

7. **sp³d³ hybridization** Here one s, three p and three d orbital (d_{xy}, d_{yz}, d_{xz}) intermix to form seven new sp³d³ hybrid orbitals.

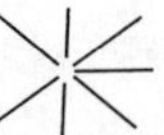

   - Shape of the species is pentagonal bipyramidal and bond angle is between 72° to 90°.

For example, IF_7, XeF_6.

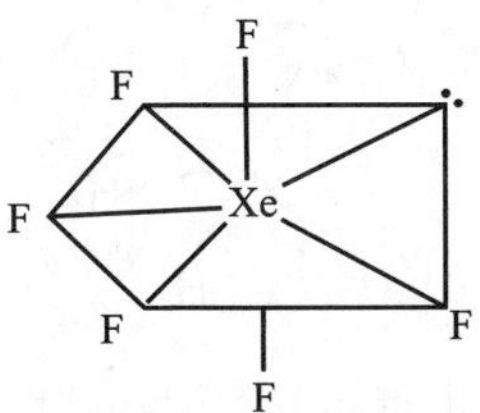

Rules to Find the Type of Hybridization

For covalent compounds and ions

1. Count the total number of valence electrons and (±) charge, to find a particular value. For example, in the PO_4^{-3} number of valence elctrons is $5 + 4 \times 6 + 3 = 32$. For NH^+_4, this number is $5 + 4 - 1 = 8$
2. Now divide the total value of electrons to get the quotient X (number of bond pair electrons)
   - If total value of X is between 2 to 8, divide it by 2
   - If total value is between 10 to 56 divide it by 8
   - If total value is 58 or more, divide it by 18.
3. If any remainder is left, divide again as above to get another quotient Y (number of lone pair electrons)
4. If X or X + Y = 2 = sp

   $3 = sp^2$

   $4 = sp^3$

   $5 = sp^3d$

   $6 = sp^3d^2$

   $7 = sp^3d^3$

   For example,

   $PCl_5 = 5 + 5 \times 7 = 40/8 = 5$ that is, the hybridizationis sp^3d

   $SF_6 = 6 + 6 \times 7 = 48/8 = 6$ that is, sp^3d^2 hybridization is present.

Rule to find the geometry of covalent compounds The shape or geometry of a molecule or ion can be found out by finding the type of hybridization, number of bond pairs and lone pair of electrons using the following relation.

$$P = ½ (V + M - C + A)$$

Here, P= total numbers of pairs of electrons around the central atom which gives the present hybridization as calculated above.

A = Charge on anion

C = Charge on cation

M = Number of monovalent atoms

V = Number of electrons in the valence shell of the central atom.

To find lone pair of electrons

Lone pair = P – N

Here,

P = Total numbers of pairs of electrons around the central atom which gives hybridization as above.

N = Number of atoms surrounding the central atom or number of bond pairs of electrons.

For example,

SO_2

$P = ½ (6 + 0 - 0 + 0) = 3(sp^3)$

$lp = (3 - 2) = 1$

SF_6

$P = ½ (6 + 6 - 0 + 0) = 6(sp^3d^2)$

$lp = (6 - 6) = 0$

Hybridization in complexes Coordination number of ligands is used to find the hybridization.

TABLE 4.1 Hybridization

Coordination Number	Hybridization		
2	sp		
3	sp^2		
4	sp^3	Or	dsp^2
5	sp^3d	Or	dsp^3
6	sp^3d^2	Or	d^2sp^3
	When ligands are weak like		When ligands are strong like
	H_2O, F, Cl, Br, I		CO, CN, NH_3

For example, in $[Fe(CN_6)]^{-3}$ the coordination number is 6 and ligand is strong, so the hybridization is d^2sp^3.

Similary, in $[Fe(H_2O)_6]^{+3}$ the coordination number is 6 and ligand is weak, so hybridization is sp^3d^2.

VSEPR (Valence Shell Electron Pair Repulsion Theory)

- Valence shell electron pair repulsion theory was introduced by Nyholm and Gillispie to predict the shape of polyatomic molecules and ions.
- According to this theory, beside hybridization, the nature of electrons around the central atom also decide the shape of molecule.

- There may be two types of electrons around the central atom, that is, bond pair or lone pair type.
- These electrons undergo electron–electron repulsion and the decreasing order of electronic repulsion is lp–lp. lp–bp > bp–bp
- Due to this electronic repulsion, the shape of the molecule becomes distorted and the bond angle changes.

Distortion in shape α e^-–e^- repulsion

Distortion in shape or change in bond angle α electronic repulsion

For example,

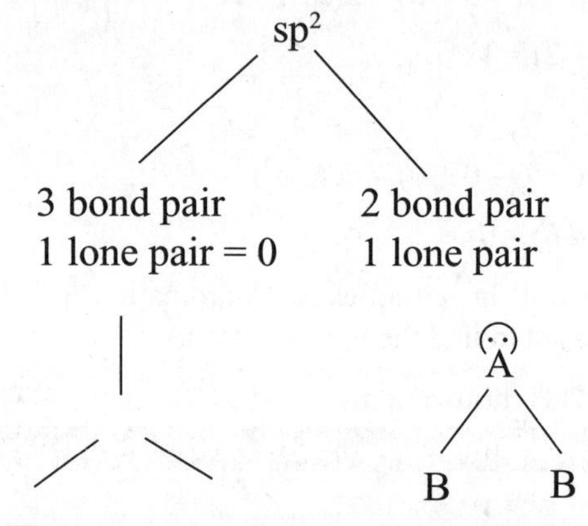

Trigonal shape due to bond pair of e^-

Angular or bent shape due to lone pair of e^- ($l_p - b_p$ repulsion)

For example, BX_3, BH_3, SO_3

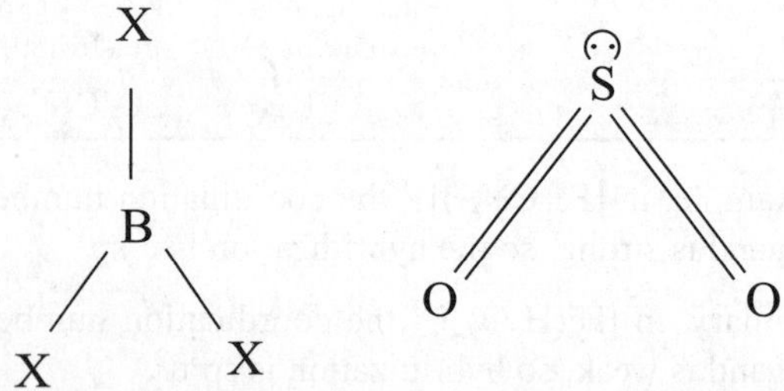

Normal trigonal shape

Bent shape

Here, S atom has two bond pairs and one lone pair of electron, so lp–bp type of repulsion distorts shape, that is, it bends and changes the bond angle and the shape becomes angular. Same holds true for $SnCl_2$ and $PbCl_2$.

sp³ Hybridization

1. When the central atom has four bond pairs of electrons, the shape will be normal with normal bond angle that is tetrahedral and 109° 28. For example, CH_4, CCl_4, $SiCl_4$, NH_4^+, BX_4^-.

2. When the central atom has 3 bond pairs and 1 lone pair of electron, there will be lp–bp type of repulsion, which distorts shape and changes bond angle, that is, the shape is pyramidal and the bond angles are less than 109° 28.

   For example, $\ddot{N}H_3$, PH_3, NX_3, PX_3

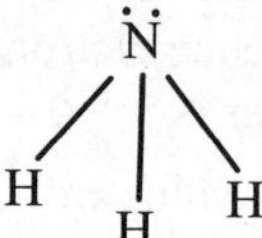

In ammonia, the bond angle is 107º.

3. When the central atom has 2 lone pair and 2 bond pair of electrons, there will be lp–lp type of electronic repulsion, so the shape will be distorted and it will be angular or bent. For example, H_2O, H_2S, OF_2, SCl_2

4. When the central atom has 3 lone pairs and 1 bond pair of electron, there will be lp–lp type of electronic repulsion. So shape is highly distorted and it becomes linear. For example, I-Cl, ClF.

sp³d Hybridization

1. When the central atom has 5 bond pair of electrons, the shape will be normal with normal bond angle, that is, trigonal bipyramidal and bond angle of 90°, 120°. As only bp–bp type of electronic repulsion occurs, so there is no distortion in shape and no change in bond angle.

   For example, PCl_5, AsF_5 and PF_5.

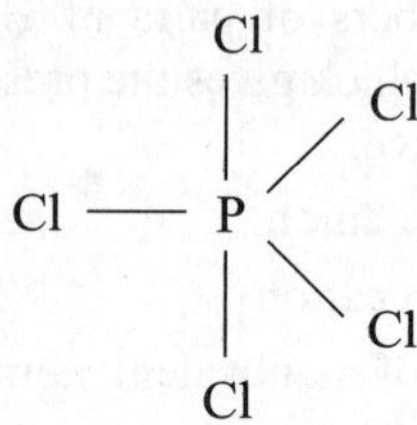

2. When the central atom has 4 bond pair and 1 lone pair of electrons, the shape will be distorted and it will possess a see-saw like structure. For example, $SeCl_4$, $TeCl_4$, SF_4,

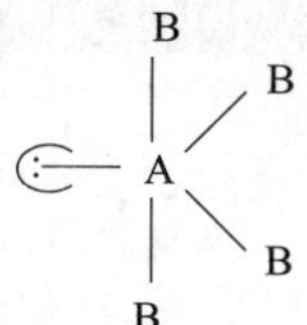

3. When the central atom has 3 bond pairs and 2 lone pair of electrons, the shape will be distorted and it will be a T-shape like structure. For example, ClF_3, BrF_3

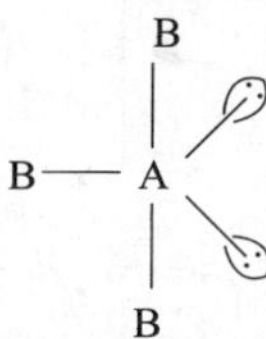

4. When the central atom has 2 bond pair and 3 lone pair of electrons, the shape will be distorted and the shape will be linear. For example, XeF_2, I_3^-, ICl_2^-

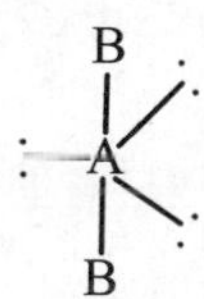

sp³d² Hybridization

1. When the central atom has 6 bond pairs of electrons, the shape will be normal with normal bond angles that is octahedral (90°, 120°). As only bp–bp type of electronic repulsion occurs, so there is no distortion in shape or change in bond angle. For example, SF_6, $TeCl_6$

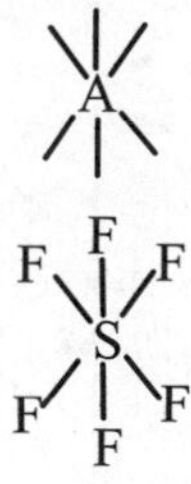

2. When the central atom has 5 bond pair and 1 lone pair of electrons, the shape will be distorted and it will be square pyramidal. For example, BrF_5, IF_5

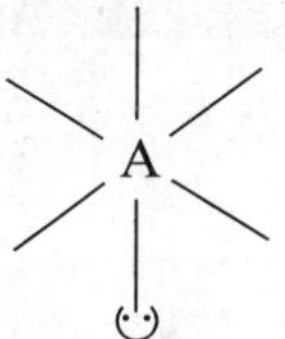

3. When the central atom has 4 bond pair and 2 lone pair of electrons, the shape will be distorted and it will be square planar. For example, XeF_4

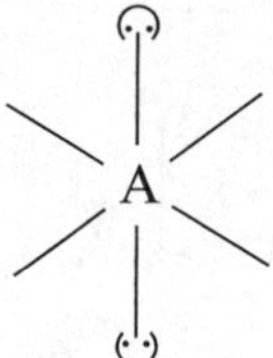

sp³d³ Hybridization

1. When the central atom has 7 bond pair of electron, the shape will be normal with normal bond angles, that is, pentagonal bipyramidal. As bp–bp type of electronic repulsion occurs, so there is no distortion in shape and no change in bond angle. For example, IF_7

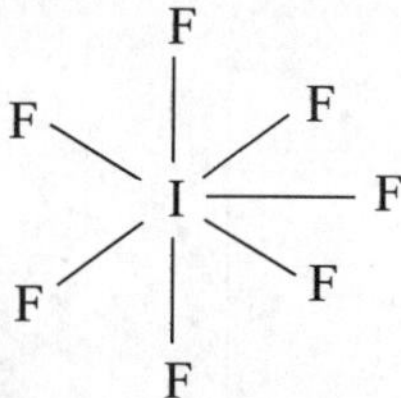

2. When central atom has 6 bond pair and 1 lone pair of electrons, the shape will be distorted and the shape will be distorted pentagonal bipyramidal. For example, XeF_6.

Hybridization and Shapes of Some Simple Molecules

Number of Bonds	Number of Lone Pairs	Number of Charge Clouds	Molecular Geometry and Shape	Examples
2	0	2	Linear	O = C = O
3	0	3	Trigonal planar	$H_2C=O$
2	1	3	Bent	SO_2
4	0	4	Tetrahedra	CH_4
3	1	4	Trigonal pyramidal	NH_3
2	2	4	Bent	H_2O
5	0	5	Trigonal bipyramidal	PCl_5
4	1	5	Seesaw	SF_4
3	2	5	T-shaped	ClF_3
2	3	5	Linear	$[I_3]^-$

Number of Bonds	Number of Lone Pairs	Number of Charge Clouds	Molecular Geometry and Shape	Examples
6	0	6	Octahedral	F, F, F, S, F, F, F
5	1	6	Square Pyramidal	[Cl, Cl, Cl, Sb, Cl, Cl]$^{2-}$
4	2	6	Square Planar	F, F, Xe, F, F

Dipole Moment

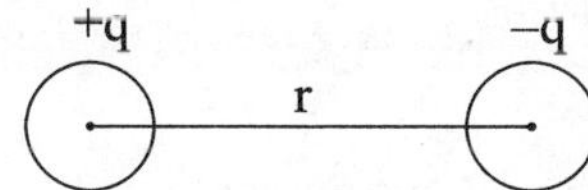

Dipole moment is used to measure the polarity in a molecule. It is denoted by μ. Mathematically, it is given as:

$\mu = q \times r$ coulomb metre

$\mu = e \times d$ esu cm

1 debye $= 1 \times 10^{-18}$ esu cm.

It is represented by (⟼) from electropositive to electronegative species or less electronegative to more electronegative species. For example, BX_3

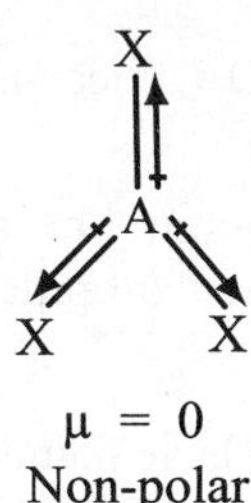

$\mu = 0$

Non-polar

Calculation of Resultant Bond Moments

Let AB and AC be two polar bonds inclined at an angle θ; their dipole moments are μ_1 and μ_2

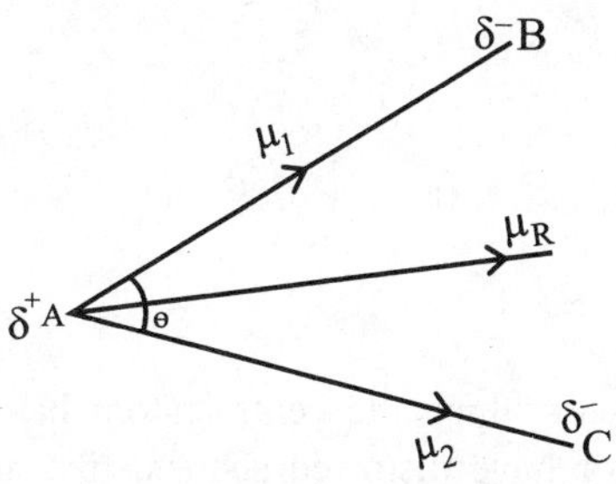

Resultant dipole moment may be claculated using vectorial method.

$$\mu_R = \sqrt{\mu_1^2 + \mu_2^2 + 2\mu_1\mu_2 \cos\theta}$$

when θ=0, the resultant is maximum

$$\mu_R = \mu_1 + \mu_2$$

when θ=180°, the resultant is minimum

- Dipole moment ∝ Electronegativity difference. For example, HF > HCl > HBr > HI.

- Dipole moment ∝ Number of lone pair of electrons.
- For example, $HF > H_2O > NH_3$
- Fluorine has 3 lone pair, oxygen has 2 lone pair, and ammonia has 1 lone pair of electron.
- Dipole moment $\propto \frac{1}{\theta}$

  ortho > meta > para

  For example,

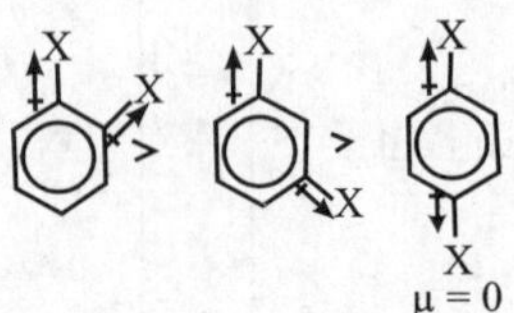

- Homo atomic molecules like X_2, N_2, O_2 and molecules having normal shapes according to hybridization like linear, trigonal, tetrahedral, will be non-polar, as for them the dipole moment is zero. For example, BX_3,

$$CH_4,\ CCl_4,\ SiCl_4,\ PCl_5$$
$$\mu = 0$$

O=C=O

Here, μ (net) = 0 as C = O bonds are in opposing directions.

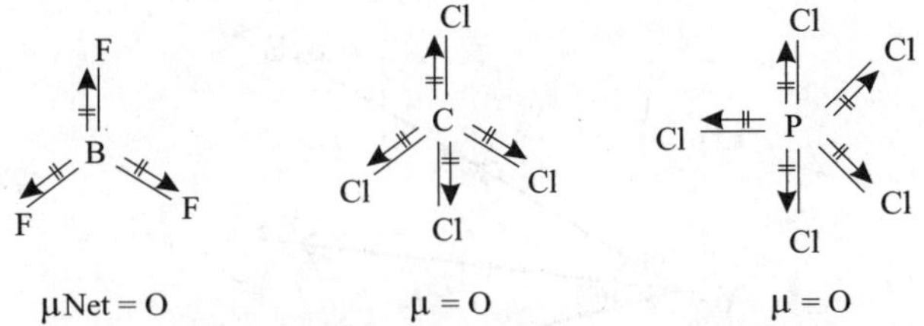

- Molecules in which the central atom has lone pair of electrons or have distorted shapes, like angular, pyramidal, sea-saw shapes will have some value of dipole moment and will be polar in nature. For example, H_2O, H_2S, OF_2, NH_3, PH_3, PCl_3, SCl_4, SO_2.

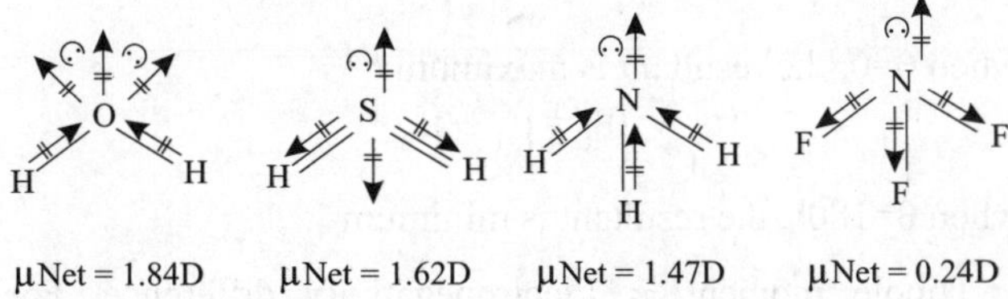

In case of SO_2, NF_3, μ_{Net} becomes less as it is substractive with μ due to lp electrons.

- Dipole moment of a cis-alkene is more than transalkene. In trans-alkenes, it is zero due to symmetry in most of the cases.

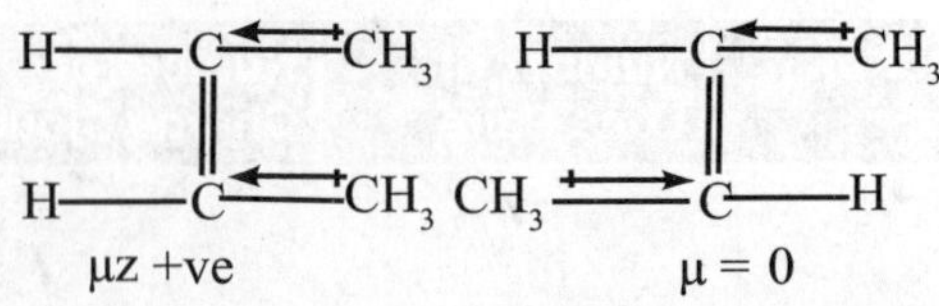

Exception Unsymmetric alkenes with odd number of carbon has some value of dipole moment. For example, trans-2-pentene.

CH_3, H, C = C, H, CH_2CH_3

μ = +ve (But less than Cis)

Specific Cases of Dipole Moment

$CH_3F < CH_3Cl > CH_3Br > CH_3I$

$CH_3Cl > CH_2Cl_2 > CHCl_3 > CCl_4$

Highly polar Non-polar

$NH_3 > NI_3 > NBr_3 > NCl_3 > NF_3$

Uses

- To find geometry of a complex.
- To find ionic character or nature

  Ionic nature % = $\frac{\mu_{\text{observed}}}{\mu_{\text{observed}}(q \times r)} \times 100$
- To distinguish cis and trans alkenes.

Molecular Orbital Theory

- Molecular orbital theory was given by Hund and Mulliken.
- It is based on LCAO (Linear Combination of Atomic Orbitals) model.
- Atomic orbitals undergo linear combination to form same number of molecular orbitals, if they fullfil the following conditions:

1. Atomic orbitals must have comparable energies.
2. Atomic orbitals must overlap linearly for enough and effective overlapping.
3. Atomic orbitals must have same symmetry along with the major molecular axis, for example, if Z axis is the main molecular axis, then only p_z–p_z orbitals will overlap and not p_x or p_y.

- Molecular orbitals are formed due to constructive and destructive interference of atomic orbitals.
- Constructive interaction of orbitals between orbital lobes having same wave function ψ produces bonding molecular orbitals like σ, π and Δ these are HoMOs (Highest occupied molecular orbitals).

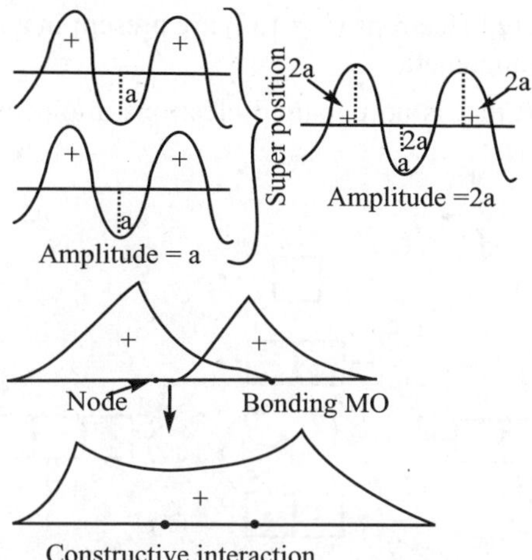

FIGURE 4.8 Depiction of Interactions Involving HoMOs

- Destructive interaction between orbitals having different sign of ψ produces anti-bonding molecular orbitals or LuMOs (Lowest unoccupied molecular orbitals). For example, σ^*, π^*, δ^*.

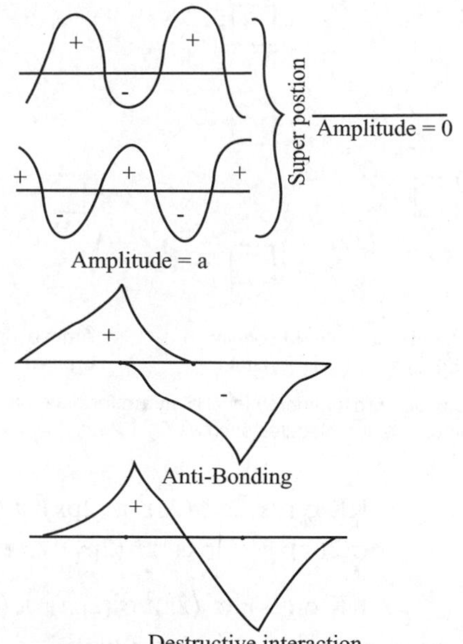

FIGURE 4.9 Interactions Involving LuMOs

Facts Related to HoMOs and LuMOs

- Energy: LuMOs > HoMOs
- Wavelength: LuMOs < HoMOs
- LuMOs have nodal planes while HoMOs may or may not have nodal planes.
- Electrons contribute force of attraction in HoMOs while they contribute repulsion in LuMOs.
- The shape of the molecular orbitals formed depend upon shape of atomic orbital from which they are formed.
- Like atomic orbitals, molecular orbitals also follow:
  1. Pauli exclusion principle—Any molecular orbital can have a maximum of two electrons with opposite spin.
  2. Hund's rule—In degenerate molecular orbital, before pairing, each molecular orbital must have one electron.
  3. Aufbau principle—Electrons are filled from molecular orbital of lower energy to higher energy.

Formation of Various Molecular Orbital

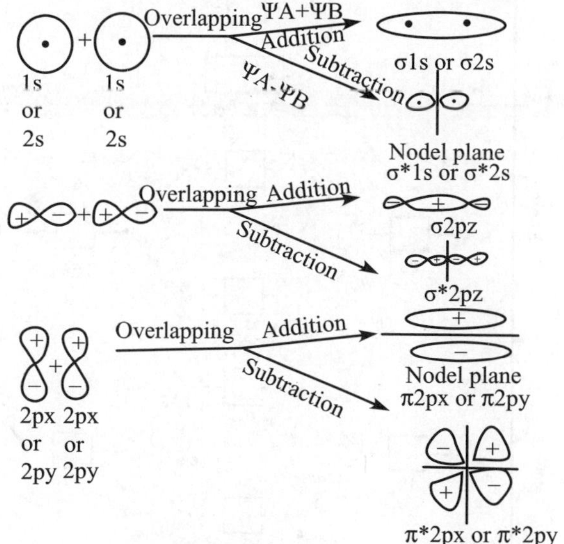

FIGURE 4.10 Molecular Orbitals

Order of Filling Electrons in Molecular Orbital

1. Molecules with N_2 configuration or 14 e^-.

   $\sigma 1s\ \sigma^* 1s,\ \sigma 2s\ \sigma^* 2s,\ \pi 2p_X \approx \pi 2p_y,\ \sigma 2p_z$

2. Molecules with O_2 configuration or more than 14 e.

   $\sigma 1s\ \sigma^* 1s\ \sigma 2s\ \sigma^* 2s,\ \sigma 2p_z\ \pi 2p_x \approx \pi 2p_y,\ \pi^* 2p_x \approx \pi^* 2p_y\ \sigma^* 2p_z$

- σ 1s is the lowest energy molecular orbital while σ^*2p_z is the highest energy molecular orbital.
- Due to intermixing of 2s and 2p orbitals in cases where the number of elecruons is more than 16, $\sigma 2p_z$ is taken after σ^*2s here.
- Bond order $\frac{n_b - n_a}{2}$

Here n_b = Number of bonding molecular orbital electrons

n_a = Number of anti-bonding molecular orbital electrons

$$\text{Bond order} \propto \text{Bond dissociation energy}$$
$$\propto \text{Bond angle}$$
$$\propto \frac{1}{\text{Bond length}}$$

- Higher the bond order, higher will be stability and shorter will be the bond length.
- If unpaired electrons (n = 1, 2) are present in a molecule it is paramagnetic.
- If n = 0 that is no unpaired electrons, molecule is diamagnetic.

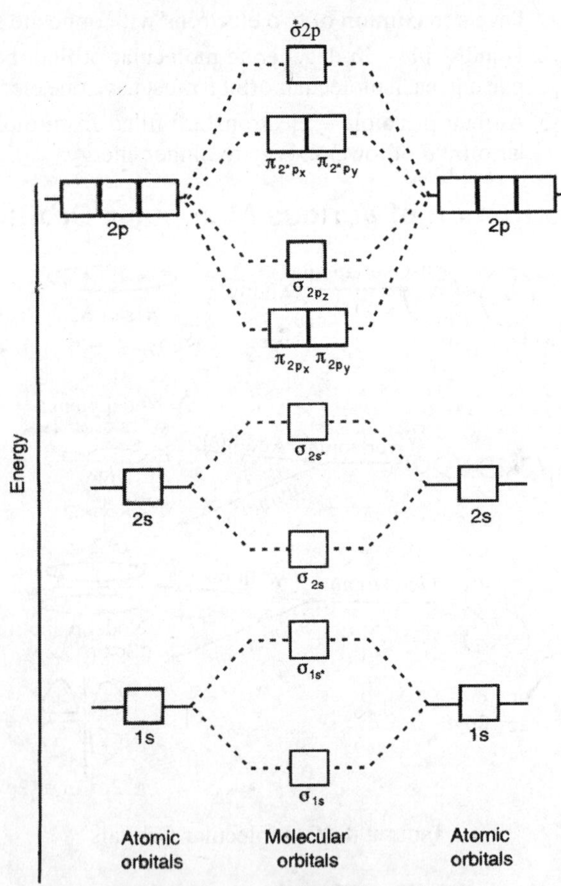

(i) Molecular energy level diagram for diatomic homonuclear molecules such as N_2, C_2 and B_2.

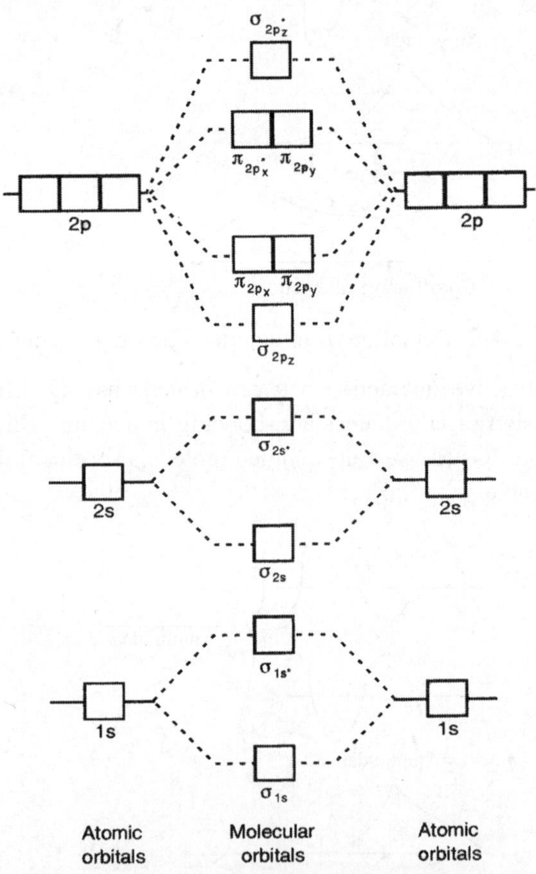

(ii) Molecular orbital energy level diagram for diatomic homonuclear molecules such as O_2, F_2 etc.

Examples:

H_2 : $\sigma(1s)^2$

H_2^+ : $\sigma(1s)^1$

H_2^- : $\sigma(1s)^2\,\sigma^*(1s)^1$

N_2^- : KK $\sigma(2s)^2\,\sigma^*(2s)^2\,\pi(2px)^2\,\pi(2py)^2\,\sigma(2pz)^2\,\pi^*(2px)^1$

N_2^{2-} : KK $\sigma(2s)^2\,\sigma^*(2s)^2\,\pi(2px)^2\,\pi(2py)^2\,\sigma(2pz)^2\,\pi^*(2px)^1\,\pi^*(2py)^1$

O_2 : KK $\sigma(2s)^2\,\sigma^*(2s)^2\,\sigma(2pz)^2\,\pi(2px)^2\,\pi(2py)^2\,\pi^*(2px)1\,\pi^*(2py)^1$

O_2^+ : KK $\sigma(2s)^2\,\sigma^*(2s)^2\,\sigma(2pz)^2\,\pi(2px)^2\,\pi(2py)^2\,\pi^*(2px)^1$

O_2^- : KK $\sigma(2s)^2\ \sigma^*(2s)^2\ \sigma(2pz)^2\ \pi(2px)^2\ \pi(2py)^2\ \pi^*(2px)^2\ \pi^*(2py)^1$

O_2^{2-} : KK $\sigma(2s)^2\ \sigma^*(2s)^2\ \sigma(2pz)^2\ \pi(2px)^2\ \pi(2py)^2\ \pi^*(2px)^2\ \pi^*(2py)^2$

F_2 : KK $\sigma(2s)^2\ \sigma^*(2s)^2\ \sigma(2pz)^2\ \pi(2px)^2\ \pi(2py)^2\ \pi^*(2px)^2 \pi^*(2py)^2\ \sigma^*(2pz)^2$

KK means $\sigma 1s^2$, $\sigma^* 1s^2$.

Some Orders Related to Molecular Orbital Theory

	O_2^{2+}	O_2^+	O_2	O_2^-	O_2^{-2}
Bond order	3	2.5	2	1.5	1

  Bond order ⟶

  Decreasing order of bond order, bond angle, bond dissociation energy.
  Increasing order of bond length.
- Magnetic Nature

  $O_2 > O_2^- \approx O_2^+ > O_2^{2+} = O_2^{2-}$

  n = 2, 1, 1, 0, 0

  Paramagnetic (O_2); Weakly Paramagnetic (O_2^-, O_2^+); Diamagnetic (O_2^{2+}, O_2^{2-})
- CO, NO^+, CN^-, N_2 (14 e^-) all have bond order 3 and are diamagnetic.
- NO, CN, N_2^-, N_2^+ all have a bond order equal to 2.5 and are paramagnetic.
- H_2, Li_2, B_2 all have a bond order equal to one and are diamagnetic except B_2.
- H_2^-, H_2^+, He_2^+ all have a bond equal to order $^1/_2$ and are paramagnetic.
- All molecules with fractional bond order are paramagnetic.
- Molecules with whole number bond order are mostly diamagnetic, except O_2, B_2, N_2^{2-}.

Facts To Remember

- When z-axis is the major molecular axis, σ molecula orbital will be formed only by s–s, p_z–p_z or s–p_z. AO and no MO is formed by p_x or p_y.

  $\sigma 1s$, $\sigma 2s$, $\sigma 2p_z$ do not have any nodal plane.

  $\sigma^* 1s$, $\sigma^* 2s$, $\sigma^* 2p_z$, $\pi 2p_x$, $\pi 2p_y$ all have one nodal plane each.
- $\pi^* 2p_x$ and $\pi^* 2p_y$ have two nodal planes each.

Some Important Features of a Bond

Bond length

- Bond length is the average distance between the centers of the nuclei of the two bonded atoms.
- It is determined by X-ray diffraction and spectroscopic methods.
- In case of ionic compounds, it is the sum of ionic radius of cation and anion, while in case of covalent compounds, it is sum of their covalent radius.

Factors affecting bond length

- Bond length ∝ Size of atom

  For example, HF < HCI < HBr < HI

  As F < CI < Br < I (Atomic size)
- $\text{Bond length} \propto \dfrac{1}{\text{Bond order or multiplicity}}$

  For example, C–C > C=C > C≡C

  1.54Å 1.34Å 1.32Å
- $\text{Bond length} \propto \dfrac{1}{s\%}$

  that is, $sp^3 > sp^2 > sp$

  25% 33% 50%

  $CH_3–CH_3 > CH_2=CH_2 > HC\equiv CH$
- Bond length ∝ Electronic repulsion

  For example, $H_2^- > H_2^+$
- Resonance and hyperconjugation also change bond length.
- For example, in benzene, C–C bond length is 1.39 Å.

Bond energy

- It is the energy needed to break one mole of bonds of a particular type, so as to separate them into gaseous atoms. It is also called **bond dissociation energy.**
- Bond energy can also be defined as the energy released during the formation of one mole of a particular bond.

Factors affecting bond energy

- Bond energy ∝ Bond order or multiplicity

  For example, C≡C > C=C > C–C
- $\text{Bond length} \propto \dfrac{1}{\text{Bond order or multiplicity}}$

For example,

$HF > HCl > HBr > HI$

- Bond energy ∝ s%

For example

$sp > sp^2 > sp^3$

50% 33% 25%

$HC \equiv CH > CH_2 = CH_2 > CH_3–CH_3$

- Bond length ∝ $\frac{1}{\text{Lone pair of electrons / electronic repulsion}}$

For example, C – C > N – N > O – O > F – F

lp e^- 0	1	2	3	
B.E. (kJ)	347	163	146	138.8

Some diatomic molecules in order of bond energy

C≡O > N≡N > C≡N > C≡C

Bond angle It is the angle between the lines representing the directions of the bonds or the orbitals having bonding pair of electrons.

Factors affecting bond angle

- Bond angle ∝ Bond order ∝ s% ∝ $\frac{1}{\text{Bond length}}$
- For example,

–C≡C– > >C=C< > –C–C–

180° 120° 109° 28°

- Bond angle is also affected by electronic repulsion. (see VSEPR theory)

For example, NH_4^+ > NH_3 > NH_2^-

no lp 1 lp 2 lp e^-

- Bond angle ∝ $\frac{1}{\text{Size of terminal atom}}$

For example, $I_2O > Br_2O > Cl_2O > OF_2$

- Bond angle ∝ $\frac{1}{\text{Size of terminal atom /electronegativity}}$

Normally, bond angle decreases down the group, as electronegativity decreases.

For example, $NH_3 > PH_3 > AsH_3 > BiH_3$

$H_2O > H_2S > H_2Se > H_2Te$

$BF_3 > PF_3 > ClF_3$

Bond angle ∝ Electronegativity of terminal atom

For example, $PF_3 < PCl_3 < PBr_3 < PI_3$

Facts To Remember

PF_3 has greater bond angle as composed to PH_3 due to resonance in PF_3, a double bond character develops.

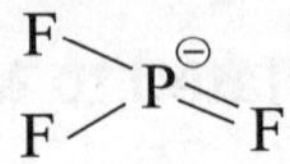

- Bond angle ∝ $\frac{1}{\text{No. of lpe-}}$

eq	$NH_4^+ > NH_3 > \overset{\ominus}{N}H_2$
lp	0 1 2

Van der waals interactions These are very weak and short-lived interactions of following types:

Dipole–dipole interactions These are the strongest interactions present between two oppositely charged poles of two dipoles. These are also known as **Keemer forces.**

For example, SO_2, HX, NH_3.

Diplole-Dipole Interactions

Dipole-induced dipole interactions A non-polar molecule may also became induced dipole in presence of permanent dipole. The interaction between them is called dipole-induced dipole interaction.

Diplole-Induced Diplole Interactions

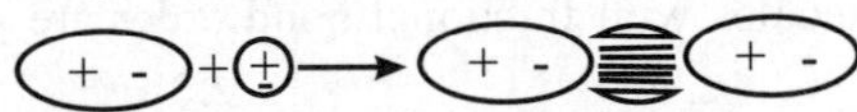

For example, inert gases in H_2O. (Increases down the group) Ion-dipole interaction Such interaction is present in between any ion and oppositely charged end of a dipole.

Ion-Dipole Interactions

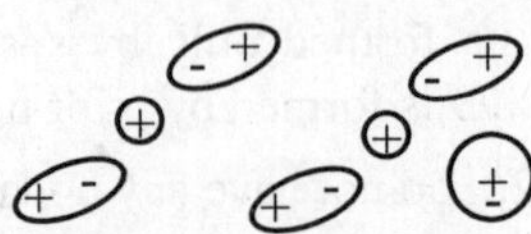

- The momentary dipole induced dipole attractions are called **london** or **dispersive forces.**

London forces ∝ Size or molar mass or surface area

Points To Remember

- **Formyl Charge** It is equal to $V - N - {}^1/_2 B$ where

  V = Total number of valence electrons in the free atom

  N = Total number of lone pair of electrons

  B = Total number of shared electrons that is, bonded electrons

  For example, in case of NH_4^+

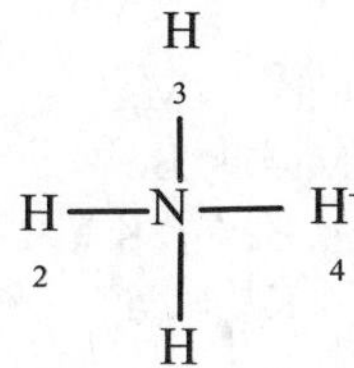

  Here, FC on N-atom $= 5 - 0 - {}^1/_2 \times 8 = +1$

  FC on 1, 2, 3 H-atoms $= 1 - 0 - {}^1/_2 \times 2 = 0$

  FC on H^+ (4 H-atom) $= 0 - 0 - {}^1/_2 \times 2 = -1$
- CO_3^{2-}, NO_3 - involve only $p\pi - p\pi$ bonding.
- SO_4^{2-}, PO_4^{2-}, ClO_4^- involve $p\pi - d\pi$ bonding.
- **In a hypervalent**, for example, SO_4^{-2}, PO_4^{-3}, ClO_4^-, at least one atom has more than 8 electrons.
- Molecular solids have low heat of fusion.
- Breaking of covalent bonds occurs during melting of SiO_2.
- **Banana bonding** is shown by boron hydride, that is, diborane B_2H_6.
- **Band Theory of Metallic Bonds** It is based on molecular orbital theory. The highest occupied energy band is called the valence band while the lowest occupied energy band is called as conduction band. The difference in the energy between top of valence bond and bottom of conduction bond is called **energy gap.**

For example,

1. When energy gap is very very small, conduction occurs (as in metals)
2. When energy gap is small, less conduction occurs (as in semi-conductor)
3. When energy gap is large, no conduction occurs (as in insulators).

- NO_2

  Also same case in –OH, $-NH_2$ groups.
- C_2 molecule has no s bond but has π bond.
- Bond energy: $Cl_2 > F_2 > Br_2 > I_2$.
- μ_{obs} of CO is greater than $\mu_{expected}$ due to presence of a coordinate bond.
- AgX, $BaSO_4$, $PbSO_4$, are nearly insoluble or insoluble in H_2O as H_{Hyd} > Lattice energy

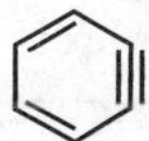

  In benzyne, the triple bond (≡) is partial due to less effective overlapping. All C atoms are sp^2 hybridized.
- Bond length of CO > CO^+.

⇒ NH_4X, NH_4NO_3, $K_4Fe(CN)_6$, $CuSO_4 \cdot 5H_2O$ Contain all 3-types of bonds.

⇒ $NO^+_2 > NO_2 > NO^-_2$ [D. O of Bond angle]

⇒ Hydrogen bonding increases enolisation e.g.,

⇒ In Benzene, bond order is 1.5 while in CO_3^{2-} it is 1.33.

CHAPTER-END EXERCISES

PRACTICE QUESTIONS – I

1. Bond angle of 109°28' is found in:
   (a) NH_3 (b) H_2O
   (c) $\overset{+}{C}H_3$ (d) $^{+}NH_4$

2. Hybridization of the underlined atom changes in which of the following transitions?
   (a) $\underline{Al}H_3$ changes to AlH_4^-
   (b) $H_2\underline{O}$ changes to H_3O^+
   (c) $\underline{N}H_3$ changes to NH_4^+
   (d) in all cases

3. The number of lone pairs on Xe in XeF_2, XeF_4 and XeF_6 respectively are:
   (a) 3, 2, 1 (b) 2, 4, 6
   (c) 1, 2, 3 (d) 6, 4, 2

4. A square planar complex is formed by hybridizationof the following atomic orbitals:
   (a) s, p_x, p_y, p_z
   (b) s, p_x, p_y, p_z, d
   (c) $d_{x^2-y^2}$, s, p_x, p_y
   (d) s, p_x, p_y, p_z, d_{z^2},

5. Which of the following hydrogen bonds is the strongest?
   (a) F–H....F (b) O–H....O
   (c) O–H....F (d) O–H....N

6. Which one of the following pairs of species have the same bond order?
   (a) CN^- and NO^+
   (b) CN^- and CN^+
   (c) $O2^-$ and CN^-
   (d) NO^+ and CN^+

7. Which of the following species exhibits the diamagnetic behaviour?
   (a) O_2^+ (b) O_2
   (c) NO (d) O_2^{2-}

8. In which of the following compounds will the bond angle be maximum?
   (a) NH_3 (b) NH_4^+
   (c) PCl_3 (d) SCl_2

9. The charge/size ratio of a cation determines its polarizing power. Which one of the following sequences represents the increasing order of the polarizing power of the cationic species, K^+, Ca^{2+}, Mg^{2+}, Be^{2+}?
   (a) $Be^{2+} < K^+ < Ca^{2+} < Mg^{2+}$
   (b) $K^+ < Ca^{2+} < Mg^{2+} < Be^{2+}$
   (c) $Ca^{2+} < Mg^{2+} < Be^{2+} < K^+$
   (d) $Mg^{2+} < Be^{2+} < K^+ < Ca^{2+}$

10. Which of the following is an electron deficient molecule?
   (a) C_2H_6 (b) PH_3
   (c) B_2H_6 (d) SiH_4

11. In which of the following molecules/ions all the bonds are unequal?
   (a) SF_4 (b) SiF_4
   (c) XeF_4 (d) BF_4^-

12. Among the following pairs, the one in which the two species are not isostructural is:
   (a) SiF_4 and SF_4
   (b) IO_3^- and XeO_3
   (c) BH_4^- and NH_4^+
   (d) PF_6^- and SF_6

13. In which of the following ionization processes, the bond order has increased and the magnetic behaviour has changed?
   (a) $NO \rightarrow NO^+$ (b) $O_2 \rightarrow O_2^+$
   (c) $N_2 \rightarrow N_2^+$ (d) $C_2 \rightarrow C_2^+$

14. Number of P-O bonds in P_4O_{10} is:
   (a) 17 (b) 16
   (c) 15 (d) 6

15. Of the following sets which does not contain isoelectronic species?
   (a) SO_3^{2-}, CO_3^{2-}, NO_3^-
   (b) CN^-, N_2, C_2^{2-}
   (c) BO_3^{3-}, CO_3^{2-}, NO_3^-
   (d) PO_4^{3-}, SO_4^{2-}, ClO_4^-

16. The pair of species having identical shapes for molecules of both species is:
(a) CF_4, SF_4 (b) XeF_2, CO_2
(c) BF_3, PCl_3 (d) PF_5, IF_5

17. In which of the following molecules all the bonds are not equal?
(a) AlF_3 (b) NF_3
(c) ClF_3 (d) BF_3

18. Which of the following molecules/ions does not contain unpaired electrons?
(a) O_2^{2-} (b) B_2
(c) N_2^+ (d) O_2

19. Which of the following is an electron deficient molecule?
(a) C_2H_6 (b) NH_3
(c) $BeCl_2$ (d) $SiCl_4$

20. The number and type of bonds between two carbon atoms in calcium carbide are:
(a) One sigma, two pi
(b) Two sigma, two pi
(c) One sigma, one pi
(d) Two sigma, one pi

21. The ratio of π and σ bonds in benzene is:
(a) 1 : 2 (b) 1 : 4
(c) 1 : 6 (d) 1 : 6

22. The bond length between C-C bond in sp3 hybridized molecule is:
(a) 1.2 Å (b) 1.39 Å
(c) 1.33 Å (d) 1.54 Å

23. The planar structure of BF_3 can be explain by the fact that BF_3 is:
(a) sp hybridized
(b) sp^2 hybridized
(c) sp^3 hybridized
(d) sp_3 hybridized

24. Dipole moment is shown by:
(a) 2-methyl propane
(b) Trans-2-butene
(c) Cis-2-butene
(d) Carbon tetrachloride

25. O_2^{2+} has a bond order of:
(a) 1 (b) 2
(c) 3 (d) 4

26. In 1, 3-butadiene, the carbon is hybridized as:
(a) sp (b) sp^2
(c) sp^3 (d) sp^2 and sp^3

27. The most polar covalent species among the ones listed is:
(a) C–S (b) C–O
(c) C–F (d) C–Br

28. Correct order of polarizing power is:
(a) $Cs^+ < K^+ < Mg^{2+} < Al^{3+}$
(b) $K^+ < Cs^+ < Mg^{2+} < Al^{3+}$
(c) $Cs^+ < K^+ < Al^{3+} < Mg^{2+}$
(d) $K^+ < Cs^+ < Al^{3+} < Mg^{2+}$

29. H-bond is strongest in:
(a) C_2H_5OH (b) H–F
(c) H_2O (d) CH_3COCH_3

30. Hybridization of 1 and 2 carbon atom in $^1CH_2={}^2C=CH_2$ are:
(a) sp, sp (b) sp^2, sp^2
(c) sp^2, sp (d) sp^3, sp^2

31. Which of the following molecules contains one lone pair of electrons on the central atom?
(a) NH_3 (b) CH_4
(c) $CHCl_3$ (d) Cl_2

32. The molecule having highest bond energy is:
(a) N–N (b) F–F
(c) C–C (d) O–O

33. The halide having the highest melting point is:
(a) NaF (b) NaCl
(c) NaBr (d) NaI

34. Born-Haber cycle is used to determine:
(a) Electron affinity (b) Lattice energy
(c) Crystal energy (d) All the above

35. The number of s and π-bonds present in pent-1-ene-4-yne is:
(a) 3, 10 (b) 9, 4
(c) 4, 9 (d) 10, 3

36. The ion which is not tetrahedral in shape is:
(a) BF_4^- (b) NH_4^+
(c) $[Ni(CN)_4]^{2-}$ (d) $NiCl_4^{2-}$

37. The paramagnetic species among the following is:
(a) KO_2 (b) SiO_2
(c) TiO_2 (d) BaO_2

38. Intermolecular hydrogen bonding is strongest in:
(a) Methylamine
(b) Phenol
(c) Formaldehyde
(d) Methanol

39. In $[Ag(CN)_2]^-$ the number of π bonds is:
(a) 2 (b) 3
(c) 4 (d) 6

40. $BaSO_4$ is insoluble in water because of:
(a) Low hydration energy
(b) Low lattice energy
(c) High lattice energy
(d) Both (a) and (c)

41. Acetylene has .... bonds:
(a) 2σ, 2π (b) 2σ, 3π
(c) 3σ, 2π (d) 3σ, 3π

42. Which of the following statement is true?
(a) Hybridization of the central atom in NH_3 and CH_4 is sp^2
(b) $BeCl_2$ has "V" shape while SO_2 is linear
(c) SF_6 is octahedral and F–S–F bond angle is 90°
(d) CO_2 has dipole moment

43. Both BF_3 and NF_3 are covalent but BF_3 molecule is non-polar while NF_3 is polar because:
(a) Atomic size of boron is smaller than nitrogen
(b) BF_3 is planar but NF_3 is pyramidal
(c) Boron is a metal while nitrogen is gas
(d) BF bond has no dipole moment while NF bond has dipole

44. The boiling point of alcohol is high due to:
(a) Hydrogen bonding
(b) Large size of alcohol
(c) Presence of –OH group
(d) High molecular weight

45. The hybridization of I in IF_3 is:
(a) sp^3d (b) sp^3
(c) sp^3d^2 (d) sp^3d^3

46. What is the number of sigma and pi bonds present in a molecule of sulphuric acid?
(a) 6σ, 2π (b) 6σ, 0π
(c) 2σ, 4π (d) 2σ, 2π

47. The shape of IF_7 molecule is:
(a) Pentagonal bipyramidal
(b) Trigonal pyramidal
(c) Tetrahedral
(d) Square planar

48. The highest dipole moment is of:
(a) CF_4 (b) CH_3OH
(c) CO_2 (d) CH_3F

49. The most polar bond is:
(a) O–H (b) C–O
(c) C–Br (d) C–C

50. Which one of the following is the correct set with reference to molecular formula, hybridization of central atom and shape of the molecule?
(a) CO_2, sp^2, bent (b) H_2O, sp^2, bent
(c) $BeCl_2$, sp, linear (d) H_2O, sp^3, linear

51. What is the hybridization state of the central atom in the conjugate base of NH_4^+ ion?
(a) sp (b) sp^3
(c) sp^3 (d) dsp^2

52. Which one of the following molecules contains both ionic and covalent bonds?
(a) CH_2Cl_2 (b) K_2SO_4
(c) $BeCl_2$ (d) SO_2

53. Which one of the following is a correct set with respect to molecule, hybridization and shape?
(a) $BeCl_2$, sp^2, linear
(b) $BeCl_2$, sp^2, triangular planar
(c) BCl_3, sp^2, triangular planar
(d) BCl_3, sp^3, tetrahedral

54. Which of the following shows isostructural species?
(a) NH_4^+ and NH_2^-
(b) CH_3^- and CH_3^+
(c) SO_4^{2-}, PO_4^{3-} and $[BF_4^-]$
(d) NH_4^+ and NH_3

55. Which one of the following is a planar molecule?
(a) NH_3 (b) H_3O^+
(c) BCl_3 (d) PCl_3

56. The molecule having largest dipole moment among the following is:
(a) CHI_3 (b) CH_4
(c) $CHCl_3$ (d) CCl_4

57. Compound X is an anhydride of sulphuric acid. The number of sigma bonds and the number of pi bonds present in X are respectively:

(a) 3, 3 (b) 4, 2
(c) 2, 4 (d) 4, 3

58. Which one of the following is a correct set?

(a) H_2O, sp^3, angular
(b) H_2O, sp^2, linear
(c) NH_4^+, dsp^2, square planar
(d) CH_4, dsp^2, tetrahedral

59. Sulphur reacts with chlorine in 1 : 2 ratio and forms X. Hydrolysis of X gives a sulphur compound Y. What is the hybridization state of central atom in the anion of Y?

(a) sp^3 (b) sp
(c) sp^2 (d) sp^3d

60. How many sigma and pi bonds are present in toluene?

(a) $3\pi + 15\sigma$ (b) $6\pi + 6\sigma$
(c) $3\pi + 6\sigma$ (d) $3\pi + 8\sigma$

61. Which of the following molecules has a zero dipole moment?

(a) $BeCl_2$ (b) HCl
(c) NH_3 (d) H_2O

62. Which one of the following molecules has the smallest bond angle?

(a) H_2O (b) NH_3
(c) H_2Se (d) H_2S

63. In which of the following molecules the central atom does not follow the octet rule?

How many kinds of water are possible?

(a) CO_2 (b) H_2S
(c) BF_3 (d) PCl_3

64. Which one of the following statements is true?

(a) The dipole moment of NF_3 is more than NH_3
(b) The dipole moment of NF_3 is less than NH_3
(c) The dipole moment of NH_3 is zero
(d) The dipole moment of NF_3 is equal to NH_3

65. Bond angle in water is:

(a) 90° (b) 104.5°
(c) 109.5° (d) 120°

66. Which of the following is a linear molecule?

(a) SO_2 (b) CH_4
(c) H_2O (d) $BeCl_2$

67. Which of the following represents the Lewis structure of N_2 molecule?

(a) ${}^{x}_{x}N \equiv N^{x}_{x}$ (b) ${}^{x}_{x}\overset{xx}{\underset{xx}{N}} \equiv N^{x}_{x}$
(c) ${}^{x}_{x}\overset{xx}{\underset{xx}{N}} \equiv \overset{xx}{\underset{xx}{N}}{}^{x}_{x}$ v (d) ${}^{x}_{x}N — {}^{x}_{x}\overset{xx}{\underset{xx}{N}}{}^{x}_{x}$

68. Hydrogen bond is strongest in:

(a) S–H—O (b) O–H—S
(c) F–H—F (d) O–H—N

69. Which of the following has the highest bond order?

(a) N_2 (b) O_2
(c) He_2 (d) H_2

70. Which of the following is diamagnetic?

(a) H_2^+ (b) O_2
(c) Li_2 (d) He_2^+

71. Among the following compounds, the one that is polar and has the central atom with sp^2 hybridization is:

(a) SiF_4 (b) BF_3
(c) $HClO_2$ (d) H_2CO_3

72. Iodine pentafluoride has which of the following hybridization?

(a) d^2sp^3 (b) dsp^3
(c) sp^3d (d) sp^3d^2

73. The number of lone pairs of electrons present in central atom of ClF_3 is:

(a) 0 (b) 1
(c) 2 (d) 3

74. Bond angle in XeO_3 is:

(a) 107° (b) 119°
(c) 92° (d) 103°

75. Which of the following molecule is planar?

(a) CH_4 (b) NH_3
(c) C_2H_4 (d) $SiCl_4$

76. The paramagnetism of O_2^+ is due to the presence of an odd electron in the MO

(a) σ^*2s (b) $\pi 2py$
(c) σ^*2px (d) π^*2py

77. Among the following compound which one has maximum number of lone pairs of electrons on central atom:

(a) $[ClO_3]^-$ (b) XeF_4
(c) SF_4 (d) $[I_3]^-$

78. The acid having O–O bond is:

(a) $H_2S_2O_3$ (b) $H_2S_2O_6$
(c) $H_2S_2O_8$ (d) $H_2S_4O_6$

79. The species having tetrahedral shape is:
(a) $[PdCl_4]^{2-}$
(b) $[Ni(CN)_4]^{2-}$
(c) $[Pd(CN)_4]^{2-}$
(d) $[NiCl_4]^{2-}$

80. Total number of lone pair of electrons in $XeOF_4$ is:
(a) 0 (b) 1
(c) 2 (d) 3

81. The correct order of hybridization of the central atom in the following species NH_3, $PtCl_4^{-2}$, PCl_5 and BCl_3 is:
(a) dsp^2, sp^3d, sp^2 and sp^3
(b) sp^3, dsp^2, sp^3d, sp^2
(c) dsp^2, sp^2, sp^3, sp^3d
(d) dsp^2, sp^3, sp^2, sp^3d

82. The hybridization of atomic orbitals of nitrogen inNO_2^+, NO_3^- and NH_4^+ are:
(a) sp^2, sp^3 and sp^2 respectively
(b) sp, sp^2 and sp^3 respectively
(c) sp^2, sp and sp^3 respectively
(d) sp^2, sp^3 and sp respectively

83. Amongst H_2O, H_2S, H_2Se and H_2 Te the one with the highest boiling point is:
(a) H_2O because of hydrogen bonding
(b) H_2Te because of higher molecular weight
(c) H_2S because of hydrogen bonding
(d) H_2Se because of lower molecular weight.

84. Which one of the following compounds has sp^2 hybridization?
(a) CO_2 (b) SO_2
(c) N_2O (d) CO

85. Which contains both polar and non-polar bonds?
(a) NH_4Cl (b) HCN
(c) H_2O_2 (d) CH_4

86. Among the following compounds, the one that is polar and has the central atom with sp2 hybridization is:
(a) H_2CO_3 (b) SiF_4
(c) BF_3 (d) $HClO_2$

87. Which one of the following has zero dipole moment?
(a) ClF (b) PCl_3
(c) SiF_4 (d) $CFCl_3$

PRACTICE QUESTIONS – II

88. Arrange the following compounds in order of decreasing dipole moment.:

(1) Cl Cl Cl

(2) 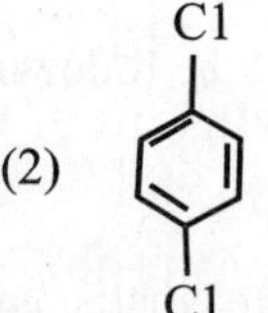

(3) 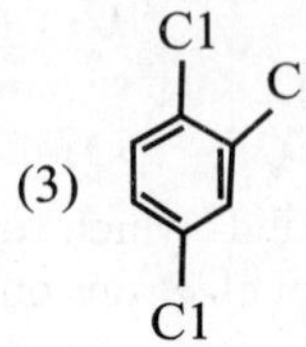

(4) 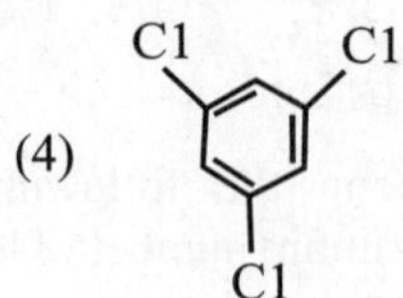

(a) 1 > 3 > 2 = 4
(b) 3 > 1 > 2 = 4
(c) 1 > 2 > 3 > 4
(d) 4 > 3 > 2 > 1

89. Which among the following pairs represent isovalent resonating forms?

(1) R—C(=O)O⁻ ⟷ R—C(O⁻)=O

(2) C(=O)(⁻O)(O⁻) ⟷ C(O⁻)(=O)(O⁻)

(3) 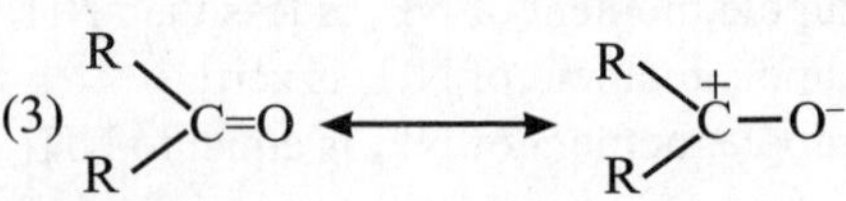

(4) 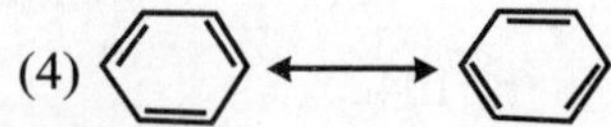

(a) 1, 2, 4
(b) 2, 3, 4
(c) 3, 4
(d) 1, 3

90. In which of the following species, intramolecular Hbonding occurs?

1. Acetate ion 2. Salicylate ion
3. Propanoic acid 4. O-nitrophenol

(a) 1, 3, 4 (b) 1, 2, 4
(c) 2, 4 (d) 4 only

91. The correct order of bond angles (smallest first) in H_2S, NH_3, BF_3 and SiH_4 is:

(a) $H_2S < SiH_4 < NH_3 < BF_3$
(b) $NH_3 < H_2S < SiH_4 < BF_3$
(c) $H_2S < NH_3 < SiH_4 < BF_3$
(d) $H_2S < NH_3 < BF_3 < SiH_4$

92. In X–Hi—-Y, where both X and Y are electronegative elements:

(a) Electron density of X will increase and the eletron density on H will decrease
(b) On both species electron density will increase
(c) On both species electron density will decrease
(d) On X the electron density will decrease and on H it will increases

93. The molecular shapes of SF_4, CF_4 and XeF_4 are:

(a) The same with 2, 0 and 1 lone pairs of electrons on the central atom, respectively
(b) The same with 1, 1 and 1 lone pair of electrons on the central atom, respectively
(c) Different with 0, 1 and 2 lone pairs of electrons on the central atom, respectively
(d) Different with 1, 0 and 2 lone pairs of electrons on the central atom, respectively

94. Which of the following statements is not correct for σ and π bonds which is formed between two carbon atoms?

(a) Free rotation of atoms about s bond is allowed but not in case of a π bond
(b) σ bond determines the direction between carbon atoms but a π bond has no primary effect in this regard
(c) σ bond is stronger than a π bond
(d) Bond energies of σ bond and π bond are of the order of 264 kJ/mol and 347 kJ/mol, respectivel

95. The bond order in NO is 2.5 while that in NO^+ is 3. Which of the following statements is true for these two species?

(a) bond length in NO^+ is greater than in NO
(b) bond length in NO is greater than in NO^+
(c) bond length in NO^+ is equal than in NO
(d) bond length is unpredictable

96. H_2O is dipolar, whereas BeF_2 is not. It is because:

(a) The electronegativity of F is greater than that of O
(b) H_2O involves hydrogen bonding whereas BeF_2 is a discrete molecule
(c) H_2O is linear and BeF_2 is angular
(d) H_2O is angular and BeF_2 is linear

97. Which one of the following pairs of molecules will have permanent dipole moments for both members?

(a) SiF_4 and NO_2
(b) NO_2 and CO_2
(c) NO_2 and O_3
(d) SiF_4 and CO_2

98. Using MO theory predict which of these species has the shortest bond length?

(a) O_2^+ (b) O_2^{2-}
(c) O_2^- (d) O_2^{2+}

99. Among the following species the ones having square planar geometry for central atom are:

1. XeF_4 2. SF_4
3. $[NiCl_4]^{2-}$ 4. $[PdCl_4]^{2-}$

(a) 1 and 4 (b) 1 and 2
(c) 2 and 3 (d) 3 and 4

100. Stability of Ge^{2+}, Sn^{2+} and Pb^{2+} is in order:

(a) $Ge^{2+} > Sn^{2+} > Pb^{2+}$
(b) $Sn^{2+} > Ge^{2+} > Pb^{2+}$
(c) $Pb^{2+} > Sn^{2+} > Ge^{2+}$
(d) $Sn^{2+} > Pb^{2+} > Ge^{2+}$

101. Which of the following statement is not correct regarding the properties of ionic compounds?

(a) Ionic compounds have high melting and boiling points
(b) Their reaction velocity in aqueous medium is very high.
(c) Ionic compounds in their molten and aqueous solutions do not conduct electricity.
(d) They are highly soluble in polar solvents.

102. Identify the correct sequence of increasing number of π bonds in the structures of the following molecules:

1. $H_2S_2O_6$ 2. H_2SO_3
3. $H_2S_2O_5$

(a) 1, 2, 3 (b) 2, 3, 1
(c) 2, 1, 3 (d) 1, 3, 2

103. If the molecule of HCl was totally polar, the expected value of dipole moment was 6.12 D but the experimental value of dipole moment calculated was 1.03 D. Calculate the percentage ionic character.

(a) 0
(b) 17
(c) 50
(d) 90

104. The correct order of the lattice energies for the following ionic compounds is:

(a) $Al_2O_3 > CaO > MgBr_2 > NaCl$
(b) $MgBr_2 > Al_2O_3 > CaO > NaCl$
(c) $Al_2O_3 > MgBr_2 > CaO > NaCl$
(d) $NaCl > MgBr_2 > CaO > Al_2O_3$

105. Which of the following is correct?

(a) The rate of ionic reactions are very slow
(b) The number of electrons present in the valence shell of S in SF_6 is 12
(c) According to VSEPR theory $SnCl_2$ is a linear molecule
(d) The correct order of stability to form ionic compoundsamong Na^+, Mg^{2+} and Al^{3+} is $Al^{3+} > Mg^{2+} > Na^+$

106. Bond order of nitric oxide is:

(a) 1
(b) 2.5
(c) 2
(d) 1.5

107. The correct order of bond angle of NO^+_2, NO_2 and NO_2^- is:

(a) $NO_2^+ < NO_2 < NO_2^-$
(b) $NO_2^+ = NO_2^- < NO_2$
(c) $NO_2^+ > NO_2^- > NO_2$
(d) $NO_2^+ > NO_2 < NO_2^-$

108. The decreasing order of the boiling points of the following hydrides is:

1. NH_3 2. PH_3
3. AsH_3 4. SbH_3
5. H_2O

(a) 5 > 4 > 1 > 3 > 2
(b) 5 > 1 > 2 > 3 > 4
(c) 2 > 4 > 3 > 1 > 5
(d) 4 > 3 > 1 > 2 > 5

109. The sequence that correctly describes the relative bond strength pertaining to oxygen molecule and its cation or anion is:

(a) $O_2^{2-} > O_2^- > O_2 > O_2^+$
(b) $O_2 > O_2^+ > O_2^- > O_2^{2-}$
(c) $O_2^+ > O_2 > O_2^{2-} > O_2^-$
(d) $O_2^+ > O_2 > O_2^- > O_2^{2-}$

110. Which of the following are isoelectronic and isostructural? NO_3^-, CO_3^{2-}, ClO_3^-, SO_3

(a) NO_3^-, CO_3^{2-} (b) SO_3, NO_3^-
(c) ClO_3^-, CO_3^{2-} (d) CO_3^{2-}, SO_3

111. Among the following, the molecule with the highest dipole moment is:

(a) CH_3Cl (b) CH_2Cl_2
(c) $CHCl_3$ (d) CCl_4

112. The nodal plane in the π bond of ethene is located in:

(a) The molecular plane.
(b) A plane parallel to the molecular plane.
(c) A plane perpendicular to the molecular plane which bisects. The carbon–carbon s bond at right angle.
(d) A plane perpendicular to the molecular plane which contains the carbon–carbonbond.

113. Specify the coordination geometry and hybridiza-tion of N and B atoms in a 1 : 1 complex of BF_3 and NH_3.

(a) N: tetrahedral, sp ; B: tetrahedral, sp^3
(b) N: pyramidal, sp^3; B: pyramidal, sp^3
(c) N: pyramidal, sp^3; B: planar, sp^2
(d) N: pyramidal, sp^3; B: tetrahedral, sp^3

114. The common features among the species CN^-, CO and NO^+ are:

(a) Bond order three and isoelectronic
(b) Bond order three and weak field ligands
(c) Bond order two and π-acceptors
(d) Isoelectronic and weak field ligands.

115. The correct order of increasing C–O bond length of CO, CO_3^{2-} , CO_2 is

(a) $CO_3^{2-} < CO_2 < CO$
(b) $CO_2 < CO_3^{2-} < CO$
(c) $CO < CO_3^{2-} < CO_2$
(d) $CO < CO_2 < CO_3^{2-}$

116. KF combines with HF to form KHF_2. The compound contains the species

(a) K^+, F^- and H^+ (b) K^+, F^- and HF
(c) K^+ and $[HF_2]^-$ (d) $[KHF]^+$ and F_2

117. The following compounds are to be arranged in order of their increasing thermal stabilities. Identify the correct order.

1. K_2CO_3 2. $MgCO_3$
3. $CaCO_3$ 4. $BeCO_3$

(a) 1 < 2 < 3 < 4 (b) 4 < 2 < 3 < 1
(c) 4 < 2 < 1 < 3 (d) 2 < 4 < 3 < 1

118. Among the following species, identify the isostructural pairs NF_3, NO_3^-, BF_3, H_3O^+, HN_3.

(a) [NF_3, NO_3^-] and [BF_3, H_3O^+]
(b) [NF_3, HN_3] and [NO_3^-, BF_3]
(c) [NF_3, H_3O^+] and [NO_3^-, BF_3]
(d) [NF_3, H_3O^+] and [HN_3, BF_3]

119. Which type of bond is not present in HNO_2 molecule?

(a) Covalent
(b) Coordinate
(c) Ionic
(d) Ionic as well as coordinate

120. The molecules that will have dipole moment:

(1) 2, 2-dimethylpropane
(2) Trans-1-pent-2-ene
(3) Cis-hex-3-ene
(4) 2, 2, 3, 3-tetramethylbutane

(a) 2, 3 (b) 1, 3
(c) 1, 2, 3 (d) all of these

121. The linear structure is seen in:

(1) $SnCl_2$ (2) NCO^-
(3) NO_2^+ (4) CS_2

(a) (1), (2) and (3)
(b) (2), (3) and (4)
(c) (1), (3) and (4)
(d) None of these

122. Which bond angle θ, would result in the maximum dipole moment for the triatomic molecule XY_2 shown below?

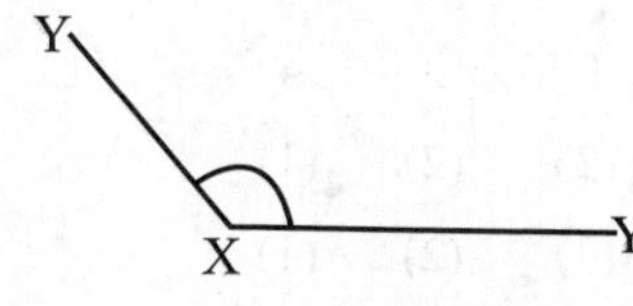

(a) θ = 120° (b) θ = 90°
(c) θ = 145° (d) θ = 175°

123. Correct order of dipole moment is:

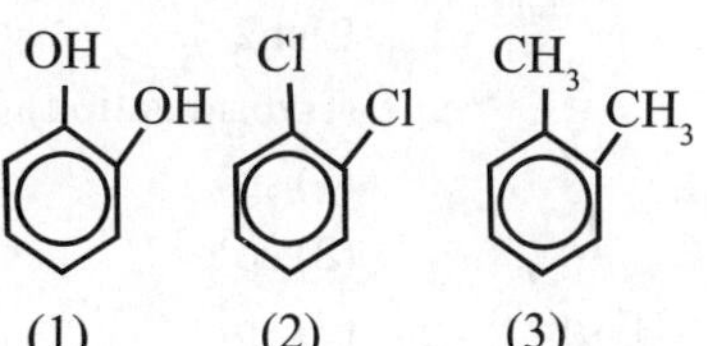

(a) 1 = 2 = 3 (b) 3 < 2 < 1
(c) 1 < 2 < 3 (d) 2 < 3 < 1

124. Match the following:

List 1 (Hydridization)	List 2 (Geometry of the molecule)
1. sp^2	(1) Trigonalbipyramidal
2. sp^3	(2) Planar trigonal
3. sp	(3) Octahedral
4. sp^3d	(4) Tetrahedral
	(5) Linear

The correct matching is:

	1	2	3	4
(a)	(2)	(4)	(5)	(1)
(b)	(3)	(5)	(1)	(4)
(c)	(4)	(3)	(1)	(2)
(d)	(4)	(2)	(5)	(3)

125. Consider the following statements:

1. The bond order of NO is 2.5
2. The bond order of NO^+ is 3
3. The bond order of O_2 is 1.5
4. The bond order of CO is 3

Which of these statements are correct?

(a) 1, 2 and 3 (b) 2, 3 and 4
(c) 1, 3 and 4 (d) 1, 2 and 4

126. Match the following:

List 1	List 2
1. ClF_3	(1) planar
2. NF_3	(2) pyramidal
3. BF_3	(3) T-shaped
4. $[CrF_3]x$	

The correct matching is:

	1	2	3	4
(a)	(–)	(3)	(2)	(1)
(b)	(3)	(2)	(1)	(–)
(c)	(3)	(1)	(2)	(4)
(d)	(3)	(2)	(1)	(2)

127. Match the following:

List 1	List 2
(Species)	(Hybridization)
1. BCl_3	(1) sp^3
2. NH_3	(2) sp^3d^2
3. $[Cu(NH_3)_4]^{2+}$	(3) sp^2
4. $[Fe(H_2O)_6]^{2+}$	(4) sp^2d

The correct matching is:

	1	2	3	4
(a)	(3)	(1)	(4)	(2)
(b)	(1)	(3)	(2)	(4)
(c)	(1)	(3)	(4)	(2)
(d)	(3)	(1)	(2)	(4)

128. Match the following:

List 1	List 2
(Compounds)	(Structures)
1. CS_2	(1) Bent
2. SO_2	(2) Linear
3. BF_3	(3) Trigonal planar
4. NH_3	(4) Tetrahedral
	(5) Trigonal pyramidal

The correct matching is:

	1	2	3	4
(a)	(1)	(2)	(4)	(5)
(b)	(2)	(1)	(3)	(5)
(c)	(1)	(2)	(5)	(4)
(d)	(2)	(1)	(5)	(4)

129. Which of the following types of bonds are present in $CuSO_4.5H_2O$?

(1) electrovalent (2) covalent
(3) coordinate

Select the correct answer using the code given below.

(a) 1 and 2 only
(b) 1 and 3 only
(c) 1, 2 and 3
(d) 2 and 3 only

130. In which one of the following pairs, molecules/ions have similar shape?

(a) CCl_4 and $PtCl_4$
(b) NH_3 and BF_3
(c) BF_3 and t-butyl carbonium ion
(d) CO_2 and H_2O

PRACTICE QUESTIONS – III

131. Which of the following are non-polar?

(1) SiF_4 (2) XeF_4
(3) SF_4 (4) BF_3
(5) NF_3

Select the correct answer using the code given below:

(a) 1, 2 and 4 (b) 3, 4 and 5
(c) 2, 3 and 4 (d) 1, 3 and 4

132. Consider the following statements:

(1) Ionic bond
(2) Covalent bond
(3) Coordinate
(4) Hydrogen bond

Which of the statements given above are correct?

(a) 1 and 3 (b) 2 and 4
(c) 1 and 2 (d) 1, 2 and 3

133. Match the following:

List 1	List 2
(Molecule)	(Bond angle)
1. BF_3	(1) 88°
2. NF_3	(2) 96°
3. PF_3	(3) 103°
4. ClF_3	(4) 110°
	(5) 120°

The correct matching is:

	1	2	3	4
(a)	(4)	(3)	(2)	(1)
(b)	(5)	(3)	(2)	(1)
(c)	(4)	(3)	(5)	(2)
(d)	(3)	(4)	(1)	(2)

134. Which are the species in which sulphur undergoes sp3 hybridization?

(1) SF_4 (2) SCl_2
(3) SO_4^{2-} (4) H_2S

Select the correct answer using the code given below:

(a) 1 and 2 (b) 2, 3 and 4
(c) 1, 3 and 4 (d) 1, 2 and 3

135. Consider the following type of energies and decide the correct order of magnitude.

1. Covalent single bond energy
2. Average translational kinetic energy of gases at room temperature
3. Rotational barrier energy in ethane between eclipsed and staggered forms
4. Ionization energy of hydrogen atom

(a) 4 > 1 > 3 > 2 (b) 4 > 1 > 2 > 3
(c) 2 > 1 > 4 > 3 (d) 4 > 3 > 2 > 1

136. Which of the following species have undistorted octahedral structures?

(1) SF_6 (2) PF_6^-
(3) SiF_6^{2-} (4) XeF_6

Select the correct answer using the codes given below:

(a) 1, 3 and 4 (b) 1, 2 and 3
(c) 1, 2 and 4 (d) 2, 3 and 4

137. Consider the given statements about the molecule $(H_3C)_2$ CH–CH=CH–C≡C–CH=CH_2.

1. Three carbon atoms are sp^3 hybridized
2. Three carbon atoms are sp^2 hybridized
3. Two carbon atoms are sp hybridized Of three statements

(a) 1 and 2 are correct
(b) 1 and 3 are correct
(c) 2 and 3 are correct
(d) 1, 2 and 3 are correct

138. Match the following:

List 1	**List 2**
(Species)	(Bond order)
1. O_2^{2+}	(1) 1.0
2. O_2	(2) 2.0
3. F_2	(3) 2.5
4. O_2^+	(4) 3.0

The correct matching is:

	1	2	3	4
(a)	(4)	(1)	(2)	(2)
(b)	(2)	(3)	(1)	(4)
(c)	(4)	(2)	(1)	(3)
(d)	(3)	(4)	(1)	(2)

139. The halogen form compound among themselves with the formula AA, AA_3, AA_5 and AA_5, where A is the heavier halogen. Which of the following pairs representing their structures and being polar and non-polar are correct?

1. AA, linear, polar
2. AA_3, T-shaped, polar
3. AA_5, square pyramidal, polar
4. AA_7, pentagonal bipyramidal, non-polar

(a) 1 and 3 are correct
(b) 1, 2 and 3 are correct
(c) 2, 3 and 4 are correct
(d) 1, 2, 3 and 4 are correct

140. Consider the following molecules or ions.

(1) CH_2Cl_2 (2) NH_4^+
(3) SO_4^{2-} (4) ClO_4^-
(5) NH_3

sp^3 hybridization is involved in the formation of

(a) (1), (2), (5) only
(b) (1), (2) only
(c) (1), (2), (3), (4)
(d) (1), (2), (3), (4), (5)

141. Consider the following halogen containing compounds

(1) $CHCl_3$ (2) CCl_4
(2) CH_2Cl_2 (4) CH_3Cl
(5) $C_6H_5Cl_2$

The compounds with a net zero dipole moment are

(a) (2) and (5) only (b) (3) only
(c) (3) and (4) only (d) (4) only

142. The electronegativity difference between N and F is greater than that between N and H, yet the dipole moment of NH_3 (1.5 D) is larger than that of NF_3 (0.2 D).

This is because:

(a) In NH_3 as well as NF_3 the atomic dipole and bond dipole are in opposite directions
(b) In NH_3 the atomic dipole and bond dipole are in the opposite directions whereas in NF_3 these are in the same direction
(c) In NH_3 as well as in NF_3 the atomic dipole and bond dipole are in same direction
(d) In NH_3 the atomic dipole and bond dipole and in the same direction whereas in NF_3 these are in opposite directions

143. According to molecular orbital theory which of the following statement about the magnetic character and bond order is correct regarding O_2^{+}?

(a) paramagnetic and bond order < O_2
(b) paramagnetic and bond order > O_2
(c) diamagnetic and bond order < O_2
(d) diamagnetic and bond order > O_2

144. Pick out the isoelectronic structures from the following:

1. CH_3 2. H_3O^+
3. NH_3 4. CH_2

(a) 1 and 2 (b) 3 and 4
(c) 1 and 3 (d) 2, 3 and 4

145. Arrange the following compounds in order of increasing dipole moment:

1. Toluene,
2. m-dichlorobenzene,
3. O-dichlorobenzene,
4. P-dichlorobenzene

(a) 1 < 4 < 2 < 3 (b) 4 < 1 < 2 < 3
(c) 4 < 1 < 3 < 2 (d) 4 < 2 < 1 < 3

146. $(CH_3{-}CH_2)(M(CO)_3)C{=}C(C{\equiv}N)_2$ Contains σ and π bonds respectively as?

(a) 19, 10 (b) 19, 11
(c) 17, 11 (d) 17, 10

147. Which of the following are correct here?

(1) Stability in decreasing order $N_2 > N_2^{+} > N_2^{2-}$
(2) Bond length in decreasing order $H_2^{-} > H_2^{+} > H_2$
(3) No of Lone pair e^- in decreasing order $Cl_2O > XeF_2 > XeO_3$
(4) C is but 2-ene > trans pent 2-ene > trans but 2-ene Dipole moment in decreasing order

(a) 1, 2 (b) 2, 3
(c) 1, 2, 3 (d) 1, 2, 4

148. If dipole moment of HX is 0.76×10^{-18} esu c.m and inter atomic distance is 1.41 A°. The ionic nature percentage in HX is nearly.

(a) 11.7% (b) 15.6%
(c) 23.6% (d) 5.05%

149. The value of bond order in IO_4^- is?

(a) 1.75 (b) 1.5
(c) 1.25 (d) 1.33

150. The correct order of hybridization of central atom in these species can be given as

XeO_3, $PdCl_4^{2-}$, $Ni(CO)_4$, XeF_4

(a) sp^3, dsp^2, dsp^2, sp^3d^2
(b) dsp^2, dsp^2, sp^3, sp^3d^2
(c) sp^3, dsp^2, sp^3, sp^3d^2
(d) sp^3, sp^3, dsp^2, sp^3d^2

151. Increasing order of bond strength of O_2, O_2^-, O_2^{2-} and O_2^+ is

(a) $O_2^+ < O_2 < O_2^- < O_2^{2-}$
(b) $O_2^- < O_2 + < O_2^- < O_2^{2-}$
(c) $O_2^- < O_2^{2-} < O_2^+ < O_2$
(d) $O_2^{2-} < O_2^- < O_2 < O_2^+$

152. Select correct statement:

(a) When a covalent bond is formed, transfer of electrons takes place
(b) Pure H_2O does not contain any ion
(c) A bond is formed when attractive forces overcome repulsive forces
(d) HF is less polar than HBr

153. Which one of the following compounds has the smallest bond angle in its molecule?

(a) SO_2 (b) OH_2
(c) SH_2 (d) NH_3

ANSWER KEYS

1. (d)	2. (a)	3. (a)	4. (c)	5. (a)	6. (a)	7. (d)	8. (b)	9. (b)	10. (c)
11. (a)	12. (a)	13. (a)	14. (b)	15. (a)	16. (b)	17. (c)	18. (a)	19. (c)	20. (c)
21. (b)	22. (d)	23. (b)	24. (c)	25. (c)	26. (b)	27. (c)	28. (a)	29. (b)	30. (c)
31. (a)	32. (a)	33. (d)	34. (d)	35. (d)	36. (c)	37. (a)	38. (d)	39. (c)	40. (d)
41. (c)	42. (c)	43. (b)	44. (a)	45. (a)	46. (a)	47. (a)	48. (d)	49. (a)	50. (c)
51. (c)	52. (b)	53. (c)	54. (c)	55. (c)	56. (c)	57. (a)	58. (a)	59. (a)	60. (a)
61. (a)	62. (c)	63. (c)	64. (b)	65. (b)	66. (d)	67. (a)	68. (c)	69. (a)	70. (c)
71. (d)	72. (d)	73. (c)	74. (d)	75. (c)	76. (d)	77. (d)	78. (d)	79. (d)	80. (b)
81. (b)	82. (b)	83. (a)	84. (b)	85. (c)	86. (a)	87. (c)	88. (a)	89. (a)	90. (c)
91. (c)	92. (b)	93. (d)	94. (d)	95. (a)	96. (d)	97. (c)	98. (d)	99. (a)	100. (c)
101. (c)	102. (b)	103. (b)	104. (a)	105. (b)	106. (b)	107. (c)	108. (a)	109. (d)	110. (a)
111. (a)	112. (a)	113. (a)	114. (a)	115. (d)	116. (c)	117. (b)	118. (c)	119. (d)	120. (a)
121. (b)	122. (b)	123. (d)	124. (a)	125. (d)	126. (b)	127. (d)	128. (b)	129. (c)	130. (c)
131. (a)	132. (c)	133. (b)	134. (b)	135. (a)	136. (b)	137. (b)	138. (c)	139. (d)	140. (d)
141. (a)	142. (d)	143. (b)	144. (d)	145. (b)	146. (b)	147. (d)	148. (a)	149. (a)	150. (c)
151. (d)	152. (c)	153. (c)							

HINTS AND EXPLANATIONS FOR SELECTIVE QUESTIONS

1. As it is sp^3d hybridized with tetrahedral shape.

2. Here hybridization changes from sp^2 to sp^3.

4. As square planar geometry has dsp2 hybridization.

6. As both are isoelectronic so have the same bond order (3).

7. O_2^{2-} has 18 electrons, which are filled in such a way that all molecular orbitals are fully filled, so diamagnetic.
$\sigma 1s^2\ \sigma^* 1s^2, \sigma 2s^2\ \sigma^* 2s^2, \sigma 2p_z^2, \pi 2p_x^2 = \pi 2p_y^2, \pi^* 2p_x^2 = \pi^* 2p_y^2$

8. As bond angles of NH_3 is 107°, for NH_4^+ it is 109.5°, PCl_3 is 100°, therefore, bond angle of NH_4^+ is maximum.

9. Higher the charge/size ratio, more is the polarizing power i.e.,
$K^+ < Ca^{2+} < Mg^{2+} < Be^{2+}$

11. In SF_4 the bonds are not equal as it has see-saw shape and have 2 axial and 2 equatorial bonds.

12. As SiF_4 is tetrahedral in shape while SF_4 has a see-saw structure.

15. As SO_3^{2-}, CO_3^{2-} and NO_3^- has 42, 32, 32 electrons respectively, that is, SO_3^{2-} is not isoelectronic.

16. As both are linear in shape.

17. Chlorine atom in ClF_3 is sp^3d hybridized. so the bonds are not equal as it has a distorted T-shape.

18. O_2 has two unpaired electrons but are paired in O_2^{2-}

20. Calcium carbide is ionic carbide having $[:C \equiv C:]^{2-}$

21. As in benzene there are 12 sigma and 3 pi bonds.

28. As polarizing power of cation ∝ charge on cation.

29. As F is more electronegative than nitrogen.

31. Here nitrogen atom has three bp and one lp electron.

35. CH_2=CH–CH_2–C≡CH
It has 10 sigma and 3 pi bonds.

36. As it is square planar in shape.

37. As in it O_2 is in O_2^- form so have one unpaired electron

39. As –C≡N– thus two cyanide ions have 4π bonds.

51. The conjugate base of NH_4^+ is NH_3 which is sp^3 hybridized.

54. All these are tetrahedral in shape.

57. A hydride of sulphuric acid is SO_3. which has three sigma and three pi bonds.

61. As it is a linear molecule, so it has no dipole moment.

62. Here bond angle is as follows:
$NH_3 > H_2O > H_2S > H_2Se$

67. Lewis structure of N_2 molecule X
$^{X}_{X}N \equiv N^{X}_{X}$

68. Fluorine, having the highest value of electronegativity forms the strongest hydrogen bonds.

69. As bond order for N_2, O_2, He_2, H_2 are 3, 2, 0, 1 respectively.

75. Ethylene is a planar molecule in which carbon atom is sp^2 hybridized.

78. Only $H_2S_2O_8$ has O—O bond

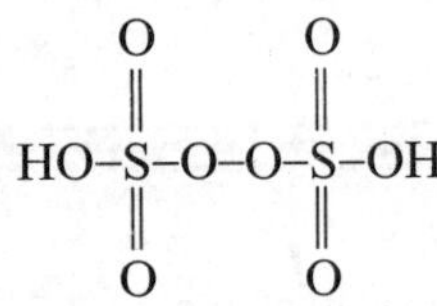

83. H_2O has the highest boiling point due to H-bonding. The order is $H_2S < H_2Se < H_2Te < H_2O$

84. In SO_2 S atom is sp^2 hybridized.

85. The O–O bond is non polar while O–H bond is polar.

87. SiF_4 has a symmetrical tetrahedral structure so individual bond moments cancel each other resulting in zero dipole moment.

91.

Species	lp	bp	VSEPR	Bond angle
H_2S	2	2	lp–lp lp–bp	90°
NH_3	1	3	bp–bp lp–bp bp–bp	107°
BF_3	0	3	bp–bp	120°
SiH_4	0	4	bp–bp	109° 28°

So, bond angle $H_2S < NH_3 < SiH_4 < BF_3$

92. In X–H——Y, X and Y both are electronegative elements then electron density on X will increase and on H will decrease.

93. SF_4 (sp^3d, trigonal bipyramidal with one equatorial position occupied by one lone pair), CF_4 (sp^3, tetrahedral, no lone pair), XeF_4 (sp^3d^2, square planar, two lone pairs).

94. Sigma bond is stronger than the pi bond, so it must be having higher bond energy than pi bond.

95. Bond length is inversely proportional to bond order.
Bond order in $NO^+ = 3$
$NO = 2.5$
So, bond length in $NO > NO^+$.

96. In a linear symmetrical molecule like BeF_2, the bond angle between the three atoms is 180°, hence the polarity due to one bond is cancelled by the equal polarity due to other bond, while it is not so in angular molecules, like H_2O.

98. As for O_2^{+2} bond order is maximum so it has minimum bond length.
Bond order of $O_2 = 2$
Bond order of $O_3 = 1.5$
Bond order of $H_2O_2 = 1$
So, $O_2 < O_3 < H_2O_2$

99. In XeF_4 : sp^3d^2 hybridization. Shape is square planar instead of octahedral due to presence of two lone pair of electrons on Xe atom.
SF_4 : sp^3d hybridization

100. Due to inert pair effect in case of lower oxidation state, stability increase down the group.

101. As they show conductivity in this state due to presence of free ions.

103. Ionic percentage $\dfrac{\mu_{calculated}}{\mu_{expected}} \times 100$
$$= \frac{1.03}{6.12} \times 100 = 17\%$$

104. Lattice energy $\propto$ charge on ion.

106. Molecular orbital configuration of NO is
$\sigma 1s^2\ \sigma^* 1s^2\ \sigma 2s^2\ \sigma^* 2s^2\ \pi 2px^2\ \pi 2py^2\ \sigma 2pz^2\ \pi^* 2px^1$
Bond order $= \frac{1}{2}(10 - 5) = 1.5.$

109. As bond strength $\propto$ bond order.

110. Both NO_3^- and CO_3^{-2} have same number of electrons (32 electrons) and sp^2 hybridized central atom. (isostructural)

111. Methane molecule has zero dipole moment value because of symmetrical nature. Dipole moment increases with the replacement of one of the H by Cl atom. Replacement of second H atom by Cl atom decreases the dipole moment value because bond angle increases due to lp–lp repulsions between the two Cl atoms. When third Cl atom is introduced bond angle further increase. When fourth Cl atom is introduced, the molecule (CCl_4) again is symmetrical in nature and has zero dipole moment value. Thus, CH_3Cl has maximum dipole moment.

112. The nodal plane in the π–bond of ethene is located in the molecular plane.

113. In the 1 : 1 complex

$H_3N : \rightarrow BF_3$, both N and B have sp^3 hybridized and tetrahedral geometry.

114. All the species are isoelectronic since each one of them has 14 e^- and bond order = 3.

115. As Bond length $\propto \dfrac{1}{\text{Bond order}}$

	Bond order
$C^- \equiv O^+$	3
O=C=O	2
CO_3^{-2}	1.33

Since the bond length increases as the bond order decreases that is $CO < CO_2 < CO_3^{-2}$

116. $KHF_2 \rightarrow K^+ + HF_2^-$

HF_2^- ion exists due to hydrogen boding between

$\overset{-}{[}\text{F}\cdots\cdots\overset{+\delta}{\text{H}}\text{—}\overset{-\delta}{\text{F}}]$.

118. NF_3, H_3O^+ are pyramidal in shape while NO_3^-, BF_3 are planar molecules.

119. H–O–N=O only covalent bonds are present.

121. $SnCl_2$ has sp^2 hybridization and angular structure. In CS_2 carbon is sp hybridized and is linear. NCO^- and NO_2^+ being isoelectronic with CS_2 have same type of shape.

122. $\mu = \sqrt{(\mu_1^2 + \mu_2^2 + 2\,\mu_1\mu_2 \cos\theta)}$ if $\theta = 90°$, μ is maximum.

123. In (1) Hydrogen bonding decreases $\theta < 60°$

In (2) dipole-dipole repulsion increases $\theta > 60°$

In (3) θ remains 60°

Hence, the correct order is 2 < 3 < 1.

125. Bond order of NO = (10 – 5)/2 = 2.5

Bond order of NO^+ = (10 – 4)/2 = 3

Bond order of CO = (10 – 4)/2 = 3

Bond order of O_2 = 2

129. $CuSO_4.5H_2O$

$$Cu^{++}\ .\ O=\overset{\overset{O}{\|}}{\underset{\underset{O}{\|}}{S}}-O^-\ .\ 5H_2O$$

Here, ionic and covalent bonds are present in $CuSO_4$ while H_2O molecules are attached by coordinate bonds.

130. BF_3 and t-butyl carbonium ion

Both of these have sp^2 hybridization and trigonal planar shape.

131.

SiF_4	sp^3	tetrahedral non-polar
XeF_4	sp^3d^2	square planar non-polar
SF_4	sp^3d	square pyramidal polar
BF_3	sp^2	trigonal planar non-polar
NF_3	sp^2	pyramidal polar

132. $NaHCO_3 \rightarrow Na^+ + HCO_3^-$

Ionic bond

$$H-O-\underset{\underset{O}{\|}}{C}-O^-$$

Covalent bond

135. As ionization energy of hydrogen atom is nearly 1310 kJ mol^{-1}, covalent single bond energies are of the order of a few hundred kJ mol^{-1}, molecular translation energy of gases is equal to 3/2 RT, which is nearly 3.7 kJ mol^{-1}, at 27°C and the rotational barrier energy is nearly between eclipsed and staggered forms of ethane and is about 12 kJ mol^{-1}.

136. XeF_6 has a distorted octahedral structure due to the presence of one lone pair of electrons on the Xe atom. SF_6, PF_6^- and SiF_6^{2-} all have undistorted octahedral structures.

137. This molecule has four carbon atoms (that is, –CH=CH– and $CH=CH_2$) sp^2 hybridized.

143. For O_2

Bond order = ½(Nb – Na) = ½(10 – 6) = 2 for O_2^+

Bond order = ½(10 – 5) = 2.5

Thus, bond order of O_2^+ > bond order of O_2. Further as there is one unpaired electron present, so the ion is paramagnetic.

144. 1. CH_3^+ ; 6 +3 – 1 = 8 (electrons)
2. H_3O^+ ; 8 + 3 – 1 = 10
3. NH_3 ; 7 + 3 = 10
4. CH_3^- ; 6 + 3 + 1 = 10
Thus 2, 3 and 4 are isoelectronic structures.

147. As the correct order of lone pair of electrons is as follows:

	XeF_2 >	Cl_2O >	XeO_3
lpe– →	3	2	1

148. Ionic nature % = $\frac{\mu_{obs}}{q \times r} \times 100$

$= \frac{0.76 \times 10^{-18}}{4.8 \times 10^{-10} \times 1.41 \times 10^{-8}}$

= 11.7 %

149. $O=I(=O)-O^-$ (with two I=O bonds) Bond order $= \frac{2+2+2+1}{4}$

= 1.75

150. As XeO_3 (sp^3), $PdCl_4^{2-}$ (dsp^2), $Ni(CO)_4$ (sp^3), XeF_4 (sp^3d^2)

151. As bond strength ∝ bond order

153.

Molecule	Hybridization	Repulsion	Bond angle
SO_2	sp^2	lp–bp bp–bp	119°
OH_2	sp^3	lp–lp bp–lp bp–lp	104.5°
SH_2	sp^3	– do –	90°
NH_3	sp^3	lp–bp bp–bp	107°

Previous Years' Questions

1. Which one of the following is not paramagnetic?
   (a) NO (b) N_2^+
   (c) CO (d) O_2^-
   [2000]

2. Cation and anion combines in a crystal to form following type of compound.
   (a) Ionic (b) Metallic
   (c) Covalent (d) Dipole-dipole
   [2000]

3. Among the following the electron deficient compound is:
   (a) BCl_3 (b) CCl_4
   (c) PCl_5 (d) CH_4
   [2000]

4. The relationship between the dissociation energy of N_2 and N_2^+ is:
   (a) Dissociation energy of N_2^+ > dissociation energy of > dissociation energy of N_2
   (b) Dissociation energy of N_2 = dissociation energy of N_2^+
   (c) Dissociation energy of N_2 > dissociation energy of N_2^+
   (d) Dissociation energy of N_2 can either be lower or higher than the dissociation energy of N_2^+
   [2000]

5. Which one of the following molecules will form a linear polymeric structure due to hydrogen bonding?
   (a) NH_3 (b) H_2O
   (c) HCl (d) HF
   [2000]

6. Among the following ions the pπ-dπ overlap could be present in:
   (a) NO_2^- (b) NO_3^-
   (c) PO_4^{3-} (d) CO_3^{2-}
   [2000]

7. Nitrogen forms N_2, but phosphorous is converted into P_4 from P, the reason is:
   (a) Triple bond is present between phosphorous atom
   (b) pπ – pπ bonding is strong
   (c) pπ – pπ bonding is weak
   (d) Multiple bond is formed easily
   [2001]

8. Main axis of a diatomic molecule is z, molecule orbitals px and py overlap to form, which of the

following orbital?

(a) π –molecular orbital
(b) σ–molecular orbital
(c) δ–molecular orbital
(d) No bond will form

[2001]

9. In X – H — Y, X and Y both are electronegative elements:

(a) Electron density of X will increase and on H will decrease
(b) In both electron density will increase
(c) In both electron density will decrease
(d) On X electron density will decrease and on H increases

[2001]

10. Which of the following two are isostructural?

(a) XeF_2, IF_2^- (b) NH_3, BF_3
(c) CO_3^{2-}, SO_3^{2-} (d) PCl_5, ICl_5

[2001]

11. In which of the following bond angle is maximum:

(a) NH_3 (b) NH_4^+
(c) PCl_3 (d) SCl_2

[2001]

12. In NO_3^- ion, number of bond pair and lone pair of electron on nitrogen atom are:

(a) 2,2 (b) 3,1
(c) 1,3 (d) 4,0

[2002]

13. Which of the following has pπ – dπ bonding?

(a) NO^{3-} (b) SO_3^{2-}
(c) BO_3^{3-} (d) CO_3^{2-}

[2002]

14. Which of the following statements is not correct for σ– and π– bonds formed between two carbon atoms?

(a) Free rotation of atoms about σ-bond is allowed but not in case of a π–bond
(b) σ–bond determines the direction between carbon atoms but a π-bond has no primary effect in this regard
(c) σ-bond is stronger than a π-bond
(d) Bond energies of σ-bond and π-bond are of the order of 264 kJ/mol and 347 kJ/mol, respectively.

[2003]

15. BrF_3 molecule, the lone pairs occupy equatorial position to minimize:

(a) Lone pair-bond pair repulsion only
(b) Bond pair-bond pair repulsion only
(c) Lone pair-lone pair repulsion and lone pair-bond pair repulsion
(d) Lone pair-lone pair repulsion only

[2004]

16. Among the following the pair in which the two species are not isostructural is:

(a) SiF_4 & SF_4 (b) IO_3^- & XeO_3
(c) BH_4^- & NH_4^+ (d) PF_6^- & SF_6

[2004]

17. In an octahedral structure, the pair of d orbitals involved in d^2sp^3 hybridisation is:

(a) $dx^2 - y^2$, dz^2
(b) dxz , $dx^2 - y^2$
(c) dz^2 , dxz
(d) dxy , dyz

[2004]

18. H_2O is dipolar, whereas BeF_2 is not. It is because:

(a) The electronegativity of F is greater than that of O
(b) H_2O involves hydrogen bonding whereas BeF_2 is a discrete molecule
(c) H_2O is linear and BeF_2 is angular
(d) H_2O is angular and BeF_2 is linear

[2004]

19. In a regular octahedral molecule, MX_6 the number of X – M – X bonds at 180° is:

(a) Three (b) Two
(c) Six (d) Four

[2004]

20. The correct order in which the O – O bond length increases in the following is

(a) $O_2 < H_2O_2 < O_3$
(b) $O_3 < H_2O_2 < O_2$
(c) $H_2O_2 < O_2 < O_3$
(d) $O_2 < O_3 < H_2O_2$

[2005]

21. Which of the following molecules has trigonal planar geometry?

(a) BF_3 (b) NH_3
(c) PCl_3 (d) IF_3

[2005]

22. Which of the following would have a permanent dipole moment?

(a) SiF_4 (b) SF_4
(c) XeF_4 (d) BF_3

[2005]

23. Which of the following is the electron deficient molecule?

(a) C_2H_6 (b) B_2H_6
(c) SiH_4 (d) PH_3

[2005]

24. The correct sequence of increasing covalent character is represented by:

(a) $LiCl < NaCl < BeCl_2$
(b) $BeCl_2 < LiCl < NaCl$
(c) $NaCl < LiCl < BeCl_2$
(d) $BeCl_2 < NaCl < LiCl$

[2005]

25. Which of the following species has a linear shape?

(a) NO_2^+ (b) O_3
(c) NO_2^- (d) SO_2

[2006]

26. In which of the following molecules are all the bonds not equal?

(a) AlF_3 (b) NF_3
(c) ClF_3 (d) BF_3

[2006]

27. Which of the following is not isostructural with $SiCl_4$?

(a) PO_4^{3-} (b) NH_4^+
(c) SCl_4 (d) SO_4^{2-}

[2006]

28. The electronegativity difference between N and F is greater than that between N and H yet the dipole moment of NH_3 (1.5 D) is larger than that of NF_3 (0.2 D). This is because

(a) In NH_3 as well as NF_3 the atomic dipole and bond dipole are in opposite directions
(b) In NH_3 the atomic dipole and bond dipole are in the opposite directions whereas in NF_3 these are in the same direction
(c) In NH_3 as well as in NF_3 the atomic dipole and bond dipole are in same direction
(d) In NH_3 the atomic dipole and bond dipole and in the same direction whereas in NF_3 these are in opposite directions

[2006]

29. Which of the following is not a correct statement?

(a) Every AB_5 molecule does infact have square pyramid structure
(b) Multiple bonds arc always shorter than corresponding single bonds
(c) The electron deficient molecules can act as lewis acids
(d) The canonical structures have no real existence

[2006]

30. The number of unpaired electrons in a paramagnetic diatomic molecule of an element with atomic number 16 is:

(a) 4 (b) 1
(c) 2 (d) 3

[2006]

31. In which of the following pairs, the two species are iso structural?

(a) SF_4 and XeF_4 (b) SO_3^{2-} and NO_3^-
(c) BF_3 and NF_3 (d) BrO_3 and XeO_3

[2007]

32. The sequence of ionic mobility in aqueous solution is:

(a) $Rb^+ > K^+ > Cs^+ > Na^+$
(b) $Na^+ > K^+ > Rb^+ > Cs^+$
(c) $K^+ > Na^+ > Rb^+ > Cs^+$
(d) $Cs^+ > Rb^+ > K^+ > Na^+$

[2008]

33. Four diatomic species are listed below in different sequences. Which of these presents the correct order of their increasing bond order?

(a) $C_2^{2-} < He_2^+ < NO < O_2^-$
(b) $He_2^+ < O_2^- < NO < C_2^{2-}$
(c) $O_2^- < NO < C_2^{2-} < He_2^+$
(d) $NO < C_2^{2-} < O_2^- < He_2^+$

[2008]

34. The angular shape of ozone molecule (O_3) consist of:

(a) 1 sigma and 1 pi bonds
(b) 2 sigma and 1 pi bonds
(c) 1 sigma and 2 pi bonds
(d) 2 sigma and 2 pi bonds

[2008]

35. The correct order of increasing bond angles in the following triatomic species is:

(a) $NO_2^+ < NO_2 < NO_2^-$
(b) $NO_2^+ < NO_2^- < NO_2$
(c) $NO_2^- < NO_2^+ < NO_2$
(d) $NO_2^- < NO_2 < NO_2^+$

36. According to MO theory, which of the following ranks the nitrogen species in terms of increasing bond order?

(a) $N_2^{2-} < N_2^- < N_2$
(b) $N_2 < N_2^{2-} < N_2$

(c) $N_2^- < N_2^{2-} < N_2$
(d) $N_2^- < N_2 < N_2^2$

[2009]

37. In which of the following molecules/ions BF_3, NO_2^-, NH_2^- and H_2O, the central atoms is sp^{2-} hybridised?

(a) NH_2^-, H_2O (b) NO_2^-, H_2O
(c) BF_3, NO_2^- (d) NO_2^-, NH_2^-

[2009]

38. In which one of the following species the central atom has the type of hybridisation which is not the same as present in other three?

(a) PCl_5 (b) SF_4
(c) I_3 (d) $SbCl^{2-}{}_5$

[2010]

39. Which has minimum bond length?

(a) O_2^+ (b) O_2
(c) O_2^{2-} (d) O_2^-

[2011]

40. Which of the two ions from the list given below that have the geometry that is explained by the same hybridization of orbitals NO_2^-, NO_3^-, NH_2^-, NH_4^+, SCN^-?

(a) NO_2^- and NO_3^- (b) NO_2^- and NH_2^-
(c) SCN^- and NH_2^- (d) NH_4^+ and NO_3^-

[2011]

41. The pairs of species of oxygen and their magnetic behaviours are noted below. Which of the following represents the correct description.

(a) O_2^+, O_2 $\rightarrow$ Both Paramagnetic
(b) O_2, O_2^{2-} $\rightarrow$ Both Paramagnetic
(c) O_2^-, O_2^{2-} $\rightarrow$ Both diamagnetic
(d) O_2^+, O_2^{2-} $\rightarrow$ Both Paramagnetic

[2011]

42. The pair of species with the same bond order is?

(a) O_2^{2-}, B_2 (b) N_2, O_2
(c) NO^+, CO (d) O_2^+, NO^+

[2013]

43. Bond order 1.5 is shown by?

(a) O_2^- (b) O_2^+
(c) O_2^{2-} (d) O_2

[2013]

44. Which of the following is paramagnetic?

(a) CO (b) $\bar{C}N$
(c) O_2^- (d) NO^+

[2013]

45. Which of the following is a polar molecule?

(a) BF_3 (b) SF_4
(c) SiF_4 (d) XeF_4

[2013]

46. Which one of the following molecules contain no π-bond.

(a) CO_2 (b) SO_2
(c) NO_2 (d) H_2O

[2013]

47. Which one of the following species contains three bond pairs and one lose pair around the central atem?

(a) PCl_3 (b) BF_3
(c) H_2O (d) NH_2

[2013]

48. Which one of the following Species has plane triangular shape?

(a) N_3^- (b) NO_2^-
(c) CO_2 (d) NO_3^-

[2014]

49. Which of the following molecule has the maximum dipole moment?

(a) CH_4 (b) NH_3
(c) NF_3 (d) CO_2

[2014]

50. Be_2^{2+} is isoelectronic with Which of the following ions?

(a) H^+ (b) Li^+
(c) Na^+ (d) Mg^{2+}

[2014]

51. Which of the following species contains equal number of σ and π-bonds?

(a) HCO_3^- (b) XeO_4
(c) $CH_2(CN)_2$ (d) $(CN)_2$

[2015]

52. Maximum bond angle at nitrogen atam is present in which of the following?

(a) NO_2 (b) NO_2^-
(c) NO_2^+ (d) NO_3^-

[2015]

53. Which of the following options represents the correct bond order?

(a) $O_2^- < O_2 < O_2^+$ (b) $O_2^- > O_2 > O_2^+$
(c) $O_2^- < O_2 > O_2^+$ (d) $O_2^- > O_2 < O_2^+$

[2015]

54. In Which of the following pairs, both the species are not iso structural?

(a) NH_3, PH_3 (b) XeF_4, XeO_4
(c) $SiCl_4$, PCl_4^+ (d) Diamond, carbide

[2015]

55. Decreasing order of stability O_2, O_2^-, O_2^+ and O_2^{2-} is?

(a) $O_2^- > O_2^{2-} > O_2^+ > O_2$
(b) $O_2 > O_2^+ > O_2^{2-} > O_2^-$
(c) $O_2^{2-} > O_2^- > O_2 > O_2^+$
(d) $O_2^+ > O_2 > O_2^- > O_2^{2-}$

[2015]

56. Which one of the following compounds shows the presence of intramolecular hydrogen bond?

(a) H_2O_2 (b) HCN
(c) Cellulose (d) Concentrated acetic acid

[2016]

57. The hybridisation of atamic orbitals of nitrogen in NO_2^+, NO_3^- and NH_4^+ respectively are?

(a) Sp, Sp2 and Sp3 (b) Sp, Sp3 and Sp2
(c) Sp2, Sp3 and Sp (d) Sp2, Sp and Sp3

[2016]

58. Which of the following pairs of ions is isoelectronic and iso structural?

(a) SO_3^{2-}, NO_3^- (b) ClO_3^-, SO_3^{2-}
(c) CO_3^{2-}, NO_3^- (d) ClO_3^-, CO_3^{2-}

[2016]

59. Match the compound given in coloumn 1 with The hybridisation and shape gives coloumn 2 and mark the correct option

Coloumn 1	Coloumn 2
(a) XeF_6	(1) Distorted octanedtral
	(2) Square planar
(b) XeO_3	(3) Pyramidal
(c) $XeOF_4$	(4) Square pyramidal
(d) XeF_4	

Codes-	A	B	C	D
(A)	1	2	4	3
(B)	1	3	4	2
(C)	4	1	2	3
(D)	4	3	1	2

60. Consider the molecules CH_4, NH_3, and H_2O which of the given statement is false?

(a) The H-O-H bondangle in H_2O is larger than the H-C-H bond angle in CH_4
(b) The H-O-H bondangle in H_2O is smaller than the H-N-H bond angle in NH_3
(C) The H-C-H bond angle in CH_4, the H-N-H bond angle in NH_3 and the H-O-H bondangle in water are all greater than 90°
(d) The H-C-H bond angle in CH_4 is larger than the H-N-H bond angle in NH_3

[2016]

Answer Keys

1. (c)	2. (a)	3. (a)	4. (c)	5. (d)	6. (c)	7. (c)	8. (d)	9. (b)	10. (a)
11. (b)	12. (d)	13. (b)	14. (b)	15. (a)	16. (a)	17. (a)	18. (d)	19. (a)	20. (d)
21. (a)	22. (b)	23. (b)	24. (c)	25. (a)	26. (c)	27. (c)	28. (d)	29. (a)	30. (c)
31. (d)	32. (d)	33. (b)	34. (b)	35. (d)	36. (a)	37. (c)	38. (d)	39. (a)	40. (a)
41. (a)	42. (c)	43. (a)	44. (c)	45. (b)	46. (d)	47. (a)	48. (d)	49. (b)	50. (b)
51. (b)	52. (c)	53. (c)	54. (b)	55. (d)	56. (c)	57. (a)	58. (b,c)	59. (b)	60. (a)

Hints and Explanations

1. Here CO is not paramagnetic as does not have any unpaired electron.

2. It is ionic bond and the compound will be ionic in nature.

3. Boron in BCl_3 has 6 electron in its outer most orbital. Hence BCl_3 is an electron pair deficient compound.

4. As dissociation energy is directly proportional to bond order and here N_2 has bond order three while N_2^+ has 2.5 so bond dissociation energy of $N_2 > N_2^+$.

5. HF forms linear polymer structure due to hydrogen bonding.

6. In P – O bond, π bond is formed by the sidewise overlappiing of d-orbital of P and p-orbital of oxygen. Hence it is formed by pπ - dπ overlapping.

7. Nitrogen forms N_2 (i.e., N ≡ N) but phosphorous forms P_2, it is at a time convert in P_4, because in P_2, pπ - pπ bonding is present which is a weaker bonding.

8. Here no molecular orbital can be formed as effective overlapping is not possible.

9. In X - H --- Y, X and Y both are electrogative elements then electron density on X will increase and on H will decrease.

10. As both are linear.

11. As bond angles of NH_3 = 107°, NH_4^+ = 109.5°, PCl_3 = 100°. Therefore bond angle of NH_4^+ is maximum.

12. In NO_3^-, nitrogen atom has only four bond pair of electrons and no lone pair electron.

$$O{=}N\begin{cases}\rightarrow O\\ O^-\end{cases}$$

13. Here SO_3^{2-} can have pπ-dπ overlapping (bonding) as sulphur atom can have d-orbitals which is invovled in pi-bond formation with the vacant p-orbital of oxygen atom.

14. As sigma bond is stronger than the pi bond, so it must be having higher bond energy than pi bond.

15. As lp-lp type of repulsion is more than lp-bp of type repulsion. So lp electron pair occupies equatorial position to minimize electronic repulsions.

16. As SiF_4 is tetrahedral in shape while SF_4 has See-saw shape sp they are not identical in shape.

17. In sp^3d^2 hybridisation the d-orbitals used are $dx^2 - y^2$, dz^2 since they can form effectively overlapping with three p-orbitals.

18. In a linear symmetrical molecule like BeF_2, the bond angle between three atoms is 180°, hence the polarity due to one bond is cancelled by the equal polarity due to other bond, while it is not so in angular molecules. like H_2O.

19. As in MX_6 the shape is octahedral in which three X – M – X can have 180° angle.

20. Bond length $\alpha \frac{1}{\text{Bond order}}$ order
B.O. of $O_2 = 2$
B.O. of $O_3 = 1.5$
B.O. of $H_2O_2 = 1$
So the bond length increase as follows.
$O_2 < O_3 < H_2O_2$

21. BF_3 is sp^2 hybridised so it is trigonal planner in shape.

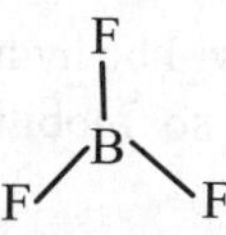

22. SF_4 has a permanent dipole moment so it is polar because it has distorted see-saw shape.

$$\ddot{-}S\begin{cases}F\\F\\F\\F\end{cases}$$

23. The compound of which central atom is not having octet state is known as electron deficient compound. Here B_2H_6 is an electron deficient compound.

24. As difference of electronegativity increases % ionic character increases and covalent character decreases i.e., negativity difference decreases covalent character increases.

Na is more +ve than Li and
Li is more +ve than Be.

25. $O = \overset{+}{N} = O$
Here N has no lone pair electron.

26. Chlorine atom in ClF_3 is sp3d hybridized so in it bonds are not equal as it has distorted T-shape.

27. Here SCl_4 is See-saw in shape while rest are tetrahedral.

28.

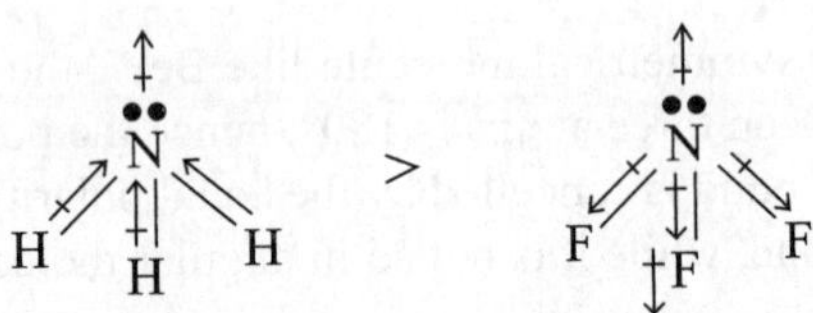

Ammonia has more dipole moment than NF_3 as in ammonia μ(net) is in the direction of lone pair electrons i.e., it is additive while in NH_3 μ (net) is opposite to lone pair i.e., substrative.

29. Since square pyramidal shape in AB_5 molecule is not possible as AB_5 is sp^3d-hybridised so the shapes can be trigonal bipyramidal, see-saw, T-shape and linear depending on number of lone pair and bond pair electrons.

30. S_2 molecule is paramagnetic like O_2 as both have two unpaired electrons.

31. As both are trigonal pyramidal in shape.

32. As smaller the size of cation, higher will be hydration and its effective size will increase so mobility in aqueous solution will decrease.

33. $He_2^+ < O_2^- < NO < C_2^{2-}$

Bond order 0.5 1.5 2.5 3.0

These bond orders are on the basis of their molecular orbital configurations using molecular orbital theory.

$He_2^+ : \sigma(ls)^2\ \sigma^*\ (ls)^1$

$$\text{Bond order} = \frac{2-1}{2} = 0.5$$

$O_2^- : KK\ \sigma\ (2s)^2\ \sigma^*(2s)^2\ \sigma(2pz)^2\ \pi\ (2px)^2\ \pi\ (2py)^2\ \pi^*\ (2px)^2\ \pi^*(2py)^1$

$$\text{Bond order} = \frac{10-7}{2} = 1.5$$

NO: $KK\ \sigma(2s)^2\ \sigma^*(2s)^2\ \sigma\ (2pz)^2\ \pi\ (2px)^2\ \pi\ (2py)^2\ \pi^*(2px)^1$

$$\text{Bond order} = \frac{10-5}{2} = 2.5$$

$C_2^{2-} : \sigma(ls)^2\ \sigma^*(ls)^2\ \sigma(2s)^2\ \sigma^*(2s)^2, \sigma(2pz)^2\ \pi(2px)^2 \approx \pi\ (2py)^2$

$$\text{Bond order} = \frac{10-4}{2} = 3$$

34. The angular shape of ozone molecule consists of 2 sigma and 1 pi bond.

$$\overset{..}{O} \quad O \quad O$$

35. $NO_2 > NO_2^+ > NO_2^-$
132° 130° 115° (Bond angles)
It is decided on the basis of presence of lone pair of electrons on nitrogen atom.
As bond angle ∝ 1/lp electrons

36. $N_2^{2-} < N_2^- < N_2$
B.O 2 < 2.5 < 3

37. B-atom is BF_3 and N-atom in NO_2^- is sp^2-hybridzed.

38. In $SbCl_5^{2-}$ hybridisation is sp^3d^2 while in rest three it is sp^3d.

39. As Bond length × $\frac{1}{\text{Bond order}}$
Here O_2^+ has 2.5 Bond order value so it has least bond length.

40. Both NO_2^- and NO_3^- are sp^2 hybridized with trigonal planar geometry.

41. Both O_2^+ and O_2 are paramagnetic due to 1, 2 un-powered e^- respectively.

42. For NO^+ and $CO(14)e^- \rightarrow$

$$\text{B.O.} = \frac{10-4}{2} = 3$$

43. $O_2^-\ (17)\ e^- \rightarrow n_B = 10$
$n_A = 7$

$$\text{B.O.} = \frac{10-7}{2} = 1.5$$

44. $O_2^-\ (17e^-) \rightarrow \sigma 1s^2 \overset{*}{\sigma} 1s^2\ \sigma 2s^2 \overset{*}{\sigma}\ 2s^2$
$\sigma 2p_z^2\ \pi 2p_x^2 = \pi 2p_y^2\ {}^*\pi 2p_x^2 = \pi^* 2p_y^1$
n = 1 Paramagnetic

45. 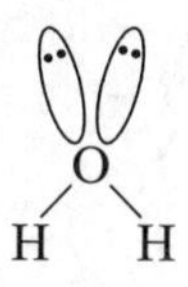
SF_4 is a polar molecule as for it μ_{Net} is positive.

46. 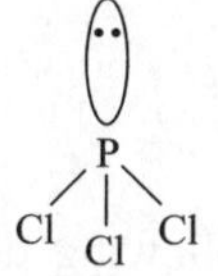
It has only 2 σ bonds and no-π-bond. Rest have π-bonds.

47. It has 3σp and 1 *l*p

48. Here NO_3^- has Trigonal planer shape while N_3^- CO_2, are NO_2^- are linear

$N=N^+=N^-$ $O=C=O$

$O—N=O$

49. Here μ (Dipole moment) decreases as follows.

	NH_3 >	NF_3 >	$CH_4 = CO_2$	
μ →	1.47D	0.9 D	0	0

50. Be_2^{2+} ($2e^-$) is iso electronic with Li^+.

$H^+ \rightarrow$ No e–

$H^+ \rightarrow$ No e–

Na^+, $Mg^{2+} \rightarrow$ 10 e–

51. 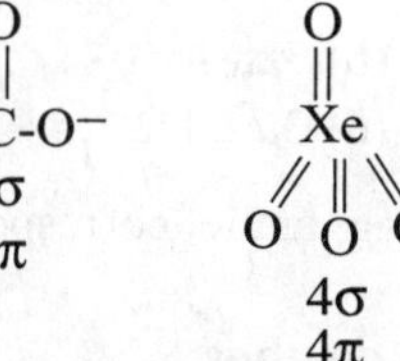

H-O-C-O⁻: 1–σ, 1–π

XeO_4: 4σ, 4π

C≡N–C≡N: 3σ, 4π

$CH_2(CN)_2$: 6σ, 4π

Hence XeO_4 has same number of σ and π-bonds.

52.

NO_2^+	NO_2	NO_2^-	NO_3^-
sp	sp	sp^2	sp^2
180°	<120°	115.4°	120°

Hence NO_2^+ has maximum bond angle

53. Using M.O.T the bond order is as follows:

$O_2^- < O_2 < O_2^+$

B.O → 1.5   2   2.5

54. NH_3, $PH_3 \rightarrow$ Pyramidal

$SiCl_4$, $PCl_4^+ \rightarrow$ tetrahedral

Diamond, carbide → tetrahedral

$XeF_4 \rightarrow$ square planar

$XeO_4 \rightarrow$ tetrahedral

Hence XeF_4, XeO_4 both differ in shape

55. Using M.O.T

Stability and Bond order

$O_2^+ > O_2 > O_2^- > O_2^{2-}$

B.O. 2.5   2   1.5   1

As B.O. ↓ so stability ↓

56. CH_3COOH, HCN, H_2O, can have intermolecular H-bonding, cellulose has Intra molecular Hydrogen bonding.

57. $O=\overset{+}{N}=O$ (sp) NO_3^- (sp^2) $[NH_4]^+$ (sp^3)

58. SO_3^{2-}, ClO_3^- have $42e^-$ and Pyramidal structure. CO_3^{2-}, NO_3^- have $32e^-$ and Triangular planar structure.

59. 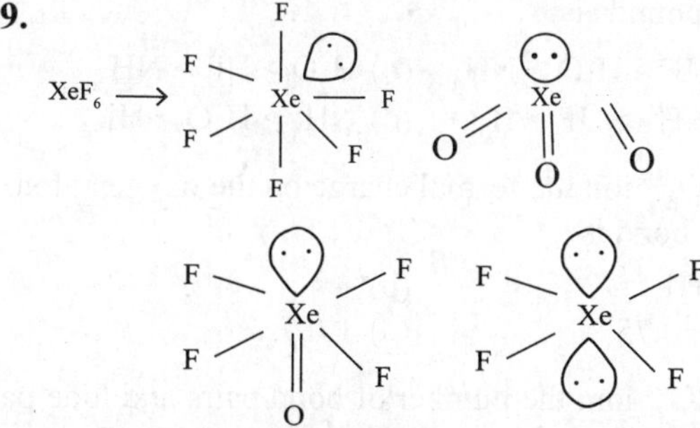

Hence option B is Correct.

60. The Bond angle in these molecules are:

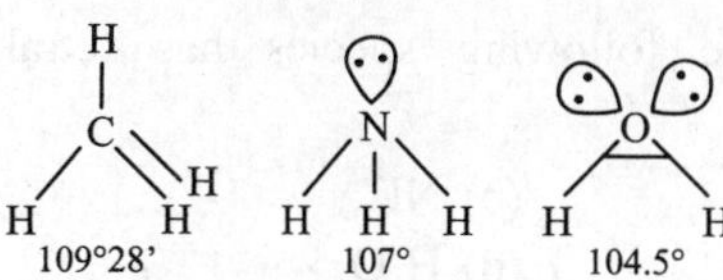

Hence option a is false.

NCERT EXEMPLAR

1. Isostructural species are those which have the same shape and hybridisation. Among the given species identify the isostructural pairs.
   (a) [NF_3 and BF_3] (b) [BF_4^- and NH_4^+]
   (c) [BCl_3 and $BrCl_3$] (d) [NH_3 and NO_3^-]

2. Polarity in a molecule and hence the dipole moment depends primarily on electronegativity of the constituent atoms and shape of a molecule. Which of the following has the highest dipole moment?
   (a) CO_2 (b) HI
   (c) H_2O (d) SO_2

3. The types of hybrid orbitals of nitrogen in NO^+_2, NO_3^- and NH_4^+ respectively are expected to be
   (a) *sp*, *sp*^{3} and *sp*^{2} (b) *sp*, *sp*^{2} and *sp*^{3}
   (c) *sp*^{2}, *sp* and *sp*^{3} (d) *sp*^{2}, *sp*^{3} and *sp*

4. Hydrogen bonds are formed in many compounds e.g., H_2O, HF, NH_3 . The boiling point of such compounds depends to a large extent on the strength of hydrogen bond and the number of hydrogen bonds. The correct decreasing order of the boiling points of above compounds is:
   (a) HF > H_2O > NH_3 (b) H_2O > HF > NH_3
   (c) NH_3 > HF > H_2O (d) NH_3 > H_2O > HF

5. In PO_4^{3-} ion the formal charge on the oxygen atom of P–O bond is
   (a) + 1 (b) – 1
   (c) – 0.75 (d) + 0.75

6. In NO_3^- ion, the number of bond pairs and lone pairs of electrons on nitrogen atom are:
   (a) 2, 2 (b) 3, 1
   (c) 1, 3 (d) 4, 0

7. Which of the following species has tetrahedral geometry?
   (a) BH_4^- (b) NH_2^-
   (c) CO^2_{3-} (d) H_3O_+

8. Number of π bonds and σ bonds in the following structure is–

   (naphthalene structure with H atoms)

   (a) 6, 19 (b) 4, 20
   (c) 5, 19 (d) 5, 20

9. Which molecule/ion out of the following does **not** contain unpaired electrons?
   (a) N_2^+ (b) O_2
   (c) O_2^{2-} (d) B_2

10. In which of the following molecule/ion all the bonds are **not** equal?
    (a) XeF_4 (b) BF_4^-
    (c) C_2H_4 (d) SiF_4

11. In which of the following substances will hydrogen bond be strongest?
    (a) HCl (b) H_2O
    (c) HI (d) H_2S

12. If the electronic configuration of an element is $1s^2$ $2s^2$ $2p^6$ $3s^2$ $3p^6$ $3d^2$ $4s^2$, the four electrons involved in chemical bond formation will be_____.
    (a) $3p^6$ (b) $3p^6$, $4s^2$
    (c) $3p^6$, $3d^2$ (d) $3d^2$, 4s2

13. Which of the following angle corresponds to *sp*^{2} hybridisation?
    (a) 90° (b) 120°
    (c) 180° (d) 109°

The electronic configurations of three elements, A, B and C are given below.

Answer the questions 14 to 17 on the basis of these configurations.

A	$1s^2$	$2s^2$	$2p^6$		
B	$1s^2$	$2s^2$	$2p^6$	$3s^2$	$3p^3$
C	$1s^2$	$2s^2$	$2p^6$	$3s^2$	$3p^5$

14. Stable form of **A** may be represented by the formula :
    (a) A (b) A_2

(c) A_3 (d) A_4

15. Stable form of **C** may be represented by the formula:

(a) C (b) C_2
(c) C_3 (d) C_4

16. The molecular formula of the compound formed from **B** and **C** will be:

(a) BC (b) B_2C
(c) BC_2 (d) BC_3

17. The bond between B and C will be:

(a) Ionic (b) Covalent
(c) Hydrogen (d) Coordinate

18. Which of the following order of energies of molecular orbitals of N_2 is correct?

(a) $(\pi 2p_y) < (\sigma 2p_z) < (\pi^* 2p_x) \approx (\pi^* 2p_y)$
(b) $(\pi 2p_y) > (\sigma 2p_z) > (\pi^* 2p_x) \approx (\pi^* 2p_y)$
(c) $(\pi 2p_y) < (\sigma 2p_z) > (\pi^* 2p_x) \approx (\pi^* 2p_y)$
(d) $(\pi 2p_y) > (\sigma 2p_z) < (\pi^* 2p_x) \approx (\pi^* 2p_y)$

19. Which of the following statement is **not** correct from the view point of molecular orbital theory?

(a) Be_2 is not a stable molecule.
(b) He_2 is not stable but He^+_2 is expected to exist.
(c) Bond strength of N_2 is maximum amongst the homonuclear diatomic molecules belonging to the second period.
(d) The order of energies of molecular orbitals in N_2 molecule is $\sigma 2s < \sigma^* 2s < \sigma 2p_z < (\pi 2p_x = \pi 2p_y) < (\pi^* 2p_x = \pi^* 2p_y) < \sigma^* 2p_z$

20. Which of the following options represents the correct bond order:

(a) $O_2^- > O_2 > O_2^+$ (b) $O_2^- < O_2 < O_2^+$
(c) $O_2^- > O_2 < O_2^+$ (d) $O_2^- < O_2 > O_2^+$

21. The electronic configuration of the outer most shell of the most electronegative element is:

(a) $2s^2 2p^5$ (b) $3s^2 3p^5$
(c) $4s^2 4p^5$ (d) $5s^2 5p^5$

22. Amongst the following elements whose electronic configurations are given below, the one having the highest ionisation enthalpy is

(a) $[Ne]3s^2 3p^1$ (b) $[Ne]3s^2 3p^3$
(c) $[Ne]3s^2 3p^2$ (d) $[Ar]3d^{10} 4s^2 4p^3$

Answer Keys

1. (b)	2. (c)	3. (b)	4. (b)	5. (b)	6. (d)	7. (a)	8. (c)	9. (c)	10. (c)
11. (b)	12. (d)	13. (b)	14. (a)	15. (b)	16. (d)	17. (b)	18. (a)	19. (d)	20. (b)
21. (a)	22. (b)								

Hints and Explanations for Selective Questions

1. NF_3 is pyramidal and BF_3 is planar triangular

BF_4^- and NH_4^+ both are tetrahedral

BCl_3 is triangular planar and $BrCl_3$ is pyramidal

NH_3 is pyramidal and NO_3^- is triangular planar

2. Due to maximum difference in electronegativity of H and O atoms, H_2O will have highest dipole moment.

3. Us formula H= ½[V+M-C+a]

10. C_2H_4 has one double bond and four single bonds. Bond length of double bond (C = C) is smaller than single bond (C – H)

10. C_2H_4 has one double bond and four single bonds. Bond length of double bond (C = C) is smaller than single bond (C – H)

12. in transition elements (n-1) d and ns orbitals take part in bond formation.

14. octet of A is complete which shows A is a noble gas i.e neon with atomic number 10.

15. stable form is (Cl_2)

21. electronic configuration representation:

(a) $2s^2 2p^5$ = fluorine = most electronegative element
(b) $3s^2 3p^5$ = chlorine
(c) $4s^2 4p^5$ Bromine
(d) $5s^2 5p^5$ Iodine

22. (b) and (d) have exactly half filled p-orbitals but due to smaller in size it has highest ionization enthalpy.

AIIMS ESSENTIALS

ASSERTION AND REASON

In the following questions, two statements (Assertion) A and Reason (R) are given. Mark

(a) If A and R both are correct and R is the correct explanation of A;

(b) If A and R both are correct but R is not the correct explanation of A;

(c) A is true but R is false;

(d) A is false but R is true,

(e) A and R both are false.

1. (A) The compound $(CF_3)_3$ N shows almost no basic behaviour even though $(CH_3)_3$ N does.

(R) There is not hydrogen bonding in $(CF_3)_3N$.

2. (A) Water is a good solvent for ionic compounds but poor one for covalent compounds.

(R) Hydration energy of ions released is sufficient to overcome lattice energy and to break hydrogen bonds in water while covalent bonded compounds interact so weakly that even Van der Waal's force between molecules of covalent compounds cannot be broken.

3. (A) Dipole moment of cis-pent-2- ene is higher than that of trans-pent-2-ene.

(R) trans-pent-2-ene is more stable than cis-pent-2-ene.

4. (A) All F – S – F angles in SF_4 are greater than 90° but less than 180°.

(R) The lone pair-bond pair repulsion is weaker than bond pair-bond pair repulsion.

5. (A) In solid state hybridisation of Be atom in $BeCl_2$ is found to be sp^3.

(R) As two chlorine atoms formed two coordinate bonds with Be atom which already have two covalent bonds also.

6. (A) Covalent bond in hydrogen molecule is non-polar.

(R) Most of the covalent bonds are polar.

7. (A) Inter-electronic repulsion between bond pair-bond pair, bond pair-lone pair and lone pair-lone pair in a molecule follows the order: bond pair-bond pair < bond-pair-lone pair < lone pair-lone pair.

(R) Bond pair electrons are found in between two nuclei whereas lone pair electrons are attached with only one nucleus and occupy more space.

8. (A) The electronic structure of O_3 is:

$$O{=}\overset{\oplus}{\ddot{O}}{-}O^{\ominus}$$

(R) $O{=}\ddot{O}{=}O$ structure is not allowed because octet around O cannot be expanded.

9. (A) LiCl is predominantly a covalent compound.

(R) Electronegativity difference between Li and Cl is too small.

10. (A) Ionic compounds tend to be non-volatile.

(R) Intermolecular forces in these compounds are weak.

11. (A) Bond order in a molecule can assume any value, positive or negative, integral or fractional including zero.

(R) It depends upon the number of electrons in the bonding and antibonding orbitals.

12. (A) The atoms in a covalent molecule are said to share electrons, yet some covalent molecules are polar.

(R) Bond pair electrons are found in between two nuclei whereas lone pair electrons are attached with only one nucleus and occupy more space.

13. (A) Nitrogen is unreactive at room temperature but becomes reactive at elevated temperatures (on heating or in the presence of catalysis).

(R) In nitrogen molecule, there is extensive delocalization of electrons.

14. (A) The solubility of MgF_2 is more than that of $MgCl_2$.
(R) As $MgCl_2$ has less lattice energy than MgF_2.

15. (A) The bond angle of PBr_3 is greater than PH_3 but the bond angle of NBr_3 is less than NH_3.
(R) Electronegativity of phosphorous atom is less than that of nitrogen.

16. (A) Double bonds in C_2 consists of both pi bonds.
(R) The four electrons are present in the 2π molecular orbitals.

17. (A) N_2 and NO^+ both are diamagnetic substances.
(R) NO^+ is isoelectronic with N_2.

18. (A) AlF_3 is a high melting solid while SiF_4 is a gas.
(R) AlF_3 is an ionic compound while SiF_4 is a polar covalent compound.

19. (A) NO_3^- is planar while NH_3 is pyramidal.
(R) N in NO_3^- is sp^2 and in NH_3 it is sp^3 hybridized.

20. (A) Na_2SO_4 is soluble in water while $BaSO_4$ is insoluble.
(R) Lattice energy of $BaSO_4$ exceeds its hydration energy.

21. (A) The energy splitting between $\sigma2p$ and $\pi2p$ orbitals is quite large.
(R) The overlap of p-orbitals is more when they are oriented along the axis to form sigma orbitals than when they are oriented two overlap side base to form π-orbitals.

22. (A) Bond order can assume any value including zero.
(R) Higher the bond order, shorter is the bond length and greater is the bond energy.

23. (A) The dipole moment helps to predict whether a molecule is polar or non-polar.
(R) The dipole moment helps of predict the geometry of molecules.

24. (A) p-Hydroxybenzoic acid has lower boiling point than o-hydroxybenzoic acid.
(R) o-Hydroxybenzoic acid has intramolecular hydrogen bonding.

[2007]

25. (A) Boron always forms covalent bond.
(R) The small size of B^{3+} favours formation of covalent bond.

[2007]

Answer Keys

1. (b)	2. (a)	3. (b)	4. (c)	5. (a)	6. (b)	7. (b)	8. (a)	9. (c)	10. (c)
11. (a)	12. (a)	13. (c)	14. (d)	15. (b)	16. (a)	17. (b)	18. (c)	19. (a)	20. (a)
21. (a)	22. (b)	23. (a)	24. (d)	25. (a)					

CHAPTER 5

States of Matter

Chapter Outline

■ General Properties ■ Gas Laws ■ Kinetic Theory of Gases ■ Ideal and Real Gases ■ Van Der Waals Equation ■ Critical Phenomenon and Liquefaction of Gases ■ General Properties.

Matter is a substance having mass and volume. It has three states—Solid, Liquid and Gas. Plasma is supposed to be it's fourth state. $S \rightleftharpoons L \rightleftharpoons G$.

Thermal Energy increases
Molecular Interaction Energy increases

General Properties

- Only non-metals form gases. (F_2, Cl_2 etc)
- Volume of a gas is equal to the volume of the container in which it is taken as it tries to occupy all the space. It is expressed in litres or m^3 or mL or cm^3.
- Gases show high diffusibility, thermal expansion and high compressibility due to lare empty spaces between molecules.
- Gases always form a homogeneous mixture.
- Gaseous pressure is measured by a **manometer**, while atmospheric pressure is measured by a **barometer**.
- Gaseous pressure is measured in newton per m^2 or mm Hg or atmosphere or torr.

  1 atm = 760 mm Hg or torr

  $= 1.01 \times 10^5$ N/m^2
- At standard condition, temperature is taken as 298 K and pressure of gas is taken as 1 atm. This is called **Standard** or **Normal Temperature and Pressure State** (STP or NTP).
- The molecules of gases have maximum energy and minimum force of attraction.
- Gases have low densities. The density of a gas with respect to hydrogen is called its **relative density**.
- A gas below critical temperature is a vapour. Vapour is the gaseous form of a solid or liquid at room temperature.
- Vapour state can be liquefied by increasing pressure while gaseous state can be liquefied by increasing pressure and decreasing temperature.

Triple Point

Solid, liquid and gas, all phases exist together at a particular temperature and pressure, it is known as triple point. For water, the triple point occurs at a temperature of 0.0098°C and a pressure of 4.58 mm Hg.

Gas Laws

As gases show their dependency on pressure, volume and temperature so the relationship of these factors can be explained through gas laws as discussed below.

Boyle's Law

According to Boyle's law, "At constant temperature, the volume of a given mass of a gas is inversely proportional to its pressure."

$$V \propto \frac{1}{P}.$$

PV = Constant (At constant temperature)

$P_1V_1 = P_2V_2$ (At constant temperature)

Initial Conditions   Final Conditions

The differential form of Boyle's law can be written as

$$(dP/dV)_T = -\frac{K}{V^2}.$$

Plots between pressure and volume

These plots are called **isotherms.**

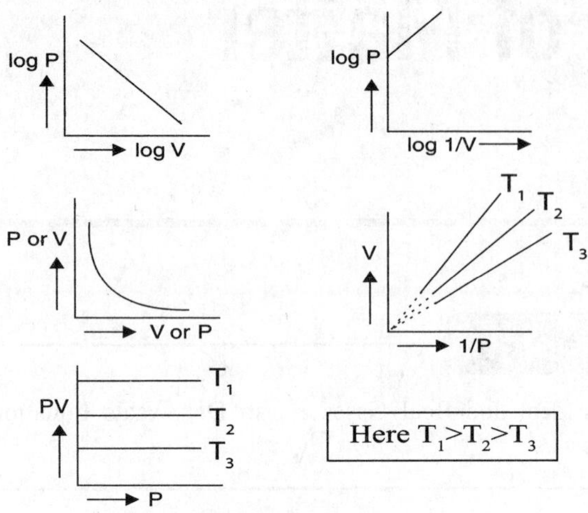

Figure 5.1 Isotherms

The size of weather balloons becomes larger as it rises to higher altitude as the external pressure decreases.

- Air at the sea level is dense as it is compressed by the mass of air present above it but the density and pressure decrease with the increase in altitude.

Charles Law

According to Charles law, "At constant pressure, the volume of a given mass of a gas increases or decreases by 1/273 of its volume at 0°C for every one degree centigrade rise or fall in temperature."

$$V_t = V_0 + \frac{V_0}{273} \times T \text{ (At constant temperature)}$$

$$= V_0 (1 + T/273) = V_0 (273 + T/273)$$

$$= V_0 \frac{T}{273}$$

Therefore

$V_t \propto T$ or $V \propto T$ (At constant pressure)

At constant pressure, the volume of a given mass of a gas is directly proportional to its temperature in Kelvin.

$$\frac{V_1}{T_1} = \frac{V_2}{T_2} \text{ (At constant pressure)}$$

$V \propto 1/d$ (d = Density)

dT = Constant

Hence $d_1T_1 = d_2T_2$

Charles law is not applicable for liquids.

Plots between volume and temperature

These plots are called **isobars.**

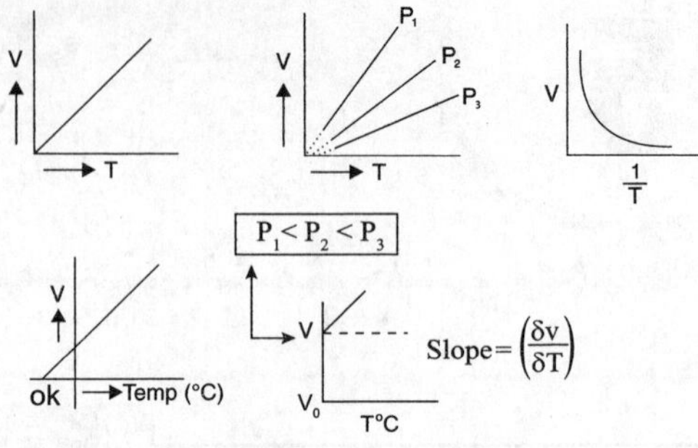

Figure 5.2 Isobars

Absolute scale of temperature and absolute zero

According to Charles law, "If the temperature of a gas is lowered upto –273°C, the volume of the gas becomes zero."

- This temperature is known as absolute zero and this zero point on the scale is known as Kelvin scale. The absolute value of this temperature is –273.15°C.
- At absolute zero, the pressure, kinetic energy and heat content of the gas is also zero.
- It does not indicate that volume of the gas is zero and at this temperature there is no significance of Gas Laws as gas phase is not observed as it is either liquified or solidified.
- When a graph is plotted between the volume of the gas against this temperature at constant pressure it is referred to as an isobar and it is always a straight line.

Pressure–Temperature Law (Gay-Lussac's Law)

According to Gay-Lussac's law, "At constant volume, the pressure of a given mass of a gas increases or decreases by 1/273 of its pressure at 0°C for every 1°C rise or fall in temperature."

$$P_t = P_0 + \frac{P_0}{273} \times T$$

$$= P_0 \frac{(273 + T)}{273} = P_0 \frac{T}{273}$$

$P \propto T$ Hence, $\frac{P}{T}$ = Constant

$$\frac{P_1}{T_1} = \frac{P_2}{T_2} \text{ (At constant volume)}$$

Plots between pressure and temperature

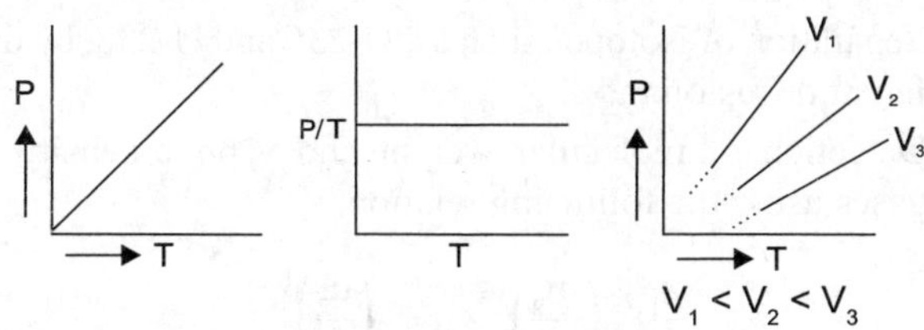

FIGURE 5.3 Isochors

Here curves are called **isochors** and the slope will be greater for lesser volumes.

Avogadro's Law

According to Avogadro's law, "At constant temperature and pressure, equal volume of gases will have same number of molecules or moles."

$$V \propto N \text{ or } n$$

$$\frac{V}{n} = \text{Constant}$$

$$\frac{V_1}{n_1} = \frac{V_2}{n_2}$$

or $\dfrac{V_1}{N_1} = \dfrac{V_2}{N_2}$

Here n = Number of moles
N = Number of molecules

This law is applied only at high temperature and low pressure.

Molar volume 1 mole of a gaseous substance occupies a volume of 22.4 litre at NTP that is molar volume.

Avogadro number 1 mole of a substance will have 6.02×10^{23} number of molecules.

1 g atom of any element will have 6.02×10^{23} atoms.

Combined gas law and ideal gas equation On combining Boyle's, Charles and other gas laws we get the following:

As $V \propto \frac{1}{P}$

$$V \propto T$$

$$V \propto n$$

$$V \propto \frac{nT}{P}$$

or $PV \propto nT$

or $PV = nRT = nST$

$$PV = \frac{w}{M}RT$$

$$P = \frac{w}{mV}RT$$

$$P = CRT$$

$$P = \frac{d}{m}RT$$

- R (or S) is the molar gas constant or **universal gas constant**.

$$R = 0.0821 \text{ L atm K}^{-1} \text{ mol}^{-1}$$
$$= 8.314 \text{ joule K}^{-1} \text{ mol}^{-1}$$
$$= 8.314 \times 10^{7} \text{ erg K}^{-1} \text{ mol}^{-1}$$
$$= 2 \text{ cal K}^{-1} \text{ mol}^{-1}$$

- For a single molecule, gas constant is known as **Boltzmann constant (k)**.

$$k = R/N_0$$
$$= 1.38 \times 10^{-23} \text{ J/deg abs/molecule}$$
$$= 1.38 \times 10^{-16} \text{ erg/deg abs/molecule}$$

As $\dfrac{PV}{T} = \text{Constant}$

$$\frac{P_1V_1}{T_1} = \frac{P_2V_2}{T_2}$$

Dalton's Law of Partial Pressure

According to Dalton's law, "Total pressure of a mixture of non-reacting gases is equal to the sum of partial pressure of these gases at constant temperature and constant volume."

$$P_{mix} = P_1 + P_2 + P_3 \ldots\ldots$$

Here P_{mix} = Pressure of the mixture
P_1, P_2, P_3 = Partial pressure of gases

$$\text{Partial pressure of any gas} = \frac{\%\text{ of gas}}{100} \times P_{mix}$$

- Partial pressure of any component A is given as

$$P_A = \frac{\text{Moles of A}}{\text{Total moles}} \times P$$

- Total pressure of a gaseous mixture having different components is given as

$$P_{mix} = (n_1 + n_2 + n_3 \ldots\ldots)\frac{RT}{V}$$

$$P_{mix} = \left(\frac{w_1}{m_1} + \frac{w_2}{m_2} + \frac{w_3}{m_3}\right)\frac{RT}{V}$$

Here w_1, w_2, w_3 = weight of component or non-reacting gases and m_1, m_2, m_3 are their molar masses.
T = Temperature in Kelvin
V = Volume in litre

- When a gas is collected over water, it mixes with water vapours so the correct pressure of moist gas is given as

$$P_{\text{moist gas}} = P_{\text{dry gas}} + P_{\text{V.P. of water}}$$

$$P_{\text{dry gas}} = P_{\text{moist gas}} - P_{\text{V.P. of water}}$$

Aqueous tension = Partial pressure of water vapour in moist gas.

- Vapour pressure of water varies with temperature. For example, at 0°C it is 4.6 torr while at 25°C it is 23.8 torr.
- Dalton's law is not applicable for a mixture of reacting gases like N_2 and O_2.

Graham's Law of Diffusion

According to Graham's law, "At constant pressure and temperature the rate of diffusion of a gas is inversely proportional to the square root of its density or molecular weight."

$$r \propto \frac{1}{\sqrt{M}} \text{ or } \frac{1}{\sqrt{d}}$$

It is applicable only at low pressure.

- If pressures are different then

$$\frac{r_1}{r_2} = \frac{P_1}{P_2} \cdot \sqrt{\frac{M_2}{M_1}} \text{ or } \sqrt{\frac{d_2}{d_1}}$$

r = Rate of diffusion or effusion of a gas or liquid.
M and d are molecular weight and density respectively.

- For any two gases, the ratio of rate of diffusion at constant pressure and temperature can be shown as

$$\frac{r_1}{r_2} = \sqrt{\frac{M_2}{M_1}} \text{ or } \sqrt{\frac{d_2}{d_1}}$$

- r = V/t (volume/time)

$$\therefore \frac{V_1 t_2}{V_2 t_1} = \sqrt{\frac{M_2}{M_1}}$$

- $r = \frac{n}{t} = \frac{d}{t}$

$$\therefore \frac{n_1 t_2}{n_2 t_1} = \sqrt{\frac{M_2}{M_1}}$$

$$= \frac{d_1 t_2}{d_2 t_1} = \sqrt{\frac{M_2}{M_1}}$$

Here n and d are number of moles and distance traveled for a particular gas respectively.

Diffusion
It is a process in which gaseous molecules spread out to completely occupy a given volume/space.
Effusion
It is a process in which gaseous molecules in a container, escape though an orifice (pinhole) in the same.

Uses of diffusion

- Detecting the presence of marsh gas in mines.
- Separation of isotopes such as, U-235 and U-238, by different diffusion rates.
- Detection of molecular weight and vapour density of gases using the following relation.

$$\frac{r_1}{r_2} = \left(\frac{m_2}{m_1}\right)^{1/2} \text{ or } \left(\frac{d_2}{d_1}\right)^{1/2}$$

Kinetic Theory of Gases

- Kinetic theory of gases was put forward by Bernoulli and was developed by Clausius and Kelvin.
- It was explained by Maxwell and Boltzmann.
- It is also known by other names, such as dynamic particle model and microscopic model.

Main Features

- Every gas is composed of a large number of tiny particles called molecules. These molecules are identical in mass size etc.
- The volume of gaseous molecules is negligible as compared to the volume of the gas.
- Gas molecules show random motion in straight lines in all possible directions but at a constant speed.
- Molecules undergo collisions which are perfectly elastic, that is, there is no change in energy.
- The force of attraction between molecules is negligible thus gas molecules can move freely and independently of each other.
- The effect of gravity on molecular motion is negligible.
- Pressure of the gas is due to collisions of molecules with the walls of the container.
- At any particular instance, molecules have different speeds so they have different kinetic energies. However, the average kinetic energy of these molecules is directly proportional to absolute temperature. Hence, K.E. ∝ absolute temperature.

Kinetic Gas Equation

On the basis of above points, it can be given as

$$PV = \frac{1}{3} mnu_{rms}^2$$

Here P = Pressure of gas
V = Volume of gas
n = Number of molecules of gas
u_{rms} = RMS velocity

$$(K.E.)_{AV} = \frac{3}{2}\frac{RT}{N}$$

$$(K.E.)_{AV} = \frac{3}{2}kT$$

$$k = R/N$$

Here k = Boltzmann constant

$= 1.38 \times 10^{-16}$ ergs

Molecular Speed Graph or Maxwell's Distribution of Velocities

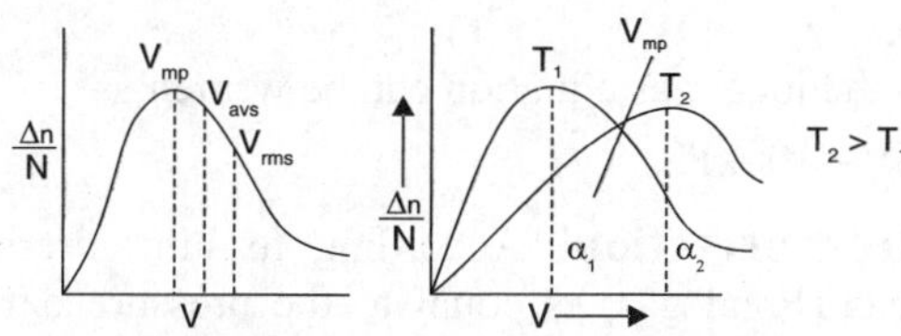

FIGURE 5.4 Molecular Speed Graph

- Molecules have different speeds due to collisions.
- Only a small fraction of the total molecules have either very high or very low speeds.
- None of the molecules have zero velocity.
- After Vmp (most probable velocity) velocity decreases.
- All types of velocities increase with the increase in temperature.

Different Velocity Terms

1. **Average velocity** ($\bar{u}$) It is the average of different velocities possessed by the molecules.

$$u_{av} = \frac{u_1 + u_2 + u_3}{n} \ldots.$$

$$u_{av} = \frac{n_1u_1 + n_2u_2 + n_3u_3}{n_1 + n_2 + n_3}$$

- Here n_1, n_2, n_3 are the number of molecules having u_1, u_2, u_3 velocities respectively.
- Relation between u_{av}, temperature and molar mass is

given as $u_{av} = \sqrt{\left(\frac{8RT}{\pi M}\right)} = \sqrt{\frac{8PV}{\pi M}}$

- It is also denoted by $\bar{u}$.

2. **Root mean square velocity (u_{rms})** It is the square root of the mean of the square of the velocities of different molecules.

$$u_{rms} = \sqrt{\frac{n_1u_1^2 + n_2u_2^2}{n_1 + n_2} + \text{------}}$$

$$u_{rms} = \sqrt{\left(\frac{3RT}{M}\right)}$$

$$u_{rms} = \sqrt{\left(\frac{3PV}{M}\right)}$$

3. **Most probable velocity** (α) It is the velocity possessed by maximum number of molecules.

$$u_{mp} = \sqrt{\left(\frac{2RT}{M}\right)}$$

Relation between α, $\bar{u}$ and u_{rms}

$\alpha:\bar{u}:u_{rms} = 1:1.128:1.224$, which means

$u_{rms} > \bar{u} > \alpha$

$\bar{u}_{av} = u_{rms} \times 0.9213$

$\alpha\ (u_{mp}) = 0.8164 \times u_{rms}$

IDEAL AND REAL GASES

Kinetic theory of gases makes certain assumptions about the volume of gaseous molecules which hold true at STP but not at higher temperature or pressure.

Real gases behave differently from the ideal gases on the basis of differences between the two given below.

Ideal Gases

- Ideal gases obey gas laws under all conditions of temperature and pressure.
- No gas is ideal in reality.
- Force of attraction is absent between their molecules.
- Volume of molecules is negligible as compared to the total volume of the gas.

Real Gases

- Real gases obey gas laws only at high temperature and low pressure.
- All the gases are real.
- Here force of attraction between molecules cannot be neglected at high pressure and low temperature.

Deviation of real gas from ideal gas behaviour

- The extent of deviation of a real gas from ideal gas behaviour is expressed in terms of compressibility factor Z. It is an empirical correction for the non ideal behaviour of real gases which allows the simple form of the combined gas laws to be retained. It is given as:

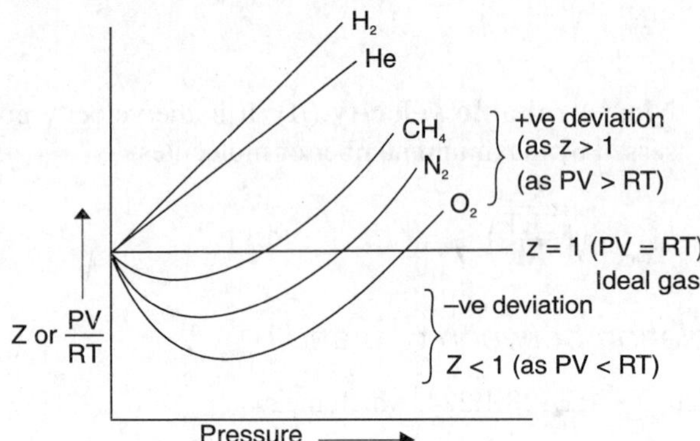

FIGURE 5.5 Deviation of Real Gases from Ideal Gas Behaviour

$$Z = \frac{PV}{nRT}$$

When Z = 1, gas exhibits ideal behaviour
When Z < 1, negative deviation is seen
When Z > 1, positive deviation is seen
When Z < 1, gas is more compressible (V < 22.4 L)
When Z > 1, gas is less compressible (V > 22.4)

- For He and H_2, Z > 1 as PV > RT (as $a/V^2 = 0$) that is, positive deviations.
- At very low pressure, PV ≈ RT (as a/V^2 and b are neglected) that is, Z ≈ 1 so nearly ideal gas behaviour.
- At low pressure, PV < RT
  Z < 1 so negative deviation.
- At moderate pressure, PV = RT
  Z = 1 so ideal gas behaviour.
- At high pressure, PV > RT
  (as b cannot be neglected), that is, Z > 1 so positive deviation.
- With increase in temperature, gases show a decrease in deviation from ideal behaviour.

VAN DER WAALS EQUATION

Van der Waals equation modified the kinetic theory of gases refuting following two points of the kinetic theory:

- The force of attraction between gaseous molecules is negligible.
- The volume of gaseous molecules is negligible as compared to the total volume of the gas.

Van der Waals made two corrections, which are explained below.

Volume correction According to van der Waals, the actual space available inside the vessel for the movement of gas molecules is not the real volume of the gas, actually it is given as

$$V_{\text{real gas}} = V - b$$

Here V is the volume of the container, b is volume occupied by gas molecules. It is called **co-volume** or **excluded volume.**

Now the ideal gas equation can be written as

$$P(V - b) = RT$$

Pressure correction According to him, pressure of a real gas (Preal gas) is equal to the pressure developed due to collisions (P), along with pressure loss due to attraction (p′)

$$P_{\text{real gas}} = P + p'$$

As $p' \propto n^2$ [n^2 is number of molecules attracting or attracted]

$$p' \propto n^2 \propto d^2 \propto \frac{1}{V^2}$$

$$\text{So,} \quad p' = \frac{a}{V^2}$$

Here 'a' is van der Waals force of attraction constant.

Hence, $P_{\text{real gas}} = P + \frac{a}{V^2}$

Now ideal gas equation can be written after correction of pressure and volume as

$$\left(P + \frac{a}{V^2}\right).(V - b) = RT \text{ (For one mole)}$$

$$\left(P + n^2\frac{a}{V^2}\right).(V - nb) = nRT \text{ (For n number of moles)}$$

Units of a and b

- $a = L^2\ mol^{-2}\ atm$
  or $cm^4\ mol^{-2}$ dyne
  or $m^4\ mol^{-2}$ Newton
- $b = L\ mol^{-1}$
  or $cm^3\ mol^{-1}$
  or $m^3\ mol^{-1}$

Significance of a and b: 'a' gives the magnitude of attraction forces present between gas molecules.

$$\text{Liquefication of gases} \propto a$$

The constancy in value of 'b' shows that gas molecules are incompressible. b = 4NVm

$$b = 4N\left(\frac{4}{3}\pi r^3\right)$$

The values of a and b are 0.1 to 0.01 and 0.01 to 0.001 respectively

At high pressure V.W. equation becomes (PV = RT + Pb)

At low pressure V.W equation becomes

$$(PV = RT - \frac{a}{V})$$

Boyle's temperature It is the temperature at which real gases obey gas laws or the temperature above which a real gas behaves like an ideal gas. It is denoted by T_b.

$$T_b = \frac{a}{Rb}$$

Easily liquefiable gases have high value of T_b, for example, T_b of O_2 is 40 K.

Inversion temperature (T_i) It is the temperature at which a gas shows neither cooling nor heating effect in Joule–Thomson adiabatic expansion. ($\mu_{JT} = 0$)

- Below the inversion temperature, cooling effect occurs.

$$(\mu_{JT} = -ve)$$

- Above inversion temperature, heating effect occurs.

$$(\mu_{JT} = +ve)$$

$$T_i = \frac{2a}{Rb}$$

Mean free path It is the average distance travelled by a particle between any two successive collisions. It is denoted by λ.

$$\lambda \propto \frac{T}{P}$$

$$\lambda = \frac{1}{\sqrt{2}\pi\sigma^2 n}$$

Here

λ = Mean free path

σ = Collision diameter

n = Number of molecules per unit volume

Collision frequency (Z) It is the number of collisions which take place in one second among the molecules present in 1 cm³ (1cc) of the gas.

$$Z = \frac{1}{\sqrt{2\sigma^2 \mu_{rms} \eta^2}}$$

$Z \propto P^2$ (At particular temperature)

$Z \propto T^{3/2}$ (At particular pressure)

$Z \propto T^{1/2}$ (At particular volume)

Critical Phenomenon and Liquefaction of Gases

Critical temperature (T_c) It is the temperature above which the gas cannot be liquified with the help of high pressure alone. It is denoted by T_c. Higher the value of T_c easier is the liquation of a gas.

$$T_c = \frac{8a}{27Rb}$$

Each gas has a definite critical temperature, for example, T_c for CO_2, O_2 and H_2 are 31.1°C, –118.8°C and –240°C respectively. (Here CO_2 is most easily liquified)

Critical volume (V_c) It is the volume occupied by one mole of the gas at critical temperature and critical pressure. It is denoted by V_c.

$$V_c = 3b$$

Critical pressure (P_c) It is the minimum pressure required to liquify any gas at critical temperature. It is denoted by P_c.

$$P_c = \frac{a}{27\,b^2}$$

Relation between T_c, V_c and P_c

$$P_cV_c = 3/8\ RT_c$$

Liquefaction of gases A gas can be liquefied by decreasing the temperature or increasing the pressure. When a gas is compressed at any temperature the intermolecular distance decreases. As a result, intermolecular forces become effective and decreases kinetic energy, this causes liquefaction.

Liquefaction ∝ a (Intermolecular force)

Increasing order of liquefaction of some gases:

He < H_2 < O_2 < N_2 <
(0.034) (0.244) (1.36) (1.39)

CO_2 < HCl < NH_3 < SO_2
(3.59) (3.8) (4.17) (6.71)

Methods of liquefaction of gases

1. Linde's process, by adiabatic expansion of compressed gas, that is, Joule–Thomson effect.
2. Claude's process, by adiabatic expansion of compressed gas involving mechanical work.
3. Through adiabatic demagnetization.
4. Cooling by freezing mixtures, for example, NaCl and ice (251 K), $CaCl_2$ and ice (218 K).

Liquid State - General Properties

- Liquids have no definite shape but have a definite volume.
- As compared to gases, density of liquids is higher but is lower than that of solids.
- As compared to solids, their compressibility is greater but is less than that of gases as small voids are present between liquid molecules.
- The diffusion rate is slower than that of gases but more than that of solids due to small voids and slow molecular motion by liquids.
- The distance of separation between the molecules of a liquid is in the range of 10^{-7} to 10^{-9} m.
- Liquids diffuse slowly as the molecules of liquids undergo large number of collisions with the neighbouring molecules.
- A liquid resembles a gas near critical temperature of the gas and a solid near the melting point of solid.

Facts to Remember

- The attraction and repulsion present between molecules of liquids can be explained by Lennard–Jones 6–12 potential model.
- H. Eyring and T. Ree introduced Hole theory or Vacancy theory of liquids is done on the system and W is positive.

Vapour Pressure

- Vapour pressure is the pressure, at a particular temperature, exerted by vapours over liquid surface when vapours are in equilibrium with liquid. Vapour pressure increases with increase in temperature. The variation of vapour pressure of liquid with temperature is given as

$$\log P = \frac{-A}{T} + 1$$

A = Constant

P = Vapour pressure of liquid

T = Temperature

- The plot of log P vs $\frac{1}{T}$ will be a straight line.

At boiling point, vapour pressure = atmospheric pressure

The vapour pressure of H_2O at 373 K is 76 cm of Hg.

- At critical temperature, the meniscus between liquid and vapour disappears.
- Vapour pressure ∝ extent or strength of H-bonding.

For example, H – F > H – O – H > NH_3 (N bonded to three H atoms)

- Heat of vaporization ∝ extent or strength of H-bonding, for example, HF > H_2O > NH_3
- The amount of heat needed to convert one gram of a liquid into its vapours at its boiling point is known as heat or enthalpy or latent heat of vaporization.

Trouton's law

$$\frac{\Delta H_{vap}}{T_b} = 21 \text{ cal K}^{-1} \text{ mol}^{-1}$$

ΔH_{vap} = Heat of vaporization

T_b = Boiling point of a liquid

Clausius–Clapeyron equation It explains the effect of temperature on vapour pressure of a liquid and also the effect of pressure on the boiling point of a liquid.

$$\log \frac{P_2}{P_1} - \frac{\Delta H_{vap}}{2.303\,R} \cdot \frac{(T_2 - T_1)}{(T_1 \cdot T_2)}$$

Facts to Remember

- Bubbles are formed in a liquid due to vapour pressure.
- Rate of evaporation does not change with time at a particular temperature. However, rate of condensation increases with time.
- At equilibrium, rate of condensation and rate of evaporation are equal.

Difference Between Evaporation and Boiling Processes

- During evaporation, cooling occurs as the average kinetic energy of the liquid decreases since molecules with higher kinetic energy leave the surface to form vapours.

- Boiling occurs only at a specific temperature while evaporation can take place at any temperature.
- Boiling occurs below the surface while evaporation occurs only at surface.

Surface Tension

- Surface tension is the force at right angles to the surface of a liquid along one cm or one metre (unit) length of the surface.
- Its units are: Newton metre^{-1} or [N m^{-1}], dyne cm^{-1}.
- Due to surface tension the surface area of the liquid decreases to minimum, for example, falling drops are spherical. Sphere is the minimum surface area for a given volume.
- Due to surface tension a liquid rises in the capillary tube, water moves upwards in soil and some insects are able to walk on the surface of water.
- Surface tension $\propto \dfrac{1}{\text{Temperature}}$
- At critical temperature, surface tension is zero.
- The effect of temperature is given by **Eotvos equation.**

$\gamma = K (T_c - T) (d/M)^{2/3}$

K = Constant

d = Density

M = Molar mass

Measurement of surface tension To measure surface tension, a stalagmometer is used. It is based on the principle that surface tension is directly proportional to the amount or weight of the spherical drop falling from the vertically kept capillary tube.

$$\gamma \propto M \quad \text{so} \quad \frac{r_1}{r_2} = \frac{M_1}{M_2}$$

Surface Energy

Surface energy is work done in ergs required to increase the surface area by 1 square centimetre. Its units are erg/cm^2 or Joule/m^2.

FACTS TO REMEMBER

- Soap, detergent, alcohol, cholesterol are some surface active substances or **surfactants** which decrease surface tension.
- Due to force of adhesion, water drops stick to a glass surface.

Viscosity

Viscosity is the internal resistance of a liquid to flow. It increases with increase of temperature.

Viscosity coefficient (η) Viscosity coefficient is the force of friction needed to maintain a velocity difference of 1 cm sec^{-1} between any two parallel layers of 1 cm^2 area and which are 1 cm apart.

$$\eta = \frac{f\,.\,x}{A\,.\,v} = \frac{\text{dynes} \times \text{cm}}{\text{cm}^2 \times \text{sec}^{-1}} = \text{dyne cm}^{-2}\ \text{sec}$$

$$= 1 \text{ poise}$$

f = Force

a = Area

v = Velocity difference

x = Distance between two layers

1 Poise = 1 g cm^{-1} sec^{-1}

Since dyne = g × cm × sec^{-2}

1 Poise = $\frac{1}{10}$ Newton metre^2 sec^{-1}

or Pas or kg m^{-1} s^{-1}

Effect of temperature on viscosity It can be shown by Arrhenius equation as follows:

$$\eta = Ae^{E_a/RT}$$

Here, T = Temperature

R = Universal gas constant

E_a = Activation energy

Measurement of viscosity

- When the same volume of two liquids say X and Y are flowing from the same height and through the same capillary then

$$\frac{\eta_x}{\eta_y} = \frac{d_x t_x}{d_y t_y}$$

d_x = Density of liquid x

d_y = Density of liquid y

t_x = Time for flow of liquid x

t_y = Time for flow of liquid y.

- Ostwald viscometer is used to measure viscosity.

Fluidity

It is the reciprocal of viscosity coefficient of a liquid denoted by ø.

$$ø = \frac{1}{\eta}$$

Facts to Remember

- In case of superfluid liquids, the resistance to flow becomes almost zero. This is called super fluidity.
- Helium at 2.18 K becomes a super fluid liquid.
- Viscosity ∝ H-bonding ∝ Molecular weight
Glycerol > Glycol > Ethanol > CH_3OH

Points to Remember

1. Kinetic energy does not depend upon type of gas molecules.
2. The order of kinetic energy for different states is solid < liquid < gas.
3. Kinetic energy at absolute zero temperature is zero.
4. Boyle's Law in terms of kinetic energy
$PV = 1/2\ Mu^2_{rms}$
5. Charles Law in terms of kinetic energy
$V = (M/3P)\ u^2_{rms}$
6. Gay-Lussac's law
$P = (M/3V)\ u^2_{rms}$
7. **Loschmidt Number** It is the number of molecules persent in 1 cm^3 of a gas or vapour at STP (value = 2.7678×10^{19} per cc).
8. At constant P, V/T = constant, hence plot of V versus T is a line passing through the origin.
9. Gas constant is work done per degree per mole.
10. Aqueous tension depends only on temperature.
11. Real gases show deviation from ideal behaviour at low temperature and high pressure.
12. The van der Waals constant b is a measure of the size of the molecules, $b = 4 \times v$ (v = actual volume). b is called co-volume or excluded volume.
13. With increase in temperature, most probable velocity increases but the fraction possessing it decreases.
14. Hydrogen and helium show heating effect on adiabatic expansion (Joule–Thomson effect) because their inversion temperatures are low.
15. $C_p - C_v = R$ for 1 mole of an ideal gas.
16. Beckmann thermometer is used only for the measurement of small differences in temperature, like depression in freezing point.
17. **Plasma State** It is the gaseous mixture of electrons and cations existing at high temperature or internal electrical fields in discharge tubes.
18. **Escape Velocity (Ve)** It is the minimum velocity needed by an object to escape from the gravitational field of another body.
$$V_e = \sqrt{2}\ gr$$

Important Formula

1. **Boyle's Law**
$P_1V_1 = P_2V_2$ (At constant temperature)
Initial condition — Final condition
2. **Charles Law**
$\frac{V_1}{T_1} = \frac{V_2}{T_2}$ (At constant pressure)
3. **Gay-Lussac's Law**
$P = \left(\frac{M}{3V}\right) u^2_{rms}$
4. **Pressure–Temperature Law (Amonton's Law)**
$\frac{P_1}{T_1} = \frac{P_2}{T_2}$ (At constant volume)
5. **Avogadro's Law**
$\frac{V_1}{n_1} = \frac{V_2}{n_2}$
Here n = Number of moles
N = Number of molecules
6. **Ideal Gas Equation**
$PV = nRT$
$PV = \frac{w}{M} RT$
$P = \frac{w}{mV} RT$
Value of R is 0.0821 L atm K^{-1} mol^{-1}
= 8.314 joule K^{-1} mol^{-1}
= 8.314×10^7 erg K^{-1} mol^{-1}
= 2 cal K^{-1} mol^{-1}
7. **Dalton's Law of Partial Pressure**
$P_{mix} = P_1 + P_2 + P_3$ ......
Here P_{mix} = Pressure of the mixture
P_1, P_2, P_3 = Partial pressure of gases
8. **Graham's Law of Diffusion**
$r_1/r_2 = \sqrt{(M_2/M_1)}$ or $\sqrt{(d_2/d_1)}$
$\frac{V_1t_2}{V_2t_1} = \sqrt{(M_2/M_1)}$

9. $r = n/t = d/t$

$$\frac{n_1t_2}{n_2t_1} = \sqrt{(M_2/M_1)}$$

$$= \frac{d_1t_2}{d_2t_1} = \sqrt{\frac{M_2}{M_1}}$$

Here n and d are number of moles and distance traveled for a particular gas.

10. If pressure are different, then

$$\frac{r_1}{r_2} = \frac{P_1}{P_2} = \sqrt{(M_2/M_1)} \text{ or } \sqrt{(d_2/d_1)}$$

11. **Kinetic Gas Equation**

$$PV = \frac{1}{3} mnu^2_{rms}$$

12. **Average Speed (v)**

$$u_{av} = \frac{n_1u_1 + n_2u_2 + n_3u_3}{n_1 + n_2 + n_3}$$

$$u_{av} = \sqrt{\left(\frac{8RT}{pM}\right)}$$

13. **Root Mean Square Velocity (u)**

$$u_{rms} = \sqrt{\left(\frac{3RT}{M}\right)}$$

14. **Most Probable Velocity (α)**

$$u_{mp} = \sqrt{\left(\frac{2RT}{M}\right)}$$

15. $u_{av} = u_{rms} \times 0.9213$

$\alpha\ (u_{mp}) = 0.8164 \times u_{rms}$

16. **Van der Waals Equation**

$$\left(P + n^2\frac{a}{V^2}\right)(V - nb) = nRT$$

17. **Vapour Pressure**

$$\log P = \frac{-A}{T} + 1$$

Here A = Constant

P = Vapour pressure of liquid

T = Temperature

18. **Trouton's Law**

$$\frac{\Delta H_{vap}}{T_b} = 21 \text{ cal K}^{-1} \text{ mol}^{-1}$$

Here ΔH_{vap} = Heat of vaporization

T_b = Boiling point of a liquid

19. **Clausius–Clapeyron Equation**

$$\log\frac{P_2}{P_1} = \frac{\Delta H_{vap}}{2.303\,R}\cdot\frac{(T_2 - T_1)}{(T_1\cdot T_2)}$$

20. **Surface Tension**

$$\gamma = K\,(T_C - T)\,(d/M)^{2/3}$$

Here K = Constant

d = Density

M = Molar mass

21. **Measurement of Surface Tension**

$$\frac{r_1}{r_2} = \frac{M_1}{M_2}$$

22. **Viscosity Coefficient (η)**

$$\eta = \frac{f\,.\,x}{A\,.\,v} = \frac{\text{dynes} \times \text{cm}}{\text{cm}^2 \times \text{cm sec}^{-1}} = \text{dyne cm}^{-2}\text{ sec}$$

= 1 poise

Here f = Force

a = Area

v = Velocity difference

x = Distance between two layers

23. **Effect of Temperature on Viscosity**

$$\eta = Ae^{E_a/RT}$$

Here T = Temperature

R = Universal gas constant

E_a = Activation energy

24. **Fluidity**

$$\text{ø} = \frac{1}{\eta}$$

SOLVED NUMERICALS

Gas Laws

1. A flask is of a capacity of one litre. What volume of air will escape from the flasks it is heated from 27°C and 37°C? Assume pressure to be constant.

Solution:

T = 27°C = 300 K

T_1 = 37°C = 310 K

V = 1 L

V_1 = ?

$\frac{V}{T} = \frac{V_1}{T_1}$ (At constant pressure)

$\frac{1}{300} = \frac{V_1}{310}$

$V_1 = \frac{310}{300} = 1.0333$ L

As capacity of flask is 1 L.

So, volume of air escaped out = 1.033 – 1
= 33.3 mL.

2. How large a balloon could be filled with 4.0 g of He gas at 22°C and 720 mm of Hg?

Solution:

P = 720/760 atm

T = 295 K

w = 4 g

m = 4 (He)

$PV = \frac{m}{w} RT$

$V = \frac{wRT}{m \times P} = \frac{4 \times 0.0821 \times 295 \times 760}{4 \times 720}$

= 25.565 L

3. Calculate the density of CO_2 at 100°C and 800 mm Hg pressure.

Solution:

P = 800/760 atm

T = 373 K

$PV = \frac{w}{m} RT$

$P = \frac{d}{m} RT$ (As w/v = Density)

$\frac{800}{760} = \frac{d}{44} \times 0.0821 \times 300$

d = 1.5124 g L^{-1}.

4. Calculate the total pressure in a 10 L cylinder which contains 0.4 g of helium, 1.6 g of oxygen and 1.4 g of nitrogen at 27°C. Also calculate the partial pressure of helium gas in the cylinder. Assume ideal behaviour of gases. Given R = 0.082L atm K^{-1} mol^{-1}.

Solution:

Given V = 10 L

Mole of He = 0.4/4 = 0.1

Mole of O_2 = 1.6/32 = 0.05

Mole of N_2 = 1.4/28 = 0.05

Total moles = 0.1 + 0.05 + 0.05 = 0.2

As PV = nRT

$P_{total} = \frac{n\,RT}{V} = \frac{0.2 \times 0.082 \times 300}{10}$

= 0.492 atm

P_{He} = Mole fraction of He × Total pressure

$= \frac{0.1}{0.2} \times 0.492 = 0.246$ atm

5. A gas occupies 300 mL at 27°C and 730 mm pressure. What would be its volume at STP?

Solution:

$V_2 = \frac{300}{1000}$ L, $P_2 = \frac{730}{760}$ atm

T_2 = 27 + 273 = 300 K

At STP

V_1 = ?,

P_1 = 1 atm

T_1 = 273 K

$\frac{P_1V_1}{T_1} = \frac{P_2V_2}{T_2}$

$\frac{1 \times V_1}{273} = \frac{730 \times 300}{760 \times 1000 \times 300}$

V_1 = 0.2622 L

V (at STP) = 262.2 mL.

6. Calculate the volume occupied by 7g N_2 at 27°C and 750 mm of Hg.

Solution:

w = 7 g

T = 27 + 273 = 300 K

P = 750/760 atm

m = 28

$$PV = \frac{w}{m}RT$$

$$\frac{750}{760} \times V = \frac{7}{28} \times 0.0821 \times 300$$

$V = 6.239$ L.

7. Find the density of CO_2 at 100°C and 800 mm Hg pressure.

Solution:

P = 800/760 atm

T = 273 + 100 = 373 K

$$CO_2 = \frac{w}{V}$$

For CO_2

$$PV = \frac{w}{m}RT$$

$$P = \frac{d}{m}RT \text{ [as } d = w/V]$$

$$\frac{800}{760} = \frac{d}{44} \times 0.0821 \times 373 \quad [m \text{ for } CO_2 = 44]$$

$d = 1.5124$ g L^{-1}

8. In Victor Meyer's experiment, 0.23 g of a volatile solute displaced air which measures 112 mL at NTP. Find the vapour density and mol. wt. of substance.

Solution:

Volume occupied by solution at NTP

= Volume of air displaced at NTP = 112 mL

$$PV = \frac{w}{m}RT$$

At NTP, P = 1 atm, T = 273 K

$$1 \times \frac{112}{1000} = \frac{0.23}{m} \times 0.0821 \times 273$$

Molecular weight = 46.02

$$\text{Vapour density} = \frac{\text{Mol. wt.}}{2} = \frac{46.02}{2} = 23.01.$$

9. Find the total pressure in a mixture of 4 g of O_2 and 2 g H_2 confined in a bulb of 1 L of 0°C.

Solution:

$$\text{Moles of } O_2 \text{ in mixture} = \frac{\text{Wt}}{\text{Mol. wt.}} = \frac{4}{32}.$$

$$\text{Moles of } H_2 \text{ in mixture} = \frac{2}{4} = 1$$

$$\text{Moles of } O_2 \text{ in mixture} = \frac{4}{32}$$

$$\text{Total moles present in mixture} = 1 + \frac{4}{32} = \frac{36}{32}$$

T = 273 K, V = 1 L

PV = nRT

$$P \times 1 = \frac{36}{32} \times 0.0821 \times 273$$

P = 25.215 atm

10. A certain amount of an ideal gas occupies a volume of 10 m^3 at a given temperature and pressure. What would be its volume after reducing its pressure to half the initial value and raising the temperature twice the initial value?

Solution:

$P_2 = 0.5\,P_1$

$T_2 = 2\,T_1$

$$\frac{P_1V_1}{T_1} = \frac{P_2V_2}{T_2}$$

$$\frac{P_1V_1}{T_1} = \frac{0.5P_1V_2}{2T_1}$$

$V_2 = 4V_1$

$V_1 = 1\ m^3$

$V_2 = 4.0\ m^3$

11. 6.02×10^{22} oxygen molecules are present at –13°C in a 1500 mL vessel. What would be the pressure of the gas? (Avogadro's number = 6.02×10^{23}, R = 0.0821 atm mol^{-1} K^{-1}).

Solution:

$$\text{Number of moles of } O_2 = \frac{6.02 \times 10^{22}}{6.02 \times 10^{23}} = 0.1$$

n = 0.1

V = 15 L

T = –13 + 273 = 260 K

$$P = \frac{nRT}{V} = \frac{0.1 \times 0.0821 \times 260}{15} = 1.42 \text{ atm}$$

12. 20 dm^3 of SO_2 diffuses through a porous partition in 60 sec. What volume of O_2 will diffuse under similar conditions in 30 seconds?

Solution:

$$r_{SO_2} = \frac{V_{SO_2}}{t_{SO_2}} = \frac{20}{60} = 0.333 \text{ dm}^3\text{/sec}$$

$$r_{O_2} = \frac{V_{O_2}}{t_{O_2}} = \frac{V}{30} \text{ dm}^3\text{/sec}$$

$$\frac{r_{SO_2}}{r_{O_2}} = \sqrt{\left(\frac{M_{O_2}}{M_{SO_2}}\right)}$$

$$0.333/V/30 = \sqrt{\left(\frac{32}{64}\right)}$$

$V = 14.14\ dm^3$

13. 20% of N_2O_4 molecules are dissociated in a sample of gas at 27°C and 760 torr. Calculate the density of the equilibrium mixture.

Solution:

$N_2O_4 \rightleftharpoons 2NO_2$

1 mol	0	At t = 0
(1 – 0.2) mol	0.4	At equilibrium

Total moles = 0.8 + 0.4 = 1.2

PV = nRT

$1 \times V = 1.2 \times 0.0821 \times 300$

V = 29.556 L

$$d_{N_2O_4} = \frac{m}{V} = \frac{0.8 \times 92}{29.556} = 2.490$$

$$d_{NO_4} = \frac{m}{V} = \frac{0.4 \times 46}{29.556} = 0.6225$$

$d_{mix} = 2.490 + 0.6225 = 3.113$ g/L.

14. One mole of nitrogen gas at 0.8 atm takes 38 seconds to diffuse through a pin hole, whereas one mole of unknown compound of xenon with fluorine at 1.6 atm takes 57 seconds to diffuse through the same hole. Calculate the molecular formula of the compound.

Solution:

$$\frac{r_1}{r_2} = \sqrt{\left(\frac{M_2}{M_1}\right)} \times \frac{P_1}{P_2}$$

$$\frac{n_1}{t_1} \times \frac{t_2}{n_2} = \sqrt{\left(\frac{M_2}{M_1}\right)} \times \frac{P_1}{P_2}$$

$$\frac{n_1}{t_1} \times \frac{t_2}{n_2} = \sqrt{\left(\frac{M_2}{M_1}\right)} \times \frac{P_1}{P_2}$$

$$M_g = \left(\frac{57}{38} \times \frac{1.6}{0.8}\right)^2 \times 28$$

On solving

$M_g = 252$

Thus, compound is XeF_6 because it can have only one xenon atom.

Kinetic Theory of Gases and Various Velocities

15. Calculate the temperature at which the root mean square velocity of SO_2 molecules is the same as that of O_2 at 27°C.

Solution:

$$u = \sqrt{\left(\frac{3RT_{SO_2}}{M_{SO_2}}\right)} = \sqrt{\left(\frac{3RT_{O_2}}{M_{O_2}}\right)}$$

$$\sqrt{\left(\frac{3RT_{O_2}}{64}\right)} = \sqrt{\left(3R \times \frac{300}{32}\right)}$$

$$T_{SO_2} = 600 \text{ K}$$

16. Two flasks A and B have equal volumes. Flask A contains H_2 at 27°C while B contains equal mass of C_2H_6 at 627°C. In which flask and by how many times are molecules moving faster? Assume ideal gas nature for both.

Solution:

For H_2, Average velocity

$$= \sqrt{\frac{8RT}{\pi M}} = \sqrt{\left(\frac{8R \times 300}{\pi \times 2}\right)}$$

For C_2H_6 Average velocity

$$= \sqrt{\frac{8RT}{\pi M}} = \sqrt{\left(\frac{8R \times 900}{\pi \times 2}\right)}$$

$$\frac{\text{Average velocity } H_2}{\text{Average velocity } C_6H_6} = \frac{\sqrt{(300 \times 30)}}{\sqrt{(900 \times 2)}}$$

= 2.236:1

17. A 4:1 molar mixture of He and CH_4 is contained in a vessel at 20 bar pressure. Due to a hole in the vessel the gas mixture leaks out. What is the composition of the mixture effusing out initially?

Solution:

$$\text{Rate of effusion} = K\frac{P}{\sqrt{M}}$$

Here P is partial pressure of the gas

M = Molar mass, M of He = 4,

M of CH_4 = 16

$$\frac{\text{Rate of effusion of He}}{\text{Rate of effusion of } CH_4} = \frac{KP_{He}/\sqrt{M_{He}}}{KP_{CH_4}/\sqrt{M_{CH_4}}}$$

$$= \frac{P_{He}}{P_{CH_4}} \times \sqrt{\left(\frac{M_{CH_4}}{M_{He}}\right)}$$

$$= \frac{4}{1}\sqrt{\left(\frac{16}{4}\right)} = 8$$

18. Find out the kinetic energy of 10 mol of gas at 200°C.

Solution:

$$K.E. = \frac{3}{2}RT$$

$R = 8314 \times 10^7$ erg/K/mol

T = 200 + 273 = 473 K

n = 10 moles of gas

$$K.E. = \frac{3}{2} \times 10 \times 8314 \times 10^7 \times 473$$

$= 58987.83 \times 10^7$ erg.

19. Find out the rms velocity of O_2 molecules at NTP (density of O_2 = 0.001429 g/cc at NTP, density of Hg = 13.6 g/cc, g = 981 cm/sec^2).

Solution:

h = 76 cm

$$u_{rms} = \sqrt{\left(\frac{3P}{d}\right)} = \sqrt{\left(3 \times h \times d \times \frac{g}{d}\right)}$$

$$= \frac{\sqrt{(3 \times 76 \times 13.6 \times 981)}}{\sqrt{0.001429}}$$

$= 46137.6\ cm\ sec^{-1}$

20. Find out the rms velocity of a CO_2 molecule at 1000°C.

Solution:

T (absolute temp. of CO_2) = 1000 + 273

= 1273 K

$R = 8.314 \times 10^7$ erg/K/mole

Mol. wt. of CO_2 = 44

$$u_{rms} = \frac{\sqrt{(3 \times 8.314 \times 10 \times 1273)}}{\sqrt{44}}$$

= 84,948 cm/sec.

Van der Waals equation

21. 2 moles of ammonia occupied a volume of 5 litres at 27°C. Calculate the pressure if the gas obeyed van der Waals equation. ($a = 4.17$ atm $L^2\ mol^{-2}$, $b = 0.0371\ L\ mol^{-1}$).

Solution:

n = 2 , V = 5 litres,

T = 273 + 27 = 300 K

$R = 0.082$ L atm $deg^{-1}\ mole^{-1}$

$P + n^2 \frac{a}{V^2} (V - nb) = nRT$

$[P + (2^2 \times 4.17/5^2)]\ [5 - 2 \times 0.037]$

$2 \times 0.082 \times 300$

P = 9.33 atm

22. Using van der Waals equation, calculate the constant 'a' when two moles of a gas confined in a four litre flask exerts a pressure of 11.0 atmospheres at a temperature of 300 K. The value of 'b' is $0.05\ L\ mol^{-1}$.

Solution:

For n moles of a gas

$$\left(P + n^2 \frac{a}{V^2}\right).\ (V - nb) = nRT$$

$[1 + (2^2a/4^2)]\ [4 - 2 \times 0.05] = 2 \times 0.082 \times 300$

$a = 6.46$ atm $L^2\ mol^{-2}$.

CHAPTER-END EXERCISES

PRACTICE QUESTIONS - I

1. Which one of the following statements is wrong for gases?
- (a) Gases do not have a definite shape and volume
- (b) Volume of the gas is equal to volume of container the confining the gas
- (c) Confirmed gas exerts uniform pressure on the walls of its container in all directions
- (d) Mass of gas cannot be determined by weighing a container in which it is enclosed

2. Which of the following mixture of gases does not obey Dalton's law of partial pressure?
- (a) O_2 and CO_2
- (b) N_2 and O_2
- (c) Cl_2 and SO_2
- (d) NH_3 and HCl

3. In which of the following pairs, the critical temperature of latter gaseous species is higher than the first?
- (a) CO_2, H_2
- (b) H_2, NH_3
- (c) NH_3, He
- (d) CO_2, He

4. The dimensions of pressure are same as that of:
- (a) Energy
- (b) Energy per unit volume
- (c) Force per unit area
- (d) Force per unit volume

5. Ideal gas obeying kinetic theory of gases can be liquefied if:
- (a) $T > T_c$
- (b) $P > P_c$
- (c) $P > P_c$ and $T < T_c$
- (d) It cannot be liquefied at any value of P and T.

6. Based on kinetic theory of gases, which of the following laws can be proved?
(a) Boyle's law (b) Charles law
(c) Avogadro's law (d) All of these

7. According to the kinetic theory of gases, in an ideal gas, between two successive collisions the gas molecules travels:
(a) In a circular path
(b) In a wavy path
(c) In a straight line path
(d) With an accelerated velocity

8. Which of the following law leads us to arrive at the conclusion that 1 g-molecule of each gas at STP occupies a volume of 22.4 L?
(a) Dalton's law
(b) Law of combining volumes
(c) Avogadro's law
(d) Boyle's law

9. 0.1 mole of gas absorbs 41.75 J of heat, the rise in temperature occurs 20°C, the gas must be:
(a) Triatomic (b) Diatomic
(c) Monoatomic (d) (b) and (c) both

10. Which pair of the gases diffuses with the same rate at same temperature and pressure?
(a) CO and NO (b) NO_2 and CO_2
(c) NH_3 and PH_3 (d) NO and C_2H_6

11. Van der Waals real gas, act as an ideal gas, at which condition?
(a) High temperature, low pressure
(b) Low temperature, high pressure
(c) High temperature, high pressure
(d) Low temperature, low pressure

12. The compressibility factor of an ideal gas is:
(a) 1 (b) 2
(c) 4 (d) 0

13. The temperature below which a gas can be made to liquefy by variation in pressure is called its:
(a) Inversion temperature
(b) Critical temperature
(c) Neutral temperature
(d) Curie point

14. By the ideal gas law the pressure of 0.60 mol NH_3 gas in a 3.00 litre vessel at 25°C is:
(a) 48.9 atm (b) 4.89 atm
(c) 0.489 atm (d) 489 atm

15. At what centigrade temperature will be the volume of a gas at 0°C double of itself, when pressure remains constant?
(a) 0°C (b) 273°C
(c) 273 K (d) 546 K

16. Pressure of a mixture of 4 g of O_2 and 2 g of H_2 confined in a bulb of 1.0 L capacity at 0°C is:
(a) 25.18 atm (b) 31.205 atm
(c) 40.215 atm (d) 15.210 atm

17. A closed container contains equal number of oxygen and hydrogen molecules at a total pressure of 740 mm. If oxygen is removed from the system then pressure will:
(a) Become double of 740 mm
(b) Become half of 740 mm
(c) Become 1/9 of 740 mm
(d) Remain unchanged

18. The volume occupied by 9.0 g of nitrogen gas at 300 K and 750 mmHg pressure is:
(a) 5.854 (b) 6.7432
(c) 8.8462 (d) 8.0225

19. What are the conditions under which the relation between volume (V) and number of moles (n) of gas is plotted? (P = pressure; T = temperature):
(a) Constant P and T (b) Constant T and V
(c) Constant P and V (d) Constant n and V

20. The rms velocity of CO_2 at a temperature T (in Kelvin) is $\times$ cm sec^{-1}. At what temperature (in Kelvin), the rms velocity of nitrous oxide would be 4 $\times$ cm sec^{-1}? (Atomic weights of C, N and O are respectively 12, 14 and 16)
(a) 16 T (b) 2T
(c) 4T (d) 32 T

21. 'n' moles of an ideal gas at temperature T (in Kelvin) occupy 'V' litres of volume, exerting a pressure of 'P' atomospheres. What is its concentration (in mol L^{-1})? (R = gas constant):
(a) P/RT (b) PT/R
(c) RT/P (d) R/PT

22. 7.5 g of gas occupy 5.6 litres of volume at STP. The gas is ....
(Atomic weight of C, N, and O are 12, 14 and 16 respectively):
(a) NO (b) N_2O
(c) CO (d) CO_2

23. A gas diffuses four times as quickly as oxygen. The molar weight of gas is:

(a) 2 (b) 4
(c) 8 (d) 16

24. Compressibility factor for 1 mol of a van der Waals gas at 0°C and 100 atmospheric pressure is found to be 0.5, the volume of gas molecule is:

(a) 2.0224 L (b) 1.4666 L
(c) 0.8542 L (d) 0.1119 L

25. The density of air is 0.001293 g/cc at STP. Its vapour density is:

(a) 0.001293 (b) 8.2786
(c) 14.48 (d) 6.2706

26. At 27°C, a closed vessel contains a mixture of equal weights of helium (mol. wt. = 4), methane (mol. wt. = 16) and sulphur dioxide (mol. wt. = 64). The pressure exerted by the mixture is 210 mm. If the partial pressures of helium, methane and sulphur dioxide are P_1, P_2 and P_3 respectively, which one of the following is correct?

(a) $P_3 > P_2 > P_1$ (b) $P_1 > P_2 > P_3$
(c) $P_1 > P_3 > P_2$ (d) $P_2 > P_3 > P_1$

27. 4 g of an ideal gas occupies 5.6035 litres of volume at 546 K and 2 atm pressure. What is its molecular weight?

(a) 4 (b) 16
(c) 32 (d) 64

28. The kinetic energy of 4 moles of nitrogen gas at 127°C is ……. Kcals. ($R = 2$ cal mol^{-1} K^{-1}):

(a) 4400 (b) 3200
(c) 4800 (d) 1524

29. If a gas contains only three molecules that move with velocities of 100, 200, 500 ms^{-1}, what is the rms velocity of the gas is ms^{-1}?

(a) $100\sqrt{8/3}$ (b) $100\sqrt{30}$
(c) $100\sqrt{10}$ (d) 800 / 3

30. A gaseous mixture contains 56 g of N_2, 44 g of CO_2 and 16 g of CH_4. The total pressure of mixture is 720 mm of Hg. The partial pressure of methane is:

(a) 75 mm (b) 160 mm
(c) 180 mm (d) 215 mm

31. At 25°C and 730 mm pressure, 380 mL of dry oxygen was collected. If the temperature is constant, what volume will the oxygen occupy at 760 mm pressure?

(a) 365 mL (b) 2 mL
(c) 10 mL (d) 20 mL

32. The density of an ideal gas is 0.03 g cm^{-3}, Its pressure is 106 g cm^{-1} sec^{-2}. What is its rms velocity (in cm sec^{-1})?

(a) 10^3 (b) 3×10^4
(c) 10^8 (d) 10^4

33. At a certain pressure, volume of a gas at 27°C is 20 litre. If the pressure and temperature are doubled, its volume will be:

(a) 20 litre (b) 40 litre
(c) 8.2 litre (d) 10.9 litre

34. A and B are ideal gases. The molecular weights of A and B are in the ratio of 1:4. The pressure of a gas mixture containing equal weights of A and B is P atm. What is the partial pressure (in atm) of B in the mixture?

(a) P/5 (b) P/2
(c) P/2.5 (d) 3P/4

35. A gas has double the average velocity of SO_2 gas at any temperature. The gas may be:

(a) CO_2 (b) C_2H_4
(c) CH_4 (d) O_2

36. If two moles of ideal gas at 540 K has volume 44.8 L, then its pressure will be:

(a) 1 atm (b) 2 atm
(c) 3 atm (d) 4 atm

37. The densities of two gases are in the ratio of 1:16. The ratio of their rates of diffusion is:

(a) 16:1 (b) 4:1
(c) 1:4 (d) 1:16

38. The critical temperature of water is higher than that of O_2 because the H_2O molecule has:

(a) Fewer electrons than O_2
(b) Two covalent bonds
(c) V-shape
(d) Dipole moment

39. If pressure of 2 mol of an ideal gas at 546 K having volume 44.8 litre is:

(a) 2 atm (b) 3 atm
(c) 4 atm (d) 1 atm

40. A gas behaves most like an ideal gas under conditions of:

(a) High pressure and low temperature
(b) High temperature and high pressure
(c) Low pressure and high temperature
(d) Low pressure and low temperature

41. At constant volume and temperature conditions, the rates of diffusion D_A and D_B of gases A and B having densities ρ_A and ρ_B are related by the expression:

(a) $D_A = [D_B \rho_A/\rho_B]^{1/2}$
(b) $D_A = [D_B \rho_B/\rho_A]^{1/2}$
(c) $D_A = D_B [\rho_A/\rho_B]^{1/2}$
(d) $D_A = D_B [\rho_B/\rho_A]^{1/2}$

42. The term that accounts for intermolecular force in van der Waals equation for non-ideal gas is:
(a) RT (b) V – b
(c) $(P + \alpha/V^2)$ (d) $(RT)^{-1}$

43. A bottle of dry ammonia and a bottle of dry hydrogen chloride connected through a long tube are opened simultaneously at both ends, the white ammonium chloride ring first formed will be:
(a) At the centre of the tube
(b) Near the hydrogen chloride bottle
(c) Near the ammonia bottle
(d) Throughout the length of the tube

44. If a gas expands at constant temperature:
(1) The pressure decreases
(2) The kinetic energy of the molecules remains the same
(3) The kinetic energy of the molecules decreases
(4) The number of molecules of the gas increase
(a) 1, 2 (b) 1, 2, 3
(c) 1, 2, 4 (d) 2, 3

45. The temperature at which real gases obey the ideal gas laws over a wide range of pressure is called:
(a) Critical temperature
(b) Boyle's temperature
(c) Inversion temperature
(d) Reduced temperature

Practice Questions - II

46. As the temperature is raised from 20°C to 40°C, the average kinetic energy of neon atoms changes by a factor of which of the following?
(a) 1/2 (b) 2
(c) $\sqrt{313/293}$ (d) 313/293

47. According to Boyle's law:
(a) $(dP/dV)_T = -K/P^2$ (b) $(dP/dV)_T = K/V$
(c) $(dV/dP)_T = -K/P^2$ (d) $(dV/dP)_T = -K/P$

48. The values of van der Waals constant 'α' for the gases O_2, N_2, NH_3 and CH_4 are 1.360, 1.390, 4.170 and 2.253 L atm. mol^{-2} respectively. The gas which can most easily be liquefied is:
(a) O_2 (b) N_2
(c) NH_3 (d) CH_4

49. To which of the following gaseous mixtures is Dalton's law not applicable?
(a) $Ne + He + SO_2$ (b) $NH_3 + HCl$
(c) $O_2 + N_2 + CO_2$ (d) $N_2 + H_2 + O_2$

50. The ratio of rate of diffusion of helium with respect to methane under similar conditions of constant temperature and pressure:
(a) 2 (b) 0.5
(c) 16 (d) 4

51. The density of neon will be highest at:
(a) STP (b) 0°C, 2 atm
(c) 273°C, 1 atm (d) 273°C, 2 atm

52. At constant volume for a fixed number of a moles of gas, the pressure of the gas increases with the rise in temperature due to:
(a) Increase in average molecular speed
(b) Increase in rate of collisions
(c) Increase in molecular attraction
(d) Increase in mean free path

53. Four one litre flasks are separately filled with the gases O_2, F_2, CH_4 and CO_2 under same conditions. The ratio of the number of molecules in these gases are:
(a) 2:2:4:3 (b) 1:1:1:1
(c) 1:2:3:4 (d) 2:2:3:4

54. An ideal gas expands according to PV = constant. On expansion, the temperature of gas:
(a) Will rise
(b) Will drop
(c) Will remain constant
(d) Cannot be determined because the external pressure is not known

55. Which of the following is true about u_{rms}, $\bar{u}$, α?
(a) $u_{rms} > \bar{u} > \alpha$ (b) urms $< \bar{u} < \alpha$
(c) $u_{rms} > \bar{u} < \alpha$ (d) urms $< \bar{u} > \alpha$

56. According to the kinetic theory of gases:
(a) The pressure exerted by a gas is proportional to mean square velocity of the molecules
(b) The pressure exerted by the gas is proportional to the root mean square velocity of the molecules

(c) The root mean square velocity is inversely proportional to the temperature
(d) The mean translational K.E. of the molecule is directly proportional to the absolute temperature.

57. 6.4 g SO_2 at 0°C and 0.99 atm pressure occupies a volume of 2.241 L. Predict which of the following is correct?
(a) The gas is ideal
(b) The gas is real with intermolecular attraction
(c) The gas is real without intermolecular repulsion
(d) The gas is real with intermolecular repulsion greater than intermolecular attraction

58. V vs T curves at constant pressure P_1 and P_2 for an ideal gas are shown below.

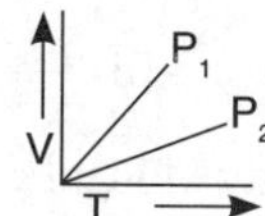

Which of the following is correct?
(a) $P_1 > P_2$ (b) $P_1 < P_2$
(c) $P_1 = P_2$ (d) All of above

59. If increase in temperature and volume of an ideal gas is two times, then the initial pressure P changes to:
(a) 4P (b) 2P
(c) P (d) 3P

60. What is the ratio of kinetic energies of 3 g of hydrogen and 4 g of oxygen at T (K)?
(a) 12:1 (b) 6:1
(c) 1:6 (d) 24:1

61. The van der Waals equation of state is:

$$P + \frac{(V - nb)}{V^2} = nRT$$

The pressure exerted by individual gas molecules on the walls of the container depends upon the
(a) Frequency of the collisions of the molecules with the walls as well as the momentum imparted by the molecules to the walls
(b) Frequency of molecular collision
(c) Mean free path of the molecules
(d) Momentum and critical pressure of the gas molecules

62. The density of a gas is 1.964 g dm^{-3} at 273 K and 76 cm Hg. The gas is:
(a) CH_4 (b) C_2H_6
(c) CO_2 (d) Xe

63. Two gas bulbs A and B are connected by a tube having a stopcock. Bulb A has a volume of 100 mL and contains hydrogen. After opening the gas from A to the evacuated bulb B, the pressure falls down by 40%. The volume (mL) of B must be:
(a) 75 (b) 150
(c) 125 (d) 200

64. At 27°C, 500 mL of helium diffuses in 30 minutes. What is the time (in hours) taken for 1000 mL of SO_2 to diffuse under same experimental conditions?
(a) 240 (b) 340
(c) 200 (d) 440

65. If two moles of an ideal gas at a temperature 546 K, occupy a volume of 44.8 litres its pressure must be:
(a) 4 atm (b) 3 atm
(c) 2 atm (d) 1 atm

66. The following graph illustrates:

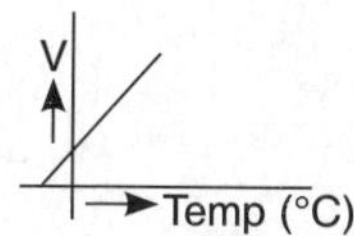

(a) Boyle's law (b) Charles law
(c) Dalton's law (d) Gay-Lussac's law

67. Containers A and B have same gases. Pressure, volume and temperature of A are all twice that of B, then the ratio of number of molecules of A and B are:
(a) 1:2 (b) 2
(c) 1:4 (d) 4

68. Which one of the following statement is not true about the effect of an increase in temperature on the distribution molecular speeds in a gas?
(a) The most probable speed increases
(b) The fraction of the molecules with the most probable speed increases
(c) The distribution becomes broader
(d) The area under the distribution curve remains the same as the under the lower temperature

69. A general form of equation of state for gases is PV = RT [A + B/V + C/V² + ...], where V is the molar volume of the gas and A, B, C .... are constants for the gas. The values of A and B, if the gas obeys van der Waals equation, are respectively:
(a) $1, \frac{a}{27\,b}$ (b) $1, b - \frac{a}{RT}$
(c) $\frac{a}{RT}, b$ (d) $\frac{2a}{RT}, b$

70. If the rms velocity of a gas at 100 K is 10^4 cm sec^{-1}, what is the temperature (in °C) at which the rms velocity will be 3×10^4 cm sec^{-1}?
(a) 900 (b) 627
(c) 327 (d) 1217

71. For which of the following parameters the structural isomers C_2H_5OH and CH_3OCH_3 would be expected to have the same values? (assume ideal behaviour):
(a) Gaseous densities at the same temperature and pressure
(b) Heat of vaporization
(c) Boiling points
(d) Vapour pressure at the same temperature

72. Which curve does not represent Boyle's law?

(a) 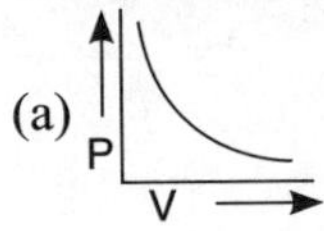

(b) 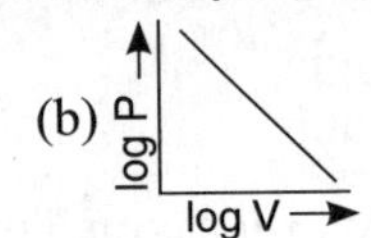

(c) 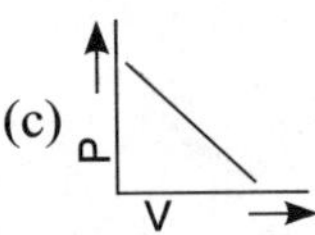

(d) 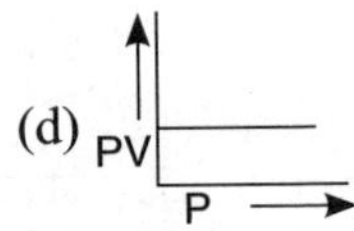

73. At what temperature, the rate of diffusion of N_2 would be 1.625 times the rate of effusion of SO_2 at 50°C?
(a) 110 K (b) 173 K
(c) 373 K (d) 273 K

74. 0.24 g of a volatile gas upon vaporization gives 45 mL vapour at NTP. What will be the vapour density of the substances? (density of $H_2 = 1$):
(a) 95.39 (b) 5.973
(c) 95.93 (d) 59.73

75. A monoatomic ideal gas undergoes a process in which the ratio of P to V at any instant is constant and equals to 1. What is the molar heat capacity of the gas?
(a) 4R/2 (b) 3R/2
(c) 5R/2 (d) 0

76. X mL of H_2 has effused through a hole in a container in 5 seconds. The time taken for the effusion of the same volume of the gas specified below under identical conditions is:
(a) 10 seconds:He (b) 20 seconds:O_2
(c) 25 seconds:CO (d) 55 seconds:CO_2

77. The ratio between the root mean square velocity of H_2 at 50 K and that of O_2 at 800 K is:
(a) 4 (b) 2
(c) 1 (d) $\frac{1}{4}$

78. If C_1, C_2, C_3 .... represents the speed of n_1, n_2, n_3, ... molecules, then the root mean square of speed is:
(a) $\left(\frac{n_1C_1^2 + n_2C_2^2 + n_3C_3^2 + \ldots}{n_1 + n_2 + n_3 + \ldots}\right)^{1/2}$
(b) $\left(\frac{n_1C_1^2 + n_2C_2^2 + n_3C_3^2 + \ldots}{n_1 + n_2 + n_3 + \ldots}\right)^2$
(c) $\frac{(n_1C_1^2)^{1/2}}{n_1} + \frac{(n_2C_2^2)^{1/2}}{n_2} + \frac{(n_3C_3^2)^{1/2}}{n_3} + \ldots$
(d) $\left[\frac{(n_1C_1 + n_2C_2 + n_3C_3 + \ldots)^2}{n_1 + n_2 + n_3 + \ldots}\right]^{1/2}$

79. The root mean square velocity of an ideal gas at constant pressure varies with density (d) as:
(a) d^2 (b) d
(c) $\sqrt{d}$ (d) $\frac{1}{\sqrt{d}}$

80. The average velocity of ideal gas molecules at 27°C is 0.3 m/sec. The average velocity at 927°C will be:
(a) 0.6 m/sec (b) 0.3 m/sec
(c) 0.9 m/sec (d) 3.0 m/sec

81. If the rate of effusion of helium gas at a pressure of 1000 torr is 10 torr min^{-1}. Find the rate of effusion of hydrogen gas at a pressure of 2000 torr at the same temperature.
(a) 20 torr min^{-1} (b) 10 torr min^{-1}
(c) $30\sqrt{2}$ torr min^{-1} (d) $20\sqrt{2}$ torr min^{-1}

82. The rate of diffusion of methane at a given temperature is twice that of a gas X. The molecular weight of X is
(a) 64.0 (b) 32.0
(c) 4.0 (d) 8.0

83. An ideal gas obeying kinetic gas equation can be liquefied if:
(a) It cannot be liquefied at any value of P and T
(b) Its temperature is more than Boyle's temperature
(c) Its temperature is more than critical temperature
(d) Its pressure is more than critical pressure

84. Equal masses of methane and hydrogen are mixed in an empty container at 25°C. The fraction of the total pressure exerted by hydrogen is:
(a) 1/2 (b) 8/9
(c) 1/9 (d) 16/17

85. Helium atom is two times heavier than a hydrogen molecule. At 298 K, the average kinetic energy of helium atom is:
(a) Same as that of a hydrogen molecule
(b) Half that of a hydrogen molecule

(c) Two times that of a hydrogen molecule
(d) Four times that of a hydrogen molecule

86. The correct representation of Charle's law is given in:

(a) 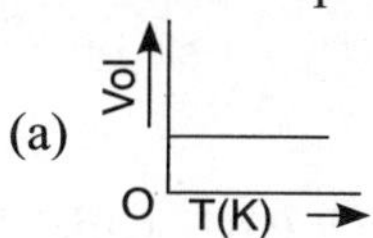

(b) 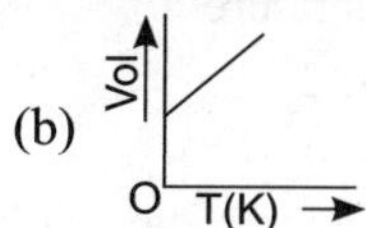

(c) 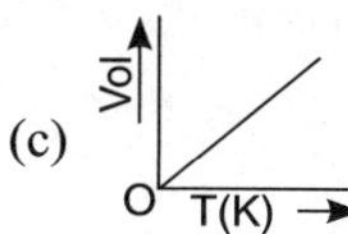

(d) 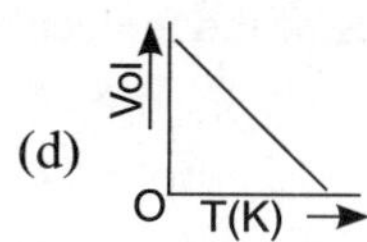

87. A 2.24 L cylinder of oxygen at NTP is found to develop a leakage. When the leakage was plugged the pressure dropped to 570 mm of Hg. The number of moles of gas that escaped will be:
(a) 0.050 (b) 0.025
(c) 0.075 (d) 0.01

88. Distribution of fraction of molecules with velocity is represented in the figure.
The velocity corresponding to point X is

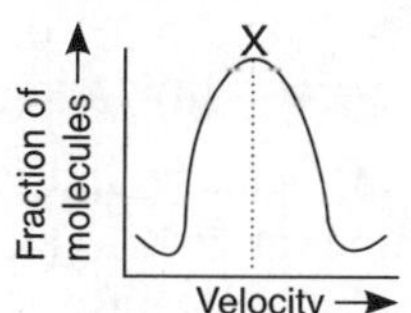

(a) $\sqrt{\frac{2RT}{M}}$ (b) $\sqrt{\frac{3RT}{M}}$
(c) $\sqrt{\left(\frac{8RT}{\pi M}\right)}$ (d) None of the above

89. A balloon having weight 50 kg is filled with 685.2 kg of helium gas at 760 mm pressure and 25°C. What will be its pay load if it displaces 5108 kg of air?
(a) 4372.8 kg (b) 4392.6 kg
(c) 4444.4 kg (d) 3482.9 kg

90. For non-zero value of force of attraction between gas molecules, gas equation will be:
(a) $PV = nRT - \frac{n^2}{V}a$ (b) $PV = nRT + nbP$
(c) $P = \frac{nRT}{V-b}$ (d) $PV = nRT$

91. Four rubber tubes are respectively filled with H_2, He, N_2 and O_2. Which tube will be reinflated first?
(a) H_2 filled tube (b) N_2 filled tube
(c) He filled tube (d) O_2 filled tube

92. A gas has the van der Waals constants, a = 1.49 L^2 atm mol^{-2} and b = 0.04 L mol^{-1}. Its Boyle's temperature is nearly:
(a) 50°C (b) 354°C
(c) 454 K (d) 408 K

93. A monoatomic gas 'A' and a diatomic gas 'B', both initially at the same temperature and pressure are compressed adiabatically from a volume V to V/2. The gas which has higher temperature is:
(a) B
(b) A
(c) Both have same temperature
(d) Can not be said

94. One litre of gas A at 2 atm pressure at 27°C and two litres of gas B at 3 atm pressure at 127°C are mixed in a 4 litre vessel. The temperature of the mixture is maintained at 327°C. What is the total pressure of the gaseous mixture?
(a) 3.93 atm (b) 3.25 atm
(c) 4.25 atm (d) 6.25 atm

95. Pressure vs density curve for an ideal gas at two different temperatures T_1 and T_2 is shown below. Which is the correct relation here?

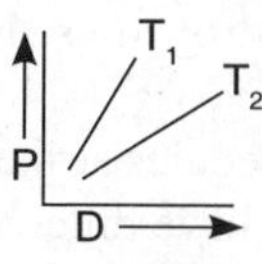

(a) $T_1 . T_2$ (b) $T_1 < T_2$
(c) $T_1 = T_2$ (d) Cannot be said

Practice Questions - III

96. The maximum number of molecules is present in:
(a) 15 L of H_2 gas at STP
(b) 5 L of N_2 gas at STP
(c) 0.5 g of H_2 gas
(d) 10 g of O_2 gas

97. The root mean square velocity of one mole of a monoatomic gas having molar mass M is u_{rms}. The relation between the average kinetic energy (E) of the gas and u_{rms} is:
(a) $u_{rms} = \sqrt{(3E/2M)}$
(b) $u_{rms} = \sqrt{(2E/3M)}$
(c) $u_{rms} = \sqrt{(2E/M)}$
(d) $u_{rms} = \sqrt{(E/3M)}$

98. Positive deviation from ideal behaviour takes place because of:

(a) Molecular interaction between atoms and $\frac{PV}{nRT} > 1$

(b) Molecular interaction between atoms and $\frac{PV}{nRT} < 1$

(c) Finite size of atoms and $\frac{PV}{nRT} > 1$

(d) Finite size of atoms and $\frac{PV}{nRT} < 1$

99. At a certain temperature for which RT = 25 L atm. mol^{-1}, the density of a gas, in g L^{-1}, is d = 2.00 P + 0.020 P^2, where P is the pressure in atmosphere. The molecular weight of the gas in g mol–1 is:

(a) 60 (b) 75
(c) 50 (d) 35

100. At 100°C and 1 atm, if the density of liquid water is 1.0 g cm^{-3} and that of water vapour is 0.0006 g cm^{-3}, then the volume occupied by water molecules in 1 litre of steam at that temperature:

(a) 6 cm^3 (b) 60 cm^3
(c) 0.6 cm^3 (d) 0.06 cm^3

101. The compressibility factor of a gas is less than unity at STP. Therefore:

(a) $V_m > 22.4$ L (b) $V_m < 22.4$ L
(c) $V_m = 22.4$ L (d) $V_m = 44.8$ L

102. A closed vessel contains equal number of molecules of O_2 and H_2. Consider the following statements:

1. The average speed of hydrogen molecules will be greater.
2. Weight of hydrogen is 1/8th of the weight of oxygen.
3. Hydrogen molecules strike the walls more often.
4. The two gases have different average energies.

The statements are wrong in:

(a) 2 and 4 (b) 2 and 3
(c) 1 and 4 (d) 1 and 3

103. The rms velocity of hydrogen is $\sqrt{7}$ times the rms velocity of nitrogen. If T is the temperature of the gas:

(a) $T(H_2) = T(N_2)$ (b) $T(H_2) > T(N_2)$
(c) $T(H_2) < T(N_2)$ (d) $T(H_2) = \sqrt{7}\,T(N_2)$

104. Which of the following statement (s) is (are) correct?

(1) A plot of log KP versus 1/T is linear
(2) A plot of log (X) versus time is linear for a first order reaction $X \longrightarrow P$
(3) A plot of log P versus 1/T is linear at constant volume
(4) A plot of P versus 1/V is linear at constant temperature.

(a) 1, 2 (b) 2, 4
(c) 2, 3 (d) 1, 4

105. When the critical volume of gas is 0.072 L mol^{-1}. The radius of the molecule will be, in cm:

(a) $(4\pi/3 \times 10^{-20})^{1/3}$
(b) $(3/4\pi \times 10^{-23})^{1/3}$
(c) $(3/4\pi \times 10^{-17})$
(d) $(3\pi/4 \times 10^{-33})^{1/3}$

106. Equal weights of ethane and hydrogen are mixed in an empty container at 25°C. The fraction to total pressure exerted by hydrogen is:

(a) 1:2 (b) 1:1
(c) 1:16 (d) 15:16

107. The ratio of root mean square velocity to average velocity of a gas molecule at a particular temperature is:

(a) 1:1.086 (b) 2:1.086
(c) 1.086:1 (d) 1.086:2

108. Equal weights of methane and oxygen are mixed in an empty container at 25°C. The fraction of the total pressure exerted by oxygen is:

(a) 1/2 (b) 2/3
(c) 1/3 × 273/298 (d) 1/3

109. An L.P.G. cylinder contains 15 kg of butane gas at 27°C and 10 atmospheric pressure. It was leaking and its pressure fell down to 8 atmospheric pressure after one day. The gas leaked in 5 days is:

(a) 10 kg (b) 3 kg
(c) 15 kg (d) 12 kg

110. The partial pressure of oxygen in a flask containin 16 g O_2 and 32 g SO_2 is:

(a) 1/16 of total pressure
(b) 1/2 of total pressure
(c) 2/3 of total pressure
(d) None of the above

111. Match the following:

List I	List II
1. Critical temperature	(1) a/R_b
2. Boyle's temperature	(2) $2a/R_b$
3. Inversion temperature	(3) T/T_c
4. Reduced temperature	(4) $8a/27\,R_b$

The correct matching is

	1	2	3	4
(a)	(2)	(4)	(1)	(3)
(b)	(4)	(1)	(2)	(3)
(c)	(3)	(2)	(1)	(4)
(d)	(4)	(3)	(2)	(1)

112. 10 mL of propane are mixed with 70 mL of oxygen and exploded. What would be the volume of residual gases after explosion and after exposure of the residual gases to alkali? All volume measurements are made at the same temperature and pressure.

(a) 55 mL, 25 mL (b) 40 mL, 15 mL
(c) 50 mL, 20 mL (d) 35 mL, 40 mL

113. The most probable speeds of the molecules of gas A at T_1K and gas B at T_2K are in the ratio 0.715:1. The same ratio for gas A at T_2K and gas B T_1K is 0.954. Find the ratio of molar masses M_A:M_B.

(a) 1.965 (b) 1.0666
(c) 1.987 (d) 1.466

114. A gas cylinder has 370 g of oxygen at 298 K and 30 atm pressure. If the cylinder was heated upto 348 K then the valve were held open until the gas pressure was 1 atm and the temperature remains 348 K. What mass of oxygen would escape in this condition?

(a) 349 g (b) 359 g
(c) 329 g (d) 339 g

115. A 200 mL flask having oxygen at 220 mm and a 300 mL flask having nitrogen at 100 mm are connected in such a way that O_2 and N_2 may combine in their volumes, if temperature is kept constant. Find the total pressure of the gaseous mixture.

(a) 158 mm (b) 138 mm
(c) 148 mm (d) 168 mm

116. Equal masses of methane and oxygen are mixed in an empty container at 25°C. The fraction of the total pressure exerted by oxygen is:

(a) 1/3 × 273/298 (b) 1/3
(c) 1/2 (d) 2/3

117. In a mixture of CO_2 and H_2, the P_{CO_2} is 0.4 atm and P_{Total} is 2 atm. The percentage composition of the mixture by volume can be given as:

(a) CO_2 = 20%, H_2 = 80%
(b) CO_2 = 40%, H_2 = 60%
(c) CO_2 = 80%, H_2 = 20%
(d) CO_2 = 60%, H_2 = 40%

118. Find kinetic energy of 0.05 mole of an ideal gas at 273 K.

(a) 1702.2 J (b) 170.22 J
(c) 17.022 J (d) 34.44 J

119. Van der Waal equation for CH_4 at low pressure can be given as:

(a) $PV = RT - Pb$ (b) $PV = RT + \frac{a}{V}$
(c) $PV = RT - \frac{a}{V}$ (d) $PV = RT + Pb$

120. An open vessel having air is heated from 27°C to 127°C. The fraction of air which goes out with respect to originally present is:

(a) 2/3 (b) 1/3
(c) 3/4 (d) 1/4

Answer Keys

1. (d)	2. (d)	3. (b)	4. (c)	5. (d)	6. (d)	7. (c)	8. (c)	9. (d)	10. (d)
11. (a)	12. (a)	13. (b)	14. (b)	15. (b)	16. (a)	17. (b)	18. (d)	19. (a)	20. (a)
21. (a)	22. (a)	23. (a)	24. (d)	25. (c)	26. (b)	27. (b)	28. (c)	29. (c)	30. (c)
31. (a)	32. (d)	33. (a)	34. (a)	35. (c)	36. (b)	37. (b)	38. (d)	39. (a)	40. (c)
41. (d)	42. (c)	43. (b)	44. (a)	45. (b)	46. (d)	47. (c)	48. (c)	49. (b)	50. (a)
51. (b)	52. (a)	53. (b)	54. (c)	55. (a)	56. (d)	57. (a)	58. (b)	59. (c)	60. (a)
61. (a)	62. (c)	63. (b)	64. (a)	65. (c)	66. (b)	67. (b)	68. (b)	69. (b)	70. (b)
71. (a)	72. (c)	73. (c)	74. (d)	75. (a)	76. (b)	77. (c)	78. (a)	79. (d)	80. (a)
81. (d)	82. (a)	83. (a)	84. (b)	85. (a)	86. (c)	87. (b)	88. (a)	89. (a)	90. (a)
91. (a)	92. (c)	93. (b)	94. (b)	95. (a)	96. (a)	97. (c)	98. (b)	99. (c)	100. (c)
101. (b)	102. (a)	103. (c)	104. (b)	105. (b)	106. (d)	107. (c)	108. (d)	109. (c)	110. (c)
111. (b)	112. (c)	113. (d)	114. (b)	115. (c)	116. (b)	117. (a)	118. (b)	119. (c)	120. (d)

Hints and Explanations for Selective Questions

2. NH_3 and HCl react to form solid NH_4Cl.

4. By definition, pressure is force per unit area.

5. Ideal gas cannot be liquefied.

8. Avogadro's law helps us to arrive at this conclusion.

9. As the condition of constant pressure, or constant volume is not mentioned so the gas may be monoatomic or diatomic.

11. At high temperature and low pressure, a real gas acts as an ideal gas.

14. $P = \frac{nRT}{V} = \frac{0.6 \times 0.082 \times 298}{3} = 4.89$ atm

16. Total mol of gases $= \frac{4}{32} + \frac{2}{2} = \frac{9}{8}$

$P = \frac{nRT}{V} = \frac{9 \times 0.082 \times 273}{8}$

$= 25.18$ atm.

18. $V = \frac{W\,RT}{M\,P} = \frac{9 \times 0.0821 \times 300 \times 760}{28 \times 750}$

$= 8.0225$ L

20. As molecular weight is same, so rms $\infty \sqrt{T}$, that is, for 4 ×, temperature must be 16 times.

22. $PV = \frac{W}{M} RT$

$M = \frac{WRT}{PV} = \frac{7.5 \times 0.082.62 \times 273}{1 \times 5.6} = 29.98$

that is, NO (14 + 16 = 30)

24. $Z = \frac{PV}{RT}$

So, $V = \frac{ZRT}{P}$

$= \frac{0.5 \times 0.082 \times 273}{100} = 0.1119$ L.

27. $M = \frac{WRT}{PV}$

$= \frac{4 \times 0.082 \times 546}{2 \times 5.6035} = 16.$

28. $KE = \frac{3}{2}\ nRT$

$= \frac{3}{2} \times 4 \times 2 \times 400 = 4800$ K.

30. $P(CH_4) = \frac{nCH_4}{\text{Total moles}} \times 720$

$= \frac{1}{4} \times 720 = 180$ mm.

31. $V_2 = \frac{P_1V_1}{P_2}$

$= \frac{730 \times 380}{760} = 365$ mL.

33. $\frac{P_1V_1}{T_1} = \frac{P_2V_2}{T_2}$

$\frac{P \times 20}{300} = \frac{2P \times V_2}{600}$

$V_2 = 20$ L.

35. $V_{av} \cdot \propto \sqrt{(1/M)}$

$\frac{V\,(SO_2)}{V\,(gas)} = \sqrt{[M(gas)/M(SO_2)]}$

On solving

M (gas) = 16 that is, CH_4.

36. $P = \frac{nRT}{V} = \frac{2 \times 0.082 \times 540}{44.8}$

$= 1.97$ atm ≈ 2 atm.

37. $r_1 : r_2 = \sqrt{(16:1)}$ or 4:1

rates are in the inverse ratio of square root of their densities

38. More the dipole moment more is the critical temperature.

39. P = nRT/V

$P = \frac{2 \times 0.082 \times 546}{44.8}$

$= 2$ atm.

43. As ammonia will have a higher diffusion rate than NH_3 due to its lower molecular weight.

44. As temperature is constant, pressure decreases during expansion while kinetic energy remains same.

45. The temperature at which a real gas behaves like an ideal gas is called Boyle's temperature or Boyle's point.

46. As average $K.E = \frac{3}{2}\frac{RT}{N_0}$

that is, (K.E) $\propto$ T

Hence $(K.E)_{313}/(K.E)_{293} = 313/293$

48. As liquification of gas $\propto$ 'a'

that is, $NH_3 > CH_4 > N_2 > O_2$.

51. As d ∝ pressure

∝ 1/temperature

53. Follow Avogadro's law.

57. 6.4 g of SO_2 at NTP (that is, at 0°C temperature and 1 atm pressure) occupies a volume of 2.24 L.

64 g of SO_2 that is, 1 mole SO_2 at NTP occupies a volume of 22.41 L.

Now 1 mole of an ideal gas occupies 22.4 L at NTP condition.

Therefore SO_2 acts as an ideal gas.

59. $\frac{PV}{T} = \frac{P'V'}{T'}$.

Here V′ = 2V, T′ = 2T

$\frac{PV}{T} = \frac{P' \times 2V}{2T}$ or P = P′

that is, pressure remains same.

60. Moles of $H_2 = 3/2 = 1.5$.

Moles of $O_2 = 4/32 = 1/8$

kinetic energy ∝ n (moles of gas)

$\frac{K.E.(H_2)}{K.E.(O_2)} = \frac{1.5}{1.8} = 12{:}1.$

62. $M = \frac{w}{V} \cdot \frac{RT}{P} = \frac{dRT}{P}$

that is, $M = \frac{1.964 \times 10^{-3} \times 82.1 \times 273}{1}$

= 44, that is, the gas is CO_2.

63. According to Boyle's law,

$P_A \times V_A = 0.4P_A \times V_B$

$100 \times P_A = 0.4\, P_A \times V_B$

$V_B = 250$ cc

Before opening the stopcock, volume of gas in bulb B must be (250 – 100) = 150 mL.

64. $\frac{V_1T_2}{V_2T_1} = \sqrt{(M_2/M_1)}$

$\frac{500 \times T_2}{1000 \times 30} = \sqrt{64/4}$

$T_2 = \frac{1000 \times 30 \times 4}{500} = 240.$

65. Given number of moles of an ideal gas (n) = 2; temperature of the gas (T) = 546 volume of gas (V) = 44.8 L

we know from the ideal gas equation the pressure (P)

$= \frac{nRT}{V}$

$= \frac{2 \times 0.0821 \times 546}{44.8} = 2$ atm

(where R = gas constant equal to 0.0821 litre atm/K/atm)

67. For A.

$P_1 = 2P, V_1 = 2V, T_2 = 2T$

For B

$P_2 = P, V_2 = V, T_2 = T$

According to ideal gas equation,

$P_1V_1/n_1RT_1 = P_2V_2/n_2RT_2$

$2P \times 2V/n_1R \times 2T = P \times V/n_2RT$

$2/n_1 = 1/n_2$

$n_1/n_2 = 2{:}1$

68. Most probable velocity increases and fraction of molecule possessing most probable velocity decreases.

70. $V_1/V_2 = \sqrt{(T_1/T_2)}$

$10^4/3 \times 10^4 = \sqrt{(100/T_2)}$

$1/9 = 100/T_2$

$T_2 = 9 \times 100 = 900.$

71. Density $= \frac{\text{Mass}}{\text{Volume}}$

Due to ideal behaviour at a given temperature and pressure volume and molar mass are same. so they have same vapour density.

73. r ∝ u

$u = \sqrt{(3RT/M)}$ or $r_1/r_2 = \sqrt{(T_1M_2/T_2M_1)}$

$r(N_2)/r(SO_2) = \sqrt{(T_1 \times 64/323 \times 28)} = 1.625$

or

$T_1 = \frac{(1.625)^2 \times 323 \times 28}{64} = 373$ K.

76. Using Graham's law, To_2 is 20 seconds.

77. $\mu_1/\mu_2 = \sqrt{(T_1M_2/M_1T_2)}$

$= \sqrt{(50 \times 32/800 \times 2)} = 1$

79. As $u = \sqrt{(3PV/m \times N_0)} = \sqrt{(3P/N_0)} \cdot (1/d)$

So, $u \propto 1/\sqrt{d}$

80. $V_1/V_2 = \sqrt{(T_1/T_2)}$

$0.3/V_2 = \sqrt{(300/1200)}$

$V_2 = 0.6$ m/sec.

81. At a given temperature, rate of effusion ∝ P and α $1/\sqrt{M}$.

Rate of effusion of hydrogen gas

$= 10 \text{ torr min}^{-1} \times \frac{2000}{1000} \times \sqrt{(4/2)}$

$= 20\sqrt{2} \text{ torr min}^{-1}$

82. $r_{CH_4}/r_x = \sqrt{(M_x/M_{CH_4})}$

$2 = \sqrt{(M_x/16)}$

$M_x = 64.$

84. Applying Dalton's law.

$P_{H_2} = \frac{W/2}{W/2 + W/16} \times P_{total}$

On solving,

$P_{H_2} = 8/9 \, P_{total}.$

85. Average kinetic energy depends only on temperature and does not depend upon the nature of the gas.

87. P (drop in pressure) = 760 – 570

$= 190 \text{ mm} = \frac{190}{760} \text{ atm}$

PV = nRT

$n = \frac{\text{P (drop in pressure) V}}{\text{RT}}$

$= \frac{190 \times 2.24}{760 \times 0.821 \times 273} = 0.0249$

= 0.025 moles.

88. It is most probable velocity given by $\sqrt{2}$ RT/M.

89. Mass of the filled balloon

= 50 + 685.2 = 735.2 kg

Pay load = Mass of displaced air – Mass of balloon = 5108 – 735.2 = 4372.8 kg.

90. For n mol of a real gas, van der Waals equation is

$P + \frac{n^2a}{V^2}(V - nb) = nRT$

When $\frac{n^2a}{V^2} \neq 0$

$P + \left(\frac{n^2a}{V^2}\right) V = nRT$

$PV + \frac{n^2a}{V^2} = nRT$

$PV = nRT - \frac{n^2a}{V^2}.$

91. This is because rate of diffusion of H_2 is maximum ($r \propto 1/\sqrt{M}$)

92. $T_B = \frac{a}{Rb} = \frac{149}{0.0812. \times 0.04} = 453.7 \text{ K}$

= 454 K.

94. For gas A, the volume increases from one litre to four litre. This would reduce the pressure but the temperature increases from 300 K to 600 K.

$p_A = 2 \times \frac{1}{4} \times \frac{600}{300} = 1 \text{ atm}$

Similarly,

$p_B = 3 \times \frac{2}{4} \times \frac{600}{400} = \frac{4}{9} \text{ atm}$

So, total pressure is (1 + 2.25) atm = 3.25 atm.

96. In 15 L of H_2 gas at STP, the number of molecules.

$= \frac{6.023 \times 10^{23}}{22.4} \times 15 = 4.033 \times 10^{23}$

In 5 L of N_2 gas at STP

$= \frac{6.023 \times 10_{23} \times 5}{22.4} = 1.344 \times 10^{23}$

In 0.5 g of H_2 gas

$= \frac{6.023 \times 10_{23} \times 0.5}{2} = 1.505 \times 10^{23}$

In 10 g of O_2 gas

$= \frac{6.023 \times 10^{23}}{32} = 1.882 \times 10^{23}$

Hence, maximum molecules are in 15 litre H_2 at STP.

97. Rms of one mole of monoatomic gas is

$u_{rms} = \sqrt{(3RT/M)}$

Average K.E. is $E = \frac{3}{2}RT$ or 2E = 3RT

$u_{rms} = \sqrt{(2E/M)}$

98. For positive deviation

PV/nRT > 1

PV = nRT + nPb

PV/nRT = 1 + Pb/RT

PV/nRT > 1.

99. For ideal gases,

$PV = nRT = \frac{w}{M} RT$

$P = \frac{RT}{M}\frac{(w)}{V} = \frac{RT}{M} d$

M = RT (d/P)

Given d = 2.00 + 0.020 P^2 (for a real gas)

d/P = 2.00 + 0.020 P

$\underset{P \to 0}{lt}$ d/P = 2.00

which is d/P for an ideal gas.

So, M = RT × 2 = 25 × 2 = 50 g mol^{-1}

100. $1L = 1000\ mL = 1000\ cm^3$

mass = density × volume

$= (0.0006\ g\ cm^{-3}) \times (1000\ cm^3)$

$= 0.6\ g$

18 g of water = 18 cm^3

0.6 g of water = 0.6 cm^3

Hence, actual volume occupied by molecules

$= 0.6\ cm^3$.

101. $Z = PV/nRT$

Given $Z < 1$

$PV/nRT < 1$ or $PV < nRT$

$(1\ atm) \times V < 1\ mol \times (0.0821\ L\ atm\ K^{-1}\ mol^{-1}) \times (273\ K)$

$V_m < 0.0821 \times 273\ L$

$V_m < 22.4\ L$.

102. At the same temperature, oxygen and hydrogen molecules will have the same average energy; weight of H_2 molecules is 1/16 of O_2 molecules.

103. $\mu H_2/\mu N_2 = \sqrt{[(T_{H_2}/M_{H_2}) \times (M_{N_2}/T_{N_2})]}$

$\sqrt{7} = \sqrt{(T_{H_2} \times 28/T_{N_2} \times 2)} = T_{H_2} \times 14/T_{N_2}$

$T_{N_2} = 2T_{H_2}$

So, $T_{N_2} > T_{H_2}$

104. At constant temperature

$V = K/P$ (Boyle's law)

So, $P = K/V$

As such, a graph plotted between P and 1/V is linear at a constant temperature.

105. $V_c = 3b$, assuming the gas to obey van der Waals equation.

$$b \text{ (the covolume)} = \frac{0.072}{3} = 0.024\ L\ mol^{-1}$$

$$b = \frac{24\ cm^3}{6 \times 10^{23}} \text{ per molecule, } (N_A = 6 \times 10^{23})$$

$$b = 4 \times 10^{-23}\ cm^3 \text{ per molecule} = 4 \times \frac{4}{3}\pi r^3.$$

$$\frac{4}{3}\pi r^3 = 10^{-23}$$

$$r^3 = \frac{3}{4\pi} \times 10^{-23}$$

$$r = (3/4\pi \times 10^{-23})^{1/3}\ cm.$$

106. Applying Dalton's law

$$P_{H_2} = \frac{W/2}{W/2 + W/30} \times P_{total}$$

On solving,

$P_{H_2} = 15/16\ P_{total}$

107. $u_{rms} : u_{av}$

$\sqrt{(3RT/M)} : \sqrt{(8RT/\pi M)}$

$\sqrt{3} : \sqrt{(8/\pi)} = 1.086:1$

108. Pressure exerted by oxygen will proportional to mole fraction of O_2.

$$\text{Mole fraction of } O_2 = \frac{W/32}{W/16 + W/32} = 1/3$$

109. According to ideal gas equation, $PV = nRT$

$$\frac{P_1}{P_2} = \frac{n_1}{n_2} \text{ (V and T are constant)}$$

$$\text{or } \frac{P_1}{P_2} = \frac{w_1}{w_2} \qquad \text{as } n = \frac{w}{M}$$

Hence, $10/8 = 15/w_2$ or $w_2 = 12$ kg

Gas leaked in one day = 15 – 12 = 3 kg

Gas leaked in 5 days = 3 × 5 = 15 kg

110. $\text{Moles of } O_2 = \frac{33}{32} = 1$

$\text{Moles of } SO_2 = \frac{32}{64} = 0.5$

Total moles = 1 + 0.5 = 1.5

Let total pressure = P

Partial pressure of $O_2 = XO_2 \times P$

(where XO_2 = mole fraction of oxygen)

$$= \frac{1}{1.5} \times P = \frac{P}{3/2} = \frac{2}{3}P$$

112. $C_3H_8 + 5O_2 \longrightarrow 3CO_2 + 4H_2O\ (l)$

1 mL 5 mL 3 mL

In this reaction, 6 mL of reactants give 3 mL of products.

So, 10 mL of propane + 50 mL of oxygen would yield 30 mL of CO_2.

Unreacted O_2 = (70 – 50) mL = 20 mL

So, volume after explosion = 30 + 20 = 50 mL

After exposure to alkali only O_2 remains = 20 mL.

113. Most probable speed = $\sqrt{[2RT/M]}$

So $\sqrt{[2RT_1/M_1]} : \sqrt{[2RT_2/M_2]} = 0.715:1$

$T_1/M_1 \times M_2/T_2 = (0.715)^2$

$T_2 . M_1 \times M_2/T_1 = (0.954)^2$

So $(M_2/M_1)^2 = (0.715)^2 \times (0.954)^2$

$$M_1/M_2 \text{ that is, } M_A/M_B = \frac{1}{0.715 \times 0.954}$$

$= 1.466.$

114. Initial moles of $O_2 = \frac{370}{32} = 11.6$

$$V = \frac{nRT}{P} = \frac{11.6 \times 0.082 \times 298}{30}$$

$= 9.43$ L

As volume is same and P is 1 atm so at 348 K moles of O_2 (n) $= \frac{PV}{RT}$

$$= \frac{1 \times 9.43}{0.082 \times 348} = 0.33$$

So, weight of O_2 left $= 0.33 \times 32 = 10.6$ g

Hence O_2 escaped $= 370 - 10.6 = 358.4$ g ≈ 359 g

116. Let the mass of methane and oxygen is w

$$\text{Mole fraction of oxygen} = \frac{w/32}{w/32 + w/16}$$

$$= \frac{1/32}{1/32 + 1/16} = \frac{3/32}{3/32} = 1/3$$

Let the total pressure be P.

The pressure exerted by oxygen (partial pressure) $= X_{O_2} \times P_{total} = P \times 1/3$

117. Mole fraction of CO_2 (X_{CO_2}) $= \frac{P_{CO_2}}{P_{Total}} = \frac{0.4}{2} = 0.2$

Mole fraction of H_2(X_{H_2}) $= 1 - X_{CO_2} = 1 - 0.2 = 0.8$

Hence % $Co_2 = 0.2 \times 100 = 20\%$

% $H_2 = 0.8 \times 100 = 80\%$

118. $K.E = \frac{3}{2} nRT = \frac{3}{2} \times 0.05 \times 8.314 \times 273$

$= 170.22$ J

119. As at low pressure volume is high

So $V - b = V$

Now V.W. equation can be given as

$$\left(P + \frac{a}{V^2}\right)(V) = RT$$

$$PV + \frac{a}{V} = RT$$

$$PV = RT - \frac{a}{V}$$

PREVIOUS YEARS' QUESTIONS

1. Which of the following expressions correctly represents the relationship between the average molar kinetic energy, K.E. of CO and N_2 molecules at the same temperature?

(a) $\overline{KE}_{CO} < \overline{KE}_{N2}$

(b) $\overline{KE}_{CO} > \overline{KE}_{N_2}$

(c) $\overline{KE}_{CO} = \overline{KE}_{N_2}$

(d) Cannot be predicted unless volumes of the gases are given.

[2000]

2. The beans are cooked earlier in pressure cooker, because:

(a) Boiling point increase with increasing pressure
(b) Boiling point decrease with increasing pressure
(c) Extra pressure of pressure cooker, softens the beans
(d) Internal energy is not lost while cooking in pressure cooker.

[2001]

3. Van der Waal's real gas, act as an ideal gas, at which condition?

(a) High temperature, low pressure
(b) Low temperature, high pressure
(c) High temperature, high pressure
(d) Low temperature, low pressure

[2002]

4. In Haber's process, 30 L of dihydrogen and 30 L of dinitrogen were taken for reaction which yielded only 50% of expected product. What is the composition of the gaseous mixture under afore–said conditions in the end:

(a) 20 L NH_3, 25 L N_2, 15 L H_2
(b) 20 L NH_3, 20 L N_2, 20 L H_2
(c) 10 L NH_3, 25 L N_2, 15 L H_2
(d) 20 L NH_3, 10 L N_2, 30 L H_2

[2003]

5. The maximum number of molecules is present in:

(a) 15 L of H_2 gas at STP
(b) 5 L of N_2 gas at STP

(c) 0.5 g of H_2 gas
(d) 10 g of O_2 gas

[2004]

6. If a gas expands at constant temperature, it indicates that:

(a) Kinetic energy of molecules remains the same.
(b) Number of the molecules of gas increases.
(c) Kinetic energy of molecules decreases.
(d) Pressure of the gas increases.

[2008]

7. What volume of oxygen gas (O_2) measured at 0ºC and 1 atm, is needed to burn completely one litre of propane gas (C_3H_8) measured under the same conditions?

(a) 5 L (b) 10 L
(c) 7 L (d) 6 L

[2008]

8. What is the dominant intermolecular force that must be overcome in converting liquid CH_3OH into gas?

(a) Dipole-dipole intraction
(b) Covalent bond
(c) London dispersion forces
(d) Hydrogen bonding

[2009]

9. By what factor does the volume of a gas increase when temperature in Kelvin is doubled?

(a) 1.4 (b) 2
(c) 2.8 (d) 4

[2011]

10. A gaseous mixture was prepared by taking equal moles of CO and N_2, If the total pressure of the mixture was found 1 atmosphere, the partial pressure of N_2 in the mixture is:

(a) 1 atm (b) 0.5 atm
(c) 0.8 atm (d) 0.9 atm

[2011]

11. 50 ml each of gas A and of gas B takes 150 and 200 seconds respectively for effusing through a pin hole under the similar conditions. If molar mass of gas B is 36, the molar mass of gas A will be:

(a) 32 (b) 20.25
(c) 64 (d) 96

[2012]

12. A certain gas takes three times as long to effuse out as helium. Its molecular mass will be:

(a) 36 u (b) 64 u
(c) 27 u (d) 9 u

[2012]

13. For real gases, van der Waals equation is written:

As $\left(p = \frac{an^2}{V^2}\right)(V - nb) = nRT$

where *a* and *b* are van der Waals constants
Two sets of gases are

(I) O_2, CO_2, H_2 and He
(II) CH_4, O_2 and H_2

The gases given in set-I in increasing order of *b* and gases given in set-II in decreasing order of *a*, are arranged below. Select the correct order from the following:

(a) (I) $He < H_2 < CO_2 < O_2$ (II) $CH_4 > H_2 > O_2$
(b) (I) $O_2 < He < H_2 < CO_2$ (II) $H_2 > O_2 > CH_4$
(c) (I) $H_2 < He < O_2 < CO_2$ (II) $CH_4 > O_2 > H_2$
(d) (I) $H_2 < O_2 < He < CO_2$ (II) $O_2 > CH_4 > H_2$

[2012]

14. Maximum devation from Ideal gas is expected from?

(a) $H_{2(g)}$ (b) $N_{2(g)}$
(c) $CH_{4(g)}$ (d) $NH_{3(g)}$

[2013]

15. Equal masses of H_2, O_2, and CH_4 have been taken in a container of valume 'V' at temperature 270°c in identical conditions. The ratio of the volumes of gases $H_2 : O_2 : CH_4$ would be?

(a) 8:1:2 (b) 16:1:2
(c) 8:16:1 (d) 16:8:1

[2014]

16. A gas such as Carbon monoxide would be most likely to obey the ideal gas law at:

(a) Low temperature, high pressure
(b) High temperature, low pressure
(c) High temperature, high pressure
(d) Low temperature, low pressure

[2015]

17. Equal moles of hydrogen and oxygen gases are placed in container with a pin hole through which both can escape what traction of the oxygen escapes. What fraction of the oxygen escapes in this time required for one half of the hydrogen to escape?

(a) $\frac{1}{4}$ (b) $\frac{1}{2}$
(c) $\frac{1}{8}$ (d) $\frac{3}{8}$

[2016]

Answer Keys

1. (c)	**2.** (a)	**3.** (a)	**4.** (c)	**5.** (a)	**6.** (a)	**7.** (a)	**8.** (d)	**9.** (a)	**10.** (b)
11. (b)	**12.** (a)	**13.** (c)	**14.** (d)	**15.** (b)	**16.** (b)	**17.** (c)			

Hints and Explanations

1. K.E. is proportional directly to $\sqrt{T}$. If T is same, K.E. is also same.

2. The beans are cooked earlier in pressure cooker because boiling point increases with increasing pressure.

3. At high temperature and low pressure real gas act as an ideal gas.

4. $3H_2 \quad + \quad N_2 \quad \rightarrow \quad 2NH_3$

given 30 L 30 L –

H_2 is limiting reagent

Expected volume of NH_3 = 20 L

Actual volume of NH_3 = 10 L

H_2 consumed = 15 L ; H_2 left = 15 L

N_2 consumed = 5 L ; N_2 left = 25 L

NH_3 formed = 10 L.

5. In 15 L of H_2 gas at STP, the number of molecules.

$$= \frac{6.023\times10^{23}}{22.4} \times 15 = 4.033 \times 10^{23}$$

In 5 L of N_2 gas at STP

$$= \frac{6.023\times10^{23}\times5}{22.4} = 1.344 \times 10^{23}$$

In 0.5 g of H_2 gas

$$= \frac{6.023\times10^{23}\times0.5}{2} = 1.505 \times 10^{23}$$

In 10 g of O_2 gas

$$= \frac{6.023\times10^{23}}{32} = 1.882 \times 10^{23}$$

Hence, maximum molecules are in 15 litre H_2 at STP.

6. The average translational K.E. of one molecule of an ideal gas is as follows:

$$K_t = \frac{K.E.}{N_A} = \frac{3/2\pi RT}{N_A} = \frac{3}{2}KT$$

When R/N_A = Boltzmann constant

i.e., $E_t \propto T$

Thus, at constant temperature K.E. of molecules remains same.

7. $C_3H_8 + 5O_2 \rightarrow 3CO_2 + 4H_2O$

1 5 3 4 volume

So 1 volume or 1 litre of propane requires to 5 volume or 5 litre of O_2 to burn completely.

8. As H-bonds break when liquid CH_3OH changes into vapour phase.

11. As $\frac{V_A}{V_B}\frac{B}{A} \sqrt{\frac{M_B}{M_A}}$ or $\frac{200}{150} = \sqrt{\frac{36}{M_A}}$

on solving $M_A = 20.25$

12. Using Graham's Law

$$\frac{r_1}{r_2} = \frac{V_1+2}{V_2+1} = \sqrt{\frac{M_2}{M_1}}$$

As $V_1 = V_2$ so

$$\frac{t_2}{t_1} = \frac{3}{1} = \sqrt{\frac{M_2}{4}}$$

on solving $M_2 = 36$

14. The extent of deviation can be explained by compressibility factor 'Z'. Gases undergoing liquification easily are more deviated.

$NH_3 \quad SO_3 \quad CH_4 \quad N_2 \quad H_2$ $\longrightarrow$

Dec. Order of deviation

15. Using Avogadro Law

$$V \propto n \propto \frac{w}{m}$$

$$V_{H_2} : V_{O_2} : V_{CH_4} = \frac{w}{2} : \frac{w}{32} = \frac{w}{16}$$

$= 16 : 1 : 2$

16. At high temperature and low pressure any gas (CO) can show ideal gas behaviour.

17. Using Grahams Law

$$\frac{n_{O_2}}{n_{H_2}} = \sqrt{\frac{M_{H_2}}{M_{O_2}}}$$

$$\frac{n_{O_2}}{0.5} = \sqrt{\frac{2}{32}} = \frac{1}{4}$$

$$n_{O_2} = \frac{1}{8}$$

NCERT EXEMPLAR

1. A person living in Shimla observed that cooking food without using pressure cooker takes more time. The reason for this observation is that at high altitude:
   (a) Pressure increases
   (b) Temperature decreases
   (c) Pressure decreases
   (d) Temperature increases

2. Which of the following property of water can be used to explain the spherical shape of rain droplets?
   (a) Viscosity
   (b) Surface tension
   (c) Critical phenomena
   (d) Pressure

3. A plot of volume (*V*) versus temperature (*T*) for a gas at constant pressure is a straight line passing through the origin. The plots at different values of pressure are shown in Fig. 5.6. Which of the following order of pressure is correct for this gas?

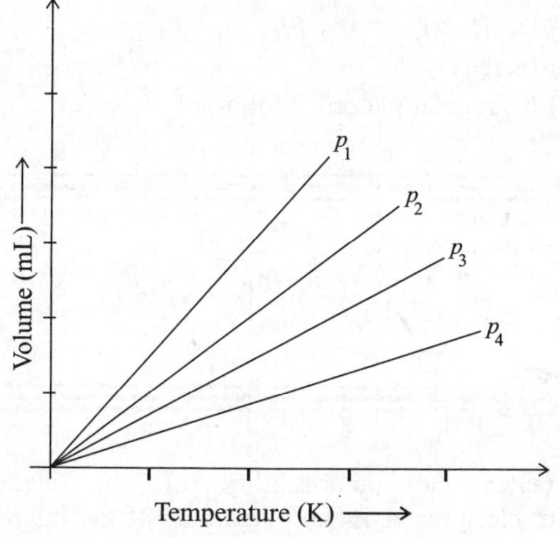

FIGURE 5.6

   (a) $p_1 > p_2 > p_3 > p_4$
   (b) $p_1 = p_2 = p_3 = p_4$
   (c) $p_1 < p_2 < p_3 < p_4$
   (d) $p_1 < p_2 = p_3 < p_4$

4. The interaction energy of London force is inversely proportional to sixth power of the distance between two interacting particles but their magnitude depends upon
   (a) Charge of interacting particles
   (b) Mass of interacting particles
   (c) Polarisability of interacting particles
   (d) Strength of permanent dipoles in the particles

5. Dipole-dipole forces act between the molecules possessing permanent dipole. Ends of dipoles possess 'partial charges'. The partial charge is
   (a) More than unit electronic charge
   (b) Equal to unit electronic charge
   (c) Less than unit electronic charge
   (d) Double the unit electronic charge

6. The pressure of a 1:4 mixture of dihydrogen and dioxygen enclosed in a vessel is one atmosphere. What would be the partial pressure of dioxygen?
   (a) 0.8×10^5 atm
   (b) 0.008 Nm^{-2}
   (c) 8×10^4 Nm^{-2}
   (d) 0.25 atm

7. As the temperature increases, average kinetic energy of molecules increases. What would be the effect of increase of temperature on pressure provided the volume is constant?
   (a) Increases
   (b) Decreases
   (c) Remains same
   (d) Becomes half

8. Gases possess characteristic critical temperature which depends upon the magnitude of intermolecular forces between the particles. Following are the critical temperatures of some gases.

Gases	H_2	He	O_2	N_2
Critical temperature in Kelvin	33.2	5.3	154.3	126

From the above data what would be the order of liquefaction of these gases? Start writing the order from the gas liquefying first

(a) H_2, He, O_2, N_2
(b) He, O_2, H_2, N_2
(c) N_2, O_2, He, H_2
(d) O_2, N_2, H_2, He

9. What is SI unit of viscosity coefficient (η)?

(a) Pascal
(b) Nsm^{-2}
(c) km^{-2} s
(d) N m^{-2}

10. Atmospheric pressures recorded in different cities are as follows:

Cities	Shimla	Bangalore	Delhi	Mumbai
p in N/m^2	1.01×10^5	1.2×10^5	1.02×10^5	1.21×10^5

Consider the above data and mark the place at which liquid will boil first.

(a) Shimla
(b) Bangalore
(c) Delhi
(d) Mumbai

11. Which curve in Fig. 5.7 represents the curve of ideal gas?

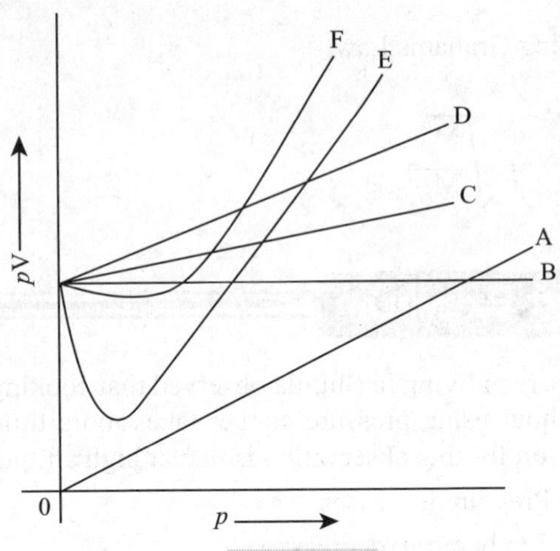

FIGURE 5.7

(a) B only
(b) C and D only
(c) E and F only
(d) A and B only

12. Increase in kinetic energy can overcome intermolecular forces of attraction. How will the viscosity of liquid be affected by the increase in temperature?

(a) Increase
(b) No effect
(c) Decrease
(d) No regular pattern will be followed

13. How does the surface tension of a liquid vary with increase in temperature?

(a) Remains same
(b) Decreases
(c) Increases
(d) No regular pattern is followed

ANSWER KEYS

1. (c)	2. (b)	3. (c)	4. (c)	5. (c)	6. (c)	7. (a)	8. (d)	9. (b)	10. (a)
11. (a)	12. (c)	13. (b)							

HINTS AND EXPLANATIONS FOR SELECTIVE QUESTIONS

1. At high altitude pressure is low so boiling takes place at low temperature and therefore cooking takes more time.

2. Due to surface tension the surface of water drops is under tension and tends to take spere share in order to reduce tension.

3. Greater the polarisability of the ineracting particles,greater is the magnitude of the interacting energy.

7. At constant volume the temperature and presuure gets increased.

10. Shimla has lowest atmospheric pressure,hence liquid will boil first in the city.

13. Surface tension of a liquid decreases with increase in temperature due to less force of attraction between molecules.

AIIMS ESSENTIALS

Assertion and Reason

In the following questions two statements (Assertion) A and Reason (R) are given, Mark.

(a) If A and R both are correct and R is the correct explanation of A;
(b) If A and R both are correct but R is not the correct explanation of A;
(c) A is true but R is false;
(d) A is false but R is true,
(e) A and R both are false.

1. (A) : The viscosity of an ideal gas is independent of pressure at constant temperature.
   (R) : As the pressure is increased, the effect of the increase in number density of molecules in compensated by a proportionate decrease in the mean free path.

2. (A) : Considering Van der Waal's equation of state for a real gas $(P + n^2a/V^2)(V - nb) = nRT$; The constant 'a' for O_2 is less than that for H_2O (g).
   (R) : The molar mass of O_2 is almost twice that of H_2O

3. (A) : In the Joule-Thomson expansion of hydrogen and helium, the temperature of the gases increases.
   (R) : These gases have very high inversion temperature.

4. (A) : The heat capacities of all the diatomic molecules attain a limiting value at higher temperatures.
   (R) : At room temperature , the translational and rotational degrees of freedom contribute to the heat capacities of gases, while the vibrational degree of freedom becomes active only at higher temperatures.

5. (A) : All molecules of an ideal gas move with the same speed.
   (R) : There is no attraction between the molecules in an ideal gas.

6. (A) : Considering the Van der Waal's quation of state $(P + a/V^2) + (V - b) = R$ for ammonia (NH_3) and nitrogen (N_2), the value of a for NH_3 is larger than that of N_2.
   (R) : Ammonia has a lower molecular weight than nitrogen.

7. (A) : The hydrogen gas warms up during the Joule-Thomson expansion.
   (R) : The temperature at which the Joule Thomson coefficient becomes negative is called Joule-Thomson inversion temperature.

8. (A) : The gases He and CO_2 are very different in their behaviour at any temperature and pressure, but their compressilibility factors are nearly the same.
   (R) : The 'Law of corresponding states' states that the compressibility factor of any gas can be written as a universal function of the reduced variables.

9. (A) : In Van der Waal's equation $(P + a/V^2)(V - b) = RT$ pressure correction (a/V^2), is due to force of attraction between molecules.
   (R) : Due to force of attraction, volume of molecules cannot be neglected.

10. (A) : A lighter gas diffuses more rapidly than a heavier gas.
    (R) : At a given temperature, the rate of diffusion of a gas is inversely proportional to the square root of its density.

11. (A) : When temperature of an ideal gas increases from 27°C to 127°C at constant P, volume increases by 100 L.

(R) : V α T at constant P.

12. (A) : Van der Waal's constant 'a' is a measure of force of attraction between molecules.

(R) : This constant is the volume correction.

13. (A) : Cp – Cv = R for an ideal gas

(R) : R is the work done when temperature of one mol of an ideal gas is increased by 1°.

14. (A) : Van der Waal's 'b' is expressed in the unit of atm L^2 mol^{-2}.

(R) : 'b' is pressure correction due to force of attraction.

15. (A) : Cp – Cv = R for an ideal gas

(R) : $(\partial E/\partial V)_T = 0$ for an ideal gas

16. (A) : The Joule-Thomson coefficient for an ideal gas is zero.

(R) : There are no intermolecular attractive forces in an ideal gas.

17. (A) : H_2 and He show some ideal gas behaviour.

(R) : All real gases deviate from ideal gas behaviour at low temperature and high pressure.

18. (A) : An ideal gas cannot be liquefied even by attaining 0.003 K with the help of adiabatic demagnetization.

(R) : The Van der Waal's constant 'a' is negligible for an ideal gas and molecules are assumed as point mass.

19. (A) : 8 g of methane occupies 11.207 litres of volume at 273 K and 1 atm pressure.

(R) : one mole of any gas at STP occupies 22.414 litres of volume.

20. (A) : The heat capacities of all the diatomic molecules attain a limiting value at higher temperature.

(R) : At room temperature , the translational and rotational degrees of freedom become active only at higher temperature .

21. (A) : The triple point of a pure substance having three phases in equilibrium has zero degree of freedom.

(R) : None of the variables like temperature and pressure can be changed without change in the number of phases.

22. (A) : Whatever be the temperature, liquid CO_2 does not exist at normal atmospheric pressure.

(R) : The triple point of CO_2 lies above 1.0 atm.

23. (A) : A value of Van der Waal's constant 'a' is larger for NH_3 than PH_3.

(R) : Hydrogen bonding is present in NH_3

24. (A) : Absolute zero is lowest possible temperature

(R) : A lower temperature would correspond to negative value of volume of gas which is physically meaningless.

25. (A) : H_2 when allowed to expand at room temperature, it causes heating effect.

(R) : H_2 has inversion temperature much below room temperature.

Answer Keys

1. (a)	2. (b)	3. (c)	4. (a)	5. (d)	6. (b)	7. (c)	8. (e)	9. (c)	10. (a)
11. (d)	12. (c)	13. (a)	14. (e)	15. (a)	16. (a)	17. (b)	18. (a)	19. (a)	20. (d)
21. (a)	22. (a)	23. (a)	24. (a)	25. (a)					

CHAPTER 6

Thermodynamics and Chemical Energetics

Chapter Outline

■ Thermodynamics ■ Thermodynamic Process ■ Thermodynamic Equilibrium ■ Heat ■ Work ■ Internal Energy ■ First Law of Thermodynamics or Law of Conservation of Energy ■ Second Law of Thermodynamics ■ Entropy ■ Zeroth Law of Thermodynamics ■ Third Law of Thermodyanmics ■ Gibbs Free Energy (G) ■ Thermochemistry ■ Heat or Enthalpy of Reaction ■ Sources of Energy ■ Conservation of Energy

Thermodynamics

Thermodynamics is a branch of science which deals with the quantitative relationships between different forms of energy; the study of interconversion of heat and work.

Main Objectives

- To decide the feasibility of a given transfromation.
- To derive various energy changes and their inter relations.
- To derive laws like Phase rule, Law of mass action.
- Presentations of experimental data in a correct manner.

Limitations

- It is not applicable when a substance is present in a small amount.
- It gives no explanation about the rate of reaction.
- It gives no explanation of system far from equation.
- It does not deal with internal structure of atoms and molecules.

Terms Related to Thermodynamics

Surrounding It consists of the region excluding the system.

System It is the region or space to be investigated or the Region at which study of pressure, temperature etc. are to be made. Systems are of two types:

Homogenous It is uniform all over, that is, made up of one phase only. For example, pure liquid or pure solid or pure gas.

Heterogenous It is non-uniform as it consists of two or more phases. For example, ice and water, solid in contact with a liquid.

Systems are further divided into three parts:

Open system In such a system, both matter and energy can be exchanged with the surroundings. For example, a beaker containing boiling water, lime kiln, ice in an open beaker.

Closed system Such a system exchanges energy with its surroundings. For example, heating of liquid in a sealed tube or ice in a closed beaker.

Isolated system In such a system, there is no exchange of matter or energy with the surroundings. For example, liquid in a sealed thermos flask or ice in a thermos flask.

Properties of a System or State Variables

State variables or functions

- The state of a system is fixed by its macroscopic properties. If any of its macroscopic property is changed the state of a system also changes. These properties are therefore referred to as state variables.
- The state variables determine the state of a system. These are fundamental properties or thermodynamic parameters. These depend only upon the initial and final states. Pressure, temperature, volume, mass and composition are important state variables. Some otherstate variables being enthalpy, free energy, internal energy and entropy. These are of the following two types, *viz*., intensive and extensive properties.

Intensive properties Such properties remain same on any division in system, that is, they do not depend upon amount of substance present in the system. For example, temperature, pressure, concentration, density, viscosity, surface tension, specific heat, refractive index, pH, EMF of a dry cell, vapour pressure, dipole moment.

Extensive properties These properties depend on the amount of substance, that is, their values are different in the divided system than in the entire system. For example, mass, volume, energy, work, internal energy, entropy, enthalpy, heat capacity, length.

- An extensive property can be made intensive by specifying it in unit amount of matter. (By dividing two extensive properties only) For example,

$$\text{Density} = \left(\frac{\text{Mass}}{\text{Volume}}\right)$$

THERMODYNAMIC PROCESS

Thermodynamic process is the path along which a change of state occurs.

Types of Processes

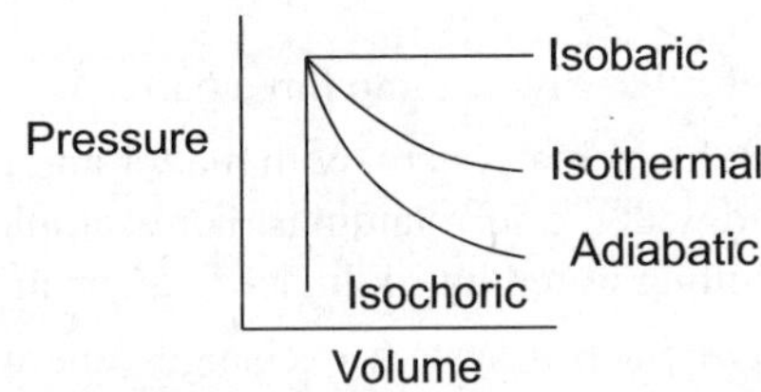

FIGURE 6.1 DIFFERENT TYPES OF PROCESSES

Isothermal process Here temperature is kept constant, that is, $\Delta T = 0$, $\Delta E = 0$.

- It is achieved by using a thermostatic control.
- Heat can be absorbed or evolved here, that is, it can be exchanged with the surroundings. For example, freezing, melting, evaporation, condensation.

Isobaric process Here pressure is kept constant, that is, $\Delta P = 0$.

Example: Expansion of a gas in an open system.

- Vaporization and heating of water up to its boiling point occurs at the same atmospheric pressure.

Isochoric process Here volume is kept constant, that is, $\Delta V = 0$.

For example, heating of a substance in a closed vessel (system) or in a non-expanding chamber.

Adiabatic process Here no exchange of heat takes place between the system and the surroundings. $Q = 0$.

- It is achieved by insulating the system or in closed and insulated containers (thermos).

Cyclic process Here the system undergoes a series of changes but finally comes back to the initial state.

- $\Delta E = 0$, $\Delta H = 0$.

Reversible or Quasi-static Process

- It is carried out in such a way that the system remains in a state of equilibrium.
- All changes occurring at any particular stage of the process will be exactly reversed when change is carried out in the opposite direction.
- It involves slow changes during operation.
- This process may occur in any direction.
- It gives rise to maximum work.
- Here driving force and opposing force differ from each other by a small value.

Irreversible Process

- Here direction of change can not be reversed by small changes in variables. All processes occurring naturally are irreversible.
- It involves fast changes during operation.
- It is a unidirectional process.
- It gives rise to network.

$$W_{rev} > W_{irr}$$

- Here, driving and opposing forces differ by a large amount.

THERMODYNAMIC EQUILIBRIUM

Thermodynamic equilibrium is of three types, namely, mechanical, thermal and chemical equilibriums.

Mechanical equilibrium There is no flow of matter from one part to another, which means that pressure is constant.

Thermal equilibrium There is no flow of heat from one part to another, that is, temperature is constant.

Chemical equilibrium There is no change in composition of any part of the system with time.

HEAT

- Heat is the energy transfer due to difference in temperature.
- Heat is a form of energy which the system can exchange with the surroundings, if they are at different temperatures; heat flows from higher temperature to lower temperature.

- Heat is expressed as Q.
  Heat absorbed by the system = Q positive
  Heat evolved by the system = Q negative

WORK

Amount of energy transferred by a force. It is the energy transfer due to difference in pressure, that is, mode of energy transfer.
W = Intensity factor × Capacity factor
It is of following types:

1. Mechanical Work = Force × Displacement
2. Electrical Work = Potential difference × Charge flow
   $$V \times Q = EnF$$
3. Expansion Work = $P \times \Delta V = -P_{ex.}(V_2 - V_1)$
   P = External pressure
   ΔV = Increase or decrease in volume
4. Gravitational Work = mgh
   Here m = Mass of body
   g = Acceleration due to gravity
   h = Height moved

Unit of work is dyne cm or erg (CGS) or newton metre or joule (SI).

FACTS TO REMEMBER

- If the gas expands, $[V_2 > V_1]$ then work is done by the system and W is negative.
- If the gas contracts, $[V_2 < V_1]$ then work is done on the system and W is positive.

Different Types of Work and Their Formulas

1. Work done in a reversible isothermal process
   $$w = -2.303\ nRT \log_{10} \frac{V_2}{V_1}$$
   $$w = -2.303\ nRT \log_{10} \frac{P_1}{P_2}$$
2. Work done in an irreversible isothermal process
   $W = -P_{ext}(V_2 - V_1)$ or,
   $W = -P\,\Delta V$

INTERNAL ENERGY

- The energy stored within a substance is called its internal energy. The absolute value of internal energy cannot be determined.
- It is the total energy of a substance depending upon its chemical nature, temperature, pressure, and volume, amount of substrate. It does not depend upon path by which the final state is achieved.
  $E = E_t + E_r + E_v + E_e + E_n + E_{PE} + E_B$
  E_t = Transitional energy
  E_r = Rotational energy
  E_v = Vibrational energy
  E_e = Electrical energy
  E_n = Nuclear energy
  E_{PE} = Potential energy
  E_B = Bond energy
- The exact measurement of interal energy is not possible so it is determined as ΔE, that is, change in internal energy is measured as:
  $$\Delta E = \Sigma E_p - \Sigma E_R$$
- It is an extensive property.
- Internal energy is a state property.
- The change in the internal energy does not depend on the path by which the final state is reached.
- For a cyclic process, ΔE is zero (E = state function)
  $$E \alpha T$$
- At constant volume (Isochoric) $Q_v = \Delta E$
- In adiabatic expansion, the gas cools as ΔE decreases.
- For exothermic process, ΔE is negative as $E_R > E_p$
- For endothermic process, ΔE is positive as $E_R < E_p$
- Internal energy for an ideal gas is a function of temperature only it means when temperature is kept constant, ΔE is zero.
- For an ideal gas, it is totally kinetic energy as there is no molecular interaction.
- It is determined by using Bomb Calorimeter
  $$\Delta E = \frac{Z \times \Delta T \times m}{w}$$
  Z = Heat capacity of Bomb Calorimeter
  ΔT = Rise in temperature
  w = Weight of substrate (amount)
  m = Molar mass of substrate

FIRST LAW OF THERMODYNAMICS OR LAW OF CONSERVATION OF ENERGY

- First law of thermodynamics was introduced by Helmholtz.
- This law states that energy can neither be created nor destroyed but can be converted from one form to another.
- The total energy of universe is constant
  ΔE = Q – W (work done by the system)

$\Delta E = Q + W$ (work done on the system) $Q = \Delta E - W$

$\delta E = \delta Q - \delta W$ (work done by the system or in expansion)

$\delta E = \delta Q + \delta W$ or $\delta E = \delta Q + P\delta V$

(Work done on the system or in compression).

Q = + ve: Heat is absorbed by the system.

Q = – ve: Heat is evolved by the system.

- For a reversible cyclic process $\Delta E = 0$,

$$Q = -W_{max} = P\Delta V = 2.303nRT \log_{10} \frac{V_2}{V_1} \text{ or } \frac{P_1}{P_2}$$

- For an isochoric process $\Delta E = Q$ ($\Delta V = 0$)
- For an adiabatic process $\Delta E = W$ (as $q = 0$)
- During adiabatic compression

$$W = \frac{(P_1V_1 - P_2V_2)}{(\gamma - 1)}$$

- It fails to explain the expansion of gases invaccum when $\Delta E = O$, $\Delta H = O$, $W = O$.

Enthalpy Change

- Enthalpy change is the amount of heat absorbed or evolved by a system at constant pressure or constant volume.
- It is not a state function.
- It is denoted by H at constant pressure.
- It is denoted by E at constant volume.
- It is the sum of internal energy and pressure volume work.
- $H = E + PV$ or, $\Delta H = \Delta E + P\Delta V$
- Enthalpy of the elements in gaseous, liquid or solid or cobined state is taken as zero.

Relation between Δ H and Δ E

$$\Delta H = \Delta E + \Delta n_g RT = \Delta E + P \cdot \Delta V$$
$$= \Delta E + \Delta(PV)$$

Here ΔH = change in enthalpy in constant pressure

ΔE = change in enthalpy in constant volume

Δn_g = difference in the moles of gaseous reactants and product

Joule–Thomson Effect

The phenomenon of change of temperature when a gas is made to expand adiabatically, from a region of high pressure into a region of low pressure, is known as Joule–Thomson effect.

Joule–Thomson coefficient (μ) It is the number of degrees of temperature change produced per atmospheric drop in pressure under constant enthalpy conditions when the gas is allowed to expand through a porous plug.

$$\mu_{JT} = \left(\frac{\delta T}{\delta P}\right)_n = \frac{-\left(\frac{\delta T}{\delta P}\right)_T}{C_P}$$

- $\mu_{JT} = 0$ for an ideal gas as van der Waals forces of attraction are negligible.

SECOND LAW OF THERMODYNAMICS

- It is not possible to convert heat into work without compensation.
- Work can always be converted into heat but the conversion of heat into work does not take place under all conditions.
- It is impossible to construct a machine that is able to transfer heat by a cyclic process, from a colder to a hotter body, unless work is done on the machine by some outside agency.
- Whenever a spontaneous process takes place it is accompanied by an increase in the total entropy of the universe.

Carnot Cycle

- Carnot cycle demonstrates the maximum convertibility of heat into work without any compensation.
- It is a theoretical and an imaginary cycle

$$\eta = \frac{W}{Q_2} = \frac{T_2 - T_1}{T_2} = \frac{Q_2 - Q_1}{Q_2}$$

η = Hermodynamic efficiency

Q_2 = Heat absorbed from the source at T_2

Q_1 = Heat lost to sink at T_1

When T_2 is much greater than T_1 η is more but less than one.

ENTROPY

- The disorder or randomness in a system is measured in terms of entropy (S). It is a state function and an extensive property.
- The absolute value of S is not determined, so mostly change in entropy ΔS is measured.

$$\Delta S = \frac{Q_{rev}}{T} = -2.303nR \log \frac{V_2}{V_1}$$

As, $Q_{rev} = -W_{rev} = -2.303 \text{ nR} \log \frac{V_2}{V_1}$

Unit of entropy is J K^{-1} mol^{-1}

Randomness α Entropy

Facts About Entropy

- In nature, all processes are irreversible followed by an increase in entropy. Entropy of the universe tends towards a maximum.

$$\Delta S = 2.303 \times C_P \log\left(\frac{T_2}{T_1}\right)$$

C_p = Molar heat capacity at constant pressure.

- For a spontaneous process, entropy increases.

ΔS = Positive

At 298K and 1 atm $\Delta S° = \Sigma S°_p - \Sigma S°_R$

Spontaneous Process

The chemical or physical process which occurs in a particular set of conditions, either on its own or after proper initiation is known as spontaneous process. All natural processes are spontaneous.

- $\Delta S_{sys} + \Delta S_{surr} = 0$ for reversible processes
- $\Delta S_{sys} + \Delta S_{surr} > 0$ for irreversible processes

Entropy Changes in Different Processes

1. Isothermal reversible expansion of an ideal gas

$$\Delta S = \text{nR} \log_e\left(\frac{V_2}{V_1}\right) \text{ or } \left(\frac{P_1}{P_2}\right)$$

2. Adiabatic reversible expansion

$Q = 0, \Delta S = 0$

3. Isobaric process

$$\Delta S = C_P \log\frac{T_2}{T_1}$$

4. Isochoric process

$$\Delta S = C_V \log\frac{T_2}{T_1}$$

5. Entropy changes during phase transition

$$\Delta S = S_2 - S_1 = \frac{Q_{rev}}{T} = \frac{\Delta H}{T}$$

$$\Delta S_{fusion} = \frac{\Delta H_{fusion}}{T_{fusion \text{ or } MP}}$$

$$\Delta S_{vap} = \frac{\Delta H_{vap}}{T_{vap \text{ or } BP}}$$

Unavailable energy = Entropy × Temperature

- When a rubber band is stretched, entropy decreases because the macro molecules get uncoiled and hence arranged in a more ordered manner, which means that randomness decreases.
- When an egg is boiled, the entropy increases because denaturation occurs resulting into a change of proteins from helical form into random coiled form.
- Molecules kept in a large volume container have a high value of entropy.
- Cases of increase in entropy
  1. Dissolution of a solute in water
  2. Decomposition of a compound
  3. Vaporization and fusion
  4. Expansion of ideal gas from one container to an evacuated chamber.
- Decrease in entropy occure during crystallization and combination.
- Entropy is directly proportional to atomic weight. For example, $I_2 > Br_2 > Cl_2$
- Entropy is directly proportional to number of bonds For example, ethane > ethene > ethyne

ZEROTH LAW OF THERMODYNAMICS

According to the zeroth law, "Two systems in thermal equilibrium with a third system are also in thermal equilibrium with each other."

THIRD LAW OF THERMODYANMICS

At the absolute zero of temperature, the entropy of any substance may become zero but does become zero in case of perfectly crystalline structure.

$$\int_{\lim T \to 0} S = 0$$

- In case of CO and NO molecules in solid state, entropy or randomness is not zero even at absolute zero temperature (0 K).

GIBBS FREE ENERGY (G)

- Free energy of a system is defined as the maximum amount of energy available to a system during a process that can be converted into useful work. It is a state function and an extensive property.

- It is denoted by G or $\Delta G°$

$$G = H - TS$$
$$\Delta G = \Delta H - T\Delta S$$
$$\Delta G° = \Delta H° - T\Delta S°$$

This is called Gibbs–Helmholtz equation.

- It is a state function and an extensive property.
- During a spontaneous change, ΔG decreases.
- ΔG = – ve (spontaneous process)
- ΔG = + ve (non-spontaneous process)
- ΔG = 0 (process in equilibrium)
- In exergonic reaction, ΔG = –ve.
- In endergonic reaction, ΔG = +ve.

TABLE 6.1 Gibbs Free Energy: Spontaneous and Non-spontaneous Processes

ΔH	ΔS	ΔG = ΔH – T ΔS	Remarks
–	+	Always –ve	Spontaneous
+	–	Always +ve	Non-spontaneous
+	+	+ve at low temp	Non-spontaneous
		–ve at high temp	Spontaneous
–	–	–ve at low temp	Spontaneous
		+ve at high temp	Non-spontaneous

Some Relations

$$\Delta G^0 = -n\,FE^\circ$$

Here

n = Number of mole of electron

F = Faraday constant

E° = Standard emf

$$\Delta G^\circ = -2.303\,RT\log K$$

K = Equilibrium constant

$$\Delta G^\circ = \Sigma G^\circ_P - \Sigma G^\circ_R$$

Heat Capacity

- Heat capacity of a system is defined as the quantity of heat required to raise the temperature of the system by one degree.
- The specific heat capacity is denoted by C and can be calculated with the help of the formula

  $$Q = C \times m \times \Delta T$$

  Here,

  Q = Heat required to raise the temperature by 1 degree

  C = Specific heat capacity

  m = Mass

  ΔT = Temperature change

- Let a very small quantity of heat dQ be given to a system so that the temperature of the system rises by dT.

  Thus, heat capacity $= \dfrac{dQ}{dT}$

- At constant volume, Q = ΔE

  So, heat capacity at constant volume,

  $$C_v = \left(\frac{\partial E}{\partial T}\right)_v$$

- At constant pressure, Q = ΔE + PΔV = ΔH

  So, heat capacity at constant pressure,

  $$C_p = \left(\frac{\partial H}{\partial T}\right)_p$$

- For 1 mole of a gas, heat capacities at constant volume and constant pressure are denoted by C_v and C_p respectively. These are termed as **molar heat capacities**. Thus, for 1 mole of a gas,

  $$C_v = \left(\frac{\partial E}{\partial T}\right)_v$$

  and $C_p = \left(\frac{\partial H}{\partial T}\right)_p$

- $C_P - C_V = P\Delta V = R$

Importance of C_p and C_v From the ratio of C_p and C_v we get an idea about the atomicity of a gas.

- For monoatomic gas, C_p = 5 calories; C_v = 3 calories

  hence $\gamma = \dfrac{C_P}{C_V} = \dfrac{5}{3} = 1.67$

- For diatomic gas, C_p = 7 calories; C_v = 5 calories

  hence $\gamma = \dfrac{C_P}{C_V} = \dfrac{7}{5} = 1.40$

- For triatomic gas, C_p = 8 calories; C_v = 6 calories

  hence $\gamma = \dfrac{C_P}{C_V} = \dfrac{8}{6} = 1.33$

Phase transition The change of matter from one state (solid, liquid or gas) to another state is called Phase transition.

- Such changes occur at definite temperatures, such as melting point (solid to liquid), boiling point (liquid to vapours) etc. and are accompanied by absorption or evolution of heat.

Heat engine A machine which can do work by using heat that flows out spontaneously from a high temperature source to a low temperature sink is called a heat engine.

Driving force The overall force responsible for reaction to occur.

THERMOCHEMISTRY

Thermochemistry deals with the heat changes during chemical reactions. It is called energetics and is based on First Law of Thermodynamics.

Exothermic Reaction

- Heat is evolved in this case.
- ΔH is –ve as $H_R > H_p$
- $\Delta H = H_p - H_R = $ –ve (At constant pressure) $\Delta E = (E_p - E_R) = $ –ve (At constant volume) $E_R > E_P$

Endothermic Reaction

- Heat is absorbed in this case.
- ΔH or ΔE = +ve for endothermic reaction
- $\Delta H = H_P - H_R$ = +ve (At constant pressure)
  As $H_P > H_R$
- $\Delta E = (E_P - E_R)$ = +ve (At constant volume)
  As $E_P > E_R$

Heat or Enthalpy of Reaction

- Heat of reaction is the change in enthalpy (amount of heat evolved or absorbed) when the number of g moles of the substance, indicated by a chemical reaction, have undergone complete reaction.
- It is determined by water and bomb calorimeters.
  Mathematically, it is given as:
  $$\Delta H = \Sigma H_P - \Sigma H_R$$
- It can be express by $[\Delta H]_p$ or $[\Delta E]_V$

Factors Affecting Heat of Reaction

Nature or physical state of reactant and products

$$H_2(g) + \frac{1}{2} O_2(g) \rightleftharpoons H_2O(g)\ \Delta H = -57.8\text{kcal}$$

$$H_2(g) + \frac{1}{2} O_2(g) \rightleftharpoons H_2O(l)\ \Delta H = -68.32\text{kcal}$$

When steam condenses heat is evolved.

Allotropic forms of the element

$$C_{diamond} + O_2\ (g) \longrightarrow CO_2\ (g)\ \Delta H = -94.3\ \text{kcal}$$

$$C_{amorphous} + O_2\ (g) \longrightarrow CO_2\ (g)\ \Delta H = -97.6\ \text{kcal}$$

$$C_{diamond} \longrightarrow C_{amorphous}\quad \Delta H = 3.3\ \text{kcal}$$

This heat is called **heat** of transition.

Enthalpies of solution

$$H_2S(g) + I_2(g) \longrightarrow 2HI + S;\ \Delta H = -17.3\ \text{kcal}$$

$$H_2S(sol) + I_2(sol) \longrightarrow 2HI + S;\ \Delta H = -21.93\ \text{kcal}$$
sol.

Temperature

It is given by Kirchoff equation.

Effect of temperature on heat of reaction is given by Kirchoff s relation.

$$\frac{\Delta H_2 - \Delta H_1}{T_2 - T_1}\ \Delta C_p$$

$$\frac{\Delta E_2 - \Delta E_1}{T_2 - T_1} = \Delta C_v$$

Pressure and volume

$$\Delta H = \Delta E + \Delta n^* RT$$

$$\Delta H = \Delta E + P\Delta V$$

$$\Delta H = \Delta E + P\Delta V$$

$$\Delta H = \Delta E + (P_f V_f - P_i V_i)$$

Δn^* = number of gaseous product – number of reactant molecules.

Heat of Combustion

- Heat of combustion is the change in enthalpy, when one mole of a substance is completely oxidized.
- ΔH is –ve as heat in always evolved in this case, that means, it is an exothermic process.
- Heat of combustion is used to calculate calorific value of food and fuels.
- It is also useful in confirming structure of organic molecules having C, H, O, N etc.
- Enthalpy change accompanied by combustion of 1 g solid or 1 g liquid or 1 cc gas is called **calorific value**.

$$\text{C.V.} = \frac{\text{Heat of combustion}}{\text{Molecular weight}}$$

$$C + O_2 \longrightarrow CO_2\ \Delta H = -\ 94\ \text{kcal}$$

$$C + O_2 = -\frac{94}{12} = -7.83\ \text{kcal}$$

$$C_{12}H_{22}O_{11} + 12O_2 \longrightarrow 12CO_2 + 11H_2O$$

$$\Delta H = -\ 1349\ \text{kcal}$$

$$\text{C.V.} = -\frac{1349}{342} = -3.94$$

$$\Delta H(\text{heat of reaction}) = -\Sigma\Delta H_P^\circ - \Sigma\Delta H_R^\circ$$

Calorimetry

Heat exchange = Heat capacity × Temperature change (heat exchange—when no chemical change or state change occurs)

Heat of Formation

The amount of heat evolved or absorbed, or change in enthalpy when 1 mole of a substance is obtained from its constituents or free elements as in following example,

$$N_2 + 3H_2 \rightleftharpoons 2NH_3; \quad \Delta H = -22 \text{ kcal}$$

$$N_2 + O_2 \rightleftharpoons 2NO; \quad \Delta H = +42 \text{ kcal}$$

- Heat of formation of NH_3 is –11 kcal
  Heat of formation of NO is +21 kcal
- Heat of formation at 25°C and 1 atm pressure is called **standard heat of formation** ($\Delta H°$).
- Standard heat of formation of a free element is taken as zero. ($G_{graphite}$, $S_{monoclinic}$)
- Heat of formation may be +ve or –ve.
- If ΔH is –ve, compound formation is exothermic.
- If ΔH is +ve, compound formation is endothermic.
- Stability of a compound formed by exothermic reaction is more than that of a compound formed by endothermic reaction.
- Greater the liberated energy, greater is the stability of the compound.

Following compounds are placed in the observed order of stability. HF > HCl > HBr > HI.

Heat of Neutralization

- Heat of neutralization is the enthalpy change accompanied during neutralization of 1 g mole of an acid and base. It is also known as heat of formation of water.

  $H^+ Cl^- + Na^+ OH^- \longrightarrow NaCl + H_2O + 13.7$ kcal/mole or 57.3 kJ/mole

  $H^+ + OH^- \longrightarrow H_2O + 13.7$ kcal or 57.3 kJ/mole
  $\Delta H = -13.7$ kcal or –57.3 kJ/mole
- It is always exothermic. $\Delta H = -ve$
- In laboratory, it is measured by using polythene on polystyrene bottles.
- Heat of neutralization of strong acid and strong base is always –13.7 kcal/mole or –57.3 kJ/mole.
- It is independent of the nature of strong acid or strong base.
- If one electrolyte is weak then ΔH will be less than –13.7 kcal as some amount of heat will be absorbed in ionization of weak electrolyte, as shown here.

  $CH_3COOH + NaOH \longrightarrow CH_3COONa + H_2O +$ 55.2 kJ/mole or 13.4 kcal/mole

  $HCl + NH_4OH \rightleftharpoons NH_4Cl + H_2O$

  $\Delta H = -12.3$ kcal, that is heat of ionization of CH_3COOH is 0.3 kcal or 1.9 kJ/mole
- If both the electrolytes are weak then

  $CH_3COOH + NH_3 \longrightarrow CH_3COONH_4 + H_2O$ + 50.4 kJ/mole

  $CH_3COOH + NH_4OH \longrightarrow CH_3COONH_4 + H_2O$ $\Delta H = -11.9$ kcal
- In case of HF, ΔH is more due to increased hydration of F^- ions.

Heat of Hydration

The enthalpy change that occurs during hydration of one g mole substance is called heat of hydration.

$CuSO_4 + 5H_2O \longrightarrow CuSO_4.5H_2O + 78.21$ kJ/mole or 18.7 kcal/mole

$\Delta H = -18.7$ kcal/mole

- It is mostly exothermic

  $\Delta H_{sol} = \Delta H_{hy}$ – Lattice energy
- If $\Delta H_H >$ L.E. then solute dissolve in water
- $\Delta H_H <$ L.E. then solute does not dissolve in water
- $\Delta H \approx$ L.E. then solute is in equilibrium with solvent in water

Bond Energy

Bond energy may be defined as, "The quantity of heat evolved when a bond is formed between two free atoms in a gaseous state to form a molecular product in a gaseous state." It is also known as enthalpy of formation of the bond.

It can also be defined as, "The average quantity of heat required to break (dissociate) bonds of that type present in one mole of the compound."

Making use of the above defined defination in the case of

$C(g) + 4H(g) \longrightarrow CH_4(g); \quad \Delta H = -398$ kcal

the average bond energy per mole of C–H bond in this example $= -\frac{398}{4} = -99.5$ kcal

When a reaction is carried out in the gas phase at constant pressure then enthalpy of the reaction is given as:

$$\Delta H = \Sigma BE_R - \Sigma BE_P$$

or

ΔH = (Energy needed to break the bonds in the molecules of the reactants) – (Energy released when bonds in the molecules of the products are formed).

For example, in the case of formation of HCl (g) from H_2 (g) and Cl_2 (g), $\Delta H = -185$ kJ/mol

$E_{H-H} = 437$ kJ

$E_{C-Cl} = 244$ kJ

$E_{H-Cl} = 433$kJ

So, $\Delta H = E_{H-H} + E_{C-Cl} - (2E_{H-Cl})$

$= 437 + 244 - (2 \times 433)$

$= -185$ kJ

It is help ful in finding heat of reaction and *resonance* energy.

Heat of Atomization

It is the enthalpy change (heat required) when bonds of one mole of a substance are broken down completely to obtain atoms in the gaseous phase (isolated) or, it is the enthalpy change when one mole of atoms in the gas phase are formed from the corresponding element in its standard state.

In case of diatomic molecules it is also called bon**d dissociation enthalpy**.

It is denoted by ΔH_a or ΔH.

For example, $H_2\,(g) \longrightarrow 2H\,(g) - 435$ kJ

$\Delta H = +435$ kJ/mol

$CH_4\,(g) \longrightarrow C\,(g) + 4H\,(g) + 1665$ kJ

$\Delta H = +1665$ kJ/mol

Enthalpy of Dissociation or Ionization

It is defined as, "The quantity of heat absorbed when one mole of a substance is completely dissociated into its ions."

For example, in case of $H_2O\,(l) \longrightarrow H^+ + OH^-$, the value of $\Delta H = 13.7$ kcal

Phase Transition and Transition Energy

- The change of matter from one state (solid, liquid or gas) to another state is called phase transition.
- Such changes occur at definite temperatures such as melting point (solid to liquid), or boiling point (liquid to vapours), and are accompanied by absorption or evolution of heat. The enthalpy change during such phase transitions is called heat of transition or transition energy.

  For example,

  C (diamond) $\longrightarrow$ C (amorphous)

  $\Delta H = 3.3$ kcal

Heat of Dilution

Heat of dilution is the change in enthalpy that occurs when a solution containing one mole of solute is diluted from one concentration to another concentration.

KCl $\longrightarrow$ KCl

$(X\,H_2O)(Y\,H_2O)$

$HCl + 10H_2O \longrightarrow HCl.10H_2O$

$\Delta H = -69$ kJ/m

It's value depands upon the concentration of solution and amount of solvent used.

Heat of Solution

Heat of solution is change in enthalpy, that occurs when one mole of a solute is dissolved in excess of solvent.

- $\Delta H = +$ve for (hydrated salts)
- $\Delta H = -$ve for (anhydrous salts)

  $KCl + H_2O \longrightarrow KCl\,(aq)$

  $\Delta H = -18.6$ kJ/mole

  $H_2SO_4 + H_2O \longrightarrow H_2SO_4\,(aq)$

  $\Delta H = -20.2$ kcal

Heat of Fusion

Heat of fusion is the change in enthalpy, that occurs in converting one mole of solid into the liquid state at its melting point. It is equal tolatent heat of fusion per gram multiplied by the molar mass.

- It is always +ve as heat is needed to overcome the intermolecular forces between constituents particles of solid.

  $H_2O(s) \longrightarrow H_2O(l)$; $\Delta H = 0.6$ kJ/mole

Heat of Vaporization

Heat of vaporization is the change in enthalpy, that occurs in converting one mole of liquid into the gaseous state at its boiling point. For example,

$H_2O(l) \longrightarrow H_2O(g)$; $\Delta H = 42$ kJ/mole

Heat of Sublimation

Heat of sublimation is the change in enthalpy that occurs in converting one mole of a solid directly into its vapour state at a temperature below its melting point.

$I_2(s) \longrightarrow I_2(g)$; $\Delta H = 62.4$ kJ/mole

- $\Delta H_{sub} = \Delta H_{vap} + \Delta H_{fus}$

Hess's law This rule is a *consequence* of I[st] Law of Thermodynamics and according to the Hess's law, the enthalpy change for a process is independent of the path it follows.

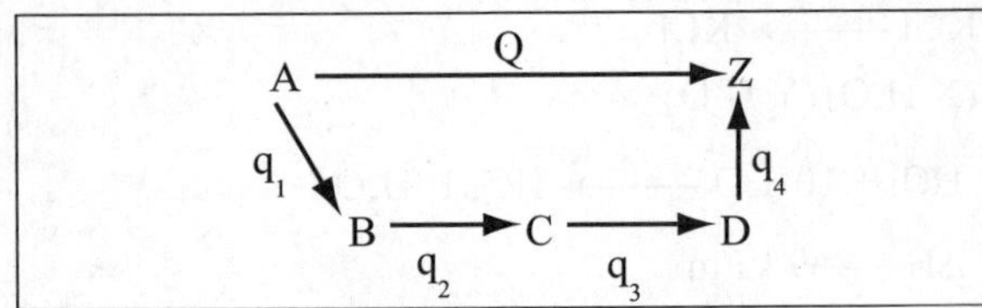

FIGURE 6.2 Hess's Law

Examples:

$C + O_2 \longrightarrow CO_2$; $\Delta H = -94.3$ kcal

$C + \frac{1}{2} O_2 \longrightarrow CO$; $\Delta H = -68.3$ kcal

$CO + \frac{1}{2} O_2 \longrightarrow CO_2$; $\Delta H = -26$ kcal

$Q = Q_1 + Q_2 = -94.3$

- ΔH or ΔE is not proportional to the path of a reaction.
- ΔH or ΔE is a state function which depends only on initial and final states (first law of thermodynamics).
- ΔH or ΔE is an extensive property.

SOURCES OF ENERGY

Energy is needed in every field of life and its importance is increasing day by day. Major sources of energy are given below:

Fossil fuels Some major fuels are natural gas, coal and petroleum products, that is, carbon containing species (major sources).

These undergo combustion to release heat in the form of energy, and this energy is further used according to need. For example, during electricity manufacture this released heat converts water into steam, which in turn runs the turbine, and thus generates electricity.

Some Major Chemical Fuels

Natural gas: It is mainly composed of methane and some other gaseous hydrocarbons like ethane, propane and butane.

Petroleum: It contains hundred of hydrocarbons and many other organic compounds having S, O, N. For example, kerosene oil, petrol, diesel, LPG (it is liquefied petroleum gas having mainly butane and isobutane).

Coal: It is a solid and a crude form of carbon having higher hydrocarbons and some other organic compounds having S, O and N.

Wood and dung The major constituent in these is methane.

Falling water Flowing or falling water has mechanical energy which can be used to run the turbine which in turn generaes electricity (hydroelectric power).

Nuclear energy It is the latest major source of energy. Here energy is obtained by the fission of fissile material like U-238 and Pu-241.

Geothermal sources, wind energy, ocean currents may also be good sources of energy.

The Sun as the Primary Source of Energy

Sun is the ultimate source of energy. Its energy is called solar energy. (Fig. 6.3)

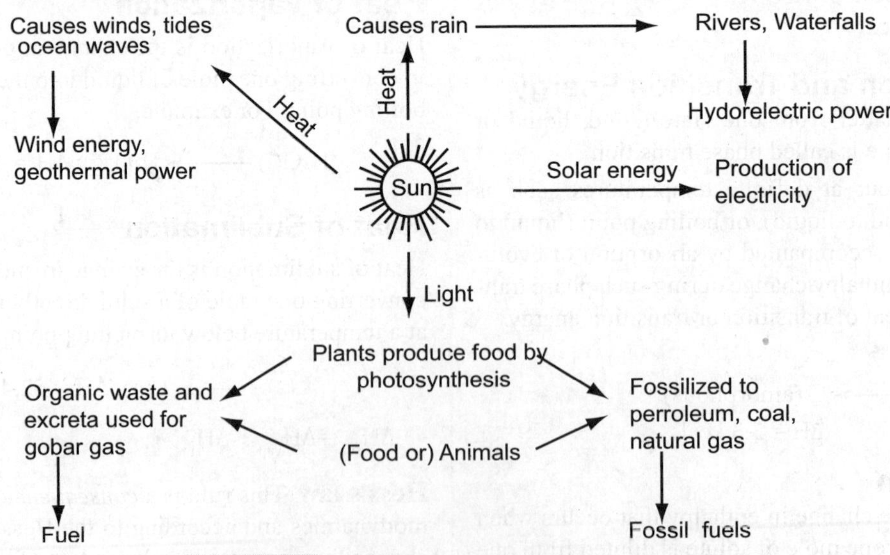

FIGURE 6.3 Sun: The Ultimate Source of Energy

Conservation of Energy

Eighty per cent of the energy requirements are fulfilled by fossil fuel. Their stock, however, is limited and these are depletable or non-renewable sources of energy. Hence, their conservation is a must.

Conservation of energy is possible in the following ways:

- We must use most efficient type of fuel, preferably from a renewable source.
- We must use high efficiency devices for the combustion of such fuel. We must use solar home appliances instead of electrical and fuel appliances.
- Deforestation must be avoided.
- Electrical appliances and vehicles must be regularly serviced.
- We must shut off the engine of the vehicles when stopping for a long period.
- Excessive use of electric appliances and vehicles should be avoided.

Points to Remember

Lavosier and laplace law The amount of heat required to break a compound into its element is equal to the heat evolved during its formation.

$$\Delta H_D = -\Delta H_f$$

A calorimeter It is used to find enthalpy of reaction

$$\Delta H = \frac{Z \times \Delta T \times m}{w}$$

1. It is not applicable when (i) reaction is slow (ii) ΔH is low (iii) reaction does not complete.
2. ΔH does not change with temperature appreciably.

Resonance Energy

Resonance energy = Expected heat of hydrogenation – Calculated heat of hydrogenation

For example, resonance energy of CO_2 is 32 calorie per mole.

For adults 2500–3000 kcal energy per day is needed.

Trouton's rule law

$$\frac{\Delta H_{vap}}{T_b} = \text{Joule/K mol}$$

Important Formula

Work

Electrical Work = Potential difference × Charge flow
= $V \times Q = EnF$

Expansion Work = $P \times \Delta V$
P = External pressure
ΔV = Increase or decrease in volume.

Different types of works and the formulas

1. Work done in a reversible isothermal process

$$W = -2.303\ nRT \log \frac{V_2}{V_1}$$

$$W = -2.303\ nRT \log \frac{P_1}{P_2}$$

2. Work done in an irreversible isothermal process

$W = -P_{ext}(V_2 - V_1)$

that is, $W = -P\,\Delta V$

- $\Delta E = \Sigma E_p - \Sigma E_R$

Bomb calorimeter

$$\Delta E = \frac{Z \times T \times m}{w}$$

First law of thermodynamic or Law of conservation of energy

$\Delta E = Q - W$ (that is, work done by the system)
$\Delta E = Q + W$ (that is, work done on the system)
or
$Q = \Delta E - W$
$\delta E = \delta Q - \delta W$
(Work done by the system or in expansion) or
$\delta E = \delta Q + \delta W$ or $\delta E = \delta Q + P\delta V$

For a reversible cycle

$\Delta E = 0,$

$$Q = -W_{max} = P\Delta V = 2.303\ nRT \log \frac{V_2}{V_1} \text{ or } \frac{P_1}{P_2}$$

For an isochoric process

$\Delta E = Q(\Delta V = 0)$

During adiabatic compression

$$W = \frac{(P_1V_1 - P_2V_2)}{(\gamma - 1)}$$

Enthalpy change

$\Delta H = \Delta E + P\Delta V$

Relation between ΔH and ΔE

$\Delta H = \Delta E + \Delta n_g RT$

Entropy

$$\Delta S = \frac{Q_{rev}}{T} = -2.303\ nR \log \frac{V_2}{V_1}$$

$$\Delta S = 2.303 \times C_P \log \frac{T_2}{T_1}$$

$$\Delta S^\circ = \Sigma S^\circ{}_P - \Sigma S^\circ{}_R$$

Entropy changes in processes

Isothermal reversible expansion of an ideal gas

$$\Delta S = R \log_e \frac{V_2}{V_1} \text{ or } \frac{P_1}{P_2}$$

Adiabatic reversible expansion

$Q = 0, \Delta S = 0$

Isobaric process

$$\Delta S = C_P \log \frac{T_2}{T_1}$$

Isochoric process

$$\Delta S = C_V \log \frac{T_2}{T_1}$$

Phase Transition

$$\Delta S = S_2 - S_1 = \frac{Q_{rev}}{T} = \frac{\Delta H}{T}$$

$$\Delta S_{fusion} = \frac{\Delta H_{fusion}}{T_{fusion\ or\ Mp}}$$

$$\Delta S_{vap} = \frac{\Delta H_{vap}}{T_{vap\ or\ BP}}$$

Gibbs free energy

- $G = H - TS$
- $\Delta G^\circ = \Delta H^\circ - T\Delta S^\circ$

Some relations

- $\Delta G^\circ = -n\ F\ E^\circ$
- n = Number of mole of electron
- E° = Standard emf
- $\Delta G^\circ = -2.303\ RT \log K$

  Here K = Equilibrium constant
- $\Delta G^\circ = \Sigma G^\circ{}_P - \Sigma G^\circ{}_R$
- At constant pressure, $Q = \Delta E + P\Delta V = \Delta H$
- $C_P - C_V = P\Delta V = R$

Heat or enthalpy of reaction

Mathematically, $\Delta H = \Sigma H_P - \Sigma H_R$

- It can be expressed by $[\Delta H]_P$ or $[\Delta E]_V$

Kirchoff equation

$$\frac{\Delta H_2 - \Delta H_1}{T_2 - T_1} = \Delta C_P$$

$$\frac{\Delta E_2 - \Delta E_1}{T_2 - T_1} = \Delta C_V$$

Pressure and volume

$\Delta H = \Delta E + \Delta nRT$

$\Delta H = \Delta E + P\Delta V$

$\Delta H = \Delta E + \Delta(PV)$

Δn = Number of gaseous product – Number of reactant molecules

Calorimetry

Heat exchange = Heat capacity × temperature change (heat exchange—when no chemical change or state change occur)

Heat of hydration

$\Delta H_{sol} = \Delta H_{hy}$ – Lattice energy

Bond energy

- $\Delta H = \Sigma BE_R - \Sigma BE_P$
- $\Delta H_{sub} = \Delta H_{vap} + \Delta H_{fus}$

Hess's law $Q = Q_1 + Q_2 + Q_3 + Q_4 \ldots\ldots$

Resonance energy

Resonance energy = Expected heat of hydrogenation – Calculated heat of hydrogenation

SOLVED NUMERICALS

Work and Energy

1. lg H_2 gas at S.T.P. is expanded so that volume is doubled. Find the work done in Joules. (1 litre atm = 101.3 J)

Solution:

1 g H_2 at S.T.P. = 0.5 mol = 11.2 L

$W = P(V_2 - V_1) = P(2V_1 - V_1)$

$= 1 \times 11.2$ L atm

$= 11.2 \times 101.3$ J.

$= 1134.56$ J.

2. Calculate the work done when 2 moles of hydrogen expand isothermally and reversibly at 25°C from 20 to 60 litres.

Solution:

$$W = -2.303\ nRT \log_{10} \frac{V_2}{V_1}$$

$$= -2.303 \times 2 \times 2 \times 298 \times \log_{10} \frac{60}{20}$$

$= -309.7$ calories

3. A gas absorbs 100 J heat and is simultaneously compressed by a constant external pressure of 1.50 atm from 8 L to 2 L in volume. Here ΔE will be:

Solution:

By first law,

$\Delta E = Q - W = Q - P\Delta V$

$Q = +100$ J

$W = P\Delta V = 1.50$ atm $(2 - 8) = -9.00$ L atm

$$= -900\text{L atm} \times \frac{8.314\ \text{J mol}^{-1}\text{K}^{-1}}{0.0821\ \text{L atm mol}^{-1}\text{L}^{-1}}$$

$= -911.4$ kJ

$\Delta E = 100 + 911.4 = 1011.4$ J

4. Calculate the minimum work necessary to compress 64 g of O_2 from 10 to 5 litres at 300 K. How much heat is evolved in this process?

Solution:

$$W = -2.303\ nRT \log_{10} \frac{V_2}{V_1}$$

$$W = -2.303 \times \frac{64}{32} \times 2 \times 300 \log \frac{5}{10}$$

$= -831.8$ calories

$Q = W = 831.8$ calories

5. 1 mole of NH_3 ($\gamma = 1.33$) gas at 300 K is expanded under adiabatic condition to increase the volume 8 times. Find the final temperature and work done here.

Solution:

For an adiabatic change

$$T_2 = T_1 \left(\frac{V_1}{V_2}\right)^{\gamma-1} = 300 \left(\frac{1}{8}\right)^{1.33-1}$$

$$= 300 \left(\frac{1}{8}\right)^{1/2} = 150\text{K}$$

$W = -nCv(T_2 - T_1)$

$= -1 \times 3 \times 2\ (150 - 300) = 900$ cal

Entropy

6. Calculate the entropy change involved in the conversion of one mole of water at 373 K to vapour at the same temperature. (Latent heat of vaporization of water = 2.257 kJ g^{-1})

Solution:

$$\Delta S_V = \frac{\Delta H_V}{T_b}$$

$\Delta H_V = 2.257 \times 18 \times 10^3$ J mol^{-1}

$T_b = 373$ K

$$\Delta S_V = \frac{2.257 \times 18 \times 10^2}{373} = 108.9\ \text{J K}^{-1}\ \text{mol}^{-1}$$

7. Calculate the entropy change in melting 1 g of ice at 0°C in SI units. Latent heat of fusion of ice is 80 cal g^{-1}.

Solution:

$$\Delta S_f = \frac{H_1}{T_1} = \frac{80 \times 4.184 \times 1}{273} = 1.226\ \text{JK}^{-1}\ \text{g}^{-1}$$

8. Calculate the standard entropy change associated with the following reaction at 298K.

$P_4\ (s) + SO_2\ (g) \longrightarrow P_4O_{10}\ (s)$

At 298 K; S° for $P_4 = 41.1$ JK^{-1} mol^{-1}

$O_2 = 205.0$ JK^{-1} mol^{-1} and

$P_4O_{10} = 231.0$ JK^{-1} mol^{-1}

Solution:

$\Delta S° = \Sigma\Delta S°_P - \Sigma\Delta S°_R$

$\Delta S° = [\Delta S°\ P_4O_{10} - (\Delta S°\ P_4 + 5\Delta S°\ O_2)\ ]$

$= 231 - (41.1 + 5 \times 205.0)$

$= 231 - 1066.1 = -835.1$ JK^{-1} mol^{-1}

9. 5 moles of an ideal gas expands reversibly, from a volume of 8 dm^3 to 80 dm^3 at a temperature of 27°C. Calculate the change in entropy.

Solution:

$\Delta S = 2.303\ nR \log_{10} \frac{V_2}{V_1}$

$\Delta S = 2.303 \times 5 \times 8.314 \times \log_{10} \frac{80}{8}$

$= 95.736 \times 1$

$= 95.736\ JK^{-1}$

10. Find ΔS°_f at 25°C for PCl_5 (g) If ΔH°_f is –374.9 kJ/mole and ΔG°_f is –305 kJ/mol.

Solution:

$\Delta G^\circ = \Delta H^\circ - T\Delta S^\circ$

$T\Delta S^\circ = \Delta H^\circ + \Delta S^\circ$

$\Delta S^\circ = \frac{\Delta H^\circ - \Delta G^\circ}{T}$

$= \frac{-374.9 - (-305)}{298}$

$= \frac{-374.9 + 305}{298} = \frac{-69.9}{298}$

$= -0.2346$ kJ/mol K

$= -234.56$ J/mol K

11. The heat liberated on complete combustion of 7.8g benzene is 327 kJ. This heat has been measured at constant volume and at 27°C. Calculate heat of combustion of benzene at constant pressure at 27°C. (R = 8.314 J/mol/K)

Solution:

$C_6H_6\ (l) + 7^{1/2}\ O_2\ (g) \longrightarrow 6CO_2\ (g) + 3H_2O(l)$

$\Delta n = 6 - 7^{1/2} = -3/2$

$\Delta H = \Delta E - \Delta nRT$

(ΔE per mole for benzene = (327 × 78)/7.8 = –327 kJ)

$= -3270 \times 10^3 + (-3/2) \times 8.314 \times 300$

$= -32737413 = -3273.74$ kJ

12. Calculate the standard internal energy change for the following reaction at 25°C.

$2H_2O_2\ (l) \longrightarrow 2H_2O\ (l) + O_2\ (g)$

$H_2O_2(l) = -188.0\ kJ\ mol^{-1}$

$H_2O\ (l) = -286.0\ kJ\ mol^{-1}$

$(R = 8.314\ JK^{-1}mol^{-1})$

Solution:

$2H_2O_2\ (l) \longrightarrow 2H_2O\ (l) + O_2\ (g)$

$\Delta H^\circ = \Delta H^\circ_f$ (products) – ΔH°_f (reactants)

$= (2 \times 286 + 0.0) - (2 \times -188)$

$= -196$ kJ

$\Delta n = 3 - 2 = 1$

$\Delta E^\circ = \Delta H^\circ - \Delta nRT$

$= -196 \times 10^3 - 1 \times 8.314 \times 298$

$= -193.58$ kJ

13. When glucose is oxidized in the body, about 40% of the energy evolved in the reaction is available for molecular activity. How much energy is obtained from the oxidation of 4.0 g of glucose?

Solution:

$C_6H_{12}O_6\ (s) + 6O_2\ (g) \longrightarrow 6CO_2\ (g) + 6H_2O(l)$

$\Delta H = -2817$ kJ

As 1 mole of glucose (180 g) gives 2817 kJ of energy.

So 4 g of glucose gives $= \frac{2817 \times 4}{180} = 62.6$ kJ

Amount of energy available for muscular activity

$= \frac{62.6 \times 40}{100} = 25.04$ kJ

(As 40% of the energy is available)

Various Enthalpies

14. The lattice energy of NaCl is –780 kJ mol^{-1} The enthalpies of hydration of Na^+ (g) and Cl^- (g) ions are –406 kJ mol^{-1} and –364 kJ mol^{-1} respectively. The enthalpy of solution of NaCl (s) is

Solution:

$NaCl\ (s) \longrightarrow Na^+\ (g) + Cl^-\ (g)\quad \Delta H = +780$ kJ

$Na^+\ (g) + aq \longrightarrow Na^+(aq)\quad \Delta H = -406$kJ

$Cl^-\ (g) + aq \longrightarrow Cl^-\ (aq)\quad \Delta H = -364$ kJ

The net reaction is dissolution of NaCl (s).

$\Delta H_{solution} = 780 - (406 + 364) = 10\ kJ\ mol^{-1}$

15. Calculate enthalpy change of the following reactions: $CH_2 = CH_2\ (g) + H_2\ (g) \longrightarrow CH_3 - CH_3\ (g)$ The bond energy of C – H = 414 kJ/mol

C–C = 347 kJ/mol,

C=C = 615kJ/mol and

H–H = 435kJ/mol

Solution:

$CH_2 = CH_2\ (g) + H_2\ (g) \longrightarrow CH_3 - CH_3\ (g)$

ΔH = Sum of BE of reactants – Sum of BE of products

= (1C=C + 4C–H + 1H–H) – (1C–C + 6C–H)

$= (615 + 4 \times 414 + 435) - (347 + 6 \times 414)$

$= 2706 - 2831 = -125$ kJ

16. Calculate the resonance energy of N_2O from the following data:

$N \equiv N$ 946 kJ mol^{-1}

$N = N$ 418 kJ mol^{-1}

$O = O$ 498 kJ mol^{-1}

$N = O$ 607 kJ mol^{-1}

ΔH°_f of N_2O is 82.0 kJ mol^{-1}.

Solution:

$N_2 (g) + 1/2\ O_2 (g) \longrightarrow N_2O (g)$,

$\Delta H_{exp} = 82$ kJ

$\Delta H_{calc} = (1N{\equiv}N + \frac{1}{2} O{=}O) - (1N{=}N{+}1N{=}O)$

$= (946 + \frac{1}{2} \times 498) - (418 + 607)$

$= 1195 - 1025 = +170$ kJ

Resonance energy

$= \Delta H_{f_{N_2O}}$ (exp.) – ΔH, N_2O (calc.)

$= 82 - 170 = -$ 88kJmol^{-1}.

17. Using bond energy data, calculate heat of formation of isoprene. Given BE of C–H = 98.8 kcal, H–H = 104 kcal, C–C = 83 kcal, C=C = 147 kcal and $C(s) \longrightarrow C(g) = 171$ kcal.

Solution:

$5C(s) + 4H_2 (g) \longrightarrow H_2C{=}C(CH_3){-}CH{=}CH_2$; ΔH =?

$\Delta H = \Sigma BE$ (reactants) – ΣBE (products)

$= [5BE_{C(s)\to C(g)} + 4BE_{H-H}] - [2BE_{C-C} + 2BE_{C=C} + 8BE_{C-H}]$

$= [5 \times 171 + 4 \times 104] - [2 \times 83 + 2 \times 147 + 8 \times 98.8]$

$= 20.6$ kcal.

18. Calculate the enthalpy of hydrogenation of C_2H_2 (g) to C_2H_4 (g). Given bond energies:

C–H = 414.0 kJ mol^{-1}

C≡C = 827.6 kJ mol^{-1}

C=C = 606.0 kJ mol^{-1}

H–H = 430.5 kJ mol^{-1}

Solution:

$H{-}C{\equiv}C{-}H + H{-}H \longrightarrow H{-}C(H){=}C(H){-}H$

$\Delta H = [2\Delta H_{C-H} + \Delta H_{C\equiv C} + \Delta H_{H-H}] - [4\Delta H_{C-H} + H_{c=c}]$

$= [2 \times 414.0 + 827.6 + 430.5] - [4 \times 414 + 606]$

$= 2086.1 - 2262 = -175.9$ kJ mol^{-1}.

19. If the bond energies of C=C and C–C at 298 K are 590 and 331 kJ mol^{-1} respectively then enthalpy of polymerization per mole of ethylene is:

Solution:

Polymerization reaction occurs as:

$nCH_2{=}CH_2 \longrightarrow (-CH_2{-}CH_2-)_n$

One mole of C = C bond is broken and two moles of C–C bonds are formed per mole of ethylene.

$\Delta H = 590 - 2 \times 331$

$= 590 - 662 = -72$ kJ per mole of ethylene

20. From the following given data find the enthalpy of hydration of anhydrous Al_2Cl_6 (s) in kcal.

(1) $2Al(s) + 6HCl (aq) = Al_2Cl_6(aq) + 3H_2(g)$

$\Delta H = -$ 239760 cal

(2) $H_2(g) + Cl_2(g) = 2HCl(g)$

$\Delta H = -$ 44000 cal

(3) HCl (g) + aq = HCl (aq)

$\Delta H = -17315$ cal

(4) $Al_2Cl_6(s) + aq = Al_2Cl_6(aq)$

ΔH =?

(5) $2Al(s) + 3Cl_2(g) = Al_2Cl_6(s)$

$\Delta H = -$ 321960 cal

Solution:

(1) $2Al(s) + 6HCl (aq) = Al_2Cl_6(aq) + 3H_2(g)$

$\Delta H = -$ 239760 cal

3 × (ii) $3H_2 (g) + 3Cl_2(g) = 6HCl (g)$

$\Delta H = -$ 132000 cal

(4) $Al_2Cl_6 (s) = 2Al (s) + 3Cl_2 (g)$

$\Delta H = +321{,}960$ cal

6 × (iii) 6HCl (g) + aq = 6HCl (aq)

$\Delta H = -$ 103,890 cal

Add: $Al_2Cl_6 (s) + aq = Al_2Cl_6 (aq)$

$\Delta H = -$ 153,690 cal or –153.69 kcal.

CHAPTER-END EXERCISES

PRACTICE QUESTIONS – I

1. In the exothermic reaction the enthalpy of reaction is always:

(a) Zero (b) Positive
(c) Negative (d) None of these

2. The enthalpy change of a reaction does not depend on:

(a) Initial and final enthalpy change of reaction
(b) State of reactants and products
(c) Different intermediate reactions
(d) Nature of reactants and products

3. For an ideal gas Joule-Thomson coefficient is:

(a) Positive
(b) Negative
(c) Zero
(d) Dependent on molecular weight

4. If the value of ΔH in a reaction is positive, then the reaction is called:

(a) Exothermic (b) Endothermic
(c) Polymorphic (d) Polytropic

5. Heat required to raise the temperature of 1 mol of a substance by 1° is called:

(a) Specific heat (b) Molar heat capacity
(c) Water equivalent (d) Specific gravity

6. Plants and living beings are examples of:

(a) Isolated system (b) Adiabatic system
(c) Open system (d) Closed system

7. In thermodynamics, a process is called reversible when:

(a) The surroundings are always in equilibrium with the system
(b) There is no boundary between system and sur roundings
(c) The surroundings and system change into each other
(d) The system changes into the surroundings sponta neously

8. Which of the following is an extensive property?

(a) Molarity of substance
(b) Number of moles
(c) Molar volume of substance
(d) Mole fraction of substance

9. If a gas at constant temperature and pressure expands, then its:

(a) Internal energy decreases
(b) Entropy increases and then decreases
(c) Internal energy increases
(d) Internal energy remains constant

10. If a gas absorbs 200 J of heat and expands by 500 cm^3 against a constant pressure of 2×10^s Nm^{-2}, then change in internal energy is:

(a) – 200 J (b) – 100J
(c) + 100J (d) + 300J

11. The internal energy of a substance does not depend upon:

(a) Translational energy
(b) Vibrational energy
(c) Energy due to gravitational pull
(d) Rotational energy

12. Internal energy of a substance/system is a:

(a) State function (b) Path function
(c) Work function (d) None

13. Which one of the following statements is false?

(a) Temperature is a state function
(b) Work is a state function
(c) Change in the state depends upon initial and final state
(d) Work appears at the boundary of the system

14. A heat engine absorbs heat Q_1 from a source at temperature T_1 and heat Q_2 from a source at temperature T_2. Work done is found to be J (Q_1 + Q_2). This is in accordance with:

(a) First law of thermodynamics
(b) Second law of thermodynamics
(c) Joules equivalent law
(d) None of these

15. For which of the following processes will the entropy increase?

(a) Reaction of magnesium with oxygen to form magnesium oxide
(b) Reaction of nitrogen and hydrogen to form ammonia
(c) Sublimation of dry ice
(d) Condensation of steam

16. Which of the following is correct equation?
(a) $\Delta U = \Delta Q - W$ (b) $\Delta W = \Delta U + \Delta Q$
(c) $\Delta U = \Delta W + \Delta Q$ (d) None of these

17. For a reaction to occur spontaneously:
(a) $(\Delta H - T\Delta S)$ must be negative
(b) $(\Delta H + T\Delta S)$ must be negative
(c) ΔH must be negative
(d) ΔS must be negative

18. The internal energy change when a system goes from state A to B is 40 kJ/mol. If the system goes from A to B by a reversible path and returns to state A by an irreversible path what would be the net change in internal energy?
(a) 40 kJ (b) > 40 kJ
(c) < 40 kJ (d) zero

19. In an irreversible process taking place at constant T and P and in which only pressure–volume work is being done, the change in Gibbs free energy (dG) and change in entropy (dS), satisfy the criteria:
(a) $(dS)_{V,E} < 0, (dG)_{T,P} < 0$
(b) $(dS)_{V,E} > 0, (dG)_{T,P} < 0$
(c) $(dS)_{V,E} = 0, (dG)_{T,P} = 0$
(d) $(dS)_{V,E} = 0, (dG)_{T,P} > 0$

20. The enthalpy change for a reaction does not depend upon the:
(a) Physical state of reactants and products
(b) Use of different reactants for the same product
(c) Nature of intermediate reaction steps
(d) Difference in initial or final temperatures of involved substances

21. (ΔH–ΔU) for the formation of carbon monoxide (CO) from its elements at 298 K is:
(a) $-1238.78\ J\ mol^{-1}$
(b) $1238.78\ J\ mol^{-1}$
(c) $-2477.57\ J\ mol^{-1}$
(d) $2477.57\ J\ mol^{-1}$

22. Standard molar enthalpy of formation of CO_2 is equal to:
(a) Standard molar enthalpy of combustion of carbon (graphite)
(b) Standard molar enthalpy of combustion of gaseous carbon
(c) Sum of standard molar enthalpies of formation of CO and O_2
(d) Zero

23. Which one of the following has ΔS° greater than zero?
(a) $CaO\ (s) + CO_2\ (g) \rightleftharpoons CaCO_3\ (g)$
(b) $NaCl\ (aq) \rightleftharpoons NaCl\ (s)$
(c) $NaNO_3\ (s) \rightleftharpoons Na^+\ (aq) + NO_3^-\ (aq)$
(d) $N_2(g) + 3H_2(g) \rightleftharpoons 2NH_3(g)$

24. For a spontaneous process, the entropy:
(a) Increases (b) Decreases
(c) Unchanged (d) Not clear

25. For a spontaneous process, the correct statement is:
(a) Entropy of the system always increases
(b) Free energy of the system always increases
(c) Total entropy change is always negative
(d) Total entropy change is always positive

26. Which one of the following has the maximum entropy of vaporization?
(a) Water (l) (b) Toluene (l)
(c) Diethyl ether (l) (d) Acetone (l)

27. Which of the following is not an endothermic reaction?
(a) Combustion of methane
(b) Decomposition of water
(c) Dehydrogenation of ethane or ethylene
(d) Conversion of graphite to diamond

28. Which of the following statements is false?
(a) Work is a state function.
(b) Temperature is a state function.
(c) Change of state is completely denned when initial and final states are specified.
(d) Work appears at the boundary of the solution.

29. Molar heat capacity of water in equilibrium with ice at constant pressure is:
(a) Zero (b) Infinity
(c) $40.45\ J\ K^{-1}\ mol^{-1}$ (d) $75.48\ J\ K^{-1}\ mol^{-1}$

30. Standard molar enthalpy of formation of CO_2 is equal to:
(a) Zero
(b) The standard molar enthalpy of combustion of gaseous carbon.
(c) The sum of standard molar enthalpies of formation Of CO and CO_2
(d) The standard molar enthalpy of combustion of carbon (graphite)

31. For which of the following changes $\Delta H \neq \Delta E$?
(a) $H_2 + I_2 \rightleftharpoons 2HI$
(b) $HCl + NaOH \longrightarrow NaCl + H_2O$
(c) $C(s) + O_2\ (g) \rightleftharpoons CO_2\ (g)$
(d) $N_2 + 3H_2 \longrightarrow 2NH_3$

32. $H_2S(g) \longrightarrow HS(g) + H(g), \Delta H° = x_1,$
$\Delta H_f^\circ [H_2S(g)] = x_2, \Delta H_f^\circ [H(g)] = x_3$
hence, ΔH_f° (HS) is:

(a) $x_1 + x_2 - x_3$ (b) $x_3 - x_1 - x_2$
(c) $x_1 - x_2 - x_3$ (d) $x_3 - x_1 + x_2$

33. Classify each of the following processes as spontaneous or non-spontaneous.

I. $H_2O(1) \longrightarrow H_2O(g)$, T = 25°C vessel open to atomsphere with 50% relative humidity.

II. $H_2O(s) \longrightarrow H_2O(l)$, T = 25°C, P=l atm

(a) I and II are both non-spontaneous
(b) I and II are both spontaneous
(c) I is non-spontaneous and II is spontaneous
(d) I is spontaneous and II is non-spontaneous

34. The Gibbs free energy is defined as:

(a) G = H – T.S (b) G = H + T.S
(c) G = E – T.S (d) G = E + T.S

35. Which of the following statement is true for ΔG?

(a) It is always proportional to ΔH
(b) It may be less than or greater than or equal to ΔH
(c) It is always greater than ΔH
(d) It is always less than ΔH

36. The free energy change (ΔG°) is negative when:

(a) The surroundings do no electrical work on the system
(b) The surroundings do electrical work on the system
(c) The system does electrical work on the surround ings
(d) The system does no electrical work on the surroundings

37. The amount of heat required to raise the temperature of a body through 1°C is called its:

(a) Molar heat (b) Specific heat
(c) Entropy (d) Thermal capacity

38. The amount of heat required to raise the temperature of one mole of the substance through 1 K is called, its:

(a) Molar heat (b) Entropy
(c) Thermal capacity (d) Specific heat

39. Bond energy of N–H, H–H, and N≡N bonds are Q_1, Q_2 and Q_3; ΔH of $N_2 + 3H_2 \longrightarrow 2NH_3$ is:

(a) $Q_3 + 3Q_2 - 2Q_1$ (b) $2Q_1 - Q_3 - 2Q_2$
(c) $Q_3 + 3Q_2 - 6Q_1$ (d) $Q_1 + Q_2 - Q_3$

40. Which of the following gas molecule has the maximum specific heat at constant pressure?

(a) Helium (b) Argon
(c) Nitrogen (d) Oxygen

Practice Questions – II

41. An ideal gas is allowed to expand both reversibly and irreversibly in an isolated system. If T_i is the initial temperature and T_f is the final temperature, which of the following statements is correct?

(a) $(T_f)_{irrev} > (T_i)_{rev}$
(b) $T_f > T_i$ for reversible process but $T_f = T_i$ for irreversible process
(c) $(T_f)_{irrev} = (T_i)_{rev}$
(d) $T_f = T_i$ for both reversible and irreversible processes

42. For the reaction of one mole of Zn dust with one mole of H_2SO_4 in a bomb calorimeter, ΔU and w corresponds to:

(a) ΔU < 0, w = 0 (b) ΔU < 0, w < 0
(c) ΔU > 0, w = 0 (d) ΔU > 0, w > 0

43. For a phase change:

$H_2O\ (l) \longrightarrow H_2O\ (s)$

(a) ΔG = 0 (b) ΔS = 0
(c) ΔH = 0 (d) ΔU = 0

44. Asuming that water vapour is an ideal gas, the internal energy change (ΔU) when 1 mol of water is vapourized at 1 bar pressure and 100°C, (Given: Molar enthalpy of vaporization of water at 1 bar and 373 K = 41 kJ mol^{-1} and R = 8.3 J $mol^{-1}K^{-1}$) will be:

(a) 3.7904 kJ mol^{-1} (b) 37.904 kJ mol^{-1}
(c) 41.00 kJ mol^{-1} (d) 4.100 kJ mol^{-1}

45. For the un-balanced chemical reaction given below that occurs in an automobile engine

$$C_8H_{18} + O_2 \rightarrow CO_2 + H_2O$$

The sign of ΔH, ΔS, and ΔG for the reaction are?

(a) + – + (b) –, +, +
(c) – – – (d) –, +, –

46. For a reversible a diabatic ideal gas expansion $\frac{dP}{p}$ is equal to:

(a) $\frac{dV}{V}$ (b) $-\gamma\frac{dV}{V}$

(c) $\left(\frac{\gamma}{\gamma-1}\right)\frac{dV}{V}$ (d) $\gamma\frac{dV}{V}$

47. Find the value of $\Delta H^{\circ}_{Reaction}$ for the reaction given below by using the given data.

$$\underset{'s'}{3CO} + \underset{g}{2O_2} \rightarrow \underset{'s'}{CO_3O_4}$$

$$\underset{'s'}{2CO} + \underset{g}{O_2} \rightarrow \underset{'s'}{2COO}\ \Delta H^0_1 = -475.8\ kJ$$

$$\underset{'s'}{6COO} + \underset{g}{O_2} \rightarrow \underset{'s'}{2CO_3\ O_4}\ \Delta H^0_2 = -355\ kJ$$

(a) –445.6 kJ (b) +891.2 kJ
(c) +445.6 kJ (d) –891.2 kJ

48. 2 mole of an ideal gas at 27°C temperature is expanded reversibly from 2 L to 20 L. Find entropy change in cal. (R = 2 cal/mol K):

(a) 92.1 (b) 0
(c) 4 (d) 9.2

49. The molar heat capacity of water at constant pressure is 80 $JK^{-1}mole^{-1}$ when 1 KJ of heat is supplied to 100 g of water which is free to expand the increase in temperature of water is:

(a) 2.4 K (b) 2.25 K
(c) 3.25 K (d) 2.1 K

50. What would be the work done obtained from an isothermal reversible expansion of 1 mol of Cl_2 from $1dm^3$ to 50 dm^3 at 273K?

(a) –8.88 KJ/mole (b) 8.88 KJ/m
(c) –4.44 KJ/m (d) –17.76 KJ/m

51. What per cent T_1 is of T_2 for a heat engine whose efficiency is 15%?

(a) 15% (b) 85%
(c) 30% (d) 70%

52. 1 mole of an ideal gas at 300 K is expanded isother-mally and reversible from an initial volume of 1 litre to 10 litre. The work in this process is (R = 2 cal mol^{-1} K^{-1}):

(a) 163.7 cal (b) zero
(c) –1381.8 cal (d) 9 1itreatm

53. Standard state Gibbs free energy change for isomerization reaction, cis-2-pentene $\rightleftharpoons$ trans-2-pentene is –3.67 kJ/mol at 400 K. if more trans-2-pentene is added to the reaction vessel, then:

(a) More cis-2-pentene is formed
(b) Additional trans-2-pentene is formed
(c) Equilibrium remains unaffected
(d) Equilibrium is shifted in the forward direction

54. The standard enthalpy of formation ($\Delta_f H^{\circ}$) at 298 K for methane, CH_4(g) is –74.8 kJ mol–1, the additional information required to determine the average energy for C–H bond formation would be:

(a) The dissociation energy of H_2 and enthalpy of sublimation of carbon
(b) Latent heat of vaporization of methane
(c) The first four ionization energies of carbon and electron gain enthalpy of hydrogen
(d) The dissociation energy of hydrogen molecule, H_2

55. Which plot represent an exothermic process?

(a)
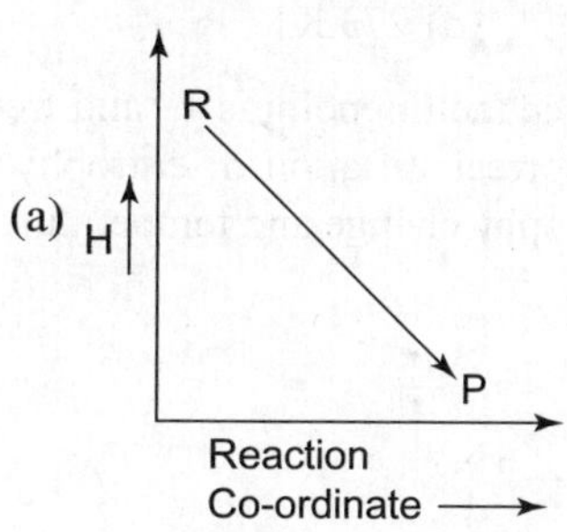

(b)
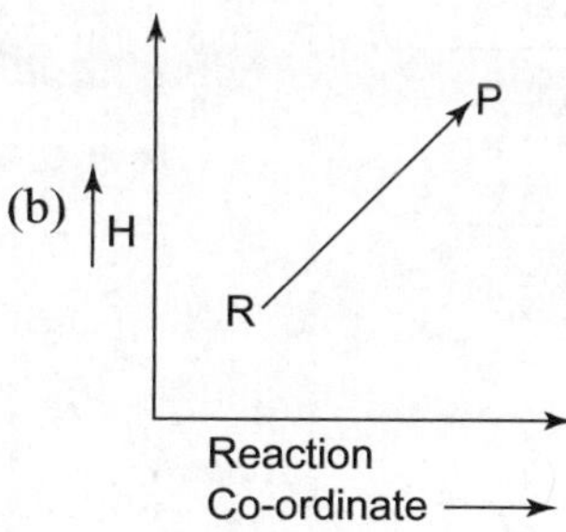

(c)
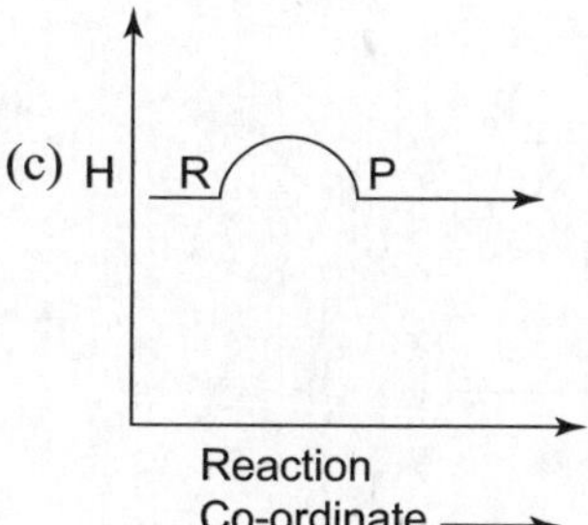

(d)
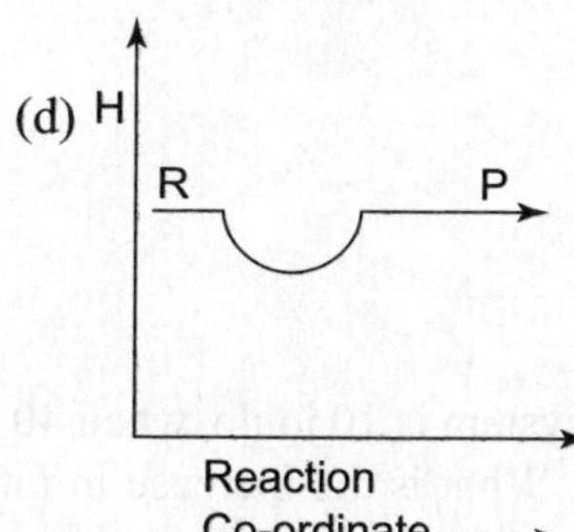

56. Consider the reaction $N_2 + 3H_2 \rightleftharpoons 2NH_3$ carried out at constant temperature and pressure. If ΔH and ΔU are the enthalpy and internal energy changes for the reaction, which of the following expressions is true?

(a) $\Delta H = 0$ (b) $\Delta H = \Delta U$
(c) $\Delta H < \Delta U$ (d) $\Delta H > \Delta U$

57. The enthalpy and entropy change for the reaction Br_2 (l) + Cl_2 (g) ⟶ 2BrCl (g) are 30 kJ mol^{-1} and 105 J K^{-1} mol^{-1} respectively. The temperature at which the reaction will be in equilibrium is:

(a) 450 K (b) 300 K
(c) 285.7 K (d) 273 K

58. For a given substance meltingpoint is T_B and freezing point is T_A these correct variation of entrophy is by graph between entrophy change and temperature is:

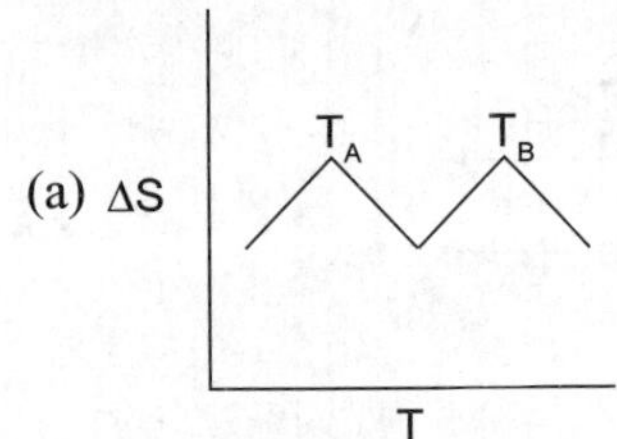

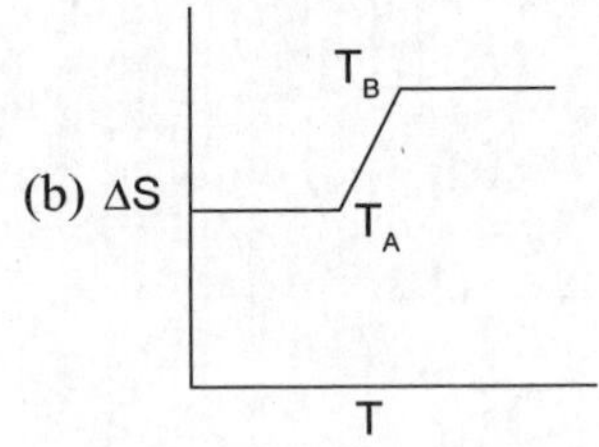

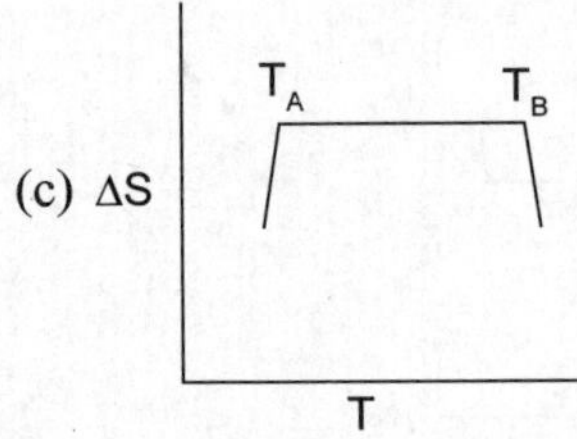

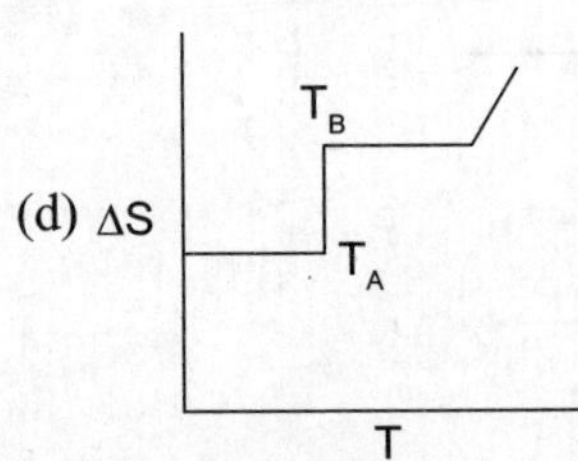

59. The work done by a system is 10 joule, when 40 joule heat is supplied to it. What is the increase in internal energy of system?

(a) 30 J (b) 50 J
(c) 40 J (d) 20 J

60. The standard entropies of CO_2 (g), C (s) and O_2 (g) are 213.5, 5.74 and 205 J K^{-1} respectively. The standard entropy of the formation of CO_2 (g) is:

(a) 1.16 JK^{-1} (b) 2.76 JK^{-1}
(c) 1.86 JK^{-1} (d) 2.12 JK^{-1}

61. For a reaction at 300 K, enthalpy and entropy changes are -11.5×10^3 J mol^{-1} and -105 JK^{-1} mol^{-1} respectively. What is the change in Gibbs free energy?

(a) 25 kJ mol^{-1} (b) 30 kJ mol^{-1}
(c) 15 kJ mol^{-1} (d) 20 $kJmol^{-1}$

62. The standard enthalpy of decomposition of N_2O_4 to NO_2 is 58.04 kJ and standard entropy of this reaction is 176.7 J K^{-1}. The standard free energy change for this reaction at 25°C, is:

(a) 5.39 kJ (b) –5.39 kJ
(c) 539 kJ (d) 53.9 kJ

63. The entropy values in J K^{-1} mol^{-1} of H_2 (g) = 130.6, Cl_2 (g) = 223 and HC1 (g) = 186.7 at 298 K and 1 atm pressure. Then entropy change for the reaction: H_2 (g) + Cl_2 (g) ⟶ 2HC1 (g) is

(a) +540.3 (b) +727.3
(c) –166.9 (d) +19.8

64. ΔH°_f (298) of methanol is given by chemical equation:

(a) CH_4 (g) + 1/2 O_2 (g) ⟶ CH_3OH (g)
(b) CO (g) + $2H_2$ (g) ⟶ CH_3OH (l)
(c) C (graphite) + 1/2 O_2 (g) + $2H_2$ (g) ⟶ CH_3OH(l)
(d) C (diamond) + 1/2 O_2 (g) + $2H_2$ (g) ⟶ CH_3OH(l)

65. 2 moles of an ideal gas is expanded isothermally and reversibly from 1 litre of 10 litre at 300 K. The enthalpy change (in kJ) for the process is:

(a) 11.4 kJ (b) –11.4 kJ
(c) 0 kJ (d) 4.8 kJ.

66. The enthalpy of vaporization of a liquid is 30 kJ mol^{-1} and entropy of vaporization is 5 J mol^{-1} K. The boiling point of the liquid at 1 atm is:

(a) 250 K (b) 400 K
(c) 450 K (d) 600 K

67. Which of the following reaction defines ΔH°_f?

(a) C (diamond) + O_2 (g) ⟶ CO_2 (g)
(b) $\frac{1}{2}H_2$ (g) + $\frac{1}{2}F_2$ (g) ⟶ HF (g)
(c) N_2 (g) + $3H_2$ (g) ⟶ $2NH_3$ (g)
(d) CO (g) + $\frac{1}{2}O_2$ (g) ⟶ CO_2 (g)

68. If the standard entropies of CH_4, O_2, CO_2 and H_2O are 186.2, 205.3, 213.6 and 69.96 J K^{-1} mol^{-1} respectively, then standard entropy change for the reaction:

$CH_4\ (g) + 2O_2\ (g) \longrightarrow CO_2\ (g) + 2H_2O\ (l)$ is:

(a) –215.6 J K^{-1} mol^{-1}
(b) –243.3 J K^{-1} mol^{-1}
(c) –130.5 J K^{-1} mol^{-1}
(d) –85.6 J K^{-1} mol^{-1}

69. The standard entropy change for the reaction:

$SO_2\ (g) + \frac{1}{2} O_2\ (g) \longrightarrow SO_3\ (g)$ is (where S° for SO_2 (g), O_2 (g) and SO_3 (g) are 248.5, 205 and 256.2 J K^{-1} mol^{-1} respectively)

(a) 198.2 J K^{-1} mol^{-1} (b) –192.8 J K^{-1} mol^{-1}
(c) –94.8 J K^{-1} mol^{-1} (d) 94.8 J K^{-1} mol^{-1}

70. In thermodynamics, a process is called reversible when:

(a) Surroundings and system change into each other
(b) There is no boundary between system and surroundings
(c) The surroundings are always in equilibrium with the system
(d) The system changes into the surroundings spontaneously

71. Under the same conditions how many mL of 1 M KOH and 0.5 MH_2SO_4 solutions, respectively, when mixed to form total volume of 100 ml, produces the maximum rise in temperature?

(a) 67, 33 (b) 33, 67
(c) 40, 60 (d) 50, 50

72. Identify the state function among the following:

(a) Q (b) Q – w
(c) Q/w (d) Q + w

73. For a reaction at 300 K, enthalpy and entropy changes are $-11.5 \times 10^3\ J\,mol^{-1}$ and $-105\ JK^{-1} mol^{-1}$ respectively. What is the change in Gibbs free energy?

(a) 25 kJ mol^{-1} (b) 30 kJ mol^{-1}
(c) 15 kJ mol^{-1} (d) 20 kJ mol^{-1}

74. For the reaction $H_2\ (g) + \frac{1}{2} O_2\ (g) \longrightarrow H_2O(l)$, the value of ΔH = –285.8 kJ mol^{-1} and ΔS = 0.163 JK^{-1} mol^{-1}. The free energy change at 300 K. for the reaction, is:

(a) –289.6 kJ mol–^{1} (b) 437.5 kJ mol^{-1}
(c) –334.7 kJ mol^{-1} (d) –291.6 kJ mol^{-1}

75. For an endothermic reaction, where ΔH represents the enthalpy of the reaction in kJ/mol, the minimum value for the energy of activation will be:

(a) Less than ΔH (b) Zero
(c) More than ΔH (d) Equal to ΔH

76. Which of the following equations represent standard heat of formation of C_2H_4?

(a) 2C (diamond) + $2H_2$ (g) $\longrightarrow$ C_2H_4 (g)
(b) 2C (graphite) + $2H_2$ (g) $\longrightarrow$ C_2H_4 (g)
(c) 2C (diamond) + 4H (g) $\longrightarrow$ C_2H_4 (g)
(d) 2C (graphite) + 4H (g) $\longrightarrow$ C_2H_4 (g)

77. The difference between heats of reaction at constant pressure and constant volume for the reaction:

$2\ C_6H_6(l) + 15\ O_2\ (g) \longrightarrow 12\ CO_2\ (g) + 6H_2O(1)$ at 25°C in kJ is:

(a) –7.43 (b) +3.72
(c) –3.72 (d) + 7.43

78. The enthalpy change (ΔH) for the reaction, $N_2\ (g) + 3H_2\ (g) \longrightarrow 2NH_3\ (g)$ is –92.38 kJ at 298 K. The internal energy change ΔU at 298 K is:

(a) –92.38 kJ (b) –87.42 kJ
(c) –97.34 kJ (d) –89.9 kJ

79. The work done by a system is 10 joule, when 40 joule heat is supplied to it. What is the increase in internal energy of system?

(a) 30 J (b) 50 J
(c) 40 J (d) 20 J

80. The increase in internal energy of the system is 100 J when 300 J of heat is supplied to it. What is the amount of work done by the system?

(a) – 200 J (b) + 200 J
(c) – 300 J (d) – 400 J

81. What is the value of ΔE, when 64 g oxygen is heated from 0°C to 100°C at constant volume? (C_v on an average is 5 JK^{-1} mol^{-1}):

(a) 1500 J (b) 1800 J
(c) 2000 J (d) 2200 J

82. To calculate the amount of work done in joules during a reversible isothermal expansion of an ideal gas, the volume must be expressed in:

(a) dm^3 only (b) m^3 only
(c) cm^3 only (d) Any one of them

83. If 0.75 mole of an ideal gas is expanded isothermally at 27°C from 15 litres to 25 litres, then work done by the gas during this process is (R = 8.314 J K^{-1} mol^{-1}):

(a) –1054.2 J (b) –896.4 J
(c) –954.2 J (d) –1254.3 J

84. The entropy change when 36 g of water evaporates at 373 K is ($\Delta H = 40.63$ kJ mol^{-1}):

(a) 218 J K^{-1} (b) 150 J K^{-1}
(c) 118 J K^{-1} (d) 200 J K^{-1}

85. Heat energy absorbed by a system in going through a cyclic process shown below is?

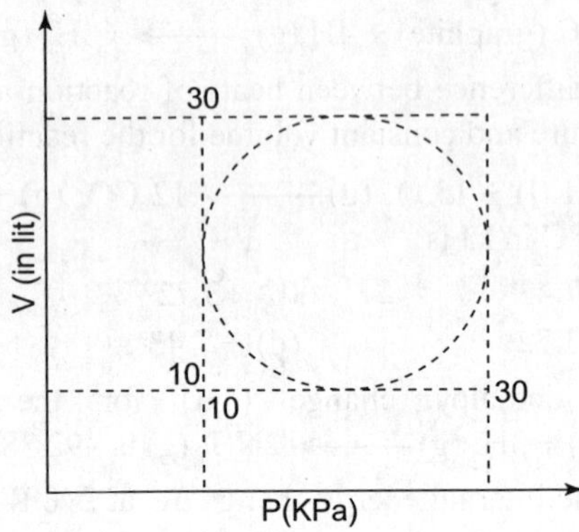

(a) 10πJ (b) 100πJ
(c) 1000πJ (d) $10,000\pi$J

86. If the standard entropies of CH_4 (g), H_2O(g), CO_2 (g) and H_2 (g) are 186.2, 188.2, 197.6 and 130.6 JK^{-1} mol^{-1} respectively, then the standard entropy change for the reaction:

$CH_4\ (g) + H_2O\ (g) \longrightarrow CO_2\ (g) + 3H_2\ (g)$ is:

(a) 215 JK^{-1} mol^{-1} (b) 225 JK^{-1} mol^{-1}
(c) 145 JK^{-1} mol^{-1} (d) 285 JK^{-1} mol^{-1}

87. Two moles of an ideal gas are compressed at 300 K from a pressure of 1 atm to a pressure of 2 atm. The change in free energy is:

(a) 5.46 kJ mol^{-1} (b) 2.46 kJ mol^{-1}
(c) 3.46 kJ mol^{-1} (d) 8.46 kJ mol^{-1}

88. In monoatomic gases, ratio of specific heat at constant pressure to that at constant volume is:

(a) 3/5 (b) 5/3
(c) 7/5 (d) 4/5

89. The standard entropies of H_2 (g), I_2 (s) and HI (g) are 130.6, 116.7 and 206.3 J K^{-1} mol^{-1} respectively. The change in standard entropy in the reaction $H_2\ (g) + I_2\ (s) \longrightarrow 2HI\ (g)$ is:

(a) 185.6 JK^{-1} mol^{-1} (b) 170.5 JK^{-1} mol^{-1}
(c) 165.9 JK^{-1} mol^{-1} (d) 165.9 JK^{-1} mol^{-1}

90. In the reaction: $CO\ (g) + \frac{1}{2} O_2\ (g) \longrightarrow CO_2\ (g)$, the change in ΔS^o is (given S^o for CO, O_2 and CO_2 are 197.6,205.3 and 213.6 JK^{-1} mol^{-1} respectively):

(a) –78.6 JK^{-1} mol^{-1} (b) –50 JK^{-1} mol^{-1}
(c) –86.5 JK^{-1} mol^{-1} (d) –30 JK^{-1} mol^{-1}

PRACTICE QUESTIONS – III

91. One mole of an ideal gas is allowed to expand reversibly and adiabatically from a temperature of 27°C. If work done during the process is 3 kJ, then final temperature of the gas is ($C_v = 20$ JK^{-1} mol^{-1})

(a) 150 K (b) 200 K
(c) 175 K (d) 225 K

92. The change in entropy, in the conversion of one mole of water at 373 K to vapour at the same temperature is (Latent heat of vaporization of water = 2.257 kJ g^{-1})

(a) 99 JK^{-1} (b) 129 JK^{-1}
(c) 89 JK^{-1} (d) 109 JK^{-1}

93. The direct conversion of A to B is difficult, hence it is carried out by the following path:

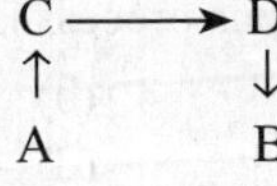

Given

$\Delta S(A \longrightarrow C) = 50$ e.u.

$\Delta S(C \longrightarrow D) = 30$ e.u.

$\Delta S(B \longrightarrow D) = 20$ e.u.

where e.u. is entropy unit then $\Delta S(A \longrightarrow B)$ is:

(a) +100 e.u. (b) +60 e.u.
(c) –100 e.u. (d) –60 e.u.

94. One mole of monatomic ideal gas at T (K) is expanded from 1L to 2L adiabatically under a constant external pressure of 1 atm the final temperature of the gas in Kelvin is:

(a) T (b) $\frac{T}{2^{5/3-2}}$

(c) $T - \frac{2}{3 \times 0.0821}$ (d) $T + \frac{3}{2 \times 0.0821}$

95. One mole of a non-ideal gas undergoes a change of state (2.0 atm, 3.0 L, 95 K) $\longrightarrow$ (4.0 atm, 5.0 L, 245 K) with a change in internal energy, $\Delta U = 30.0$ L atm. The change in enthalpy (ΔH) of the process in L atm is:

(a) 40.0
(b) 42.3
(c) 44.0
(d) Not defined, because pressure is not constant

96. The ΔH_f° for $CO_2(g)$, CO(g) and $H_2O(g)$ are –393.5, –110.5 and –241.8 kJ mol^{-1} respectively. The standard enthalpy change (in kJ) for the reaction:

$CO_2\,(g) + H_2\,(g) \longrightarrow CO(g) + H_2O(g)$ is

(a) 524.1 (b) 41.2
(c) –262.5 (d) –41.2

97. Calculate the enthalpy change for the combustion of cyclopropane at 298 K, if the enthalpy of formation CO_2 (g), H_2O(1) and propene (g) are –393.5, –385.8 and 20.42 kJ mol^{-1} respectively. The enthalpy of isomerization of cyclopropane to propene is –33.0 kJ mol^{-1}.

(a) 1802 kJ mol^{-1} (b) 2091 kJ mol^{-1}
(c) 2196 kJ mol^{-1} (d) None

98. The enthalpy change involved in the oxidation of glucose is –2880 kJ mol^{-1}. Twenty five per cent of this energy is available for muscular work. If 100 kJ of muscular work is needed to walk one kilometre, what is the maximum distance that a person will be able to walk after consuming 120 g of glucose?

(a) 7.9 km (b) 9.7 km
(c) 4.8 km (d) 8.4 km

99. Anhydrous $AlCl_3$ is covalent. From the data given below, predict whether it would remain covalent or become ionic in aqueous solution (ionization energy of Al = 5137kJ mol^{-1} $\Delta H_{hydration}$ for Al^{+3} = –4665 kJ mol^{-1}, $\Delta H_{hydration}$ for Cl^- = –381 kJ mol^{-1})

(a) Ionic (b) Covalent
(c) Both (a) and (b) (d) None of these

100. The standard molar enthalpies of formation of cyclohexane (1) and benzene (1) at 25°C are –156 and +49 kJ mol^{-1} respectively. The standard enthalpy of hydrogenation of cyclohexene (1) at 25°C is –119 kJ/mol. Find resonance energy of benzene.

(a) –152 kJ mol^{-1} (b) –159 kJ mol^{-1}
(c) +152 kJ mol^{-1} (d) +159 kJ mol^{-1}

101. An athlete is given 100 g of glucose of energy equivalent to 1560 kJ. He utilizes 50% of this gained energy in the event. In order to avoid storage of energy in the body, calculate the mass of water he would need to perspire. Enthalpy of H_2O for evaporation is 44 kJ mol^{-1}.

(a) 346 g (b) 316 g
(c) 323 g (d) 319 g

102. The standard enthalpy of combustion at 25°C of H_2, C_6H_{10} and cyclohexane (C_6H_{12}) are –241, –3800 and –3920 kJ mol^{-1} respectively. Calculate heat of hydrogenation of cyclohexane (C_6H_{10}).

(a) –161 kJ mol^{-1} (b) –131 kJ mol^{-1}
(c) –121 kJ mol^{-1} (d) None

103. The standard heats of formation of CH_4 (g), CO_2 (g) and H_2O (g) are –76.2, –394.8 and –241.6 kJ mol^{-1} respectively. Calculate the amount of heat evolved by burning 1 m^3 methane measured under normal conditions.

(a) 162897.3 kJ (b) 64078.2 kJ
(c) 35182 kJ (d) None

104. Calculate the resonance energy of N_2O from the following data: ΔH_f of N_2O = 82 kJ mol^{-1}. Bond energies of N≡N, N=N, O=O and N=O bonds are 946, 418, 498 and 607 kJ mol^{-1} respectively.

(a) –88 kJ mol^{-1}
(b) –170 kJ mol^{-1}
(c) –82 kJ mol^{-1}
(d) –258 kJ mol^{-1}

105. Calculate ΔH_f° for chloride ion from the following data:

$$\frac{1}{2}H_2\,(g) + \frac{1}{2}Cl_2\,(g) \longrightarrow HCl(g), \quad \Delta H_f^\circ = -92.4\text{kJ}$$

$$HCl\,(g) + nH_2O\,(l) \longrightarrow H^+\,(aq) + Cl^-\,(aq) \quad \Delta H_{208} = -74.8\text{ kJ}$$

$\Delta H_f^\circ [H^+] = 0.0$kJ

(a) – 189 kJ (b) – 167 kJ
(c) +167 kJ (d) –191 kJ

106. 0.16 g of methane is subjected to combustion at 27°C in a bomb calorimeter system. The temperature of the calorimeter system (including water) was found to rise by 0.5°C. Calculate the heat of combustion of methane at constant volume. The thermal capacity of the calorimeter system is 177 kJ K^{-1} (R = 8.314 J K^{-1} mol^{-1}):

(a) –695 kJ mol^{-1} (b) –1703 kJ mol^{-1}
(c) –890 kJ mol^{-1} (d) –885 kJ mol^{-1}

107. At 300 K, the standard enthalpies of formation of C_6H_5COOH (s), CO_2 (g) and H_2O (s) are –408, –393 and –286 kJ mol^{-1} respectively. Calculate the heat of combustion of benzoic acid at constant volume.

(a) –3296 kJ mol^{-1}
(b) –3200 kJ mol^{-1}
(c) –3201 kJ mol^{-1}
(d) –3603 kJ mol^{-1}

108. The heat liberated on complete combustion of 7.8 g benzene is 327 kJ. This heat was measured at constant volume and at 27°C. Calculate the heat of combustion of benzene at constant pressure (R = 8.3 J mol^{-1} K^{-1}).

(a) –3274 kJ mol^{-1}
(b) –1637 kJ mol^{-1}
(c) –3270 kJ mol^{-1}
(d) –3637 kJ mol^{-1}

109. The enthalpies of solution of $BaCl_2$ (s) and $BaCl_2.2H_2O$ (s) are –20.6 and 8.8 kJ mol^{-1} respectively. The enthalpy change for the hydration of $BaCl_2$ (s) is:

(a) 29.8 kJ (b) –11.8 kJ
(c) –20.6 kJ (d) –29.4 kJ.

110. For the reaction,
A (g) + 2B (g) $\longrightarrow$ 2C (g) + 3D (g)
The value of ΔH at 27°C is 19.0 kcal. The value of ΔE for the reaction would be (given R = 2.0 cal K^{-1} mol^{-1})

(a) 20.8 kcal (b) 19.8 kcal
(c) 18.8 kcal (d) 17.8 kcal

111. Following reaction occurs at 25°C.
2NO(g, 1×10^{-5} atm) + Cl_2 (g, 1×10^{-2} arm) $\rightleftharpoons$ 2NOCl(g, 1×10^{-2}atm) ΔG° is:

(a) –45.65 kJ (b) –66.53 kJ
(c) –22.82 kJ (d) –35.65 kJ

112. Determine ΔH and ΔE for reversible isothermal evaporation of 90 g of water at 100°C. Assume that water vapour behaves as an ideal gas and heat of evaporation of water is 540 cal g^{-1} (R = 2.0 cal mol^{-1} K^{-1}).

(a) 48600 cal, 44870 cal
(b) 43670 cal, 47700 cal
(c) 47700 cal, 43670 cal
(d) 44870 cal, 48670 cal

113. ΔG° for the reaction, x + y $\rightleftharpoons$ z is –4.606 kcal. The value of equilibrium constant of the reaction at 227°C is:

(a) 0.01 (b) 100
(c) 2 (d) 10

114. The standard heat of combustion of Al is –837.8 kJ mol^{-1} at 25°C. If Al reacts with O_2 at 25°C, which of the following releases 250 kcal of heat?

(a) The reaction of 0.312 mol of Al
(b) The formation of 0.624 mol of Al_2O_3
(c) The reaction of 0.712 mol of Al
(d) The formation of 0.615 mol of Al_2O_3

115. The dissociation energies of CH_4 and C_2H_6 to convert them into gaseous atoms are 360 and 620 kcal mol respectively. The bond energy of C – C bond is:

(a) 280 kcal mol^{-1} (b) 240 kcal mol^{-1}
(c) 160 kcal mol^{-1} (d) 80 kcal mol^{-1}

116. Calculate Q and W for the isothermal reversible expansion of one mole of an ideal gas from an initial pressure of 1.0 bar to a final pressure of 0.1 bar at a constant temperature of 273 K.

(a) 5.22 kJ, –5.22 kJ
(b) –27.3 kJ, 27.3 kJ
(c) 27.3 kJ, –27.3 kJ
(d) –5.22 kJ, 5.22 kJ

117. If at 298 K the bond energies of C–H, C–C, C=C and H–H bonds are respectively 414, 347, 615 and 435 kJ mol^{-1}, the value of enthalpy change for the reaction $H_2C = CH_2$ (g) + H_2 (g) $\longrightarrow$ $H_3C - CH_3$ (g) at 298 K will be:

(a) +250 kJ (b) –250 kJ
(c) +125 kJ (d) –125 kJ

118. An ideal gas expands in volume from 1×10^{-3} m^3 to 1×10^{-2} m^3 at 300.K against a constant pressure of 1×10^5 Nm^{-2}. The work done is:

(a) –900 kJ (b) –900 J
(c) 270 kJ (d) 940 kJ

119. The enthalpies of combustion of carbon and carbon monoxide are –393.5 and –283 kJ mol^{-1} respectively. The enthalpy of formation of carbon monoxide per mole is:

(a) –676.5 kJ (b) –110.5 kJ
(c) 110.5 kJ (d) 676.5 kJ

120. If the bond dissociation energies of XY, X_2 and Y_2 are in the ratio of 1:1:0.5 and ΔH_f for the formation of XY is –200 kJ/mole. The bond dissociation energy of X_2 will be:

(a) 100 kJ/mole (b) 400 kJ/mole
(c) 600 kJ/mole (d) 800 kJ/mole

121. A schematic plot of In K_{eq} versus inverse of temperature for a reaction is shown in the figure:

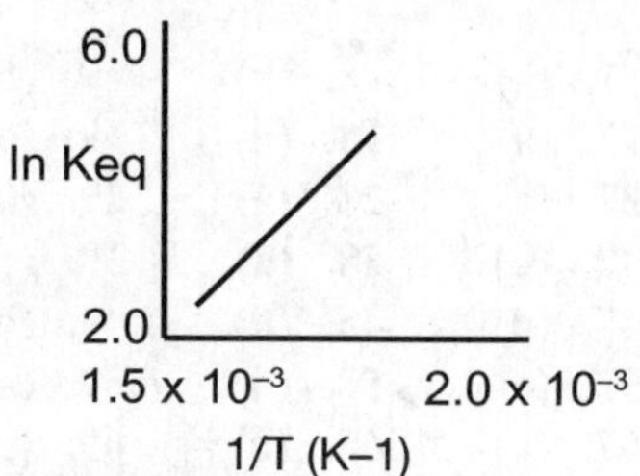

The reaction must be:
(a) Exothermic
(b) Endothermic
(c) One with negligible enthalpy change
(d) High spontaneous at ordinary temperature

122. The enthalpy changes for the following processes are listed below.

Cl_2 (g) = 2C1 (g); 242.3 kJ mol^{-1}
I_2 (g) = 21 (g); 151.0 kJ mol^{-1}
ICl (g) = I (g) + Cl(g); 211.3 kJ mol^{-1}
I_2 (s) = I_2 (g); 62.76 kJ mol^{1}

Given that the standard states for iodine and chlorine are I_2 (s) and Cl, (g), the standard enthalpy of formation for ICl (g) is:
(a) –14.6 kJ mol^{-1} (b) –16.8 kJ mol^{-1}
(c) +16.8 kJ mol^{-1} (d) +244.8 kJ mol^{-1}

123. In the conversion of lime stone to lime,

$CaCO_3$ (s) $\longrightarrow$ CaO (s) + CO_2 (g)

The values of ΔH° and ΔS° are +179.1 kJ mol^{-1} and 160.2 J/K respectively at 298 K and 1 bar. Assuming that ΔH° and ΔS° do not change with temperature, temperature above which conversion of limestone to lime will be spontaneous is:
(a) 1200 K (b) 845 K
(c) 1118 K (d) 1008 K

124. Standard entropy of X_2, Y_2 and XY_3 are 60, 40 and 50 JK^{-1} mol^{-1}, respectively. For the reaction:

1/2 X_2 + 3/2 Y_2 $\longrightarrow$ XY_3, ΔH = –30 kJ, to be at equilibrium, the temperature will be:
(a) 1250 K (b) 500 K
(c) 750 K (d) 1000 K

125. Oxidizing power of chlorine in aqueous solution can be determined by the parameters indicated below:

$\frac{1}{2}Cl_2(g) \xrightarrow{1/2\ \Delta_{diss}\ H^-} Cl(g) \xrightarrow{\Delta_{eg}\ H^-}$

$Cl^-(g) \xrightarrow{\Delta_{hyd.}\ H^-} Cl^-(aq)$

The energy involved in the conversion of $\frac{1}{2}$ Cl_2 (g) to Cl^- (g)
(Using the data, Δ_{diss} H^-Cl_2 = 240 kJ mol^{-1}, Δ_{eg} H^-Cl = –349 kJ mol^{-1}, Δ_{hyd} H^-Cl = –381 kJ mol^{-1}) will be:
(a) +152 kJ mol^{-1} (b) –610 kJ mol^{-1}
(c) –850 kJ mol^{-1} (d) +120 kJ mol^{-1}

126. 10 mole of an ideal gas expand isothermally and reversibly from a pressure of 10 atm to 1 atm at 27°C. What will be the largest mass that be lifted through a height of 100 metre?
(a) 5.855 Kg (b) 58.55 Kg
(c) 585.5 Kg (d) 29.28 Kg

127. If $\Delta H°_{formation}$ for O_3, CO_2, NH_3 and HI can be given as 142.1, –393.2, –46.2 and + 25.8 KJ/mole. The order of their stability in increasing order can be given as:
(a) $NH_3 < CO_2 < HI < O_3$
(b) $CO_2 < NH_3 < HI < O_3$
(c) $O_3 < HI < NH_3 < CO_2$
(d) $O_3 < HI < NH_3 < CO_2$

128. Which of the following are endothemic reactions?
(1) Decomposition of H_2O_2
(2) Conversion of graphite into diamond
(3) Combustion of ethane
(4) Dehydration of ethanol into ethane
(a) 1, 2 (b) 2, 4
(c) 1, 2, 4 (d) 1, 4

129. 36 ml of pure water takes 100 sec to evaporate from a vessel and heater connected to an electric source which delivers 806 watt. The $\Delta H_{Vap.}$ of water is:
(a) 40.3 KJ/mole (b) 406 KJ/mole
(c) 80.6 KJ/mole (d) 20.15 KJ/mole

130. Five moles of a gas is put through a series of changes as shown graphically in a cyclic process. The processes A → B, B → C and C → A are respectively

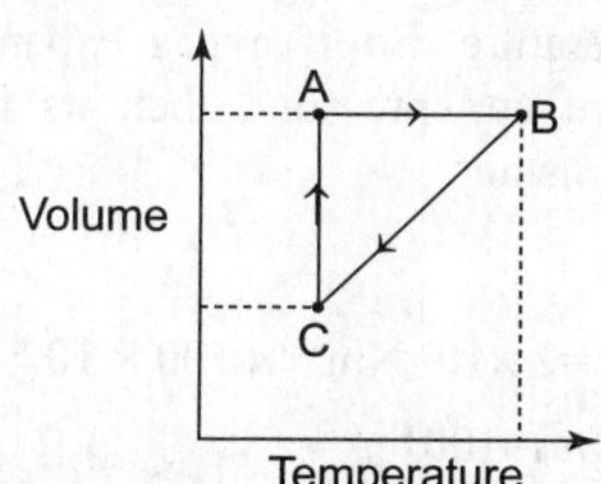

(a) Isochoric, isobaric, isothermal
(b) Isobaric, isocharic, isothermal
(c) Isochoric, isothermal, isobaric
(d) Isobaric, isothermal, isochoric

Answer Keys

1. (c)	2. (c)	3. (c)	4. (b)	5. (b)	6. (c)	7. (a)	8. (b)	9. (d)	10. (c)
11. (c)	12. (a)	13. (b)	14. (c)	15. (a)	16. (a)	17. (a)	18. (d)	19. (d)	20. (c)
21. (b)	22. (a)	23. (c)	24. (a)	25. (d)	26. (d)	27. (a)	28. (a)	29. (b)	30. (d)
31. (c)	32. (a)	33. (b)	34. (a)	35. (b)	36. (c)	37. (b)	38. (a)	39. (c)	40. (d)
41. (a)	42. (a)	43. (a)	44. (b)	45. (d)	46. (b)	47. (d)	48. (d)	49. (b)	50. (a)
51. (b)	52. (c)	53. (a)	54. (a)	55. (a)	56. (c)	57. (c)	58. (d)	59. (a)	60. (b)
61. (d)	62. (a)	63. (d)	64. (c)	65. (c)	66. (b)	67. (b)	68. (b)	69. (c)	70. (c)
71. (c)	72. (d)	73. (d)	74. (c)	75. (d)	76. (a)	77. (a)	78. (b)	79. (a)	80. (b)
81. (c)	82. (d)	83. (c)	84. (d)	85. (b)	86. (a)	87. (c)	88. (b)	89. (d)	90. (c)
91. (a)	92. (d)	93. (b)	94. (c)	95. (c)	96. (b)	97. (b)	98. (c)	99. (a)	100. (a)
101. (d)	102. (c)	103. (c)	104. (a)	105. (b)	106. (d)	107. (b)	108. (a)	109. (d)	110. (d)
111. (a)	112. (a)	113. (b)	114. (b)	115. (d)	116. (a)	117. (d)	118. (b)	119. (b)	120. (d)
121. (a)	122. (c)	123. (c)	124. (c)	125. (b)	126. (b)	127. (d)	128. (c)	129. (a)	130. (a)

Hints and Explanations for Selective Questions

3. If a gas is ideal, there is no force of attraction between its molecules, hence expansion takes place without loss of energy.

4. If the value of ΔH is positive, then the heat is absorbed by the system during the reaction.

6. Plants and living beings are capable of exchanging both, matter and energy with their surroundings, that is, they are examples of open system.

7. In thermodynamics a process is called reversible when the surroundings are always in equilibrium with the system.

8. Number of moles depends upon the quantity of a substance present in a system, so it is an extensive property of a substance.

9. The internal energy of a gas depends upon its pressure and temperature. So if a gas expands at constant temperature and pressure, then its internal energy remains constant.

10. $Q = 200$ J

$W = P\Delta V = 2 \times 10^5 \text{ Nm}^{-2} \times 500 \times 10^{-6} \text{ m}^3$

$= 100\text{Nm} = 100\text{J}$

$\Delta E = Q - W = 200 - 100 = 100\text{J}$

11. Internal energy does not depend upon energy due to gravitational pull.

12. Internal energy is a state function.

13. A state function is independent of the manner in which it is performed. Since work is path dependent, it is not a state function.

18. ΔE = 0, in a cyclic process.

21. $C(s) + \frac{1}{2}O_2(g) \longrightarrow CO\ (g);\ (\Delta n = 1 - \frac{1}{2}) = \frac{1}{2}$

$\Delta H - \Delta U = \Delta nRT = +\frac{1}{2} \times 8.314 \times 298$

$= +\ 1238.78 \text{ J mol}^{-1}$

22. The standard molar enthalpy of formation of CO_2 and standard molar enthalpy of combustion of carbon (graphite) refer to the same chemical equation, so are equal.

23. As on moving from solid state to liquid state entropy increases.

25. $\Delta S_{total} = \Delta S_{system} + \Delta S_{surroundings}$

For a spontaneous process ΔS_{total} must be positive that is, $\Delta S_{total} > 0$

26. As here least volatile is H_2O (1), hence the maximum heat is required to vaporize it. and $\Delta S = \frac{\Delta H}{T}$, hence the maximum entropy.

27. As combustion is always exothermic ΔH = –ve

28. Statement A is false, as work is a path function

30. $C\ (graphite) + O_2\ (g) \longrightarrow CO_2\ (g)$

The standard molar enthalpy of formation of CO_2 is equal to the standard molar enthalpy of combustion of carbon (graphite).

31. $\Delta H \neq \Delta E$ when $\Delta ng \neq 0$

32. $\Delta H° = \Delta H_f^\circ\text{(products)} - \Delta H°\text{(reactants)}$

$\Delta H° = \Delta H_f^\circ(HS) + \Delta H_f^\circ(H) - \Delta H_f^\circ(H_2S)$

$x_1 = x + x_3 - x_2$

$x = (x_1 + x_2 - x_3)$

34. $G = H - T.S$

It is a single valued function of thermodynamic state of the system.

35. The values of change in Gibbs free energy (ΔG) and change in enthalpy (ΔH), depend upon entropy change (ΔS) for a thermochemical reaction.

36. If the change in free energy (ΔG) is negative, then the electrical work is done by the system on the surroundings.

39. ΔH = (BE) of all bonds in reactants – (BE) of all bonds in products

$= Q_3 + 3Q_2 - 6Q_1$ (each NH_3 has three N–H bonds)

40. The oxygen molecule has the maximum specific heat.

41. Work done is reversible process is maximum. Thus $T_2 << T_1$ in reversible process or $T_{2rev} << T_{2\ irr}$

42. Bomb calorimeter is commonly used to find the heat of combustion of organic substances which consists of a sealed combustion chamber, called a bomb. If a process is run in a sealed container then no expansion or compression is allowed, so w = 0 and

$\Delta U = Q.$

$\Delta U < 0, w = 0.$

43. $\Delta G° = -RT \ln K$

$\Delta G°$ = standard free energy change, K = equilibrium constant, if a substance is in equilibrium between two phases at constant temperature and pressure , its chemical potential must have the same value in both the phases. So, $AG° = 0$

45. As for any spontaneous change the best condition is:

$\Delta H = -ve$

$\Delta S = +ve, \Delta G = -ve$

46. As PV^γ = Constant

$P \cdot \gamma V^{\gamma-1} dV + V^\gamma \cdot dP = 0$

So $\frac{dP}{p} = -\gamma \frac{dV}{V}$

47. $\Delta H° = \frac{3}{2} \times \Delta H°_1 + \frac{\Delta H°_2}{2}$

$= -713.7 - 177.5$

$= -891.2$ kJ

51. $\eta = 1 - \frac{T_1}{T_2}$

$\frac{T_1}{T_2} = 1 - \eta$

$= 1 - 0.15 = 0.85 = 85\%$

$T_1 = 85\%\ T_2$

52. $W = -2.303\ nRT \log \frac{V_2}{V_1}$

$= -2.303 \times 1 \times 2 \times 300 \times \log \frac{10}{1}$

$= -1381.8$ cal.

54. $C + 2H_2 \longrightarrow CH_4;\ \Delta H = -74.8$ kJ

$\Delta H = [HC_{(s\to g)} + 2 \times E_{H-H}] - 4 \times E_{C-H}$

55. $\Delta G = \Delta H - T\Delta S$

$= -(2.5 \times 10^3) - (298 \times 7.4) = -4705.2$ J.

As ΔG is –ve, so the reaction is spontaneous

56. $N_2 + 3H_2 \longrightarrow 2NH_3$

$\Delta n = 2 - 4 = -2$

$\Delta H = \Delta U + \Delta nRT = \Delta U - 2RT$

$\Delta H < \Delta U$

57. At equilibrium $\Delta G = 0$. So $\Delta H = T\Delta S$

So $T = \Delta H/\Delta S = 30 \times 10^3/105$

$= 285.7$ K

58. As for a pure substance T_A and T_B represent the same temperature so, option d is correct here.

59. The change in internal energy

$\Delta E = \Delta Q - \Delta W = 40 - 10 = 30$ J.

60. $\Delta S°(CO_2) = S(CO_2) - S(C) + S(O_2)$

$= 213.5 - [5.74 + 205] = 2.76\ J\ K^{-1}$

61. $\Delta G = \Delta H - T\Delta S$

$= -(11.5 \times 10^3) - [300(-105)]$

$= 11500 - (-31500) = 20000\ J\ mol^{-1}$

$= 20\ kJ\ mol^{-1}$

62. $\Delta G° = \Delta H° - T\Delta S°$

$= 58.04 - [298 \times (176.7 \times 10^{-3})]$

$= 58.04 - 52.65 = 5.39$ kJ.

63. As $\Delta S° = \Sigma S°_p - \Sigma S°_R$
So, $\Delta S° = 2S°\ HC1 - (S°\ H_2 + S°\ Cl_2)$
$= 2 \times 186.7 - (130.6 + 223.0)$
$= 19.8\ JK^{-1}\ mol^{-1}$

64. Methanol can be prepared synthetically by heating carbon monoxide and hydrogen gases under pressure in the presence of a catalyst.

$$C\ (graphite) + \frac{1}{2} O_2 \longrightarrow CO\ (g) \quad (1)$$
$$CO(g) + 2H_2(g) \longrightarrow CH_3OH(l) \quad (2)$$
$$C(graphite) + \frac{1}{2} O_2(g) + 2H_2(g) \longrightarrow CH_3OH(l)$$

65. As $\Delta H = nC_p\ \Delta T$
As for isothermal process, $\Delta T = 0$,
Hence $\Delta H = 0$

66. $\Delta S = \frac{\Delta Q_{rev}}{T}$

So $T = \frac{\Delta Q}{\Delta S} = \frac{30 \times 1000}{75} = 400\ K$

68. $\Delta S° = \Sigma S°_p - \Sigma S°_R$
$= [(213.6 + 2 \times 69.96) - (186.2 + 2 \times 205.3)]$
$= -243.3\ J\ K\ mol^{-1}$.

69. $\Delta S° = \Sigma S°_p - \Sigma S°_R$
$= [256.2 - (248.5 + \frac{1}{2} \times 205)]$
$= -94.8\ J\ K^{-1}\ mol^{-1}$

70. The driving force and the opposing force differ by infinitesimally small amount. So the system and surroundings always remain in equilibrium state with one another.

71. It gives the neutralization of 1 mole H^+ and 1 mole of OH^- ions.

72. $Q = \Delta E - W$ (first law of thermodynamics)
$Q + W = \Delta E$
Now $Q + W$ is a state function as it is equal to ΔE (state function).

73. $\Delta G = \Delta H - T\Delta S$
$= -(11.5 \times 10^3) - [300\ (-105)]$
$= 11500 - (-31500) = 20000\ J\ mol^{-1}$
$= 20\ kJ\ mol^{-1}$

74. $\Delta G = \Delta H - T\Delta S$
$= -285.8 - (300 \times 0.163)$
$= -285.8 - 48.9 = -334.7\ kJ\ mol^{-1}$

76. Standard heat of formation of C_2H_4 is the heat change when 1 mole of C_2H_4 is formed from its elements in the standard state [carbon as graphite and hydrogen as H_2].

77. $\Delta H - \Delta E = \Delta ng\ RT$
$= (12 - 15) \times \frac{8.314}{1000} \times 298$
$= -7.433\ kJ\ mol^{-1}$

78. $\Delta H = \Delta U + \Delta nRT$
(here Δn = number of molecules of products – number of molecules of reaction)
$-92.38 \times 1000 = \Delta U - 2 \times 8.314 \times 298$
$\Delta U = -87424\ J = -87.424\ kJ$.

79. The change in internal energy
$\Delta E = \Delta Q - \Delta W = 40 - 10 = 30\ J$.

80. $\Delta W = \Delta Q - \Delta E = 300 - 100 = +200J$.

81. $n = \frac{\text{Weight of oxygen}}{\text{Mol. wt. of oxygen}} = \frac{64}{16} = 4$
$\Delta E = nCv(T_2 - T_1)$
$= 4 \times 5 \times (373 - 273) = 4 \times 5 \times 100 = 2000\ J$.

82. $W = 2.3\ n\ RT\ \log (V_1/V_2)$
Since the relation involves ratio of V_1 and V_2, so the unit of volume of gas may be expressed in any one of the given units that is, m^3, dm^3 or cm^3.

83. $W = 2.303\ nRT\ \log (V_1/V_2)$
$= 2.303 \times 0.75 \times 8.314 \times 300\ \log (15/25)$
$= 4302\ \log 0.6 = 4302 \times (-0.2218) = -954.2\ J$.

84. Mol. wt. of water = 18
Number of moles in 36 g of water
$= \frac{\text{Weight of water}}{\text{Mol. wt. of water}} = \frac{36}{18} = 2$

$\Delta S = \frac{\Delta H}{T} = \frac{40630}{373} = 109\ JK^{-1}\ mol^{-1}$

Entropy change for the evaporation of 36 g or 2 moles of water = $109 \times 2 = 218\ JK^{-1}$

86. $\Delta S° = \Sigma S°_p - \Sigma S°_R$
$= [(197.6 + 3 \times 130.6) - (186.2 + 188.2)]$
$= 215\ J\ K^{-1}\ mol^{-1}$

87. $\Delta G = 2.303 \times nRT\ \log (P_2/P_1)$
$= 2.303 \times 2 \times 8.314 \times 300\ \log (2/1)$
$= 11500\ \log 2 = 11500 \times 0.3010$
$= 3460\ J\ mol^{-1} = 3.46\ kJ\ mol^{-1}$

88. For monoatomic gases
$C_p = 5/2\ R.\ C_v = 3/2\ R$

$\gamma = C_p/C_v = \frac{(5/2)R}{(3/2)R} = \frac{5}{3}$

89. $\Delta S° = \Sigma S°_p - \Sigma S°_R$
$= [(2 \times 206.3) - (130 + 116.7)]$
$= 165.9 \text{ J K}^{-1} \text{ mol}^{-1}$

90. $\Delta S° = \Sigma S°_P \Sigma S°_R$
$= 213.6 -(197.6 + \frac{1}{2} \times 205.3)$
$= -86.5 \text{ J K}^{-1} \text{ mol}^{-1}$

91. $W = nC_v(T_1 - T_2)$ or $T_1 - T_2 = \frac{W}{nC_v}$
$T_2 = T_1 - \frac{W}{nC_v} = 300 - \frac{3000}{1 \times 20} = 150 \text{ K}$

92. Latent heat of vaporization of water in kJ mol^{-1}
$= (2.257 \times 18) = 40.63 \text{ kJ mol}^{-1}$
Change in entropy, $\Delta S = \frac{\Delta H}{T} = \frac{40.63}{373}$
$= 0.109 \text{ kJK}^{-1} \text{ mol}^{-1} = 109 \text{ JK}^{-1} \text{ mol}^{-1}$
so change in entropy of one mole
$= 109 \text{ JK}^{-1} \text{ mol}^{-1} \times 1 \text{ mole} = 109 \text{ JK}^{-1}$.

93. $\Delta S(A \longrightarrow B)$
$= \Delta S(A \longrightarrow C) + \Delta S(C \longrightarrow D)$
$- \Delta S(B \longrightarrow D$
$= 50 + 30 - 20 = 60$e.u.

94. During adiabatic irreversible process,
$W_{surr} = \Delta E = - P\Delta V$, $nC_V \Delta T = - P\Delta V$
$1 \times \frac{R}{\gamma - 1} \times (T_f - T) = - P(V_2 - V_1)$
$T_f = T - \frac{(\gamma - 1)}{R} P(V_2 - V_1)$
$T_f = \frac{T(5/3 - 1) \times 1 \times (2 - 1)}{0.0821}$
$T_f = T - \frac{2}{3 \times 0.0821}$

95. $\Delta H = \Delta U + \Delta(PV)$
$= \Delta U + (P_2V_2 - P_1V_1)$
$= 30 + (4 \times 5 - 2 \times 3)$
$= 30 + 14 = 44$ atm

96. $\Delta H = \Delta H°_f$ (products) $- \Delta H°_f$ (reactants)
$= [\Delta H°_f(CO) + \Delta H°_f(H_2O)] - [\Delta H°_f(CO_2) + \Delta H°_f(H_2)]$
$= [-110.5 - 241.8] - [-393.5 + 0]$
$= -352.3 + 393.5 = +41.2 \text{ kJ}$

98. As energy available for muscular work
$\frac{2880 \times 25}{100} = 720 \text{ kJ mol}^{-1}$
So the energy available for muscular work for 120 g of glucose $= 720 \times \frac{120}{180} = 480 \text{ kJ}$
As 100 kJ is used for walking = 1 km So 480 kJ is used for walking
$= 1 \times \frac{480}{100} = 4.8 \text{ km}$

99. If AIC1, is present in ionic state in aqueous solution (Al^{+3} and three Cl^{-1} ions), then standard heat of hydration of Al^{3+} and three Cl^- ions.
$\Delta H°$ (hydration) $= -4665 + (3 \times -381)$
$= -5808 \text{ kJ mol}^{-1}$.
As hydration energy is greater than ionization energy of Al, so $AlCl_3$ would be ionic in solution.

100. Given
(1) $C_6H_{10} + H_2 \longrightarrow C_6H_{12}$; $\Delta H = -119 \text{ kJ mol}^{-1}$
This reaction involves breaking up of one double bond and addition of one H_2 molecule.
$C_6H_{10} + 3H_2 \longrightarrow C_6H_2$;
$\Delta H = 3 \times -119 = -357 \text{ kJ}$
(because, this reaction involves breaking up of one double bond and addition of three H_2 molecules). Also given
$6C + 6H_2 \longrightarrow C_6H_{12}$; $\Delta H = -156 \text{ kJ}$
$C_6H_6 + 3H_2 \longrightarrow C_6H_{12}$; $\Delta H = - 357 \text{ kJ}$
$6C + 3H_2 \longrightarrow C_6H_6$; $\Delta H = +201 \text{ kJ}$
Thus, resonance energy = difference between ΔH_f (obs) $- \Delta H_f$ (calc) $= 49 - 201 = -152$ kJ

101. Amount of energy left unused
$= \frac{1560}{2} = 780 \text{ kJ}$
Amount of water perspired
$= \frac{\text{Energy left unused}}{\text{Enthalpy of evaporation of water}}$
$= \frac{780 \text{ kJ}}{44 \text{ kJ mol}^{-1}} = \frac{780}{44} \text{ mole}$
$= \frac{780 \times 18}{44} = 319.14 \text{ g}$

102. Given that:
(1) $H_2(g) + \frac{1}{2} O_2(g) \longrightarrow H_2O$ (l);
$\Delta H = -241 \text{ kJ}$
(2) $C_6H_{10} + 7/2O_2 \longrightarrow 6CO_2(g) + 5H_2O$;
$\Delta H = -3800 \text{ kJ}$
(3) $C_6H_{12} + 9O_2 \; 6CO_2(g) + 6H_2O$ (l);
$\Delta H = -3920 \text{ kJ}$
Our aim is to get
$C_6H_{10} + H_2 \longrightarrow C_6H_{12}$; $\Delta H = ?$
So add (1) and (2) and then subtract (3), to get $\Delta H = -121 \text{ kJ mol}^{-1}$'

103. $CH_4 + 2O_2 \longrightarrow CO_2 + 2H_2O$

$\Delta H = \sum \Delta H°_f \text{(products)} - \sum \Delta H°_f \text{(reactants)}$

$[\Delta H_f (CO_2) + 2\ \Delta H_f (H_2O)] - [\Delta H_f (CH_4) + 2\Delta H_f (CO_2)]$

$= [-394.8 + 2 \times -241.6] - [-76.2 + 0]$

$= -878.0 + 76.2 = 802.2\ kJ\ mol^{-1}$

Now 22.4 L (or $22.4 \times 10^{-3}\ m^3$) of CH_4 on combustion gives heat = $802.2\ kJ\ mol^{-1}$

$1\ m^3 = 802.2/22.4 \times 10^{-3}$

$= 35812\ kJ\ mol^{-1}$

104. $N \equiv N + \frac{1}{2} O = O \longrightarrow N = N = O$

$\Delta H°_f = \sum BE$ of reactants $- \sum BE$ of products

$= [BE (N{\equiv}N) + \frac{1}{2} BE(O{=}O)] - [BE(N{=}N) + BE(N{=}O)]$

$= (946 + \frac{1}{2} \times 498) - (418 + 607) = 170$ kJ resonance energy

$= \Delta H°_f \text{(observed)} - \Delta H°_f \text{(calculated)} = 82 - 170$

$= -88\ kJ\ mol^{-1}$

105. $\Delta H°_f$ is to be calculated for the following reaction:

$\frac{1}{2} Cl_2 (g) + (aq) \longrightarrow Cl^- (aq);\ \Delta H = ?$

Given is:

(1) $\frac{1}{2} H_2 (g) + \frac{1}{2} Cl_2 (g) \longrightarrow HCl (g)$ $\quad \Delta H = -92.4\ kJ$

(2) $HCl (g) + (aq) \longrightarrow H^+ (aq) + Cl^- (aq)$ $\quad \Delta H = -74.8\ kJ$

(3) $\frac{1}{2} H_2 (g) + (aq) \longrightarrow H^+ (aq);\ \Delta H = 0.0\ kJ$

Operating (1) + (2) – (3), we obtain

$\Delta H°_f$ of $Cl^- (aq) = -167.2\ kJ$

106. Heat produced by the combustion of 0.16 g of CH = 17.7 × 0.5 = 8.85 kJ

Heat produced by the combustion of

1.0 mole of CH_4 (16 g) $= \frac{8.85 \times 16}{0.16} = 885$ kJ

As combustion has been carried in bomb calorimeter (closed vessel), the heat produced is Q_v, that is, Q = –885 kJ

107. The reaction is:

$C_6H_5COOH (s) + 15/2\ O_2 (g) \longrightarrow 7\ CO_2 (g) + 3H_2O (1)$

$\Delta H = \sum \Delta H°_f \text{(products)} - \sum \Delta H°_f \text{(reactants)}$

$= [7 \times \Delta H_f (CO_2) + 3 \times \Delta H_f (H_2O)]$

$- [\Delta H_f (C_6H_5COOH) + 15/2\ \Delta H_f (CO_2)]$

$= [7 \times (-393) + 3 \times (-286)] - [-408 + 0]$ kJ

$= -3201$ kJ

Now $\Delta E = \Delta H - \Delta n_g RT$

$\Delta n_g = 7 - 15/2 = -1/2$;

R = 8.314/1000 kJ mol^{-1} Putting the values, we get

$\Delta E = -3199.9\ kJ = -3200\ kJ$

108. ΔE per mole $= \frac{327}{7.8} \times 78 = -3270$ kJ

T = 27 + 273 = 300K,

R = 8.3/1000 kJ $mol^{-1}\ K^{-1}$

$C_6H_6 (l) + 15/2\ O_2 (g) \longrightarrow 6CO_2 (g) + 3H_2O(l)$

$\Delta n_g = n_p - n_r = 6 - 15/2 = -3/2$

$\Delta H = \Delta E + \Delta n_g RT$

$\Delta H = -3270 + [(-3/2) \times (8.3/1000) \times 300]$

$\Delta H = -3274$ kJ

109. Given:

(1) $BaCl_2 (s) \longrightarrow BaCl_2 (aq);\ \Delta H_1 = -20.6\ kJ$

(2) $BaCl_2.2H_2O (s) \longrightarrow BaCl_2 (aq);\ \Delta H_2 = 8.8\ kJ$

Now subtract (2) from (1), we obtain

$BaCl_2 (s) \longrightarrow BaCl_2.2H_2O (s)$;

$\Delta H = \Delta H_1 - \Delta H_2 = -29.4kJ$

110. $\Delta n_g = n_p - n_r = 5 - 3$

$\Delta H = \Delta E + \Delta n_g RT$

$\Delta E = \Delta H - \Delta n RT$

$= 19 - (2 \times 2 \times 10^{-3} \times 300)$

$= 19 - 1.2 = 17.8$ kcal

111. $2NO (g) + Cl_2 (g) \rightleftharpoons 2NOCl (g)$

$K_P \frac{P_{2NOCl}}{P_{2NOCl_2}} = 1 \times 10^8$

K_P is thermodynamics equilibrium constant (relative to 1 atm) and is considered as unitless quantity

$\Delta G^0 = -2.303\ RT \log K_p$

$= -2.303 \times 8.314 \times 10^{-3} \times 298 \log 10^8$ kJ

$= -45.65$ kJ

112. $\Delta H = m \times T_b$

$= 90 \times 540 = 48600$ cal

$\Delta H = \Delta E + P\Delta V$

$\Delta H = \Delta E + P(V_g - V_l)$

Since volume of liquid is negligible as compared to volume of its vapour, so

$V_g >>> V_1$

$\Delta H = \Delta E + P.V_g$

$\Delta H = \Delta E + nRT \qquad (PV = nRT)$

$\Delta E = \Delta H - nRT$

$= 48600 - (90/18 \times 2 \times 373)$

$= 44870$ cal.

113. $\Delta G^0 = -2.303\ RT \log K$

$$-4.606 = -2.303 \times \frac{2}{100} \times 500 \log K$$

$$\log K = \frac{-4.606}{-2.303 \times 2/100 \times 500} = 2$$

$K = \text{antilog } (2) = 100$

114. $2Al + 3/2\ O_2 \longrightarrow Al_2O_3$

2mol 0.5 mol O_2 1 mol Al_2O_3

1 mol 0.75 mol O_2 0.5 mol Al_2O_3

$\Delta H^0 = -837.8\ kJ\ mol^{-1}$

837.8 kJ mol^{-1} = combustion of 1 mol of Al or formation of 0.5 mol Al_2O_3

$$250\ kcal = \frac{250 \times 8.314\ kJ}{2}$$

$= 1.24$ mol Al or 0.62 mol of Al_2O_3

115. $CH_4\ (g) \longrightarrow C(g) + 4H(g)$,

$\Delta H = 360$ kcal for four (C – H) bonds

$$BE\ (C-H) = \frac{360}{4} = 90\ kcal$$

$C_2H_6 \longrightarrow 2C(g) + 6H\ (g)$

$\Delta H = 620$ kcal for six (C – H) bonds and one (C – C) bond.

6 BE (C – H) + BE (C – C) = 620

6 × 90 + BE (C – C) = 620

BE(C – C) = 80 kcal mol^{-1}

116. n = 1, P_1 = 1 bar,

$P_2 = 0.1$ bar, T = 273K

$W = -2.303\ nRT \log P_1/P_2$

$= -2.303 \times 1 \times 8.314 \times 273 \log 1/0.1$

$= -2.303 \times 1 \times 8.314 \times 273 \times 1$

$= -5227$ J

$= -5.227$ kJ

Now, $Q = \Delta E - W$

Since temperature is constant, $\Delta E = 0$

$Q = -W = +5.227$ kJ

117. $CH_2 = CH_2 + H_2 \longrightarrow CH_3–CH_3$

ΔH = (BE)reactants – (BE)products

$= 4(BE)_{C-H} + (BE)_{C=C}{}^{+} + (BE)_{H-H}$

$[6(BE)_{C-H} + (BE)_{C-C}]$

$= -125$kJ

118. Work done due to change in volume against constant pressure is:

$W = -P\ (V_2 - V_1) = -1 \times 10^5\ Nm^{-2}$

$(1 \times 10^{-2} - 1 \times 10^{-3})m^3$

$= -900$ Nm

$= -900$ J.

119. (1) $C\ (s) + O_2\ (g) \longrightarrow CO_2\ (g)$

$\Delta H = -393.5$ kJ

(2) $CO\ (g) + 1/2\ O_2\ (g) \longrightarrow CO_2\ (g)$

$\Delta H = -283.0$ kJ

(1) and (2) gives

(3)$C\ (s) + 1/2\ O_2\ (g) \longrightarrow CO\ (g)$,

$\Delta H = -110.5$ kJ

This equation (iii) also represents the formation of one mole of CO and so enthalpy change is the heat of formation of CO (g).

120. Let the bond dissociation energy of XY, X_2 and Y_2 be × kJ mol^{-1}, × kJ mol^{-1} and 0.5x kJ mol^{-1} respectively.

$1/2\ X_2 + 1/2\ Y_2 \longrightarrow XY;\ \Delta H_f = -200\ kJ\ mol^{-1}$

ΔH reaction = (sum of bond dissociation energy of all reactants) – (sum of bond dissociation energy of product)

$= [1/2\ \Delta H\ X_2 + 1/2\ \Delta H\ Y_2 - \Delta H\ XY]$

$$= \frac{x}{2} + \frac{0.5x}{2} - x = -200$$

$$x = \frac{200}{0.25} = 800\ kJ\ mol^{-1}$$

121. $$\text{In} \frac{K_2}{K_1} = \frac{\Delta H}{R} = \frac{1}{T_1} - \frac{1}{T_2}$$

$$\text{In} \frac{6}{2} = \frac{\Delta H}{R} [1.5\ ] \times 10^{-3} - 2 \times 10^{-3}$$

Since ΔH of reaction wil be negative, so reaction is exothermic.

122. $I_2\ (s) + Cl_2\ (g) \longrightarrow 2ICl\ (g);\ \Delta H$

$$\Delta H = E_{Cl_2} + E_{\underset{s}{I_2} - \underset{g}{I_2}} + E_{\underset{g}{I_2}} - 2E_{ICl}$$

$= [242.3 + 62.76 + 151.0] - 2 \times [211.3]$

$= 33.46$

$$\Delta H = \frac{33.46}{2} = 16.83\ kj\ mol^{-1}$$

124. $1/2\ X_2 + 3/2\ Y_2 \longrightarrow XY_3$

$\Delta S_{reaction} = 50 - (3/2 \times 40 + 1/2 \times 60)$

$= -40\ J\ mol^{-1}$

$\Delta G = \Delta H - T\Delta S$

As at equilibrium $\Delta G = 0$.

So $\Delta H = T\Delta S$

$30 \times 103 = T \times 40$

On solving, we get

$T = 750$ K.

125. For this process $1/2\ Cl_2\ (g) \longrightarrow Cl^-\ (aq)$

$\Delta H = 1/2\ \Delta H_{diss}$ of $Cl_2 + \Delta_{eq} Cl + \Delta_{hyd} Cl^-$

$= +\frac{240}{2} - 349 - 381$

$= 120 - 349 - 381$

$= -610$ kJ/mole

126. $W = -nRT \log \frac{P_1}{P_2}$

$= -2.303\ nRT \log_{10} \frac{P_1}{P_2}$

$= -2.303 \times 10 \times 8.314 \times 300 \times \log \frac{10}{1}$

$= -2.303 \times 10 \times 8.314 \times 300$

$= -23.03 \times 8.314 \times 300$

$= -57441.426$

As $w = -mgh$

$m = \frac{w}{gh} = \frac{57441.426}{9.8 \times 100}$

$= 58.55$kg

127. As more the –ve value of ΔH_F° more is the stability.

$O_3 < HI < NH_3 < CO_2$

ΔH_y° 142.1, 25.8, –46.2, –393.2 kJ/mole

129. 1 watt = 1J/S

Total heat supplied for 36 ml

$H_2O = 806 \times 100 = 80600$J

$\Delta H = \frac{80600}{36} \times 18$

$= 403005$ J/m

$= 40.3$ kJ/m

130. As in process A → B volume is constant so it is isochoric.

In B → C, pressure remains constant so it isobaric.

In C → A, temperature remains constant so it isothermal.

Previous Years' Questions

1. The entropy change in the fusion of 1 mol of a solid melting at 27°C (Latent heat of fusion, 2930 J mol^{-1}) is:

(a) 9.77 JK^{-1} mol^{-1} (b) 10.73 JK^{-1} mol^{-1}
(c) 2930 JK^{-1} mol^{-1} (d) 108.5 JK^{-1} mol^{-1}

[2000]

2. For the reaction:

$C_2H_5OH\ (l) + 3O_2\ (g) \rightarrow 2CO_2\ (g) + 3H_2O\ (l)$

Which one is true:

(a) $\Delta H = \Delta E - RT$ (b) $\Delta H = \Delta E + 2RT$
(c) $\Delta H = \Delta E - 2RT$ (d) $\Delta H = \Delta E + RT$

[2000]

3. The factor of ΔG values is important in metallurgy. The ΔG values for the following reactions at 800°C are given as:

$S_2\ (s) + 2O_2\ (g) \rightarrow 2SO_2\ (g)$; $\Delta G = -544$ kJ

$2Zn\ (s) + S_2\ (s) \rightarrow 2ZnS\ (s)$; $\Delta G = -293$ kJ

$2\ Zn\ (s) + O_2\ (g) \rightarrow 2\ ZnO\ (s)$; $\Delta G = -480$ kJ

the ΔG for the reaction,

$2ZnS\ (s) + 3O_2\ (g) \rightarrow 2\ ZnO\ (s) + 2SO_2\ (g)$

will be:

(a) –357 kJ (b) –731 kJ
(c) –773 kJ (d) –229 kJ

[2000]

4. When 1 mol gas is heated at constant volume temperature is raised from 298 to 308 K. Heat supplied to the gas is 500 J. Then which statement is correct?

(a) q = W = 500 J, ΔU = 0
(b) q = ΔU = 500 J, W = 0
(c) q = W = 500 J, ΔU = 0
(d) ΔU = 0, q = W = –500 J

[2001]

5. Change in enthalpy for reaction,

$2H_2O_2\ (l) \rightarrow 2H_2O\ (l) + O_2\ (g)$

if heat of formation of H_2O_2 (l) and H_2O (l) are –188 and –286 kJ/mol respectively is:

(a) –196 kJ/mol (b) +196 kJ/mol
(c) +948 kJ/mol (d) –948 kJ/mol

[2001]

6. Enthalpy of $CH_4 + 1/2\ O_2 \rightarrow CH_3OH$ is negative. If enthalpy of combustion of CH_4 and CH_3OH are X and Y respectively. Then which relation is correct?

(a) $X > Y$ (b) $X < Y$
(c) $X \geq Y$ (d) $X = Y$

[2001]

7. $PbO_2 \rightarrow PbO$ $\Delta G_{298} < 0$
$SnO_2 \rightarrow SnO$ $\Delta G_{298} > 0$
Most probable oxidation state of Pb and Sn will be:
(a) Pb^{4+}, Sn^{4+} (b) Pb^{4+}, Sn^{2+}
(c) Pb^{2+}, Sn^{2+} (d) Pb^{2+}, Sn^{4+}

[2001]

8. 2 mol of an ideal gas at 27°C temperature is expanded reversibly from 2 L to 20 L. Find entropy change in cal. (R = 2 cal/mol K):
(a) 92.1 (b) 0
(c) 4 (d) 9.2

[2002]

9. Unit of entropy is:
(a) $JK^{-1}\ mol^{-1}$ (b) $J\ mol^{-1}$
(c) $J^{-1}\ K^{-1}\ mol^{-1}$ (d) $JK\ mol^{-1}$

[2002]

10. In a closed insulated container, a liquid is stirred with a paddle to increase the temperature which of the following is true?
(a) $\Delta E = W \neq 0, q = 0$
(b) $W = 0, \Delta E = q \neq 0$
(c) $\Delta E = 0, W = q \neq 0$
(d) $\Delta E = W = q \neq 0$

[2002]

11. Heat of combustion ΔH° for C(s), H_2 (g) and CH_4 (g) are –94, –68 and –213 Kcal/mol, then ΔH° for $C(s) + 2H_2\ (s) \rightarrow CH_4\ (g)$ is:
(a) –117 kcal (b) –170 Kcal
(c) –17 Kcal (d) –87 Kcal

[2002]

12. The molar heat capacity of water at constant pressure, C, is 75 $JK^{-1}\ mol^{-1}$. When 1.0 kJ of heat is supplied to 100 g of water which is free to expand, the increase in temperature of water is:
(a) 4.8 K (b) 6.6 K
(c) 1.2 K (d) 2.4 K

[2003]

13. What is the entropy change (in $JK^{-1}\ mol^{-1}$) when 1 mol of ice is converted into water at 0°C? (The enthalpy change for the conversion of ice to liquid water is 6.0 kJ mol^{-1} at 0°C):
(a) 2.198 (b) 21.98
(c) 20.13 (d) 2.013

[2003]

14. For which one of the following equations $\Delta H°_{react}$ equal to $\Delta H°_f$ for the product?
(a) $N_2\ (g) + O_3\ (g) \rightarrow N_2O_3\ (g)$
(b) $CH_4(g) + 2Cl_2(g) \rightarrow CH_2Cl_2(l) + 2HCl(g)$
(c) $Xe\ (g) + 2F_2\ (g) \rightarrow XeF_4\ (g)$
(d) $2CO\ (g) + O_2\ (g) \rightarrow 2CO_2\ (g)$

[2003]

15. The densities of graphite and diamond at 298 K are 2.25 and 3.31 g cm^{-3} respectively. If the standard free energy difference (ΔG°) is equal to 1895 J mol^{-1}, the pressure at which graphite will be transformed into diamond at 298 K is:
(a) 7.92×10^8 Pa
(b) 9.92×10^8 Pa
(c) 9.92×10^{10} Pa
(d) 9.92×10^{18} Pa

[2003]

16. For the reaction,
$C_3H_8\ (g) + 5O_2\ (g) \rightarrow 3CO_2\ (g) + 4H_2O\ (l)$
at constant temperature, ΔH – ΔE is:
(a) + 3RT (b) –RT
(c) +RT (d) –3RT

[2003]

17. Standard enthalpy and standard entropy changes for the oxidation of ammonia at 298 K are –382.64 kJ mol^{-1} and –145.6 $JK^{-1}\ mol^{-1}$, respectively. Standard Gibbs energy change for the same reaction at 268 K is:
(a) –221.1 kJ mol^{-1} (b) –339.3 kJ mol^{-1}
(c) –439.3 kJ mol^{-1} (d) –523.2 kJ mol^{-1}

[2004]

18. If the bond energies of H – H, Br – Br, and H – Br are 433, 192 and 364 kJ mol^{-1} respectively, then ΔH° for the reaction:
$H_2\ (g) + Br_2\ (g) \rightarrow 2HBr\ (g)$ is
(a) –261 kJ (b) +103 kJ
(c) +261 kJ (d) –103 kJ

[2004]

19. The work done during the expansion of a gas from a volume of 4 dm^3 to 6 dm^3 against a constant external pressure of 3 atm is:
(a) – 6 J (b) –608 J
(c) +304 J (d) –304 J

[2004]

20. Considering entropy (S) as a thermodynamic parameter, the criterion for the spontaneity of any process is:

(a) $\Delta S_{system} + \Delta S_{surroundings} > 0$
(b) $\Delta S_{system} - \Delta S_{surroundings} > 0$
(c) $\Delta S_{system} > 0$ only
(d) $\Delta S_{surroundings} > 0$ only

[2004]

21. Which of the following pairs of a chemical reaction is certain to result in a spontaneous reaction?

(a) Exothermic and increasing disorder
(b) Exothermic and decreasing disorder
(c) Endothermic and increasing disorder
(d) Endothermic and decreasing disorder

[2005]

22. A reaction occurs spontaneously if:

(a) $T\Delta S < \Delta H$ and both ΔH, ΔS are +ve
(b) $T\Delta S > \Delta H$ and ΔH = +ve, ΔS = –ve
(c) $T\Delta S > \Delta H$ and both ΔH, ΔS are +ve
(d) $T\Delta S = \Delta H$ and both ΔH, ΔS are +ve

[2005]

23. The absolute enthalpy of neutralization of the reaction $MgO(s) + 2HCl(aq) \rightarrow MgCl_2(aq) + H_2O(l)$ will be:

(a) –57.33 kJ/mole
(b) >–57.33 kJ/mole
(c) <–57.33 kJ/mole
(d) 57.33 kJ/mole

[2005]

24. Assume each reaction is carried out in an open container. For which reaction will $\Delta H = \Delta U$?

(a) $PCl_5(g) \rightarrow PCl_3(g) + Cl_2(g)$
(b) $2CO(g) + O_2(g) \rightarrow 2CO_2(g)$
(c) $H_2(g) + Br_2(g) \rightarrow 2HBr(g)$
(d) $C(s) + 2H_2O(g) \rightarrow 2H_2(g) + CO_2(g)$

[2006]

25. Identify the correct statement for change of Gibb's energy for a system (ΔG_{system}) at constant temperature and pressure:

(a) If $\Delta G_{system} = 0$, the system is still moving in a particular direction
(b) If ΔG_{system} = –ve, the process is not spontaneous
(c) If ΔG_{system} = +ve, the process is spontaneous
(d) If $\Delta G_{system} = 0$, the system has attained equilibrium

[2006]

26. The enthalpy and entropy change for the reaction $Br_2(l) + Cl_2(g) \rightarrow 2BrCl(g)$ are 30 kJ mol^{-1} and 105 J K^{-1} mol^{-1} respectively. The temperature at which the reaction will be in equilibrium is:

(a) 450 K (b) 300 K
(c) 285.7 K (d) 273 K

[2006]

27. The enthalpy of hydrogenation of cyclohexene is –119.5 kJ mol^{-1}. If resonance energy of benzene is –150.4 kJ mol^{-1}, its enthalpy of hydrogenation would be:

(a) –269.9 kJ mol^{-1} (b) –358.5 kJ mol^{-1}
(c) –508.9 kJ mol^{-1} (d) –208.1 kJ mol^{-1}

[2006]

28. Consider the following reactions:

(1) $H^+(aq) + OH^-(aq) = H_2O(l)$
$\Delta H = -X_1$ kJ mol^{-1}
(2) $H_2(g) + 1/2\,O_2(g) = H_2O(l)$
$\Delta H = -X_2$ kJ mol^{-1}
(3) $CO_2(g) + H_2(g) = CO(g) + H_2O(l)$
$\Delta H = -X_3$ kJ mol^{-1}
(4) $C_2H_2(g) + 5/2\,O_2(g) = 2CO_2(g) + H_2O(l)$
$\Delta H = +X_4$ kJ mol^{-1}

Enthalpy of formation of $H_2O(l)$ is:

(a) $-X_2$ kJ mol^{-1} (b) $+X_3$ kJ mol^{-1}
(c) $-X_4$ kJ mol^{-1} (d) $+X_1$ kJ mol^{-1}

[2007]

29. Given that bond energies of H–H and Cl–Cl are 430 kJ mol^{-1} and 240 kJ mol^{-1} respectively and ΔH_f for HCl is –90 kJ mol^{-1}, bond enthalpy of HCl is:

(a) 290 kJ mol^{-1} (b) 380 kJ mol^{-1}
(c) 425 kJ mol^{-1} (d) 245 kJ mol^{-1}

[2007]

30. Bond dissociation enthalpy of H_2, Cl_2 and HCl are 434, 242 and 431 kJ mol^{-1} respectively. Enthalpy of formation of HCl is:

(a) –93 kJ mol^{-1} (b) 245 kJ mol^{-1}
(c) 93 kJ mol^{-1} (d) –245 kJ mol^{-1}

[2008]

31. For the gas phase reaction,

$PCl_5(g) \rightleftharpoons PCl_3(g) + Cl_2(g)$

Which of the following conditions are correct?
(a) $\Delta H < 0$ and $\Delta S < 0$
(b) $\Delta H > 0$ and $\Delta S < 0$
(c) $\Delta H = 0$ and $\Delta S < 0$
(d) $\Delta H > 0$ and $\Delta S > 0$

[2008]

32. Which of the following are not state functions?
(1) $q + w$ (2) q
(3) w (4) $H - TS$
(a) (1), (2) and (3) (b) (2) and (3)
(c) (1) and (4) (d) (2), (3) and (4)

[2008]

33. Standard free energies of formation (in kJ/mol) at 298 K are –237.2, –394.4 and –8.2 for H_2O (l), CO_2 (g) are pentane (g) respectively. The value of $E°_{cell}$ for the pentane-oxygen fuel cell is:
(a) 1.0968 V (b) 0.0698 V
(c) 1.968 V (d) 2.0968 V

[2008]

34. The values of ΔH and ΔS for the reaction
$$\underset{\text{graphite}}{C} + \underset{\text{'g'}}{CO_2} \rightarrow \underset{\text{'g'}}{2\,CO} \text{ are}$$
170 kJ and 170 JK^- respectively. This reaction will be spontaneous at:
(a) 910 K (b) 1110 K
(c) 710 K (d) 510 K

[2009]

35. Standrad entropies of X_2, Y_2 and XY_3 are 60, 40 and 50 JK^{-1} mol^{-1} respectively. For the reaction,
$\frac{1}{2}X_2 + \frac{3}{2}Y_2 \leftrightarrow XY_3$, $\Delta H = -30$ kJ to be at equilibrium, the temperature should be:
(a) 500 K (b) 750 K
(c) 1000 K (d) 1250 K

[2010]

36. For an endothermic reaction, energy of activation is E_a and enthalpy of reaction is ΔH (both of these in kJ/mol). Minimum value of E_a will be:
(a) Equal to zero (b) Less than ΔH
(c) Equal to ΔH (d) More than ΔH

[2010]

37. For vaporization of water at 1 atmospheric pressure, the values of ΔH, and ΔS are 40.63 kJ $mole^{-1}$ and 108.8 Jk^{-1} mol^{-1}. The temperature when Gibbs free energy change (ΔG) for transformation will be zero is?
(a) 273.4 K (b) 293.4 K
(c) 373.4 K (d) 393.4 K

[2010]

38. Which of the following is the correct option for free expansion of an ideal gas under adiabatic conditions?
(a) $q = 0, \Delta T < 0, W \neq 0$
(b) $q = 0, \Delta T \neq 0, W = 0$
(c) $q \neq 0, \Delta T = 0, W = 0$
(d) $q = 0, \Delta T = 0, W = 0$

[2011]

39. If the enthalpy change for the transition of liquid water to steam is 30 kJ mol^{-1} at 27° C the entropy change for the process would be:
(a) 1.0 J mol^{-1} K^{-1} (b) 0.1 J mol^{-1} K^{-1}
(c) 10 J mol^{-1} K^{-1} (d) 100 J mol^{-1} K^{-1}

[2011]

40. Equal volumes of two monatomic gases A and B at same temperature and pressure are mixed. The ratio of specific heats $\left(\frac{C_P}{C_V}\right)$ of the mixture will be:
(a) 1.5 (b) 1.67
(c) 0.83 (d) 3.3

[2012]

41. In which of the following reactions, standard reaction entropy change (ΔS°) is positive and standard Gibb's energy change (ΔG°) decreases sharply with increasing temperature?
(a) $\frac{1}{2}\underset{\text{graphite}}{C} + \frac{1}{2}\underset{\text{'g'}}{O_2} \rightarrow \frac{1}{2}\underset{\text{'g'}}{CO_2}$
(b) $\underset{\text{graphite}}{C} + \frac{1}{2}\underset{g}{O_2} \rightarrow \underset{g}{CO}$
(c) $\underset{\text{'g'}}{CO} + \frac{1}{2}\underset{\text{'g'}}{O_2} \rightarrow \underset{\text{'g'}}{CO_2}$
(d) $\underset{\text{'s'}}{Mg} + \frac{1}{2}\underset{g}{O_2} \rightarrow \underset{\text{'s'}}{MgO}$

[2012]

42. Standard enthalpy of vapourization $dH°_{vap}$ for water at 100°C is 40.66 kJ/mole. The internal energy of vapourization of water at 100°C (in kJ mol^{-1}) is:
(a) +37.56 (b) +40.66
(c) –43.76 (d) 43.76

[2012]

43. The enthalpy of fusion of water is 1.435 k. cal/mole. The molar entropy change for the melting of ice at 0°C is:

(a) 0.526 cal mol^{-1} K^{-1}
(b) 5.26 cal mol^{-1} K^{-1}
(c) 10.52 cal mol^{-1} K^{-1}
(d) 21.04 cal mol^{-1} K^{-1} **[2012]**

44. The Gibbs' energy for the decomposition of Al_2O_3 at 500°C is as follows?

$$\frac{2}{3}Al_2O_3 \rightarrow \frac{4}{3}Al + O_2 \qquad \Delta G = 960 \text{ kJ/mol}$$

The potential difference needed for the electrolytic reduction for Al_2O_3 at 500°C is at least?

(a) 2.5 V (b) 3.0V
(c) 4.5 V (d) 5.0V

[2012]

45. A reaction having equal energies of activation for forward and reverse reactions has:

(a) $\Delta G = 0$
(b) $\Delta H = 0$
(c) $\Delta H = \Delta G = \Delta S = 0$
(d) $\Delta S = 0$

[2013]

46. For the reaction $X_2O_{4\ell} \longrightarrow 2XO_{2'g'}$
$\Delta U = 2.1$ k.cal $\Delta S = 20$ cal k^{-1} at 300 k
Hence ΔG is?

(a) 2.7 k.cal (b) –2.7 k.cal
(c) 9.3 k.cal (d) –9.3 k.cal

[2014]

47. Which of the following statement is correct for the spontaneous absorption of a gas?

(a) ΔS is negative and therefore ΔH should be highly positive.
(b) ΔS is negative and therefore ΔH should be highly negative.
(c) ΔS is positive and therefore ΔH should be negative.
(d) ΔS is positive and therefore ΔH should be highly positive.

[2014]

48. Which one is not equal to zero for an ideal solution?

(a) ΔH_{mix} (b) ΔS_{mix}
(c) ΔV_{mix} (d) $\Delta P_{obs.} - P_{Raoult}$

[2015]

49. The heat of combustion of carbon to CO_2 is –393.5 KJ/Mole. The heat released upon the formation of 35.2 g CO_2 from carbon and oxygen gas is:

(a) –315 KJ (b) +315 KJ
(c) –630 KJ (d) –3.15 KJ

[2015]

50. The correct thermodynamic condition for the spontaneous reaction at all temperatures is?

(a) $\Delta H > 0$ and $\Delta S < 0$
(b) $\Delta H < 0$ and $\Delta S > 0$
(c) $\Delta H < 0$ and $\Delta S < 0$
(d) $\Delta H < 0$ and $\Delta S = 0$

[2016]

51. For a sample of perfect gas when its pressure is changed isothermally from P_i to P_f, the entropy change is given by?

(a) $\Delta S = nR\,\ell n\left(\frac{P_i}{P_f}\right)$ (b) $\Delta S = nR\,\ell n\left(\frac{P_f}{P_i}\right)$

(c) $\Delta S = nRT\,\ell n\left(\frac{P_f}{P_i}\right)$ (d) $\Delta S = RT\,\ell n\left(\frac{P_i}{P_f}\right)$

[2016]

Answer Keys

1. (a)	2. (a)	3. (b)	4. (b)	5. (a)	6. (b)	7. (d)	8. (d)	9. (a)	10. (a)
11. (c)	12. (d)	13. (b)	14. (c)	15. (b)	16. (d)	17. (b)	18. (d)	19. (b)	20. (a)
21. (a)	22. (c)	23. (c)	24. (c)	25. (d)	26. (c)	27. (d)	28. (a)	29. (c)	30. (a)
31. (d)	32. (b)	33. (a)	34. (b)	35. (b)	36. (d)	37. (c)	38. (d)	39. (d)	40. (b)
41. (b)	42. (a)	43. (b)	44. (a)	45. (b)	46. (b)	47. (b)	48. (b)	49. (a)	50. (b)
51. (a)									

HINTS AND EXPLANATIONS

1. As $\Delta S = \frac{Q}{T}$

$= \frac{2930}{300} = 9.77 \text{ J K}^{-1} \text{ mol}^{-1}$

2. As $\Delta n = 2 - 3 = -1$

$\Delta H = \Delta E + \Delta nRT$

So $\Delta G = \Delta E - RT$

3. $2ZnS \rightarrow 2Zn + S_2$; $\Delta G° = +293$ J (1)

$2Zn + O_3 \rightarrow 2ZnO$; $\Delta G° = -616$ J (2)

$S_2 + 2O_2 \rightarrow 2SO_2$; $\Delta G° = -408$ J (3)

ΔG for the reaction,

$2ZnS\ (s) + 3O_2\ (g) \rightarrow 2\ ZnO\ (s) + 2SO_2\ (g)$ can be obtained by adding equation (1), (2) and (3)

So $\Delta G° = 293 - 616 - 408 = -731$ J.

4. For an isochoric system, $\Delta V = 0$

As $q = \Delta U$ and when $q = \Delta U$, hencc 'W' will be zero according to first law of thermodynamics

$q = \Delta U - W$

5. $\Delta H°_f = \Sigma H°_f$ (product) $- \Sigma H°_f$ (reactant)

$2H_2O_2\ (l) \rightarrow 2H_2O\ (l) + O_2(g)$

$\Delta H°_f = 2 \times \Sigma H°_f (H_2O) - 2 \times \Sigma H°_f (H_2O_2)$

$= 2 \times -286 - 2 \times (-188)$

$= -196 \text{ kJ mol}^{-1}$

6. $CH_4 + 1/2\ O_2 \rightarrow CH_3OH$

$\Delta H = X - Y$

So $X - Y < 0$

7. Since for Pb, ΔG is negative, hence the reaction is spontaneous so the probable oxidation number is +2 for lead. Since for Sn, ΔG is positive, hence the reaction is not spontaneous so the probable oxidation number is +4.

8. $\Delta S = \frac{\Delta Q}{T} = 2.303\ n\ R \log_{10} \frac{V_2}{V_1}$

$= 2.303 \times 2 \times 2 \times \log_{10} \frac{20}{2}$

$= 9.2$

9. As $\Delta S = \frac{\Delta Q}{T} \quad \frac{J}{K\ mol}$

10. As $q = 0$

So change in internal energy = work done

$\Delta E = W \neq$ zero

11. $C\ (s) + O_2 \rightarrow CO_2$ $\Delta H = -94$ (1)

$H_2 + 1/2\ O_2 \rightarrow H_2O$ $\Delta H = -68$ (2)

$CH_4 + 2O_2 \rightarrow CO_2 + 2H_2O$ $\Delta H = -213$ (3)

Required equation

$= (1) + 2 \times (2) - 3$

$= -94 - (68 \times 2) + 213$

$= -17$ Kcal

12. $C = \frac{q}{m(T_2 - T_1)}$

$75 = \frac{1000}{5.55(\Delta T)}$ $(m = 100/8 = 5.55)$

$\Delta T = \frac{1000}{5.55 \times 75} = 2.4K$

13. $\Delta S = \frac{\Delta H_f}{T} = \frac{6.0}{273} = 0.02198 \text{ kJ K}^{-1} \text{ mol}^{-1}$

$(T = 0°C + 273 = 273\ K)$

$= 0.02198 \times 1000 \text{ JK}^{-1} \text{ mol}^{-1}$

$= 21.98 \text{ JK}^{-1} \text{ mol}^{-1}$

14. For the reaction, $\Delta H°$ (react) $= \Delta H°_f$

$Xe\ (g) + 2F_2\ (g) = XeF_4(g)$

As in this reaction one mole of XeF_4 is formed from its constituent elements.

15. The pressure at which graphite will be transformed into diamond at 298 K is 9.92×10^8 Pa.

16. For the reaction:

$C_3H_8\ (g) + 5O_2 \rightarrow 3CO_2\ (g) + 4H_2O\ (l)$

$\Delta n = n_P - n_R$

$= 3 - 6 = -3$

$\Delta H = \Delta E + \Delta nRT$

$\Delta H - \Delta E = \Delta nRT = -3RT$

17. Given that:

$\Delta H = -382.64 \text{ kJ mol}^{-1}$

$\Delta S = -145.6 \text{ J K}^{-1} \text{ mol}^{-1} = -145.6 \times 10^{-3} \text{ kJ K}^{-1}$

$\Delta G = \Delta H - T\Delta S$

So $\Delta G = -382.64 - (298 \times -145.6 \times 10^{-3})$

$= -339.3 \text{ kJ mol}^{-1}$

18. H_2 (g) + Br_2 (g) → 2HBr (g), ΔH° = ?

ΔH = –[2 × bond energy of HBr – (bond energy of H_2 + bond energy of Cl_2)]

ΔH = –[2 × 364 – (433 + 192)]

= –[728 – 625] = –103 kJ

19. Work done = $-P_{ext}(V_2 - V_1)$

= –3 × (6 – 4) = –6 litre-atm

= –6 × 101.32 J

(As 1 litre-atm = 101.32 J)

= –607.92 = –608 J

20. In irreversible process

$\Delta S_{system} + \Delta S_{surroundings} > 0$

So in the spontaneous process ΔS is positive.

21. For a spontaneous reaction ΔH = –ve, ΔS = +ve spontaneity depends upon both critical minimum energy and maximum randomness/disorder.

22. ΔG = ΔH – TΔS

ΔG = –ve for spontaneous reaction when ΔS = +ve, ΔH = –ve and TΔS > ΔH ⇒ ΔG = –ve

When randomness favours and energy factor do not favour then spontaneity depends upon the temperature dependent means at high temperature.

TΔS >ΔH, ΔG = –ve and reaction is feasible.

23. MgO is the oxide of weak base and we know that heat of neutralization of 1 equation of strong acid with strong base is –57.33 kJ/mol.

⇒ With weak base some heat is absorbed in dissociation of weak base.

Heat if neutralization of weak base with strong acid will be –57.33 kJ/mol.

24. As Δn = 0 so ΔH = ΔU

25. At equilibrium ΔG = 0

26. At equilibrium ΔG = 0. So ΔH = TΔS

So T = ΔH/ΔS = $30 \times 10^{-3}/105$

= 285.7 k

27. ΔH_{obs} = ΔH + Energy needed to disturb resonance

= 3 × –119.5 + 150.4 = –208.1 kJ mol^{-1}

28. In this set of reactions the enthalpy of formation of H_2O (1) is given by equations (2), so it is $-X_2$ kJ/mol.

29. HCl → 1/2 H_2 + 1/2 Cl_2

ΔH = Σ B.E. (products) – Σ B.E. (reactants)

= 1/2 [B.E. (H_2) + B.E. (Cl_2)] – B.E. (HCl)

= 1/2 (430 + 240) – (–90)

= 1/2 × (670 + 90)

= 335 + 90 = 425 kJ mol^{-1}.

30. $H_2 + Cl_2 \rightarrow 2HCl$

$\Delta H_{reactions} = \Sigma(B.E)_{reactant} - \Sigma(B.E)_{product}$

= [(B.E)H – H + (B.E)Cl – Cl] – [2(B.E)H – Cl]

= 434 + 242 – (431) × 2

= –186 kJ

As $\Delta H_{reaction}$ = –186 kJ

So Enthalpy of formation of HCl

$= \frac{-186}{2}$ kJ = –93 kJ mol^{-1}

31. PCl_5 (g) ⇌ PCl_3 (g) + Cl_2 (g)

ΔH = ΔE + ΔnRT

Here, Δn = Change in number of moles of product and reactant

As Δn +ve, so ΔH = +ve

$\Delta S = S_{Product} - S_{Reactant}$

ΔS = +ve

32. State functions or state variables depend only on the state of the system.

Here ‘w’ represents work done and ‘q’ represents amount of heat so both of these are not state functions.

33. C_5H_{12} (g) + $8O_2$ (g) → $5CO_2$ (g) + $6H_2O$ (l)

ΔG° = [(–394.4 × 5) + (–237.2 × 6)] – [(–8.2) + (8 × 0)]

= –3387.5 kJ.

The standard free energy change of elementary substances is taken as zero.

For the fuel cell, the complete cell reaction is:

C_5H_{12} (g) + $8O_2$ (g) → $5CO_2$ (g) + $6H_2O$ (l)

This reaction is the combination of the following two half reactions:

C_5H_{12}(g) + $10H_2O$(l) → $5CO_2$ + $32H^+$ + 32e

$8O_2$ (g) + 32 H^+ + 32e → $16H_2O$ (l)

As the number of electrons exchanged is 32 here, so n = 32

ΔG° = –nFE°

–3387.5 × 10^3 J = –32 × 96500 J/Volt × E°

On solving, we get

E° = 1.09698 V.

34. For a reaction to spontaneous it is needed that:

$\Delta G = -ve$

i.e., $\Delta H - T.\Delta S <$ zero

or

$\Delta H < T.\Delta S$

It means $T > \dfrac{\Delta H}{\Delta S}$

$T > \dfrac{170 \times 1000}{170} = T > 1000$ K

It means at a temperature more than 1000 K process is spontaneous i.e., at 1110 K here.

35. $\Delta S = \Sigma S_P - \Sigma S_R$

$\Delta S = 50 - \left(\dfrac{1}{2} \times 60 + \dfrac{3}{2} \times 40\right)$

$= -40$ JK^{-1} mol^{-1}

$\Delta G = \Delta H - T\Delta S$

At equilibrium $\Delta G = 0$

$T = \dfrac{\Delta H}{\Delta S} = \dfrac{-30 \times 10^3}{-40}$

$T = 750$ K

37. $\Delta G = \Delta H - T\,\Delta S$

As $\Delta G = 0$ so

$T = \dfrac{\Delta H}{\Delta S} = \dfrac{40.63 \times 1000}{108.8}$

$= 373.4$ K

38. For an ideal gas for free expansion

$\Delta T = 0, q = 0, w = 0$

39. As $\Delta S = \dfrac{\Delta H_{vap}}{T}$

$= \dfrac{30 \times 10^3}{300}$

$= 100$ J mol^{-1} K^{-1}

40. As $\gamma = \dfrac{C_p}{C_v} = 1.67$ for monatomic gases here volume of both gases are equal to γ is constant. Hence value is 1.67.

41. $\underset{\text{Graphite}}{C} + \dfrac{1}{2}\underset{g}{O_2} \rightarrow \underset{g}{CO}$

$\Delta ng = +\dfrac{1}{2}$ (Increase in volume)

Hence ΔS is positive

42. As $\Delta H = \Delta U + \Delta ng\, RT$

$40.660 = \Delta U + 1 \times 8.314 \times 10^{-3} \times 373$

On Solving

$\Delta U = +37.56$ kJ/mole

43. $\Delta S_{fusion} = \dfrac{DH_{fusion}}{T} = \dfrac{1.435 \times 10^3}{273}$

$= 5.260$ cal mol^{-1} K^{-1}

44. $\Delta G° = -nFE°$

$960 \times 1000 = -4 \times 96500 \times \varepsilon°$

On solving $E° = -2.48$ V

$\Delta E = 0 - (-2.48) = 2.48$ V

$= 2.5$ V

45. When $E_{a(F)} = E_{a(b)}$

$\Delta H = 0$

46. As $\Delta n_g = \Sigma n_P - \Sigma n_R = 2 - 0 = 2$

$\Delta U = 2.1$ kcal $= 2.1 \times 10^3$ cal

$\Delta H = \Delta U + \Delta n_g RT$

$= (2.1 \times 10^3) + (2 \times 2 \times 300)$

$= 3300$ cal

$\Delta G = \Delta H - T.\Delta 5$

$= 3300 - (300 \times 20)$

$= -2700$ cal

$= -2.7$ k.cal

47. As $\Delta G = \Delta H - T\Delta S$

As adsorption of a gas means $\Delta S = -ve$ hence in order to keep ΔG as $-ve$

ΔH must be highly $-ve$

$(\Delta H > T\Delta S)$

48. As during mixing there is no change in volume

$\Delta V_{mix} = 0$ so $\Delta H_{mix} = 0$

Hence $\Delta S_{mix} \neq$ zero

49. $C + O_2 \longrightarrow CO_2 \quad \Delta H_f = -393.5$ KJ/M

As heat released for 44 g $CO_2 = -393.5$

So heat released for 35.2 g $CO_2 = \dfrac{-393.5 \times 35.2}{44}$

$= -315$ KJ/Mole

50. Using $\Delta G = \Delta H - T.\Delta S$

When $\Delta H = -ve$, $\Delta S = +ve$

$\Delta G = -ve$

Similarly when

$\Delta H = -ve$

$\Delta S = 0$

$\Delta G = -ve$

51. (a) $\Delta S = nR\ \ell n\left(\frac{P_i}{P_f}\right)$ (b) $\Delta S = nR\ \ell n\left(\frac{P_f}{P_i}\right)$

(c) $\Delta S = nRT\ \ell n\left(\frac{P_f}{P_i}\right)$ (d) $\Delta S = RT\ \ell n\left(\frac{P_i}{P_f}\right)$

NCERT EXEMPLAR

1. Thermodynamics is **not** concerned about______.
   (a) Energy changes involved in a chemical reaction.
   (b) The extent to which a chemical reaction proceeds.
   (c) The rate at which a reaction proceeds.
   (d) The feasibility of a chemical reaction.

2. Which of the following statements is correct?
   (a) The presence of reacting species in a covered beaker is an example of open system.
   (b) There is an exchange of energy as well as matter between the system and the surroundings in a closed system.
   (c) The presence of reactants in a closed vessel made up of copper is an example of a closed system.
   (d) The presence of reactants in a thermos flask or any other closed insulated vessel is an example of a closed system.

3. The state of a gas can be described by quoting the relationship between___.
   (a) Pressure, volume, temperature
   (b) Temperature, amount, pressure
   (c) Amount, volume, temperature
   (d) Pressure, volume, temperature, amount

4. The volume of gas is reduced to half from its original volume. The specific heat will be ______.
   (a) Reduce to half
   (b) Be doubled
   (c) Remain constant
   (d) Increase four times

5. During complete combustion of one mole of butane, 2658 kJ of heat is released. The thermochemical reaction for above change is In the following questions, two statements (Assertion) A and (Reason) R are given. Mark
   (a) $2C_4H_{10}(g) + 13O_2(g) \rightarrow 8CO_2(g) + 10H_2O(l)$
   CH = –2658.0 kJ mol^{-1}
   (b) $C_4H_{10}(g) + 13O_2(g) \rightarrow 4CO_2(g) + 5H_2O(g)$
   CH = –1329.0 kJ mol^{-1}
   (c) $C_4H_{10}(g) + 13O_2(g) \rightarrow 4CO_2(g) + 5H_2O(g)$
   CH = –2658.0 kJ mol^{-1}
   (d) $C_4H_{10}(g) + 13O_2(g) \rightarrow 4CO_2(g) + 5H_2O(g)$
   CH = +2658.0 kJ mol^{-1}

6. $_fU^V$ of formation of $CH_4(g)$ at certain temperature is –393 kJ mol^{-1}. The value of $_fH^V$ is:
   (a) zero (b) $< {_fU^V}$
   (c) $> {_fU^V}$ (d) equal to $_fU^V$

7. In an adiabatic process, no transfer of heat takes place between system and surroundings. Choose the correct option for free expansion of an ideal gas under adiabatic condition from the following.
   (a) $q = 0, T \neq 0, w = 0$
   (b) $q \neq 0, T = 0, w = 0$
   (c) $q = 0, T = 0, w = 0$
   (d) $q = 0, T < 0, w \neq 0$

8. The pressure-volume work for an ideal gas can be calculated by using the

   expression w= $-\int_{vi}^{vf}$ pexdV. The work can also be

   calculated from the pV– plot

   by using the area under the curve within the specified limits. When an ideal gas is compressed (a) reversibly or (b) irreversibly from volume Vi to Vf .choose the correct option.
   (a) w (reversible) = w (irreversible)
   (b) w (reversible) < w (irreversible)
   (c) w (reversible) > w (irreversible)
   (d) w (reversible) = w (irreversible) + $p_{ex.}$ V

9. The entropy change can be calculated by using the expression $S = T^{rev}$.

   When water freezes in a glass beaker, choose the correct statement amongst the following:
   (1) S (system) decreases but S (surroundings) remains the same.
   (2) S (system) increases but S (surroundings) decreases.

(3) S (system) decreases but S (surroundings) increases.
(4) S (system) decreases and S (surroundings) also decreases.

10. On the basis of thermochemical equations (1), (2) and (3), find out which of the algebric relationships given in options (a) to (d) is correct.
(1) C (graphite) + O_2 (g) → CO_2 (g); rH = x kJ mol^{-1}
(2) C (graphite) + 12O_2 (g) → CO_2 (g); rH = y kJ mol^{-1}
(3) C (graphite) + 12O_2 (g) → CO_2 (g); rH = z kJ mol^{-1}

(a) $z = x + y$ (b) $x = y - z$
(c) $x = y + z$ (d) $y = 2z - x$

11. Consider the reactions given below. On the basis of these reactions find out which of the algebric relations given in options (a) to (d) is correct?
(1) C (g) + 4H (g) → CH_4 (g); rH = x kJ mol^{-1}
(2) C (graphite,s) + 2H (g) → CH_4 (g); rH = y kJ mol^{-1}

(a) $x = y$ (b) $x = 2y$
(c) $x > y$ (d) $x < y$

12. The enthalpies of elements in their standard states are taken as zero. The enthalpy of formation of a compound
(a) Is always negative
(b) Is always positive
(c) May be positive or negative
(d) Is never negative

13. Enthalpy of sublimation of a substance is equal to:
(a) Enthalpy of fusion + enthalpy of vapourisation
(b) Enthalpy of fusion
(c) Enthalpy of vapourisation
(d) Twice the enthalpy of vapourisation

14. Which of the following is **not** correct?
(a) G is zero for a reversible reaction
(b) G is positive for a spontaneous reaction
(c) G is negative for a spontaneous reaction
(d) G is positive for a non-spontaneous reaction

Answer Keys

1. (c)	**2. (c)**	**3. (d)**	**4. (c)**	**5. (c)**	**6. (b)**	**7. (c)**	**8. (b)**	**9. (c)**	**10. (c)**
11. (c)	**12. (c)**	**13. (a)**	**14. (b)**						

Hints and Explanations for Selective Questions

1. Thermodynamics is not concerned by the reaction rate.

2. No matter is exchanged between the system and the surroundings but texchange of energy takes place for a closed vessel of copper.

4. the specific heat of substance is the heat required to raise the temperature of 1 gram of a substance by one degree.

7. applying first law of thermodynamics if there occur no change of internal energy,the temperature will also remain constant.

11. same bonds are formed in reactions (a) and (b) but bonds between reactant molecules are broken only in reaction (b).In broken bond energy realeased in reaction (b) Is greater than reaction (a)

13. enthalpy of sublimation of a subsatnace is equal to enthalpy of fusion + enthalpy of vapourisation.

14. ΔG <0 process is spontaneous
. ΔG >0 process is non spontaneous
. ΔG =0 process is equilibrium.

AIIMS ESSENTIALS

ASSERTION AND REASON

(a) If A and R both are correct and R is the correct explanation of A;
(b) If A and R both are correct but R is not the correct explanation of A;
(c) A is true but R is false;
(d) A is false but R is true;
(e) Both A and R are false.

1. (A) : The enthalpy of formation of H_2O (l) is greater than that of H_2O (g).
(R) : Enthalpy change is –ve for the condensation reaction, $H_2O\ (g) \rightarrow H_2O\ (l)$

2. (A) : $C_P - C_V = R$ for ideal gas
(R) : $(dU/dV)_T = 0$ for ideal gas

3. (A) : All combustion reactions are exothermic
(R) : Products are more stable than reactants in exothermic process.

4. (A) : Chlorine when tried to be solidified does not have zero entropy when at absolute zero.
(R) : Chlorine contains a mixture of isotopes and is difficult to solidify.

5. (A) : Heat of neutralization of HF is more than that of HCl by NaOH.
(R) : HCl is stronger acid than HF.

6. (A) : The thermodynamic function which determines the spontaneity of a process is the free energy, for a process to be spontaneous, the change in free energy must be negative.
(R) : The change in free energy is related to the change in enthalpy and change in entropy. The change in entropy for a process must be always positive if it is spontaneous.

7. (A) : All exothermic reactions are spontaneous at room temperature.
(R) : Endothermic reaction is spontaneous at all temperatures.

8. (R) : In ($\Delta G = \Delta H - T\Delta S$), ΔG becomes negative and negative sign of ΔG indicates spontaneous reaction.
(R) : ΔG is negative when $T\Delta S > \Delta H$

9. (A) : As temperature increases, heat of reaction also increases for exothermic as well for endothermic reactions.
(R) : ΔH varies with temperatures as given by
$\Delta H_2\ (\text{at } T_2) = \Delta H_1\ (\text{at } T_1) + \Delta C_P\ (T_2 - T_1)$

10. (A) : Efficiency of a reversible engine is maximum (100%) when temperature of the sink is –273°C (T_1)
(R) : η (efficiency) $= \dfrac{T_2 - T_1}{T_2}$

11. (A) : As solid changes to liquid and then to vapour state, entropy increases.
(R) : As going from solid to liquid and then to vapour state, disorder increases.

12. (A) : In the case of an ideal gas, the changes in Gibbs and Helmholtz free energies are equal to each other ($\Delta G = \Delta A$) for isothermal reversible processes.
(R) : There is no change in internal energies and enthalpies for ideal gases at constant temperature.

13. (A) : When H_2O is added to CaO, heat is liberated.
(R) : Reaction between CaO and H_2O is exothermic.

14. (A) : In the following reaction
$C(s) + O_2\ (g) \rightarrow CO\ (g)$,
$\Delta H = \Delta E - RT$
(R) : ΔH is related to ΔE by equation
$\Delta H = \Delta E + \Delta n_g RT$

15. (A) : In the following reaction:
$CO + 1/2\ O_2 \rightarrow CO_2$, $\Delta H = -68$ kcal
and thus $\Delta H_f\ (CO_2) = -68$ kcal mol^{-1}
(R) : 1 mol of CO_2 is formed and enthalpy change is the heat of formation of CO_2.

16. (A) : Heat of neutralization of HF (aq), a weak acid with NaOH (aq) is less than 13.7 kcal, in an exothermic reaction.
(R) : Some heat is lost in the ionization of a weak acid

17. (A) : The standard enthalpy of formation of graphic is taken as zero but of diamond is not zero but it is equal to 1.816 kJ/mol.

(R) : Among carbon allotropes diamond is taken as standard state

18. (A) : In case of some glassy solids having mixture of isotopes, crystals of CO, N_2O, NO etc. entropy is not zero even at absolute zero temperature

(R) : These kind of solids don't have perfect order even at absolute zero temperature

19. (A) : Enthalpy changes are positive when $Na_2SO_4 \cdot 10H_2O$, $CuSO_4 \cdot 5H_2O$ and salts like NaCl, KCl etc. which do not form hydrates are dissolved in water. But enthalpy changes are negative when anhydrous salts capable of forming hydrates are dissolved in water.

(R) : The difference in the behaviour is due to large differences in the molecular weights of hydrated and anhydrous salts. The substances with larger molecular weights usually show positive enthalpy change on dissolution.

20. (A) : ΔE is state function of the system.

(R) : As it depends upon the final and initial state of the system.

21. (A) : In any reversible cycle process, the net increase in entropy of the system is zero.

(R) : As $\Delta S_{universe} = \Delta S_{system} + \Delta Ssurr$ as in case of reversible cyclic process $\Delta Ssystem$ is positive, but of surrounding is negative in same amount. That is why $\Delta S_{universe} = 0$

22. (A) : The enthalpy of neutralization of equivalent of HF and 1 equivalent NaOH is 14 kcal/mol. which is higher than HCl and NaOH?

(R) : It is because the enthalpy of hydration of F^- is higher that of is heat enthalpy of dissociation.

23. (A) : The entropy change in the reaction.

$$\overset{\ominus}{H-O} + H-CH_2-CH_2 \longrightarrow$$
$$\ominus\, Br$$

$$H_2O + Br + CH_2 = CH_2$$

is positive.

(R) : Entropy increase when number of particles or (i.e., molecules or ions) of the product or, in the product directions increases.

24. (A) : The amount of work done in the isothermal expansion is greater than work done in the adiabatic system for same final volume.

(R) : In the adiabatic expansion of a gas temperature and pressure both decrease due to decrease in internal energy of the system.

25. (A) : There is a natural asymmetry between converting work to heat and converting heat to work.

(R) : No process is possible in which the sole result in the absorption of heat from a reservation and its complete conversion into work.

26. (A) : ΔH and ΔE are almost the same for the reason $\underset{g}{N_2} + \underset{g}{O_2} \rightleftharpoons \underset{g}{2NO}$

(R) : All reactants and products are gases

As → (R)

[2013]

Hint - $\Delta H = \Delta E$ as $\Delta n_g = 0$

27. (A) : The increase in internal energy (ΔE) for the evaporation of one mole of water at 1 atm and 373 k is zero.

(R) : For all isothermal procers $\Delta E = 0$

[2013]

Answer Keys

1. (a)	2. (b)	3. (a)	4. (a)	5. (b)	6. (a)	7. (a)	8. (a)	9. (d)	10. (a)
11. (a)	12. (a)	13. (b)	14. (d)	15 (e)	16. (e)	17. (c)	18. (a)	19. (a)	20. (a)
21. (c)	22. (a)	23. (a)	24. (a)	25. (a)	26. (b)	27. (a)			

CHAPTER 7A

Chemical Equilibrium

CHAPTER OUTLINE

■ Types of Reactions ■ Equilibrium

TYPES OF REACTIONS

A chemical reaction is a process in which the reactants undergo changes and produce substances, called products, which are different from them.

Chemical reactions are classified as irreversible and reversible reactions.

Irreversible Reactions

Irreversible reactions occur in one direction only and go on to completion. For example,

1. When unreactive products are formed

$$NH_4NO_2 \xrightarrow{\Delta} N_2 + 2H_2O$$

2. All precipitate reactions are irreversible

$$BaCl_2(aq) + H_2SO_4(aq) \longrightarrow BaSO_4(ppt.) + 2HCl(aq)$$

3. Neutralization reaction

$$H_2SO_4 + 2KOH \longrightarrow K_2SO_4 + 2H_2O$$

4. Redox reactions

$$2FeCl_3 + SnCl_2 \longrightarrow SnCl_4 + 2FeCl_2$$

5. Combustion reactions

$$2Mg + O_2 \longrightarrow 2MgO$$

Reversible Reactions

Reversible reactions occur in both forward and backward directions and therefore never go on to completion.

For example,

$$N_2 + O_2 \rightleftharpoons 2NO$$

$$3Fe + 4H_2O \rightleftharpoons Fe_3O_4 + 4H_2$$

In an open flask, vaporization of water as well as decomposition of calcium carbonate are irreversible reactions and in closed flasks both, the reactions become reversible.

Active Mass or Molar Concentration

$$C_A \text{ or } [A] = \frac{w}{m \times V\,(\text{liters})} = \gamma.\ \text{molarity}$$

Here γ is activity coeficient (for dilute solution it is taken as unity).

1. Unit is mol L^{-1}
2. At constant temperature

   Partial pressure $\propto$ Active mass
3. Active mass of pure solid or pure liquid is assumed to be one.
4. Active mass of any gas at NTP is 1/22.4

EQUILIBRIUM

The word “Equilibrium” in physical sense is explained as the “No change of state of the body”.

When the two opposing processes (reactions) occur simultaneously with equal rates, the system will be in the state of equilibrium.

Equilibrium is classified as shown below:

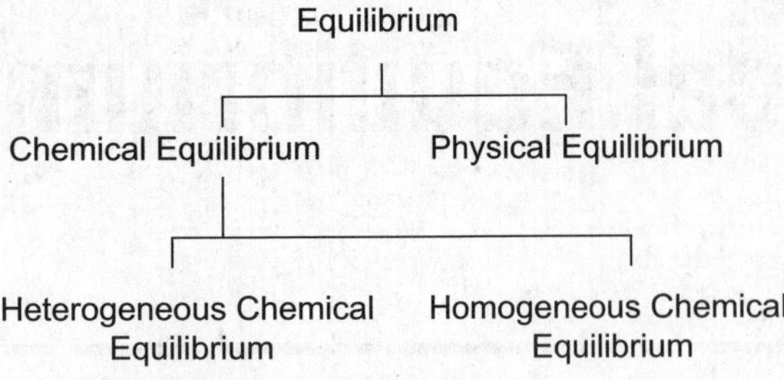

FIGURE 7.1 Classification of Equilibrium

Physical Equilibrium

When an equilibrium exists between same chemical species, it is called **physical equilibrium.** In it, the opposing process involves only physical changes.

$$\underset{\text{Ice}}{H_2O\ (s)} \underset{1\ \text{atm}}{\overset{0°C}{\rightleftharpoons}} \underset{\text{Water}}{H_2O\ (l)}$$

FACTS TO REMEMBER

- Physical equilibrium may be of these types:

(i) Solid $\rightleftharpoons$ liquid

e.g., Ice $\rightleftharpoons$ water

(ii) Liquid $\rightleftharpoons$ Gas

e.g., $\underset{'l'}{\text{Water}} \rightleftharpoons \underset{'g'}{\text{Steam}}$

(iii) Solid $\rightleftharpoons$ Gas

e.g., $\underset{'s'}{I_2} \rightleftharpoons \underset{'g'}{I_2}$ By Sublimation

Here rate of evaporation and condensation are same.

(iv) Solid $\rightleftharpoons$ Solution

e.g., $\underset{\text{Solid}}{\text{Sugar}} \overset{\text{In water}}{\rightleftharpoons}$ Sugar Solution

Here rate of dissolution and precipitation are same

(v) Gas $\rightleftharpoons$ Solution

e.g., $\underset{g}{O_2} \rightleftharpoons \underset{\text{aq. soln.}}{O_2}$

Chemical Equilibrium

- When an equilibrium exists between different chemical species, it is called chemical equilibrium.

$N_2(g) + 3H_2(g) \rightleftharpoons 2NH_3(g)$

$CaCO_3\ (s) \rightleftharpoons CaO\ (s) + CO_2\ (g)$

- If a chemical equilibrium has only one phase, it is called **homogeneous** and if it has more than one phase it is called **heterogeneous**.

$H_2\ (g) + I_2\ (g) \rightleftharpoons 2HI\ (g)$

Number of phases = 1 (Homogeneous chemical equilibrium)

$2HgO\ (s) \rightleftharpoons 2Hg\ (l) + O_2\ (g)$

Number of phases = 3 (Heterogeneous chemical equilibrium)

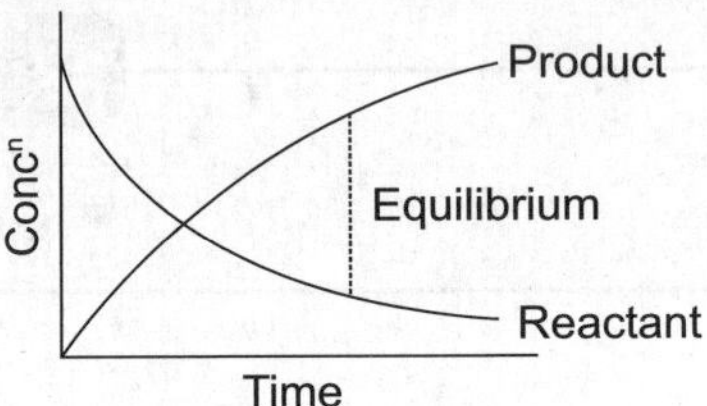

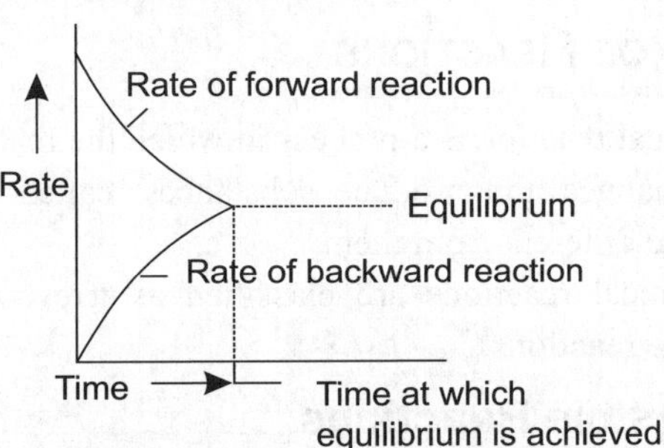

FIGURE 7.2 Chemical Equilibrium

- Chemical equilibrium is the stage of a reversible reaction at which the active masses of the reactants and the products become constant in mixture and do not change with time.
- It is the state of a reversible reaction at which measurable properties like colour, density, pressure and concentration are nearly unchangeable.

Features

- Rate of forward reaction is equal to rate of backward reaction.
- It is obtained only when reversible reaction is carried out in a closed space.
- At constant temperature it is characterized by properties like colour, density, pressure.
- It is possible from both the directions.
- It is dynamic in nature. It means the reaction is not going to be ceased as reaction occurs in both the directions with equal rates.
- A positive catalyst can set up equilibrium in less time but cannot change it.

- At equilibrium, ΔG is equal to zero.

As $\Delta G = \Delta H - T\Delta S$

And $\Delta H = T\Delta S$

So $\Delta G = O$

Types of chemical equilibrium Chemical equilibrium can be classified into following two types, *viz*., homogeneous and heterogenous.

Homogeneous Here reactants and products are present in the same phase or physical states.

Example: $N_2(g) + O_2(g) \rightleftharpoons 2NO(g)$

$$\underset{'l'}{R\text{–}COOH} + \underset{'l'}{R'\text{–}OH} \longrightarrow \underset{'l'}{R\text{–}COOR'} + \underset{'l'}{H_2O}$$

Heterogeneous Here reactants and products are present in different phases or physical states.

Example: $3Fe(s) + 4H_2O(l) \rightleftharpoons Fe_3O_4(s) + 4H_2(g)$

Law of Mass Action

- Law of mass action was introduced by Guldberg and Waage.
- It states that the rate at which a substance reacts is directly proportional to its active mass and the rate at which different substances react together is directly proportional to the product of their active masses.
- $A + B \underset{V_2}{\overset{V_1}{\rightleftharpoons}} C + D$

  If active masses of and D are [A], [B], [C], [D] respectively, then

  Rate of reaction of A ∝ [A]

  Rate of reaction of B ∝ [B]

  $V_1 \propto [A]\,[B] = K_1\,[A]\,[B]$

  $V_2 \propto [C][D] = K_2\,[C]\,[D]$
- Law of mass action is not applicable for solids and for them, active mass is always one. For example,

  $KClO_3(s) \longrightarrow 2KCl + 3O_2$

Equilibrium Constant

- Equilibrium constant is the ratio of the rate of forward and backward reactions at a particular temperature, or in other words it is the ratio of active masses of the reactants to that of active masses of products at a par ticular temperature.
- It is denoted by K_c when the concentration is taken into consideration and by K_p when the partial pressure is taken into consideration.

For a reaction

$$A + B \underset{V_2}{\overset{V_1}{\rightleftharpoons}} C + D$$

$V_2 \propto [A]\,[B]$ or $V_1 = K_1[A][B]$

$V_2 \propto [C]\,[D]$ or $V_2 = K_2[C][D]$

We know that at equilibrium $V_1 = V_2$

So, $K_1\,[A]\,[B] = K_2\,[C]\,[D]$

$$\frac{K_1}{K_2} = \frac{[C]\,[D]}{[A]\,[B]}$$

$$K_c = \frac{[C]^{n_1}[D]^{n_2}}{[A]^{m_1}[B]^{m_2}} \text{ or } \frac{[\text{product}]}{[\text{reactant}]} \text{ at constant temperature}$$

or

For $m_1A + m_2B \rightleftharpoons n_1C + n_2D$

$$K_c = \frac{[C]^{n_1}[D]^{n_2}}{[A]^{m_1}[B]^{m_2}}$$

- Completion of reaction $\propto K_c$

  Stability of reactant $\propto 1/K_c$

  Stability of product $\propto K_c$

K_c does not depend upon

- Initial concentration or pressure of reactants and products involved.
- The presence of a catalyst.
- Presence of an inert gas.
- The direction from which equilibrium has been set up.

K_c depends upon

(1) Temperature

- The variation of equilibrium constant with temperature is given by van't Hoff equation as follows:

  $K = Ae^{-\Delta H°/RT}$

  $$\log_{10}\frac{K_2}{K_1} = \frac{\Delta H°}{2.303\,R}\,\frac{(T_2 - T_1)}{T_1.T_2}$$

  $K \propto T \quad (\Delta H = +ve)$

  $K \propto 1/T \quad (\Delta H = -ve)$
- When a graph is plotted between $\log_{10}K$ and 1/T, a straight line will be formed, the slope of which is equal to $\dfrac{-\Delta H°}{2.303\,R}$

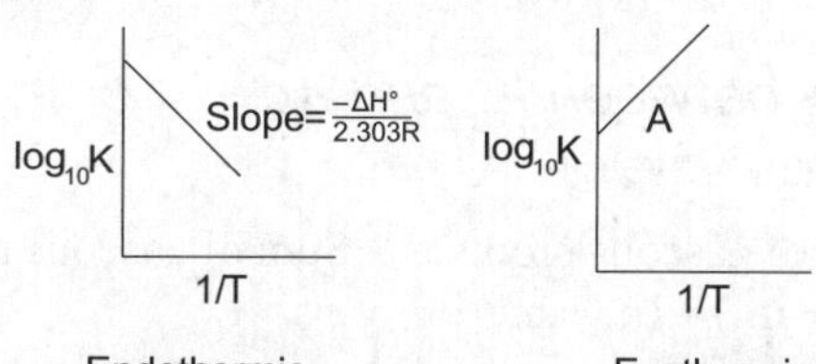

FIGURE 7.3 Graphs for Endothermic and Exothermic Reactions

(2) *Mode of representation of reaction*

Let us consider a reaction

$A + B \rightleftharpoons C + D$

$K_c = \frac{[C]\,[D]}{[A]\,[B]}$

Now if we write the above reaction as follows:

$C + D \rightleftharpoons A + B$

$K'_c = \frac{[A]\,[B]}{[C]\,[D]}$

Hence, $K_c = 1/K'_c$

(3) *Stoichiometric representation of chemical equation*

Let us consider a reaction

$A + B \rightleftharpoons C + D$

$K_c = \frac{[C]\,[D]}{[A]\,[B]}$

If we write the above reaction as follows:

$m_1A + m_2B \rightleftharpoons n_1C + n_2D$, then

$K_c = \frac{[C]^{n_1}\,[D]^{n_2}}{[A]^{m_1}\,[B]^{m_2}}$

Examples:

1. $N_2 + 2O_2 \rightleftharpoons 2NO_2$

   Here $K'_c = \frac{[NO_2]^2}{[N_2][O_2]^2}$

   $\frac{1}{2} N_2 + O_2 \rightleftharpoons NO_2$

   Here $K'_c = \frac{[NO_2]^2}{[N_2]^{1/2}[O_2]} = \sqrt{K_c}$

2. $N_2 + 3H_2 \rightleftharpoons 2NH_3$ and

   $NH_3 \rightleftharpoons \frac{1}{2} N_2 + 3/2\ H_2$ then

   $K'_c = 1/\sqrt{k_c}$

Equilibrium constant in terms of partial pressure

If partial pressures of A, B, C, D are given as P_A, P_B, P_C, P_D respectively, then for the reaction, K_p is given as

$m_1A + m_2B \rightleftharpoons n_1C + n_2D$

$K_p = \frac{[P_C]^{n_1}\,[P_D]^{n_2}}{[P_A]^{m_1}\,[P_B]^{m_2}}$

Relation between K_p and Kc

$K_p = K_c.(RT)^{(m_1 + m_2) - (n_1 + n_2)}$

Δn = Sum of gaseous products – Sum of gaseous reactants

$\Delta n = (m_1 + m_2) - (n_1 + n_2)$

$K_p = K_c.(RT)^{\Delta n}$

- If Δn is zero $K_p = K_c$, as in

  $N_2 + O_2 \rightleftharpoons 2NO$

- If Δn is positive, then $K_p > K_c$, as in

  $PCl_5 \rightleftharpoons PCl_3 + Cl_2$

  $2NH_3 \rightleftharpoons N_2 + 3H_2$

- If Δn is negative, then $K_c > K_p$, as in

  $2SO_2 + O_2 \rightleftharpoons 2SO_3$

Units of equilibrium constant (K)

$K_c = (mol/L)^{\Delta n}$

$Kp = (atm)^{\Delta n}$

Equilibrium constant for a multi-step reaction

$K = K_1 . K_2 . K_3$

For example, for a multistep reaction like

$H_2O_2 + 2H^+ \longrightarrow 2I^- \longrightarrow I_2 + 2H_2O$

We follow the steps as given below:

$H_2O_2 + I^- \longrightarrow OH^- + HOI\ (K_1)$

$H^+ + OH^- \longrightarrow H_2O\ (K_2)$

$HOI + H^+ + I^- \longrightarrow H_2O + I_2\ (K_3)$

Here $K = K_1 . K_2 . K_3$

Reaction or concentration quotient: (Q)

It may be defined in a same manner to equilibrium constant ($K_{eq.}$) however the molar concentrations in case of Q are not necessarily equilibrium values.

$m_1A + m_2B \rightleftharpoons n_1C + n_2D$

$Q = \frac{[C]^{n_1}[D]^{n_2}}{[A]^{m_1}[B]^{m_2}}$

It helps in predicting the direction of reaction as follows:

(1) If $Q = k_{eq}$, The reaction is at equilibrium.
(2) If $Q > k_{eq}$, Backward reaction is favourable.
(3) If $Q < k_{eq}$ forward reaction is favourable.

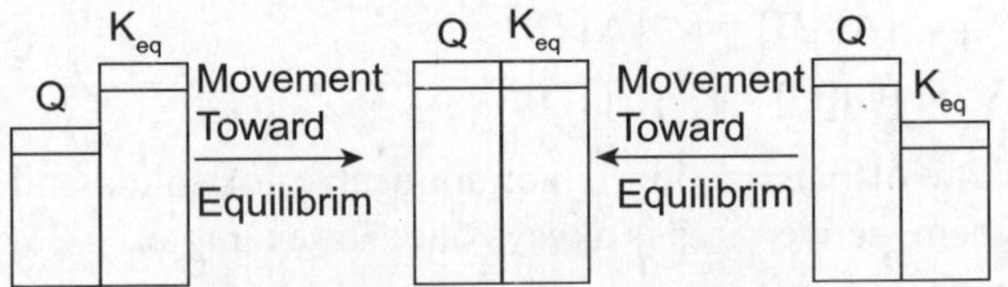

FIGURE 7.4 Reaction Quotient

Relation between standard gibbs free energy and equilibrium constant

The relation between the two is given by the equation mentioned below:

$\Delta G = \Delta G° + 2.303\ RT\ \log_{10} Q$

As $\Delta G = 0$, Q = K. So,

$\Delta G° = -2.303\ RT\ \log_{10} K$

Here if $\Delta G° > 0$ (+ve), then < 1, making the backward reaction feasible.

If $\Delta G° < 0$ (–ve), then > 1, making the forward reaction feasible.

Expressions of K_c and K_p for Some Reactions

1. In case of HI or NO formation

$$K_c = \frac{4x^2}{(a-x)(b-x)}$$

$$K_p = \frac{4x^2}{(a-x)(b-x)}$$

2. In case of NH_3 formation

$$K_c = \frac{4x^2V^2}{(a-x)(b-3x)^3}$$

3. In case of PCl_5 dissociation

$$K_c = \frac{x^2}{(a-x)V}$$

$$K_p = \frac{Px^2}{(1-x^2)}$$

4. Dissociation of N_2O_4

$$K_c = \frac{4\alpha^2 a}{(1-\alpha)V}, K_p = \frac{4\alpha^2 p}{(1-\alpha^2)}$$

Here a and b represent the initial concentration of the reactants.

Degree of Dissociation (α)

Degree of dissociation is the ratio of number of molecules dissociated at equilibrium, to the initial number of molecules.

It is given as:

$$\alpha = \frac{D-d}{d(n-1)} = \frac{1}{(n-1)}\left(\frac{D-d}{d}\right)$$

Here $D = \frac{\text{Molecular weight}}{2}$

= density before dissociation

d = vapour density

= density after dissociation

n = number of species formed after dissociation

Le Chatelier's Principle

- Le Chatelier's principle describes the effect of change in concentration, pressure and temperature on a reversible system.
- If the system at equilibrium is subjected to a change of concentration or temperature or pressure, the system adjusts itself in such a way that the effect of these changes can be neglected or minimized.

Effect of concentration

- An increase in concentration of any substance favours the reaction in which it is used up, that is, in opposite direction.
- An increase in concentration of reactant favours formation of more product, that is, forward reaction.
- Increase in concentration of product favours backward reaction, that is, its continuous removal is essential for its greater formation.

Effect of pressure

- High pressure is favourable for the reaction in which there is a decrease in volume or $n_r > n_p$. [As $P \propto \frac{1}{V}$]
- Low pressure is favourable for the reaction, in which there is an increase in volume or $n_r < n_p$.
- Pressure is kept constant when volume is constant or $n_r = n_p$.

  Here n_r = Moles of gaseous reactant

  n_p = Moles of gaseous product

Effect of temperature

- Increase in temperature favours the forward reaction in endothermic reaction (ΔH = +ve).
- Decrease in temperature favours the forward reaction in exothermic reaction (ΔH = –ve).

Applications of Le Chatelier's principle

1. *Ice–water system (melting of ice)*

   $$\text{Ice} \rightleftharpoons \text{Water} - \text{Heat } (\Delta H = +\text{ve})$$

   Since this is an endothermic process and there is a decrease in volume, the favourable conditions for melting of ice are high temperature and high pressure.

2. *Water-steam system (evaporation of water)*

   $$\text{Water} \rightleftharpoons \text{Steam} - \text{Heat } (\Delta H = +\text{ve})$$

   Since this is an endothermic process and there is an increase in volume, so the favourable conditions for evaporation of water are high temperature and low pressure.

3. *Solubility of gases in liquids* When a gas dissolves in a liquid, there is a decrease in its volume, so an increase in pressure will favour the dissolution of a gas in liquid, that is, solubility of a gas is directly proportional to partial pressure of the gas (Henry's law).

On opening a carbonated water bottle, CO_2 comes out as its pressure decreases.

4. *Effect of temperature on solubility.*
   - Exothermic substances like CaO, $Ca(OH)_2$, KOH are more soluble in cold water than in hot water, which means that low temperature favours their solubility.
   - Endothermic substances like sugar and urea are more soluble in hot water than in cold water, which implies that high temperature favours their solubility.
5. *Formation of nitric oxide.*

   $N_2\,(g) + O_2\,(g) \rightleftharpoons 2NO\,(g) - 43200$ calories

   - Here favourable conditions for formation of NO are: higher concentration of N_2, O_2, constant pressure (as $n_p = n_r$) and high temperature (as the forward reaction is endothermic).
6. *Formation of ammonia.*

   $N_2\,(g) + 3H_2\,(g) \rightleftharpoons 2NH_3\,(g) + 22400$ calories

   - Favourable conditions for NH_3 formation are: higher concentration of N_2, H_2, high pressure (as $n_p < n_r$) and low temperature (as it is exothermic).
7. Formation of SO_3:

   $$2SO_2 + O_2 \rightleftharpoons 2SO_3$$

   $$\Delta H = -193.2 \text{ kJ}$$

   The favourable conditions for SO_3 formation are low temperature, high pressure, more conc^n of SO_2, O_2
8. Formation of H_2 by Bosch method:

   $\underbrace{H_2 + CO}_{\text{Water gas}} + \underset{g}{H_2O} \rightleftharpoons \underset{g}{2H_2} + \underset{g}{CO_2}\ \Delta H = +42$ kJ

   The favourable conditions are high temperature, constant pressure and more conc^n of H_2, CO, steam.
   - Conversion of a solid into liquid and of liquid into gas is favoured by high temperature in most cases.
   - Melting of solids is accompanied by increase of volume so low pressure favours their melting while high pressure favours solidification.

Effects of Adding an Inert Gas on the Equilibrium

- When n_p is equal to n_r, adding an inert gas either at constant volume or pressure, does not have any effect. (In case of HI or NO formation)
- When $n_p \neq n_r$, there is no effect of adding an inert gas at constant volume.
- When $n_p \neq n_r$, at constant pressure, on adding inert gas, equilibrium will shift towards the side which has more volume, that is, the dissociation of ammonia will be more at constant pressure by adding inert gas like argon (Ar).

$$2NH_3 \rightleftharpoons N_2 + 3H_2$$

$$n_R \quad < \quad n_p$$

$$2 \qquad\quad 4$$

POINTS TO REMEMBER

Van't Hoff reaction isotherm It shows the free energy change of a reaction at any given temperature, pressure and composition of the reacting system as follows:

$$\Delta G = \Delta G° + RT \text{ In } J$$

Since at equilibrium, ΔG is 0 so $\Delta G^0 = -RT \text{ In } J_{eq}$

Here J stands for reaction quotient of partial pressures of products and reactants.

$$J = \frac{\text{Concentration terms of products}}{\text{Concentration terms of reactants}}$$

J_{eq} means the partial pressure of the products and the reactants at the equilibrium that is J_{eq} can be replaced by K_p.

So, $\Delta G° = -RT \text{ In } K_p$

- The equilibrium equation can not inform us how long will it take to attain equilibrium during a reaction.
- Le-Chatelier's principle is not applicable for solids.

  for e.g., $\underset{s}{Fe} + \underset{\text{'s'}}{S} \rightarrow \underset{\text{'s'}}{FeS}$

Important formula

Active mass or molar concentration

$$C_A \text{ or } [A] = \frac{w}{m \times V\,(\text{litre})}$$

Equilibrium constant

For $m_1A + m_2B \rightleftharpoons n_1C + n_2D$

$$K_c = \frac{[C]^{n_1}[D]^{n_2}}{[A]^{m_1}[B]^{m_2}}$$

Equilibrium constant in terms of partial pressure

$$K_p = \frac{[P_C]^{n_1}[P_D]^{n_2}}{[P_A]^{m_1}[P_B]^{m_2}}$$

Relation between K_p and K_c

$K_p = K_c.(RT)^{(n_1+n_2)-(m_1+m_2)}$

$\Delta n_g = (n_1 + n_2) - (m_1 + m_2)$

$K_p = K_c.(RT)^{\Delta n}$

Degree of dissociation (α)

$\alpha = \frac{D-d}{d(n-1)}$

Here $D = \frac{\text{Molecular weight}}{2}$ d = Vapour density

Reaction quotient or concentration quotient (Q)

For $m_1A + m_2B \rightleftharpoons n_1C + n_2D$, then

$Q = \frac{[C]^{n_1}[D]^{n_2}}{[A]^{m_1}[B]^{m_2}}$

- when $[C]^{n_1}[D]^{n_2} = [A]^{m_1}[B]^{m_2}$

  $Q = K_{eq}$ (reaction is at equilibrium).

Solved Numericals

K_p and K_c

1. K_c for the reaction

$N_2(g)\ 3H_2(g) \rightleftharpoons 2NH_3(g)$

is 0.5 mole^{-2} litre^2 at 400 K. Find K_p. Given R = 0.82 litre arm degree^{-1} mole^{-1}.

Solution:

$K_p = K_c.(RT)^{\Delta n}$

$\Delta n = 2 - 4 = -2$

$K_c = 0.5$ mole^{-2} litre^2

T = 400 K

R = 0.82 litre arm degree^{-1} mole^{-1}

$K_p = 0.5 \times (0.082 \times 400)^{-2} = 4.636 \times 10^{-4}$ arm^{-2}

2. For the reaction:

$A + B \rightleftharpoons 2C$

2 moles of A and 3 moles of are allowed to react. If the equilibrium constant is 4 at 673 K. Find out the moles at that equilibrium.

Solution:

A +	B ⇌	$2C$	
2	3	0	Initial moles
$(2-x)$	$(3-x)$	$2x$	Moles at equilibrium
$\frac{(2-x)}{v}$	$\frac{(3-x)}{v}$	$\frac{2x}{v}$	Molar conecentration at equilibrium

$K_c = \frac{[C]^2}{[A][B]} = \frac{4x^2}{(2-x)(3-x)} = 4$

$x^2 = 6 - 5x + x^2$

On solving:

$x = 1.2$

Moles of at equilibrium = $2x$

$= 2 \times 1.2 = 2.4.$

3. For the reaction $H_2 + I_2 \rightleftharpoons 2HI$. The equilibrium concentration of H_2, I_2 and HI are 8.0, 3.0 and 28.0 moles per litre respectively. Calculate the equilibrium constant of the reaction.

Solution:

$H_2(g)$ +	$I_2(g)$ ⇌	$2HI(g)$	
8.0	3.0	28.0	Molar concentration at equilibrium

$K_c = \frac{[H]^2}{[H_2][I_2]} = \frac{(28.0)^2}{8.0 \times 3.0} = 32.66$

4. Find the value of K_p for dissociation of:

$2HI \rightleftharpoons H_2 + I_2$,

if the equilibrium concentration of HI, H_2 and I_2 are 3.53, 0.4789 and 0.4789 mole litre^{-1} respectively.

Solution:

$2HI$ ⇌	H_2 +	I_2	
3.53	0.4789	0.4789	Molar concentration at equilibrium

$K_c = \frac{[H_2][I_2]}{[HI]^2} = \frac{0.4789 \times 0.4789}{[3.53]^2} = 1.84 \times 10^{-2}$

$K_p = K_c(RT)^{\Delta n}$

Here $\Delta n = 0$

$K_p = K_c = 1.814 \times 10^{-2}$

5. A gaseous mixture contains 0.30 mol CO, 0.10 mole H_2 and 0.2 mole H_2O, plus an unknown amount of CH_4 in each litre. This mixture is in equilibrium at 1200 K.

$CO (g) + 3H_2 (g) \rightleftharpoons CH_4 (g) + H_2O (g)$
What is the concentration of CH_4 in this mixture? The equilibrium constant, $K_c = 3.92$.

Solution:

$CO (g) + 3H_2 (g) \rightleftharpoons CH_4 (g) + H_2O (g)$

0.30	0.10	x	0.2	Eq. moles
0.30	0.10	x	0.2	Eq. conc.
1	1	1	1	

$$K_c = \frac{[CH_4][H_2O}{[CO][H_2]^3}$$

$$3.92 = \frac{x \times 0.2}{0.30 \times (0.10)^3}$$

On solving

$x = 5.88 \times 10^{-3}$ mole litre1^{-1}

6. The K_P value for the reaction:

$H_2 + I_2 \rightleftharpoons 2HI$

at 460° C is 49. If the initial partial pressure of H_2 and I_2 is 0.5 atm respectively. Determine the partial pressure of each gas at equilibrium.

Solution:

	H_2 (g) + I_2 (g) $\rightleftharpoons$		2HI (g)
At t = 0	0.5 atm	0.5 atm	0
At equilibrium	$(0.5 - x)$	$(0.5 - x)$	$2x$ atm

$$K_p = \frac{(P_{HI})^2}{P_{H2} \times P_{I2}} = \frac{(2x)^2}{(0.5 - x)0.5 - x}$$

$$49 = \frac{(2x)^2}{(0.5 - x)^2}$$

$$7 = \frac{2x}{(0.5 - x)^2}$$

$x = 0.389$ atm

At equilibrium

$P_{H_2} = 0.5 - 0.389 = 0.111$ atm

$P_{I_2} = 0.5 - 0.389 = 0.111$ atm

$P_{HI} = 2 \times 0.389 = 0.778$ atm

Degree of Dissociation

7. 20% of N_2O_4 molecules are dissociated in a sample of gas at 27°C and 760 torn. Calculate the density of the equilibrium mixture.

Solution:

N_2O_4	$\rightleftharpoons$	$2NO_2$	
1		0	Initial moles
(1 – 0.2) = 0.8		2 × 0.2 = 0.4 (as x = 20/100)	Moles at equilibrium

Total number of moles at equilibrium = 0.8 + 0.4 = 1.2

From gas equation $V = \frac{nRT}{P}$

$$= \frac{1.2 \times 0.0821 \times 300 \times 760}{760} = 29.56 \text{ litre}$$

Density (d) = $\frac{M}{V}$

(Molecular weight of $N_2O_4 = 2 \times 14 + 4 \times 16 = 92$)

$$= \frac{92}{29.56} = 3.11 \text{ g liter}^{-1}$$

8. The vapour density of N_2O_4 at a certain temperature is 30. Calculate, the percentage dissociation of N_2O_4 at this temperature.

Solution:

Normal molecular mass of N_2O_4 = 28 + 64 = 92

Normal vapour density (VD before dissociation)

$$d = \frac{\text{Normal molecular mass}}{2} = \frac{92}{2} = 46$$

Vapour density after dissociation (d) = 30.

$N_2O_4(g) \rightleftharpoons 2NO_2 (g)$

$$\text{Degree of dissociation } (\alpha) = \frac{D - d}{d(n - 1)}$$

$$= \frac{46 - 30}{30\ (2 - 1)} = \frac{16}{30 \times 1} = 0.535$$

$\alpha\% = 53.5$

Standard Free Energy Change

9. The value of K_p at 298 for the reaction

$\frac{1}{2}N_2 (g) + \frac{3}{2}H_2 (g) \rightleftharpoons NH_3 (g)$

is found to be 826.0, partial pressures being measured in atmospheric units. Calculate ΔG° at 298 K. (R = 1.98 calories)

Solution:

$\frac{1}{2}N_2 (g) + \frac{3}{2}H_2 (g) \rightleftharpoons NH_3 (g)$

$\Delta G° = -2.303\ RT \log K_p$

$= -2.303 \times 1.98 \times 298 \times \log 826$

$= -1358.86 \times 2.9169 = -3963.66$ calories

10. Find the equilibrium concentration ratio of M, if two mole of each of and L react at 27°C.

ΔG° = 460 cal

$K + L \rightleftharpoons M + N$

Solution:

$-\Delta G^\circ = 2.303\ RT \log K$

that is, $\log K = \dfrac{-\Delta G^\circ}{2.303\ RT}$

$= \dfrac{-460}{2.303 \times 2 \times 300} = -0.333$

So, K = 0.464

At equilibrium

[K] = [L] and [M] = [N]

So, $K = \dfrac{[M][N]}{[K][L]} = \dfrac{[M]^2}{[K^2]} = 0.464$

[M]:[K] = 0.682

CHAPTER-END EXERCISES

PRACTICE QUESTIONS – I

1. The reaction $PC1_5\ (g) \rightleftharpoons PC1_3\ (s) + Cl_2\ (g)$ is an example of:
(a) Backward reaction
(b) Forward reaction
(c) Irreversible reaction
(d) Reversible reaction

2. Partial pressure of O_2 in the reaction:
$2\underset{'s'}{Ag_2O} \rightleftharpoons 4\underset{'s'}{Ag} + \underset{g}{O_2}$ is
(a) K_p (b) $\sqrt{K_P}$
(c) $3\sqrt{K_P}$ (d) $2\ K_P$

3. The value of K_p in the reaction:
$MgCO_3\ (s) \rightleftharpoons MgO\ (s) + CO_2\ (g)$ is
(a) $K_p = P(CO_2)$
(b) $K_p = \dfrac{P(MgCO_3)}{P(CO_2) \times P(MgO)}$
(c) $K_p = \dfrac{P(CO_2) \times P(CO_2) \times P(MgO)}{P(MgCO)_3}$
(d) $K_p = \dfrac{P(CO_2) \times P(MgO)}{P(MgCO)_3}$

4. One of the following equilibria is not affected by change in volume of the flask:
(a) $PCl_5\ (g) \rightleftharpoons PCl_3\ (g)\ Cl_2\ (g)$
(b) $N_2\ (g) + 3H_2\ (g) \rightleftharpoons 2NH_3\ (g)$
(c) $N_2\ (g) + O_2\ (g) \rightleftharpoons 2NO\ (g)$
(d) $SO_2Cl_2\ (g) \rightleftharpoons SO_2\ (g) + Cl_2\ (g)$

5. At equilibrium total number of moles for the reaction $2HI \rightleftharpoons H_2 + I_2$, if α is degree of dissociation, are:
(a) 2 (b) 2 – α
(c) 1 (d) I – α

6. If equilibrium constant for the reaction:
$N_2 + 3H_2 \rightleftharpoons 2NH_3$ is K_c, then the equilibrium constant for the reaction $NH_3 \rightleftharpoons \frac{1}{2}N_2 + \frac{3}{2}H_2$ will be.
(a) $\dfrac{1}{K_c}$ (b) $\dfrac{1}{K_c^2}$
(c) $\sqrt{K_c}$ (d) $\dfrac{1}{\sqrt{K_c}}$

7. $A_2\ (g) + B_2\ (g) \rightleftharpoons 2AB\ (g)$; ΔH = +ve, it:
(a) Increase by pressure
(b) It occurs at 1000 atm pressure
(c) It occurs at high temperature
(d) It occurs at high pressure and temperature

8. For a reversible reaction, the concentration of the reactants are doubled, then the equilibrium constant:
(a) Becomes one-fourth
(b) Is doubled
(c) Is halved
(d) Remains same

9. $2SO_2\ (g) + O_2\ (g) \rightleftharpoons 2SO_3\ (g)$,
in the above reaction K_p and K_c are related as:
(a) $K_p = K_c \times (RT)$ (b) $K_p = K_c \times (RT)^{-1}$
(c) $K_c = K_p \times (RT)^2$ (d) $K_p = K_c \times (RT)^{-2}$

10. In the reaction, $N_2 + 3H_2 \rightleftharpoons 2NH_3$ + heat, relationship between K_p and K_c is:
(a) $K_p = K_c(RT)^{-2}$ (b) $K_p = K_c\ (RT)^2$
(c) $K_p = K_c(RT)^{-3}$ (d) $K_c = K_p\ (RT)^3$

11. Which of the following change will shift the reaction in forward direction:
$I_2\ (g) \rightleftharpoons 21\ (g)$ Take ΔH° = +150 kJ
(a) Increase in concentration of I
(b) Increase in total pressure
(c) Decrease in concentration of I_2
(d) Increase in temperature

12. A reversible reaction is said to have attained equilibrium, when:
(a) Backward reaction stops
(b) Both backward and forward reactions take place at equal speed
(c) Both backward and forward reactions stop
(d) Concentration of each of the reactants and products becomes equal

13. In what manner will increase of pressure affect the following equation:
$C(s) + H_2O\ (g) \rightleftharpoons CO\ (g) + H_2\ (g)$
(a) Shift in the reverse direction
(b) Shift in the forward direction
(c) Increase in the yield of hydrogen
(d) No effect

14. The equilibrium between water and its vapour, in an open vessel:
(a) Can be achieved
(b) Depends upon pressure
(c) Cannot be achieved
(d) Depends upon temperature

15. Which of the following equilibrium, in gaseous phase, would be unaffected by an increase in pressure:
(a) $N_2 + 3H_2 \rightleftharpoons 2NH_3$
(b) $N_2 + O_2 \rightleftharpoons 2NO$
(c) $N_2O_4 \rightleftharpoons 2NO_2$
(d) $CO_2 + \frac{1}{2} O_2 \rightleftharpoons CO_2$

16. Which of these reactions are having homogeneous equilibrium:
(a) $\underset{\text{'g'}}{N_2O_4} \rightleftharpoons \underset{\text{'g'}}{2NO_2}$
(b) $\underset{\text{'g'}}{H_2} + \underset{\text{'g'}}{Cl_2} \rightleftharpoons \underset{\text{'g'}}{2HCl}$
(c) $\underset{\text{'s'}}{CaCO_3} \rightleftharpoons \underset{\text{'s'}}{CaO} + \underset{\text{'g'}}{CO_2}$
(d) Both (a) and (b)

17. The law of chemical equilibrium was first given by:
(a) Guldberg (b) Boyle
(c) Waage (d) Both (a) and (c)

18. A saturated solution of non-radioactive sugar was taken and a little radioactive sugar was added to it. A small amount of it gets dissolved in solution and an equal amount of sugar was precipitated. This proves:
(a) The equilibrium has been established in the solution
(b) Radioactive sugar can displace non-radioactive sugar from its solution.
(c) Equilibrium is dynamic in nature
(d) None of the above

19. Which of the following graphs represents an exothermic reaction:

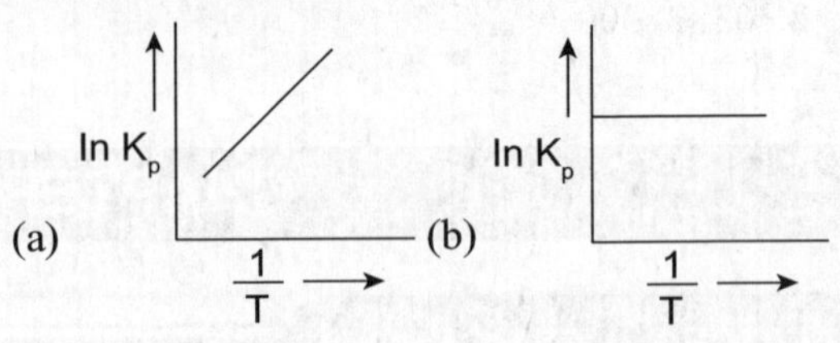

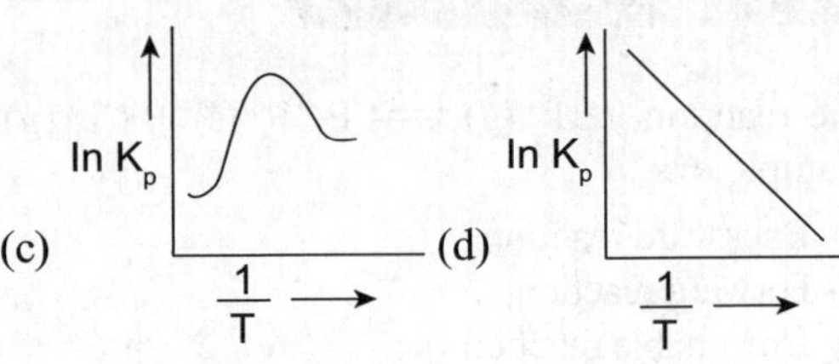

20. The relation between K_p and K_c for the reaction $2NO\ (g) + Cl_2\ (g) \rightleftharpoons 2NOCl\ (g)$ is:
(a) $K_p = K_c\ (RT)^{-1}$
(b) $K_p = K_c$
(c) $K_p = K_c/(RT)^2$
(d) $K_p = K_c/RT$

21. When two reactants A and are mixed to give products and D, the reaction quotient Q, at the initial stages of the reaction:
(a) Is zero
(b) Decreases with time
(c) Is independent of time
(d) Increases with time

22. When a catalyst is added to a reversible reaction in equilibrium state, the value of equilibrium constant:
(a) Increases
(b) Decreases
(c) Does not change
(d) Becomes zero

23. A vessel at equilibrium, contains SO_3, SO_2 and O_2. Now some helium gas is added, so that total pressure increases while temperature and volume remain constant. According to Le Chatelier's Principle, the dissociation of SO_3:
(a) Decreases
(b) Remains unaltered
(c) Increases
(d) Change unpredictably

24. In a reversible reaction, the catalyst:

(a) Decreases activation energy of forward reaction
(b) Increases activation energy of forward reaction
(c) Decreases activation energy of both forward and backward reactions
(d) Increases activation energy of backward reaction

25. A chemical reaction is catalysed by a catalyst X. Hence, X:

(a) Increases activation energy of the reaction
(b) Does not affect equilibrium constant of the reaction
(c) Decreases rate constant of the reaction
(d) Reduces enthalpy of the reaction

26. The chemical equilibrium of a reversible reaction is not influenced by:

(a) Temperature
(b) Pressure
(c) Catalyst
(d) Concentration

27. For equilibrium reaction:

$2NO_2\,(g) \rightleftharpoons N_2O_4\,(g) + 14.6$ J, increase in temperature would

(a) Favour the formation of N_2O_4
(b) Stop reaction
(c) Favour the decomposition of N_2O_4
(d) No alter the equilibrium

28. Which of the following favours the backward reaction in a chemical equilibrium:

(a) Decreasing the concentration of one of the reactants
(b) Increasing the concentration of one of the reactants
(c) Increasing the concentration of one or more of the products
(d) Removal of at least one of the products at regular intervals

29. If an inert gas is added in the reaction:

$N_2 + 3H_2 \rightleftharpoons 2NH_3$ at constant volume, then its equilibrium

(a) Remains unaffected
(b) Favours the backward reaction
(c) Favours the forward reaction
(d) Increases the dissociation of reactants

30. In the reaction $H_2 + I_2 \rightleftharpoons 2HI$ at equilibrium, some I_2 is added. What happens to the equilibrium:

(a) It gets shifted to the right
(b) It remains unchanged
(c) It gets shifted to the left
(d) First (b) then (c)

31. The role of a catalyst in a reversible reaction is to:

(a) Alter the equilibrium constant of the reaction
(b) Increase the rate of forward reaction
(c) Allow the equilibrium to be achieved quickly
(d) Decrease the rate of backward reaction

32. Le Chatelier's Principle is applicable to:

(a) Heterogenous reaction
(b) Homogenous reaction
(c) Irreversible reaction
(d) System in equilibrium

33. $N_2(g) + 3H_2\,(g) \rightleftharpoons 2NH_3 + 22.4$ kcal

formation of NH_3 by above reaction shows:

(a) Cyanamide process
(b) Serpeck's process
(c) Haber process
(d) None of these

34. For reaction:

$H_2 + I_2 \rightleftharpoons 2HI, = 47.6$, if the initial number of moles of each reactant and product is 1 mole, then at equilibrium:

(a) $[I_2] = [H_2], [I_2] > [HI]$
(b) $[I_2] < [H_2], [I_2] = [HI]$
(c) $[I_2] = [H_2], [I_2] < [HI]$
(d) $[I_2] > [H_2], [I_2] = [HI]$

35. Which of the following reaction will be favoured at low pressure:

(a) $N_2 + 3H_2 \rightleftharpoons 2NH_3$
(b) $H_2 + I_2 \rightleftharpoons 2HI$
(c) $PCl_5 \rightleftharpoons PCl_3 + Cl_2$
(d) $N_2 + O_2 \rightleftharpoons 2NO$

36. For the chemical reaction:

$3X\,(g) + Y\,(g) \rightleftharpoons X_3Y\,(g)$, the amount of X_3Y at equilibrium is affected by

(a) Temperature and pressure
(b) Temperature only
(c) Pressure only
(d) Temperature, pressure and catalyst

37. At constant temperature, the equilibrium constant (K_p) for the decomposition reaction:

$N_2O_4 \rightleftharpoons 2NO_2$ is expressed by

$K_p = (4x^2P)/(1 - x^2)$, where P = pressure, x = extent of decomposition. Which one of the following statements is true?

(a) K_p increases with increase of P
(b) K_p increases with increase of x
(c) K_p increases with decrease of x
(d) K_p remains constant with change in P and x

38. For the reaction:
$PCl_5(g) \rightleftharpoons PCl_3(g) + Cl_2(g)$
the forward reaction at constant temperature is favoured by
1. Introducing an inert gas at constant volume
2. Introducing chlorine gas at constant volume
3. Introducing an inert gas at constant pressure
4. Increasing the volume of the container
5. Introducing $pc1_5$ at constant volume

(a) 1,2,3
(b) 4, 5
(c) 2, 3,5
(d) 3,4, 5

39. For the reaction:
$CO(g) + H_2O(g) \rightleftharpoons CO_2(g) + H_2(g)$
at a given temperature, the equilibrium amount of CO_2 (g) can be increased by
(a) Adding a suitable catalyst
(b) Adding an inert gas
(c) Decreasing the volume of the container
(d) Increasing the amount of CO (g)

40. In a reaction $A_2(g) + 4B_2(g) \rightleftharpoons 2AB_4(g)$; $\Delta H < 0$.
The formation of AB_4 (g) will be favoured by:
(a) Low temperature and high pressure
(b) High temperature and high pressure
(c) Low temperature and low pressure
(d) High temperature and low pressure

41. In what manner will increase of pressure affect the following equation:
$C(s) + H_2O(g) \rightleftharpoons CO(g) + H_2(g)$
(a) Shift in the reverse direction
(b) Shift in the forward direction
(c) Increase in the yield of hydrogen
(d) No effect

42. The solubility of a solute in water varies with temperature and is given as $S = Ae^{-\Delta H/RT}$.
Here, ΔH is enthalpy of the solution. For the given solute variation of $\log_e S$ with temperature can be shown by the figure given below.

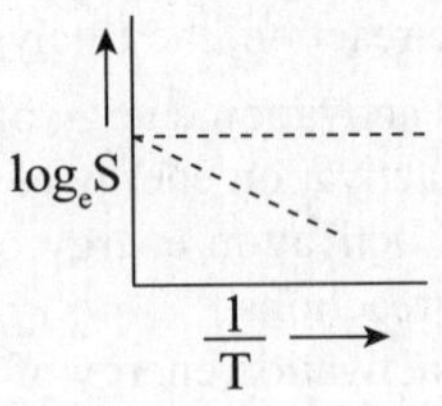

This solute may be:
(a) $CaSO_4$
(b) $CuSO_4$
(c) $CuSO_4 5H_2O$
(d) MgO

43. In which of the following cases does the reaction go farthest to completion:
(a) K = 1 (b) K = 10
(c) $K = 10^{-2}$ (d) $K = 10^2$

44. Consider the following reactions:
1. $AB_2(g) + \frac{1}{2}B_2(g) \rightleftharpoons AB_3(g)$
2. $2AB_3(g) \rightleftharpoons 2AB_2 + B_2(g)$

If K_1 and K_2 are the equilibrium constants at 27 °C of reactions 1 and 2 respectively, then K_1 and K_2 are related as:
(a) $K_1^2 = K_2$ (b) $K_2 K_2^2 = 1$
(c) $K_1 = 2K_2$ (d) $K_1 K_2^2 = 1$

45. For the dissociation of $PC1_5$ into $PC1_3$ and Cl_2 in gaseous phase reaction, if d is the observed vapour density and D the theoretical vapour density with 'a' as degree of dissociation. Variation of D/d with 'a' is given by which graph:

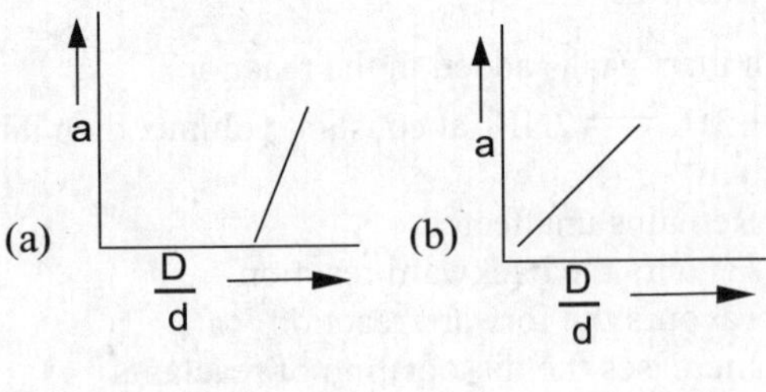

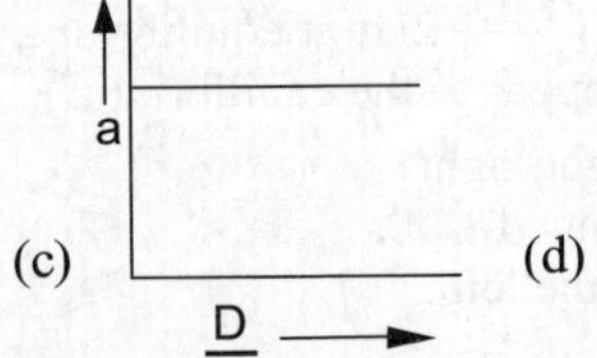

(d) None of these

PRACTICE QUESTIONS – II

46. At 700 K, the equilibrium constant K_p for the reaction $2SO_3(g) \rightleftharpoons 2SO_2\ (g) + O_2\ (g)$ is 1.80×10^{-3}. What is the numerical value in mole per litre of equilibrium constant K_c for this reaction at the same temperature:

(a) 8.1×10^{-8} (b) 9.1×10^{-9} mol L^{-1}
(c) 3.1×10^{-7} (d) 6.1×10^{-7} mol L^{-1}

47. For the reaction,

$2NO_2\ (g) \rightleftharpoons 2NO\ (g) + O_2\ (g)$
($K_c = 1.8 \times 10^{-6}$ at 184°C)
(R = 0.0831 kJ/(mol K))
when K_p and K_c are compared at 184°C it is found that:

(a) K_p is greater than K_c
(b) K_p is less than K_c
(c) $K_p = K_c$.
(d) Whether K_p is greater than, less than or equal to K_c depends upon the total gas pressure

48. For the following three reactions A, B and C, equilibrium constants are given:

(a) $CO\ (g) + H_2O\ (g) \rightleftharpoons CO_2(g) + H_2(g);\ K_1$
(b) $CH_4\ (g) + H_2O(g) \rightleftharpoons CO(g) + 3H_2(g);\ K_2$
(c) $CH_4(g) + 2H_2O(g) \rightleftharpoons CO_2(g) + 4H_2(g);\ K_3$

Which of the following relation is correct?

(a) $K_1\sqrt{K_2} = K_2$ (b) $K_2K_3 = K_1$
(c) $K_3 = K_1K_2$ (d) $K_3.K_2^3 = K_1^2$

49. For the $N_2 + 3H_2 \rightleftharpoons 2NH_3$, the initial mole ratio of $N_2:H_2$ is 1:3. If at equilibrium only 50% has reacted and equilibrium pressure is P. Find the value of P_{NH_3} at equilibrium.

(a) $\frac{P}{3}$ (b) $\frac{P}{5}$
(c) $\frac{P}{9}$ (d) $\frac{P}{6}$

50. Consider an endothermic reaction $X \longrightarrow Y$ with the activation energies E_b and E_f for the backward and forward reactions, respectively. In general:

(a) $E_b < E_f$ (b) $E_b > E_f$ (c) $E_b = E_f$
(d) There is no definite relation between E_b and E_f

51. The exothermic formation of ClF_3 is represented by the equation:

$Cl_2\ (g) + 3F_2\ (g) \rightleftharpoons 2ClF_3\ (g);\ \Delta H = -329$ kJ. Which of the following will increase the quantity of ClF_3 in an equilibrium mixture of Cl_2, F_2 and ClF_3?

(a) Increasing the temperature (b) Removing Cl_2
(c) Increasing the volume of the container
(d) Adding F_2

52. For the reaction equilibrium,

$N_2O_4\ (g) \rightleftharpoons 2NO\ (g)$

the concentrations of N_2O_4 and NO_2 at equilibrium are 4.8×10^{-2} and 1.2×10^{-2} mol L^{-1} respectively. The value of K_c for the reaction is:

(a) 3.3×10^2 mol L^{-1} (b) 3×10^{-1} mol L^{-1}
(c) 3×10^{-3} mol L^{-1} (d) 3×10^3 mol L^{-1}

53. If K_{eq} for the reaction is 81 $P + Q \rightleftharpoons 2R$

If we start with 1 mole each of P and Q. What is the mole fraction of R at equilibrium:

(a) $\frac{1}{9}$ (b) $\frac{11}{9}$
(c) $\frac{4}{9}$ (d) $\frac{9}{11}$

54. For the reaction, $H_2 + I_2 \rightleftharpoons 2HI$ the equilibrium concentration of H_2, I_2 and HI are 8.0, 3.0 and 28.0 mole/litre, respectively, the equilibrium constant is:

(a) 28.34 (b) 32.66
(c) 34.78 (d) 38.88

55. In a chemical equilibrium rate constant of forward reaction is 7.5×10^{-4} and the equilibrium constant is 1.5. The rate constant of backward reaction is:

(a) 2.5×10^4 (b) 5×10^{-4}
(c) 2.5×10^{-4} (d) 5×10^4

56. $A + 2B \rightleftharpoons 2C + D$, initial concentration of was 1.5 times that of A, but the equilibrium concentration of A and are found to be equal. The equilibrium constant for the reaction is:

(a) 4 (b) 8
(c) 12 (d) 16

57. The equilibrium constant for the reaction: $SO_3\ (g) \rightleftharpoons SO_2\ (g) + \frac{1}{2}O_2\ (g)$ is $K_c = 4.9 \times 10^{-2}$. The value of K_c for the reaction $2SO_2\ (g) + O_2\ (g) \rightleftharpoons 2SO_3\ (g)$ will be:

(a) 416 (b) 2.40×10^{-3}
(c) 9.8×10^{-2} (d) 4.9×10^{-2}

58. The value of K_p for the reaction:

$2H_2S(g) \rightleftharpoons 2H_2(g) + S_2(g)$ is 1.2×10^{-2} at 1065°C. The value for K_c is

(a) $< 1.2 \times 10^{-2}$ (b) $> 1.2 \times 10^{-2}$
(c) 1.2×10^{-2} (d) 0.12×10^{-2}

59. 1 mole of CH_3COOH and 1 mole of C_2H_5OH reacts to produce $\frac{2}{3}$ mole of $CH_3COOC_2H_5$. The equilibrium constant is:

(a) 2 (b) +2
(c) –4 (d) +4

60. The equilibrium constant for the following reaction will be $3A + 2B \rightleftharpoons C$:

(a) $\frac{[3A][2B]}{[C]}$ (b) $\frac{[C]}{[3A][2B]}$
(c) $\frac{[C]}{[A]^2[B]^2}$ (d) $\frac{[C]}{[A]^3[B]^2}$

61. The equilibrium constant of a reaction is 300. If the volume of reaction flask is tripled, the equilibrium constant is:

(a) 300 (b) 600
(c) 900 (d) 100

62. At 100°C the vapour density of nitrogen peroxide (N_2O_4) is 26.8. The percentage dissociation into NO_2 molecules is:

(a) 71.64% (b) 61.57%
(c) 83.56% (d) 67.39%

63. One mole of HI was heated in a sealed tube at 440°C till the equilibrium was reached. HI was found to be 22% decomposed. The equilibrium constant for dissociation reaction, $2HI \rightleftharpoons H_2 + I_2$ is:

(a) 1.99 (b) 0.282
(c) 0.01988 (d) 0.0796

64. In the reaction $N_2 (g) + 3H_2 (g) \rightleftharpoons 2NH_3 (g)$, when 100 mL of N_2 has reacted, the volumes of H_2 and NH_3 are:

(a) 300 mL of H_2 and 300 mL of NH_3
(b) 100 mL of H_2 and 200 mL of NH_3
(c) 300 mL of H_2 and 200 mL of NH_3
(d) 100 mL of H_2 and 100 mL of NH_3

65. If a mixture containing 3 moles of hydrogen and 1 mole of nitrogen is converted completely into ammonia, the ratio of volumes of reactants and products at the same temperature and pressure would be:

(a) 2:1 (b) 1:2
(c) 1:3 (d) 3:1

66. Equilibrium constant for the reaction
$2A (g) \rightleftharpoons B (g) + C (g)$ at 780 and 10 atm pressure is 3.52. The equilibrium constant of this reaction at 780 and 20 atm pressure is:

(a) 5.72 (b) 11.44
(c) 28.6 (d) 3.52

67. The ratio of K_p/K_c for the reaction:
$CO (g) + \frac{1}{2} O_2 (g) \rightleftharpoons CO_2 (g)$ is

(a) 1 (b) RT
(c) $(RT)^{1/2}$ (d) $(RT)^{-1/2}$

68. For the reaction, $A + B \rightleftharpoons C + D$, the rate constants for the forward and backward reactions are found to be 4.2×10^{-2} and 3.36×10^{-3} mol L^{-1} s^{-1} respectively. What is the equilibrium constant for the reaction:

(a) 11.5 (b) 12.5
(c) 8.0 (d) 6.0

69. The vapour density of N_2O_4 at a certain temperature is 30. What is the percentage dissociation of N_2O_4 at this temperature:

(a) 53.3 (b) 106.6
(c) 26.7 (d) None of these

70. The rate constants for the forward and backward reactions of hydrolysis of ester are 1.1×10^{-2} and 1.5×10^{-3} mol L^{-1} s^{-1} respectively. The equilibrium constant of the reaction, $CH_3COOC_2H_5 + H^+ \rightleftharpoons CH_3COOH + C_2H_5OH$ is:

(a) 6.53 (b) 7.34
(c) 7.75 (d) 8.33

71. In the reaction $2H_2S (g) \rightleftharpoons 2H_2 (g) + S_2 (g)$ the concentration of H_2S is 0.5 mol L^{-1} and concentration of H_2 is 0.1 mol L^{-1} while concentration of S_2 is 0.4 mol L^{-1} in one litre vessel. The value of equilibrium constant of the reaction is:

(a) 0.016 (b) 0.013
(c) 0.020 (d) 0.030

72. If equilibrium constants of reaction:
$N_2 + O_2 \rightleftharpoons 2NO$ is K_1, and $\frac{1}{2} N_2 + \frac{1}{2} O_2 \rightleftharpoons NO$ is K_2 then

(a) $K_1 = K_2$ (b) $K_1 = 2K_2$
(c) $K_2 = \sqrt{K_1}$ (d) $K_1 = \frac{1}{2} K_2$

73. At constant temperature, the equilibrium constant for the decomposition reaction $N_2O_4 \rightleftharpoons 2NO_2$ is expressed by $K_p = (4x^2p)(1 - x^2)$, where P = pressure, x = extent of decomposition. Which one of the following statement is true:

(a) K_p remains constant with change in P and x
(b) K_p increases with decrease of x
(c) K_p increases with increase of x
(d) K_p increase with increase of P

74. For a gaseous reaction $2A + B \rightleftharpoons C + D$, the partial pressures of A, B, C and D at equilibrium are 0.5, 0.8, 0.7 and 1.2 atm. The value of K_p for this reaction is:

(a) 2.4 atm (b) 6.2 arm ^{2}
(c) 4.2 atm^{-1} (d) 8.4 atm^{-3}

75. For the reaction $C(s) + CO_2(g) \rightleftharpoons 2CO(g)$, the partial pressures of CO_2 and CO are 4 and 8 atm respectively. The value of K_p for this reaction is:
(a) 14 atm (b) 16 atm
(c) 18 atm (d) 12 atm

76. In which of the following gaseous reaction, K_p and K_c have the same values:
(a) $2H1 \rightleftharpoons H_2 + I_2$
(b) $N_2 + 3H_2 \rightleftharpoons 2NH_3$
(c) $2SO_2 + O_2 \rightleftharpoons 2SO_3$
(d) $PCI_5 \rightleftharpoons PCI_3 + Cl_2$

77. In which of the following gaseous reaction, the value of K_p is less than K_c:
(a) $PC1_5 \rightleftharpoons PCI_3 + Cl_2$
(b) $2SO_2 + O_2 \rightleftharpoons 2SO_3$
(c) $2HI \rightleftharpoons H_2 + I_2$
(d) $N_2 + 3H_2 \rightleftharpoons 2NH_3$

78. The value of K_c for the reaction $N_2 + 3H_2 \rightleftharpoons 2NH_3$ at 400 is 0.5. The value of K_p for the reaction is (R = 0.0821 Latm mol^{-1} K^{-1}):
(a) 25.6×10^{-2} (b) 4.6×10^{-4}
(c) 46.6×10^{-4} (d) 16.6×10^{-8}

79. The value of K_p for the reaction, $2SO_2 + O_2 \rightleftharpoons 2SO_3$ at 700 is 1.3×10^{-3} atm^{-1}. The value of K_c at same temperature will be:
(a) 1.4×10^{-2} (b) 7.4×10^{-2}
(c) 5.2×10^{-2} (d) 3.1×10^{-2}

80. The ratio of Kp/Kc for the reaction:
$SO_2(g) + \frac{1}{2}O_2(g) \rightleftharpoons SO_3(g)$ is
(a) $(RT)^{-1/2}$ (b) $(RT)^{1/2}$
(c) RT (d) 1

81. For the reaction $N_2 + 3H_2 \rightleftharpoons 2NH_3$, the value of K_p is 6×10^{-3} at 500°C at a total pressure of 0.5 atm. If we take equal number of moles of N_2, H_2 the mole fraction of NH_3 is:
(a) 0.01 (b) 0.001
(c) 3×10^{-3} (d) 1×10^{-4}

82. If K_p of the reaction $A(g) + 2B(g) \rightleftharpoons 2C(g) + D(g)$ is 0.05 at 1000 K, then the value K_c of the reaction:
(a) 20000 R (b) 0.02 R
(c) 5×10^{-5}/R (d) 5×10^{-5} R

83. The reaction $PCl_5(s) \rightleftharpoons PCl_3(g) + Cl_2(g)$ is in equilibrium. If the equilibrium concentration of PCI_3 (g) is doubled, then concentration of Cl_2 (g) would become:
(a) 1/2 of its initial value
(b) 1/4 of its initial value
(c) Four times of its initial value
(d) Two times of its initial value

84. 9.2 g of N_2O_4 (g) is taken in a closed one litre vessel and heated till the following equilibrium is reached $N_2O_4(g) \rightleftharpoons 2NO_2(g)$. At equilibrium, 50% of N_2O_4 (g) is dissociated. What is the equilibrium constant (in mol L^{-1})? (molecular weight of N_2O_4 is 92):
(a) 0.1 (b) 0.2
(c) 0.4 (d) 2

85. Equilibrium constant for the reaction:
$H_2O(g) + CO(g) \rightleftharpoons H_2(g) + CO_2(g)$ is 81. If the velocity constant of the forward reaction is 162 L mol^{-1} sec^{-1}, what is the velocity constant (in L mol^{-1} $sec.^{-1}$) for the backward reaction:
(a) 13122 (b) 2
(c) 261 (d) 243

86. One mole of A (g) is heated to 300°C in a closed one litre vessel till the following equilibrium is reached.
$A(g) \rightleftharpoons B(g)$
The equilibrium constant of this reaction at 300°C is 4. What is the concentration of B (in mol L^{-1}) at equilibrium:
(a) 0.2 (b) 0.6
(c) 0.75 (d) 0.1

87. One mole of A(g) is heated to 200°C in a one litre closed flask, till the following equilibrium is reached.
$A(g) \rightleftharpoons B(g)$
The rate of forward reaction at equilibrium is 0.02 mol L^{-1} min^{-1}. What is the rate (in mol L^{-1}, min^{-1}) of the backward reaction at equilibrium:
(a) 0.04 (b) 0.01
(c) 0.02 (d) 1

88. In the reaction $PC1_5(g) \rightleftharpoons PCl_3(g) + Cl_2(g)$, the equilibrium concentrations of $PC1_5$ and $PC1_3$ are 0.4 and 0.2 mole/litre respectively. If the value of K_c is 0.5, what is the concentration of Cl_2 in mole/litre:
(a) 2.0 (b) 1.5
(c) 1.0 (d) 0.5

89. Consider the following reaction equilibrium:
$N_2(g) + 3H_2(g) \rightleftharpoons 2NH_3(g)$
Initially, 1 mole of N_2 and 3 mole of H_2 are taken in a 2 litre flask. At equilibrium state, if the number of

union of N_2 in 0.6, what is the total number of moles of all gases present in the flask:

(a) 0.8 (b) 1.6
(c) 3.2 (d) 6.4

90. In which of the following reactions, the concentration of reactant is equal to concentration of product at equilibrium (K = equilibrium constant):

(a) $A \rightleftharpoons B$; K = 0.01
(b) $R \rightleftharpoons P$; K = 1
(c) $X \rightleftharpoons Y$; K =10
(d) $L \rightleftharpoons J$; = 0.025

91. In which of the following reactions, the concentration of product is higher than the concentration of reactant at equilibrium? = equilibrium constant):

(a) $A \rightleftharpoons B$; K = 0.001
(b) $M \rightleftharpoons N$; K = 10
(c) $X \rightleftharpoons Y$; K = 0.005
(d) $R \rightleftharpoons P$; K = 0.01

92. At Kp for the following reaction is 1 atm $X(g) \rightleftharpoons Y(g) + Z(g)$

At equilibrium, 50% of X (g) is dissociated. The total pressure of the equilibrium system is 'P' atm. what is the partial pressure (in atm) of X (g) at equilibrium:

(a) 1 (b) 4
(c) 2 (d) 0.5

93. At 550 K, the K_c for the following reaction is 10^4 mol^{-1} lit

$X(g) + Y(g) \rightleftharpoons Z(g)$

At equilibrium, it was observed that

$[X] = \frac{1}{2}[Y] = \frac{1}{2}[Z]$

What is the value of [Z] (in mol L^{-1}) at equilibrium:

(a) 2×10^{-4} (b) 10^{-4}
(c) 2×10^4 (d) 10^4

94. 4 moles each of SO_2 and O_2 gases are allowed to react to form SO_3 in a closed vessel. At equilibrium 25% of O_2 is used up. The total number of moles of all the gases at equilibrium is:

(a) 6.5 (b) 7.0
(c) 8.0 (d) 2.0

95. For the reversible reaction,

$N_2(g) + 3H_2(g) \rightleftharpoons 2NH_3(g)$

At 500°C, the value of K_p is 1.44×10^{-5} when partial pressure is measured in atmospheres. The corresponding value of K_c, with concentration in mole L^{-1}, is:

(a) $1.44 \times 10^{-5}/(0.082 \times 500)^{-2}$
(b) $1.44 \times 10^{-5}/(8.314 \times 773)^{-2}$
(c) $1.44 \times 10^{-5}(0.082 \times 773)^{2}$
(d) $1.44 \times 10^{-5}/(0.082 \times 773)^{-2}$

96. For the reaction $PQ_{2\,'g'} \rightleftharpoons PQ_g + Q_g$ the degree of dissociation (α) of PQ_2 can be related with P or V as:

(a) $\alpha \propto \frac{1}{\sqrt{P}}$ (b) $\alpha \propto \frac{1}{P}$
(c) $\alpha \propto \frac{1}{\sqrt{V}}$ (d) $\alpha \propto V$

97. If the equilibrium constant for the reaction, $N_2(g) + 3H_2(g) \rightleftharpoons 2NH_3(g)$ at 750 is 49, then the equilibrium constant for the reaction, $NH_3(g) \rightleftharpoons 1/2N_2(g) + 3/2\,H_2(g)$ at the same temperature will be:

(a) 1/49
(b) 49
(c) 1/7
(d) 49^2

98. For the reaction:

$N_2(g) + 3H_2(g) \rightleftharpoons 2NH_3(g)$, $\Delta H = -93.6$ kJ mol^{-1}, the concentration of H_2 at equilibrium can be increased by:

(1) Lowering the temperature
(2) Increasing the volume of the system
(3) Adding N_2 at constant volume
(4) Adding H_2 at constant volume

(a) (ii) and (iv) are correct
(b) Only (ii) is correct
(c) (i), (ii) and (iii) are correct
(d) (iii) and (iv) are correct

99. For which of the following reaction, $K_p = K_c$:

(a) $2NOC1(g) \rightleftharpoons 2NO(g) + Cl_2(g)$
(b) $N_2(g) + 3H_2(g) \rightleftharpoons 2NH_3(g)$
(c) $H_2(g) + Cl_2(g) \rightleftharpoons 2HCl(g)$
(d) $PCl_3(g) + Cl_2(g) \rightleftharpoons PCl_5(g)$

100. In which of the following reactions, equilibrium is independent of pressure:

(a) $N_2(g) + O_2(g) \rightleftharpoons 2NO(g)$; $\Delta H = +ve$
(b) $2SO_2 + O_2(g) \rightleftharpoons 2SO_3(g)$; $\Delta H = -ve$
(c) $3H_2(g) + N_2(g) \rightleftharpoons 2NH_3(g)$; $\Delta H = -ve$
(d) $PC1_5(g) \rightleftharpoons PC1_3(g) + Cl_2(g)$; $\Delta H = +ve$

PRACTICE QUESTIONS – III

101. In an equilibrium reaction, $2SO_2\ (g) + O_2\ (g) \rightleftharpoons 2SO_3\ (g)$, the partial pressure SO_2, O_2 and SO_3 are 0.662, 0.101 and 0.331 atm respectively. What should be the partial pressure of oxygen if the equilibrium concentration of SO_3 and SO_2 becomes equal:

(a) 0.4 atm^{-1} (b) 0.6 atm^{-1}
(c) 0.12 atm^{-1} (d) 0.8 atm^{-1}

102. The equilibrium constant of mutarotation of α-D-glucose to β-D-glucose is 1.8. What per cent of the α-form remains under equilibrium:

(a) 35.7 (b) 64.3
(c) 55.6 (d) 44.4

103. The equilibrium constant for the reaction: $H_2\ (g) + S\ (g) \rightleftharpoons H_2S\ (g)$ is 18.5 at 925 and 9.25 at 1000 respectively. What is the enthalpy of the reaction:

(a) –142.16 kJ/mole (b) –71.08 kJ/mole
(c) –35.54 kJ/mole (d) None of these

104. For a gaseous equilibrium: $2A\ (g) \rightleftharpoons 2B\ (g) + C\ (g)$, K_p has a value of 1.8 at 700 K. What is the value of K_c for the equilibrium $2B\ (g) + C\ (g) \rightleftharpoons 2A$ at the same pressure:

(a) 0.031 (b) 1.3×10^{-3}
(c) 44.4 (d) 38

105. For the decomposition of solid calcium bicarbonate if the total pressure is 0.2 barat 420K.
Find ΔG° for this decomposition:

(a) 32 KJ/mole (b) 16 KJ/mole
(c) –16 KJ/m (d) –32 KJ/m

106. The equilibrium constant for the reaction:
$H_3PO_4 \rightleftharpoons H^+ + H_2PO_4^-$ is K_1,
for reaction $H_2PO_4 \rightleftharpoons H^+ + HPO_4^{2-}$ is K_2 and
for reaction $HPO_4^{2-} \rightleftharpoons H^+ + PO_4^{3-}$ is K_3.
The equilibrium constant (K) for
$H_3PO_4 \rightleftharpoons 3H^+ + PO_4^{3-}$ will be:

(a) $K_1, \times K_2 \times K_3$ (b) $K_1/K_2\ K_3$
(c) $K_2/K_1\ K_3$ (d) $K_1, + K_2 + K_3$

107. For the reaction $\underset{g}{2SO_2} + \underset{g}{O_2} \rightleftharpoons \underset{g}{2SO_3}$

If we start with 2 mol. SO_2 and 1mol. O_2 in 1L flask, the mixture needs 0.4 mol MnO_4^- in acidic medium for the complete oxidation of SO_2. The value of K_c is:

(a) 1/2 (b) 2
(c) 1 (d) 0.6

108. HI was heated in sealed tube at 400°C till the equilibrium was reached. HI was found to be 22% decomposed. The equilibrium constant for dissociation is:

(a) 1.99 (b) 0.0199
(c) 0.0796 (d) 0.282

109. In a 0.5 litre capacity vessel, CO and Cl_2 are mixed to form $COC1_2$. At equilibrium, it contains 0.2 mole of $COC1_2$ and 0.1 mole each of CO and Cl_2. The equilibrium constant (K_c) for reaction: $CO + Cl_2 \rightleftharpoons COC1_2$ is

(a) 15 (b) 5
(c) 20 (d) 10

110. One mole of N_2O_4 (g) at 300 is kept in a closed container under one atmosphere. It is heated to 600 when 20% by mass of N_2O_4 (g) decomposes to NO_2 (g). The resultant pressure is:

(a) 1.2 atm (b) 2.4 atm
(c) 2.0 atm (d) 1.0 atm

111. An equilibrium mixture for the reaction, $2H_2S\ (g) \rightleftharpoons 2H_2\ 9g) + S_2\ (g)$ had 1 mole of H_2S, 0.2 mole of H_2 and 0.8 mole of S_2 in a 2 litre flask. The value of K_c in mol L^{-1} is:

(a) 0.08 (b) 0.016
(c) 0.004 (d) 0.160

112. 1.25 moles of NOCl were placed in a 2.50 L reaction chamberat 427°C. After equilibrium was reached, 1.10molesofNOClremained. Calculatetheequilibrium constant K_c for the reaction, $2NOC1\ (g) \rightleftharpoons 2NO\ (g) + Cl_2\ (g)$:

(a) 1.6×10^{-3} (b) 5.6×10^{-4}
(c) 2.6×10^{-3} (d) 4.6×10^{-4}

113. $A + B \rightleftharpoons C + D$ is 100 at 25°C. Consider the following statements in this regard. If the initial concentration of all the four species were 1.0 M each, then the equilibrium concentration of:

1. A would be 0.182 mol/L
2. C would be 0.818 mol/L
3. D would be 1.818 mol/L

Of these statements:

(a) 1 and 3 are correct (b) 1 and 2 are correct
(c) 2 and 3 are correct (d) 1, 2 and 3 are correct

114. What is the correct sequence of active masses in increasing order in gaseous mixture, containing one gram per litre of each of the following:

1. NH_3 2. N_2
3. H_2 4. O_2

Select the correct answer using the codes given below:

(a) 3, 1, 4, 2 (b) 3, 4, 2, 1
(c) 2, 1, 4, 3 (d) 4, 2, 1, 3

115. If equilibrium constant for the reaction,
$XO^- (aq) + H_2O (l) \rightleftharpoons HXO (aq) + OH^- (aq)$
is 0.36×10^{-6} then find the value of dissociation constant (K_a) for HXO:

(a) 0.36×10^{-8} (b) 2.8×10^{-8}
(c) 2.8×10^{-10} (d) 0.36×10^{-6}

116. The equilibrium constant for the reaction,
$N_2 (g) + O_2 (g) \rightleftharpoons 2NO (g)$
At temperature T is 4×10^{-4}. The value of K_c for the reaction $NO (g) \rightleftharpoons \frac{1}{2} N_2 (g) + \frac{1}{2} O_2 (g)$ at the same temperature is:

(a) 4×10^{-6} (b) 2.5×10^2
(c) 0.02 (d) 50

117. An amount of solid NH_4HS in placed in a flask already containing ammonia gas at a certain temperature and 0.50 atm pressure. Ammonium hydrogen sulphide decomposes to yield NH_3 and H_2S gases in the flask. When the decomposition reaction reaches equilibrium, the total pressure is the flask rises to 0.84 atm, the equilibrium constant for NH_4HS decomposition at this temperature is:

(a) 0.30 (b) 0.18
(c) 0.17 (d) 0.11

118. Phosphorous pentachloride dissociates as follows, in a closed reaction vessel.
$PCl_5 (g) \longrightarrow PCl_3 (g) + Cl_2 (g)$
If total pressure at equilibrium of the reaction mixture is P and degree of dissociation of PCl_5 is x, the partial pressure of PCl_3 will be:

(a) $\left(\frac{x}{(x+1)}\right)P$ (b) $\left(\frac{2x}{(x-1)}\right)P$
(c) $\left(\frac{x}{(x-1)}\right)P$ (d) $\left(\frac{x}{(1-x)}\right)P$

119. The equilibrium constant for the reaction:
$SO_3(g) \rightleftharpoons SO_2(g) + \frac{1}{2}O_2 (g)$
is $K_c = 4.9 \times 10^{-2}$. the value of K_c for the reaction $2SO_2 (g) + O_2 (g) \rightleftharpoons 2SO_3 (g)$ will be

(a) 416 (b) 2.40×10^{-3}
(c) 9.8×10^{-2} (d) 4.9×10^{-2}

120. The equilibrium constants K_{P_1} and K_{P_2} for the reactions $X \rightleftharpoons 2Y$ and $Z \rightleftharpoons P + Q$, respectively are in the ratio of 1:9. If the degree of dissociation of X and Z be equal then the ratio of total pressure at these equilibria is:

(a) 1:36 (b) 1:1
(c) 1:3 (d) 1:9

121. For the chemical equilibrium:
$CaCO_{3\,'s'} \rightleftharpoons CaO_{'s'} + CO_{2\,g}$
ΔH_r° can be determind from which of the following plot:

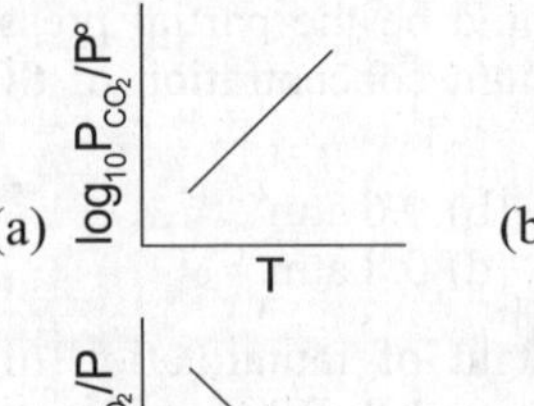

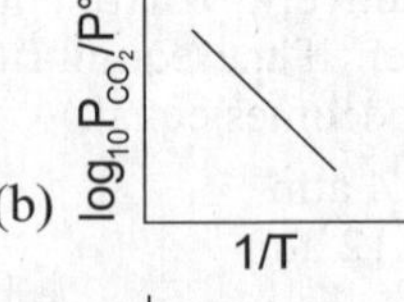

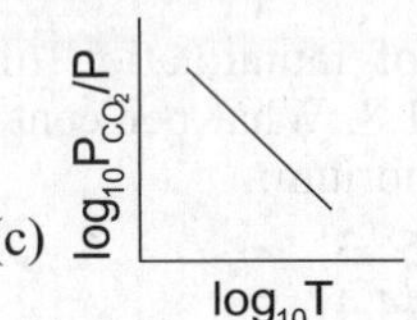

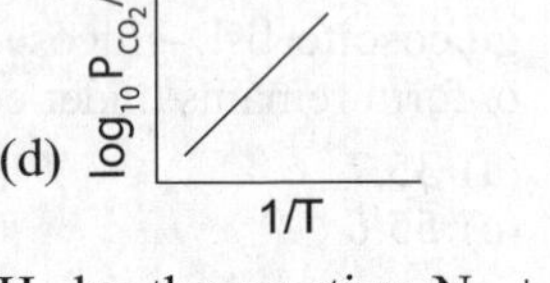

122. For the synthesis of NH_3 by the reaction $N_2 + 3H_2 \rightleftharpoons 2NH_3$ in the Haber process the attainment of equilibrium is correctly predicted by the curve:

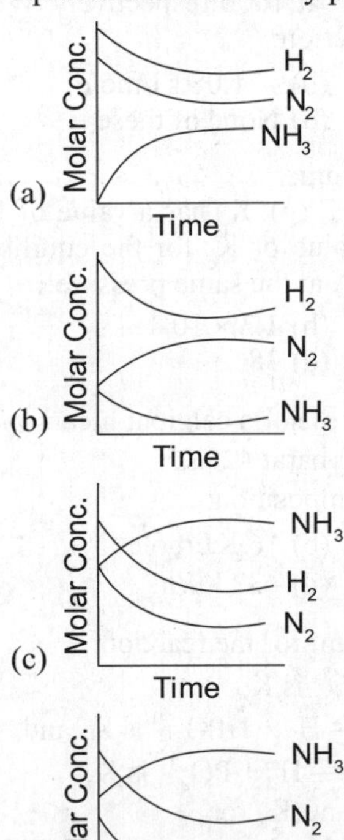

123. For the hypothetical reactions, the equilibrium constant (K) values are given:
$A \rightleftharpoons B \quad K_1 = 2$
$B \rightleftharpoons C \quad K_2 = 4$
$C \rightleftharpoons D \quad K_3 = 3$
The equilibrium constant (K) for the reaction: $A \rightleftharpoons D$ is

(a) 3 (b) 6
(c) 12 (d) 24

124. One mole of a compound AB reacts with one mole of compound CD according to the equation:

$AB + CD \rightarrow AD + CB$

When equilibrium had been established it was found that 3/4 mole each of reactant AB and CD has been converted to AD and CB. There is no change in volume. The equilibrium constant of the reaction is?

(a) 9 (b) $\frac{1}{9}$

(c) 9/16 (d) 16/9

125. At a certain temperature $PQ_{2\,(g)}$ dissociates according to the reaction:

$2PQ_{2(g)} \rightleftharpoons 2PQ_{(g)} + Q_{2(g)}$

if degree of dissociation α is quite small as compared to unity. The expression of K_P in terms of α and initial pressure P can be given as:

(a) $\frac{P\alpha^3}{2}$ (b) $\frac{P\alpha^3}{3}$

(c) $\frac{P\alpha^2}{2}$ (d) $\frac{P\alpha^2}{3}$

126. The dissociation of ClF_3 occurs as follows:

$2ClF_{3(g)} \rightleftharpoons Cl_{2(g)} + 3F_{2(g)}\ \Delta H = +ve$

It is favoured by:

(1) Addition of more fluorine

(2) Increase of temperature

(3) Increase of volume of the container

(4) Adding of inertgas at constant pressure

(a) I, II (b) II, III

(c) I, II, III (d) II, III, IV

Answer Keys

1. (d)	2. (a)	3. (a)	4. (c)	5. (a)	6. (d)	7. (c)	8. (d)	9. (b)	10. (a)
11. (d)	12. (b)	13. (a)	14. (c)	15. (b)	16. (d)	17. (d)	18. (c)	19. (a)	20. (a)
21. (d)	22. (c)	23. (b)	24. (c)	25. (b)	26. (c)	27. (c)	28. (d)	29. (a)	30. (a)
31. (c)	32. (d)	33. (c)	34. (c)	35. (c)	36. (a)	37. (d)	38. (b)	39. (d)	40. (a)
41. (a)	42. (c)	43. (d)	44. (b)	45. (a)	46. (c)	47. (a)	48. (c)	49. (a)	50. (a)
51. (d)	52. (c)	53. (d)	54. (b)	55. (b)	56. (a)	57. (a)	58. (a)	59. (d)	60. (d)
61. (a)	62. (a)	63. (c)	64. (c)	65. (a)	66. (d)	67. (d)	68. (b)	69. (a)	70. (b)
71. (a)	72. (c)	73. (a)	74. (c)	75. (b)	76. (a)	77. (d)	78. (b)	79. (b)	80. (a)
81. (a)	82. (c)	83. (a)	84. (b)	85. (b)	86. (c)	87. (c)	88. (c)	89. (c)	90. (b)
91. (b)	92. (a)	93. (a)	94. (a)	95. (d)	96. (a)	97. (c)	98. (a)	99. (c)	100. (a)
101. (a)	102. (a)	103. (b)	104. (c)	105. (b)	106. (a)	107. (b)	108. (b)	109. (d)	110. (b)
111. (b)	112. (b)	113. (a)	114. (d)	115. (b)	116. (d)	117. (d)	118. (a)	119. (a)	120. (a)
121. (b)	122. (a)	123. (d)	124. (a)	125. (a)	126. (d)				

Hints and Explanations For Selective Questions

2. As both Ag, Ag_2O are solids so $K_p = P_{O_2}$

3. As active mass of solid substance is always unity, they cannot be included in the equilibrium constant expression.

Thus equilibrium constant in terms of partial pressure

$(K_p) = P_{(co_2)}$

11. As the value of ΔH^0 is positive, therefore heat is absorbed by the system during the process. So on increasing the temperature, the reaction will shift in the forward direction.

12. A reversible reaction is said to have attained equilibrium, when both the backward and forward reactions take place at equal speed.

13. It is according to Le Chatelier's Principle.

As $n_p > n_r$

14. Evaporation process continues, till the entire water changes into vapour form. So in an open vessel, the water vapour equilibrium can not be achieved.

15. $N_2 + O_2 \rightleftharpoons 2NO$

As number of product molecules is equal to number reactant molecules, therefore this reaction will be unaffected by an increase in pressure.

16. In Homogeneous Equilibrium reactions and products are in same physical states, which is in both A and B here.

18. This is because equilibrium is dynamic in nature.

20. $K_p = K_c(RT)^{\Delta n} = Kc\ (RT)^{-1}$

$\Delta n = 2 - 3 = -1$

21. As the concentration of products increases with time, So reaction quotient Q will increase with time.

$$Q = \frac{[C][D]}{[A][B]}$$

23. As if a reaction takes place at constant volume and temperature, then addition of an inert gas (nitrogen, helium, argon) does not change the molar concentration of the reactants and products.

25. It is clear that a catalyst increase the rates of both forward and backward reactions equally. So the catalyst X does not affect the equilibrium constant of the reaction.

26. As a catalyst only increases the rate of a reaction, so it does not affect the chemical equilibrium.

27. The given reaction is endothermic, so increase of tem perature would favour the decomposition of N_2O_4.

28. An increase in the concentration of product always favours the backward reaction.

29. When an inert gas is added in a reversible reaction at constant volume, it is unable to change the molar concentration of the reactants and products. So the state of equilibrium remains unaffected.

30. When the concentration of any reactant is added in the reversible reaction, it favours the forward reaction.

35. As the number of moles of product is greater than the number of moles of reactant, therefore this reaction is favoured at low pressure.

36. Temperature and pressure both affect equilibrium but catalyst has no effect (Le Chatelier's Principle).

37. K_p (equilibrium constant) is independent of pressure and concentration.

38. This is according to Le Chatelier's Principle.

39. Catalyst has no effect on equilibrium. Decreasing the volume or pressure has no effect as n_r is equal to n_p therefore, the equilibrium shifts in the forward direction by increasing the concentration of reactants CO_2 (g).

40. As the reaction is exothermic, the formation of AB_4 is favoured by low temperature . As the forward reaction is accompanied by decrease in the number of moles. So it is favoured by high pressure.

41. It is according to Le Chatelier's principle.

As $n_p > n_r$.

42. As the slope of the graph is negative, so ΔH is negative. As dissolution of hydrated salt is endothermic, so the solute is $CuSO_4.5H_2O$

43. A large value of K_c or K_p shows that forward reaction occurs to a large extent.

44. Equation (2) is obtained by reversing the equation (1) and then multiplied by two.

So, $K_2 = (1/K_1)^2$ or $K_2K_1^2 = 1$

45. $(1 + a) = D/d$

When $a = 0$, $D/d = 1$

When 'a' increases, (1 + a) increases, so D/d also increases.

46. $$(K_c) = \frac{Kp}{(RT)^{\Delta n}} = \frac{1.8 \times 10^{-3}}{(8.134 \times 700)^1}$$

$$= 3.1 \times 10^{-7}$$

47. As $K_p = K_c(RT)^{\Delta n}$

Here $\Delta n = 3 - 2 = 1$

So $K_p = K_c\ (0.0831 \times 457)^1$.

Hence $K_p > K_c$ here.

48. As equation (c) = equation (a) + equation (b)

So $K_3 = K_1 K_2$

49.

N_2 +	$3H_2$	$\rightleftharpoons$	$2NH_3$
1	3		0
1–0.5	3–1.5		1

$n_T = 0.5 + 1.5 + 1 = 3$

$$P_{NH_3} = \frac{n_{NH_3}}{n_T} \times P = \frac{P}{3}$$

50. For endothermic reaction, $\Delta H = +ve$

$\Delta H = E_f - E_b$, it means $E_b < E_f$

51. $Cl_2 (g) + 3F_2 (g) \rightleftharpoons 2ClF_3 (g); \Delta H = -329$ kJ.

As Favourable conditions for ClF_3 formation are

1. Decrease in temperature
2. Addition of reactants (F_2 and Cl_2)
3. Increase in pressure, that is, decrease in volume

52. $K_c = \dfrac{[NO_2]^2}{[N_2O_4]} = \dfrac{(1.2 \times 10^{-2})^2}{4.8 \times 10^{-2}} = 3 \times 10^{-3}$ mol L^{-1}

53. $P + Q \rightleftharpoons 2R$

P	Q	R
1	1	0
1–x	1–x	2x

$n_T = 2$

$K = \dfrac{(R)^2}{(P)(Q)} = \dfrac{(2x)^2}{(1-x)(1-x)}$

$\sqrt{81} = \dfrac{2x}{1-x}$, on solving $x = \dfrac{9}{11}$

54. $KC = \dfrac{[HI]^2}{[H_2][I_2]} = \dfrac{28 \times 28}{8 \times 3} = 32.66$

55. $K_b = K_f/K_c$

$= 7.5 \times 10^{-4}/1.45 = 5 \times 10^{-4}$

57. $K_{c_1} = 4.9 \times 10^{-2} = \dfrac{[SO_2][O_2]^{1/2}}{[SO_3]}$

$\dfrac{[SO\ ]}{[SO_2]\ [O_2]}$

$K_{c_2} = (1/K_{c_1})^2 = (1/4.9 \times 10^{-2})^2 = 416.5$

58. As $\Delta n_g = 3 - 2 = 1$ so $K_p > K_c$

Hence $K_c < 1.2 \times 10^{-2}$

59. $CH_3COOH + C_2H_5OH \rightleftharpoons CH_3COOC_2H_5 + H_2O$

CH_3COOH	C_2H_5OH	$CH_3COOC_2H_5$	H_2O
1	1	0	0
1–α	1–α	α	α
$1-\frac{2}{3}$	$1-\frac{2}{3}$	$\frac{2}{3}$	$\frac{2}{3}$

$K_c = \dfrac{\frac{2}{3} \times \frac{2}{3}}{1 - \frac{2}{3} \times 1 - \frac{2}{3}} = 4$

62. $\alpha = \dfrac{D-d}{(n-1) \times d} = \dfrac{46 - 26.8}{26.8} = \dfrac{19.2}{26.8} = 0.7164$

$\alpha\% = 71.64$

64. As 1 mole of nitrogen reacts with 3 moles of hydrogen to form 2 moles of ammonia, so 100 mL of nitrogen (N_2) reacts, with $3 \times 100 = 300$ mL of H_2.

Therefore volume of ammonia produced $= 2 \times 100 = 200$ mL.

65. $N_2 (g) + 3H_2 (g) \rightleftharpoons 2NH_3 (g)$,

As 4 moles of reactants give 2 moles of product, so ratio of volumes of reactants and products at the same temperature and pressure will be 2:1 or 4:2.

66. As in this reversible reaction, the temperature is same in both cases, therefore its equilibrium constant will also remain same that is 3.52.

67. $K = K_c (RT)^{\Delta n}$

$\dfrac{K_p}{K_c} = (RT)^{\Delta n} = (RT)^{1-3/2} = (RT)^{-1/2}$

As $[\Delta n = n_p - n_r = 1 - 3/2 = -1/2]$

68. The equilibrium constant of the given reaction

$K_c = \dfrac{K_f}{K_b} = \dfrac{4.2 \times 10^{-2}}{3.36 \times 10^{-3}} = 12.5$

69. Vapour density, $D = \dfrac{\text{Molecular Weight}}{2} = \dfrac{92}{2} = 46$

$d = 30$

$\alpha = \dfrac{D-d}{d} = \dfrac{46-30}{30} = \dfrac{16}{30} = 0.533 = 53.3\%$

70. $K_c = \dfrac{K_f}{K_b} = \dfrac{1.1 \times 10^{-2}}{1.5 \times 10^{-3}} = 7.34$

71. $K_c = \dfrac{[H_2]^2[S_2]}{[H_2S]} = \dfrac{(0.1)^2\,(0.4)}{(0.5)^2} = 0.016$

72. The equilibrium constant $(K_2) = \dfrac{[NO]^2}{[N_2][O_2]}$

Similarly, equilibrium constant $(K_2) = \dfrac{[NO]}{[N_2]^{1/2}[O_2]^{1/2}}$

$= \left(\dfrac{[NO]^2}{[N_2][O_2]}\right)^{\frac{1}{2}} = (K_1)^{\frac{1}{2}} = \sqrt{K_1}$

73. As the equilibrium constant of reaction in terms of partial pressure (K_p) depends upon concentration and not on pressure (P) or extent of decomposition (x) so K_p remains constant with change in P and x here.

74. The equilibrium constant for the reaction in terms of partial pressure

$K_p = \dfrac{(P_c) \times (P_D)}{(P_A)^2 \times (P_B)} = \dfrac{0.7 \times 1.2}{(0.5)^2 \times 0.8} = 4.2$ atm^{-1}

75. $K_p = \dfrac{(P_{CO})^2}{(P_{CO2}) \times (P_C)} = \dfrac{(8^2)}{4 \times 1} = 16$ atm

76. In the reaction $2HI \rightleftharpoons H_2 + I_2$

Value of $\Delta n = (1 + 1) - (2) = 0$

So $K_p = K_c$

77. Here $\Delta n = 2 - 4 = -2$.
So, the value of K_p is less than K_c.

78. $K_p = K_c(RT)^{\Delta n}$
$= 0.5 \times (0.0821 \times 400)^{-2} = 4.6 \times 10^{-4}\ atm^{-2}$

79. $K_c = \frac{K_p}{(RT)^{\Delta n}} = \frac{1.3 \times 10^{-3}}{(0.0821 \times 700)^{-1}} = 7.4 \times 10^{-2}$

80. $K_c = \frac{K_p}{(RT)^{\Delta n}}$ or $\frac{K_p}{K_c} = (RT)^{-1/2}$

81. $N_2 + 3H_2 \rightleftharpoons 2NH_3$

1	1	0
1–x	1–3x	2x

$n_{Total} = 1–x + 1–3x + 2x$
$= 2\ (1–x)$

Now $K_p = \frac{16x^2(1-x)^2}{(1-3x)^3 \times 0.25}$

on solving
$x = 0.01$

So $\frac{x}{1-x} = 0.01$

82. $K_c = \frac{K_p}{(RT)^{\Delta n}} = \frac{0.05}{(1000 \times R)^1} = \frac{5 \times 10^{-5}}{R}$

(Here $\Delta n = 4 - 3 = 1$).

83. $K_c = \frac{[PCl_3][Cl_2]}{[PCl_5]}$

As the value of $[PCl_5]$ is 1 (due to its solid form), therefore $(K_c) = [PCl_3][Cl_2]$
When the concentration of PCl_3 is doubled, then concentration of Cl_2 will become $\frac{1}{2}$ in order to maintain the equilibrium constant.

86. $A_g \longrightarrow B_g$
(1–x) x

$K_c = \frac{x}{(1-x)}$

$x = 4 - 4x$
$3x = 4$
$x = \frac{3}{4} = 0.75$

88. $K_c = \frac{[PCl_3][Cl_2]}{[PCl_5]}$

$[Cl_2] = \frac{K_c \times [PCl_5]}{[PCl_3]}$

$= \frac{0.5 \times 0.4}{0.2} = 1.0$

89. $N_2 + 3H_2 \rightleftharpoons 2NH_3$

1	3	0
1–α	3–3α	2α
0.6	3–3(0.4)	2 × 0.4

$n_T = 0.6 + 1.8 + 0.8 = 3.2$

94. $2SO_2 + O_2 \rightleftharpoons 2SO_3$
Initials moles 4 4 0
Moles at equlibrium 4(1 – 0.25) 4(1 – 0.25) 0.25 × 2
Total number of moles
$= 4(1 - 0.25) + 4(1 - 0.25) + 0.5$
$= 4 - 1 + 4 - 1 + 0.5 = 6.5$

95. $\Delta n = 2 - 4 = -2$
$K_p = K_c.(RT)^{\Delta n}$
$K_c = K_p/(RT)^{\Delta n}$
$= \frac{1.44 \times 10^{-5}}{(0.0821 \times 773)^{-2}}$

99. $K_p = K_c$ when $\Delta n = 0$
$\Delta n = 2 - 2 = 0$
So, the correct option is:
$H_2\ (g) + Cl_2\ (g) \rightleftharpoons 2HCl\ (g)$

100. In (a), as $n_p = n_r$, pressure has no effect on equilibrium.

101. $K_p = \frac{(P_{SO_3})^2}{(P_{SO_2})^2} = \frac{(0.331)^2}{(0.662)^2 \times (0.101)} = 2.5\ atm^{-1} = 2.5\ atm^{-1}$
$= 2.5\ atm^{-1}$
Since concentration of SO_3 is equal to the concentration of SO_2, so $P_{SO_3} = P_{SO_2}$
Thus,
$K_p = \frac{1}{P_{O_2}}$ or $P_{O_2} = \frac{1}{K_p} = \frac{1}{2.5} = 0.4\ atm^{-1}$

102. α – D Glucose $\rightleftharpoons$ β – D Glucose
Initial concentration 1 0
Concentration at equlibrium, $1 - x$ x
$K = \frac{\beta\text{-D-glucose}}{\alpha\text{-D-glucose}}$
$1.8 = \frac{x}{1-x}$
$1.8\ (1 - x) = x$
$1.8 - 1.8x = x$ or $1.8 = 2.8x$
$x = 1.8/2.8 = 0.643 = 0.643 \times 100 = 64.3\%$
β-D-glucose = 64.3%
α-D-glucose remains at equilibrium
= 100 – 64.3 = 35.7%

103. $\log\frac{K_2}{K_1} = \frac{\Delta H}{2.303\ R} \times \frac{[T_2 - T_1]}{T_1T_2}$

$\log\frac{9.25}{18.5} = \frac{\Delta H}{2.303 \times 8.314} \times \frac{75}{925 \times 1000}$

$$-0.0310 = \frac{\Delta H \times 75}{2.303 \times 8.314 \times 925 \times 1000}$$

$\Delta H = -71080.57$ J/mole $= -71.08$ kJ/mole

104. For $2B(g) + C(g) \rightleftharpoons 2A(g)$ at 700°C

$K_p = \frac{1}{1.8}$

$K_p = K_c (RT)^{\Delta n}$

$\Delta n = 2 - (2+1) = -1$

$$Kc = \frac{Kp}{(RT)^{\Delta n}} = \frac{1}{8.8} \times \frac{1}{(0.821 \times 973)^{-1}}$$

$$= \frac{1}{1.8} \times 0.0821 \times 973 = 44.379$$

105. $\underset{\text{'s'}}{Ca(HCO_3)_2} \xrightarrow{\Delta} \underset{\text{'s'}}{CaCO_3} + \underset{g}{CO_2} + \underset{g}{H_2O}$

Here $P_{Total} = 0.2$ bar

At Equilibrium $P_{CO_2} = P_{H_2O} = 0.1$ bar

$K_p = P_{CO_2} P_{H_2O} = 0.1 \times 0.1 = 0.01$

$\Delta G° = -2.303\ RT \log_{10} K_{eq}$

$= -2.303 \times 8.314 \times 420 \times \log 0.01$

$= 16083$ J $= 16.08$ KJ/M

106. The equilibrium constant

$$K_1 = \frac{[H^+][H_2PO_4^-]}{[H_3PO_4]}$$

$$K_2 = \frac{[H^+][HPO_4^{2-}]}{[H_2PO_4^-]}$$

$$K_3 = \frac{[H^+][PO_4^{3-}]}{[HPO_4^{2-}]}$$

$$K = \frac{[H^+]^3[PO_4^{3-}]}{[H_3PO_4]}$$

The product of $K_1 \times K_2 \times K_3$

$$= \frac{[H^+][H_2PO_4^-]}{[H_3PO_4]} \times \frac{[H^+][HPO_4^{2-}]}{[H_2PO_4^-]} \times \frac{[H^+][PO_4^{3-}]}{[HPO_4^{2-}]}$$

$$= \frac{[H^+]^3[PO_4^{3-}]}{[H_3PO_4]} = K$$

Thus, $K = K_1 \times K_2 \times K_3$

108.

$2HI \longrightarrow$	H_2	+	I_2
1	0		0
$(1-x)$	$\left(\frac{x}{2}\right)$		$\left(\frac{x}{2}\right)$
$1 - 0.22$	0.11		0.11
$= 0.78$			

$x = 22\%$ of 1 mole $= \frac{22}{100} \times 1 = 0.22$

$$K = \frac{[H_2][I_2]}{[HI]^2} = \frac{0.11 \times 0.11}{(0.78)^2} = 0.0199$$

$= 0.0199$

109. In this reaction,

Number of moles of $COC1_2 = 0.2$

Number of moles of CO = 0.1

Number of moles of Cl = 0.1

The concentration of $COC1_2$

$$= \frac{\text{Number of moles}}{\text{Volumes in litres}} = \frac{0.2}{0.5} = 0.4 \text{ mol L}^{-1}$$

Concentration of CO $= \frac{0.1}{0.5} = 0.2$ mol L^{-1}

Concentration of Cl $= \frac{0.1}{0.5} = 0.2$ mol L^{-1}

The equilibrium constant of the reaction

$$K_c = \frac{[COCl_2]}{[CO][Cl_2]} = \frac{0.4}{0.2 \times 0.2} = 10$$

110.

$N_2O_4 \rightleftharpoons$	$2NO_2$
1	0
$1 - 0.2$	0.4
$= 0.8$	

Number of moles at 300 = 1

Number of moles at 600 = 0.8 + 0.4 = 1.2

V is constant

$$\frac{P_2}{P_1} = \frac{n_2T_2}{n_1T_1}$$

$$\frac{P_2}{1} = \frac{1.2 \times 600}{1 \times 300} = 2.4 \text{ atm}$$

111. $K = \frac{[H_2]^2[S_2]}{[H_2S]^2}$

$[H_2] = \frac{2 \times 0.2}{2} = 0.2$ mol/L

$[S_2] = 0.8/2 = 0.4$ mol/L

$[H_2S] = 1$ mol/L

$K_c = \frac{[0.2]^2[0.4]}{1} = 0.016$ mol/L

112. As final (equlibrium) amount of NOCl = 1.10 mol So, change in the amount of NOCl

$= 1.25 - 1.10 = 0.15$ mol

	$2NOCl(g) \rightleftharpoons$	$2NO(g)$ +	$Cl_2(g)$
Initial moles	1.25	–	–
Change (moles)	–0.15	+0.15	+0.15/2
Final (moles)	1.10	0.15	0.075
Final cone.	1.10/2.50	0.15/2.50	0.075/2.50

$$K_c = \frac{[NO(g)]^2[Cl_2(g)]}{[NOCl(g)]^2}$$

$$= \frac{(0.15/2.50)^2(0.075/2.50)}{(1.10/2.50)^2}$$

$$= \frac{(0.15)^2 \times 0.075}{(1.10)^2 \times 2.50} = 5.578 \times 10^{-4} = 5.6 \times 10^{-4}$$

113. Let X = change in concentration of A

	A +	B	C +	D
Initial conc.	1.0 M	1.0 M	1.0 M	1.0 M
Equation conc.	$1-x$	$1-x$	$1+x$	$1+x$

$K = \frac{(1+x)(1+x)}{(1-x)(1-x)} = 100$

$= \frac{1+x}{1-x} = 10$

$= 1 + x = 10 - 10x = 11x = 9$

$x = 9/11 = 0.818$

So, [A] = 1 – 0.818 = 0.182 M

[C] = 1 + 0.818 = 0.818 M

[D] = 1.818 M

114. Active masses of:

$NH_3 = \frac{1 \text{ g}/17 \text{ g mol}}{1 \text{ L}} = \frac{1}{17}$ mol/L

$N_2 = \frac{1 \text{ g}/28 \text{ g mol}}{1 \text{ L}} = \frac{1}{28}$ mol/L

$H_2 = \frac{1 \text{ g}/2 \text{ g mol}}{1 \text{ L}} = \frac{1}{2}$ mol/L

$O_2 = 1 \text{ g}/32 \text{ g mol} = \frac{1}{32}$ mol/L

Increasing order of active masses is:

1/32 < 1/28 < 1/17 < 1/2 mol/L

$O_2 < N_2 < NH_3 < H_2$, that is, 4 < 2 < 1 < 3.

115. $HXO \rightleftharpoons H^+ + XO^-$

$K_a = \frac{[H^+][XO]}{[HXO]}$

$XO^- + H_2O \rightleftharpoons HXO + OH^-$

$K_{eq} = \frac{[HXO][OH^-]}{[XO^-]}$

$K_a K_{eq} = [H^+][OH^-] = K_w = 1 \times 10^{-4}$

$K_a = \frac{K_w}{K_{eq}} = \frac{1.0 \times 10^{-14}}{0.36 \times 10^{-6}} = 2.8 \times 10^{-8}$

116. $N_2(g) + O_2(g) \rightleftharpoons 2NO\ (g)$

$K_c = \frac{[NO]^2}{[N_2][O_2]} = 4 \times 10^{-4}$

$NO \rightleftharpoons \frac{1}{2} N_2\ (g) + \frac{1}{2} O_2\ (g)$

$K_c = \frac{[N_2]^{1/2}[O_2]^{1/2}}{[NO]}$

$= \sqrt{(1/K_c)} = \sqrt{(1/4 \times 10^{-4})} = 50$

117. $NH_4HS\ (s) \rightleftharpoons NH_3\ (g) + H_2S\ (g)$

	NH_4HS (s)	NH_3 (g)	H_2S (g)
Initial pressure	0	0.5	0
At equilibrium	0	0.5 + x	x

Total pressure = 0.5 + 2x = 0.84

So, x = 0.17 atm.

$K_p = P_{NH_3} \times P_{H_2S} = 0.67 \times 0.17 = 0.11\ atm^2$

118. $PCl_5 \rightleftharpoons PCl_3 + Cl_2$

1   0   0

1–x x x

P_{PCl_5} = Total pressure × mole fraction of $PCl_3 = P\left(\frac{x}{1+x}\right)$

119. $K_{C_1} = 4.9 \times 10^{-2} = \frac{[SO_2][O_2]1/2}{[SO_3]}$

$K_{C_2} = -\frac{[SO_3]^2}{[SO_2]^2[O_2]}$

$K_{C_2} = (1/K_{C_1})^2 = (1/4.9 \times 10^{-2})^2 = 416.5$

120. $X \rightleftharpoons 2Y$

1   0

(1 – x)   2x

$K_{P_1} = \frac{(2x)^2}{(1-x)} (P_1/1+x)^1$

$Z \rightleftharpoons P + Q$

1   0   0

(1 – x)   x   x

$K_{P2} = \frac{(x)^2}{(1-x)} (P_2/1+x)^1 \quad \frac{4 \times P_1}{P_2} = \frac{1}{9}$

On solving, we get

$P_1 : P_2 = 1:36$

121. For $CaCO_3 \overset{\Delta}{\rightleftharpoons} CaO + CO_2$ ('s' 's' g)

$K_P = P_{CO_2} = \frac{P_{CO_2}}{P^o}$

P° represents reference pressure i.e.,/Atm.

As according to van't Hoff equation

$\frac{d \ln K_P}{dT} = \frac{\Delta H^o}{RT^2}$

$\ln K_P = \frac{-\Delta H^o}{RT} + 1$

$2.303 \log_{10} K_p = \frac{-\Delta H^o}{RT} + 1$

$2.303 \log_{10} \frac{P_{CO_2}}{P^o} = \frac{-\Delta H^o}{RT} + 1$

It means plot of log $\frac{P_{CO_2}}{P^o}$ versus $\frac{1}{T}$ must be a straight line with a negative slope.

122. Here option (A) is correct as conc^n of N_2, H_2 decrease while NH_3 increases with time and after equilibrium state these become almost constant.

123. As for a multistep wise reaction

$K = K_1 \cdot K_2 \cdot K_3 = 2 \times 4 \times 3 = 24$

124. As after equilibrium

$(AB) = \frac{1}{4V}, (CD) = \frac{1}{4V}$

$(AD) = \frac{3}{4V}$ $(CB) = \frac{3}{4V}$

$$\text{As } K = \frac{[AD][CD]}{[AB][CD]} = \frac{\frac{3}{4V}\times\frac{3}{4V}}{\frac{1}{4V}\times\frac{1}{4V}} = 9$$

125. $\underset{g}{2PQ_2} \rightleftharpoons \underset{g}{2PQ} + \underset{g}{Q_2}$

$C(1-\alpha)$ $C\alpha$ $\frac{C\alpha}{2}$

$$K_P = \frac{(P_{PQ})^2(P_{Q_2})}{(P_{PQ_2})^2}$$

$$K_P = \frac{\frac{C\alpha}{2}\times(C\alpha)^2\times P_T}{[C(1-\alpha)][C(1+\frac{\alpha}{2})]}$$

$$K_P = \frac{\alpha^3 P_T}{2(1-\alpha)^2(1+\frac{\alpha}{2})}$$

As $1 >>> \alpha$ so

$$K_P = \frac{\alpha^3 P_T}{2} = \frac{\alpha^3 P}{2}$$

126. As addition of more F_2 favours the backword reaction i.e formation of ClF_3.

PREVIOUS YEARS' QUESTIONS

1. The value of Kp in the reaction $MgCO_3$ (s) $\leftrightarrow$ MgO (s) + CO_2 (g) is:
(a) $K_p = P_{CO_2}$
(b) $K_p = \frac{P_{MgCO_3}}{P_{CO_2} \times P_{MgCO_3}}$
(c) $K_p = P_{CO_2} \times \frac{P_{CO_2} \times P_{MgCO}}{P_{MgCO_3}}$
(d) $K_p = \frac{P_{CO_2} \times P_{MgCO}}{P_{MgCO_3}}$

[2000]

2. In which of the following case, the value of K_p is less than K_c?
(a) $N_2 + O_2 \leftrightarrow 2NO$
(b) $H_2 + Cl_2 \leftrightarrow 2HCl$
(c) $2SO_2 + O_2 \leftrightarrow 2SO_3$
(d) $PCl_5 \leftrightarrow PCl_3 + Cl_2$

[2000]

3. For a reversible reaction, if the concentrations of the reactants are doubled, at constant temperature the equilibrium constant will be:
(a) One-fourth (b) Halved
(c) Doubled (d) The same

[2000]

4. Reaction BaO_2 (s) $\leftrightarrow$ BaO (s) + O_2 (g); ΔH = +ve. In equilibrium condition, pressure of O_2 depends on:
(a) Increase mass of BaO_2
(b) Increase mass of BaO
(c) Temperature of equilibrium
(d) Mass of BaO_2 and BaO both

[2002]

5. Reaction quotient for the reaction: N_2 (g) + $3H_2$ (g) $\leftrightarrow$ $2NH_3$ (g) is given by $Q = \frac{[NH_3]^2}{[N_2][H_2]^3}$ the reaction will proceed from right to left, if Kc is equilibrium constant
(a) Q < Kc (b) Q = 0
(c) Q > Kc (d) Q = Kc.

[2003]

6. For the reaction $H_2 + I_2 \leftrightarrow 2HI$, the equilibrium concentrations of H_2, I_2 and HI are 8, 3 and 28 mol L^{-1} respectively. Equilibrium constant of the reaction is:
(a) 32.67 (b) 31.67
(c) 34.67 (d) 36.67

[2003]

7. The following equilibria are given:
$N_2 + 3H_2 \leftrightarrow 2NH_3$ K_1
$N_2 + O_2 \leftrightarrow 2NO$ K_2
$H_2 + \frac{1}{2}O_2 \leftrightarrow H_2O$ K_3
The equilibrium constant of the reaction
$2NH_3 + 5/2\ O_2 \leftrightarrow 2NO + 3H_2O$
in terms of K_1, K_2 and K_3 is:
(a) $K_1 K_3^2/K_2$ (b) $K_2 K_3^3/K_1$
(c) $K_1 K_2 K_3$ (d) $K_1 K_2/K_3$

[2003, 07]

8. Equilibrium constants K_1 and K_2 for the following equilibria:
NO (g) + $\frac{1}{2}O_2 \xleftrightarrow{K_1}$ NO_2 (g) and
2NO (g) $\xleftrightarrow{K_2}$ 2NO (g) + O_2

are related as:

(a) $K_2 = 1/K_1^2$ (b) $K_2 = K_1^2$
(c) $K_2 = 1/K_1$ (d) $K_2 = K_1/2$

[2005]

9. For the reaction:

$CH_4\ (g) + 2O_2\ (g) \leftrightarrow CO_2\ (g) + 2H_2O\ (l)$
$\Delta_r H = -170.8\ kJ\ mol^{-1}$
Which of the following statements is not true?

(a) Addition of CH_4 (g) or O_2 (g) at equilibrium will cause a shift to the right
(b) The reaction is exothermic
(c) At equilibrium, the concentrations of CO_2 (g) and H_2O (l) are not equal
(d) The equilibrium constant for the reaction is given by $Kp = \frac{[CO_2]}{[CH_4][O_2]}$

[2006]

10. The enthalpy and entropy change for the reaction:

$Br_2\ (l) + Cl_2\ (g) \rightarrow 2BrCl\ (g)$
are 30 kJ mol^{-1} and 105 J K^{-1} mol^{-1} respectively. The temperature at which the reaction will be in equilibrium is

(a) 450 K (b) 300 K
(c) 285.7 K (d) 273 K

[2006]

11. The value of equilibrium constant of the reaction:

$HI\ (g) \rightleftharpoons \frac{1}{2} H_2\ (g) + \frac{1}{2} I_2$ is 8.0
The equilibrium constant of the reaction
$H_2\ (g) + I_2\ (g) \rightleftharpoons 2HI\ (g)$ will be

(a) 16 (b) 1/8
(c) 1/16 (d) 1/64

[2008]

12. If the concentration of OH^- ions in the reaction:

$Fe(OH)_3\ (s) \rightleftharpoons Fe^{3+}\ (aq) + 3OH^-\ (aq)$ is decreased by one-fourth times, then equilibrium concentration of Fe^{3+} will increase by

(a) 64 times (b) 4 times
(c) 8 times (d) 16 times

[2008]

13. The dissociation equilibrium of a gas AB_2 can be shown as:

$2AB_{2\,g} \rightleftharpoons 2AB_g + B_{2\,g}$

The degree of dissociation is 'X' and is small compared to 1. The expression relating the degree of dissociation 'X' with equilibrium constant K_P and total pressure 'P' is:

(a) $\frac{2K_P}{P}$ (b) $\left(\frac{2K_P}{P}\right)^{\frac{1}{3}}$
(c) $\left(\frac{2K_P}{P}\right)^{\frac{1}{2}}$ (d) $\left(\frac{K_P}{P}\right)$

[2008]

14. The dissociation constant for acetic acid and HCN at 25°C are 1.5×10^{-5} and 4.5×10^{-10} respectively. The equilibrium constant for the equilibrium

$\bar{C}N + CH_3COOH \rightleftharpoons HCN + CH_3COO^-$

(a) 3×10^{-5} (b) 3×10^{-4}
(c) 3×10^{4} (d) 3×10^{5}

[2009]

15. In which of the following equilibrium K_c and K_p are not equal?

(a) $2C_{(s)} + O_{2(g)} \rightleftharpoons 2CO_{2(g)}$
(b) $2NO_{(g)} \rightleftharpoons N_{2(g)} + O_{2(g)}$
(c) $SO_{2(g)} + NO_{2(g)} \rightleftharpoons SO_{3(g)} + NO_{(g)}$
(d) $H_{2(g)} + I_{2(g)} \rightleftharpoons 2HI_{(g)}$

[2010]

16. The reaction:

$2A_{(g)} + B_{(g)} \rightleftharpoons 3C_{(g)} + D_{(g)}$
has begun with the concentrations of A and B both at an initial value of 1.00 M. When equilibrium is reached, the concentration of D is measured and found to be 0.25 M. The value for the equilibrium constant for this reaction is given by the expression

(a) $[(0.75)^3(0.25)]/[(1.00)^2(1.00)]$
(b) $[(0.75)^3(0.25)]/[(0.50)^2(0.75)]$
(c) $[(0.75)^3(0.25)]/[(0.50)^2(0.25)]$
(d) $[(0.75)^3(0.25)]/[(0.75)^2(0.25)]$

[2010]

17. The value of DH for the reaction

$X_{2\,(g)} + 4Y_{2\,(g)} \rightleftharpoons 2\ XY_{4\,(g)}$

is less then zero formation of XY_4 will be favoured at?

(a) High temperature and high pressure
(b) High pressure and low temperature
(c) High temperature and low pressure
(d) Low pressure and low temperature

[2011]

18. Given that the equilibrium constant for the reaction:

$2SO_2\ (g) + O_2\ (g) \rightleftharpoons 2SO_{2(g)}$

has a value of 278 at a particular temperature. What is the value of the equilibrium constant for the following reaction at the same temperature?

$SO_3\,(g) \rightleftharpoons SO_2\,(g) + 1/2O_2\,(g)$

(a) 1.3×10^{-5} (b) 1.6×10^{-3}
(c) 6.0×10^{-2} (d) 3.6×10^{-3}

[2012]

19. Given the reaction between 2 gases represented by A_2 and B_2 to give the compound $AB_{(g)}$.

$A_2\,(g) + B_2\,(g) \rightleftharpoons 2AB\,(g)$

At equilibrium, the concentration
of $A_2 = 3.0 \times 10^{-3}M$
of $B_2 = 4.2 \times 10^{-3}M$
of $AB = 2.8 \times 10^{-3}M$

If the reaction takes place in a sealed vessel at 527°C, then the value of Kc will be:

(a) 2.0 (b) 4.5
(c) 0.62 (d) 1.9

[2012]

20. $KMnO_4$ can be prepared from K_2MnO_4
As per reaction
$3MnO_4^{2-} + 2H_2O \rightleftharpoons 2MnO_4^- + MnO_2 + 4OH^-$
The reaction can go to completion by removing OH^- ions by adding ?

(a) KOH (b) HCl
(c) CO_2 (d) SO_2

[2013]

21. Using the Gibb's energy change:
$\Delta G = +63.3$ KJ for the following reaction
$AgNO_{3(s)} \rightleftharpoons 2Ag^+_{(aq)} + CO_3^{2-}{}_{(aq)}$
The K_{SP} of $Ag_2CO_{3(s)}$ in water at 25°C is (R = 8.314 JK^{-1} $mole^{-1}$)

(a) 3.2×10^{-26} (b) 8×10^{-12}
(c) 2.9×10^{-3} (d) 7.9×10^{-2}

[2014]

22. For the reversible reaction:
$N_{2(g)} + 3H_{2(g)} \rightleftharpoons 2NH_{3(g)} + \text{Heat}$
The equilibrium shifts in forward direction
(a) By increasing the concentration of $NH_{3(g)}$
(b) By decreasing the pressure
(c) By decreasing the concentation of $N_{2(g)}$ and $H_{2(g)}$
(d) By increasing pressure and decreasing temperature

[2014]

23. If the equilibrium constant for:
$N_{2(g)} + O_{2(g)} \rightleftharpoons 2NO_{(g)}$ is K,
The equilibrium constant for
$\frac{1}{2}N_{2(g)} + \frac{1}{2}O_{2(g)} \rightleftharpoons NO_{(g)}$ will be

(a) $K^{1/2}$ (b) $\frac{1}{2}K$
(c) K (d) K^2

[2015]

24. If the value of an equilibrium constant for a particular reaction is 1.6×10^{12}, then at equilibrium the system will contain:
(a) Mostly reactants
(b) Mostly products
(c) All reactants
(d) Similar amount of reactants and products

[2015]

25. If E°_{cell} for a given reaction has a negative value, which of the following gives correct relationship for the value of ΔG° and K_{eq} ?
(a) $\Delta G^\circ > 0, K_{eq} > 1$
(b) $\Delta G^\circ > 0, K_{eq} < 1$
(c) $\Delta G^\circ < 0, K_{eq} < 1$
(d) $\Delta G^\circ < 0, K_{eq} > 1$
(d) Similar amount of reactants and products

[2016]

ANSWER KEYS

1. (a)	2. (c)	3. (d)	4. (c)	5. (c)	6. (a)	7. (b)	8. (a)	9. (d)	10. (c)
11. (d)	12. (a)	13. (b)	14. (c)	15. (a)	16. (b)	17. (b)	18. (c)	19. (c)	20. (a)
21. (b)	22. (d)	23. (a)	24. (b)	25. (b)					

Hints and Explanations

1. $MgCO_3$ (s) ↔ MgO (s) + CO_2 (g)
MgO and $MgCO_3$ are solid and they do not exert any pressure and hence only pressure exerted in by CO_2 so $Kp = Pco_2$

2. In the $2SO_2 + O_2 \leftrightarrow 2SO_3$,
$\Delta n = 2–3 = –1$.
In this reaction Δn_g is negative so the value of Kp is less than K_c.

3. As equilibrium constant is independent of concentration or pressure so on doubling concentration it will remain unchanged.

4. Reaction
BaO_2 (s) ↔ BaO (s) + O_2 (g); ΔH = +ve. In equilibrium $Kp = Po_2$
Hence, the equilibrium is based upon partial pressure of O_2 (Po_2) and it is based upon increasing of temperature, formation of O_2 (g) increases.

5. Forward reaction, $Kc = \frac{[NH_3]^2}{[N_2][H_2]^3}$ for backward reaction $Q = \frac{[NH_3]^2}{[N_2][H_2]^3}$ for reaction of proceed from right to left
Q (backward rate) > Kc (forward rate)

6. The equilibrium constant of the reaction

$$Kc = \frac{[HI]^2}{[H_2][I_2]} = \frac{(28)^2}{(8)\times(3)} = 32.67$$

7. For equilibrium

$$N_2 (g) + 3H_2 (g) \rightleftharpoons 2NH_3 (g)\ K_1 = \frac{[NH_3]}{[N_2][H_2]^3} \quad (1)$$

$$N_2 (g) + O_2 (g) \rightleftharpoons 2NO (g)\ K_2 = \frac{[NO]^2}{[N_2][O_2]} \quad (2)$$

$$H_2 (g) + \frac{1}{2} O2 (g) \rightleftharpoons H_2O (g)\ K_3 \frac{[H_2O]^2}{[H_2][O_2]^{1/2}} \quad (3)$$

For reaction

$$2NH_3 (g) + 5/2\ O_2 (g) \rightleftharpoons 2NO (g) + 3H_2O (g)\ K = \frac{[NO]^2[H_2O]^3}{[NH_3]^2[O_2]^{5/2}} \quad (4)$$

From equation (1), (2) and (3)

$$K = \frac{K_2 \times K_3}{K_1}$$

8. $NO (g) + \frac{1}{2} O_2 \xleftrightarrow{K_1} NO_2$ (g) and
$2NO (g) \xleftrightarrow{K_2} 2NO (g) + O_2$

$$K_1 = \frac{[NO_2]}{[NO][O_2]^{1/2}}$$

$$K_1^2 = \frac{[NO_2]^2}{[NO]^2[O_2]} \quad (1)$$

$$K_2 = \frac{[NO]^2[O^2]}{[NO^2]^2} \quad (2)$$

from equation (1) and (2)

$K_2 = 1/K_1^2$

9. As the equilibrium constant for the reaction is given by $Kp = \frac{[CO_2]}{[CH_4][O_2]}$

10. At equilibrium ΔG = 0.
So ΔH = TΔS
T = ΔH/ΔS
= 30 × 10^{-3}/105
= 285.7 K

11. $HI (g) \rightleftharpoons \frac{1}{2} H_2 (g) + \frac{1}{2} I_2 (g)$

$$K = \frac{[H_2]^{1/2}[I_2]^{1/2}}{[HI]} = 8$$

$H_2 (g) + I_2 (g) \rightleftharpoons 2HI (g)$

$$K' = \frac{[HI]^2}{[H_2][I_2]} = (1/8)^2$$

K’ = 1/64.

12. $Fe(OH)_3 (S) \rightleftharpoons Fe^{3+} (aq) + 3OH^- (aq)$

$$K = \frac{[Fe^{3+}][OH^-]^3}{[Fe(OH)_3]}$$

$K = [Fe^{3+}] [OH^-]^3$

(as activity of solid is taken unity)
Concentration OH^- ion in the reaction is decreased by one-fourth times then equilibrium concentration of Fe^{3+} will be increased by 64 times in order to keep the value of K constant.

13. $2AB_2 \rightleftharpoons 2AB + B_2$ (g, g, g)

	$2AB_2$	$2AB$	B_2
Initially	2 moles	0	0
At equilibrium	2 – 2x	2x	x

Total moles = 2 – 2x + 2x + x

= (2 + x)

$$P_{AB_2} = \frac{2-2x}{2+x}.P$$

$$P_{AB} = \frac{2x}{2+x}, P$$

$$P_{B_2} = \frac{x}{2+x}.P$$

$$K_P = \frac{(P_{AB})^2.P_{B_2}}{(P_{AB_2})^2}$$

$$K_P = \frac{\left(\frac{2x}{2+x}P\right)^2\left(\frac{x}{2+x}P\right)}{\left(\frac{2-2x}{2+x}P\right)^2} = \frac{4\times 3}{(2+x)\,(2-2x)^2}P$$

As 2 >>> x so

$$K_P = \frac{4\times 3}{2.(2)^2}.P$$

$$x^3 = \frac{2K_P}{P}$$

$$x = \left(\frac{2K_P}{P}\right)^{\frac{1}{3}}$$

14. As

$$CH_3COOH \rightleftharpoons CH_3CHOO^- + H^+$$

$$K_1 = 1.5 \times 10^{-15} = \frac{[CH_3COO^-][H^+]}{[CH_3COOH]}$$

$$HCN \rightleftharpoons H^+ + CN^-$$

$$K_2 = 4.5 \times 10^{-10} = \frac{[H^+][CN^-]}{[HCN]}$$

Now $CN^- + CH_3COOH \rightleftharpoons CH_3COO^- + HCN$

$$K = \frac{[HCN][CH_3COO]}{(CN^-)\,(CH_3COOH)}$$

$$K = \frac{1}{K_2}.K_1$$

$$= \frac{1}{4.5\times 10^{-10}} \times 1.5\times 10^{-5}$$

$$= 3.3 \times 104^4$$

18. $K' = \left(\frac{1}{K}\right)^{1/2} = \left(\frac{1}{278}\right)^{1/2} = 6\times 10^{-2}$

19. $A_2 + B_2 \rightleftharpoons 2AB$

$$K_c = \frac{[AB]^2}{[A_2][B_2]} = \frac{2.8\times 2.8\times 10^{-6}}{3\times 10^{-3}\times 4.2\times 10^{-3}} = 0.62$$

20. Here $\bar{OH}$ are from very weakly acidic H_2O so must be removed by using weak acid CO_2 as any strong acid like HCl can easily reverse the reaction KOH will increase $\bar{OH}$ concentration so again shifts the reaction back work.

21. $\Delta G° = -2.303\ RT \log_{10} K_{SP}$

$63.3 = -2.303 \times 8.314 \times 10^{-3} \times 298, \log_{10} K_{SP}$

On solving $\log_{10} K_{SP} = -11.09$

K_{SP} = Antilog (–11.09)

$K_{SP} = 8 \times 10^{-12}$

22. $N_2 + 3H_2 \rightleftharpoons 2NH_3$ + Heat

Using Lechatlier's principle

As $n_p(2) < n_p(4)$ that is

Volume is decreasing, So high pressure favours forward reaction

As ΔH = –ve, i.e exothermic reaction so low temperature is favourable.

23. $N_2 + O_2 \rightleftharpoons 2NO$

$$K = \frac{(NO)^2}{(N_2)\,(O_2)}$$

For reaction $\frac{1}{2}N_2 + \frac{1}{2}O_2 \rightleftharpoons NO$

$$K_C = \frac{(NO)}{(N_2)^{1/2}\,(O_2)^{1/2}} = \sqrt{K}$$

24. Amount of product $\propto K_c$ as K_e is 1.6×10^{12} a very high value so reaction is almost completed and there is almost 100% product.

25. As $\Delta G° = -n\,F\,E°_{cell}$ As $E°_{cell}$ = –ve

It means $\Delta G°$ > zero or (+ve)

Now $\Delta G° = -2.303\ nRT \log_{10} K_{eq}$

As $\Delta G°$ > zero or +ve

So $K_{eq} < 1$

AIIMS ESSENTIALS

Assertion and Reason

In the following questions, two statements (Assertion) A and Reason (R) are given. Mark

(a) If A and R both are correct and R is the correct explanation of A;
(b) If A and R both are correct but R is not the correct explanation of A;
(c) A is true but R is false;
(d) A is false but R is true.
(e) both A and R are false.

1. (A) : For a reaction at equilibrium, the free energy for the reaction is minimum.
(R) : The free energy for both reactants and products decreases and become equal.

2. (A) : The endothermic reactions are favored at lower temperature and the exothermic reactions are favoured at higher temperature.
(R) : When a system in equilibrium is disturbed by changing the temperature, it will tend to adjust itself so as to overcome the effect of change.

3. (A) : The reaction
$2NO\ (g) + O_2\ (g) \leftrightarrow 2NO_2$ is favoured in the forward direction with increase of pressure.
(R) : The reaction is exothermic

4. (A) : Concentration of the reactant and product does not change with time at equilibrium for a chemical reaction.
(R) : The rate of a reaction is zero at equilibrium.

5. (A) : In the Haber process,
$N_2 + 3H_2 \leftrightarrow 2NH_3$
Pressure is about 200 atm in presence of catalyst and temperature is kept 500°C even the reaction is exothermic.
(R) : Energy needed for this reaction is easily obtained at this temperature .

6. (A) : When $CaCO_3$ (s) is heated, the loss of CO_2 (g) from the system causes the reaction to go almost to completion to leave a residue of CaO (s).
(R) : Heating causes gas particles to move with more energy.

7. (A) : When the following equilibrium is studied in a vessel of twice the volume of the earlier one,
$A\ (g) \leftrightarrow B\ (g) + C\ (g)$
equilibrium constant is decreased.
(R) : Equilibrium constant

$$Kc = \frac{x^2}{(1-X)\,V}$$

8. (A) : For the reaction
$A + B \leftrightarrow C$, Kc = 2. If we start with 1 mol each of A, B, and C in 5 L flask, then at equilibrium, molar concentration of A and B decreases and that of C increases.
(R) : Reaction quotient Q > Kc with given quantity of A, B, and C hence reaction is reversed.

9. (A) : The equilibrium constant for a reaction having positive ΔH° increases with increase of temperature.
(R) : The temperature dependence of the equilibrium constant is related to ΔH° and not ΔH° for the reaction.

10. (A) : The value of K gives us a relative idea about the extent to which a reaction proceeds.
(R) : The value of K is independent of the stoichiometry of reactants and products at the point of equilibrium.

11. (A) : On opening a sealed soda bottle dissolved carbon dioxide gas escapes.
(R) : Gas escapes to each the new equilibrium condition of lower pressure.

12. (A) : The equilibrium constant is fixed and a characteristic for any given chemical reaction at a specified temperature.
(R) : The composition of the final equilibrium mixture at a particular temperature depends upon the starting amount of reactants.

13. (A) : Ice melts slowly at higher altitudes.
(R) : The melting of ice is favoured at high pressure because ice → water shows decrease in volume.

14. (A) : When $Q = K_c$, reaction is at equilibrium
(R) : At equilibrium ΔG is 0.

15. (A) : Dissociation of PCl_5 decreases in presence of Cl_2
(R) : An increase in concentration of Cl_2 favors backward reaction to decrease dissociation of PCl_5

16. (A) : Water boils at high temperature in pressure cooker
(R) : Inside the pressure cooker pressure is less than 1 atm.

17. (A) : Some hydrated salts like blue vitriol dissolve in water with absorption of heat
(R) : Le Chatellier principle is not valid for solid – solid heterogeneous systems.

18. (A) : For every chemical reaction at equilibrium standard Gibbs energy of reaction is zero.
(R) : At constant temperature and pressure, chemical reactions are spontaneous in the direction of deceasing Gibbs energy.

Answer Keys

1. (a)	**2. (d)**	**3. (b)**	**4. (c)**	**5. (b)**	**6. (b)**	**7. (d)**	**8. (a)**	**9. (c)**	**10. (c)**
11. (a)	**12. (c)**	**13. (a)**	**14. (b)**	**15. (a)**	**16. (c)**	**17. (b)**	**18. (d)**		

CHAPTER

7B

Ionic Equilibrium

Chapter Outline

■ Types of Substances ■ Arrhenius Theory ■ Ostwald Dilution Law ■ Ionic Product of Water ■ Buffer Solution ■ Solubility and Solubility Product ■ Salt Hydrolysis ■ Acid and Base

Types of Substances

Substances can be classified into following two types, *viz.*, non-eletrolytes and electrolytes.

Non-electrolytes

Non-electrolytes do not conduct electricity in aqueous solution or molten state, for example, solution of urea, glucose, sugar, glycerine.

Electrolytes

Electrolytes conduct electricity in aqueous solution or molten state, for example, NaCl. These can be further divided into following types:

Strong electrolytes These get strongly ionized in water, hence they exhibit high conduction, for example, strong acids like HCl, H_2SO_4, HNO_3, strong bases like KOH, NaOH and salt of strong acid or strong base like NaCl, CH_3COONa, NH_4X.

Weak electrolytes These are ionized in water less strongly, hence they show less conduction, for example, weak acids like CH_3COOH, HCN, H_3PO_4, H_2CO_3, weak bases like NH_4OH and their salts like NH_2CN, CH_3COONH_4.

Arrhenius Theory

Arrhenius theory is also known as introduction of ionic theory or dissociation of electrolytes theory. The main postulates of this theory are:

- Electrolytes on dissolving in a solvent, ionize into ions.

$$AB \underset{H_2O}{\overset{\text{Solvent}}{\rightleftharpoons}} \underset{\text{Ions}}{A^+ + B^-}$$

[Ionization]

- The process of ionization is reversible and the ionization constant (K) is given as

$$K = \frac{[A^+]\,[B^+]}{[AB]}$$

- The ions are always discharged in equivalent amount, no matter what their relative speeds are during electrolysis.
- Ions act like molecules towards elevating the boiling point, depressing the freezing point, lowering vapour pressure and osmotic pressure.

Evidence in Support of Ionic Theory

- Presence of ions in solid electrolytes can be noticed by X-ray diffraction method.
- Ions obey Ohm's law (I = E/R)
- Ionic reactions, for example,

$$AgNO_3 + NaCl \longrightarrow Ag^+\,Cl^- \downarrow + Na^+NO_3^-$$

- Colour of some solutions is due to the presence of ions, for example, $CuSO_4$ solution is blue due to Cu^{2+}.

Degree of Ionization

Degree of ionization is the extent to which an electrolyte gets ionized in a solvent. It is denoted by α.

$$\alpha = \frac{\text{Number of molecules dissociated}}{\text{Total number of molecules}}$$

Degree of ionization or α depends on:

- Nature of solute and solvent. For strong electrolytes α is more than that for weak electrolytes.
- $\alpha \propto$ Dielectric constant of solvent
  Greater the dielectric constant of a solvent, more will be ionization of electrolyte in it.
- The degree of dissociation of weak electrolyte is directly proportional to dilution, that is, α is maximum at infinite dilution.
- $\alpha \propto \frac{1}{\text{Concentration}}$
- $\alpha \propto$ Temperature

Ostwald Dilution Law

- Ostwald dilution law is the law of mass action for weak electrolytes and dilute solutions and gives the relationship between the dissociation of a weak electrolyte.
- It is not applicable for strong electrolytes or highly concentrated solutions.
  Consider a weak electrolyte AB in V litres of a solution.

	$AB \rightleftharpoons$	A^+ +	B^-
Initial moles	1	0	0
Moles at equilibrium	$1-\alpha$	α	α
Concentration at equilibrium	$(1-\alpha)/V$	(α/V)	(α/V)

$$K = \frac{[B^-][A^+]}{[AB]}$$

$$K = \frac{(\alpha/V)\times(\alpha/V)}{(1-\alpha)/V}$$

$$\text{So, } K = \frac{\alpha^2/V^2}{(1-\alpha)/V} = \frac{\alpha^2}{(1-\alpha)/V}$$

As $1 >>> \alpha$ so $(1-\alpha) = 1$

$$K = \frac{\alpha^2}{V} \text{ or } \alpha^2 = KV \quad \text{or } \alpha = \sqrt{KV}$$

$$\alpha \propto \sqrt{V}$$

Hence, at constant temperature, degree of dissociation is directly proportional to the square root of its dilution.

- If C is the concentration then

$$AB \rightleftharpoons A^+ + B^-$$

$$C(1-\alpha) \quad C\alpha \quad C\alpha$$

$$K = \frac{C\alpha^2}{(1-\alpha)}$$

$K = C\alpha^2$ or $\alpha^2 = K/C$

$\alpha = \sqrt{(K/C)}$

that is, $\alpha \propto \sqrt{(1/C)}$

Hence, at constant temperature, degree of dissociation is inversely proportional to the square root of its concentration.

- $\alpha = \frac{\lambda_V}{\lambda_\infty}$ or $\frac{\Lambda_V}{\Lambda_\infty}$
  Here λ_v or Λ_v is equivalent conductivity at V dilution. λ_∞ or Λ_∞ is equivalent conductivity at infinite dilution.
- $\lambda_v = K_v \times V$
  Here K_v = Specific conductivity
  V = Dilution
- $\lambda_\infty = \lambda_c + \lambda_\alpha$
  Here λ_c and λ_α are the ionic mobilities of cation and anion respectively.

Common Ion Effect

The value of degree of dissociation for a weak electrolyte is decreased by the addition of a strong electrolyte having a common ion.

- As a result of this effect, the concentration of the uncommon ion of the weak electrolyte decreases.
  Examples:
  (1) $NH_4OH \rightleftharpoons NH_4^+ + OH^-$ (Weak)
  $NH_4Cl \rightleftharpoons NH_4^+ + Cl^-$ (Common ion)
  Here the value of α for NH_4OH will be decreased by NH_4Cl.
  (2) $CH_3COOH \rightleftharpoons CH_3COO^- + H^+$
  $CH_3COONa \rightleftharpoons CH_3COO^- + Na^+$ (Common ion)
  Here the value of α for CH_3COOH will be decreased by CH_3COONa.
- Solubility of a partially soluble salt decreases due to common ion effect, for example, presence of $AgNO_3$ or KCl, decreases the solubility of AgCl in water.
- Salting out of soap by addition of NaCl.
- Purification of NaCl by passing HCl gas.

Isohydirc Solution

These are solutions having same concentration of common ions.

Ionic Product Of Water

Ionic product of water is the product of the molar concentrations of H^+ or H_3O^+ and OH^- ions. It is denoted by K_w.

$H_2O + H_2O \rightleftharpoons H_3^+O + OH^-$
$K_w = [H_3O^+] . [OH^-]$
or $[H^+]. [OH^-]$
Here K_w = Ionic product of water

- $K_w = K_a . K_b$
- $pK_w = -\log_{10} K_w$
- $K_w = K_a \times K_b$
- $pK_w = pK_a + pK_b$
- At 25°C $K_w = 1 \times 10^{-14}$
  $pK_w = 14$
- The value of K_w increases with increase in temperature, for example, at 98°C K_w is 1×10^{-12}.

pH Scale

- pH scale was introduced by Sorenson, to measure acidity or basicity of a solution.
- pH stands for *potentiel de-H*$^+$ or concentration of H^+.
  It is given as:
  $[H^+] = 10^{-pH}$
  $pH = -\log_{10} [H^+]$
  pH of aqueous solution is equal to negative logarithm of H^+ (H_3^+O) concentration in mole/litre.
  $pH = \log_{10} \frac{1}{[H^+]}$

pH of weak acid

CH_3COOH	$\rightleftharpoons$	CH_3COO^- +	H^+
1		0	0
$C(1-x)$		Cx	Cx

$K = Cx^2$
$x = \sqrt{(K_a/C)}$
$[H^+] = Cx = C\sqrt{(K_a/C)} \approx \sqrt{(K_a/C)}$
$pH = -\log_{10} Cx$
$pH = -\log_{10} \sqrt{(K_a/C)}$
Here C = Molar concentration of acid
x = Degree of dissociation
K_a = Dissociation constant of acid

pH of weak base

NH_4OH	$\rightleftharpoons$	NH_4^+	+	OH^-
1		0		0
$(1-x)$		Cx		Cx

$[OH^-] = Cx = \sqrt{(K_b.C)}$
$pOH = -\log_{10} Cx$
$pOH = -\log_{10} \sqrt{(K_b.C)}$
Here K_b = Dissociation constant of the base

pOH

$[OH^-] = 10^{-pOH}$
$pOH = -\log_{10} [OH^-]$ or $\frac{1}{\log_{10}[OH^-]}$

- $pH + pOH = 14$
- $pH + pOH = pK_w$
- $pK_a = -\log_{10} K_a$
- $pK_a \propto \frac{1}{K_a} \propto \frac{1}{\text{Acidic strength}}$
- $pK_b = -\log_{10} K_b$
- $pK_b \propto \frac{1}{K_b} \propto \frac{1}{\text{Basic strength}}$
- pK = −ve log of dissociation constant
- A weak acid has high value of pK_a
- A weak base has high value of pK_b

Facts to Remember

- pH of strong acid or base does not depend upon temperature.
- pH of weak acid decreases with increase in temperature, due to increase in ionization.
- pH of weak base increases with increase in temperature, due to increase in ionization or $[OH^-]$ ion concentration.

pH Value and Nature of a Solution

- If $[H^+] > 10^{-7}$, pH is less than 7 and the solution is acidic.
- If $[H^+] = 10^{-7}$, pH is 7 and the solution is neutral.
- If $[H^+] < 10^{-7}$, pH is more than 7 and the solution is basic.

Table 7.1 pH Range of Some Substances

Substance	pH range
Gastric juice	1–3.0
Soft drink	2–4.0
Lemon	2.2–3.4
Vinegar	2.2–2.4
Urine	4.8–8.4
Milk	6.3–6.6
Saliva	6.5–7.5
Blood	7.3–7.5
Sea water	8.5
Tears	7.4

Limitations of pH scale

- pH value of a solution does not instantaneously give us an idea of the relative strength of the solution.
- pH is zero for 1N solution of strong acid.
- pH is negative for concentrations 2N, 3N, 10N of strong acids.
- At higher concentrations, in place of pH, **Hammelt acidity functions** are used.

Buffer Solution

- Buffer solution is the solution whose pH, on addition of a small amount of strong acid or a base, does not change much or noticeably.
- It is also called **reserve acidity** or **basicity** of the solution, as it resists change in pH value and this action to resists pH change is called **buffer action**.

Features

- It has a definite pH, that is a definite reserve acidity or basicity.
- Its pH does not change over a long period time.
- Its pH does not change on dilution.
- Its pH changes slightly (unnoticeable) on the addition of small quantity of a strong acid or base.

Buffer Capacity

Buffer capacity is the ratio of number of moles of the acid or base added in one litre of the solution, so as to change its pH by unity. It is denoted by φ.

$$\varphi = \frac{\text{Number of moles of acid/base added to 1 L of solution}}{\text{Change in pH}}$$

Types of Buffer Solutions

1. **Acidic buffer** It is a solution of a weak acid and its salt with a strong base. For example,
   - $CH_3COOH + CH_3COONa$
   - Phthalic acid + potassium phthalate
   - Boric acid + borax
   - $H_2CO_3 + NaHCO_3$
   - Citric acid + sodium citrate
2. **Basic buffer** It is a solution of a weak base and it salt with a strong acid. For example,
   - $NH_4OH + NH_4Cl$
   - Glycerine + Glycerine hydrogen chloride
3. **Ampholytes** These are compounds that contain both acidic and basic groups and therefore exist as **zwitter-ions** at a certain pH (isoelectric point). Hence, proteins and amino acids also act like a buffer solution.

Note: Zwitterions are ions that contain both positive and negative charges and are thus amphotieric.

4. A mixture of acidic salt and normal salt of a polybasic acid. For example,
   $Na_2HPO_4 + Na_3PO_4$
5. Salt of weak acid and weak base (in water).
   For example, CH_3COONH_4, NH_4CN

Uses of Buffer solutions

- To determine pH value of unknown solution with the help of an indicator.
- $NH_4Cl + NH_4OH$ and $(NH_4)_2CO_3$ buffer solutions are used to precipitate carbonates of group V elements of the periodic table group.
- CH_3COOH and CH_3COONa buffer is used to remove PO_4^{-3} in qualitative inorganic analysis after IInd group.
- CH_3COOH and CH_3COONa buffer is used to precipitate lead chromate quantitatively in gravimetric analysis.
- They are also used in paper, dye, printing ink, dairy product.

Solubility and Solubility Product

Solubility

- Solubility is the weight of solute in grams, present in 100 mL of solvent.
- It is denoted by s and is expressed in mole per litre or gram per litre.
- Solubility (s) $\propto \dfrac{1}{\text{Concentration of common ions}}$
  ar number of common ions

Solubility increases due to complex ion formation. For example, the solubility of AgCl in water, in presence of $AgNO_3$.

Here, AgCl has more solubility in ammonia, due to complex formation.

$$AgCl + 2NH_3 \longrightarrow Ag(NH_3)_2Cl$$

$HgCl_2$ has more solubility in KI due to complex formation (Nesseler's reagent).

$$HgI_2 + 2KI \longrightarrow K_2HgI_4$$

In the analysis of IInd group elements, SnS, Sb_2S_3, As_2S_3 are soluble in $(NH_4)_2S$ due to complex ion formation.

$$SnS + S^{-2} \longrightarrow SnS_2^{-2}$$

$$Sb_2S_3 + 3S^{-2} \longrightarrow 2SbS_3^{-3}$$

$$As_2S_5 + 3S^{-2} \longrightarrow 2AsS_4^{-3}$$

Simultaneous solubility It is solubility of a solution of two electrolytes having common ions.

Examples:

1. AgBr + AgSCN
2. $CaF_2 + SrF_2$
3. $MgF_2 + CaF_2$

Solubility Product

Solubility product is the product of the molar concentrations of ions of an electrolyte in saturated solution at a particular temperature. It is denoted by K_{sp} or S.

For a binary electrolyte AB

$$\underset{\text{Solid}}{AB} \rightleftharpoons \underset{\text{Un-ionized}}{AB} \rightleftharpoons A^+ + B^-$$

$$\text{So, } K = \frac{[A^+][B^-]}{[AB]}$$

or $K.[AB] = [A^+]\,[B^-]$

$K_{sp} = [A^+][B^-]$ (At constant temperature)

General representation

$$A_xB_y \rightleftharpoons xA^{+y} + yB^{-x}$$

$$K_{sp} = [A^{+y}]^x\,[B^{-x}]^y$$

Examples:

1. $CaF_2 \rightleftharpoons Ca^{+2} + 2F^-$

   $K_{sp} = [Ca^{+2}]\,[F^-]^2$

2. $Ag_2CrO_4 \rightleftharpoons 2Ag^+ + CrO_4^{-2}$

   $K_{sp} = [Ag^+]^2\,[CrO_4^{-2}]$

Relation Between Solubility (s) and Solubility Product (K_{sp})

A_xB_y	$\rightleftharpoons$	xA^{+y}	+	yB^{-x}
a		0		0
a – s		x s		y s

$K_{sp} = (x\,s)^x\,(y\,s)^y$

or

$K_{sp} = x^x\,y^y\,(s)^{x+y}$

If α is given:

$K_{sp} = x^x\,y^y\,(\alpha\,s)^{x+y}$

For example, $Cu_2Cl_2 \rightleftharpoons \underset{2s}{2Cu^+} + \underset{2s}{2Cl^-}$

So $K_{sp} = 2^2\,2^2\,(s)^{2+2} = 16\,s^4$

$Fe(OH)_3 \rightleftharpoons \underset{s}{Fe^{+3}} + \underset{3s}{3OH^-}$

$K_{sp} = 1^1\,3^3\,(s)^{1+3} = 27\,s^4$

$Al_2(SO_4)_3 \rightleftharpoons \underset{2s}{2Al^{+3}} + \underset{3s}{3SO_4^{-2}}$

$K_{sp} = 2^2\,3^3\,(s)^{2+3} = 108\,s^5$

$Na_3Li_3(AlF_6)_2 \rightleftharpoons 3Na^+ + 3Li^{+3} + 2AlF_6^{-3}$

$K_{sp} = 3^3\,3^3\,2^2\,(s)^{3+3+2} = 2916\,s^8$

Solubility Product and Precipitation

- If $K_{sp} \approx$ ionic product, the solution is saturated, and for precipitation, more solute should be added.
- If ionic product > K_{sp}, the solution is super saturated, and therefore gets easily precipitated.
- If ionic product < K_{sp}, the solution is unsaturated and therefore no precipitation takes place.

Salt Hydrolysis

- When a salt is added to water, ions of the salt interact with water to cause acidity or basicity in aqueous solution. This ionic interaction is called **salt hydrolysis**.
- Interaction of cation is cationic hydrolysis and interaction of anion is anionic hydrolysis.
- Hydrolysis is reverse of neutralization and is an endothermic process.
- If hydrolysis constant is K_h and neutralization constant is K_n, then $K_n = 1/K_h$.
- A solution of the salt of strong acid and weak base is acidic and for it pH < 7 or $[H^+] > 10^{-7}$, for example, $FeCl_3$ (salt of a weak base + strong acid). Here, the solution is acidic and involves cationic hydrolysis.
- A solution of the salt of strong base and weak acid is basic and for it, pH > 7 or $[H^+] < 10^{-7}$, for example, KCN (salt of a strong base + weak acid). Here, the solution is basic and involves anionic hydrolysis.
- A solution of the salt of a weak acid and a weak base is:
  1. Acidic, if $K_a > K_b$
  2. Basic, if $K_a < K_b$
  3. Neutral, if $K_a = K_b$ is neutral
- CH_3COONH_4 (salt of a weak acid + weak base) Here, the solution is neutral and involves both cationic and anionic hydrolysis.

- A solution of the salt of strong acid and strong base is neutral or pH = 7 or $[H^+] = 10^{-7}$
- A salt of a strong acid and a strong base is never hydrolyzed.

  However, ions are hydrated, for example K_2SO_4 (salt of a strong base + strong acid)

Degree of Hydrolysis

It is the amount or fraction of the salt which is hydrolyzed.
It is denoted by h.

$$h = \frac{\text{Number of moles of salt hydrolyzed}}{\text{Total moles of the salt taken}}$$

Various Expressions for K_h, h and pH for Different Type of Salts

1. For the salt of weak acid and strong base like KCN,

$$K_h = \frac{K_w}{K_a}$$
$$h = \sqrt{(K_h/C)}$$
$$h = \sqrt{(K_w/K_a \cdot C)}$$
$$pH = \frac{1}{2}[pK_w + pK_a + \log C]$$
$$pOH = \frac{1}{2}[pK_w - pk_a - \log C]$$

2. For the salt of strong acid and weak base like $FeCl_3$,

$$K_h = \frac{K_w}{K_b}$$
$$h = \sqrt{(K_h/C)}$$
$$h = \sqrt{(K_w/K_b \cdot C)}$$
$$pH = \frac{1}{2}[pK_w - pK_b - \log C]$$

3. For the salt of weak acid and weak base like CH_3COONH_4,

$$K_h = \frac{K_w}{K_a K_b}$$
$$h = \sqrt{(K_w/K_a K_b)}$$
$$pH = \frac{1}{2}[pK_w + pK_a - pK_b]$$

Acid and Base

Arrhenius Concept of Acid and Base

According to arrhenius concept, "Acids are H^+ ion donor in water and bases are OH^- ion donor in water."

Acid

- $HA + H_2O \longrightarrow A^- + H_3O^+$
  where H_3O^+ hydronium ion or hydrated proton.
- Water can accept H^+ to form hydronium ion (H_3O^+) due to hydrogen bonding. For example, $[H_5O_2^+, H_7O_3^+]$
- H^+ can hold water molecule by hydrogen bond, as it has high heat of hydration. For example, HNO_3, H_2SO_4, HX
- H_3O^+ has sp^3 hybridization and is trigonal pyramidal in shape.
- Number of H^+ donated = Basicity or protosity of the acid

  Examples:

  1. H_3PO_4 (tribasic)

     O
     ↑
     H–O–P–O–H
     ½
     O–H

  2. H_3PO_3 (dibasic)

     H
     ½
     H–O–P–O–H
     ↓
     O

  3. H_3PO_2

     O
     ↑
     H–P–H
     ½
     O–H

Base

- $B.OH + H_2O \longrightarrow B^+ + H_3O_2^-$ [or $H_5O_3^-$]
  For example, NaOH, KOH.

Strength of Acid and Base

1. Acidic strength $\propto K_a$
   K_a = Dissociation constant of the acid
2. Base strength $\propto K_b$
   K_b = Dissociation constant of the base
3. Relative strength = $\sqrt{(K_1/K_2)}$

$$H_3PO_4 \overset{K_1}{\rightleftharpoons} H_2PO_4^- + H^+ \text{ or } H_3O^+$$
$$H_2PO_4^- \overset{K_2}{\rightleftharpoons} HPO_4^{-2} + H^+$$
$$HPO_4^- \overset{K_3}{\rightleftharpoons} PO_4^{-3} + H^+$$

Here $K_1 > K_2 >> K_3$ so the correct acidic strength order is $H_3PO_4 > H_2PO_4^- > HPO_4^{-2}$

4. As negative charge develops the removal of H^+ becomes more and more difficult so acidic nature decreases.

FACTS TO REMEMBER

- This concept fails to explain acidic and basic nature of $AlCl_3$, BX_3, NH_3.

Bronsted Lowery Concept or Proton Concept

Acid

Acids are proton or H^+ donor.

$$\underset{\text{Acid}}{HA} \longrightarrow \underset{\text{Conjugate base}}{A^- + H^+}$$

For example, $HX \longrightarrow H^+ + \underset{\text{Conjugate base}}{X^-}$

$$\underset{\text{Acid}}{H_2SO_4} \longrightarrow H^+ + \underset{\text{Conjugate base}}{HSO_4^-}$$

$$HNO_3 \longrightarrow H^+ + NO_3^-$$

Base

Bases are proton or H^+ acceptor

$$\text{Base} + \underset{\text{Proton given by acid}}{H^+} \longrightarrow \underset{\text{Conjugate base}}{(\text{Base H})^+}$$

For example, $\underset{\text{Base}}{OH^-} + H^+ \longrightarrow \underset{\text{Conjugate base}}{H_2O}$

Amphoteric or Ampholyte Substances

Such a substance can behave both like an acid and a base that is, it can donate or accept H^+ or proton.

Examples:

1. $\underset{\text{Conjugate acid}}{NH_4^+} \xleftarrow{+H^+} \underset{\text{Ampholyte}}{NH_3} \xrightarrow{-H^+} \underset{\text{Conjugate base}}{NH_2^-}$

2. $\underset{\text{Conjugate acid}}{H_3O^+} \xleftarrow{+H^+} \underset{\text{Ampholyte}}{H_2O} \xrightarrow{-H^+} \underset{\text{Conjugate base}}{OH^-}$

Some other examples are HSO_4^-, HCO_3^-, $H_2PO_4^-$, HPO_4^{2-}, $H_2PO_3^-$, HS^- and $HC_2O_4^-$

FACTS TO REMEMBER

- $H_2PO_2^-$ and HPO_3^{2-} are only bronsted bases. As H_3PO_2 and H_3PO_3 are monobasic and dibasic acid respectively, they can release only one and two H^+ ions respectively.

Lewis Concept of Acid and Base

Lewis acids are electron deficient species which can accept an electron pair. Following are the species which act as Lewis acids:

1. Cations
   $-\overset{|}{C}^+-$, NO_2^+, X^+ (Halonium ion)
2. Electron deficient central atoms

BeX_2,	BX_3,	FeX_3,	AlX_3,	$SnCl_2$,	$ZnCl_2$
↓	↓	↓	↓	↓	↓
4	6	6	6	4	4

3. Central atom with multiple bonds only
   O=C=O, O=S=O, S=C=S, SO_3
4. If in a molecule, central atom has vacant d orbitals, it can have more than eight electrons, that is, an expansion of octet state occurs and it can behave as a Lewis acid.
   For example,
   (1) $\underset{\text{Lewis acid}}{PX_5} + X^- \longrightarrow \underset{\text{Lewis base}}{PX_6^-}$
   (2) $\underset{\text{Lewis acid}}{SiX_4} + 2X^- \longrightarrow \underset{\text{Lewis base}}{SiX_6^{-2}}$

- Some other examples of Lewis acids are: SnX_4, PX_3, GeX_4, SF_4, SeF_4, $TeCl_4$
- Elements with an electron sextet (for example O, S)

Strength of Lewis Acid

1. Lewis acid strength ∝ Electronegativity difference
   Example:
   AlF_3 $AlCl_3$ $AlBr_3$ AlI_3 ⟶
   Acidic nature increases
2. In case of boron halide (BX_3)
   BI_3 > BBr_3 > BCl_3 > BF_3
   No back bonding; Less back bonding; Maximum back bonding
3. In BF_3 fluorine shifts its electron pairs back to boron atom. In F-atom, due to its small size and more electron density ($2p^5$), there is more e^-–e^- repulsion. But in case of B, the atom size is large and 2p orbital is almost vacant.
4. The decreasing order of strength of some Lewis acids is given below:
   BX_3, AlX_3, FeX_3, GaX_3, SbX_5, InX_3, SnX_4, AsX_5, ZnX_2, HgX_2

Lewis bases These are species which donate lone pair of electrons.

- Any anion or molecule with octate state and central atom with lone pair of electrons, is a Lewis base as it can donate its lone pair.

  For example, X^-, OH^-, CN^-

  $\ddot{N}$ (p), $\ddot{N}H_3$, $R-\ddot{N}H_2$, $R_2\ddot{N}H$ (s), $R_3\ddot{N}$ (t)

  $\ddot{P}H_3$ $H-\ddot{O}-H$ $R-\ddot{O}-H$ $R-\ddot{O}-R$

- Some multiple bonded molecules which form complexes with transition metals. For example, CO, NO, C_2H_4.

Strength of Lewis Base

$$\text{Lewis base} \propto \frac{1}{\text{Electronegativity difference}}$$

$NH_3 > NI_3 > NBr_3 > NCl_3 > NF_3$

Least electronegativity difference

Facts to Remember

- CO, $RCH{=}CH_2$, $RC{\equiv}CH$ are border line Lewis bases.
- H_3O^+, $\overset{+}{N}H_4$ are neither Lewis acid nor Lewis bases.

Factors Affecting Acidic Strength

Effect of electronegativity difference

- Acidic strength ∝ Electronegativity difference
  1. $HF > H_2O > NH_3 > CH_4$
     As electronegativity of F > O > N > C.
     $HNO_3 > H_2CO_3 > H_3BO_3$
     As electronegativity of N > C > B.
  2. HClO > HBrO > HIO
     $HClO_3 > HBrO_3 > HIO_3$
     As electronegativity of Cl > Br > I.
- Acidic strength ∝ Size of central atom or ease of removal of H^+

  HF < HCl < HBr < HI

  $H_2O < H_2S < H_2Se$

  $NH_3 < PH_3 < AsH_3 < SbH_3$

- Strength of oxyacids ∝ Oxidation number of central atom
  1. $\overset{+1}{HOX} < \overset{+3}{HXO_2} < \overset{+5}{HXO_3} < \overset{+7}{HXO_4}$
     Increasing order of acidic nature →
  2. $N^{-3}H_3 < H_2N_2^{+1}O_2 < HN^{+3}O_2 < HN^{+5}O_3$

Facts to Remember

- Order of acidic strength: $H_3PO_2 > H_3PO_3 > H_3PO_4$

Effects of a solvent on acidic strength A solvent also effects acidic strength and it is called **levelling effect**. For example,

1. CH_3COOH
   - CH_3COOH behaves like a weak acid in water, but in liquid ammonia, it is a strong acid as ammonia has a greater tendency to accept H^+ than water.

     For example,

     $$\underset{\text{Acid}}{CH_3COOH} + \underset{\text{Base}}{H_2O} \rightleftharpoons \underset{\text{Conjugate base}}{CH_3COO^-} + \underset{\text{Conjugate acid}}{H_3O^+}$$

     $$\underset{\text{Acid}}{CH_3COOH} + \underset{\text{Base}}{NH_3} \rightleftharpoons \underset{\text{Conjugate base}}{CH_3COO^-} + \underset{\text{Conjugate acid}}{NH_4^+}$$

   - CH_3COOH in HF behaves like a base, as HF is stronger acid than CH_3COOH.

     $$\underset{\text{Base}}{CH_3COOH} + \underset{\text{Acid}}{HF} \longrightarrow \underset{\text{Conjugate acid}}{CH_3COO^+H_2} + \underset{\text{Conjugate base}}{F^-}$$

2. HNO_3
   HNO_3 is a strong acid in water but a weak acid in CH_3COOH and acts as a base in HF.
3. $HClO_4$
   $HClO_4$ (the strongest acid) behaves as a weak acid in HF.

Strength of Lewis Base

$HClO_4$, $HClO_3$, HI, H_2SO_4, HBr, HCl, H_3O^+, HNO_3, HF, HSO_4^-, H_3PO_4, HCOOH, CH_3COOH, H_2CO_3, H_2S, NH_4^+, HCN, C_6H_5OH, H_2O, R–OH, HC≡CH, NH_3, $H_2C{=}CH_2$, CH_4, C_2H_6, C6H_{12} (cyclohexane)

Corresponding Conjugate Bases in Increasing Order of Acidic Strength

ClO_4^-, ClO_3^-, I^-, H_2O, NO_3^-, F^-, SO_4^{2-}, $H_2PO_4^-$, $HCOO^-$, CH_3COO^-, HCO_3^-, HS^-, NH_3, CN^-, $C_6H_5O^-$, OH^-, OR^-, $C^- \equiv CH$, NH_2^-, $CH^- = CH_2$, CH_3^-, $C_2^-H_5$, $C_6^-H_{11}$

Strongest conjugate base

Nature of Oxides

Lux–flood concept

- An acidic oxide takes up oxygen while a basic oxide gives up oxygen.

  Example:

  $CaO + CO_2 \longrightarrow CaCO_3$

  Basic Acidic

  $6Na_2O + P_4O_{10} \longrightarrow 4Na_3PO_4$

- An acidic oxide produces an acid with water while a basic oxide produces a base with water.

  Example:

  $SO_3 + H_2O \longrightarrow H_2SO_4$

  Acidic

  $Na_2O + H_2O \longrightarrow 2NaOH$

  Basic

Some Acidic Oxides

Cl_2O_7, P_4O_{10}, SO_2, CO_2, SiO_2, SO_3, N_2O_3, B_2O_3, CrO_3, WO_3, MoO_3, Mn_2O_7.

Some Basic Oxides

M_2O (here M = IA metal), MO (Here M = IIA metal), Sc_2O_3, TiO_2, ZrO_2.

Some Amphoteric Oxides

BeO, CrO, Al_2O_3, Sb_2O_3, SnO, PbO, ZnO, Cr_2O_3.

Acid–Base Indicators

- Acid–base indicators are weak organic acids or bases which exhibit a sharp colour change at the end point of a titration.
- Such indicators exhibit different colours in acidic and basic medium. For example, methy1 orange exhibits red colour in acidic medium while shows yellow colour in basic medium.
- At the end point of an acid–base titration, a sharp change of pH occurs, so colour of indicator changes.
- The range of pH value between lower and upper limit at which the colour of indicator changes, is called **pH range.**
- This pH range depends upon the nature and concentration of acid and base used in the titration. For example, phenolphthalein or methy1 red or methyl orange indicator can be used in the range of titration of strong acid versus strong base.
- Phenolphthalein can also be used in titration between weak acid with a strong base, but it does not work in the titration of a weak base with a strong acid titration.
- While titrating a base with a strong acid, both methy1 orange and methy1 red can be used.
- In the titration of a weak acid with weak base, there is no suitable indicator, as pH change is not sharp at the end point.
- When several indicators are mixed in a definite proportion, a universal indicator is formed which exhibits characteristics colours at different pH values.
- An indicator exhibits dark colour in quinoid form, that is when benzoid form changes into quinoid form, a sharp colour change occurs.
- At the end point in acid–base titration, acid is present in slight excess.
- The colour of an indicator is determined by the ratio of $[In^-]/[HIn]$.

Hard and Soft Acids and Bases (HSAB Principle)

Hardness is the property to retain valence electrons very strongly. On the basis of this concept, hard and soft acids or bases are described as follows:

Hard acids Acceptor atom is small sized, has high charge and has no electron. For example, H^+, Li^+, Na^+, Be^{2+}, Mg^{2+}, Al^{3+}, Cr^{3+}, Fe^{3+}, BF_3, $AlCl_3$.

Soft acids Acceptor atom is large sized, has a low charge and possesses electrons. For example, Pb^{2+}, Cd^{2+}, Hg^{2+}, Cu^+, Ag^+, Hg^{2+}.

Hard bases These are Lewis bases that hold electrons strongly. For example, H_2O, NH_3, R–OH, RNH_2, OH^-, F^-, RO^-.

Soft bases These are Lewis bases from which, electron can be easily removed. For example, R_2S, RS^-, I^-, CN^-, CO, R^-.

TABLE 7.2 Acid Base Titrations

S.No.	Type of titration	Example	pH change at the end point
1	Strong acid–weak base	0.1 M HCl and 0.1 M NH_4OH	4–6.3
2	Strong base–weak acid	0.1 M NaOH and 0.1 M CH_3COOH	7.7–9.7
3	Weak acid–weak base	0.1 M CH_3COOH and 0.1 M NH_4OH	No sharp change is observed
4	Strong acid–strong base	0.1 M HCl and 0.1 M NaOH	5.5–8.5

TABLE 7.3 Some Indicators and Their Colour Changes with the Respective pH Range

Indicator	Colour in acidic medium	Colour in basic medium	pH range
Phenolphthalein	Colourless	Pink	8.2–10
Thymol blue	Yellow	Blue	8–9.6
Phenol red	Yellow	Red	6.8–8.4
Methyl red	Red	Yellow	4.2–6.3
Methyl orange	Red	Yellow	3.1–4.4

POINTS TO REMEMBER

1. According to Ingold concept, an **electrophile** is an acid while a **nucleophile** is a base. For example, $AlCl_3$ (electrophile) is an acid and NH_3 (nucleophile) is a base.
2. In the reaction $I_2 + I^- \longrightarrow I_3^-$, the Lewis base is I^-.
3. $[Fe(H_2O)_6]^{3+} + H_2O \longrightarrow [Fe(H_2O)_5OH]^{2+} + H_3O^+$
   Acid, Base, Conjugate base, Conjugate acid
4. $[Zn(H_2O)_5OH]^{2+} + H_3O^+ \longrightarrow [Zn(H_2O)_6]^{3+} + H_2O$
   Base, Acid, Conjugate acid, Conjugate base
5. The reaction of a cation or anion with water accompanied by cleavage of O–H bond, is called **salt hydrolysis**.
6. A change of 0.2 unit in pH of blood may cause death also.
7. The buffer system present in blood is H_2CO_3 + $NaHCO_3$.
8. In case of strong acid with 10^{-8}m concentration pH is 6.95 and in case of strong base with same conc^n pH is 7.0414.
   (H^+) from acid $= 1 \times 10^{-8}$
   (H^+) from $H_2O = 1 \times 10^{-7}$
   Net $(H^+) = 1.1 \times 10^{-7}$
   $pH = -\log [1.1 \times 10^{-7}] = 6.95$
   Similarly, for base $[\bar{O}H]$ Net $= 1.1 \times 10^{-8}$
   pOH = 6.95
   pH = 14–6.95 = 7.05

Important formula

Degree of Ionization

$$\alpha = \frac{\text{Number of molecules dissociate}}{\text{Total number of molecules}}$$

Ostwald dilution law

For a binary weak electrolyte

$$K = \frac{\alpha^2}{V} \quad \text{or} \quad \alpha^2 = KV \quad \text{or} \quad \alpha = \sqrt{(KV)}$$

- If 'C' is the concentration, then

$$K = \frac{C\alpha^2}{(1-\alpha)}$$

$$K = C\,\alpha^2 \quad \text{or} \quad \alpha^2 = K/C$$

$$\alpha^2 = (K/C)$$

- $\alpha = \frac{\lambda_v}{\lambda_\infty}$ or $\frac{\Lambda_v}{\Lambda_\infty}$
- Here λ_v or Λ_v is equivalent conductivity at V dilution. λ_∞ or Λ_∞ is equivalent conductivity at infinite dilution.
- $\lambda_v = K_v \times V$
  Here K_v = specific conductivity
  V = dilution
- $\lambda_\infty = \lambda_c + \lambda_\alpha$
  Here λ_c and λ_α are ionic mobilities of cations and anions.

Ionic product of water

- $K_w = K_a \times K_b$
- $pK_w = pK_a + pK_b$

pH scale

- $[H^+] = 10^{-pH}$
- $pH = -\log_{10} [H^+]$
- $pH = \log_{10} \dfrac{1}{[H^+]}$

pOH

$[OH^-] = 10^{-OH}$

$pOH = -\log[OH^-]$ or $\dfrac{1}{\log[OH^-]}$

Solubility product

$K_{sp} = [A^+] [B^-]$ (At constant temperature.)

$A_xB_y \rightleftharpoons xA^{+y} + yB^{-x}$

$K_{sp} = [A^{+y}]^x [B^{-x}]^y$

Relation between solubility (s) and solubility product (K_{sp})

$K_{sp} = (x\ s)^x . (y\ s)^y$

or

$K_{sp} = x^x\ y^y\ (s)^{x+y}$

Degree of hydrolysis

$$h = \frac{\text{Number of moles of salt hydrolyzed}}{\text{Total moles of the salt taken}}$$

Various expressions for K_h, h and pH fordifferent types of salts

For salt of weak acid and strong base (KCN)

$K_h = \dfrac{K_w}{K_a}$

$h = \sqrt{(K_h/C)}$

$h = \sqrt{(K_w/K_a \cdot C)}$

$pH = \frac{1}{2}[pK_w + pK_a + \log C]$

$pOH = \frac{1}{2}[pK_w - pK_a - \log C]$

For salt of strong acid and weak base ($FeCl_3$)

$K_h = \dfrac{K_w}{K_b}$

$h = \sqrt{(K_h/C)}$

$h = \sqrt{(K_w/K_b \cdot C)}$

$pH = \frac{1}{2}[pK_w - pK_b - \log C]$

For salt of weak acid and weak base

$K_h = \dfrac{K_w}{K_a K_b}$

$h = \sqrt{(K_w/K_a.K_b)}$

$pH = \frac{1}{2}[pK_w + pK_a - pK_b]$

SOLVED NUMERICALS

Dissociation Constant, Degree of Dissociation

1. The dissociation constant of CH_3COOH at 298 K is 1.75×10^{-5}. What will be the hydrogen ion concentration of 0.1 M solution of acetic acid?

Solution

$CH_3COOH + H_2O \rightleftharpoons CH_3COO^- + H_3O^+$

$C(1 - \alpha) \qquad C.\alpha \qquad C.\alpha$

$\alpha = \sqrt{(K\alpha/C)} = \sqrt{(1.75 \times 10^{-5}/0.1)}$

$= 0.01323$

$[H_3O^+] = [H^+] = C.\ \alpha$

$= 0.1 \times 0.01323$

$= 1.323 \times 10^{-3}$ mole/litre.

2. The dissociation constants of formic acid and acetic acid are 1.77×10^{-4} and 1.75×10^{-5}, respectively. Find out the relative strength of two acids.

Solution

K (formic acid) $= 1.77 \times 10^{-4}$

K (acetic acid) $= 1.75 \times 10^{-5}$

$$\frac{\text{Strength of formic acid}}{\text{Strength of acetic acid}} = \frac{\sqrt{K(\text{formic acid})}}{\sqrt{K(\text{acetic acid})}}$$

$= \sqrt{(1.77 \times 10^{-4}/1.75 \times 10^{-5})}$

$= 3.18$

3. The ionization constant of NH_4^+ in water is 5.6×10^{-10} at 25°C. The rate constant for the reaction of NH_4^+ and OH^- to form NH_3 and H_2O at 25°C is 3.4×10^{-10} litre mol^{-1} sec^{-1}. Find the rate constant for proton transfer from water to NH_3.

Solution

$NH_4^+ + HOH \rightleftharpoons NH_4OH + H^+$

$$K_a = 5.6 \times 10^{-10}$$

$$NH_3 + HOH \underset{K_2}{\overset{K_1}{\rightleftharpoons}} NH_4^+ + OH^-$$

$$K_b = 3.4 \times 10^{10}$$

$$K(NH_3) = \frac{K_1}{K_2} = \frac{K_w}{K_a(NH_4^+)}$$

$$\frac{K}{3.4\times10^{10}} = \frac{1\times10^{-14}}{5.6\times10^{-10}}$$

$$K_1 = 6.07 \times 10^5$$

4. At what dilution benzoic acid would be 10% dissociated? The dissociation constant of the acid is 6.6×10^{-5} at 25°C.

Solution

Let the dilution be V litre per mole then

$$K = \frac{\alpha^2}{V(1-\alpha)}$$

$$\alpha = \frac{10}{100} = 0.1$$

$$V = \frac{\alpha^2}{K(1-\alpha)} = \frac{0.1\times0.1}{66\times10^{-5}\times0.9}$$

$$\frac{10^5}{66\times9} = 168 \text{ litre}$$

pH Scale

5. How many grams of NaOH must be dissolved in one litre of solution to give it a pH value of 11?

Solution

Given pH of solution = 11

$pH = -\log [H^+]$

$11 = -\log [H^+]$

$[H^+] = 10^{-11}$ M

$K_w = [H^+][OH^-] = 1 \times 10^{-14}$

$$[OH^-] = \frac{1\times10^{-14}}{10^{-11}} = 1 \times 10^{-3} \text{ M}$$

$[OH^-] = 1 \times 10^{-3}$ M

Strength = Molarity × mol. weight

$= 10^{-3} \times 40 = 0.04$ g/litre.

6. Find the pH of 0.004 M NH_4OH having 3.2% dissociation.

Solution

$NH_4OH \longrightarrow$	NH_4^+	+	OH^- (weak base)
C	0	0	conc. before ionization
C(1 – α)	C. α	C. α	conc. after ionization)

$$[H^+] = C.\alpha = 3 \times 10^{-3} \times \frac{4.3}{100}$$

$= 1.29 \times 10^{-4}$ M

$pH = -\log [H^+] = -\log [1.29 \times 10^{-4}]$

$pH = 3.8894$

7. Find the pH of solution obtained by mixing 20 mL of 0.1 M HCl and 30 mL of 0.2 M H_2SO_4.

Solution

Milli equivalent of H^+ from HCl = 20 × 0.1 = 2

Milli equivalent of H^+ from H_2SO_4
= 30 × 0.2 × 2 = 12

Total milli equivalent of H^+ in solution = 2 + 12 = 14

Total volume of solution = 20 + 30 = 50

$$[H^+] = \frac{M_{eq}}{C_{in\,ml}} = \frac{14}{50} = 0.28$$

$pH = -\log [H^+] = -\log (0.28) = 0.5528$

8. Find the pH and $[OH^-]$ in 0.2 M aqueous solution of sodium butyrate. K_a for butyric acid = 2.0×10^{-5}.

Solution

$NaBu + HOH \rightleftharpoons BuH + NaOH$

$[OH^-] = C.x = C\sqrt{(K_h/C)} = \sqrt{(K_h.C)}$

$= \sqrt{(K_w \times C/K_a)}$

$= \sqrt{(1 \times 10^{-14} \times 0.2/2.0 \times 10^{-5})}$

$= 1 \times 10^{-5}$

$pOH = -\log [OH^-] = -\log 10^{-5} = 5$

$pH = 14 - pOH = 14 - 5 = 9$

9. Calculate the pH of a solution containing 2.0 g of NaOH per litre of water.

Solution

As the solution contains 2.0 g of NaOH per litre of water, its normality $= \frac{2}{40} = \frac{1}{20}$ N.

$[OH^-] = 1/20$ g ion/litre

$[H^+][OH^-] = 10^{-14}$

$[H^+][1/20] = 10^{-14}$

$[H^+] = 20 \times 10^{-14} = 2 \times 10^{-13}$ g ion/litre

$pH = -\log [H^+]$

$= -\log [2 \times 10^{-13}]$

$= -[\log 2 - 13 \log 10]$

$= -[0.3010 - 13] = 12.699$

10. Calculate the pH of a solution obtained by mixing 50 mL of 0.01 M $Ba(OH)_2$ solution with 50 mL water.

Solution

$N_1V_1 = N_2V_2$

$Ba(OH)_2$ mixture

[Normality of solution of $Ba(OH)_2$ = 2 × its molarity, as its mol. wt. = 2 × eq. wt.]

$0.01 \times 2 \times 50 = N_2 \times (50 + 50)$

(as it is assumed as 100% ionized).

$N_2 = \frac{0.01 \times 100}{100} = 0.01$ N

$[OH^-] = \alpha N = 1 \times 0.01 = 10^{-2}$

$pOH = -\log [OH^-] = -\log 10^{-2} = 2 \log 10$

pOH = 2

pH + pOH = 14

pH + 2 = 14

pH = 12

11. Will the pH of water be same at 4°C and 25°C? Explain.

Solution

At 25°C, Ionic product of water = 1×10^{-14}.

So, $pK_a = 14$

$[H^+] [OH^-] = 1 \times 10^{-14}$

$-\log [H^+] - \log [OH^-] = 14$

pH + pOH = 14

It is neutral compound so, its pH and pOH are 7 at 25°C. As on decreasing the temperature K_w decreases so, pK_w increases. Hence, at 4°C pK_w is higher than at 25°C and at higher pK_w, pH will be higher, i.e., at 4°C pH of water is higher than 7.

Buffer Solutions

12. 0.15 mole of pyridinium chloride has been added into 500 mL of 0.2 M pyridine solution. Find the pH and $[OH^-]$ ion concentration in the resulting solution. $K_b = 1.5 \times 10^{-9}$ M (assuming no change in volume).

Solution

$$pH = 14 - pK_b - \log \frac{[Salt]}{[Base]}$$

$pK_b = -\log K_b = -\log (1.5 \times 10^{-9}) = 8.8239$

[Salt] = [pyridinium chloride]

$$= \frac{0.15 \times 1000}{500} = 0.30 \text{ mole litre}^{-1}$$

[Base] = [Pyridine] = 0.20 mole $litre^{-1}$

$$pH = 14 - 8.8239 - \log_{10} \frac{0.30}{0.20}$$

$= 14 - 9 = 5$

pOH = 14 – 5 = 9

$[OH^-] = 10^{-9}$

13. Find the pH of a solution containing 0.10 M H_3BO_3 and 0.18 M NaH_2BO_3. Given $K_{a1} = 7.3 \times 10^{-10}$.

Solution

$$pH = pK_a + \log_{10} \frac{[Salt]}{[Acid]}$$

$$pH = -\log_{10} K_a + \log_{10} \frac{[0.18]}{[0.10]}$$

$= -\log (3.3 \times 10^{-7}) + \log 1.8 = 9.39$

14. Calculate the pH of a 0.01 M NH_4OH solution containing 10 mol/litre of NH_4Cl? ($pK_b = 4.76$).

Solution

[Conjugate base] = 0.10 M NH_3 in H_2O

[Conjugate acid] = 1.0 M NH_4^+

$$pOH = pK_b + \log \frac{[Salt]}{[Base]}$$

$$= 4.76 + \log \frac{1.0}{0.1}$$

$= 4.76 + \log 10 = 5.76$

pH = 14 – pOH

= 14 – 5.76 = 8.24

Solubility Product

15. The solubility of Ag_2CrO_4 at 25°C is 8.0×10^{-5} mole/litre. Find its solubility product.

Solution

$$Ag_2CrO_4 \rightleftharpoons 2Ag^+ + CrO_4^{-2}$$

8.0×10^{-5} $2 \times 8.0 \times 10^{-5}$ 8.0×10^{-5}

$K_{sp} = [Ag^+]^2 [CrO_4^{-2}]$

$= 2 [8.0 \times 10^{-5}]^2 [8.0 \times 10^{-5}]$

$= 2.048 \times 10^{-12}$

16. The solubility product of SrF_2 in water is 8×10^{-10}. Find its solubility in 0.1 M NaF aqueous solution.

Solution

Let the solubility of SrF_2 be s mol $litre^{-1}$.

$SrF_2 (s) \rightleftharpoons Sr^{2+} (aq) + 2F^- (aq)$

$K_{sp} = [Sr^{2+}] [F^-]^2 = 4s^3$

$s = [K_{sp}/4]^{1/3}$

$= [8 \times 10^{-10}/4]^{1/3} = 5.85 \times 10^{-4}$

In 0.1 M NaF

$K_{sp} = [s]\,[0.1 + s]^2$

$0.01\ s = 8 \times 10^{-10} \quad (s <<< 0.01)$

$s = \dfrac{8\times10^{-10}}{0.01} = 8 \times 10^{-8}$ M

17. Find the pH at which $Mg(OH)_2$ begin to precipitate from a solution containing 0.10 M Mg^{2+} ions. K_{sp} for $Mg(OH)_2 = 10^{-11}$.

Solution

Precipitate of $Mg(OH)_2$ starts when

$[Mg^{2+}]\,[OH^-]^2 = K_{sp}\ Mg(OH)_2$

$[0.1]\,[OH^-]^2 = 10^{-11}$

$[OH^-]^2 = 10^{-5}$

pOH = 5

pH = 14 – 5 = 9

Salt Hydrolysis

18. Calculate the hydrolysis constant of the salt containing NO_2^- ions. K_a for $HNO_2 = 4.5 \times 10^{-10}$ and $K_w = 1 \times 10^{-14}$.

Solution

$K_h = \dfrac{K_w}{K_a} = \dfrac{1\times10^{-14}}{4.5\times10^{-10}} = 2.2 \times 10^{-5}$

19. Find the percentage of hydrolysis in 0.003 M aqueous solution of NaOCN. Given K_a for HOCN = 3.33×10^{-4} and $K_w = 1 \times 10^{-14}$.

Solution

$NaOCN + HOH \rightleftharpoons HOCN + NaOH$

$x = \sqrt{(K_h/C)} = \sqrt{(K_w/K_a.C)}$

$= \sqrt{(1 \times 10^{-14}/3.33 \times 10^{-14} \times 0.003)}$

$= 1 \times 10^{-4}$

% Hydrolysis $= \dfrac{1\times10^{-4}}{100} = 10^{-2} = 0.01\%$

20. Find the pH of an aqueous solution of 1.0 M ammonium formate assuming complete dissociation. (pK_a of formic acid = 3.8 and pK_b of NH_3 = 4.8.)

Solution

$HCOONH_4 + HOH \rightleftharpoons HCOOH + NH_4OH$

pH for a salt of weak acid and weak base

$pH = \frac{1}{2}\,[\log_{10} K_b - \log_{10} K_a - \log K_w]$

$= \frac{1}{2}\,[pK_a + pK_w - pK_b]$

$= \frac{1}{2}\,[3.8 + 14 - 4.8]$

$= 6.5.$

CHAPTER-END EXERCISES

PRACTICE QUESTIONS – I

1. Ostwald's dilution law is applicable for:

(a) Strong electrolyte
(b) Weak electrolyte
(c) Both strong and weak electrolytes
(d) None of these

2. The conjugate base of OH^- is:

(a) H_2O (b) O^{2-}
(c) O^- (d) O_2

3. The ionization constant of phenol is higher than that of ethanol because:

(a) Phenoxide ion is bulkier than ethoxide
(b) Phenoxide ion is stronger base than ethoxide
(c) Phenoxide ion is stabilized through delocalization
(d) Phenoxide ion is less stable than ethoxide

4. One of the following species acts as both Bronsted acid and base:

(a) $H_2PO_2^-$ (b) HPO_3^{2-}
(c) HPO_4^{2-} (d) All of the above

5. Which of the following is not true?

(a) The conjugate base of $H_2PO_4^-$ is HPO_4^{2-}.
(b) pH + pOH = 14 for all aqueous solutions.
(c) The pH of 1×10^{-8} M HCl is 8.
(d) 96500 coulombs of electricity when passed through a $CuSO_4$ solution deposit 1 g equivalent of copper at the cathode.

6. The conjugate base of $H_2PO_4^-$ is:

(a) HPO_4^{2-} (b) H_3PO_4
(c) PO_4^{3-} (d) P_2O_5

7. When rain is accompanied by a thunderstorm, the collected rain water will have a pH value:
(a) Slightly lower than that of rain water without thunderstorm
(b) Slightly higher than that when the thunderstorm in not there
(c) Uninfluenced by occurrence of thunderstorm
(d) Which depends on the amount of dust in air

8. The molar solubility (in mol L^{-1}) of a sparingly soluble salt MX_4 is 's'.The corresponding solubility product is Ksp.s is given in terms of Ksp by the relation:
(a) $s = (Ksp/256)^{1/5}$ (b) $s = (128Ksp)^{1/4}$
(c) $s = (Ksp/128)^{1/4}$ (c) $s = (256Ksp)^{1/5}$

9. The H^+ ion concentration of a solution is 1 mol L^{-1}. Its pH is:
(a) 0.1 (b) 1.5
(c) 1.0 (d) 0

10. The pH of 10^{-10} M NaOH solution is:
(a) 10 (b) 7.01
(c) 6.99 (d) 4

11. Which of the following is a tribasic acid?
(a) H_3PO_4 (b) H_3PO_3
(c) H_3PO_2 (d) HPO_3

12. In which of the following acid–base titration, pH is greater than 8 at equivalence point?
(a) Acetic acid vs ammonia
(b) Acetic acid vs sodium hydroxide
(c) Hydrochloric acid vs ammonia
(d) Hydrochloric acid vs sodium hydroxide

13. Which one of the following is not a buffer solution?
(a) 0.8 M H_2S + 0.8 M KHS
(b) 2M $C_6H_5NH_2$ + 2M $C_6H_5NH_3^+Br^-$
(c) 3M H_2CO_3 + 3M $KHCO_3$
(d) 0.05 M $KClO_4$ + 0.05 M $HClO_4$

14. The pH of pure water at 80°C will be:
(a) = 7 (b) < 7
(c) > 7 (d) None of these

15. A centinormal solution of a monobasic acid is 100% ionized. Its pH is:
(a) 2 (b) 4
(c) 3 (d) 1

16. When a buffer solution of CH_3COONa and CH_3COOH is diluted with water then:
(a) H^+ ion concentration increases
(b) H^+ ion concentration decreases
(c) H^+ ion concentration remains constant
(d) CH_3COO^- ion concentration increases

17. Which of the following would produce a buffer solution when mixed in equal volume?
(a) 1 M CH_3COOH and 0.5 M NaOH
(b) 1 M CH_3COOH and 0.5 M HCl
(c) 1 M NH_4OH and 0.5 M NaOH
(d) 1 M NH_4Cl and 0.5 M HCl

18. At 25°C the pH value of a solution is 6, the solution is:
(a) Neutral (b) Acidic
(c) Alkaline (d) Basic

19. The pH of N/10 HCl is:
(a) 3 (b) 4
(c) 2 (d) 1

20. Which one of the following statement is correct?
(a) Bronsted-Lowry theory could not explain the acidic nature of BCl_3
(b) The pH of 0.01 M NaOH solution is 2
(c) The ionic product of water at 25°C is 10^{-10} mol^2 L^{-2}
(d) The pH of a solution can be calculated using the equation pH = log [H^+]

21. Which one of the following is not correct?
(a) pH = 1/log[H^+] (b) pH = log 1/[H^+]
(c) [H^+] = 10–pH (d) pH = –log [H^+]

22. Which one of the following statements is not correct?
(a) Cl^- is a lewis acid.
(b) The pH of 10^{-8} M HCl solution is less than 7.
(c) The ionic product of water at 25°C is 10^{-14} mol^2 L^{-2}.
(d) Bronsted–Lowry theory could not explain the acidic character of $AlCl_3$.

23. NH_4Cl solution is:
(a) Neutral (b) Acidic
(c) Basic (d) Amphoteric

24. In the titration of iodine against hype the indicator used is:
(a) Starch
(b) Potassium ferricyanide
(c) Methyl orange
(d) Methyl red

25. The indicator used for titration of weak base and strong acid is:
(a) Thymol blue (b) Methyl orange
(c) Phenolpthalein (d) Fluorescein

26. The correct order of relative acidity is:
(a) $HClO > HClO_2 > HClO_3 > HClO_4$
(b) $HClO_4 > HClO_3 > HClO_2 > HClO$
(c) $HClO > HClO_4 > HClO_2 > HClO_3$
(d) $HClO_3 > HClO_2 > HClO_4 > HClO$

27. CH_3COOH is weaker acid than H_2SO_4. It is due to:
(a) More ionization (b) Less ionization
(c) Covalent bond (d) Electrovalent bond

28. Which one of the following pair show buffer's solution?
(a) NaCl + NaOH
(b) $CH_3COONa + CH_3COOH$
(c) $CH_3COOH + CH_3COONH_4$
(d) $H_2SO_4 + CuSO_4$

29. The pH of a solution of H_2O_2 is 6.0. Some chlorine gas is bubbled into this solution. Which of the following is correct?
(a) The pH of resultant solution becomes 8.0
(b) Hydrogen gas is liberated from resultant solution
(c) The pH of resultant solution becomes less than 6.0 oxygen gas is liberated
(d) Cl_2O is formed in the resultant solution

30. Which of the following is correct?
(a) The pH of one litre solution containing 0.49 g of H_2SO_4 is 2.0.
(b) The conjugate base of H_2S is S^{2-}.
(c) BF_3 is lewis base.
(d) Phenolphthalein is colourless in basic medium.

31. At 90°C, pure water has $[H_3O^+]$ 10^{-6} mole $litre^{-1}$. What is the value of K_w at 90°C?
(a) 10^{-8} (b) 10^{-6}
(c) 10^{-12} (d) 10^{-14}

32. Of the given anions, the strongest bronsted base is:
(a) ClO_4^- (b) ClO_3^-
(c) ClO_2^- (d) ClO^-

33. The pH of a 10^{-8} molar solution of HCl in water is:
(a) –8 (b) 8
(c) Between 6 and 7 (d) Between 7 and 8

34. A acidic buffer solution can be prepared by mixing the solutions of:
(a) Sodium chloride and sodium hydroxide
(b) Sulphuric acid and sodium sulphate
(c) Ammonium chloride and ammonium hydroxide
(d) Ammonium acetate and acetic acid

35. Molten sodium chloride conducts electricity due to the presence of:
(a) Free ions
(b) Free molecules
(c) Free electrons
(d) Atoms of sodium and chlorine

36. For the reactions given below the correct relations is:
$$HCl + H_2O \rightleftharpoons H_3O^+ + Cl^c \ (K_a)$$
$$Cl^- + H_2O \rightleftharpoons HCl + OH^- \ (K_b)$$
(a) $K_b = K_a$ (b) $K_w = K_a.K_b$
(c) $\frac{K_a}{K_b} = K_w$ (d) $K_b = \frac{1}{K_a}$

37. The range of pH for acidic and basic buffers is:
(a) From $pH = pK_a + 1$ to $pH = pK_b - 1$
(b) From $pH = pK_a + 1$ to $pH = pK_b + 1$
(c) From $pH = pK_a$ to $pH = pK_b + 2$
(d) From $pH = pK_a \pm 1$ to $pK_b \pm 1$

38. A liquid is in equilibrium with its vapour at its boiling point. On the average, the molecules in the two phases have equal:
(a) Kinetic energy
(b) Total energy
(c) Inter-molecular forces
(d) Potential energy

39. The best indicator for detection of end point in titration of a weak acid and strong base is:
(a) Methyl red (5 to 6)
(b) Methyl orange (3 to 4)
(c) Phenolphthalein (8 to 9.6)
(d) Bromothymol blue (6 to 7.5)

40. The compound that is not a Lewis acid is:
(a) $SnCl_4$ (b) $BeCl_2$
(c) BF_3 (d) $AlCl_3$

41. The conjugate acid of NH_2^- is:
(a) NH_4^+ (b) NH_3
(c) NH_2OH (d) N_2H_4

42. The pH of the neutralization point of 0.1 N ammonium hydroxide with 0.1 NHCl is:
(a) 1 (b) 6
(c) 9 (d) 7

43. The compound whose 0.1 M solution is basic is:
(a) Ammonium acetate
(b) Ammonium sulphate
(c) Sodium acetate
(d) Ammonium chloride

44. Amongst the following hydroxides, the one which has the lowest value of K_{sp} at ordinary temperature (about 25°C) is:
(a) $Mg(OH)_2$ (b) $Ca(OH)_2$
(c) $Ba(OH)_2$ (d) $Be(OH)_2$

45. Fear or excitement, generally causes one to breathe rapidly and it results in the decrease of CO_2 concentration in blood. In what way will it change the pH of blood?
(a) pH will increase (b) pH will decrease
(c) No change (d) pH will adjust to 7

46. A buffer solution can be prepared from a mixture of:
1. Sodium acetate and acetic acid in water.
2. Sodium acetate and hydrochloric acid in water.
3. Ammonia and ammonium chloride in water.
4. Ammonia and sodium hydroxide in water.

(a) 1, 3, 4 (b) 2, 3, 4
(c) 1, 2, 4 (d) 1, 2, 3

47. Which among the following acts as a basic buffer solution?
(a) $NH_4Cl + NH_4OH$
(b) $NaCl + NaOH$
(c) $NH_4Cl + CH_3COOH$
(d) $CH_3COONa + CH_3COOH$

48. Conjugate base of hydrazoic acid is:
(a) HN_3^- (b) N_2^-
(c) Azide ion (d) Nitride ion

49. Which of the following are Lewis acids?
(1) BF_3 (2) H_2O
(3) HSO^-_4 (4) SO_3
(a) (1) and (3) (b) (1) and (2)
(c) (1) and (4) (d) (3) and (4)

50. Out of Ca^{2+}, Al^{3+}, Cr_3, Mg^{2+}, and Zn^{2+}, the reagents NH_4Cl and aqueous NH_3 will precipitate:
(a) Ca^{2+}, Al^{3+} (b) Al^{3+}, Cr^{3+}
(c) Bi^{3+}, Mg^{2+} (d) Mg^{2+}, Zn^{2+}

51. Which among the following is the least soluble?
(a) MnS ($K_{sp} = 7 \times 10^{-16}$)
(b) FeS ($K_{sp} = 4 \times 10^{-19}$)
(c) PtS ($K_{sp} = 8 \times 10^{-73}$)
(d) NiS ($K_{sp} = 3 \times 10^{-12}$)

52. The ratio of dissociation constant of two weak acids Hx and Hy is 4. At what molar concentration ratio, the two acids can have same pH value?
(a) 1 (b) 0.75
(c) 0.25 (d) 4

53. Which one of the following is true for any diprotic acid, H_2X?
(a) $K_{a_2} = K_{a_1}$ (b) $K_{a_2} > K_{a_1}$
(c) $K_{a_1} > K_{a_2}$ (d) $K_{a_2} = 1/K_{a_1}$

54. An aqueous solution of IM NaCl and IM HCl is:
(a) Not a buffer but pH < 7
(b) Not a buffer but pH > 7
(c) A buffer with pH < 7
(d) A buffer with pH > 7

55. For sparingly soluble salt A_pB_q, the relationship of its solubility product (L_s) with its solubility (s) is:
(a) $L_s = s^{p+q}\, p^p\, q^q$ (b) $L_s = s^{p+q}\, p^q\, q^p$
(c) $L_s = s^{pq}\, p^p\, q^q$ (d) $L_s = s^{pq}\, (pq)^{p+q}$

56. When NaOH is titrated with HCl, variation of conductance (x-axis) with addition of HCl (y-axis) can be shown by:

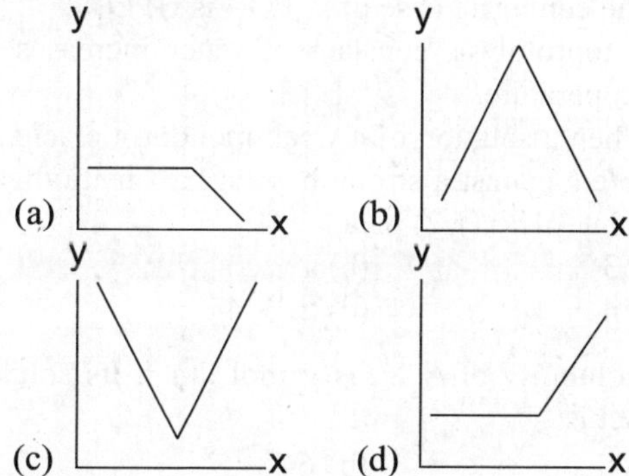

57. Which of the following curve represents the graph of pH during the titration of NaOH with HCl here?

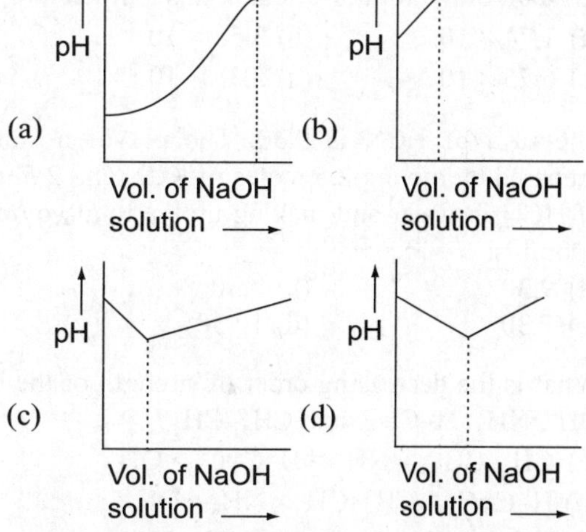

58. The solubility of A_2X_3 is y mol dm^{-3}, its solubility product is:
(a) $6\, y^4$ (b) $64\, y^4$
(c) $36\, y^5$ (d) $108\, y^5$

59. pKa values of three acids A, B and C are 4.3, 3.3 and 5.5 respectively. Which of the following represents the correct order of acid strength?
(a) A > B > C (b) C > A > B
(c) B > A > C (d) C > B > A

60. Which of the following on reaction with H_2S does not produce metallic sulphide?
(a) $CdCl_2$ (b) $ZnCl_2$
(c) $CoCl_2$ (d) $CuCl_2$

PRACTICE QUESTIONS – II

61. The pH 0.1 M solution the following salts increases in the order:
(a) NaCl < NH_4Cl < NaCN < HCl
(b) HCl < NH_4Cl < NaCl < NaCN
(c) NaCN < NH_4Cl < NaCl < HCl
(d) HCl < NaCl < NaCN < NH_4Cl

62. Which of the following statement is correct?
1. The pH of 1.0×10^{-8} M solution of HCl is 8.
2. The conjugate base of $H_2PO_4^-$ is HPO_4^{2-}.
3. Autoprotolysis constant of water increases with temperature.
4. When a solution of a weak monoprotic acid is titrated against a strong base at half neutralization point, pH = (1/2) pKa.

(a) 2, 3 (b) 1, 2, 3
(c) 3, 4 (d) 2, 3, 4

63. The solubility of A_2X_3 is y mol dm^{-3}. Its solubility product is:
(a) $6y^4$ (b) $64\ y^4$
(c) $36\ y^5$ (d) $108\ y^5$

64. If pKb for fluoride ion at 25°C is 10.83, the ionization constant of hydrofluoric acid at this temperature is:
(a) 1.74×10^{-5} (b) 3.52×10^{-3}
(c) 6.75×10^{-4} (d) 5.38×10^{-2}

65. The pKa pf HCN is 9.30. The pH of a solution prepared by mixing 2.5 moles of KCN and 2.5 moles of HCN in water and making up the total volume to 500ml is:
(a) 9.30 (b) 8.30
(c) 7.30 (d) 10.30

66. What is the decreasing order of strength of the bases OH^-, NH_2^-, $H–C≡C^-$ and $CH_3–CH_2^-$?
(a) $CH_3–CH_2^-$, > NH_2^- > $H–C≡C^-$ > OH^-
(b) $H–C≡C^-$ > $CH_3–CH_2^-$ > NH_2^- > OH^-
(c) OH^- > NH_2^- > $H–C≡C^-$ > $CH_3–CH_2^-$
(d) NH_2^- > $H–C≡C^-$ > OH^- > $CH_3–CH_2^-$

67. The following equilibrium is established when hydrogen chloride is dissolved in acetic acid:

$HCl + CH_3COOH \leftrightarrow Cl^- + CH_3COOH_2^+$

The set that characterizes the conjugate acid base pair is:
(a) (HCl, CH_3COOH) and ($CH_3COOH_2^+$, Cl^-)
(b) (HCl, $CH_3COOH_2^+$) and (HCl, $CH_3COOH_2^+$) and (CH_3COOH, Cl^-)
(c) ($CH_3COOH_2^+$, HCl) and (Cl^-, CH_3COOH)
(d) (HCl, Cl^-) and ($CH_3COOH_2^+$, CH_3COOH)

68. The pKa of acetyl salicylic acid (aspirin) is 3.5. The pH of gastric juice in human stomach is about 2–3 and pH in the small intestine is about 8. Aspirin will be:
(a) Ionized in the small intestine and almost unionized in the stomach.
(b) Unionized in the small intestine and in the stomach.
(c) Completely ionized in the small intestine and in the stomach.
(d) Ionized in the stomach and almost unionized in the small intestine.

69. An example of a reversible reaction is:
(a) KNO_3 (aq) + NaCl (aq) = KCl (aq) + $NaNO_3$ (aq)
(b) 2Na(s) + H_2O (l) = 2NaOH (aq) + H_2(g)
(c) $AgNO_3$(aq) + HCl (aq) = AgCl (s) + $NaNO_3$ (aq)
(d) $Pb(NO_3)_2$ (aq) + 2NaI (aq) = PbI_2 (s) + 2$NaNO_3$ (aq)

70. A certain weak acid has a dissociation constant of 1.0×10^{-4}. the equilibrium constant for its reaction with a strong base is:
(a) 1.0×10^{10} (b) 1.0×10^{14}
(c) 1.0×10^{-10} (d) 1.0×10^{-4}

71. A certain buffer solution contains equal concentration of X^- and HX. The Kp for X^- is 10^{-10}. The pH of the buffer is:
(a) 6 (b) 10
(c) 4 (d) 14

72. The precipitate of CaF_2 (Ksp = 1.7×10^{-10}) is obtained when equal volumes of the following are mixed:
(a) 10^{-2} M Ca^{2+} + 10^{-3} MF^-
(b) 10^{-3} M Ca^{2+} + 10^{-5} MF^-
(c) 10^{-4} M Ca^{2+} + 10^{-4} MF^-
(d) 10^{-5} M Ca^{2+} + 10^{-3} MF^-

73. Liquid NH_3 can ionize as follows:

$2NH_3 \rightleftharpoons NH_4^+ + NH_2^-$

A –50°C, the value of $K_{NH_3} = [NH_4^+][NH_2^-]$ is 10^{-30}. The number of amide ions present in per cm^3 of NH_3 (Pure) are:

(a) 6×10^5 (b) 6×10^{23}
(c) 6×10^8 (d) 6×10^{-8}

74. Hydrogen ion concentration in mol/L in solution of pH 5.4 will be:

(a) 3.88×10^6 (b) 3.98×10^{-6}
(c) 3.98×10^8 (d) 3.68×10^{-6}

75. Which has highest pH?

(a) CH_3COOK (b) Na_2CO_3
(c) NH_4Cl (d) $NaNO_3$

76. The correct order of increasing H_3O^+ in the following aqueous solution is:

(a) 0.01M H_2S < 0.01M H_2SO_4 < 0.01M NaCl < 0.01M $NaNO_2$
(b) 0.01M NaCl < 0.01M $NaNO_2$ < 0.01M H_2S < 0.01M H_2SO_4
(c) 0.01M H_2S < 0.01M $NaNO_3$ < 0.01M NaCl < 0.01M H_2SO_4
(d) 0.01M $NaNO_2$ < 0.01M NaCl < 0.01M H_2S < 0.01M H_2SO_4

77. pH of 0.005 M calcium acetate (pKa of CH_3COOH = 4.74) is:

(a) 7.37 (b) 9.37
(c) 9.26 (d) 8.37

78. Solubility of a M_2S salt is 3.5×10^{-6} then find out solubility product:

(a) 1.7×10^{-6} (b) 1.7×10^{-16}
(c) 1.7×10^{-18} (d) 1.7×10^{-12}

79. The solubility in water of a sparingly soluble salt AB_2 is 1.0×10^{-5} mol L^{-1}. Its solubility product number will be:

(a) 4×10^{-15} (b) 4×10^{-10}
(c) 1×10^{-15} (d) 1×10^{-10}

80. What is the pH of 0.01M Glycerine solution?

For glycine $K_{a_1} = 4.5 \times 10^{-3}$ and $K_{a_2} = 1.7 \times 10^{-10}$ at 298K:

(a) 3 (b) 7.2
(c) 6.1 (d) 10

81. An acid solution of 0.005 M has a pH of 5. The degree of ionization of acid is:

(a) 0.1×10^{-2} (b) 0.2×10^{-2}
(c) 0.5×10^{-4} (d) 0.6×10^{-6}

82. The solubility product of $BaCl_2$ is 4×10^{-9}. Its solubility in mol L^{-1} is:

(a) 4×10^{-3} (b) 4×10^{-9}
(c) 1×10^{-3} (d) 1×10^{-9}

83. pKa value of four acids are given below. The strongest acid is:

(1) 4.0 (2) 3.5
(3) 2.5 (4) 2

(a) 1 (b) 2
(c) 3 (d) 4

84. A solution has hydrogen ion concentration 0.0005 M, its pOH is:

(a) 8.2798 (b) 10.6990
(c) 12.7854 (d) 13.3344

85. At 25°C the pH of solution containing 0.10 M sodium acetate and 0.03 M acetic acid is [pKa value of CH_3COOH = 4.57]:

(a) 3.24 (b) 4.59
(c) 5.09 (d) 6.67

86. At 80°C, distilled water (H_3O^+) concentration is equal to 1×10^{-6} mol/litre. At the same temperature the value of Kw is:

(a) 1×10^{-3} (b) 1×10^{-6}
(c) 1×10^{-9} (d) 1×10^{-12}

87. When 10 mL of 0.1 M acetic acid (pKa = 5.0) is titrated against 10 mL of 0.1 M ammonia solution (pKb = 5.0) the equivalence point occurs at pH:

(a) 5.0 (b) 6.0
(c) 9.0 (d) 7.0

88. 40 ml of 0.1 M ammonia solution is mixed with 20 mol of 0.1 M HCl. What is the pH of the mixture? (pKb of ammonia solution is 4.74):

(a) 4.74 (b) 2.26
(c) 9.26 (d) 5.00

89. The solubility of AgCl in 0.2 M NaCl is Ksp AgCl = 1.8×10^{-10}:

(a) 1.8×10^{-11} M (b) 9×10^{-10} M
(c) 6.5×10^{-12} M (d) 5.6×10^{-11} M

90. Ionization constant of acetic acid is 1.8×10^{-5}. The concentration of H^+ ions in 0.1 M solution is:

(a) 1.8×10^{-3} M (b) 1.8×10^{-5} M
(c) 1.3×10^{-3} M (d) 1.34×10^{-3} M

91. The dissociation constant of a weak acid is 4.9×10^{-8}, its percentage ionization at 0.1 M is:
(a) 0.07% (b) 0.007%
(c) 0.7% (d) 0.0007%

92. The pKa of a weak acid is 4.8. What should be the ratio of [acid]/[salt], if a buffer of pH = 5.8 is required?
(a) 0.1 (b) 10
(c) 1 (d) 2

93. The pH of a 0.1 M aqueous solution of a weak acid (HA) is 3. What is its degree of dissociation?
(a) 1% (b) 10%
(c) 50% (d) 25%

94. The pH range if methyl red indicator is:
(a) 4.2 to 6.3 (b) 8.3 to 10.0
(c) 8.0 to 9.6 (d) 6.8 to 8.4

95. 75 ml of 0.2 M HCl is mixed with 25 ml of M HCl. To this solution, 300 ml of distilled water is added. What is the pH of the resultant solution?
(a) 1 (b) 2
(c) 4 (d) 0.2

96. The dissociation constant of two acids HA_1 and HA_2 are 3.0×10^{-4} and 1.8×10^{-5} respectively. The relative strengths of the acids is:
(a) 1:16 (b) 1:4
(c) 4:1 (d) 16:1

97. If the ionic product of water (Kw) is 1.96×10^{-14} at 35°C, what is its value at 10°C?
(a) 1.96×10^{-14} (b) 3.92×10^{-14}
(c) 2.95×10^{-15} (d) 1.96×10^{-13}

98. 0.005 M acid solution has 5 pH. The percentage ionization of acid is:
(a) 0.8% (b) 0.6%
(c) 0.4% (d) 0.2%

99. 100 ml of 0.015 M HCl solution is mixed with 100 ml of 0.005 M HCl. What is the pH of the resultant solution?
(a) 2.5 (b) 1.5
(c) 2 (d) 1

100. The solubility product of A_2X_3 is 1.08×10^{-23}. Its solubility will be:
(a) 1.0×10^{-3} M (b) 1.0×10^{-4} M
(c) 1.0×10^{-5} M (d) 1.0×10^{-6} M

101. M(OH) × has Ksp = 4×10^{-12} and solubility 10^{-4} M, × is:
(a) 1 (b) 2
(c) 3 (d) 4

102. The pH values of 1 M solutions of CH_3COOH (1), CH_3COONa (2), CH_3COONH_4 (3), and KOH (4) will be in the order:
(a) 4 > 3 > 2 > 1 (b) 4 > 2 > 3 > 1
(c) 1 > 3 > 2 > 4 (d) 2 > 1 > 3 > 4

103. For preparing a buffer solution of pH 6 by mixing sodium acetate and acetic acid, the ratio of the concentration of salt and acid should be (Ka = 10^{-5}):
(a) 1:10 (b) 10:1
(c) 100:1 (d) 1:100

104. The strongest Bronsted base among the following ions is:
(a) CH_3O^- (b) $C_2H_5O^-$
(c) $(CH_3)_2CHO^-$ (d) $(CH_3)_3CO^-$

105. The degree of ionization of water was found to be 1.8×10^{-9}. The Ka (ionization constant) of water is:
(a) 1.8×10^{-9} (b) 1.8×10^{-16}
(c) 1.0×10^{-14} (d) Cannot be calculated

106. The decreasing order of acidic nature of H_2SO_4 (1), H_3PO_4 (2), and $HClO_4$ (3) is:
(a) 1 > 2 > 3 (b) 1 > 3 > 2
(c) 3 > 1 > 2 (d) 3 > 2 > 1

107. The number of H^+ ions present in 1cm³ of a solution whose pH is 10 is:
(a) 10^{-10} (b) 10^{-13}
(c) 6.02×10^{10} (d) 6.02×10^{13}

108. The Ka value of formic acid and acetic acid are respectively 1.77×10^{-4} and 1.75×10^{-5}. the ratio of the acid strength of 0.1 N acids is:
(a) 0.1 (b) 0.3
(c) 3. 178 (d) 100

109. If 0.1 M of a weak monobasic acid is taken and its percentage degree of ionization is 1.34%, then calculate its ionization constant:
(a) 0.8×10^{-5} (b) 1.79×10^{-5}
(c) 0.182×19^{-5} (d) None of these

110. A weak monobasic acid is half neutralized by a strong base. If the pH of the solution is 5.4, its pKa is:
(a) 6.8 (b) 2.7
(c) 5.4 (d) 10.8

111. The solubility of AgCl in moles per litre when its solubility product is 1.56×10^{-10} at 25°C is:
(a) 0.576×10^{-8} mol/litre
(b) 1.056×10^{-4} mol/litre
(c) 1.249×10^{-5} mol/litre
(d) 1.478×10^{-6} mol/litre

112. If the solubility of sodium hexafluoroaluminate is 'a' mol/litre, its solubility product is:

(a) a^8 (b) 27 a^4
(c) 180 a^3 (d) 2916 a^8

113. If the solubility of $BaSO_4$ (mol. wt. 233) is 2.33×10^{-4} g/100 mL then the solubility product of $BaSO_4$ is:

(a) 1×10^{-5} mol L^{-1}
(b) 1×10^{-10} mol L^{-1}
(c) 1×10^{-4} mol L^{-1}
(d) 1×10^{-8} mol L^{-1}

114. The solubility product of calcium fluoride is 3.2×10^{-11} M^3. Its solubility in saturated solution is:

(a) 8×10^{-12} M (b) 2×10^{-4} M
(c) 4×10^{-12} M (d) 1×10^{-4} M

115. Equal volumes of the following Ca^{2+} and F^- solutions are mixed. In which of the solutions will precipitation occurs?

[Ksp of $CaF_2 = 1.7 \times 10^{-10}$]

1. 10^{-2} M Ca^{2+} + 10^{-5} M F^-
2. 10^{-3} M Ca^{2+} + 10^{-3} M F^-
3. 10^{-4} M Ca^{2+} + 10^{-2} M F^-
4. 10^{-2} M Ca^{2+} + 10^{-3} M F^-

Select the correct answer using the codes given below:

(a) In 4 only (b) In 1 and 2
(c) In 3 and 4 (d) In 2, 3 and 4

116. For which of the following sparingly soluble salt, the solubility (S) and solubility produce (Ksp) are related by the expressions S = $(Ksp/4)^{1\backslash 3}$?

(a) $BaSO_4$ (b) $Ca_3(PO_4)_2$
(c) Hg_2Cl_2 (d) Ag_2PO_4

117. ZnS is not precipitated by passing H_2S in acidic medium but CuS is precipitated. The reason for it is:

(a) Ksp CuS << Ksp ZnS
(b) Ksp CuS >> Ksp ZnS
(c) Ksp CuS = Ksp ZnS
(d) None of these

118. Consider of following acids:

1. HCN 2. HCOOH
3. CH_3COOH 4. Cl–CH_2COOH

Correct order of acid strength is:

(a) 2 > 3 > 1 > 4 (b) 4 > 2 > 3 > 1
(c) 4 > 3 > 2 > 1 (d) 3 > 2 > 4 > 1

119. The hydrogen ion concentration is 0.2 M ethanoic acid ($K_C = 2 \times 10^{-5}$ mol dm^{-3}) is approximately:

(a) 10^{-4} (b) 2×10^{-2}
(c) 2×10^{-6} (d) 2×10^{-3}

120. The pH of solution made by mixing 50 mL of 0.01 M barium hydroxide solution with 50 mL of H_2O is:

(a) 3.0 (b) 6.0
(c) 12.0 (d) 15.0

PRACTICE QUESTIONS – III

121. Find the molar solubility of $Fe(OH)_3$ in a buffer solution that 0.10 M in NH_4Cl and 0.10 M in NH_3. If K_b $(NH_3) = 1.8 \times 10^{-5}$ and Ksp $[Fe(OH)_3] = 2.6 \times 10^{-39}$:

(a) 4.458×10^{-25} M (b) 3.458×10^{-25} M
(c) 2.229×10^{-24} M (d) 4.458×10^{-22} M

122. The acid ionization (hydrolysis) constant of Zn^{2+} is 1.0×10^{-9}. Which of the following statement are correct?

(1) The basic dissociation constant of $Zn(OH)^+$ is 1.0×10^5.
(2) The pH of 0.001 M $ZnCl_2$ solution is 6.
(3) The basic dissociation constant of $Zn(OH)^+$ is 1.0×10^{-5}.
(4) The $[H_3O^+]$ ion concentration in 0.001 M $ZnCl_2$ solution is 1.0×10^{-4}.

(a) 1, 2 (b) 2, 3
(c) 3, 4 (d) 1, 2, 4

123. A $(OH)_2$ is a partially soluble substance. Its Ksp value is 4×10^{-12}, which of the following statement is correct?

(1) The solubility is unaffected by pH of the medium
(2) Its solubility has been decreased in a buffered medium at pH at 1 L
(3) Its solubility has been increased in a buffered medium having pH at 9
(4) Its saturated solution has pH is equal to 10.3

(a) 1, 2, 3 (b) 3 and 4
(c) 2 and 3 (d) 2, 3 and 4

124. Find the minimum volume of H_2O needed to dissolve 1g $CaSO_4$ at 248K [K_{SP} of $CaSo_4 = 9.1 \times 10^{-6}$]:

(a) 2.43L (b) 1.215L
(c) 4.86L (d) 24.3L

125. Ionization constant of CH_3COOH is 1.7×10^{-5} and concentration of H^+ ion is 3.4×10^{-4}. Then initial concentration of CH_3COOH is:

(a) 3.4×10^{-4} (b) 3.4×10^{-3}
(c) 6.8×10^{-4} (d) 6.8×10^{-3}

126. Which of the following is not a buffer solution?

(a) 0.8M H_2S + 0.8M KHS
(b) 2M $C_6H_5NH_2$ + 2M$C_6H_5\overset{+}{N}H_3\overset{-}{Br}$
(c) 3M H_2CO_3 + 3M $KHCO_3$
(d) 0.05 $KClO_4$ + 0.05M $HClO_4$

127. Which of the following is a set of amphiprotic species?

(a) H_3O^+, HCO_3^-, HSO_4^-
(b) H_2O, $H_2PO_2^-, HCO_3^-$
(c) H_2O, HPO_3^{2-}, HPO_4^{2-}
(d) H_2O, $H_2PO_3^-, H_2PO_4^-$

128. One litre of 0.5 M KCI solution is electrolysed for one minute in a current of 16.08 mA. Considering 100% efficiency, the pH of resulting solution will be:

(a) 7 (b) 9
(c) 8 (d) 10

129. $Ag^+ + NH_3 \rightleftharpoons [Ag(NH_3)^+]$; $K_1 = 3.5 \times 10^{-3}$
$[Ag(NH_3)]^+ + NH_3 \rightleftharpoons [Ag(NH_3)_2]^+$; $K_2 = 1.7 \times 10^{-3}$ Then the formation constant of $[Ag(NH_3)_2]^+$ is:

(a) 6.08×10^{-6} (b) 6.08×10^{6}
(c) 6.08×10^{-9} (d) None

130. When 0.1 mole of CH_3NH_2 (ionization constant $K_b = 5 \times 10^{-4}$) is mixed with 0.08 mol HCl and the volume is made up of 1 litre. Find the $[H^+]$ of resulting solution:

(a) 8×10^{-2} (b) 2×10^{-11}
(c) 1.23×10^{-4} (d) 8×10^{-11}

131. Which of the following solutions will have pH close to 1.0?

(a) 100 ml of (M/10) HCl + 100 ml of (M/10) NaOH
(b) 55 ml of (M/10) HCl + 45 ml of (M/10) NaOH
(c) 10 ml of (M/10) HCl + 90 ml of (M/10) NaOH
(d) 75 ml of (M/10) HCl + 25 ml of (M/5) NaOH

132. If $Ag^+ + 2NH_3 \rightleftharpoons Ag(NH_3)_2^+$; $K_1 = 1.8 \times 10^7$ $Ag^+ + Cl^- \rightleftharpoons AgCl$; $K_2 = 5.6 \times 10^9$

Then for

$AgCl + 2NH_3 \rightleftharpoons [Ag(NH_3)_2]^+ + Cl^-$,

Equilibrium constant will be:

(a) 0.32×10^{-2} (b) 3.11×10^{2}
(c) 10.08×10^{16} (d) 1.00×10^{-17}

133. There sparingly soluble salts A_2X, AX and AX_3 have the same solubility product. Their solubilities will be in the order:

(a) $AX_3 > AX > A_2X$
(b) $AX_3 > A_2X > AX$
(c) $AX > AX_3 > A_2X$
(d) $AX > A_2X > AX_3$

134. Which of the following solution(s) have pH between 6 and 7?

1. 2×10^{-6} M NaOH
2. 2×10^{-6} M HCl
3. 10^{-8} M HCl
4. 10^{-13} M NaOH

(a) 1, 2 (b) 2, 3
(c) 3, 4 (d) 2, 3, 4

135. What is the pH value at which $(MgOH)_2$ begins to precipitate from a solution containing 0.10 M Mg^{+2} ion? Ksp of $Mg(OH)_2$ is 1×10^{-11}:

(a) 3 (b) 6
(c) 9 (d) 11

136. 50 litres of 0.1 M HCl are mixed with 50 litres of 0.2 M NaOH. The POH of the resulting solution is:

(a) 12.70 (b) 12.34
(c) 8.7 (d) 4.2

137. The pH of 0.05 M aqueous solution of diethylamine is 12. Its Kb is:

(a) 2×10^{-3} (b) 2.5×10^{-3}
(c) 3×10^{-3} (d) 4.5×10^{-3}

138. Ka for HCN is 5×10^{-10} at 25°C. For maintaining a constant pH of 9, the volume of 5 M KCN solution required to be added to 10 mL of M HCN solution is:

(a) 2 mL (b) 3 mL
(c) 4.2 mL (d) 5.6 mL

139. The approximate pH of a solution formed by mixing equal volumes of solutions of 0.1 M sodium propionate and 0.1 M propanoic acid (the dissociation constant of propanoic acid is 1.3×10^{-5} mol dm^{-3}) will be:

(a) 2.45 (b) 4.89
(c) 5.98 (d) 6.89

140. The dissociation constant of acetic acid is 1.6×10^{-5}. The degree of dissociation (α) of 0.01 M acetic acid in the presence of 0.1 M HCl is equal to:

(a) 0.4 (b) 0.026
(c) 1.6 (d) 0.016

141. If the equilibrium constant of the reaction of weak acid HA with strong base is 10^9, then pH of 0.1 M NaA is:
(a) 3 (b) 9
(c) 7 (d) 6

142. If Ksp of $Al(OH)_3$ is 1.0×10^{-15}.M. Find at what pH does 1.0×10^{-3}.M Al^{3+} precipitate on the addition of buffer of NH_4Cl and NH_4OH solution:
(a) 10 (b) 10.5
(c) 11 (d) 12

143. Acetic acid and aq. NH_3 are weak monobasic acid and weak monobasic base respectively and Ka of acetic acid is equal to K_b of aq. NH_3. Which of the following statements are incorrect?
(1) If acetic acid is exactly neutralized by aqueous NH_3 then pH of resulting solution is equal to $\frac{1}{2}$ pkw.
(2) All the above mixing would result solution having pH = 7 at 25°C.
(3) If acetic acid is exactly half neutralized by NaOH, then pH of resulting solution is equal to pKa.
(4) If aq. NH_3 is exactly half neutralized by HCl, then pOH of resulting solution is equal to pK_b.
(a) 2 and 4 (b) 2 and 3
(c) 1 and 3 (d) 2 only

144. Four species are listed below:
(1) HCO_3^- (2) H_3O^+
(3) HSO_4^- (4) HSO_3F
Which one of the following is the correct sequence of their acid strength?
(a) 4 < 2 < 3 < 1 (b) 2 < 3 < 1 < 4
(c) 1 < 3 < 2 < 4 (d) 3 < 1 < 4 < 2

145. The solubility product of a salt having general formula MX_2 in water is, 4×10^{-12}.
The concentration of M^{2+} ions in the aqueous solution of the salt is:
(a) 1.6×10^{-4} M (b) 2.0×10^{-6} M
(c) 1.0×10^{-4} M (c) 4.0×10^{-10} M

146. The pKa of a weak acid, HA is 4.80. The pK_b of a weak base, BOH is 4.78. The pH of an aqueous solution of the corresponding salt, BA will be:
(a) 9.58 (b) 4.79
(c) 7.01 (d) 9.22

147. In a saturated solution of AgCl NaCl is added gradually. The concentration of Ag^+ is plotted against the concentration of Cl^-. The graph can be represented as:

(a)
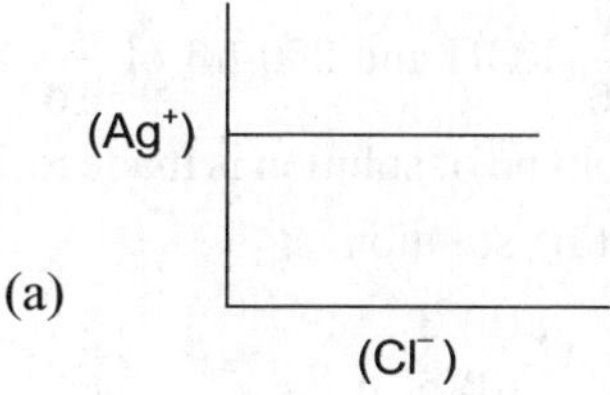

(b)
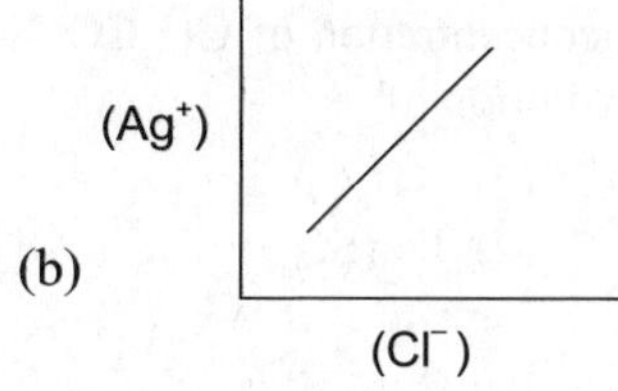

(c)
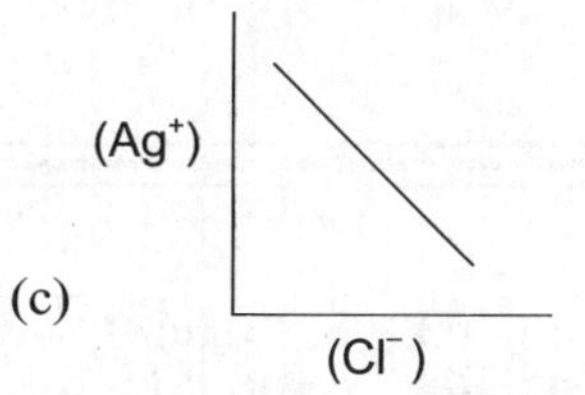

(d)
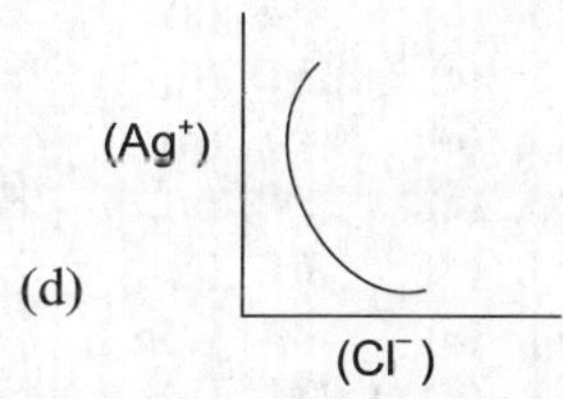

148. What is the minimum pH needed to prevent the precipitation of ZnS in a solution that is 0.01 M $ZnCl_2$ and saturated with 0.1 M H_2S?
$[K_{SP} = 10^{-21}, K_{a_1} \times K_{a_2} = 10^{-20}]$
(a) 0 (b) 1
(c) 3 (d) 5

149. The two bronsted bases in the following reaction are
$HC_2O_4^- + PO_4^{3-} \rightleftharpoons HPO_4^{2-} + C_2O_4^{2-}$:
(a) $HC_2O_4^-$ and HPO_4^{2-} (b) HPO_4^{2-} and $C_2O_4^{2-}$
(c) PO_4^{3-} and $C_2O_4^{2-}$ (d) $HC_2O_4^-$ and PO_4^{3-}

150. When in a 100 ml aqueous HNO_3 solution of pH = 1, 900 ml of distilled water is added, the pH of the resultant solution becomes:
(a) 2 (b) 4
(c) 8 (d) 0.5

151. Solubility of AgCl in H_2O, 0.02 M $MgCl_2$, 0.02 M NaCl, and 0.05 M $AgNO_3$ are S_1, S_2, S_3, and S_4 respectively then:
(a) $S_1 > S_3 > S_5 > S_4$ (b) $S_1 > S_3 > S_2 > S_4$
(c) $S_1 > S_2 = S_3 > S_4$ (d) $S_4 > S_3 > S_1 > S_2$

152. When 200 ml of $\frac{W}{20}$ KOH and 200 ml of $\frac{N}{10}$ HNO_3 are mixed and the volume of solution is made to 1 litre. The pH of the resulting solution is:

(a) 2 (b) 3
(c) 4 (d) 5

153. The hydronium ion concentration of CH_3COOH can be increased by the addition of:

(1) NaCN (2) NH_4Cl
(3) $AgNO_3$ (4) $Fe_2(SO_4)_3$

(a) 2, 3 (b) 2, 3, 4
(c) 1, 3, 4 (d) 1, 2, 4

154. Which of the following solutions will have no effect on pH on dilution?

(1) 0.2M KHS
(2) 0.8M NH_4NO_3
(3) 0.01M NaCl
(4) 1M CH_3COONH_4

(a) 1, 3 (b) 1, 4
(c) 2, 3 (d) 2, 4

Answer Keys

1. (b)	2. (b)	3. (c)	4. (c)	5. (c)	6. (b)	7. (a)	8. (a)	9. (d)	10. (b)
11. (a)	12. (b)	13. (d)	14. (b)	15. (a)	16. (c)	17. (a)	18. (b)	19. (d)	20. (a)
21. (a)	22. (a)	23. (b)	24. (a)	25. (b)	26. (b)	27. (b)	28. (b)	29. (c)	30. (a)
31. (c)	32. (d)	33. (c)	34. (d)	35. (a)	36. (b)	37. (d)	38. (b)	39. (c)	40. (a)
41. (b)	42. (b)	43. (c)	44. (d)	45. (c)	46. (d)	47. (a)	48. (c)	49. (c)	50. (b)
51. (c)	52. (c)	53. (c)	54. (a)	55. (a)	56. (d)	57. (a)	58. (d)	59. (c)	60. (c)
61. (b)	62. (a)	63. (d)	64. (c)	65. (a)	66. (a)	67. (d)	68. (a)	69. (a)	70. (a)
71. (c)	72. (a)	73. (a)	74. (b)	75. (b)	76. (d)	77. (d)	78. (b)	79. (a)	80. (c)
81. (b)	82. (c)	83. (d)	84. (b)	85. (c)	86. (d)	87. (d)	88. (c)	89. (b)	90. (c)
91. (a)	92. (a)	93. (a)	94. (a)	95. (a)	96. (c)	97. (c)	98. (d)	99. (c)	100. (c)
101. (b)	102. (b)	103. (b)	104. (d)	105. (b)	106. (c)	107. (c)	108. (c)	109. (b)	110. (c)
111. (c)	112. (d)	113. (b)	114. (b)	115. (c)	116. (c)	117. (a)	118. (b)	119. (d)	120. (c)
121. (a)	122. (b)	123. (d)	124. (a)	125. (d)	126. (d)	127. (d)	128. (c)	129. (a)	130. (d)
131. (d)	132. (a)	133. (d)	134. (b)	135. (c)	136. (a)	137. (b)	138. (a)	139. (b)	140. (d)
141. (b)	142. (a)	143. (d)	144. (a)	145. (c)	146. (c)	147. (d)	148. (b)	149. (c)	150. (a)
151. (b)	152. (a)	153. (b)	154. (b)						

Hints and Explanations for Selective Questions

2. Conjugate base of OH^-

$$OH^- \rightleftharpoons O^{2-} + H^+$$

4. $HPO_4^{2-} + H_2O \rightleftharpoons H_2PO_4^{2-} + OH^-$

$HPO_4^{2-} + H_2O \rightleftharpoons PO_4^{3-} + H_3O^+$

$H_2PO_2^-$ is a conjugate base of H_3PO_2 (a monobasic acid and does not give H^+). HPO_3^{2-} is a conjugate base of $H_2PO_3^-$ and does not ionize further so H_3PO_3 is a dibasic acid.

5. 1×10^{-8} M HCl is acidic.

pH < 7

So false.

6. H_3PO_4 is a tribasic acid, thus ionizing in three steps:

I. $H_3PO_4 \rightleftharpoons H^+ + H_2PO_4^-$

II. $H_2PO_4^- \rightleftharpoons H^+ + HPO_4^{2-}$

III. $HPO_4^{2-} \rightleftharpoons H^+ + PO_4^{3-}$

Conjugate base is formed when an acid loses its proton. Thus HPO_4^{2-} is the conjugate base of $H_2PO_4^-$ (which is an acid in step II, but is the conjugate base of H_3PO_4 in step I).

7. During thunderstorm, there is formation of 2NO which changes to NO_2 and ultimately to $2HNO_3$ (acid rain)?

$$N_2 + O_2 \longrightarrow 2NO \xrightarrow{O_2} NO_2$$

$$N_2O_5 \xrightarrow{H_2O} 2HNO_3 \ (pH < 7)$$

8. For the solute Ax By $\rightleftharpoons xA^{+y} + yB^{-x}$

$Ksp = x^x\, y^y\, (s)^{x+y}$

$MX_4 \rightleftharpoons M^{4+} + 4X^-$

$x = 1, y = 4$

$Ksp = (4)^4 (1)^1 (s)^5 = 256\, s^5$

$s = (Ksp/256)^{1/5}$

15. As $[H^+] = c\,\alpha\, \frac{1}{100} \times 1 = 10^{-2}$ so pH = 2.

22. As Cl^- is a conjugate base of HCI.

31. For pure water, $[H_3O^+] = [OH^-]\ Kw = 10^{-6} \times 10^{-6} = 10^{-12}$

32. Among the given oxyacids HClO is the weakest as acidity depends on the oxidation state (+1 in HClO, which is lowest). Hence its conjugate base (ClO^-) will be the strongest.

33. Molar conc. of HCl = 10^{-8} pH = 8. But this cannot be possible as pH of an acidic solution can not be more than 7. so we have to consider $[H^+]$ coming from H_2O.

35. In molten state the cations and anions become free and flow of current is due to migration for these ions in opposite directions in the electric field.

36. $HCl + H_2O \rightleftharpoons H_3O^+ + Cl^- \ (K_a)$

$Cl^- + H_2O \rightleftharpoons HCl + \overline{O}H \ (K_b)$

As we know that $K_w = [H^+]\,[\overline{O}H]$

$= K_a.K_b$

37. In case of acidic buffer pH = $pK_a \pm 1$ and for a basic buffer pH = $pK_b \pm 1$.

39. The pH of the solution at the equivalence point will be greater than 7 due to salt hydrolysis. So an indicator giving colour on the basic side will be suitable.

40. Sn in $SnCl_4$ has its octet completed while the central atoms in other molecules have incomplete octet.

41. NH_2^- (based) + $H^+ \longrightarrow NH_3$ (conjugate acid)

42. $NH_4OH + HCl \longrightarrow NH_4Cl$

NH_4Cl is a salt of weak base and strong acid, so it gives the acidic solution with pH < 7.

43. CH_3COONa give basic solution as it is a salt of weak acid and strong base.

44. Solubility of hydroxide of group 2 elements increase down the group. Therefore $Be(OH)_2$ has lowest solubility and hence lowest solubility product.

45. pH of blood remains same because it has a buffer solution of H_2CO_3/HCO_3^-.

46. In all the 1, 2, 3 buffer solution is prepared. In choice 2, when CH_3COONa reacts with HCl it forms CH_3COOH + NaCl.

47. Weak base (NH_4OH) + salt of a strong acid (NH_4Cl).

48. $HN_3 \longrightarrow H^+ + N_3^-$

Hydrazoic acid — conjugate base — (Azide ion)

49. BF_3 and SO_3 are Lewis acids because they can accept electron pairs.

50. $NH_4Cl + NH_4OH$ precipitate radical of group III *viz.* Al^{3+} and Cl^{3+}.

51. Solubility product α solubility PtS having least Ksp is least soluble.

52. $K_{x_1} = \frac{[H^+][x^-]}{[Hx]}$

As $[H^+] = [x^-]$

$K_{a_1} = \frac{[H^+]^2}{[Hx]}$

Similarly, $K_{a_2} = \frac{[H^+][y^-]}{[Hy]} = \frac{[H^+]^2}{[Hy]}$

As $[H^+]$ is same for the two acids so

$\frac{Ka_1}{K_{a_1}} = \frac{[Hy]}{[Hx]} = \frac{4}{1}$

$\frac{[Hx]}{[Hy]} = \frac{1}{4} = 0.25$

53. $H_2X \rightleftharpoons H^+ + HX^- \ (Ka_1)$

$HX^- \rightleftharpoons H^+ + H^{2-} \ (Ka_2)$

So $Ka_1 > Ka_2$

55. $Ap\,Bq \rightleftharpoons pA^{q+} + B^{p-}$

pS qS

Ksp or Ls = $[A^{q+}]^p . [B^{p-}]^q = (pS)^p.(qS)^q = p^p . S^p . q^q . S^q$

$Ls = S^{p+q}. p^p. q^q$

59. As acidic strength is inversely proportional to pKa.

61. HCl: Strong acid

NH_4Cl: Salt of weak base + strong acid $\longrightarrow$ it gives acidic solution.

NaCl: Salt of strong base + strong base $\longrightarrow$ it gives neutral solution with pH = 7.

NaCl: Salt of strong base + weak acid $\longrightarrow$ it gives basic solution with pH > 7.

Thus the increasing order of pH is:

$HCl < NH_4Cl < NaCl < NaCN$

63. $A_2B_3 \rightleftharpoons 2A^{+3} + 3B^{-2}$

$\quad\quad\quad\quad 2S \quad 3S$

$Ksp = [A^{+3}]2\ [B^{-2}]^3 = (2S)^2\,(3S)^3$

$Ksp = 108\ S^5$

64. pKa = 14 – pKb = 14 – 10.83 = 3.17

Ka = Antilog (–pKa)

Ka = Antilog (–3.17) = 6.75×10^{-4}

65. In the Henderson's equation, pH = pKa + log [Salt]/[Acid] when [Salt] = [Acid] pH = pKa = 9.30

66. As the strength of conjugate base α 1/acidic strength As acidic strength is

$HOH > HC \equiv CH > NH_3 > CH_3 - CH_3$ So

$CH_3 - CH_2^- > NH_2^- > H - C \equiv C^- > OH^-$

67. $HCl + CH_3COOH \rightleftharpoons Cl^- + CH_3COOH_2^+$

Acid Base Conj.base Conj.acid

68. Gastric juice is acidic. So the ionization of aspirin is suppressed whereas small intestine has basic medium therefore aspirin is more ionized.

69. As all the reactants and products are present in aqueous form in

$KNO_3\,(aq) + NaCl\,(aq) = KCl\,(aq) + NaNO_3\,(aq)$

So it is a reversible reaction. In others either solid or gas is generated which is insoluble or volatile and hence makes the reaction unidirectional.

70. The equilibrium constant for the neutralization of a weak acid with a strong base,

$$K = \frac{Ka}{Kw} = \frac{1.0\times10^{-4}}{1.0\times10^{-14}} = 1.0 \times 10^{10}$$

71. For a basic buffer, pH = 14 – pKb – log $\frac{[salt]}{[base]}$ = 14 – (–log 10^{-10}) –log 1, pH = 4

72. Calculate the ionic product $[Ca^{2+}][F^-]^2$ in each case and see that ionic product must exceed Ksp for precipitation. Concentration of each ion is halved due to dilution effect (volume doubled).

73. $2NH_3 \rightleftharpoons NH_4^+ + NH_2^{\ominus}$

For this reaction k = 10^{-30}

$[\overset{\ominus}{N}H_2] = \sqrt{k} = \sqrt{10^{-30}} = 10^{-15}$

As No. of $NH_2^{\ominus}$ in 10^3 cm³ = $10^{-15} \times 6.02 \times 10^{23}$

As No. of $NH_2^{\ominus}$ in 10^3 cm³ = $\frac{10^{-15} \times 6.02 \times 10^{23}}{10^3}$

$= 6 \times 10^5$ ions

74. $[H^+]$ = antilog of pH = antilog of 5.4

$= 3.98 \times 10^{-6}$

75. Na_2CO_3 is salt of weak acid H_2CO_3 and strong base NaOH, therefore, has pH more than 7. Also, it dissociate to give two moles of NaOH.

$Na_2CO_3 + 2H_2O \longrightarrow 2NaOH + 2H_2CO_3$

76. The correct order of H_3O^+ is as follows:

$\underset{\text{Basic}}{NaNO_2} < \underset{\text{Neutral}}{NaCl} < \underset{\text{Weak acid}}{H_2S} < \underset{\text{Strong acid}}{H_2SO_4}$

77. 0.005 M calcium acetate $(CH_3CaOO)_2$ Ca

As $[CH_3COO^-] = 0.01$ M

$CH_3COO^- + H_2O \rightleftharpoons CH_3COOH + OH^-$

alkaline

$$pH = 7 + \frac{pKa}{2} + \frac{\log C}{2} = 7 + 2.37 + \frac{\log 0.01}{2}$$

$= 7 + 2.37 - 1 = 8.37$

78. Solubility product = $(2x)^2\,(x) = 4x^3$

$= 4\,(3.5 \times 10^{-6})^3 = 1.7 \times 10^{-16}$

79. $AB_2 \rightleftharpoons A^{2+} + 2B^-$

$\quad\quad\quad\quad S \quad\quad 2S$

$Ksp = [A^{2+}]\,[B^-]^2$

$= (S)\,(2S)^2 = 4S^3$

$= 4\,(1 \times 10^{-5})^3 = 4 \times 10^{-15}$

80. Glycine $\left(\begin{matrix} CH_2-COOH \\ | \\ NH_2 \end{matrix}\right)$ is more acidic than basic so

$K = Ka_1 \cdot Ka_2$

$= 4.5 \times 10^{-3} \times 1.7 \times 10^{-10} = 7.65 \times 10^{-13}$

$[H^+] = \sqrt{KC}$

$= \sqrt{7.65 \times 10^{-3} \times 0.01} = 0.87 \times 10^{-7}$ M

pH = –log H^+

= –log $[0.87 \times 10^{-7}]$ = 6.07

81. pH = 5.

So $[H^+] = 1 \times 10^{-5}$

As $[H^+] = C\alpha$

$1 \times 10^{-5} = 0.005 \times \alpha$

or $\alpha = 1 \times 10^{-5}/0.005$

$= 0.2 \times 10^{-2}$

82. $s = (Ksp/4)^{1/3} = (4 \times 10^{-9}/4)^{1/3} = 1 \times 10^{-3}$

83. As acidic strength α 1/pKa

84. As $[H^+] = 0.0005 = 5 \times 10^{-4}$

So $[OH^-] = \dfrac{1\times10^{-14}}{5\times10^{-4}}$

$= 2 \times 10^{-11}$

pOH $= -\log 2 \times 10^{-11}$

$= 11 - 0.3010 = 10.6990$

85. $pH = pKa + \log_{10}\dfrac{[Salt]}{[Acid]}$

$= 4.57 + \log_{10}\dfrac{0.1}{0.03}$

$= 5.57 - 0.4771$

$= 5.09$

87. As here complete neutralization occurs due to same milliequivalents so pH will be 7.

90. $[H^+] = \sqrt{KaC}$

$= \sqrt{(1.8 \times 10^{-5} \times 0.1)}$

$= 1.3 \times 10^{-3}$

91. As $\alpha = \sqrt{Ka/C}$

$= \sqrt{(4.9 \times 10^{-8}/0.1)}$

$= 7 \times 10^{-4}$

so α% = 0.07.

93. As pH = 3

So $[H^+] = 1 \times 10^{-3}$

$\alpha = [H^+]/C$

$= 1 \times 10^{-3}/0.1$

$= 0.01$

α% = 1%.

95. $M_R = \dfrac{M_1V_1 + M_2V_2}{\text{Total volume}}$

$= \dfrac{0.2\times75+1\times25}{400} = 1/10$

$= 0.1$

so net $[H^+] = 0.1$ hence pH = 1.

96. $HA_1/HA_2 = \sqrt{(Ka_1/Ka_2)}$

$= \sqrt{(3.0 \times 10^{-4}/1.8 \times 10^{-5})}$

$= \sqrt{(300/18)}$

$\approx 4:1$

98. As pH = 5

So $[H^+] = 1 \times 10^{-5}$

$\alpha = [H^+]/C$

$= 1 \times 10^{-5}/0.005$

$= 0.002$

= so α% = 0.2%

99. $M_R = \dfrac{M_1V_1 + M_2V_2}{\text{Total volume}}$

$= \dfrac{100\times0.015+100\times0.005}{200} = 1/100$

i.e., $[H^+] = 1/100 = 0.01$

so pH = 2

100. $A_2X_3 \rightleftharpoons 2A^{+3} + 3X^{-2}$

2S 3S

$Ksp = [A^{+3}]^2 [X^{-2}]^3$

$= (2S)^2 (3S)^3 = 108\ S^5$

$S = \dfrac{Ksp^{1/5}}{108} = \dfrac{1.08\times10^{-23\ 1/5}}{108}$

$S = [10^{-25}]^{1/5} = 10^{-5}$

101. $M(OH)x \rightleftharpoons M^{+x} + x\ OH^-$

10^{-4} $10^{-4} \times x\ 10^{-4}$

$Ksp = [M^{+x}] [OH^-]^x = 4 \times 10^{-12}$

$[10^{-4}] [x \times 10^{-4}]^x = 4 \times 10^{-12}$

$[x \times 10^{-4}]\ x = 4 \times 10^{-8}$

It holds good if x = 2

102. Solution of I is acidic (pH < 7), solution of II is basic (ph > 7), solution of III is neutral (pH = 7) while solution of IV is strongly basic (pH > > 7).

Thus the order of pH is IV > II > III > I

103. $pH = pKa + \log\dfrac{[Salt]}{[Acid]}$

$\log\dfrac{[Salt]}{[Acid]} = pH - pKa = 6 - 5 = 1$

$[pKa = -\log Ka = -\log 10^{-5} = 5]$

$\dfrac{[Salt]}{[Acid]} = \text{antilog}(1) = 10$

104. The corresponding acids are:
(a) CH_3OH (b) C_2H_5OH
(c) $(CH_3)_2CHOH$ (d) $(CH_3)_3COH$.

Among 1°, 2° and 3° alcohols, the order of acidity is 1°> 2° > 3°. A weak acid has strong conjugate base.

Therefore (d) is strongest conjugate base.

105. $H_2O \rightleftharpoons H^+ + OH^-$

$$C = \frac{1000}{18} = 55.55 \text{ M}$$

$$Ka = \frac{[H^+][OH^-]}{[H_2O]} = \frac{C\alpha\ C\alpha}{C} = C\alpha^2$$

$$Ka = 55.55 \times (1.8 \times 10^{-9})^2 = 1.8 \times 10^{-16}$$

106. Acidic strength of oxyacids increases with the increase in electronegativity of the non-metal and its oxidation number. Thus, the order is:
$HClO_4$ (3) > H_2SO_4 (1) > H_3PO_4 (2)

107. pH = 10 means

$[H^+] = 10^{-10}$ mol $L^{-1} = 10^{-13}$ g ions/cc
$= 10^{-13} \times (6.02 \times 10^{23})$ ions/cc
$= 6.02 \times 10^{10}$ ions/cc

108. $HCOOH/CH_3COOH$
$= \sqrt{(\text{Ka for formic acid/Ka for acetic acid})}$
$= \sqrt{(1.77 \times 10^{-4}/1.75 \times 10^{-5})}$
$= \sqrt{(10/1)}$
$= 3.178/1$
$= 3.178$

117. The one with lower value of Ksp gets precipitated first. So Ksp of CuS will be less than Ksp of ZnS.

118. Acidity constants of HCN = 7.2×10^{-10};
HCOOH = 17.7×10^{-5}
$CH_3COOH = 1.75 \times 10^{-5}$
$Cl - CH_2COOH = 1.36 \times 10^{-5}$
Hence, the correct order is $CH_2ClCOOH$ > HCOOH > CH_3COOH > HCN

119. $CH_3COOH\ (aq) \rightleftharpoons CH_3COO^- + H^+$

Initial conc.	C	–	–
Eq. conc.	C – X	X	X

$$Ka = \frac{X.X}{C.X} = 2 \times 10^{-5}$$

$$\frac{X^2}{0.2 - X} = 2 \times 10^5$$

Assuming 0.2 – X = 0.2
$X^2/0.2 = 2 \times 10^{-5}$
$[H^+] = X = \sqrt{(4 \times 10^{-6})} = 2 \times 10^{-3}$ mol dm^{-3}

120. The volume of solution is made double by the addition of 50 mL of water, therefore, the molarity of dilute:

$$Ba(OH)_2 \text{ solution} = \frac{0.01}{2} \text{ M} = 0.005 \text{ M}$$

$$[OH^-] = 0.005 \frac{\text{mol } Ba(OH)_2}{L} \times \frac{2\text{mol } OH^-}{1\text{mol } Ba(OH)_2} = 0.01 \text{ M}$$

pOH = –log $[OH^-]$ = –log (0.01) = 2
pH = 14 – 2 = 12.

121. $NH_3 + H_2O \rightleftharpoons NH_4^+ + OH^-$

$$K_b = 1.8 \times 10^{-5} = \frac{[NH_4^+][OH^-]}{[NH_3]} = \frac{0.1[OH^-]}{0.1}$$

$[OH^-] = 1.8 \times 10^{-5}$
$Fe(OH)_3\ (s) \rightleftharpoons Fe^{3+}\ (aq) + 3OH^-\ (aq)$
$Ksp = 2.6 \times 10^{-39} = [Fe^{3+}]\ (1.8 \times 10^{-5})^3$

$$[Fe^{3+}] = \frac{2.6 \times 10^{-39}}{5.832 \times 10^{-15}}$$

So $[Fe^{3+}] = 4.458 \times 10^{-25}$ M

124. $CaSO_4 \rightleftharpoons Ca^{+2} + SO_4^{2-}$

$K_{SP} = s^2$ or $s = \sqrt{K_{SP}} = \sqrt{9.1 \times 10^{-6}}$
$= 3.02 \times 10^{-3}$ m/L
$= 3.02 \times 10^{-3} \times 136 = 0.411$g/L

As for dissolving 0.411g $CaSO_4$ H_2O needed = 1L

So for dissolving 1g $CaSO_4$ H_2O needed $= \frac{1}{0.411}$
= 2.43L

125. $CH_3COOH \rightleftharpoons CH_3COO^- + H^+$

$$Ka = \frac{[CH_3COO^-][H+]}{[CH_3COOH]}$$

Given here,
$[CH_3COO^-] = [H^+] = 3.4 \times 10^{-4}$ M
Ka for $CH_3COOH = 1.7 \times 10^{-5}$
As CH_3COOH is a weak acid, so in it (CH_3COOH) is equal to initial concentration. Hence

$$1.7 \times 10^{-5} = \frac{(3.4 \times 10^{-4})(3.4 \times 10^{-4})}{[CH_3COOH]}$$

$[CH_3COOH] = 3.4 \times 10^{-4} \times 3.4 \times 10^{-4}\ 1.7 \times 10^{-5}$
$= 6.8 \times 10^{-3}$ M

126. Here, $KClO_4 + HClO_4$ cannot act like a buffer as it is a combination of strong acid and it's salt with strong base.

127. Here H_2O, $H_2PO_3^-$ and $H_2PO_4^-$ represents set of amphiprotic species.

128. The redox changes are:

$K^+ + e \longrightarrow K$ [at cathode]

$2H_2O \longrightarrow 4H^+ + O_2 + 4e^-$ [at anode]

W/E = it/96500

Eq. of K^+ lost = equivalent of H^+ formed

$$\frac{It}{96500} = \frac{16.08\times10^{-3}\times1\times60}{96500} = 10 \times 10^{-6}$$

$$[H^+] = \frac{10\times10^{-6}}{10^3} = 10 \times 10^{-9} = 10^{-8}$$

pH = 8.

130. $CH_3NH_2 + HCl \rightleftharpoons CH_3NH_3^+Cl^-$

0.1 0.08 0

0.02 0 0.08

As it is a basic buffer solution.

$$pOH = pK_b + \log\frac{0.08}{0.02}$$

$= -\log 5 \times 10^{-4} + \log 4$

$= 3.30 + 0.602 = 3.902$

pH = 14 − 3.902 = 10.09

$[H^+] = 7.99 \times 10^{-11} = 8 \times 10^{-11}$ M

131. $$MH^+ = \frac{M_1V_1 - M_2V_2}{V_1 + V_2}$$

$$= \frac{0.1\times75 - 0.2\times25}{100} = \frac{7.5 - 5.0}{100}$$

$MH^+ = 2.5/100 = 2.5 \times 10^{-2}$ M

$= 25 \times 10^{-3}$ M

$pH = -\log [H^+] = -\log (25 \times 10^{-3})$

$= -1.398 + 3 = 1.602$

i.e., in this case, pH is close to 1.

132. $$K_1 = \frac{[Ag(NH_3)_2]^+}{[Ag^+][NH_3]^2}$$

$$K_2 = \frac{[Ag(Cl)]}{[Ag^+][Cl^-]}$$

$$K = \frac{[Ag(NH_3)_2]^+[Cl^-]}{[AgCl][NH_3]^2} = \frac{K_1}{K_2}$$

$$\underline{K_1} = \frac{[Ag(NH_3)_2]^+}{[Ag^+][NH_3]^2} \times \frac{[Ag^+][Cl^-]}{[AgCl]}$$

$$K_2 = \frac{1.8\times10^7}{5.6\times10^9} = 0.32 \times 10^{-2}$$

133. $A_2X \rightleftharpoons 2A^+ + X^{-2}$

2s s

$Ksp = (2s)^2 (s) = 4s^3$

$s = [K_{sp}/4]^{1/3}$

$AX \rightleftharpoons A^+ + X^-$

s_1 s_1

$Ksp = (s_1)^2$

$s_1 = (K_{sp})^{1/2}$

$AX_3 \rightleftharpoons A^{+3} + 3X^-$

s_2 $3s_2$

$K_{sp} = (s_2)(3s_2)^3 = 27s_2^4$

$s_2 = [K_{sp}/27]^{1/4}$

Thus, it is clear that $s_1 > s > s_2$ so

$AX > A_2X > AX_3$

134. pH of 10^{-8} M HC1 is not 8 but it is less than 7 because in this case contribution of H^+ from water is not neglected.

Total $H^+ = 10^{-8}$ (from acid) + 10^{-7} (from water)

$= 10^{-8}(1 + 10)$

$= 11 \times 10^{-8}$ M

$pH = -\log [H^+] = -\log [11 \times 10^{-8}]$

$= -[\log 11 + \log 10^{-8}]$

$= -[1.0414 - 8]$

$= 6.9586 \simeq 6.96$

135. When $Mg(OH)_2$ starts undergoing precipitation, then

$[Mg^{+2}][OH^-]^2$ = Ksp of $Mg(OH)_2$

$[0.1][OH^-]^2 = 1 \times 10^{-11}$

$[OH^-] = 10^{-5}$ M

Now $pOH = -\log [OH^-] = -\log 10^{-5} = 5$

pH = 14 − pOH = 14 − 5 = 9

136. $$M(H^+) = \frac{M_1V_1(\text{base}) - M_2V_2(\text{acid})}{V_1 + V_2}$$

$$= \frac{0.2\times50 - 0.1\times50}{100} = 5 \times 10^{-2}$$

$pH = -\log [H^+] = -\log (5 \times 10^{-2})$

= 1.301

Now, pH + pOH = 14

pOH = 14 − pH = 14 − 1.301 = 12.69

137. pH = 12 $[H^+] = 10^{-12}$

Now $[OH^-] = \frac{10^{-14}}{[H^+]} = \frac{10^{-14}}{10^{-12}} = 10^{-2}$ M

= 0.01 M

$(C_2H_5)_2NH + H_2O \rightleftharpoons (C_2H_5)_2NH^+ + OH^-$

initial molar conc	0.05	0	0
At equation	0.05 − 0.01	0.01	0.01

$K_b = \frac{[OH^-]^2}{[(C_2H_5)_2NH]} = \frac{(0.01)^2}{0.04}$

$= \frac{10^{-4} \times 10^2}{4} = \frac{10^{-2}}{4}$

$= 2.5 \times 10^{-3}$

138. Let X = mL of KCN solution to be added.

(10 + X) mL = Total volume of solution after mixing the two solutions.

Molarity (M_2) of HCN in the final solution:

$M_2 = \frac{M_1V_1}{V_2} = \frac{2M \times 10\,mL}{(10+X)\,mL} = \frac{20M}{(10+X)}$

Molarity (M'_2) of KCN in the final solution:

$M'_2 = \frac{M_1V_1}{V_2} = \frac{5M \times X\,mL}{(10+X)\,mL} = \frac{5XM}{(10+X)}$

Now using Henderson equation, we get:

$pH = -\log(Ka) + \log\frac{[KCN]}{[HCN]}$

$9.00 = -\log(5 \times 10^{-10}) + \log\frac{[5X/(10+X)]M}{[20/(10+X)]M}$

9.00 = 9.30 + log (X/4)

log (X/4) = −0.30

X/4 = 0.50

X = 2.0 mL

139. As CH_3CH_2COONa is a strong electrolyte.

So $CH_3CH_2COONa \longrightarrow CH_3CH_2COO^- + Na^+$

0 M 0.1 M 0.1 M

Now consider solution of 0.1 M propanoic acid having initially 0.1 M $CH_3CH_2COO^-$.

$CH_3CH_2COONa \longrightarrow CH_3CH_2COO^-(aq) + Na^+(aq)$

Initial conc	0.1 M	0.1 M	–
Final conc.	(0.1 − X)M	(0.1 + X)M	XM

$Ka = \frac{[CH_3CH_2COO^-(aq)][H^+]}{[CH_3CH_2COOH]} = 1.3 \times 10^{-5}$

$\frac{(0.1+X)(X)}{(0.1-X)} = 1.3 \times 10^{-5}$

Now assuming that 0.1 + X = 0.1

0.1 − X = 0.1, one gets

$\frac{(0.1)(X)}{0.1} = 1.3 \times 10^{-5}$ or $X = 1.3 \times 10^{-5} = [H^+]$

So $pH = -\log(1.3 \times 10^{-5}) = 4.89$

140. $HCl(aq) \longrightarrow H^+(aq) + Cl^-(aq)$

0.1 M 0.1 M

$CH_3COOH(aq) + H_2O(l) \rightleftharpoons CH_3COO^-(aq) + H_3O^+(aq)$

Initial conc.	0.01			0.1
Change	−X		+X	+X
Final conc.	(0.01 − X)		X	(0.1 + X)

$Ka = \frac{[CH_3COO^-][H_3O^+]}{[CH_3COOH]} = \frac{X(0.1+X)}{(0.01-X)}$

Now assuming that (0.1 + X) = 0.1 and (0.01 – X) = 0.01

$1.6 \times 10^{-5} = \frac{X \times 0.1}{0.01}$

$X = 1.6 \times 10^{-4}$

So degree of dissociation $\alpha = \frac{X}{C}$

$= \frac{1.6 \times 10^{-4}}{0.01} = 0.016 = 1.6\%$

141. $HA + OH^- \rightleftharpoons H_2O + A^-$

$K = \frac{[A^-]}{[HA][OH^-]}$

$10^9 = \frac{[A^-][H^+]}{[HA][OH^-][H^+]} = \frac{K_a}{K_w} = \frac{Ka}{1 \times 10^{-14}}$

$K_a = 1 \times 10^{-5}$

pKa = 5

NaA is a salt of weak acid and strong base

$pH = \frac{1}{2}[14 + pKa + \log_C]$

= 7 + 2.5 − 0.5 = 9

142. $Al(OH)_3(s) \rightleftharpoons Al(aq) + 3OH^-(aq)$

$Ksp = [Al^{3+}][OH^-]^3$

$Al(OH)_3$ precipitates when

$[Al^{3+}][OH^-]^3 > Ksp$

$(1 \times 10^{-3})[OH^-]^3 > 1.0 \times 10^{-15}$

$[OH^-]^3 > 1 \times 10^{-12}$

$[OH^-] > 1 \times 10^{-4}$ M

$pOH = -\log 1 \times 10^{-4} = 4$

$pH = 14 - 4 = 10$

144. Here the correct order is IV < II < III < I.

145. $MX_2\,(s) \rightleftharpoons M^{2+}\,(aq) + 2X^-\,(aq)$

$\qquad\qquad\qquad s \qquad\quad 2s$

$Ksp = s.\,(2s)^2 = 4s^3$

$4 \times 10^{-12} = 4s^3$

$s^3 = 1 \times 10^{-12}$

$s = 1 \times 10^{-4}$ M

$[M^{2+}] = 1 \times 10^{-4}$ M.

146. It is a salt of weak acid and weak bases.

$[H^+] = \sqrt{[(K_w \times Ka)]/K_b]}$

On solving, we get

pH = 7.01.

148. $K_{SP} = (Zn^{+2})\,[S^{-2}]$

$$[S^{-2}] = \frac{10^{-21}}{0.01} = 10^{-19}$$

$$\text{As } K_{a_1} \times K_{a_2} = \frac{[H^+]^2[S^{-2}]}{[H_2S]}$$

$$[H^+]^2 = \left[\frac{10^{-20} \times 0.1}{10^{-19}}\right]$$

$[H^+]^2 = 0.01$

$[H^+] = 0.1$

pH = 1

152. $$N_R = \frac{N_AV_A - N_BV_B}{V_T} = \frac{200 \times \frac{1}{10} - 200 \times \frac{1}{20}}{1000}$$

$$N_R = \frac{1}{100} = (0.01)$$

$pH = -\log_{10}[H^+]$

$= -\log[0.01]$

$= 2$

153. Here except NaCN rest will increase H^+ concentration as its aqueous solutions basic.

154. As I, IV are buffer solutions so they will have no effect of dilution on their pH values.

Previous Years' Questions

1. Which one of the following is true for any diprotic acid, H_2X?

(a) $Ka_2 = Ka_1$ (b) $Ka_2 > Ka_1$

(c) $Ka_1 > Ka_2$ (d) $Ka_2 = \frac{1}{Ka_1}$

[2000]

2. A base dissolved in water yields a solution with a hydroxyl ion concentration of 0.05 mol litre^{-1}. The solution is:

(a) Basic (b) Acid

(c) Neutral (d) Either (b) or (c)

[2000]

3. Conjugate acid of NH_2^- is:

(a) NH_4^+ (b) NH_3

(c) NH_2 (d) NH

[2000]

4. Which of the following statements about pH and H^+ ion concentration is incorrect?

(a) Addition of one drop of concentrated HCl in NH_4OH solution decreases pH of the solution.

(b) A solution of the mixture of one equivalent of each of CH_3COOH and NaOH has a pH of 7

(c) pH of pure neutral water is not zero

(d) A cold and concentrated H_2SO_4 has lower H^+ ion concentration than a dilute solution of H_2SO_4

[2000]

5. The ionization constant of phenol is higher than that of ethanol because:

(a) Phenoxide ion is bulkier than ethoxide

(b) Phenoxide ion is stronger base than ethoxide

(c) Phenoxide ion is stabilized through delocalization

(d) Phenoxide ion is less stable than ethoxide

[2000]

6. The reaction $PCl_5(g) \leftrightarrow PCl_3\,(s) + Cl_2(g)$ is an example of:

(a) Backward (b) Forward

(c) Irreversible (d) Reversible

[2001]

7. Ionization constant of CH_3COOH is 1.7×10^{-5} and concentration of H^+ ion is 3.4×10^{-4}. Then initial concentration of CH_3COOH is:
(a) 3.4×10^{-4} (b) 3.4×10^{-3}
(c) 6.8×10^{-4} (d) 6.8×10^{-3}

[2001]

8. In HS^-, I^-, RNH_2 and NH_3 order of proton accepting tendency will be:
(a) $I^- > NH_3 > RNH_2 > HS^-$
(b) $HS^- > RNH_2 > NH_3 > I^-$
(c) $RNH_2 > NH_3 > HS^- > I^-$
(d) $NH_3 > RNH_2 > HS^- > I^-$

[2001]

9. Solution of 0.1 N NH_4OH and 0.1 N NH_4Cl has pH 9.25, then find out pKb of NH_4OH:
(a) 9.25 (b) 4.75
(c) 3.75 (d) 8.25

[2002]

10. Which has highest pH?
(a) CH_3COOK (b) Na_2CO_3
(c) NH_4Cl (d) $NaNO_3$

[2002]

11. Solubility of MX_2 type electrolyte is 0.5×10^{-4} mol L^{-1}. Then find out Ksp of electrolytes:
(a) 5×10^{-12} (b) 25×10^{-10}
(c) 1×10^{-13} (d) 5×10^{-13}

[2002]

12. Which one of the following orders of acid strength is correct?
(a) $RCOOH > HC \equiv CH > HOH > ROH$
(b) $RCOOH > ROH > HOH > HC \equiv CH$
(c) $RCOOH > HOH > ROH > HC \equiv CH$
(d) $RCOOH > HOH > HC \equiv CH > ROH$

[2003]

13. The solubility product of AgI at 25°C is 1.0×10^{-16} $mol^2\ L^{-2}$. The solubility of AgI in 10^{-4} N solution of KI at 25°C is (in mol L^{-1}):
(a) 1.0×10^{-10} (b) 1.0×10^{-8}
(c) 1.0×10^{-16} (d) 1.0×10^{-12}

[2003]

14. Which one of the following compounds is not a protonic acid?z
(a) $SO_2(OH)_2$ (b) $B(OH)_3$
(c) $PO(OH)_3$ (d) $SO(OH)_2$

[2003]

15. The rapid change of pH near the stoichiometric point of an acid-base titration is the basis of indicator detection. pH of the solution is related to the concentrations of the conjugate acid (HIn) and base (In^-) forms of the indicator by the expression:

(a) $\log\frac{[In^-]}{[HIn]} = pK_{In} - pH$

(b) $\log\frac{[HIn]}{[In^-]} = pK_{In} + pH$

(c) $\log\frac{[HIn]}{[In^-]} = pH - pK_{In}$

(d) $\log\frac{[In^-]}{[HIn]} = pH - pK_{In}$

[2004]

16. The solubility product of a sparingly soluble salt AX_2 is 3.2×10^{-11}. Its solubility (in mol L^{-1}) is:
(a) 5.6×10^{-6} (b) 3.1×10^{-4}
(c) 2×10^{-4} (d) 4×10^{-4}

[2004]

17. At 25°C, the dissociation constant of a base, BOH is 1.0×10^{-12}, the concentration of hydroxyl ions in 0.01 M aqueous solution of the base would become:
(a) 2.0×10^{-6} mol L^{-1} (b) 1.0×10^{-5} mol L^{-1}
(c) 1.0×10^{-6} mol L^{-1} (d) 1.0×10^{-7} mol L^{-1}

[2005]

18. What is the correct relationship between the pHs of isomolar solutions of sodium oxide (pH_1), sodium sulphide (pH_2), sodium selenide (pH_3) and sodium telluride (pH_4)?
(a) $pH_1 > pH_2 \approx pH_3 > pH_4$
(b) $pH_1 < pH_2 < pH_3 < pH_4$
(c) $pH_1 < pH_2 < pH_3 \approx pH_4$
(d) $pH_1 > pH_2 > pH_3 > pH_4$

[2005]

19. H_2S gas when passed through a solution of cations containing HCl precipitates the cations of second group of qualitative analysis but not those belonging to the fourth group. It is because:
(a) Solubility product of group II sulphides is more than that of group IV sulphides
(b) Presence of HCl increases the sulphide ion concentration
(c) Sulphides of group IV cations are unstable in HCl.
(d) Presence of HCl decreases the sulphide ion concentration

[2005]

20. The hydrogen ion concentration of a 10^{-8} M HCl aqueous solution at 298 K ($Kw = 10^{-14}$) is:

(a) 9.525×10^{-8} M (b) 1.0×10^{-8} M
(c) 1.0×10^{-6} M (d) 1.0525×10^{-7} M

[2006]

21. Which of the following pairs constitutes buffer?

(a) HNO_3 and NH_4NO_3 (b) HCl and KCl
(c) HNO_2 and $NaNO_2$ (d) NaOH and NaCl

[2006]

22. Calculate the pOH of a solution at 25°C that contains 1×10^{-10} M of hydronium ions, i.e., H_3O^+:

(a) 7.000 (b) 4.000
(c) 9.000 (d) 1.000

[2007]

23. A weak acid, HA has a Ka of 1.00×10^{-5}. If 0.100 mol of this acid is dissolved in one litre of water, the percentage of acid dissociated at equilibrium is closest to:

(a) 99.0% (b) 1.00%
(c) 99.9% (d) 0.100%

[2007]

24. Equal volumes of three acid solutions of pH 3, 4, and 5 are mixed in a vessel. What will be the H^+ ion concentration in the mixture?

(a) 3.7×10^{-3} M (b) 1.11×10^{-3} M
(c) 1.11×10^{-4} M (d) 3.7×10^{-4} M

[2008]

25. Equimolar solutions of the following were prepared in water separately. Which one of the solutions will record the highest pH?

(a) $MgCl_2$ (b) $BaCl_2$
(c) $SrCl_2$ (d) $CaCl_2$

[2008]

26. Which of the following molecule acts as a Lewis acid?

(a) $(CH_3)_2$ (b) $(CH_3)_3B$
(c) $(CH_3)_3P$ (d) $(CH_3)_3N$

[2009]

27. The ionization constant of NH_4OH is 1.77×10^{-5} at 298 K. Hydrolysis constant of ammonium chloride is:

(a) 5.65×10^{-10} (b) 5.65×10^{-12}
(c) 5.65×10^{-13} (d) 6.5×10^{-12}

28. In a buffer solution containing equal concentration of B^- and HB the K_b for B^- is 10^{-10}. The pH of buffer solution is:

(a) 4 (b) 0
(c) 7 (d) 6

[2010]

29. What is $[H^+]$ in mol/L of a solution that is 0.20 M in CH_3COONa and 0.10 M in CH_3COOH? K_a for $CH_3COOH = 1.8 \times 10^{-5}$.

(a) 9.0×10^{-6} (b) 3.5×10^{-4}
(c) 1.1×10^{-5} (d) 1.8×10^{-5}

[2010]

30. Which of the following molecular hydride acts as a lewis acid?

(a) CH_4 (b) B_2H_6
(c) CH_3 (d) H_2O

[2010]

31. If pH of a saturated solution of $Ba(OH)_2$ is 12, the value of its $K_{(sp)}$ is:

(a) 5.00×10^{-7} M^3
(b) 4.00×10^{-6} M^3
(c) 4.00×10^{-7} M^3
(d) 5.00×10^{-6} M^3

[2010]

32. Which of the following is least likely to behave as Lewis base?

(a) H_2O (b) NH^3
(c) BF_3 (d) OH^-

[2011]

33. A buffer solution is prepared in which the concentration of NH_3 is 0.3 M and the concentration of NH_4^+ is 0.2 M. If the equilibrium constant K_b for NH_3 is 1.8×10^{-5}, what is the pH of the solution?

(a) 8.56
(b) 10.72
(c) 9.04
(d) 9.43

[2011]

34. Buffer solutions have constant acidity and alkalinity because:

(a) They have fixed value of pH
(b) These give unionized or acid or base on reaction with added acid or alkali
(c) They have large excess of H^+ or OH^- ions
(d) Acids and alkalies in these solutions are shielded from attack by other ions

[2012]

35. pH of saturated solution of $Ba(OH)_2$ is 12. The value of solubility product (K_{sp}) of $Ba(OH)_2$ is:

(a) 3.3×10^{-7}
(b) 5.0×10^{-7}
(c) 4.0×10^{-6}
(d) 5.0×10^{-6}

[2012]

36. Equimolar solutions of the following substances were prepared separately. Which one of these will record the highest pH value?

(a) $BeCl_2$
(b) $AlCl_3$
(c) LiCl
(d) $BaCl_2$

[2012]

37. Which of these is least likely to act as a Lewis base?

(a) CO (b) PF_3
(c) BF_3 (d) F^-

[2013]

38. Which of the following is the strongest acid?

(a) H_2SO_4 (b) $HClO_3$
(c) $HClO_4$ (d) H_2SO_3

[2013]

39. Which of the following salts will give highest PH in water?

(a) NaCl (b) Na_2CO_3
(c) KCl (d) $CuSO_4$

[2014]

40. The K_{SP} of Ag_2CrO_4, AgCl, AgBr and AgI are respectively 1.1×10^{-12}, 1.8×10^{-10}, 5×10^{-13}, 8.3×10^{-17}. Which of the following precipitate last if $AgNO_3$ solution is added to the solution containing equal moles of NaCl, NaBr, NaI and Na_2SO_4:

(a) AgCl
(b) AgBr
(c) AgI
(d) Ag_2CrO_4

[2015]

41. Which of the following pairs of solution is not an acidic buffer?

(a) $HClO_4$ and $NaClO_4$
(b) CH_3COOH and CH_3COONa
(c) H_3PO_4 and Na_3PO_4
(d) H_2CO_3 and Na_2CO_3

[2015]

42. What is the PH of the resulting solution when equal volumes of 0.1 M NaOH and 0.01 M HCl are mixed?

(a) 1.04 (b) 2.0
(c) 7 (d) 12.65

[2015]

43. Among the following correct order of acidity is?

(a) $HClO_4 < HClO_2 < HClO < HClO_3$
(b) $HClO < HClO_2 < HClO_3 < HClO_4$
(c) $HClO_3 < HClO_4 < HClO_2 < HClO$
(d) $HClO_2 < HClO < HClO_3 < HClO_4$

[2016]

44. MY and NY_3, two nearly insoluble salts have the same K_{SP} values of 6.2×10^{-13} at room temperature. Which statement would be true in regard to MY and NY_3?

(a) The molar solubility of MY in water is less than that of NY_3
(b) The salts MY and NY_3 are more soluble in 0.5 m KY than in pure water
(c) The addition of the salt of KY to solution of MY and NY_3 will have no effect on their solubilities
(d) The molar solubilities of MY and NY_3 in water are identical

[2016]

45. The solubility of $AgCl_{(s)}$ with solubility product 1.6×10^{-10} in 0.1 M NaCl Solution would be?

(a) 1.26×10^{-6} M (b) 1.6×10^{-9} M
(c) 1.6×10^{-11} M (d) Zero

[2016]

46. The percentage of pyridine (C_5H_5N) that forms pyridinium ion ($C_5H_5N^+H$) in 0.1 M aqueous pyridine solution is ($K_B = 1.7 \times 10^{-9}$):

(a) 0.006 % (b) 0.013 %
(c) 0.77 % (d) 1.6 %

[2016]

Answer Keys

1. (c)	2. (a)	3. (b)	4. (b)	5. (c)	6. (d)	7. (d)	8. (c)	9. (b)	10. (b)
11. (d)	12. (c)	13. (d)	14. (b)	15. (d)	16. (c)	17. (d)	18. (d)	19. (d)	20. (d)
21. (c)	22. (b)	23. (b)	24. (d)	25. (b)	26. (b)	27. (a)	28. (a)	29. (a)	30. (b)
31. (a)	32. (c)	33. (d)	34. (b)	35. (c)	36. (c)	37. (c)	38. (c)	39. (b)	40. (d)
41. (a)	42. (d)	43. (b)	44. (a)	45. (b)	46. (b)				

Hints and Explanations

1. $H_2X \leftrightarrow H^+ + HX^-$ (Ka_1)

$HX^- \leftrightarrow H^+ + X^{2-}$ (Ka_2)

So $Ka_1 > Ka_2$

2. $[H^+][OH^-] = 1 \times 10^{-14}$

$$[H^+] = \frac{1 \times 10^{-14}}{0.05} = 2 \times 10^{-13} \text{ mol L}^{-1}$$

$pH = -\log[H^+] = -\log[2x10^{-13}] = -\log 2 - \log 10^{-13}$

$= -\log 2 - (-13)\log 10$

$= -0.3010 + 13.0000 = 12.6990.$

Since the value of pH > 7, so the solution is basic.

3. As NH_3 after losing a proton (H^+) gives NH_2^-

$NH_3 + H_2O \leftrightarrow NH_2^- + H_3O^+$

4. CH_3COOH is weak acid while NaOH is strong base, so one equivalent of NaOH can not be neutralized with one equivalent of CH_3COOH. Therefore the solution of one equivalent of each does not have pH value as 7.

5. Phenoxide ion is stabilized through delocalization of 6 π electrons or by resonance.

6. The dissociation of PCl_5 is an example of a reversible reaction

$PCl_5 \rightleftharpoons PCl_3 + Cl_2$

7. $CH_3COOH \leftrightarrow CH_3COO^- + H^+$

$$Ka = \frac{[CH_3COO^-][H^+]}{[CH_3COOH]}$$

Given

$[CH_3COO^-] = [H^+] = 3.4 \times 10^{-4}$ M

Ka for $CH_3COOH = 1.7 \times 10^{-5}$

Since CH_3COOH is weak acid, so in it (CH_3COOH) is equal to initial concentration. Hence

$$1.7 \times 10^{-5} = \frac{(3.4\times10^{-4})(3.4\times10^{-4})}{[CH_3COOH]}$$

$$[CH_3COOH] = \frac{3.4 \times 10^{-4} \times 3.4 \times 10^{-4}}{1.7 \times 10^{-5}}$$

$= 6.8 \times 10^{-3}$ M

8. As a strong base has higher tendency to accept the proton hence increasing order of acidic strength is:

$RNH_3^+ < NH_4^+ < H_2S < I^-$

So increasing order of base is:

$RNH_2 > NH_3 > HS^- > I^-$

9. $pOH = pK_b + \log [Salt]/[Base]$

$pOH + pH = 14$

$pOH = 14 - pH$

$14 - pH - \log [salt]/[Base] = pK_b$

$14 - 9.25 - \log (0.1/0.1) = pK_b$

$14 - 9.25 - 0 = pK_b$

$pK_b = 4.75$

10. Na_2CO_3 is salt of weak acid H_2CO_3 and strong base NaOH, therefore, it has a pH more than 7. Also, It dissociates to give two moles of NaOH (strong base).

$Na_2CO_3 + 2H_2O \rightarrow 2NaOH + 2H_2CO_3$

11. $s = 0.5 \times 10^{-4}$ moles/lit

for MX_2, $Ksp = s \times (2s)^2 = 4s^3$

$[MX_2 \leftrightarrow M^{2+} + 2X^-]$

$Ksp = 4 \times (0.5 \times 10^{-4})^3 = 4 \times 0.125 \times 10^{-12}$

$= 0.5 \times 10^{-12} = 5 \times 10^{-13}$

12. $RCOOH > HOH > ROH > HC \equiv CH$

Depending upon the rate of donation of proton it is the deceasing order.

13. Ksp for AgI $= 1 \times 10^{-16}$

In solution of KI, I^- would be due to both AgI and KI, 10^{-4} KI would provide $= 10^{-4}$ I^- AgI would provide, say $= xI^-$ (x is solubility of AgI)

Total $I^- = (10^{-4} + x)$, Ksp of AgI

$= (10^{-4} + x)x$ Ksp $= 10^{-4}x + x^2$

As x is very small so x^2 can be ignored i.e.,

$10^{-4}x = 10^{-16}$

Or

$$x \text{ (solubility)} = \frac{10^{-16}}{10^{-4}} = 10^{-12} \text{ (mol}^{-1})$$

14. $B(OH)_3$ does not provide H^+ ions in water instead it accepts OH^- ion and hence it is Lewis acid.

$$B(OH)_3 + H_2O \leftrightarrow [B(OH)_4]^- + H^+$$

15. For acid-base indicator

$HIn \leftrightarrow H^+ + In^-$

$$KIn = \frac{[H^+][In^-]}{[HIn]} \text{ or } [H^+] = KIn \times \frac{[HIn]}{[In^-]}$$

$$\text{Or } \log H^+ = \log KIn + \log \frac{[HIn]}{[In^-]}$$

Taking negative on both sides

$$pH = pKIn + \log \frac{[In^-]}{[H\,In]}$$

16. For AX_2; Ksp $= 4S^3$

So $3.2 \times 10^{-11} = 4S^3$

On solving

$S = 2 \times 10^{-4}$ m/L

17. Given Kb $= 1.0 \times 10^{-12}$

[BOH] = 0.01 M

	BOH	$\leftrightarrow$	B^+	+	OH^-
Initial	c		0		0
At eq.	c(1 – x)		cx		cx

$$Kb = \frac{c^2x^2}{c(1-x)} = \frac{cx^2}{(1-x)}.$$

$$1.0 \times 10^{-12} = \frac{0.01\,x^2}{0.01\,(1-x)}.$$

On solving

$x = 1.0 \times 10^{-7}$ mol L^{-1}

18. As higher the basic character, higher is the pH so the correct order is:

$pH_1 > pH_2 > pH_3 > pH_4$

19. H_2S being weak acid does not dissociate appreciably in presence of HCl and thus $[S^{2-}]$ becomes less due to common ion effect (H^+).

20. $HCl \rightarrow H^+ + Cl^-$

10^{-8}

H_2O	$\leftrightarrow$	H^+	+	OH^-
		$10^{-8} + a$		a

$Kw = [H^+]\,[OH^-]$

$10^{-14} = (10^{-8} + a) \times a$

$a = 0.95 \times 10^{-7}$

$[H^+] = 10^{-8} + 0.95 \times 10^{-7} = 1.05 \times 10^{-7}$

21. The mixture containing a weak acid (HNO_2) and its conjugate base (NO_2^-) acts as acidic buffer.

22. As $H^+ = 1 \times 10^{-10}$

So pH = 10

At 25°C pH + pOH = 14

So pOH = 14 – 10 = 4

23. As for a weak acid degree of dissociation (α) is given as:

$\alpha = \sqrt{(Ka/C)}$

$= \sqrt{(1 \times 10^{-5}/0.1)}$

$= 0.01$

So $\alpha\% = 0.01 \times 100 = 1\%$

24. $pH = -\log_{10} [H^+]$

$[H^+] = 10^{-pH}$

$[H^+]$ of solution 1 $= 10^{-3}$

$[H^+]$ of solution 2 $= 10^{-4}$

$[H^+]$ of solution 3 $= 10^{-5}$

Total concentration of $[H^+]$

$= 10^{-3}\ (1 + 1 \times 10^{-1} + 1 \times 10^{-2})$

$= 10^{-3}\ \left(\frac{1}{1} + \frac{1}{10} + \frac{1}{100}\right)$

$= 10^{-3}\ \left(\frac{111}{100}\right)$

$= 1.11 \times 10^{-3}$

Therefore, H^+ ion concentration in mixture of equal volume of these acid solutions:

$$= \frac{1.11 \times 10^{-3}}{3} = 3.7 \times 10^{-4} \text{ M}$$

25. Equimolar solutions of the given chlorides when prepared in water forms their respective hydroxides. $Be(OH)_2$ is amphoteric, but the hydroxides of other alkaline earth metals are basic. The basic strength increases down the group. Hence higher the basic character higher will be the pH.

26. Here $(CH_3)_3B$ is lewis acid as it is electron deficient.

27. $$K_h = \frac{K_w}{K_a} = \frac{1 \times 10^{-14}}{1.7 \times 10^{-5}} = 5.65 \times 10^{-10}$$

28. For acidic buffer solution

$$pH = pKa + \log \frac{[salt]}{[acid]}$$

Given $[B^-] = [HB]$

And K_b for $B^- = 10^{-10}$

So $K_a = 10^{-4}$ for HB

$pH = pKa = 4$

29. For acidic buffer solution

$$[H^+] = \frac{Ka[CH_3COOH]}{[CH3COO^-]}$$

$$= \frac{1.8 \times 10^{-5} \times 0.10}{0.20} = 9 \times 10^{-6} M$$

30. Here B_2H_6 acts as a lewis acid

31. At 25 C pH + pOH = 14

pOH = 2

[OH] = 10^{-2} M

Now let solubility of $Ba(OH)_2$ be S

$Ba(OH)_2 \rightarrow Ba^{+2} + 2OH^{-2}$

S S 2S

$[OH^-] = 2s = 10^{-2}$

$$[\text{Solubility of } Ba(OH)_2]S = \frac{10^{-2}}{2} = 5 \times 10^{-3} \text{ mol/L}$$

Now Ksp for $Ba(OH)_2 = 4s^3$

$= 4 (5 \times 10^{-3})^3 = 5 \times 10^{-7} M^3$

32. They give unionized acid or base on reaction with added acid or alkali.

35. pH = 12, pOH = 2, $[OH^-] = 10^{-2}$

$K_{sp} = [Ba^{+2}] [OH^-]^2$

$$= \frac{10^{-2}}{2}[10^{-2}]^2$$

$= 5 \times 10^{-7}$

36. $BaCl_2$ is made up of $Ba(OH)_2$ + HCl.

$Ba(OH)_2$ is strongest base among the given options.

37. BF_3 is an e deficient molecule so can't act like Lewi's base.

38. Here acidic nature decreases as follows:

$HClO_4 > H_2SO_4 > HClO_3 > H_2SO_3$

39. Na_2CO_3 will make aqueous solution more basic here due to more concentration of $\bar{O}H$ ions.

$Na_2CO_3 \rightleftharpoons 2Na^+ + CO_3^{2-}$

$2H.OH \rightleftharpoons 2OH^- + 2H^+$

2NaOH H_2CO_3

40. For Ag_2CrO_4 $s = \sqrt[3]{\frac{K_{SP}}{4}} = \sqrt[3]{\frac{1.1 \times 10^{-12}}{4}} = 0.65 \times 10^{-4}$

AgCl $s = \sqrt{K_{SP}} = \sqrt{1.8 \times 10^{-10}} = 1.34 \times 10^{-5}$

AgBr $s = \sqrt{K_{SP}} = \sqrt{5 \times 10^{-13}} = 0.71 \times 10^{-6}$

AgI $s = \sqrt{K_{SP}} = \sqrt{8.3 \times 10^{-17}} = 0.9 \times 10^{-8}$

Hence solubility is as follows:

$Ag_2CrO_4 > AgCl > AgBr > AgI$

41. Acidic buffer is a mixtured of W.A. and its salt with S.B.

$CH_3COOH + CH_3COONa$
$H_3PO_4 + CH_3PO_4$ } Acidic buffer
$H_2CO_3 + Na_2CO_3$

As $HClO_4$ is a S.A. with salt $NaClO_4$ (S.A + S.B) so it can't be a buffer.

42. When $N_AV_A < N_BV_B$ resultant Solution will be basic

$N_BV_B - N_AV_A = NV$

$0.1 \times 1 - 0.01 \times 1 = N \times 2$

AS $N_BV_B > N_AV_B$, So Solution is basic

$N = \frac{0.09}{2} = 0.045$

$(\bar{O}H) = 0.045$

$P^{OH} = -\log (0.045) = 1.35$

$P^H = 14 - 1.35 = 12.65$

43. In case of Oxy acids

Acidic nature ∝ number. of central atom

+1 +3 +5 +7

$HClO < HClO_2 < HClO_3 < HClO_4$

44. For MY $\rightarrow s = \sqrt{K_{SP}} = \sqrt{6.2 \times 10^{-13}} = 7.874 \times 10^{-7}$

$$NY_3 \rightarrow s = \left(\frac{K_{SP}}{27}\right)^{1/4} = \left(\frac{6.2 \times 10^{-13}}{27}\right)^{1/4} = 3.89 \times 10^{-4}$$

It means molar solubility of MY is less than that of NY_3.

In 0.5 MKY solubility of both due to common ions.

45. $AgCl \rightleftharpoons \underset{s}{Ag^+} + \underset{s}{Cl^-}$

$NaCl \rightleftharpoons \underset{0.1}{Na^+} + \underset{0.1}{Cl^-}$

$K_{sp} = s \cdot (s + 0.1)$

$K_{sp} = s \cdot (0.1)$

$$s = \frac{K_{sp}}{0.1} = \frac{1.6 \times 10^{-10}}{0.1}$$

$$= 1.6 \times 10^{-9}$$

NCERT EXEMPLAR

1. We know that the relationship between K_c and K_p is $K_p = K_c\,(RT)^{\Delta n}$

What would be the value of Δn for the reaction:

$NH_4Cl\ (s) \rightleftharpoons NH_3\ (g) + HCl\ (g)$

(a) 1 (b) 0.5
(c) 1.5 (d) 2

2. For the reaction $H_2\ (g) + I_2(g) \rightleftharpoons 2HI\ (g)$, the standard free energy is $\Delta G^\ominus > 0$. The equilibrium constant (K) would be ________.

(a) $K = 0$ (b) $K > 1$
(c) $K = 1$ (d) $K < 1$

3. Which of the following is not a general characteristic of equilibria involving physical processes?

(a) Equilibrium is possible only in a closed system at a given temperature.
(b) All measurable properties of the system remain constant.
(c) All the physical processes stop at equilibrium.
(d) The opposing processes occur at the same rate and there is dynamic but stable condition.

4. PCl_5, PCl_3 and Cl_2 are at equilibrium at 500K in a closed container and their concentrations are 0.8×10^{-3} mol 1^{-3}, 1.2×10^{-3} mol L^{-1} and 1.2×10^{-3} mol L^{-1} respectively. The value of K_c for the reaction $PCl_5\ (g) \rightleftharpoons PCl_3\ (g) + Cl_2\ (g)$ will be

(a) 1.8×10^3 mol L^{-1} (b) 1.8×10^{-3}
(c) 1.8×10^{-3} L mol^{-1} (d) 0.55×104

5. Which of the following statements is incorrect?

(a) In equilibrium mixture of ice and water kept in perfectly insulated flask mass of ice and water does not change with time.
(b) The intensity of red colour increases when oxalic acid is added to a solution containing iron (III) nitrate and potassium thiocyanate.
(c) On addition of catalyst the equilibrium constant value is not affected.
(d) Equilibrium constant for a reaction with negative ΔH value decreases as the temperature increases.

6. When hydrochloric acid is added to cobalt nitrate solution at room temperature, the following reaction takes place and the reaction mixture becomes blue. On cooling the mixture it becomes pink. On the basis of this information mark the correct answer.

$\underset{\text{(pink)}}{[Co\,(H_2O)_6]^{3+}\,(aq)} + 4Cl^-\,(aq) \rightleftharpoons \underset{\text{(blue)}}{[CoCl_4]^{-2}\,(aq)} + 6H_2O\,(l)$

(a) $\Delta H > 0$ for the reaction
(b) $\Delta H < 0$ for the reaction
(c) $\Delta H = 0$ for the reaction
(d) The sign of ΔH cannot be predicted on the basis of this information.

7. The pH of neutral water at 25°C is 7.0. As the temperature increases, ionization of water increases, however, the concentration of H^+ ions and OH^- ions are equal. What will be the pH of pure water at 60°C?

(a) Equal to 7.0 (b) Greater than 7.0
(c) Less than 7.0 (d) Equal to zero

8. The ionisation constant of an acid, K_a, is the measure of strength of an acid. The K_a values of acetic acid, hypochlorous acid and formic acid are 1.74×10^{-3}, 3.0×10^{-8} and 1.8×10^{-4} respectively. Which of the following orders of pH of 0.1 mol dm^{-3} solutions of these acids is correct?

(a) Acetic acid > hypochlorous acid > formic acid
(b) Hypochlorous acid > acetic acid > formic acid
(c) Formic acid > hypochlorous acid > acetic acid
(d) Formic acid > acetic acid > hypochlorous acid

9. Ka_1, Ka_2 and Ka_3 are the respective ionization constants for the following reactions.a_1

$H_2S \rightleftharpoons H^+ + HS^-$
$HS^- \rightleftharpoons H^+ + S^{2-}$
$H_2S \rightleftharpoons 2H^+ + S^{2-}$

The correct relationship between Ka_1, Ka_2 and Ka_3 is

(a) $Ka_3 = Ka_1 \times Ka_2$ (b) $Ka_3 = Ka_1 + Ka_2$
(c) $Ka_3 = Ka_1 - Ka_2$ (d) $Ka_3 = Ka_1 / Ka_2$

10. Acidity of BF_3 can be explained on the basis of which of the following concepts?

(a) Arrhenius concept
(b) Bronsted Lowry concept

(c) Lewis concept
(d) Bronsted Lowry as well as Lewis concept.

11. Which of the following will produce a buffer solution when mixed in equal volumes?
(a) 0.1 mol dm^{-3} NH_4OH and 0.1 mol dm^{-3} HCl
(b) 0.05 mol dm^{-3} NH_4OH and 0.1 mol dm^{-3} HCl
(c) 0.1 mol dm^{-3} NH_4OH and 0.05 mol dm^{-3} HCl
(d) 0.1 mol dm^{-3} CH_4COONa and 0.1 mol dm^{-3} NaOH

12. In which of the following solvents is silver chloride most soluble?
(a) 0.1 mol $dm^{-3}$$AgNO_3$ solution
(b) 0.1 mol dm^{-3}HCl solution
(c) H_2O
(d) Aqueous ammonia

13. What will be the value of pH of 0.01 mol dm^{-3} CH_3COOH ($K_a = 1.74 \times 10^{-5}$)?
(a) 3.4 (b) 3.6
(c) 3.9 (d) 3.0

14. Ka for CH_3COOH is 1.8×10^{-5} and Kb for NH_4OH is 1.8 × 10-5. The pH of ammonium acetate will be:
(a) 7.005 (b) 4.75
(c) 7.0 (d) Between 6 and 7

15. Which of the following options will be correct for the stage of half completion of the reaction $A \rightleftharpoons B$.
(a) $\Delta G^\ominus = 0$ (b) $\Delta G^\ominus > 0$
(c) $\Delta G^\ominus < 0$ (d) $\Delta G^\ominus$ = -RTIn2

16. On increasing the pressure, in which direction will the gas phase reaction proceed to re-establish equilibrium, is predicted by applying the Le Chatelier's principle. Consider the reaction.
$N_2(g) + 3H_2(g) \rightleftharpoons 2NH_3(g)$
Which of the following is correct, if the total pressure at which the equilibrium is established, is increased without changing the temperature?
(a) *K* will remain same
(b) *K* will decrease
(c) *K* will increase
(d) *K* will increase initially and decrease when pressure is very high

17. What will be the correct order of vapour pressure of water, acetone and ether at 30°C.? Given that among these compounds, water has maximum boiling point and ether has minimum boiling point?
(a) Water < ether < acetone
(b) Water < acetone < ether
(c) Ether < acetone < water
(d) Acetone < ether < water

18. At 500 K, equilibrium constant, Kc, for the following reaction is 5. $12H_2(g) + 12I_2(g) \rightleftharpoons HI(g)$
What would be the equilibrium constant Kc for the reaction $2HI(g) \rightleftharpoons H_2(g) + I_2(g)$
(a) 0.04 (b) 0.4
(c) 25 (d) 2.5

19. In which of the following reactions, the equilibrium remains unaffected on addition of small amount of argon at constant volume?
(a) $H_2(g) + I_2(g) \rightleftharpoons 2HI(g)$
(b) $PCl_5(g) \rightleftharpoons PCl_3(g) + Cl_2(g)$
(c) $N_2(g) + 3H_2(g) \rightleftharpoons 2NH_3(g)$
(d) The equilibrium will remain unaffected in all the three cases.

Answer Keys

1. (d)	2. (d)	3. (c)	4. (b)	5. (b)	6. (a)	7. (c)	8. (d)	9. (a)	10. (c)
11. (c)	12. (d)	13. (a)	14. (c)	15. (a)	16. (a)	17. (b)	18. (a)	19. (d)	

Hints and Explanations for Selective Questions

2. $\Delta G° > 0$ means $\Delta G°$ is +ve. This can be so only if In K is -ve,

4. $K_C = \dfrac{[PCl_3][Cl_2]}{[PCl_5]}$

12. Silver chloride forms a soluble complex with aqueous ammonia.

15. $\Delta G^\ominus = -RT \ln K$
At the stage of half completion of reaction [A] = [B],
Therefore, K = 1.
Thus, $\Delta G^\ominus > 0$

16. According to Le-Chatelier's principle, at constant temperature, the equilibrium composition will change but K will remain same.

AIIMS ESSENTIALS

Assertion and Reason

In the following questions, two statements (Assertion) A and Reason (R) are given. Mark

(a) If A and R both are correct and R is the correct explanation of A;
(b) If A and R both are correct but R is not the correct explanation of A;
(c) A is true but R is false;
(d) A is false but R is true,
(e) A and R both are false.

1. (A) : The pH of a buffer solution containing equal moles of acetic acid and sodium acetate is 4.8(pKa of acetic acid is 4.8)
   (R) : The ionic product of water at 25ºC is 10^{-14} mol^2 lit^{-2}

2. (A) : In the titration of Na_2CO_3 with HCl using methyl orange indicator, the volume of the acid required at the equivalence point is twice that of the acid required using phenolphthalene indicator.
   (R) : Two moles of HCl are required for complete neutralization of one mole of Na_2CO_3.

3. (A) : Solubility of $CaSO_4$ in 0.1 M K_2SO_4 is 10^{-8} M hence its K_{sp} is 10^{-16}
   (R) : Since for $CaSO_4$ $K_{sp} = (s)^2$

4. (A) : According to principle of common ion effect, the solubility of HgI_2 is expected to be less in an aqueous solution of KI than in water. But HgI_2 dissolves in an aqueous solution of KI to form a clear solution.
   (R) : Iodide ion, I^- is highly polarisable.

5. (A) : On dissolving AgCl in 0.1 M solution of NaCl, $CaCl_2$ and NH_3 solution the solubility is lowest in NH_3.
   (R) : AgCl form complex with NH_3 in aqueous solution.

6. (A) : The addition of a small amount of a 'neutral' electrolyte (one that does not share a common ion) such as NaCl to a dilute solution of acetic acid, will cause an increase in the degree of dissociation of the acid.
   (R) : Due to the increased ionic strength, the mean ionic activity coefficient of H_3O^+ and CH_3COO^- will increase.

7. (A) : On addition of NH_4Cl to NH_4OH solution, pH decreases but remains greater than 7.
   (R) : Addition of NH_4^+ ion decreases ionization of NH_4OH thus, $[OH^-]$ is decreased, hence, pH decreases.

8. (A) : The pH range of any indicator is form $pK_{In} - 1$ to $pK_{In} + 1$
   (R) : As the pH of the indicator can be given as follows:

$$pH = pK_{In} + \log\frac{[In^-]}{[HIn]}$$

   As pH range depends upon the ratio of $[In^-]$ to

9. (A) : 0.33 M solution of KCN is more basic than 0.33 M solution of KF.
   (R) : 0.33 M solution of KCN is more basic than 0.33 M solution of CH_3COOK.

10. (A) : On mixing equal volumes of 1M HCl and 2 M CH_3COONa, an acidic buffer solution is formed.
    (R) : Resultant mixture contains CH_3COOH and CH_3COONa which are parts of acidic buffer.

11. (A) : H_3PO_3 is a dibasic acid.
    (R) : There are two hydrogen atoms directly attached to phosphorus.

12. (A) : Sodium carbonate can be titrated against sulphuric acid by using either phenolphthalene or methyl orange as indicator.
    (R) : The volume of sulphuric acid required to produce colour change for the two indicators is different.

13. (A) : A mixture of the solutions of a weak acid and its sodium salt acts as a good buffer.
    (R) : The ratio of the salt to the acid in the mixture does not change substantially when small amount of acid or base is added to the buffer.

14. (A) : pH of an amphiprotic anion always independent of concentration.
    (R) : pH of amphiprotic anion can be given by 1/2 $[p^{k_1} + p^{k_2}]$. Here p^{k_1} and p^{k_2} are the dissociation constants of the acid in which amphiprotic anion is formed.

15. (A) : In the volumetric estimation of iron (II) in acid solution in the presence of excess chloride by titration with permaganate solution, excess phosphoric acid and manganese (II) sulphate are added to the solution containing iron (II) before the titration.

(R) : Phosphoric acid complexes the iron (III) produced by the oxidation, and manganese (III) depresses the reduction potential of permanganate.

16. (A) : When 0.1 M diprotic weak acid H_2X dissociates with its dissociation constants $K_{a_1} = 10^{-4}$ and $K_{a_2} = 10^{-12}$, then $[X^{-2}]$ is almost equal to 10^{-12} M.

(R) : Since $K_{a_2} << K_{a_1}$ for 0.1 M so H_2X so $[X^{-2}]$ is negligible w.r.t. $[HX^-]$. Hence $[X^{-2}] \approx K_{a_2}$

17. (A) : On mixing 500 mL of 10^{-6} M Ca^{2+} ion and 500 mKL of 10^{-6} MF^- ion, the precipitate of CaF_2 will be obtained.

$K_{sp}\ (CaF_2) = 10^{-18}$

(R) : If Ksp is greater than ionic product, precipitate will be obtained.

Answer Keys

1. (b)	2. (b)	3. (d)	4. (b)	5. (d)	6. (a)	7. (a)	8. (d)	9. (b)	10. (a)
11. (a)	12. (d)	13. (a)	14. (a)	15. (a)	16. (a)	17. (d)			

CHAPTER

8

Hydrogen and its Component

Chapter Outline

■ Rogue Element or Hydrogen (^1_1H) ■ Resemblance of Hydrogen with Alkali Metals (Ia) ■ Resemblance with Halogens ■ Types of Hydrogen ■ Methods of Preparation of Dihydrogen ■ Hydrides ■ Water ■ Hard and Soft Water ■ Heavy Water or Deuterium Oxide D_2O ■ Hydrogen Peroxide (Auxochrome) H_2O_2

Rogue Element or Hydrogen (^1_1H)

Introduction and occurrence

- It was discovered by Henry Cavendish and named by Lavoisier ('Hydro' means water and 'gen' means creating.)
- It is the most abundant element in the universe (70%) and the third most abundant element on the surface of the globe.
- It is the first, lightest, smallest, non-metallic element in the periodic table.
- It does not have any 'neuron'.
- It has maximum specific heat.
- It exists in H^+ (Hydronium) and H^- (Hydride) states.

H — H^+ : Exists in H_2O as H_3^+O stable in gaseous state

H — H^- : (Metal hydride)

- In case of H_2^+ and H_2^-, bond order is half (½) .
- Inversion temperature of H_2 is –80°C.
- Sun's atmosphere has 90 % hydrogen.
- H_2 is dried by P_2O_5 but concentrated H_2SO_4 cannot be used as H_2 catches fire with it.

Position in periodic table

Since an atom of hydrogen contains only one electron in its K-orbit, it has an equal tendency to loose or to gain this one electron to get a stable configuration thus, hydrogen neither belongs to the first group nor to the seventh group. Therefore its position is not well certain.

Resemblance of Hydrogen with Alkali Metals (IA)

(1) Electronic configuration

It has same electronic configuration like alkali metals as it has only one valence electron (ns^1).

H (Z = 1) : $1s^1$

Li (Z = 3) : $1s^2\ 2s^1$

(2) Formation of monopositive ion

Hydrogen also forms cation (H^+) like M^+ formed by alkali metals by loosing one electron.

$H \rightarrow H^+ + e^-$

$Li \rightarrow Li^+ + e^-$

(3) Electropositive character

It is also electropositive in nature like these metals.

(4) Valency and oxidation state

Hydrogen is monovalent like alkali metals and in most of its compounds it exists in +1 oxidation state like alkali metals.

Example,

$\overset{+1}{H}Cl, \overset{+1}{Li}Cl$

(5) Affinity for non-metals

Like alkali metals hydrogen has a great affinity for non metals and it combines with them to form a number of compounds like oxides, halides etc.,

$H_2 + Cl_2 \rightarrow 2HCl$

$2Na + Cl_2 \rightarrow 2NaCl$

(6) Reducing property

It is also a good reducing agents like these metals and can reduce many compounds. As they can be easily oxidized by loosing their valence electrons so they can act as good reducing agents.

Example,

$CuO + H_2 \xrightarrow{\text{Heat}} Cu + H_2O$

$B_2O_3 + 6K \xrightarrow{\text{Heat}} 2B + 3K_2O$

Resemblance with Halogens

(1) Electronic configuration

Hydrogen has similar electronic configuration to like halogens as it needs just one electron to get the next inert gas configuration like halogens.

$H : 1s^1$

$F : 1s^2\ 2s^2\ 2p^5$

(2) Non-metallic nature

It also shows non metallic nature like halogens and it is a bad conductor of heat and electricity.

(3) Atomicity

It can also exist in diatomic state (H_2) like halogens (X_2).

(4) Formation of mononegative ions

It forms mononegative ion by gaining an electron like halogens.

$H + e^- \rightarrow H$

$F + e^- \rightarrow F^-$

(5) Ionization energy

It has a close value in ionization energy to halogens.

Example,

element	H	F	Cl	Br
I.E.	1310	1681	1255	1121 kJ mol^{-1}

(6) Oxidation state

Like halogens it can show –1 oxidation state or it can form H^- ion like X^- formed by halogens.

$\overset{+1}{Na}\ \overset{-1}{H}$ Sodium hydride

$\overset{+1}{Na}\ \overset{-1}{Cl}$ Sodium chloride

(7) Reaction with metals

Hydrogen reacts with alkali and alkaline earth metals to form hydrides similar to halogens which react with these metals and form halides example, NaH and NaCl, CaH_2 and $CaCl_2$ etc.

(8) Reaction with non-metals

Hydrogen reacts with non-metals like C, Si, Ge etc. to form covalent compounds like halogens.

Hydrides:	CH_4	SiH_4	GeH_4
Halides:	CCl_4	$SiCl_4$	$GeCl_4$

(9) Liberation at anode during electrolysis

Hydrogen is released at anode when compounds like LiH, NaH, CaH_2 etc., are subjected to electrolysis in the molten state. It is similar to halogens which also released at anode during the electrolysis of their compounds like NaX, KX, PbX_2 etc.

Isotopes of Hydrogen

Hydrogen has three isotopes. It is easily revealed by the mass spectrographic studies of hydrogen and these are as follows:

$_1H^1$	$_1H^2$ or D	$_1H^3$ or T
Protium	Deuterium	Tritium
n = 0	n = 1	n = 2

In ordinary hydrogen the ratio of $_1H^1$: $_1H^2$ is 6400 : 1.

Urey discovered heavy hydrogen or deuterium as follows:

$$_7N^{14} + _0n^1 \rightarrow _6C1^3 + \underset{\text{Deuterium}}{_1H^2}$$

Oilphant, Harteck, Rutherford prepared tritium in the following manner:

$$\underset{\text{Deutron}}{_1H^2} + _1H^2 \rightarrow _1H^3 + _1H^1$$

Isotopic Effect

The difference in the physical properties and in the rates of reactions of the isotopes due to a difference in mass number is called isotopic effect. Due to isotopic effect the rate of reaction is slower in case of heavier isotope.

Physical Properties of Dihydrogen and Dideuterium

Property	H_2	D_2
Molecular mass	2.016	4.028
Boiling point (K)	20.39	23.67
Enthalpy of dissociation (kJ mol^{-1}) (298.2 K)	435.88	443.35
Enthalpy of fusion (kJ mol^{-1})	0.117	0.197
Enthalpy of vaporization (kJ mol^{-1})	0.904	1.226
Internuclear distance (pm)	74.14	74.14
Melting point (K)	13.957	18.73

Types of Hydrogen

(1) Active or Atomic Hydrogen

It is obtained as follows:

$$\underset{\text{Molecular}}{H_2} \xrightarrow{\text{elec. Arc}} \underset{\text{Atomic}}{2H} - 104.5 \text{ kcal}$$

Here high temperature and low pressure are favourable conditions for its formation. Half life of active hydrogen is 0.3 second.

(2) Nascent hydrogen

It is formed in a contact of a substance to be reduced

$$Zn + \text{dil. } H_2SO_4 \rightarrow ZnSO_4 + 2[H]$$

$$FeCl_3 + [H] \rightarrow FeCl_2 + HCl$$

It is more reactive and stronger reducing agent than ordinary hydrogen. Nascent hydrogen can not be isolated like atomic hydrogen. It has less reducing power than atomic hydrogen.

(3) Occluded Hydrogen

Some metals (Pd, Pt, Au, Ni) can adsorb hydrogen on their surface, it is called occlusion. Decreasing order of occlusion property of metals is

Colloidal Pd > Pd > Pt > Au > Ni

(4) Ortho and Para Hydrogens (Nuclear spin Isomers)

Dihydrogen has two nuclear spin isomers known as ortho and para dihydrogen.

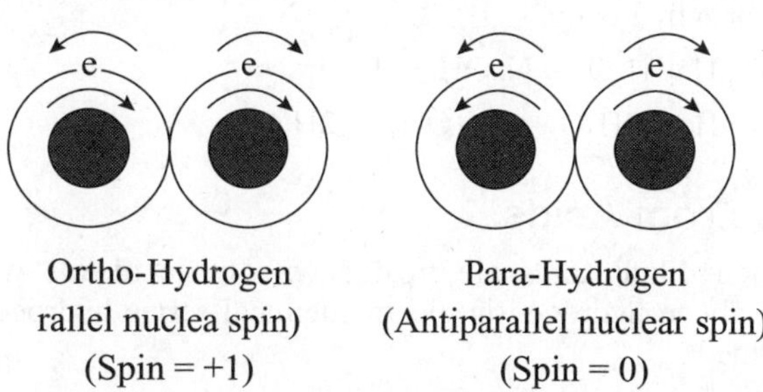

Ortho-Hydrogen rallel nuclea spin) (Spin = +1)   Para-Hydrogen (Antiparallel nuclear spin) (Spin = 0)

In ortho hydrogen the nuclear spin is in same direction but in para hydrogen it is in opposite direction. Para hydrogen has lower energy i.e., more stability so it is favoured at low temperature but at room temperature ortho hydrogen is favourable due to its higher molecular energy.

At ordinary temperature the ratio of ortho and para hydrogen is 3 : 1 but on decreasing temperature percentage of para form increases.

The ortho and para hydrogen have nearly same physical properties. However, the thermal conductivity of para is 50% more than that of ortho and the melting point of para hydrogen is 0.15 K less than that of ortho hydrogen.

Methods of Preparation of Dihydrogen

(1) With Metals

Many metals react with H_2O at different temperatures to release hydrogen:

(i) With very active metals

Metals like Na, K, Ca react with H_2O to form hydrogen.

Example,

$$2Na + 2H_2O \rightarrow 2NaOH + H_2$$

(ii) With less active metals

Metals like Mg, Al, Zn etc., react with H_2O on heating to form hydrogen.

$$Mg + H_2O \xrightarrow{\Delta} MgO + H_2$$

(iii) When steam is passed over metals like Fe, Co, Ni, Sn etc. hydrogen is formed.

$$\underset{\text{Steam}}{3Fe + 4H_2O} \xrightleftharpoons{\text{High temp.}} Fe_3O_4 + 4H_2$$

(2) By the Action of Water on Ionic Hydrides

Hydrides of alkali and alkaline earth metals react with water to form hydrogen.

$$NaH + H_2O \rightarrow NaOH + H_2$$

$$CaH_2 + 2H_2O \rightarrow Ca(OH)_2 + 2H_2$$

(3) From Acids

When dilute acids are treated with those metals which are above hydrogen in electrochemical series hydrogen is formed.

Example,

$$Zn + H_2SO_4\,(aq.) \rightarrow ZnSO_4 + H_2\,(g)$$

$$Fe + 2HCl\,(aq) \rightarrow FeCl_2 + H_2\,(g)$$

Facts To Remember

Active metals like Zn, Mg, Al, Fe can not produce hydrogen with dilute HNO_3 as HNO_3 being an oxidant oxidizes H_2 into H_2O.

(4) From Alkalies

When metals like Zn, Al, Sn are treated with Alkalies like NaOH, KOH etc., hydrogen is formed.

Example,

$$2Al + 2NaOH + 2H_2O \rightarrow \underset{\text{Sodium meta aluminate}}{2NaAlO_2} + 3H_2$$

$$Zn + 2NaOH \rightarrow \underset{\text{Sodium zincate}}{Na_2ZnO_2} + H_2$$

(5) Lab Method

When diluted sulphuric acid is passed over granulated zinc even at room temperature hydrogen is obtained which is collected by the downward displacement of water.

$$Zn + \text{dil. } H_2SO_4 \rightarrow ZnSO_4 + H_2\,(g)$$

Facts To Remember

Here pure zinc is not used as it is non-porous in nature so impure granulated zinc is used which becomes porous due to the presence of some impurities.

Here concentrated sulphuric acid can not be used as it gives SO_2 and not hydrogen.

$$Zn + \text{conc. } H_2SO_4 \rightarrow ZnSO_4 + SO_2 + 2H_2O$$

(6) By Electrolysis

Highly pure hydrogen can be prepared by the electrolysis of water having small amount of an acid or a base.

$$2H_2O \xrightarrow{\text{Electrolysis}} 2H_2 + O_2$$

$$H_2O \rightleftharpoons H^+ + OH^-$$

At Cathode:

$$H^+ + e \rightarrow H$$

$$H + H \rightarrow H_2\,(g)$$

At Anode:

$$4OH^- \rightarrow 4OH + 4e$$

$$4OH^- \rightarrow 2H_2O + O_2\,(g)$$

(7) Industrial Methods

Bosch, Lane methods are used to prepare hydrogen.

(i) Bosch Method

Here water gas mixed with steam is passed over the catalyst heated up to 450°C to get hydrogen as follows:

$$\underset{\text{Water gas}}{CO + H_2} + H_2O \xrightarrow[450°C]{Fe_3O_4 + Cr_2O_3} 2H_2 + CO_2\uparrow$$

(ii) Lane is Method

Here super heated steam is passed over iron at high temperature to get hydrogen. Here iron can be regenerated by passing water gas.

$$3Fe + 4H_2O \xrightarrow[1000°C]{\Delta} Fe_3O_4 + 4H_2 \uparrow$$

This process is called gassing and it is continued for nearly 10 minutes till the surface of iron is covered by a layer of the iron oxide and the reaction begins to slow down. This iron oxide can be further reduced into iron if hydrogen supply is cut off and water gas is circulated. The circulation of water gas is called reduction period and it is nearly for 20 minutes.

$$Fe_3O_4 + H_2 \rightarrow 3FeO + H_2O$$

$$FeO + H_2 \rightarrow Fe + H_2O$$

$$Fe_3O_4 + CO \rightarrow 3FeO + CO_2$$

$$FeO + CO \rightarrow Fe + CO_2$$

(iii) Uyeno Method

$$2Al + 2KOH + 2H_2O \rightarrow 2KAlO_2 + 3H_2$$

(iv) Water Gas Shift Reaction

$$CO + H_2O \xrightarrow[\text{Fe and CO oxides}]{673\ K} CO_2 + H_2$$

(v) From Hydrocarbons

When steam is passed over heated hydrocarbons, hydrogen is obtained and the phenomenon is called steam reforming of hydrocarbons.

$$\underset{\text{Hydrocarbons}}{C_nH_{2n+2}} + nH_2O \xrightarrow[\text{Catalyst}]{1170\ K} \underset{\text{Water gas}}{CO + (2n+1)\ H_2}$$

Example,

$$CH_4\ (g) + H_2O\ (g) \xrightarrow[\text{Catalyst}]{1270\ K} CO\ (g) + 3H_2\ (g)$$

$$C_3H_8\ (g) + 3H_2O\ (g) \xrightarrow[\text{Catalyst}]{1270\ K} 3CO\ (g) + 7H_2\ (g)$$

(vi) From Natural Gas

Natural gas (CH_4) when mixed with steam and then passed over nickel catalyst heated to 1173 K, hydrogen is formed.

$$CH_4 + H_2O \xrightarrow{\Delta\ Ni} CO + 2H_2$$

Physical Properties

(1) It is a colourless, tasteless and odourless gas.

(2) It is the lightest among all the known substances and has a density of 0.0899 gm/cm^3.

(3) It is less soluble in water.

(4) It is highly inflammable and should be handled with care.

(5) Its melting and boiling points are 13.957 K and 20.39 K respectively.

Chemical Properties

It is not much reactive at ordinary temperature due to its high bond dissociation energy (436 kJ mol^{-1}) but at higher temperatures it becomes more reactive.

(1) Combustion

Hydrogen undergoes combustion in air or O_2 (with almost invisible pale blue flame) giving water by an exothermic reaction.

$$2H_2 + O_2 \rightarrow 2H_2O + 485\ kJ$$

(2) Reaction with Halogens

It reacts with halogens and form HX as follows:

$$\underset{(g)}{H_2} + \underset{(g)}{F_2} \xrightarrow{\text{Low temperature}} \underset{(g)}{2HF}$$

(Violent reaction)

$$\underset{(g)}{H_2} + \underset{(g)}{Cl_2} \xrightarrow{\text{Light}} \underset{(g)}{2HCl}$$

$$\underset{(g)}{H_2} + \underset{(g)}{Br_2} \xrightarrow[\text{Catalyst}]{620\ K} \underset{(g)}{2HBr}$$

$$\underset{(g)}{H_2} + \underset{(g)}{I_2} \xrightarrow[\text{Fe}]{620\ K} \underset{(g)}{2HI}$$

(3) Reaction with Nitrogen

Hydrogen combines with nitrogen in 3 : 1 ratio at 490°C and 200 atmosphere pressure in presence of iron catalyst to produce ammonia.

$$N_2\ (g) + 3H_2\ (g) \rightleftharpoons 2NH_3\ (g)$$

(4) Reaction with Non-metals of Oxygen Family

It reacts with oxygen, sulphur, selenium and tellurium to form their corresponding compounds under favourable conditions.

$$2H_2 + O_2 \xrightarrow{654°C} \underset{\text{Water}}{2H_2O}$$

$$H2 + S \xrightarrow[\text{Pressure}]{427°C} H_2S$$

$$H2 + Se \xrightarrow[\text{Pressure}]{\text{High temp}} H_2Se$$

(5) Reaction with Carbon

It combines with carbon at 1147°C to form methane.

$$C\,(s) + 2H_2\,(g) \xrightarrow{1147°C} \underset{\text{Methane}}{CH_4\,(g)}$$

However in presence of an electric arc it gives acetylene.

$$2C\,(s) + H_2\,(g) \xrightarrow[3027°C]{\text{Electric arc}} \underset{\text{Acetylene}}{C_2H_2\,(g)}$$

(6) Reaction with Carbon Monoxide

It reacts with carbon monoxide at 427°C and at 200 atmosphere pressure in presence of catalyst ZnO/Cr_2O_3 to give methanol.

$$CO + 2H_2 \xrightarrow[\text{700 K, 200 atm}]{ZnO\,/\,Cr_2O_3} \underset{\text{Methanol}}{CH_3OH}$$

(7) Reaction with Metals

It reacts with many metals at high temperature to form metal hydrides.

Example,

$2Li + H_2 \rightarrow 2LiH$

$Ca + H_2 \rightarrow CaH_2$

(8) Reducing Properties

It can reduce metal oxides into metals.

Example,

$Fe_3O_4 + 4H_2 \rightarrow 3Fe + 4H_2O$

$Ag_2O + H_2 \rightarrow 2Ag + H_2O$

(9) Hydrogenation Reactions

Hydrogen is used for these reactions in presence of catalysts like Ni, Pd, Pt, etc.

Example,

$$\text{Vegetable oil} + H_2 \xrightarrow{\Delta,\,Ni} \text{Vegetable ghee}$$

$$\underset{\text{Alkyne}}{-C\equiv C-} \xrightarrow{H_2/Ni} \underset{\text{Alkene}}{\text{H}>C=C<\text{H}} \xrightarrow{H_2} \underset{\text{Alkane}}{\text{H}_2C-C\text{H}_2}$$

Uses

1. In airships and balloons a mixture of 15 % hydrogen and 85 % helium is used as hydrogen is highly inflammable.
2. It is used to prepare methyl alcohol, ammonia, synthetic petrol etc.
3. It is also used to produce low temperature i.e., as a cryogenic fluid.
4. Hydrogen is used in the manufacture of synthetic petrol by heating with coal and heavy oils under very high pressure in the presence of a catalyst.
5. It is used in oxy-hydrogen torch in hydrogen welding.
6. Liquid hydrogen when mixed with liquid oxygen can be used as a fuel for rockets.

HYDRIDES

Hydrogen can form binary hydrides of MHx or MmHn types with nearly all elements except noble gases and metals of VII, VIII, IX groups (hydride gap) .

Types of Hydrides

Gibb's classified hydrides into following types:

(1) Saline or Ionic Hydride (MHn)

These are ionic or salt like hydrides which are formed by alkali metals and alkaline earth metals except Be and Mg. Be and Mg, due to their small size, polarize H^- ions and develop covalent nature. Infact, these are formed by the elements of very low electronegativity which can easily transfer their electrons to hydrogen atoms and form ionic bonds.

During the formation of such hydrides some part of energy is liberated which makes hydrogen molecules hyperactive.

These are non volatile, non conducting crystalline solids having rock salt like structure.

Preparation

The alkali and alkaline earth metals can combine directly with hydrogen on heating and form solid saline hydrides of the type MH and MH_2 respectively

Example,

$$2Li + H_2 \xrightarrow{700°–800°C} 2LiH$$
(In molten state) Lithium hydride

$$2Na + H_2 \xrightarrow{250°–400°C} 2NaH$$

$$Ca + H_2 \xrightarrow{1073\ K} CaH_2$$

$$Sr + H_2 \xrightarrow{1073\ K} SrH_2$$

Facts To Remember

During the preparation of such hydrides, the metal should be hydrogenated in the presence of a surface active agent in kerosene or some form of agitation should be used to prevent the decomposition of surface film of hydride on the metal which leads to incomplete conversion.

Properties

(1) These hydrides are white crystalline solids and are frequently soluble in polar solvents.

(2) These are more dense than the parent metal and are volatile in nature.

(3) The thermal stability of these hydrides decreases down the group i.e., from Li to Cs and from Ca to Ba.

(4) The lattice energy also decreases down the group i.e., from LiH to CsH. For LiH it is equal to 904 kJ, while for CsH is 653 kJ.

(5) The hydrides of Li, Sr and Ca require a higher temperature for decomposition as compared to other hydrides.

(6) All these hydrides are strong reducing agents.

(7) The chemical reactivity of hydrides increases down the group i.e., from LiH to CsH and from CaH_2 to BaH_2.

(8) Among the alkali metal hydrides, LiH is most stable and least reactive due to smaller size of Li^+ ion which is capable of polarizing H^- ion.

(9) These hydrides react with water at room temperature to form corresponding hydroxides and hydrogen.

$$2LiH + 2H_2O \rightarrow 2LiOH + 2H_2$$

$$2CsH + 2H_2O \rightarrow 2CsOH + 2H_2$$

$$CaH_2 + 2H_2O \rightarrow Ca(OH)_2 + 2H_2$$

The reaction of alkali metal hydrides is increasingly violent from LiH to CsH. The reaction is less violent for the alkaline earth metal hydrides.

(10) LiH reacts with nitrogen at high temperature to form $LiNH_2$ which further changes into Li_2NH and finally into Li_3N. Calcium hydride also reacts with N_2 at about 500°C forming Ca_3N_2. However other hydrides do not react with nitrogen.

Structure

These hydrides are ionic in nature. Since an ionic bond is non-directional, the molecules of these hydrides have no geometry. The typical gathering of these molecules in the crystal gives a particular type of lattice. The alkali metal hydrides have face centered cubic lattice, while alkaline earth metal hydrides have orthorhombic lattice.

Facts To Remember

- BeH_2, MgH_2, LiH have covalent nature and covalent polymeric structure.
- Thermal stability of such hydrides decreases down the group.
- CaH_2 is known as hydrolith.

(2) Molecular or Covalent Hydrides

These hydrides are formed by all p-block non metallic elements except zero group elements.

Example, CH_4, NH_3, HF etc.

Such hydrides have a general formula MH_{8-n} (n = number of valence electrons).

Example, CH_4, NH_3, HX etc.

Preparation

These are prepared by the following methods

(1) By Direct Combination of Elements with Hydrogen

Many elements combine with hydrogen at elevated temperatures to form such hydrides. The combination may also be possible at room temperature if nascent hydrogen is used in place of hydrogen gas.

$H_2 + F_2 \rightarrow 2HF$

$N_2 + 3H_2 \rightarrow 2NH_3$

(2) By Reduction of Compounds with Nascent Hydrogen

Some compounds like halides on reduction with nascent hydrogen give these hydrides.

Example,

$GeCl_4 + 8\,[H] \rightarrow GeH_4 + 4HCl$

$AsCl_3 + 6\,[H] \rightarrow AsH_3 + 3HCl$

(3) By Hydrolysis

Some compounds like borides, carbides, silicides, nitrides etc., on hydrolysis either by water or by a dilute mineral acid give these hydrides.

Example,

$Mg_3B_2 + 6HCl \rightarrow 3MgCl_2 + B_2H_6$

$CaC_2 + 2H_2O \rightarrow Ca(OH)_2 + C_2H_2$

(4) By Reduction of Halides with $LiAlH_4$

Some halides like SiH_4, GeH_4 etc. on reduction with $LiAlH_4$ give these hydrides.

Example,

$SiCl_4 + LiAlH_4 \rightarrow LiCl + AlCl_3 + SiH_4$

$GeCl_4 + LiAlH_4 \rightarrow LiCl + AlCl_3 + GeH_4$

(5) By Electric Discharge

Higher boron hydrides and some other hydrides can be obtained by passing electric discharge between the metal electrodes in the hydrogen atmosphere.

Properties

The properties of such hydrides are not homologous and irregular variations are observed from one group to another group. However in general these are either gases or liquids having low boiling points and volatile nature. These are poor conductors of electricity even in the liquid state also. These are not very stable so decompose on heating.

FACTS TO REMEMBER

Hydrides of III group (AlH_3, B_2H_6) being electrophiles behave like lewis acids.

(3) Interstitial or Metallic Hydrides

These are formed by transition and inner transition elements as at elevated temperatures they absorb hydrogen which easily fits in the interstitial sites of their lattices. example, ScH_2, LaH_2, VH etc. These hydrides have metallic appearance and almost similar properties to the parent metal atom. Such hydrides are good conductors of electricity in solid state and also used to store hydrogen in fuel cells.

Metallic hydrides are also non-stoichiometric in nature.

Example, $TiH_{1.73}$, $ZrH_{1.92}$,

Preparation

These are prepared by the following methods:

(1) By Occlusion

Many metals like Pd, Pt, Ni etc., occlude hydrogen at moderate or high temperatures and form these hydrides example, Pd can adsorb 900 volumes of hydrogen.

(2) Electrolytic Method

During an electrolytic process which involve liberation of hydrogen at cathode, it can be obtained by using a particular metal as cathode. The hydrogen liberated at cathode gets occluded in the metal cathode and as a result the metallic hydride is formed.

Example, Pd when used as cathode occludes nearly 1000 time of its volume of hydrogen.

(3) By Reduction

Copper hydride can be obtained by reduction of Cu(II) solution with the sodium salt of hypophosphorous acid at about 343 K. It can also be obtained by the reduction of Cu_2Cl_2 by lithium boron hydride.

(4) By the Action of Atomic Hydrogen

Hydrides of silver and gold can be obtained by passing atomic hydrogen over these metals at room temperature.

(5) By Heating with Hydrogen

Uranium hydride can be obtained by heating the metal with H_2 very rapidly at 523-573 K.

$$2U + 3H_2 \xrightarrow{523\text{ - }573K} 2UH_3$$

Properties

As metallic hydrides do not have definite compositions so their properties vary from one compound to another. Some common properties of such hydrides are as follows:

(1) These are usually grey to blackish in colour with a metallic lustre.

(2) They are hard and good conductors of electricity like metals.

(3) The reactivity of such hydrides towards common reagents is more than that of the parent metals.

(4) Most of these hydrides are decomposed by acids and show reducing properties.

(5) As on heating they emit the occluded hydrogen and act as strong reducing agents e.g., Uranium hydride precipitates silver from silver nitrate solution and it also reduces cupric salts into cuprous salts.

(6) These hydrides are susceptible to atmospheric oxidation although this tendency varies widely. Finely divided UH_3 ignites spontaneously in air, whereas hydrides of Ti, Zr and Hf oxidize rapidly on being heated to 673-873K.

(7) Most of these hydrides are not attacked by water upto at least 373 K.

Structure

Presently, there are two separate theories to explain the nature of bonding in these hydrides.

According to the first theory, it is believed that hydrogen atoms lose their electrons to the conduction band of the metal. Complete transfer of an electron to leave H^+ is not possible on energy grounds. Therefore, it is believed that there is a transfer of electron density, leaving hydrogen with a fractional positive charge. This view is supported by the fact that on passing electric current through palladium hydride, hydrogen sets free at the cathode.

According to the second theory, H^- ions are formed by the loss of electrons from the metal to hydrogen atoms in such a hydride. This view is supported by the fact that metallic properties of such hydrides decrease as the hydrogen to metal ratio increases.

(4) Polymertic Hydrides

These are also known as polynuclear hydrides. These are formed by some elements having electronegative values between 1.4 and 2.0. In such hydrides, hydride molecules are linked together in two and three dimensions through hydrogen bridges e.g., (BeH_2) n, (MgH_2) n, (AlH_3) n etc.

In beryllium hydride, BeH_2 molecules are linked together through hydrogen bonding.

Uses

(1) Saline hydrides are used as a source of hydrogen and as reagents in organic chemistry.

(2) Diborane (covalent hydride) is used as a high energy fuel and propellent.

(3) Metallic hydrides are used as reducing agents e.g., $LiAlH_4$ is widely used in organic chemistry as a reductant.

(4) Metallic hydrides are used in the manufacture of vacuum tubes and in some metallurgical operations.

WATER

(H_2O)

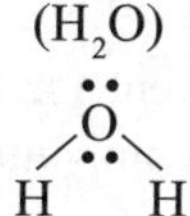

Water is the most abundant compound in the biosphere. In our body 70% is water. It is an oxide of hydrogen with a molecular formula H_2O having one part of hydrogen and eight parts of oxygen by weight.

Structure of Water Molecule

A water molecule consist of two hydrogen atoms joined to an oxygen atom by covalent bonds. It has an angular or bent shape. In water the central atom oxygen is sp^3 hybridized and due to the presence of two lone pairs of electrons on it, the shape of water molecule becomes angular or bent with a bond angle of 104.5° due to lp-lp electron repulsion.

In water molecule, O – H bond length is 95.7 pm. The bond *moments* of the two O – H bonds makes the molecule a permanent dipole and the dipole *moment* value of H_2O molecule is 1.85D.

In liquid state water molecules are held together by intermolecular hydrogen bonding and each water molecule can have hydrogen bonding with four other water molecules. In the solid state (ice) of water, the water molecules come more closer and get arranged in tetrahedral manner. The structure of ice is normally hexagonal in which each oxygen atom of water is tetrahedrally surrounded by four others oxygen atoms through a hydrogen atom. Each hydrogen atom is covalently bonded to one oxygen atom and is linked to another oxygen atom by a hydrogen bond. This type of packing gives ice and open cage structure with large open spaces. This makes the density of ice less than that of water so it floats on water. The floating ice prevents or delays the freezing of underlying water. This property helps in saving the lives of aquatic animals even in winter.

Physical Properties

(1) *Boiling point*: It has a boiling point of 100°C and a dipole moment of 1.85 D.

(2) *Density*: Maximum density of H_2O is at 4°C.

(3) *Stability*: Water molecule has a very high thermal stability due to high negative heat of formation ($\Delta H^\circ_f = -285.9$ kJ mol^{-1}) e.g., at 1500 K, it dissociates only to an extent of less than 0.02 % and at 2270 K and 1 atmospheric pressure it dissociates only 0.6 %.

(4) *Ionization*: Water is not so much ionized as it has a small but measurable electrical conductivity.

$$H_2O\ (l) + H_2O\ (l) \rightleftharpoons H_3O^+\ (aq) + OH^-\ (aq)$$

$$K_w = 1.008 \times 10^{-14}\ mol^2\ L^{-2} \text{ at } 298\ K.$$

(5) Water is a good solvent for the ionic compounds but bad for the covalent compounds.

(6) It is a universal solvent due to its high dielectric constant, high liquid range and ability to dissolve most of the compounds.

(7) AgX, $BaSO_4$, CaF_2, AlF_3 are insoluble in H_2O due to very strong forces of attraction in their lattice i.e., higher lattice energy.

(8) R – OH, R – COOH and carbohydrate (oses) are soluble in H_2O due to H-bonding

(9) One H_2O molecule can form H – bonds with four other H_2O molecules
Density of H_2O > Density of Ice

(10) Molarity of pure water is 55.5 M.

Chemical Properties

Water can behave as an acid, as a base, as an oxidizing agent, as a reducing agent and as a ligand to the metal ions. Some of the important chemical properties of water are as follows:

(2) Amphoteric Nature

Although water is neutral towards litmus (pH = 7 at 298 K), it acts both as an acid and a base and hence shows amphoteric nature.

It can act as a base towards acids stronger than itself and as an acid towards base stronger than itself.

Example,

$$\underset{\text{Base}}{H_2O\ (l)} + \underset{\text{acid}}{HX\ (aq)} \rightleftharpoons \underset{\text{acid}}{H_3O^+\ (aq)} + \underset{\text{base}}{X^-\ (aq)}$$

$$\underset{\text{Acid}}{H_2O\ (l)} + \underset{\text{base}}{NH_3\ (aq)} \rightleftharpoons \underset{\text{acid}}{NH_4^+\ (aq)} + \underset{\text{base}}{OH^-\ (aq)}$$

(3) Hydrolysing Nature

Water can hydrolysed many halides, nitrides, phosphides, carbides etc.

Example,

$$\underset{\text{Silicon tetra chloride}}{SiCl_4} + 2H_2O \rightarrow H_4SiO_4 + 4HCl$$

$$Al_4C_3 + 12H_2O \rightarrow 4Al(OH)_3 + 3CH_4$$

$$AlN + 3H_2O \rightarrow NH_3 + Al(OH)_3$$

$$Ca_3P_2 + 6H_2O \rightarrow 2PH_3 + 3Ca(OH)_2$$

(4) Oxidising and Reducing Nature

Water acts as an oxidizing as well as a reducing agent. It acts as an oxidizing agent when it reacts with active metals.

$$2Na + 2H_2O \rightarrow 2NaOH + H_2$$

$$Ca + H_2O \rightarrow CaO + H_2$$

Water acts as a reducing agent when it reacts with highly electronegative elements.

$$2F_2 + 2H_2O \rightarrow 4HF + O_2$$

$$2Cl_2 + 2H_2O \xrightarrow{\text{Sun light}} 4HCl + O_2$$

(5) Formation of Hydrates

Water reacts with many metal salts to form their hydrates. Depending upon the mode of linkage of water molecules, there are three types of hydrates which are as follows:

(i) In some hydrates, water molecules get attached to certain oxygen containing anions through hydrogen bonds e.g., $CuSO_4.5H_2O$. In this hydrate four water molecules are coordinated to central Cu^{2+} ion while the fifth water molecule is attached to the sulphate group by hydrogen bonds. Some other examples are $FeSO_4.7H_2O$, $MgSO_4.7H_2O$.

(ii) In some hydrates, water molecules act as ligands and get attached to a metal ion by coordinate bonds and form complex ions
Example,

$$[Cr(H_2O)_6]^{3+}\ (Cl^-)_3$$

$$[Ni(H_2O)_6]^{2+}\ (NO_2^-)_2$$

(iii) In some hydrates, water molecules occupy the interstitial sites (void) in the crystal lattice e.g., $BaCl_2.2H_2O$.

Hard and Soft Water

Soft Water

It gives lather easily with soap. e.g., Rain water, distilled water etc.

Hard Water

It does not give lather with soap. e.g., Sea water, water of rivers, lakes etc.

Reason of Hardness in Water

It is due to the presence of insoluble salts of Ca^{+2}, Mg^{+2}, Fe^{+2} in water.

Due to the presence of these salts water does not produce lather with soap easily because soaps are sodium or potassium salts of higher fatty acids like sodium palmitate ($C1_5H_{31}COONa$), sodium stearate ($C_{17}H_{35}COONa$). When hard water having Ca^{2+}, Mg^{2+} ions is treated with soap solution a precipitate of calcium or magnesium salts of the higher fatty acids is obtained which are insoluble in water.

$$\underset{\text{Sodium palmitate (soap) soluble in water}}{Ca^{2+} + 2C_{15}H_{31}COONa} \rightarrow \underset{\text{Calcium palmitate (insoluble in water)}}{(C_{15}H_{31}COO)\,2Ca\downarrow} + 2Na^+$$

$$\underset{\text{Sodium stearate (soap) soluble in water}}{Mg^{2+} + 2C_{17}H_{35}COONa} \rightarrow \underset{\text{Magnesium Stearate (insoluble in water)}}{(C_{17}H_{35}COO)_2\,Mg\downarrow} + 2Na^+$$

Hence no lather can be produced till all the calcium or magnesium ion get precipitated.

Hardness is expressed in ppm of $CaCO_3$ as follows:

$$\text{Degree of hardness} = \frac{\text{wt. of } CaCO_3 \text{ gms}}{10^6 \text{ gm of } H_2O}$$

$$1\,CaCO_3 \equiv 1\,MgCl_2 \equiv 1\,MgSO_4 \equiv 1CaCl_2$$

$$1\,CaSO_4$$

Types of Hardness

(1) Simple or Temporary Hardness

It is due to bicarbonates of Mg^{+2} Ca^{+2} and the removal of hardness is called Softening of water. It is possible as follows:

(a) By Boiling

When water is boiled for nearly 15 minutes the soluble bicarbonates decomposes into their insoluble carbonates and carbon dioxides which can be removed by filtration or decantation thus making water soft.

$$\underset{\text{Calcium bicarbonate (soluble)}}{Ca(HCO_3)_2} \xrightarrow{\Delta} \underset{\text{Calcium carbonate (Insoluble)}}{CaCO_3\downarrow} + CO_2 + H_2O$$

$$\underset{\text{Magnesium bicarbonate (soluble)}}{Mg(HCO_3)_2} \xrightarrow{\Delta} \underset{\text{Magnesium carbonate (Insoluble)}}{MgCO_3\downarrow} + CO_2 + H_2O$$

(b) By Clark's method

Here a calculated quantity of $Ca(OH)_2$ is added to remove temporary hardness in water. Here insoluble carbonates are formed which get precipitated and can be easily removed by filtration as follows:

$$Ca(HCO_3)_2 + Ca(OH)_2 \leftrightarrow 2CaCO_3\downarrow + 2H_2O$$

$$Mg(HCO_3) + 2Ca(OH)_2 \leftrightarrow 2CaCO_3\downarrow + \underset{\text{Insoluble}}{Mg(OH)_2\downarrow} + 2H_2O$$

(2) Permanent Hardness

It is due to the presence of soluble chlorides and sulphates of Ca^{+2} and Mg^{+2}. Permanent hardness is removed by the following methods:

(a) Soda Lime Method

Permanent hardness of calcium ions can be removed by washing soda but not of magnesium salts. When a calculated amount of washing soda is added in hard water the insoluble carbonates are formed which get precipitated and can be removed by filtration.

$$CaSO_4 + Na_2CO_3 \leftrightarrow \underset{\text{Insoluble}}{CaCO_3\downarrow} + Na_2SO_4$$

(b) By Calgon Method (Masking or Hiding of Ions)

Here calcium and magnesium ions are hiden or masked by calgon reagent i.e., sodium hexa meta phosphate ($Na_2[Na_4(PO_3)_6]$) or sodium poly meta phosphate formula's must be insameline not in broken from ($(NaPO_3)$ x. ($X > 100$).

$$2CaCl_2 + Na_2\,[Na_4(PO_3)_6] \leftrightarrow Na_2[Ca_2(PO_3)_6] + 4NaCl$$

$$2MgCl_2 + Na_2[Na_4(PO_3)_6] \leftrightarrow Na_2[Mg_2(PO_3)_6] + 4NaCl$$

These complex salts remain dissolved in water and do not cause hardness as the ions Ca^{2+} and Mg^{2+} are not free to cause any hardness.

(c) Ion Exchange Method

It is a modern method for softening of water. Here Ca^{2+} and Mg^{2+} ions are exchanged by those ions which do not cause hardness. For this purpose ions exchangers are used as follows:

(i) By Ion Exchange Resin (Organic ion Exchangers)
Here cation exchange resins (RCOOH or R – SO_2OH) remove Ca^{2+} and Mg^{2+} ions from hard water while anion exchange resin (OH^-, NH_2^-) remove Cl^-, SO_4^{-2} from hard water. Resin which exchanges cation for H^+ ion is known as cation exchanger and is shown as resin-H^+ while the resin which exchanges anion is known as anion exchanger and shown resin-OH^-.

$$Ca^{2+} + 2\text{ resin-}H^+ \rightarrow Ca\text{ (resin)}_2 + 2H^+$$

$$Mg^{2+} + 2\text{ resin-}H^+ \rightarrow Mg\text{ (resin)}_2 + 2H^+$$

Cation exchanger

$$Cl^- + \text{resin-}OH^- \rightarrow \text{resin} - Cl^- + OH^-$$

$$SO_4^{2-} + 2\text{ resin-}OH^- \rightarrow (\text{resin})_2 - SO_4^{2-} + 2OH^-$$

Anion exchanger

These H^+ and OH^- combines to form water molecules which is called deionised or demineralised water.

$$H + OH^- \rightarrow H_2O$$

(ii) Inorganic Cation Exchangers (Permutit Process)
Some complex inorganic salts have the property of exchanging Ca^{2+} and Mg^{2+} ions from hard water for Na^+ ions to remove hardness e.g, Permutit is artificial zeolite or Sodium alumino ortho silicate [$Na_2Al_2\ Si_2\ O_8 \cdot xH_2O$] which exchange calcium and magnesium ions from hard water by giving Na_+ ion to hard water.

$$\underset{\text{Sodium zeolite (permutit)}}{Ca^{2+} + Na_2\ Al_2\ Si_2\ O_8 \cdot xH_2O} \rightarrow \underset{\text{Calcium zeolite}}{Ca\ Al_2\ Si_2\ O_8 \cdot xH_2O} + 2Na^+$$

$$Mg^{2+} + Na_2\ Al_2\ Si_2\ O_8 \cdot xH_2O \rightarrow \underset{\text{Magnesium zeolite}}{Mg\ Al_2\ Si_2\ O_8 \cdot xH_2O} + 2Na^+$$

HEAVY WATER OR DEUTERIUM OXIDE D_2O

It was discovered by Urey and it is deuterium oxide i.e., D_2O. Nearly 6000 parts of H_2O has one part D_2O.

Theoretically six different varieties of heavy water are possible in terms of three different isotopes of oxygen

(i) $H - O^{16} - D$ (ii) $H - O^{17} - D$

(iii) $H - O^{18} - D$ (iv) $D - O^{16} - D$

(v) $D - O^{17} - D$ (vi) $D - O^{18} - D$

It is non radioactive in nature.

It is prepared by fractional distillation of ordinary water and exhaustive electrolysis of ordinary water having N/2 NaOH solution in seven stages. Here 99% pure D_2O is formed.

In the preparation of heavy water from simple water's electrolysis. Cathode is of steel vessel while canoed is of cylindrical nickel sheet.

Physical property	H_2O	D_2O
1. Boiling point °C	100	101.42
2. Freezing point °C	0	3.82
3. Viscosity at 20°C	10.09	12.6
4. Density at 20°C (g/ml)	0.997	1.017
5. Maximum density at °C	4	11.6
6. Ionic product Kw at 25°C	1×10^{-14}	3×10^{-15}
7. Dielectric constant at 20°C	82	80.5
8. Solubility of NaCl per 100 gm of water at 25°C	35.9	30.5
9. Molecular weight	18	22

Most of the physical properties of heavy water are more than simple water. It is due to greater nuclear mass and greater degree of association of deuterium.

Chemical Properties

The reactivity of D_2O is less than H_2O as D – O is stronger than H – O bond however it shows nearly all the reactions of water. Some of the reactions of heavy water are as follows:

(1) Electrolysis

Electrolysis of D_2O gives D_2 at cathode and O_2 at anode.

$$2D_2O \xrightarrow{\text{electrolysis}} 2D_2 + O_2$$

(2) With Active Metals

It reacts with active metals like alkali and alkaltine metals and releases heavy hydrogen.

$2Na + 2D_2O \rightarrow D_2 + 2NaOD$

$Ca + 2D_2O \rightarrow D_2 + Ca(OD)_2$

(3) Reaction with Metal Oxides

It reacts with these oxides to give their deuterohydroxides.

Example,

$Na_2O + D_2O \rightarrow 2NaOD$

$CaO + D_2O \rightarrow Ca(OD)_2$

(4) With Non-Metal Oxides

It reacts with these oxides to give their deuteroacids.

Example,

$SO_3 + D_2O \rightarrow \underset{\text{Deuterosulphuric acid}}{D_2SO_4}$

$N_2O_5 + D_2O \rightarrow \underset{\text{Deuteronitric acid}}{2DNO_3}$

(5) With Carbides

It reacts with metal carbides to give hydrocarbons

Example,

$CaC_2 + 2D_2O \rightarrow \underset{\text{Deuteroacetylene}}{C_2D_2} + 2Ca(OD)_2$

$Al_4C_3 + 12D_2O \rightarrow 4Al(OD)_3 + \underset{\text{Deuteromethane}}{3CD_4}$

(6) With Metal Nitrides

It reacts with metal nitrides to form deuteroammonia.

Example,

$Mg_3N_2 + 6D_2O \rightarrow 3Mg(OD)_2 + 2ND_3$

$AlN + 3D_2O \rightarrow Al(OD)_3 + ND_3$

(7) Deuterolysis

Hydrolysis of salts in heavy water is called deutrolysis.

$AlCl_3 + 3D_2O \rightarrow Al(OD)_3 + 3DCl$

(8) Exchange Reaction

It reacts with many compounds having hydrogen to exchange their hydrogen with deuterium. These are called exchange reactions.

Example,

$H_2O_2 + 2D_2O \rightarrow D_2O_2 + 2HOD$

$NH_4Cl + 4D_2O \rightarrow ND_4Cl + 4HOD$

Uses

(1) It is used to prepare D_2, CD_4, C_2D_2, ND_3, PD_3 etc.

(2) It is used as a moderator to slow down velocity of neutrons in nuclear reactors.

(3) It is used as a tracer to study reaction mechanism of biochemical process, electrophillic substitution etc.

(4) The exchange reaction of D_2 with H_2 is used to determine the structure of oxy acids. For example H_3PO_3 is dibasic while H_3PO_2 is monobasic.

Biological & Physiological action of Heavy water

It checks plant growth, stops seed germination and disturbs aquatic life.

HYDROGEN PEROXIDE (AUXOCHROME) H_2O_2

It was discovered by Thenard. It is also called oxygenated water and perhydrol (30 % H_2O_2) also.

Methods of Preparation

It is prepared by following methods:

(1) From True Peroxide

True peroxides are peroxides of IA and IIA groups like Na_2O_2, BaO_2

$3BaO_2 + 2H_3PO_4 \rightarrow Ba_3(PO_4)_2 + 3H_2O_2$

HNO_3 can not be taken as it oxidises H_2O_2 into O_2

BaO_2 is the best choice as $BaSO_4$ can be easily removed

H_3PO_4 is better than H_2SO_4 as it does not react with H_2O_2 and prevents reverse reaction.

BaO_2 should be hydrated but not in excess.

(2) Merck Method

Here CO_2 is passed through ice cold BaO_2 solution to get H_2O_2.

$BaO_2 + H_2O + CO_2 \rightarrow BaCO_3 \downarrow + H_2O_2$

(3) By the Electrolysis of (50%) or Moderate Concentrated H_2SO_4

Here 30 % H_2O_2 is obtained at anode (Pt) by the electrolysis of 50 % H_2SO_4

$2H_2SO_4 \rightarrow 2H^+ + 2HSO_4^-$

$2HSO_4^- \rightarrow H_2S_2O_8 + 2e^-$ (at anode)

$2H^+ + 2e^- \rightarrow H_2$ (at cathode)

FACTS TO REMEMBER

Here in place of 50% H_2SO_4, NH_4HSO_4 dissolved in excess of H_2SO_4 can also be used for electrolysis.

$$(NH_4)_2S_2O_8 + 2H_2O \rightarrow 2NH_4.HSO_4 + H_2O_2$$

Distillation of Per di sulphuric acid also gives H_2O_2.

$$[H_2S_2O_8 \xrightarrow[0°C]{2H_2O} H_2O_2 + 2H_2SO_4]$$
Pt

(4) By Redox reaction

2-Ethyl anthraquinol on oxidation by air gives H_2O_2. It is a cyclic process.

$$\text{2-Ethyl-anthraquinol} \underset{\text{red}^n \; H_2\,Pd}{\overset{O_2 \text{ oxi}}{\rightleftharpoons}} \text{2-Ethylanthra-quinone} + H_2O_2$$

2-Ethyl -anthraquinol   2-Ethylanthra -quinone

When a mixture of ammonium sulphate and sulphuric acid is taken in 1 :1 ratio H_2O_2 is obtained.

Concentration of H_2O_2

- It is concentrated below 70°C by vacuum distillation or by distillation under reduced pressure.
- When dilute solution of H_2O_2 is evaporated on a water
- bath at 152°C low pressure 20 – 30°C H_2O_2 is obtained.
- When 20 – 30°C H_2O_2 undergoes reduced pressure distillation 90 % H_2O_2 is obtained.
- When 90 % H_2O_2 is crystallized by freezing mixture of solid CO_2 and ether 100% H_2O_2 is obtained.

Storage of H_2O_2

- It is stored in wax-lined, wax amber colour bottles to avoid decomposition due to alkali metal oxides present in glass.
- To avoid decomposition of H_2O_2 in it negative catalyst like H_3PO_4, glycerol, acetanilide, sodium stannate, sodium pyrophosphate are added to it in small amount .
- Dilute solution of H_2O_2 is stable in presence of alcohol or ether or in acidic medium.
- The presence of sand, MnO_2, Fe, Al_2O_3 increase decomposition (positive catalysis).

Physical Properties

(1) It is a colourless, odourless, syrupy liquid and highly soluble in water.

(2) In H_2O_2 oxidation state of oxygen is –1.

(3) H_2O_2 is more hydrogen bonded than is water and so has a higher boiling point (425K). It is more dense than water (density 1.4 g cm^{-3}).

Chemical Properties

(1) Decomposition of H_2O_2

Pure H_2O_2 is not very stable and undergo composition which is an example of disproportion reaction or auto oxidation reaction.

$$2H_2O_2 \xrightarrow{MnO_2} 2H_2O + O_2$$

(2) Here positive catalysts are MnO_2, Pt, Fe etc. and negative catalysts are H_3PO_4, glycerol etc.

(3) Oxidizing Properties

$H_2O_2 \rightarrow H_2O + [O]$

Due to nascent oxygen it can act not only as an oxidant but also as a permanent bleaching agent.

Oxidation and reduction by H_2O_2 in acidic medium is generally slow while it is rapid in alkaline medium.

In acidic medium:

$H_2O_2 + 2H^+ + 2e^- \rightarrow 2H_2O$

In basic medium:

$H_2O_2 + 2e^- \rightarrow 2OH^-$

Example,

$$2FeSO_4 \xrightarrow[+H_2O_2]{H_2SO_4} Fe_2(SO_4)_3 + 2H_2O$$

Ferrous sulphate   Ferric sulphate

$$2K_4Fe(CN)_6 \xrightarrow{H_2O_2} 2K_3Fe(CN)_6 + 2KOH$$

Potassium ferro cyanide   Potassium ferricyanide

It oxidizes sulphites to sulphates example,

$$H_2O_2 \rightarrow H_2O + [O]$$
$$Na_2SO_3 + [O] \rightarrow Na_2SO_4$$

$$Na_2SO_3 + H_2O_2 \rightarrow Na_2SO_4 + H_2O$$

$$MNO_2 + H_2O_2 \xrightarrow{\quad} MNO_3 + H_2O$$
Metal nitrite — Metal nitrate

It oxidizes arsenites to arsenates example,

$$H_2O_2 \rightarrow H_2O + [O]$$
$$Na_3AsO_3 + [O] \rightarrow Na_3AsO_4$$
$$\overline{Na_3AsO_3 + H_2O_2 \rightarrow Na_3AsO_4 + H_2O}$$

$$2KI + H_2O_2 \xrightarrow{\quad} 2KOH + I_2$$

$$HCHO + H_2O_2 \xrightarrow{\text{Pyrogallol}} HCOOH + H_2O$$

$$PbS + 4H_2O_2 \xrightarrow{\quad} PbSO_4 + 4H_2O$$

$$Hg + H_2O_2 \xrightarrow{\quad} HgO + H_2O$$

When a precipitate chromium hydroxide suspended in sodium hydroxide is treated with hydrogen peroxide, it gets oxidized into chromate and a yellow solution of sodium chromate is obtained.

Example,

$$2Cr(OH)_3 + 4NaOH + 3H_2O_2 \rightarrow \underset{\text{Sodium chromate}}{2Na_2CrO_4} + 8H_2O$$

- It gives blue chromium peroxide (CrO_5) with acidic $K_2Cr_2O_7$.

  $H_2Cr_2O_7 + 4[O]\ 2CrO_5 + H_2O$
- CrO_5 has butterfly structure with two peroxy bonds and in it oxidation number of chromium is +6 in it.
- CrO_5 is stable in ether layer forming a blue layer having $CrO(O_2)_2$.

(4) Bleaching Action of H_2O_2

It is due to nascent oxygen. It has permanent bleaching action on human hair, silk, wool etc.

Coloured substance + [O] → Colourless substance

(5) Reducing Properties

H_2O_2 can act as a weak reducing agent in both acidic and basic mediums.

In acidic medium:

$$H_2O_2 \rightarrow 2H^+ + O_2 + 2e^-$$

In basic medium:

$$H_2O_2 + 2OH \rightarrow 2H_2O + O_2 + 2e^-$$

Example,

$$H_2O_2 \xrightarrow{\text{acid}} 2H^+ + O_2 + 2e^- \qquad E^o = -0.67\ V$$

$$H_2O_2 + 2OH \xrightarrow{\quad} 2H_2O + O_2 + 2e^- \qquad E^o = +0.08\ V$$

Examples,

$$X_2 + H_2O_2 \xrightarrow{\quad} 2HX + O_2$$

$$2KMnO_4 + 3H_2SO_4 + 5H_2O_2 \xrightarrow{\quad} K_2SO_4 + 2MnSO_4 + 8H_2O + 5O_2$$

$$3K_3[Fe(CN)_6] + 2KOH + H_2O_2 \xrightarrow{\quad} 2K_4[Fe(CN)_6] + 2H_2O + O_2$$

$$PbO_2 \xrightarrow{H_2O_2} PbO + H_2O + O_2$$

$$Ag_2O \xrightarrow{H_2O_2} 2Ag + H_2O + O_2$$

$$MXO \xrightarrow{H_2O_2} MX + H_2O + O_2$$

It dissolves manganese dioxide (black) suspended in dilute H_2SO_4 due to its reduction into pale ink manganese sulphate which is soluble.

$$MnO_2 + H_2SO_4 + H_2O_2 \rightarrow MnSO_4 + 2H_2O + O_2$$

(6) Acidic Properties

Pure H_2O_2 is a very weak acid (Ka = 1.55×10^{-12} at 25°C). It can also act as a dibasic acid.

$$H_2O_2 \rightleftharpoons H^+ + \underset{\text{(Hydroperoxide ion)}}{HO_2^-}$$

$$HO_2^- \rightleftharpoons H^+ + \underset{\text{(Peroxide ion)}}{O_2^{2-}}$$

Example,

$$2NaOH + H_2O_2 \rightarrow Na_2O_2 + 2H_2O$$
$$Ba(OH)_2 + H_2O_2 \rightarrow BaO_2 + 2H_2O$$

(7) Addition Reaction

It is capable to show addition reaction and forms addition compounds.

Example,

$$\begin{array}{c} CH_2 \\ \| \\ CH_2 \end{array} + \begin{array}{c} HO \\ | \\ HO \end{array} \rightarrow \begin{array}{c} CH_2OH \\ | \\ CH_2OH \end{array}$$

Ethylene glycol

It can also form sodium perborate ($NaBO_2.H_2O_2.3H_2O$), $(NH_4)_2SO_4.H_2O_2$, $NH_2CONH_2.H_2O_2$.

Test of H_2O_2

- It gives I_2 with KI which turns starch paper blue.

$$2KI + H_2O_2 \rightarrow 2KOH + I_2$$

$$\text{Starch} + I_2 \rightarrow \text{Blue complex}$$

- With TiO_2:

$$TiO_2 + H_2SO_4 + H_2O_2 \rightarrow TiO_2.nH_2O$$

or

$H_2[TiO_2(SO_4)_2]$
Orange red
(Per titanic acid)

Structure of H_2O_2

The structure of H_2O_2 is given as,

H – O – O – H
(Baeyer Structure)

↕

$H_2O \rightarrow O$
(Kingzett Structure)

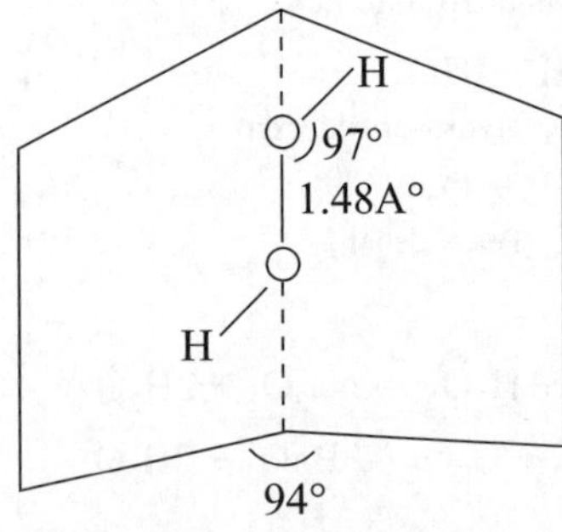

Open Book Structure

The X-ray and dipole moment studies of H_2O_2 confirms that it has a non-planar structure which is shown as below. In this structure the dihedral angle is 111.5° in gas phase, but in solid state due to hydrogen bonding it reduces to 90.2°. Here the two oxygen atoms are linked by a peroxide linkage. Here O – O bond length is 1.48 Å while O – H is 0.95 Å and the O – O – H bond is nearly 97°.

Uses of H_2O_2

(1) H_2O_2 is used to bleach hair, teeth etc.

(2) As a fuel in rockets.

(3) As an antiseptic and germicides in wounds under the name Perhydrol (30 % H_2O_2).

(4) To clear paintings by converting PbS (black) into $PbSO_4$ (white).

(5) The addition compound of H_2O_2 with urea is called Hyperal.

Strength of H_2O_2

It can be given by-

(I) Volumetric strength (e.g., 10 volume, 20 volume, 30 volume)

X-volume H_2O_2 means that one ml of it gives X-ml of O_2 at STP.

Example,

15 volume of H_2O_2 solution means 1 ml of this solution on decomposition releases 15 ml of O_2 at NTP.

(II) Percentage by Weight: The concentration of H_2O_2 in a solution can be expressed as percentage of H_2O_2 in solution (W/V) .

example, 20 % solution of H_2O_2 means 20 gm of H_2O_2 are present in 1000 ml of water.

(III) Concentration (gm / litre) , normality and molarity

FACTS TO REMEMBER

- Concentration or strength of 10 volume H_2O_2 solution = 30.35 gm / lit.
  Volume strength = 5.5 x Normality
  Volume strength = 11.2 x Molarity
- 10 vol H_2O_2 = 3.036 % H_2O_2 by wt. = 1.786 N = 0.893 M

  or

  11.2 vol H_2O_2 = 3.4 % H_2O_2 by wt. = 2N = 1M

Use of Liquid Hydrogen as a Fuel

The liquefied hydrogen is commonly called liquid hydrogen. It has been used as a fuel for rockets (The Saturn V rocket).

Liquid hydrogen is considered as an ideal rocket fuel due to its following unique properties:

- Hydrogen is the lightest known gas having very low mass.
- It is quite abundant in the form of water on earth.
- It is combustible and have a high enthalpy of combustion.
- Water is the only product of its combustion.
- It is non-polluting and no harmful pollutant like CO, CO_2, SO_2, nitrogen oxides etc. are produced during the combustion of hydrogen.

Hydrogen Economy

The use of liquid hydrogen as a fuel is possibly one of the best solution to the energy crisis that the modern world might have to face after a few years. Our present sources of energy like coal, petroleum etc., are depletable so it is must to find out new sources. Hydrogen economy is a good solution to this problem.

Hydrogen economy is related to the transportation and storage of energy in the form of liquid or gaseous hydrogen.

Many efforts are being made to reduce the cost of production of hydrogen. Better and more safe methods of transportation and storage of hydrogen are also being worked out. If the problems of safe storage and high cost of product of hydrogen are solved, hydrogen will prove an ideal fuel not only for the rockets but also for vehicles and domestic uses.

Advantage of the use of hydrogen as an automobile fuel

The use of liquid hydrogen as an automobile fuel has the following advantages:

- It releases a much more amount of energy per unit weight of the fuel.
- The combustion of hydrogen does not produce polluting emissions like CO, CO_2, oxides of nitrogen and sulphur, aldehydes etc. Here combustion product of hydrogen is water having few traces of nitrogen oxides only. Hence the use of hydrogen is more safe and does not cause any pollution.
- The technology needed to modify internal combustion engines for the use of hydrogen as a fuel is simple and does not involve much cost.
- The fuel cells involving the catalytic combustion of hydrogen are already in successful use and have an efficiency of 70-85%.

Hydrates or Hydrated Salts

- These salts are associated with water molecules and water molecules are called water of crystallization. example,

Cationic Hydrate ($MgCl_2 . 6H_2O$)

- Water molecules are held by cations.

Anionic Hydrate ($MgSO_4 . 7H_2O$)

- Water molecules are held by anions and by cations also.

Lattice Hydrate ($Na_2CO_3 . 10H_2O$)

- Water molecules are held in particular sites of the crystal.

Hygroscopic Substances

- Such substances absorb moisture from air. example, NH_4NO_3, $NaClO_4$

Deliquescent Substances

Such substances absorb a lot of water from air. example, NaOH, $Mg(ClO_4)_2 . 6H_2O$.

Efflorescent substances

Such substances lose water of crystallization. example, $CuSO_4 . 5H_2O$, $FeSO_4 . 7H_2O$.

CHAPTER-END EXERCISES

PRACTICE QUESTIONS – I

1. Hydrogen will not reduce:
(a) Heated cupric oxide
(b) Heated ferric oxide
(c) Heated stannic oxide
(d) Heated aluminium oxide

2. Tritium is a radioactive isotope of hydrogen. It emits:
(a) α-particles (b) β-particles
(c) γ-rays (d) Neutrons

3. Hydrogen can be prepared by:
(a) Electrolysis of pure water
(b) Kolbe's electrolysis
(c) Lane's Process
(d) All of these

4. Which of the following is reduced by hydrogen peroxide in acid medium?
(a) $FeSO_4$
(b) Potassium iodide
(c) Potassium ferrocyanide
(d) Potassium permanganate

5. A metal which does not liberate H_2 (g) from acids is:
(a) Fe (b) Zn
(c) Mn (d) Ag

6. Para and ortho hydrogen differ in:
(a) Atomic mass
(b) Atomic number
(c) Number of neutrons
(d) Spins of protons

7. Hydrogen from HCl can be prepared by
(a) P (b) Mg
(c) Hg (d) Cu

8. Heavy water is obtained by:
(a) Prolonged electrolysis of H_2O
(b) Heating H_2O
(c) Boiling water
(d) Fractional distillation of H_2O

9. A hydride of nitrogen which is acidic is:
(a) NH_3 (b) N_3H
(c) N_2H_2 (d) N_2H_4

10. Which one of the following compounds undergoes hydrolysis during distillation to yield hydrogen peroxide?
(a) HNO_3 (b) $H_4P_2O_7$
(c) $H_2S_2O_8$ (d) $H_2S_2O_6$

11. Which of the following is a true peroxide?
(a) SO_2 (b) BaO_2
(c) NO_2 (d) MnO_2

12. D_2O is used in/as:
(a) Insecticide (b) Industry
(c) Medicine (d) Nuclear reactor

13. A variety of water which contains soluble salts of Ca and Mg is known as:
(a) Soft water
(b) Heavy water
(c) Conductivity water
(d) Hard water

14. The boiling point of water is exceptionally high because:
(a) There is covalent bond between H and O
(b) Water molecule is not linear
(c) Water molecule is linear
(d) Water molecules associate due to hydrogen bonding

15. The process used for the removal of hardness of water is:
(a) Baeyer (b) Hoope
(c) Calgon (d) Serpeck

16. The metal that cannot displace hydrogen from diluted HCl is:
(a) Fe (b) Cu
(c) Zn (d) Mg

17. Hydride ion:
(a) Is a strong conjugate base of H_2
(b) Is a strong conjugate acid of H_2
(c) Is a strong conjugate base of H^-
(d) Is a strong conjugate acid of H^+

18. Aluminium reacts with concentrated HCl and concentrated NaOH to liberate the gases ...... respectively.

(a) H_2 and H_2 (b) O_2 and O_2
(c) O_2 and H_2 (d) H_2 and O_2

19. NH_3 and H_2O form NH_4OH by:
(a) Covalent bond
(b) Coordinate bond
(c) Electrovalent bond
(d) None of these

20. The formula of exhausted permutit is:
(a) $K_2Al_2Si_2O_8.xH_2O$
(b) $CaAl_2Si_2O_8.xH_2O$
(c) $Na_2Al_2Si_2O_8.xH_2O$
(d) $CaB_2Si_2O_8.xH_2O$

21. Reaction of N_2O_5 with D_2O gives:
(a) NO (b) DNO_3
(c) DNO_2 (d) NO_2

22. The H – O – H angle in water molecule is about:
(a) 90° (b) 105°
(c) 135° (d) 180°

23. Which group element of d-block do not form hydride at all?
(a) 7 (b) 9
(c) 8 (d) All of these

24. The reaction of H_2O with X does not liberate gaseous product. Which of the following is X?
(a) PbO_2 (b) $KMnO_4$ / H^+
(c) PbS (d) Cl_2

25. D_2O will have maximum density at:
(a) 20°C (b) 9°C
(c) 11.5°C (d) 15.5°C

26. Heavy water (D_2O) freezes at:
(a) 3.8°C (b) 38°C
(c) 0°C (d) –3.8°C

27. Action of water or dilute mineral acids on metals can give:
(a) Tritium (b) Dihydrogen
(c) Trihydrogen (d) Mono hydrogen

28. Water is oxidized to oxygen by:
(a) H_2O_2 (b) $KMnO_4$
(c) ClO_2 (d) Fluorine

29. What is the formula of Calgon?
(a) $MgSO_4$ (b) Na_3PO_4
(c) $Mg_3(PO_4)_2$ (d) $(NaPO_3)_6$

30. The products formed when heavy water is reacted with magnesium nitride, are:
(a) ND_3, $Mg(OD)_2$ (b) ND_3, $Mg(OH)_2$
(c) NH_3, $Mg(OH)_2$ (d) NH_3, $Mg(OD)_2$

31. Reaction of potassium with water is:
(a) Hydrolysis (b) Absorption
(c) Exothermic (d) Endothermic

32. The reagent commonly used to determine hardness of water titrimetrically is:
(a) Disodium salt of EDTA
(b) Sodium thiosulphate
(c) Sodium citrate
(d) Oxalic acid

33. In the following compound H is covalent bonded in case of:
(a) CaH_2 (b) NaH
(c) SiH_4 (d) BaH_2

34. In which of the following substances, hydrogen bonding is absent?
(a) H_2O (b) HF
(c) Salicylaldehyde (d) CCl_4

35. Pure water can be obtained from sea water by:
(a) Plasmolysis (b) Sedimentation
(c) Centrifugation (d) Reverse osmosis

36. The reagent used in the detection of fluoride in water is:
(a) Phenolphthalein
(b) Zirconium-alizarin S
(c) Soda lime
(d) Bayer's reagent

37. Water softening by Clarke's process uses:
(a) Potash alum
(b) Calcium bicarbonate
(c) Calcium hydroxide
(d) Sodium bicarbonate

38. Which one of the following is used for reviving the exhausted "permutit"?
(a) 10 % NaCl solution
(b) 10 % $MgCl_2$ solution
(c) 10 % $CaCl_2$ solution
(d) HCl solution

39. The chemical formula of zeolite is:
(a) $Na_2Al_2Si_2O_8.xH_2O$
(b) $Na_2(Na_4(PO_3)_6$

(c) $Ca_2Al_2Si_2O_8$
(d) $K_2Al_2Si_2O_8.xH_2O$

40. What is formed when calcium carbide reacts with heavy water?

(a) CaD_2 (b) C_2D_2
(c) Ca_2D_2O (d) CD_2

41. By which of the following process permanent hardness of water can be removed?

(a) Washing soda (b) Soda lime
(c) Sodium chloride (d) Sodium bicarbonate

42. What is false about H_2O_2?

(a) Pale blue liquid
(b) Can be oxidized by O_3
(c) Act as both oxidizing and reducing agent
(d) Two –OH bonds lie in the same plane

43. The H – O – O bond angle in H_2O_2 is:

(a) 97° (b) 106°
(c) 104.5° (d) 109.28°

44. H_2O_2 exists as ..... in alkaline medium

(a) $HO_2^{\oplus}$ (b) $H_2O^{\oplus}$
(c) HO_2^- (d) O_2^{2-}

45. The value of .... is less for D_2O compared to that of H_2O_2:

(a) Boiling point
(b) Latent heat of vapourization ($J.g^{-1}$)
(c) Density (g/ml^{-1}) at 20°C
(d) Dielectric constant at 20°C

46. Which of the following is not correct regarding the electrolytic preparation of H_2O_2?

(a) Sulphuric acid undergoes oxidation
(b) Hydrogen is liberated at anode
(c) 50 % H_2SO_4 is used
(d) Lead is used as cathode

47. For making H_2O_2 in the laboratory:

(a) Na_2O_2 is added to boiling water
(b) PbO_2 is added to an acidified solution of $KMnO_4$
(c) BaO_2 is added to CO_2 bubbling through cold water
(d) MnO_2 is added to dilute cold H_3PO_4

48. Which of the following cannot be oxidized by H_2O_2:

(a) O_3 (b) PbS
(c) Na_2SO_3 (d) KI + HCl

49. When H_2O_2 reacts with chlorine the product obtained is:

(a) H_2 (b) O_2
(c) HOCl (d) ClO_2

50. In which of the following reactions, H_2O_2 acts as a reducing agent?

(a) $PbO_2(s) + H_2O_2(aq) \rightarrow PbO(s) + H_2O(l) + O_2(g)$
(b) $KNO_2(aq) + H_2O_2(aq) \rightarrow KNO_3(aq) + H_2O(l)$
(c) $2KI(aq) + H_2O_2(aq) \rightarrow 2KOH(aq) + I2(s)$
(d) $Na_2SO_3(aq) + H_2O_2(aq) \rightarrow Na_2SO_4(aq) + H_2O(l)$

51. Speed of decomposition of H_2O_2 is reduced by:

(a) Pt (b) Alcohol
(c) Na_2CO_3 (d) NaOH

52. A commercial sample of hydrogen peroxide is labeled as 10 volume, its percentage strength is:

(a) 1 % (b) 3 %
(c) 10 % (d) 90 %

53. H_2O_2 acts as an oxidizing agent in:

(a) Acidic medium
(b) Alkaline medium
(c) Acidic and alkaline medium
(d) Neutral medium

54. Which one of the following reactions does not form gaseous product?

(a) $PbS + H_2O_2 \rightarrow$
(b) $Cl_2 + H_2O_2 \rightarrow$
(c) $PbO_2 + H_2O_2 \rightarrow$
(d) Acidified $KMnO_4 + H_2O_2 \rightarrow$

55. Structure of H_2O_2 is:

(a) Planar (b) Linear
(c) Non planar (d) Three dimensional

56. What is the gas liberated when alkaline formaldehyde solution is treated with H_2O_2?

(a) CH_4 (b) H_2
(c) CO_2 (d) O_2

57. Electrolysis of X gives Y at anode. Vacuum distillation of Y gives H_2O_2. The number of peroxy (O – O) bonds present in X and Y respectively are:

(a) 1, 1 (b) 1, 2
(c) Zero, 1 (d) Zero, Zero

58. Phosphoric acid is preferred to sulphuric acid in the preparation of H_2O_2 from barium peroxide because:

(a) Phosphoric acid is available at low cost than sulphuric acid
(b) Phosphoric acid acts as a preservator by retarding decomposition of H_2O_2
(c) Sulphuric acid reacts with explosion
(d) All of the above

59. In comparison to water, heavy water:

(a) Has lower density
(b) Has lower boiling point
(c) Has lower dissociation constant
(d) In general, reacts faster.

60. Hydrogen peroxide can not:

(a) Give silver peroxide with moist silver oxide
(b) Turn the mixture of aniline, $KClO_3$ and dil. H_2SO_4 violet
(c) Liberate iodine from KI
(d) Turn the titanium salt yellow

PRACTICE QUESTIONS – II

61. When zeolite is treated with hard water, the hardness is reduced due to exchange of:

(a) Ca^{2+} ions of water with Na^+ ions
(b) Ca^{2+} ions of water with H^+ ions
(c) Cl^- ions of water with OH^- ions.
(d) SO_4^{2-} ions of water with OH^- ions.

62. All of the following physical properties of D_2O are considered to be related to isotope effects except D_2O, which has a:

(a) Higher melting point than H_2O
(b) Smaller equilibrium dissociation constant than H_2O
(c) Higher boiling point than H_2O
(d) Smaller density than H_2O

63. There are three isotopes of hydrogen and three naturally occurring isotopes of oxygen (O^{16}, O^{17} and O^{18}).

How many kinds of water are possible?

(a) 9 (b) 16
(c) 8 (d) 18

64. If the molar mass of monoatomic deuterium (d) is 2.0141 g then what is the density of diatomic deuterium (D_2) gas at 25°C and 1.00 atmospheric pressure?

(a) 0.165 g / lit (b) 5.125 g / lit
(c) 1.565 g / lit (d) 3.698 g / lit

65. Which of the properties of interstitial hydrides is correct?

(a) They give rise to metals fit for fabrication
(b) They generally form non-stoichiometric species
(c) They can be used as hydrogenation catalysts
(d) The hydrogen dissolved in titanium improves its mechanical properties.

66. Hydrogen peroxide when added to a solution of $KMnO_4$ acidified with H_2SO_4:

(a) Acts as an oxidizing agent
(b) Acts as a reducing agent
(c) Reduces H_2SO_4
(d) Forms water only

67. Which of the following could act as a propellant for rockets?

(a) Liquid hydrogen + liquid nitrogen
(b) Lliquid nitrogen + liquid oxygen
(c) Liquid hydrogen + liquid oxygen
(d) Liquid oxygen + liquid argon

68. How many litres of hydrogen gas can be generated by reacting 6.25 g of barium hydride with water at 20°C and 755 mm Hg pressure according to the following chemical equation:

$BaH_2\ (s) + 2H_2O\ (l) \rightarrow Ba(OH)_2\ (aq) + 2H_2\ (g)$

(a) 2.17 litre (b) 3.58 litre
(c) 1.07 litre (d) 4.57 litre

69. Which one of the following processes will produce hard water?

(a) Saturation of water with $CaCO_3$
(b) Addition of Na_2SO_4 to water
(c) Saturation of water with $MgCO_3$
(d) Saturation of water with $CaSO_4$

70. Which are the compounds, which can be oxidized and reduced by H_2O_2 in acidic and basic medium respectively?

(a) $\{K_3[Fe(CN)_6], K_4[Fe(CN)_6]\}$
(b) $\{K_4[Fe(CN)_6], K_3[Fe(CN)_6]\}$
(c) $[K_2O, NaAlO_2]$
(d) $[KMnO_4, K_2MnO_4]$

71. The molarity of pure water at 277 K is:

(a) 1 M (b) 5 M
(c) 55.5 M (d) 2.5 M

72. The oxidation states of the most electronegative element in the products of the reaction, BaO_2 with diluted H_2SO_4 are:

(a) 0 and –1 (b) –1 and –2
(c) –2 and 0 (d) –2 and +1

73. One mole of magnesium nitride on the reaction with an excess of water gives:

(a) Two moles of ammonia
(b) Two moles of nitric acid
(c) One mole of ammonia
(d) One mole of nitric acid

74. Which property of heavy water is more in magnitude as that compared with normal water?

(1) Ionisation constant
(2) Boiling point
(3) Molecular mass
(4) Density

(a) 1, 2, 3 (b) 2, 3, 4
(c) 1, 3, 4 (d) 1, 2, 4

75. Commercial 11.2 volume H_2O_2 solution has a molarity of:

(a) 0.5 (b) 1.0
(c) 0.75 (d) 11.2

76. The normality of 30 volume H_2O_2 solution is:

(a) 5.86 (b) 7.53
(c) 5.36 (d) 3.36

77. The volume strength of 1.5 N H_2O_2 solution is:

(a) 4.8 (b) 8.4
(c) 4.0 (d) 8.0

78. Out of LiH, CaH_2 and CuH

(a) All are covalent hydrides
(b) LiH is ionic, CaH_2 and CuH are intermediate hydrides
(c) All are ionic hydrides
(d) LiH, CaH_2 are ionic and CuH is covalent hydride

79. 2 g of aluminium is treated separately with excess of dilute H_2SO_4 and excess of NaOH. The ratio of the volumes of hydrogen evolved is:

(a) 1:1 (b) 1:2
(c) 2:1 (d) 2:3

80. For decolourization of one mole of $KMnO_4$ the moles of H_2O_2 required is:

(a) 1/2 (b) 3/2
(c) 5/2 (d) 7/ 2

81. Which compounds are considered to be binary ionic hydrides?

(a) MgH_2, AlH_3, SiH_4, H_2S
(b) BeH_2, B_2H_6, CH_4, NH_3
(c) MgH_2, SrH_2, AlH_3, SiH_4
(d) NaH, KH, CaH_2, BaH_2

82. When two ice cubes are pressed over each other, they unite to form one cube. Which of the following forces is responsible to hold them together?

(a) Van der Waals forces
(b) Covalent attraction
(c) Ionic interaction
(d) Hydrogen bond formation

83. When zeolite, which is hydrated sodium aluminium silicate, is treated with hard water the sodium ions are exchanged with:

1. H^+ ions 2. Ca^{2+} ions
3. SO_4^{-2} ions 4. Mg^{2+} ions

(a) 2, 4 (b) 1, 2, 3
(c) 2, 4, 5 (d) All are correct

84. Which of the following reactions does not produce hydrogen?

(a) $CH_4 \xrightarrow{\text{Cracking } 1000°C}$
(b) $CH_4 + \text{steam} \xrightarrow{\text{Ni-Cr, } 820°C}$
(c) $C + H_2O \rightarrow$
(d) Water gas + steam $\rightarrow$

85. Softening of hard water is done using sodium aluminium silicate (Zeolite). This causes:

(a) Adsorption of Ca^{2+} and Mg^{2+} ions of hard water replacing Al^{3+} ions
(b) Adsorption of Ca^{2+} and Mg^{2+} ions of hard water replacing Na^+ ions
(c) Both of these
(d) None of these

86. How many grams of calcium hydride are required to produce 2.28 litre of hydrogen gas at 25°C and 0.975 atm pressure according to the chemical equation shown below?

$CaH_2 (s) + 2H_2O (l) \rightarrow Ca(OH)_2 (aq) + 2H_2 (g)$

(a) 2.98 (b) 0.91
(c) 3.48 (d) 1.91

87. Polyphosphates are used as water softening agents because they:

(a) Form soluble complexes with cationic species
(b) Precipitate cationic species
(c) Precipitate anionic species
(d) form soluble complexes with anionic species

88. Indicate the coefficient in front of H_2O_2 when the following redox equation is balanced in an acidic medium.

$H_2O_2 (aq) + Cl^- (aq) \rightarrow H_2O (l) + Cl_2 (aq)$

(a) 4 (b) 3
(c) 2 (d) 1

89. Which compounds are considered to be binary covalent hydrides?

(a) MgH_2, SrH_2, AlH_3, SiH_4
(b) BeH_2, B_2H_6, CH_4, NH_3
(c) MgH_2, AlH_3, SiH_4, H_2S
(d) NaH, KH, CaH_2, BaH_2

90. Find the degree of hardness of a sample of water containing 12 mg of magnesium sulphate (molar mass 120) per kg of water.

(a) 0.1 ppm (b) 1 ppm
(c) 10 ppm (d) 100 ppm

PRACTICE QUESTIONS – III

91. Consider the following statements:

1. Atomic hydrogen is obtained by passing hydrogen through an electric arc.
2. Hydrogen gas will not reduce heated aluminium oxide.
3. Finely divided palladium absorbs large volume of hydrogen gas.
4. Pure nascent hydrogen is best obtained by reacting Na with C_2H_5OH.

Which of the above statements is/are correct?

(a) 1 alone (b) 2 alone
(c) 1, 2, and 3 (d) 2, 3 and 4

92. Indicate the coefficient in front of H_2O_2 when the following redox equation is balanced in an acidic medium.

$H_2O_2\ (aq) + MnO_4^-\ (aq) \rightarrow O_2\ (g) + Mn^{2+}\ (aq)$

(a) 4 (b) 5
(c) 6 (d) 7

93. Which of the following is correct statements?

(1) Temporary hardness of water is due to the presence of bicarbonates of calcium and magnesium in it
(2) Permutit is artificial zeolite
(3) H_2O_2 acts as an oxidizing agent in the following reaction:
$Cl_2 + H_2O_2 \rightarrow O_2 + 2HCl$
(4) H_2O_2 is used an bleaching agent for delicate textiles.

(a) 1, 2, 3 (b) 2, 3, 4
(c) 1, 2, 4 (d) 1, 3, 4

94. Acetone exhibits keto- enol tautomerism :

$$(CH_3)_2C=O \rightleftharpoons CH_3-\underset{}{\overset{\displaystyle OH}{\overset{|}{C}}}=CH_2$$

Which of the following products is obtained when acetone is treated with an excess of D_2O for a sufficient time in the presence for a small amount of a dilute NaOH solution?

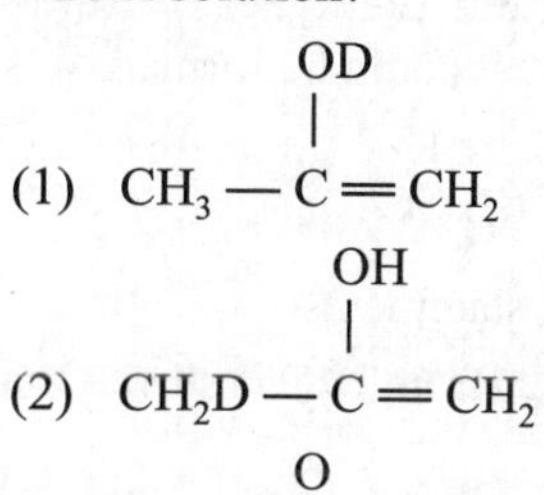

(3) $CD_3-\overset{\displaystyle O}{\overset{\|}{C}}-CD_3$

(4) $CH_2=\underset{\displaystyle OD}{\underset{|}{C}}-CH_3$

(a) 1, 2, 3 (b) 1, 3, 4
(c) 2, 3, 4 (d) 1, 2, 3, 4

95. Select the correct statements:

(1) H_2 is thermally stable
(2) Ionisation of CH_3COOH is slower than that of CH_3COOD
(3) Kinetic isotopic effect is observed when there is retardation in the rate when H_2O is replaced by D_2O
(4) H^+ can exist as $H_4O_4^+$ in water

(a) 1, 2 and 3 (b) 1, 2 and 4
(c) 1, 3 and 4 (d) 2, 3 and 4

96. Point out the correct statement:

(1) Tritium is formed continuously in nuclear reactions induced by cosmic rays.
(2) Deuterium is non-radioactive in contrast to tritium

(3) The kinetic study of reaction rates show in many cases, so called isotopic effect in which hydrogen reacts faster than deuterium.
(4) The para forms of both H_2 and D_2 are stable at low temperature while the ortho forms are stable at higher temperatures.

(a) 1, 2, 3 (b) 1, 2, 4
(c) 2, 3, 4 (d) 2, 4

97. HCl is added to the following oxides, which of them will give H_2O_2?

(1) Na_2O_2 (2) BaO
(3) BaO_2 (4) MnO_2
(a) 1, 2 (b) 1, 3
(c) 1, 2, 3 (d) I only

98. There are three different isotopes of hydrogen ^1H, ^2H, ^3H and three different isotopes of oxygen ^{16}O, ^{17}O, ^{18}O. Indicate the number of different types of isotopically substituted water, HO that could be formed.

(a) 12 (b) 15
(c) 18 (d) 21

99. Consider the following statements:

1. Rate of transfer of D^+ from D_2O is slower than that of H^- from H_2O.
2. Ka for $CH_3COOH \leftrightarrow CH_3COO^- + H^+$ is similar that that of Ka for
3. $^1H_3$ is a radioactive isotope.

Here correct statements are:

(a) 1, 2, 3 (b) 2, 3
(c) 1, 2 (d) 1, 3

100. Which one of the following statements about the zeolite is true?

(1) They have open structure which enables them to take up small molecules.
(2) They are used as cation exchangers
(3) Zeolites are aluminosilicates having three dimensional network.
(4) None of the SiO_4^{4-} units are replaced by AlO_4^{5-} and AlO_6^{9-} ions in zeolites

(a) 1, 2, 3 (b) 1, 2, 4
(c) 2, 3, 4 (d) 3, 4

101. Select the correct statement.

(1) Para hydrogen has a lower internal energy than that of ortho hydrogen.
(2) Para hydrogen is more stable at lower temperature.
(3) Ortho and para hydrogen are different due to difference in their electron spins.
(4) Ortho and para hydrogen are different due to difference in their nuclear spins.

(a) 1, 2, 3 (b) 1, 2, 4
(c) 2, 3, 4 (d) 3, 4

102. In which of the following reactions. Hydrogen peroxide acts as a oxidising agent?

(1) $2KI + H_2O_2 + H_2SO_4 \rightarrow K_2SO_4 + I_2 + H_2O$
(2) Ag_2O (moist) $+ H_2O_2 \rightarrow 2Ag + H_2O + O_2$
(3) $Na_2SO_3 + H_2O_2 \rightarrow Na_2SO_4 + H_2O$
(4) $PbS + 4H_2O_2 \rightarrow PbSO_4 + 4H_2O$

(a) 1, 2, 3 (b) 2, 3, 4
(c) 1, 2, 4 (d) 1, 3, 4

103. Write a balanced net ionic equation for the reaction of bromine with water.

(a) $2Br_2(g) + 2H_2O(l) \rightarrow O_2(g) + 4H^+(aq) + 4Br^-(aq)$
(b) $Br_2(l) + 2H_2O(l) \rightarrow 2HOBr(aq) + H_2(g)$
(c) $Br_2(l) + H_2O(l) \rightarrow HOBr(aq) + H^+(aq) + Br^-(aq)$
(d) $2Br_2(l) + 2H_2(l) \rightarrow O_2(g) + 4HBr(aq)$

104. The percentage weight of hydrogen in H_2O_2 is:

(a) 6.50 (b) 5.88
(c) 25 (d) 50

105. 11.2 volumes of H_2O_2 is about:

(a) 0.34 % (b) 0.034 %
(c) 34 % (d) 3.4 %

106. Consider the following statements about ortho and para hydrogen.

(1) In ortho hydrogen, the spins of the proton are in the same direction.
(2) Ortho hydrogen is more stable than para hydrogen at the ambient condition.
(3) At ordinary temperature, ordinary hydrogen is a mixture of about 75 % para and 25 % ortho forms.
(4) The two forms have similar chemical properties, but differ in physical properties like specific heat and thermal conductivity.

Which of the statements are correct?

(a) 1, 2 and 3 (b) 1, 2 and 4
(c) 1, 3 and 4 (d) 2, 3 and 4

107. 10- volume H_2O_2 is equivalent to:

(1) 10% (2) 30 g/ L
(3) 1.786 N (4) 3 % H_2O_2 (W / V)
(a) 1, 2, 3 (b) 1, 2, 4
(c) 2, 3, 4 (d) 2, 4

108. The dipole moment of H_2O is 1.85 D. The O–H bond moment would be (given the H–O–H bond angle as 104.5° and cos 104.5° = –0.25)

(a) 1.51 D (b) 2.41 D
(c) 1.86 D (d) 1.91 D

109. Which of the following is correct?
(1) $NH_4^+ + D_2O \rightleftharpoons NH_3D^+ + HDO$
(2) $PH_3 + D_2O \rightleftharpoons PD_3 + H_2O$
(3) $CH_3OH + D_2O \rightleftharpoons CH_3OD + HDO$
(4) $CH_3NH_2 + D_2O \rightleftharpoons CH_3NHD + HDO$
(a) 1, 2, 3 (b) 2, 3, 4
(c) 1, 2, 4 (d) 1, 3, 4

110. A compound 'x' is formed by slowly adding calculated quantities of sodium peroxide to ice cold 20% solution of the sulphuric acid. Which of the following incorrectly describes the compound 'x'?
(1) It is a mild bleaching agent and its bleaching action is due to reduction.
(2) When shaken with potassium dichromate in sulphuric acid and ether, a blue colour is obtained in the ether layer.
(3) It is a strong oxidizing agent in acidic medium and has no oxidizing power in alkaline medium.
(4) It is a weak monobasic acid.
(a) 2, 3, 4 (b) 1, 3, 4
(c) 1, 2, (d) 1, 2, 3

Answer Key

1. (d)	2. (b)	3. (d)	4. (d)	5. (d)	6. (d)	7. (b)	8. (a)	9. (b)	10. (c)
11. (b)	12. (d)	13. (d)	14. (d)	15. (c)	16. (b)	17. (a)	18. (a)	19. (c)	20. (b)
21. (b)	22. (b)	23. (d)	24 (c)	25. (c)	26. (a)	27. (b)	28. (d)	29. (d)	30. (a)
31. (c)	32. (a)	33. (c)	34. (d)	35. (d)	36. (b)	37. (c)	38. (a)	39. (a)	40. (b)
41. (b)	42. (d)	43. (a)	44. (c)	45. (d)	46. (c)	47. (c)	48. (a)	49. (b)	50. (a)
51. (b)	52. (c)	53. (c)	54. (a)	55. (c)	56. (b)	57. (c)	58. (b)	59. (c)	60. (a)
61. (a)	62. (d)	63. (d)	64. (a)	65. (b)	66. (d)	67. (c)	68. (a)	69. (d)	70. (b)
71. (c)	72. (b)	73. (c)	74. (b)	75. (b)	76. (c)	77. (b)	78. (b)	79. (a)	80. (c)
81. (d)	82. (d)	83. (a)	84. (c)	85. (b)	86. (d)	87. (a)	88. (d)	89. (b)	90. (c)
91. (c)	92. (b)	93. (c)	94. (b)	95. (c)	96. (a)	97. (b)	98. (c)	99. (a)	100. (a)
101. (b)	102. (d)	103. (c)	104. (b)	105. (d)	106. (b)	107. (c)	108. (a)	109. (d)	110. (b)

Hints and Explanations for Selective Questions

11. $BaO_2 + H_2SO_4 \rightarrow BaSO_4 + H_2O_2$

16. Cu is below hydrogen in the electrochemical series and unable to displace hydrogen from HCl.

19. $NH_3 + H_2O \rightarrow NH_4^+ + OH^-$

25. D_2O has maximum density at 11.5°C

28. Fluorine oxidizes water to O_2

$2F_2 + 2H_2O \rightarrow O_2 + 4HF$

31. Potassium reacts with H_2O liberating H_2 and heat.

37. In Clarke's process quick lime is added to hard water which produces $Ca(OH)_2$ which further converts bicarbonates into insoluble carbonates

40. $CaC_2 + 2D_2O \rightarrow Ca(OD)_2 + C_2D_2$

49. $H_2O_2 + Cl_2 \rightarrow 2HCl + O_2$

50. $\overset{+4}{PbO_2}$ is reduced to $\overset{+2}{PbO}$.

51. Alcohol retards decomposition of H_2O_2 (negative catalyst).

58. It liberates H_2O_2 and also acts as a preservator by retarding decomposition.

59. Dissociation constant: Water = 1.0×10^{-14}

and heavy water = 0.3×10^{-14}

61. When zeolite is treated with hard water the hardness is reduced due to exchange of Ca^{2+} or Mg^{2+} ions with Na^+ ions.

$Na_2Al_2Si_2O_8.\times H_2O + Ca^{2+} \rightarrow$

$CaAl_2Si_2O_8.\times H_2O + 2Na^+$

or

$Na_2Al_2Si_2O_8.\times H_2O + Mg^{2+} \rightarrow$

$MgAl_2Si_2O_8.\times H_2O + 2Na^+$

66. Acidified $KMnO_4$ solution is reduced by H_2O_2 to Mn^{2+} ions.

67. Mixture of liquid hydrogen and liquid oxygen is used as a propellant for rockets.

69. $MgCO_3$ and $CaCO_3$ are water insoluble. $CaSO_4$ dissolves in water adding Ca^{2+} ions which are responsible for producing hardness in water.

71. Density of water at 277 K is 1 g cm^{3-}.

So molarity $= \frac{1000}{18}$

72. $BaO_2 + H_2SO_4 \rightarrow BaSO^{2-}{}_4 + H_2O^{-1}{}_2$

73. $Mg_3N_2(s) + 6H_2O(l) \rightarrow 3Mg(OH)_2 + 2NH_3(g)$

1 mol 2 mol

75. Molarity $= \frac{\text{Volume Strength}}{11.2}$

$= \frac{11.2}{11.2} = 1.0$

76. $30 = N \times 5.6$

$N = \frac{30}{5.6} = 5.36$

77. Volume strength of H_2O_2 = Normality × 5.6

Volume strength of 1.5 NH_2O_2 = 1.5 × 5.6

= 8.4

79. $2Al + 3H_2SO_4 \rightarrow Al_2(SO_4)_3 + 3H_2$

$2Al + 2NaOH + 2H_2O \rightarrow 2NaAlO_2 + 3H_2$

So the ratio of volumes of hydrogen evolved is 1 : 1.

82. Hydrogen bonds are formed between water molecules in ice.

83. $Na_2Al_2Si_2O8$ $.xH_2O$ is written as Na_2Z, Na^+ ions exchange Ca^+ and Mg^{2+} ions of hard water.

$Na_2Z + Mg^{2+} \rightarrow 2Na^+ + MgZ$

$Na_2Z + Ca^{2+} \rightarrow 2Na^+ + CaZ$

84. (a) $C + H_2O \rightarrow \underset{\text{water gas}}{CO + H_2}$

(b) $\underset{\text{water gas}}{CO + H_2} + H_2O \xrightarrow{\text{Catalyst}} CO_2 + 2H_2$

(c) $\underset{(g)}{CH_4} + \underset{(g)}{H_2O} \xrightarrow{\text{Ni-Cr,820°C}} CO + 3H_2$

(d) $CH_4\ (g) \xrightarrow{\text{Cracking 1000°C}} C + 2H_2$

85. $\underset{\text{Zeolite}}{n\ Na_2O.Al_2O_3.SiO_2.yH_2O} + CaCl_2 \rightarrow$

$CaO.Al_2O_3.SiO_2.yH_2O + 2NaCl$

87. Polyphosphates are used as water softener because these form soluble complexes with cationic species (Ca^{+2} and Mg^{+2}) present in hard water. The complex calcium and magnesium ions do not form any ppt. with soap and hence water readily produce lather with soap solution.

$2Ca^{+2} + Na_2[Na_4(PO_3)_6] \rightarrow Na_2[Ca_2(PO_3)_6] + 4Na^+$

$2Mg^{+2} + Na_2[Na_4(PO_3)_6] \rightarrow \underset{\text{Soluble complex}}{Na_2[Mg_2(PO_3)_6]} + 4Na^+$

90. 12 mg of $MgSO_4$ is present in 103 g of water thus 106 g of H_2O contains

= 12000 mg of $MgSO_4$

120 g of $MgSO_4 = \frac{100}{120} \times \frac{12000}{1000}$ g of $CaCO_3$

= 10 g of $CaCO_3$

So hardness of water = 10 ppm

104. $\underset{34\text{ g}}{H_2O_2} \rightarrow \underset{2\text{ g}}{2H};$

34 g of H_2O_2 gives 2 g of hydrogen

100 g of H_2O_2 will give hydrogen

$= \frac{2}{34} \times \frac{100}{1} = 5.88$ g

105. 11.2 = Normality × 5.6

Normality $= \frac{11.2}{5.6}$

Strength (g/l) $= \frac{11.2}{5.6} \times 17$

% Strength $= \frac{11.2 \times 17}{5.6 \times 10} = 3.4\%$

107. $\underset{68\text{ g}}{2H_2O_2} \rightarrow \underset{22400\text{ mL at stp}}{2H_2O + O_2}$

10- vol. H_2O_2 means that 1 vol. of H_2O_2 ≡ 10 vol. of O_2 at stp

or 1 mL of H_2O_2 ≡ 10 mL of O_2 at stp

22400 mL of O_2 at stp = 68 g of H_2O_2
As, 10 mL of O_2 at stp = 3×10^{-2} g of H_2O_2
So 1 mL of H_2O_2 sol. = 3×10^{-2} of H_2O_2
100 mL of H_2O_2 sol. = 3 g of H_2O
So, 3% W / V.

108. The resultant dipole moment is as follows:
$\mu = \sqrt{(\mu_1^2 + \mu_2^2 + 2\mu_1 \mu_2 \cos\theta)}$
$1.85 = \sqrt{(2\mu^2 + 2\mu^2 \cos 104.5°)}$
$1.85 = \sqrt{(3/2)} \times \mu$
On solving, we get
$\mu = 1.51$ D.

109. PH_3 does not contain labile hydrogen. The electronegativity values of P and H are nearly the same.

110. The compound is H_2O_2
$Na_2O_2 + H_2SO_4 \rightarrow Na_2SO_4 + H_2O_2$
The reaction described is a test for H_2O_2

Previous Years' Questions

1. A solid compound 'X' on heating gives CO_2 gas a residue. The residue mixed with water forms 'Y'. On passing an excess of CO_2 through 'Y' in water, a clear solution 'Z' is obtained. On boiling:

(a) $Ca(HCO_3)_2$ (b) Na_2CO_3
(c) $CaCO_3$ (d) K_2CO_3

[2004]

2. Which of the following compound is a Peroxide?

(a) NO_2 (b) KO_2
(c) BaO_2 (d) MnO_2

[2010]

3. Select the correct statements about heavy water?

(1) Heavy water is used as a moderator in nuclear reactor
(2) Heavy water is more associated than ordinary water
(3) Heavy water is more effective solvent than ordinary water

(a) 1 and 2 (b) 1, 2 and 3
(c) 2 and 3 (d) 1 and 3

[2010]

4. Which of the following statement about hydrogen is in correct ?

(a) Hydrogen never acts as cation in ionic salts
(b) Hydronium ion H_3O^+ exists frecly in solution
(c) Di hydrogen does not act as a reducing agent
(d) Hydrogen has there isotopes of which tritium is the most common

[2016]

Answer Keys

1. (c) **2.** (c) **3.** (a) **4.** (c)

Hints and Explanations

1. Here the compound (X) is $CaCO_3$ and sequence of reactions is as follows:

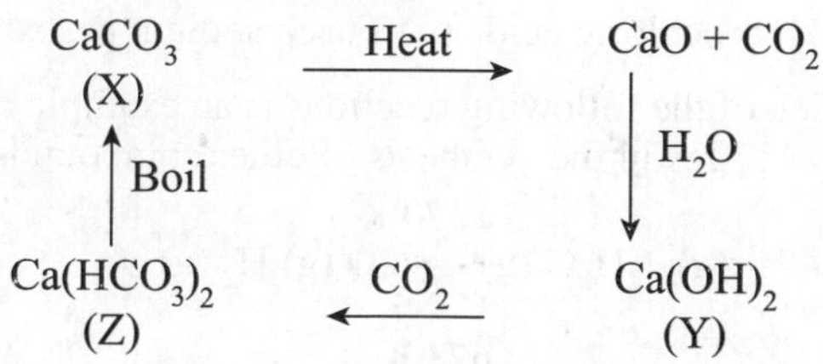

2. Here BaO_2 is peroxide, KO_2 is superoxide and NO_2, MnO_2 are Oxides.

3. Heavy water is used as a moderator in nuclear reactor and also it is more associated than ordinary water.

4. H_2 can act as reductant and out of three isotopes of Hydrogen ^1_1H (Protonium) is most abundant while ^3_1H is least.

NCERT EXEMPLAR

1. Hydrogen resembles halogens in many respects for which several factors are responsible. Of the following factors which one is most important in this respect?
   (a) Its tendency to lose an electron to form a cation.
   (b) Its tendency to gain a single electron in its valence shell to attain stable electronic configuration.
   (c) Its low negative electron gain enthalpy value.
   (d) Its small size.

2. Why does H^+ ion always get associated with other atoms or molecules?
   (a) Ionisation enthalpy of hydrogen resembles that of alkali metals.
   (b) Its reactivity is similar to halogens.
   (c) It resembles both alkali metals and halogens.
   (d) Loss of an electron from hydrogen atom results in a nucleus of very small size as compared to other atoms or ions. Due to smal size it cannot exist free.

3. Metal hydrides are ionic, covalent or molecular in nature. Among LiH, NaH, KH, RbH, CsH, the correct order of increasing ionic character is:
   (a) LiH > NaH > CsH > KH>RbH
   (b) LiH < NaH < KH < RbH < CsH
   (c) RbH > CsH > NaH > KH > LiH
   (d) NaH > CsH > RbH > LiH > KH

4. Which of the following hydrides is electron-precise hydride?
   (a) B_2H_6 (b) NH_3
   (c) H_2O (d) CH_4

5. Radioactive elements emit α, β and γ rays and are characterised by their half-lives. The radioactive isotope of hydrogen is:
   (a) Protium (b) Deuterium
   (c) Tritium (d) Hydronium

6. Consider the reactions:
   (A) $H_2O_2 + 2HI \rightarrow I_2 + 2H_2O$
   (B) $HOCl + H_2O_2 \rightarrow H_3O^+ + Cl^- + O_2$
   Which of the following statements is correct about H_2O_2 with reference to these reactions? Hydrogen perioxide is ________.
   (a) An oxidising agent in both (A) and (B)
   (b) An oxidising agent in (A) and reducing agent in (B)
   (c) A reducing agent in (A) and oxidising agent in (B)
   (d) A reducing agent in both (A) and (B)

7. The oxide that gives H_2O_2 on treatment with dilute H_2SO_4 is —
   (a) PbO_2 (b) $BaO_2 .8H_2O + O_2$
   (c) MnO_2 (d) TiO_2

8. Which of the following equations depict the oxidising nature of H_2O_2?
   (a) $2MnO_4^- + 6H^+ + 5H_2O_2 \rightarrow 2Mn^{2+} + 8H_2O + 5O_2$
   (b) $2Fe^{3+} + 2H^+ + H_2O_2 \rightarrow 2Fe^{2+} + 2H_2O + O_2$
   (c) $2I^- + 2H^+ + H_2O_2 \rightarrow I_2 + 2H_2O$
   (d) $KIO_4 + H_2O_2 \rightarrow KIO_3 + H_2O + O_2$

9. Which of the following equation depicts reducing nature of H_2O_2?
   (a) $2[Fe(CN)_6]^{4-} + 2H^+ + H_2O_2 \rightarrow 2[Fe(CN)_6]^{3-} + 2H_2O$
   (b) $I_2 + H_2O_2 + 2OH^- \rightarrow 2I^- + 2H_2O + O_2$
   (c) $Mn^{2+} + H_2O_2 \rightarrow Mn^{4+} + 2OH^-$
   (d) $PbS + 4H_2O_2 \rightarrow PbSO_4 + 4H_2O$

10. Hydrogen peroxide is ________.
   (a) An oxidising agent
   (b) A reducing agent
   (c) Both an oxidising and a reducing agent
   (d) Neither oxidising nor reducing agent

11. Which of the following reactions increases production of dihydrogen from synthesis gas?
   (a) $\underset{g}{CH_4} + \underset{g}{H_2O} \xrightarrow[1200\,K]{Ni} \underset{g}{CO} + \underset{g}{3H_2}$
   (b) $\underset{s}{C} + \underset{g}{H_2O} \xrightarrow{1270\,K} \underset{g}{CO} + \underset{g}{H_2}$
   (c) $\underset{g}{CO} + \underset{g}{H_2O\,(g)} \xrightarrow[Catalyst]{673\,K} CO_2 + H_2$
   (d) $C_2H_6 + 2H_2O \xrightarrow[Ni]{1270\,K} 2CO + 5H_2$

12. When sodium peroxide is treated with dilute sulphuric acid, we get ________.
   (a) Sodium sulphate and water
   (b) Sodium sulphate and oxygen
   (c) Sodium sulphate, hydrogen and oxygen
   (d) Sodium sulphate and hydrogen peroxide

13. Hydrogen peroxide is obtained by the electrolysis of ________.
   (a) Water (b) Sulphuric acid
   (c) Hydrochloric acid (d) Fused sodium peroxide

14. Which of the following reactions is an example of use of water gas in the synthesis of other compounds?
   (a) $CH_4\,(g) + H_2O\,(g) \xrightarrow[Ni]{1270\,K} CO\,(g)\,H_2\,(g)$
   673 K

(b) $CO (g) + H_2O (g) \rightarrow CO (g) + H (g)$
Catalyst
1270 K

(c) $C_nH_{2n+2} + nH_2O (g) \rightarrow nCO (2n+1) + H$
Ni

(d) $CO (g) + 2H_2 (g) \rightarrow CH OH (l)_3$
Cobalt
Catalyst

15. Which of the following ions will cause hardness in water sample?

(a) Ca^{2+} (b) Na^+
(c) Cl^- (d) K^+

16. Which of the following compounds is used for water softening?

(a) $Ca_3 (PO_4)_2$ (b) Na_3PO_4
(c) $Na_6P_6O_{18}$ (d) Na_2HPO_4

17. Elements of which of the following group(s) of periodic table do not form hydrides.

(a) Groups 7, 8, 9 (b) Group 13
(c) Groups 15, 16, 17 (d) Group 14

18. Only one element of ________ forms hydride.

(a) Group 6 (b) Group 7
(c) Group 8 (d) Group 9

Answer Keys

1. (b)	**2. (d)**	**3. (b)**	**4. (d)**	**5. (c)**	**6. (b)**	**7. (b)**	**8. (c)**	**9. (b)**	**10. (c)**
11. (c)	**12. (d)**	**13. (b)**	**14. (d)**	**15. (a)**	**16. (c)**	**17. (a)**	**18. (a)**		

Hints and Explanations for Selective Questions

2. H^+ has very small size as compared to normal atomic and ionic sizes of 50 to 220pm.

3. Ionic character increases as size of atom increases.

5. generally for radioactive nucleides with neutron proton ratio > 1.5

9. H_2O_2 act as a reducing agent as I_2 is reduced to I^-

AIIMS ESSENTIALS

Assertion and Reason

In the following questions, two statements (Assertion) A and Reason (R) are given. Mark

(a) If A and R both are correct and R is the correct explanation of A;

(b) If A and R both are correct but R is not the correct explanation of A;

(c) A is true but R is false;

(d) A is false but R is true.

(e) Both A and R are false.

1. (A) : Atomic hydrogen is more reactive than nascent hydrogen

(R) : The energy content of atomic hydrogen is more than ascent hydrogen

2. (A) : Calcium oxide is basic oxide and H_2O_2 is amphiprotic.

(R) : H_2O_2 acts as oxidizing as well as reducing agent.

3. (A) : H_2O_2 liberates O_2 when it reacts with acidified $KMnO_4$ solution.

(R) : $KMnO_4$ oxidises H_2O_2 to O_2.

4. (A) : Ortho – hydrogen is less stable than Para hydrogen at 0°C.

(R) : At 0°C, highest % of para- hydrogen is present.

5. (A) : The O – O bond length in H_2O_2 is longer than that of O_2F_2.

(R) : H_2O_2 is ionic as well as covalent in nature.

6. (A) : Crystals of $NaHCO_3$ and $KHCO_3$ show hydrogen bonds of different kinds.

(R) : In $NaHCO_3$, the bicarbonate ions are linked in an infinite chain while in $KHCO_3$ a dimeric chain is formed.

7. (A) : The molarity of 20 volume H_2O_2 is 3.58 M.

(R) : Volume strength = 5.6 × M.

8. (A) : Complete hydrolysis of one mole of peroxydisulsphuric acid gives one mole of H_2O_2 and two moles of H_2SO_4.

(R) : Peroxydisulphuric acid has one peroxy linkage.

9. (A) : Para-hydrogen with lower energy is favoured at high temperatures.

(R) : The thermal conductivity of parahydrogen is lesser than that of ortho-hydrogen.

10. (A) : Permanent hardness of water is due to the presence of chlorides or sulphates of Ca or Mg.

(R) : Permanent hardness is removed by boiling.

11. (A) : Hydrogen gas is liberated by the action of aluminium with a concentrated solution of NaOH.

(R) : Aluminium with NaOH forms sodium metaborate.

12. (A) : ScH_2 is an example of ionic hydride.

(R) : All metals form ionic hydride.

13. (A) : Hydrogen shows resemblance with alkali metals as well as halogens.

(R) : Hydrogen exists in atomic form only at high temperature.

14. (A) : Zinc hydroxide dissolve in excess of NaOH solution to evolve dihydrogen gas.

(R) : Zinc hydroxide is amphoteric in nature

15. (A) : Water is an amphoteric substance.

(R) : Water has a tendency to accept and donate a proton easily.

Answer Keys

1. (a)	2. (d)	3. (a)	4. (b)	5. (b)	6. (a)	7. (d)	8. (a)	9. (d)	10. (c)
11. (b)	12. (d)	13. (b)	14. (d)	15. (d)					

CHAPTER 9

The s-block Elements

Chapter Outline

■ s-Block Elements ■ Alkali Metals [Ia] ■ Sodium ($_{11}Na^{23}$) ■ Compounds of Sodium Sodium Oxide (Na_2O) ■ Sodium Peroxide (Na_2O_2) ■ Caustic Soda or Sodium Hydroxide (NaOH) ■ Sodium Chloride (NaCl) ■ Sodium Carbonate or Washing Soda ($Na_2CO_3.10H_2O$) ■ Sodium Bicarbonate or Baking Soda ($NaHCO_3$) ■ Micro Cosmic Salt ($Na(NH_4)\ HPO_4.4H_2O$) ■ Potassium ($_{19}K^{39}$) ■ Potassium Super Oxide (KO_2) ■ Potassium Hydroxide (KOH) ■ Potassium Chloride (KCl) ■ Potassium Carbonate (K_2CO_3) ■ Potassium Iodide (KI) ■ Potassium Sulphate (K_2SO_4) ■ Potassium Bicarbonates ($KHCO_3$) ■ Biological Role of Sodium (Na) and Potassium (K) ■ Alkaline Earth Metals and their Compounds [IIa (ns^2)] ■ Chemical Properties and Compounds ■ Magnesium ($_{12}Mg^{24}$) ■ Compounds of Magnesium Magnesia (MgO) ■ Magnesium Chloride ($MgCl_2$) ■ Magnesium Sulphate or Epsom Salt or Epsomite [$MgSO_4.7H_2O$] ■ Magnesium Hydroxide $Mg(OH)_2$ ■ Magnesium Carbonate ($MgCO_3$) ■ Magnesium Bicarbonate $Mg(HCO_3)_2$ ■ Calcium ($_{20}Ca^{40}$) ■ Calcium Oxide Or Quick–Lime (CaO) ■ Calcium Hydroxide or Slaked Lime $Ca(OH)_2$ ■ Calcium Oxide or Marble or Lime Stone ($CaCO_3$) ■ Calcium Sulphate Dihydrate or Gypsum ($CaSO_4.2H_2O$) ■ Plaster of Paris ($CaSO_4$.½ H_2O) or [$(CaSO_4)_2\ H_2O$] ■ Bleaching Powder $CaOCl_2$ or Ca(OCl)Cl ■ Cement ■ Biological Role of Mg and Ca

s-Block Elements

It includes alkali metals (I A) and alkaline earth metals (II A). In these elements the differentiating electron is filled in the s-sub shell of the valence orbit.

Alkali Metals [Ia]

It includes six elements i.e., Li, Na, K, Rb, Cs, Fr.

Alkali metals have a general electronic configuration ns^1. These metals are known as alkali metals as their hydroxides are soluble in water and their aqueous solution is strongly alkaline in nature. These metals are always found in combined states and not in free state due to their high reactivity. Out of these metals Na and K are quite abundant in nature and occupy 7th and 8th position in abundance by weight in earth crust. Out of these metals francium (Fr) is least abundant and it is a radioactive element with a very short half life of 21 minutes.

General Features

The general features of these elements are as follows:

Electronic Configuration

All these elements have a general electronic configuration ns^1 i.e., last valence electron is in s-orbital and the inner orbits are complete. Due to same electronic configuration they have a great resemblance in their properties. Here lithium shows some abnormalities in the properties as its electronic configuration is little different from the other alkali metals.

Example,

$_3Li : 1s^2, 2s^1$

$_{11}Na : 1s^2, 2s^2\ 2p^6, 3s^1$

$_{19}K : 1s^2, 2s^2\ 2p^6, 3s^2\ 3p^6\ 4s^1$

$_{37}Rb : 1s^2, 2s^2\ 2p^6, 3s^2\ 3p^6\ 3d^{10}, 4s^2\ 4p^6, 5s^1$

$_{55}Cs : 1s^2, 2s^2\ 2p^6, 3s^2\ 3p^6\ 3d^{10}, 4s^2\ 4p^6\ 4d^{10}, 5s^2\ 5p^6, 6s^1$

$_{87}Fr : 1s^2, 2s^2\ 2p^6, 3s^2\ 3p^6\ 3d^{10}, 4s^2\ 4p^6\ 4d^{10}\ 4f^{14}, 5s^2, 5p^6\ 5d^{10}, 6s^2\ 6p^6, 7s^1$

As they have only one valence s-electron out side the noble gas core so their electronic configuration can also be given as [noble gas] ns^1.

Example,

Li : [He] $2s^1$

Na : [Ne] $3s^1$

Physical Properties

(1) Physical State

These are soft, silvery white solids with metallic lustre and having high malleability and ductility. They are *cuttable* with the help of knife except lithium and on cutting their metallic lustre fades rapidly due to surface oxidation by atmospheric air. These metals have silvery lustre due to the presence of highly mobile electrons in their metallic lattices. As they have only one valence electron so metallic bonding is not so strong, this makes them soft and this softness increases with the increase of atomic number. The metallic bond strength decreases with the increase in atomic size.

Li, Na, K, Rb, Cs ⟶

Decreasing order of hardness

(2) Atomic and Ionic Radii

The atoms of these elements are largest in size in their periods and the ionic radii of these metals are smaller than the atomic radii of the metals atoms i.e., $M > M^+$.

Explanation: M is greater than M^+ because during the formation of M^+ the outermost orbit is destroyed and effective nuclear charge increases as number of protons are more than electrons.

Size α Number of orbitals α 1 / Effective nuclear charge (Zeff)

	Na	→	Na^+	$+ e^-$
	$1s^2\ 2s^2\ 2p^6\ 3s^1$		$1s^2\ 2s^2\ 2p^6$	
Orbit:	3		2	

The value of atomic and ionic radii increases on moving down the group i.e. from Li to Fr due to increase in number of orbits and decrease in the value of effective nuclear charge.

	Li	Na	K	Rb	Cs
Atomic radii (Å)	1.23	1.57	2.03	2.16	2.35
Ionic Radii (Å)	0.60	0.95	1.33	1.48	1.67

Physical properties of group 1 elements (Alkali Metals)

Property	Element					
	Li	Na	K	Rb	Cs	Fr
Abundance in earth's crust (ppm)	18	22700	18400	78.12	2.6	
Atomic mass	6.94	22.99	39.10	85.47	132.91	223
Boiling point (K)	1615	1156	1032	961	944	
Density (g cm^{-3}) (at 293 K)	0.53	0.97	0.86	1.53	1.90	
Electronegativity (pauling scale)	0.98	0.93	0.82	0.82	0.79	
First ionization enthalpy (kJ mol^{-1})	520	496	419	403	376	
Hydration energy (kJ mol^{-1})	-499	-390	-305	-300	-277	
Ionic radius (M^+) (pm)	76	102	138	152	167	
Melting point (K)	454	371	336	312	302	
Metallic radius (pm)	152	186	227	248	265	375

(3) Density and Atomic Volume

These are light metals having low densities due to their larger sizes of atoms. The density increases on moving from Li to Cs due to increase in atomic mass. The effect of large size is overcome however, the density of potassium is less than that of sodium, it is due to unusual increase in atomic size from sodium to potassium.

	Li	Na	K	Rb	Cs
Density (g mL)	0.54	0.97	0.86	1.53	1.87

The value of atomic volume for these elements increases with the increase in atomic number i.e. from Li to Cs.

	Li	Na	K	Rb	Cs
Atomic Volume (ml)	13	23.7	44.4	55.8	69.3

(4) Melting and Boiling Point

These metals have low values of melting and boiling points as they have only one valence electron so the energy binding their atoms in their crystal lattices is relatively low. On moving from Li to Cs the value M.P. and B.P. decreases as the strength of metallic bond decreases.

	Li	Na	K	Rb	Cs
M.P. (ºC)	181	98	63	39	28.5
B.P. (ºC)	1347	880	766	688	705

(5) Ionization Energy

Alkali metals have lowest value of ionization energy in each period as due to their large size and poor effective nuclear charge the removal of the valence electron is very easy. The value of ionization energy decreases from Li to Cs due to increase in size and decrease in effective nuclear charge the removal of valence electron becomes easier. Due to very low ionization energy K, Cs are used in photoelectric cells as photoelectric effect is easily possible in them.

	Li	Na	K	Rb	Cs
I. E. (eV)	5.4	5.1	4.3	4.2	3.9

Decreases →

The value of second ionisation energy is quite high infact largest energy gap between first and second ionization energies values as it is very difficult to remove another electron from M^+ ions due to their stable inert gas configuration.

Value of II I.P decreases from Li^+ to Cs^+

$$\xrightarrow{Li^+,\ Na^+,\ K^+,\ Rb^+,\ Cs^+}$$

(6) Electropositive and Metallic Nature

Due to their low ionization energies these are highly electropositive metals with high reactivity and these properties increases from Li to Cs as ionization energy decreases.

(7) Electronegativity

As these metals are highly electropositive so they have very low tendency to attract electrons hence they have low value of electronegativity. The value of electronegativity decreases from Li to Cs as electropositive nature increases.

(8) Valency and Oxidation States

These metals can easily loose their only one valence electron to form M^+ ion so they are univalent in nature and exhibit only +1 oxidation state.

$$M - e^- \rightarrow M^+$$

Here M^+ is colourless and diamagnetic in nature as it has no unpaired electron.

These metals do not show higher oxidation states as M^+ has stable inert gas configuration so removal of further electrons is not easy under ordinary conditions.

(9) Reducing Nature

Alkali metals are powerful reducing agents as they have large size and low ionization energies so they can easily loose their valence electron and get oxidized. In free state, this nature increases from Li to Cs as the removal of electron becomes easier.

Their reducing character increases in NH3 due to the availability of ammoniated electron.

Example,

$$M + NH_3 \rightarrow (M^+).\ NH_3 \quad + \quad [e\ NH_3]^-$$

$(M^+).\ NH_3$	$[e\ NH_3]^-$
Conducts electricity	ammoniated
Paramagnetic	anion e^- gives
H_2 not evolved	Deep blue colour

The tendency of an element to loose an electron in solution is directly proportional to its standard oxidation potential as these metals have high value of $E^o_{O.P.}$ so they are powerful reducing agent.

	Li	Na	K	Rb	Cs
$E^o_{O.P.}$ (V)	+3.05	+2.71	+2.93	+2.99	+2.99

Li is the strongest reducing agent in aqueous solution while Cs is the strongest reducing agent in gaseous state.

(10) Hydration of Ions and Hydration Energy

$$M^+ (g) + aq \rightarrow \underset{\text{Hydrated cation}}{M^+ (aq)}$$

The alkali metal cations have a very strong tendency to get hydrated by water molecules and this tendency decreases as the size of M^+ ions increases so the degree of hydration of M^+ ion decreases from Li^+ to Cs^+. The process of hydration is exothermic in nature and the energy released when one gram mole of an M^+ ion is dissolved in water is called hydration energy. It also decreases from Li^+ to Cs^+.

- The radii of the hydrated ion also decreases from Li^+ to Cs^+ however ionic conductance increases.

(11) Heat of Atomization and Specific Heat

The values of heat of atomization and specific heat decreases from Li to Cs as metallic bond strength decreases from Li to Cs.

	Li	Na	K	Rb	Cs
Heat of at.(kJ mol^{-1})	163.6	104.2	83.2	79.5	75.7

	Li	Na	K	Rb	Cs
Sp. heat at 0°C	0.94	0.29	0.17	0.08	0.05

(12) Conductivity

These metals are good conductors of heat and electricity due to the presence of loosely held valence electron which can freely move throughout the structure.

(13) Flame Test

These metals and their salts impart a characteristic colour during flame test. It is due to the fact that electron goes into excited state and when it returns to the original state energy is released in the form of light in the visible range of the spectrum. This released energy increases from Li to Cs so the frequency of the coloured light also increases.

Due to this effect alkali metals are used in fire works

Li	Na	K	Rb	Cs
crimson red	golden yellow	pale violet	violet	violet

Properties decrease on moving from Li to Cs

Li Na K Rb Cs ⟶

Metallic bond strength, hardness, hydration energy, specific heat, heat of atomization, melting and boiling points, ionization energy, electronegativity, electron affinity and reducing nature in solution.

Properties increase on Moving From Li to Cs

Li Na K Rb Cs ⟶

Atomic and ionic size, electropositive nature, density, reactivity, metallic nature, reducing nature in gaseous state and ionic conductance.

Chemical Properties of Alkali Metals

(1) Effect of Air

Alkali metals (except lithium) in air get tarnished at once and give oxides, hydroxides and carbonates so they are kept in inert solvents like kerosene oil, paraffin oil to prevent their reaction with air.

$$M \xrightarrow{O_2} M_2O \xrightarrow{H_2O} MOH$$

$$\xrightarrow{CO_2} M_2CO_3$$

$$M_2O \xrightarrow{O_2} M_2O_2$$

- Reactivity of IA metals towards air is

$$Cs > Rb > K > Na > Li$$

(2) Reaction With O_2

Alkali metals on combustion with oxygen give oxides (M_2O), peroxide (M_2O_2) and superoxide (MO_2).

$$M \xrightarrow{O_2} M_2O \xrightarrow{O_2} M_2O_2$$

$$\xrightarrow{O_2} M_2O_2$$

The formation and stability of these type of oxides by alkali metals can be expressed on the basis of their ionic radii and positive field around it so it can stabilize only a smaller anion while Na^+ being larger in size can stabilize a more large anion and so on. The strong positive field around Li^+ can attract the negative charge very firmly so O_2^- ion can not form O_2^{2-} ion that's why only Li_2O is possible but in case of other M^+, O^{2-} can further combine with oxygen so other oxides are also possible as follows:

$$O^{2-} \xrightarrow{\frac{1}{2}[O_2]} O_2^{2-} \xrightarrow{[O]_2} 2O_2^-$$

Monoxide ion — Peroxide ion — Super oxide ion

- Li gives only Li_2O, Na forms true peroxide and K, Rb, Cs form MO_2 (super oxides).

$$\xrightarrow[\text{White} \quad \text{pale Yellow} \quad \text{bright yellow} \quad \text{orange}]{Li_2O \quad Na_2O \quad K_2O \quad Rb_2O \quad Cs_2O}$$

Li_2O, Na_2O, K_2O and Rb_2O have antifluorite crystal structures while Cs_2O has an anti $CdCl_2$ layer structure.

On moving from Li_2O to Cs_2O, basic nature increases.

M_2O_2 are colourless, diamagnetic in nature and strong oxidizing agents. MO_2 are coloured, paramagnetic and stronger oxidizing agents than M_2O_2.

$$Na_2O_2 + H_2SO_4 \rightarrow Na_2SO_4 + H_2O_2$$

$$4KO_2 + 2H_2SO_4 \rightarrow 2K_2SO_4 + 2H_2O + 3O_2$$

$$2KO_2 + 2H_2O \rightarrow 2KOH + H_2O_2 + O_2$$

KO_2 is used in space capsules for artificial respiration.

$$KO_2 + 2CO_2 \rightarrow 2K_2CO_3 + 3O_2\uparrow$$

(3) Reaction with Water

Alkali metals react with water to form metal hydroxide which are strongly alkaline in nature.

$$2M + 2H_2O \rightarrow 2MOH + H_2\uparrow$$

Reactivity order towards water

$$Cs > Rb > K > Na > Li$$

$$\xrightarrow{LiOH,\ NaOH,\ KOH,\ RbOH,\ CsOH}$$

Basic nature, solubility in H_2O, thermal stability increase from LiOH to CsOH.

(4) With Hydrogen

Alkali metals react with hydrogen to form metal hydride which is ionic in nature.

$$2M + H_2 \rightarrow 2MH$$

$$\xrightarrow{LiH,\ NaH,\ KH,\ RbH,\ CsH}$$

Reducing nature, reactivity, ionic nature increase but stability decreases.

These hydrides react with water and liberate hydrogen.

$$MH + H_2O \rightarrow MOH + H_2$$

These hydrides are strong reductant specially at higher temperatures Example,

$$SiCl_4 + 4NaOH \rightarrow SiH_4 + 4NaCl$$

(5) With Halogen

Alkali metals react with halogens to form metal halides, which are ionic in nature.

$$2M + X_2 \rightarrow 2MX + \text{heat}$$

Reactivity order of Alkali Metals: Reactivity order towards halogens is

$$Cs > Rb > K > Na > Li$$

$$\xrightarrow{LiX,\ NaX,\ KX,\ RbX,\ CsX}$$

Ionic nature, solubility in H_2O increases while lattice energy decreases.

These halides are colourless however on heating they become coloured due to non-stoichiometry and crystal defects.

FACTS TO REMEMBER

- LiF is ionic but insoluble in H_2O due to very high lattice energy while LiCl, LiBr, LiI are covalent.
- Other MX are crystalline solids with high melting and boiling points.
- Halides of K, Rb, Cs can also form polyhalides example,

$$KI + I_2 \rightarrow KI_3$$

(6) Metal Carbonates (M_2CO_3)

Alkali metals form M_2CO_3 type of carbonates which are thermally very stable and readily soluble in water. Their solubility in water, thermal stability increase from Li_2CO_3 to Cs_2CO_3 as electropositive nature increases down the group.

These carbonates do not decompose on heating except Li_2CO_3 which decompose on heating.

$$M_2CO_3 \xrightarrow{\Delta} XXX$$

$$Li_2CO_3 \xrightarrow{\Delta} Li_2O + CO_2\uparrow$$

Li_2CO_3 is insoluble in water and decompose on heating as Li^+ ion being smallest cation has maximum polarizing power so it distorts the electron cloud of the neighbour oxygen atoms of the carbonate ion, as a result C – O becomes weak and Li – O becomes stronger, so Li_2O is formed on decomposition.

(7) Metal Bicarbonates ($MHCO_3$)

Alkali metals form $MHCO_3$ type of bicarbonates and their solubility in H_2O, thermal stability increase from $NaHCO_3$ to $CsHCO_3$.

While $LiHCO_3$ is insoluble in water.

On heating bicarbonates decompose as follows

$$2MHCO_3 \xrightarrow{\Delta} M_2CO_3 + CO_2 + H_2O$$

(8) Metal Nitrates

These elements form MNO_3 type of nitrates which are colourless and soluble in water except lithium nitrate all other metal nitrates decompose as follows:

$$2MNO_3 \xrightarrow{\Delta} 2MNO_2 + O_2$$

$$4LiNO_3 \xrightarrow{\Delta} 2Li_2O + 4NO_2 + O_2$$

(9) Metal Sulphates

These elements form M_2SO_4 type of sulphates and these are soluble in water except Li_2SO_4. These sulphates on fusion with carbon form sulphides as follows:

$$M_2SO_4 + 4C \rightarrow M_2S + 4CO$$

(10) Formation of Amalgam

These metals react with mercury to form metal amalgams. The reactivity of metals to form amalgam decreases down the group here.

$$M + Hg \longrightarrow \underset{\text{Metal amalgam}}{M.Hg} + Heat$$

(11) Formation of Complexes

These metals have very low tendency to form complexes and this tendency decreases down the group.

(12) Heating Effect of Alkali Metal Salts

Carbonates

$$\underset{\text{(IA)}}{M_2CO_3} \xrightarrow{\Delta} \text{No gas}$$

$$Li_2CO_3 \xrightarrow{\Delta} Li_2O + CO_2 \uparrow$$

$$\text{Rest metal carbonates} \xrightarrow{\Delta} MO + CO_2 \uparrow$$

$$MgCO_3 \xrightarrow{\Delta} MgO + CO_2$$

Nitrates

$$\underset{\text{I.A}}{2M-NO_3} \xrightarrow{\Delta} \underset{\text{metal nitrate}}{2M-NO_2} + O_2 \uparrow$$

$$\text{Rest Metal Nitrates} \xrightarrow{\Delta} \text{Rest metal oxide} + NO_2 + O_2 \uparrow$$

Example,

$$2Pb(NO_3)_2 \xrightarrow{\Delta} 2PbO + 4NO_2 + O_2 \uparrow$$

SODIUM ($_{11}Na^{23}$)

Metallic sodium was first isolated by Davy by the electrolysis of fused caustic soda. The most common compound of sodium is NaCl i.e., Rock salt or Lake salt and it has F.C.C unit cell and co-ordination number is 6 for both Na^+ and Cl^- in it.

Occurrence

It is present in earth crust as alumino silicate (2.27 %). As it is a very reactive metal so it is not present in free state in nature and found in combined state in the form of these ores.

Important Ores

Soda feldspar – $Na_2OAl_2O_3.6SiO_2$

Cryolite – Na_3AlF_6

Chile saltpetre – $NaNO_3$

Glauber's salt – $Na_2SO_4.10H_2O$

Borox – $Na_2B_4O_7 . 10H_2O$

Extraction

Sodium is a very strong reductant so it can not be isolated by the reduction of its oxides hence it is normally extracted by the electrolytic reduction of its chloride.

Sodium is extracted by electrolysis of fused NaCl by Down's method. Here, an aqueous solution of NaCl cannot be used as the formed sodium at cathode can react at once with water to form NaOH. As the melting point of NaCl is quite high so a mixture of $CaCl_2$ and KF is used to lower the melting point of NaCl from 800° to 600°C.

Here, the Down's cell has a steel tank lined with fire bricks. It has a circular graphite anode in the centre of it which is surrounded by a ring shaped steel cathode. The anode and the cathode are separated by a steel gauze diaphragm. This diaphragm also prevents the contact of sodium metal and chlorine to form NaCl again. Here anode is covered by a conical hood which provides the outlet for the escape of Cl_2 gas. The cathode is provided with a circular trough attached with a pipe connected to a reservoir to collect the produced molten sodium. The sodium obtained by this method is nearly 99.8% pure. It has nearly 1% impurity of calcium which can be easily removed by cooling sodium metal.

Reactions

$NaCl \rightleftharpoons Na^+ + Cl^-$

At Cathode (Iron Rod)

$Na^+ + e^- \rightarrow Na$

At Anode (Graphite Rod)

$2Cl^- + 2e^- \rightarrow Cl_2 + 2e^-$

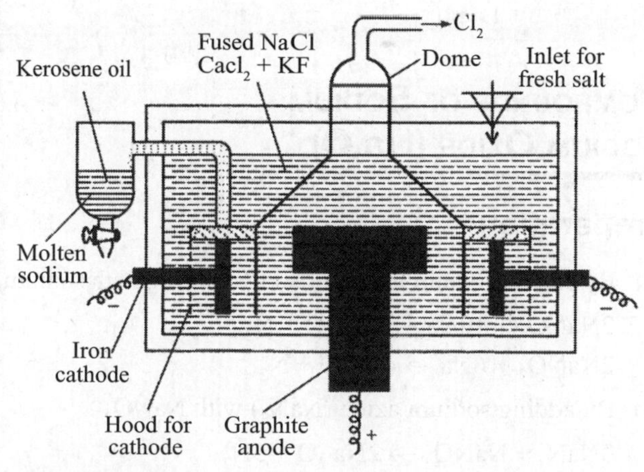

FIGURE 9.1 Down's Cell for the Extraction of Sodium

Castner Process

Here electrolysis of NaOH is made at 330°C using iron cathode and nickel anode.

$2NaOH \leftrightarrow 2Na^+ + 2OH^-$

At Cathode

$2Na^+ + 2e^- \rightarrow 2Na(s)$

At Anode

$4OH^- + 2e^- \rightarrow 2H_2O + O_2 + 4e^-$

During electrolysis, oxygen and water are produced. Water formed at the anode gets partly evaporated and is partly broken down and hydrogen is discharged at cathode.

$H_2O \leftrightarrow H^+ + OH^-$

At Cathode

$2H^+ + 2e^- \rightarrow 2H \rightarrow H_2 \uparrow$

Physical Properties

(1) It is a light silvery white soft metal cuttable with a knife.

(2) It has a density of 0.97 gm per cm3.

(3) It is soluble in liquid ammonia and its solution is of deep blue colour.

(4) Its solution has high conductivity and very strong reducing nature due to the presence of ammoniated electrons.

$Na + (x + y) NH_3 \rightarrow Na^+ (NH_3)x + (NH_3)y\ e^-$

Ammoniated electron

Chemical Properties

It is highly a reactive metal and some of its chemical properties are as follows:

(1) Action of Air

In air, it gets tarnished due to the formation of a layer having oxide, hydroxide and carbonate on its surface. Due to this, it looses its shining. To prevent sodium by the action of air it is kept under kerosene.

$4Na + O_2 \rightarrow 2Na_2O$

$Na_2O + H_2O \rightarrow 2NaOH$

(moisture)

$2NaOH + CO_2 \rightarrow Na_2CO_3 + H_2O$

(from air)

(2) With O_2

Sodium of heating with oxygen forms sodium peroxide in a vigorous manner.

$2Na + O_2 \rightarrow Na_2O_2$

(3) With H_2O

Sodium reacts with water violently to give sodium hydroxide and hydrogen.

$2Na + 2H_2O \rightarrow 2NaOH + H_2$

(4) With Halogens

Sodium reacts with halogens to form its ionic crystalline solid halides.

$2Na + X_2 \rightarrow 2NaX$
$2Na + Cl_2 \rightarrow 2NaCl$
$2Na + Br_2 \rightarrow 2NaBr$

(5) With Ammonia

Sodium on heating with ammonia gives sodamide as follows:

$2Na + 2NH_3 \xrightarrow{297\text{-}397^{\circ}C} 2NaNH_2 + H_2$
Sodamide

(6) With Non-Metals

Sodium can react with some non-metals like hydrogen, phosphorous as follows:

$2Na + H_2 \xrightarrow{360^{\circ}C} 2NaH$
$3Na + P \rightarrow Na_3P$
Sodium phosphide

(7) Reduction Action

Being a strong reductant sodium can reduce many oxides and halides as follows:
$3CO_2 + 4Na \rightarrow 2Na_2CO_3 + C$
$SiO_2 + 4Na \rightarrow Si + 2Na_2O$
$Al_2O_3 + 6Na \rightarrow 2Al + 3Na_2O$
$BeCl_2 + 2Na \rightarrow Be + 2NaCl$

Uses

(1) It is used in the preparation of many sodium compounds like Na_2O_2, $NaNH_2$, NaCN etc.

(2) Sodium amalgam is used extensively in the preparation of many organic compounds.

(3) Being strong reductant it is used in the extraction of non metals like boron and silicon.

(4) Liquid sodium or its alloy with potassium can be used as a coolant in the nuclear reactors.

(5) Sodium lead alloy is used in the preparation of famous anti knocking compound tetra ethyl lead. $[(C_2H_5)_4 Pb]$.

(6) As sodium has high thermal conductivity so it is used for filling hollow valves of air craft engines.

(7) Sodium is also used in the production of sodium vapour lamps.

Summery of Important Reactions of Sodium

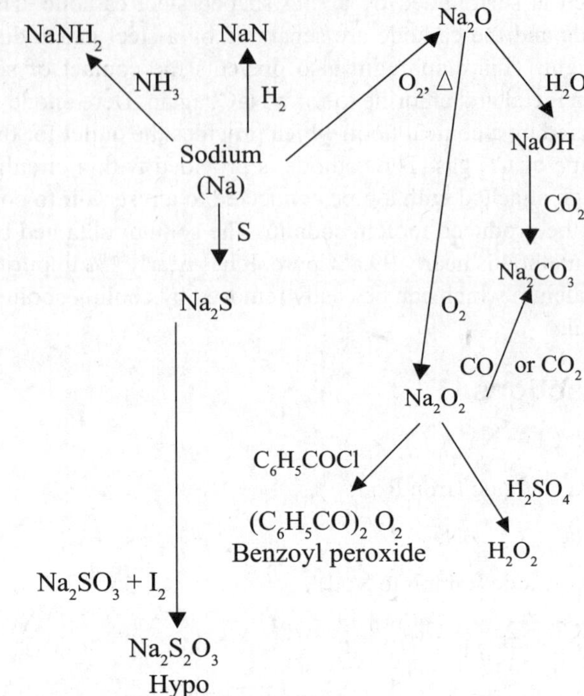

Compounds of Sodium

Sodium Oxide (Na_2O)

Preparation

(1) By adding sodium nitrate or sodium nitrite with sodium.
$2NaNO_3 + 10Na \rightarrow 6Na_2O + N_2$
$2NaNO_3 + 6Na \rightarrow 4Na_2O + N_2$

(2) By adding sodium azide (NaN_3) with $NaNO_2$
$3NaN_3 + NaNO_2 \rightarrow 2Na_2O + 5N_2$
(Pure)

Physio-Chemical Properties

(1) It is white amorphous powder.

(2) *Reaction with Water*

It dissolves in water with evolution of much heat.

$Na_2O + H_2O \rightarrow 2NaOH$

(3) *Effect of Heat*

$2Na_2O \xrightarrow{400^{\circ}C} Na_2O_2 + 2Na$

(4) *Reaction with Liquid Ammonia*

$Na_2O + NH_3 \rightarrow NaOH + NaNH_2$
Sodamide

SODIUM PEROXIDE (Na_2O_2)

Preparation

It is prepared by heating sodium with excess of dry air, free from CO_2 on aluminium tray as follows:

$$2Na + O_2 \xrightarrow{350^\circ C} Na_2O_2$$

Physio-Chemical Properties

(1) It is slightly yellow powder due to the presence of small amount of sodium oxide but in pure form it is colourless.

(2) Effect of Air

It turns white on exposure to air due to the formation of NaOH and Na_2CO_3.

$$2Na_2O_2 + 2H_2O \rightarrow 4NaOH + O_2$$
$$2NaOH + CO_2 \rightarrow Na_2CO_3 + H_2O$$

(3) Reaction with Water

When it is treated with water it gives NaOH and oxygen.

$$2Na_2O_2 + 2H_2O \rightarrow 4NaOH + O_2$$

(4) With dilute Acids

It reacts with cold and dilute acids to give hydrogen peroxide while on heating it gives water and oxygen as follows :

$$Na_2O_2 + H_2SO_4 \ Na_2SO_4 + H_2O_2$$
$$2Na_2O_2 + 2H_2SO_4 \xrightarrow{\Delta} 2Na_2SO_4 + 2H_2O + O_2$$

(5) As a Powerful Oxidising Agent

Being a strong oxidizing agent it oxidizes chromic compounds into chromates and sulphides into sulphates etc.

$$3Na_2O_2 + 3H_2O \ 6NaOH + 3[O]$$
$$\underline{2Cr(OH)_2 + 4NaOH + 3[O] \rightarrow 2Na_2CrO_4 + 5H_2O}$$

$$2Cr(OH)_3 + 3Na_2O_2 \rightarrow 2Na_2CrO_4 + 2NaOH + 2H_2O$$

$$MnSO_4 + 2Na_2O_2 \rightarrow Na_2MnO_4 + Na_2SO_4$$

$$2C_6H_5COCl + Na_2O_2 \rightarrow (C_6H_5CO)_2O_2 + 2NaCl$$

Benzoyl chloride Benzoyl peroxide

(6) Action of CO and CO_2

It reacts with carbon monoxide and carbon dioxide to form sodium carbonate as follows:

$$Na_2O_2 + CO \rightarrow Na_2CO_3$$
$$2Na_2O_2 + 2CO_2 \rightarrow 2Na_2CO_3 + O_2$$

Uses

(1) As it readily combines with CO and CO_2 so it is used for the purification of air in submarines, diving bells and other confined spaces.

(2) It is used as an oxidizing agent.

(3) It is also used for bleaching straw, silk etc.

(4) It is used in the preparation of dyes and some other chemicals like benzoyl peroxide, sodium per borate etc.

CAUSTIC SODA OR SODIUM HYDROXIDE (NaOH)

Methods of Preparation

It is prepared by following methods:

(1) Gossage or Causticization Method

In this method a suspension of lime (CaO + $Ca(OH)_2$) is treated with sodium carbonate to get NaOH as follows:

$$\underset{Aq.}{Na_2CO_3} + \underset{Aq.}{Ca(OH)_2} \rightarrow \underset{Aq.}{2NaOH} + CaCO_3$$

From here calcium carbonate can be easily separated out and caustic soda solution can be easily drained out and evaporated to dryness to get in crystalline form. This NaOH is not pure and have impurities of $CaCO_3$, Na_2CO_3, $Ca(OH)_2$ etc.

(2) Lowig's Method

Here a mixture of sodium carbonate and ferric oxide is heated in a revolving furnace upto redness to get sodium-ferrite which is first of all cooled and hydrolysed by hot water into NaOH solution and in soluble ferric oxide. The solution is filtered and evaporated upto dryness to get the flacks of NaOH.

$$Na_2CO_3 + Fe_2O_3 \xrightarrow[-CO_2\uparrow]{\text{Fussion}} 2NaFeO_2$$
$$2NaFeO_2 \xrightarrow[\Delta]{H_2O} 2NaOH + Fe_2O_3$$

- In both Gossage and Lowig's methods, the initial material sodium carbonate.

(3) Modern Method or By Electrolysis of Brine Solution (aq. NaCl)

For the electrolysis of aqueous NaCl electrolytic cells like Nelson, Castner-Kelner, Salvay cells are used. Here NaOH, Cl_2, H_2 are formed.

Reactions

$$NaCl \leftrightarrow Na^+ + Cl^-$$

$$H_2O \leftrightarrow H^+ + OH^-$$

At Anode

$$2Cl^- \rightarrow Cl_2 + 2e^-$$

At Cathode

$$2Na^+ + 2e^- \rightarrow 2Na$$

$$2Na + 2H.OH \rightarrow 2NaOH + H_2\uparrow$$

As here chlorine can react with NaOH solution even in cold so it is necessary that it must be kept away from NaOH by using a porous diaphragm or by using a mercury cathode so that this reaction can be checked:

$$2NaOH + Cl_2 \rightarrow NaClO + NaCl + H_2O$$

Description of Cells

Nelson etc.

Physical Properties

(1) It is a white crystalline, deliquescent soapy touch solid with a M.P. of 591.4 K.

(2) It is highly soluble in water but less in alcohol.

(3) It is corrosive in nature and bitter in taste.

Chemical Properties

(1) Effect of Atmosphere

It absorbs moisture and CO_2 from atmosphere and changes into Na_2CO_3 so it can not be kept in atmosphere.

$$2NaOH + CO_2 \rightarrow Na_2CO_3 + H_2O$$

(2) Basic Nature

It is a strong base so it reacts with acid or acidic oxide to give salts as follows:

$$NaOH + HNO_3 \rightarrow NaNO_3 + H_2O$$

$$2NaOH + CO_2 \rightarrow Na_2CO_3 + H_2O$$

$$NaOH + HCl \rightarrow NaCl + H_2O$$

$$2NaOH + H_2SO_4 \rightarrow Na_2SO_4 + 2H_2O$$

It can also react with amphoteric oxide like Al_2O_3, ZnO, SnO, PbO etc., to form sodium metalates as follows:

$$Al_2O_3 + 2NaOH \rightarrow 2NaAlO_2 + H_2O$$

Sodium meta aluminate

$$ZnO + 2NaOH \rightarrow Na_2ZnO_2 + H_2O$$

Sodium zincate

$$SnO + 2NaOH \rightarrow Na_2SnO_2 + H_2O$$

Sodium stannite

$$PbO + 2NaOH \rightarrow Na_2PbO_2 + H_2O$$

Sodium plumbite

(3) With Metals

It can react with metals like Zn, Sn, Pb and Al to give sodium metalates but it can not react with alkali metals.

IA metal + NaOH → No reaction

Rest metals + NaOH → $Na_2MO_2 + H_2\uparrow$

Sodium metalate

M may be Zn, Sn , Pb, Be

Example,

$$2NaOH + Zn \rightarrow Na_2ZnO_2 + H_2$$

Sodium zincate

(4) With Non-Metals

It can react with non-metals like B, Si, P, S and halogens as follows:

With Boron

$$2B + 6NaOH \rightarrow 2Na_3BO_3 + 3H_2$$

Sodium borate

With Silicon

$$2NaOH + Si + H_2O \rightarrow Na_2SiO_3 + 2H_2$$

Sodium silicate

With White Phosphorous

$$P_4 + 3NaOH + 3H_2O \rightarrow 3NaH_2PO_2 + PH_3$$

Sodium hydrogenphosphite

With Sulphur

$$4S + 6NaOH \rightarrow Na_2S_2O_3 + 2Na_2S + 3H_2O$$

Hypo

With Halogens

$$2NaOH + X_2 \rightarrow NaXO + NaX + H_2O$$
Dil & cold Sodium hypo halite

Example,

$$2NaOH + Cl_2 \rightarrow NaOCl + NaCl + H_2O$$

$$6NaOH + 3X_2 \rightarrow 5NaX + NaXO_3 + 3H_2O$$
Hot & conc. Sodium halate

Example,

$$6NaOH + 3Br_2 \rightarrow 5NaBr + NaBrO_3 + 3H_2O$$

(5) With Salts

It can react with metallic salts to form metallic hydroxides which may be either insoluble or further soluble in NaOH to give oxyacid salts.

(a) Salts of Fe, Cr, Cu form insoluble hydroxides with NaOH as follows:

$$FeCl_3 + 3NaOH \rightarrow \underset{\text{Red ppt.}}{Fe(OH)_3} + 3NaCl$$

$$CrCl_3 + 3NaOH \rightarrow \underset{\text{Green ppt.}}{Cr(OH)_3} + 3NaCl$$

$$CuSO_4 + 2NaOH \rightarrow \underset{\text{Blue colour ppt.}}{Cu(OH)_2} + Na_2SO_4$$

(b) With those salts whose insoluble hydroxides further dissolve in excess of NaOH to give sodium metalates:

Example,

When NaOH is added in $AlCl_3$ sodium meta aluminate is formed as follows:

$$3NaOH + AlCl_3 \rightarrow \underset{\text{White ppt.}}{Al(OH)_3} + 3NaCl$$

$$NaOH + Al(OH)_3 \rightarrow \underset{\text{Sodium meta aluminate}}{NaAlO_2} + 2H_2O$$

When NaOH is added in the given compounds their sodium matalates are formed.

$$SnCl_2 + 2NaOH \rightarrow Sn(OH)_2 + 2NaCl$$

$$Sn(OH)_2 + 2NaOH \rightarrow Na_2SnO_2 + 2H_2O$$

$$ZnCl_2 + 2NaOH \rightarrow Zn(OH)_2 + 2NaCl$$

$$Zn(OH)_2 + 2NaOH \rightarrow Na_2ZnO_2 + 2H_2O$$

(c) Unstable Hydroxides: Some salts give stable hydroxides with NaOH Example,

$$2AgNO_3 + 2NaOH \rightarrow 2AgOH + 2NaNO_3$$

$$2AgOH \rightarrow \underset{\text{Brown}}{Ag_2O} + H_2O$$

$$HgCl_2 + 2NaOH \rightarrow Hg(OH)_2 + 2NaCl$$

$$Hg(OH)_2 \rightarrow \underset{\text{Yellow}}{HgO} + H_2O$$

(6) With Ammonium Salts

It liberates ammonia

$$NH_4Cl + NaOH \xrightarrow{\Delta} NH_3\uparrow + HCl$$

(7) Formation of HCOOH

It reacts with carbon monoxides 150 – 200ºC under pressure to form sodium formate which undergoes hydrolysis to give formic acid.

$$NaOH + CO \xrightarrow[5\text{-}10\text{ atm 'p'}]{150\text{-}200^\circ} HCOONa \xrightarrow{H.OH} HCOOH$$

Uses of NaOH

(1) It is used in the hydrolysis of ester and saponification.

(2) It is used in the preparation of paper, soap, rayon, dyes, drugs and a number of chemicals. `

(3) It is used as a reagent in the laboratory.

(4) It is used in the refining of Bauxite (by Baeyer's method), petroleum and vegetable oils etc.

(5) It is used to prepare $H_2\uparrow$ [uyeno method]

$$2Al + 2NaOH + 2H_2O \rightarrow 2NaAlO_2 + 3H_2\uparrow$$

Sodium Chloride (NaCl)

It is called common salt or rock salt or sea salt or table salt. Sea water has nearly 2.95% NaCl. 28% aqueous NaCl solution is called Brine.

Preparation

It is mainly manufactured from sea water by evaporation in sun. As it contains impurities of $MgCl_2$, $CaCl_2$, $CaSO_4$ so it is further purified by passing HCl gas where due to common ion effect pure NaCl gets precipitated.

Physical Properties

(1) It is a white crystalline solid and hygroscopic in nature.

(2) It melts at 1081 K and boils at 1713 K.

(3) It dissolves in water and the process of dissolution is endothermic.

(4) Its solubility is 36 g per 100 g of water at 273 K. The solubility does not increase much with increase in temperature.

Summary of Some Important Reactions

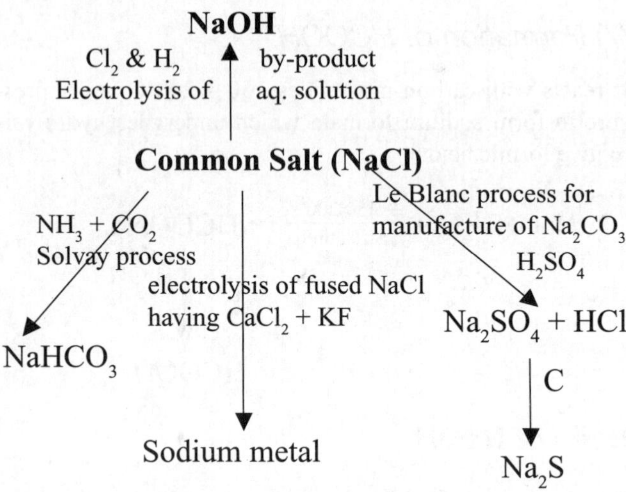

Uses

(1) It is an essential constituent of our food.

(2) It is used in the manufacture of sodium, sodium hydroxide, washing soda, hydrogen chloride, chlorine etc.

(3) It is used in freezing mixture.

(4) It is used for the preparation of soap.

(5) It is used for regenerating ion exchange resins.

Sodium Carbonate or Washing Soda ($Na_2CO_3.10H_2O$)

Na_2CO_3 soda-ash

$Na_2CO_3 .H_2O$ crystal carbonate

Methods of Preparation

It is prepared by following methods:

(1) Le-BlancProcess

Here the raw material used to prepare Na_2CO_3 are NaCl, $CaCO_3$, Coke and H_2SO_4

$$2NaCl + H_2SO_4 \rightarrow \underset{\text{Salt cake}}{Na_2SO_4} + 2HCl$$

$$Na_2SO_4 + 4\text{'C'} \rightarrow \underset{\text{Black salt}}{Na_2S} + 4CO \uparrow$$

$$Na_2S + CaCO_3 \rightarrow \underset{\text{Black ash}}{Na_2CO_3 + CaS}$$

- This method is used to prepare Washing soda, Salt cake, Black salt, black ash

(2) Solvay Ammonia Soda Process

Na_2CO_3 is industrially prepared by Solvay method using NaCl, $CaCO_3$ and NH_3 as follows:

$$2NH_3 + CO_2 + H_2O \rightarrow (NH_4)_2 CO_3$$

$$2NaCl + (NH_4)_2CO_3 \rightarrow Na_2CO_3 + 2NH_4Cl$$

$$Na_2CO_3 + CO_2 + H_2O \rightarrow 2NaHCO_3$$

$$2NaHCO_3 \xrightarrow{\Delta} Na_2CO_3 + CO_2 + H_2O$$

Here Na_2CO_3 , $NaHCO_3$, $CaCl_2$ are obtained as biproducts. $NaHCO_3$ is precipitated due to its low solubility product. $KHCO_3$ cannot be precipitated due to its high solubility product so it can not be prepared by this method.

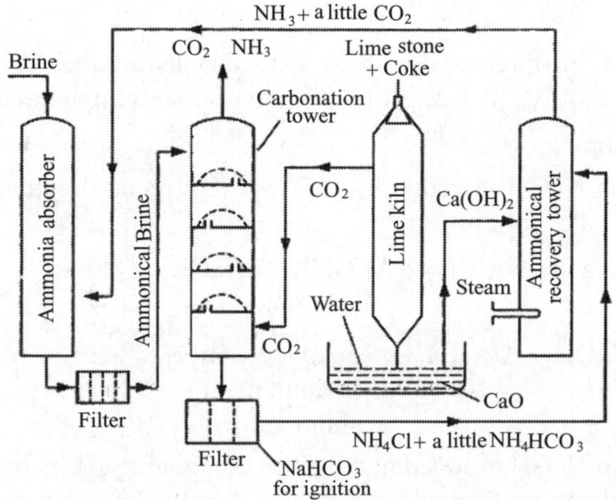

FIGURE 9.2 Solvay Ammonia Soda Process

Details of the Process

This process is carried out in following steps one by one.

(1) Saturation of Brine Solution by Ammonia

It is done in a saturated tank (ammonia absorber) by passing ammonia gas through 30 % brine solution. Here the possible impurities of calcium, magnesium and iron salts present in the brine can be removed as carbonates by carbon dioxide present in ammonia. These impurities can be easily removed by filtration in the form of precipitates and the clear liquid is allowed to enter the carbonation tower.

$$2NH_3 + CO_2 + H_2O \rightarrow (NH_4)_2CO_3$$

$$CaCl_2 + (NH_4)_2CO_3 \rightarrow CaCO_3 \downarrow + 2NH_4Cl$$

$$MgCl_2 + (NH_4)_2CO_3 \rightarrow MgCO_3 \downarrow + 2NH_4Cl$$

(2) Carbonation

The clear ammonical brine solution is cooled and then allowed to enter the carbonation tower. It flows down slowly in the tower and when it comes in contact with the carbon dioxide stream, the sparingly soluble sodium bicarbonate is formed alongwith ammonium chloride.

$$NH_3 + H_2O + CO_2 \rightarrow NH_4HCO_3$$

$$NaCl + NH_4HCO_3 \rightarrow NaHCO_3 + NH_4Cl$$

Here CO_2 is prepared by heating lime stone in a lime kiln and the lime formed in the kiln is dissolved in water to get $Ca(OH)_2$ which is transferred to ammonia recovery tower.

$$CaCO_3 \xrightarrow{\Delta} CaO + CO_2$$

$$CaO + H_2O \rightarrow Ca(OH)_2$$

(3) Filtration

The milky solution having $NaHCO_3$ as a fine suspension obtained after carbonation is filtered with the help of a rotary vacuum filter. From here the filtrate of $NaHCO_3$ is separated out and the remaining solution having NH_4Cl (with little NH_4HCO_3) is pumped into ammonia recovery tower.

(4) Calcination

Sodium bicarbonate obtained after filtration is strongly heated in specially designed cylindrical vessels for ignition to get sodium carbonate.

$$2NaHCO_3 \xrightarrow{\Delta} Na_2CO_3 + CO_2 + H_2O$$

(5) Recovery of Ammonia

The filtrate from the vacuum filter having NH_4Cl is mixed with slaked lime and steam is passed. Here ammonia is formed again as follows:

$$2NH_4Cl + Ca(OH)_2 \rightarrow 2NH_3 + H_2O + CaCl_2$$

$$NH_4HCO_3 \xrightarrow{\Delta} NH_3 + H_2O + CO_2$$

From here the mixture of ammonia and carbon dioxide is again passed into the saturating tower so that the process goes on.

- Sodium carbonate obtained by this process is quite pure and the byproduct obtained in this process is calcium chloride and it is a very cheap process also.

Physical Properties

(1) It is a white crystalline solid partially soluble in water and its aqueous solution is basic.

(2) *Efflorescence*

It is an efflorescent substance and looses its water of crystallization to form its mono hydrate in open air as follows:

$$Na_2CO_3.10H_2O \xrightarrow{air} \underset{\substack{\text{'Powder form'} \\ \text{(loss of weight)}}}{Na_2CO_3.H_2O} + 9H_2O$$

(3) Heating Effect

On heating, it changes into anhydrous sodium carbonate as follows:

$$Na_2CO_3.10H_2O \xrightarrow[-10H_2O]{} Na_2CO_3 \xrightarrow{\Delta \text{ strongly}} \text{XXX No Reaction}$$

(4) With Acids

It is easily decomposed by acids as follows:

$$Na_2CO_3 + HCl \rightarrow NaHCO_3 + NaCl$$

$$NaHCO_3 + HCl \rightarrow NaCl + H_2O + CO_2\uparrow$$

(5) With CO_2

When CO_2 is passed through the concentrated solution of sodium carbonate, sodium bicarbonate gets precipitated.

$$Na_2CO_3 + H_2O + CO_2 \rightarrow 2NaHCO_3$$

(6) With Silica

When it is treated with silica it gives sodium silicate as follows:

$$Na_2CO_3 + SiO_2 \rightarrow Na_2SiO_3 + CO_2\uparrow$$

Sodium silicate is known as water glass or soluble glass as it dissolves in water.

(7) With Sulphur and Sulphur dioxide

When aqueous solution of sodium carbonate is treated with sulphur dioxide and sulphur sodium thiosulphate is formed.

$Na_2CO_3\ (aq) + SO_2 \rightarrow Na\ _2SO_3 + CO_2$

$NaSO_3 + S \rightarrow NaS_2O_3$

(8) With Salts of Non-Alkali Metals

It reacts with salts of non-alkali metals to form insoluble normal or basic carbonates as follows:

(a) With Lead Acetate

When it is treated, with lead acetate, it gives basic lead carbonate as follows:

$3(CH_3COO)_2Pb + 3Na_2CO_3 + H_2O \rightarrow$

$2PbCO_3\ Pb(OH)_2 + CO_2 + 6CH_3COONa$
Basic lead carbonate

(b) Reaction with Copper Sulphate

When it is treated with copper sulphate, it gives basic copper carbonate as follows:

$2CuSO_4 + 2Na_2CO_3 + H_2O \rightarrow$

$Cu(OH)_2.\ CuCO_3 + CO_2 + Na_2SO_4$
Basic copper carbonate

(c) With Zinc Sulphate

When it is treated with zinc sulphate, it gives basic zinc carbonate as follows:

$5ZnSO_4 + 5Na_2CO_3 + 4H_2O \rightarrow$

$[3Zn(OH)_2.\ 2ZnCO_3].H_2O + 3CO_2 + 5Na_2SO_4$
Basic zinc carbonate

(9) Carbonates of some metals like Fe, Al, Sn undergo hydrolysis at once into hydroxides as follows:

$Fe_2(SO_4)_3 + 3Na\ _2CO_3 \rightarrow Fe_2(CO_3)_3 + 3Na\ _2SO_4$

$Fe_2(CO_3)_3 + 3H_2O \rightarrow 2Fe(OH)_3 + 3CO_2$

Uses

(1) It is used for softening hard water, refining of petrol and in the manufacture of glass, borax etc.

(2) $Na_2CO_3 + K_2CO_3$ is a fusion mixture so it is used in quantitative and qualitative analysis.

(3) It is used for washing purpose in laundry.

(4) It is used in paper, textile and paint industries.

(5) Na_2CO_3 gives pink colour with HPH, yellow colour with MeOH or methyl red and blue colour with red litmus.

Sodium Bicarbonate or Baking Soda ($NaHCO_3$)

Preparation

It is prepared by passing carbon dioxide through a saturated solution of sodium carbonate as follows:

$Na_2CO_3 + CO_2 + H_2O \rightarrow 2NaHCO_3$

The sodium bicarbonate can be easily precipitated from here as it is less soluble in water.

Properties

(1) It is a white crystalline solid partially soluble in water and its aqueous solution is basic. $NaHCO_3$ gives yellow colour with methyl red or methyl orange but no colour with HPH i.e., phenolphthalein.

$HCO_3^- + H_2O \rightarrow H_2O + CO_2 + OH^-$

(2) Heating Effect

On heating at 100°C, it decomposes into Na_2CO_3 with the evolution of CO_2.

$2NaHCO_3 \xrightarrow{100°C} Na_2CO_3 + CO_2 + H_2O$

(3) With zinc sulphate it forms normal zinc carbonate.

$ZnSO_4 + 2NaHCO_3 \rightarrow$
$ZnCO_3 + Na_2SO_4 + H_2O + CO_2$

Uses

(1) It is used to remove acidity in stomach and in the making of baking powder (Sodium bicarbonate + potassium hydrogen tartarate).

(2) It is used in fire extinguishers.

(3) It is used in medicines as a mild antiseptic for skin infections.

(4) It is used in making effervescent drinks.

Micro Cosmic Salt ($Na(NH_4)\ HPO_4.4H_2O$)

Preparation

It is prepared by dissolving ammonium chloride and dihydrogen phosphate in equimolar amounts in hot water as follows:

$$NH_4Cl + Na_2HPO_4 + 4H_2O \rightarrow Na(NH_4)HPO_4.4H_2O + NaCl$$

The partially soluble micro cosmic salt can be easily separated out and filtered.

Properties

(1) It is colourless crystalline solid, partially soluble in water.

(2) Heating Effect

On heating it melts into a clear transparent mass of sodium meta phosphate. Sodium metal phosphate can combine with metallic oxides to form colored ortho phosphates i.e. why micro cosmic salt is used for the detection of coloured ions like borax bead test.

$$Na(NH_4)HPO_4 \rightarrow \underset{\text{Sodium meta phosphate}}{NaPO_3} + NH_3 + H_2O$$

$$NaPO_3 + CuO \rightarrow \underset{\text{Blue mass}}{NaCuPO_4}$$

Uses

It is used in the detection of coloured ions and for testing silica.

POTASSIUM ($_{19}K^{39}$)

Like sodium it is not present in free state but it is present in combined state in nature in plants' salt bed, in rocks and in sea water.

Minerals

The main minerals of potassium are:

Carnalite	$KCl.MgCl_2.6H_2O$
Feldspar	$K_2O.Al_2O_3.6SiO_2$ (clay)
Kainite	$KCl.MgSO_4.MgCl_2.3H_2O$
Sylvine	KCl
Polyhalite	$K_2SO_4.MgSO_4.CaSO_4.6H_2O$
Indian Saltpetre	KNO_3

Extraction of Potassium

Potassium is not obtained by the electrolysis of fused KCl as K has lower boiling point (1039 K) than the melting point of KCl (1063 K) and hence vapourises. Therefore potassium metal is extracted by using following methods:

(1) By the Electrolysis of Fused KOH

Here the reaction are as follows:

$$KOH \rightarrow K^+ + OH^-$$

At cathode: $K^+ + e^- \rightarrow K$

At anode: $4OH^- \rightarrow 2H_2O + O_2 + 4e^-$

(2) Modern Method

Here potassium is obtained by the reduction of molten KCl with metallic sodium in stainless steel vessel at 1120 – 1150 K.

$$KCl\ (l) + Na\ (g) \xrightarrow{1120\text{-}1150K} NaCl + K\ (g)$$

By the Reduction of KF

It is prepared by reducing potassium fluoride with CaC_2 at 1270 K in steel cylinders as follows:

$$2KF + CaC_2 \rightarrow 2K + CaF_2 + 2C$$

Compounds of Potassium

Potassium forms oxides, hydroxides, carbonate and halides as follows:

Oxides

Potassium forms following three oxides:

K_2O (Potassium mono oxide)

KO_2 (Potassium super oxide)

K_2O_3 (Potassium sesquioxide)

POTASSIUM SUPER OXIDE (KO_2)

Preparation

(1) KO_2 is prepared by burning potassium with oxygen (free from moisture) as follows:

$$K + O_2 \rightarrow KO_2$$

(2) KO_2 can also be prepared by treating KOH with ozone as follows:

$$2KOH + O_3 \rightarrow 2KO_2 + H_2O$$

Properties

(1) It is a chrome yellow powder which easily dissolves inwater and gives H_2O_2 as follows:

$$2KO_2 + 2H_2O \rightarrow 2KOH + H_2O_2 + O_2$$

(2) It can react with CO and CO_2 as follows:

$$2KO_2 + CO \rightarrow K_2CO_3 + O_2$$

$$2KO_2 + CO_2 \rightarrow K_2CO_3 + 3/2\ O_2$$

(3) It reacts with sulphur on heating to form K_2SO_4.

$$2KO_2 + S \rightarrow K_2SO_4$$

Uses

(1) It is used as air purifier in space capsules and breathing mask as it can not only produce oxygen but also remove CO_2. It is also used as an oxidizing agent.

- K_2O_3 is prepared by passing oxygen through potassium dissolved in liquid ammonia.
- $4K + 3O_2 \rightarrow 2K_2O_3$
- K_2O is a white hygroscopic solid while KO_2 is chrome yellow powder.
- K_2O and KO_2 on hydrolysis form KOH.
- $K_2O + H_2O \rightarrow 2KOH$

Potassium Hydroxide (KOH)

It is mainly obtained by the electrolysis of an aqueous solution of KCl just like NaOH.

It is also obtained by the action of soda lime of potassium carbonate.

Its aqueous solution is known as potash lye.

It is used for the absorption of gases like CO_2, SO_2 etc.

Alcoholic KOH is used as a dehydrating agent for the dehydration of alkyl halides in organic chemistry.

It is also used in making soaps.

Potassium Chloride (KCl)

Occurrence

It occurs as sylvine (KCl) and as carnalite (KCl.$MgCl_2$.$6H_2O$).

Preparation

(1) From Sylvine (KCl)

It is a mixture of KCl and NaCl. When the boiling hot saturated solution of the mixed salts in water is cooled, KCl separates out and NaCl is left behind in the solution.

(2) From Carnalite (KCl.$MgCl_2$.$6H_2O$)

It is always found mixed with NaCl and $MgSO_4$. To remove NaCl and $MgSO_4$, the ore is ground and extracted with a hot 20% solution of $MgCl_2$. Carnalite dissolves while NaCl and $MgSO_4$ remain undissolved. These are filtered off and the solution which contains only carnalite is crystallized so that cubic crystals of KCl separate out leaving behind $MgCl_2$ in the solution.

Properties

(1) It is a colourless crystalline solid having cubic crystals.

(2) Its melting point and boiling point are 768°C and 1411°C respectively.

(3) It is extremely soluble in water and closely resembles NaCl in most of its properties, except that its solubility increases rapidly with the temperature and is more readily fusible.

Uses

(1) It is used in the manufacture of potassium and its compounds.

(2) It is used as a potassium fertilizer (K-type fertiliser), since it supplies potassium (as K_2O) to the soil.

Potassium Carbonate (K_2CO_3)

It is also called pearl ash.

Preparation

It is prepared by the following methods:

(1) Leblanc Process

Here KCl is treated with H_2SO_4 to form K_2SO_4 which on further heating with $CaCO_3$ and carbon gives K_2CO_3 as follows:

$$KCl + H_2SO_4 \rightarrow KHSO_4 + HCl$$

$$KCl + KHSO_4 \rightarrow K_2SO_4 + HCl$$

$$K_2SO_4 + 2C + CaCO_3 \rightarrow \underbrace{K_2CO_3 + CaS}_{\text{Black ash}} + 2CO_2$$

(2) Precht Process

$$2KCl + 3[MgCO_3.3H_2O] + CO_2 \xrightarrow{20^{\circ}C} 2[KHCO_3.MgCO_3.4H_2O] + MgCl_2$$

Potassium magnesium hydrogen carbonates

$$2\,[KHCO_3.MgCO_3.4H_2O] \xrightarrow{140^{\circ}C} K_2CO_3 + 2MgCO_3 + CO_2 + 9H_2O$$

Properties

(1) It is a white deliquescent solid.

(2) It is highly soluble in water and due to hydrolysis, it forms an alkaline solution.

$$K_2CO_3 + H_2O \rightleftharpoons KHCO_3 + K^+ + OH^-$$

(3) Melts at lower temperature (700°C) when mixed with Na_2CO_3, thus it forms fusion mixture (Na_2CO_3 + K_2CO_3).

(4) Action of CO_2

$$K_2CO_3 + H_2O + CO_2 \rightarrow 2KHCO_3$$

Uses

It is used in the manufacturing of hard glass.

Its mixture with Na_2CO_3 can be used as a fusion mixture in lab.

Potassium Iodide (KI)

Preparation

(1) When ferroso ferric iodide and potassium carbonate are treated as follows, potassium iodide is formed.

$$4K_2CO_3 + Fe_3I_8 + 4H_2O \rightarrow 8KI + 4CO_2 + Fe(OH)_2.2Fe(OH)_3$$

From here the precipitate can be easily filtered off and the solution on crystallization gives the crystals of KI.

(2) When KOH or K_2CO_3 are treated with HI, KI is formed as follows:

$$KOH + HI \rightarrow KI + H_2O$$

$$K_2CO_3 + 2HI \rightarrow 2KI + CO_2 + H_2O$$

(3) When iodine is heated with hot and concentrated. solution of KOH, KI and KIO_3 solution is formed which is evaporated to dryness and the obtained solid residue is ignited with powdered charcoal to obtained KI as follows:

$$3I_2 + 6KOH \rightarrow 5KI + KIO_3 + 3H_2O$$

$$KIO_3 + 3C \rightarrow KI + 3CO$$

Physio-Chemical Properties

(1) It is a white crystalline solid which is highly soluble in water and alcohol.

(2) It dissolves free iodine and forms KI_3

$$KI + I_2 \rightarrow KI_3$$

Unstable

(3) With H_2SO_4

Here sulphuric acid decomposes KI as follows:

$$2KI + 2H_2SO_4 \rightarrow K_2SO_4 + I_2 + SO_2 + 2H_2O$$

(4) As a Reducing Agent

Being a strong reducing agent it can reduce $KMnO_4$, $K_2Cr_2O_7$, $CuSO_4$, etc., as follows:

It reduces $KMnO_4$ into $MnSO_4$ as follows:

$$2KMnO_4 + 10\,KI + 8H_2SO_4 \rightarrow 6K_2SO_4 + 2Mn(SO_4) + 5I_2 + 8H_2O$$

It reduces $K_2Cr_2O_7$ into chromium sulphate as follows:

$$6KI + K_2Cr_2O_7 + 7H_2SO_4 \rightarrow 4K_2SO_4 + Cr_2(SO_4)_3 + 3I_2 + 7H_2O$$

It reduces HNO3 into nitrogen dioxide as follows:

$$2KI + 4HNO_3 \rightarrow 2KNO_3 + 2NO_2 + I_2 + 2H_2O$$

It reduces copper sulphate into cuprous iodide as follows:

$$2CuSO_4 + 4KI \rightarrow Cu_2I_2 + 2K_2SO_4 + I_2$$

Cuprous iodide

(5) Formation of Insoluble Iodides

It forms insoluble iodides with lead acetate and silver nitrate etc. as follows:

$$Pb(CH_3COO)_2 + 2KI \rightarrow PbI_2 + 2CH_3COOK$$

Yellow

$$AgNO_3 + KI \rightarrow AgI + KNO_3$$

Yellow

(6) With Mercurous Chloride

When it is treated with $HgCl_2$ a red precipitate of HgI_2 is formed which further dissolves in excess of KI giving potassium mercuric iodide. Its alkaline solution is called Nesseler's reagent which is used for the detection and estimation of ammonium ions.

$$HgCl_2 + 2KI \rightarrow HgI_2 + 2KCl$$

$$HgI_2 + 2KI \rightarrow K_2HgI_4$$

Uses

(1) It is used for making Nesseler's reagent.

(2) It is used as a solvent of iodine as a reagent in laboratory.

(3) It is used in medicine and photography also

Potassium Sulphate (K_2SO_4)

Preparation

(1) By heating naturally occurring mineral, Schonite

$$\underset{\text{Schonite}}{K_2SO_4.MgSO_4.6H_2O} + 2KCl \xrightarrow{\Delta} 2K_2SO_4 + MgCl_2 + 6H_2O$$

(2) By Treating KCl with H_2SO_4

$$2KCl + H_2SO_4 \rightarrow K_2SO_4 + 2HCl$$

(3) By Heating dry KO_2 with Sulphur

$$2KO_2 + S \xrightarrow{\Delta} K_2SO_4$$

Physio-Chemical Properties

(1) It is a white crystalline solid.

(2) It does not form hydrates.

(3) When heated with carbon, it is reduced to potassium sulphide.

(4) It resembles Na_2SO_4.

Potassium Bicarbonates ($KHCO_3$)

Preparation

By Passing CO_2 through cold saturated solution of K_2CO_3.

$$K_2CO_3 + H_2O + CO_2$$

Properties

(1) It is in the form of white powder.

(2) Chemically it resembles $NaHCO_3$ except that it is more soluble in water.

Biological Role of Sodium (Na) and Potassium (K)

- It is macronutrient and taken as NaCl.
- Their requirements are 5-15 gm for adults, 1-3 gm for childrens and less than one gm for infants per day.

Functions

- It maintains normal osmotic pressure between tissue fluids and cells and controls permeability of cells.
- It takes part in absorption of water, glucose, galactose and amino acids.
- It helps in conduction of nerve impulse and helps to initiate and regulate heart beat.
- It helps in CO_2 transport as $NaHCO_3$.

Deficiency

Its deficiency leads sweating, vomiting, diarrhea, poor renal functions, muscular pain, low blood pressur, poor growth of bones etc.

Potassium (K)

- It is a major mineral element and required about 4 gm by adults, 2-3 gm by childrens per day.

Function

- It maintains osmotic pressure, helps in absorption of H_2O and essential for conduction of nerve impulse.
- It is essential component for protein synthesis.
- It acts as a co-factor for pyruvic acid kinase.

Deficiency

- Its deficiency causes vomiting, diarrhea, muscular weakness, tachycardia and paralysis.

Alkaline Earth Metals and their Compounds [IIa (ns^2)]

These are called alkaline earth metals as their oxides are present in earth crust and are alkaline in nature.

Be, Mg, Ca, Sr, Ba, Ra

Electronic Configuration

These elements have two electrons in their valence orbit and their general electronic configuration is ns^2 type. Except beryllium rest of these metals have eight electrons in the penultimate shell.

$_4Be: 1s^2\ 2s^2$

or $[He]\ 2s^2$

$_{12}Mg: 1s^2\ 2s^2\ 2p^6\ 3s^2$

or $[Ne]\ 3s^2$

$_{20}Ca: 1s^2\ 2s^2\ 2p^6\ 3s^2\ 3p^6\ 4s^2$

or $[Ar]\ 4s^2$

$_{38}Sr: 1s^2\ 2s^2\ 2p^6\ 3s^2\ 3p^6\ 3d^{10}\ 4s^2\ 4p^6\ 5s^2$

or $[Kr]\ 5s^2$

$_{56}Ba: 1s^2\ 2s^2\ 2p^6\ 3s^2\ 3p^6\ 3d^{10}\ 4s^2\ 4p^6\ 4d^{10}\ 5s^2\ 5p^6\ 6s^2$

or $[Xe]\ 6s^2$

$_{88}Ra: 1s^2\ 2s^2\ 2p^6\ 3s^2\ 3p^6\ 3d^{10}\ 4s^2\ 4p^6\ 4d^{10}\ 4f^{14}\ 5s^2\ 5p^6\ 5d^{10}\ 6s^2\ 6p^6\ 7s^2$

or $[Rn]\ 7s^2$

Physical Properties

(1) Physical State

These are silvery grayish white metals which are soft in nature but harder than alkali metals due to stronger metallic bond. These metals are less malleable and ductile than alkali metals. Their hardness decreases down the group.

(2) Atomic and Ionic Radii

The atomic and ionic radii of these elements are quite large but smaller than alkali metals due to higher nuclear charge than alkali metals. Values of atomic and ionic radii increase down the group from Be to Ra due to increase in number of orbit and decrease of effective nuclear charge successively.

	Be	Mg	Ca	Sr	Ba	Ra
Atomic radii (Å)	1.12	1.60	1.97	2.15	2.22	-
Ionic radii (Å)	0.31	0.65	0.99	1.13	1.35	1.40

(3) Density and Atomic Volume

These metals have more density than alkali metals due to more close packed arrangement of atoms. The value of density decreases first upto calcium and then increases it is due to the difference in the crystals structures of these elements.

Atomic volume increases from Be to Ra as atomic number increases.

	Be	Mg	Ca	Sr	Ba	Ra
Density (g ml–1)	1.84	1.74	1.55	2.63	3.62	5.5

	Be	Mg	Ca	Sr	Ba	Ra
At. vol. (ml)	4.90	13.97	25.9	35.54	36.7	38.0

- Density of calcium is lower than magnesium due to more atomic volume of calcium.

(4) Melting and Boiling Points

These elements have higher values of melting and boiling points than alkali metals due to more stronger metallic bonds as they have two valence electrons. The value of m.p. and b.p. suffer any regular change down the group as follows:

	Be	Mg	Ca	Sr	Ba	Ra
m.p. (°C)	1277	650	838	763	714	700
b.p.	2770	1107	1440	1380	1640	1140

(5) Ionisation Energy

The ionization energies of these elements are quite low but more than the alkali metals due to their smaller size and more effective nuclear charge than alkali metals. On moving down the group from Be to Ra ionization energy decreases as size increases and effective nuclear charge decreases.

The second ionization energy of these elements is nearly double when the first ionization energy however the value is less than second ionization energy for alkali metals.

	Be	Mg	Ca	Sr	Ba	Ra
First I.E	9.32	7.64	6.11	5.70	5.2	5.3
Sec. I.E.	18.21	15.03	11.87	11.0	10.0	10.1

(6) Electronegativity

These elements have very low values of electronegativity but more than alkali metals. It is due to their large size and less effective nuclear charge. On moving down the group electronegativity decreases.

	Be	Mg	Ca	Sr	Ba	Ra
E.N	1.47	1.23	1.04	1.0	0.97	0.97

(7) Electropositive Nature

These elements are highly electropositive in nature as they have low ionization energy so they can easily loss their two valence electrons to form M^{2+} cations and their electropositive nature increases down the group as size increases and ionization energy decreases. Their electropositivity is less than that of corresponding alkali metal value due to presence of one extra electron than alkali metals.

$M - 2e \rightarrow M^{2+}$

(8) Nature to form Dipositive Ion (M^{2+})

These elements can form M^{2+} ion in both the solid state as well as in aqueous solution states. As M^{2+} has no unpaired electron so it is colourless and diamangnetic dipositive ion.

(9) Oxidation State

All these elements show +2 oxidation states in their compounds. It is their most common and most stable oxidation state as M^{2+} ions has noble gas configuration so removal of further electron is extremely difficult.

(10) Hydration of M^{2+} Ions and Hydration Energy

The M^{2+} ion is extensively hydrated and gives $[M(H_2O)_x]^{2+}$ with the release of a huge amount of energy i.e., hydration energy.

The value of hydration energy and degree of hydration decreases from Be^{2+} to Ba^{2+} as ionic size increases. However, the ionic mobilities increases from Be^{2+} to Ba^{2+}.

$M^{2+} + xH_2O \rightarrow [M(H_2O)_x]^{2+}$ + hydration energy

	Be^{2+}	Mg^{2+}	Ca^{2+}	Sr^{2+}	Ba^{2+}
Heat of hydration	570	460	395	355	305

- Hydration energy of IIA is five times more than IA.

(11) Reducing Nature

These metals are strong reducing agents as they can easily loose their two valence electrons as follows:

$M \rightarrow M^{2+} + 2e^-$

Their reducing nature increases down the group as the value of oxidation potential increases from Be to Ba.

Their reducing nature is less than alkali elements due to higher values of heat of sublimation and ionization energies.

(12) Flame Test

Except Be and Mg the other elements of this group and their salts impart characteristic colours during the flame test as the electrons get excited to higher energy levels and return back to their original state after some time.

Be, Mg,	Ca,	Sr,	Ba,	Ra
Colourless	Brick Red	crimson red	green	crimson red

Be and Mg are colourless due to their small sizes and high excitation energies. The excitation of electrons is quite dificult and not possible in this test.

Trends in Various Properties

Be, Mg, Ca, Sr, Ba, Ra →

Increases from Be to Ra

Solubility of compounds or salt (except sulphate), ionic size or atomic size, reducing power, reactivity, density and electropositive nature.

FACTS TO REMEMBER

- IIA have greater hardness, M.P, B.P, I.E densities etc., than IA due to their small atomic sizes.
- Be forms covalent compounds (due to its small size) and not peroxides.

Decreases from Be to Ra

Electronegativity, ionization energy, electron affinity, hydration energy, polarizing power.

Chemical Properties and Compounds

These metals are highly reactive but less reactive than alkali metals. Their high reactivity is due to their low ionization energies and it increases from Be to Ra.

(1) Action of Air

In air these metals get tarnished however Be and Mg are less effected.

(2) Reaction with Oxygen

These metals react with oxygen to form MO type of oxides and MO_2 type peroxides and this tendency increases down the group.

(a) Oxides (MO)

These metals react slowly with oxygen to give MO type oxides.

$$2M + O_2 \rightarrow 2MO$$

Example,

BeO	MgO	CaO	BaO
Covalent Amphoteric	magnesia	quicklime	baryta

$\longrightarrow$ Basic Nature increases

These oxides are white crystalline solids and quite stable except BeO (covalent) rest are ionic oxides having face centered cubic structure. Except BeO rest oxides react with water to form basic hydroxides as follows:

$$MO + H_2O \rightarrow M(OH)_2 + \text{Heat}$$

From BeO to BaO

Ionic nature, solubility in water and basic nature increase while lattice energy and stability decrease.

(b) Peroxides (MO_2)

Ba, Sr also form peroxides due to their high electropositivity as follows:

$$2BaO + O_2 \xrightarrow{\Delta} 2BaO_2$$

$$2SrO + O_2 \xrightarrow{\Delta} 2SrO_2$$

All MO_2 are white solid ionic compounds.

- Be metal is relatively less reactive and does not react below 880 K while magnesium burns with white dazzling light.

(3) Formation of Hydroxides $M(OH)_2$

Except Be all metals react with water to give hydroxides and the decreasing order of reactivity towards water is as follows: Ba > Sr > Ca > Mg

$$M + 2H_2O \longrightarrow M(OH)_2 + H_2\uparrow$$

- Beryllium does not react with water even at higher temperature due to its lower oxidation potential while magnesium reacts with hot water.

$Be(OH)_2$, $Mg(OH)_2$, $Ca(OH)_2$, $Sr(OH)_2$, $Ba(OH)_2$ $\longrightarrow$

Basic nature and solubility in water increases

- $Be(OH)_2$ is amphoteric in nature while rest are basic and their basicity is less than that of alkali metal hydroxides as in these metal hydroxides the M – O bonds are stronger than in alkali metal hydroxides.
- Solubility increases from $Be(OH)_2$ to $Ba(OH)_2$ or down the group as hydration energy becomes more than lattice energy as lattice energy decreases much more than hydration energy here.
- Aq. Solution of $Ca(OH)_2$ is lime water
- Aq. Solution of $Ba(OH)_2$ is Baryta water

(4) Formation of Hydrides (MH_2)

Except Be all these metals combine directly with hydrogen to form MH_2 type of hydrides.

$$M + H_2 \rightarrow MH_2$$

BeH_2, MgH_2, CaH_2, SrH_2, BaH_2 $\longrightarrow$

Covalent ... ionic or saline

Ionic nature increases but Solubility in water decreases

These MH_2 are highly reactive and can react with water to form hydroxides but this reaction is violent.

$$MH_2 + 2H_2O \rightarrow M(OH)_2 + 2H_2$$

Be becomes passive due to formation of oxide layer but BeH_2 can be formed as follows

$$2BeCl_2 + LiAlH_4 \rightarrow 2BeH_2 + LiCl + AlCl_3$$

BeH_2 gets polymerized by hydrogen bridges and have a three centric banana shaped structure.

CaH_2 is known as Hydrolith and it is used to prepare hydrogen.

$$CaH_2 + 2H_2O \rightarrow Ca(OH)_2 + 2H_2$$

(5) Formation of Halides (MX_2)

All these metals combine directly with halogens at higher temperature to form MX_2 type of halides.

$$M + X_2 \xrightarrow{\Delta} MX_2$$

These halides can also be obtained by the action of HX on MO or $M(OH)_2$ or MCO_3.

$$MO + 2HX \rightarrow MX_2 + H_2O$$

$$M(OH)_2 + 2HX \rightarrow MX_2 + 2H_2O$$

$$MCO_3 + 2HX \rightarrow MX_2 + H_2O + CO_2$$

These halides are ionic, hygroscopic in nature, high melting, non-volatile solids and form hydrates Example, $CaCl_2.6H_2O$.

BeX_2 MgX_2 CaX_2 SrX_2 BaX_2

Ionic nature increases

→

Solubility decreases

(Except in fluorides which are insoluble)

$BeCl_2$ is covalent polymeric halide due to small size of Be^{2+} and its more polarizing power. It is a low melting volatile solid.

$BeCl_2$ fumes in moist air as it is very easily hydrolysed by water as follows:

$$BeCl_2 + 2H_2O \rightarrow Be(OH)_2 + 2HCl$$

Structure of $BeCl_2$: Its structure is just like BeH_2 which is polymeric as $BeCl_2$ is electron deficient compound.

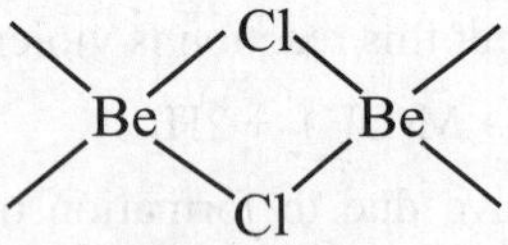

Cl

Be Be

Cl

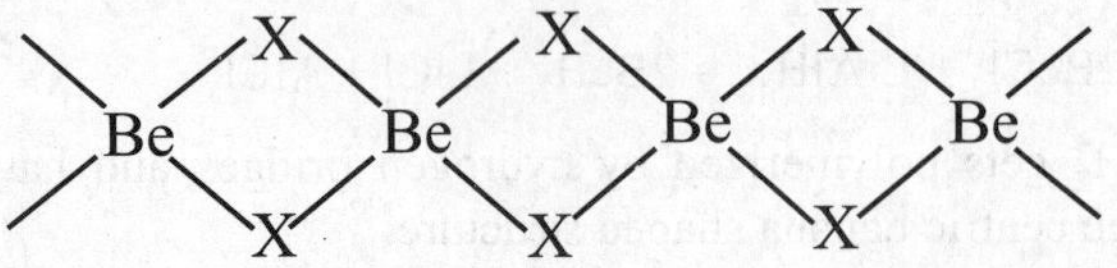

X X X

Be Be Be Be

X X X

(6) Formation of Nitrides (M_3N_2)

All these metals combine directly with nitrogen to form M_3N_2 type of nitrides at high temperature. Their stability, ease of formation decrease from Be to Ba.

$$3M + N_2 \rightarrow M_3N_2$$

These nitrides liberate ammonia on hydrolysis.

$$M_3N_2 + 6H_2O \rightarrow 2NH_3 + 3M(OH)_2$$

Here Be_3N_2 is covalent and volatile while other M_3N_2 are ionic and non-volatile.

(7) Formation of Carbides

Except Be all these metals combine with carbon to form MC_2 type of carbides on heating.

$$M + 2C \xrightarrow{\Delta} MC_2$$

All these carbides are ionic in nature and have a distorted sodium chloride type of structure except Be_2C which has an antifluorite structure.

These carbides on hydrolysis give hydrocarbons.

Example,

$$CaC_2 + 2H_2O \rightarrow C_2H_2 + Ca(OH)_2$$

Ethyne

$$Be_2C + 4H_2O \rightarrow CH_4 + 2Be(OH)_2$$

Methane

$$Mg_2C_3 + 4H_2O \rightarrow C_3H_4 + 2Mg(OH)_2$$

Propyne

(8) Formation of Carbonates (MCO_3)

These metals form MCO_3 type of carbonates which are insoluble in water but dissolve in $CO_2 + H_2O$ and their solubility decreases down the group.

$$MCO_3 + H_2O + CO_2 \rightarrow M(HCO_3)_2$$

Thermal stability, ionic nature increase from $BeCO_3$ to $BaCO_3$ as the polarizing power of M^{2+} ion decreases with the increase in size from Be^{2+} to Ba^{2+}.

The increase of thermal stability can also be explained by the decomposition temperatures of these carbonates

	$BeCO_3$	$MgCO_3$	$CaCO_3$
Decomposition temp (K)	< 300	813	1173
	$SrCO_3$	$BaCO_3$	
	1563	1633	

- IIA metals do not form solid bicarbonates.
- $SrCO_3$ is used in the manufacture of glass of picture tube for colour T.V.

(9) Formation of Sulphates (MSO_4)

These metals form MSO_4 type of sulphates by the action of dilute H_2SO_4 as follows:

$$M + \text{diluted } H_2SO_4 \rightarrow MSO_4 + H_2 \uparrow$$

$$MO + H_2SO_4 \rightarrow MSO_4 + H_2O$$

$$M(OH)_2 + H_2SO_4 \rightarrow MSO_4 + 2H_2O$$

$$MCO_3 + H_2SO_4 \rightarrow MSO_4 + H_2O + CO_2$$

The solubility of these sulphates in water decreases from $BeSO_4$ to $BaSO_4$ as their solubility product and heat of hydration of solvation decreases.

	$BeSO_4$	$MgSO_4$	$CaSO_4$
Ksp	very high	10	2.4×10^{-5}
	$SrSO_4$	$BaSO_4$	
	7.6×10^{-7}	1.5×10^{-9}	

These sulphates are thermally very stable due to their high lattice energies but on strong heating give metal oxides as follows:

$$2MSO_4 \xrightarrow{\text{Heat}} 2MO + 2SO_2 + O_2$$

Their stability of the sulphates increases down the group or from $BeSO_4$ to $MgSO_4$ as the thermal decomposition and basic nature of the metal increases.

	$BeSO_4$	$MgSO_4$	$CaSO_4$	$SrSO_4$
Decomposition temp (K)	773	1168	1422	1647

- $BaSO_4$ is insoluble in H_2O so used to detect stomach problem (barium meal).
- $MgSO_4\ .7H_2O$ or $[Mg(H_2O)_6\ SO_4\ .H_2O]$ is Epsom salt (mild laxative).
- $K_2SO_2\ .\ MgSO_4\ .6H_2O$ is potash magnesia (a fertilizer).

(10) Formation of Metal Nitrates $[M(NO_3)_2]$

The metal nitrates of these elements are $M(NO_3)_2$ types and they can be prepared as follows:

$$\underset{\text{diluted}}{MO + 2HNO_3} \rightarrow M(NO_3)_2 + H_2O$$

$$\underset{\text{diluted}}{MCO_3 + 2HNO_3} \rightarrow M(NO_3)_2 + H_2O + CO_2$$

Facts To Remember

- Beryllium can form basic beryllium nitrate also as follows:
- $Be(NO_3)_2 \xrightarrow{400K} \underset{\text{Basic beryllium nitrate}}{[Be_4O(NO_3)_6]}$
- Magnesium nitrate can be crystallized as $Mg(NO_3)_2\ .6H_2O$.
- Stroncium and barium nitrates can be used in pyrotechniques to provide red and green flames.

These nitrates are soluble in water and they undergo decomposition on heating to give metal oxides.

$$2M(NO_3)_2 \rightarrow 2MO + 4NO_2 + O_2$$

(11) Formation of Oxalates

These metals form MC_2O_4 type of oxalates and their solubility is as follows:

BeC_2O_4,	MgC_2O_4,	CaC_2O_4, SrC_2O_4, BaC_2O_4
Highly soluble	Less soluble	Partially soluble

(12) Formation of Amalgams

All these metals dissolve in Hg to form Amalgams.

(13) Dissolution in Liquid Ammonia

These metals dissolve in liquid ammonia to form coloured solutions. In dilute solutions the colours is bright blue due to solvated electron while concentrated solutions are bronze in colour.

$$M \rightarrow M^{2+} + 2e^-$$

$$2NH_3 + 2e^- \rightarrow 2NH_2^- + H_2$$

$$M^{2+} + 2NH_2^- \rightarrow M(NH_2)_2$$

$$M(NH_3)_6 \rightarrow M(NH_2)_2 + 4NH_3 \uparrow + H_2 \uparrow$$

(14) Complex Formation

Except beryllium and magnesium none of these metals have normally any tendency to form complexes. Beryllium and magnesium can form stable complexes due to their smaller size and more charged density.

Beryllium can form many stable complexes like $[BeF_4]^{2-}$, $[Be(H_2O)_4]^{2+}$, $[Be_4O(A)_6]$ here A may be NO_3 –, $RCOO^-$ etc.

The famous complex of magnesium is chlorophyll in-which magnesium is bonded to the four heterocyclic nitrogen atoms. It plays an important in photosynthesis.

Ca, Sr, Ba can also form complexes with some strong complexing agents like EDTA, acetylacetone etc.

Abnormal Behaviour of Be

Be has abnormal behaviour due to its small size, high electronegativity, high polarising power of Be^{+2}.

(1) Be is the hardest element in the group with highest melting and boiling points which is effected by atmosphere.

(2) It does not react with water, hydrogen and acids directly like other metals of this group.

(3) It sublimes on heating and no conductance of electricity in molten state occurs in it.

(4) It dissolve in alkali and releases hydrogen.

$$\underset{\text{Sodium beryllate}}{Be + 2NaOH + 2H_2O \rightarrow Na_2BeO_2.2H_2O + H_2}$$

(5) Be salts have low solubility due to high hydration energy.

(6) Be forms complexes due to its small size, positive charge.

Example, $\underset{sp^2}{[BeF_3]^-}$ $\underset{sp^3}{[BeF_4]^{-2}}$ $\underset{sp^3}{[Be(H_2O)_4]^{+2}}$

(7) Be shows similar properties to aluminium due to diagonal relationship.

(8) Its oxide BeO and hydroxide $Be(OH)_2$ are amphoteric while other MO and $M(OH)_2$ are basic.

(9) It shows maximum coordination number of four while others can show a coordination number six also as it does not have d-orbitals.

(10) Its carbide on hydrolysis gives methane while rest carbides give acetylene on hydrolysis.

Diagonal Relationship between Be and Al

Beryllium shows diagonal relationship with aluminium as both have nearly same charge / radius ratio, electronegativity and $E^o_{O.P.}$ Some similar properties are given as follows:

(1) Both the elements exist together in nature in the mineral beryl ($3BeO.Al_2O_3.6SiO_2$).

(2) Both the elements are not affected by the atmosphere and are not attacked by acids due to the formation of an oxide film on their surface.

(3) With NaOH

Both these elements dissolves in NaOH as follows:

$$Be + 2NaOH \rightarrow \underset{\text{Sodium beryllate}}{Na_2BeO_2} + H_2$$

$$2Al + 2NaOH + 2H_2O \rightarrow \underset{\text{Sodium aluminate}}{2NaAlO_2} + 3H_2$$

(4) Amphoteric Nature of Oxides

Their oxides are hard, non-volatile having high melting points and amphoteric in nature. They dissolve in acids and alkalis as follows:

$$BeO + 2HCl \rightarrow BeCl_2 + H_2O$$

$$Al_2O_3 + 6HCl \rightarrow 3AlCl_3 + 3H_2O$$

$$BeO + 2NaOH \rightarrow \underset{\text{Sodium beryllate}}{Na_2BeO_2} + H_2O$$

$$Al_2O_3 + 2NaOH \rightarrow \underset{\text{Sodium aluminate}}{2NaAlO_2} + H_2O$$

(5) Amphoteric Nature of Hydroxides

They form amphoteric oxides which are insoluble in water and on heating gives metal oxides.

(6) Formation of Covalent Compounds

Both the metals form covalent compounds by reacting with non-metals Example,

$$BeO + C + Cl_2 \rightarrow BeCl_2 + CO$$

$$Al_2O_3 + 3C + 3Cl_2 \rightarrow 2AlCl_3 + 3CO$$

Both $BeCl_2$ and $AlCl_3$ are covalent halides with low melting points. They behave like lewis acid and dissolve in organic solvents.

(7) Carbides of both the metals on reaction with water gives methane.

$$Be_2C + 4H_2O \rightarrow 2Be(OH)_2 + CH_4$$

$$Al_4C_3 + 12H_2O \rightarrow 4Al(OH)_3 + 3CH_4$$

(8) Formation of Fluoro Complex Anions Both these metals form fluoro complex anions like BeF_4^{2-}, AlF_6^{3-}.

MAGNESIUM ($_{12}Mg^{24}$)

Occurrence

It is not present in free state in nature but it is present in the form of ores as follows :

Carnalite ($KCl.MgCl_2.6H_2O$)

Magnesite ($MgCO_3$)

Dolomite ($MgCO_3 . CaCO_3$)

Kieserite ($MgSO_4. H_2O$)

Asbestos [$CaMg_3 (SiO_3)_4$] etc.

It is an essential constituent of chlorophyll the green colouring matter in plants, involved in photosynthesis.

It is also present 0.13 % extent in sea water.

Extraction

As it is a highly electropositive metal and a strong reductant so it can not be obtained in the metallic form by normal reduction methods. It can also not be obtained by the electrolysis of aqueous solutions its salts as it has a great affinity with water so it is extracted by the electrolysis of its fused anhydrous salts mainly as follows:

(1) By the Electrolysis of anhydrous $MgCl_2$

Anhydrous $MgCl_2$ can be obtained from carnalite or from sea water.

(a) From Carnalite

Carnalite is a dehydrated compound and it can not be dehydrated by heating only as the water of crystallization reacts with $MgCl_2$ to give magnesium oxide as follows:

$$\underset{\text{Hexahydrate}}{MgCl_2.6H_2O} \xrightarrow{\Delta} \underset{\text{Dihydrate}}{MgCl_2.2H_2O} \xrightarrow{\Delta}$$

$$\underset{\text{Oxychloride}}{Mg_2OCl_2} \xrightarrow{\text{strongly heated}} MgO$$

To make carnalite anhydrous it is first treated in air and then in a current of HCl gas.

$$MgCl_2.6H_2O \xrightarrow[\text{of HCL gas}]{\text{heated in presence}} MgCl_2 + 6H_2O$$

(b) From Sea Water

Anhydrous $MgCl_2$ can also be obtained from sea water having magnesium compounds. When sea water is treated with $Ca(OH)_2$, $Mg(OH)_2$ get precipitated. The precipitate is dissolved in dilute HCl to get the solution and which is concentrated and finally crystallized to get anhydrous $MgCl_2$.

$$\underset{(MgCl_2 \text{ or } MgSO_4)}{\text{Sea water}} + Ca(OH)_2 \rightarrow \underset{\text{ppt.}}{Mg(OH)_2}$$

$$Mg(OH)_2 + 2HCl \rightarrow MgCl_2.6H_2O$$

$$MgCl_2.6H_2O \xrightarrow[HCl(g)]{\text{heated}} \underset{\text{Anhydrous}}{MgCl_2} + 6H_2O$$

Electrolysis

The anhydrous $MgCl_2$ is fused with NaCl and anhydrous calcium chloride in the ratio of 35 %, 50 % and 15 % respectively. This mixture is subjected to electrolysis at 973 – 1023 K in presence of an inert gas in an electrolytic cell. Here NaCl and anhydrous $CaCl_2$ are used to lower the fusion temperature and to increase the conductivity of the fused mass.

After electrolysis magnesium gets discharged at cathode in molten state. It being lighter than the electrolyte floats over the surface of the fused mass so can be easily removed with perforatory ladles it is 99.9 % pure. It can be further purified by remelting it by the flux of anhydrous $MgCl_2$ and NaCl.

$$MgCl_2 \rightleftharpoons Mg^{2+} + 2Cl^-$$

At cathode:

$$Mg^{2+} + 2e^- \rightarrow Mg$$

At anode:

$$2Cl^- \rightarrow Cl_2 + 2e$$

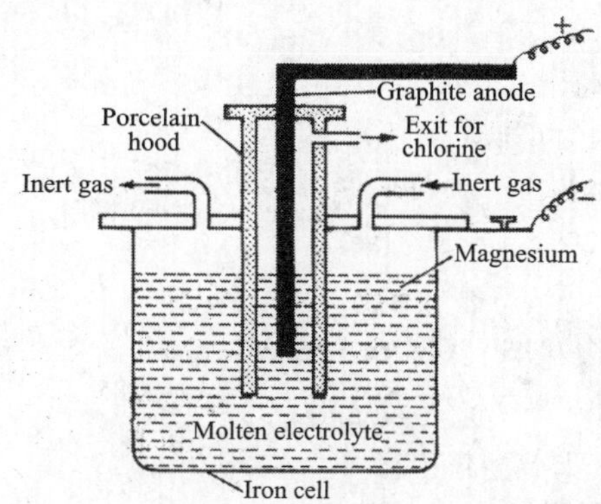

(2) By Thermal Reduction of MgO

Here MgO is obtained by heating magnesite. The reduction of MgO is made at 2000ºC in vacuum by reducing agent likes C or Si or Al.

$$MgCO_3 \rightarrow MgO + CO_2$$

$$2MgO + Si \rightarrow 2Mg + SiO_2$$

$$3MgO + 2Al \rightarrow 3Mg + Al_2O_3$$

(3) From MgO.CaO

When this mixed oxide is heated with ferrosilicon above 1000ºC under reduced pressure Mg is obtained.

- This mixed oxide is obtained by heating dolomite.

Properties

(1) It is a hard malleable, ductile, silvery white metal.

(2) Mg burns in N_2, O_2, CO_2, SO_2

$$2Mg + O_2 \xrightarrow{\Delta} 2MgO$$

(white dazzling light so used in flash light photography)

$$2Mg + CO_2 \xrightarrow{\Delta} 2MgO + \text{'C'}$$

Carbon black [proves CO_2 has 'C']

(3) Mg dissolves in dilute acids with the removal of H_2.

$$Mg + H_2SO_4 \rightarrow MgSO_4 + H_2$$

(4) Magnesium acts as a strong reducing agent.

Example,

$$B_2O_3 + 3Mg \rightarrow 3MgO + 2B$$

$$3Mg + KClO_3 \xrightarrow{\Delta} KCl + 3MgO$$

burn with explosion

(5) It react with halides to form Grignard reagent which is used in various organic synthesis.

$$R\text{-}X + Mg \xrightarrow{\text{ethers}} R - Mg - X$$

Grignard reagent

Facts To Remember

- Mg-fire cannot be extinguished by CO_2.
- Mg is refined by subliming at 600ºC and 1mm Hg Pressure.
- Being more electropositive magnesium prevents ships, pipes from corrosion.

Uses

(1) Agnalium [15% Mg + 85-99% Al] is used in making parts of aeroplanes and motors etc.

(2) Electron metal (95% Mg, 4.5% Zn, 0.5% Cu) is used in the manufacture of air crafts.

(3) Mg is used in the extraction of B, Si, Ti etc.

(4) Mg is used as a deoxidizer in Cu, steel etc.

Property	Element					
	Be	Mg	Ca	Sr	Ba	Ra
Abundance in earth's crust (ppm)	2	27640	46668	384	390	10–6
Atomic mass	9.01	24.31	40.08	87.62	137.33	226.03
Boiling point (K)	2745	1363	1767	1655	2078	(1973) (uncertain)
Density (g cm^{-3}) (at 293 K)	1.84	1.74	1.55	2.63	3.59	(5.5)
Electronegativity (pauling scale)	1.57	1.31	1.00	0.95	0.89	0.9
E° (V) at 298 K [for M^{2+} (aq) + $2e^- \rightarrow M$ (s)]	–1.97	–2.36	–2.84	–2.89	–2.92	–2.92
First ionization enthalpy (kJ mol^{-1})	899	737	590	549	503	509
Hydration energy (kJ mol^{-1})	–2385	–1920	–1650	–1480	–1380	–
Ionic radius (M^{2+}) (pm)	31	72	100	118	135	148
Melting point (K)	1562	924	1124	1062	1002	973
Metallic radius (pm)	112	160	197	215	222	–

Compounds of Magnesium

Magnesia (MgO)

Preparation

It is prepared by heating following compounds as follows:

$$Mg(OH)_2 \xrightarrow{\Delta} MgO + H_2O$$

$$MgCO_3 \xrightarrow{\Delta} MgO + CO_2$$

$$2Mg(NO_3)_2 \xrightarrow{\Delta} 2MgO + 4NO_2 + O_2$$

$$2Mg(s) + O_2 \xrightarrow{\Delta} 2MgO\ (s)$$

$$MgCl_2.6H_2O \xrightarrow[\text{heating}]{\text{strong}} MgO\ (s)$$

Properties

(1) It is an amorphous white powder and slightly soluble in water as follows.

$$MgO + H_2O\ Mg(OH)_2$$

(2) It is quite stable and fuses at 2800 K so it is used as a refractory material for lining electric furnaces.

(3) It is a basic oxide and forms salt with acids.

$$MgO + H_2SO_4 \rightarrow MgSO_4 + H_2O$$

$$MgO + 2HCl \rightarrow MgCl_2 + H_2O$$

Uses

(1) It is used as a refractory material in furnace due to its high melting point and basic flux nature.

(2) It is used in medicine as an antacid.

(3) It is also used in making soral cement.

(4) It is used for filing rubber.

(5) When it is mixed with asbestos it can be used as an insulator for steam pipes and boilers.

Magnesium Chloride ($MgCl_2$)

Occurrence

It occurs as carnalite ($KCl.MgCl_2.6H_2O$) and bischofite ($MgCl_2.H_2O$). It is found in sea water and mineral springs etc.

Preparation

(1) From Carnalite

It is extracted from carnalite by fractional crystallisation here the mineral is fused and cooled to 175°C, when practically all KCl is separated out only $MgCl_2$ is remained in the fused state. This fused mass on cooling gives the crystals of $MgCl_2.6H_2O$ as follows:

$$MgCl_2 + 6H_2O \xrightarrow{} MgCl_2.6H_2O$$
In vaccum

(2) From Sea Water

When sea water is treated with lime the magnesium ions present in it can be precipitated in the form of magnesium hydroxide. When the precipitate of $Mg(OH)_2$ is dissolved in HCl and the solution formed is crystallized the crystals of $MgCl_2.6H_2O$ are obtained.

$$Mg^{2+} + Ca(OH)_2 \rightarrow Mg(OH)_2 \downarrow + Ca^{2+}$$
(from sea water)

$$Mg(OH)_2 + 2HCl \rightarrow MgCl_2 + H_2O$$

(3) Lab Method

It can be prepared by dissolving magnesium oxide or magnesium carbonate in dilute hydrochloric acid.

$$MgO + 2HCl \rightarrow MgCl_2 + H_2O$$

$$MgCO_3 + 2HCl \rightarrow MgCl_2 + H_2O + CO_2$$

This solution on concentration and cooling gives crystals of $MgCl_2.6H_2O$.

(4) Anhydrous $MgCl_2$ cannot be obtained by heating hydrated form of $MgCl_2$ so it is prepared by heating magnesium metal or hydrated magnesium chloride in a current of dry HCl or chlorine gas.

$$Mg + Cl_2 \rightarrow MgCl_2$$
(anhydrous)

Properties

(1) It is a colourless, crystalline, deliquescent solid which is highly soluble in water.

(2) Heating Effect: On heating it decomposes into magnesium oxide as follows:

$$MgCl_2.6H_2O \xrightarrow{\Delta} Mg(OH)Cl + HCl + 5H_2O$$

$$Mg(OH)Cl \xrightarrow{\Delta} MgO + HCl$$

Uses

(1) It is used for the preparation many magnesium compounds like $Mg(OH)_2$, MgO, $MgCO_3$ etc.

(2) When hydrated magnesium chloride is mixed with magnesium oxide it becomes quite hard mass paste and known as Soral magnesia i.e., $MgCl_2.5MgO\ x\ H_2O$ which is used as a cement in dental filling.

(3) It is used in lubricating cotton thread in spinning.

Magnesium Sulphate or Epsom Salt or Epsomite [$MgSO_4.7H_2O$]

It occurs in the form of minerals like Epsom salt ($MgSO_4.7H_2O$), Kieserite ($MgSO_4.H_2O$), kainite ($KCl.MgSO_4.3H_2O$)

Preparation

(1) From Magnesite

It is prepared from magnesite ore by reacting them with dil. H_2SO_4 as follows:

$$MgCO_3 + H_2SO_4 \rightarrow MgSO_4 + H_2O + CO_2$$

(2) From Dolomite

When powdered dolomite is boiled with sulphuric acid, it is obtained as the calcium sulphate being less soluble can be easily precipitated while $MgSO_4$ remains in the solution.

$$\underset{\text{Dolomite}}{MgCO_3.CaCO_3} + 2H_2SO_4 \longrightarrow MgSO_4 + CaSO_4\downarrow + 2CO_2 + 2H_2O$$

(3) From Keiserite ($MgSO_4.H_2O$)

When the powdered ore is dissolved in water, a solution is obtained. When the solution is concentrated and cooled the crystals of magnesium sulphate is obtained.

$$MgSO_4.H_2O + 6H_2O \rightarrow MgSO_4.7H_2O$$

(4) Lab method

It can be prepared by dissolving MgO or $MgCO_3$ into dilute H_2SO_4 and evaporating the solution to get the crystals of magnesium sulphate.

$$MgO + H_2SO_4 \rightarrow MgSO_4 + H_2O$$

$$MgCO_3 + H_2SO_4 \rightarrow MgSO_4 + H_2O + CO_2$$

Properties

(1) It is a colourless crystalline solid which is soluble in water. It is isomorphous with $ZnSO_4.7H_2O$

(2) It is an efflorescent compound and loses water of crystallization on exposure to air.

(3) Heating effect: It undergoes decomposition on heating as follows:

$$MgSO_4.7H_2O \xrightarrow[-H_2O]{150^\circ C} MgSO_4.H_2O$$

$$\underset{\text{Kieserite}}{MgSO_4.H_2O} \xrightarrow{200^\circ C} MgSO_4$$

$$MgSO_4 \xrightarrow{\Delta} MgO + SO_3$$

(4) Reduction by Lamp Black

$$2MgSO_4 + \text{'C'} \rightarrow 2MgO + 2SO_2 + CO_2$$

Uses

(1) It is used as purgative (medicine), in dying and tanning, ceremic cement.

(2) $Mg(ClO_4)_2$ is anhydrous magnesium Per chlorate, a famous dehydrating agent.

(3) It is used in manufacturing of paints, fire proofing farics etc.

Facts To Remember

Magnesia mixture is a solution of $MgCl_2$ and NH_4Cl in ammonia which is used to detect phosphate or arsenate ions.

Magnesium Hydroxide $Mg(OH)_2$

$Mg(OH)_2$ occurs in nature as brucite.

Preparation

(1) **From MgO:** It is prepared by dissolving MgO into water as follows:

$$MgO + H_2O \rightarrow Mg(OH)_2$$

(2) **By Treating Magnesium Salt with Alkali**

$$MgCl_2 + Ca(OH)_2 \rightarrow Mg(OH)_2 + CaCl_2$$

$$MgCl_2 + 2NaOH \rightarrow Mg(OH)_2 + 2NaCl$$

Physical-Chemical Properties

(1) It is a white powder and partially soluble in water.

(2) It is a basic oxide and forms salts with acids.

(3) On heating $Mg(OH)_2$ decomposes to form MgO.

(4) It dissolves in ammonium chloride solution forming a complex ion. It partly explains why magnesium is not precipitate with the hydroxides of Al, Cr and Fe in IIIrd group of qualitative analysis.

$$Mg(OH)_2 + 2NH_4Cl\ MgCl_2 + 2NH_4OH$$

Uses

(1) $Mg(OH)_2$ is an aqueous suspension used in medicines as an antacid, called 'Milk of magnesia'.

MAGNESIUM CARBONATE ($MgCO_3$)

$MgCO_3$ occurs in nature as Dolomite $MgCO_3.CaCO_3$ and as Magnesite $MgCO_3$.

Preparation

(1) By treating an aqueous solution of Magnesium salt with $NaHCO_3$

$$MgSO_4 + 2NaHCO_3 \rightarrow MgCO_3 + Na_2SO_4 + H_2O + CO_2$$

(2) By passing CO_2 through suspension of MgO in water

$$MgO + H_2O + 2CO_2 \rightarrow Mg(HCO_3)_2$$
$$Mg(HCO_3)_2 + MgO \rightarrow 2MgCO_3 + H_2O$$

(3) From Magnesium Sulphate and Sodium Carbonate

It can not be directly obtained by the reaction of these two as a white precipitate of basic magnesium carbonate or magnesium alba is obtained first which is suspended in water and CO_2 is passed into it to get magnesium bicarbonate (fluid magnesium) whose solution on boiling gives magnesium carbonate.

$$2MgSO_4 + 2Na_2CO_3 + H_2O \rightarrow MgCO_3.Mg(OH)_2 + 2Na_2SO_4 + CO_2$$
$$MgCO_3.Mg(OH)_2 + 3CO_2 + H_2O \rightarrow 2Mg(HCO_3)_2$$
$$Mg(HCO_3)_2 \rightarrow MgCO_3 + CO_2 + H_2O$$

Physical-Chemical Properties

(1) It is a white solid powder which is insoluble in water.

(2) On suspension in water and by passing CO_2 it forms $Mg(HCO_3)_2$

$$MgCO_3 + H_2O + CO_2\ Mg(HCO_3)_2 \rightarrow$$
Magnesium bicarbonate

(3) With Acids

It dissolves in acids giving salts as follows:

$$MgCO_3 + H_2SO_4 \rightarrow MgSO_4 + H_2O + CO_2$$
$$MgCO_3 + 2HCl \rightarrow MgCl_2 + H_2O + CO_2$$

(4) Heating Effect

$$MgCO_3 \xrightarrow{\Delta} MgO + CO_2$$

Uses

(1) $(MgCO_3)X$, $[Mg(OH)_2].3H_2O$ (Magnesium Alba) is used in tooth powder as an antacid and laxative.

(2) It is also used in the manufacturing of glass, ceramics etc.

(3) It is also used as a filler for paper rubber and pigments.

MAGNESIUM BICARBONATE $Mg(HCO_3)_2$

Preparation

By passing CO_2 when $MgCO_3$ is suspended in water.

$$MgCO_3 + H_2O + CO_2 \rightarrow Mg(HCO_3)_2$$

Physical-Chemical Properties

(1) It is found only in liquid form.

(2) On Boiling

$$Mg(HCO_3)_2 \rightarrow MgCO_3 + H_2O + CO_2$$

(3) A solution having 12 grams of $MgCO_3$ per 100 cc. of water containing dissolved CO_2 is known as 'fluid magnesia'.

(4) The precipitate obtained by reacting Epsom salt ($MgSO_4.7H_2O$) with sodium carbonate is known as magnesia alba. It has got a variable composition

$$Mg(CO_3)X\ [Mg(OH)_2]Y.2H_2O.$$

CALCIUM ($_{20}Ca^{40}$)

Occurrence

$Ca_3(PO_4)_2$ is present in bones, teeth etc. $CaCO_3$ is present in protective shells of marine animals.

Important Ores and Minerals

$CaCO_3$ (Limestone, marble, chalk or calcite)

$CaSO_4.2H_2O$ (Gypsum)

$CaCO_3\ .MgCO_3$ (Dolomite)

CaF_2 (Fluorspar)

$Ca_3(PO_4)_2$ (Phosphorite)

$3Ca_3(PO_4)_2.Ca(OH)_2$ (Hydroxypatite)

Extraction

(1) It is obtained by electrolysis of a fused mass having 6 parts $CaCl_2$ and 1 part CaF_2 at nearly 973 K in an electrolytic cell made of graphite. Graphite act an anode while cathode is of water cooled iron cathode. On passing a current calcium is discharged at cathode and easily solidified. Here CaF_2 is added to decrease its melting point.

$$CaCl_2 \xrightleftharpoons{700^\circ C} Ca + Cl_2$$

At cathode

At cathode: $Ca_2+ + 2e \rightarrow Ca$

At anode : $2Cl \rightarrow Cl_2 + 2e$

It must be removed quickly by screw rearrangement as calcium is soluble in fused $CaCl_2$ at this temperature.

(2) Thermite or Gold Schmidt Method

$3CaO + 2Al\ Ca + Al_2O_3$

Physio-Chemical Properties

(1) It is a silvery white, malleable and ductile metal with a specific gravity of 1.5.

(2) It is a good conductor of heat and electricity.

(3) Effect of Air or Moisture

On exposure to air it is covered by a white powder having $Ca(OH)_2$ and Ca_3N_2.

$$2Ca \xrightarrow{O_2} 2CaO$$

$$3Ca \xrightarrow{N_2} Ca_3N_2$$

$$Ca + 2H_2O \rightarrow Ca(OH)_2 + H_2$$

(4) Reaction with Non-Metal

$$Ca + H_2 \xrightarrow{\Delta} CaH_2 \text{ [Hydrolith]}$$

$$Ca + 2C \xrightarrow{\Delta} CaC_2$$

$$Ca + Cl_2 \xrightarrow{\Delta} CaCl_2$$

(5) With NH_3

It dissolves in liquid ammonia to give calcium ammoniate which on heating gives calcium amide which on further on heating gives calcium nitride.

$$Ca \xrightarrow{6NH_3} Ca(NH_3)_6 \xrightarrow[-4NH_3,\ H_2]{D} Ca(NH_2)_2$$

Calcium amide

$$3Ca(NH_2)_2 \xrightarrow{\Delta} Ca_3N_2 + 4NH_3$$

(6) Reducing Action

It can reduce metal oxides

Example,

$$V_2O_5 + 5Ca \rightarrow 2V + 5CaO$$

$$Cr_2O_3 + 3Ca \rightarrow 2Cr + 3CaO$$

Uses

(1) It is drying agent in the preparation of absolute alcohol.

(2) It is used in the preparation of CaH_2 and metallic vanadium.

(3) It is used in metal casting as deoxidant.

(4) It is used to remove sulphur during petroleum refining.

(5) It is used as a chemical means to remove traces of gases from vacuum tube.

Calcium Oxide or Quick–Lime (CaO)

It is called burnt lime or lime also.

Preparation

(1) Industrial Preparation

It is prepared by heating lime stone in a rotatory kiln at 800 – 1000°C.

As it is a reversible reaction so the evolved CO_2 must be removed so that the equilibrium may shift in the forward direction.

$$CaCO_3 \xrightarrow[1000^\circ]{\Delta} CaO \xrightarrow{H_2O} Ca(OH)_2$$

$-CO_2$ Slaking of lime

Physical Properties

(1) It is a white amorphous solid with a melting point of 2870 K.

(2) On exposure to atmosphere, it absorbs moisture and CO_2.

(3) On heating in an oxy-hydrogen flame, it becomes incandescent and releases a bright white light called lime light.

Chemical Properties

(1) Basic Nature

Being a basic oxide it react with acidic oxides to form salts as follows:

$$CaO + CO_2 \rightarrow CaCO_3$$

$$CaO + SiO_2 \rightarrow CaSiO_3$$

(2) Reaction with Water

On reaction with water it forms calcium hydroxide by giving a hissing sound it is an exothermic reaction and known as slaking of lime.

$$CaO \xrightarrow{\text{Heat}} Ca(OH)_2 + 15 \text{ Kcal}$$

$[Ca(OH)_2 + H_2O]$ is milk of lime

(3) With Carbon

When it is heated with carbon at 3000ºC, calcium carbide is formed.

Calcium carbide on reaction with water gives acetylene with nitrogen gives $CaCN_2$.

Imege

$$CaO \xrightarrow[3000°C]{3'C'} CaC_2 + CO\uparrow$$

Calcium carbide

$CaC_2 \xrightarrow{H_2O} C_2H_2 + Ca(OH)_2$

$CaC_2 \xrightarrow{N_2\ (\Delta)} CaCN_2 + C$

Nitrolim (fertilizer)

Uses

(1) It is used in the manufacture of glass, cement, bleaching powder, soda-lime etc.

(2) It is used as a basic flux.

(3) It is used as a drying agent for gases and alcohols (NH_3, C_2H_5OH).

(4) It is used in the refining of sugar.

(5) It is also used as a disinfectant and germicide.

(6) It is used for white washing.

Calcium Hydroxide or Slaked Lime $Ca(OH)_2$

Preparation

It is prepared by the reaction of lime with water as follows:

$$CaO + H_2O \longrightarrow Ca(OH)_2 + 15 \text{ Kcal}$$

Physical Properties

(1) It is a white amorphous powder and sparingly soluble in water. Its suspension in water is called milk of lime while the clear solution is called lime water but chemically both are same things.

Chemical Properties

(1) Reaction with CO_2

It slowly absorbs CO_2 from air forming insoluble calcium carbonate's milky solution. If more CO_2 is passed in this milky solution soluble calcium bicarbonate is formed and milkyness is disappeared.

$$Ca(OH)_2\ (aq) + CO_2 \rightarrow CaCO_3 + H_2O$$

$$CaCO_3 + H_2O + CO_2 \rightarrow Ca(HCO_3)_2$$

Calcium bicarbonate (soluble)

(2) Reaction with Chlorine

When it is treated with chlorine in cold milk of lime form, calcium hypo chlorite is formed but when it is hot, calcium chlorate is formed.

$$2Ca(OH)_2 + 2Cl_2 \longrightarrow CaCl_2 + 2H_2O + Ca(OCl)_2$$

Calcium hypo chlorite

$$6Ca(OH)_2 + 6Cl_2 \longrightarrow 5CaCl_2 + 6H_2O + Ca(ClO_3)_2$$

Calcium chrolate

On heating with chlorine slaked lime gives bleaching powder.

$$Ca(OH)_2 \xrightarrow[30°C]{Cl_2} CaOCl_2$$

bleaching powder

When it is heated with chlorine upto redness or with NH_4Cl, calcium chloride is formed.

$$Ca(OH)_2 \xrightarrow{Cl_2 \text{ red hot}} CaCl_2$$

$$Ca(OH)_2 \xrightarrow{NH_4Cl} CaCl_2 + NH_3$$

Uses

(1) It is used in softening of water, purification of coal gas, sugar etc.

(2) It is used in the manufacturing of bleaching powder, sodium carbonate etc.

(3) It is also used in making of mortar and plaster used as building materials.

Calcium Oxide or Marble or Lime Stone ($CaCO_3$)

In nature it occurs as lime stone, ice land spar, marble and shells of sea animals.

Preparation

Lab Method

In lab it is prepared by passing CO_2 through lime water or by adding sodium carbonate solution into calcium chloride as follows:

$$Ca(OH)_2 + CO_2 \rightarrow CaCO_3 + H_2O$$

$$CaCl_2 + Na_2CO_3 \rightarrow CaCO_3 + 2NaCl$$

Physio-Chemical Properties

(1) It is a white solid and which is almost insoluble in water.

(2) *Heating Effect*

$$CaCO_3 \xrightarrow{1000^\circ C} CaO + CO_2$$

(3) *With dilute Acids*

$$CaCO_3 + H_2SO_4 \rightarrow CaSO_4 + H_2O + CO_2 \uparrow$$

$$CaCO_3 + 2HCl \rightarrow CaCl_2 + H_2O + CO_2 \uparrow$$

Uses

(1) It is used in the preparation of cement, washing soda ($NaHCO_3$ by Solvay method)

(2) In the extraction of many metals like iron.

(3) Marble in used as a building material.

(4) Precipitated chalk is used in the manufacture of paints, medicines and toothpaste etc.

Calcium Sulphate Dihydrate or Gypsum ($CaSO_4 . 2H_2O$)

Calcium sulphate occurs as anhydride ($CaSO_4$) and gypsum ($CaSO_4.2H_2O$). Naturally occurring calcium sulphate is called Alabaster.

Preparation

In lab it is prepared by the reaction of calcium carbonate and calcium chloride with dilute acids as follows:

$$CaCO_3 + H_2SO_4 \rightarrow CaSO_4 + H_2O + CO_2 \uparrow$$

$$CaCl_2 + H_2SO_4 \rightarrow CaSO_4 + 2HCl$$

$$CaCl_2 + Na_2SO_4 \rightarrow CaSO_4 + 2NaCl$$

Physio-Chemical Properties

(1) It is a white crystalline solid which is partially soluble in water and its dissolution in water is exothermic.

(2) It also dissolves in dilute acids, ammonium sulphate etc.

(3) *Heating Effect*

On heating gypsum gives Plaster of Paris (Calcium sulphate hemihydrate). Plaster of Paris when mixed with water, gives a hard mass with light expansion. Gypsum on heating at 200°C gives anhydrous calcium sulphate known as dead burnt plaster .

$$2CaSO_4 .2H_2O \underset{\text{Setting } (\Delta H = -Ve)}{\overset{120^\circ C}{\rightleftarrows}} \underset{\text{plaster of paris}}{(CaSO_4)_2 H_2O + 3H_2O}$$

$$\xrightarrow[200^\circ C]{\Delta} \underset{\text{Dead burnt plaster}}{2CaSO_4 + H_2O}$$

On very strong heating gypsum decomposes into CaO.

$$2CaSO_4 \xrightarrow{\text{strong heating}} 2CaO + 2SO_2 + O_2$$

(4) With Carbon

On strong heating with carbon it forms calcium sulphide.

$$CaSO_4 + 4C \rightarrow CaS + 4CO$$

Uses

(1) Gypsum is used in the manufacture of plaster of paris, cement, black board chalk etc.

(2) It is also used in manufacturing of mortar and cement.

Plaster of Paris ($CaSO_4 . \frac{1}{2} H_2O$) or [$(CaSO_4)_2 H_2O$]

It is known as calcium sulphate hemihydrate.

Preparation

On heating gypsum at 120°C Plaster of Paris is formed as follows:

$$2CaSO_4 .2H_2O \underset{\text{Setting}(\Delta H = -Ve)}{\overset{120^\circ C}{\rightleftarrows}} \underset{\text{Plaster of paris}}{(CaSO_4)_2 H_2O + 3H_2O}$$

Physio-Chemical Properties

(1) It is a white powder.

(2) *Setting of Plaster of Paris*

When plaster of paris is mixed with a sufficient amount of water a paste is formed which on standing for some time changes into a hard mass with more volume. Here gypsum is formed with the evolution of heat.

$$(CaSO_4)_2.H_2O \xrightarrow[H_2O]{\text{Setting}} \underset{\text{(orthorhombic)}}{\underset{\text{Gypsum}}{CaSO_4.2H_2O}}$$

$$\xrightarrow{\text{Hardening}} \underset{\text{Gypsum (monoclinic)}}{CaSO_4.2H_2O}$$

Setting of plaster of paris may be catalysed with the help of NaCl while retarded by adding borax or alum.

Plaster of paris with alum is called Keene cement.

(3) *Heating Effect*

When plaster of paris is heated at 200°C it gives anhydrous Calcium sulphate (dead plaster or dead burnt). Which does not have any setting property.

$$CaSO_4)_2\,H_2O \xrightarrow[200°C]{\Delta} \underset{\text{Dead burnt plaster}}{2CaSO_4 + H_2O}$$

Uses

(1) Plaster of paris is used for plaster of broken bones, making statues, toys, chalks etc.

(2) It is also used as building material.

(3) It is used for making moulds for casting.

Bleaching Powder $CaOCl_2$ or Ca(OCl)Cl

It is also known as Calcium chloro hypochlorite or chloride of lime.

Preparation

It is prepared by passing a current of chlorine over dry slaked lime. The methods used are Bachmann method, Hasenclever method.

$$Ca(OH)_2 + Cl_2 \longrightarrow \underset{\text{Bleaching powder}}{CaOCl_2} + H_2O$$

Physio-Chemical Properties

(1) It is a mixture of calcium hypo chlorite ($CaOCl_2.4H_2O$) and basic calcium chloride ($CaCl_2.\ Ca(OH)_2.4H_2O$)

(2) It is an oxidizing agent and a bleaching agent.

(3) Cl_2 (available chlorine) is generally obtained in less amount (35-40%) from bleaching powder by using CO_2 $CaOCl_2 + CO_2 \longrightarrow CaCO_3 + Cl_2\uparrow$

(4) $CoCl_2$ acts as a catalyst to decompose bleaching powder to liberate oxygen

$$2CaOCl_2 \xrightarrow{CoCl_2} 2CaCl_2 + O_2\uparrow$$

Uses

(1) It is used to purify water, to prepare $CHCl_3$, germicide etc.

(2) It is used for bleaching cotton, wood pulp etc.

Cement

It is used as a building material. It was called Portland cement by Mason Joseph Aspdin. As on setting, it becomes as hard as Portland rock found in England.

Composition of Cement

Composition of cement is given as:

CaO,	SiO_2,	Al_2O_3,	MgO,	Fe_2O_3,	SO_3
62%	23%	7%	2.5%	2.5%	1.5%

It contains dicalcium silicate (Ca_2SiO_4 , 26 %) , tricalcium silicate (Ca_3SiO_3 , 51 %), and tricalcium aluminate (Ca_3 Al_2O_6 , 11.7 %), Gypsum etc (2-3 %).

- Tricalcium Aluminate ($3CaO.Al_2O_3$) is the fastest setting component in cement while Dicalcium Silicate (2CaO. SiO_2) is the slowest setting component.

Manufacture

The essential raw materials required for the manufacture of cement are lime stone and clay. Lime stone provides lime while clay provides silica, alumina and iron oxide. Gypsum is also required at the final stage of its manufacture.

It is manufactured in the following steps:

(1) Mixing of Raw Materials

First of all the raw materials (lime stone and clay) are mixed either by using dry process or wet process.

(a) Dry Process

It is used when raw materials are hard and dry. Here lime stone is broken into small pieces and mixed with clay in proper proportion. This mixture is finely powdered and pulverised to get a uniform mixture called raw meal.

(b) Wet Process

It is used when the raw materials are soft and damp. First clay is washed with water and then mixed with crushed lime stone in a proper proportion. The mixture is finely ground and homogenized to get the product slurry.

(2) Burning in Rotary Kiln

The raw meal and slurry obtained from the above processes are charged from one end of the rotary kiln. The rotary kiln has a long steel cylinder resting on rollers and made to revolve slowly. The kiln is heated by a blast of burning coal blown from the otherside. Due to the revolution of the kiln, the charge moves from one end to the other and gets heated upto 1400 – 1600 K.

Reactions in Kiln

$$CaCO_3 \xrightarrow{\text{Heat}} CaO + CO_2$$

$$2CaO + SiO_2 \rightarrow \underset{\text{Dicalcium silicate}}{2CaO.SiO_2}$$

$$3CaO + SiO_2 \rightarrow \underset{\text{Tri calcium silicate}}{3CaO.SiO_2}$$

$$3CaO + Al_2O_3 \rightarrow \underset{\text{Tri calcium aluminate}}{3CaO.Al_2O_3}$$

$$4CaO + Al_2O_3 + Fe_2O_3 \rightarrow 4CaO.Al_2O_3.Fe_2O_3$$

The resting product obtained in the form of greenish black or grey coloured hard balls and is called cement clinker. It is taken to a revolving tube called as cooler to make it cooled.

(3) Grinding with Gypsum

The cement clinker is mixed with 2-3 % gypsum and ground to an exceedingly fine powder in grinding machines.

Here gypsum is added to regulate the setting time of the cement. The fine powdered mixture is called Portland cement. It is packed in airtight bags and sold in the market.

Setting of Cement

When mixed with water, cement sets to a very hard stone like mass. It is believed that cement reacts with water to form a gelatinous mass. This undergoes dehydration and forms three dimensional cross links between –Si – O – Si– and –Si – O – Al– chains. This results in its setting into a very hard mass. Setting of cement is an exothermic process.

$$3CaO.Al_2O_3 + 6H_2O \xrightarrow{\text{Hydration}} \underset{\text{Colloidal gel}}{3CaO.Al_2O_3.6H_2O}$$

$$2CaO.SiO_2 + xH_2O \xrightarrow{\text{Hydration}} 2CaO.SiO_2.xH_2O$$

$$4CaO.Al_2O_3.Fe_2O_3 + 6H_2O \xrightarrow{\text{Hydration}} 3CaO.Al_2O_3.6H_2O + Fe_2O_3.CaO$$

$$3CaO.SiO_2 + H_2O \xrightarrow{\text{Hydration}} Ca(OH)_2 + 2CaO.SiO_2$$

$$3CaO.Al_2O_3 + 6H_2O \xrightarrow{\text{Hydration}} 3Ca(OH)_2 + 2Al(OH)_3$$

When gypsum is added the setting of cement slows down due to the following reaction:

$$3CaO.Al_2O_3 + 3CaSO_4 + 2H_2O \rightarrow 3CaO.Al_2O_3.3CaSO_4.2H_2O$$

Concrete

Concrete is the mixture of cement, sand, gravel and water. It is very hard mass which is used for the construction purposes

Reinforced Concrete (RCC)

When cement concrete is filled in and around a wire netting of iron rods for setting, the resulting structure is called reinforced concrete. Which are very strong and used in the construction of bridges, beams, pillars etc.

- Fly ash, a waste material from the steel industry can also be added in cement for reducing the cost.
- Rice husk can also be used in cement manufacture as it has very high silica content.
- Cement without iron is white in colour.

Biological Role of Mg and Ca

Magnesium (Mg)

It is a major mineral and about 70% of it is found in bones as magnesium phosphate.

About 200-300 mg of it is required for adults and 150- 200 mg for children per day.

Function

- It is a important component of bones and teeth.
- It is essential for binding ribosomal subunits during protein synthesis.
- Mg^{2+} ions along with Ca^{2+} ions transmit electrical impulses along the nerve fibre and help in the contraction of muscles.
- It is important element for nerve function and activates enzymes of carbohydrate metabolism e.g., for Hexokinase, Phosphofructokinase, Euolase in glycolysis and isocitric dehydrogenase in Kreb-cycle.
- It is an important constituent of chlorophyll which is the green colouring pigment present in plants and is responsible for initiating the process of photosynthesis. The process of photosynthesis can be represented as follows:

$$6H_2O + 6CO_2 \xrightarrow[\text{Sun light}]{\text{Chlorophyll}} C_6H_{12}O_6 + 6O_2$$

Deficiency

- Due to its deficiency nerves become more irritable and muscles become weak.
- Its deficiency causes cardiac arrhythmia and peripheral vasodialation.

Calcium (Ca)

- Ca^{2+} ions are present as phosphates in the bones of both human beings and animals.
- It is a major mineral. It absorbs from intestine and controlled by Vitamin D, PTH.
- Its excess is stored in bones. About 800 mg of it is required by adults, 1.2 gm for pregnant lady, 350-550 mg for infants per day. 99 % calcium is present in bones and teeth.
- In blood its concentration is regulated by calcitonin and PTH hormones.

Function

- It is a essential element for the formation bones and teeth.
- It is a essential element for blood clotting and muscle contraction.
- It controls permeability of cell membranes and helps in nerve conduction.
- It activates enzymes like ATPase, arginine kinase, succinic dehydrogenase etc.,
- It acts as chemical messenger (IIIrd messenger)
- Ca^{2+} ions along with Mg^{2+} ions catalyse the formation of pyrophosphate linkages which control various biological systems. The hydrolysis of pyrophosphates liberate energy is controlled by Ca^{2+} ions.

Deficiency

- Rickets is caused by the deficiency of calcium in childrens and infants.
- Its deficiency causes Osteomalacia a bone disorder of adults and Osteophresis in adults specially ladies in which bones becomes soft so fractures occur easily.
- Low Ca^{++} level increases blood clotting time.
- Its deficiency leads to muscular tetany.

Unforgettable Guidelines

- Al and BaO_2 is an ignition mixture.
- $V + Al_2O_3$ is activated alumina.
- Thermite mixture is $[Fe_2O_3 + Al] + BaO_2$

  3 : 1
- Ultramarine is $Na_5Al_3\ Si_3\ S_3O_{12}$ (blue) used in Calico or Blue printing.
- Damp Al_2S_3 gives the smell of H_2S (foul smell)
- An alloy of Na and K is a liquid at room temperature. It is used in special thermometers for recording temperature above the B.P of mercury (357ºC).
- Sodium sulphate or Glauber's Salt is $Na_2SO_4.10H_2O$ and it is used as purgative and to prepare craft paper, Hypo, Soda glass etc.
- K^+ can be detected in lab by using picric acid (yellow ppt formation) or tartaric acid (white ppt.).
- Sodium sesquicarbonate ($Na_2CO_2 . NaHCO_3 . 2H_2O$) is neither deliquescent nor efflorescent and used in wool washing.
- Sodium bicarbonate is present in Seildlitz powder.

  Polyhalite is $K_2SO_4 .MgSO_4 . 2CaSO_4 . 2H_2O$

POINT TO REMEMBER

- $BeCO_3$ can be kept only in the atmosphere of CO_2
- $BeCl_2$ is most covalent among (MCl_2).
- Ca has lowest density, Mg has lowest M.P, Be has no peroxide.
- Be and Mg crystallize in H.C.P., Ca and Sr in C.C.P. and Ba in B.C.C. structures.
- Amongst alkaline earth metals, M.P. of Mg is lowest while density of Ca is the lowest.
- Magnesium burns with dazzling light even in CO_2 and N_2.
- $CaCl_2.6H_2O$ is widely used for melting ice on roads, particularly in very cold countries, because a 30% eutectic mixture of $CaCl_2 / H_2O$ freezes at – 55ºC as compared with $NaCl / H_2O$ at –18ºC.
- Hydroxyapatite, $Ca_5(PO_4)_3OH$ is the main component of tooth enamel. Cavities in your teeth are formed when acids decompose the weakly basic apalite coating.
- The ions Na^+, K^+, Mg^{2+} and Ca^{2+} are the most abundant metal ions in biochemical systems.
- Most of the kidney stones consist of calcium oxalate, $CaC_2O_4.H_2O$ which dissolves in dilute strong acids but remains insoluble in bases.
- $BaSO_4$ being insoluble in H_2O and opaque to X – rays is used under the name barium meal to scan the X-ray of the human digestive system.
- Only Mg displaces hydrogen from a very dilute HNO_3.
- The ionization energy of radium is higher than that of barium.
- Albaster is ($CaSO_4.H_2O$) used in making ornaments.
- Calcium magnesium silicate is Asbestos Ca [Mg_3 (SiO_3)$_4$].
- Beryllium is amphoteric as it reacts both with acid and bases.

  $$Be + 2NaOH + 2H_2O \rightarrow \underset{\text{Sodium beryllate}}{Na_2[Be(OH)_4]} + H_2$$

- K_2SO_4 . $MgSO_4$. $6H_2O$ is potash magnesia (a fertilizer).

CHAPTER-END EXERCISES

PRACTICE QUESTIONS – I

1. The oxidation state of sodium in sodium amalgam is:

(a) –1 (b) +1
(c) +2 (d) 0

2. The number of electrons in $[19K^{40}]^{-1}$ are:

(a) 18 (b) 20
(c) 30 (d) 40

3. Sodium amalgam is useful as:

(a) Catalyst (b) Oxidizing agent
(c) Bleaching agent (d) Reducing agent

4. Which of the following has the least ionization potential?

(a) He (b) Li
(c) Zn (d) N

5. The correct order of increasing ionic character is:

(a) $BeCl_2 < MgCl_2 < BaCl_2 < CaCl_2$
(b) $BeCl_2 < MgCl_2 < CaCl_2 < BaCl_2$
(c) $BaCl_2 < CaCl_2 < MgCl_2 < BeCl_2$
(d) $BeCl_2 < BaCl_2 < MgCl_2 < CaCl_2$

6. The increasing order of atomic radius for the elements Na, Rb, K and Mg is:

(a) Mg < Na < K < Rb
(b) Rb < K < Mg < Na
(c) Na < K < Mg < Rb
(d) Na < Mg < K < Rb

7. The pair of elements that have similar chemical properties are:

(a) Beryllium and boron
(b) Lithium and magnesium
(c) Carbon and nitrogen
(d) Aluminium and magnesium

8. Lithium is strongest reducing agent among alkali metals due to which of the following factor?

(a) Hydration energy (b) Electron affinity
(c) Lattice energy (d) Ionization energy

9. The alkali metal that reacts with nitrogen directly to form nitride is:

(a) K (b) Na
(c) Li (d) Rb

10. The number of covalent bonds formed by beryllium is:

(a) 4 (b) 5
(c) 2 (d) 3

11. The pair of compounds which cannot exist together is:

(a) $NaHCO_3$ & $NaOH$
(b) Na_2CO_3 & $NaHCO_3$
(c) Na_2CO_3 & $NaOH$
(d) $NaHCO_3$ & $NaCl$

12. Which one of the following compounds in aqueous solution gives a white precipitate with perchloric acid?

(a) NaCl (b) $MgCl_2$
(c) KCl (d) $FeCl_3$

13. As the alkaline earth metals (except Be) tend to lose their valence electrons readily, they act as:

(a) Bases
(b) Strong reducing agents
(c) Weak oxidizing agents
(d) Weak reducing agents

14. In view of their low ionization energies, the alkali metals are:

(a) Strong oxidizing agents
(b) Weak oxidizing agents
(c) Strong reducing agents
(d) Weak reducing agents

15. When sodium is treated with sufficient oxygen/air, the product obtained is:

(a) NaO_2 (b) NaO
(c) Na_2O (d) Na_2O_2

16. Which of the following has lowest melting point?

(a) Cs (b) Na
(c) Li (d) K

17. The ionic mobility of alkali metal ions in aqueous solution is maximum for:

(a) K^+ (b) Rb^+
(c) Li^+ (d) Na^+

18. As the nuclear charge increases from neon to calcium, the orbital energies:

(a) Increase very slow
(b) Increase very rapidly
(c) Fall
(d) Increase

19. Which of the following has maximum ionization energy?

(a) $Ca \rightarrow Ca^{2+} + 2e^-$ (b) $Mg \rightarrow Mg^{2+} + 2e^-$
(c) $Ba \rightarrow Ba^+ + e^-$ (d) $Be \rightarrow Be^+ + e^-$

20. Which of the following does not precipitate in the Solvay's process, for the manufacture of Na_2CO_3?

(a) CO_2 (b) H_2SO_4
(c) NH_3 (d) NaCl solution

21. Mg is an important component of which biomolecule occurring extensively in living world?

(a) Chlorophyll (b) Haemoglobin
(c) ATP (d) Florigen

22. Mortar is a mixture of:

(a) Slaked lime + silica + H_2O
(b) $CaCO_3$ + silica + H_2O
(c) Slaked lime + plaster of paris + H_2O
(d) Plaster of paris + silica

23. The reaction of slaked lime with Cl_2 gas gives:

(a) A mixture of $Ca(OCl)_2$, $Ca(OH)_2$, $CaCl_2$ and H_2O
(b) Quick lime
(c) Baryta water
(d) Only $Ca(OCl)_2$

24. Aqueous sodium hydroxide reacts with white phosphorous to form phosphine and:

(a) P_2O_3 (b) P_2O_5
(c) NaH_2PO_2 (d) Na_3PO_3

25. Metals having ns^1 as the valence electronic configuration:

(a) Have a first ionization potential of more than 10 eV/ atom
(b) Are highly electropositive
(c) Are highly electronegative
(d) Act as strong oxidizing reagents

26. An aqueous solution of sodium carbonate is alkaline because sodium carbonate is a salt of:

(a) Weak acid and weak base
(b) Weak acid and strong base
(c) Strong acid and strong base
(d) Strong acid and weak base

27. In solvay process of manufacture of Na_2CO_3 , the by products are:

(a) $CaCl_2$, CO_2, NH_3 (b) Na_2CO_3, CO_2
(c) NH_4Cl, CaO (d) CaO, Na_2CO_3

28. Aqueous solution of Carnalite gives positive test for:
(a) Potassium, magnesium and chloride ions
(b) Potassium ions only
(c) Potassium and chloride ions only
(d) Chloride ions only

29. Which of the following set of raw materials are used in the manufacture of Na_2CO_3 by Solvay process?
(a) $CaCl_2$, NH_3, CO_2
(b) NaOH, NH_3, CO_2
(c) NaCl, NH_3, CO_2
(d) $Ca(OH)_2$, NH_3, CO_2

30. A burning strip of magnesium is introduced into a jar containing a gas. After sometimes the walls of the container are coated with carbon. The gas in the container is:
(a) H_2O (b) CO_2
(c) O_2 (d) N_2

31. Which of the following covalent molecules in an exception of octet rule?
(a) CH_4 (b) CO_2
(c) $BeCl_2$ (d) H_2O

32. In a reaction sodium is heated with $BeCl_2$, in this reaction, sodium acts as:
(a) Precipitating agent
(b) Hydrolysis agent
(c) Reducing agent
(d) Oxidizing agent

33. Which is insoluble in water?
(a) CaF_2 (b) H_2O
(c) $HgCl_2$ (d) $Ca(NO_3)_2$

34. What are the products formed when an aqueous solution of magnesium bicarbonate is boiled?
(a) Mg, CO_2, H_2O
(b) $Mg(OH)_2$, H_2O
(c) $MgCO_3$, H_2O, CO_2
(d) $Mg(HCO_3)_2$, H_2O

35. Which of the following is calcium superphosphate?
(a) $Ca(H_2PO_4)_2H_2O + CaSO_4$
(b) $CaSO_4 + CaO$
(c) $Ca_3(PO_4)_2 + 2CaSO_4$
(d) $Ca(H_2PO_4)_2H_2O + 2(CaSO_2.2H_2O)$

36. What is the molecular formula of plaster of pairs?
(a) $CaSO_4.2H_2O$ (b) $2CaSO_4.H_2O$
(c) $2CaSO_4.2H_2O$ (d) $CaSO_4.3H_2O$

37. The metal that dissolves in liquid ammonia, giving a dark blue coloured solution is:
(a) Lead (b) Tin
(c) Silver (d) Sodium

38. What is the molecular formula of gypsum?
(a) $CaSO_4.2H_2O$ (b) $CaSO_4.3H_2O$
(c) $CaSO_4.H_2O$ (d) $2CaSO_4.H_2O$

39. In the Castner's process of extraction of sodium cathode is:
(a) Nickel rod (b) Iron rod
(c) Graphite rod (d) Copper rod

40. Plaster of paris is used:
(a) In dentistry and surgery.
(b) As a plaster for walls.
(c) As a drying agent.
(d) In metallurgical process.

41. Which one is used as an air purifier in spacecraft?
(a) Potassium super oxide
(b) Anhydrous $CaCl_2$
(c) Quick lime
(d) Slake lime

42. Which of the following is correct pair?
(a) NH_3, linear (b) BF_3, octahedral
(c) $BeCl_2$, liner (d) CO_2, tetrahedral

43. The active constituent of bleaching powder is:
(a) $Ca(ClO_2)_2$ (b) Ca(OCl)Cl
(c) $Ca(OCl)_2$ (d) Ca(OCl)Cl

44. What is the crystal structure of cesium chloride?
(a) Tetrahedral (b) Octahedral
(c) Face centred cubic (d) Body centerd cubic

45. In organic reactions, sodium in liquid ammonia is used as ………:
(a) Oxidizing agent (b) Reducing agent
(c) Hydrolyzing agent (d) Precipitating agent

46. Sodium metal reacts with Al_2O_3 at high temperature to give a sodium compound X. X reacts with carbon dioxide in water to form Y. Y is ……..:
(a) Na_2CO_3 (b) Na_2O_2
(c) $NaAlO_2$ (d) Na_2O

47. Aqueous NaCl solution is electrolysed using platinum electrodes. What is the product formed at cathode?
(a) O_2 (b) Cl_2
(c) H_2 (d) Na

48. Oxalic acid reacts with concentrated H_2SO_4 to give a mixture of two gases. When this mixture is passed through caustic potash, one of the gases is absorbed. What is the product formed by the absorbed gas with caustic potash?
(a) K_2HCO_3 (b) K_2CO_3
(c) KOH (d) K_2SO_4

49. Solution of $MgCl_2$ in water has pH value:
(a) 7 (b) >7
(c) < 7 (d) 14.2

50. The compound formed when gypsum is dissolved in aqueous ammonium sulphate solution is:
(a) $CaSO_4.NH_3Cl. H_2O$
(b) $CaSO_4.(NH_4)_2SO_4.H_2O$
(c) $CaCl_2.NH_4Cl._2H_2O$
(d) $CaCl_2. (NH_4)_2SO_4.H_2O$

51. Alkali metals have high oxidation potential and hence they behave as:
(a) Electrolytes (b) Lewis bases
(c) Oxidizing agents (d) Reducing agents

52. "Fluorosis" disease is caused due to the reaction of .... with excess of fluoride in the body.
(a) Mg (b) Fe
(c) K (d) Ca

53. Which one of the following electrolysis is used in Down's process of extracting sodium metal?
(a) NaCl (b) NaCl + NaOH
(c) NaOH + KCl + KF (d) NaCl + KCl + KF

54. Which one of the following reactions occur at the anode, in the Castner process of extracting sodium metal:
(a) $Na^+ + e^- \rightarrow Na$
(b) $4OH^- \rightarrow 2H_2O + O_2 + e^-$
(c) $H_2 \rightarrow 2H^+ + 2e^-$
(d) $2Cl^- \rightarrow Cl_2 + 2e^-$

55. In curing cement plasters, water is sprinkled from time to time. This helps in:
(a) Hydrated sand gravel mixed with cement
(b) Keeping it cool
(c) Converting sand into silicic acid
(d) Developing interlocking needle like crystals of hydrated silicates

56. Smallest among these species is:
(a) Hydrogen (b) Lithium
(c) Lithion ion (d) Helium

57. In the extractive of sodium by Down's process cathode and anode are respectively:
(a) Nickel and chromium
(b) Iron and graphite
(c) Copper and nickel
(d) Copper and chromium

58. Silica reacts with Mg to form a magnesium compound X. X reacts with dilute HCl and forms Y. Y is:
(a) $MgCl_2$ (b) MgO
(c) $SiCl_4$ (d) $MgSiO_3$

59. Oxone is:
(a) Na_2O_2 (b) $NaBO_3$
(c) N_2O (d) CaO

60. One of the elements present in Carnalite shows flame colouration. The colour of the flame is:
(a) Green (b) Lilac
(c) Yellow (d) Orange

61. .... process is used for the removal of hardness of water.
(a) Hoope (b) Baeyer
(c) Serpeck (d) Calgon

62. Brine solution on electrolysis will not give:
(a) O_2 (b) H_2
(c) Cl_2 (d) NaOH

63. Mg and Li are similar in their properties due to:
(a) Same ionic potential
(b) Same electron affinity
(c) Same e/m ratio
(d) Same group

64. The molecular formula of potash alum is:
(a) $K_2Al_2S_2H_{48}O_{39}$ (b) $KAl_2S_2H_{48}O_{40}$
(c) $K_2AlS_2H_{48}O_{40}$ (d) $K_2Al_2S_2H_{48}O_{40}$

65. Which one of the following is dead burnt plaster?
(a) Anhydrous $CaSO_4$ (b) $CaSO_3$
(c) $(CaSO_4)_2.H_2O$ (d) $CaSO_4.2H_2O$

66. Photoelectric effect is maximum in:
(a) K (b) Cs
(c) Na (d) Li

67. Which of the following is not a Mg ore?
(a) Magnesite (b) Gypsum
(c) Carnalite (d) Dolomite

68. An element M reacts with chlorine to form a compound X. The bond angle in X is 120°. What is M?

(a) B (b) N
(c) Be (d) Mg

69. When washing soda is heated:

(a) CO_2 is released
(b) Water vapour is released
(c) CO_2 is released
(d) $CO + CO_2$ is released

70. In which of the following reactions, MgO is not formed?

(a) $Mg + B_2O_3 \rightarrow$ (b) $Mg + NO_2 \rightarrow$
(c) $Mg + \text{dil. } HNO_3 \rightarrow$ (d) $Mg + CO_2 \rightarrow$

71. An important ore of magnesium is:

(a) Malachite (b) Cassiterite
(c) Carnalite (d) Galena

72. The most unstable carbonate is:

(a) Cs_2CO_3 (b) Li_2CO_3
(c) Na_2CO_3 (d) K_2CO_3

73. Initial setting of cement is mainly due to:

(a) Hydration and hydrolysis
(b) Dehydration and gel formation
(c) Dehydration and dehydrolysis
(d) Hydration and gel formation

74. $MgCl_2.6H_2O$, when heated, gives:

(a) Magnesium hydride
(b) Magnesium chloride
(c) Magnesium oxide
(d) Magnesium oxychloride

75. What are the products formed when Li_2CO_3 undergoes decomposition?

(a) $LiO_2 + CO$ (b) $Li_2O + CO$
(c) $Li_2O_2 + CO$ (d) $Li_2O + CO_2$

76. On heating sodium metal in a current of dry ammonia, the compound formed is:

(a) Sodium hydride (b) Sodium amide
(c) Sodium azide (d) Sodium nitride

77. Sodium nitrate decomposes above ~ 800°C to give:

(a) N_2 (b) O_2
(c) NO_2 (d) Na_2O

78. What is the reaction occurring at the anode in Down's process for the extraction of sodium?

(a) $2Cl^- \rightarrow Cl_2 + 2e^-$
(b) $NaOH \rightarrow Na+ + OH^-$
(c) $4OH^- \rightarrow 2H_2O + O_2 + 4e^-$
(d) $Na^+ + e^- \rightarrow Na$

79. Magnesium can be obtained by:

(a) Reducing magnesium salt solution with Fe
(b) Electrolysis of fused magnesium salt
(c) Electrolysis of $Mg(NO_3)_2$ solution
(d) Reducing MgO with coke

80. The solubilities of carbonates decrease down the magnesium group due to a decrease in:

(a) Inter ionic interaction
(b) Lattice energies of solids
(c) Entropy of solution formation
(d) Hydration energies of cations

81. Magnesium does not decompose in:

(a) Hot water (b) Steam
(c) Semi - hot water (d) Cold water

82. Li_2O_2 is formed when:

(a) LiOH is heated in an atmosphere of O_2..
(b) Li is heated strongly is excess of air.
(c) H_2O_2 and alcohol are added to a solution of LiOH in water.
(d) Both (a) and (c).

83. Sodium is heated in air at 300°C to form X. X absorbs CO_2 and forms Na_2CO_3 and Y. Which of the following is Y?

(a) O_2 (b) O_3
(c) H_2 (d) H_2O_2

84. Which of the following ore contains both calcium and magnesium?

(a) Lime stone (b) Dolomite
(c) Epsom salt (d) Carnallite

85. The product obtained on fusion of $BaSO_4$ and Na_2CO_3 is:

(a) BaO (b) $BaCO_3$
(c) $BaHSO_4$ (d) $Ba(OH)_2$

86. The hydration energy of Mg^{2+} ions is higher than that of:

(a) Al^{3+} (b) Be^{2+}
(c) Na^+ (d) none of these

87. Baking soda is:

(a) $NaHCO_3$ (b) Na_2CO_3
(c) $NaHCO_3.6H_2O$ (d) $Na_2CO_3.10H_2O$

88. Which of the following exists in polymeric form?

(a) B_2H_6 (b) $AlCl_3$
(c) $BeCl_2$ (d) SiC

89. Alkali metals present in their compounds are always:

(a) Monovalent (b) Bivalent
(c) Zerovalent (d) None of these

90. Crown ethers and cryptans are:

(a) Salts of alkali metals
(b) Complexes of alkali metals
(c) Organic salts of alkali metals
(d) Hydroxides of alkali metals used for inorganic quantitative analysis

91. Which of the following is the composition of carnallite?

(a) $MgCl_2.KCl.6H_2O$ (b) Fe_3O_4
(c) Na_3AlF_6 (d) $Na_2Al_2O_3$

92. Which of the following is produced when magnesium reacts with hot concentrated H_2SO_4?

(a) H_2 (b) S
(c) MgO (d) SO_2

93. Which of the following process is used in the extractive metallurgy of magnesium?

(a) Self reduction
(b) Fused salt electrolysis
(c) Thermite reduction
(d) Aqueous solution electrolysis

94. Which of the following alkali metal ions has lowest ionic mobility in aqueous solution?

(a) Na^+ (b) Li^+
(c) Rb^+ (d) Cs^+

95. NaOH is prepared by the method:

(a) Castner cell (b)Down's cell
(c) Castner Kellner cell (d)Solvay process

96. For the preparation of sodium thiosulphate by "Springs reaction" the reactants used are:

(a) $Na_2S + SO_2$
(b) $Na_2SO_3 + S$
(c) $Na_2S + Na_2SO_3 + I_2$
(d) $Na_2S + Na_2SO_3 + Cl_2$

97. Brine solution on electrolysis will not give:

(a) H_2 (b) O_2
(c) NaOH (d) Cl_2

98. In the Castner process for the extraction of sodium, the anode is made of .... metal.

(a) Nickel (b) Iron
(c) Copper (d) Sodium

99. Carnallite on electrolysis gives:

(a) Na and CO_2 (b) Mg and Cl_2
(c) Al and Cl_2 (d) Ca and Cl_2

100. Which of the following does not give flame colouration?

(a) Mg (b) Ca
(c) Sr (d) Ba

101. Several blocks of magnesium are fixed to the bottom of a ship to:

(a) Keep away the sharks
(b) Prevent action of water and salt
(c) Make the ship lighter
(d) Prevent puncturing by under sea rocks

102. Lithium is similar to magnesium in many properties. This is because

(a) Both have nearly the same size.
(b) The ratio of the charge to size is nearly the same
(c) Both have similar electronic configurations
(d) Both have nearly the same N /P ratio.

103. Which of the following has lowest thermal stability?

(a) $BeCO_3$ (b) $MgCO_3$
(c) $CaCO_3$ (d) $BaCO_3$

104. The ionic conductance of following cations in a given concentration is in the order:

(a) $Li^+ > Na^+ < K^+ < Rb^+$
(b) $Li^+ = Na^+ < K^+ < Rb^+$
(c) $Li^+ > Na^+ > K^+ > Rb^+$
(d) $Li^+ < Na^+ < K^+ < Rb^+$

Practice Questions – II

105. A sodium salt on treatment with $MgCl_2$ gives white precipitate only on heating. The anion of the sodium salt is:

(a) HCO_3^- (b) CO_3^{2-}
(c) NO_3^- (d) SO_4^{2-}

106. Gas 'A' is bubbled through slaked lime when a white precipitate is formed, on prolonged bubbling the precipitategets dissolved. On heating the resultant solution, the white precipitate reappeared with solution of gas 'B'. The gases 'A' and 'B' respectively are:

(a) CO & CO_2 (b) CO_2 & CO_2
(c) CO & CO (d) CO_2 & CO

107. Anhydrous mixture of KF and HF contains which type of ions?

(a) KH^+, F^- (b) $(KF)^+ (HF)^-$
(c) K^+,H^+, F^- (d) K+, HF_2^-

108. When K_2O is added to water, the solution is basic because it contains a significant concentration of:

(a) OH^- (b) K^+
(c) O^{3-} (d) O_2^{2-}

109. Which of the following statement is correct for $CsBr_3$?

(a) It is a covalent compound
(b) It contains Cs^{3+} and Br^- ions
(c) It contains Cs^+ and Br_3^- ions
(d) It contains Cs^+, Br^- and lattice Br_2 molecule

110. A mixture of sodium oxide and calcium oxide are dissolved in water and saturated with excess carbon dioxide gas. The resulting solution is .... It contains .....

(a) Acidic NaOH and $CaCO_3$
(b) Basic Na_2CO_3 and $CaCO_3$
(c) Neutral Na_2CO_3 and $CaCO_3$
(d) Basic NaOH and $Ca(OH)_2$

111. A certain metal M is used to prepare an antacid, which is used as a medicine for acidity. This metal accidently catches fire, which can no be extinguished by using CO_2 based extinguishers. The metal M is:

(a) Mg (b) Ba
(c) Be (d) C

112. Consider the following abbreviations for hydrated alkali ions:

$X = [Li(H_2O)_n]^+$ $Y = [K(H_2O)_n]^+$
$Z = [Cs(H_2O)_n]^+$

Which of the correct order of size of these hydrated alkali ions?

(a) Z > X > Y (b) X = Y = Z
(c) X > Y > Z (d) Z > Y > X

113. Out of the following metals that cannot be obtained by electrolysis of the aqueous solution of their salts are:

1. Ag 2. Mg
3. Cu 4. Al

(a) 2, 4 (b) 2, 3
(c) 1, 4 (d) All

114. The compound insoluble in acetic acid is:

(a) Calcium oxide (b) Calcium carbonate
(c) Calcium oxalate (d) Calcium hydroxide

115. Consider the following reactions:

1. $2Na + 2H_2O \rightarrow 2NaOH + H_2$
2. $2NaOH + Cl_2 \rightarrow NaCl + NaOCl + H_2O$
3. $4OH^- \rightarrow O_2 + 2H_2O + 4e^-$
4. $2Cl^- \rightarrow Cl_2 + 2e^-$

in the diaphragm cell used for the electrolysis of brine, the reactions that occur would include

(a) 2, 3, 4 (b) 1, 3, 4
(c) 1, 2, 3 (d) 1, 2, 4

116. Sodium sulphate is soluble in water whereas barium sulphate is sparingly soluble because:

(a) The hydration energy of sodium sulphate is more than its lattice energy .
(b) The lattice energy of barium sulphate is less than its hydration energy.
(c) The lattice energy has no role to play in solubility.
(d) The hydration energy of sodium sulphate is less than its lattice energy..

117. In electrolysis of NaCl when Pt electrode is taken then H_2 is liberated at cathode while with Hg cathode, it forms sodium amalgam. The reason for this is:

(a) more voltage is required to reduce H^+ at Hg than at Pt.
(b) concentration of H^+ ions is larger when Pt ele trode is taken.
(c) Na is dissolved in Hg while it does not dissolve in Pt.
(d) Hg is more inert than Pt.

118. The metallic lusture exhibited by sodium is explained by:

(a) Diffusion of sodium ions
(b) Oscillation of loose electrons
(c) Excitation of free electrons
(d) Existence of body centred cubic lattice.

119. When sodium chloride is electrolysed in Nelson's cell, hydrogen is liberated at cathode and not sodium, because:

(a) H^+ does not react with water
(b) H^+ has lower electrode potential
(c) H^+ has greater mobility than Na^+
(d) H^+ is smaller than Na^+

120. When zeolite, which is hydrated sodium aluminium silicate, is treated with hard water the sodium ions are exchanged with:

1. H^+ ions 2. Ca^{2+} ions
3. SO_4^{-2} ions 4. Mg^{2+} ions
(a) 2, 4 (b) 1, 2, 3
(c) 2, 4,5 (d) all are correct

121. On heating sodium metal in a current of dry ammonia, the compound formed is:

(a) Sodium hydride (b) Sodium amide
(c) Sodium azide (d) Sodium nitride

122. In NaCl, the chloride ions occupy the place in a fashion of:

(a) bcc (b) fcc
(c) both (d) none

123. Halides of alkaline earth metals form hydrates such as $MgCl_2.6H_2O$, $CaCl_2.6H_2O$, $BaCl_2.2H_2O$ and $SrCl_2.2H_2O$. This shows that halides of group 2 elements:

(a) Can absorb moisture form air
(b) Act as dehydrating agents
(c) Are hydroscopic in nature
(d) All of the above

124. Water softening by Clarke's process uses:
(a) Potash alum
(b) Calcium bicarbonate
(c) Calcium hydroxide
(d) Sodium bicarbonate

125. Consider the following statements:

1. Cs^+ ion is more highly hydrated than other alkali metal ions
2. Among the alkali metals Li, Na, K and Rb, Li has the higher melting point
3. Among the alkali metals, only Li form a stable nitride by direct combination

(a) 1 and 3 are correct
(b) 2 and 3 are correct
(c) 1, 2 and 3 are correct
(d) 1 and 2 are correct.

126. The reaction of slaked lime with Cl_2 gas gives:
(a) A mixture of $Ca(OCl)_2$, $Ca(OH)_2$, $CaCl_2$ and H_2O
(b) Quick lime
(c) Baryta water
(d) Only $Ca(OCl)_2$

127. Halides of alkaline earth metals form hydrates such as $MgCl_2.6H_2O$, $CaCl_2.6H_2O$, $BaCl_2.2H_2O$ and $SrCl_2.2H_2O$. This shows that halides of group 2 elements:

(a) Can absorb moisture from air
(b) Act as dehydrating agents
(c) Are hydroscopic in nature
(d) All of the above

128. The relative thermal stabilities of alkali metal halides are such that:

(a) CsCl > RbCl > KCl < NaCl > LiCl
(b) CsCl > RbCl < KCl > NaCl < LiCl
(c) LiCl > NaCl > KCl > RbCl > CsCl
(d) CsCl > RbCl > KCl > NaCl > LiCl

129. A deliquescent white crystalline hydroxide X reacts with a nitrate Y to form another hydroxide which decomposes to give a insoluble brown layer of its oxide. X is a powerful cautery and breaks down then proteins of skin flesh to a pasty mass. X and Y are:

(a) NaOH, $Zn(NO_3)_2$ (b) NaOH, $AgNO_3$
(c) $Ca(OH)_2.HgNO_3$ (d) NaOH, $Al(NO_3)_3$

130. The major role of fluorspar (CaF_2) which is added in small quantities in the electrolytic reduction of alumina dissolved in fused cryolite (Na_3AlF_6) is

1. As a catalyst
2. To make the fused mixture very conducting
3. To lower the temperature of melt
4. To decrease the rate of oxidation of carbon at the anode

(a) 2, 3 (b) 1, 2
(c) 2, 3 ,4 (d) 3, 4

131. Solution of sodium metal in liquid ammonia is strongly reducing due to the presence in solution of the following:

(a) Solvated electrons (b) Sodium atoms
(c) Sodium hydride (d) Sodium amide

132. Metal X on heating in nitrogen gas gives Y. Y on treatment with H_2O gives a colourless gas which when passed through $CuSO_4$ solution gives a blue colour. Y is:

(a) MgO (b) $Mg(NO_3)_2$
(c) Mg_3N_2 (d) NH_3

133. Bleaching powder loses its powder on keeping for a long time because:
(a) It absorbs moisture
(b) It changes into calcium chloride and calcium chlorate
(c) It changes into calcium hypochlorate
(d) It changes into calcium chloride and calcium hydroxide

134. On heating quick lime with coke in an electric furnace, we get

(a) CaC_2 (b) $CaCO_3$
(c) CaO (d) Ca and CO_2

135. Limestone is not used in which of the following manufacturing processes?

(a) Portland cement
(b) Phosphorus from phosphorite
(c) Iron from haematite
(d) Solvay process of sodium carbonate

136. Excess of Na^+ ions in human system causes:

(a) Anaemia
(b) High blood pressure
(c) Diabetes
(d) Low blood pressure

137. Crystalline sodium chloride is a bad conductor of electricity because:

(a) It contains free molecules
(b) The ions present in it are not free to move
(c) It does not possess ions
(d) It contains only molecules

138. Which one of the following reactions is not associated with the Solvay process of manufacture of sodium carbonate?

(a) $NaCl + NH_4HCO_3 \rightarrow NaHCO_3 + NH_4Cl$
(b) $2NaOH + CO_2 \rightarrow Na_2CO_3 + H_2O$
(c) $2NaHCO_3 \rightarrow Na_2CO_3 + H_2O + CO_2$
(d) $NH_3 + H_2CO_3 \rightarrow NH_4HCO_3$

139. KO_2 (potassium super oxide) is used in oxygen cylinders in space and submarines because it:

(a) Produces ozone
(b) Absorbs CO_2
(c) Absorbs CO_2 and increase O_2 contents
(d) Eliminates moisture

140. A metal M readily forms water soluble sulphate MSO_4, water insoluble hydroxide $M(OH)_2$ and oxide MO which becomes inert on heating. The hydroxide is soluble in NaOH. The M is:

(a) Mg (b) Sr
(c) Be (d) Ca

141. The metallic sodium dissolves in liquid ammonia to form a deep blue coloured solution. The deep blue colour is due to formation of:

(a) Solvated electron, $e(NH_3)x^-$
(b) Solvated atomic sodium, $Na(NH_3)y$
(c) $[Na^+ + Na^-]$
(d) $NaNH_2 + H_2$

142. Identify the correct order of acidic strengths of CO_2, CuO, CaO, H_2O:

(a) $CaO < H_2O < CuO < CO_2$
(b) $H_2O < CO_2 < CaO < CuO$
(c) $H_2O < CuO < CaO < CO_2$
(d) $CaO < CuO < H_2O < CO_2$

143. Carnallite on electrolysis gives:

(a) Na and CO_2 (b) Mg and Cl_2
(c) Al and Cl_2 (d) Ca and Cl_2

144. The substance not likely to contain $CaCO_3$ is:

(a) Sea shells (b) Calcined gypsum
(c) Dolomite (d) A marble statue

145. The paramagnetic species is:

(a) KO_2 (b) SiO_2
(c) TiO_2 (d) BaO_2

146. On dissolving moderate amount of sodium metal in liquid NH_3 at low temperature, which one of the following does not occur?

(a) Na^+ ions are formed in the solution.
(b) Liquid NH_3 solution remains diamagnetic.
(c) Liquid NH_3 solution becomes good conductor of electricity.
(d) Blue coloured solution is obtained.

147. Beryllium and aluminium exhibit many properties which are similar. But the two elements differ in:

(a) Forming polymeric hydrides
(b) Forming covalent halides
(c) Exhibiting maximum covalency in compounds
(d) Exhibiting amphoteric nature in their oxides.

148. One mole of magnesium nitride on the reaction with an excess of water gives:

(a) Two moles of ammonia
(b) Two moles of nitric acid
(c) One mole of ammonia
(d) One mole of nitric acid

149. Based on lattice energy and other considerations which one of the following alkali metal chlorides is expected to have the highest melting point?

(a) LiCl (b) NaCl
(c) KCl (d) RbCl

150. The following compounds have been arranged in the increasing solubilities. Identify the correct order.

1. $Be(OH)_2$ 2. $Mg(OH)_2$
3. $Ca(OH)_2$ 4. $Sr(OH)_2$
5. $Ba(OH)_2$

(a) $2 < 3 < 1 < 4 < 5$ (b) $1 > 2 > 3 > 4 > 5$
(c) $1 < 2 < 5 < 4 < 3$ (d) $1 < 2 < 3 < 4 < 5$

151. $MgSO_4$ on reaction with NH_4OH and Na_2HPO_4 forms a white crystalline precipitate. What is its formula?

(a) $Mg(NH_4)PO_4$ (b) $Mg_3(PO_4)_2$
(c) $MgCl_2.MgSO_4$ (d) $MgSO_4$

152. Among the alkali metals, cesium is the most reactive because:

(a) It has a single electron in the valence shell.
(b) Its incomplete shell is nearest to the nucleus.
(c) The outermost electron is more loosely bound than the outermost electron to the other alkali metals.
(d) It is the heaviest alkali metal.

153. The increasing order of ionic character of CsF, LiI, NaBr and KCl is:

(a) CaF < KCl < NaBr < LiI
(b) NaBr < KCl < LiI < CsF
(c) LiI < KCl < CsF < NaBr
(d) LiI < NaBr < KCl < CsF

154. A greenish yellow gas reacts with an alkali metal hydroxide to form a halate which can be used in fire works and safety matches. The gas and halate respectively are:

(a) Br_2, $KBrO_3$ (b) Cl_2, $KClO_3$
(c) I_2, $NaIO_3$ (d) Cl_2, $NaClO_3$

Practice Questions – III

155. Consider the following statements:

1. Cs^+ ion is more highly hydrated than other alkali metal ions.
2. Among the alkali metals Li, Na, K and Rb, Li has the higher melting point.
3. Among the alkali metals, only Li form a stable nitride by direct combination.

(a) 1 and 3 are correct
(b) 2 and 3 are correct
(c) 1, 2 and 3 are correct
(d) 1 and 2 are correct.

156. Alkali metals are characterized are:

1. Good conductor of heat and electricity
2. High oxidation potentials
3. High melting point
4. Solubility in liquid ammonia

(a) 1, 2 (b) 2, 3, 4
(c) 1, 2, 4 (d) all

157. $A + C + Cl_2 \xrightarrow[\text{Of about 1000 K}]{\text{High temperature}} B + CO$

$B + 2H_2O \rightarrow C + 2HCl$

Compound B is found in polymeric chain structure and is an electron deficient molecule. B must be:

(a) $BeCl_2$ (b) $BeO.Be(OH)_2$
(c) $Be(OH)_2$ (d) BeO

158. In the following sequence of reactions, identify the end product (D):

(1) $Na_2CO_3 \xrightarrow{SO_2}$ (B) $\xrightarrow{Na_2CO_3}$
(C) $\xrightarrow{\text{Element S, }\Delta}$ (D) $\xrightarrow{I_2}$

(a) Na_2SO_4 (b) $Na_2S_4O_6$
(c) Na_2S (d) $Na_2S_2O_3$

159. Identify the correct statements.

(1) Beryllium liberates H_2 from HCl or H_2SO_4.
(2) Beryllium is rendered passive on treatment with conc. HNO_3.
(3) $BeCl_2$ has a polymeric structure in the solid state but exists as a dimmer in the vapour state.
(4) $BeCl_2$ has a dimeric structure in the solid state but a polymeric structure in the vapour state.

(a) 1, 2 (b) 2, 3
(c) 1, 2, 3 (d) 2, 3, 4

160. Pick out the statement(s) which is(are) not true about the diagonal relationship of Li and Mg.

(1) Polarising powers of Li^+ and Mg^{2+} are almost same.
(2) Like Li, Mg decomposes water very fast.
(3) LiCl and $MgCl_2$ are deliquescent.
(4) Like Li, Mg readily reacts with liquid bromine at ordinary temperature.

(a) 1 only (b) 2 only
(c) 1 and 4 (d) 2 and 4

161. (P) + $H_2O \rightarrow NaOH$

(P) $\xrightarrow{O_2\text{, 400°C}}$ (Q) $\xrightarrow{H_2O\text{, at 25°C}}$ $NaOH + O_2$

(Q) is used for oxygenating in submarine. (P) and (Q) are:

(a) Na_2O_2 and O_2 (b) Na_2O and O_2
(c) Na_2O_2 and Na_2O (d) Na_2O and Na_2O_2

162. Highly pure dilute solution of sodium in liquid ammonia:

1. Shows blue colour.
2. Exhibits electrical conductivity.
3. Produces sodium amide.
4. Produces hydrogen gas.

(a) 1, 2 (b) 1, 2, 3
(c) 2, 3, 4 (d) all

163. Which of the following is incorrect?

(1) Sodium reduces CO_2 to carbon.
(2) In the Castner's process of sodium extraction, NaCl is used as an electrolyte.
(3) Magnalium is an alloy of Mg and Zn.
(4) Mg reacts with cold water and liberate hydrogen gas.

(a) 1, 2, 3 (b) 2, 3, 4
(c) 2, 3 (d) 1, 3, 4

164. The major role of fluorspar (CaF_2) which is added in small quantities in the electrolytic reduction of alumina dissolved in fused cryolite (Na_3AlF_6) is:

1. As a catalyst
2. To make the fused mixture very conducting
3. To lower the temperature of melt
4. To decrease the rate of oxidation of carbon at the anode

(a) 2, 3 (b) 1, 2
(c) 2, 3 ,4 (d) 3, 4

165. $CO_2 + NH_3 + H_2O \rightarrow X$

$X + NaCl \rightarrow Y + NH_4Cl$

$2Y \rightarrow Z + H_2O + CO_2$. 'Z' is

(a) Na_2CO_3 (b) $(NH_4)_2CO_3$
(c) NH_4HCO_3 (d) $NaHCO_3$

166. Out of the following metals that cannot be obtained by electrolysis of the aqueous solution of their salts are:

1. Ag 2. Mg
3. Cu 4. Al

(a) 2, 4 (b) 2, 3
(c) 1, 4 (d) All

167. Which of the following is correct?

(1) Mg burns in air releasing dazzling light rich in U.V. rays.
(2) $CaCl_2.6H_2O$ when mixed with ice gives freezing mixture.
(3) Mg cannot form complexes.
(4) Be can form complexes due to its very small size.

(a) 1, 2 (b) 2, 3
(c) 1, 2, 3 (d) 2, 3, 4

168. In the electrolysis of aqueous NaCl solution, side reactions taking place are:

(1) $2OH^- + Cl_2 \rightarrow 2OCl^- + H_2$
(2) $2Na + 2H_2O \rightarrow 2NaOH + H_2$
(3) $4OH^- \rightarrow O_2 + 2H_2 + 4e^-$

Select the correct alternate:

(a) 1 and 3 (b) 2 and 3
(c) 1 and 2 (d) 1, 2 and 3

169. $Rb[ICl_2]$ on heating gives:

(a) $RbI + Cl_2$
(b) $2Rb[ICl] + Cl_2$ (from two molecules)
(c) $RbCl + ICl$
(d) $2Rb + 2Cl_2 + I_2$ (from two molecules)

170. Which of the following pairs can be distinguished by the action of heat?

(1) Na_2CO_3 and $CaCO_3$
(2) $MgCl_2.6H_2O$ and $CaCl_2.6H_2O$
(3) $Ca(NO_3)_2$ and $NaNO_3$

(a) (1) and (2) (b) (1), (2) and (3)
(c) (1) and (3) (d) (1) only

171. Following statements regarding the periodic trends of chemical reactivity of the alkali metals and the halogens are given. Which of these statements give the incorrect picture?

(1) The reactivity decreases in the alkali metals but increases in the halogens with increase in atomic number down the group.
(2) In both the alkali metals and the halogens the chemical reactivity decreases with increase in atomic number down the group.
(3) Chemical reactivity increases with increase in atomic number down the group in both the alkali metals and halogens.
(4) In alkali metals the reactivity increases but in the halogens it decreases with increase in atomic number down the group.

(a) 1, 2 (b) 2, 3
(c) 1, 2, 3 (d) 2, 3, 4

172. Which of the following reactions are correct?

(1) $B + NaOH \rightarrow 2Na_3BO_3 + H_2$
(2) $P_4 + NaOH + H_2O \rightarrow NaH_2PO_2 + PH_3$
(3) $S + NaOH \rightarrow Na_2S_2O_3 + Na_2S + H_2O$

(a) 1 only (b) 3 only
(c) 2 and 3 (d) 1, 2 and 3

173. The pair of compounds which cannot exist together in aqueous solution is:

1. NaH_2PO_4 and Na_2HCO_3
2. Na_2CO_3 and $NaHCO_3$
3. NaOH and NaH_2PO_4
4. $NaHCO_3$ and NaOH

(a) 1, 2, 3 (b) 3, 4
(c) 1, 4 (d) 2, 3

174. Highly pure dilute solution of sodium in liquid ammonia:

1. Shows blue colour
2. Exhibits electrical conductivity
3. Produces sodium amide
4. Produces hydrogen gas

(a) 1, 2 (b) 1, 2, 3
(c) 2, 3, 4 (d) all

175. Consider the following reactions:

$$X + HCl \xrightarrow[\text{(addition)}]{\text{anhyd. } AlCl_3} C_2H_5Cl \xrightarrow[\text{(substitution)}]{\text{anychd. } ZnCl_2\text{ /HCl}} Y$$

Y can be converted to X on heating with …….. at ……… temperature.

(a) Cu, 300°C (b) Al_2O_3, 350°C
(c) NaOH / I2, 60°C (d) $Ca(OH)_2 + CaOCl_2$, 60°C

176. Which of the following elements occur as their chlorides in earth crust?

(1) Na (2) Cu
(3) Ag (4) K
(a) 1, 2 (b) 2, 3, 4
(c) 1, 4 (d) 1, 3, 4

177. A deliquescent white crystalline hydroxide X reacts with a nitrate Y to form another hydroxide which decomposes to give a insoluble brown layer of its oxide. X is a powerful cautery and breaks down then proteins of skin flesh to a pasty mass. X and Y are:

(a) NaOH, $Zn(NO_3)_2$ (b) NaOH, $AgNO_3$
(c) $Ca(OH)_2$.$HgNO_3$ (d) NaOH, $Al(NO_3)_3$

178. In the reaction: $Al_2(SO_4)_3.18H_2O \xrightarrow[-18H_2O]{\text{heat}} A \xrightarrow{800°C} B + C$. The product A, B and C are respectively:

(a) $Al_2(SO_4)_3$, Al_2O_3, SO_3
(b) Al_2O_3, $Al_2(SO_4)_3$, SO_3
(c) Al_2SO_4, Al_2O_3, SO_3
(d) $Al_2(SO_4)_3$, Al_2O_3, SO_2

179. Consider the following reactions:

1. $2Na + 2H_2O \rightarrow 2NaOH + H_2$
2. $2NaOH + Cl_2 \rightarrow NaCl + NaOCl + H_2O$
3. $4OH^- \rightarrow O_2 + 2H_2O + 4e^-$
4. $2Cl^- \rightarrow Cl_2 + 2e^-$

In the diaphragm cell used for the electrolysis of brine, the reactions that occur would include

(a) 2, 3, 4 (b) 1, 3, 4
(c) 1, 2, 3 (d) 1, 2, 4

180. Metal X on heating in nitrogen gas gives Y. Y on treatment with H_2O gives a colourless gas which when passed through $CuSO_4$ solution gives a blue colour. Y is:

(a) MgO (b) $Mg(NO_3)_2$
(c) Mg_3N_2 (d) NH_3

Answer Keys

1. (d)	2. (b)	3. (d)	4. (b)	5. (b)	6. (a)	7. (b)	8. (a)	9. (c)	10. (c)
11. (a)	12. (c)	13. (b)	14. (c)	15. (d)	16. (a)	17. (b)	18. (c)	19. (b)	20. (b)
21. (a)	22. (a)	23. (a)	24. (c)	25. (b)	26. (b)	27. (a)	28. (a)	29. (c)	30. (b)
31. (c)	32. (c)	33. (a)	34. (c)	35. (d)	36. (b)	37. (d)	38. (a)	39. (b)	40. (b)
41. (a)	42. (c)	43. (c)	44. (d)	45. (b)	46. (a)	47. (c)	48. (b)	49. (c)	50. (b)
51. (d)	52. (d)	53. (d)	54. (b)	55. (d)	56. (c)	57. (b)	58. (a)	59. (a)	60. (b)
61. (d)	62. (a)	63. (a)	64. (d)	65. (a)	66. (b)	67. (b)	68. (a)	69. (b)	70. (c)
71. (c)	72. (b)	73. (d)	74. (b)	75. (d)	76. (b)	77. (b)	78. (a)	79. (b)	80. (d)
81. (d)	82. (c)	83. (a)	84. (b)	85. (b)	86. (c)	87. (a)	88. (c)	89. (a)	90. (b)
91. (a)	92. (d)	93. (b)	94. (c)	95. (c)	96. (c)	97. (b)	98. (a)	99. (b)	100. (a)
101. (c)	102. (b)	103. (a)	104. (d)	105. (a)	106. (b)	107. (d)	108. (a)	109. (c)	110. (b)
111. (a)	112. (c)	113. (a)	114. (c)	115. (b)	116. (a)	117. (d)	118. (b)	119. (c)	120. (a)
121. (b)	122. (b)	123. (d)	124. (c)	125. (b)	126. (a)	127. (d)	128. (c)	129. (b)	130. (a)
131. (a)	132. (c)	133. (b)	134. (a)	135. (b)	136. (b)	137. (b)	138. (b)	139. (c)	140. (c)
141. (a)	142. (d)	143. (b)	144. (b)	145. (a)	146. (b)	147. (a)	148. (c)	149. (b)	150. (d)
151. (a)	152. (c)	153. (d)	154. (b)	155. (b)	156. (c)	157. (a)	158. (b)	159. (b)	160. (d)
161. (d)	162. (a)	163. (b)	164. (a)	165. (a)	166. (a)	167. (c)	168. (a)	169. (c)	170. (b)
171. (c)	172. (d)	173. (b)	174. (a)	175. (b)	176. (d)	177. (b)	178. (a)	179. (b)	180. (c)

Hints and Explanations for Selective Questions

1. Sodium amalgam is an alloy of Na and Hg. It is a mixture so sodium is in metallic state. Hence its oxidation state is zero.

2. The number of electrons in $K^{-1} = 19 + 1 = 20$

3. It is a reducing agent because of the presence of sodium.

4. Alkali metals because of their larger sizes have lowest values of ionization potential.

7. Lithium and magnesium shows diagonal relationship due to same electronegativity.

8. Li due to highest hydration energy among the alkali metals is strongest reducing agent.

9. $6Li + N_2 \rightarrow 2\ Li_3N$.

10. Since Be has two valence electrons, so it forms only two covalent bonds.

13. Alkaline earth metals act as strong reducing agents.

15. Sodium in excess of oxygen forms Na_2O_2.

18. fter neon the next electron enter into 4s orbital which is lower in energy than 3d orbital.

19. IE_2 of Mg is maximum.

20. Main ingredients are NH_3; NaCl solution and CO_2 only.

23. $Ca(OH)_2 + Cl_2 \rightarrow \underset{\text{Bleaching powder}}{CaOCl_2} + H_2O$

Bleaching powder is a mixed salt of calcium hypochlorite, $Ca(OCl)_2$ and basic calcium chloride, $CaCl_2.Ca(OH)_2.H_2O$.

25. ns^1 electronic configuration represents alkali metals i.e., IA group, alkali metals are highly electropositive.

26. Na_2CO_3 is a salt of weak acid and strong base. So its aqueous solution is alkaline due to hydrolysis of CO_3^{2-} ions.

27. In ammonia soda process, the by products obtained are $CaCl_2$, CO_2 and NH_3.

28. KCl $MgCl_2$ $6H_2O \rightarrow K^+, Mg^{2+}$, 3Cl So K^+, Mg^{2+}, Cl^- ions give positive tests.

30. Magnesium wire reduce CO_2 to C

$2Mg + CO_2 \rightarrow 2MgO + C$

31. In $BeCl_2$ the Be atom has only four electrons around it.

32. IA group elements are good reducing agents.

35. Mixture of calcium dihydrogen phosphate and gypsum is known as calcium super phosphate.

37. Sodium dissolves in liquid ammonia to give dark blue solution.

40. Plaster of paris, $CaSO_4 . \frac{1}{2} H_2O$ is used in dentistry and surgery.

49. $MgCl_2$ hydrolyses to give HCl. So the pH of the solution will be < 7.

56. Li^+ because of greater effective nuclear charge has the smallest size.

59. Na_2O_2 and HCl mixture is called oxone.

65. Dead burnt plaster is anhydrous $CaSO_4$

69. $Na_2CO_3.10H_2O \xrightarrow[-9H_2O]{\Delta} Na_2CO_3.H_2O$

$\xrightarrow{\Delta} Na_2CO_3 + H_2O$

71. Carnalite is $KCl.MgCl_2.6H_2O$. it is an ore of Mg.

76. $2Na + 2NH_3 \xrightarrow{\Delta} 2NaNH_2 + H_2$

77. $2NaNO_3 \xrightarrow{\Delta} 2NaNO_2 + O_2$

85. $BaSO_4 + Na_2CO_3 \rightarrow BaCO_3 + Na_2SO_4$

86. Mg^{2+} has smaller size than Na_+ and thus has higher hydration energy than Na_+.

89. Alkali metals always show monovalency in their compounds.

94. Li^+ ion due to excessive hydration has lowest ionic mobility in aqueous solution.

99. Carnalite ($MgCl_2$ $KCl.6H_2O$) is an ore of Mg which on electrolysis gives Mg at the cathode and Cl_2 at the anode.

102. The ratio of charge to size i.e., Polarising power is nearly the same for lithium and magnesium.

103. As the basicity of metal hydroxide increases down the group from Be to Ba, the thermal stability of their carbonates also increases in the same order, since the carbonates can be regarded as salts of weak carbonic acids with the metal hydroxides. $BeCO_3$ decomposes at 373 K and $BaCO_3$ decomposes at 1633 K.

106. 'A' & 'B' are CO_2 and CO_2.

$Ca(OH)_2 \xrightarrow[\Delta]{(A)\ CO_2} CaCO_3 \downarrow \xrightarrow[H_2O]{CO_2}$

$Ca(HCO_3)_2 \xrightarrow{\Delta} CaCO_3 + CO_2 \uparrow$

(soluble) (ppt.) (B)

107. $KF + HF \rightarrow K+ [HF_2]^-$.

In HF_2^- there exists a hydrogen bonding such as F^- ….. H – F.

108. $K_2O + H_2O \rightarrow 2KOH$.

109. $CsBr_3$ contains Cs+ and Br_3^- ($Br^- + Br_2$) ions.

113. Mg and Al are more electropositive than hydrogen and hence can not be obtained by electrolysis of the aqueous solution of their salts.

114. Calcium oxalate does not dissolve in acetic acid. All others [CaO, $CaCO_3$ and $Ca(OH)_2$] are bases and hence dissolve in acetic acid.

116. When H.E. > L.E. the salt is soluble in water.

117. H^+ ions are discharged at a higher potential when Hg cathode is used than that using Pt cathode.

118. These loose electrons absorb radiation and then radiate it back.

119. H^+ has greater mobility than Na^+, because of its smaller size as compared with Na^+, therefore electrode potential (or reduction potential) of H^+ (0.00 V) is greater than that of Na^+ (-2.71 V). hence H+ is reduced in preference to Na^+.

Cathode:

$2H^+ (aq) + 2e^- \rightarrow H_2 (g)$

120. $Na_2Al_2Si_2O_8.xH_2O$ is written as Na_2Z, Na^+ ions exchange Ca^+ and Mg^{2+} ions of hard water.

$Na_2Z + Mg_2^+ \rightarrow 2Na^+ + MgZ$

$Na_2Z + Ca_2^+ \rightarrow 2Na^+ + CaZ$

121. $2Na + 2NH_3 \xrightarrow{\Delta} 2NaNH_2 + H_2$

122. In NaCl (rock salt) structure, Cl^- ions form fcc type of lattice.

123. Halides of group 2 elements are hygroscopic and act as dehydrating agents.

124. In Clarke's process quick lime is added to hard water which produces $Ca(OH)_2$ which further converts bicarbonates into insoluble carbonates.

126. $Ca(OH)_2 + Cl_2 \rightarrow CaOCl_2 + H_2O$

Bleaching powder

Bleaching powder is a mixed salt of calcium hypochlorite, $Ca(OCl)_2$ and basic calcium chloride, $CaCl_2 . Ca(OH)_2 . H_2O$.

127. Halides of group 2 elements are hygroscopic and act as dehydrating agents.

130. Fluorspar is added in small quantities in the electrolytic reduction of alumina dissolved in fused cryolite (Na_3AlF_6)

(1) To make the fused mixture more conducting so alumina in a bad conductor of electricity.

(2) To lower the m.p. of fused mixture to 1140 K, alumina has a high m.p. of 2323 K.

131. $Na + (x + y) NH_3 \rightarrow Na(NH_3)x^+ + e\,(NH_3)^-_y$.

Thus, due to solvated (ammoniated) electron, solution of Na metal in liquid ammonia is highly reducing in nature.

132. $\underset{(X)}{3Mg + N_2} \xrightarrow{\Delta} \underset{(Y)}{Mg_3N_2}$

$\underset{(Y)}{Mg_3N_2} + 6H_2O \rightarrow 3Mg(OH)_2 + \underset{\text{colourless}}{2NH_3}$

$CuSO_4 + 4NH_3 \rightarrow \underset{\text{Blue complex}}{[Cu(NH_3)_4]SO_4}$

134. $CaO + 3C \xrightarrow{3300K} CaC_2 + CO$

135. Limestone ($CaCO_3$) is not required in the manufacture of P from phosphorite.

136. High blood pressure is caused by taking excess of Na^+ ions in the diet.

137. In NaCl crystal, Na^+ and Cl^- ions due to strong attractive forces are not free to move.

138. $2NaOH + CO_2 \rightarrow Na_2CO_3 + H_2O$

This reaction is not associated with solvay process.

139. $4KO_2 + 2CO_2 \rightarrow 2K_2CO_3 + 3O_2$

140. M is Be which forms water soluble $BeSO_4$, water insoluble $Be(OH)_2$ and $BeO.Be(OH)_2$ is soluble in NaOH giving sodium beryllate Na_2BeO_2.

143. Carnalite ($MgCl_2$ KCl.$6H_2O$) is an ore of Mg which on electrolysis gives Mg at the cathode and Cl_2 at the anode.

144. Calcined gypsum does not contain $CaCO_3$ where as all other materials contain $CaCO_3$ in them.

145. As in KO_2, O_2 is in O_2^- form so have one unpaired electron.

$O_2^-\,(17e^-) \rightarrow \sigma 1s^2\, \sigma^* 1s^2\, \sigma 2s^2\, \sigma^* 2s^2,\, \sigma 2p_z^2\, \pi 2px^2 = \pi 2py^2,\, \pi^*\, 2px^2 = \pi^*\, 2py^1$

146. As on dissolving moderate amount of Na metal in liquid NH_3, the solution is blue in colour and paramagnetic but not diamagnetic.

147. Be (Z = 4) has maximum covalency of 4 while Al (Z = 13) has maximum covalency of 6.

148. $\underset{1\text{ mol}}{Mg_3N_2(s)} + 6H_2O(l) \rightarrow 3Mg(OH)_2 + \underset{2\text{ mol}}{2NH_3(g)}$

149. In case of halides of alkali metals, melting point decreases going down the group because lattice enthalpies decreases as size of alkali metal increases. But LiCl has lower melting point in comparison to NaCl due to covalent nature. So, NaCl is expected to have the highest melting point among given halides.

150. If the OH^- ions and group 2 divalent cations are of comparable size, the lattice energy of the group 2 hydroxides decreases more rapidly than hydration energy. Rapidly decreasing lattice energy favours increased solubility.

151. Test of Mg^{2+} ion

$Mg^{2+} + NH_4OH + Na_2HPO_4 \rightarrow Mg(NH_4)PO_4$

154. The gas is Cl_2 and the halate used in fire works and safety matches is $KClO_3$.

$\underset{\text{greenish yellow gas}}{3Cl_2} + 6KOH \rightarrow \underset{\text{pot. chlorate (Halate)}}{KClO_3} + 5KCl + 3H_2O$

156. Melting point of alkali metals are lower.

157. $\underset{(A)}{BeO + C + Cl_2} \xrightarrow{\text{About 1000K}} \underset{(B)}{BeCl_2} + CO$

$\underset{(B)}{BeCl_2} + 2H_2O \longrightarrow \underset{(C)}{Be(OH)_2} + 2HCl$

$BeCl_2$ is found in polymeric chains due to it's electron deficiency.

158. $Na_2CO_3 + SO_2\ 2NaHSO_3 + CO_2$

(A)

$2NaHSO_3 + Na_2CO_3 \rightarrow$

(A)

$2Na\ 2SO_3 + H_2O + CO_2$

(B)

$Na\ 2SO_3 + S \xrightarrow{\Delta} Na\ 2S_2O_3$

(B) (C)

$2Na_2S_2O_3 + I_2 \rightarrow Na_2S_4O_6 + 2NaI$

(D)

159. Beryllium does not liberate H_2 from HCl or H_2SO_4. Concentrated. HNO_3 renders Be passive due to the formation of oxide film on the surface. In the solid state $BeCl_2$ has polymeric structure but in vapour state, dimmer is present.

162. Highly pure dilute solution of sodium in liquid ammonia is blue in colour due to ammoniated electron. This solution is conducting due to both ammoniated cation and ammoniated electron.

$Na^+ (x + y) NH_3 \rightarrow [Na(NH_3)x]^+ + [e (NH_3)y]^-$

164. Fluorspar is added in small quantities in the electrolytic reduction of alumina dissolved in fused cryolite (Na_3AlF_6)

(1) To make the fused mixture more conducting so alumina in a bad conductor of electricity.

(2) To lower the m.p. of fused mixture to 1140 K, alumina has a high m.p. of 2323 K.

165. $CO_2 + NH_3 + H_2O \rightarrow NH_4HCO_3$ (X)

$NH_4HCO_3 + NaCl \rightarrow NaHCO_3 + NH_4Cl$

(Y)

$2NaHCO_3 \rightarrow Na_2CO_3$ (Z) $+ H_2O + CO_2$

166. Mg and Al are more electropositive than hydrogen and hence can not be obtained by electrolysis of the aqueous solution of their salts.

169. As the electronegativity difference increases stability of ionic lattice increases so,

$Rb[ICl_2] \xrightarrow{\Delta} RbCl + ICl$

170. (1) $Na_2CO_3 \xrightarrow{\Delta}$ no effect

$CaCO_3 \xrightarrow{\Delta} CaO + CO_2$

(2) $MgCl_2.6H_2O \xrightarrow{\Delta} Mg(OH)Cl.5H_2O + HCl$

$CaCl_2.6H_2O \xrightarrow{\Delta} 2CaCl_2 + 6H_2O$

(3) $2Ca(NO_3)_2 \xrightarrow{\Delta} 2CaO + 2NO_2 + 3O_2$

$2NaNO_3 \xrightarrow{\Delta} 2NaNO_2 + O_2$

173. Acidic and basic salts do not exist together in aqueous solution because they react.

$2NaOH + NaH_2PO_4 \rightarrow Na_3PO_4 + 2H_2O$

$NaHCO_3 + NaOH \rightarrow Na_2CO_3 + H_2O$

174. Highly pure dilute solution of sodium in liquid ammonia is blue in colour due to ammoniated electron. This solution is conducting due to both ammoniated cation and ammoniated electron.

$Na + (x + y) NH_3 \rightarrow [Na(NH_3)x]^+ + [e (NH_3)y]^-$

176. NaCl (Rock salt), KCl (Sylvine), AgCl (Horn silver).

178. $Al_2(SO_4)_3.18H_2O \xrightarrow[-18H_2O]{Heat} Al_2(SO_4)_3 \xrightarrow{800^\circ C} Al_2O_3 + 3SO_3$

B C

180. $3Mg + N_2 \xrightarrow{\Delta} Mg_3N_2$

(XI) (Y)

$Mg_3N_2 + 6H_2O \rightarrow 3Mg(OH)_2 + 2NH_3$

(Y) colourless

$CuSO_4 + 4NH_3 \rightarrow [Cu(NH_3)_4]SO_4$

Blue complex

PREVIOUS YEARS' QUESTIONS

1. Which one of the following statements about the zeolite is false?

(a) They have open structure which enables them to take up small molecules.
(b) They are used as cation exchangers
(c) Zeolites are aluminosilicates having three dimensional network.
(d) None of the SiO_4^{4-} units are replaced by AlO_4^{5-} and AlO_6^{9-} ions in zeolites

[2004]

2. The correct order of the mobility of the alkali metal ions in aqueous solution is:

(a) $K^+ > Rb^+ > Na^+ > Li^+$
(b) $Rb^+ > K^+ > Na^+ > Li^+$
(c) $Li^+ > Na^+ > K^+ > Rb^+$
(d) $Na^+ > K^+ > Rb^+ > Li^+$

[2006]

3. The correct order of increasing thermal stability of K_2CO_3, $MgCO_3$, $CaCO_3$ and $BeCO_3$ is:

(a) $BeCO_3 < MgCO_3 < K_2CO_3 < CaCO_3$
(b) $BeCO_3 < MgCO_3 < CaCO_3 < K_2CO_3$
(c) $MgCO_3 < BeCO_3 < CaCO_3 < K_2CO_3$
(d) $K_2CO_3 < MgCO_3 < CaCO_3 < BeCO_3$

[2007]

4. In which of the following the hydration energy is higher than the lattice energy?

(a) $BaSO_4$ (b) $MgSO_4$
(c) $RaSO_4$ (d) $SrSO_4$

[2007]

5. The alkali metals form salt-like hydrides by the direct synthesis at elevated temperature. the thermal stability of these hydrides decreases in which of the following orders?

(a) NaH > LiH > KH > RbH > CsH
(b) LiH > NaH > KH > RbH > CsH
(c) CsH > RbH > KH > NaH > LiH
(d) KH > NaH > LiH > CsH > RbH

[2008]

6. In the case of alkali metals, the covalent character decreases in the order :

(a) MI > MBr > MCl > MF
(b) MCl > MI > MBr > MF
(c) MF > MCl > MBr > MI
(d) MF > MCl > MI > MBr

[2009]

7. Among the following which one has the highest cation to onion size ratio?.

(a) CsI (b) CsF
(c) LiF (d) NaF

[2010]

8. Which of the following alkaline earth metal sulphate has hydration energy higher than the latice energy?

(a) $SrSO_4$ (b) $CaSO_4$
(c) $BeSO_4$ (d) $BeSO_4$

[2010]

9. Which of the following compound has the lowest melting point?

(a) $CaCl_2$ (b) CaF_2
(c) CaI_2 (d) $CaBr_2$

[2011]

10. Which of the following statement is incorrect?

(a) Al reacts with excess of NaOH to give $Al(OH)_3$
(b) $NaHCO_3$ on heating gives Na_2CO_3.
(c) Pure Na-metal dissoives in liq NH_3 to give blue solutions
(d) NaOH reacts with glass to give sodium silicate.

[2011]

11. The function of sodium pump is a biological procers operating in each and every cell of all animal. Which of the following biologically important ions is also a constituent of this pump?

(a) Mg^{2+} (b) K^+
(c) Ca^{2+} (d) Fe^{2+}

[2015]

12. Solubility of the alkaline earth's metal sulphates in water decreases in the sequence?

(a) Ca>Sr>Ba>Mg (b) Ba>Mg>Sr>Ca
(c) Sr>Ca>Mg>Ba (d) Mg>Ca>Sr>Ba

[2015]

13 On heating which of the following releases CO_2 most easily?

(a) KCO_3^+ (b) $NaCO_3$
(c) $MgCO_3$ (d) $CaCO_3$

[2015]

14. Which of the following statements is false?
(a) Mg^{2+} forms a complex with ATP
(b) Mg^{2+} ions are important in the green parts of plants
(c) Ca^{2+} ions are important in blood clotting
(d) Ca^{2+} ions are not important in maintaing the heart.
[2016]

15. In context with beryllium, which one of the following statements is incorrect?
(a) It forms Be_2C (b) Its salts rarely hydrolyse
(c) It is rendered passive by nitric acid
(d) Its hydride is electron and polymeric
[2016]

Answer Keys

1. (d)	2. (b)	3. (b)	4. (b)	5. (b)	6. (a)	7. (b)	8. (c)	9. (c)	10. (a)
11. (b)	12. (d)	13. (c)	14. (d)	15. (b)					

Hints And Explanations

1. Zeolites are aluminosilicates having three dimensional open structure in which four or six membered rings predominates. So due to open chain structure, they have cavities and can take up water and other? small molecules and in them SiO_4^{4-} units are replaced by AlO_4^{5-} and AlO_6^{9-} ions.

2. As ionic mobility in aqueous solution decrease with decrease in size so the decreasing order of ionic mobility is:
$Rb^+ > K^+ > Na^+ > Li^+$

3. Since thermal stability of IA is more than IIA and on moving down the group thermal stability decreases, so the correct increasing order of thermal stability is as follows:
$BeCO_3 < MgCO_3 < CaCO_3 < K_2CO_3$

4. As Mg_2^+ ion is smallest in size so have maximum hydration energy here.
Hydration energy α 1 / Size

5. The ionic character of the bonds in hydrides increases from LiH to CsH so thermal stability of these hydrides decreases as follows:
LiH > NaH > KH > RbH > CsH

6. Bigger the anion (I^- ion), larger is the covalent character (Fajan's rules). Hence, the covalent character of the alkali metal halides decreases as given in option (a).

7. Since Cs^+ is largest cation and F^- is the smallest anion.

8. $BeSO_4$ has more hydration energy than lattice energy.

9. Here CaI_2 has the Lowest M.P.

10. As here final Product is Na AlO_2 that is sodium meta aluminate.

11. Na^+ and K^+ are constituent ions of sodium pump.

12. Solubility of II gp. Sulphates decreases down the group as ΔH Decreases from Be^{2+} to Ba^{2+}.

13. Thermal stubility of Co_3^{2-} of IA> IIA it also (↑) down the group so.
$K_2CO_3 > NaCO_3 > CaCO_3 > MgCO_3$
$MgCO_3 \rightarrow MgO + CO_2$ least stable

14. As Ca^{2+} are very important for maintaining heart beat of the heart.

15. Due to very small size of Be^{2+} its salts are easily hydrolysed due to higher hydration energy.
$BeCl_2 + 2H_2O \rightarrow Be(OH_2) + 2HCl$

NCERT EXEMPLAR

1. The alkali metals are low melting. Which of the following alkali metal is expected to melt if the room temperature rises to 30°C?
   (a) Na (b) K
   (c) Rb (d) Cs

2. Alkali metals react with water vigorously to form hydroxides and dihydrogen. Which of the following alkali metals reacts with water least vigorously?
   (a) Li (b) Na
   (c) K (d) Cs

3. The reducing power of a metal depends on various factors. Suggest the factor which makes Li, the strongest reducing agent in aqueous solution.
   (a) Sublimation enthalpy
   (b) Ionisation enthalpy
   (c) Hydration enthalpy
   (d) Electron-gain enthalpy

4. Metal carbonates decompose on heating to give metal oxide and carbon dioxide. Which of the metal carbonates is most stable thermally?
   (a) $MgCO_3$ (b) $CaCO_3$
   (c) $SrCO_3$ (d) $BaCO_3$

5. Which of the carbonates given below is unstable in air and is kept in CO_2 atmosphere to avoid decomposition.
   (a) $BeCO_3$ (b) $MgCO_3$
   (c) $CaCO_3$ (d) $BaCO_3$

6. Metals form basic hydroxides. Which of the following metal hydroxide is the least basic?
   (a) $Mg(OH)_2$ (b) $Ca(OH)_2$
   (c) $Sr(OH)_2$ (d) $Ba(OH)_2$

7. Some of the Group 2 metal halides are covalent and soluble in organic solvents. Among the following metal halides, the one which is soluble in ethanol is:
   (a) $BeCl_2$ (b) $MgCl_2$
   (c) $CaCl_2$ (d) $SrCl_2$

8. The order of decreasing ionisation enthalpy in alkali metals is:
   (a) Na > Li > K > Rb
   (b) Rb < Na < K < Li
   (c) Li > Na > K > Rb
   (d) K < Li < Na < Rb

9. The solubility of metal halides depends on their nature, lattice enthalpy and hydration enthalpy of the individual ions. Amongst fluorides of alkali metals, the lowest solubility of LiF in water is due to:
   (a) Ionic nature of lithium fluoride
   (b) High lattice enthalpy
   (c) High hydration enthalpy for lithium ion.
   (d) Low ionisation enthalpy of lithium atom

10. Amphoteric hydroxides react with both alkalies and acids. Which of the following Group 2 metal hydroxides is soluble in sodium hydroxide?
    (a) $Be(OH)_2$ (b) $Mg(OH)_2$
    (c) $Ca(OH)_2$ (d) $Ba(OH)_2$

11. In the synthesis of sodium carbonate, the recovery of ammonia is done by treating NH_4Cl with $Ca(OH)_2$. The by-product obtained in this process is:
    (a) $CaCl_2$ (b) NaCl
    (c) NaOH (d) $NaHCO_3$

12. When sodium is dissolved in liquid ammonia, a solution of deep blue colour is obtained. The colour of the solution is due to:
    (a) Ammoniated electron
    (b) Sodium ion
    (c) Sodium amide
    (d) Ammoniated sodium ion

13. By adding gypsum to cement:
    (a) Setting time of cement becomes less.
    (b) Setting time of cement increases.
    (c) Colour of cement becomes light.
    (d) Shining surface is obtained.

14. Dead burnt plaster is:
    (a) $CaSO_4 1$ (b) $CaSO_4 . 2\,H_2O$
    (c) $CaSO_4.H_2O$ (d) $CaSO_4.2H_2O$

15. Suspension of slaked lime in water is known as:
    (a) Lime water
    (b) Quick lime
    (c) Milk of lime
    (d) Aqueous solution of slaked lime

16. Which of the following elements does not form hydride by direct heating with dihydrogen?
    (a) Be (b) Mg
    (c) Sr (d) Ba

17. The formula of soda ash is:
(a) $Na_2CO_3.10H_2O$ (b) $Na_2CO_3.2H_2O$
(c) $Na_2CO_3.H_2O$ (d) Na_2CO_3

18. A substance which gives brick red flame and breaks down on heating to give oxygen and a brown gas is
(a) Magnesium nitrate
(b) Calcium nitrate
(c) Barium nitrate
(d) Strontium nitrate

19. Which of the following statements is true about $Ca(OH)_2$?
(a) It is used in the preparation of bleaching powder
(b) It is a light blue solid
(c) It does not possess disinfectant property.
(d) It is used in the manufacture of cement.

20. A chemical A is used for the preparation of washing soda to recover ammonia. When CO_2 is bubbled through an aqueous solution of A, the solution turns milky. It is used in white washing due to disinfectant nature. What is the chemical formula of A?
(a) $Ca\ (HCO_3)_2$ (b) CaO
(c) $Ca(OH)_2$ (d) $CaCO_3$

21. Dehydration of hydrates of halides of calcium, barium and strontium i.e., $CaCl_2 6H_2O$, $BaCl_2.2H_2O$, $SrCl_2.2H_2O$, can be achieved by heating. These become wet on keeping in air. Which of the following statements is correct about these halides?
(a) Act as dehydrating agent
(b) Can absorb moisture from air
(c) Tendency to form hydrate decreases from calcium to barium
(d) All of the above

Answer Keys

1. (d)	2. (a)	3. (c)	4. (d)	5. (a)	6. (a)	7. (a)	8. (c)	9. (b)	10. (a)
11. (a)	12. (a)	13. (b)	14. (a)	15. (c)	16. (a)	17. (d)	18. (b)	19. (a)	20. (c)
21. (d)									

Hints and Explanations for Selective Questions

1. Melting point decreases with the strength of metallic bonding decreases with increasing size of atom.

4. Thermal stability of metal carbonates increases as the electropositive character of metal increases. As basicity of metal hydroxide increases from $Be(OH)_2$ to $Ba(OH)_2$.

7. $BeCl_2$ is covalent and hence most soluble due to its small size. High electronegativity and high ionization enthalpy.

13. Setting of cement is an exothermic process and involves hydration of calcium aluminates and silicates.

16. Be has high ionization enthalpy and small size.

21. Extent of hydration decreases from Mg to Ba. Chlorides of alkaline earth metals are hydrated salts.

AIIMS ESSENTIALS

Assertion and Reason

In the following questions, two statements (Assertion) A and Reason (R) are given. Mark

(a) If A and R both are correct and R is the correct explanation of A;

(b) If A and R both are correct but R is not the correct explanation of A;

(c) A is true but R is false;

(d) A is false but R is true.

(e) Both A and R are false.

1. (A) Sodium metal can not be obtained by the electrolysis is of its salt in aqueous solution.

(R) Sodium is above hydrogen in electro chemical series and it reacts with water to produce sodium hydroxide and hydrogen.

2. (A) LiF is practically insoluble in water.

(R) LiF has very high lattice energy.

3. (A) $BaSO_4$ is highly insoluble in water but readily dissolves in a solution of the sodium salt of ethylene diamine tetra acetic acid.

(R) Ba^{2+} forms a very stable complex with the anion of the sodium salt of ethylene diamine tetra acetic acid which is soluble in water

4. (A) $Al(OH)_3$ is amphoteric in nature. (R) Al – O and O – H bonds can be broken with equal ease in $Al(OH)_3$

5. (A) Among the alkali metals, lithium salts exhibit the least electrical conductance in aqueous solutions

(R) Smaller the radius of the hydrated cation, lower is the electrical conductance in aqueous solution

6. (A) LiCl is predominantly a covalent compound.

(R) Electronegativity difference between Li and Cl is **too** small.

7. (A) The stability of alkali metal peroxides increases with increasing atomic number

(R) Bigger cations form more more stable lattice with bigger anions.

8. (A) $AlCl_3$ forms dimmer Al_2Cl_6 but it dissolves in water forming $[Al(H_2O)_6]_3^+$ and Cl^- ions.

(R) Aqueous solution of $AlCl_3$ is acidic due to hydrolysis.

9. (A) Sodium ions are discharged in preference to hydrogen ions at a mercury cathode.

(R) The nature of cathode can affect the order of discharge of cations.

10. (A) Mg gets oxidized when heating in CO_2 atmosphere

(R) Mg has a strong affinity for oxygen.

11. (A) Anhydrous magnesium chloride cannot be obtained by heating $MgCl_2.6H_2O$;

(R) $MgCl_2.6H_2O$ is highly stable and is unaffected by heat.

12. (A) $BaCO_3$ is more soluble in HNO_3 than in plain water.

(R) Carbonate is a weak base and reacts with the H^+ from the strong acid causing the barium salt to dissociate.

13. (A) The alkali metals can form ionic hydrides which contains the hydride ion.

(R) The alkali metals have low electronegativity, their hydrides conduct electricity when fused and liberate hydrogen at the anode.

14. (A) Sodium reacts with oxygen to form Na_2O_2 whereas potassium reacts with oxygen to form KO_2

(R) Potassium is more reactive than sodium.

15. (A) Crystals of $NaHCO_3$ and $KHCO_3$ show hydrogen bonds of different kinds.

(R) In $NaHCO_3$, the bicarbonate ions are linked in an infinite chain while in $KHCO_3$, a dimeric chain is formed.

16. (A) MgO is used for lining of steel making furnace

(R) It is a acidic flux and helps in removing basic impurities.

17. (A) When hot and concentrated NaOH reacts with chlorine, NaCl and NaClO are formed.

(R) It is a case of disproportion or auto redox reaction.

18. (A) $BeCl_2$ fumes in moist air (R) $BeCl_2$ reacts with moisture to form HCl gas.

19. (A) Mg is not present in enamel of human teeth.

(R) Mg is an essential element in biological functions of human.

20. (A) Magnesium is extracted by the electrolysis of fused mixture of $MgCl_2$, NaCl and $CaCl_2$.

(R) Calcium chloride acts as a reducing agent.

21 (A) Alkali metals dissolve in liquid ammonia to give blue solutions.

(R) Alkali metals in liquid ammonia give solvated species of the type $[M(NH_3)_n]$+(M = alkali metals)

22. (A) Sodium cannot be obtained by chemical reduction of its ore.

(R) Sodium is one of the strongest reducing agents.

23. (A) Al becomes passive in conc. HNO_3.

(R) conc. HNO_3 has no action on aluminium metal.

24. (A) Magnesium metal burns in air to give a white ash and this ash gives smell of ammonia in contact with water.

(R) The ash contains magnesium nitride also which is hydrolysed by water and ammonia is evolved.

25. (A) $MgCO_3$ is soluble in water when a current of CO_2 is passed.

(R) The solubility of $MgCO_3$ is the due to the formation of $Mg(HCO_3)_2$

ANSWER KEYS

1. (d)	2. (a)	3. (a)	4. (a)	5. (a)	6. (c)	7. (a)	8. (b)	9. (a)	10. (a)
11. (c)	12. (a)	13. (a)	14. (b)	15. (a)	16. (c)	17. (d)	18. (a)	19. (b)	20. (c)
21. (b)	22. (a)	23. (c)	24. (a	25. (a)					

HINTS AND EXPLANATIONS FOR SELECTIVE QUESTIONS

1. Sodium metal can be obtained by electrolysis of fused salt.

4. As the size of the ion increases, the tendency to rupture the O⁻ H bond decreases and hence acidic nature decreases i.e., basic nature increases.

5. Hydration energy decreases down the group i.e., Li is most hydrated the least conducted in aqueous solution.

14. K^+ being larger in size than Na^+ has a weaker positive field around it which cannot prevent the conversion of peroxide ion (O_2^{2-}) to superoxide ion $(O)^{2-}$.

16. As it is a basic flux so removes acidic impurities.

18. $BeCl_2 + 2H_2O \rightarrow Be(OH)_2 + 2HCl$

20. NaCl and $CaCl_2$ are added to provide conductivity to the electrolyte and also to lower the fusion temperature of anhydrous $MgCl_2$.

21. Blue colour is due to solvated electrons

23. Al metal is rendered passive when treated with conc. HNO_3.

CHAPTER 10A

The p-Block Element Boron Family

CHAPTER OUTLINE

■ IIIa ($ns^2 np^1$) ■ Boron ■ Compounds of Boron Borax or Tincal ($Na_2B_4O_7.10H_2O$) ■ Boric Acid or Ortho Boric Acid (H_3BO_3 or $B(OH)_3$) ■ Hydrides of Boron ■ Diborane (B_2H_6) ■ Boron Nitride (BN) ■ Aluminium ($_9Al^{27}$) ■ Aluminium Chloride $AlCl_3$ or Al_2Cl_6 ■ Aluminium Oxide or Alumina Al_2O_3 ■ Alums ■ Potash Alum ($K_2SO_4Al_2(SO_4)_3$ 24H_2O) ■ Ultra Marine ($Na_5Al_3Si_3S_3O_{12}$)

IIIa ($ns^2 np^1$)

Out of these elements, aluminium is the most abundant element. It is in fact the most abundant metal and the thirdmost abundant element in the earth crust. Boron is rare however it occurs as borax, kernite etc. Ga is more abundant than boron while In and Tl are less abundant. Ga, In and Tl are found in the form of their sulphides in traces in sulphide ores of zinc and lead.

Electronic Configuration

Their general electronic cohfiguration is $ns^2 np^1$ i.e., they have three valence electrons. The electronic configuration of these elements can be written as:

$_5$B: $1s^2, 2s^2 2p^1$
or [He] $2s^2, 2p^1$

$_{13}$Al: $1s^2, 2s^2 2p^6, 3s^2 3p^1$
or [Ne] $3s^2 3p^1$

$_{31}$Ga: $1s^2, 2s^2 2p^6, 3s^2 3p^6 3d^{10}, 4s^2 4p^1$
or [Ar] $3d^{10} 4s^2 4p^1$

$_{49}$In: $1s^2, 2s^2 2p^6, 3s^2 3p^6 3d^{10}, 4s^2 4p^6 4d^{10}, 5s^2 5p^1$
or [Kr] $4d^{10} 5s^2 5p^1$

$_{81}$Tl: $1s^2, 2s^2 2p^{6,} 3s^2 3p^6 3d^{10}, 4s^2 4p^6 4d^{10} 4f^{14}, 5s^2 5p^6 5d^{10}, 6s^2 6p^1$
or [Xe] $4f^{14}, 5d^{10}, 6s^2 6p^1$

Physical Properties

(1) Metallic Nature

The elements of this group are less metallic as compared to s-block elements.

B, Al, Ga, In, Tl
Non-metal metals

- Boron is often considered as a semi metal also as it has very high ionization energy.

(2) Atomic and Ionic Radii

They have lower value of atomic and ionic radii than that of II A group elements due to more effective nuclear charge and more number of valence electrons. On moving down the group atomic and ionic radius increase except that of galanium.

	B <	Ga <	Al <	In <	Tl.
At. radii (Å)	0.8	1.24	1.24	1.5	1.55
	B^{3+} <	A^{3+} <	Ga^{3+} <	In^{3+} <	Tl^{3+}
Ionic radii (Å)	0.2	0.52	0.6	0.8	0.95

The abnormal increase in case of aluminium is due to increase in screening effect due to availability of eight penultimate shell electrons while in case of boron there are only two penultimate shell electrons. In case of Ga, In, the screening effect is not so much effecting as it is by d-electrons present in the penultimate shell.

- The existence of B^{3+} and its value of ionic radii are estimated only.

(3) Boiling Point and Melting Point

There is no regular change in the values of melting points for these elements on moving down the group. Melting point first decreases from B to Ga and then increases. While the boiling points decreases down the group in a normal way as the bond strength holding the atoms in liquid state decreases.

Order of B.P

B	>	Al	>	Ga	>	In	>	Tl
2823		2723		2513		2323		1743 K

Order of M.P

B	>	Al	>	Tl	>	In	>	Ga
2573		933		576		430		303 K

- Boron has very high melting point due to its giant covalent polymeric structure in both solid and liquid states while Ga has low melting point as it consist of only Ga_2 molecules and have an unusual structure.

(4) Density

These elements have more value of densities than that of s-block elements (IA and IIA) and density increases on moving from B to Al as follows:

B	Al	Ga	In	Tl
2.4	2.7	5.9	7.3	11.9 g/ml

(5) Electropositive Nature

These are less electropositive elements than that of IA and IIA elements. In group electropositive nature is as follows :

$B < Al > Ga > In > Tl$

Electropositive nature decreases from Al onwards due to uneffective screening effect by d- and f- orbital electrons.

(6) Oxidation State

Boron shows only +3 oxidation state while rest elements show +1 , +3 oxidation states. The stability in these oxidation states are as follows:

Decreasing order of stability.

$B^{+3} > Al^{+3} > Ga^{+3} > In^{+3} > Tl^{+3}$

$Tl^{+} > In^{+} > Ga^{+} > Al^{+} > B^{+}$ (i.e $B^{+3} > B^{+}$)

Most stable　　　　　　　　Least

Boron does not show inert pair effect while this effect is maximum in Tl so due to inert pair effect stability increases down the group in lower oxidation state. Tl^{+3} is an oxidizing agent as it changes into its more stable Tl^{+} state. Al is a better reducing agent than carbon as Al has great affinity for O_2 than carbon.

(7) Reducing Nature

Except boron rest of the elements of the group have high values of $E^{o}_{O.P.}$ (Standard oxidation potential) so they are very strong reducing agents. Aluminium is more strong reductant than carbon as it has low ionization energy.

(8) Ionization Energy

These elements have lower value of first ionization energy than that of II A elements due to their less stable electronic configuration as removal of electron in easier. In this group ionization energy changes as follows

$B > Al < Ga > In < Tl$

Ga has more ionization energy than Al due to ineffective screening effect or more Zeff value same for Tl.

Chemical Properties

(1) Nature of Compounds

The compounds of these elements are supposed to be covalent as –

(1) Size of M^{3+} is small with high charge density i.e., more polarizing power.

(2) The sum of ionization energies is very high.

Physical properties of group 13 elements					
Property	Elements				
	B	Al	Ga	In	Tl
Atomic mass	10.81	26.98	69.72	114.82	204.38
Atomic radius (pm)	85	143	135	167	170
Ionic radius M^{3+} (pm)	27	53.5	62.0	80.0	88.5
Ionic radius M^+ (pm)	–	–	120	140	150
Density (g cm^{-3})	2.35	2.70	5.90	7.31	11.85
Melting point (K)	2453	933	303	430	576
Boiling point (K)	3923	2740	2676	2353	1730
First ionization energy (kJ mol^{-1})	800	577	578	558	590
Electronegativity (pauling scale)	2.0	1.5	1.6	1.7	1.8
E° (V) at 298 K for $M^{3+}(aq) + 3e^- \rightarrow M(s)$	–	–1.66	–0.56	–0.34	+1.26
E° (V) at 298 K for $M^+(aq) + e^- \rightarrow M(s)$	–	0.55	–	–0.18	–0.34
Abundance in earth's crust (pm)	9	83000	19	0.24	0.7

Boron can form only covalent compounds due to its very high ionization energy. The tendency to form ionic compounds increases down the group as size of M^{3+} increases and ionization energy decreases.

(2) Complex Formation

These elements have more tendency to form complexes than s-block elements due to their smaller size, more charge and presence of empty d-orbitals.

(3) Reaction with Water

Pure boron does not react with water but reacts with steam at red hot. Aluminium can reacts with cold water if there is no oxide layer formation on its surface. Ga, In and Tl can react with water in presence of oxygen or moist air.

$$2B + 3H_2O \xrightarrow{\text{Red hot}} B_2O_3 + 3H_2$$

$$2Al + 3H_2O \rightarrow Al_2O_3 + 3H_2$$

$$4Tl + 2H_2O + O_2 \rightarrow 4TlOH$$

(4) Reaction with Air

Boron in pure form is less reactive while impure boron burns in air to give B_2O_3.

$$4B + 3O_2 \rightarrow 2B_2O_3$$

Aluminium or thallium can react with air to form a stable oxide layer on its surface which protects it from further reactivity. That is why Aluminium is used in anodising (electrolytic process). Ga, In have no action of air.

(5) With Non-Metals

Boron and aluminium can combine with carbon and nitrogen on heating to give carbides and nitrides.

$$4B + C \rightarrow B_4C$$

$$4Al + 3C \rightarrow Al_4C_3$$

$$2B + N_2 \rightarrow 2BN$$

$$2Al + N_2 \rightarrow 2AlN$$

All of the elements of this group can react with halogens to form MX_3 type halides.

(6) With Acids

Except boron all other elements dissolve in dilute acids to form trivalent salts however, with concentrated acids all these elements can react except Al and Ga which have no reaction with concentrated HNO_3 as a protective oxide film is formed on their surfaces as follows:

$$2Al + 6HCl \rightarrow 2AlCl_3$$
(dil.)

$$B + 3HNO_3 \rightarrow H_3BO_3 + 3NO_2$$
(conc.)

$$2B + 3H_2SO_4 \rightarrow 2H_3BO_3 + 3SO_2$$
(conc.)

$$2Al + 6H_2SO_4 \rightarrow Al_2(SO_4)_3 + 3SO_2 + 6H_2O$$

(7) With Alkalies

Except In, Tl all these elements can react with alkalies example,

$$2B + 6NaOH \rightarrow 2Na_3BO_3 + 3H_2$$

$$2Al + 2NaOH + 2H_2O \rightarrow 2NaAlO_2 + 3H_2$$
Sodium meta aluminate

Compounds

The main compounds of these elements are as follows:

(1) Hydrides

These elements can not combine directly with hydrogen however except thallium all these elements can form hydrides. Boron forms stable covalent hydrides which are called Boranes (B_nH_{n+4} and B_nH_{n+6}).

Aluminium can form polymeric hydride $(AlH_3)_n$ i.e., Alane. Ga can form a dimeric hydride (Ga_2H_6) i.e., Digallane. Indium can form $(InH_3)_n$.

B, Al, Ga can form complex anionic hydrides also as follows:

$$MH_3 + H^- \rightarrow [MH_4]^-$$
Complex hydride
sp³ (tetrahedral)

$[MH_4]^-$ are powerful reducing agents due to H^-. e.g., $LiAlH_4$, $NaBH_4$

(2) Halides

These elements can form MX_3 type of halides mainly which are quite stable and can act like Lewis acids.

These are formed as follows:

$$B_2O_3 + 3C + 3Cl_2 \rightarrow 2BCl_3 + 3CO$$

$$Al_2O_3 + 3C + 3Cl_2 \rightarrow 2AlCl_3 + 3CO$$

BX_3 , AlX_3 , GaX_3 , InX_3 , TlX_3 ⟶

Stability decreases, Lewis acid nature increases as size of M^{+3} increases

In case of BX_3, lewis acid nature decreases as follows:

$BBr_3 > BCl_3 > BF_3$ ⟶

Back bonding occurs more in BX_3 with more electronegative X example, BF_3 is least strong lewis acid here due to maximum back bonding.

In back bonding, electrons shift from halogen atom to boron atom as B-atom has less e^- density.

BX_3 are always monomers while rest may be dimmers in vapour state or in non-polar solvents like Al_2Cl_6. In water dimerisation is not possible due to very high hydration energy.

BX_3 can form complex halides $(BX_4)^-$ in which boron can have a coordination number four also.

MF_3 are ionic while MX_3 (X = Cl, Br, I) are covalent and ionic nature increases from Ga to Tl. They can form complex halides of $(MX_6)^{3-}$ also as M has empty d-orbitals so it can have a coordination number six also.

- These elements can also form di and mono halides in gaseous state which are highly unstable like B_2X_4, $Ga^+[GaCl_4]^-$ etc.

(3) Oxides

These elements form M_2O_3 type of oxides and basic nature of oxides increases down the group as the size of M^{3+} ions increases so their interaction with water becomes very low.

B_2O_3 ,	Al_2O_3 ,	Ga_2O_3 , In_2O_3 ,	Tl_2O_3
Acidic	Amphoteric	Basic	Most basic

These are formed as follows:

$$4B + 3O_2 \xrightarrow{700^\circ C} 2B_2O_3$$
Boric anhydride

$$4Al + 3O_2 \xrightarrow{800^\circ C} 2Al_2O_3$$
Alumina

B_2O_3 being acidic can react with bases to form basic compounds.

example,

$$B_2O_3 + 2NaOH \rightarrow 2NaBO_2 + H_2O$$

$$CuO + B_2O_3 \rightarrow Cu(BO_2)_2$$

Al_2O_3 and Ga_2O_3 being amphoteric can react with acids as well as alkalies example,

$$Al_2O_3 + 6HCl \rightarrow 2AlCl_3 + 3H_2O$$

$$Al_2O_3 + 2NaOH \rightarrow \underset{\text{Sodium meta aluminate}}{2NaAlO_2} + H_2O$$

- Tl can also form a stable oxide Tl_2O which is just like an alkali metal oxide.

$$Tl_2O + H_2O \rightarrow 2TlOH$$

(4) Hydroxides

These elements form $M(OH)_3$ type of hydroxides and basic nature of these hydroxides increases down the group as the size of M^{3+} ions increases so their interaction with water becomes very low.

$B(OH)_3$,	$Al(OH)_3$,	$Ga(OH)_3$, $In(OH)_3$, $Tl(OH)_3$	
Acidic	Amphoteric	Basic	Most basic

These are formed as follows:

$$2Al(Hg) + 6H_2O \rightarrow Al(OH)_3 + 3H_2 + Hg$$

$$B_2O_3 + 3H_2O \rightarrow 2B(OH)_3 \text{ or } 2H_3BO_3$$

$B(OH)_3$ is distinctly acidic (monobasic) and acts as lewis acid by accepting OH- from H_2O.

$$B(OH)_3 + H_2O \leftrightarrow [B(OH)_4]^- + H^+$$

Boron

Occurrence

Boron is not present in free state in nature. In the combined state, it is present in the form of the salts of boric acid. Some of the important minerals of boron are as follows:

Borox – $Na_2B_4O_7.10H_2O$
Colemanite – $Ca_2B_6H_{11}.5H_2O$
Kernite (Rasorite) – $Na_2B_4O_7.4H_2O$
Boracite – $2Mg_3B_8O_{15}.MgCl_2$
Boric acid – H_3BO_3
Boronatrocalcite – $CaB_4O_7.NaBO_2.8H_2O$

Extraction of Boron

(1) From Borax or Colemanite

Boron can be extracted from borax or colemanite ores in two steps:

(i) Preparation of B_2O_3

(a) *From Borax:* First of all the finely powdered borax is converted into B_2O_3 by heating it with conc. HCl or H_2SO_4 as follows:

$$Na_2B_4O_7 + 2HCl \rightarrow 2NaCl + \underset{\text{Tetraboric acid}}{H_2B_4O_7}$$

$$Na_2B_4O_7 + H_2SO_4 \rightarrow Na_2SO_4 + H_2B_4O_7$$

$$H_2B_4O_7 + 5H_2O \rightarrow \underset{\text{Ortho boric acid}}{4H_3BO_3}$$

$$2H_3BO_3 \xrightarrow{\Delta} B_2O_3 + 3H_2O$$

or

$$Na_2[B_4O_5(OH)_4].8H_2O \xrightarrow{\text{acid}} \underset{\text{Orthoboric acid}}{H_3BO_3} \xrightarrow[-H_2O]{\text{heat}} B_2O_3$$

(b) *From Colemanite:* The powdered form of colemanite is fused with sodium carbonate and the fused mass is treated with hot water in which calcium carbonate is insoluble so it gets precipitated. The filtrate solution having borax and sodium meta borate is crystallized to get the crystals of borax first. When in the remaining solution, CO_2 is passed, sodium metaborate changes into borax. The obtained borax can be converted into B_2O_3 as in above process (a).

$$Ca_2B_6O_{11} + 2Na_2CO_3 \rightarrow 2CaCO_3 + Na_2B_4O_7 + 2NaBO_2$$

$$4NaBO_2 + CO_2 \rightarrow Na_2B_4O_7 + Na_2CO_3$$

- Colemanite can be converted into B_2O_3 by adding the mineral in water and passing SO_2 gas in the suspension as follows:

$$Ca_2B_6O_{11} + 11H_2O + 4SO_2 \rightarrow 2Ca(HSO_3)_2 + 6H_3BO_3$$

$$2H_3BO_3 \xrightarrow{\Delta} B_2O_3 + 3H_2O$$

(ii) Reduction of B_2O_3

Amorphous boron of low purity (called Moissan boron) can be obtained by reducing B_2O_3 with Mg or Na at a high temperature. It is nearly 95 – 98% pure, and is black in colour.

$$B_2O_3 \xrightarrow{\text{Mg/Na/K}} B + MgO \text{ or } Na_2O \text{ or } K_2O$$

(2) Modern Method (Electrolytic Method)
Boron can be obtained by the electrolysis of a fused mixture having boric anhydride, MgO and MgF_2 at 1373 K. The electrolysis is carried out in a carbon crucible which acts as an anode and an iron rod is used as cathode here. The magnesium that is discharged at the cathode reduces B_2O_3 into boron as follows:

$$2MgO \rightarrow 2Mg + O_2$$

$$B_2O_3 + 3Mg \rightarrow 2B + 3MgO$$

To remove the impurities, boron is heated in a vacuum at 1373 K electrically where the impurities are removed and pure boron is obtained.

- Crystalline form of boron can be obtained by reducing B_2O_3 with Al powder. This aluminium can be recovered by heating the fused with sodium hydroxide solution.

(3) Thermal decomposition of diborane or other boron hydrides also gives boron.

$$B_2H_6 \xrightarrow{\text{heat}} 2B\ (g) + 3H_2\ (g)$$

(4) By the reduction of volatile boron compounds with dihydrogen at high temperature (1543°C).

$$2BCl_3\ (g) + 3H_2\ (g) \xrightarrow{1543^\circ C} 2B\ (s) + 6HCl\ (g)$$

(5) Pyrolysis of BI3 (Van Arkel method) also gives boron.

$$2BI_3\ (g) \xrightarrow{\text{tungsten}, \Delta} 2B + 3I_2\ (g)$$

(6) By heating Potassium Fluoborate (KBF_4) with Magnesium or Potassium

$$2KBF_4 + 3Mg \xrightarrow{\Delta} 2B + 2KF + 3MgF_2$$

$$KBF_4 + 3K \xrightarrow{\Delta} B + 4KF$$

Physical Properties

(1) The two allotropic forms of boron are amorphous boron (dark brown) and crystalline boron (black).
(2) It has two isotopic forms, namely $_5B^{10}$ and $_5B^{11}$.
(3) It is a very hard substance with melting point higher than 2177°C and boiling point is about 2552°C. It is non-fusible under ordinary pressure.
(4) It is a non conductor of electricity.

Chemical properties

Boron is not much reactive so it can not react with many chemical reagents at ordinary temperature. Some of the chemical properties of boron are as follows:

(1) Action of Air

Amorphous boron burns in air at 973 K with a reddish flame giving boron oxide (B_2O_3) and boron nitride (BN).

(2) Reaction with Water

Boron is not attacked by water under normal conditions however, when steam is passed over red hot boron, hydrogen is liberated as follows:

$$2B + 3H_2O \rightarrow B_2O_3 + 3H_2 \uparrow$$

(3) Action of Alkalies

Boron reacts with fused KOH and NaOH to give borates.

$$2B + 6KOH \rightarrow \underset{\text{Potassium borate}}{2K_3BO_3} + 3H_2$$

(4) Reducing Nature

It is a powerful reducing agent.

$$3CO_2 + 4B \rightarrow 2B_2O_3 + 3C$$
$$3SiO_2 + 4B \rightarrow 2B_2O_3 + 3Si$$

(5) Reaction with Acids

Concentrated H_2SO_4 or HNO_3 oxidises boron into boric acid however it does not react with non oxidizing acids.

$$2B + 3H_2SO_4 \xrightarrow{\Delta} \underset{\text{Boric acid}}{2H_3BO_3} + 3SO_2$$

$$2B + \underset{\text{Conc.}}{6HNO_3} \rightarrow 2H_3BO_3 + 6NO_2 \uparrow$$

(6) Reaction with Metals

Boron combines with metals upon heating giving borides which are normally very hard and have high melting points

$$2Mg + 2B \rightarrow \underset{\text{Magnesium boride}}{Mg_3B_2}$$

$$Cr + B \rightarrow CrB$$

(7) Reaction with Non-metals

When it is heated with carbon in an electric furnace it gives an extremely hard substance, boron carbide (B_4C). Boron gives B_2S_3 on heating with sulphur. It also reacts with Cl and Br_2 but at higher temperatures.

$$2B + 3Cl_2 \xrightarrow{\Delta} 2BCl_3$$

Uses

(1) It is used as a semi conductor.

(2) Boron steel or boron carbide rods can be used for controlling nuclear reactions as boron has a very high cross section for capturing the neutrons. Moreover, it can absorb neutron to create another boron having even number of neutrons.

$${}_5B^{10} + {}_0n^1 \rightarrow {}_5B^{11}$$

(3) It is used in glass industry for making specific types of glass.

(4) Boron filaments are used in making light and composite material for air crafts.

Anomalous Behaviour of Boron

Boron being the first element of its group differs from other elements of its own group in many properties. The anomalous behaviour of boron is due to:

(1) its small size
(2) high ionization energy
(3) high electronegativity
(4) absence of vacant d-orbitals in its valence shell.

Some of the important properties in which boron differs from other members of its group are as follows:

(1) Boron is a non-metal while rest of the members of this group are metals.

(2) The melting and boiling points of boron are much higher than those of aluminium and the other members of this group.

(3) It can show only +3 oxidation state while other elements of the group can show +1 and +3 oxidation states.

(4) The trihalides of boron exist as discrete monomeric molecules while those of the other elements of its group can exist as dimmers having halogen-bridged structures.

(5) Boron shows allotropy and exists in two forms – crystalline and amorphous while other elements of its group do not show allotropy.

(6) It can not decompose water or steam like other members of the group which can decompose water.

(7) Boron is a non-conductor of electricity while the other elements of its group are good conductors.

(8) It can combine with metals to form borides while rest of the elements of the group can not do so.

(9) Boron shows a maximum covalence of four in its compounds (e.g., BH_4^-). Aluminium and other members of this family can show a covalence of even six due to presence of d-orbitals in them e.g., $[Al(OH)^{3-}]$.

(10) Boron forms only covalent compounds, while aluminium forms both covalent as well as ionic compounds. The other members of this group form mainly ionic compounds.

(11) The oxide and hydroxide of boron are weakly acidic, while those of aluminium and the other members of this group are either amphoteric or basic in nature.

(12) The oxide of boron i.e., B_2O_3 is soluble in water while the oxides of other elements are insoluble in water.

(13) Boron forms a number of series of hydrides called boranes. The boranes are stable covalent compounds and are volatile in nature. The hydrides of aluminium and the other elements of this group are non-volatile solids and are less stable and decompose on heating.

Diagonal Relationship between Boron and Silicon

Boron (III group) is placed diagonally opposite to silicon (IV group) in the periodic table. Therefore boron shows diagonal relationship with silicon.

The diagonal relationship between boron and silicon is due to following factors:

(1) The atomic radius of boron (85 pm) is close to that of silicon (118 pm).

(2) Both have almost similar values of electronegativity (B = 2.0, Si = 1.8) and ionization energy (B = 800 kJ mol^{-1}, Si = 786 kJ mol^{-1}).

Some of the important properties in which boron resembles silicon are as follows:

(1) Occurrence

Both boron and silicon do not occur in free state in nature. But they are always found in the combined state.

(2) Allotropy

Both boron and silicon can show allotropy and exist in two allotropic forms i.e., Amorphous and Crystalline.

(3) Non-metallic Character

Both boron and silicon are typical non-metals with high melting and boiling points. Both are non-conductors of electricity.

(4) Semi-Conducting Nature

Both of these are semi conductors at high temperature.

(5) Due to high ionization energies both boron and silicon can not form cations easily.

(6) Both boron and silicon can not combine with carbon to form carbides i.e., B_4C, SiC which are very hard in nature and used for cutting and abrasing purposes.

(7) Formation of Hydrides

Both boron and silicon can form a large number of hydrides known as boranes and silanes respectively

example,
Silane, SiH_4
Disilane, Si_2H_6 etc.

(8) Reaction with Alkalies

Both boron and silicon dissolve in alkalies to form borates and silicates respectively with the evolution of hydrogen. Boron dissolves in fused alkalies, whereas silicon dissolves both in fused as well as aqueous alkalies.

$$2B + 6NaOH \rightarrow \underset{\text{Sodium borate}}{2Na_2BO_3} + 3H_2$$

$$Si + 2NaOH + 2H_2O \rightarrow \underset{\text{Sodium silicate}}{Na_2SiO_3} + 2H_2$$

(9) Formation of Halides

These can form halides when heated with respective halogens and halides, are covalent in nature and hygroscopic and volatile. They can be easily hydrolyzed by water.

$$SiCl_4 + 4H_2O \rightarrow \underset{\text{Silicic acid}}{H_4SiO_4} + 4HCl$$

$$BC13 + 3H_2O \rightarrow \underset{\text{Boric acid}}{H_3BO_3} + 3HCl$$

Due to release of HCl, their chlorides fume in moist air.

(10) Formation of Oxides

These can form oxides when heated in oxygen and their oxides (B_2O_3, SiO_2) are high melting solids and acidic in nature and dissolve in alkalies to form borates and silicates respectively.

$$B_2O_3 + 6NaOH \rightarrow 2Na_3BO_3 + 3H_2O$$

$$SiO_2 + 2NaOH \rightarrow Na_2SiO_3 + H_2O$$

(11) Formation of Binary Compounds with Metals

Both can react with many metals forming binary compounds called borides and silicides respectively.

example,

$$2B + 3Mg \xrightarrow{\Delta} \underset{\text{Magnesium boride}}{Mg_3B_2}$$

$$Si + 2Mg \xrightarrow[\text{(In absence of air)}]{\Delta} \underset{\text{Magnesium silicide}}{Mg_2Si}$$

Some of these borides and silicides on hydrolysis results boranes and silanes respectively.

Compounds of Boron Borax or Tincal ($Na_2B_4O_7.10H_2O$)

It is called Sodium tetra borate decahydrate. It is also called Tincal or Suhaga as tincal has nearly 45 % borax.

Preparation

(1) From Colemanite

It is prepared from powdered ore of Colemanite [$Ca_2B_6O_{11}$] by boiling it with sodium carbonate followed by filtration as follows:

$$Ca_2B_6O_{11} + 2Na_2CO_3 \rightarrow Na_2B_4O_7 + 2NaBO_2 + 2CaCO_3$$

Sodium meta borate can be further converted into borax by passing CO_2 through it.

$$4NaBO_2 + CO_2 \rightarrow Na_2B_4O_7 + Na_2CO_3$$

(2) From Boric Acid

Boric acid on treatment with sodium carbonate gives borax as follows:

$$4H_3BO_3 + Na_2CO_3 \rightarrow Na_2B_4O_7 + 6H_2O + CO_2$$

Physio-Chemical Properties

(1) It exists in three forms:

(1) Prismatic Borax i.e., decahydrate form ($Na_2B_4O_7.10H_2O$): It can be obtained by the crystallization of borax solution at coordinate temperature also. It is less soluble in cold water but highly soluble in hot water.

(2) Octahedral Borax i.e., pentahydrate form ($Na_2B_4O_7.5H_2O$): It can be obtained by crystallizing borax solution at 333 K.

(3) Anhydrous Borax or Borax Glass ($Na_2B_4O_7$): It can be obtained heating borax above its melting point (1013 K). This type of borax is colourless glassy mass which abosorbs moisture and readily changes into decahydrate form.

(2) Hydrolysis

On hydrolysis it gives basic aqueous solution as follows:

$$Na_2B_4O_7 + 7H_2O \xrightarrow{\Delta} 2Na^+ + 2\,OH^- + 4H_3BO_3$$

(3) Heating Effect of Borax

Borax on heating swells up as water molecules are eliminated when it is further heated at a high temperature a glassy transparent solid mass is obtained.

$$\underset{\text{Prismatic Borax}}{Na_2B_4O_7 . 10H_2O} \xrightarrow[\text{(In absence of air)}]{\Delta} \underset{\text{Anhydrous borax}}{Na_2B_4O_7}$$

$$Na_2B_4O_7 \xrightarrow[740^\circ C]{\Delta} \underset{\text{Sodium meta borate (Glassy mass)}}{2NaBO_2} + B_2O_3$$

(4) B_2O_3 (acidic, glass like) is used in Borax bead test for detecting the presence of basic radicals like

$[Co^{+3}, Cu^{+2}, Cr^{+2}, Ni^{+2}, Fe^{+2}]$

Blue ↓ brown ↓

Green yellow

$$\underset{\text{Bead}}{B_2O_3} + \text{Pt-wire} + \text{Metal peroxide} \rightarrow \text{Meta Borates}$$

$$CuSO_4 + B_2O_3 \rightarrow \underset{\text{Copper meta borate}}{CuO.B_2O_3} + SO_3$$

(5) With Ethyl Alcohol and concentraded H_2SO_4

When it is heated with ethyl alcohol and conc. H_2SO_4 it gives volatile vapours of triethyl borate which burn with a green edged flame (test of borate ions).

$$Na_2B_4O_7 + H_2SO_4 + 5H_2O \rightarrow Na_2SO_4 + 4H_3BO_3$$

$$H_3BO_3 + 3C_2H_5OH \rightarrow \underset{\text{Tri ethyl borate}}{B(OC_3H_5)_3} + 3H_2O$$

Structure of Borax

It can be shown as $[Na_2 [B_4O_7 . 2H_2O]. 8H_2O]$. It has two tetrahedral and two triangular units attached to each other as shown in the figure given below:

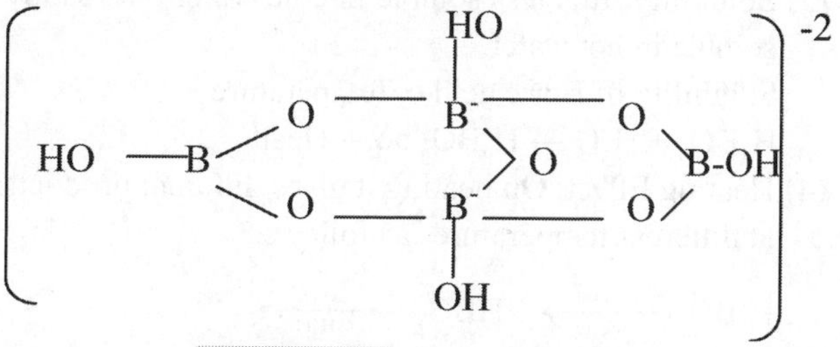

$$\left[\begin{array}{c} \text{HO–B(–O–B}^-\text{(OH)–O–B}^-\text{(OH)–O–)B–OH, bridged by O} \end{array}\right]^{-2}$$

FIGURE 10.1 Structure of Borax

Uses

(1) It is used in borax bead test for detecting basic radicals.
(2) It can be used in the softening of water as it can form insoluble calcium and magnesium borates.
(3) It is used in making optical and hard borosilicate glasses.
(4) It is used to make enamels and glazes.
(5) It can be used as a flux and as an antiseptic also.

Boric Acid or Ortho Boric Acid (H_3BO_3 or $B(OH)_3$)

Preparation

(1) From Borax

When aqueous solution of borax is acidified with conc. HCl or H_2SO_4, boric acid solution is formed which can be easily crystallised into crystals of boric acid as follows.

$$Na_2B_4O_7 + 2HCl + 5H_2O \rightarrow 4H_3BO_3 + 2NaCl$$

$$Na_2B_4O_7 + H_2SO_4 + 5H_2O \rightarrow 4H_3BO_3 + Na_2SO_4$$

(2) From Colemanite

When SO_2 is passed through a suspension of colemanite in boiling water, boric acid is formed which can be easily crystallized.

$$Ca_2B_6O_{11} + 4SO_2 + 11H_2O \rightarrow 6H_3BO_3 + 2Ca(HSO_3)_2$$

(3) From Boron trichloride

$$BCl_3 + 3H_2O \rightarrow H_3BO_3 + 3HCl$$

Physio-Chemical Properties

(1) It is white, soft, needle like crystalline solid with a soapy touch.

(2) Acidic Nature: It behaves like a weak monobasic acid. It also behaves like a Lewis acid.

$H_3BO_3 + H_2O \rightarrow H_3O^+ + H_2BO_3^-$

$H_3BO_3 + H_2O \rightarrow [B(OH)_4]^- + H^+$

(3) Solubility: It is less soluble in cold water but readily soluble in hot water.

Solubility of Boric acid α Temperature

$H_3BO_3 + H_2O \rightarrow H_3BO_3$ aq. – Heat

(4) Heating Effect: On heating it gives different products at different temperatures as follows:

$$H_3BO_3 \xrightarrow[-H_2O]{373K} HBO_2 \xrightarrow[-H_2O]{433K}$$

Meta boric acid

$$H_2B_4O_7 \xrightarrow[-H_2O]{\text{strongly heatig}} B_2O_3$$

Tetraboric acid Boron trioxide

(5) With Ethyl Alcohol: When it is heated with ethyl alcohol it gives volatile vapours of tri ethyl borate which burn with a green edged flame.

$$H_3BO_3 + 3C_2H_5OH \rightarrow B(OC_2H_5)_3 + 3H_2O$$

Tri ethyl borate

(6) With NaOH:

$$H_3BO_3 + NaOH \rightarrow NaBO_2 + 2H_2O$$

(7) With NaOOH (Sodium Hydroperoxide)

$$2H_3BO_3 + 2NaOOH \rightarrow Na_2B_2O_6 . 2H_2O + 2H_2O$$

Dihydrated sodium Peroxy Borate (Crystals)

Structure

It has a slippery layer structure in which planar BO_3^{3-} units are linked together by hydrogen bonds. In this structure each hydrogen atom can act as a bridge between the two oxygen atoms of the two adjacent BO_3^{3-} units. It forms a covalent bond with the oxygen atom of one BO_3^{3-} unit and a hydrogen bond with the oxygen atom of another BO_3^{3-} unit. In boric acid boron atom is sp^2 hybridised with planar trigonal shape.

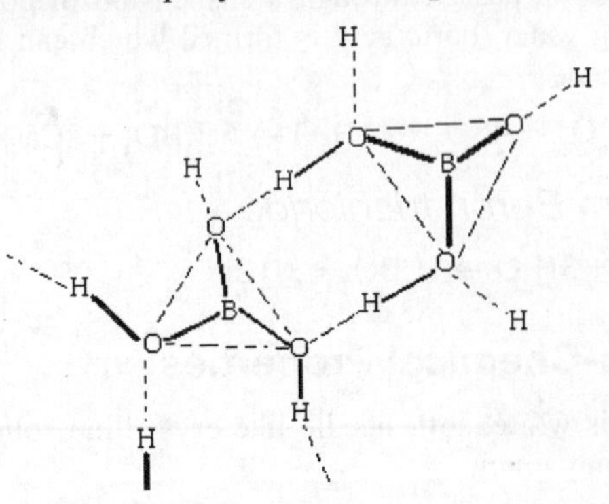

or

$$2Na^+ \left[\begin{matrix} HO \\ HO \end{matrix} > B < \begin{matrix} O-O \\ O-O \end{matrix} > B < \begin{matrix} OH \\ OH \end{matrix} \right]^{-2}$$

FIGURE 10.2 Structure of Boric Acid

Uses

(1) It is used as a antiseptic and a eye lotion (boric lotion).

(2) It is also used as a food preservative.

(3) H_3BO_3 decreases thermal expansion of glass.

(4) It is used in the manufacture of enamels and glazes for pottery.

Hydrides of Boron

Boron reacts with dihydrogen to give a number hydrides. These hydrides are known as boranes and have a general formula B_nH_{n+4} (eg- B_2H_6) and B_nH_{n+6} (eg- B_4H_{10})

Diborane (B_2H_6)

Preparation

B_2H_6 is prepared as follows:

(1) Lab Method

In lab it is prepared by the reaction of sodium borohydride with iodine in a high boiling polyether solvent $(CH_3OCH_2CH_2)_2O$

$$2NaBH_4 + I_2 \xrightarrow{\text{polyether}} B_2H_6 + 2NaI + H_2$$

(2) Industrial Method

$$2BF_3 + 6LiH \xrightarrow{450K} B_2H_6 + LiF$$

(3) From Boron Halides

$$4BX_3 + 3LiAlH_4 \rightarrow 2B_2H_6 + 3LiX + 3AlX_3$$

example,

$$4BCl_3 + 3LiAlH_4 \rightarrow 2B_2H_6 + 3LiCl + 3AlCl_3$$

(4) When boron halides and excess of hydrogen undergo reaction by passing silent electric discharge at low pressure, B_2H_6 is formed.

$$2BX_3 + 6H_2 \xrightarrow[\text{discharge}]{\text{Silentelectric}} B_2H_6 + 6HCl$$

Physio-Chemical Properties

(1) It is a *colourless gas* which is stable only at low temperatures.

(2) Heating Effect

When it is heated in between 100 – 300°C, higher boranes are formed like B_3H_9, B_4H_{10}, B_5H_{11} etc. while at red hot temperature it decomposes into boron.

$$B_2H_6 \xrightarrow{\text{Red hot}} 2B + 3H_2$$

(3) With Oxygen

All boranes catch fire in the presence of oxygen and liberate a lot of heat energy so they can be used as high energy fuels also.

$$B_2H_6 + 3O_2 \rightarrow 2B_2O_3 + 3H_2O$$
$$\Delta H = -2008 \text{ kJ/mol}$$

(4) With Water

Boranes are readily hydrolysed by water

$$B_2H_6 + 6H_2O \rightarrow 2H_3BO_3 + 6H_2$$

(5) Formation of Borohydrides

Boranes are used in the formation of hydroborates or borohydrides like $LiBH_4$ or $NaBH_4$, which are extensively used as reducing agents in many organic synthesis.

$$2LiH + B_2H_6 \rightarrow 2Li^+ [BH_4]^-$$

$$2NaH + B_2H_6 \rightarrow 2Na^+ [BH_4]^-$$

(6) Formation of Borazine

Diborane (1 vol) on reaction with NH_3 (2 vol) at 450 K gives borazine or borazol or Inorganic benzene.

$$3B_2H_6 + 6NH_3 \xrightarrow{450K} 2B_3N_3H_6 + 12H_2$$

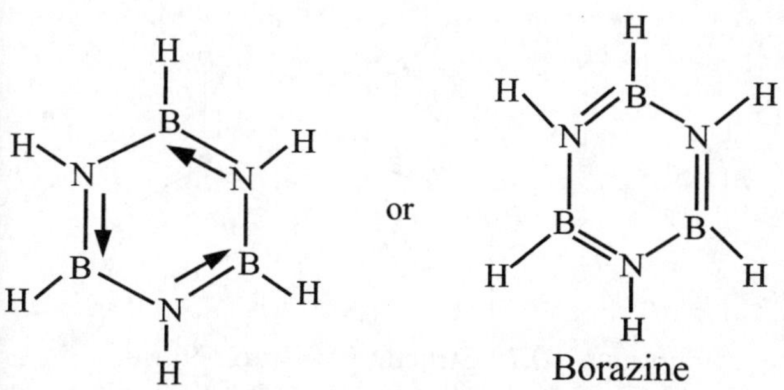

FIGURE 10.3 Structure of Barazine

Structure of Diborane

It is an electron deficient molecule as it does not have enough number of electrons needed for the formation of normal covalent bonds. Its structure was proposed by Dilthey as a bridge structure shown in the figure given below. It is also confirmed by electron diffraction studies.

In this structure of diborane, boron atom is sp^3 hybridised having one empty and three half filled hybrid orbitals i.e., four hybrid orbitals of equal energy.

$$B^* \ 1s^2 \ 2s^1 \ 2p^1{}_x \ 2p^1{}_y \ 2p^0{}_z$$

The structure of borane is of great interest. They are all electron- deficient compounds i.e., they do not posses enough valency electrons to form the expected number of covalent bonds example, in diborane there are not enough electron (there are only twelve electrons) three from each boron atom and six from six hydrogen atom to form the seven covalent bonds that an ordinary Lewis structure would required $H_3B.BH_3$.

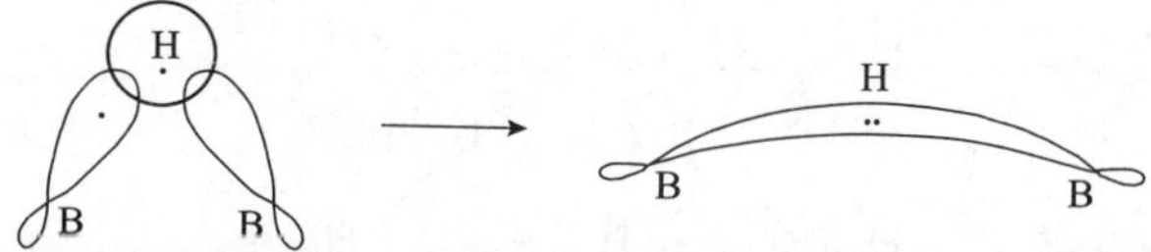

FIGURE 10.4 Formation of three Centre Bond (B-H-B)

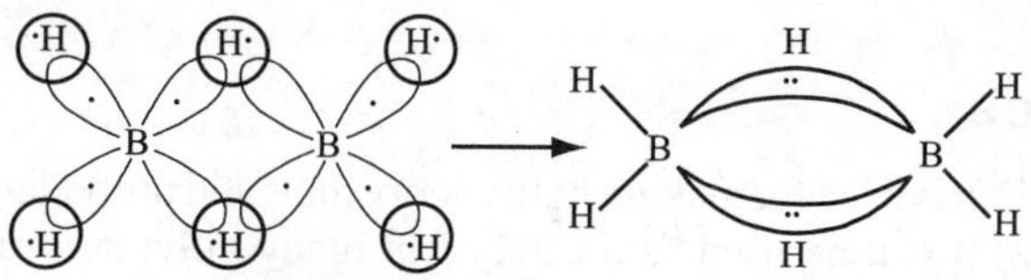

FIGURE 10.5 Structure of Diborane

These 12 valence electron of boron forms 6 covalent bonds with these 6 hydrogen. Amongst these 6 hydrogen atoms, 2H are termed as bridging hydrogen whereas 4H are termed as terminal hydrogen atoms.

When diborane is methylated then only 4 hydrogen atoms are susceptible to methylation and these 4H-atoms are entirely different from remaining 2H-atoms. These 2Hatoms are known as bridge hydrogen. Hydrogen bridge is also known as Banana bond or tau bond or 3 centre 2 electron (3C – 2e) bond.

Bond lengths:

B – B = 1.77 Å
B – H = 1.19 Å (Normal covalent bond)
B – H = 1.33 Å 3 centre 2 electron bond.

H – B – H bond angle = 121.5° ± 7.5°
H – H – B bond angle (3C – 2e) = 100°

The boron atom undergoes sp^3 hybridisation in which 1s orbital of hydrogen atom overlaps the sp^3 hybrid orbital of boron and a delocalized orbital is formed. Two boron atoms and four hydrogen atoms lie in same plane. Two Hatoms forming bridges, one above and other below, lie in the plane perpendicular to the rest of molecule.

Diborane has two type of BH bonds

(1) $B - H_t$: (Here H_t = terminal hydrogen) It is a normal covalent bond (two centre electron pair bond i.e., 2c – 2e bond).

(2) $B - H_b$: (Here H_b = bridge hydrogen). It is a bond between three atoms, $B - H_b - B$, (three centre-two electron pair bond i.e., 3c – 3e or banana bond).

Such a bond is formed by the overlaping of empty sp^3–hybridised orbital of one boron atom, 1s-orbital of hydrogen atom (H_b) and half filled sp^3 –hybridised orbital of the other boron atom.

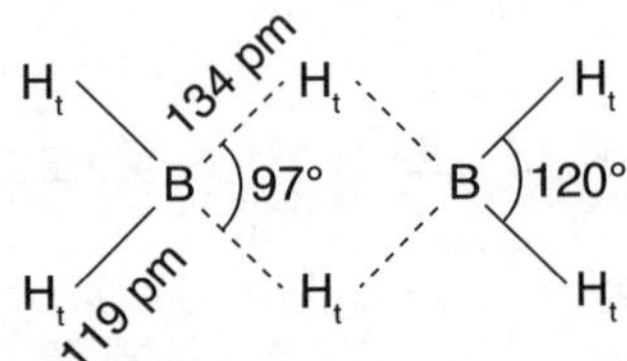

FIGURE 10.6 Structure of Diborane

Uses

(1) It is used for making high energy fuels and propellants.
(2) It can be used as a catalyst in many polymerisation reactions.
(3) It can be used as a reductant in many organic reactions like hydroboration.

BORON NITRIDE (BN)

Preparation

It is prepared by the following methods:

(1) By heating boron in the atmosphere of nitrogen or ammonia as follows:

$2B + N_2 \rightarrow 2BN$

$2B + 2NH_3 \rightarrow 2BN + 3H_2$

(2) When borax is heated with ammonium chloride at very high temperature, boron nitride is formed.

$Na_2B_4O_7 + 2NH_4Cl \rightarrow 2BN + 2NaCl + B_2O_3 + 4H_2O$

Here, the resulting mass is treated with diluted HCl, as a result B_2O_3 and NaCl dissolve while BN is left insoluble.

(3) When NH_3 reacts with BCl_3, pure BN is formed.

$BCl_3 + 6NH_3 \rightarrow \underset{\text{Boronamide}}{B(NH_2)_3} + 3NH_4Cl$

$2B(NH_2)_3 \xrightarrow{\text{Heat}} \underset{\text{Boronimide}}{B_2(NH)_3} + 3NH_3$

$B_2(NH)_3 \rightarrow 2BN + NH_3$

(4) When B_2O_3 is heated with $Hg(CN)_2$, boron nitride is formed.

$B_2O_3 + Hg(CN)_2 \rightarrow 2BN + CO + CO_2 + Hg$

Physio-Chemical Properties

(1) It is a white powder with a very high melting point of 3000°C and insoluble in water.

(2) Heating Effect: On heating in steam under pressure, it gives ammonia.

$2BN + 3H_2O \rightarrow B_2O_3 + 2NH_3$

(3) It is chemically an inert compound and it is unaffected by mineral acids. However, it dissolves in HF giving ammonium borofluoride.

$BN + 4HF \rightarrow NH_4BF_4$

(4) It does not react with alkalis. However when fused with K_2CO_3, it gives potassium cyanate and potassium metaborate.

$K_2CO_3 + BN \rightarrow KCNO + KBO_2$

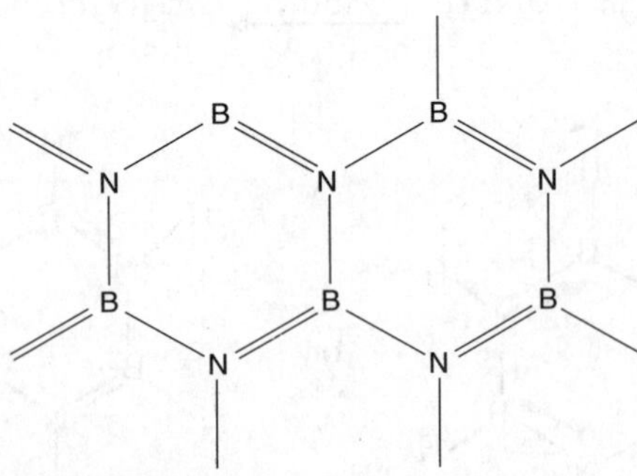

FIGURE 10.7 Structure of Briton Nitride

Structure

Structure of boron nitride is similar to the structure of graphite.

Recently an another crystalline form of boron nitride has been synthesized by application of high temperature and pressure. It is known as borazon. It is harder than diamond and used as an abrasive.

ALUMINIUM ($_9Al^{27}$)

Occurrence

It is the most abundant metal but 3rd most abundant element in earth crust (O, Si, Al, Fe). It is an important constituent of many silicate rocks.

Ores

Some important ores are as follows:

Alunite or alum stone, $K_2SO_4.Al_2(SO_4)_3.4Al(OH)_3$

Bauxite $Al_2O_3.\ 2H_2O$

Corundum, Al_2O_3

Cyrolite, Na_3AlF_6

Diaspore, $Al_2O_3.H_2O$

Feldspar, $KAlSi_3O_8$

Kaolinite, $Al_2O_3.2SiO_2.2H_2O$

Mica, $K_2O.3Al_2O_3.6SiO_2.\ 2H_2O$

Extraction of Aluminium

Aluminium is extracted from Bauxite ($Al_2O_3.2H_2O$) mainly which is associated with the impurities of Fe_2O_3 and silica. It is necessary to remove these impurities before the extraction of aluminium as these impurities make aluminium brittle and liable to corrosion.

Steps Involving in the Metallurgy of Aluminium

It involves following three steps:

(1) Purification of Bauxite

For the purification of bauxite mainly these methods are employed:

(a) Baeyer's Process

It is used when the ore has mainly the impurity of Fe_2O_3 which makes ore red. Here the powdered ore is first of all roasted to convert FeO into Fe_2O_3. Now the roasted ore is treated with concentrated caustic soda solution at 150°C under pressure for many hours. As a result Al_2O_3 dissolves in NaOH giving solution of sodium meta aluminate and the insoluble impurities of Fe_2O_3 and SiO_2 settled down. Sodium meta aluminate solution is now changed into aluminium hydroxide by hydrolysis and the precipitate of $Al(OH)_3$ is washed out and dried, which on heating gives alumina.

- Here the solution of NaOH is concentrated and can be used again.

$$\underset{\text{Bauxite}}{Al_2O_3.2H_2O} + 2NaOH \rightarrow \underset{\text{Sodium meta aluminate}}{2NaAlO_2} + 3H_2O$$

$$NaAlO_2 + 2H_2O \rightarrow Al(OH)_3\downarrow + NaOH$$

- **Summary**

$$\text{Bauxite (fine powdered)} \xrightarrow[FeO\rightarrow Fe_2O_3]{\text{roasted}}$$

$$\text{Roasted ore} \xrightarrow[\text{High pressure (150°C, 80atm) Filtered, } Fe_2O_3 \text{ as residue}]{\text{+ caustic soda solution}}$$

$$\text{Filtrate (sodium aluminate)} \xrightarrow{\text{filtered } CO_2} \underset{\text{Ppt.}}{Al(OH)_3}$$

(b) Hall's Process

When finely powdered bauxite is fused with sodium carbonate, Al_2O_3 combines with sodium carbonate giving sodium meta aluminate (fused mass). This fused mass is extracted with H_2O where Fe_2O_3 and SiO_2 remain insoluble in the residue.

$$Al_2O_3.2H_2O + Na_2CO_3 \rightarrow 2NaAlO_2 + CO_2 + 2H_2O$$

The solution having sodium meta aluminate is first warmed to 323 – 333 K and then CO_2 is circulated through it as a result $Al(OH)_3$ separates out as precipitate.

$$2NaAlO_2 + 3H_2O + CO_2 \xrightarrow{323\text{-}333K} 2Al(OH)_3\downarrow + Na_2CO_3$$

This precipitate is filtered, washed and dried. The solution of Na_2CO_3 is concentrated and can be used again.

- **Summary**

Bauxite (finely powdered) $\xrightarrow[\text{Fused, extracted with water, Residue } Fe_2O_3]{+ Na_2CO_3}$

solution $\xrightarrow[\text{50-60 °Cand filtered filtrate } (N_2CO_3]{CO_2}$

precipitate $Al(OH)_3 \longrightarrow$ Pure Al_2O_3

(c) Serpeck's Process

It is used when traces of silica is present in bauxite ore. When finely powdered bauxite is mixed with coke and heated at 2000 K in presence of nitrogen, aluminium nitride is formed. Silica is reduced to silicon which volatilizes off at this temperature.

$$Al_2O_3.2H_2O + N_2 + 3C \xrightarrow{2000K} 2AlN + 3CO\uparrow + 2H_2O$$

$$SiO_2 + 2C \rightarrow Si\uparrow + 2CO\uparrow$$

AlN is hydrolysed with water to get aluminium hydroxide. The advantage of this process is that ammonia is obtained as a valuable by-product here.

$$AlN + 3H_2O \rightarrow Al(OH)_3\downarrow + NH_3$$

- **Summary**

Bauxite (finely powdered) $\xrightarrow[\text{Heated to 1800°C}]{+ \text{coke} + N_2}$

Silica reduced to Si which volatalises	+	alumina form aluminium nitride AlN	$\xrightarrow{\text{hydrolysis}}$

Calcination of Aluminium Hydroxide

The precipitate of aluminium hydroxide is calcined at 1773 K in a rotary kiln to get pure alumina (Al_2O_3).

$$2Al(OH)_3 \xrightarrow{\text{heated}} \text{Pure } Al_2O_3 + 3H_2O$$

Electrolysis of Fused Alumina (Hall and Heroult's Method)

Pure alumina is mixed with a molten mixture of cryolite (Na_3AlF_6) and fluorspar (CaF_2) in 20 : 24 : 20 ratio for electrolysis Here the molten mixture is used to decrease the melting point of alumina from 2000°C – 900°C and to increase the conductivity of the mixture.

Here, the electrolysis takes place in an iron box having carbon lining inside which acts as cathode.

Carbon rods dipped in fused mixture of the electrolyte act as anode. The fused electrolyte is covered with a layer of coke to avoid glow as a safety point for eyes.

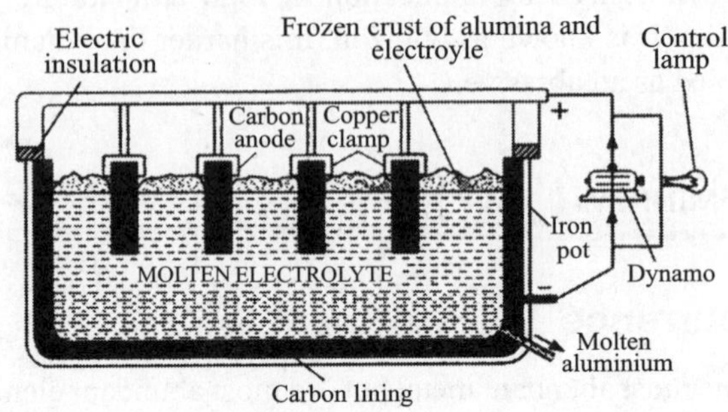

FIGURE 10.8 Electrolysis of Al_2O_3

The mechanism of electrolysis is given by two methods:

(1) $Na_3AlF_6 \rightleftharpoons 3NaF + AlF_3$

$AlF_3 \rightleftharpoons Al^{3+} + 3F^-$

At cathode: $Al^{3+} + 3e^- \rightarrow Al$

At anode: $2F^- \rightarrow F_2 + 2e^-$

The liberated fluorine reacts with alumina to give AlF_3 and O_2 as follows:

$$2Al_2O_3 + 6F_2 \rightarrow 4AlF_3 + 3O_2$$

$$2C + O_2 \rightarrow 2CO$$

$$2CO + O_2 \rightarrow 2CO_2$$

(2) $Al_2O_3 \rightleftharpoons \underset{\text{Cathode}}{Al^{3+}} + \underset{\text{Anode}}{AlO_3^{3-}}$

At cathode: $Al^{3+} + 3e^- \rightarrow Al$

At anode:

$4AlO_3^{3-} \rightarrow 2Al_2O_3 + 3O_2 + 12e^-$

The net chemical reaction taking place during electrolysis is,

$$2Al_2O_3 \rightarrow 4Al + 3O_2$$

From this process, 99.8 % pure aluminium is formed.

Refining of Aluminium (Hoope's Electrolytic Method)

Here, the electrolytic cell has an iron box having carbon coating inside. The cell has three layers having different specific gravities in it the upper layer having pure aluminium act as cathode. The middle layer having a mixture of the fluorides of Al, Ba and Na act as electrolyte. The lowest layer having impure aluminium acts as anode. The graphite rods dipped in pure aluminium and Cu-Al alloy rod at the bottom act as conductors. During electrolysis aluminium gets deposited at cathode from the middle layer and the same amount of aluminium comes from lower layer to middle layer i.e., pure aluminium moves from lower to upper layer and impurities are left behind. Aluminium obtained by Hoope's method is nearly 99.98 % pure.

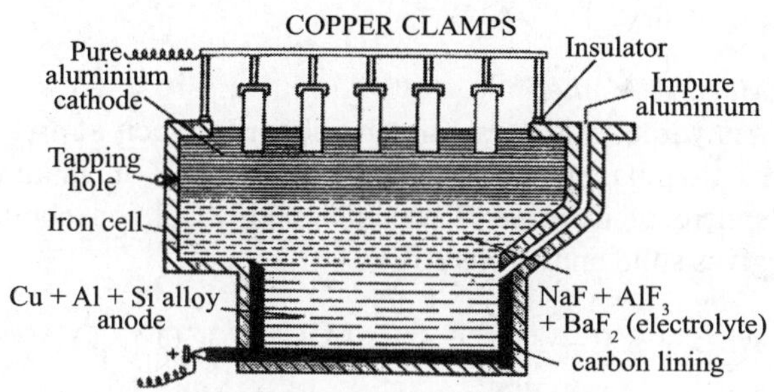

FIGURE 10.9 Hoopes Net Nod

Physical Properties

(1) It is a bluish white lustrous metal with a melting point of 660°C and a boiling point of 1800°C.

(2) It has a density of 2.7 g / ml.

(3) It is malleable, ductile and good conductor of heat and electricity.

(4) It can be welded and cast however difficult to solder

Chemical Properties

(1) Action of Air and Oxygen

It is uneffected by dry air however in moist air a thin film of Al_2O_3 is formed on its surface. On burning in oxygen it gives a bright light and this reaction is highly exothermic, so a huge amount of heat is evolved. This heat can be used in thermite process for the reduction of metal oxides like Cr_2O_3, Fe_2O_3, etc.,

$$4Al + 3O_2 \rightarrow 2Al_2O_3$$

(2) Reaction with Water

Pure aluminium does not react with water while impure aluminium can be easily corroded by water or sea water. When aluminium is heated with water hydrogen is evolved.

$$2Al + 6H_2O \rightarrow 2Al(OH)_3 + 3H_2 \uparrow$$

(3) Reaction with Non-Metals

Except oxygen it can react with non-metals directly e.g., when it is heated with nitrogen, aluminium nitride is formed while with sulphur, aluminum sulphide is formed.

$$2Al + N_2 \xrightarrow{\Delta} 2AlN$$

$$2Al + 3S \xrightarrow{\Delta} Al_2S_3$$

$$2Al + 3X_2 \xrightarrow{\Delta} 2AlX_3$$

(4) Displacement Reactions

Aluminium can displace less electropositive metals like Cu, Zn, Pb etc., from their salt solutions due to its more reactivity and electropositive nature.

$$3CuSO_4 + 3Al \rightarrow Al_2(SO_4)_3 + 3Cu$$

$$3ZnSO_4 + 3Al \rightarrow Al_2(SO_4)_3 + 3Zn$$

(5) Reaction with Alkalies

It reacts with alkali (NaOH) as follows:

$$2Al + 2NaOH + 2H_2O \rightarrow \underset{\text{Sodium meta aluminate}}{2NaAlO_2} + 3H_2 \uparrow$$

$$2Al + 6NaOH \rightarrow \underset{\text{Sodium aluminate}}{2Na_3AlO_3} + 3H_2 \uparrow$$

(6) Reaction with Acids

Being strongly electropositive, high reactivity and reduction action, it can react with dilute acids evolving hydrogen.

$$2Al + 6HCl \rightarrow 3AlCl_3 + 3H_2 \uparrow$$

$$2Al + 3H_2SO_4 \rightarrow Al_2(SO_4)_3 + 3H_2 \uparrow$$

$$2Al + \underset{\text{concentrated}}{6H_2SO_4} \rightarrow Al_2(SO_4)_3 + 3SO_2 + 6H_2O$$

- Aluminium is not affected by HNO_3 as a stable oxide layer is formed on its surface.

(7) Reducing Nature

It can act like a reducing agent to reduce some metallic and non metallic oxides

example,

$$Cr_2O_3 + 2Al \rightarrow 2Cr + Al_2O_3 + \text{heat}$$

$$3Mn_3O_4 + 8Al \rightarrow 9Mn + 4Al_2O_3 + \text{heat}$$

$$Fe_2O_3 + 2Al \rightarrow 2Fe + Al_2O_3 + \text{heat}$$

$$4Al + 3CO_2 \rightarrow 2Al_2O_3 + 3C$$

$$4Al + 3SiO_2 \rightarrow 2Al_2O_3 + 3Si$$

Uses

(1) Bronze [Cu +Al], Magnalium (Mg + Al) are used to make aeroplane parts etc.

(2) Al is used to make container for conc. HNO_3, surgical parts etc.

(3) Al –foil is used in packing of cigarettes etc.

(4) It is used in the reduction of metal oxides (Gold schimidt alumino thermite process).

(5) In is used in Thermite welding.

(6) $LiAlH_4$ is a reducing agent.

(7) A mixture of aluminium powder and aluminium nitrate is called Ammonal (an explosive used in making bombs).

Some Important Alloys of Aluminium

1. Magnalium: It is an alloy of 90% Al and 10% Mg. It is used in making balance beams.
2. Duralumin: It is an alloy of Al (95.5%), Copper (4%), Mg (0.5%) and Mn (0.5%). It is used in making aeroplane parts.
3. Alnico: It is an alloy of steel (77%), nickel (2%), aluminium (20%) and cobalt (1%). It is used in making permanent magnets.
4. Aluminium bronze: It is an alloy of 90% copper and 10% aluminium. It is used in making coins, trays and picture frames.
5. Nickeloy: It is an alloy of 95% aluminium, 4% copper and 1% nickel. It is used to make air craft parts.
6. γ-Alloy: It is an alloy of 93% aluminium, 4% copper, 2% nickel and 1% magnesium. It is used to make pistons and machinery parts mainly.

Some Stones of Aluminium

Ruby (red) Al_2O_3 + Cr

Topaz (yellow) FluoroSilicate of Al

Sapphire (blue Lapis Lazule) : Na-Al-Silicate

Emerald Be –Al Silicate

REMEMBER

- Al, Fe, Cr, Ni, Co, become inactive with conc. HNO_3, $HClO_4$, H_2CrO_4 [oxidizing agents], anode etc. due to stable oxide layer formation over metal surface that is why conc. HNO_3 is stored in Al-container.
- Al is a very good conductor. [Ag > Cu > Au > Al]
- Al – Hg is a reducing agent.

Aluminium Chloride $AlCl_3$ or Al_2Cl_6

Preparation

It is prepared as follows:

(1) By passing dry HCl or Cl_2 over heated aluminium in absence of air, anhydrous $AlCl_3$ can be obtained.

$$2Al + 3Cl_2 \rightarrow 2AlCl_3$$

$$2Al + \underset{\text{dry}}{6HCl} \rightarrow 2AlCl_3 + 3H_2$$

(2) Macah's Method

Anhydrous $AlCl_3$ can also be obtained when a mixture of alumina and carbon is heated at 1273 K in an atmosphere of chlorine. The vapours of $AlCl_3$ on cooling gives solid anhydrous $AlCl_3$.

$$Al_2O_3 + 3\text{'C'} + \underset{\text{dry}}{3Cl_2} \xrightarrow{\Delta} 2AlCl_3 + 3CO \uparrow$$

(3) Hydrated $AlCl_3$

When aluminium or aluminium hydroxide is dissolved in dilute HCl followed by crystallization hydrated aluminium chloride is obtained.

$$2Al + \underset{\text{dry}}{6HCl} \rightarrow 2AlCl_3 + 3H_2$$

$$Al(OH)_3 + 3HCl \rightarrow AlCl_3 + 3H_2O$$

Physical Properties

(1) Anhydrous $AlCl_3$ is covalent and lewis acid in nature.

(2) It is a white deliquescent solid and it is soluble in organic solvents like alcohol, ether etc.

(3) Dimer of $AlCl_3$ i.e Al_2Cl_6 has both covalent and coordinate bonds.

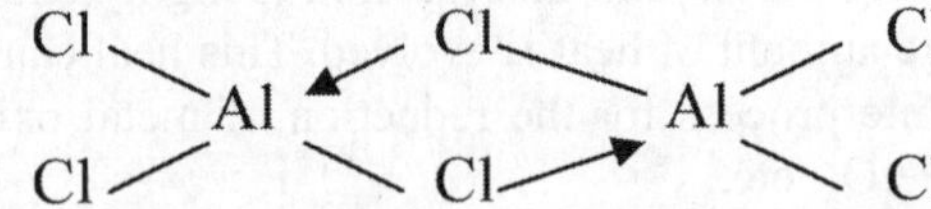

(4) In $AlCl_3$, Al is sp^2 while in Al_2Cl_6 Al is sp^3 hybridised.

Chemical Properties

(1) Hydrolysis

During hydrolysis in moist air $AlCl_3$ gives fumes of HCl and on dissolving in water it changes into hydrated form.

$$2AlCl_3 + 6H_2O \rightarrow 2Al(OH)_3 + 6HCl$$
Excess

$$2AlCl_3 + 12H_2O \rightarrow 2[Al(H_2O)_6]^{3+}.3Cl^-$$

(2) With NaOH

When sodium hydroxide is added slowly and gently into $AlCl_3$ solution a white gelatinous precipitate is formed which dissolves in excess of sodium hydroxide to give sodium meta aluminate.

$$AlCl_3 + 3NaOH \rightarrow Al(OH)_3 + 3NaCl$$

$$Al(OH)_3 + NaOH \rightarrow NaAlO_2 + 2H_2O$$

(3) With NH_4OH

When ammonium hydroxide is added in a solution of $AlCl_3$, a white gelatinous precipitate of $Al(OH)_3$ is formed which does not dissolve in NH_4OH.

$$AlCl_3 + 3NH_4OH \rightarrow Al(OH)_3 + 3NH_4Cl$$

(4) With Ammonia

Anhydrous $AlCl_3$ forms an adduct with gaseous ammonia.

$$Al_2Cl_6 + 12NH_3 \rightarrow 2[AlCl_3.6NH_3]$$

Uses

(1) $AlCl_3$ is used as a catalyst to generate electrophile in Friedal craft's reaction (also in gasoline)
(2) It is used as a mordant in dyeing.
(3) It can also be used giving the manufacture of gasoline (by cracking of high boiling fractions of petroleum).

Aluminium Oxide or Alumina Al_2O_3

Alumina (aluminium oxide, Al_2O_3) is most stable compound and exists both in the anhydrous and hydrated forms.

(1) Anhydrous Al_2O_3 occurs in nature as colourless crystals in corundum and in combination with different coloured oxides such as ruby (red), sapphire (blue), emerald (green), amethyst (violet) and topaz (yellow).
(2) Hydrate alumina occurs as bauxite.

Preparation

Al_2O_3 can be prepared by igniting aluminium hydroxide, aluminium nitrate, aluminium sulphate or ammonium alum as follows:

(1) From Aluminium Chloride

$$AlCl_3 + 3NH_4OH \rightarrow Al_2O_3 + 3NH_4Cl$$

$$2Al(OH)_3 \rightarrow Al_2O_3 + 3H_2O$$

(2) By Heating Aluminium Sulphate or Ammonium Salt

$$Al_2(SO_4)_3 \xrightarrow{\Delta} Al_2O_3 + 3SO_3$$

$$\underset{\text{Ammonium salt}}{(NH_4)_2SO_4.Al_2(SO_4)_3.24H_2O} \xrightarrow{\Delta} Al_2O_3 + 2NH_3 + 4SO_3 + 25H_2O$$

(3) By Heating mixture of AlF_3 and B_2O_3

$$2AlF_3 + B2O_3 \rightarrow Al_2O_3 + 2BF_3$$

(4) Al_2O_3 can also be prepared by H.Gold Schmidt's aluminothermic process. In this process Al powder mixed with ferric oxide (thermite) in the ratio of 1:3 respectively and ignited oxides are reduced to the metal and heat is liberated.

$$Fe_2O_3 + 2Al \rightarrow 2Fe + Al_2O_3$$

(5) $4Al + 3O_2 \rightarrow 2Al_2O_3$
This reaction involves oxidation and the process of anodizing will favour formation of Al_2O_3.

Physio-Chemical Properties

(1) It is a white crystalline, very stable and infusible compound.
(2) It is insoluble in water and almost unreactive.
(3) Al_2O_3 is not a good conductor of electricity. It is dissolved in cryolite to make it a good conductor.
(4) It has a melting point of 2323 K and a boiling point of 2523 K (due to high affinity of Al for oxygen the M.P. of aluminium oxide is very high)
(5) Electrolytic reduction of pure alumina is not possible because it melts at very high temperature.
(6) Amphoteric Nature: Al_2O_3 is an amphoteric oxide because it reacts with acids as well as bases to form salts.

$$Al_2O_3 + 6HCl \rightarrow 2AlCl_3 + 3H_2O$$

$$Al_2O_3 + 2NaOH \xrightarrow{\Delta} 2NaAlO_2 + H_2O$$

(7) Heating Effect: When alumina is heated above 800°, an exothermal change takes place, after which it becomes almost insoluble in acids.
(8) On Heating with Carbon at High Temperature

$$2Al_2O_3 + 9C \xrightarrow{2000^\circ C} Al_4C_3 + 6CO$$

$$Al_4C_3 + 12H_2O \rightarrow 4Al(OH)_3 + 3CH_4$$

Uses

(1) Alumina is used as refractory material, medium in chromatographic separation, bauxite cement (bauxite + lime).
(2) Fused bauxite called aluminium, is used as an abrasive, in the manufacture of Al and its salts and in the preparation of artificial precious gems.
(3) Alumina has different capacities for adsorbing other substances on the surface from solution. This property is used in separating mixtures into their components by chromatography.
(4) Artificial gems are prepared by directing oxyhydrogen flame on a rod of Al and finely powdered mixture of alumina, fluorspar and little colouring matter (e.g., Cr_2O_3 for rubies, 5 % TiO_2 and 1.5 % Fe_2O_3 for sapphires).

ALUMS

These are double salts of M^+ and M^{+3} having a general formula :

$M_2^I SO_4, M_2^{III}(SO_4)_3 .24H_2O$
or
$M^I M^{III} (SO_4)_2 12H_2O$

Here M^I may be K^+, Na^+, Cs^+, NH_4^+, Rb^+, Ag (Li^+ is not possible due to its small size it can not have coordination number 6) and M^{III} may be Fe^{+3}, Cr^{+3}, Mn^{+3}
Common alum is Potash Alum,
i.e, $K_2SO_4 Al_2 (SO_4)_3 24 H_2O$

Naming of Alums

If M^{+3} is not Al^{+3}, name of alum is given according to M^{+3}.
e.g., $Cs_2SO_4 . Cr_2(SO_4)_3 . 24H_2O$ i.e Chrome alum.
If M^{+3} is Al^{+3} name of alum is given according to M^+ (e.g., Rubidium alum) but not with K^+.
If neither K^+ nor Al^{+3} are present than name of alum is given according to both the cations.

example,- $(NH_4)_2SO_4. Fe_2 (SO_4)_3 24H_2O$
Ferric ammonium alum

Pseudo Alums

When in an alum mono valent element (M^I) is substituted by a bivalent element (M^{II}) it is called pseudo alum. It is not isomorphous with an alum.
example,

$MgSO_4 .Al_2(SO_4)_3 .24H_2O$
$FeSO_4 .Al_2(SO_4)_3 .24H_2O$

Some Facts about Alums

(1) These can be obtained by mixing equimolar solutions of the constituent sulphates followed by crystallization.
(2) These are crystalline compounds and all alums are isomorphous.
(3) In alum crystals, 6 water molecules are held by monovalent ion, 6 water molecules are held by trivalent ion, 12 water molecules are held in the crystal structure.
(4) Aqueous solutions of alum are acidic due to cationic hydrolysis of trivalent cation.
(5) Feature alum or Hair-salt $Al_2SO_4.18H_2O$ is a native form of aluminium sulphate.
(6) On heating an alum at high temperature the alum swells up into a porous mass which is called burnt alum.
(7) Alums act as coagulants as they are effective in precipitating colloids.
(8) Alums have germicide properties.

POTASH ALUM ($K_2SO_4Al_2(SO_4)_3$ 24H_2O)

Preparation

(1) From Bauxite or Aluminium Sulphate
When bauxite is boiled with sulphuric acid solution of aluminium sulphate is obtained. In it K_2SO_4 is added in a calculated amount followed by concentration and cooling to get the crystals of alum.

$$Al_2O_3 + 3H_2SO_4 \rightarrow Al_2(SO_4)_3 + 3H_2O$$

$$Al_2(SO_4)_3 + K_2SO_4 + 24H_2O \rightarrow \underset{\text{Potash alum}}{K_2SO_4.Al_2 (SO_4)_3. 24H_2O}$$

(2) From Alunite or Alum Stone
It is boiled with dilute sulphuric acid and a calculated amount of K_2SO_4 is added in the boiled solution the resultant solution on cooling gives the crystals of alum.

$$\underset{\text{Alum stone}}{K_2SO_4. Al_2(SO_4)_3.4Al(OH)_3} + 6H_2SO_4 \rightarrow K_2SO_4 + 3Al_2(SO_4)_3 + 12H_2O$$

$$K_2SO_4 + Al_2 (SO_4)_3 + 24H_2O \rightarrow K_2SO_4. Al_2(SO_4)_3 .24H_2O$$

Physio-Chemical Properties

(1) It is a white crystalline solid compound.

(2) It is soluble in water and the aqueous solution is acidic due to cationic hydrolysis of $Al_2(SO_4)_3$. The aqueous solution has K^+, Al^{3+} and SO_4^{2-}.

(3) Heating Effect: On heating it swells up due to elimination of water molecules as follows:

$$K_2SO_4.Al_2(SO_4)_3.14H_2O \xrightarrow{\Delta} K_2SO_4.Al_2(SO_4)_3 + 24H_2O$$

$$K_2SO_4.Al_2(SO_4)_3 \xrightarrow{\text{Red hot}} K_2SO_4 + Al_2O_3 + 3SO_3$$

Uses of Alums

(1) As mordent in dyeing.

(2) As a syptic to stop bleeding

(3) In tanning of leather

(4) In purification of water (negative impurities in water)

(5) It is used in sizing of cheap quality paper.

Ultra Marine ($Na_5Al_3Si_3S_3O_{12}$)

(1) It is an artificial form of blue coloured mineral Lapislazuli. It is a complex silicate of sodium and aluminium having nearly 12 % of sulphur as Na_2S.

(2) It can be obtained when a mixture of kaolin, soda ash, charcoal and sulphur (in excess) are heated upto redness.

(3) The blue ultra marine is most common form which can be changed into violet form by heating it in a current of dry chlorine.

(4) Blue ultra marine is used in making blue paint, wall paper, blue printed paper.

(5) It is also used in calico-printing and in laundry.

Gold Schmidt Alumino Thermic Process

Being highly electropositive in nature and having a great affinity towards oxygen at high temperatures, aluminium can displace many elements from their oxides which are less electropositive than it. This reaction is discovered by Gold Schmidt and commonly known as Gold Schmidt Alumino Thermic process.

$$\text{Metal oxide} + Al \rightarrow \underset{\text{Molten state}}{Al_2O_3 + \text{Metal}} + \text{Heat}$$

This process is helpful in the extraction of some non metals like B, Si and metals like Cr, Mn etc., from their oxides.

example,

$$Cr_2O_3 + 2Al \rightarrow 2Cr + Al_2O_3 + \text{heat}$$
$$3Mn_3O_4 + 8Al \rightarrow 9Mn + 4Al_2O_3 + \text{heat}$$
$$Fe_2O_3 + 2Al \rightarrow 2Fe + Al_2O_3 + \text{heat}$$
$$2Al + B_2O_3 \rightarrow Al_2O_3 + 2B + \text{heat}$$
$$4Al + 3SiO_2 \rightarrow 2Al_2O_3 + 3Si + \text{heat}$$

It is also helpful in thermite welding of metals e.g., welding of steel.

Thallium is also called Duckbill platypus due to its similarities with many other elements.

CHAPTER-END EXERCISES

Practice Questions – I

1. The compounds of boron and hydrogen are collectively called

   (a) Boranes (b) Boracits
   (c) Borazoles (d) Diboranes

2. The alum used for purifying water is

   (a) Chrome alum (b) Ammonium alum
   (c) Potash alum (d) Ferric alum

3. BCl_3 is a planar molecule because in this molecule boron is

   (a) Sp^2-hybridised (b) Sp-hybridised
   (c) Sp^3-hybridised (d) Sp^3d-hybridised

4. Boron when heated with carbon forms

   (a) B_4C (b) B_2C_3
   (c) BC_2 (d) B_4C_3

5. Heating an aqueous solution of aluminium chloride to dryness will give

   (a) $AlCl_3$ (b) Al_2Cl_3
   (c) $Al(OH)Cl_2$ (d) Al_2Cl_6

6. The chemical formula of zeolite is .....

   (a) $Na_2Al_2Si_2O_8.xH_2O$ (b) $Na_2(Na_4(PO_3)_6$
   (c) $Ca_2Al_2Si_2O_8$ (d) $K_2Al_2Si_2O_8.xH_2O$

7. Three centred bond is present in

(a) NH_3 (b) $AlCl_3$
(c) B_2H_6 (d) BCl_3

8. Boron nitride on reacting with caustic alkali gives

(a) NO_2 (b) N_2O
(c) Na_2BO_2 (d) NH_3

9. Which of the following does not give borax bead test?

(a) $Cu(NO_3)_2$ (b) $(CH_3COO)_2Co$
(c) $Al_2(SO_4)_3$ (d) $NiSO_4$

10. Aluminium vessels should not be washed with materials containing washing soda since

(a) Washing soda is easily decomposed.
(b) Washing soda reacts with aluminium to form insoluble aluminium oxide.
(c) Washing soda is expensive.
(d) Washing soda reacts with aluminium to form soluble aluminate.

11. Which one of the following compounds liberates CO_2 from aqueous $NaHCO_3$?

(a) $CHCl_3$ (b) CH_3Cl
(c) Anilinium chloride (d) CCl_4

12. The state of hybridization of 'B' BCl_3 is

(a) sp (b) sp^2
(c) sp^3 (d) sp^3d^2

13. Borax is not used

(a) As a flux in soldering
(b) In making enamel and pottery glazes
(c) As a styptic to stop bleeding
(d) In making optical glasses

14. The first ionization enthalpies of Li, Be, B and C are given below. Which of these corresponds to Boron?

(a) 520 kJ mol^{-1} (b) 1086 kJ mol^{-1}
(c) 800 kJ mol^{-1} (d) 899 kJ mol^{-1}

15. BCl_3 does not exist as dimmer because

(a) B has low electronegativity
(b) B has small size
(c) B has no vacant orbitals
(d) B has low electron affinity

16. Which of the following compound react with BF_3?

(a) Al_2Cl_6 (b) NH_3
(c) Ethers (d) All of these

17. What is the molecular formula of borazole?

(a) $B_3N_3H_3$ (b) $B_3N_3H_6$
(c) B_2H_6 (d) $B_6N_6H_6$

18. Which of the following is not possible

(a) BeC_2 (b) B_2H_6
(c) BeF_4^{2-} (d) BF_4^-

19. Which of the following sets of elements does not belong to transitional element set?

(a) Cu, Ag, Au (b) Tl, Zr, Hf
(c) Ga, In, Tl (d) Fe, Co, Ni

20. Which of the following is used in the preparation of ammonal?

(a) NaOH + CaO (b) Al powder
(c) Fe_2O_3 (d) $CaCN_2$ + C (graphite)

21. Among the halides of the elements of group-13 the one which is most acidic is

(a) $AlCl_3$ (b) BBr_3
(c) BF_3 (d) BCl_3

22. Which one of the following boron compounds is optically active?

(a) Boron anhydride
(b) Sodium tetraborate
(c) Borosalicylic acid
(d) Boron trifluoride

23. From B_2H_6, all the following can be prepared except

(a) $B_2(CH_3)_6$ (b) $NaBH_4$
(c) B_2O_3 (d) H_3BO_3

24. Inorganic benzene contains

(a) C, N ,H (b) C, H, Al
(c) C, H, B (d) B, N, H

25. Boric acid when burnt with ethyl alcohol gives a green edged flame due to

(a) Metaboric acid (b) Boric anhydride
(c) Orthoboric acid (d) Ethyl borate

26. Which one is metaboric acid?

(a) $B(OH)_3$ (b) H_3BO_3
(c) $H_2B_4O_7$ (d) HBO_2

27. Amorphous boron on burning in air forms

(a) Mixture of B_2O_3 and $(BN)_x$
(b) $B(OH)_3$
(c) Only $(BN)_x$
(d) Only B_2O_3

28. Which of the following is an electron deficient compound?

(a) NaCl (b) B_2H_6
(c) KCl (d) NaH

29. What is the formula of diaspore?

(a) $AlPO_4$ (b) $Al_2O_3.H_2O$
(c) Al_2O_3 (d) $Al_2O_3.2H_2O$

30. $BCl_3 + H_2O \rightarrow X$
The products formed in above reaction are

(a) $B_2H_6 + HCl$ (b) $B_2O_3 + HOCl$
(c) $H_3BO_3 + HCl$ (d) No reaction

31. On strong heating, boric acid yields

(a) B_2H_6 (b) B
(c) BO_2 (d) B_2O_3

32. In Hall's process, the main reagent is mixed with

(a) AlF_3 (b) NaF
(c) Na_3AlF_3 (d) None of these

33. Orthoboric acid (H_3BO_3) and metaboric acid (HBO_2) differ in respect of

(a) Structure (b) Basicity
(c) Melting point (d) All of these

34. The electronic configuration of group III elements is

(a) $ns^1\ np^2$ (b) $ns^1\ np^3$
(c) $ns^2\ np^1$ (d) $ns^2\ np^2$

35. When a nickel salt is heated with boric anhydride, brown coloured bead is formed which is due to the formation of

(a) NiO (b) $Ni(BO_2)_2$
(c) $Ni_2B_4O_7$ (d) None of these

36. Boric acid is a very weak acid but in presence of certain organic compounds, it acts as a strong acid. Which one of the following organic compounds can affect such change?

(a) Acetic acid (b) Glycerol
(c) Ethylene (d) Ethyl alcohol

37. Which of the following is the electron deficient molecule?

(a) B_2H_6 (b) SiH_4
(c) PH_3 (d) C_2H_6

38. Which of the following mineral contain Al?

(a) Cryolite (b) Mica
(c) Felspar (d) All of these

39. During the Serpeck's process Silica is eliminated as

(a) Si (solid) (b) Si (vapour)
(c) SiO_2 (vapour) (d) SiO_2 (vapour)

40. In diborane the two H – B – H angles are nearly

(a) 60º, 120º (b) 95º, 150º
(c) 95º, 120º (d) 120º, 180º

41. Which one of the following is used to remove silicon dioxide in the Serpeck's process of purification of bauxite?

(a) Coke (b) nickel
(c) Na_2CO_3 (d) CaO

42. During the electrolytic reduction of alumina, the reaction at cathode is ……

(a) $3F^- \rightarrow 3F + 3e^-$
(b) $2H^+ + 2e^- \rightarrow H_2$
(c) $2H_2O \rightarrow O_2 + 4H^+ + 4e^-$
(d) $Al^{3+} + 3e^- \rightarrow Al$

43. Boron compounds behave as lewis acids because of their

(a) Ionisation energy
(b) Electron deficient nature
(c) Covalent nature
(d) Ionisation energy

44. Boron halides behave as Lewis acids because they

(a) Have only six electron in the valence shell
(b) Have a lone pair of electrons on the B atom
(c) Are covalent compounds
(d) Are ionic compounds

45. Reactivity of borazole is greater than that of benzene because

(a) It has localized electron in it.
(b) Borazole is electron deficient compound
(c) Borazole is polar compound
(d) Borazole is non-polar compound

46. Aluminium reacts with hot concentrated H_2SO_4 to liberate

(a) S (b) SO_3
(c) SO_2 (d) H_2S

47. Which of the following is not a Lewis acid?

(a) $AlCl_3.6H_2O$ (b) $SnCl_4$
(c) $FeCl_3$ (d) $AlCl_3$

48. Aluminium oxide is not reduced by chemical reactions since
(a) Reducing agent contaminates
(b) The process pollutes the environment
(c) Aluminium oxide is highly stable
(d) Aluminium oxide is stable

49. 'Thermite' is a mixture of X parts of ferric oxide and Y parts of aluminium powder, X and Y respectively are
(a) 2, 3 (b) 3, 1
(c) 3, 2 (d) 1, 1

50. The pair of amphoteric hydroxide is
(a) $Al(OH)_3$, $Be(OH)_2$
(b) $Ni(OH)_2$, $Zn(OH)_2$
(c) $Be(OH)_2$, $Mg(OH)_2$
(d) $Al(OH)_3$, LiOH

51. Alum is water purifier because it
(a) Gives taste
(b) Softens hard water
(c) Coagulates the impurities
(d) Destroys the pathogenic bacteria

52. Borax is
(a) $Na_2[B_4O_5(OH)_4].8H_2O$
(b) $Na_2[B_4O_2(OH))_{10}].5H_2O$
(c) $Na_2[B_4O_6(OH)_6].7H_2O$
(d) $Na_2[B_4O_3(OH)_8].6H_2O$

53. During the electrolysis of cryolite, aluminium and fluorine are formed in ….. molar ratio.
(a) 1 : 3 (b) 1 : 1
(c) 1 : 2 (d) 2 : 3

54. Boric acid is polymeric due to
(a) Its geometry
(b) Its presence of hydrogen bonds
(c) Its acidic nature
(d) Its monobasic nature

55. Which of the following compounds contain boron?
(a) Colemanite (b) Cristoballite
(c) Kernite (d) Both A and C

56. In B_2H_6
(a) The structure is similar to that of C_2H_6
(b) The boron atoms are linked through hydrogen bridges
(c) All the atom lie in the same plane
(d) There is boron-boron bond.

57. Which of the following aluminium salt is used to stop bleeding
(a) Aluminium chloride
(b) Aluminium fluoride
(c) Aluminium sulphate
(d) Potash alum

58. Maximum bond angle is present in case of
(a) BCl_3 (b) BF_3
(c) BBr_3 (d) same in all

59. Which of the following ion does not exist in octahedral aqua ions?
(a) Al (b) Ga
(c) Tl (d) B

60. Which of the following does not exist in free state?
(a) BCl_3 (b) BH_3
(c) BF_3 (d) BBr_3

61. Which of the following reactions does not liberate gaseous product?
(a) $AlCl_3$ + NaOH →
(b) NaOH + P (white) + H_2O →
(c) Al + NaOH $\xrightarrow{\Delta}$
(d) Zn + NaOH $\xrightarrow{\Delta}$

62. Inorganic graphite is
(a) B_3N_3 (b) SiC
(c) P_4S_3 (d) $Fe(CO)_5$

63. Which of the following does not form ionic compound easily?
(a) Thalium (b) Aluminium
(c) Boron (d) Galium

64. BF_3 used as a catalyst in several industrial processes due to its
(a) Weak reducing action
(b) Strong reducing nature
(c) Weak lewis acid character
(d) Reducing nature

65. Boron and silicon resembles in all respects except
(a) Both form halides which are Lewis acids
(b) Their chlorides hydrolyse to their respectively acids
(c) Both form acidic oxides
(d) Their hydrates are stable

66. Which is strongest Lewis acid?
(a) BBr_3 (b) BI_3
(c) BF_3 (d) BCl_3

67. Ga is below Al in the periodic table, but atomic radius of Ga is less than Al. It is because of

(a) Lanthanoid contraction
(b) Greater screening effect
(c) Inert pair effect
(d) None of these

68. Which of the following reacts with BCl_3 to form diborane?

(a) $K_2Cr_2O_7$ (b) NH_3
(c) NaHg (d) $LiAlH_4$

69. Boric acid is a very weak acid but in presence of certain organic compounds, it acts a s a strong acid. Which one of the following organic compounds can affect such change?

(a) Fromic acid (b) Glycerol
(c) Ethyne (d) Ethyl alcohol

70. The structure of diborane (B_2H_6) contains

(a) Four 2c-2e bonds and two 3c-2e bonds
(b) Two 2c-2e bonds and four 3c-2e bonds
(c) Two 2c-2e bonds and two 3c-3e bonds
(d) Four 2c-2e bonds and four 3c-2e bonds

Practice Questions – II

71. In which of the following reaction boron does not act as reducing agent?

(a) $B + CO_2 \rightarrow$ (b) $B + Mg \rightarrow$
(c) $B + HNO_3 \rightarrow$ (d) $B + SiO_2 \rightarrow$

72. Diborane can't be obtained from

(a) $NaBH_4 + I_2$ (b) $Na_2B_4O_7 + HCl$
(c) $BF_3 + NaH$ (d) $BF_3 + LiAlH_4$

73. Borax is converted into B by following steps

$$\text{Borax} \xrightarrow{A} H_3BO_3 \xrightarrow{\Delta} B_2O_3 \xrightarrow{B} \text{Product}$$

Reagents (A) and (B) are:

(a) Acid, Fe (b) Acid, Mg
(c) Acid, Sn (d) Acid, Al

74. When strongly heated, orthoboric acid gives

(a) $H_2B_4O_7$ (b) B_2O_3
(c) HBO_2 (d) $NaBO_2 + B_2O_3$

75. In the reaction: $Al_2(SO_4)_3.18H_2O \xrightarrow[-18H_2O]{heat}$ A $\xrightarrow{800°C}$ B + C. The product A, B and C are respectively

(a) $Al_2(SO_4)_3$, Al_2O_3, SO_3
(b) Al_2O_3, $Al_2(SO_4)_3$, SO_3
(c) Al_2SO_4, Al_2O_3, SO_3
(d) $Al_2(SO_4)_3$, Al_2O_3, SO_2

76. White fumes appear around the bottle of anhydrous $AlCl_3$ due to:

(a) Hydrolysis of $AlCl_3$ liberating H_2 gas
(b) Decomposition of $AlCl_3$
(c) Hydrolysis of $AlCl_3$ liberating HCl gas
(d) Hydrolysis of $AlCl_3$ liberating Cl_2 gas

77. In the reaction $B_2H_6 + 2KOH + 2X \rightarrow 2Y + 6H_2$ X and Y are respectively

(a) HCl, KBO_3 (b) H_2, H_3BO_3
(c) H_2O, KBO_2 (d) H_2O, KBO_3

78. X reacts with aqueous NaOH solution to form Y and H_2. Aqueous solution of Y is heated to 323-333 K and on passing CO_2 into it, Na_2CO_3 and Z were formed. When Z is heated to 1200°C, Al_2O_3 is formed. X, Y and Z respectively are

(a) Al, $Al(OH)_3$, $AlCl_3$
(b) Al, $NaAlO_2$, $Al(OH)_3$
(c) Zn, Na_2ZnO_2, $Al(OH)_3$
(d) Al, $AlCl_3$, $NaAlO_2$

79. Which of the following compounds are formed when BCl_3 is treated with water?

(a) $B_2O_3 + HCl$ (b) $H_3BO_3 + HCl$
(c) $B_2H_6 + HCl$ (d) None of these

80. Borazine is the product of reaction between

(a) BF_3 (b) B_2H_6
(c) NH_3 (d) Both B, C

81. The number of isomers possible for disubstituted borazine, $B_3N_3HX_2$ is

(a) 2 (b) 3
(c) 4 (d) 5

82. The dissolution of $Al(OH)_3$ by a solution of NaOH results in the formation of:

(a) $[Al(H_2O)_6(OH)_3]$
(b) $[Al(H_2O)_4(OH)]^{2+}$
(c) $[Al(H_2O)_2(OH)_4]^-$
(d) $[Al(H_2O)_3(OH)_3]$

83. H_3BO_3 is:

(a) Monobasic and weak Lewis acid.

(b) Monobasic and weak bronsted acid.
(c) Monobasic and strong lewis acid.
(d) Tribasic and weak bronsted acid.

84. Which of the following is only acidic in nature?

(a) $B(OH)_3$ (b) $Al(OH)_3$
(c) $Be(OH)_2$ (d) $Mg(OH)_2$

85. Which of the following impart green colour to the burner flame?

(a) Na(OMe) (b) $B(OMe)_3$
(c) $Sn(OH)_2$ (d) $Al(OPR)_3$

86. Heating an aqueous solution of aluminium chloride to dryness will give

(a) $AlCl_3$ (b) Al_2Cl_3
(c) $Al(OH)Cl_2$ (d) Al_2Cl_6

87. In diborane

(a) 2-bridged hydrogens and four terminal hydrogens are present.
(b) 3-bridged hydrogens and three terminal hydrogens are present.
(c) 4-bridged hydrogens and two terminal hydrogens are present.
(d) None of the above.

88. Among the following compounds of boron, the species which also forms π-bond in addition to σ-bonds is

(a) BF_3 (b) BF_4^-
(c) B_2H_6 (d) BH_3

89. Ionisation of boric acid in aqueous medium gives which one of the following?

(a) $[BO_3]^{3-}$ (b) $[B(OH)_4]^-$
(c) $[B(OH)_2O]^-$ (d) $[B(OH)O_2]^{2-}$

90. Consider the oxides: B_2O_3 , Al_2O_3, Ga_2O_3 and In_2O_3. The basic character of these oxides increases as:

(a) $In_2O_3 < Ga_2O_3 < Al_2O_3 < B_2O_3$
(b) $Al_2O_3 < B_2O_3 < In_2O_3 < Ga_2O_3$
(c) $B_2O_3 < Al_2O_3 < Ga_2O_3 < In_2O_3$
(d) $Ga_2O_3 < B_2O_3 < In_2O_3 < Al_2O_3$

91. Borax is used as a cleansing agent because on dissolving in water it gives:

(a) Alkaline solution
(b) Acidic solution
(c) Bleaching solution
(d) Basic solution

92. Hydrogen bridge structure of diborane molecule is

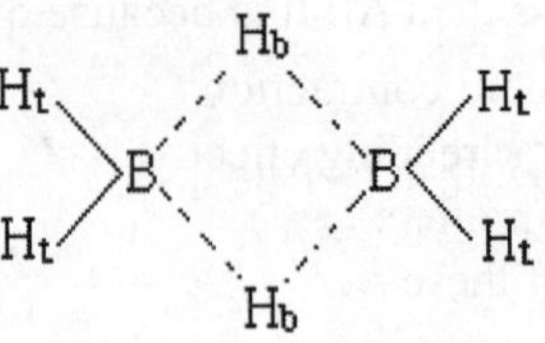

Which statement is incorrect a bont this compound?

(a) Hybridization of boron atom in the molecule is sp^3.
(b) Bond angle H_bBH_b is greater than $\angle H_tBH_t$.
(c) Bond length $B...H_b$ is greater than $B - H_t$.
(d) Bond angle $\angle H_bBH_b$ is lesser than $\angle H_tBH_t$.

93. Which of the following is incorrect with regard to the structure of diborane?

(a) Two bridged hydrogen atoms lie above and below the plane.
(b) The sp^3– hybrid orbitals of the boron atoms have one electron each.
(c) The terminal hydrogen atoms and boron atoms lie in a plane.
(d) Four terminal hydrogen atoms and two bridged |hydrogen atoms.

94. Which of the following pair shows amphoteric behaviours?

(a) Aluminium and thalium oxide
(b) Aluminium and gallium hydroxide
(c) Aluminium and thallium hydroxide
(d) Aluminium and gallium oxide

95. The oxidation state of boron family shows which of the following trend for stable +1 oxidation state

(a) Al < Ga < In < Tl
(+1 O.S. stability increases)
(b) Al < Ga < In < Tl
(stability of +3 oxidation states)
(c) Al < Ga < In < Tl
(stability +1 oxidation state)
(d) Al < Ga < In < Tl
(+3 O.S. stability increases)

96. Aluminium is more reactive than iron because its standard reduction potential is higher. Still aluminium is less easily corroded than iron because

(a) Al reacts with atmospheric carbon dioxide to form a self protective layer of Al_2O_3.
(b) it has higher reducing power and forms a self protective layer of Al_2O_3.

(c) it has higher reducing power and does not react with oxygen so easily.
(d) Both (A) and (B)

97. A solution when dilute with H_2O and boiled, it gives a white precipitate. On addition of excess NH_4Cl / NH_4OH the volume of precipitate decreases leaving behind a white gelatinous precipitate. Identify the precipitate which dissolves in NH_4OH / NH_4Cl.

(a) $Al(OH)_3$ (b) $Ca(OH)_2$
(c) $Mg(OH)_2$ (d) $Zn(OH)_2$

98. $B(OH)_3 + NaOH \leftrightarrow NaBO_2 + Na[B(OH)_4] + H_2O$
How can this reaction be made to proceed in forward direction?
(a) Addition of cis-1,2-diol
(b) Addition of borax
(c) Addition of trans-1,2-diol
(d) Addition of Na_2HPO_4

99. Which of the following statements about diborane is correct?
(1) It contains two 3-centre-2-electron bonds
(2) The B atoms in it are sp^3-hybridised
(3) The molecule is non-planar
(4) All B – H bond lengths are equal due to resonance
(a) 1, 2, 3 (b) 2, 3, 4
(c) 2, 3 (d) 1, 4

100. Aluminium chloride in acidified aqueous solution forms:
(a) Octahedral $[Al(H_2O)_6]^{+3}$ ion
(b) Tetrahedral $[A](H_2O)_6]^{+3}$ ion
(c) Octahedral $[Al(H_2O)_4]^{+3}$ ion
(d) Tetrahedral $[Al(H_2O)]^{+3}$ ion

PRACTICE QUESTIONS – III

101. The aqueous solution of an inorganic compound (X) gives white precipitate with NH_4OH which does not dissolve in excess of NH_4OH. This aqueous solution also gives white precipitate with $AgNO_3$ and the precipitate is soluble in dilute HNO_3. Here (X) is:
(a) $AlCl_3$ (b) $AlBr_3$
(c) AlN (d) $Al_2(SO_4)_3$

102. $Na_2B_4O_7 \xrightarrow[\Delta]{740°C} 2NaBO_2 + B_2O_3$
$\underbrace{X \quad + \quad Y}_{\text{Transparent}}$

$Z + CuO(s) \rightarrow Cu(BO_2)_2$
Blue Bead

The 'Z' will be:
(a) Y
(b) X
(c) Mixture of X & Y in 1 : 2 ratio
(d) Mixture of X & Y in 2 : 1 ratio

103. Which of the following is borane?
(1) B_5H_9 (2) B_5H_{10}
(3) B_5H_{11} (4) B_6H_{10}
(a) 1, 2, 3 (b) 1, 3, 4
(c) 2, 3, 4 (d) 1, 2, 4

104. Select the correct statements about diborane.
(1) Hb ….B….Hb bond angle is 122°
(2) All hydrogen in B_2H_6 lie in the same plane
(3) Each boron atom lies in sp^3 hybrid state
(4) B_2H_6 has three centered bond
(a) 1, 2, 3 (b) 1, 3, 4
(c) 2, 3, 4 (d) 1, 4

105. Which of the following statement is/are correct regarding B-F bond in BF_3?
(1) The unusual shortness and strength of the bonds may be explained by a pπ-dπ interaction between boron and fluoride.
(2) The unusual shortness and strength of the BF bond may be explained by pπ-pπ interaction between boron and fluorine atoms.
(3) All the three B-F bond lengths are equal and each of them is shorter than the sum of the covalent radii of boron and fluoride.
(4) The bond energy of the B-F bond is very high, higher than for any other single bond.
(a) 1, 2, 3 (b) 1, 3, 4
(c) 2, 3, 4 (d) 1, 4

106. Consider the following statements for diborane:
(1) Boron is approximately sp^3 hybridised.
(2) B – H – B angle is 180°.
(3) There are two terminal B – H bonds for each boron atom.
(4) There are only 12 bonding electrons available.
(a) 1, 2 and 3 are correct

(b) 2, 3 and 4 are correct
(c) 1, 3 and 4 are correct
(d) 1, 2 and 4 are correct

107. Identify the statement that is correct as far as structure of diborane is concerned.
(1) All B – H bonds in diborane are similar
(2) The hydrogen atoms are not in the same plane in diborane
(3) There are two bridging hydrogen atoms in diborane
(4) Each boron atom forms four bonds in diborane
(a) 1, 2 (b) 1, 2, 3
(c) 2, 3, 4 (d) 1, 3, 4

108. The incorrect statement regarding Aluminium & Thallium:
(1) Tl is more electropositive than Al.
(2) Al is more electropositive than Thallium.
(3) Tl^{+3} is more stable than Al^{+}.
(4) Tl^{+3} is a powerful reducing agent.
(a) 1, 2, 3 (b) 1, 3, 4
(c) 2, 3, 4 (d) 1, 4

109. 2 g of aluminium is treated separately with excess of dilute H_2SO_4 and excess of NaOH. The ratio of the volumes of hydrogen evolved is
(a) 1:1 (b) 1:2
(c) 2:1 (d) 2:3

110. Specify the co-ordination geometry around and hybridisation of N and B atoms in a 1 : 1 complex of BF_3 and NH_3.
(a) N : pyramidal, sp^3; B : pyramidal, sp^3
(b) N : tetrahedral, sp^3; B : tetrahedral, sp^3
(c) N : pyramidal, sp^3; B : tetrahedral, sp^3
(d) N : pyramidal, sp^3; B : planar, sp^3

111. Mark the correct statement regarding BF_3.
(1) It acts as a Lewis acid.
(2) It is volatile liquid at room temperature.
(3) It combines with either to form boron trifluoride etherate.
(4) It forms an addition compound with NH_3.
(a) 1, 2, 3 (b) 1, 3, 4
(c) 2, 3, 4 (d) 1, 4

112. Which one of the following is/are the incorrect statement(s)?
(1) Boric acid is a protonic acid.
(2) Beryllium exhibits coordination number of six.
(3) Chlorides of both beryllium and aluminium have-bridged chloride structures in solid phase.
(4) $B_2H_6.2NH_3$ is known as 'inorganic benzene'.
(a) 1, 2, 4 (b) 1, 2, 3
(c) 2, 3, 4 (d) 1, 3, 4

113. Which of the following is correct statement?
(1) Boron hydrides are formed when diluted HCl reacts with Mg_3B_2.
(2) All the B – H bond distances in B_2H_6 are equal.
(3) BH_3 is not a stable compound.
(4) The boron hydrides are readily hydrolysed.
(a) 1, 2 (b) 1, 2, 3
(c) 1, 3, 4 (d) 2, 3, 4

114. Which of the following statement is/are correct?
(1) $B(OH)_3$ reacts with NaOH, forming $Na[B(OH)_4]$.
(2) $B(OH)_3$ does not donate a proton and hence does not form any salt with NaOH
(3) $B(OH)_3$ partially reacts with water to form H_3O^+ and $[B(OH_4)^-]$ and behaves like a weak acid
(4) $B(OH)_3$ behaves like a strong monobasic acid in presence of sugars and this acid can be titrated against an NaOH solution using phenolphthalein as an indicator.
(a) 1, 2, 3 (b) 2, 3, 4
(c) 2, 3 (d) 1, 3, 4

115. Which of the following statement is/are correct for H_3BO_3?
(1) It has a layer structure in which BO_3 units are joined by hydrogen bonds.
(2) It is obtained by treating borax with concentrated H_2SO_4.
(3) It is mainly monobasic acid and a Lewis acid.
(4) It does not act as a proton donor but acts as an acid by accepting hydroxyl ions.
(a) 1, 2, 3 (b) 2, 3, 4
(c) 1, 3, 4 (d) All of these

116. Which of the following is correct statements?
(1) Aluminium is a strong oxidizing agent.
(2) Aluminium is extracted by the electrolysis of alumina in the presence of cryolite.
(3) Aluminium is used in the preparation of duralumin.
(4) Aluminium liberates hydrogen on treating with a base.
(a) 1, 2, 3 (b) 2, 3, 4
(c) 1, 3, 4 (d) 2, 4

117. Which of the following is incorrect?
(1) A crystalline form of boron nitride called borazon is harder than diamond.
(2) All the B – N bonds are equal in borazine.
(3) In borazole the negative charge is carried out by borons.
(4) Borazole is less reactive than benzene.

(a) 1, 2, 3 (b) 2, 3, 4
(c) 1, 3, 4 (d) All of these

118. Which of the following statement is correct?

(a) Al forms $[AlF_6]^{3-}$ ion but B does not form $[BF_6]^{3-}$ ion.
(b) Boron and aluminium halides behave as Lewis acids.
(c) The $p\pi - p\pi$ back bonding occurs in the halides of boron and not in those of aluminium.
(d) All of these

119. What is not a appropriate method for the isolation of elemental boron?

(a) Electrolytic reduction of aqueous $B(OH)_3$.
(b) Refluxing borax with sodium peroxide.
(c) High temperature reduction of B_2O_3 with magnesium.
(d) High temperature reduction of BBr_3 with hydrogen over a Ta wire.

120. A certain salt (X) gives the following tests:
On strongly heating it swells to give glassy material. When concentrated H_2SO_4 is added to a hot concentrated solution of (X), white crystals of a weak acid separate out. Identify (X)?

(a) $NaBO_2$ (b) $Na_2B_4O_7$
(c) $Na_2S_2O_3$ (d) $Ca_2B_6O_{11}$

121. Which of the following is correct?

(1) Al reacts with NaOH and liberate H_2.
(2) $AlCl_3$ is a Lewis acid.
(3) Al is used in the manufacture of electrical cables.
(4) NaOH is used during Halls process of purification of bauxite.

(a) 1, 2, 3 (b) 2, 3
(c) 2, 3, 4 (d) 1, 2, 3, 4

122. Which of the following is incorrect?

(1) Borax glass is the anhydrous form of borax.
(2) Jeweller's borax is obtained by crystallizing the solution at 25°C.
(3) Decahydrate form of borax is obtained by crystallizing the solution at 60°C.
(4) Boric acid is less soluble in hot water but more soluble in cold water.

(a) 1, 2, 3 (b) 3, 4
(c) 1, 3, 4 (d) 2, 3, 4

123. Observe the following statements regarding purification of bauxite:

1. During Hall's process, silica is removed as Si (vapour).
2. Bauxite ore contaminated with Fe_2O_3 is purified in Baeyer's process.
3. During Serpeck's process, AlN is formed.

The correct answer is
(a) 1, 2 and 3 are correct
(b) only 1 and 3 are correct
(c) only 2 and 3 are correct
(d) only 1 and 2 are correct

124. H_3BO_3 and HBO_2 differ in

(1) Basicity (2) Oxidation number
(3) Melting point (4) Structure
(a) 1, 2, 4 (b) 1, 3, 4
(c) 2, 3, 4 (d) 1, 2, 3

125. Which of the following is correct here?

(1) Diborane is a coloured gas and stable at low temperature
(2) The member of B_nH_{n+6} series are less stable than B_nH_{n+4}.
(3) The reaction of diborane with oxygen is endothermic
(4) It has banana bonds.

(a) 1, 2, 3 (b) 2, 3, 4
(c) 1, 3, 4 (d) 2, 4

126. Which one of the following methyl diboranes can exist?

(1) $B_2H_4(CH_3)_2$ (2) $B_2H_2(CH_3)_4$
(3) $B_2H_3(CH_3)_3$ (4) $B_2H(CH_3)_5$
(a) 1, 2, 3 (b) 2, 3, 4
(c) 1, 3, 4 (d) 2, 4

127. $BCl_3 + H_2O \xrightarrow{X}$

The products formed in this reaction are:
(a) $B_2H_6 + HCl$
(b) $H_3BO_3 + HClO_4$
(c) $B_2O_3 + HOCl$
(d) $H_3BO_3 + HCl$

ANSWER KEYS

1. (a)	2. (c)	3. (a)	4. (a)	5. (d)	6. (a)	7. (c)	8. (d)	9. (c)	10. (d)
11. (c)	12. (b)	13. (c)	14. (c)	15. (b)	16. (d)	17. (b)	18. (a)	19. (c)	20. (b)
21. (b)	22. (c)	23. (a)	24. (d)	25. (d)	26. (d)	27. (a)	28. (b)	29. (b)	30. (c)
31. (d)	32. (c)	33. (d)	34. (c)	35. (b)	36. (b)	37. (a)	38. (d)	39. (b)	40. (c)
41. (a)	42. (d)	43. (b)	44. (a)	45. (c)	46. (c)	47. (a)	48. (c)	49. (b)	50. (a)
51. (c)	52. (a)	53. (d)	54. (b)	55. (d)	56. (b)	57. (d)	58. (d)	59. (d)	60. (b)
61. (a)	62. (a)	63. (c)	64. (c)	65. (a)	66. (b)	67. (b)	68. (d)	69. (b)	70. (a)
71. (b)	72. (b)	73. (b)	74. (b)	75. (a)	76. (c)	77. (c)	78. (b)	79. (b)	80. (d)
81. (c)	82. (c)	83. (a)	84. (a)	85. (b)	86. (d)	87. (a)	88. (a)	89. (b)	90. (c)
91. (a)	92. (b)	93. (b)	94. (b)	95. (a)	96. (b)	97. (a)	98. (a)	99. (b)	100. (a)
101. (a)	102. (a)	103. (b)	104. (b)	105. (c)	106. (c)	107. (c)	108. (b)	109. (a)	110. (b)
111. (b)	112. (a)	113. (c)	114. (d)	115. (d)	116. (b)	117. (b)	118. (d)	119. (b)	120. (b)
121. (a)	122. (d)	123. (c)	124. (b)	125. (d)	126. (a)	127. (d)			

HINTS AND EXPLANATIONS FOR SELECTIVE QUESTIONS

2. $K_2SO_4.Al_2(SO_4)_3.\ 24H_2O$ is potash alum. It is used to purify water.

3. In BCl_3, boron is sp^2 hybridised which gives a planar structure to BCl_3 molecule.

9. $Al_2(SO_4)_3$ is colourless salt hence, will not give borax bead test.

12. Since boron in excited state has an outer shell configuration of $2s^1\ 2p^1x\ 2p^1y\ 2p_z$ it undergoes sp^2 hybridisation.

13. Boron is not used as an antiseptic to stop bleeding.

18. BF_4^- exists since maximum covalency of boron is 4. While Be, an alkaline earth metal form ionic carbides of the type BeC_2, instead Be_2C exist.

19. As Ga, In, Tl belongs to IIIA group.

20. Ammonal is a mixture of Al + NH_4NO_3

21. BBr_3 can accept the electrons most readily because it is much more electron deficient than BF_3. the decreasing order of acidity is

$BBr_3 > BCl_3 > BF_3$.

26. HBO_2 is called metaboric acid.

28. In the formation of diborane only 12 valency electrons only involved instead of 14 like in ethane.

30. $BCl_3 + 3H_2O \rightarrow B(OH)_3 + 3HCl$

Thus the products are $B(OH)_3$ i.e., H_3BO_3 and HCl.

35. Nickel metaborate, $Ni(BO_2)_2$ is brown in colour

45. Borazole is a polar compound due to electronegativity difference of the atoms

54. Boric acid consists of BO_3 units which are attached to each other by hydrogen bonding, that is why it forms polymeric structure.

56. The boron atoms are linked through hydrogen bridge in B_2H_6. the structures is not similar to ethane, there is no B – B bond and also all the atoms do not lie in the same plane.

57. Potash alum, which gives Al^{3+} ions easily which neutralize the negatively charged albuminoid substance of blood which then coagulate to form clot.

59. Al, Ga, In and Tl ions exists as octahedral aqua, ions, $[M(OH_2)_6]^{3+}$ in aqueous solution and many salts like halides, sulphates, nitrates are exist as hydrates but boron does not exist as octahedral aqua ions due to absence of vacant d-orbital.

65. Silicon due to its larger size form hydrides which are not so stable. However hydrides of boron are stable

68. $4BCl_4 + 3LiAlH_4 \rightarrow 2\ B_2H_6 + LiCl + 3AlCl_3$

74. $H_3BO_3 \xrightarrow{\text{heated}} HBO_2 + H_2O$

meta boric acid

$4HBO_2 \xrightarrow{\text{heated}} H_2B_4O_7 + H_2O$

$H_2B_4O_7 \xrightarrow{\text{strong heating}} 2B_2O_3 + H_2O$

75. $Al_2(SO_4)_3.18H_2O \xrightarrow[-18H_2O]{heat} \underset{A}{Al_2(SO_4)_3}$

$\xrightarrow{800°C} \underset{B}{Al_2O_3} + \underset{C}{3SO_3}$

78. Here X, Y and Z are Al, $NaAlO_2$ and $Al(OH)_3$ respectively.

79. BCl_3 is completely hydrolysed by water yielding boric acid and hydrochloric acid.

$BCl_3 + 3H_2O \rightarrow H_3BO_3 + 3HCl$

82. $Al(OH)_3 + OH- \rightarrow [Al(OH)_4^-]$

coordination number is 6, thus it exists as $[Al\ (H_2O)_2 (OH)_4]^-$.

84. Here $B(OH)_3$ or H_3BO_3 is acid while rest are bases.

85. The vapours of trialkyl borate, $B(OR)_3$ or R_3BO_3, burn with green edged flame.

$2BO_3^{3-} + 3H_2SO_4 \rightarrow 3SO_4^{2-} + 2H_3BO_3$
$H_3BO_3 + 3ROH \rightarrow \underset{\text{Trialkyl borate}}{R_3BO_3} + 3H_2O$

87. In B_2H_6 there are two bridged H atoms and four terminal H atoms which along with two B atoms lied in plane perpendicular to each other.

88. In BF_3, boron undergoes sp^2 hybridisation forming sigma bond with each F atom. Then there is a back $p\pi$- $d\pi$ bonding in B – F bond.

91. Borax on dissolving in water, gives alkaline solution.

$Na_2B_4O_7 + 7H_2O \rightarrow 4H_3BO_3 + 2Na^+ + 2OH^-$

94. The hydroxide of Al and Ga shows amphoteric character.

$2Al(OH)_3 + 3H_2SO_4\,(aq) \rightarrow Al_2(SO_4)_3\,(aq) + 6H_2O\,(l)$
$Al(OH)_3(s) + NaOH(aq) \rightarrow Na[Al(OH)_4]\,(aq)$

95. Oxidation states are +3 and +1, the stable oxidation state is +1 and its stability increases in the sequence
$Al < Ga < In < Tl$
(increasing stability of +1 O.S.)

97. $Al(OH)_3$ is a gelatinous white precipitate, formed by the reaction of Al^{3+} with NH_4OH in the presence of NH_4Cl.

98. Due to formation of chelated complex, the reaction moves in forward direction.

101. $AlCl_3 + 3NH_4OH \rightarrow Al(OH)_3 + 3NH_4Cl$
$Al(OH)_3 + NH_4OH \rightarrow$ No reaction
$AlCl_3 + 3AgNO_3 \rightarrow \underset{\text{White ppt.}}{3AgCl} + Al(NO_3)_3$

109. $2Al + 3H_2SO_4 \rightarrow Al_2(SO_4)_3 + 3H_2$
$2Al + 2NaOH + 2H_2O \rightarrow 2NaAlO_2 + 3H_2$
so the ratio of volumes of hydrogen evolved is 1 : 1.

111. BF_3 is not a volatile liquid but it is a colourless gas at room temperature .

112. As both $BeCl_2$ and $AlCl_3$ in solid state have bridged chloride structures.

116. Metals are reducing agents. Aluminium metal is a reducing agent but not oxidizing agent.

120. On strong heating it swells up to give a glassy mass so it may be borax. It is further confirmed as with H_2SO_4 it gives white crystals of boric acid (weak acid).

$Na_2B_4O_7.10H_2O \xrightarrow{\Delta} Na_2B_4O_7 + 10H_2O$

$Na_2B_4O_7 \xrightarrow{\Delta} 2NaBO_2 + B_2O_3$

Glassy mass

$Na_2B_4O_7 + H_2SO_4 + 5H_2O \rightarrow Na_2SO_4 + 4H_3BO_3$

Weak acid

Previous Years' Questions

1. In borax bead test which compound is formed?

(a) Double oxide (b) Tetra borate
(c) Meta-borate (d) Ortho-borate

[2002]

2. Which one of the following compounds is not a protonic acid?

(a) $PO(OH)_3$ (b) $B(OH)_3$
(c) $SO(OH)_2$ (d) $SO_2(OH)_2$

[2003]

3. Which of the following is the electron deficient molecule?
(a) B_2H_6 (b) SiH_4
(c) PH_3 (d) C_2H_6

[2005]

4. Al_2O_3 can be converted to anhydrous $AlCl_3$ by heating
(a) Al_2O_3 with HCl gas
(b) Al_2O_3 with NaCl in solid state
(c) a mixture of Al_2O_3 and carbon in dry Cl_2 gas
(d) Al_2O_3 with Cl_2 gas

[2006]

5. The stability of +1 oxidation state increases in the sequence:
(a) Al < Ga < In < Tl
(b) Tl < In < Ga < Al
(c) In < Tl < Ga < Al
(d) Ga < In < Al < Tl

[2009]

6. Tendency of BF_3, BCl_3 and BBr_3 to behave as Lewis acid decreases in the sequence?
(a) $BF_3 > BCl_3 > BBr_3$
(b) $BCl_3 > BF_3 > BBr_3$
(c) $BBr_3 > BCl_3 > BF_3$
(d) $BBr_3 > BF_3 > BCl_3$

[2010]

7. Which of the following molecular hydride acts as a Lewis acid?
(a) CH_4 (b) NH_3
(c) H_2O (d) B_2H_6

[2010]

ANSWER KEYS

1. (c) **2. (b)** **3. (a)** **4. (c)** **5. (a)** **6. (c)** **7. (d)**

HINTS AND EXPLANATIONS

1. In borax bead test metal metaborates are formed e.g.

$$Na_2B_4O_7 \xrightarrow{\Delta} \underset{\text{sod. metaborate}}{2NaBO_2} + \underset{\text{boric anhydride}}{B_2O_3}$$

$$CuO + B_2O_3 \rightarrow \underset{\text{Cupric meta borate (blue bead)}}{Cu(BO_2)_2}$$

2. Here $B(OH)_3$ or H_3BO_3 is not a protonic acid as it does not give proton on ionization directly.

3. B_2H_6 which is a dimer of BH_3 an electron deficient compound. Here B_2H_6 is Lewis acid

4. Al_2O_3 can be converted to anhydrous $AlCl_3$ by heating a mixture of Al_2O_3, carbon and dry chlorine as follows:

$$Al_2O_3 + 3C + 3Cl_2 \rightarrow 2AlCl_3 + 3CO$$

5. The stability of +1 oxidation state increases down the group due to inert pair effect. Hence the correct sequence is Al < Ga < In , Tl

6. Acidic nature decreases as Back bonding decreases in order $BBr_3 > BCl_3 > BF_3$

7. B_2H_2 is electron deficient compound formed by two BH_3 molecules. The B-H-B bonds in the centre are 3 centered 2 electron bond (also known as Banana Bond). So it tends to accept electron and react with lewis bases.

AIIMS ESSENTIALS

Assertion And Reason

In the following questions, two statements (Assertion) A and Reason (R) are given. Mark

(a) If A and R both are correct and R is the correct explanation of A;
(b) If A and R both are correct but R is not the correct explanation of A;
(c) A is true but R is false;
(d) A is false but R is true.
(e) Both A and R are false.

1. (A) AlF_3 is an ionic compound whereas BF_3 is covalent compound.
(R) BF_3 involves back π bonding.

2. (A) Al $(OH)_3$ is amphoteric in nature
(R) Al – O and O – H bonds can be broken with equal ease in $Al(OH)_3$.

3. (A) B_2H_6 contains two three centre electron pair bonds.
(R) B_2H_6 is electron deficient molecule.

4. (A) Among elements of group 13, boron has the highest melting point.
(R) Boron is metallic in nature.

5. (A) Diamond is harder than graphite.
(R) Graphite is more stable than diamond.

6. (A) Tl $(OH)_3$ is more basic than TlOH
(R) It can be explained on the basis of Fajan's rule.

7. (A) Al forms $[AlF_6]^{3-}$ but B does not form $[BF_6]^{3-}$.
(R) B does not react with F_2.

8. (A) In water, orthoboric acid behaves as a weak monobasic acid.
(R) In water, orthoboric acid acts as a proton donor.

9. (A) Gallium has higher ionization enthalpy than aluminium.
(R) This is due to imperfect shielding of outer shell electrons from the nucleus by the d- electrons of gallium.

10. A) Trihalides of boron family fume in moist air.
(R) Trihalides of group 13 elements are hygroscopic.

11. (A) BF_3 is a weaker Lewis acid than BCl_3
(R) The planar BF_3 molecule is stabilized to a greater extenet than BCl_3 by B – X π-bonding.

12. (A) Tl^{3+} acts as an oxidizing agent.
(R) Due to inert pair effect Tl^+ is more stable than Tl^{3+}.

13. (A) The compound called inorganic benzene is borazine.
(R) Borazine has a structure similar to benzene.

14. (A) Baeyer's process is used for refining of aluminium.
(R) Red bauxite contains iron oxide which is removed in Baeyer's process.

15. (A) Boric acid behaves as a weak monobasic acid.
(R) Boric acid contains hydrogen bonds in its structure.

Answer Keys

1. (b)	2. (a)	3. (b)	4. (c)	5. (b)	6. (d)	7. (b)	8. (c)	9. (a)	10. (c)
11. (a)	12. (a)	13. (a)	14. (d)	15. (b)					

CHAPTER **10B**

The p-Block Element Carbon Family

Chapter Outline

[IV (ns^2 np^2) or Group 14]

This group includes these elements:

C, Si,	Ge,	Sn, Pb
Non- metals	Metalloid	Metals

Out of these elements, silicon is the most abundant element in earth crust. In fact, it is the second most abundant element (27.2 % by weight) just after oxygen. It is present as silica and silicates. Carbon is also quite abundant (7th element in earth crust). In native state, it is found as coal, graphite and diamond and in combined state it is found as hydrocarbons, carbonates, carbohydrates, CO_2 etc. The abundance of Sn and Pb are less and nearly 2 ppm and 13 ppm respectively in nature. Ge is mainly extracted from fuel dusts obtained from the roasting of zinc and silver ores in the form of trace elements (1-5 ppm).

Electronic Configuration

The general electronic configuration of this group elements is ns^2 np^2 i.e., these elements have four electrons in their valence orbit.

$_6C$: $1s^2, 2s^2 2p^2$

or $[He]\ 2s^2 2p^2$

$_{14}Si$: $1s^2, 2s^2 2p^6, 3s^2 3p^2$

or $[Ne]\ 3s^2 3p^2$

$_{32}Ge$: $1s^2, 2s^2 2p^6, 3s^2 3p^6 3d^{10}, 4s^2 4p^2$

or $[Ar]\ 3d^{10}, 4s^2 4p^2$

$_{50}Sn$: $1s^2, 2s^2 2p^6, 3s^2 3p^6 3d^{10}, 4s^2 4p^6 4d^{10}, 5s^2 5p^2$

or $[Kr]\ 4d^{10}, 5s^2 5p^2$

$_{82}Pb$: $1s^2, 2s^2 2p^6, 3s^2 3p^6 3d^{10}, 4s^2 4p^6 4d^{10} 4f^{14}, 5s^2 5p^6 5d^{10}, 6s^2 6p^2$

or $[Xe]\ 4f^{14}, 5d^{10}, 6s^2 6p^2$

Physical Properties

(1) Atomic and Ionic Radii

The atomic and ionic radii of these elements are smaller than those of the corresponding IIIrd A group (13 gp) elements due to more effective nuclear charge. On moving down the group, atomic and ionic radii increase as effective nuclear charge decreases and number of orbits increases. However, the increase from the silicon onwards is less due to poor screening effect by d- and f-orbital electrons as follows:

	C,	Si,	Ge,	Sn,	Pb
At. radii (ppm)	77	118	12	140	146

(2) Density and Atomic Volume

On moving down the group, density increases due to more effective packing of the constituent particles in the higher element of the group. Atomic volume also increases down the group as follows:.

	C*,	Si,	Ge,	Sn,	Pb
Density (g/cm^3)	3.51	2.34	5.32	7.26	11.34
At. volume (ml)	3.4	11.4	13.6	16.3	18.26

Here C^* is for diamond and for graphite density is 2.22 g/cm^3.

(3) Melting and Boiling Points

Value of boiling and melting points of these elements are much more than those of boron family. On moving down the group, melting and boiling point decrease as follows

	C,	Si,	Ge,	Sn,	Pb
m.p.(K)	4373	1693	1218	505	600
b.p. (K)	-	3550	3123	2896	2024

- Carbon and silicon have high values of b.p. and m.p. due to their three dimensional network covalent bonded structures in which the constituent particles have strong interatomic forces of attraction. On moving down the group this interatomic force of attraction decrease so b.p. and m.p. also decrease.

(4) Ionization Enthalpy

These elements have higher ionization energy than those of corresponding IIIA group elements due to their smaller sizes and more effective nuclear charges. On moving down the group, ionization enthalpy decreases due to increase in size and decrease in effective nuclear charge as follows:

	C,	Si,	Ge,	Sn,	Pb
I.E. (kJ/mole)	1086	786	761	708	715

- After silicon the decrease of ionization energy is less due to poor screening effect by d- or f-orbital electrons.

(5) Electronegativity

These elements have higher values of electronegativities than those of IIIA group elements. On moving down the group, electronegativity decreases as follows:

C,	Si,	Ge,	Sn,	Pb
2.5	1.8	2	1.7	1.9

The electronegativity of these elements are nearly same due to filling of d- and f-orbitals

(6) Allotropy

All the elements of this group show allotropy except Pb.

Example, Carbon exists as crystalline (diamond, graphite) and amorphous (coal, coke, charcoal and carbon black etc).

Si exists in both crystalline as well as amorphous forms while Ge exists in two crystalline forms.

Tin exists in three forms as grey tin, white tin and rhombic tin.

(7) Metallic and Electropositive Nature

These are less metallic and electropositive elements than IIIrd A group elements due to their higher ionization energies and smaller sizes. On moving down the group, metallic and electropositive nature increase as size increases and ionization energy decreases.

C, Si,	Ge,	Sn, Pb
Non- metals	Metalloid	Metals

$\longrightarrow$

Increasing order of metallic and electropositive nature

(8) Catenation

It is the tendency of an element to form long chains or cyclic rings by linking its atoms with one another through covalent bonds. This tendency is shown by all of elements of this group and it decreases on moving down the group as the bond energy for catenation or strength of A – A bond decreases down the group.

C >> Si > Ge ≈ Sn >> Pb

$\longrightarrow$

Decreasing order of catenation

Bond:	C – C	Si – Si	Ge – Ge	Sn – Sn
Bond energy ($kJ\ mol^{-1}$)	348.4	222	167	155

(9) Covalency and Maximum Covalency

All these elements show a covalency of four i.e., they are tetravalent in nature.

It is explained on the basis of shifting of one of the paired ns-electron to the vacant empty orbital in the same orbit giving four sp^3 hybrids which results in a tetrahedral structure.

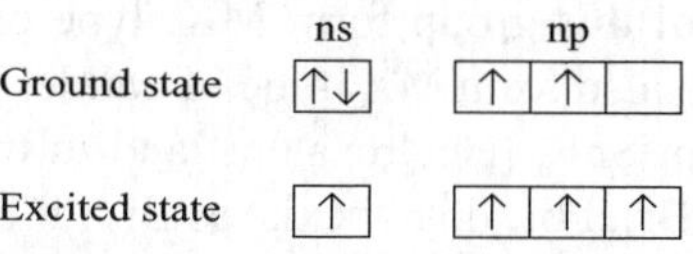

Carbon has maximum covalency four as it has no vacant d-orbitals to accommodate extra electrons while other elements of the group have vacant d-orbitals so they can show maximum covalency six also as the vacant d-orbitals permit the formation of coordinate bonds with other atoms or ions having lone pair of electrons example, silica, Ge, Sn, Pb can exhibit a covalency of six in $[SiF_6]^{2-}$, $[GeCl_6]^{2-}$, $[Sn(OH)_6]^{2-}$, $[PbCl_6]^{2-}$ etc.

(10) Oxidation States and Nature of Compounds

All the elements of this group show +4, +2 oxidation states mainly.

On moving down the group the stability of +4 state and covalent nature decrease while the stability of +2 state and ionic nature increase down the group.

Reason: These elements have ns^2 np^2 type of configuration so they can get octate state either by loosing four electrons or by gaining four electrons. This means they can form not only M^{4+} but also M^{4-} ions. But the stability of M^{4+} and M^{4-} are not much due to their higher ionization energies and electron affinities respectively.

example, In Be_2C, Al_4C_3, carbon exists as C^{4-}.

These elements specially carbon and silica have more tendency to form covalent compounds by showing tetracovalency as explained above in covalency.

- In lower oxidation state (+2) compounds are ionic while in higher oxidation state (+4) compounds are covalent (by applying Fajan's rule). On moving down the group +2 oxidation state becomes prominent and stable one due to inert pair effect (the last ns^2 electron pair do not take part in bonding)

example,- $PbCl_2$ < $PbCl_4$

Ionic Covalent

$C^{+2} < Si^{+2} < Ge^{+2} < Sn^{+2} < Pb^{+2}$

[Increasing order of stability]

[compounds are ionic]

$C^{+4} > Si^{+4} > Ge^{+4} > Sn^{+4} > Pb^{+4}$

Decreasing order of stability oxidising agent. (compounds are covalent)

(11) Tendency to Form Multiple Bonds ($p\pi - p\pi$ and $d\pi - p\pi$)

These elements have a tendency to form multiple bonds by $p\pi - p\pi$ and $d\pi - p\pi$ bonding. Carbon can form multiple bonds only by $p\pi - p\pi$ bonding while the other elements can form multiple bonds by $d\pi - p\pi$ bonding.

(i) *$p\pi - p\pi$ Multiple Bonds:* Carbon can form stable multiple bonds not only with itself but also with some other non metals like N, O, S by the lateral overlapping of the p-orbitals of these atoms through $p\pi - p\pi$ bonding.

This tendency of carbon to form $p\pi - p\pi$ multiple bonds is reflected in the structure of graphite. In which carbon atoms are sp^2-hybridised. The planar hexagonal layers of carbon atoms in graphite are held together by weak Van der Waal's forces. With in such a layer a carbon atom is linked to three other carbon atoms by sp^2 hybrid bonds and the additional pelectron from each carbon can form an extended delocalized pi-bonding system encompassing the entire layer.

- Silica can not have graphite like structure but have only diamond like structure it means silica has no tendency to form $p\pi - p\pi$ multiple bonds.

(ii) *$d\pi - p\pi$ Multiple Bonds:* Silica and other elements of the group can form $d\pi - p\pi$ multiple bonds as they have d-orbitals.

This bonding is prominent than Si is attached to oxygen or nitrogen atom. example, Trimethyl amine$[(CH_3)_3N]$ is pyramidal in shape while trisilylamine $[(SiH_3)_3N]$ is planar in shape. It can be explained as follows:

Carbon does not have d-orbitals so it can not form $d\pi - p\pi$ type of multiple bonds while Si can form these bonds due to the presence of d-orbitals.

In trimethyl amine, nitrogen atom is sp^3 hybridised and have a lone pair of electron so its shape is pyramidal while in trisilylamine, nitrogen atom is sp^2 hybridised and the lone pair of nitrogen atom is transferred to an empty d-orbital of silicon atom by $d\pi - p\pi$ bonding which is formed by the overlapping of unhybridised p-orbital of nitrogen atom with a d-orbital of silicon atom. This gives a planar shape to this molecule.

Trimethyl amine (pyramidal)

Trisilylamine (planar)

Chemical Properties

This group elements are not very much reactive however their reactivity increases down the group from C to Sn. Here Pb is more unreactive than acceptance it is due to coating of an oxide layer on its surface and the high over potential for the reduction of H^+ ion into H_2 at the lead surface.

(1) Hydrides

All the elements of this group form MH_4 type of covalent hydrides however the number of hydrides, their ease of formation and thermal stability decrease down the group. Besides MH_4, carbon can form a number of cyclic and acyclic hydrides i.e., hydrocarbons like alkanes, alkenes, alkynes, cycloalkanes, arenes etc.

Si, Ge can form less number of hydrides than carbon Their common hydrides have a general formula $M_n H_{2n+2}$ which are respectively called silanes and germanes. Sn and Pb can form only SnH_4 and PbH_4 type of hydrides.

example,

CH_4, SiH_4, GeH_4, SnH_4, PbH_4

Not hydrolysed due to absence of d-orbital — stannate — Plumbane

$\longrightarrow$

Increasing order of hydrolysis , reducing nature Thermal stability, volatility decrease

$$SiH_4 \xrightarrow[\text{Decomp}]{\Delta} Si + 2H_2\uparrow$$

'pure' (semiconductor)

Silanes are more reactive than the alkanes. It is mainly due to the larger size of Si atom and its less electronegativity which makes attack of nucleophiles easier moreover silica has low energy d-orbitals so it can easily form intermediate compounds and it decreases activation energy of the process.

On moving down the group stability decreases as the strength of M – H decreases due to increase in the sizeof M.

Mono-Silane

It is thc most important type of silane which can be prepared by the reduction of $SiCl_4$ with $LiAlH_4$.

$$SiCl_4 + LiAlH_4 \rightarrow SiH_4 + LiCl + AlCl_3$$

SiH_4 is a colourless inflammable gas. It can be used in the preparation of highly pure silicon used in the semi conductors. When SiH_4 is pyrolysed in absence of air pure silicon is obtained.

(2) Halides

All the elements of this group form MX_4 type of halides. Si, Ge, Sn, Pb, can also form MX_2 type of halides. All MX_4 are covalent except SnF_4 (electrovalent) and in them M is sp^3 hybridised. $PbBr_4$, PbI_4 do not exist as Br^- , I^- are strong reducing agents and Pb^{+4} is a strong oxidizing agent

The thermal stability of these halides decreases with the increase in molecular mass of the tetrahalides or down the group and these can be,

example,– $CC_{14} > SiCl_4 > GeCl_4 > SnCl_4 > PbCl_4$

$\longrightarrow$

Decreasing order of thermal stability

Increasing order of hydrolysis

All the MX_4 are easily hydrolysed by water except CCl_4 which is not hydrolysed due to absence of d-orbital Example,

$$SiCl_4 + 4H_2O \rightarrow Si(OH)_4 + 4HCl$$

CCl_4 can not be hydrolysed due to absence of d-orbital in the carbon atom however CCl_4 on heating with water gives phosgene as follows:

$$CCl_4 + H_2O \xrightarrow{500°c} COCl_2 + 2HCl$$

Steam — Phosgene

- Among carbon halides the thermal stability and volatile nature decrease as follows:

$CF_4 > CCl_4 > CBr_4 > CI_4$

Most thermally stable, volatile — Least volatile stable

Except CCl_4, rest of the MX_4 can form hexahalo complexe showing Lewis acidic nature due to the availability of vacant d-orbitals. example,

$$SiF_4 + F_2 \text{ or } 2F^- \rightarrow [SiF_6]^{2-}$$

$$SnCl_4 + Cl_2 \text{ or } 2Cl^- \rightarrow [SnCl_6]^{2-}$$

Dihalides

Ge, Sn, Pb can form MX_2 type of dihalides which are more ionic in nature than MX_4 their stability increases down the group.

- **Stability Order:**

  $GeX_2 < GeX_4$
  $SnX_2 < SnX_4$
  $PbX_2 > PbX_4$

(3) Oxides

The elements of this group can form two type of oxides i.e., MO, MO_2 type of oxides.

(a) Monoxide (MO) Type

All elements of this group can form monoxides of MO type. example,

CO, SiO,	GeO, SnO,	PbO
Neutral	Amphoteric	Basic

⟶ Basic nature increases

(b) Dioxides (MO_2) Type

All element of this group can form dioxides of MO_2 type.

CO_2, SiO_2,	GeO_2, SnO_2,	PbO_2
Acidic	Amphoteric	Basic

⟶ Acidic nature decreases

Here CO_2 is a gas due to weak Vanderwaal's force while rest are solid due to strong covalent bonds.

- At room temperature CO_2 is a gaseous linear molecule while SiO_2 (silica) is a giant network solid having a continuous lattice of Si and O atoms connected by covalent bonds in a tetrahedral manner.

Property	CO_2	SiO_2
Physical state	Gas	Solid
Structure	It is a monomeric linear molecules with two carbonoxygen double bonds O = C = O	It has a giant network structure having a continuous lattice of silicon and oxygen atoms in which each silicon atom is tetrahedrally bonded to four oxygen atoms through covalent bonds.
Solubility in water	Soluble	Insoluble
Acidic character	Weakly acidic	Less acidic
Reduction	On passing through red hot coke, it gets reduced to carbon monoxide. $CO_2 + C \rightarrow 2CO$	On heating with carbon in an electric furnace, it is reduced to silicon carbide which is an abrasive. $SiO_2 + 3C \rightarrow SiC + 2CO$

- PbO_2 can be used as a powerful oxidizing agent as it can dissolve in acids to give oxygen.

  $PbO_2 + 2HNO_3 \rightarrow Pb(NO_3)_2 + H_2O + ½\ O_2$

CARBON

Anomalous Behaviour of Carbon

Carbon shows abnormal behaviour due to its small size, high electronegativity and absence of d-orbitals such as:

(i) Carbon has maximum catenation power, due to maximum bond energy for catenation.

$C > Si > Ge \approx Sn > Pb$

(ii) It can not form hexavalent compounds or complex compounds (CF_6) due to absence of d-orbital while Si, Ge, Sn, Pb can form complexes like

$[SiF_6]^{-2}$, $[SnCl_6]^{-2}$, $[GeCl_6]^{-2}$, $[PbF_6]^{-2}$

Sp^3d^2 {Octahedral}

(iii) It is most compact, most hardest element in its group and have highest melting and boiling points in the group.

Occurrence

In occurs in freestate in the form of coal, diamond and graphite and in combined state in the form of CO_2, carbonates and hydrocarbons. It exists in two allotropic forms: crystalline and amorphous

(1) Crystalline Form

It is of three types as follows:

(a) Diamond

It is the hardest known substance. It is the purest form of carbon with high density (3.5 g/ml), melting point (7350ºC) and refractive index (2.45).

Structure

In diamond carbon atom is sp^3 hybridised with tetrahedral structure. Here one carbon atom is attached with four other carbon atoms by covalent bonds and this gives diamond in three dimensional polymeric structure. Here C – C bond length is 1.54 Å and bond angle is 109° 28'

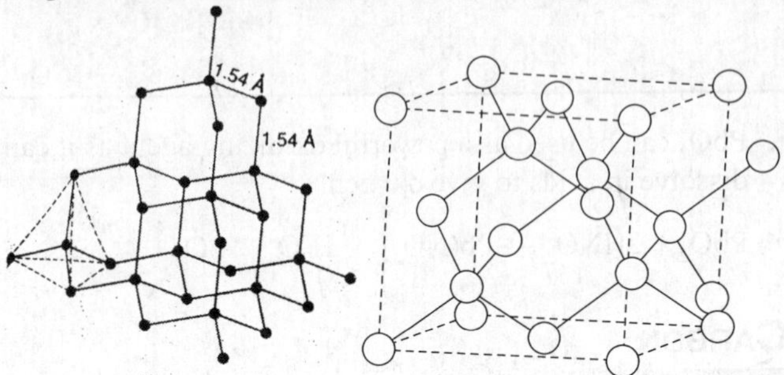

FIGURE 10.10 Structure of Diamond

Properties

(1) The high value of boiling point, melting point, density are due to its three dimensional structure which needs high energy for bond breaking. Light rays, X-rays (highest) can pass through it.

(2) It is transparent solid with a specific gravity of 3.52.

(3) Due to high refractive index on proper cutting it produces maximum total internal reflection.

(4) It is bad conductor due to absence of free electrons.

(5) It is insoluble in all solvents.

(6) Pure diamonds are colourless while impure diamond may be coloured. Kohinoor diamond is of 186 carat while Pit diamond is of 136 carat.

(7) Diamond on heating in vacuum at 1800 – 2000°C converts into graphite.

(8) Being chemically inert it does not react with acids, bases etc.

(9) When it is heated in air upto 1173 K, CO_2 is formed and on heating with fluorine upto 973 K it gives CF_4.

$$C + O_2 \rightarrow CO_2$$

$$C + 2F_2 \rightarrow CF_4$$

Uses

Diamond dyes are oftenly used in making cutting tools. It is used as an abrasive.

(b) Graphite

It is a gray coloured substance with metallic lustre and more stable and reactive than diamond.

Structure

It has a two dimensional sheet like hexagonal structure in which each carbon atom is sp^2 hybridised and it is attached with three other carbon atoms in a hexagonal planar structure method. Here C – C bond length is 1.42 Å and the distance between two successive layer is 3.35 Å. These layers are held together by weak Van der Waal's forces and slippery in nature so graphite is soft and have low density. Here C – C bond length is shorter than that of diamond because a pi bond is formed here by the fourth valence electron present in each carbon atom which is free.

Due to these pi-electrons which are able to move in these slippery structures graphite can conduct electricity.

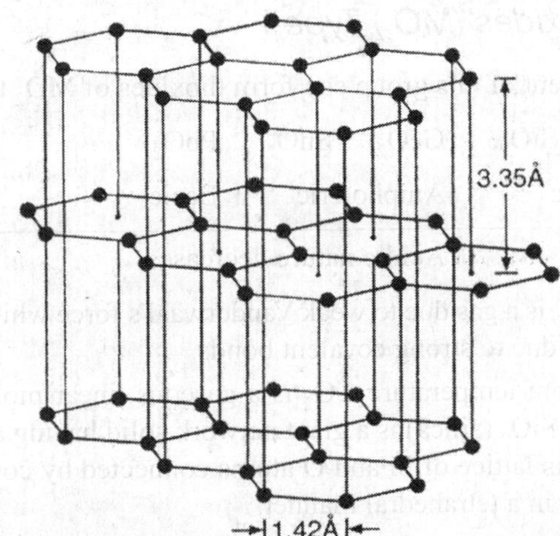

FIGURE 10.11 Structure of Graphite

Preparation

Acheson method is used to prepare graphite.

$$SiO_2 + 3C \xrightarrow{\Delta} SiC + 2CO$$

$$SiC \xrightarrow{\Delta} C + Si$$

Properties

(1) It is a soft, grayish, greasy crystalline solid with a density of 2.5 g/ml.

(2) Graphite has lowest energy among carbon allotropes.

(3) It is a conductor of electricity due to the presence of free electrons in its slippery hexagonal sheets. Its conductivity increases with increase of temperature.

(4) It is called plumbago or black lead as it leaves a black mark on paper.

(5) It is thermodynamically more stable than diamond and it has 1.9 kJ less free energy of formation than diamond.

(6) On heating at 1873 K and at a very-very high pressure (50,000-60,000 atm) it changes into diamond.

(7) It is more reactive form than diamond, however, it does not react with dilute acids or alkalies.

(8) Graphite on heating with conc. HNO_3 gives graphitic acid ($C_{11}H_4O_5$).

(9) Graphite on oxidation with alkaline $KMnO_4$ gives mellitic acid or benzene Hexacarboxylic acid [$C_6(COOH)_6$] and oxalic acid.

(10) On heating graphite with vapours some metals like K, Rb, Cs at nearly 600 K, a paramagnetic compound C_8M is obtained (M = K, Rb, Cs).

Uses

(1) It is used in pencil leads (graphite and clay) and as a lubricant in making electrodes.

(2) It is also used in electroplating, electrotyping, painting, stoves and in making refractory crucibles.

(3) In nuclear reactor it is used as moderator.

(4) It is a better lubricant on earth than Moon.

(c) Fullerenes

These are latest crystalline pure allotropes of carbon obtained by evaporation of graphite by laser. C_{60} type of fullerene was discovered by H.W. Proto, Smalley and Robert Curt (Noble prize winners) by applying laser beam on sample of graphite.

These are the only pure form of carbon as unlike diamond and graphite. These do not have dangling edge or surface bonds which are instrumental in attracting other atoms. These are large spheroidal cage like molecules of carbon having molecular formula with C_{60}, C_{70} etc.

C_{60} is called Buckminister fullerene in the honour of Robert Buckminister fuller and its structure is similar to a soccer ball. It has 12 five membered rings and 20 six membered rings. The five membered rings are connected only to six membered rings while six membered rings are connected both the type of rings. In it C – C and C = C bond lengths are 145.3 and 138.3 pm respectively.

Now C_{70} has also been discovered and referred as Buckyballs. It can be easily separated from the shoot by extraction in toluene followed by chromatographic separation over alumina.

Fullerenes are soluble in organic solvents giving coloured solutions example, C_{60} gives purple colour in toluene while C_{70} give orange colour.

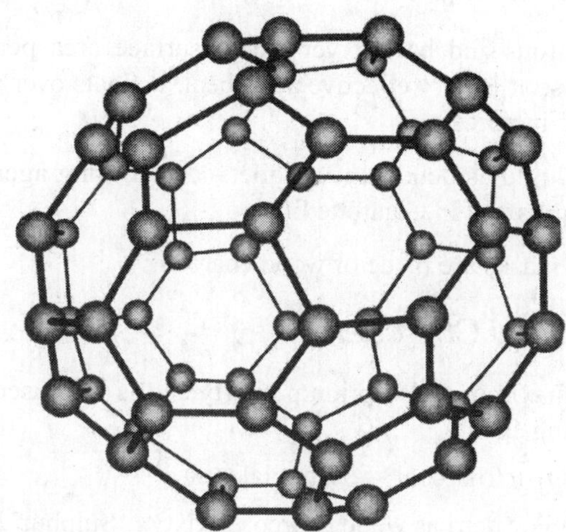

FIGURE 10.12 Structure of Buckminster Fullerene, C_{60}

(2) Amorphous Form

(a) Coal

It is crude form of carbon which is formed in nature by the slow decomposition of vegetable matter under the influence of air, heat and pressure. The main type of coal are as follows:

Type	Carbon content	Calorific value in B.T.U.
Peat coal	60%	9990
Lignite	70%	11700
Bituminous coal	80%	14900
Anthracite coal	90%	16200

Here the common coal is Bituminous which is hard like stone and burns which smoky flame.

Coal can be used as a fuel and in the manufacture of synthetic petrol, coal gas etc.

Coke

Coal – Volatile separation → Coke

On destructive distillation of coal by heating in absence of air the volatile constituents like coal gas, ammonia etc. are removed and coke is formed. Coke can be used as a fuel and as a reducing agent in iron and steel industry.

(b) Charcoal

It is the most active form of carbon which can react with many oxidizing agent acids etc. It can be obtained by burning wood, cellulose etc. in a limited supply of air. It is of following types:

(1) Wood Charcoal

It is porous and have a very large surface area per unit weight so it is very effective adsorbent. It floats over water (fills air in pores).

It is used as deodorant in toilets, decolouring agents in sugar mills and in Cigarette filter.

Gas masks are made of wood charcoal.

(2) Animal Charcoal

It has maximum adsorption property and is also used as a bleaching agent.

HNO_3 + Charcoal → Artificial tannin

- Gun Powder has wood charcoal, KNO_3, 'Sulphur'

Lamp Black or Carbon Black

It is very pure carbon having 98 – 99 % carbon. It can be obtained by the burning of substances having high content of carbon such as petroleum, kerosene oil etc. in a limited amount of air.

It is soft black powder which is used for making printing inks, black paints, ribbons of type writers and as a filler in making rubber tyre.

Properties of Carbon

(1) Formation of Carbon dioxide

When carbon (all allotropes) is heated with oxygen or air CO_2 is formed along with some amount of CO.

$$C + O_2 \rightarrow CO_2$$

$$2C + O_2 \rightarrow 2CO$$

(2) Formation of Carbides

Carbon on heating with some oxides forms carbides.

Example,

$$SiO_2 + 3C \xrightarrow{2273K} \underset{\text{Carborundum}}{SiC} + 2CO$$

$$2Al_2O_3 + 9C \xrightarrow{2273K} Al_4C_3 + 6CO$$

$$CaO + 3C \xrightarrow{2273K} \underset{\text{Calcium carbide}}{CaC_2} + CO$$

(3) Reducing Properties

Being a strong reducing agent it can reduce oxides, sulphates etc. as follows:

$$H_2O + C \rightarrow CO + H_2$$

$$Fe_2O_3 + 3C \rightarrow 2Fe + 3CO$$

$$BaSO_4 + 4C \rightarrow BaS + 4CO$$

(4) With Non-metals

It can react with some non metals like hydrogen, sulphur etc. as follows:

$$2C + H_2 \xrightarrow{\text{electric arc}} \underset{\text{Acetylene}}{C_2H_2}$$

$$\underset{\text{Red hot}}{C} + \underset{\text{Vapours}}{2S} \rightarrow \underset{\text{Carbon disulphide}}{CS_2}$$

(5) With Acids

Charcoal on dissolving in hot and dilute nitric acid gives a brown coloured artificial tannin while concentrated HNO_3 oxidises carbon into CO_2 as follows:

$$C + 4HNO_3 \rightarrow CO_2 + 4NO_2 + 2H_2O$$

When carbon is oxidized with concentrated H_2SO_4, CO_2 is formed as a major product along with mellitic acid as follows:

$$12\text{'C'} + 9\,H_2SO_4 \xrightarrow{\Delta} C_6(COOH)_6 + 6H_2O + 9SO_2$$

Mellitic acid or Benzene hexcarboxylic acid

$$C + 2H_2SO_4 \rightarrow CO_2 + 2SO_2 + 2H_2O$$

Compounds of Carbon

Carbon Monoxide

$:C \equiv O:$

It is present in some quantity in volcanic gases, exhaust gases of internal combustion engines and chimney gases.

Preparation

It can be prepared as follows:

(1) It is obtained by the incomplete combustion of carbon as follows:

$$2C + O_2 \xrightarrow[\text{Less air}]{} 2CO$$

(2) From HCOOH

When formic acid or sodium formate is heated with concentrated H_2SO_4, pure carbon monoxide is formed as follows:

$$HCOOH \xrightarrow{\text{conc. } H_2SO_4} H_2O + CO\uparrow$$

$$2HCOONa + H_2SO_4 \rightarrow Na_2SO_4 + 2H_2O + 2CO$$

(3) From Oxalic Acid

When oxalic acid is treated with concentrated sulphuric acid, carbon monoxide is formed as follows:

$$\begin{array}{l} COOH \\ | \\ COOH \end{array} \xrightarrow{\text{conc. } H_2SO_4} CO\uparrow + CO_2\uparrow + H_2O$$

(4) From Potassium Ferrocyanide

When potassium ferrocyanide is treated with sulphuric acid, carbon monoxide is formed as follows:

$$K_4Fe(CN)_6 + 6H_2SO_4 + 6H_2O \rightarrow 2K_2SO_4 + FeSO_4 + 3(NH_4)_2SO_4 + 6CO$$

(5) By Passing Steam over Red Hot Coke

When steam is passed over red hot coke, a mixture of CO and H_2 is formed.

$$C + H_2O \rightarrow \underbrace{CO + H_2}_{\text{Water gas}}$$

(6) By Heating Metal Carbonates

Carbonates of Ca, Ba, Mg on heating with zinc gives CO as follows:

$$CaCO_3 + Zn \longrightarrow CaO + ZnO + CO$$

(7) By the Reduction of Carbon dioxide

It can be reduced into CO by passing it over red hot zinc as follows:

$$Zn + CO_2 \xrightarrow{\text{heat}} ZnO + CO$$

(8) By the Reduction of Heavy Metal Oxide

These oxides on heating with carbon give CO as follows:

$$Fe_2O_3 + 3C \rightarrow 2Fe + 3CO$$

$$ZnO + C \rightarrow Zn + CO$$

Physical Properties

(1) It is a neutral, highly poisonous, colourless, odourless gas and burns with blue flame.

$$2CO + O_2 \rightarrow 2CO_2 + \text{heat}$$

(2) It is less soluble in water.

(3) Its density is nearly equal to that of air.

(4) It is a reducing agent in Mond's process (Ni).

(5) It is a linear molecule in which carbon atom is sphybridized.

(6) When it mixes with heamoglobin of blood it forms carboxy-haemoglobin which destroys oxygen carrying capacity of blood as it can absorb oxygen.

$$\text{Haemoglobin} + CO \rightarrow \underset{\text{(stable)}}{\text{Carboxy-Haemoglobin}}$$

- Here due to suffocation death may occur.

Chemical Properties

(1) Combustion

Carbon monoxide burns in air with blue flame with an exothermic reaction.

$$2CO + O_2 \rightarrow 2CO_2$$

(2) Reducing Properties

It is a strong reducing agent so it is widely used for the extraction of metals from their oxides.

example,

$$Fe_2O_3 + 3CO \xrightarrow{600\text{-}900^\circ C} 2Fe + 2CO_2$$

$$ZnO + CO \rightarrow Zn + CO_2$$

$$2Cu(OH)_2 + CO \rightarrow Cu_2O + 2H_2O + CO_2$$

$$I_2O_5 + 5CO \rightarrow I_2 + 5CO_2$$

This iodine turns $CHCl_3$ and CCl_4 layer violet so it is a test of CO.

(3) Reaction with Chlorine

It combines with chlorine in presence of sunlight to form phosgene as follows:

$$CO + Cl_2 \rightarrow \underset{\text{Phosgene}}{COCl_2}$$

(4) Reaction with Metals

Carbon monoxide reacts with metals to form metalcarbonyls.

$$M + xCO \rightarrow M[CO]_x$$

Here M may be Ni, Cr, Mo, Fe, Co etc. example,

$$Ni + 4CO \xrightarrow{325\text{–}345K} \underset{\text{Nickel carbonyl}}{Ni(CO)_4}$$

$$Fe + 5CO \xrightarrow{473K} \underset{\text{Iron penta carbonyl}}{Fe(CO)_5}$$

(5) Reaction with Dihydrogen

When it reacts with dihydrogen at 425-675K in presence of a catalyst, methanol is formed as follows:

$$CO + 2H_2 \xrightarrow[ZnO + Cu]{425\text{-}675K} \underset{\text{Methanol}}{CH_3OH}$$

Uses

(1) It is used as a reducing agent and in the preparation of metal carbonyls.

(2) It is used in war gas preparation (phosgene)

(3) It is used in the refining of nickel in Mond's process

(4) It is used in the preparation of methanol, phosgene, synthetic petrol etc.

(5) It is used in gaseous fuel as water and producer gas.

POINTS TO REMEMBER

- Carbogen (mixture of O_2 + 5-10% CO_2) is used for artificial respiration for victims of CO poisoning.

Structure: The structure of CO is represented as follows:

$$:\overset{-}{C}:::\overset{+}{O}: \quad \text{or} \quad :\overset{-}{C}\equiv\overset{+}{O}: \quad \text{or} \quad :C \overset{\leftarrow}{=} O:$$

Here carbon atom is sp hybridized and the carbon - oxygen bond length is 1.13 Å. It has lone pair of electrons on carbon atom so it can act as a ligand and can form a coordinate bond with metal atoms during complex formation.

CARBON DIOXIDE
O = C = O OR CO_2

It is present 0.03 – 0.05 % by volume in air. It is a linear non-polar molecule having zero value of dipole moment.

Preparation

It is prepared by following methods:

(1) From Carbonates and Bicarbonates

Metal carbonates and bicarbonates on heating give carbon dioxide as follows:

$$M - CO_3 \xrightarrow{\Delta} CO_2 \uparrow + MO$$

example,

$$CaCO_3 \xrightarrow{\Delta} CaO + CO_2$$

$$M - HCO_3 \xrightarrow{\Delta} M\text{-}CO_3 + CO_2 \uparrow + H_2O$$

example,

$$Ca(HCO_3)_2 \xrightarrow{\Delta} CaO + 2CO_2 + H_2O$$

(2) Lab method

When metal carbonates or bicarbonates are treated with mineral acids, CO_2 is formed as follows:

$$M - CO_3 + HCl \xrightarrow{\Delta} CO_2 + M - Cl + H_2O$$

$$MHCO_3 + HCl \rightarrow MCl + H_2O + CO_2$$

For example, When $CaCO_3$ is treated with dilute HCl, carbon dioxide is formed.

$$CaCO_3 + 2HCl \xrightarrow{\Delta} CaCl_2 + CO_2 + H_2O$$

(3) By Alcoholic Fermentation

During the fermentation of molasses into alcohol, CO_2 is also formed in good amount.

$$\underset{\text{Glucose}}{C_6H_{12}O_6} \xrightarrow{\text{enzyme}} \underset{\text{Ethanol}}{2C_2H_5OH} + \underset{\text{by product}}{2CO_2}$$

(4) By the Complete combustion of Carbon

$C + O_2 \rightarrow CO_2$

$2C + O_2 \rightarrow 2CO$

$2CO + O_2 \rightarrow 2CO_2$

(5) From Fuel Gases

Fuel gases produced during the combustion of coal or coke have a good amount of CO_2 mixed with N_2, O_2 and CO.

On passing the gaseous mixture through K_2CO_3, CO_2 is absorbed and potassium bicarbonate is formed which on heating gives CO_2 as follows:

$K_2CO_3 + CO_2 + H_2O \rightarrow 2KHCO_3$

$2KHCO_3 \rightarrow K_2CO_3 + H_2O + CO_2$

Properties

(1) It is a colourless, odourless, tasteless gas which is heavier than air.

(2) It is partially soluble in water under normal pressure but more soluble at high pressure.

(3) It is non-poisonous but it does not support life.

(4) CO_2 is acidic in nature and called Carbonic anhydride.
$H_2O + CO_2 \rightarrow H_2CO_3$

(5) Dry Ice: CO_2 gas can be easily liquefied into a mobile liquid which can be further changed into a white solid mass of CO_2. Solid carbon dioxide is known as dry ice since it does not wet the surface on which it melts. It is used as a coolant for preserving fresh articles in food industry and for making cold baths in the laboratory. Under the name dry cold it is used as a refrigerant. It can provide not only cold but also an inert atmosphere which can help in the killing of unwanted bacterias.

Property	CO	CO_2
B.P. (K)	81.5	194.5
C – O bond length (pm)	112.8	116.3
Density (g/L) at 273 K	1.250	1.977
ΔH_f (kJ mol^{-1})	110.5	-393.5
M.P. (K)	68	216.4 (at 5.2 atm)
Structure	$:C \equiv O:$	$O = C = O$

Chemical Properties

(1) Non-Combustible Nature and Non-supporter of Combustion

It is neither combustible nor supports it, however, active metals like Na, K, Mg etc. continue to burn in the atmosphere of CO_2.

$CO_2 + 4Na \rightarrow 2Na_2O + C$

$CO_2 + 2Mg \rightarrow 2MgO + C$

CO_2 can be used as a fire extinguisher except in case of active metals like Na, Mg etc.

(2) Reduction

When carbon dioxide is passed over red hot coke, it gets reduced to carbon monoxide.

$CO_2 + C \rightarrow 2CO$

Red hot coke

(3) Acidic Nature

It is an acidic oxide as it forms carbonic acid on dissolving in water and can form salts with bases as follows:

$CO_2 + H_2O \rightarrow H_2CO_3$

$2NaOH + CO_2 \rightarrow Na_2CO_3 + H_2O$

$Na_2CO_3 + CO_2 + H_2O \rightarrow 2NaHCO_3$

(4) Action on Lime Water

CO_2 turns limewater into milky in soluble calcium carbonate.

$Ca(OH)_2 + CO_2 \longrightarrow \underset{\text{Milky}}{CaCO_3\downarrow} + H_2O$

This milkiness disappears when more CO_2 is passed as soluble calcium bicarbonate is formed.

$CaCO_3 + H_2O + CO_2 \rightarrow \underset{\text{Soluble}}{Ca(HCO_3)_2}$

$Ca(HCO_3)_2 \rightarrow CaCO_3 + CO_2 + H_2O$

(5) Formation of Carbonates

It reacts with some basic oxides like Na_2O, K_2O to form their carbonates.

$K_2O + CO_2 \rightarrow K_2CO_3$

$Na_2O + CO_2 \rightarrow Na_2CO_3$

(6) Photosynthesis

Here CO_2 is changed into carbohydrates (glucose) etc., by plants in presence of sunlight and chlorophyll.

$$6CO_2 + 6H_2O \xrightarrow[\text{Sun light}]{\text{Chlorophyll}} C_6H_{12}O_6 + 6O_2$$

$$6xCO_2 + 5xH_2O \rightarrow (C_6H_{10}O_5)x + 6xO_2$$

Uses

(1) It is used in cold drinks and aerated water.

(2) CO_2 (15%) is used as fire extinguisher (except Mg-fire)

(3) O_2 + 5 – 10% CO_2 is Carbogen which is used for artificial respiration in case of CO poisoning and pneumonia patients.

(4) It is used in the manufacture of sodium carbonate by solvay method.

(5) Dry fire extinguisher is $SiO_2 + NaHCO_3$.

(6) Foamite fire extinguisher is $NaHCO_3 + Al_2(SO_4)_3$.

Structure

It is a linear molecule with zero dipole moment here C – O bond length is 1.15 Å (less than C = O bond) as it is resonance hybrid of following structures.

$$O = C = O \leftrightarrow \overset{+}{O} \equiv C - \overset{-}{O} \leftrightarrow \overset{-}{O} - C \equiv \overset{+}{O}$$

- Carbon can also form some other oxides like C_3O_2, C_5O_2 etc, which are less stable. Some graphite oxides like C_2O and C_2O_3 are also formed which are very unstable.

Carbon Suboxide (C_3O_2): It is obtained by the dehydration of malonic acid.

$$CH_2\begin{cases}COOH\\COOH\end{cases} \xrightarrow{\Delta, P_4O_{10}} O = C = C = C = O + 2H_2O$$

Some of the chemical reactions of C_3O_2 are as follows:

$$C_3O_2 + 2H_2O \longrightarrow CH_2\begin{cases}COOH\\COOH\end{cases}$$
Malonic acid

$$C_3O_2 + 2HCl \longrightarrow CH_2\begin{cases}COCl\\COCl\end{cases}$$
Malonyl chloride

$$C_3O_2 + 2NH_3 \longrightarrow CH_2\begin{cases}CONH_2\\CONH_2\end{cases}$$
Malonyl amide

Carbon Disulphide (CS_2)

Preparation

It is prepared by passing sulphur vapours over red hot coke or charcoal as follows:

$$C + 2S \rightleftharpoons CS_2\ ;\ \Delta H = 79.4\ kJ$$

Physio-Chemical Properties

(1) It is a colourless, mobile liquid with high refractive index and characteristic smell.

(2) It is highly volatile in nature with a boiling point of 319 K.

(3) It is insoluble in water but miscible with alcohol, benzene, ether etc in all proportions and soluble in sulphur, iodine, phosphorous, waxes etc.

(4) With Oxygen: It is highly inflammable and burns with a blue flame to give CO_2 and SO_2 as follows:

$$CS_2 + 3O_2 \rightarrow CO_2 + 2SO_2$$

(5) With Calcium Hydroxide When vapours of carbon disulphide are passed over calcium hydroxide, calcium carbonate and calcium thiocarbonate are formed as follows:

$$3CS_2 + 3Ca(OH)_2 \rightarrow CaCO_3 + 2CaCS_3 + 3H_2O$$

Calcium thiocarbonate

(6) With Sodium Hydroxide When vapours of CS_2 are passed over NaOH, a mixture of sodium carbonate and sodium thiocarbonate is formed.

$$3CS_2 + 6NaOH \rightarrow Na_2CO_3 + 2Na_2CS_3 + 3H_2O$$

(7) With Chlorine When chlorine is passed through boiling CS_2 having a small amount of I_2 or $AlCl_3$, CCl_4, S_2Cl_2 are formed as follows:

$$CS_2 + 3Cl_2 \rightarrow CCl_4 + S_2Cl_2$$

Sulphur monochloride

(8) With PCl_5 It reacts with PCl_5 to form thiocarbonyl chloride.

$$CS_2 + PCl_5 \rightarrow CSCl_2 + PSCl_3$$

Thiocarbonyl chloride

Uses

(1) It is used as a solvent for rubber, resins, oils, fats, sulphur, iodine etc.

(2) It is used in the manufacture of CCl_4, rayons, artificial silk etc.

(3) It is used in the extraction of oils from oil seeds and in the vulcanization of rubber.

(4) It is used as an insecticide also.

CARBIDES

These are binary compounds formed by carbon with metals or less electro negative element when carbon is heated with these elements at high temperature

Carbides are of following types:

(1) Salt Like or Ionic Carbides

These are ionic in nature and also known as electrovalent carbides. These are formed by carbon and IA, IIA, IIIA (except boron), coinage metals, Zn, Cd and some lanthanide elements. These are transparent colourless crystals and non conductors of electricity.

These undergo hydrolysis to give hydrocarbons and according to the nature of hydrocarbon produced they are of following types:

(a) Methanides

They have C^{4-} ions and give methane with water.

Example, Al_4C_3 , Be_2C

$$Al_4C_3 + 12H_2O \rightarrow 4Al(OH)_3 + 3CH_4$$

$$Be_2C + 4H_2O \rightarrow 2Be(OH)_2 + CH_4$$

(b) Acetylides

They have C_2^{2-} or $[-C \equiv C-]^{-2}$ ions and they give C_2H_2 on reaction with water

Example, CaC_2

$$CaC_2 + 2H_2O \rightarrow Ca(OH)_2 + C_2H_2$$

Metals of IA group and Cu, Ag can form M_2C_2 type carbides while IIA metals Zn and Cd can form MC_2 type carbides.

(c) Allylides

These have C_3^{4-} ions and give propyne or allylene $(CH_3 - C \equiv C- H)$ with water.

example,- Mg_2C_3

$$Mg_2C_3 + 4H_2O \rightarrow 2Mg(OH)_2 + CH_3C{\equiv}CH$$

- Ionic hydrides are normally crystalline solids the carbides of IA and IIA are colourless while rest are coloured.

(2) Metallic or Interstitial Carbides

These are formed by carbon and transition elements of group IV, V and VI. These can be prepared by heating the powdered carbon at high temperature. In such carbides small sized carbon atoms occupy interstitial positions in the crystal lattices of transition metals. These carbides are very hard and have metallic lustre, high electrical conductivity with very high melting points. Their conductivity increases with decrease of temperature. Most of these carbides are chemically inert so remain unaffected by the action of water mineral acids etc. These are interstitial and weakly paramagnetic in nature as the carbon atoms occupy the interstices in the closely packed arrays of metal atoms due to these carbons atoms hardness, stability and melting point increases.

Type

These are of two types

MC Type (M = Ti, Zr, V, Ta, Mo, W etc) and
Example, WC (Tungsten carbide), TiC (Titanium carbide)

M_2C Type (M = V, Mo, W)
example, MoC_2 (Molybdenum carbide)

(3) Covalent Carbides

These are formed by carbon with boron or silicon as they are closer in size and electronegativity to carbon atom. These are very hard, stable and decompose at very high temperature. These are chemically inert and not attacked by water dilute and concentrated acids.

example,

(1) Boron Carbide (B_4C)

It is a black crystalline very hard solid with a melting point of 2350°C. It is hardest known artificial substance.

$$2B_2O_3 + 7C \rightarrow B_4C + 6CO$$

(2) Silicon Carbide or Carborundum (SiC)

It is a colourless crystalline solid which is very hard thermally stable and chemically inert. It is prepared as follows:

$$SiO_2 + 3C \xrightarrow{2270\text{ K}} SiC + 2CO$$

$$2Si + C_2H_2 \rightarrow 2SiC + H_2$$

(4) Border Line Carbides

These are formed by Fe, Mn, Co, Ni and Cr. These are intermediate carbides between metallic and covalent carbides.

(5) Mixed Carbides

Such carbides give a mixture of hydrocarbons on hydrolysis example, UC_2, ThC_2 etc.

Uses

1. Boron carbide is used for cutting diamonds, drilling holes in rocks and for making grinding wheels. It is also used as a shield against radiation.
2. Silicon carbide is used as an abrasive and used in the glass cutting and grinding. It is also used for making metal melting crucibles.
3. Aluminium carbide is used in the preparation of methane.
4. Calcium carbide is used in the preparation of acetylene.
5. Tungsten carbide is used for making high speed cutting and drilling tools.

Fuels

Fuels are those substances which produce heat energy after their combustion. The product of heat energy formed after combustion of the fuels depends on the presence of combustible substances like C, CO, H_2 etc.

Features of a Good Fuel

A good fuel must have following features:

- It should have high calorific value.
- It should be cheap and easily available.
- Its combustion should be at a moderate speed.
- It should not leave any undesirable product or bad smell.
- It should give very little ash during its combustion.

Type of Fuels

Fuels are mainly of following types:

(a) Solid Fuels

example, wood, coal, coke, charcoal etc.

(b) Liquid Fuels

Petroleum products like petrol diesel, kerosene oil etc.

(c) Gaseous Fuels

Water gas, producer gas, natural gas, coal gas, L.P.G. etc.

Gaseous fuel is considered to be the best fuel due to following facts:

It has high calorific value and it needs no special device for its combustion.

It does not give almost any smoke and ash.

Important Gaseous Fuels

(1) Water Gas [H_2 + CO]

It is a mixture of CO and H_2 mainly. It has nearly 40 volumes of CO, 50 volumes of H_2, 5 volumes of CO_2 and 5 volumes of N_2.

It can be obtained by passing steam through the layers of red hot coal at 1000 – 1200°C.

$$\underset{\text{Red hot}}{C} + H_2O \xrightarrow{1000^\circ c} H_2 + CO - 28\text{kcal}$$

It has high calorific value 2700 Kcal / m^3 (310 B.T.U per cubic feet). Its calorific value can be further increased by adding gaseous hydrocarbons in it.

It is used in H_2 preparation by Bosch method and methyl alcohol preparation by Patart method.

It can be used for welding purposes also as its flame is short and hot.

It produces reducing medium.

(2) Producer Gas [CO + N_2]

It is a mixture of CO and N_2 mainly having 35 % CO and 65 % N_2 mainly.

It is manufactured in gas producer by the incomplete combustion of coal or coke in restricted supply of air.

$$C + \underset{\text{Hot air}}{(N_2{+}O_2)} \xrightarrow{1100^\circ c} CO + N_2$$

It is poisonous, insoluble in water and heavier than air.

Its calorific value is low (103 B.T.U per cubic feet) due to the presence of nitrogen content in excess.

It is cheaper, so maximum used in industry

It is used as a gas fuel open hearth furnace used for the manufacture of steel and glass.

It is used as fuel in motor vehicles also.

(3) Semiwater Gas

Water gas + producer gas

[CO +	H_2 +	N_2]
25	10	50
28%	12%	55%

Red 'C' + steam + air → Semi water gas

It is used in steel industries and in internal combustion engines.

(4) Coal Gas

It is a mixture of H_2, CH_4, CO, N_2, C_2H_2, C_2H_4 etc. It has H_2 (40 – 55%), CH_4 (25 – 35%), N_2 (2 –12%), CO (4 – 11%), C_2H_4 & C_2H_2 (2 – 3%).

It is a very good gaseous fuel as it has nearly 94 % combustible gases and a high calorific value of 450-660 B.T.U. per cubic feet.

It is prepared by heating coal in absence of air at a temperature greater than 1000ºC that is by the destructive distillation of coal in large fire clay retorts. Here after destructive distillation mainly coal gas, coke, coal tar and ammonical liquor are formed.

It is water insoluble gas.

Uses

It is a common gas fuel, illuminant and produces inert or reducing atmosphere during metallurgical operations.

It is also used in smelting of metals and alloys and as a reductant.

(5) Oil Gas

It is a mixture of lower hydrocarbons like CH_4, C_2H_4, C_2H_2 etc.

It is prepared by cracking of kerosene oil when a thin stream of oil is dropped on the red hot surface of the iron retort the kerosene present into this stream of oil is cracked into gaseous hydrocarbons mainly.

$$\underset{\text{(higher hydrocarbons)}}{\text{Kerosene}} \xrightarrow[\text{absence of air}]{\text{Heat}} \underset{\text{(lower hydrocarbons)}}{\text{Oil gas}}$$

$$C_{16}H_{34} \rightarrow 6CH_4 + 2C_2H_4 + C_2H_2 + 4C$$

Uses

It is used in lab in Bunsen burners but now it is replaced with petrol gas (mixture of petrol vapours + air).

(6) Natural Gas

It is a mixture of gaseous hydrocarbons like CH_4 (85%), C_2H_6 along with nitrogen. It is found mainly in petroleum abundant regions.

Its composition is as nearly follows:

CH_4	C_2H_6	C_3H_8	C_4H_{10}	N_2
85%	8%	3%	1-2%	2%

It is used mainly as a gaseous fuel and in the production of hydrogen, carbon black etc.

(7) L.P.G (Liquefied Petroleum Gas)

It is a mixture of n-butane, iso-butane, butene and propane having mainly normal butane.

It is mostly isolated from natural gas and by the cracking units of refineries.

It is colourless, non toxic, odourless gas with a density of 0.55 g/ml and an average calorific value of 55 kJ per gm.

In it some traces of ethy mercaptan (C_2H_5SH) or methyl mercaptan (CH_3SH) are added for generating smell in it, to check any leakage.

It is an important domestic fuel and direct motor fuel.

(8) Gobar Gas or Biogas

It is a mixture of CH_4 (60–70 %), CO_2, H_2, H_2S, N_2 etc. It is produced by the anaerobic fermentation of organic wastes (animal dung, human excreta, vegetable wastes etc.).

It can be used as a fuel for cooking, lightening purposes etc.

It is not only a very good alternative energy source but also its slurry (wastes left after its formation) is a good fertilizer.

SILICON AND ITS COMPOUNDS SILICON

Silicon is the second most abundant element (28 % by weight) in earth crust. It is widely present as silica (SiO_2). Three crystalline modifications of SiO_2 are quartz, cristobalite and tridymite of which first two are important. Quartz is used as a piezoelectric material.

Silica occurs as silicates mainly as Si has a great affinity for oxygen example, Aluminium silicate {Rock clay} (most widely distributed)

Preparation

(1) From Silica

Here silica is reduced by coke in an electric furnace to get silicon.

$$SiO_2 + 2C \rightarrow Si + 2CO$$

$$Si + C \rightarrow SiC$$

$$2SiC + SiO_2 \longrightarrow \underset{\text{Crystalline form}}{3Si} + 2CO$$

(2) Silica can also be reduced into silicon by magnesium powder.

$$SiO_2 + 3Mg \longrightarrow \underset{\text{Amorphous form}}{Si} + 2MgO$$

(3) From Silicon tetrachloride

Ultra pure silicon is obtained by zone refining and by the reduction of very pure $SiCl_4$.

$$SiCl_4 + 2H_2 \rightarrow Si + 4HCl$$

(4) From Silicon Hydride

$$SiH_4 \xrightarrow{402^\circ C} Si + 2H_2$$

It is purified by zone refining process.

Physio-chemical Properties

(1) It is a very hard solid element which has two allotropic forms i.e.,

(a) Amorphous

It is dark brown powder and a more reactive form.

(b) Crystalline

It is yellow crystalline solid, iso structural to diamond and less reactive.

(2) Reaction with Metals It combines with metals giving silicides

example,

$$2Mg + Si \rightarrow Mg_2Si$$

(3) Solubility It dissolves in aqua regia and alkalies. On fusion with alkali it gives a silicate

example,

$$Si + 2NaOH + H_2O \rightarrow Na_2SiO_3 + 2H_2O$$

(4) With Steam It decomposes steam as follows:

$$Si + 2H_2O \rightarrow SiO_2 + 2H_2$$

(5) Ignition Amorphous silicon can burn with O2 and F2.

$$Si + O_2 \rightarrow SiO_2$$

$$Si + 2F_2 \rightarrow SiF_4$$

Uses

(1) Silicon chips doped with P, As, Al or Ga to enhance the semi conductor properties are used for computing devices.

(2) It is used in the manufacture of many alloys with high strength, hardness, resistant power against acids like ferrosilicon, manganese silicon, bronze etc. Ferro – silicon is used to prepare acid resistant steel.

Compounds of Silicon Carborundum (Silicon Carbide) (SiC)

Preparation

It can be obtained by heating a mixture of sand, carbon, common salt and saw dust in an electric furnace.

$$SiO_2 + 3C \rightarrow SiC + 2CO$$

Here salt added, acts as flux and saw dust makes the mass porous while two carbon rods connected by a thin carbon core act as electrode in the furnace. Carborundum (SiC) is obtained round the central core of carbon. It is crushed, washed with H_2SO_4, NaOH, H_2O and finally dried.

Physio-chemical Properties

(1) In pure form it is colourless but in commercial samples it is yellow, green or blue coloured.

(2) It is very hard mass but less than diamond.

(3) It is chemically inert so resists the attack of almost all the reagents except fused caustic soda.

(4) It dissolves in the fused alkali in presence of air to give sodium silicate as follows:

$$4NaOH + SiC + 2O_2 \rightarrow Na_2SiO_3 + Na_2CO_3 + 2H_2O$$

Uses

It is used as an abrasive in place of emery. It is used in the manufacturing of grind-stones, knife sharpeners etc.,

Preparation

(1) From Silicon or Silica

It can be obtained by heating silicon or an intimate mixture of silica and coke in a stream of chlorine where it distills over as a colourless liquid as follows:

$$Si + 2Cl_2 \rightarrow SiCl_4$$

$$SiO_2 + 2C + 2Cl_2 \rightarrow SiCl_4 + 2CO$$

(2) From Magnesium Silicide

It can be prepared by passing dry chlorine over heated magnesium silicide as follows:

$$Mg_2Si + 4Cl_2 \rightarrow SiCl_4 + 2MgCl_2$$

Physio-Chemical Properties

(1) It is a colourless fuming liquid which boils at 330 K and freezes at 230 K.
(2) It is a covalent compound having electronic formula analogous to that of carbon tetrachloride.

$$Cl\overset{\bullet\times}{\underset{\bullet\times}{\times Cl \bullet}}Cl \qquad Cl\overset{\bullet\times}{\underset{\bullet\times}{\times Si \times}}Cl$$

(3) Hydrolysis

It undergoes hydrolysis to form silicic acid as follows:

$$SiCl_4 \xrightarrow[-HCl]{H_2O} Cl_3Si-OH \xrightarrow[-3HCl]{3H_2O} HO-Si(OH)_2-OH$$

Uses

(1) $SiCl_4$ alongwith ammonia is used in warfare for the production of smoke screen. In the presence of moisture, it hydrolyses to give hydrogen chloride which combines with ammonia to form dense white clouds at ammonium chloride.

Silicones

These are organo-silicon polymeric compounds having a general formula $(R_2SiO)n$. They have Si – O – Si linkages.

$$-O-\underset{R}{\overset{R}{Si}}-O-\underset{R}{\overset{R}{Si}}-O-$$

Preparation

The alkyl or aryl substituted chlorosilanes are prepared by the reaction of Grignard reagent and silicon tetrachloride.

$$\underset{\text{Grignard reagent}}{RMgCl} + SiCl_4 \rightarrow RSiCl_3 + MgCl_2$$

$$2RMgCl + SiCl_4 \rightarrow R_2SiCl_2 + 2MgCl_2$$

$$2RCl + Si \xrightarrow{Cu,\ 570\ K} R_2SiCl_2$$

$$3RMgCl + SiCl_4 \rightarrow R_3SiCl + 3MgCl_2$$

Here R may be $-CH_3$, $-C_2H_5$ or $-C_6H_5$ etc.,

Dialkyl dichloro silanes on hydrolysis gives silanols as follows:

$$R_2SiCl_2 + 2HOH \xrightarrow{-2HCl} \underset{\text{Dialkyl silandiol}}{R_2Si(OH)_2}$$

Polymerisation of dialkyl silandiol yields linear thermoplastic polymer.

$$HO-SiR_2-OH + HO-SiR_2-OH \longrightarrow HO-SiR_2-O-SiR_2-OH$$

Polymerisation continues on both the ends and thus, chain increases in length.

$RSiCl_3$ on hydrolysis gives a cross linked silicone. The formation can be explained in three steps:

$$R-SiCl_3 \xrightarrow[-3HCl]{3H_2O} R-Si(OH)_3$$

$$\underset{\text{Silicon}}{nOH-SiR(OH)-OH} \xrightarrow{-nH_2O} \text{cross-linked } -O-SiR(O-)-O-SiR(O-)-O- \text{ network}$$

Cyclic (ring) silicones are formed when water is eliminated from the terminal –OH groups of linear silicones.

```
    R   R
     \ /
     Si
    /  \
   O    O
R  |    |  R
 >Si  Si<
R   \O/    R
```

Cyclic silicon

R_3SiCl on hydrolysis forms only a dimmer.

$R_3SiOH + OH\ SiR_3 \rightarrow R_3Si - O - SiR_3$

Properties

(1) The lower silicones are oily liquids however higher silicons having long chains or ring structures are waxy and rubber like solids.

(2) These are non toxic in nature.

(3) These have very high thermal stability.

(4) Some low molecular weight silicones dissolve in organic solvents like benzene, ether etc.

(5) These are chemically inert, water repellent, insulators, lubricants and antifoaming agents etc.

(6) Viscosity of silicone oils remains the same at different temperatures.

Uses

(1) Silicones oils are used for high temperature oil baths, high vacuum pumps and low temperature lubrication.

(2) These are mixed with paints and enamels to make them resistant to the effects of high temperature , sunlight, chemicals and damp.

(3) These are used for making water proof papers, all weather lubricants etc.

(4) They are used as insulating material for electric motors etc.

(5) These are used in making vaseline like greases which can be used as lubricants in aeroplanes.

(6) Silicone-rubbers are very useful as they retain their elasticity over a range of temperatures.

(7) Silicones, resins, silicon fluids and silicon rubbers are also widely used.

SILICATES

These are alumino silicates of magnesium, calcium etc. example, Quartz, feldspar, mica and zeolites. These are present in almost all rocks, clays etc. Glass and cement are man made and highly useful silicates.

The silicates are complex network solids having the basic structural unit as silicate ion $(SiO_4)^{4-}$ which is tetrahedral in shape. On the basis of the manner in which different $(SiO_4)^{4-}$ units are linked together silicates are divided into following types:

(1) Orthosilicates

These are simple silicates and in such silicates discrete SiO_4^{4-} tetrahedral are present and there is no sharing of oxygen atoms between adjacent tetrahedral.

example,- Zircon ($ZrSiO_4$)

Foresterite or Olivine (Mg_2SiO_4)

Willemite (Zn_2SiO_4)

Phenacite (Be_2SiO_4) etc.

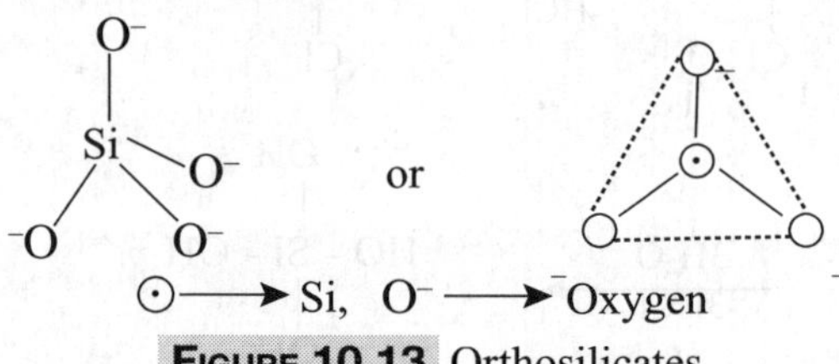

FIGURE 10.13 Orthosilicates

(2) Pyrosilicates or Islands

In such silicates the two tetrahedral units share one oxygen atom (corner) between them containing basic unit of $Si_2O_7^{6-}$ anion. Here each silica atom is surrounded by 3.5 oxygen atoms.

Example,- Thortveitite $Sc_2(Si_2O_7)$

Hemimorphtie $Zn_4(OH)_2\ Si_2O_7.H_2O$ etc.

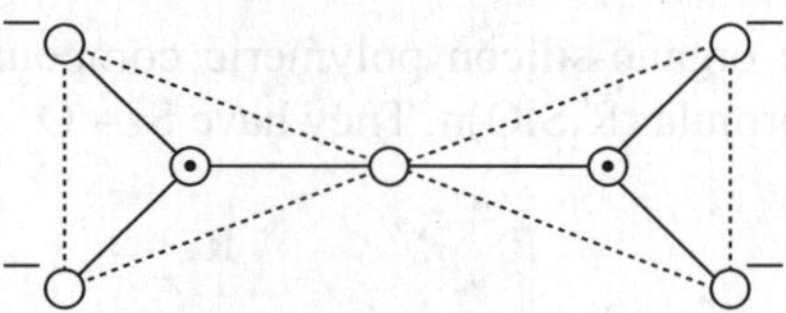

FIGURE 10.14 Islands

(3) Cyclic or Ring Silicates

In such silicates the two tetrahedral units share two oxygen atoms (two corners) per tetrahedron to form a close ring containing basic unit of $(SiO_3)_n^{2n-}$ or $(SiO_3^{2-})_n$.

Example,- Wollastonite $Ca_3(Si_3O_9)$

Benitoite $BaTi(Si_3O_9)$

Beryl $Be_3Al_2(Si_6O_{18})$

Catapleite $Na_2ZrSi_3O_9.2H_2O$ etc.

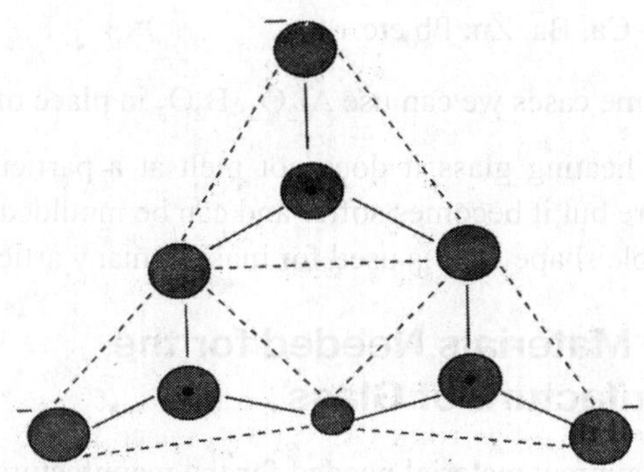

FIGURE 10.15 Ring silicates

(4) Chain Silicates

The sharing of two oxygen atoms (two corners) per tetrahedron leads to the formation of a long chain. Anions of these silicates have two general formulas $(SiO_3)_n^{2n-}$ and $(Si_4O_{11})_n^{6n-}$.

Example,- Pyroxenes and asbestos.

Jadeite $NaAl(SiO_3)_2$

Spodumene $LiAl(SiO_3)_2$

Diopside $CaMg(SiO_3)_2$ e

Tremolite $Ca_2Mg_5(Si_4O_{11})_2(OH)_2$

Enstatite $MgSiO_3$ etc.

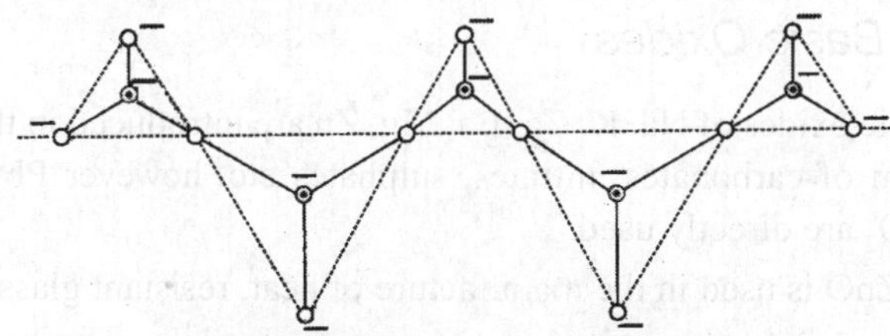

FIGURE 10.16 Chain silicates

(5) Two Dimensional Sheet Silicates

In such silicates, sharing of three oxygen atoms (three corners) by each tetrahedron unit results in an infinite two dimensional sheet of primary unit $(Si^2O_5)_n^{2n-}$. The sheets are held together by electrostatic force of the cations that lie between them.

Micas example- Muscovite $KAl_2(OH)_2(Si_3AlO_{10})$

Clays example- Talc $Mg_3(OH)_2(Si_4O_{10})$

Kaolin $Al_2(OH)_4(Si_2O_5)$ etc.

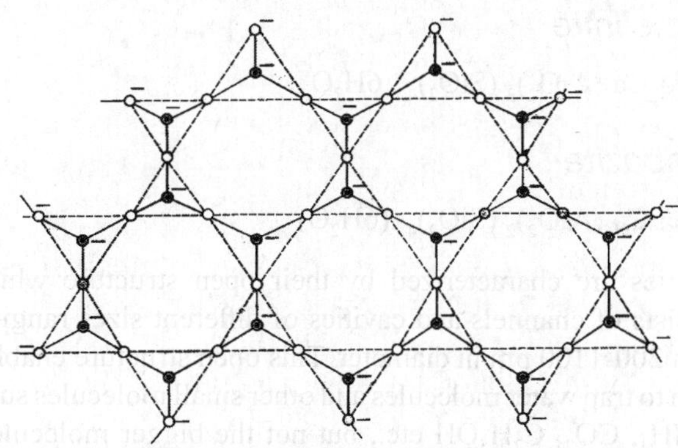

FIGURE 10.17 Sheet Silicates

(6) Three Dimensional or Frame Work Silicates

In such silicates all the four oxygen atoms (four corners) of SiO_4^{4-} tetrahedral are shared with other tetrahedral, resulting in a three dimensional network with the general formula $(SiO_2)n$.

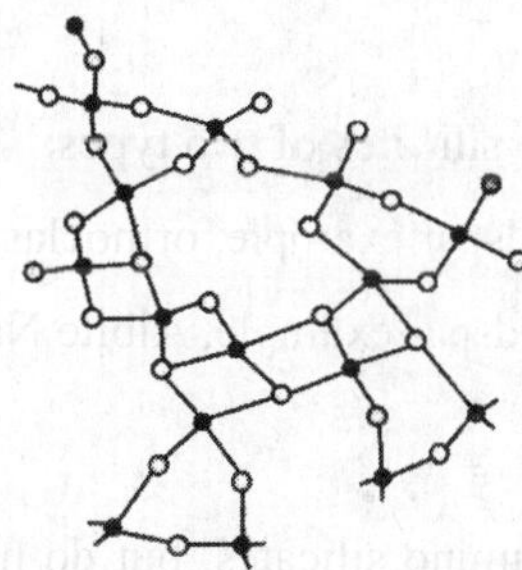

FIGURE 10.18 Frame work silicates

example,

Zeolite

These are hydrated alumino silicate having a three dimensional structure (honey comb like) with a general formula

$M_x^{+n}[(AlO_2)x(SiO_2)y]n^x.zH_2O$.

Here, M = Na, K, Ca,

'n' represents the charge on the metal ion and 'z' is the number of moles of water of hydration. Some important examples of zeolites are as follows:

Erionite

$Na_2K_2CaMg(AlO_2)_2(SiO_2)_6.6H_2O$

Gemelinite

$Na_2Ca (AlO_2)_2 (SiO_2)_4 . 6H_2O$

Chabazite

$Na_2Ca (AlO_2)_2 (SiO_2)_4 .(6H_2O)$

Zeolites are characterized by their open structure which consists of channels and cavities of different sizes ranging from 200-1100 pm in diameter. This open structure enables them to trap water molecules and other small molecules such as NH_3, CO_2, C_2H_5OH etc., but not the bigger molecules. Thus zeolites act as molecular sieves.

In zeolites, cations are free to move through the open pore structure and can be exchanged reversibly with other cations. Therefore, they act as ion-exchangers.

Zeolites act as shape-selective catalysts due to presence of cavities. They are widely used as shape selective catalysts and as ion-exchangers in water softening operations.

Feldspar

These are alumino silicates of two types:

(1) Orthoclase feldspar example, orthoclase $KAlSi_3O_8$

(2) Plagioclase feldspar example, Albite $NaAlSi_3O_8$

Ultramarines

These are also alumino silicates, but do not contain water of hydration. Most of the ultra marines are coloured and are used as pigments and in calico printing.

example, Sodalite $Na_8 (AlO_2)_6 (SiO_2)_6 Cl_2$

FACTS TO REMEMBER

- SiO_2 is silica gel (a semi permeable membrane).
- Being adsorbent it is used to remove sulphur.
- Na_2SiO_3 is used a filter in saponification to remove excess of water.

GLASS

It is an amorphous and transparent or translucent super cooled liquid having a mixture of silicates of different metals, one of which is always an alkali metal like Na or K and borates.

The general formula of an ordinary glass can be given as: $A_2O.BO.6SiO_2$

Here, A = Na or K

B = Ca, Ba, Zn, Pb etc.

- In some cases we can use Al_2O_3, B_2O_3 in place of SiO_2.

On heating glass it does not melt at a particular temperature but it becomes softer and can be moulded into any desirable shape so it is used for making many articles.

Raw Materials Needed for the Manufacture of Glass

The main raw material needed for the manufacture of glass is as follows:

(1) Acidic Oxides

Many acidic oxides like SiO_2, B_2O_3, P_2O_5 etc. can be used according to the need of the quality of the glass.

- **Silica (SiO_2):** It must not be very fine or coarse otherwise the reaction will be violent or slow respectively. It must not have any impurities of Fe_2O_3 etc.
- **Boron Trioxide (B_2O_3):** B_2O_3 in the form of borax or boric acid is used in the manufacture of glass having low coefficient of expansion.
- **Phosphorous Pentoxide (P_2O_5):** P_2O_5 in the form of calcium phosphate is used in the manufacture of opalescent glass.

(2) Basic Oxides

Basic oxides of Na, K, Ca, Ba, Mg, Zn are introduced in the form of carbonates, nitrates, sulphates etc. however PbO, PbO_2 are directly used.

- ZnO is used in the manufacture of heat, resistant glass.

(3) Colouring Matter

The addition of many transition metal compounds to the glass mixture give colours to the glasses.

Compound added	Colour imparted
Cobalt oxide (CoO)	Blue
Cuprous oxide (Cu_2O)	Red
Iron (III) compounds	Yellow
Cadmium sulphide (CdS)	Lemon yellow
Chromium trioxide (Cr_2O_3)	Green
Manganese dioxide (MnO_2)	Purple
Auric chloride ($AuCl_3$)	Ruby

Manufacture

The essential raw materials needed for the manufacture of a particular type of glass is called batch. It is mixed with broken glass pieces (cullet) to decrease the melting temperature. Now this fused mass is heated upto 1673 K in a pot or tank furnace using producer gas. Here glass is formed as a result of following reactions:

Silica decomposes the carbonates, by setting the evolution of CO_2.

$$Na_2CO_3 + SiO_2 \rightarrow Na_2SiO_3 + CO_2$$

$$CaCO_3 + SiO_2 \rightarrow CaSiO_3 + CO_2$$

At 1673 K silica dissolves in the silicates of sodium and calcium to form glass.

$$Na_2SiO_3 + CaSiO_3 + 4SiO_2 \rightarrow \underset{\text{Glass}}{Na_2SiO_3.\ CaSiO_3.4SiO_2}$$

Annealing

If glass is cooled rapidly it becomes brittle and fragile that is why the articles of glass must be cooled neither very rapidly nor very slowly. Such objects are passed through a long tunnel like furnace which is quite hot at the mouth end side and very cold at the tail end side hence these articles are heated or cooled gradually. This is called annealing.

Type of Glasses

(1) Soda Lime or Soft Glass or Ordinary Glass

It is a mixture of sodium and calcium silicates and is obtained by heating a mixture of Na_2O_3, CaO and SiO_2 (or sand) in a furnace around 1700 K. It is quite cheap and used in making glass tubes, bottles, window panes etc.

(2) Potash Glass or Hard Glass

When Na_2CO_3 is replaced by K_2CO_3 in soda glass, hard glass is produced. Its approximate composition is K_2SiO_3. $CaSiO_3$. $4SiO_2$ or $K_2O.CaO.6SiO_2$.

It is more resistant to the action of acids, water etc. and it fuses with difficulty.

It can withstand high temperatures so it is used for making combustion tubes and scientific apparatus.

(3) Flint Glass or High Refractive Index Glass or Lead Potash Glass

It is obtained by fusing $PbCO_3$, K_2CO_3, and SiO_2. Its composition is $K_2O.PbO.6SiO_2$.

This glass has high refractive index and is used for making high quality optical instruments like lenses, prisms etc.

(4) Borosil Glass or Pyrex Glass or Corning Glass

It is obtained by adding borax (or boron oxide) and Al_2O_3 to ordinary glass mixture.

It is mixture of sodium borosilicates and aluminium borosilicates have a composition, $Na_2O.B_2O_3.Al_2O_3.SiO_2$.

It is heat resistant, has low coefficient of thermal expansion and can withstand sudden changes of temperature. It is used for making laboratory glass wares, thermometers etc.

(5) Crooke's Glass

It is a special type of glass having CeO_2 (cerium oxide) which has the property to cut off ultraviolet rays which are harmful to eyes. It is used for making spectacle lenses.

(6) Water Glass

It is obtained by heating Na_2CO_3 and SiO_2 and has the composition $Na_2O,.SiO_2$.

$$Na_2CO_3 + SiO_2 \rightarrow Na_2SiO_3 + CO_2$$

(7) Jena Glass

It is a mixture of zinc and barium borosilicates and it is resistant to heat, shock etc. It is used for making high quality of glass wares.

(8) Quartz Glass

It is obtained from pure silica. It does not break even when plunged in water while red hot.

Etching of Glass

Glass can be attacked by HF. This leads to etching of glass. Here for etching the glass is coated with a thin layer of wax and an aqueous of HF is applied to its exposed part. After a few while it is kept in water and wax is removed. The marks are engraved on its exposed parts.

TIN AND ITS COMPOUNDS TIN (Sn)

Tin (Sn) is present in p-block of the periodic table. It is a member of group 14 (IV A) called carbon family and is placed after carbon, silicon and germanium in this group. It is a typical metal. The valence shell electronic configuration of the metal (Z = 50) is [Ar] $4d^{10}$ $5s^2$ $5p^2$. It exhibits +2 and +4 oxidation states in its compounds,. However, +2 oxidation state is more stable.

Occurrence

In native or free state it is found only in traces at few places but in nature it occurs mainly in combined state in the form of ores.

Ores of Sn

Cassiterite: SnO_2

Tin pyrite: $SnS_2.Cu_2S.FeS$

Extraction of Tin

It is mainly extracted from Cassiterite ore which has nearly 10% of Sn as SnO_2. The ore has silicious impurities, $FeWO_4$, $MnWO_4$, iron and copper pyrites etc. The extraction occurs as follows:

(1) Crushing and Concentration of the Ore

First of all the ore is crushed into finely powdered form. It is then concentrated by gravity process to remove the lighter silicious impurities along with gangue.

Now the ore is concentrated by magnetic separation method to remove the impurities of $FeWO_4$, $MnWO_4$ and other magnetic impurities.

(2) Roasting

The concentrated ore is roasted in a revolving furnace in the free supply of air so that the impurities of S, As, Sb can be removed from the ore in the form of their volatile oxides.

Here iron and copper pyrite also changed into their sulphate.

$$S + O_2 \rightarrow SO_2$$

$$4As + 3O_2 \rightarrow 2As_2O_3$$

$$4Sb + 3O_2 \rightarrow 2Sb_2O_3$$

$$CuS + 2O_2 \rightarrow CuSO_4$$

$$FeS + 2O_2 \rightarrow FeSO_4$$

(3) Washing with Water

The roasted ore is washed with water so that soluble sulphate salts of Cu and Fe and some other impurities can be removed. The ore obtained from here has nearly 70% of tin in the form of SnO_2 and it is called black tin.

(4) Smelting

The roasted ore (black tin) is mixed with nearly one fifth amount of its weight of powdered anthracite along with a little lime stone (flux). Now this mixture is heated in a reverberatory furnace at 1200-1300°C. As a result SnO_2 is reduced into tin metal which collects at the bottom of the furnace.

$$SnO_2 + 2C \rightarrow Sn + 2CO$$

Black tin (99.5%)

$$CaO + SiO_2 \rightarrow CaSiO_3$$

$$SnO + SiO_2 \rightarrow SnSiO_3$$

This molten or crude tin has many impurities also which can be removed in the form of slag. The slag has mainly 20-40 % $SnSiO_3$ from which Sn can be recovered easily by smelting it in the same manner.

$$SnSiO_3 + CaO + C \rightarrow Sn + CaSiO_3 + CO$$

(5) Refining

Tin obtained from smelting has the impurities of Pb, Fe, W, S, As etc., so its refining is necessary. It is possible by these methods:

(a) Liquation: Here impure Sn is heated on the sloping hearth of a reverberatory furnace at a temperature which is more than the melting point of tin. As a result pure tin melts and flows down while impurities are left on the hearth.

(b) Poling: Tin obtained on liquation is melted again and stirred with logs of green wood. Here the hydrocarbon gases evolved from the logs, agitate the metal and bring the impurities to the surface as a result these are removed due to oxidation by air in the form scum. From here 99 % pure Sn is obtained.

(c) Electrolytic Refining: Tin obtained after poling process is further purified by this method. Here a block of impure tin and a thin sheet of impure tin are used as anode and cathode respectively. In an electrolytic bath having $SnSO_4$ acidified with H_2SO_4 as electrolyte. When electric current is passed 100 % pure tin is obtained at the cathode.

PROPERTIES OF TIN

Physical Properties

(1) It exist in three allotropic forms (grey, white and rhombic) out of which white tin is the most common and most stable form.

Grey tin $\xrightleftharpoons{291\text{ K}}$	White tin $\xrightleftharpoons{440\text{ K}}$	Rhombic tin
Amorphous	Crystalline	Crystalline
powder	teragonal	rhombic
(brittle)	(malleable)	(brittle)
Density = 5.8	Density = 7.8	Density = 7.29
gm/cm^3	gm/cm^3	gm/cm^3

- The conversion of white tin into grey tin in very cold countries like Switzerland etc. is accompanied by a decrease in its density or increase in volume as a result it becomes very brittle and powdered form. it is called tin disease or tin plague or tin pest.

(2) It is a silvery white lustrous metal which is softer than zinc but harder than lead. It has a melting point of 505 K and a boiling point of 2543 K.

(3) It is malleable and ductile at 473 K. On bending it produces a peculiar sound cry of tin or tin cry which is due to the rubbing of crystal against one another.

Chemical Properties

(1) Action of Air

At ordinary temperature tin is not effected by the action of air and water. When tin is heated in air 1775-1875 K it burns with a bright flame to give SnO_2.

$$Sn + O_2 \rightarrow SnO_2$$

- Molten tin can decompose steam by releasing hydrogen.

$$Sn + 2H_2O \rightarrow SnO_2 + 2H_2 \uparrow$$

(2) With Chlorine

On heating it reacts with chlorine to give stannic chloride.

$$Sn + 2Cl_2 \rightarrow SnCl_4$$

(3) With Sulphur

On heating it reacts with sulphur to give stannic sulphide.

$$Sn + 2S \rightarrow SnS_2$$

Stannic sulphide

(4) Reaction with Alkalies

It dissolve in hot and concentrated alkalies to give stannates.

$$Sn + 2NaOH + H_2O \rightarrow Na_2SnO_3 + 2H_2$$

Sodium stannate

(5) Reaction with Acids

Tin is uneffected by the action the action of organic acids that is why it is used for tinning cooking utensils made of brass.

It dissolves slowly in dilute HCl but rapidly in concentrated HCl forming stannous chloride as follows:

$$Sn + 2HCl \rightarrow SnCl_2 + H_2$$

It dissolves in dilute and cold or hot and concentrated sulphuric acid to give $SnSO_4$ or $Sn(SO_4)_2$ as follows:

$$Sn + H_2SO_4 \rightarrow SnSO_4 + H_2O$$

Dil & cold Stannous sulphate

$$Sn + 4H_2SO_4 \rightarrow Sn(SO_4)_2 + 4H_2O + 2SO_2$$

Hot & conc. Stannic sulphate

It dissolves in both cold and dilute nitric acid as well as concentrated nitric acid to give stannic nitrate and stannic acid respectively.

$$4Sn + 10HNO_3 \rightarrow 4Sn(NO_3)_2 + NH_4NO_3 + 3H_2O$$

Cold & dil. Stannic nitrate

$$Sn + 4HNO_3 \rightarrow H_2SnO_3 + 4NO_2 + H_2O$$

Conc. Stannic acid

Uses

(1) It is used to protect iron objects from corrosion by electrolytic deposition of tin on iron objects i.e. tin plating.

(2) It is used in making house hold utensils and tinning brass utensils.

(3) Tin foils are widely used for wrapping cigarettes, confectionary and for making the tooth paste tubes etc.

(4) Tin amalgam is used in the making of mirrors.

(5) It is used for making many useful alloys.

Alloys	% Composition	Uses
Babbitmetal	Sn 90%, Cu 3%, Sb 7%	Bearing of machinery
Bell metal	Cu 80%, Sn 20%	Bells, gongs
Britannia metal	Sn 93%, Sb 5%, Cu 2%	Cheap table wares
Bronze	Cu 80%, Zn 10%, Sn 10%	Coins, statues, utensils, valves
Gun metal	Cu 90%, Sn 10%	Gum barrels, gears
Pewter	Sn 80%, Pb 20%	Utensils
Rose metal	Sn 25%, Pb 25%, Bi 50%	Electric fuses
Soft solder	Sn 50%, Pb 50%	Soldering
Type metal	Pb 70%, Sb 20%, Sn 10%	Printing type
Wood's metal	Sn 12.5%, Pb 25%, Cd 12.5%, Bi 50%	Fire alarms, fuses, automatic water sprinklers

Flow Sheet of the Metallurgy of Tin

METALLURGY OF TIN (Sn)

Cassiterite (SnO_2)

↓

Concentration by levigation and Electro Magnetic Separation

Crushed powdered ore is washed with water to remove lighter siliceous impurities and the impurities of $FeWO_4$ and $MnWO_4$ are removed by electromagnetic separation

↓

Roasting in reverberatory furnace

Here the impurities of S, As are removed as SO_2 and As_2O_3

$$S + O_2 \rightarrow SO_2$$

$$4As + 3O_2 \rightarrow 2As_2O_3$$

↓

Smelting

Roasted ore + carbon + lime stone are heated in reverberatory furnace at 1200-1300°C

$$CaCO_3 + SiO_2 \rightarrow CaSiO_3 + CO_2$$

$$SnO_2 + 2C \rightarrow Sn + 2CO$$

Black tin (99.5%)

Purification by Liquation and Poling or Electrolytic method

↓

Pure Sn

Compounds of Tin Stannous Oxide (SnO)

Preparation

(1) From $SnCl_2$ and NaOH

When NaOH is added in $SnCl_2$, stannous hydroxide is obtained. Which on heating in an atmosphere of CO_2 gives SnO.

$$SnCl_2 + 2NaOH \rightarrow Sn(OH)_2 + 2NaCl$$

$$Sn(OH)_2 \xrightarrow{\Delta, CO_2} SnO + H_2O$$

(2) From $SnCl_2$ and Na_2CO_3

When a mixture of $SnCl_2$ and Na_2CO_3 is heated in an atmosphere of CO_2, SnO is formed as follows:

$$SnCl_2 + Na_2CO_3 \xrightarrow{\Delta, CO_2} SnO + CO_2 + 2NaCl$$

(3) From Stannous Oxalate

When stannous oxalate is heated in absence of air, SnO is formed.

$$SnC_2O_4 \xrightarrow{\Delta} SnO + CO + CO_2$$

Physio-Chemical Properties

(1) It is a dark grey or blackish powder which is insoluble in water.

(2) It is an amphoteric oxide and dissolves both in acids and bases.

$$SnO + 2HCl \rightarrow SnCl_2 + H_2O$$

Stannous chloride

$$SnO + 2NaOH \rightarrow Na_2SnO_2 + H_2O$$

Sodium stannite

(3) Burning in Air: It burns in air with incandescence giving stannic oxide.

$$2SnO + O_2 \rightarrow 2SnO_2$$

Stannic Oxide (SnO_2)

In nature it is found as tin stone or cassiterite.

Preparation

(1) By Heating Tin in Air

$$Sn + O_2 \rightarrow SnO_2$$

(2) From Meta stannic Acid

$$Sn + 4HNO_3 \rightarrow H_2SnO_3 + 4NO_2 + H_2O$$

$$H_2SnO_3 \rightarrow SnO_2 + H_2O$$

Meta stannic acid

or

$$2SnO + O_2 \xrightarrow{Heat} 2SnO_2$$

$$H_2Sn_5O_{11}.4H_2O \xrightarrow{Heat} 5SnO_2 + 5H_2O$$

Properties

(1) It is a white powder which is insoluble in water but soluble in concentrated H_2SO_4 and alkalies (KOH).

$$SnO_2 + 2H_2SO_4 \rightarrow Sn(SO_4)_2 + 2H_2O$$

$$SnO_2 + 2KOH \rightarrow K_2SnO_3 + H_2O$$

Potassium stannates

Uses

(1) It is used in making pottery and glass white glazes.

(2) It is used for making milky glass and as a polishing powder (putty powder).

Stannous Chloride ($SnCl_2$)

Preparation

(1) From Stannous Oxide and Tin

$$SnO + 2HCl \rightarrow SnCl_2 + H_2O$$

$$Sn + 2HCl \rightarrow SnCl_2 + H_2$$

The solution of $SnCl_2$ on concentration and cooling gives the crystals of hydrated stannous chloride.

(2) Anhydrous $SnCl_2$ can be prepared as follows:

$$Sn + Cl_2 \rightarrow SnCl_2$$

Dry

$$Sn + HgCl_2 \rightarrow SnCl_2 + Hg$$

Physio-Chemical Properties

(1) It is a white crystalline hygroscopic solid which is soluble in water, alcohol and ether.

- It undergoes hydrolysis as follows:

$$SnCl_2 + H_2O \rightleftharpoons Sn(OH)Cl + HCl$$

(2) With Sodium Hydroxide: When NaOH is added in it sodium stannite is formed as follows:

$$SnCl_2 + 2NaOH \rightarrow \underset{\text{White ppt.}}{Sn(OH)_2} + 2NaCl$$

$$Sn(OH)_2 + 2NaOH \rightarrow \underset{\text{Sodium stannite}}{Na_2SnO_2} + 2H_2O$$

(3) With H_2S: It gives a dark brown precipitate with H_2S as follows:

$$SnCl_2 + H_2S \rightarrow \underset{\text{Brown ppt.}}{SnS} + 2HCl$$

The brown precipitate dissolves in ammonium sulphide.

$$SnS + (NH_4)_2S_2 \rightarrow \underset{\text{Ammonium thio stannate}}{(NH_4)_2\, SnS_3}$$

(4) Reducing Properties: It is a strong reducing agent

example, (i) It reduces mercuric chloride into mercury as follows:

$$2HgCl_2 + SnCl_2 \rightarrow \underset{\substack{\text{(White ppt.)}\\ \text{Mercurous chloride}}}{Hg_2Cl_2} + SnCl_4$$

$$Hg_2Cl_2 + SnCl_2 \rightarrow \underset{\text{(Grey black)}}{2Hg} + SnCl_4$$

example, (ii) It reduces ferric chloride into ferrous chloride as follows:

$$2FeCl_3 + SnCl_2 \rightarrow 2FeCl_2 + SnCl_4$$

example, (iii) It reduces auric chloride into metallic gold as follows:

$$2AuCl_3 + 3SnCl_2 \rightarrow \underset{\substack{\text{Colloidal}\\ \text{gold}}}{2Au} + 2SnCl_4$$

example, (iv) It reduces nitro compounds into amines as follows:

$$RNO_2 + 6HCl + 3SnCl_2 \rightarrow RNH_2 + 3SnCl_4 + 2H_2O$$

Uses

It is used as a reducing agent, as a mordent in dyeing and for making purple of casius.

Stannic Chloride ($SnCl_4$)

Preparation

(1) By passing dry chlorine over fused tin or stannous chloride:

$$Sn + 2Cl_2 \rightarrow SnCl_4$$

$$SnCl_2 + Cl_2 \rightarrow SnCl_4$$

(2) By distilling tin with excess of mercuric chloride:

$$2HgCl_2 + Sn \rightarrow SnCl_4 + 2Hg$$

Physio-Chemical Properties

(1) It is colourless fuming liquid with unpleasant smell.

(2) It is hygroscopic in nature and can form many crystalline hydrates having 3, 5, 6, 8 water molecules as water of crystallisation.

example, $SnCl_4.5H_2O$ is called butter of tin or oxymuriate of tin.

(3) It is soluble in water and undergoes hydrolysis as follows:

$$SnCl_4 + 4H_2O \rightarrow Sn(OH)_4 + 4HCl$$

(4) It dissolves in conc. HCl giving chloro stannic acid.

$$SnCl_4 + 2HCl \rightarrow H_2SnCl_6$$

(5) It combines with ammonia to form crystalline adduct, $SnCl_4.4NH_3$ which can be sublimed without any decomposition.

Uses

(1) It is used for fire proofing cotton and for increasing the weight of silk.

(2) Butter of tin acts as a mordent in dyeing.

Lead and its Compounds Lead or Plumbum (Pb)

It is heavier than tin. The valence shell electronic configuration of lead (Z = 82) is [Xe] $4f^{14}\ 5d^{10}\ 6s^2\ 6p^2$ and has four electrons in the valence shell. The compounds of lead exist in both +2 and +4 oxidation states. However, the compounds in +2 oxidation state are more stable.

Occurrence

It occurs in combined state in the form of ores like galena.

Ores of Pb

Galena: PbS

Anglesite: $PbSO_4$

Stolzite: $PbWO_4$

Cerrusite: $PbCO_3$

Wulfenite: $PbMnO_4$

Extraction

Lead is mainly extracted from galena in which 6-8 % lead is present. From it, it is extracted by following two methods:

(1) Air or Self reduction Process

It is employed when galena is rich in lead. Here lead is extracted as follows:

(1) Crushing and Concentration

First of all the ore is crushed into finely powdered form and concentrated by froth floatation method.

(2) Roasting

The concentrated ore is roasted in a reverberatory furnace having three doors D_1, D_2 and D_3 for proper air supply at below 1023 K. Here PbS is oxidized into PbO and $PbSO_4$.

$$2PbS + 3O_2 \rightarrow 2PbO + 2SO_2$$

$$PbS + 2O_2 \rightarrow PbSO_4$$

(3) Reduction

Here the doors are closed to reduce air supply and more galena is introduced to the furnace. On increasing the temperature of the furnace to the fusion temperature both PbO and $PbSO_4$ get reduced into metallic lead by excess of galena as follows:

$$2PbO + PbS \rightarrow 3Pb + SO_2$$

$$PbSO_4 + PbS \rightarrow 2Pb + 2SO_2$$

Molten lead can be drawn of from the bottom of the furnace easily.

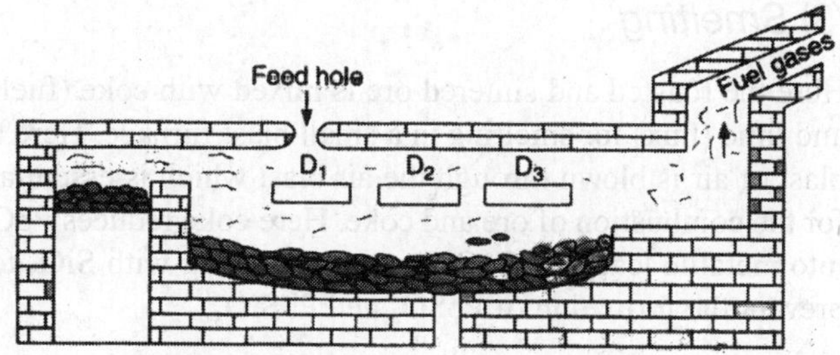

FIGURE 10.19 Reverberatory Furnace

(2) Carbon Reduction Method

It is employed when galena is poor in lead. Here lead is extracted as follows:

(1) Crushing and Concentration

First of all the ore is crushed into finely powdered form and concentrated by froth floatation method.

(2) Roasting and Sintering

The concentrated ore is mixed with lime and roasted in a specific type of furnace called Sinterer as shown in the figure. When a slow blast of air is passed from the bottom and a suction is applied at the top PbS get oxidized into PbO. Here lime being more basic than PbO prevents the formation of both $PbSO_4$ and $PbSiO_3$.

$$2PbS + 3O_2 \rightarrow 2PbO + 2SO_2$$

$$PbSiO_3 + CaO \rightarrow PbO + CaSiO_3$$

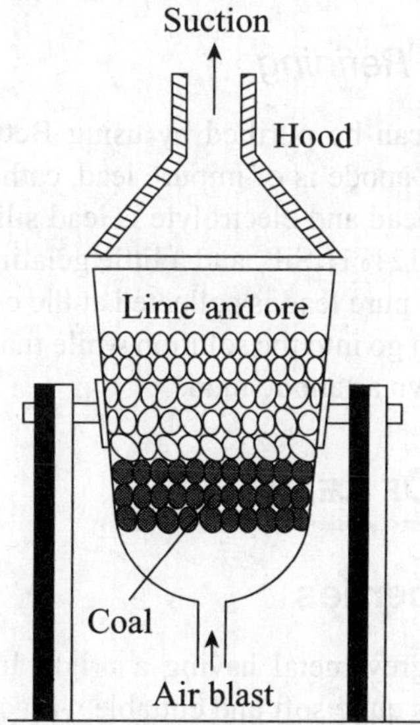

FIGURE 10.20 Sinterer

(3) Smelting

Here the roasted and sintered ore is mixed with coke (fuel) and lime (flux) for smelting in a small blast furnace. Here a blast of air is blown through the air blast which is essential for the combustion of ore and coke. Here coke reduces PbO into metallic lead and CaO forms fusible slag with SiO_2 to prevent the formation of $PbSO_4$ and $PbSiO_3$.

$$PbO + C \rightarrow Pb + CO$$

$$PbO + CO \rightarrow Pb + CO_2$$

$$CaO + SiO_2 \rightarrow CaSiO_3$$

$$PbSiO_3 + CaO \rightarrow PbO + CaSiO_3$$

Here both molten lead and slag collect in different layers at the bottom so can be easily removed by separate tapping holes.

Purification

Lead obtained by above processes has the impurities of Cu, Sb, Zn, Ag, Bi, etc. it makes it hard and brittle. It is purified by these methods:

(1) Softening Process

Here impure metal is melted on the hearth of a reverberatory furnace as a result the base metals get oxidized and can be removed as scum on the surface.

(2) Desilverisation

From impure lead, silver can be removed by Parke's method.

(3) Electrolytic Refining

Desilverised lead can be purified by using Bett's electrolytic method. Here anode is of impure lead, cathode is of a thin sheet of pure lead and electrolyte is lead silicofluoride ($PbSiF_6$) having 8-12% H_2SiF_6 and a little gelatin. On passing electric current pure lead is collected at the cathode and impurities of Fe, Sn go into the solution while that of Sb, Cu etc. are thrown down as anode mud.

Properties Of Lead

Physical Properties

(1) It is a bluish grey metal having a bright lustre when freshly cut. It is quite soft and cuttable with a knife.

(2) It is malleable and very little ductile.

(3) It is a heavy metal having a density of 11.33 gm per cm^3 and a melting point of 600 K.

Chemical Properties

(1) Effect of Air

Dry air has no action on lead however in moist air its surface is covered with a thin film of basic lead carbonate.

On heating in air at low temperature PbO and at high temperature Pb_3O_4 are obtained as follows:

$$2Pb + O_2 \rightarrow 2PbO$$

$$6PbO + O_2 \rightarrow 2Pb_3O_4$$

(2) With Water

Air free pure water has no action on lead but lead dissolves slowly in water having dissolved air giving $Pb(OH)_2$. It is called plumbosolvancy.

$$2Pb + 2H_2O + O_2 \rightarrow 2Pb(OH)_2$$

- Plumbosolvancy is more when water has presence of ammonium salts, nitrates and organic acids but the presence of phosphates, sulphates and bicarbonates decrease it.

(3) With Sulphur

Lead on heating with sulphur gives lead sulphide as follows:

$$Pb + S \rightarrow PbS$$

(4) With Chlorine

Lead on heating with chlorine gives lead chloride as follows:

$$Pb + 2Cl_2 \rightarrow PbCl_4$$

(5) With Alkalies

Lead dissolves slowly in sodium hydroxide to form sodium plumbate as follows:

$$Pb + 2NaOH \rightarrow Na_2PbO_2 + H_2$$

(6) With Acids

It dissolves slowly in dilute HCl but rapidly in concentrated HCl forming chloro plumbous acid as follows:

$$Pb + 2HCl \rightarrow PbCl_2 + H_2$$

$$PbCl_2 + 2HCl \rightleftharpoons H_2PbCl_4$$

It dissolves in hot and concentrated sulphuric acid to give $PbSO_4$ however the reaction is retarded due to the formation of $PbSO_4$ on the surface of lead.

$$Pb + 2H_2SO_4 \rightarrow PbSO_4 + SO_2 + 2H_2O$$

It dissolves in both cold and dilute nitric acid as well as moderate concentrated nitric acid to give lead nitrate as follows:

$$\underset{\text{Cold. \& dil.}}{3Pb + 8HNO_3} \rightarrow 3Pb(NO_3)_2 + 4H_2O + 2NO$$

$$\underset{\text{Moderate \& conc.}}{Pb + 4HNO_3} \rightarrow Pb(NO_3)_2 + 2H_2O + 2NO_2$$

It also dissolves in acetic acid in the presence of oxygen to form lead acetate as follows:

$$Pb + 2CH_3COOH + \frac{1}{2} O_2 \rightarrow (CH_3COO)_2Pb + H_2O$$

Uses

(1) It is used for making bullet shots, lead accumulators, lead pipes and cable covering.

(2) It is also used for making many pigments like white lead, red lead, chrome red, chrome yellow etc.

(3) It is used to prepare the famous anti knocking substance tetra ethyl lead.

(4) Lead sheets are used for making lead chambers used in the manufacture of sulphuric acid.

(5) It is also used in a number of alloys like rose metal, type metal, pewter etc.

Flow Sheet of the Metallurgy of Lead

METALLURGY OF LEAD

Galena (PbS)

Concentration by Froth Floatation Process
Reduction

Air Reduction Process

It is in reverberatory furnace

& in presence of air

$2PbS + 3O_2 \rightarrow 2PbO + 2SO_2$

$PbS + 2O_2 \rightarrow PbSO_4$

↓

Mixed it with more galena

and heat at higher temperature

in furnace

↓

PbO_{2n} & $PbSO_4$ are reduced by

PbS into lead metal (Crude lead)

$PbS + 2PbO_2 \rightarrow 3Pb + SO_2$

$PbS + PbSO_4 \rightarrow 2Pb + 2SO_2$

Purification: By Perk method & Cupellation

Carbon Reduction Process

The ore is mixed with

lime& heated in sinterer

$2PbS + 3O_2 \rightarrow 2PbO + 2SO_2$

$PbS \rightarrow PbO$

↓

Mixed it with C & CaO and

smelted in blast furnace to

get crude lead

Lead Mono Oxide or Litharge or Plumbous Oxide or Lead (II) Oxide (PbO)

Preparation

(1) By Heating Lead in Air

It can be prepared by heating lead in air or oxygen at 300°C it changes to a yellow powder known as Massicot which on further heating at 880°C becomes reddish yellow known as litharge. Chemically both are PbO.

$$2Pb \xrightarrow{\Delta\, O_2} 2PbO$$

(2) By Heating Lead Nitrate

$$2Pb(NO_3)_2 \xrightarrow{\Delta} 2PbO + 4NO_2 + O_2$$

(3) By Heating Lead Carbonate:

$$PbCO_3 \xrightarrow{\Delta} PbO + CO_2 \uparrow$$

Physio-Chemical Properties

(1) When it is a yellow powder it is called massicot and when on heating it is in buff coloured crystalline form it is called litharge.

(2) Amphoteric Nature: It is an amphoteric oxide which is insoluble in water but soluble in acids and alkalies.

$$PbO \xrightarrow{2HCl} 2PbCl_2 + H_2O$$

$$PbO \xrightarrow{2NaOH} \underset{\text{Sodium plumbate}}{Na_2PbO_2} + H_2O$$

(3) Heating Effect:

$$\underset{\text{Litharge}}{6PbO} + O_2 \xrightarrow{\Delta} \underset{\text{Red lead}}{2Pb_3O_4}$$

(4) Reduction: It can be reduced into metallic lead by heating it with H_2 or CO or with carbon.

$$PbO + H_2 \rightarrow Pb + H_2O$$

$$PbO + CO \rightarrow Pb + CO_2$$

$$PbO + C \rightarrow Pb + CO$$

Uses

(1) PbO is used in paints, utensils and flint glass.

(2) Glycerol + Massicot : A cement for glass and stone.

(3) It is used for making flint glass.

(4) It is used for glazing pottery.

Plumbic Oxide or Lead Dioxide or Lead (IV) Oxide (PbO_2)

Preparation

(1) From Litharge

Litharge on heating (fusion) with KNO_3 or $KClO_3$ gives lead dioxide.

$$PbO + KNO_3 \longrightarrow PbO_2 + KNO_2$$

$$3PbO + KClO_3 \longrightarrow 3PbO_2 + KCl$$

(2) By Heating Red Lead with dilute HNO_3:

$$\underset{\text{Red lead}}{Pb_3O_4} + 4HNO_3 \xrightarrow{\Delta} 2Pb(NO_3)_2 + 2H_2O + \underset{\text{Brown}}{PbO_2 \downarrow}$$

(3) Lab Method

In lab it is prepared by treating lead acetate with bleaching powder having slaked lime as follows:

$$(CH_3COO)_2Pb + Ca(OH)_2 \rightarrow Pb(OH)_2 + (CH_3COO)Ca$$

$$Pb(OH)_2 + \underset{\text{Bleaching powder}}{CaOCl_2} \rightarrow PbO_2 + CaCl_2 + H_2O$$

Physio-Chemical Properties

(1) It is a chocolate brown coloured powder which does not dissolve in water.

(2) Heating Effect: On heating at 300°C it changes into litharge.

$$2PbO_2 \xrightarrow{300^\circ C} 2PbO + O_2$$

As oxygen is liberated here so it can act as a strong oxidant.

(3) Amphoteric Nature: Being an amphoteric oxide it can react with both acids and bases

example, (i) It forms plumbic chloride on reaction with cold and concentrated HCl.

$$PbO_2 + 4HCl \rightarrow PbCl_4 + 2H_2O$$

example, (ii) It dissolves in concentrated HCl and H_2SO_4 to give $PbCl_2$ and $PbSO_4$ respectively.

$$PbO_2 + 4HCl \rightarrow PbCl_2 + Cl_2 + 2H_2O$$

Conc.

$$2PbO_2 + 2H_2SO_4 \rightarrow 2PbSO_4 + 2H_2O + O_2$$

example, (iii) It also dissolves in hot and concentrated solution of sodium hydroxide give sodium plumbate.

$$PbO_2 + 2NaOH \rightarrow \underset{\text{Sodium plumbate}}{Na_2PbO_3} + H_2O$$

(2) With $KMnO_4$

$$KMnO_4 + \text{conc. } HNO_3 + PbO_2 \rightarrow \underset{\substack{\text{Purple} \\ \text{(Permanganic acid)}}}{HMnO_4}$$

Uses

It is used in match industry, lead storage battery and as an oxidising agent.

Red Lead or Tri Lead Tetra-Oxide (Pb_3O_4)

Pb_3O_4 is Triplumbic tetra-oxide Minium or Sindur. It is a mixed oxide of PbO_2 and PbO in 1 : 2 ratio.

Preparation

When massicot (litharge) is heated with air at 400°C in a reverberatory furnace red lead is formed.

$$\underset{\text{Litharge}}{6PbO} + O_2 \xrightleftharpoons{470^\circ C} \underset{\text{Red lead}}{2Pb_3O_4}$$

Physio-Chemical Properties

(1) It is a red power which is insoluble in water.

(2) Heating Effect (above 470°C): On heating it turns violet or blackish and at above 470°C it decomposes into PbO and O_2.

$$2Pb_3O_4 \xrightarrow{\Delta} 6PbO + O_2$$

(3) Oxidizing Properties: Being an oxidant it can oxidize HCl into chlorine and evolve oxygen with hot and concentrated sulphuric acid as follows:

$$Pb_3O_4 + 8HCl \rightarrow 3PbCl_2 + 4H_2O + Cl_2$$

$$2Pb_3O_4 + 6H_2SO_4 \rightarrow 6PbSO_4 + 6H_2O + O_2$$

It can oxidized CO into CO_2.

$$Pb_3O_4 + 4CO \rightarrow 3Pb + 4CO_2$$

(4) With Nitric Acid: On reaction with nitric acid it gives lead nitrate and a brown precipitate of PbO_2 which suggests that it is a mixture of PbO and PbO_2.

$$Pb_3O_4 + 4HNO_3 \rightarrow 2Pb(NO_3)_2 + PbO_2 \downarrow + 2H_2O$$

Uses

(1) It is used in making flint glass, match industry and as a red pigment in making protective paints for coating iron and steel.

(2) Pb_3O_4 + linseed oil is a protective paint for iron, silver mirror etc.

Lead (II) Sulphide (PbS)

Preparation

When H_2S is passed through an acidified solution of lead nitrate or any other lead salt, the black ppt. of PbS is formed which can be filtered, washed and dried easily.

$$Pb(NO_3)_2 + H_2S \rightarrow \underset{\text{Black ppt.}}{PbS} + 2HNO_3$$

Physio-Chemical Properties

(1) It is a black solid which is insoluble in water and also in yellow ammonium sulphide.

(2) It dissolves in 50 % HNO_3 solution to give lead nitrate as follows:

$$PbS + 2HNO_3 \rightarrow \underset{\text{Lead nitrate}}{Pb(NO_3)_2} + H_2S$$

Uses

It is used in ceramics.

Lead (II) Halides or Plumbous Halides (PbX_2)

Lead can form stable dihalides of PbX_2 type example, PbF_2, $PbCl_2$, $PbBr_2$, PbI_2 which are ionic in nature.

$PbF_2 > PbCl_2 > PbBr_2 > PbI_2$

White solids Yellow solid

$\longrightarrow$

Decreasing order of ionic nature

Lead Chloride or Plumbous Chloride ($PbCl_2$)

Preparation

It is prepared by the reaction of hydrochloric acid or some soluble chloride with a lead salt solution as follows:

$$Pb(NO_3)_2 + 2HCl \rightarrow PbCl_2 + 2HNO_3$$

$$Pb(NO_3)_2 + 2NaCl \rightarrow PbCl_2 + 2NaNO_3$$

$$PbO + 2HCl \rightarrow PbCl_2 + H_2O$$

Physio-Chemical Properties

(1) It is a white crystalline solid, soluble in hot water. It is soluble in excess of hydrochloric acid giving chloro plumbous acid.

$$PbCl_2 + 2HCl \rightarrow H_2PbCl_4$$

(2) With Hot Lime Water: It reacts with hot lime water to give Pb(OH)Cl (a white pigment).

$$PbCl_2 + Ca(OH)_2 \rightarrow \underset{\text{Pattuson's white lead}}{Pb(OH)Cl} + CaO + HCl$$

(3) Heating Effect: On heating in air it forms lead oxychloride as follows:

$$2PbCl_2 + ½\, O_2 \rightarrow Pb_2OCl_2 + Cl_2$$

Lead (IV) Halides or Plumbic Halides (PbX_4)

Among PbX_4 only PbF_4 and $PbCl_4$ are known however PbF_4 has not been obtained in the pure state.

Lead Tetrach loride or Plumbic Chloride ($PbCl_4$)

Preparation

It is obtained by the reaction of lead dioxide and well cooled hydrochloric acid.

$$PbO_2 + 4HCl \xrightarrow{273\text{ K}} PbCl_4 + 2H_2O$$

Physio-Chemical Properties

(1) It is a yellow liquid compound, covalent in nature and soluble in organic solvents.

(2) It is unstable compound and decomposes easily as follows:

$$PbCl_4 \rightarrow PbCl_2 + Cl_2$$

(3) It is hydrolysed by water as follows:

$$PbCl_4 + 2H_2O \rightarrow PbO_2 + 4HCl$$

(4) It forms stable double salt with ammonium chloride, which decomposes when reacts with sulphuric acid.

$$PbCl_4 + 2NH_4Cl \rightarrow \underset{\text{Amm. hexachloro plumbate}}{(NH_4)_2\, PbCl_6}$$

$$(NH_4)_2PbCl_6 + H_2SO_4 \rightarrow (NH_4)_2SO_4 + PbCl_4 + 2HCl$$

(5) With Concentrated HCl:

$$PbCl_4 + 2HCl \rightarrow H_2[PbCl_6]$$

Lead Acetate or Sugar of Lead ($(CH_3COO)_2Pb$)

Preparation

It can be prepared by dissolving lead oxide i.e., litharge or lead carbonate or basic lead carbonate in acetic acid (45%) and crystallizing the solution as follows.

$$PbO + 2CH_3COOH \rightarrow (CH_3COO)_2Pb + H_2O$$

$$PbCO_3 + 2CH_3COOH \rightarrow (CH_3COO)_2Pb + CO_2 + H_2O$$

$$2PbCO_3.Pb(OH)_2 + 6CH_3COOH \rightarrow 3(CH_3COO)_2Pb + 2CO_2 + 4H_2O$$

Physio-Chemical Properties

(1) It is a white crystalline solid which is soluble in water. Its solution taste's sweet and called sugar of lead but it is poisonous in nature.

(2) Heating Effect: On heating, it decomposes as follows:

$$(CH_3COO)_2Pb \rightarrow PbO + CO_2 + \underset{\text{Acetone}}{CH_3COCH_3}$$

(3) With Sodium Bicarbonate: On reaction with sodium carbonate it forms a white precipitate of lead carbonate.

$$(CH_3COO)_2Pb + 2NaHCO_3 \rightarrow PbCO_3 + 2CH_3COONa + CO_2 + H_2O$$

(4) With Sodium Carbonate: On reaction with sodium carbonate solution it forms a white precipitate of basic lead carbonate.

$$3(CH_3COO)_2Pb + 3Na_2CO_3 + H_2O \rightarrow \underset{\text{Basic lead carbonate}}{2PbCO_3.Pb(OH)_2} + 6CH_3COONa + CO_2$$

(5) With Potassium Chromate: On reaction with potassium chromate it gives a yellow precipitate of lead chromate.

$$(CH_3COO)_2Pb + K_2CrO_4 \rightarrow PbCrO_4 + 2CH_3COOK$$

(6) With H_2S: When H_2S is passed over its solution a black precipitate of PbS is formed

$$(CH_3COO)_2Pb + H_2S \rightarrow \underset{\substack{\text{Lead sulphide} \\ \text{(Black)}}}{PbS} + 2CH_3COOH$$

(7) With Litharge: When a solution of lead acetate is boiled with litharge, basic lead acetate is formed.

$$(CH_3COO)_2Pb + PbO + H_2O \rightarrow \underset{\text{Basic lead acetate}}{Pb(OH)_2Pb(CH_3COO)_2}$$

(8) With dilute HCl: When dilute HCl is added to a cold solution of lead acetate, a white precipitate of lead chloride is formed which is soluble in hot water.

$$(CH_3COO)_2Pb + 2HCl \rightarrow PbCl_2 + 2CH_3COOH$$

Uses

(1) It is used as a laboratory reagent for the test of sulphide and chloride ions.

(2) It is used in medicine for curing skin diseases.

(3) It is used as a mordant in dyeing and calicoprinting.

(4) It is used in the manufacture of chrome yellow and chrome red used as lead pigments.

Basic Lead Carbonate or White Lead $2PbCO_3.Pb(OH)_2$

White lead can be manufactured by any of these three methods: Dutch process, Carter's process, Electrolytic process

(1) Dutch Process

It is a very slow method which takes about 6 to 8 weeks for the conversion of lead metal into white lead in the presence of acetic acid, carbon dioxide and moist air. Here best quality of white lead is formed.

Here following chemical reactions take place:

$$2Pb + 2H_2O + O_2 \rightarrow 2Pb(OH)_2$$

$$Pb(OH)_2 + 2CH_3COOH \rightarrow Pb(CH_3COO)_2 + 2H_2O$$

$$Pb(OH)_2 + Pb(CH_3COO)_2 \rightarrow \underset{\text{Basic lead acetate}}{Pb(OH)_2Pb(CH_3COO)_2}$$

$$3[Pb(OH)_2Pb(CH_3COO)_2] + 4CO_2 + 2H_2O \rightarrow \underset{\text{White lead}}{2[2PbCO_3.Pb(OH)_2]} + 6CH_3COOH$$

Here, white lead is removed from unchanged lead and washed with water to remove soluble lead acetate and finally it is dried.

(2) Carter's Process

It is comparatively quicker method and here the process is carried out in wooden cylinders which are rotated. Molten lead is first atomized by a jet of superheated steam or compressed air and brought in the wooden cylinders. A spray of acetic acid, carbon dioxide and air is blown into the cylinders.

After chemical reactions white lead is formed just like in Dutch process but in 5 – 12 days only.

(3) Electrolytic Process

The electrolytic cell is divided into two compartments by using a porous diaphragm. In it, the anodic compartment is filled with sodium acetate solution and have anodes of lead plates while cathodic compartment has cathodes of steel dipped into a solution of sodium carbonate. When electricity is passed, lead of anode starts dissolving in the form of lead acetate, which meets with sodium carbonate to form white lead which is washed with water and finally dried.

$$CH_3COONa \rightleftharpoons CH_3COO^- + Na^+$$

Anode

↓ dissolves lead

$$Pb(CH_3COO)_2$$

$$3Pb(CH_3COO)_2 + 4Na_2CO_3 + 2H_2O \rightarrow 2PbCO_3.Pb(OH)_2 + 2NaHCO_3 + 6CH_3COONa$$

Physio-Chemical Properties

(1) It is white, heavy amorphous powder which is poisonous in nature.

(2) It is insoluble in water but miscible with linseed oil.

(3) Heating Effect: On heating, it decomposes into red lead at 700 K.

$$2PbCO_3.Pb(OH)_2 \rightarrow \underset{\text{Red lead}}{Pb_3O_4} + H_2O + CO + CO_2$$

It is darkened in air due to the formation of black precipitate of PbS.

Uses

(1) White lead mixed with linseed oil is used as a white paint having a large covering power. It is normally mixed with $BaSO_4$ which increases its property of brushing out without affecting its covering power.

(2) It is also used for the preparation of red lead and pottery glazes.

- The main disadvantage of white lead is that it is not only poisonous but also it becomes black if H_2S is present in atmosphere due to the formation of PbS. So in place of it various substitutes like lithopone, titanium oxide, sublimated white lead etc., are used as white pigment now.

CHAPTER-END EXERCISES

PRACTICE QUESTIONS – I

1. The inert form of carbon is:
(a) Graphite (b) Coal
(c) Diamond (d) Charcoal

2. Diamond and graphite are:
(a) Allotropes (b) Isomers
(c) Polymers (d) Isotopes

3. The geometry of $(SiO_4)^{4-}$ ion is:
(a) Square planar (b) Tetrahedral
(c) Planar triangular (d) Octahedral

4. Which of the following types of forces bind together the carbon atoms in diamond?
(a) Dipolar (b) Van der Waal's
(c) Ionic (d) Covalent

5. Poisonous gas present in the exhaust fumes of an automobile is:
(a) CO_2 (b) CH_4
(c) CO (d) C_2H_6

6. Germanium is transparent in:
(a) Visible light (b) Infra violet region
(c) Infra-red region (d) Ultra violet region

7. When $SiCl_4$ is allowed to undergo hydrolysis to gives
(a) $Si(OH)_4$ – Silicic acid
(b) SiO_2 – Silicic acid
(c) $Si(OH)Cl_3$ – Silicic acid
(d) $SiCl_4$ do not undergo hydrolysis

8. Carbon is soluble in:
(a) Concentrated HNO_3
(b) Diluted HCl
(c) Concentrated H_2SO_4
(d) Concentrated HCl

9. H_2SO_4 is not used for the preparation of CO_2 from marble chips because:
(a) Calcium sulphate is sparingly soluble and gets dposited on marble chips and stops the reaction
(b) Diluted HCl
(c) It does not react
(d) Huge amount of heat is evolved

10. Which of the following is a good conductor of electricity?
(a) Coal (b) Diamondl
(c) Graphite (d) None

11. Which of the following is chemically inactive allotropic form of carbon?
(a) Coal (b) Diamondl
(c) Animal charcoal (d) Charcoal

12. Moderate electrical conductivity is shown by:
(a) Diamond (b) Carborundum
(c) Animal charcoal (d) Charcoal

13. Carbon shows tetravalency due to:
(a) Dsp^2 hybridisation (b) Sp^3 hybridisation
(c) S^2 p^2 hybridisation (d) D^3p hybridisation

14. Quartz is a crystalline variety of:
(a) Silicon (b) Silica
(c) Sodium silicate (d) Silicon carbide

15. he metal which does not form ammonium nitrate by reaction with dilute nitric acid is:
(a) Pb (b) Fe
(c) Mg (d) Al

16. CO forms a volatile compound with:
(a) Copper (b) Nickel
(c) Aluminium (d) Sodium

17. The purest form of coal is:
(a) Anthracite (b) Peat
(c) Bituminous coal (d) Lignite

18. Dry powder fire extinguisher contains:
(a) Sand and K_2CO_2
(b) Sand and baking soda
(c) Sand
(d) Sand and Na_2CO_3

19. The silicates which contain discrete tetrahedral units are:
(a) Ortho silicates
(b) Sheet silicates
(c) Three dimensional silicates
(d) Pyrosilicate

20. "CARBOGEN" is:
(a) Mixture of O_2 and CO_2
(b) Mixture of CO and CO_2
(c) $COCl_2$
(d) Pure form of carbon

21. In the ground state of carbon atom has unpaired electrons:
(a) 4 (b) 3
(c) 2 (d) No

22. Silica is soluble in:
(a) HF (b) HNO_3
(c) H_2SO_4 (d) HCl

23. Which of the following halide of carbon is used as refrigerant?
(a) CCl_4 (b) CF_4
(c) CH_2Cl_2 (d) CH_2F_2

24. Which one of the following carbides is most likely to be non-stoichiometric?
(a) SiC (b) CaC_2
(c) Fe_3C (d) None of these

25. SiO_2 is reacted with sodium carbonate. What is the gas liberated?
(a) O_2 (b) O_3
(c) CO (d) CO_2

26. The component present in greater proportion in water gas is:
(a) CO (b) H_2
(c) CO_2 (d) CH_4

27. Which can do glass etching?
(a) HIO_4 (b) HF
(c) SiF_4 (d) HNO_3

28. The substance used as a smoke screen in warfare is:
(a) PH_3 (b) Acetylene
(c) $SiCl_4$ (d) PCl_4

29. For prevention of rusting of iron, which is used in paints?
(a) PbO b) PbO_2
(c) Pb_3O_4 (d) $PbSO_4$

30. Amorphous form of silica is:
(a) Cristobalite (b) Tridymite
(c) Asbestos (d) Fumed silica

31. What is the C – C bond length (in angstroms) in diamond?
(a) 5.2 (b) 2.0
(c) 1.54 (d) 3.35

32. Which of the following oxides is amphoteric in character?
(a) CO_2 (b) SiO_2
(c) SnO_2 (d) CaO

33. Elemental silicon to be used as semiconductor is purified by:
(a) Floatation
(b) Electrolysis
(c) Zone refining
(d) Heating under vacuum

34. The type of bonding present in tetravalent compounds of group 14 elements is:
(a) Zero (b) -2
(c) +2 (d) -4

35. CO is absorbed by:

(a) CCl_4
(b) Pyrogallol
(c) Ammonical solution of cuprous chloride
(d) $CHCl_3$

36. The use of diamond as a gem depends on its:

(a) Chemical inertness
(b) Hardness
(c) Purest form of carbon
(d) High refractive index

37. Fire extinguisher contain a bottle of H_2SO_4 and:

(a) $MgCO_3$ (b) $CaCO_3$
(c) $NaHCO_3$ (d) Any carbonate

38. Which of the following is potassium ferricyanide?

(a) $K_3[Fe(CN)_3]$ (b) $K_3[Fe(CN)_4]$
(c) $K_3[Fe(CN)_6]$ (d) $K_4[Fe(CN)_6]$

39. A mixture of carbon monoxide, hydrogen and hydrocarbons is known as:

(a) Producer gas
(b) Water gas
(c) Semi water gas
(d) Carbureted water gas

40. Formula for tear gas is:

(a) N_2O (b) $COCl_2$
(c) Cl_3NO_3 (d) none of these

41. The molecule having zero dipole moment is:

(a) CCl_4 (b) H_2O
(c) HCl (d) $CHCl_2$

42. Glass reacts with HF to produce:

(a) H_2SiF_6 (b) Na_3AlF_6
(c) H_2SiO_3 (d) SiF_4

43. When steam is passed through red hot coke:

(a) CO and N_2 are formed
(b) CO_2 and H_2 are formed
(c) Petrol gas is obtained
(d) CO and H_2 are formed

44. Carbon dioxide is used for extinguishing fire because:

(a) It is a colourless gas
(b) It is neither combustile nor a supporter of combution
(c) It has a relatively high critical temperature
(d) In solid state, it is called dry ice

45. What is the compound formed when acetaldehyde is reacted with $LiAlH_4$?

(a) CH_3COOH (b) CH_3COCH_3
(c) C_2H_5OH (d) $C2H_5OC_2H_5$

46. The ability of a substances to assume two or more crystalline structures is called:

(a) Amorphism (b) Diamondl
(c) Graphite (d) None

47. The shape of tetrahalides of group 14 elements is:

(a) Tetrahedral
(b) Octahedral
(c) Trigonal bipyramid
(d) Square planar

48. Maximum ability of catenation is shown by:

(a) Sulphur (b) Nitrogen
(c) Oxygen (d) Carbon

49. Lead and tin are extracted from their chief ores by:

(a) Electrolysis and self reduction
(b) Carbon reduction and self reduction
(c) Self reduction and carbon reduction
(d) Self reduction and electrolysis.

50. Which of the following is used in the preparation of aerated water (soda)?

(a) SO_2 (b) HCl
(c) CO_2 (d) CO

51. Silicon has the characteristics of a:

(a) Non metal (b) Metalloid
(c) Metal (d) None of these

52. Graphite is a good conductor of electricity because if contains:

(a) Mobile electrons
(b) Bonded electrons
(c) Strong C = C bonds
(d) Strong C – C bonds

53. Sodium nitroprusside when added to an alkaline solution of sulphide ions produce:

(a) Purple colouration
(b) Blue colouration
(c) Red colouration
(d) Brown colouration

54. Which one of the following allotropic forms of carbon is isomorphous with crystalline silicon?
(a) Coal (b) Diamond
(c) Coke (d) Graphite

55. Coal gas:
(a) Is not a good fuel
(b) Is not used for lighting purpose
(c) Burns with non-smoky flame
(d) Burns with a smoky flame

56. An ideal mixture of producer gas contains:
(a) 65% CO, 35 % N_2
(b) 65% CO, 30 % N_2 and 5 % CO_2
(c) 35 % CO, 65 % N_2
(d) 30 % CO, 65 % N_2 and 5 % CO_2

57. One can obtain a silica garden if:
(a) Silicon salts are grown in garden
(b) Crystals of coloured cations are added to a strong solution of sodium silicate
(c) Silicon tetrafluoride is hydrolysed
(d) Sodium silicate solution is heated with base

58. Silicon is an important constituent of:
(a) Amalgams (b) Chlorophyll
(c) Haemoglobin (d) Rocks

59. Elements of group 14:
(a) Form M^{4-} and M^{4-} ions
(b) Form M^{4+} ions
(c) Exhibit oxidation state of +4
(d) Exhibit oxidation state of +4 and +2

60. What is the number of free electrons present on each carbon atom in graphite?
(a) Zero (b) 1
(c) 2 (d) 3

61. A dry ice piece is composed of:
(a) Solid C_6H_6 (b) Solid SO_2
(c) Solid CO_2 (d) Solid He

62. Which one of the following is used as an acid flux in metallurgy?
(a) SiO_2 (b) SO_2
(c) Na_2CO_3 (d) CaO

63. In laboratory burners, we use:
(a) Coal gas (b) Oil gas
(c) Gobar gas (d) Producer gas

64. Tungsten carbide is an example of:
(a) Passive solid solution
(b) Interstitial solid solution
(c) Sandwich solid solution
(d) Substitutional solid solution

65. Silicons are:
(a) Water repelling in nature
(b) With high dielectric strength
(c) With high thermal stability
(d) All of these

66. Pπ – pπ multiple bond is seen in:
(a) All carbon family member
(b) Carbon and less in other metals
(c) Boron family and hot in carbon family
(d) Sn but not in carbon

67. Which of the following pairs liberate a gas when they react with each other:
(a) Mg, B_2O_3 (b) Mg, CO_2
(c) Fused NaOH, C (d) SiO_2, Na

68. Coal gas is a mixture of:
(a) H_2 saturated and unsaturated hydrocarbons CO, CO_2, N_2 and O_2
(b) CO and H_2
(c) CO, CO_2, CH_4
(d) Saturated and unsaturated hydrocarbons:

69. Which does not exist?
(a) $[SiCl_6]^{2-}$ (b) $[CCl_6]^{2-}$
(c) $[SnCl_6]^{2-}$ (d) $[GeF_6]^{2-}$

70. SiO_2 is reacted with sodium carbonate. What is the gas liberated?
(a) O_2 (b) O_3
(c) CO (d) CO_2

71. What is the formula of carbon suboxide?
(a) C_3O_2 (b) C_2O_4
(c) CO (d) CO_2

72. In which of the following silicates, cyclic structure is present?

(a) Emerald (b) Talc
(c) Mica (d) Asbestos

73. Which one of the following is a correct set for SiO_2?

(a) Angular, basic (b) Linear, basic
(c) Linear, acidic (d) Tetrahedral, acidic

74. A pseudo solid is:

(a) Glass (b) $CaCO_3$
(c) Rock salt (d) Diamond

75. The number of carbon compound is very large because:

(a) It has strong catenation property
(b) It shows allotropy
(c) It has non metallic character
(d) Carbon is tetravalent

76. In graphite, electrons are:

(a) Spread out between the structure
(b) Localized on earth C-atom
(c) Present in anti-bonding orbital
(d) Localized on every third C-atom

77. What is X in the following reaction?

$SiCl_4 + H_2O \rightarrow X + HCl$

(a) H_4SiO_4 (b) SiH_4
(c) $(SiOOH)_2$ (d) $Si(ClO_4)_4$

78. Destructive distillation of coal does not give:

(a) Sarbides (b) Ammonia
(c) Gas carbon (d) Coke

79. Which of the following is a ferrous alloy?

(a) German silver (b) Gun metal
(c) Nichrome (d) Devarda's alloy

80. Carbon-60 contains:

(a) 12 pentagons and 20 hexagons
(b) 20 pentagons and 12 hexagons
(c) 30 pentagons and 20 hexagons
(d) 24 pentagons and 36 hexagons

Practice Questions – II

81. If CO_2 is passed in excess in to lime water, the milkines first formed disappears due to:

(a) Formation of water soluble calcium bicarbonate.
(b) The solution getting heated by exothermic reation.
(c) Reversal of the original reaction.
(d) Formation of volatile calcium derivative.

82. In group 4 A or 14 group of the extended form of the periodic table with increase in atomic number, the oxidizing power of tetravalent species increases in the order:

(a) Ge > Sn > Pb (b) Pb > Sn > Ge
(c) Pb > Ge > Sn (d) Ge > Pb > Sn

83. The species present in solution when CO_2 is dissolved in water are:

(a) CO_2, H_2CO_3, HCO_3^-, CO_3^-
(b) H_2CO_3, CO_3^{2-}
(c) CO_3^{2-}, HCO_3^-
(d) CO_2, H_2CO_3

84. The correct order of stability is:

(a) $SnF_2 > PbF_2 > GeF_2$
(b) $GeF_2 > SnF_2 > PbF_2$
(c) $GeF_2 > PbF_2 > SnF_2$
(d) $PbF_2 > SnF_2 > GeF_2$

85. The CO bond length is least in which one of the following?

(a) Acetone (b) Carbonate ion
(c) Carbon dioxide (d) Carbon monoxide

86. Which of the following reacts readily with SiO_2 to give SiF_4?

(a) C_2F_6 (b) XeF_6
(c) CaF_2 (d) NaF

87. Which of the following oxides of carbon are stable?

(a) CO_2 (b) C_3O_2
(c) CO (d) All of these

88. The correct match is:

(a) SnO, PbO - amphoteric
(b) $B(OH)_3$- basic
(c) PbO_2 – only acidic
(d) GeO_2 – basic

89. $CH_3Cl + Si\ Cu \xrightarrow[570K]{Cu\ Powder} (X) \xrightarrow{2H_2O} (Y)$
(Y) & (X) respectively are,

(a) $SiCl_4$, $Si(OH)_4$
(b) $Si(OH)_4$, $SiCl_4$
(c) $(CH_3)_2Si(OH)_2$, $(CH_3)_2SiCl_2$
(d) $(CH_3)_2SiCl_2$, $(CH_3)_2Si(OH)_2$

90. Which element-element bond has highest bond energy?

(a) SiF_4 (b) Si – Si
(c) Sn – Sn (d) Ge – Ge

91. Pb and Sn are extracted from their chief ore by:

(a) Carbon reduction and self reduction
(b) Self reduction and carbon reduction
(c) Electrolysis and self reduction
(d) Self reduction and electrolysis

92. Very pure silicon is an insulator, but becomes a p-type or an n-type semiconductor when doped with a:

(a) Group 12 and a group 16 element respectively
(b) Group 4 and group 6 element respectively
(c) Group 1 and a group 12 element respectively
(d) Group 13 and a group 15 element respectively

93 Which statement is most inconsistent with the chemistry of silicon?

(a) It is hard, gray, semiconducting solid that melts at 1410°C.
(b) It is obtained by oxidation of silica sand with coke.
(c) In nature, it is generally found combined with ox gen in SiO_2 and in various silicate minerals.
(d) It crystallizes in a diamond like structure and does not form the graphite like allotrope.

94. $[SiO_4]^{4-}$ has tetrahedral structure the silicate formed by using three oxygen has a:

(a) Two dimensional sheet structure
(b) Pyrosilicate structure
(c) Linear polymeric structure
(d) Three dimensional structure.

95. The correct match is:

(a) C_{60}- Buckminster fullerence
(b) Borazole- $B_3N_3H_6$
(c) $Na_2B_4O_7 . 4H_2O$- Kernite
(d) All of these

96. Hydrolysis of dimethyldichloro silane; $(CH_3)_2SiCl_2$ followed by condensation polymerization yields straight chain polymer of:

(A) ~O–Si(O)(O)–Si(O)(O)–O~ (each Si bearing one O above and one O below)

(B) ~Si–O–O–Si~

(C) ~O–Si(CH_3)(CH_3)–O–Si(CH_3)(CH_3)–O~

(D) ~O–Si(CH_3)(CH_3)–Si(CH_3)(CH_3)–O~

97. The soldiers of Napolean army while at Alps during freezing winter suffered a serious problem as regards to the tin buttons of their uniforms. White metallic tin buttons got converted to grey power. This transformation is related to:

(a) A change in the partial pressure of oxygen in the air
(b) An interaction with water vapour contained in the humid air.
(c) A change in the crystalline structure of tin.
(d) An interaction with nitrogen of the air at very low temperatures.

98. Oxalic acid is heated with concentrated H_2SO_4. when the resultant gases are passed over red hot carbon, X is obtained. X, is:

(a) CO (b) CO_2
(c) C_2H_4 (d) CH_4

99. Dipole moment of CO is very small (0.1 D) (inspite of the larger difference in electronegativity between C and O) and negative end of the dipole lies near the less electronegative C-atom. It is due to:

(a) Occupation of antibonding molecular orbitals
(b) Occupation of the lone pairs as well as empty do bitals
(c) Occupation of lone-pairs
(d) Occupation of empty d-orbitals

100. In carbon-60 all carbon atoms are:
(a) sp^2-hybridised with a graphite like shape
(b) sp^2-hybridised with a diamond shape
(c) sp^3-hybridised with a square antiprism shape
(d) sp^2-hybridised with a truncated icosahedron shape

101. Which one of the following has a two dimensional layer structure?
(a) Tremolite asbestos, $Ca_2Mg_5(Si_4O_{11})(OH)_2$
(b) Zircon, $ZrSiO_4$
(c) Silica, SiO_2
(d) Talc, $Mg_3(OH)_2(Si_4O_{10})$

102. Method used for obtaining highly pure silicon used as a semiconductor material, is:
(a) electrochemical (b) oxidation
(c) zone refining (d) crystallization

103. Which one of the following is correct about stability of the given ions?
(a) $Sn^{4+} > Sn^{2+}$ (b) $Pb^{4+} > Pb^{2+}$
(c) $Pb^{2+} > Pb^{4+}$ (d) $Si^{2+} > Si^{4+}$

104. $A + CO \rightarrow CO_2$
$B + CO \rightarrow CO_2$
$C + O_2 \rightarrow CO_2$
A, B & C respectively are:
(a) Fe_2O_3, ZnO, CH_4
(b) CH_4, Carbon , Fe_2O_3
(c) HCOOH, carbon, CH_4
(d) Fe_2O_3, CH_4, ZnO

105. Graphite is a:
(a) Good conductor of electricity
(b) Good conductor of heat
(c) Bad conductor of heat
(d) Both A and B

106. In silicon dioxide:
(a) there are double bonds between silicon and oxygenatoms.
(b) silicon is bonded to two silicon atoms.
(c) each silicon atom is surrounded by two oxygen atoms and each oxygen atom is bonded to two silcon atoms.
(d) each silicon atom is surrounded by four oxygen atoms and each oxygen atom is bounded to two silicon atoms.

107. In the equilibrium:
$C \text{ (s, diamond)} \rightleftharpoons C \text{ (s, graphite)} + \text{Heat}$ (density of diamond and graphite are 3.5 and 2.3 g/cm^3 respectively), the equilibrium will be shifted to the left at
(a) High temperature and low pressure
(b) Low temperature and very high pressure
(c) Low temperature and low pressure
(d) High temperature and very high pressure

108. To get an 'n' type doped semiconductor, impurity to be added to silicon should have the following number of valence electrons:
(a) 1 (b) 2
(c) 3 (d) 5

109. (1) $SiO_2 + NaOH \rightarrow ?$
(2) $SiO_2 + HF \rightarrow ?$
The products of (2) & (1) respectively are
(a) SiF_4, Na_2SiO_3
(b) H_2SiF_6, SiO_4^{4-}
(c) Na_2SiO_4, H_2SiF_6
(d) Na_2SiO_3, SiF_4

110. What is not a true statement?
(a) Compounds with C = C double bonds and Si = Si double bonds are quite common.
(b) BF_3 is a gaseous molecular halide but AlF_3 is a high melting ionic solid.
(c) BeO is amphoteric but the oxides of the other group 2A elements are basic.
(d) B differs from other elements of group 3A by forming mainly covalent molecular compounds.

111. Graphite is a soft, solid, lubricant, extremely difficult to melt. The reason for this anomalous behaviour is that graphite:
(a) Has molecules of variable molecular masses like polymers.
(b) Has carbon atoms arranged in large plates of rings of strongly bound carbon atoms with weak inte plate bonds.
(c) Is an allotropic form of diamond.
(d) Is a non crystalline substance.

112. Carbon shows strong catenation while silicon shows little or no catenation because:

(a) The Si–Si bond is stronger than the C–C bond.
(b) The C–C bond is stronger than the Si–Si bond.
(c) Silicon is a metalloid and carbon is a non-metal.
(d) Silicon forms ionic compounds whereas carbo forms covalent compounds.

113. Ge(II) compounds are powerful reducing agents, whereas Pb (IV) compounds are strong oxidants. It can be due to:

(a) Lead, the inert pair effect in more pronounced than in germanium
(b) The ionic radii of Pb^{2+} and Pb^{4+} are larger than those of Ge^{2+} and Ge^{4+}
(c) Lead is more electropositive than germanium
(d) The ionization potential of lead is less than that of germanium

114. The catenation tendency of C, Si and Ge is in the order Ge < Si < C. The bond energies (in kJ mol^{-1}) of C – C, Si – Si and Ge – Ge bonds, respectively are:

(a) 348, 180, 167 (b) 180, 167, 348
(c) 348, 167, 180 (d) 167, 180, 348

115. Which statement is inconsistent about graphite?

(a) A two dimensional sheet like structure in which each C atom uses sp^2 hybrid orbitals.
(b) Pi electrons are delocalized and free to move perpendicular to the plane of the hexagonal sheets.
(c) Carbon sheets are separated by a distance of 335 pm and are held together by weak London dispe sion forces.
(d) Electrical conductivity parallel to the planar sheets is 10^{20} times greater than the conductivity of diamond.

Practice Questions – III

116. Which statement is consistent about diamond?

(1) When an electrical potential is applied diamond becomes a valuable semiconductor
(2) The hardest known substance with the highest melting point for an element.
(3) A covalent network structure in which each C atom uses sp^3 hybrid orbitals.
(4) A tetrahedral array of sigma bonds with bond lengths of 154 pm.

(a) 1, 2 (b) 1, 2, 3
(c) 2, 3, 4 (d) 3, 4

117. Select correct statement:

(a) GeO_2, SnO_2 and PbO_2 are all acidic oxides.
(b) Oxides of carbon family (MO_2) are all networ soids with octahedral coordination.
(c) Silicon dioxide (silica) is a network solid with terahedral coordination and is a giant molecule.
(d) None of these.

118. Which of the following radicals evolve gas turning lime water milky?

(a) $C_2O_4^{2-}$ (b) CO_3^{2-}
(c) SO_3^{2-} (d) All of these

119. Which of the following statements is/are correct?

(1)The carbon atom is sp^2-hybridised in the CO_2 mol ecule as well as the molecule of its hydrate H_2CO_3.
(2)The carbon dioxide molecule behaves as a non po lar molecule even though two of its resonating structures. $^-O - C \equiv O^+$ and $O^+ \equiv C - O^-$
(3) The carbon dioxide molecule is linear because the carbon atom utilizes its sp-orbitals to form σ-bonds.
(4) Carbon dioxide is the anhydride of the unstable dibasic acid $O = C(OH)_2$.

(a) 1, 3, 4 (b) 1, 3, 4
(c) 1, 2, 4 (d) 1, 2, 3, 4

120. The correct statement is:

(a) $[SiF_6]^{2-}$ is known whereas $[SiCl_6]^{2-}$ is not.
(b) Diamond is covalent yet it has high melting point.
(c) SiO only exist at high temp.
(d) All of these.

121. Consider following statements:

1. In diamond, each carbon atom is linked tetrahe drally to four other carbon atoms by sp^3 bonds.
2. Graphite has planar hexagonal layers of carbon at oms held together by weak Van der Waal's forces.
3. Silicon exists only in diamond structure due to its tendency to form pπ – pπ bond to itself. In this:

(a) Only 1 is correct
(b) Only 1 and 2 are correct
(c) Only 2 and 3 are correct.
(d) All are correct statement here.

122. A colourless poisonous gas (A) burns with blue flame is passed through aqueous NaOH at high pressure and temperature to give a compound (B). (B) on heating gives (C) and (C) gives white precipitate (D) with $CaCl_2$ both (C) and (D) decolourise acidified $KMnO_4$. Here (A) can be:

(a) CO (b) CO_2
(c) SO_2 (d) H_2S

123. $H_2C_2O_4 \xrightarrow{\Delta}$ Gas (P) + gas (Q) + liquid (R)
Gas (P) burns with a blue flame and is oxidized to gas (Q).
Gas (P) + Cl_2 → (S) $\xrightarrow{NH_3, \Delta}$ (T)
(P), (Q), (R) and (T) are respectively
(a) CO, CO_2, $COCl_2$, $HCONH_2$
(b) CO, CO_2, H_2O, $COCl_2$
(c) CO_2, CO, H_2O, $HCONH_2$
(d) CO, CO_2, H_2O, NH_2CONH_2

124. Which statement is consistent about carbon monoxide?
(1) One of its main industrial uses is in the synthesis of methanol, CH_3OH.
(2) Toxicity of CO results from its ability to bond strongly to iron (II) atom in haemoglobin.
(3) It is formed by burning carbon or hydrocarbon in excess oxygen
(4) It is a colourless, odourless and toxic gas.
(a) 2, 4 (b) 1, 2, 4
(c) 1, 2, 3 (d) 2, 3, 4

125. A solid element Y conducts electricity and forms two chlorides YCln a colourless volatile liquid and YCl_{n-2} a colourless solid. Then Y belongs to which of the following groups of periodic table?
(a) 16 (b) 15
(c) 14 (d) 13

126. Which of the following statements is incorrect in the context of silicones?
(1) They are more stable to heat than other polymers.
(2) The Si –O bond energy is high.
(3) They are strongly water- repellent, are good elec trical insulators, and have nonsticking and anti forming properties.
(4) The Si –O bond energy is low.
(a) 1, 2, 4 (b) 1, 2, 3
(c) 2, 3, 4 (d) 2, 4

127. $(Me)_2SiCl_2$ on hydrolysis will produce:
(a) $(Me)_2Si(OH)_2$
(b) $(Me)_2Si = O$
(c) $-[-O-(Me)_2SI-O-]_n-$
(d) $Me_2SiCl(OH)$

128. Among the following substituted silanes the one which will give rise to cross linked silicone polymer on hydrolysis is:
(a) R_4Si (b) $RSiCl_3$
(c) R_2SiCl_2 (d) R_3SiCl

129. Which of the statement is not correct here:
(a) Carbon has maximum catenation power due to high bond energy.
(b) CO forms carboxyhaemoglobin with haemogl bin.
(c) Si is the most abundant element in earth crust.
(d) None is incorrect.

130. Which of the following statements is incorrect for silicon?
(1) Forms an oxide (SiO_2) that is amphoteric and has a Ga, As structure.
(2) Does not undergo coordination number expasion.
(3) Forms strong but unconjugated multiple bonds of the pπ-dπ variety, especially with O and N.
(4) Forms molecular halides that are not hydrolysed.
(a) 2, 4 (b) 1, 2, 4
(c) 1, 2, 3 (d) 2, 3, 4

131. Which one of the following is correct statement of fullerenes?
(a) Fullerenes are made by heating of graphite in an electric arc in the presence of Hydrogen.
(b) Fullerenes are the only impure form of carbon due to presence of dangling bonds.
(c) It contains twenty six- membered rings and twelve five membered rings.
(d) Both (a) and (b).

132. Which of the following statement is/are correct?
(1) C_{60} has a geodesic dome structure.
(2) Solid C_{60} has a cubic close-packed structure.
(3) The name buckminsterfullerene was given to C_{60}.
(4) The common name for C_{60} is 'bucky ball'
(a) 1, 2, 3 (b) 2, 3
(c) 2, 3, 4 (d) 1, 4, 3, 4

133. What is not a reaction commonly associated with cabondioxide?
(a) $CO_2 (g) + H_2O (l) \rightleftharpoons H+ (aq) + HCO_3- (aq)$
(b) $C_6H1_2O_6 (aq) + 6O_2 (g) \rightleftharpoons 6CO_2 (g) + 6H_2O (l)$
(c) $CaCO_3 (s) + heat \rightarrow CaO (s) + CO_2 (g)$
(d) Oxyhaemoglobin + $CO_2 \rightleftharpoons$ Carboxyhaemoglobin + O_2

134. Electrical conductivity in graphite maximized:
(a) Perpendicular to the pi frame work and perpendicular to the planar C-atom framework.
(b) Parallel to the pi framework and perpendicular to the planar C-atom framework.

(c) Parallel to the pi framework and parallel to the planar C-atom framework.
(d) Perpendicular to the pi framework and parallel to the planar C-atom framework.

135. Which of the following statements are correct?

(1) The hybridisation of C in graphite is sp^2.
(2) SiO_2 reacts with Na_2CO_3 and liberates CO.
(3) SiO_2 is used as acid flux.
(4) The distance between the layers in graphite is 3.35 x 10-3 cm.

(a) 1, 2 and 3 (b) 1, 2 and 4
(c) 1, 3 and 4 (d) 2, 3 and 4

136. Which of the following statement is correct for graphite?

(1) Its layers are very tightly packed, almost withou tany space between them.
(2) It cleaves easily between the layers which acounts for the remarkable softness of the crystals.
(3) Its density is lower than that of diamond
(4) It has a layered structure and the bonding between the layers is very weak.

(a) 2, 4 (b) 1, 2, 4
(c) 1, 2, 3 (d) 2, 3, 4

137. Which one is correct statement for zeolite?

(a) Hydrated zeolites are used as ion exchangers in hardening of soft water
(b) ZSM -5 us used to convert gasoline to alcohol
(c) They are alumino silicates
(d) All of these

138. Which of the following statement is incorrect?

(1) Graphite has such a high thermo-dynamical sta- bi ity that diamond spontaneously changes into graphite in ordinary conditions.
(2) Graphite and diamond have equal thermodynamic stability.
(3) Graphite is thermodynamically more stable than diamond.
(4) Diamond is thermodynamically more stable than graphite.

(a) 2, 4 (b) 1, 2, 4
(c) 1, 2, 3 (d) 2, 3, 4

139. An organic acid (a) reacts with concentrated H_2SO_4 to give a neutral oxide (B), acidic oxide (C) and a diatomic oxide (D). When (D) reacts with chlorine gas, a poisonous gas (E) is evolved, this gas with ammonia gives an organic compound (F). Then, compounds (A) and (F) can be found as:

(a) (A) = CH_3COOH and (F) = NH_2CONH_2
(b) (A) =$H_2C_2O_4$ and (F) = NH_2CONH_2
(c) (A) = CCl_4 and (F) = CH_3CHO
(d) (A) = $CHCl_3$ and (F) – $H_2C_2O_4$

140. The hybridization state of nitrogen atoms in the molecules$(CH_3)_3N$ and $(SiH_3)_3N$ are:

(a) Respectively sp^2 and sp^3
(b) Respectively sp^3 and sp^2
(c) Both sp^2
(d) Both sp^3

141. In graphite, the hybridization state of each carbon atom and the π–- bond order of each carbon- carbon bond are, respectively,

(a) sp^3 and 1 (b) sp^2 and 3/2
(c) sp^2 and 1/3 (d) sp and 1/2

142. A- tin $\xrightarrow{T_1}$ β- tin $\xrightarrow{T_2}$ γ- tin

Temperatures T_1 and T_2 will be:
(a) T_1 = 284 K, T_2 = 290 K
(b) T_1 = 496 K, T_2 = 310 K
(c) T_1 = 284 K, T_2 = 442 K
(d) T_1 = 296 K, T_2 = 310 K

143. The correct order of decreasing ionic character of PbF_2, $PbCl_2$, $PbBr_2$ and PbI_2 are:

(a) $PbF_2 > PbCl_2 > PbBr_2 > PbI_2$
(b) $PbF_2 > PbBr_2 > PbCl_2 > PbI_2$
(c) $PbF_2 < PbCl_2 > PbBr_2 > PbI$
(d) $PbF_2 < PbCl_2 < PbBr_2 < PbI_2$

Answer Keys

1. (c)	2. (a)	3. (b)	4. (d)	5. (c)	6. (c)	7. (a)	8. (a)	9. (a)	10. (c)
11. (b)	12. (c)	13. (b)	14. (b)	15. (a)	16. (b)	17. (a)	18. (b)	19. (a)	20. (a)
21. (c)	22. (a)	23. (d)	24. (c)	25. (d)	26. (b)	27. (b)	28. (c)	29. (c)	30. (d)
31. (c)	32. (c)	33. (c)	34. (a)	35. (c)	36. (d)	37. (c)	38. (c)	39. (d)	40. (c)
41. (a)	42. (a)	43. (d)	44. (b)	45. (c)	46. (d)	47. (c)	48. (d)	49. (c)	50. (c)
51. (b)	52. (a)	53. (a)	54. (b)	55. (c)	56. (c)	57. (b)	58. (d)	59. (d)	60. (b)
61. (c)	62. (a)	63. (b)	64. (b)	65. (d)	66. (b)	67. (c)	68. (a)	69. (b)	70. (a)
71. (a)	72. (a)	73. (d)	74. (a)	75. (a)	76. (d)	77. (a)	78. (a)	79. (c)	80. (a)
81. (a)	82. (b)	83. (a)	84. (d)	85. (d)	86. (b)	87. (d)	88. (a)	89. (c)	90. (a)
91. (b)	92. (d)	93. (b)	94. (a)	95. (d)	96. (c)	97. (c)	98. (a)	99. (a)	100. (d)
101. (d)	102. (c)	103. (c)	104. (a)	105. (d)	106. (d)	107. (d)	108. (d)	109. (a)	110. (a)
111. (b)	112. (b)	113. (a)	114. (d)	115. (b)	116. (c)	117. (c)	118. (d)	119. (a)	120. (d)
121. (d)	122. (a)	123. (d)	124. (b)	125. (c)	126. (b)	127. (c)	128. (b)	129. (c)	130 (b)
131. (c)	132. (d)	133. (d)	134. (c)	135. (c)	136. (d)	137. (c)	138. (b)	139. (b)	140 (b)
141. (c)	142. (c)	143. (a)							

Hints and Explanations for Selective Questions

1. Diamond is very hard substance. It is sp^3 hybridised, hence it is inert.

2. Carbon have many allotropes like Diamond, Graphite, Buckminsterfullerene or Bucky Balls (made by evaporation of graphite).

4. In a diamond each carbon atom is sp^3 hybridised and thus forms covalent bonds with four other carbon atoms lying at the corners of a regular tetrahedron.

8. $C + \text{Conc. } 4HNO_3 \rightarrow CO_2 + 4NO_2 + 2H_2O$

9. Marble and H_2SO_4 is not used for preparation of CO_2 because $CaSO_4$ is sparingly soluble and get deposited on marble chips and stops the reaction.

10. Graphite possesses free electron in its structure, it is responsible for the electrical conductivity.

11. Diamond is chemically inactive allotrope of carbon because of its compact structure and saturated nature.

12. Graphite show moderate conductivity due to the presence of unpaired or free fourth valence electron on each carbon atom.

15. Pb does not form ammonium nitrate with dil. HNO_3.

16. With nickel carbon monoxide forms volatile compound.

17. As anthracite contains more than 90% coal.

18. Mixture of SiO_2 (sand) and $NaHCO_3$ (baking soda) is used as a dry powder fire extinguisher.

20. 90 % O_2 and 10 % CO_2 used for reviving respiration.

21. Electronic configuration of carbon is $1s^2\ 2s^2\ 2p^2$, the unpaired electron of p-orbitals are two.

22. As it forms hydrofluorosilicate

25. $SiO_2 + Na_2CO_3 \xrightarrow{\Delta} Na_2SiO_3 + CO_2$

26. Water gas contains about 50 volume of H_2, 40 volume of CO and 5 volume of CO_2 and N_2 etc.

27. Glass is a mixture of sodium and calcium silicate. It reacts with HF forming sodium and calcium fluorosilicates respectively.

$Na_2SiO_3 + 6HF \rightarrow Na_2SiF_6 + 3H_2O$

$CaSiO_3 + 6HF \rightarrow CaSiF_6 + 3H_2O$

The etching of glass is based on these reactions.

28. $SiCl_4$ gets hydrolysed in moist air and gives white fumes which are used as a smoke screen in warfare.

31. In diamond C – C single bond in there. Therefore bond length is 1.54 Å.

32. SnO_2 is an amphoteric oxide.

33. Zone refining method is the best method for purific aion of semi conductor.

34. Tetravalent compounds of group 14 elements are covalent which are formed by mutual sharing of electrons.

36. Diamond have high refractive index. The value of $\mu = 2$, only some synthetic compound having such a high value of refractive index.

37. Fire extinguisher contain a bottle of H_2SO_4 and-$NaHCO_3$, which on reaction gives carbon dioxide.

39. $CO + H_2$ is called as water gas. $CO + H_2 +$ Hydrocarbons is known as Carburetted water gas.

41. CCl_4 is a symmetric structure, Tetrahedral structure with all identical C – Cl bonds.

42. $6HF + SiO_2 \rightarrow H_2SiF_6 + 2H_2O$

43. $C + H_2O \xrightarrow{\text{red hot}} CO + H_2$

45. $CH_3 - CHO + H_2 \xrightarrow{LiAlH_4} C_2H_5OH$

46. he phenomenon of occurrence of a substance in two or more crystalline structures is called polymorphism.

47. The shape of tetrahalides of group 14 elements are tetrahedral where the element involves sp^3 hybridisation.

49. Lead is extracted by self reduction process while tin is extracted by carbon reduction process.

51. Silicon has the characteristics of the semi-metal.

52. As carbon have mobile π-electrons which is responsible for electrical conductance.

53. Sodium nitroprusside when added to an alkaline solution of sulphide ions produce purple colouration.

55. Coal gas burns with non-smoky flame because it contains small hydrocarbons.

57. This solution gives a coloured growth of crystals after sometime known as silica garden.

58. Rocks are mainly made up of silicates.

61. Solid carbon dioxide (at very low temperature) is known as dry ice.

64. Tungsten carbide is an example of interstitial solid solution.

68. Coal gas is mixture of CO, CO_2, N_2, O_2, H_2 and saturated and unsaturated hydrocarbons.

70. $SiO_2 + Na_2CO_3 \xrightarrow{\Delta} Na_2SiO_3 + CO_2$

72. Beryl has cyclic structure. Beryl containing 1-2% Cr having grass green colour is known as Emerald.

74. Glass is a pseudo solid because it is super cooled liquid.

76. In graphite, each carbon is sp^2 –hybridised and the single occupied unhybridised p-orbitals of C-atoms overlap sidewise to give π-electron cloud which is delocalized and thus the electrons are spread out between the structure.

78. Carbides can produced by action of acetylene on metals at high temperature and pressure.

81. Calcium bicarbonate is water soluble, it is produced when excess CO_2 is passed through lime water.

83. $CO_2 + H_2O \leftrightarrow H_2CO_3 \leftrightarrow H^+ + HCO_3^- \leftrightarrow H^+ + CO_3^{2-}$

84. The stability of fluorides increases as the size of central atom increases.

The correct order is

$PbF_2 > SnF_2 > GeF_2$

86. Xenon hexafluoride readily reacts with SiO_2 and gives SiF_4. This is the basic reason that XeF_6 cannot be stored in glass vessel, that is why XeF_6 is stored in nickel alloy (monel metal vessel).

90. Si – F bond dissociation energy is highest. Its value is 540 kJ mol^{-1}.

Where as Si – Si = 215 kJ mol^{-1}

Ge – Ge = 165 kJ mol^{-1}

Sn – Sn = 152 kJ mol^{-}

97. Transition of white to grey tin occurs at any temperature below 15.2°C, it becomes rapid at -50°C.

98. $H_2C_2O_4 \xrightarrow[-H_2O]{H_2SO_4} CO + CO_2 \rightarrow CO$

CO_2 is reduced by "C" to CO.

102. Si obtained by reduction of $SiCl_4$ with H_2 is further purified by zone refining method to get Si of very high purity. Silicon is purified by zone-refining process because the impurities present in it are more soluble in the liquid phase than in the solid phase.

108. Silicon has four valence electron. When it is doped with an atom having 5 valence electrons then due to the presence of one extra electron it is called n-type.

122. $\underset{(A)}{CO} + NaOH \xrightarrow{\text{High, P\&T}} \underset{(B)}{HCOONa}$

$\underset{(B)}{HCOONa} \xrightarrow{\Delta} \underset{(C)}{(COONa)_2} + H_2$

$(COONa)_2 + CaCl_2 \rightarrow \underset{(D)}{CaC_2O_4} + 2NaCl$

$5CaC_2O_4 + 2KMnO_4 + 5H_2SO_4 \rightarrow$

$5CaSO_4 + K_2SO_4 + MnSO_4 + 8H_2O + 10CO_2$

125. The element Y belongs to group 14 of the periodic table which forms two chlorides YCl_4 a colourless volatile liquid and YCl_2 a colourless solid.

127. Si due to larger size cannot form π-bonds. The product is polymeric in nature and is known as Silicone.

128. $R-SiCl_3 \xrightarrow{H_2O} R-Si(OH)_3$

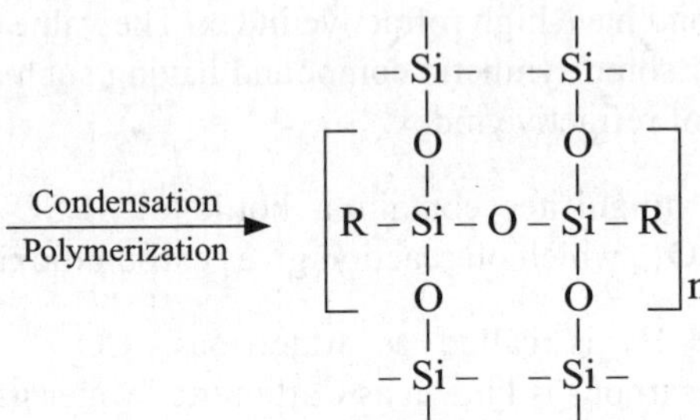

139. $\underset{(a)}{(COOH)_2} \xrightarrow{\text{conc. } H_2SO_4} \underset{(b)}{H_2O} + \underset{(c)}{CO_2} + \underset{(d)}{CO}$

$\underset{(d)}{CO} + Cl_2 \xrightarrow{b/w} \underset{\substack{(E)\\(\text{poisonous})}}{COCl_2} \xrightarrow[-2HCl]{NH_3} \underset{(F)}{NH_2CONH_2}$

Previous Years' Questions

1. Glass reacts with HF to produce
(a) H_2SiF_6 (b) Na_3AlF_6
(c) H_2SiO_3 (d) SiF_4
[2000]

2. Which of the following oxidation states are the most characteristic for lead and tin respectively:
(a) +4, +2 (b) +2, +4
(c) +4, +4 (d) +2, +2
[2007]

3. The correct order of C – O bond length among CO, CO_3^{2-} and CO_2 is
(a) $CO_2 < CO_3^{2-} < CO$
(b) $CO < CO_3^{2-} < CO_2$
(c) $CO_3^{2-} < CO_2 < CO$
(d) $CO < CO_2 < CO_3^{2-}$
[2007]

4. Which one of the following anions is present in the chain structure of silicates?
(a) $Si_2O_7^{6-}$ (b) $(Si_2O_5^{2-})n$
(c) $(SiO_3^{2-})n$ (d) SiO_4^{4-}
[2007]

5. The straight chain polymer is formed by
(a) Hydrolysis of CH_3SiCl_3 followed by condensation polymerization.
(b) Hydrolysis of $(CH_3)_4$ Si by addition polymerization.
(c) Hydrolysis of $(CH_3)_2SiCl_2$ followed by condensation polymerization.
(d) Hydrolysis of $(CH_3)_3$ SiCl followed by condensation polymerization.
[2009]

6. Which of the following oxide is amphoteric?
(a) SiO_2 (b) CO_2
(c) SnO_2 (d) CaO
[2011]

7. Name the type of the structure of silicate in which one oxygen atom of $[SiO_4]^{4-}$ is shared?

(a) Linear chain Silicate (b) Three dimensional
(c) Pyro Silicate (d) Sheet Silicate

[2011]

8. The basic structural unit of silicates is?

(a) SiO^- (b) SiO_3^{2-}
(c) SiO_4^{2-} (d) SiO_4^{4-}

[2013]

9. Which of these is not a manomer for a high molecular mass silicon polymer ?

(a) $MeSiCl_3$ (b) Me_2SiCl_2
(c) Me_3SiCl (d) $PhSiCl_3^-$

[2013]

ANSWER KEYS

1. (a)	2. (b)	3. (d)	4. (c)	5. (c)	6. (c)	7. (c)	8. (d)	9. (c)

HINTS AND EXPLANATIONS

1. Glass reacts with HF to give H_2SiF_6 as follows:

$$6\ HF + SiO_2 \rightarrow H_2SiF_6 + 2H_2O$$

2. As due to inert pair effect Pb^{2+} is more stable than Pb^{4+} so for lead +2 oxidation state is more common while +4 is more common for tin:

3. As bond length is inversely proportional to bond order or more the single bond nature in resonance hybrid, more is the bond length so the increasing bond length order is

$$CO < CO_2 < CO_3^{2-}$$

4. Anions of chain silicates are $(SiO_3^{2-})n$ and $(Si_4O_{11})n^{6-}$.

5. Hydrolysis of $(CH_3)_2SiCl_2$ will give linear polymer on hydrolysis followed by polymerization.

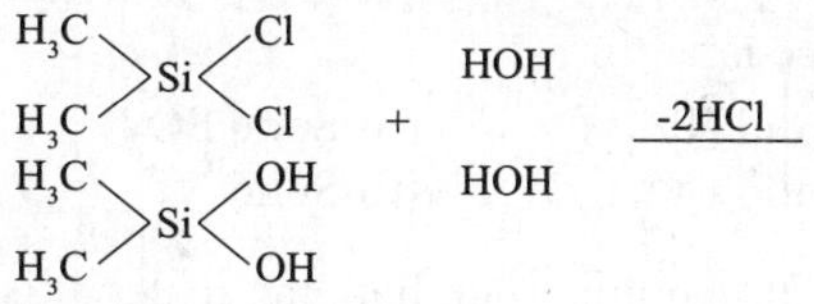

$$(H_3C)_2SiCl_2 + 2\,HOH \xrightarrow{-2HCl} (H_3C)_2Si(OH)_2$$

Dimethyldichlorosilane

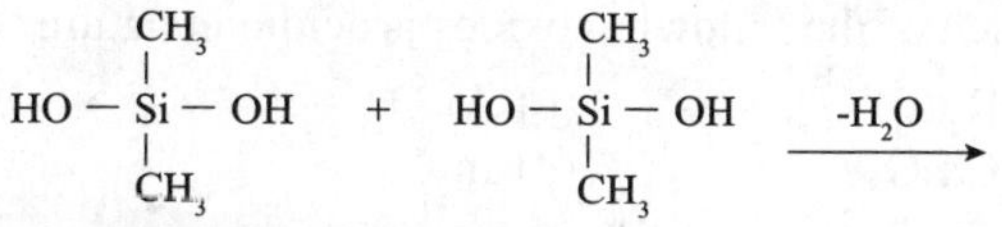

$$HO-Si(CH_3)_2-OH + HO-Si(CH_3)_2-OH \xrightarrow{-H_2O}$$

Dimethyl silanol

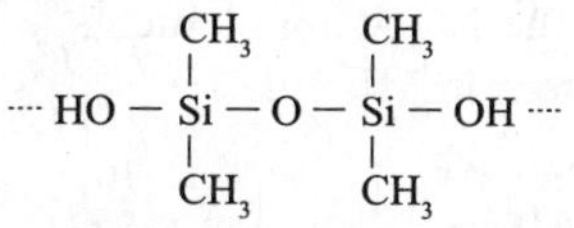

$$\cdots HO-Si(CH_3)_2-O-Si(CH_3)_2-OH \cdots$$

Linear polymer

6. Here SnO_2 is amphoteric CaO is basic while CO_2 SiO_2 acidic in nature.

7. In Pyrosilicate $[Si_2O_7]^{6-}$ O-atom is shared. [See text part].

8. SiO_4^{4-} is the basic structural unit of silicates

9. Me_3SiCl an hydrolysis gives $Me_3Si(OH)$ Which has only one reacting site so no polymerisation accurs

NCERT EXEMPLAR

1. The element which exists in liquid state for a wide range of temperature and can be used for measuring high temperature is:

(a) B (b) Al
(c) Ga (d) In

2. Which of the following is a Lewis acid?

(a) $AlCl_3$ (b) $MgCl_2$
(c) $CaCl_2$ (d) $BaCl_2$

3. The geometry of a complex species can be understood from the knowledge of type of hybridisation of orbitals of central atom. The hybridisation of orbitals of central atom in $[Be(OH)_4]^-$ and the geometry of the complex are respectively:

(a) sp^3, tetrahedral (b) sp^3, square planar
(c) $sp^3 d^2$, octahedral (d) dsp^2, square planar

4. Which of the following oxides is acidic in nature?

(a) B_2O_3 (b) Al_2O_3
(c) Ga_2O_3 (d) In_2O_3

5. The exhibition of highest co-ordination number depends on the availability of vacant orbitals in the central atom. Which of the following elements is not likely to act as central atom in MF_6^{3-} ?

(a) B (b) Al
(c) Ga (d) In

6. Boric acid is an acid because its molecule:

(a) Contains replaceable H^+ ion
(b) Gives up a proton
(c) Accepts OH^- from water releasing proton
(d) Combines with proton from water molecule

7. Catenation i.e., linking of similar atoms depends on size and electronic configuration of atoms. The tendency of catenation in Group 14 elements follows the order:

(a) C > Si > Ge > Sn
(b) C >> Si > Ge ≈ Sn
(c) Si > C > Sn > Ge
(d) Ge > Sn > Si > C

8. Silicon has a strong tendency to form polymers like silicones. The chain length of silicone polymer can be controlled by adding:

(a) $MeSiCl_3$ (b) Me_2SiCl_2
(c) Me_3SiCl (d) Me_4Si

9. Ionisation enthalpy ($_i H_1$ kJ mol^{-1}) for the elements of Group 13 follows the order.

(a) B > Al > Ga > In > Tl
(b) B < Al < Ga < In < Tl
(c) B < Al > Ga < In > Tl
(d) B > Al < Ga > In < Tl

10. In the structure of diborane:

(a) All hydrogen atoms lie in one plane and boron atoms lie in a plane perpendicular to this plane.
(b) 2 boron atoms and 4 terminal hydrogen atoms lie in the same plane and 2 bridging hydrogen atoms lie in the perpendicular plane.
(c) 4 bridging hydrogen atoms and boron atoms lie in one plane and two terminal hydrogen atoms lie in a plane perpendicular to this plane.
(d) All the atoms are in the same plane.

11. A compound X, of boron reacts with NH_3 on heating to give another compound Y which is called inorganic benzene. The compound X can be prepared by treating BF_3 with Lithium aluminium hydride. The compounds X and Y are represented by the formulas.

(a) B_2H_6, $B_3N_3H_6$ (b) B_2O_3, $B_3N_3H_6$
(c) BF_3, $B_3N_3H_6$ (d) $B_3N_3H_6$, B_2H_6

12. Quartz is extensively used as a piezoelectric material, it contains ___________.

(a) Pb (b) Si
(c) Ti (d) Sn

13. The most commonly used reducing agent is:

(a) $AlCl_3$ (b) $PbCl_2$
(c) $SnCl_4$ (d) $SnCl_2$

14. Dry ice is:

(a) Solid NH_3 (b) Solid SO_2
(c) Solid CO_2 (d) Solid N_2

15. Cement, the important building material is a mixture of oxides of several elements. Besides calcium, iron and sulphur, oxides of elements of which of the group (s) are present in the mixture?

(a) Group 2
(b) Groups 2, 13 and 14
(c) Groups 2 and 13
(d) Groups 2 and 14

Answer Keys

1. (c)	2. (a)	3. (a)	4. (a)	5. (a)	6. (c)	7. (b)	8. (c)	9. (d)	10. (b)
11. (a)	12. (b)	13. (d)	14. (c)	15. (b)					

Hints and Explanations for Selective Questions

1. Melting point of Ga = 30 ° C
Boiling point of Ga = 2240 ° C

2. Electron deficient will act as Lewis acid.

4. acidic nature decreases on moving down the group.

12. Quartz is a crystalline form of silica.

13. +4 oxidation state of Sn is more stable than +2 oxidation state.

15. cement contains elements of group 2(Ca),group 13 (Al) and group 14(Si)

AIIMS ESSENTIALS

Assertion and Reason

In the following questions, two statements (Assertion) A and Reason (R) are given. Mark

(a) If A and R both are correct and R is the correct explanation of A;

(b) If A and R both are correct but R is not the correct explanation of A;

(c) A is true but R is false;

(d) A is false but R is true

(e) Both A and R are false.

1. (A) : Lead pipes are never used for carrying water.
(R) : Lead in the presence of air is attacked by water to form soluble $Pb(OH)^2$ which provides highly poisonous Pb^{2+} ions.

2. (A) : CF_4 and NF_3 cannot be hydrolysed.
(R) : Carbon and nitrogen both do not have vacant d-orbital.

3. (A) : Silicones are hydrophobic in nature.
(R) : Si – O – Si linkages are moisture sensitive.

4. (A) : CO_2 is a gas while SiO_2 is a solid at room temperature.
(R) : CO_2 is a covalent compound and SiO_2 is an ionic solid.

5. (A) : Diamond is the hardest possible substance and is a net work covalent solid.
(R) : All the C- atoms in diamond are sp_{2-} hybridized

6. (A) : PbI_4 is a stable compound
(R) : Iodide stabilizes higher oxidation state.

7. (A) : Si – Si bonds are much stronger then Si – O bonds.
(R) : Silicon does not form double bonds with itself.

8. (A) : Adamantine silicon is obtained by heating silica with aluminium.
(R) : Admantine silicon is an alloy of aluminium and silicon.

9. (A) : Pb^{4+} compounds are stronger oxidizing agents than Sn^{4+} compounds.
(R) : The higher oxidation states for the group 14 elements are more stable for the heavier members of the group due to insert pair effect.

10. (A) : The man who fit or fixes water pipes in our homes etc. is called a plumber.
(R) : Pipes used for carrying water were made up of lead at one time.

11. (A) : When silicon is heated with magnesium in an electric furnace Mg_2Si is produced.

(R) : Magnesium reacts with Si directly at high temperature.

12. (A) : Between $SiCl_4$ and CCl_4, only $SiCl_4$ reacts with water.

(R) : $SiCl_4$ is ionic and CCl_4 is covalent.

Answer Keys

1. (a)	2. (a)	3. (b)	4. (c)	5. (c)	6. (d)	7. (d)	8. (c)	9. (c)	10. (a)
11. (a)	12. (c)								

CHAPTER 11A

Organic Chemistry - I

CHAPTER OUTLINE

■ Organic and Inorganic Compounds ■ Homologous Series ■ Hybridization ■ IUPAC System ■ Functional Group Preference Table

Organic and Inorganic Compounds

- The word organic means 'pertaining to life'.
- Compounds like sugars, fats, oils, dyes, proteins, vitamins which were isolated directly or indirectly from living organisms like animals and plants were called **organic compounds** and their study called **organic chemistry.**
- Compounds like common salt, alum, nitre, blue and green vitriol etc., which were isolated from the non-living sources such as rocks and minerals were called **inorganic compounds.**

Berzelius Hypothesis

According to this theory, which is also known as **vital force theory,** organic compounds cannot be synthesized in the laboratory because they require the presence of a 'mysterious' force, that is, a vital force which exists only in living organisms. However, this assertion was disproved by Freidrich Wohler.

Wohler's Synthesis

Wohler synthesized urea, an organic compound, in the laboratory by heating ammonium cyanate obtained by double decomposition of ammonium chloride and potassium cyanate.

$$NH_4Cl + K\,CNO \longrightarrow \underset{\text{Ammonium cyanate}}{NH_4CNO} + KCl$$

$$NH_4CNO \xrightarrow[\text{Rearrangement}]{\text{Heat}} \underset{\text{Urea}}{NH_2\text{–}\overset{\overset{\displaystyle O}{\|}}{C}\text{–}NH_2}$$

Modern Definition of Organic Chemistry

- Organic compounds may also be referred to as hydrocarbons. Hence, the study of hydrocarbons and their derivatives such as CH_3OH, $CH_3\ CN$ is called organic chemistry.
- Organic chemistry is treated as a separate branch because organic compounds:
  1. Contain a few selected elements (C, H, O, N, S)
  2. Are large in number
  3. Have complex structures
  4. Have covalent and coordinate bonds
  5. Are mainly soluble in non-polar solvents
  6. Have low melting and boiling points
  7. Are bad conductors of heat and electricity
  8. Undergo molecular reactions which are generally slow and reversible
  9. Show isomerism and homology.

Diversity of Organic Compounds

The reasons for the occurrence of a large number of organic compounds is attributed to their following properties:

Catenation The self-linking property of carbon atoms through covalent bonds to form long, straight or branched chains and rings of different sizes is called catenation.

- Amongst all elements in the periodic table carbon shows maximum catenation. This property is primarily due to its small size, electronic configuration and maximum bond energy, that is, strenght of carbon–carbon bonds for catenation.

  The catenation bond strength follows the order:

  C > Si > S > P >

  The linking in carbon atoms is possible through a single, double or triple bond as shown below:

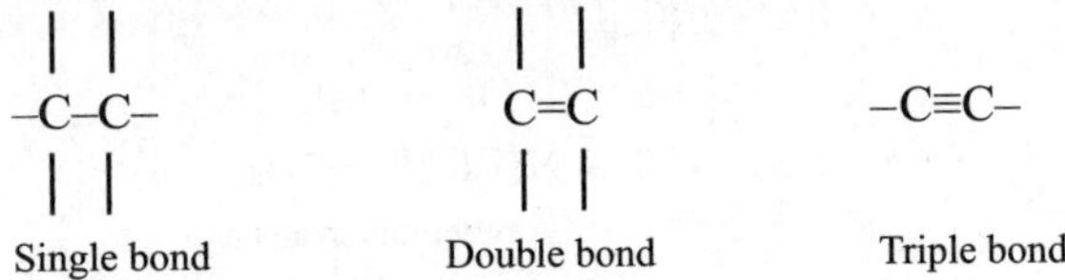

- Due to catenation property of carbon atoms, they may link to form:

  (a) an open chain which may be straight or branched

  (b) a closed chain.

  Both are shown as follows:

Open chain

–C–C–C–, –C–C=C–, –C–C≡C

Straight chain

–C–C–C– (with a –C– branch)

Branched chain

Closed chain or ring form

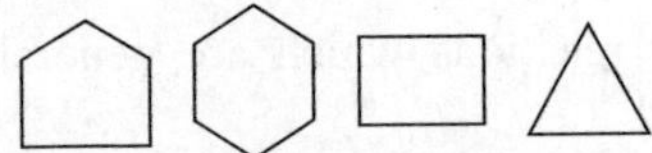

Electronegativity and strength of bonds The electronegativity of carbon (2.5) is close to a number of other elements like hydrogen (2.1), nitrogen (3.0), phosphorus (2.1), chlorine (3.0) and oxygen (3.5), so it can form strong covalent bonds with these elements.

Tendency to form multiple bonds Due to its small size, carbon atom has a strong tendency to form multiple bonds with another carbon, oxygen or nitrogen atom.

Isomerism Many organic compounds show the phenomenon of isomerism, by virtue of which a single molecular formula may be represented in two or more structures.

Tetravalency of carbon The tetravalent nature and tetrahedral structure of carbon was studied by Liebel and van't Hoff.

C 109°28'

- The carbon atom is tetravalent, that is, one carbon atom can combine with four monovalent atoms or groups of atoms or with an equivalent number of atoms or groups having other valencies. Thus, it can form various types of compounds like

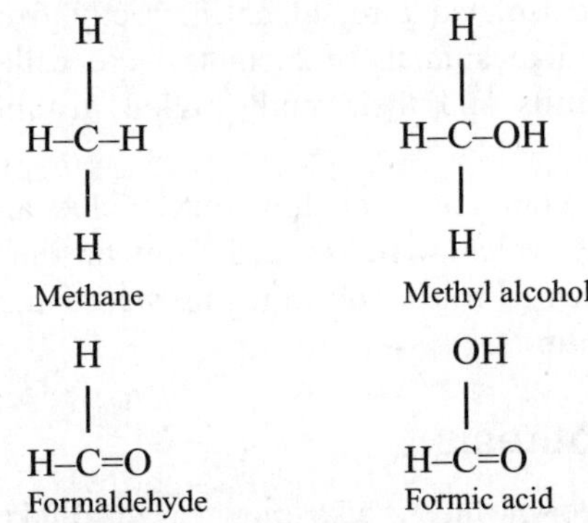

- All the four valencies of carbon are alike and symmetrical with respect to the atom, that is, they are equal and equivalent to each other.
- Methane forms only one mono-substituted isomer, that is, monochloromethane on chlorination.

Classes of Carbon Atoms

Primary carbon atom (1° or p) A carbon atom present either alone or attached to not more than one other carbon atom is called primary carbon atom.
For example,

$CH_3–CH_3$
p p

Secondary carbon atom (2° or s) A carbon atom which is attached to two other carbon atoms is called secondary carbon atom.
For example,

$$\underset{}{CH_3}–\underset{s}{CH_2}–CH_3$$

Tertiary carbon atom (3° or t) A carbon atom which is attached to three other carbon atoms is called tertiary carbon atom.
For example,

$$CH_3–\overset{t}{\underset{|\atop CH_3}{CH}}–CH_3$$

Quaternary carbon atom (4° or q) A carbon atom which is attached to four other carbon atoms is called quaternary carbon atom.
For example,

$$CH_3–\overset{CH_3}{\underset{CH_3}{\overset{|\,q}{\underset{|}{C}}}}–CH_3, \quad \overset{1°}{C}–\overset{2°}{C}–\overset{\overset{1°}{C}}{\overset{|}{\underset{3°}{C}}}–\overset{\overset{1°}{C}}{\overset{|}{\underset{4°}{\underset{|\,1°}{\underset{C}{C}}}}}–\overset{1°}{C}$$

Facts To Remember

- Hydrogen atoms attached to primary, secondary or tertiary carbon atoms are called primary, secondary or tertiary hydrogen atoms respectively.
- The number of hydrogen atoms attached to primary, secondary, tertiary and quaternary carbon atoms in alkanes are respectively three, two, one and zero.
- If a carbon atom is unsaturated it is not indicated as p, s or t carbon atom but counted for others.
  For example,

  $$CH_2{=}CH–\underset{s}{CH_2}–\underset{p}{CH_3} \quad \underset{p°}{CH_3}–C{\equiv}C–\underset{s°}{CH_2}–\underset{p°}{CH_3}$$

Classification of Organic Compounds

Organic compounds have been divided into two classes:

- Acyclic compounds (Aliphatic compounds)
- Cyclic compounds (Aromatic compounds)

Acyclic compounds The organic compounds in which all the carbon atoms are linked to one another to form open chains, either straight or branched, are called acyclic or open chain or aliphatic compounds.

For example,

$CH_3CH_2CH_2CH_3$ Butane; $CH_3–\overset{CH_3}{\overset{|}{CH}}–CH_3$ Isobutane; $CH_3CH_2CH_2OH$ Propanol

Cyclic compounds The compounds which have at least one ring or closed chain of atoms in their structure are called cyclic compounds. These are of two types:

Homocyclic compounds Such compounds contain rings which are made up of only one kind of atoms. If all the atoms in the ring are carbon atoms, they are called carbo-cyclic compounds. These are of the following two types:

Alicyclic compounds These are carbocyclic compounds which resemble aliphatic compounds in their properties.

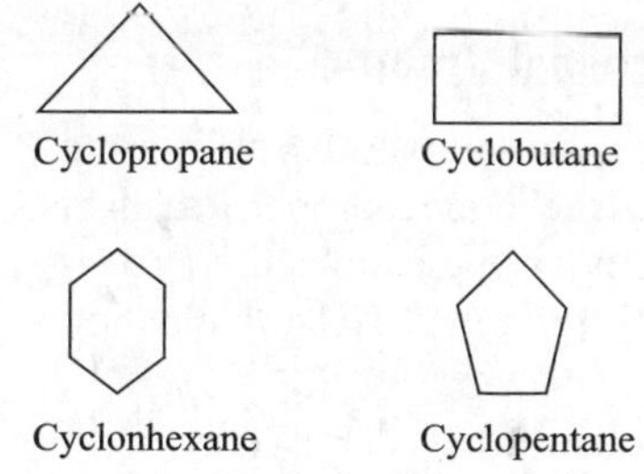

Cyclopropane Cyclobutane

Cyclonhexane Cyclopentane

Aromatic compounds The compounds containing one or more fused or isolated benzene rings are called aromatic or **benzoid** compounds.

Benzene Naphthalene Phenol

The arromatic compounds without a benzene ring but having an another unsaturated ring/s are called non-benzoid compounds

e.g.,

Tropone, cyclopenta dienylanion, Tropolone

Heterocyclic compounds: The cyclic compounds having one or more hetero atoms (other than C-atom). For example, (O, N, S) in the ring are called heterocyclic compounds. These are of following two types:

Alicyclic heterocyclic compounds Such compounds resemble with aliphatic compounds in their properties.

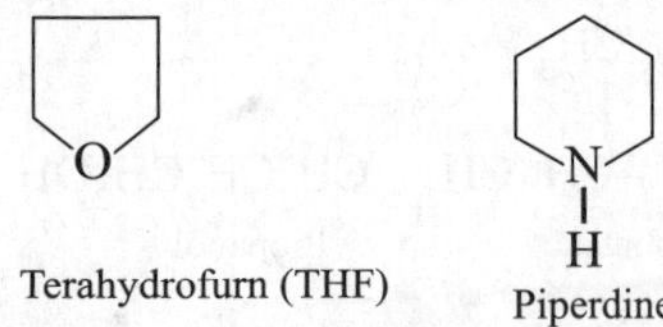

Terahydrofurn (THF)   Piperdine

Aromatic heterocyclic compounds Those compounds which resemble benzene and other aromatic compounds in most of their properties are called aromatic heterocyclic compounds. For example,

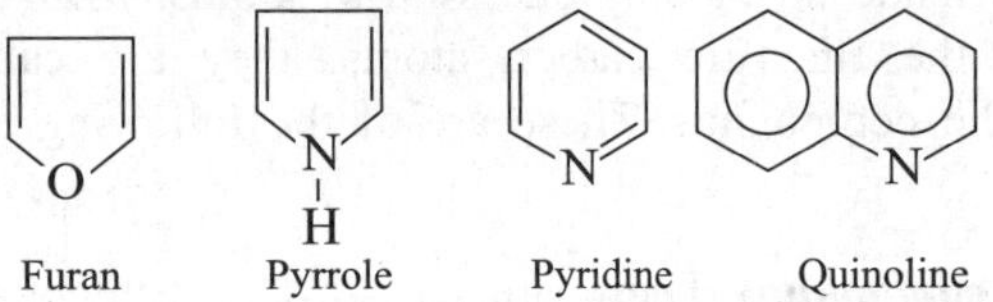

Furan   Pyrrole   Pyridine   Quinoline

Functional Group

It is an atom or a group of atoms present in a molecule which determine its chemical properties, that is, the site of chemical activity in the molecule.
For example, –OH (hydroxy), –CHO (aldehydic) – COOH (carboxylic).

HOMOLOGOUS SERIES

A series of structurally similar compounds having same functional groups is called homologous series and the phenomenon is termed as homology. It is written in the increasing order of the number of carbon atoms.

Characteristics of Homologous Series

- Any two successive members of a homologous series differ in their molecular formula by a CH_2 group and their molecular weights differ by 14.
- All the members can be prepared by a number of general methods.
- The members have many similar chemical properties.
- The physical properties of the members show a regular gradation with increase in molecular weight.
- The IUPAC names of all homologues have the same ending as –ane in alkanes, –ene in alkenes, –ol in alkanols.

For example,

CH_4 (Methane)
C_2H_6 (Ethane)
C_3H_8 (Propane)
C_4H_{10} (Butane)

Hydrocarbon radicals

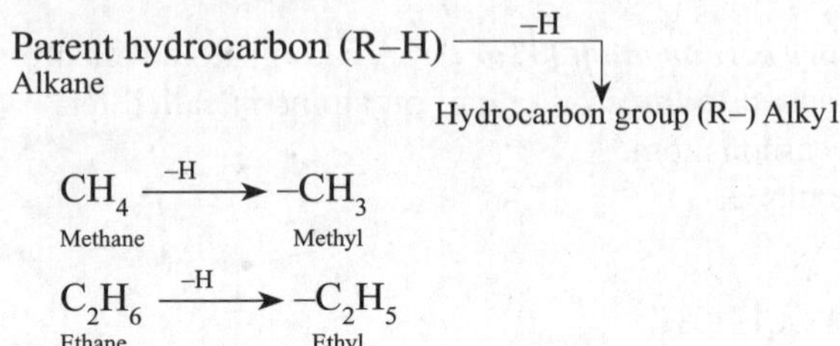

HYBRIDIZATION

Pauling and Slater introduced it to explain the shape of molecules which can be explained by VBT. It is the intermixing or re-distribution of energy among two or more half filled or full filled or incomplete or empty orbitals of comparable energies to form same number of hybrids. Hybrids have identical energies and similar shapes.

Facts about Hybridization

- Number of atomic orbitals taking part in hyrbridization is equal to number of hybrids formed.
- Electrons do not undergo hybridization.
- A hybrid bond is always a sigma bond.
- A hybrid bond is always stronger than a non-hybrid bond.
- Hybridization occurs at the time of bond formation.
- Hybridization α overlapping (for enough over lapping orbitals must be fairly apart i.e., neither very close nor very far).
- Hybridization is for increasing stability and decreasing reactivity and energy.
- Hybridization is of central atom in a molecule (NH_3, H_2O, CH_4).
- Hybridization is not meant for isolated atoms but for bonded atoms.

Types of Hybridization

(1) sp Hybridization

Here one s and one p orbital form two sp hybrids after intermixing. Shape of molecule is linear and bond angle is 180º.

e.g., X – M – X
sp

(M = Be, Zn, Hg)

H – C ≡ C – H
sp sp

Some other examples are CO_2, CS_2 etc.

Characteristic of sp-Hybrid Orbitals

(1) Shape of sp-hybrid orbital is oval.

(2) Its relative power of overlapping is 1.93 with respect to s orbital.

(3) Both sp-hybrid orbitals are completely equivalent and symmetrical.

(4) These two sp-hybrid orbitals are colinear, i.e., angle between the hybrid orbitals is 180°.

(5) Energy of sp-hybrid orbital is more than s-orbital but less than the p-orbital.

In sp-hybrid orbital, one lobe is bigger while other lobe is small. The bigger lobe is very large with respect to p-orbital, hence it has higher degree of overlapping, so it forms stronger bond. For example, in the formation of acetylene. In acetylene, hybridization of carbon is sp. One 2s orbital blends with only one 2p orbital to form two sp-hybrid orbitals. In this case, two unhybridized orbitals remain, each with one electron.

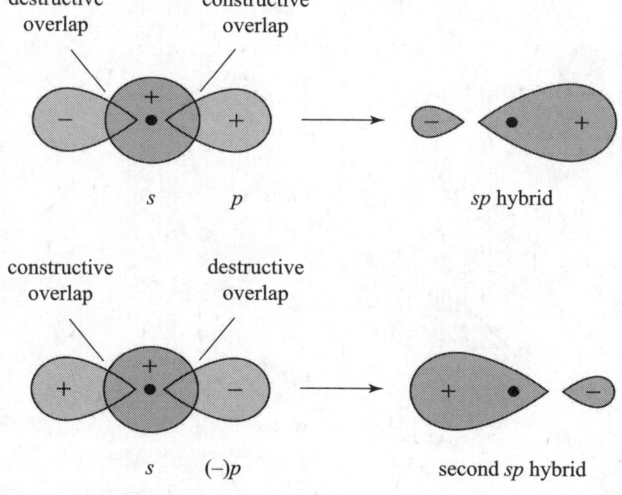

(2) sp² Hybridization

Here one s and two p orbital intermix to form three new sp^2 hybrids. Shape of the species is trigonal or co-planar and bond angle is 120°.

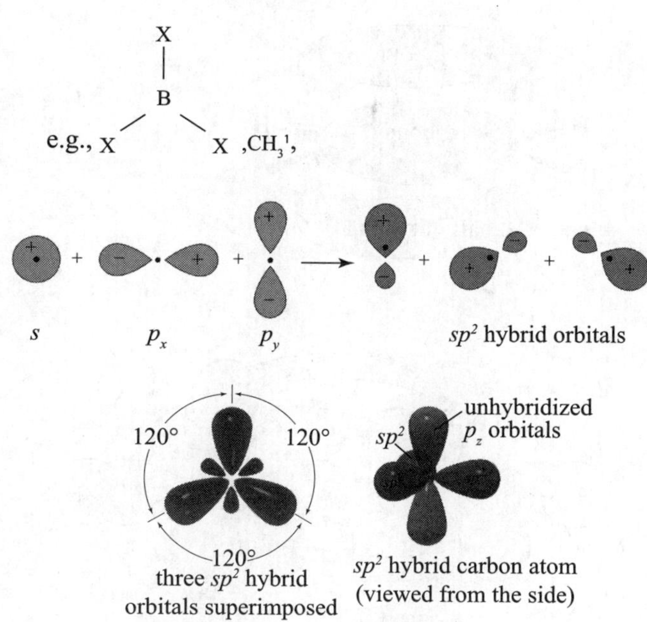

Characteristics

(1) These hybrid orbitals are planar with bond angle 120°.

(2) These sp^2-hybrid orbitals are completely equivalent and symmetrical.

(3) These are stronger than s and p orbitals. Its relative power of overlapping is 1.99 with respect to s-orbital.

(4) As in this hybridization contribution of p-orbitals is more hence it is less oval than sp-hybrid orbitals. In this case, one lobe is bigger and one lobe is smaller and it forms stronger bond.

When carbon is bonded to another atom by a double bond, the carbon atom is in the sp^2-hybrid state. Example is ethylene. In order to form sp^2 bonding orbitals, carbon hybridizes its 2s orbital with only two of its orbitals. one p-orbital remains unhybridized on the carbon atom. In ethylene, two sp^2 carbons are joined by a sigma bond formed due to the overlap of one sp^2 orbital from each carbon atom.

Both carbon atoms have also unhybridized p-orbitals, which can be oriented parallel to each other and thereby overlap. Both lobes of the p-orbitals merge above and below the sigma bond forming a π MO. Thus double bond is composed of a σ bond and a π bond. The ethylene molecules are completed when σ bonds are formed between the overlapping sp^2 orbitals of carbon and the s-orbitals of the hydrogens.

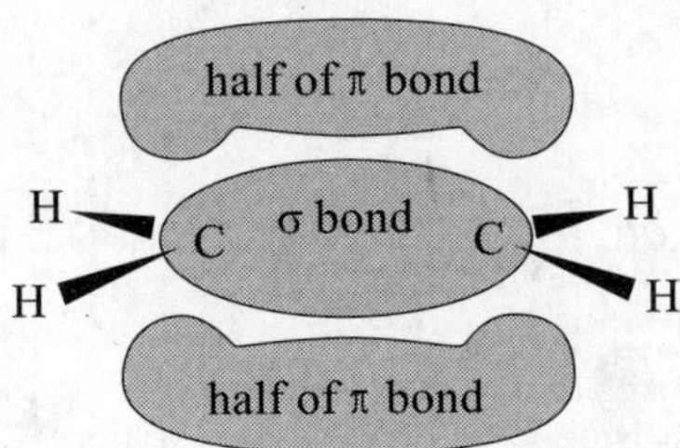

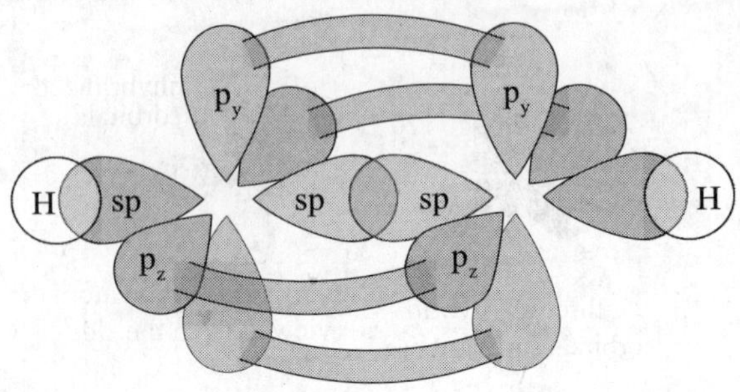

(3) sp^3 Hybridization

Here one s and three p orbitals intermix to give four new sp^3 hybrids. Shape of the species is tetrahedral and bond anglc is 109° 28'.

e.g., C_2H_6, CX_4, SiX_4, NH_4^+, BX_4^-, NH_3, PH_3, H_2O, H_2S,

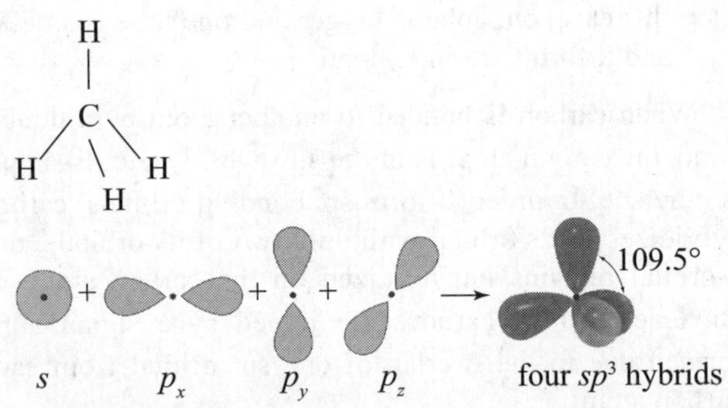

Characteristics

(1) These orbitals are directed towards the four corners of a regular tetrahedron and the angle between each pair of them is 109°28' or 109.5°.

(2) All the four sp^3 hybrid orbitals are completely equivalent and symmetrical.

(3) As in sp^3 hybridization, the contribution of p-orbitals is 75%, its shape is almost same as that of the parent p-orbitals except that the bigger lobe in sp^3-orbital is somewhat more spread and shorter in length than the pure p-orbitals.

Their relative power of overlapping is 2.00 with respect to s-orbital. This shows that sp^3-orbitals are stronger than sp^2 which is stronger than sp-orbitals.

To find hybridization in organic compound

—C— (sp^3) —C= (sp^2) —C≡ (sp) =C= (sp)

e.g., In it, all carbon atoms are sp^2 hybridized.

(1)

Me_3C (sp^2) C = C = C = C (sp^2) CMe_3 (sp, sp)

(2)

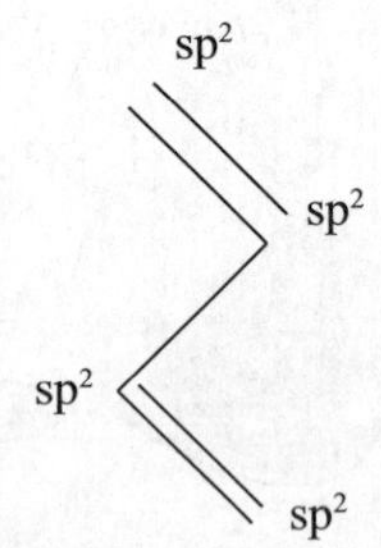

(3)

sp^3 C ≡ N (sp), sp^2, sp^2, sp^3

(4)

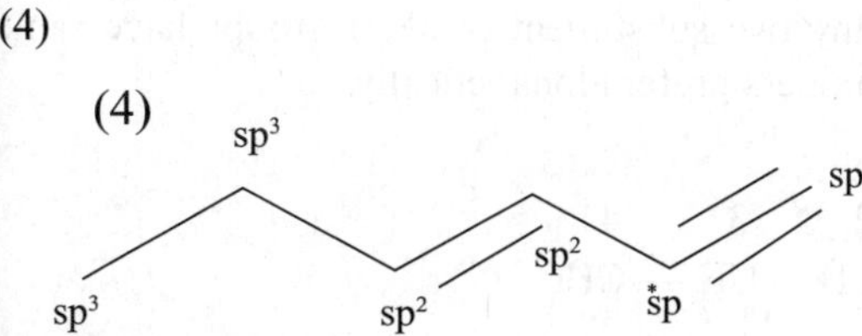

IUPAC System

"International union of pure and applied chemists"

Basic Rules of nomenclature

Carbon chain length	Root word (Alk.)	Carbon chain	Root word
C_1	Meth-	C_9	Non-
C_2	Eth-	C_{10}	Dec-
C_3	Prop-	C_{11}	Un dec-
C_4	But-	C_{12}	Do dec-
C_5	Pent-		
C_6	Hex-		
C_7	Hept-		
C_8	Oct-		

Selection of the longest chain of C-atoms

To write the IUPAC name, the first step is to select the longest chain,i.e., parent or main chain.

e.g.,

$CH_3-CH-CH_2-CH_2-CH_3$ with CH_2-CH_3 on the CH → 6C-atom chain [Longest here]; 5C-atom chain

$CH_3-CH_2-C(CH_2-CH_3)(CH_3)-CH(CH_3)-CH_2-CH_3$ → 6C-atom chain [longest here]

Rules for numbering of C-atoms of the chain

- For numbering of C-atoms of the chain priority order is given as follows:
- functional group > (=) or (≡) > substituent or alkyl group e.g.,

$\overset{1}{CH_3}-\overset{2}{CH}(OH)-\overset{3}{CH}(Br)-\overset{4}{CH_3}$ — Prefer it (as functional group is at more corner)

Numbering 2, 1 from the other end — Neglect it

FACTS TO REMEMBER

Prefer the chain having maximum or all possible functional groups, or (=) or (≡) bonds and for it longest chain rule can be neglected. For example (1)

CH_2-COOH
$CH-COOH$
CH_2-COOH

3 C-atom chain [since, it includes all the three functional groups, so it is prefered here]

example (2)

$CH_3-CH_2-C(=CH_2)-CH_2-CH_2-CH_3$ → 6C-atom[longest possible chain]; → 5C-atom chain including (=) bond so it is preferred

- If in a given compound, there are two or more chains having similar number of C-atoms than a chain with maximum side chains is preferred.

example

$CH_3-CH_2-CH-CH_2-CH_3$ with $H_3C-C(CH_3)-CH_3$ on the CH → 5C-atom chain with one-side chain; 5C-atom chain with 3-side chains so it is preferred

$CH_3-CH(CH_3)-CH(CH_2CH_3)-CH(CH_3)-CH_2-CH_3$ → 6 'C'-atom chain only two-side chains; 6 'C'-atom chain with 3-side so it is preferred here.

- In case of (=) or (≡) bond preference for numbering is given to the one which is at more corner or have lower value of locant number.

e.g.,

$\overset{1}{CH_2}=\overset{2}{CH}-\overset{3}{C}\equiv\overset{4}{C}-\overset{5}{CH_2}-\overset{6}{CH_3}$ Prefer it

Numbering 6 5 4 3 2 1 Neglect it

example,

```
 1   2     3     4   5    6     7       Prefer it
CH ≡ C — CH2—CH=CH—CH2—CH3
 7   6     5     4   3    2     1       Neglect it
```

- When both (=) and (≡) bond have same locant numbers prefer (=) bond.

example,

```
CH2 = CH — C ≡ CH
 1     2    3   4
```

- Similarly, in case of substituent and alkyl groups prefer which has lower locant number value.

example,

```
 1    2    3     4     5      Prefer it as Cl is on 2nd C-atom
CH3—CH—CH—CH2—CH3             that is, locant number is two
 5   4|   3|    2     1
     Cl   CH3
```

example,

```
 1     2      3      4      5     Prefer it, as CH3 has lower locant
CH3 — CH —CH —CH2 —CH3            number (2) while in case of Br-atom
 5     4|    3|     2      1      locant number is 3
      CH3    Br
```

- In case any two substituent or alkyl groups have same locant numbers prefer alphabetically.

example,

```
 1      2      3      4
CH3 — CH —CH — CH3
        |      |
        Br     Cl
```

2- bromo 3- chloro butane [correct]
3- bromo 2- chloro butane [incorrect]

example,

```
 6      5      4      3      2      1
CH3 — CH2 — CH — CH — CH2 — CH3
              |     |
             CH3   CH2
                    |
                   CH3
```

3-ethyl 4-methyl hexane [correct]
4-ethyl 3-methyl hexane [incorrect]

Lowest Sum Rule

When numbering of a chain is possible form more than one side, prefer numbering of the chain from the side having lowest sum of all the locant numbers.

e.g.,

```
                                               Path (II)
          1COOH       OH                           ⟶ Sum of locant numbers
           |7      6   5|   4    3     2       1   [1+8+3+5+5+6=28]
Path(I)  CH = CH — C — C ≡ C — CH2 — COOH
           2     3   4|   5    6     7         8 ⟶ Sum of locant numbers
                     NH2                                ⟵ is ⟶
                                                  1+8+2+4+4+5=24
```

Since, in path (I) the sum is 24 whereas in path (II) it is 28 hence path (I) is preferred here for numbering.

Naming of Complex Alkyl Substituents

- When a side chain further includes another side chain, it is numbered and named as,

E.g.,

```
 1      2        3      4       5      6
CH3 — CH  —  CH — CH2 — CH2 — CH3
        |      |
       CH3  1'CH —Cl
               |
            2'CH2 — Br
```

3[2'-bromo 1'-chloro] ethyl 2-methyl hexane

- Prefixes like bis, tris, tetrakis etc., can also be used to indicates the multiplicity of substituted substituent.

e.g.,

```
       CH2 — Br
        |
CH3 — C — CH2 — CH2 — CH2 — CH3
        |
       CH2 — Br
```

2, 2-bis [Bromomethyl] hexane

- Use prefixes like di, tri, tetra etc., in case of simple substituents with same times locant numbers.

e.g.,

```
  Cl   Br
  |    |
 CH — C — CH3
  |    |
  Cl   Br
```

2, 2-dibromo 1, 1 dichloro propane

- All prefixes are written before alk (root word) as in above cases howeverene, yne or main functional group names, that is, suffix names are written after root word with their locant numbers are follow.

--------- Prefix names (Alphabetically) + Alk. + ---------- + ene or yne or main functional group name

(------- means locant number and di, tri etc.)

e.g.,

```
1 COOH         Br
  |            |
  CH = CH — C — CH— CH3
  2      3   4|   5|     6
             Cl   CH3
```

4-bromo 4-chloro 5-methyl-hex-2- en 1-oic acid

```
1 CHO                  NH2
  |                    |
  CH = CH — C ≡ C — C — CH3
  2      3    4   5  6|    7
                      Br
```

6-amino 6-bromo hept 2-ene 4-yn 1-al

- If a compound has many functional groups, the main functional group is named as suffix name while rest are given prefix names.

```
1 COOH                 O
  |                    ||
  CH = CH — CH — C — CH3
  2      3    4|     5    6
               OH
```

4-hydroxy 5-keto hex-2-en 1-oic acid

Functional Group Preference Table

Prefix name	Functional Group	Suffix name
Sulpho	$-SO_3H$	Sulphonic acid
Carboxy	–COOH	oic acid
—	—C(=O)—O—C(=O)—	Alkanoic acid-Anhydride
Carbalkoxy-or alkoxy carbonyl	–COOR	oate
Chloroformyl	–COCl	oyl chloride
Carbamoyl	$-CONH_2$	amide
Cyano	–CN	nitrile
Isocyano	–NC	Isonitrile or -carbylamine
Aldo-or formyl	–CHO	al
Keto or oxo	>C=O	one
Hydroxy	–OH	ol
Mercapto	–SH	thiol
Amino	$-NH_2$	amine

*In case carbon atom of the functional group is not countable, name them as follows:
–COOH (Carboxylic acid), –COOR (carbanoate), –COCl (carbanoyl chloride), $-CONH_2$ (carbanamide or carboxamide), -CN (carbo nitrile), –CHO (carboxaldehyde).

```
CH2—CN
|
CH —CN
|
CH2—CN
```

Propan, 1, 2, 3-tri carbo nitrile

Examples of Nomenclature

Hydrocarbons: These are named as follows:

e.g.,

```
 1      2      3      4      5      6
CH3 — CH — CH2 — CH — CH2 — CH3
       |             |
      CH3           CH3
```

2, 4 di-methyl hexane

$$\underset{CH_3}{\overset{CH_3}{>}}\overset{2}{CH}-\underset{\underset{CH_3}{|}}{\overset{3}{CH}}-\overset{4}{CH}<\begin{matrix}CH_2-CH_3\\CH_3\end{matrix}$$

2, 3 , 4-tri-methyl hexane

$$CH_2=CH-\underset{\underset{CH_3}{|}}{C}=CH-CH_3$$

3-methyl penta 1, 3 di-ene

$$CH_2=CH-\underset{\underset{CH_3}{|}}{CH}=CH_2$$

2- methyl buta 1, 3 di-ene

$$CH_3-C\equiv C-\underset{\underset{CH_3}{|}}{CH}-CH_3$$

4-methyl pent 2-yne

$$\underset{5}{CH_3}-\underset{4}{C}\equiv\underset{3}{C}-\underset{2}{CH}=\underset{1}{CH_2}$$

Pent-1-ene-3-yne

$$\underset{CH_2}{\overset{\overset{1}{CH_3}}{>}}\overset{2}{C}-\overset{3}{CH_2}-\overset{4}{C}\equiv\overset{5}{CH}$$

2-methyl pent-1-ene 4-yne

Halo alkanes These are named as follows:

e.g.,

$$\underset{\underset{Br}{|}}{CH_2}-\underset{\underset{Br}{|}}{CH}-CH_2$$

1, 2-di-bromo propane

$$\overset{\overset{CCl_3}{|}}{CH_2}-CH_2-CH_3$$

Trichloro butane

or

1, 1, 1 tri-chloro butane

$$\overset{\overset{CHCl_2}{|}}{CH_2}-CH=CH_2$$

4, 4 di-chloro butene -1

Alcohols (R-OH)

These are named as alkanols.

Alkan + ol → Alkanol

e.g.,

$$\underset{CH_3}{\overset{CH_3}{>}}CH-OH$$

Propan 2-ol

$$\underset{\underset{OH}{|}}{CH_2}-\underset{\underset{OH}{|}}{CH}-\underset{\underset{OH}{|}}{CH_2}$$

Propan-1, 2, 3 tri-ol

$$CH_3-CH=CH-\underset{\underset{Cl}{|}}{CH}-CH_2OH$$

2-Chloro-pent-3-en-1-ol

$$\underset{4}{CH_3}-\underset{3}{C}\equiv\underset{2}{C}-\underset{1}{CH_2OH}$$

But-2-yn-1-ol

$$HO\,CH_2-\underset{\underset{CH_2OH}{|}}{\overset{\overset{CH_2OH}{|}}{C}}-CH_2OH$$

Tetrakis (hydroxymethyl) methane

5, 4, 3, 2, 1, OH, OH

3-Vinyl pentan-1,4-diol

CH_3, 4', 3', 2', 1', 8, 6, 3, 2, OH, HO, 7, 4, 5, Cl, 1, 1', 2'

3-(2-Chloroethyl)-4-(2-methyl)-1-butenyl) octane-1, 8-diol

Ethers [R —O —R]

Ethers are named as alkoxy alkanes.

e.g.,

Alkoxy	Alkane
less C-atom alkyl group	More C-atom alkyl group

$$CH_3-O-CH_2-CH_2-CH_3$$

Methoxy propane

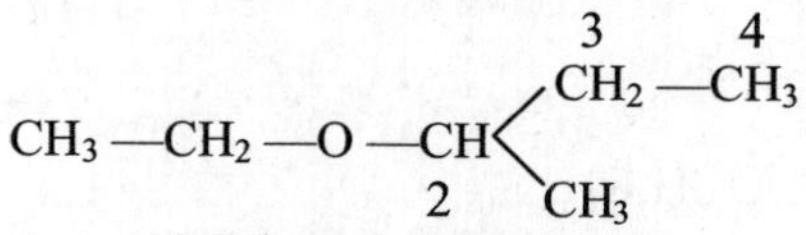

2-Ethoxy butane

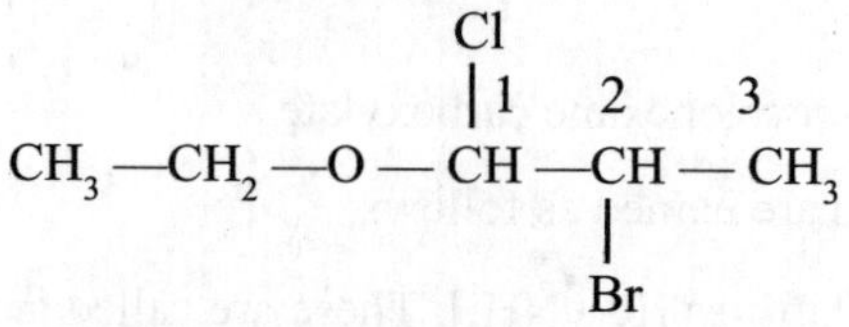

Ethoxy 2-bromo 1-chloro propane

Aldehydes [R—CHO]

These are named as a Alkanals.

Alkan + al → Alkanal

e.g.,

$CH_3—CH_2—CH_2—CHO$

Butanal

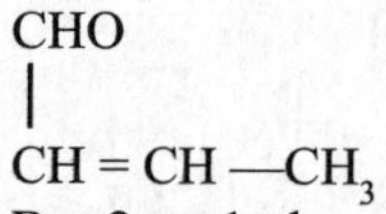

But-2-en-1-al

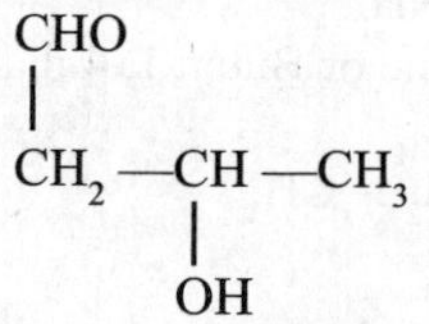

3-hydroxy butanal

$C_6H_5—CH = CH—CHO$

3-phenyl-prop-2-en-1-al

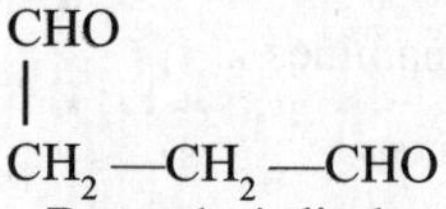

Butan 1, 4-di-al

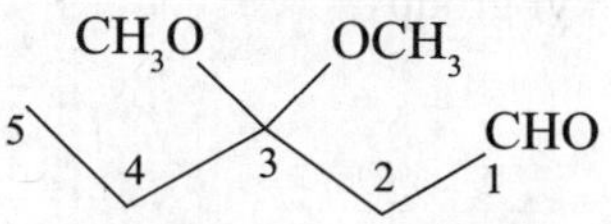

3,3-di-methoxy-pentanal

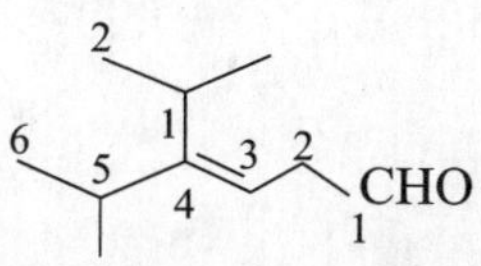

4-(1-Methylethyl)-5-methyl-3-hexenal

3. **Ketones** (R—C(=O)—R)

These are named as alkanones.

Alkane + one → Alkanone

e.g.,

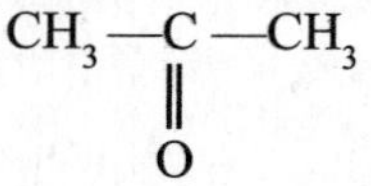

Propan-2-one

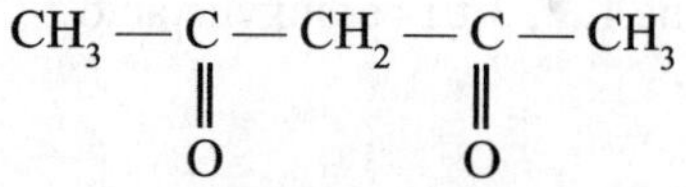

Pent 2, 4-di-one

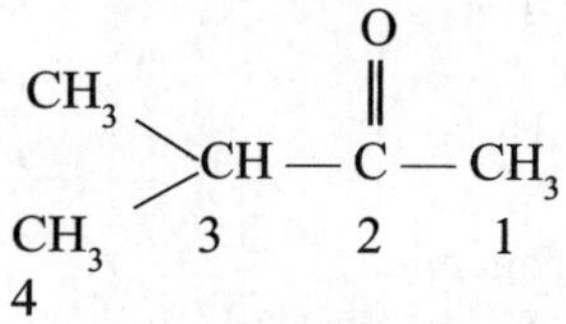

3-methyl butan-2-one

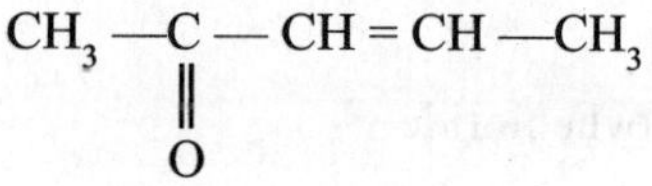

Pent-3-en-2-one

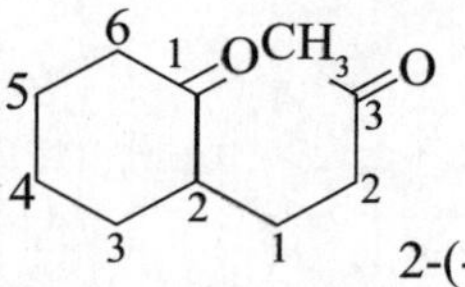

2-(-3-oxobutyl) cyclohexanone

Carboxylic Acids (R—COOH)

These are named as alkanoic acids.

Alkan + oic acid → Alkanoic acid

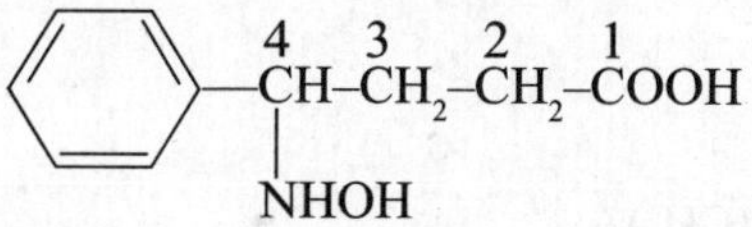

4-(Hyroxylamino)-4-phenylbutanoic acid

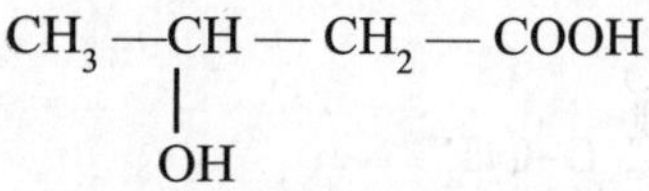

3-hydroxyl butanoic acid

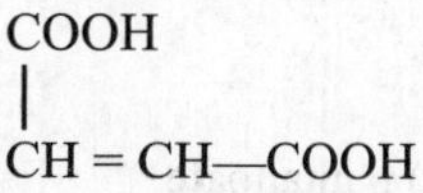

But-2-ene1, 4 di-oic acid

$$\begin{array}{l} COOH \\ | \\ CH = CH—CH_3 \end{array}$$

But 2-en-1-oic acid

$$\begin{array}{l} 1\ CH_2—COOH \\ \quad | \\ 2\ CH—COOH \\ \quad | \\ 3\ CH_2—COOH \end{array}$$

Propane 1, 2, 3-tri-carboxylic acid

Acid derivatives

$CH_3—C(=O)—N(CH_3)(C_2H_5)$

N-ethyl N-methyl ethanamide

$$\begin{array}{l} CHO \\ | \\ CH = CH—COCl \end{array}$$

4-oxo-but 2-en 1-oyl chloride

$(CH_3—CH_2—CO)_2O$

Propanoic anhydride

$\underset{4}{CH_3}—\underset{3}{C}(=O)—\underset{2}{CH_2}—\underset{1}{COOC_2H_5}$

Ethyl 3-keto butanoate

$\overset{3}{C}H_3—\overset{2}{C}H_2—\overset{1}{C}(=O)—O—C(CH_3)_3$

Tert-butyl propanoate

$\overset{5}{C}H_3–\overset{4}{C}H_2–\overset{3}{C}H_2–\overset{2}{C}H(–\overset{1}{C}(CH_3)_2–\overset{2}{C}H_3)–\overset{1}{C}(=O)–O–CH_3$

Methyl 2-(1, 1-dimethylethyl) pantanoate

Cyclohexane ring (C-1 to C-6) with $C(=O)OCH_2CH_3$ at C-1 and Cl at C-3

ethyl 3-chloro cyclohexane carboxylate

Amines Amines are named as follows:

(A) **Primary Amines** [$R—NH_2$]: These are called amino alkanes or alkanamines.

$CH_3—CH_2—NH_2$

Amino ethane or Ethanamine

$CH_3—CH(NH_2)—CH_3$

2-amino propane

$NH_2—CH_2—CH_2—CH_2—CH_2—NH_2$

1, 4-di-amino butane or Butan 1, 4 diamine

$CH_2(NH_2)—CH(NH_2)—CH_2—CH_2—CH_3$

Pentan 1, 2 diamine

(B) **Secondary amines** $\left[\begin{array}{l} R \\ R \end{array}\!\!>NH\right]$

These are named as N-alkyl alkanamines.

N-alkyl for smaller alkyl group	alkanamine for longer alkyl group

$CH_3—NH—CH_2—CH_3$

N-methyl ethanamine

$HN(CH_2—CH_3)_2$

N-ethyl ethanamine

$HN<(CH_2-CH_3)(CH_2-CH_2-CH_3)$

N-ethyl propanamine

$CH_3NH-CH_2CH_2CH_2CH_2CH_2CH_2-NHCH_3$ (carbons numbered 1–6)

N, N'-di-methyl-hexane –1, 6-di-amine

(C) **Tertiary amines (R_3N):**

R_3N

N-N di alkyl alkanamine
or
N alkyl N'-alkyl-alkan amine

e.g.,

$N(CH_3)_3$

N, N-di-methyl methanamine

$N(CH_2-CH_3)_3$

N-N di-ethyl ethanamine

$N(CH_3)(CH_2-CH_3)(CH_2-CH_2-CH_3)$

N-ethyl N'-methyl propanamine

Cynides and Isocynides

(1) NC–CH₂–CH=C(OH)–CH₂–CN (numbered 1 to 6)

3- Hydroxy hex 3-ene 1, 6-di-nitrile

(2) $CN-CH_2-CH(CN)-CH_2-CN$

Propane 1,2,3-tri-carbonitrile

(3) Cyclohexene ring with CN, OH and NH_2 substituents

3-Amino 2-hydroxyl cyclo hexene 1-carbonitrile

(4) CN–CH=C(OH)–CH=CH–CHO (numbered 1 to 5)

5-Aldo 3-hydroxy pent -3, 4 di-ene 1-nitrile

(5) $CH_3-CH_2-CH_2-NC$ (numbered 3, 2, 1)

Propane Carbyl amine

(6) $C_6H_5CH_2NC$

Benzyl carbyl amine

Nomenclature of Aromatic Compounds

(1) If the given compounds have two or more functional groups, write all secondary functional groups in prefix names in alphabetic order.

e.g.,

Benzene ring: 1-OH, 2-CH_3, 4-Br

4- Bromo -2- methylphenol

Benzene ring: 1-COOH, 3-Cl, 4-OH

3- Chloro -4- hydroxybenzoic acid

Benzene ring: 1-COCl, 3-OCH_3, 4-OH

4- Hydroxy -3- methoxybenzoyl chloride

1, 3- di bromo, 5- chlorobenzene

1- Bromo -6- chloro -3- iodo -2- nitrobenzene

N- Phenylcyclohexanecarboximide

2- Chloroethyl benzoate

5- Hydroxymethyl -2- methoxybenzonitrile

4- hyrdoxy -3- methoxybenzaldehyde

(2) In some cases when a substituent is such which when taken together with the benzene ring gives a special name to the molecule, then it is named as a derivative of that molecule with the substituent at position

For example,

4- Ethyl -2- chloro anisole

4- Acetamido phenol

4- Methoxy -2- nitro benzenesulphonic acid

(3) In case benzene ring is linked with an aliphatic compound having a functional group, it is named as phenyl derivatives of a compound.

2, 3, di hydroxy-4-phenylbutanal

(2- phenyl) octane

Ethlnylbenzene or Phenylethyne

Some Specific Examples

In case of spiro compounds in which two rings are fused at the same carbon atom, name is given as follows:

------ Prefix Spiro (X, Y), alk + ---- main suffix name/s

- Here X, Y represents no. of carbon atoms in ascending order and numbering is from smaller ring to larger ring as follows:

For example:

Spiro [2, 3] hexan -4-ol

2-Methylspiro [4, 5] deca -1, 6-diene

5-Oxaspiro [3, 4] octane

- In case of bicycle compounds in which two rings are fused at two bridge carbon atoms, numbering is made from larger ring to smaller ring and the compound is named as

------ Prefix bicyclo (X, Y, Z), alk + ---- main suffix name/s

Here X, Y, Z are no. of carbon atoms in decreasing order.

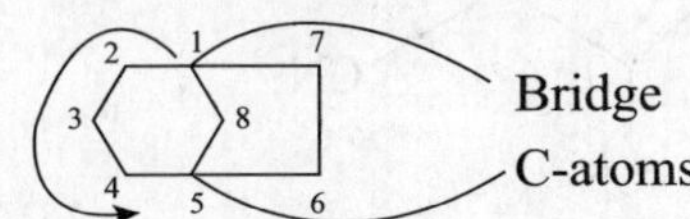

For example

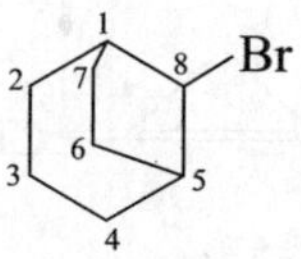

8-Bromo bicyclo [3, 2, 1] octane

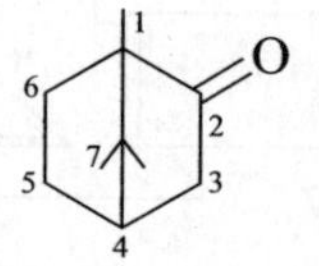

1, 7, 7-Trimethyl bicyclo [2,2,1] heptan-2-one

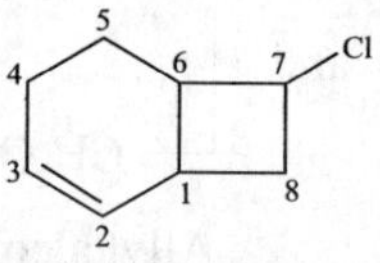

7-Chloro bicyclo [4,2,0] 2-oct–2–ene

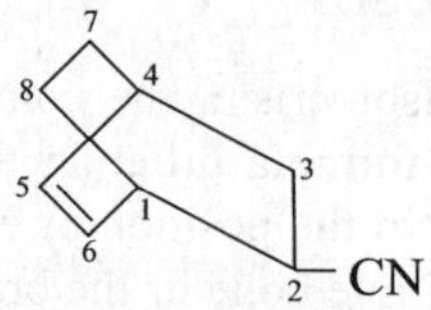

Bicyclo [2, 2, 2]oct-5-en -2-carbonitrile

Some other cases of cyclic compounds

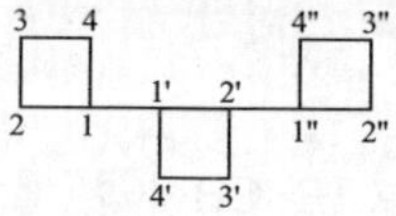

1, 1', 2', 1"-Tetcyclobutane

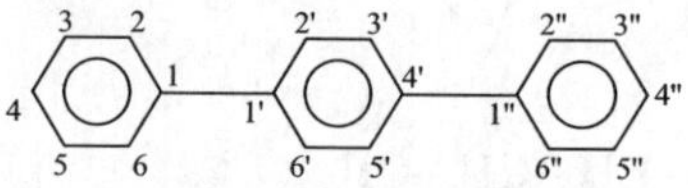

1, 1', 4', 1"-Terphenyl

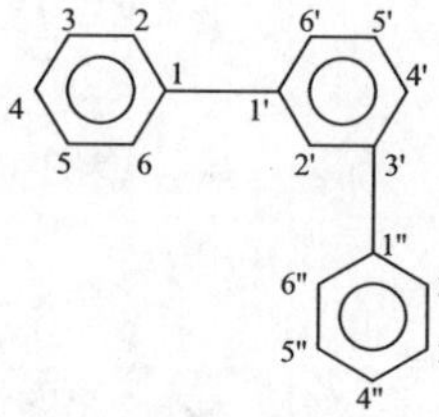

1, 1', 3', 1"-Terphenyl

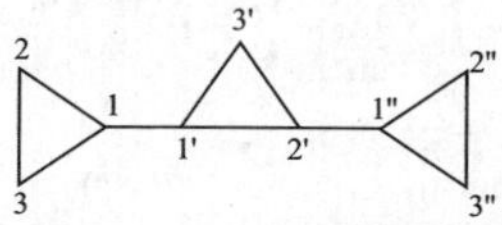

1, 1, 2', 1"- Tercyclopropane

Isomerism

The compounds which differ in their properties but have same molecular formula are called **isomers** and the phenomenon is known as isomerism. Isomerism is of following three types:

- Structural or constitutional isomerism
- Stereo or configurational isomerism
- Conformational isomerism

Structural or Constitutional Isomerism

- In structural isomerism, compounds possessing same molecular formula differ in their properties due to the difference in the linkages of atoms inside the molecule, that is, due to the difference in their structures.
- Here, isomers differ in the connectivity of carbon atoms, that is, they differ in their structural formula. For example, in the case of CH_3CH_2OH and CH_3OCH_3, both have the same molecular formula C_2H_6O but they differ in their structures.
- Structural isomers differ a lot in their physical as well as chemcial properties. Structural isomerism is of following types:

Chain or skeleton or nuclear isomerism

- In this type of isomerism, the compounds possessing same molecular formula differ in their properties due to the difference in the arrangement of carbon chain present in them.
- The isomers differ in the type of chain formed by the carbon atoms, that is, branched or unbranched chain. For example,

$CH_3–CH_2–CH_2–CH_2–CH_2–$

Linear alkane

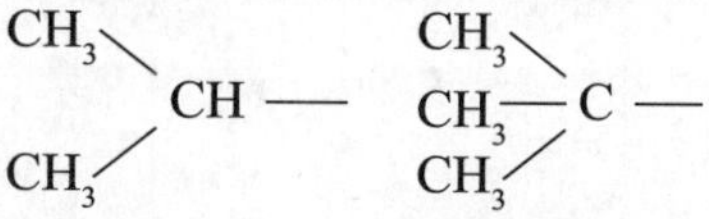

$(CH_3)_2CH—$ Iso alkane

$(CH_3)_3C—$ Neo alkane

- A chain of minimum 4 carbon atoms is necessary to show this isomerism. For example,

1. $\mathbf{C_4H_{10}}$

$CH_3–CH_2–CH_2–CH_3$

n-butane

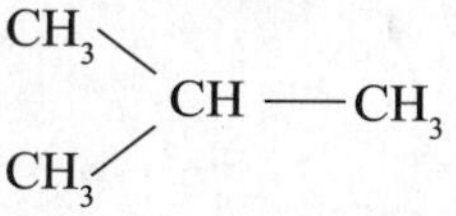

$(CH_3)_2CH—CH_3$

Isobutane (2-methyl propane)

2. $\mathbf{C_5H_{12}}$

$CH_3–CH_2–CH_2–CH_2–CH_3$

n-pentane

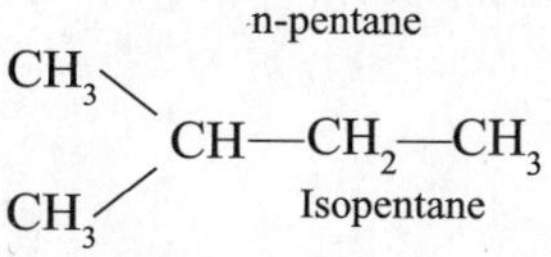

$(CH_3)_2CH—CH_2—CH_3$

Isopentane

$(CH_3)_3C—CH_3$

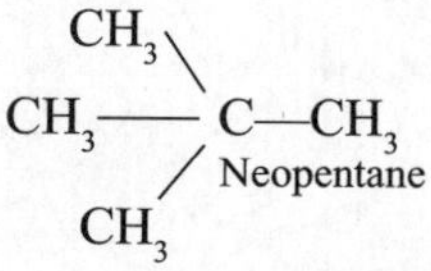

Neopentane

3. C_6H_{14} has five isomers

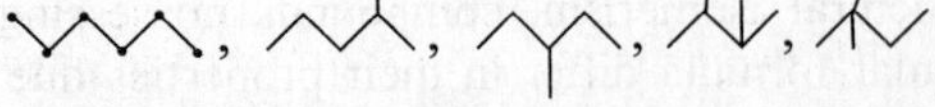

4. C_7H_{16} has 9 isomers, C_8H_{18} has 18 isomers and C_9H_{20} has 35 isomers.

Ring chain isomerism

- In ring chain isomers, mode of chain formation differs, that is, either open or close type of chain formation is observed.
- Ring chain isomers are also functional isomers.

For example,

1. Alkene and cycloalkane (C_nH_{2n})

C_4H_8

(I) $CH_3–CH_2–CH=CH_2$

(II) $(CH_3)_2C=CH_2$

(III) $CH_3–CH=CH–CH_3$ (cis and trans)

Here (I), (III) are position isomers

(IV)

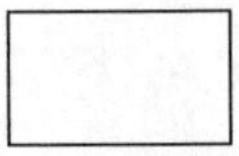

(V)

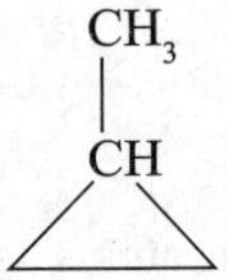

(IV), (V) Chain isomers and with respect to first 3 these are ring chain isomers (C_nH_{2n-2})

2. Cycloalkenes, alkynes and alkadienes

C_4H_6

Alkynes

$CH_3–C≡C–CH_3$

$H–C≡C–CH_2–CH_3$

Alkadienes

position

$CH_2=CH–CH=CH_2$

or

$CH_3–C(H)=C=CH_2$

Cycloalkenes

Position

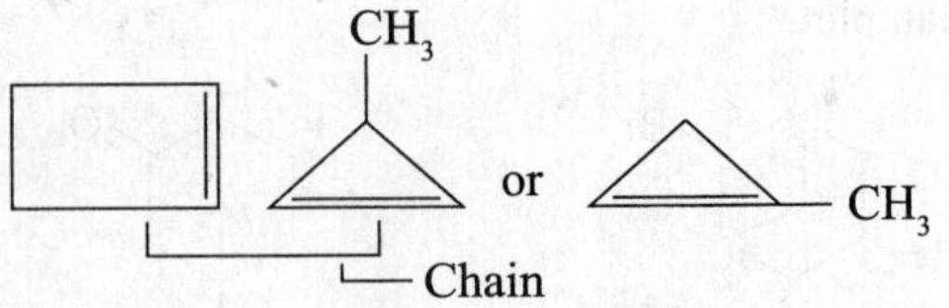

3. Cyclic ether and un-saturated Alcohol.

e.g.,

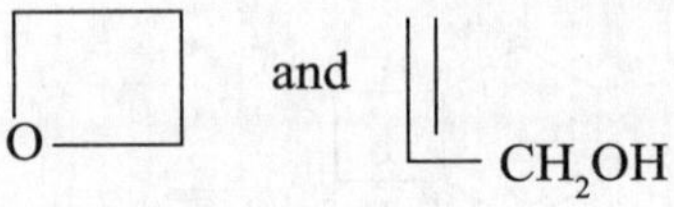

Oxa-butane　　Allyl alcohol

Position isomerism

- In this type of isomerism, the compounds possessing same molecular formula differ in their properties due to the difference in the position of either the functional group or the multiple bond or the branched chain or the substituent attached to the main carbon chain.

For example,

1. $CH_3–CH_2–C(=O)–CH_2–CH_3$

Pentan-3-one

$CH_3–C(=O)–CH_2–CH_2–CH_3$

Pentan-2-one

2. $CH_3–CH_2–CH(OH)–CH_2–CH_3$

Pentan-3-ol

$CH_3–CH(OH)–CH_2–CH_2–CH_3$

Pentan-2-ol

3. **C_4H_8**

$CH_2=CH–CH_2–CH_3$ butene-1

$CH_3–CH=CH–CH_3$

Butene-2 or But-2-ene

4. **C_8H_{10}**

CH_2—CH_3 (I) CH_3, CH_3 (II) CH_3, CH_3 (III) CH_3, CH_3 (IV)

I II III IV

Here, the structures shown as II, III and IV are positional isomers and with respect to one they are chain isomers.

Functional isomerism

- In this the type of isomerism, the compounds possessing same molecular formula differ in their properties due to the difference in their functional groups.

1. –CN (Cyanide) and –NC (Isocyanide) CH_3CN (Methylcyanide) and CH_3NC (Methylisocyanide).
2. $-NO_2$ (Nitro) and –ONO (Nitrite) $C_2H_5NO_2$ (Nitroethane) and C_2H_5ONO (Ethyl nitrite)
3. Primary (RNH_2), secondary (R_2NH) and tertiary amines (R_3N)
4. **Alcohol and ether ($C_nH_{2n+2}O$)**

- **C_2H_6O**
  CH_3–O–CH_3 (Methoxymethane)
  CH_3–CH_2–OH (Ethanol)
- **C_3H_8O**
  CH_3–O–CH_2–CH_3 (Ethyl methyl ether)
  CH_3–CH_2–CH_2OH (Propanol)
  $(CH_3)_2$CH—OH (Isopropanol)
- $C_4H_{10}O$– It has 4 alcohols and 3 ethers.

OH OH OH OH

(1) (2) (3) (4)

or 2, 3, 4 are chain isomers and with respect to (1) while (2) is position isomer of 1.

O O O

(5) (6) (7)

5 with respect to 6 or 7 is metamer while 6 and 7 are position isomers.

5. **$C_nH_{2n}O$** The isomers having this molecular formula are:

- Aldehyde
- Ketone
- Cyclic alcohol
- Cyclic ether
- Unsaturated alcohol or ether
- Epoxy ether

For example,

C_3H_6O

- CH_3–CH_2–CHO

- CH_3–C(=O)–CH_3
- CH_2=CH–CH_2–OH
- OH–CH (ring with CH_2—CH_2)
- O—CH_2 / H_2C—CH_2 (four-membered ring)
- CH_2—CH—CH_3 (with O bridging CH_2 and CH)

6. **$C_nH_{2n}O_2$** The isomers having this molecular formula are:

- acids
- esters
- aldehydes with –OH group
- ketones with –OH group
- cyclic ethers with –OH group

For example,

$C_2H_4O_2$

- CH_3–COOH
- $HCOOCH_3$
- CHO–CH_2–OH
- CH_2—CH—OH (with O bridging CH_2 and CH)

$C_3H_6O_2$

- CH_3–CH_2–COOH
- CH_3–$COOCH_3$

- $HCOOC_2H_5$
- $CH_2OH–CH_2CHO$
- $CH_2OH–COOCH_3$
- $CH_2–CH–CH_2OH$ (with O bridging the first two carbons)

$C_nH_{2n+3}N$

p°, s°, t° amines.

- E.g., $C_4H_{11}N$ has 4 primary, 3 secondary and 1 tertiary amines, that is, 8 in all.

(1) $CH_3CH_2CH_2CH_2NH_2$

(2) $CH_3–CH(CH_3)–CH_2NH_2$

(3) $CH_3CH_2–CH(CH_3)–NH_2$

(4) $CH_3–C(CH_3)_2–NH_2$

(5) $C_2H_5–N(C_2H_5)–H$

(6) $CH_3–N(CH_2CH_2CH_3)–H$

(7) $CH_3–N(CH(CH_3)–CH_3)–H$

(8) $CH_3—N(CH_3)—CH_2—CH_3$

Metamerism

- It is the type of isomerism in which the compounds possessing same molecular formula differ in their properties due to the difference in the alkyl groups present in them, that is, they have the same functional group but different alkyl groups attached to it.
- It is shown by ether, thioether, ketones, esters, secondary amines, tertiary amines and alkenes. For example,

1. $CH_3–COOCH_3$ and $HCOOC_2H_5$ are metamers
2. $C_2H_5–NH–C_2H_5$ and $CH_3–NH–CH_2CH_2CH_3$ are metamers.
3. 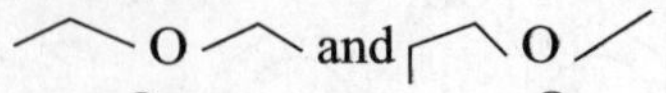 (ethyl–O–ethyl) and (propyl–O–methyl)
4. 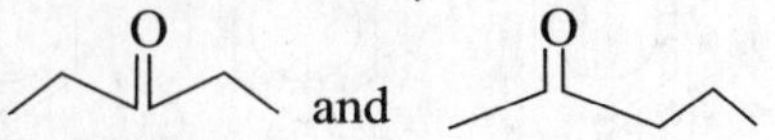 (pentan-3-one) and (pentan-2-one)

Tautomerism

- It is also known as krytomerism, allelotropism, metrotropy, keto–enol isomerism.
- It was introduced by Conard Larr.
- It is the type of isomerism in which the two functional isomers exist together in a dynamic equilibrium and the two forms existing in equilibrium are called tautomers.
- To show this isomerism C-atom must be bonded with highly El-ve elements, O, N, S with a multiple bond.
- These forms are formed due to the migration of α–H^+ or protons (cations mainly) due to which the bond position changes (desmotropism).
- The migration of H^+ or proton is called **cationtropism** or **prototropism.**
- The least stable form is called **labile form**.
- The migration of cation is possible in the following ways:

1 ***Diad system*** Here, H^+ (cation) migrates to next atom.

For example,

H–C≡N ⇌ H–N≡C

H–O–N=O ⇌ H–N(→O)=O

2 ***Triad system*** Here, H^+ migrates to next alternate atom (that is, the third atom).

For example,

$H_2C(H)–N(→O)=O \rightleftharpoons CH_2=N(→O)–O–H$

$CH_3–CH_2–C(=O)–NH_2 \rightleftharpoons CH_3–CH_2–C(OH)=N–H$

Keto–enol isomerism

- It is the most common type of tautomerism in which the two tautomeric forms are keto and enol forms. These are formed as a result of migration of α-hydrogen atoms.

- The presence of α-hydrogen atoms is a must here. The α-carbon atom, however, must not be unsaturated.

$$-\overset{O}{\overset{\|}{C}}-CH_2- \rightleftharpoons -\overset{OH}{\overset{|}{C}}=CH-$$
Keto Enol

For example,

$$CH_3-\overset{O}{\overset{\|}{C}}-C_6H_5 \rightleftharpoons CH_2=\overset{OH}{\overset{|}{C}}-C_6H_5$$
α

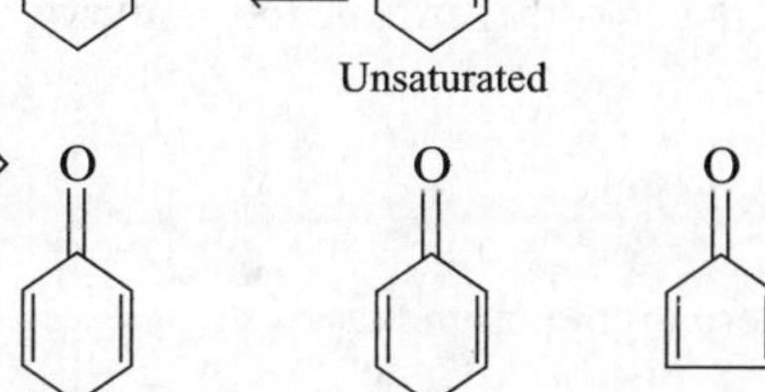

⇒ O O O

O NH

These compounds cannot show tautomerism as all α-carbon atoms are unsaturated, therefore, migration of α-hydrogen atom is not possible.

- Enol content ∝ Number of α-hydrogen atom
- Enol content ∝ Acidic nature
- Enol content ∝ Number of carbonyl groups
- Enol content in decreasing order is given as follows:

O, φ–CO–CH_2–CO–φ, φ–CO–CH_2–CO–CH_3,

CH_3–CO–CH_2–CO–CH_3

CH_3COCH_2–CHO, φ–CO–CH_2–$COOC_2H_5$, CH_3COCH_2

$COOC_2H_5$, CH_3 $COCH_3$, CH_3CHO, O O

Enolisation is favoured by presence of >C=O gp, α-H-atoms, resonance and hydrogen bonding e.g.,

e.g.,

$$\phi-\overset{O-H\cdots O}{C=CH-\overset{\|}{C}}-OC_2H_5$$

H⋯O⋯H
O O

Some Specific Cases of Tautomerism

1. Imine-enamine system

$$R-\underset{H}{\overset{a}{CH}}-CH=NH \leftrightarrow R-CH=CH-NH_2$$
Imine Enamine

Nitroso-Oximino System

$$R-\underset{H}{CH}-N=O \leftrightarrow R-CH=N-OH$$
Nitroso Oxime

Ring-Chain Tautomers

$$R-\overset{O}{\overset{\|}{C}}\diagdown CH_2 \diagup \ddot{O}-H \longrightarrow R-\overset{O-H}{\overset{|}{C}}-O \ (CH_2)$$
Keto Cyclic hemi acetal

Glucose in open structure exhibits this isomerism with both α or β -d-glucopyranose

Stereoisomerism or Space Isomerism

In stereoisomerism isomerism, the compounds possessing same molecular formula differ in their properties due to the difference in the arrangement of their atoms or groups of molecules in space.

Stereoisomerism is of the following types:

Geometrical isomerism

- It is the type of isomerism in which the compounds possessing same molecular formula differ in their properties due to the difference in their geometry, that is, due to the difference in the direction of attachment of same atoms or groups in their molecule i.e., relative spatial arrangement of atoms on groups around the double bond.
- It is not shown by compounds with single bond due to the occurrence of free rotation.

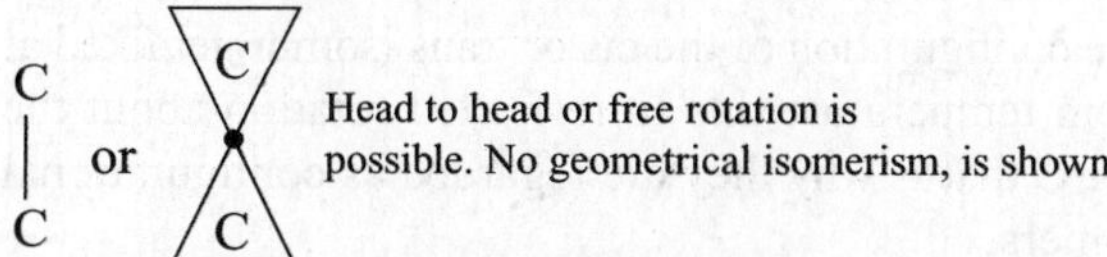

- It is shown by compounds containing [>C=C<], [>C=N–], [–N=N–] groups and by cycloalkanes.

Geometrical isomerism in alkenes

- **Case I** The given compound has same type of atoms and groups, so it does not show geometrical isomerism:

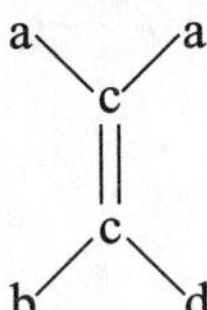

For example, butene-1, 2-methyl but-2-ene

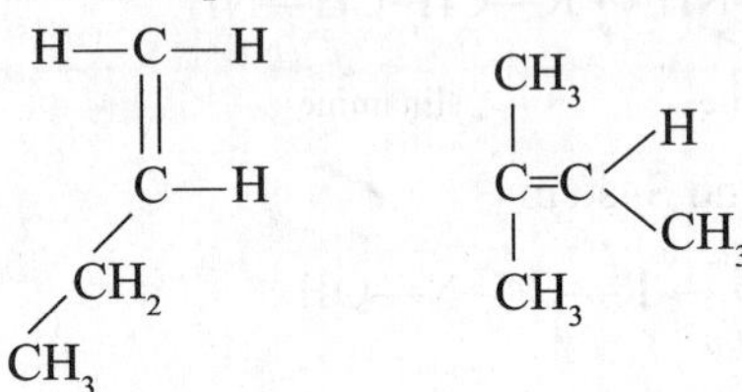

- **Case II** The given compounds exhibit geometrical isomerism:

b—C—a ‖ b—C—a (Cis) b—C—a ‖ a—C—b (Trans)

For example,

1. $C_2H_2X_2$

H—C—X ‖ H—C—X (Cis) H—C—X ‖ X—C—H (Trans)

2. $C_4H_4O_4$

H—C—COOH ‖ H—C—COOH (Cis Maleic acid) H—C—COOH ‖ HOOC—C—H (Trans Fumaric acid)

3. C_4H_8

H—C—CH_3 ‖ H_3C—C—H (trans but-2-ene) H—C—CH_3 ‖ H—C—CH_3 (Cis-but-2-ene)

FACTS TO REMEMBER

The configuration of the cis or trans isomer get fixed at room temperature due to restricted rotation about the C = C that is why they are regarded as configurational isomers.

Difference Between Cis and Trans Forms

Cis form Here same species lie on the same side of (=) bond.

- Cis form is more reactive and less stable, as same species are on the same side so the steric repulsion increases the reactivity and decreases the stability.
- The dipole moment of cis form is more than that of the trans form.
- It has lower melting point than the trans form.
- The boiling point of cis form is higher than that of trans form due to stronger dipole-2 interactions.
- The solubility, viscosity and refractive index of cis form is more than that of trans form.

Trans form Here same species lie on diagonally opposite sides.

- It is a more stable and less reactive form as here same species are present on the opposite sides.
- The dipole moment is mostly zero due to symmetry. For example,

H–C–CH_3 ‖ CH_3–C–H (zero dipole moment)

Trans-2-butene

- In case of unsymmetrical alkenes, due to a little unsymmetry there may be some dipole moment value also.

H–C–R ‖ R^1–C–H

For example, trans-2-pentene has some dipole moment value but less than cis form.
- They have higher melting point than the cis form due to more close packed structures i.e., more latice energy.
- The boiling point of trans form is less than that of cis form, but have more density than cis-form.
- The solubility, viscosity refractive index is less than that of cis form due to more stronger interactions.

Geometrical isomerism in cycloalkanes Cycloalkanes can show it due to restricted rotation of C-C bond in their rigid ring structures.

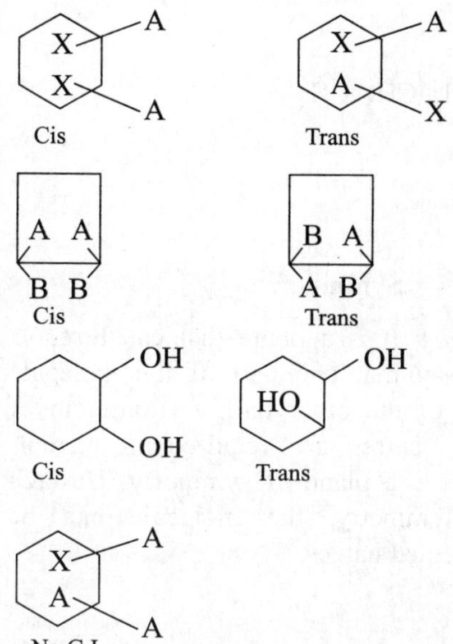

The last structure does not show geometrical isomerism, as one carbon atom has two similar species 'A'.

Geometrical isomerism in oximes and azo-compounds

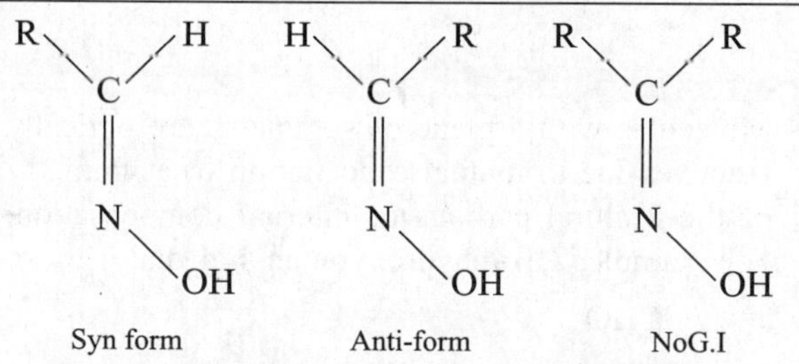

- The last structure does not show geometrical isomerism, as one carbon atom has two similar species 'R'.
- In azo compounds

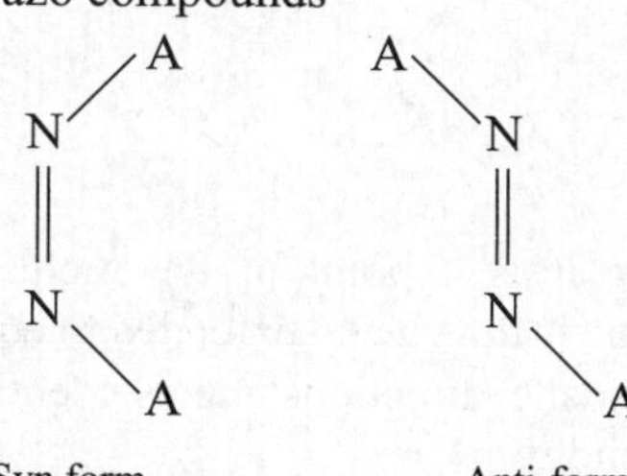

Here, the attached groups may be same or different.

Calculation of geometrical isomerism

(a) When end groups are different

Number of geometrical isomers = 2^n

Here, n = number of double bonds

For example, in CH_3–CH=CH–CH=CH–Cl

n = 2

Number of geometrical isomers = $2^n = 2^2 = 4$

(b) When end groups are same

1. when n is an even number

   Number of geometric isomers

   $= 2^{n-1} + 2^{n/2-1}$

   For example,

   X–CH=CH–CH–CH=CH–CH

   =CH–CH=CH–X

   n = 4

   Number of geometric isomers

   $= 2^{4-1} + 2^{(4/2)-1}$

   $= 2^3 + 2^1 = 10$

2. when n is an odd number

   Number of geometric isomers

   $= 2^{n-1} + 2^{(n+1)/2-1}$

   In, CH_3–CH=CH–CH=CH–CH=CH–CH_3

   n = 3

   Number of geometric isomers

   $= 2^{3-1} + 2^{(3+1)/2-1}$

   $= 2^2 + 2^1 = 6$

Optical isomerism In this type of isomerism, the compounds differ in the direction or extent of rotating plane polarized light.

Plane polarized light

- The light from an ordinary source is composed of waves vibrating in many different planes, perpendicular to the plane of its propagation.
- When such a light is passed through a nicol prism, the light coming through it is found to be vibrating in only one plane. Such a light is called as **plane polarized** light (polarizer) and here nicol prism is called as **polaroid**.

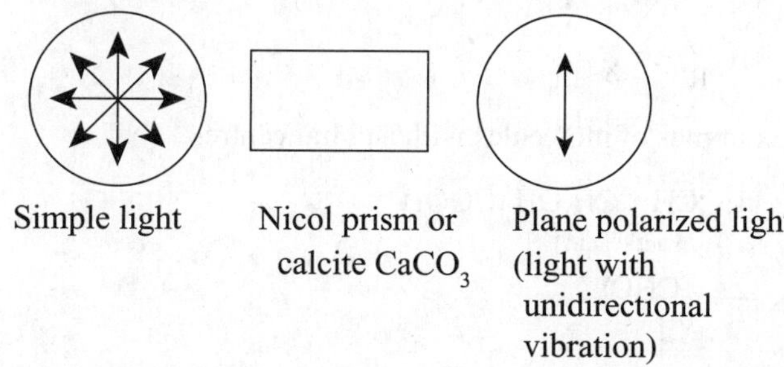

Simple light | Nicol prism or calcite $CaCO_3$ | Plane polarized light (light with unidirectional vibration)

Optical activity or optical rotation The optically active substances rotate the plane polarized light clockwise or anti-clockwise, that is, dextro and laevo rotation respectively, at an angle θ. This rotation is called **optical rotation**.

- In both the rotations, the value of θ is the same but has opposite signs, that is, same in magnitude but opposite in direction.

- Optical rotation is measured by a polarimeter in terms of specific rotation.

$$[\alpha]_{\lambda}^{t°c} = \frac{\alpha(\text{observed})}{l \times c}$$

Here, α = Specific rotation
α (observed) = Observed value of rotation
l = Path length of light travelling through the solution
c = Concentration of solution

Specific rotation depends upon:

1. nature of solute and solvent
2. wavelength of light during the experiment
3. temperature during the experiment
4. path length of light travelling through solution and concentration of solution

Reason for occurrence of optical activity or optical isomers

- The compound must be asymmetric with a non-superimposable mirror image. Examples of non-superimposable mirror images:
  The image of a hand in a mirror.
  The word Ambulance written in reverse, as seen in a mirror.
- Asymmetric molecules do not have a plane or central or axial axis of symmetry and alternating axis of symmetry.

Chiral centre or stereogenic centre

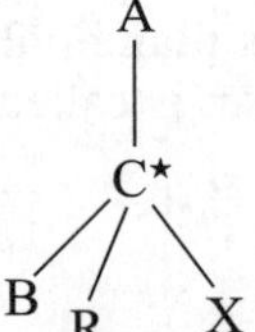

Examples of molecules with a chiral centre:

1. CH_3 *CH.OH .COOH
   Lactic acid
2. CHO
   |
   *CH.OH
   |
   *CH.OH
   |
   CHO
3. Br
   |
   CH_3–*C–CH_2–CH_2–CH_3
   |
   Cl
   2-bromo-2-chloropentane

Symmetry

Plane of symmetry It is a plane that can bisect a molecule in a way that one half of the molecule is mirror image of the other half. A minimum of two stereogenic centres are required in a molecule for it to have a plane of symmetry. Through the plane of symmetry, the molecule can be divided into two equal halves.

X
|
b–*C–a
-------- Plane of symmetry
b–*C–a
|
X

(O, R, R on ring)

- Molecules with a plane of symmetry are optically inactive, due to mutual cancellation of asymmetry of the 2 chiral carbons or internal compensation. For example, 2, 3-dihydroxybutan-1, 4-dial

CHO
|
H–*C–OH
-------- Plane of symmetry
H–*C–OH
|
CHO

Centre of symmetry It is a point in the molecule through which if a line is drawn and extended to equal distances in the opposite directions meets identical species at its either ends.

A, CO — NH, R, C, C, R, NH — CO, A

Centre of symmetry

Optical isomers differ in their optical activity, that is, the rotation of plane polarized light.

Optical isomers are of following types:

Optically active isomers: They can rotate the plane polarized light, that is, show optical activity.
For example, Lactic acid

COOH | C (H, OH, CH_3) — Dextra | COOH | C (HO, H, CH_3) — Laevo

- Dextro or d isomers show clockwise rotation of plane polarized light.
- Laevo or ℓ isomers show anti-clockwise rotation of plane polarized light.
- Both d and ℓ isomers have same value of θ but with opposite signs.
- These are non-superimpossible mirror images and called **enantiomers** or **enantiomorphs** or **antimers**.
- They have almost identical chemical properties and have nearly same physical properties.
- They have different biological properties or activities.
- d, ℓ forms cannot be separated by simple distillation etc.
- For their separation, spectrographic methods and chromatography are used.
- They can be separated by converting them into respective diastereomers, which differ in their physical properties.

Optically inactive isomers or meso form

- For the meso form, a minimum of two chiral centres are needed and they must possess a plane of symmetry.
- The meso form is optically inactive due to internal compensation.

For example,

COOH
H–C*–OH
······· Plane of symmetry
H–C*–OH
COOH

Meso tartaric acid

- Lactic acid has no meso form as it has only one chiral centre.
- Dextro or laevo and meso forms are non-superimposable mirror images of each other and are called diastereomers.
- Diastereomers have different physical properties.

Racemic mixture

- It is an equimolecular mixture of (+) and (-) (that is, d- and 1-) forms of an optically active compound.
- It is not able to rotate the plane polarized light because of external compensation. This phenomenon is called as **racemization.**
- It is also written as dl or ± form.

Some special examples of optical isomerism are:

I. Chiral centre other than carbon

O=P* (propyl, CH_3, ethyl)

II. In case of biphenyls

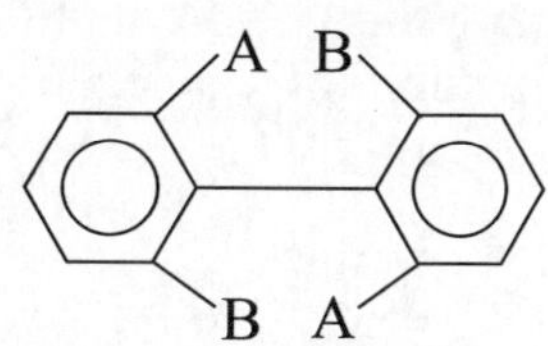

Ortho substituted

- Here, steric hinderance by bulky groups make the rotation of C–C bond difficult and causes optical activity.
- This stereo isomerism due to restricted C–C bond rotation is called **atropisomerism**.

For example,

NO_2 COOH / COOH NO_2 | COOH NO_2 / NO_2 COOH

In case of allenes

$(A)(B)C{=}C{=}C(A)(B)$ — Optically active

$(A)(B)C{=}C{=}C(A)(A)$ — Optically inactive

In case of alkylidenes

$$X,B > C_6H_{10} = C < X, A$$

- If any carbon has two identical groups, chirality is lost.

To find the number of optical isomers

Case I When the chiral molecule has no symmetry

Optically active isomer (a) = 2^n

Here, n = number of chiral centre

- Meso form (m) is absent.
- Racemic form = a/2
- Total optical isomers = a + m

For example,

$CH_3.\ ^*CH\ .\ OH\ .\ COOH$

n = 1

$a = 2^1 = 2$

m = 0

r = a/2 = 1

For example,

CH_2OH
|
$(^*CHOH)_4$ (Glucose)
|
CHO

n = 4

$a = 2^4 = 16$

m = 0

Total = 16

Case II When should the word 'Chiral' come here, the molecule has symmetry, that is, molecule is divisible.

1. When n is an even number

Optically active isomers (a) = 2^{n-1}

Meso form (m) = $2^{(2/n)-1}$

Total optical isomers = a + m

For example,

COOH
|
$(^*CHOH)_2$
|
COOH

$n = 2,\ a = 2^{n-1} = 2^1 = 2$

$m = 2^{(2/n)-1} = 2^o = 1$

Total = 3

2. When n is an odd number optically active isomer (a) = $2^{n-1} + 2^{(n-1)/2}$

Meso form (m) = $2^{(n-1)/2}$

Total optical isomers = a + m

For example,

CH_2OH
|
$(^*CHOH)_3$
|
CH_2OH

n = 3,

$a = 2^{3-1} + 2^{(3-1)/2} = 2^2 + 2^1 = 6$

$m = 2^{(3-1/2)} = 2^1 = 2$

Total optical isomers = 6 + 2 = 8

Racemization: It is the conversion of an optically active form into an optically inactive form by the action of light, heat or a chemical reagent. Half of the number of molecules change into enan-tiomers. The resulting product is, therefore, optically inactive. For example,

$$\underset{\text{2-iodobutane}}{(Et)(Me)(I)(H)\,^*C} \xrightarrow[S_{N^2}]{\text{NaI/Acetone}} \underset{\text{Optical inactive}}{(Et)(I)(Me)(H)C + (Et)(Me)(I)(H)C}$$

Racemization occurs by either acid catalysed or base catalysed mechanisms. In case of optically active halides it also occurs by SN^1 or SN^2 mechanisms.

Resolution: It is the conversion of a racemic mixture into dextro and laevo forms. It is possible by following methods:

1. **Mechanical separation method** It is possible as both forms have different crystal shapes.
2. **Biochemical separation** *Penicillium glaucum* separates dextro ammonium tartarate from the laevo form.
3. By converting enantiomers into diastereomers, as diastereomers differ in the physical properties.
4. By **chromatographic methods.**
5. By **salt formation method.**

Asymmetric synthesis: Here, an optically inactive compound is changed into an optically active compound by a suitable chemical reagent. For example,

$$\text{R(H)C=O} \xrightarrow{HCN} \text{R(H)}\overset{*}{C}\text{(OH)(CN)}$$

Optically active cyanohydrin

e.g.,

$$C_6H_5CHO \xrightarrow{HCN} C_6H_5-\overset{*}{C}H(OH)-CN \xrightarrow{\text{Hydrolysis}} C_6H_5-\overset{*}{C}H(OH)-COOH$$

(–) Mandelic acid.

If we use CH_3CHO here, product is lactic acid.

Walden or optical inversion

- As a result of a reaction, when the product formed is optically active but its configuration is reverse from that of the reactant, the process is known as optical inversion or walden inversion.
- It is the conversion of one optically active form into another optically active form with the help of strong reagents like PCl_5 or KOH.
- It involves SN^2 mechanism.

For example, (1)

$$\text{Et, Me, H, OH}-\overset{\star}{C} \xrightarrow[-POCl_3,\ -HCl]{\Delta PCl_5} \text{Et, Me, Cl, H}-\overset{\star}{C} \xrightarrow{KOH} \text{Et, Me, HO, H}-C$$

e.g., (2)

$$CH_3(CH_2)_5(H)(CH_3)C-Br \xrightarrow{^-OH} HO-\overset{\star}{C}(CH_2)_5CH_3(H)(CH_3)$$

FACTS TO REMEMBER

- Optical isomerism was first observed in quartz.
- It is also shown by $KClO_3$, $NaClO_3$, $(HCOO)_2Ba$.
- d, l notation was introduced by Emil Fischer.
- d, l form do not always represent optical isomers.

Erythro, threo system of nomenclature for isomers: In erythro form of the molecule, same species are on the same side but in threo form these are on the opposite sides.

For example,

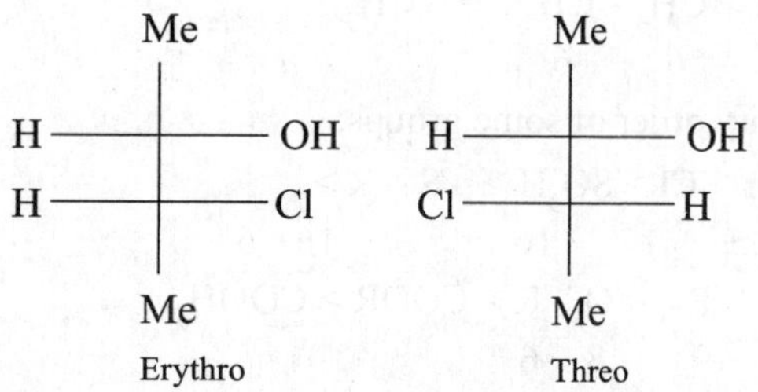

Epimers: These differ in configuration only at one chiral or stereogenic centre.

For example,

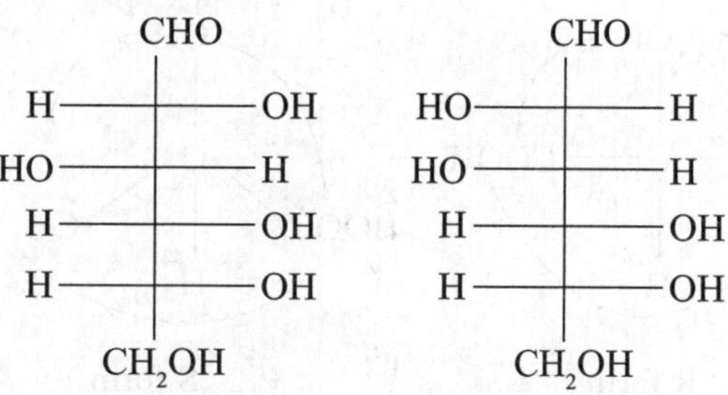

- In the given example, the configuration differs at second carbon atom.

Absolute configuration

- The actual three dimensional arrangement of groups in a molecule containing asymmetric carbon is known as absolute configuration.
- It is a three dimensional arrangement of atoms or groups in space.

R, S system of nomenclature for isomers: It was introduced by R.S. Cahn, C.K. Ingold and V adimir Prelog. R stands for rectus (means clockwise rotation), S stands for sinister (means anti-clockwise rotation)

Rules for assigning the form to molecules

- First decide priority of groups attached to the chiral centre.

For example,

$$\underset{3}{CH_3} . \underset{4}{\overset{*}{C}H} . \underset{1}{OH} . \underset{2}{COOH}$$

To decide priority of group:

1. It is decided by the atomic number (∝ atomic number) and atomic weight in case of isotopes.
2. If the first atom is same, consider the second atom and so on.

For e.g.,

$-OH > CH_3 - CH_2 - > - CH_3$
8 6 6 6

Priority order of some groups:

$I > Br > Cl > SO_3H > -S-R >$
53 35 17 16 16 6

$-SH > F > -O-R > COOR > COOH$
1 9 8 6

$> COR > CHO > CN > C_2H_5 > CH_3 > D > H$

- Here, the first irpriority must be at the top and the fourth priority must be at the bottom.

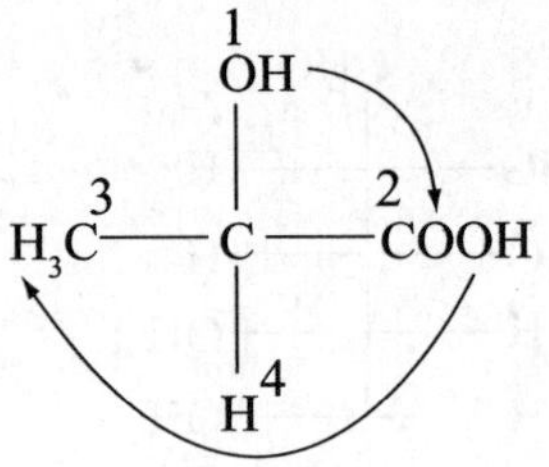

R form

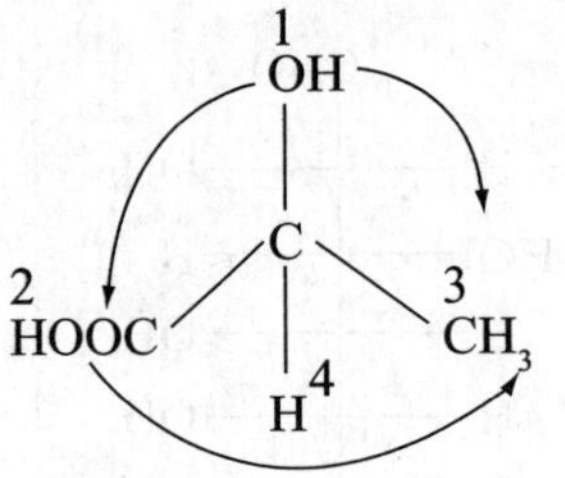

S form

- *Switching* If the fourth priority is not at the bottom, the positions are interchanged as depicted below:

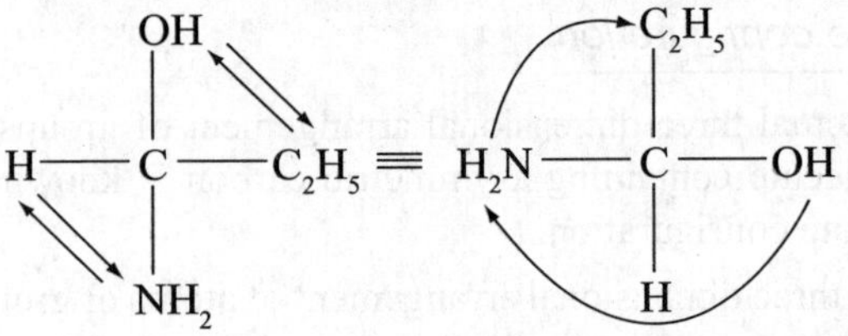

- In case of switching, the configuration in most cases, is a reverse of the given configuration (in most cases).

E, Z system of nomenclature of isomers: It is the modern way to explain geomatrical isomers as follows:
E stands for entgegen, the German word for opposite. Therefore, in this form, the species of same priorities are on the opposite side (trans).

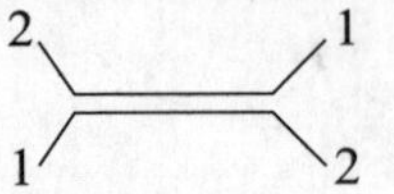

Z stands for zusammen, the German word for together. Therefore, in this form, the species of same priorities are on the same side (cis).

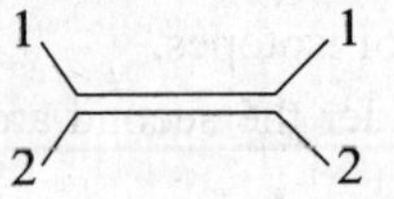

Examples

1. 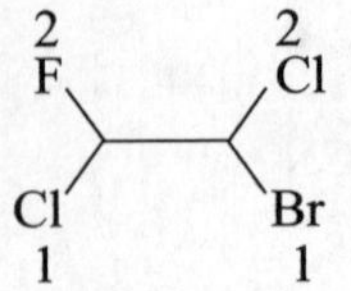 Z form
2. 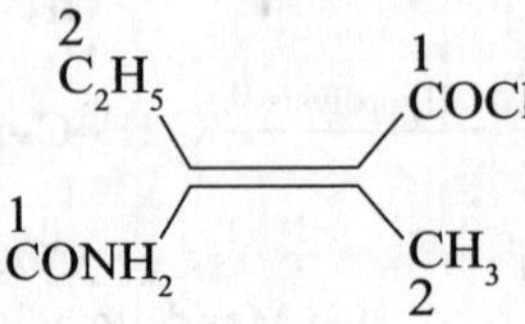 E form
3. 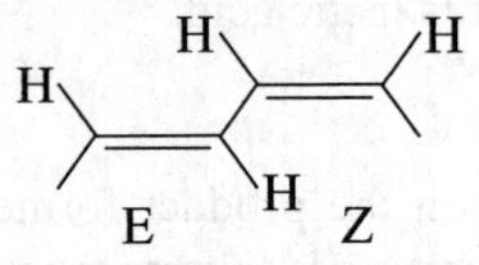 E form Z form
2'Z' 4'E' hexa 2, 4 di-ene.
4. 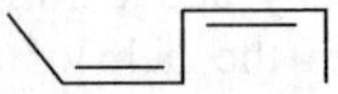 2'Z', 4'Z' hexa 2, 4 di-ene

Conformational Isomerism

- Conformational isomers are formed by the rotation around C-C bond. An alkane can have infinite number of conformations.
- Conformers differ slightly in energy. After a 360° rotation, the conformation returns to the initial form.
- The main structural work was done by Sawhorse and Newman.
- Conformers have a slight difference in their energies so the conformers cannot be separated.
- In staggered conformation, the angle between H-C-H is 60°, while in eclipsed it is less than 10°.
- Staggered form is more stable and less reactive than the eclipsed form.
- In staggered conformation, the distance between H-nuclei is 2.55 Å but in eclipsed it is 2.29 Å only.
- The rotational barrier in staggered form is 0.6 kcal/mole.
- The rotational barrier in staggered form is 2.9 kcal/mole.
- Energy barrier for free rotation, is 0.6 kcal/mole.
- For restricted rotation, the energy barrier is 0.6–16 kcal/mole.
- The energy barrier between eclipsed and eclipsed confirmation is 2.9 kcal/mole.

- **Stability orders of conformations:**
  In case of butane, stability order is
  Anti-staggered > staggered > gauche > partially eclipsed > fully eclipsed
- In case of cyclo hexane, the stability order is chair > twist boat > boat > half-chair
- Chair form is more stable than boat form, due to torsional strain or bond–bond repulsion in chair form and stearic repulsion in boat form (due to eclipsing and flag. pole–2 interaction).

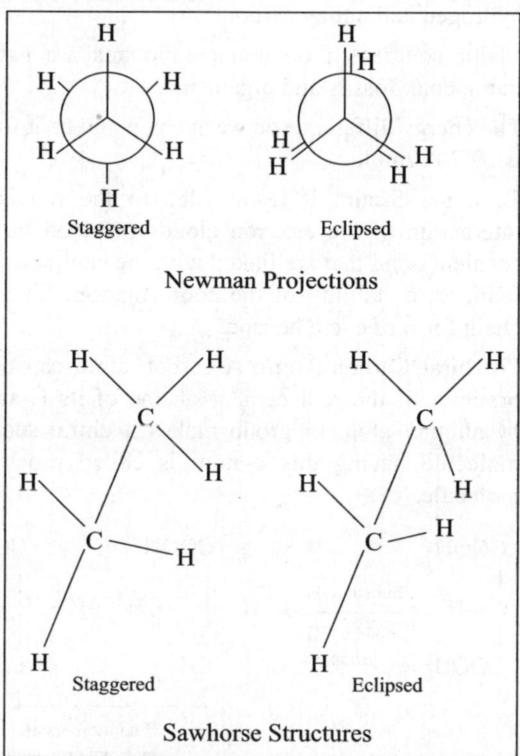

FIGURE 11.1 Ethane Conformations

(1) Anti ⇌ (60°) (2) Partially Eclipsed ⇌ (60°)

(3) Gauche Staggered ⇌ (60°) (4) Fully Eclipsed

(5) Gauche ⇌ (60°) (6) Eclipsed

FIGURE 11.2 Butane Conformations

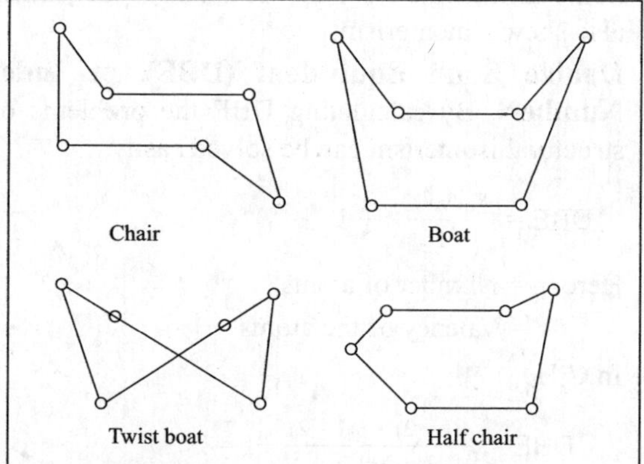

FIGURE 11.3 Conformation of Cyclohexane

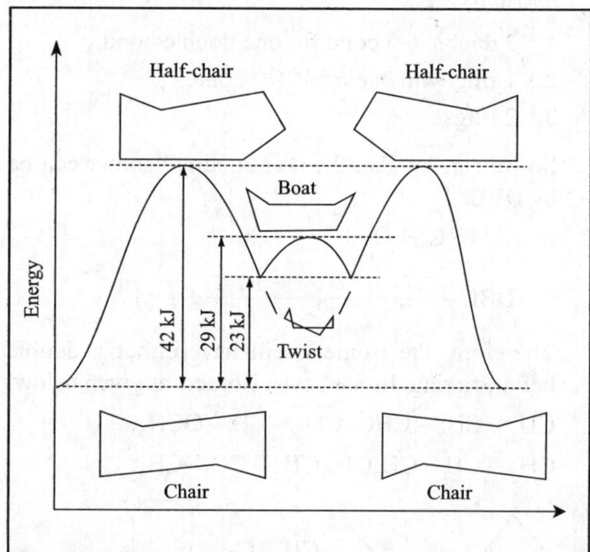

FIGURE 11.4 Conformational Energy of Cyclohexane

Points To Remember

- **Isotopic Isomerism** It arises due to the presence of different isotopes of an element in the compound.
  For example
  (i) $^{12}CH_4$ and $^{14}CH_4$
  (ii) H_2O and D_2O
- Some amides like urea, thiourea, nitrosophenol may also show tautomerism.
- **Double Bond Equivalent (DBE), or Index Number** By calculating DBE the problems on structural isomerism can be solved easily.

$$DBE = \frac{\Sigma n(V-2)}{2} + 1$$

Here, n = Number of atoms
V = Valency of the atoms

In C_4H_6

$$DBE = \frac{4(4-2)+6(1-2)}{2} + 1 = 2$$

Here, DBE is 2 which implies that the molecule may have:

1. 2 double (=) bond for one double bond,
2. 1 ring with one triple (≡) bond,
3. 2 rings

So we can say that the no structural isomer can be by DBE.

In case of C_3H_6O

$$DBE = \frac{3(4-2)+6(1-2)+1(2-2)}{2} + 1 = 1$$

Therefore, the isomers will have either a double bond or a ring. Its has seven isomers as given below:

CH_3-CH_2-CHO, $CH_2=CH-OCH_3$, $CH_2=CH-CH_2OH$, $CH_3-CO-CH_3$, (cyclopropyl)–OH, (epoxide ring)CH – CH_3, (oxetane ring with O)

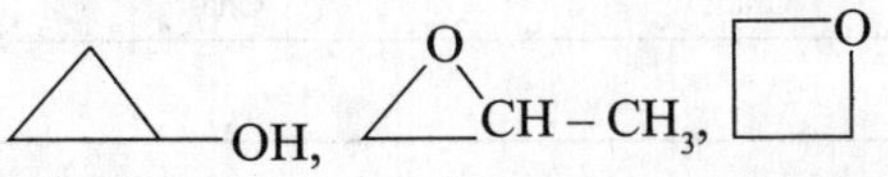

- **Homomers** Homomers are identical representations of the same compounds, that is, the molecular models of these representations are superimposable on one another.
- Regiomers differ in the position of the functional groups.

For example,

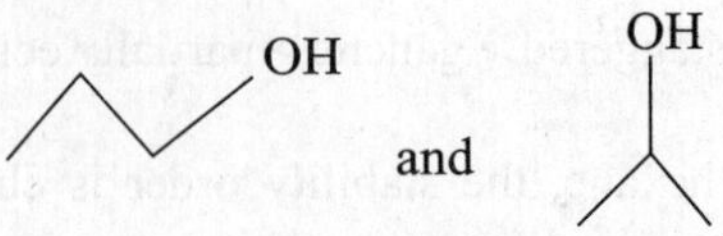

(zig-zag chain)OH and (branched chain with OH)

- Acetic acid was the first organic compound to be synthesized in laboratory.
- Methane was first synthesized artificially by Kolbe.
- Maximum number of compounds are formed by hydrogen and not by carbon.
- Major sources of organic compounds are petroleum, coal, fossils and organisms.
- The energy difference between chain and boat form is 29.7 KJ/mole.
- Torsional Strain: It results due to the repulsive interaction of the electron clouds involved in the covalent bond that are linked with the central atom. It increase stability of the conformation. for e.g., chain form of cyclo hexane.

  Prochiral Carbon Atom: A carbon atom can be a prochiral if the replacement of one of its H-atom by another atom or group makes it chiral such a molecule having this c-atom is called prochiral molecule.

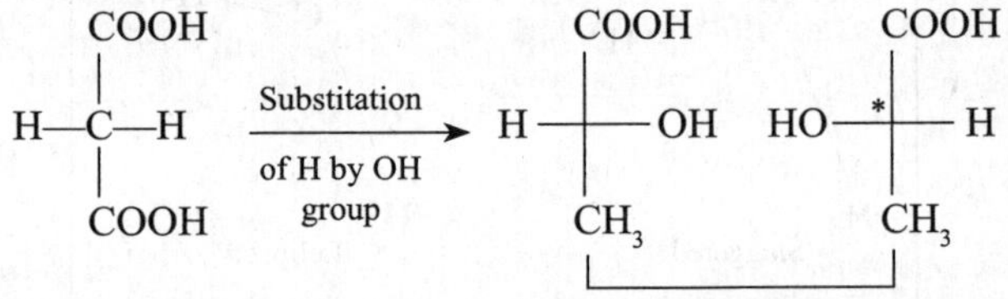

H—C(COOH)(COOH)—H —(Substitution of H by OH group)→ H—C(COOH)(CH₃)—OH and HO—C*(COOH)(CH₃)—H

Enantiomers of 2-hydroxy propanoic (Lactic acid)

R—$\overset{*}{C}H$—CH_2 (epoxide, bridged by O) is also optically active as C_2 is chiral in it.

Intra molecular H-bond

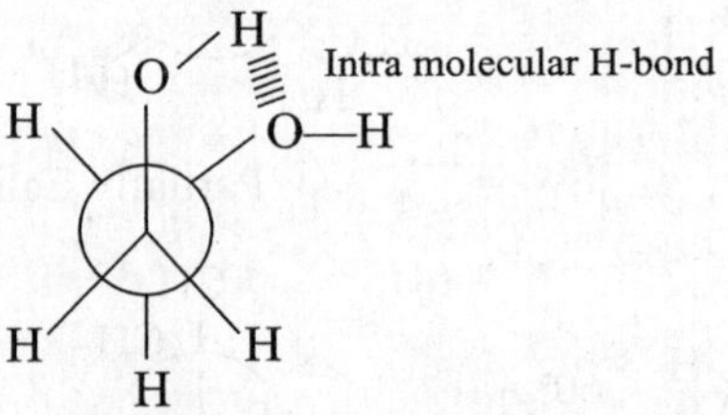

Gauche form of Ethylene glycol is stable due to intramolecular H-bonding it also gives it dipole moment.

CHAPTER-END EXERCISES

Practice Questions – I

1. The IUPAC name of C_6H_5COCl is
(a) Benzoyl chloride
(b) Benzene chloroketone
(c) Benzene carbonyl chloride
(d) Chlorophenyl ketone

2. In the reaction,
$CH_3CHO + HCN \longrightarrow CH_3CH(OH)CN$
A chiral centre is produced. This product would be
(a) Racemic mixture (b) Meso compound
(c) Dextrorotatory (d) Laevorotatory

3. Optical activity is measured by
(a) Refractometer (b) Tracer technique
(c) Spectrograph (d) Polarimeter

4. The compound having only primary hydrogen atoms is
(a) Isobutene (b) 2,3-dimethylbutene
(c) Cyclohexane (d) Propyne

5. Which of the following shows geometrical isomerism?
(a) But-1-ene (b) But-2-ene
(c) 2,3-dichlorobutane (d) Ethene

6. The number of optically active isomers of tartaric acid is
CH(OH)COOH
|
CH(OH)COOH
(a) 2 (b) 3
(c) 4 (d) 1

7. The compound in which C uses only its sp^3 hybrid orbitals for bond formation is
(a) HCOOH (b) $(NH_2)_2CO$
(c) $(CH_3)_3COH$ (d) CH_3CHO

8. Number of isomers which can be obtained theoretically from monochlorination of 2-methylbutane are
(a) 2 (b) 3
(c) 4 (d) 5

9. An organic molecule necessarily shows optical activity if it
(a) Is non-superimposable on its mirror image
(b) Is superimposable on its mirror image
(c) Contains asymmetric carbon atoms
(d) Is non-polar

10. The C–H bond distance is longest in
(a) C_2H_2 (b) C_2H_4
(c) C_2H_6 (d) $C_2H_2Br_2$

11. How many structural isomers of $C_4H_{10}O$ are ethers?
(a) 1 (b) 2
(c) 3 (d) 4

12. Which of the following has zero dipole moment?
(a) 1-butene (b) Cis-2-butene
(c) Trans-2-butene (d) 2 methyl-1-propene

13. The IUPAC name of the following compound is

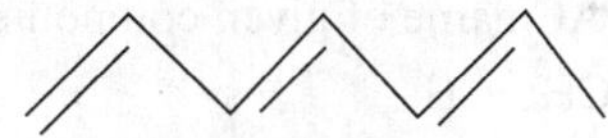

(a) 2,4,5-triheptene
(b) 2,4,6-triheptene
(c) hepta-1,3,5–triene
(d) 2,4,6-heptatriene

14. The most stable conformation of ethylene glycol is
(a) Anti (b) Gauche
(c) Fully eclipsed (d) Partially eclipsed

15. The total number of isomers for C_4H_8 are
(a) 8 (b) 7
(c) 6 (d) 5

16. $C_6H_5C\equiv N$ and $C_6H_5N\rightleftarrows C$ are which type of isomers?
(a) Tautomerism (b) Position
(c) Functional (d) Linkage

17. The IUPAC name of following compound is

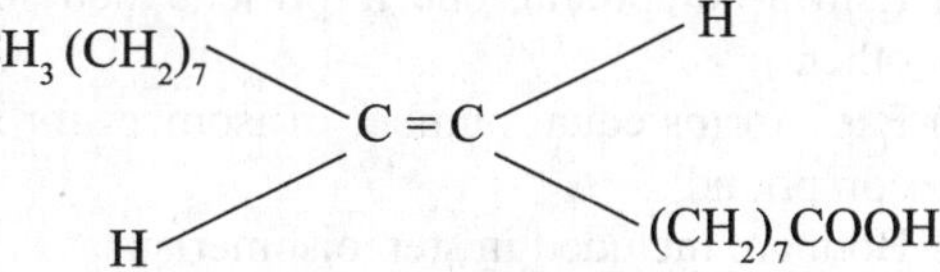

(a) Cis-octadec-9-enoic acid
(b) Trans–octadec-9-enoic acid
(c) Oleic acid
(d) None of these

18. How many stereoisomers are there for tartaric acid?
(a) 3 (b) 4
(c) 5 (d) 0

19. The number of enantiomers of the compound $CH_3CHBrCHBrCOOH$ is

(a) 1 (b) 2
(c) 3 (d) 4

20. The IUPAC name of compoud

$$\begin{array}{ccc} CH_2 - & CH - & CH_2 \\ | & | & | \\ CN & CN & CN \end{array}$$ is

(a) 1, 2, 3-tricyanopropane
(b) Propane-1, 2,3-tricarbonitrile
(c) 3-cyanopentane-1,5-dinitrile
(d) Propane-1, 2, 3-trinitrile.

21. Which of the following compounds will show metamerism?

(a) $CH_3–O–C_2H_5$ (b) $CH_3–O–CH_3$
(c) $CH_3–CO–C_2H_5$ (d) $C_2H_5–S–C_2H_5$

22. What will be the IUPAC name of given compound?

$$\begin{array}{c} CH_3 \qquad\qquad CH_2–CH_3 \\ | \qquad\qquad\quad | \\ CH_3–CH–CH–CH_2–CH–CH_3 \\ | \\ CH_2–CH_3 \end{array}$$

(a) 2,5-diethyl-4-methylhexane
(b) 3,3,6-trimethyloctane
(c) 2,5,6-trimethyloctane
(d) 3,5-dimethyl-6-ethylheptane

23. In the boat conformation of cyclohexane, the most destabilizing interaction is

(a) Eclipsing
(b) Flagpole–flagpole
(c) 1,3-diaxial
(d) 1,3-diequatorial

24. A similarity between optical and geometrical isomerism is that

(a) If in a compound, one is present then so is the other.
(b) Each forms equal number of isomers for a given compound
(c) Both are included in stereoisomerism
(d) They have no similarly

25. The IUPAC name of $CH_3COCH(CH_3)_2$ is

(a) 4-methyl isopropyl ketone
(b) 3-methylbutan-2-one
(c) isopropylmethyl ketone
(d) 2-methylbutan-3-one

26. (+)–mandelic acid has a specific rotation of +158°. What would be the observed specific rotation of a mixture containing 25 % (–)-mandelic acid and 75 % (+)-mandelic acid?

(a) + 79° (b) –118.5°
(c) –79° (d) +118.5°

27. The IUPAC name of the following compound is

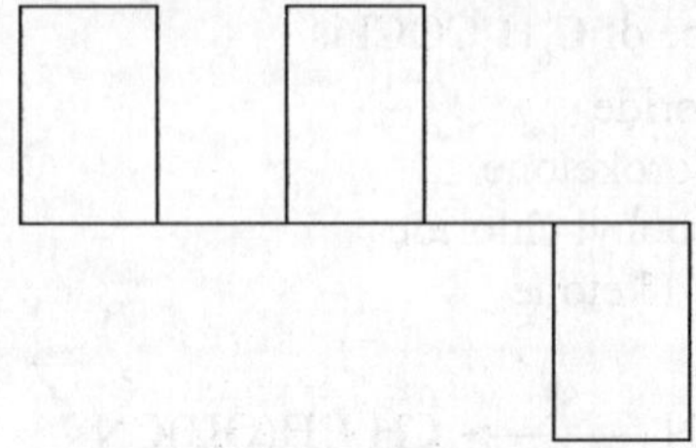

(a) Biscyclobutylcyclobutane
(b) Dicyclobutylcyclobutane
(c) 1, 1', 2', 1"-tercyclobutane
(d) Tercyclobutane

28. Among the following, the most stable compound is

(a) Cis-1, 2-cyclohexanediol
(b) Trans-1, 2-cyclohexanediol
(c) Cis-1, 3-cyclohexenediol
(d) Trans-1,3-cyclohexanediol

29. The number of isomers possible for C_7H_8O are

(a) 3 (b) 4
(c) 5 (d) 6

30. The number of geometrical isomers of $CH_3CH=CH–CH=CH–CH=CHCl$ are

(a) 4 (b) 6
(c) 8 (d) 10

31. C_3H_9N cannot represent

(a) Quaternary salt (b) 1° amine
(c) 2° amine (d) 3° amine

32. The IUPAC name of the following compound is

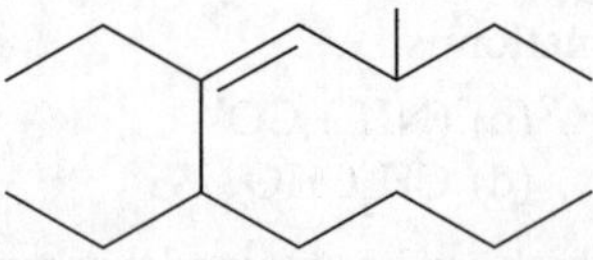

(a) 5, 6-diethyl-8-methyldec-6-ene
(b) 5, 6-diethyl-3-methyldec-4-ene
(c) 6-butyl-5-ethyl-3-methylloct-4-ene
(d) 4, 5-diethyl-8-methyldec-6-ene

33. Which of the following is optically active?

(a) Butane
(b) 2-methylpentane
(c) 4-methylpent ane
(d) 3-methylheptane

34. The hybridization of carbon atoms in C–C single bond of $HC{\equiv}C{-}CH{=}CH_2$ is
(a) sp^3–sp^3 (b) sp^2–sp^3
(c) sp–sp^2 (d) sp^3–sp

35. Number of stereoisomers of the compound, 2-chloro-4-methylhex-2-ene is
(a) 2 (b) 4
(c) 6 (d) 7

36. Number of π-electrons in cyclobutadienyl anion $(C_4H_4)^{2-}$ is
(a) 2 (b) 4
(c) 6 (d) 8

37. Racemic mixture is formed by mixing two
(a) Isomeric compounds
(b) Chiral compounds
(c) Meso compounds
(d) Enantiomers with chiral carbon

38. The number of primary amines of formula $C_4H_{11}N$ are
(a) 2 (b) 3
(c) 4 (d) 5

39. The IUPAC name of the following compound is

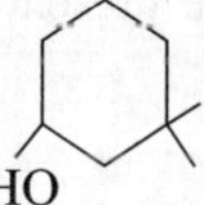

(a) 3,3-dimethyl-1-hydroxycyclohexane
(b) 1,1-dimethyl-3-hydroxycyclohexane
(c) 3,3-dimethyl-1-cyclohexanol
(d) 1,1-dimethyl-3-cyclohexanol.

40. The two optical isomers given below, namely

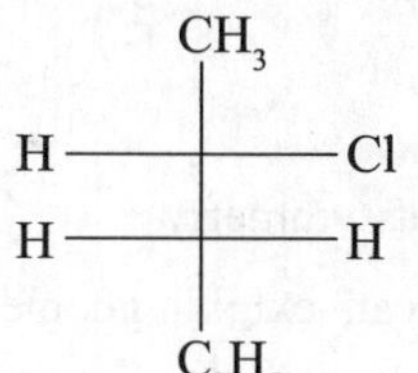

and

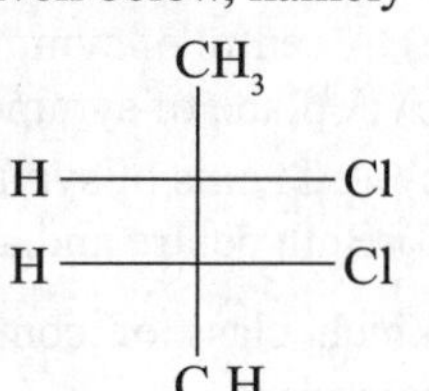

are

(a) Enantiomers
(b) Geometrical isomers
(c) Structural isomers
(d) Diastereomers

41. The C–C bond length of the following molecules are in the order
(a) $C_2H_6 > C_2H_4 > C_6H_6 > C_2H_2$
(b) $C_2H_2 > C_2H_4 > C_6H_6 > C_2H_6$
(c) $C_2H_6 > C_2H_2 > C_6H_6 > C_2H_4$
(d) $C_2H_4 > C_2H_6 > C_2H_2 > C_6H_6$

42. An isomer of ethanol is
(a) Methanol
(b) Diethyl ether
(c) Acetone
(d) Dimethyl ether

43. An organic molecule necessarily shows optical activity if it
(a) Contains asymmetric carbon atoms
(b) Is non-planar
(c) Is non-superimposable on its mirror image
(d) Is superimposable on its mirror image

44. The IUPAC name of the given compound is

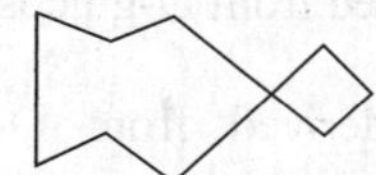

(a) 1, 1 –Cyclobutylheptane
(b) Bicyclo[6, 3, 0]nonane
(c) Sipro [3, 6] decane
(d) Spiro [3, 5] decane

45. The IUPAC name of the following compound is

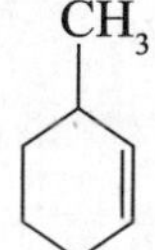

(a) 3-methyl cyclohexene
(b) 1-methyl cyclohex-2-ene
(c) 6-methyl cyclohexene
(d) 1-methyl cyclohex-5-ene

46. The maximum number of carbon atoms arranged linearly in the molecule, $CH_3 - C \equiv C - CH = CH_2$ are
(a) 3 (b) 4
(c) 5 (d) 6

47. The IUPAC name of cinnamic acid is
(a) Phenylpropanoic acid
(b) 3-phenylprop-2-enoic acid
(c) 3-phenylpropanoic acid
(d) Benzene-1, 4-dicarboxylic acid

48. IUPAC name of the given compound is

$$\begin{array}{l} CH_3 - \underset{\displaystyle CH_3}{\underset{|}{CH}} - CH_2 - \underset{\displaystyle CN}{\underset{|}{CH}} - CH_3 \end{array}$$

(a) 2-cyano- 4-methylpentane
(b) 2-methyl- 4-cyanopentane
(c) 2, 4-dimethyl pentane nitrile
(d) 2-dimethyl- 4-cyanopentane

49. The IUPAC name of the compound $CH_3 - C(CH_3)_2 - CH = C(CH_3)_2$ is
(a) 1, 1, 3, 3-tetramethyl-but-1-ene
(b) 1, 3, 3-trimethyl-pent-2-ene
(c) 2, 2, 4-trimethylbut-4-ene
(d) 2, 4, 4-trimethylpent-2-ene

50. One recently discovered allotrope of carbon (C_{60}) is commonly known as
(a) Fluorine (b) Fluorene
(c) Freon (d) Fullerene

51. The optically active tartaric acid is named as D-(+)-tartaric acid because it has a positive
(a) Optical rotation and is derived from D-glucose
(b) pH in an organic solvent
(c) Optical rotation and is derived from D-(+)-glyceraldehyde
(d) Optical rotation only when substituted by deuterium

52. The correct IUPAC name of the compound, is

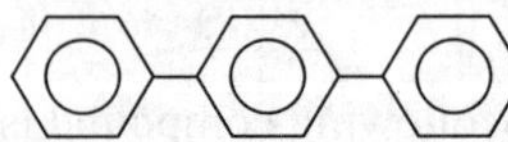

(a) p–phenyldiphenyl
(b) p, 1-diphenylbenzene
(c) 1, 1′, 4′. 1″-terphenyl
(d) terphenyl

53. Keto – enol tautomerism is observed in
(a) C_6H_5CHO
(b) $C_6H_5OCH_3$
(c) $C_6H_5COCH_2COOH$
(d) $C_6H_5COC_6H_5$

54. The IUPAC name of the following compound is

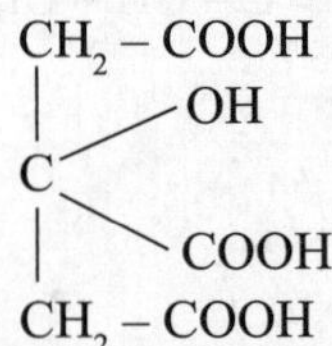

(a) 1, 2, 3-tricarboxy-2-hydroxy propane
(b) 3-carboxy-3-hydroxy-1, 5-pentane dioic acid
(c) 3-hydroxy-3-carboxy pentan1, 5-dioic acid
(d) 3-hydroxypropane-1, 2, 3-tricarboxylic acid

55. The given compound can exhibit

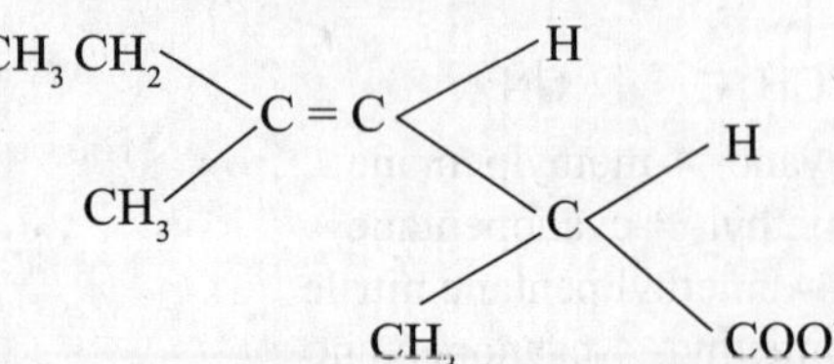

(a) Geometrical isomerism
(b) Optical isomerism
(c) Geometrical and optical isomerism
(d) Conformational isomerism

56. Maximum enol content is observed in
(a) $CH_3COCH_2COOC_2H_5$
(b)

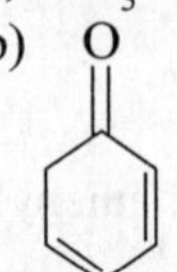

(c) $CH_3COCH_2COCH_3$
(d) CH_3COCH_3

57. Assign the IUPAC name to $[(CH_3)_2CH]_3COH$.
(a) 3-(2-methylethyl)-3(1-methylethyl) pentan-2-ol
(b) tris-(1-methylpropyl)methanol
(c) 2, 4-dimethyl-3 (1-methylethyl) pentan- 2-ol
(d) 3-(1-methylethyl)-2, 4-dimethylpentan-3-ol

58. Which will exhibit geometrical isomerism here?
(a) $A_2C{=}CAB$ (b) $A_2C{=}CB_2$
(c) $ABC{=}CAB$ (d) $ABC{=}CX_2$

59. How many tertiary carbon atoms are present in the given compound?

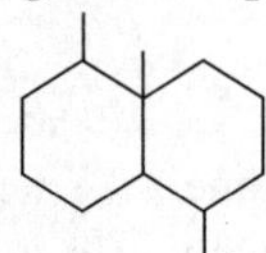

(a) 3 (b) 4
(c) 2 (d) 5

60. 2-methylpenta-2,3-diene is achiral as it has
(a) A centre of symmetry
(b) A plane of symmetry
(c) A C_2 axis of symmetry
(d) Both centre and a plane of symmetry

61. Which class of compound can exhibit geometrical isomerism?
(a) $CH_3CH{=}CHCH_3$
(b) $C_6H_5CH{=}NOH$
(c) COOH–CH
‖
COOH–CH
(d) All of these

62. How many σ and π-bonds are there in the molecule of tetracyanoethylene?
(a) 9σ and 9π (b) 5σ and 9π
(c) 9σ and 7π (d) 5σ and 8π

63. The bond between carbon atom (1) and carbon atom (2) in compound N≡C–CH=CH_2 involves the hybridization

(a) sp^2 and sp^2 (b) sp^3 and sp
(c) sp and sp^2 (d) sp and sp

64. How many secondary carbon and hydrogen atoms are present in the molecule given below respectively?

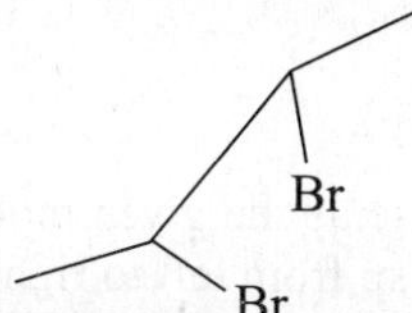

(a) 2, 3 (b) 2, 2
(c) 3, 3 (d) 2, 0

65. Which IUPAC name is correct for the given compound?

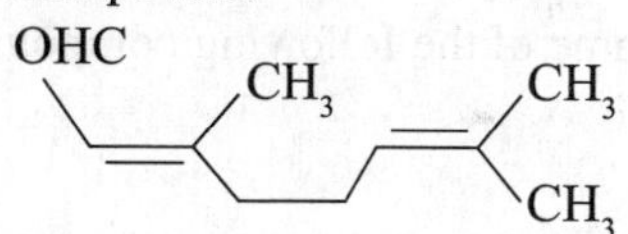

(a) 3, 7-dimethyloct-2, 6-dienal
(b) 2, 6-dimethyloct-2, 6-dienal-8
(c) 7-formyl-2, 6-dimethylhept-2, 6-diene
(d) 7-aldo-2, 6-dimethylhept-2, 6-diene

PRACTICE QUESTIONS – II

66. The compound given below has how many primary and secondary hydrogen atoms respectively?

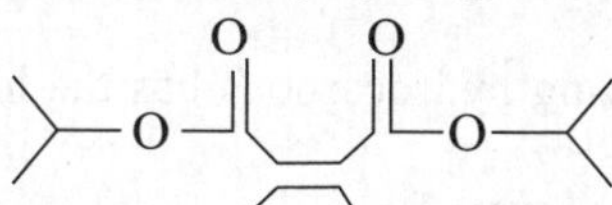

(a) 18, 1 (b) 18, 2
(c) 16 , 2 (d) 18, 4

67. The total number of optically active isomers for $CH_2OH(CHOH)_3CHO$ are

(a) 16 (b) 8
(c) 4 (d) 2

68. If optical rotation produced by compound (1) is 36° then the rotation produced by compound (2) is

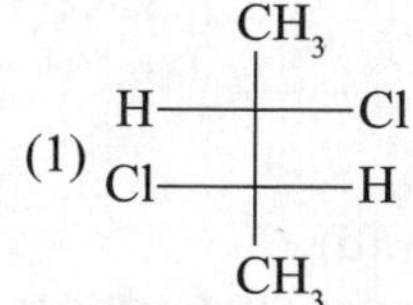

(2) with CH_3 at top, H–C–Cl, H–C–Cl, CH_3 at bottom

(a) –36° (b) 0°
(c) +18° (d) +72°

69. Total number of isomers and optically active compounds in the isomers of C_3H_5Br are

(a) 6, 4 (b) 5, 3
(c) 3, 5 (d) 8, 3

70. On monochlorination of n-pentane, the number of isomers formed is are

(a) 4 (b) 3
(c) 2 (d) 1

71. The number of possible open chain (acyclic) isomeric compounds for molecular formula C_5H_{10} would be

(a) 5 (b) 6
(c) 7 (d) 4

72. The IUPAC name of the following compound is

$H_2N–N=$ (cyclohexane ring) –COOH

(a) 4-hydrazonocyclohexane-1-carboxylic acid
(b) 4-hydrazonobenzoic acid
(c) 4-hydrazonocyclohexanoic acid
(d) 4-(n-amino) benzenecarboxylic acid

73. The compound (1) and (2) can be regarded as

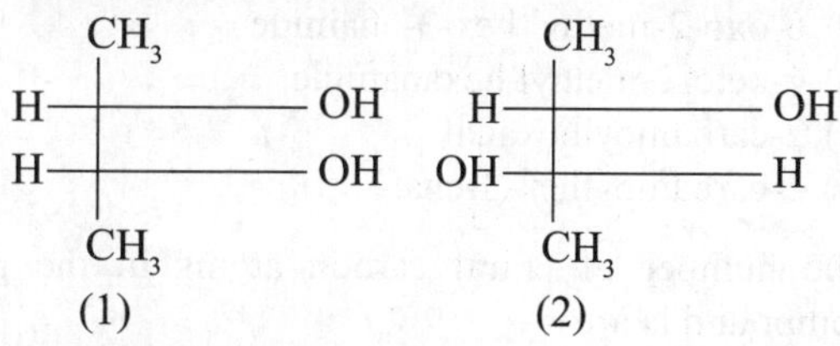

(a) Enantiomers
(b) Diastereomers
(c) Mesomers
(d) Geometrical isomer

74. Consider the following organic compound,

$\overset{1}{CH_3}–\overset{2}{CH_2}–\overset{3}{CH_2}–\overset{4}{CH_2}–\overset{5}{CH_2}–\overset{6}{CH_2}–\overset{7}{CH_3}$

To make it a chiral compound, the attack should be on carbon

(a) 1 (b) 4
(c) 3 (d) 6

75. Assign the IUPAC name for the following compound.

Cl–(benzene ring)–N=N–(benzene ring with Cl)

(a) 3, 4-dichloroazobenzene
(b) (4-chlorophenyl) (3-chlorophenyl) diazene
(c) 3, 4-bis(chlorophenyl) diazene
(d) (3-chlorophenyl) (4-chlorophenyl) diazene

76. Write the IUPAC name of the following compound.

$CO-O-C_2H_5$

$CO-Cl$

(a) Ethyl-2-(chlorocarbonyl) benzoate
(b) Ethyl-2-(chlorocarbonyl) hexanoate
(c) 2-(ethoxycarbonyl) benzoyl chloride
(d) None of these

77. Assign the IUPAC name for the following compound.

$$\begin{array}{c} CH_2-C\equiv N \\ | \\ N\equiv C-CH_2-CH_2-CH-CH_2-C\equiv N \end{array}$$

(a) 3-methyl hexane-1, 2, 3-tricarbonitrile
(b) Hexane-1,2,3-tricarbonitrile
(c) 1, 2-bis (cyanomethyl) butanenitrile
(d) 3-(cyanomethyl) hexane dinitrile

78. The IUPAC name of the compound is

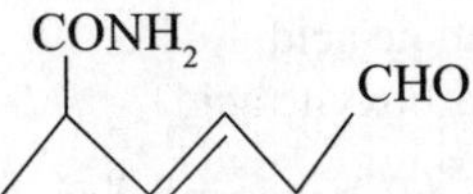

(a) 6-oxo-2-methyl hex-3-enamide
(b) 6-keto-2-methyl hexanamide
(c) 2-carbamoylhexanal
(d) 2-carbamoylhex-3-enal

79. The number of chiral carbon atoms in the given compound is are

Br

CH_2CH_3

(a) 2 (b) 3
(c) 4 (d) 1

80. The correct decreasing order of priority for the functional groups of organic compounds in the IUPAC system of nomenclature is

(a) $-COOH, -SO_3H, -CONH_2, -CHO$
(b) $-SO_3H, -COOH, -CONH_2, -CHO$
(c) $-CHO, -COOH, -SO_3H, -CONH_2$
(d) $-CONH_2, -CHO, -SO_3H, -COOH$

81. In the following reaction,

$$CH_3CHO + HCN \longrightarrow CH_3CH(OH)CN \xrightarrow{H^+/H_2O} CH_3CH(OH)COOH$$

An asymmetric centre is generated. The acid obtained would be
(a) L-isomer
(b) D-isomer
(c) 20 % D + 80% L-isomer
(d) 50% D + 50 % L-isomer

82. Which of the following represents the given mode of hybridization $sp^2 - sp^2 - sp - sp$ from left to right?
(a) $CH_2 = CH - C \equiv NCH$
(b) $CH \equiv C - C \equiv N$
(c) $CH_2 = C = C = CH_2$
(d) CH_2 / CH_2

83. Which of the following hydrocarbons has the lowest dipole moment?

(a) CH_3, CH_3, C = C, H, H

(b) $CH_3 - C \equiv C - CH_3$
(c) $CH_3\ CH_2\ CH = CH_2$
(d) $CH_2 = CH - C \equiv CH$

84. How many optically active isomers are possible for the compound shown below?

CH_3 CH_3 Br CH_2Br

(a) 4 (b) 2 (c) 8 (d) 6

85. In the compound, $CH_2{=}CH{-}CH_2{-}CH_2{-}C{\equiv}CH$, the $C_2{-}C_3$ bond is of the type
(a) $sp{-}sp^2$ (b) $sp^3{-}sp^3$
(c) $sp{-}sp^3$ (d) $sp^2{-}sp^3$

86. Which type of conformation is shown by figure 1 and 2?

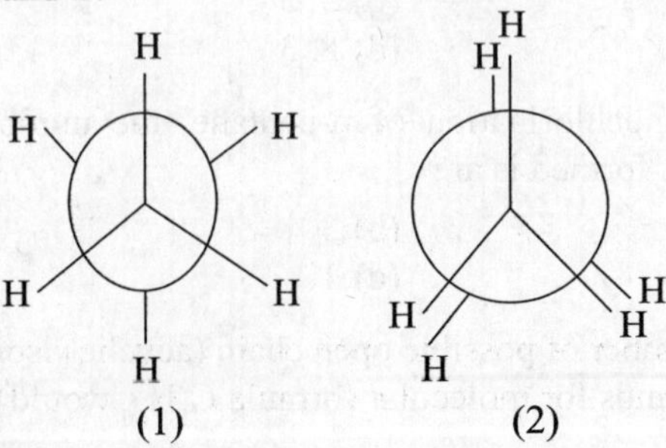

(a) 1 is eclipsed and 2 is staggered
(b) 1 is staggered and 2 is eclipsed
(c) 1 is eclipsed and 2 is gauche
(d) Both are gauche

87. An organic compound C_4H_8O is found to be optically active. Which of the following is correct structure of the given compound?

(a) $(CH_3)_2CHCHO$
(b) $CH_2 = CH - CH(OH)CH_3$
(c) $CH_3COCH_2CH_3$
(d) $CH_3CH_2CH_2CHO$

88. The number of optical isomers formed by hydrogenation of the compound, $(CH_3)_2C = CHCH_3$ are

(a) 0 (b) 1
(c) 2 (d) 3

89. Which of the following can exhibit optical isomerism?

(1) 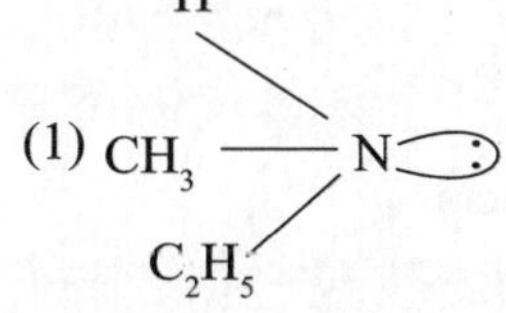

(2) $\ce{H(Cl)C=C=C=C(H)Cl}$

H╲ ╱H
C = C = C = C
Cl╱ ╲Cl

(3) 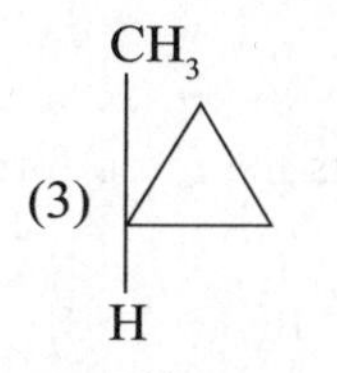

(4) 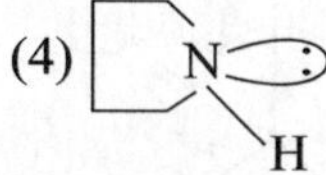

(a) 1 and 2 (b) 2 and 4
(c) 3 only (d) 2 only

90. The number and type of bonds between two carbon atoms in CaC_2 are

(a) One sigma and one π bond
(b) One sigma and two π bonds
(c) One sigma and one and a half π bonds
(d) One sigma bond

91. Which of the following has the least hindered rotation about carbon-carbon bond?

(a) Ethane (b) Ethylene
(c) Acetylene (d) Hexachloroethane

92. $CH_3-\overset{O}{\overset{\|}{C}}-CH_3$ (1) $CD_3-\overset{O}{\overset{\|}{C}}-CH_3$ (2) $CD_3-\overset{O}{\overset{\|}{C}}-CD_3$ (3)

Among these ketones which of the following is correct order of the case of their enolizations?

(a) 1 > 2 > 3 (b) 3 > 2 >1
(c) 2 > 1 > 3 (d) 3 > 1 = 2

93. Which of the following does not show geometrical isomerism?

(a) 1,1-dichloro-1-pentene
(b) 1, 2-dichloro-1-pentene
(c) 1, 3-dichloro-1-pentene
(d) 1,4-dichlroro-2-pentene

94. Ibuprofen has the following structure, this compound shows

$CH_3-\underset{}{\overset{CH_3}{\overset{|}{CH}}}-CH_2-$⟨benzene ring⟩$-\overset{CH_3}{\overset{|}{CH}}-\overset{O}{\overset{\|}{C}}-OH$

(a) Geometrical isomerism
(b) Optical isomerism
(c) Tautomerism
(d) Both (a) and (b)

95. The compound with an isopropyl group is

(a) 2, 2, 3, 3-tetramethylpentane
(b) 2, 2-dimethylpentane
(c) 2, 2, 3-trimethylpentane
(d) 2-methylpentane

96. The enolic form of acetone contains

(a) 9 sigma bonds, 1 pi bond and 2 lone pair of electrons
(b) 8 sigma bonds, 2 pi bonds and 2 lone pair of electrons
(c) 10 sigma bonds, 1 pi bond and 1 lone pair of electrons
(d) 9 sigma bonds, 2 pi bonds and 1 lone pair of electrons

97. How many chiral carbon atoms are present in 2, 3, 4-trichloropentane?

(a) Three (b) Two
(c) One (d) Four

98. The IUPAC name of $CH_3-\underset{\underset{CH_2-CH_3}{|}}{CH}-CHO$ is

(a) Butan-2-aldehyde
(b) 2-methylbutanal
(c) 3-methylisobutyraldehyde
(d) 3-methylbutanal

99. An optically active compound is

(a) 1-bromobutane
(b) β-bromobutyric acid
(c) 2-bromo-2-methylpropane
(d) 1-bromo-2-methylpropane

100. On monochlorination of 2-methyl butane, the total number of chiral compounds is

(a) 2 (b) 4
(c) 6 (d) 8

101. Which of the following compounds exhibit steroisomerism?

(a) 2-methylbutene-1
(b) 3-methylbutyne-1
(c) 3-methylbutanoic acid
(d) 2-methylbutanoic acid

102. The number of isomers for the compound with molecular formula $C_2BrClFI$ is

(a) 3 (b) 4
(c) 5 (d) 6

103. Which of the following compounds will exhibit geometrical isomerism?

(a) 1-phenyl-2-butene
(b) 3-phenyl-1-butene
(c) 2-phenyl-1-butene
(d) 1,1-diphenyl-1-propene

104. Which of the following compounds will show geometrical isomerism?

1. 2-butene 2. propene
3. 1-phenylpropene 4. 2-methylbut-2-ene

(a) 1, 2 (b) 3, 4
(c) 1, 2, 3 (d) 1, 3

105. How many optically active stereoisomers are possible for butan-2, 3-diol?

(a) 1 (b) 2
(c) 3 (d) 4

106. Choose the correct statement about the compounds A, B and C.

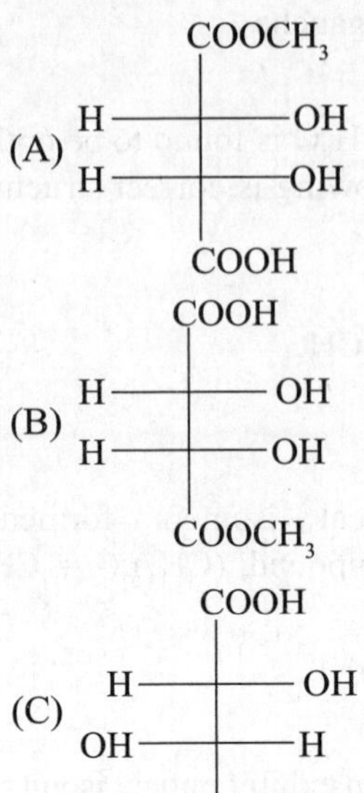

(a) A and B are identical
(b) A and B are diastereomers
(c) A and C are enantiomers
(d) A and B enantiomers

107. The number of possible enantiomeric pairs than can be produced during monochlorination of 2-methyl butane is

(a) 2 (b) 3
(c) 4 (d) 1

108. Which of the following represents the meso form of 2, 3-dibromobutane?

(a) Br; H, Br; CH$_3$, H; CH$_3$
(b) Br; H, CH$_3$; CH$_3$, H; Br
(c) Br; H, H; Br, CH$_3$; CH$_3$
(d) Br; H, Br; CH$_3$, H; H

Practice Questions – III

109. The favoured conformation (Gauche, anti or eclipsed) for the molecule 1, 2-dichloro ethene 1, 2-ethanediol and propanaldehyde will be respectively

(a) Anti, eclipsed, gauche (b) Anti, gauche, eclipsed
(c) Eclipsed, anti, gauche (d) Gauche, anti, eclipsed

110. Which is true regarding the $HOCH_2 - (CHOH)_4 - CH_2OH$?

(1) The number of racemic mixtures will be four for the compound.
(2) The total number of optical isomers is ten and number of meso isomer is two.
(3) The total number of optical isomer is sixteen and number of meso isomer is zero.
(4) The optically active isomers are eight while meso isomers are two.

(a) 1, 2 and 3 (b) 2, 3 and 4
(c) 1, 2 and 4 (d) 1 and 2 only

111. Consider the following compounds given below and select the correct statement.

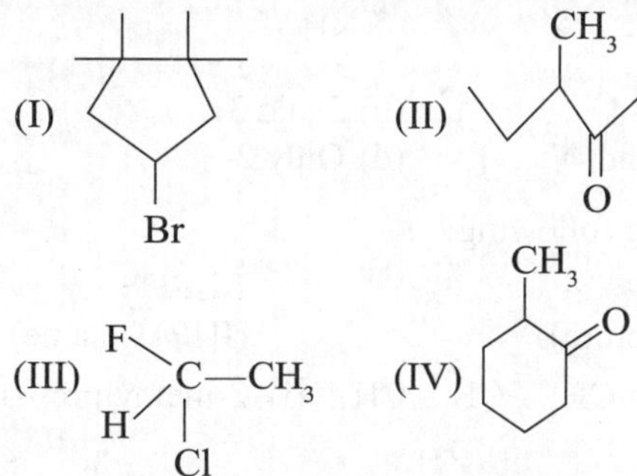

(a) Except (I) all have stereogenic centre
(b) All have stereogenic centre
(c) All have plane of symmetry
(d) Only II and III have stereogenic centre

112. Match the following:

List I	List II
1. CH_3COOH and $HCOOCH_3$	(i) Metamers
2. $CH_3–CH_2–C≡CH$ and $CH_3–C≡C–CH_3$	(ii) Position isomers
3. $CH_3CH_2CH_2NH_2$ and $CH_3–CH(NH_2)–CH_3$	(iii) Tautomers
4. CH_3CH_2OH and $(CH_3)_2O$	(iv) Functional isomer

The correct matching is:

	1	2	3	4
(a)	(iii)	(iv)	(i)	(iv)
(b)	(i)	(ii)	(iii)	(iv)
(c)	(iii)	(ii)	(i)	(iv)
(d)	(iv)	(ii)	(ii)	(iv)

113. Consider the following statements:

(1) Racemic forms are not optically active but are resolvable
(2) The presence of chiral atom is sufficient but not the necessary condition for enantiomer
(3) Meso compounds are inactive and are not resolvable
(4) Conformational enantiomers can be resolved

Now select the correct statements.

(a) 1, 2 and 3 (b) 2, 3 and 4
(c) 1, 2 and 4 (d) 1, 2, 3 and 4

114. Identify E and Z configurations of the following:

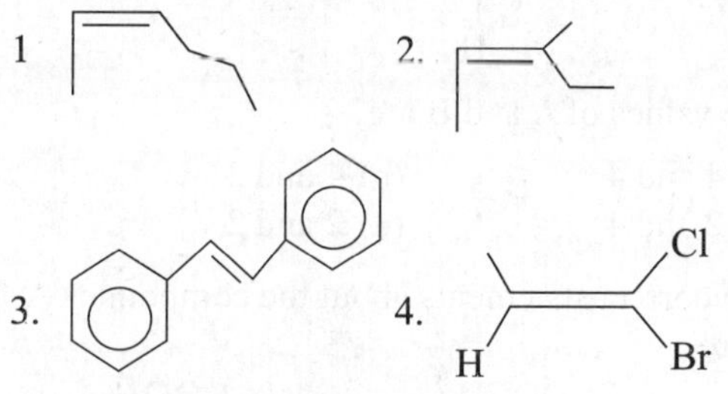

(a) 1 (E), 2 (Z), 3 (E), 4 (Z)
(b) 1 (Z), 2 (E), 3 (Z), 4 (E)
(c) 1 (Z), 2 (Z), 3 (E), 4 (E)
(d) 1 (Z), 2 (Z), 3 (E), 4 (Z)

115. Which is incorrectly matched?

Common name	IUPAC name
(a) Vanillin	4-hydroxy-3-methoxybenzaldehyde
(b) Protocatechualdehyde	2, 3-dihydroxy-benzaldehyde
(c) Adipic acid	Hexan1, 6-dioic acid
(d) Pyruvic acid	Cis-2-methyl-but-2-enoic acid

116. How many structural and geometrical isomers are possible for dimethyl cyclohexane?

(a) 3, 3 (b) 3, 6
(c) 6, 6 (d) 6, 3

117. Consider the following statements about chirality:

1. Molecules which are not superimposable on their mirror images are achiral
2. A chiral molecule can have simple axis of symmetry
3. A carbon atom to which four different species are attached is a chiral centre.
4. A compound whose molecules are achiral exhibits optical activity

Which of the statements given above are correct?

(a) 1, 2 and 4
(b) 2, 3 and 4
(c) 2 and 3
(d) 1 and 4

118. The compound, whose stereochemical formula is shown below, exhibits A-geometrical isomers and B-optical isomers

CH_3(Br)C=CH–CH(OH)–CH_3

The values of A and B are

(a) 4 and 4 (b) 4 and 2
(c) 2 and 4 (d) 2 and 2

119. The correct statements about the compounds A, Band C are

(A) $COOCH_3$; H–C–OH; H–C–OH; COOH
(B) COOH; H–C–OH; H–C–OH; $COOCH_3$
(C) COOH; H–C–OH; HO–C–H; $COOCH_3$

(a) A and B are identical
(b) A and B are enantiomers
(c) A and C are enantiomers
(d) A and B are diastereomers

120. Tautomerism is exhibited by which of the following compounds?

(1) O=⟨cyclohexane⟩=O (2) ⟨cyclohexane⟩=C=NOH
(3) cyclohexane-1,3,5-trione (4) O=⟨cyclohexa-2,5-diene⟩=O

(a) 1, 2 and 3 (b) 1 and 3
(c) 1 and 2 (d) 1, 2, 3 and 4

121. Which of the following compounds are chiral?

(1) CH_3(H)C=C=C(CH_3)CH_3
(2) CH_3(H)C=C=C(H)CH_3
(3) CH_3(CH_3)⟨cyclohexane⟩=C(CH_3)CH_3
(4) 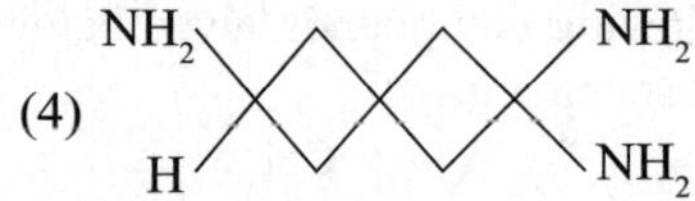

(a) 2 and 4 (b) 2 and 3
(c) 2, 3 and 4 (d) Only 2

122. Match the following:

List I (Structure)	List II (IUPAC name)
1. $CH_3 - CH_2 - CH = CH_2$	(i) 2-methylprop-1-ene
2. H_3C(H)C=C(H)CH_3	(ii) But-1-ene
3. H_3C(H_3C)C=CH_2	(iii) Cis-but-2-ene
4. H_3C(H)C=C(H)CH_3	(iv) Trans-but-2-ene

The correct matching is:

	1	2	3	4
(a)	(iii)	(ii)	(iv)	(i)
(b)	(ii)	(iii)	(i)	(iv)
(c)	(iii)	(iv)	(ii)	(i)
(d)	(ii)	(iv)	(i)	(iii)

123. Which of the following compound will exhibit optical isomerism?

(1) 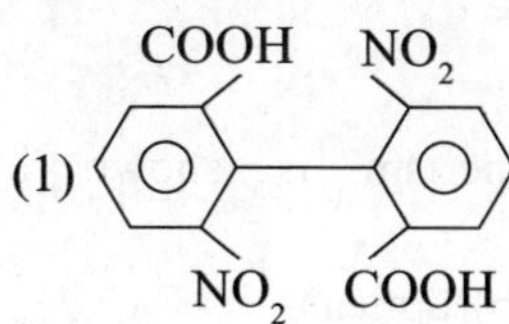

(2) H–C(OH)(COOH)–CH_3

(3) HOOC(H)C=C(H)COOH

(4) C_6H_5–CO–C_6H_5

(a) 1 and 3 (b) 1 and 2
(c) 1, 2 and 4 (d) 1 only

124. Which of the following have asymmetric carbon atoms?

1. $ClCH_2 - CH_2Br$
2. $CH_3 - CHCl_2$
3. $CH_3 - CH\,DCl$
4. $CH_2Br - CHOH - CH_3$

(a) 1, 2, 3 (b) 1, 3, 4
(c) 2, 3 (d) 3, 4

125. Tautomerism is exhibited by

1. C_6H_5–CH = CH–OH
2.
3.
4.

(a) 1 and 2 (b) 1, 3 and 4
(c) 1, 2 and 4 (d) 1, 2, 3 and 4

126. Underlined carbon is sp hybridized in

(a) $CH_3CH = CH_2$ (b) $CH_3CH_2NH_2$
(c) CH_3CONH_2 (d) CH_3CH_2CN

127. The incorrect IUPAC name is

(a) 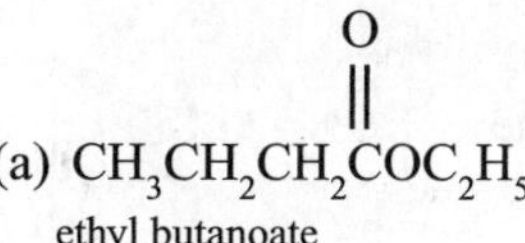

$CH_3CH_2CH_2COC_2H_5$ (C=O)
ethyl butanoate

(b) $CH_3CH(CH_3)CH_2CHO$
3-methyl butanal

(c) $CH_3CH(CH_3)COCH_2CH_3$
2-methyl-3-pentanone

(d) $CH_3CH(CH_3)CH(OH)CH_3$
2-methyl-3-butanol

128. Following types of compounds are studied under which type of isomerism?

and

(a) Chain isomerism
(b) Position isomerism
(c) Conformers
(d) Stereoisomerism

129. Among the following four structures 1 to 4

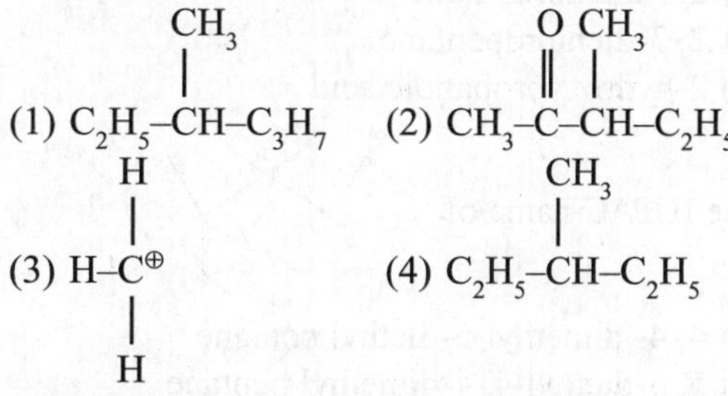

(1) C_2H_5–CH(CH_3)–C_3H_7 (2) CH_3–C(=O)–CH(CH_3)–C_2H_5
(3) H–$C^{\oplus}$(H)–H (4) C_2H_5–CH(CH_3)–C_2H_5

It is true that

(a) All four are chiral compounds
(b) Only I and II are chiral compounds
(c) Only III is a chiral compound
(d) Only II and IV are chiral compounds

130. Amongst the following compounds, the optically active alkane having lowest molecular mass is

(a) $CH_3–CH_2–CH_2–CH_3$

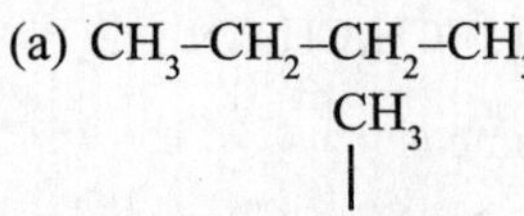

(b) $CH_3–CH_2–CH(CH_3)–CH_3$

(c) 2-cyclopropylbutane

(d) $CH_3–CH_2–C{\equiv}CH$

131. Which of the following molecules is expected to rotate the plane of plane polaized light?

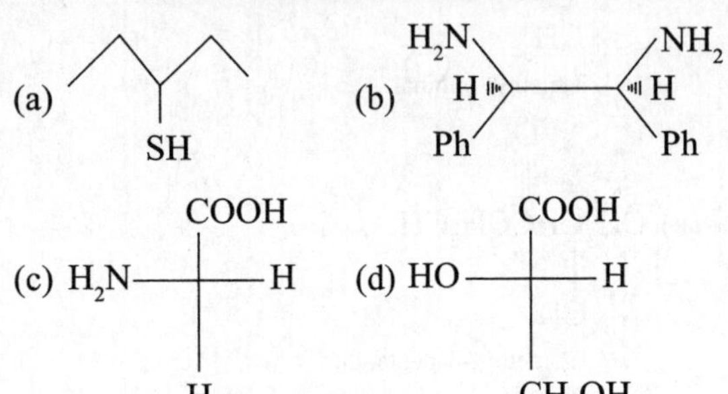

132. Which one of the following conformations of cyclohexane is chiral?

(a) Rigid (b) Chair
(c) Boat (d) Twist boat

133. The absolute configuration of

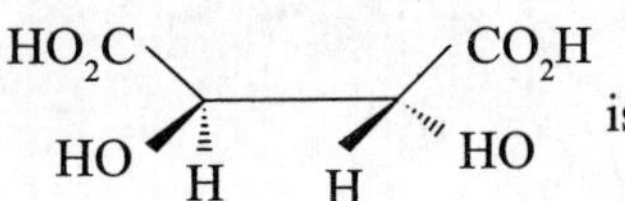

is

(a) S, S (b) R, R
(c) R, S (d) S, R

134. Which of the following will have a meso isomer also?

(a) 2-chlorobutane
(b) 2,3-dichlorobutane
(c) 2, 3-dichloropentane
(d) 2-hydroxypropanoic acid

135. The IUPAC name of is

(a) 4, 4- dimethyl-5-diethyl pentane
(b) 5, 5-diethyl-4, 4-dimethyl pentane
(c) 3-ethyl-4, 4-dimethyl heptane
(d) 1,1-diethyl-2, 2-dimethyl pentane

136. Increasing order of stability among the three main conformations (eclipse, anti, gauche) of 2-fluoroethanol is

(a) eclipse, gauche, anti
(b) gauche, eclipse, anti
(c) eclipse, anti, gauche
(d) anti, gauche, eclipse

137. Which types of isomerism is shown by 2, 3-dichlorobutane?

(a) diastereo (b) optical
(c) geometric (d) structural

138. The structure of cis bis (propenyl) ethene can be given as

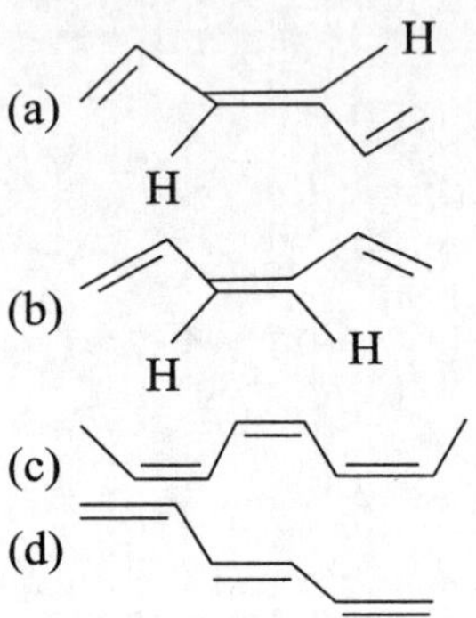

139.

H Me
H
H
Me

Hydrogenation of this compound in presence of poisoned 'Pd' catalyst gives

(a) An optically active compound
(b) A racemic mixture
(c) A diastereomeric mixture
(d) An optically inactive compound.

140. In the given conformation, C_2 is rotated about $C_2–C_3$ bond anti-clockwise by an angle of 120° then the formation obtained is

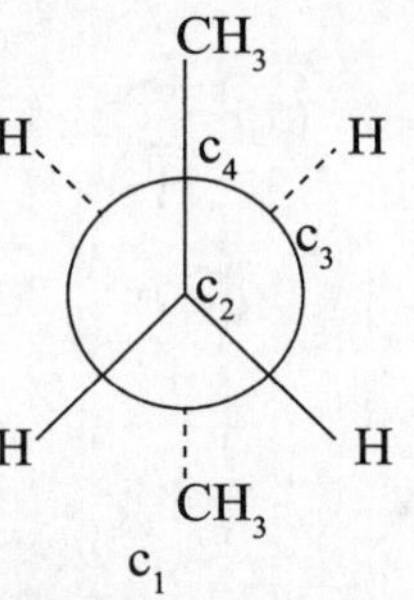

(a) Staggered conformation
(b) Gauche conformation

(c) Fully eclipsed conformation
(d) Partially eclipsed conformation.

141. The two isomers given below are

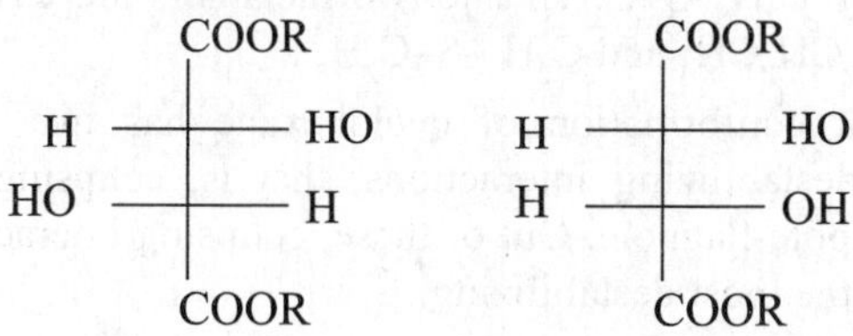

(a) Enantiomer (b) Diastereomer
(c) Mesomer (d) Epimer

142. How many isomers are possible for methyl anthracene?
(a) 1 (b) 2
(c) 3 (d) 4

143. OH NH$_2$

How many stereoisomers are possible for this compound?
(a) 2 (b) 3
(c) 4 (d) 6

144. $CH_3–\underset{\underset{NH_2}{|}}{CH}–CH = CH_2$

has geometrical and optical isomers respectively?
(a) 1, 1 (b) 2, 2
(c) 0, 1 (d) 0, 2

145. Which of the following compound can have chiral C–atom.

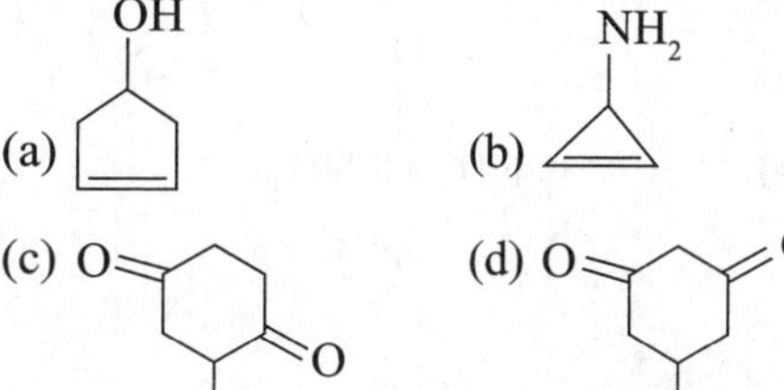

146. How many p°, s° C–atoms are present in the given compound here respectively?

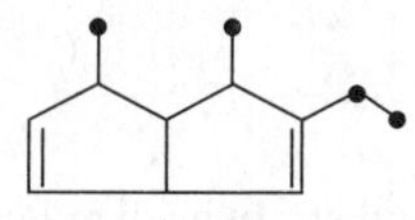

(a) 2, 2 (b) 3, 2
(c) 3, 3 (d) 3, 1

ANSWER KEYS

1. (c)	2. (a)	3. (d)	4. (b)	5. (b)	6. (a)	7. (c)	8. (c)	9. (a)	10. (c)
11. (c)	12. (c)	13. (c)	14. (b)	15. (c)	16. (c)	17. (b)	18. (a)	19. (d)	20. (b)
21. (d)	22. (b)	23. (a)	24. (c)	25. (b)	26. (a)	27. (c)	28. (d)	29. (c)	30. (c)
31. (a)	32. (b)	33. (d)	34. (c)	35. (b)	36. (c)	37. (d)	38. (c)	39. (c)	40. (d)
41. (b)	42. (d)	43. (c)	44. (c)	45. (a)	46. (b)	47. (b)	48. (c)	49. (d)	50. (d)
51. (c)	52. (c)	53. (c)	54. (d)	55. (c)	56. (b)	57. (c)	58. (c)	59. (a)	60. (c)
61. (d)	62. (a)	63. (c)	64. (b)	65. (a)	66. (b)	67. (b)	68. (b)	69. (b)	70. (b)
71. (b)	72. (a)	73. (b)	74. (c)	75. (d)	76. (a)	77. (d)	78. (a)	79. (a)	80. (b)
81. (d)	82. (a)	83. (b)	84. (a)	85. (d)	86. (b)	87. (b)	88. (a)	89. (d)	90. (b)
91. (a)	92. (a)	93. (a)	94. (b)	95. (d)	96. (a)	97. (b)	98. (b)	99. (d)	100. (a)
101. (d)	102. (d)	103. (a)	104. (d)	105. (b)	106. (d)	107. (a)	108. (b)	109. (b)	110. (c)
111. (a)	112. (d)	113. (d)	114. (c)	115. (b)	116. (b)	117. (c)	118. (b)	119. (b)	120. (a)
121. (d)	122. (b)	123. (b)	124. (d)	125. (b)	126. (d)	127. (d)	128. (a)	129. (b)	130. (c)
131. (d)	132. (d)	133. (b)	134. (b)	135. (c)	136. (c)	137. (b)	138. (c)	139. (d)	140. (b)
141. (b)	142. (c)	143. (c)	144. (d)	145. (c)	146. (d)				

Hints and Explanations for Selective Questions

2. As during the reaction, a chiral carbon is created and further since the CN^- ion can attack the planar aldehyde group both from the top and the bottom face of the aldehyde group with equal ease, a 50 : 50 mixture of the two enantiomers, that is a racemic mixture is obtained.

7.
```
   O                  O
   ‖                  ‖
 H–C–OH           NH2–C–NH2
  sp2                sp2
```

```
      CH3
      |
  CH3–C–OH (all sp3 hybridized carbons)
      |
      CH3
```

```
       O
       ‖
  H3C–C–H (sp3 and sp2 hybridized carbons)
  sp3 sp2
```

8.
```
    1     2    3     4
   CH3–CH–CH2–CH3
         |
         1CH3
```
has four types of H-atoms and gives four isomeric monochloro derivatives.

9. Non-superimposable on its mirror image.

10. The size of hybrid orbital of carbon of C–H bond decreases in the order: $sp^3 > sp^2 > sp$. Therefore, C–H bond distance decreases in the same order. Thus C–H bond distance in C_2H_6 is longest.

11. Three, that is, $CH_3OCH_2CH_2CH_3$, $CH_3–O–CH(CH_3)_2$ and $CH_3CH_2OCH_2\ CH_3$.

12. Trans-2-butene has zero dipole moment.

14. Gauche farm is highly stable due to Intra Molecular H-Bonding.

15. Six, that is, 1-butene, trans-2-butene, cis-2-butene, 2-methyl propene, cyclobutane and methyl-cyclopropane.

16. These are functional isomers.

18. Three, that is, d-tartaric acid, l-tartaric acid and meso-tartaric acid.

19. The compound $CH_3\ {}^*CHBr\ {}^*CHBrCOOH$ contains two dissimilar chiral carbon atoms and hence $2^2 = 4$ enantiomers are possible.

21. Only $C_2H_5–S–C_2H_5$. The two metamers are $CH_3–S–CH_2CH_2CH_3$ and $C_2H_5–S–C_2H_5$.

23. Boat conformation of cyclohexane has two types of destabilizing interactions, that is, eclipsing and flagpole-flagpole. Out of these, eclipsing interactions are the most destabilizing.

24. Both are included in stereoisomerism

26. Specific rotation of the mixture
$= (75/100)\ (+158°) + (25/100)\ (–158°)$
$= +79°$

29. Five and these are: benzyl alcohol, anisole, o-, m- and p-cresols (that is, methylphenols).

30. As both the carbon atoms of each of the three double bonds are differently susbstituted, therefore, $2^3 = 8$ geometrical isomers are possible.

31. C_3H_9N cannot represent a quaternary salt.

33. 3-methylheptane,
```
          CH3
          |
  CH3CH2–*CH–CH2CH2CH3
```
has a chiral carbon and it is optically active.

35.
```
  CH3–C=CH–*CH–CH2CH3
      |      |
      Cl     CH3
```
2-chloro-4-methylhex-2-ene contains a chiral carbon as well as a double bond so it will have two geometrical and two optical isomers, that is total fourstereoisomers.

38. Four primary amines are possible. These are: $CH_3CH_2CH_2CH_2NH_2$, $(CH_3)_2CH–CH_2NH_2$, $CH_3CH(NH_2)CH_2CH_3$ and $(CH_3)_3CNH_2$.

40. The two stereoisomers are not mirror images and hence, are diastereomers.

41. C–C bond length > C=C > C≡C. In benzene partial double bond is present and its bond length is in between C–C and C=C.

42. Both ethanol (C_2H_5OH) and dimethyl ether (CH_3OCH_3) have same molecular formula, that is C_2H_6O.

43. Asymmetry is essential condition for optical activity.

46. Two carbons of triple bond and the two carbons directly attached to triple bonded carbons, are linearly arranged. Hence, the total carbon atoms arranged linearly is four.

50. C_{60} is fullerene.

51. (+) sign is used for positive optical rotation. Symbol D is derived from D (+)-glyceraldehyde, that is orientation of the OH group at the second carbon of tartaric acid is the same as in D (+) glyceraldehydes, that is towards right.

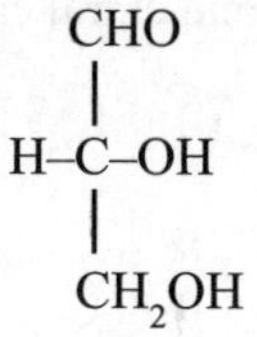

CHO
|
H–C–OH
|
CH_2OH

57. $(CH_3)_2CH–C(OH)(CH(CH_3)_2)–CH(CH_3)_2$

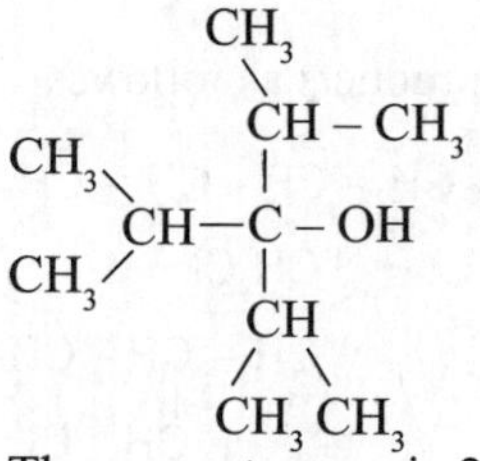

The correct name is 2, 4 dimethyl 3-(1-methyl ethyl) pentan-2-01.

60. $CH_3–C(CH_3)=C=C(H)–CH_3$

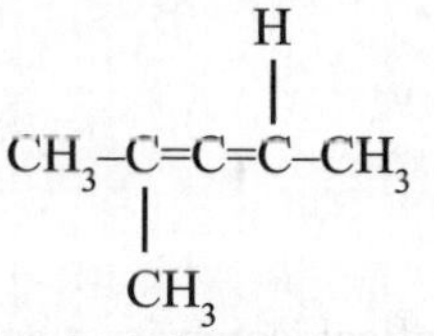

that is, it has symmetry at C_2 center from left side.

66. $(CH_3)_2CH$(s°)–O–CO–CH(CH_3)–CH(CH_3)–CO–O–CH(s°)$(CH_3)_2$

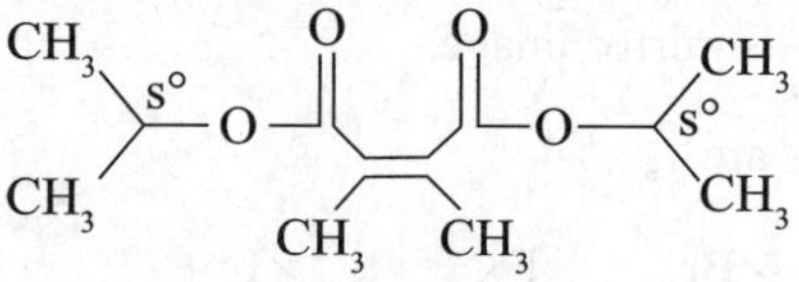

There are 6 p° c-atoms and 18 p°-H-atoms. It also contains 2 s° c-atoms and 2 s° H-atoms

67. As $a = 2^n$.

Here, n = 3

So, $a = (2)^3 = 8$

(Here, n is number of asymmetric carbon atoms and a is number of optically active isomers)

68. As it is a meso compound so it is optically inactive.

70. Here, following three isomers are formed on monochlorination

1. $CH_3CH_2CHClCH_2CH_3$
2. $CH_3CHClCH_2CH_2CH_3$
3. $CH_2ClCH_2CH_2CH_2CH_3$

71. There are six isomers, namely
1-pentene, cis-2-pentene , trans-2-pentene,
2-methyl-1-butene, 3-methyl-1-butene,
2-methyl-2-butene.

72. The prefix "hydrazono" is attached to the name of the appropriate parent hydride, which bears the principal functional group –COOH.

74. Putting a substituent at position 3 will make the molecule chiral

$\overset{1}{CH_3}–\overset{2}{CH_2}–\overset{3}{^*CH}(X)–\overset{4}{CH_2}–\overset{5}{CH_2}–\overset{6}{CH_2}–\overset{7}{CH_3}$

75. This compound is named as the derivative of 'diazene'.

77. The two –C≡N groups are attached to the unbranched chain. The name of secondary substituent, $–CH_2–C≡N$ is cyanomethyl which is linked with the parent hydride, hexane.

79. Cyclohexane ring bearing Br and CH_2CH_3 on adjacent carbons (both marked •)

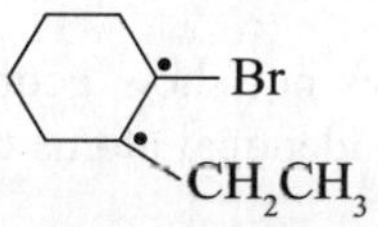

There are two chiral carbon atoms in it.

80. The correct decreasing order is as follows:
$–SO_3H$, –COOH, $–CONH_2$, –CHO

81. A racemic mixture is always produced whenever a new chiral carbon atom is generated in a reaction. Thus, 50% D and 50% L isomers are produced.

82. sp^2 sp^2 sp sp
$H_2C=CH–C≡CH$

83. Linear symmetrical structures have zero dipole moment.

84. Cyclohexane with gem-dimethyl group and $–CH_2Br$

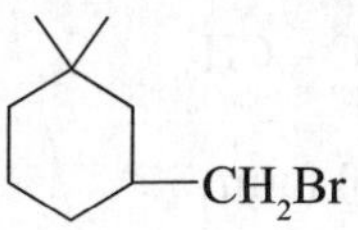

Here n = 1

so O. Isomers = $2^n = 2^1 = 2$

85. If there is a choice, the numbering is done from the end nearer to the double bond, that is

$\overset{1}{CH_2}=\overset{2}{CH}–\overset{3}{CH_2}–\overset{4}{CH_2}–\overset{5}{C}≡\overset{6}{CH}$. Thus,
$C_2–C_3$ bond is of the type $sp^2 – sp^3$.

87. The compound $CH_2{=}CH{-}CH(OH)CH_3$

$$CH_2{=}CH{-}\overset{\displaystyle H}{\underset{\displaystyle OH}{C^*}}{-}CH_3$$

has non-superimposable image, due to the presence of one asymmetric carbon atom.

88. $(CH_3)_2C = CHCH_3 \xrightarrow{H_2} (CH_3)_2CH = CH_2CH_3$ the reduced product does not contain a chiral carbon atom and hence is optically inactive.

90. CaC_2 contains C_2^{-2} unit, that is $C{\equiv}C^{-2}$, it has 1 sigma and 2 pi bonds.

91. Ethylene and acetylene have hindered rotation around C=C and C≡C bond. Ethane and hexachloroethane both have C–C bonds and have less hindered rotation but the latter has more hindered rotation because of bigger size of Cl atoms.

93. 1,2-dichloro-1-pentene

$C.(Cl)_2 = CHCH_2CH_2CH_3$ does not show geometrical isomerism since it has two identical atoms that is Cl on C_1.

94. As it has one chiral centre.

$$\overset{\displaystyle CH_3}{\overset{|}{}}\!\!CH_3{-}\overset{\displaystyle CH_3}{CH}{-}CH_2{-}C_6H_4{-}\underset{*}{\overset{\displaystyle CH_3}{CH}}{-}\overset{\displaystyle O}{\overset{\|}{C}}{-}OH$$

95. $H_3C - CH_2 - CH_2 - \overset{*}{C}H(CH_3) - CH_3$

96. $CH_2(H) - \overset{\displaystyle O}{\overset{\|}{C}} - CH_3 \rightleftharpoons CH_2{=}\overset{\displaystyle \ddot{O}-H}{C} - CH_3$

(Keto form) (Enol form)

9 sigma bonds,
1 pi bond and 2 lone pairs

97.

$$H_3C{-}\overset{\displaystyle H}{\underset{\displaystyle Cl}{^*C}}{-}\overset{\displaystyle H}{\underset{\displaystyle Cl}{C}}{-}\overset{\displaystyle H}{\underset{\displaystyle Cl}{C^*}}{-}CH_3$$

It has 2 cliral c-atoms

98 $\overset{2}{C}H_3{-}\overset{2}{C}H(\overset{3}{C}H_2{-}\overset{4}{C}H_3){-}\overset{1}{C}HO$

2-methylbutanal

99. 1-bromo-2-methylpropane has one chiral carbon.

$$H_3C{-}\overset{\displaystyle CH_3}{\underset{\displaystyle H}{^*C}}{-}\underset{\displaystyle Br}{CH_2}$$

100. It gives 2 optically active products as follows:

$$CH_3 - \underset{\displaystyle CH_3}{CH} - CH_2 - CH_3 \xrightarrow[hr]{Cl_2} CH_3 - \underset{\displaystyle CH_2Cl}{CH} - CH_2 - CH_3 +$$

$$CH_3 - \underset{\displaystyle CH_3}{CH} - \underset{\displaystyle Cl}{CH} - CH_3$$

101.

$$CH_3{-}CH_2{-}\overset{\displaystyle H}{\underset{\displaystyle CH_3}{^*C}}{-}COOH$$

exhibits stereoisomerism (optical isomerism) because it contains one asymmetric carbon and is nonsuperimposable on its mirror image.

102. Six isomers are

$(F)(Cl)C{=}C(Br)(I)$ $(F)(Cl)C{=}C(I)(Br)$

$(F)(Br)C{=}C(Cl)(I)$ $(F)(Br)C{=}C(I)(Cl)$

$(F)(I)C{=}C(Cl)(Br)$ $(F)(I)C{=}C(Br)(Cl)$

103. $H_3C{-}\underset{\displaystyle H}{C}{=}\underset{\displaystyle H}{C}{-}CH_2{-}C_6H_5$

In this, both the carbons of the double bonds carry different substituents and hence it show geometrical isomerism.

104. 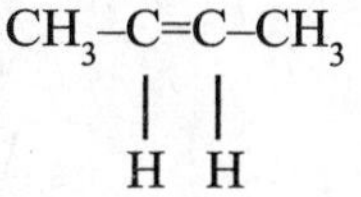 $CH_3-C=C-CH_3$ (each double-bond carbon bearing H)

and $CH_3-C=C-C_6H_5$ (each double-bond carbon bearing H)

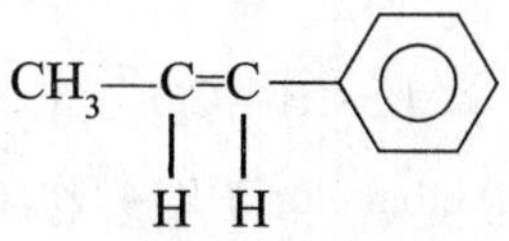

have two different groups attached to each carbon atom of the double bond.

105. 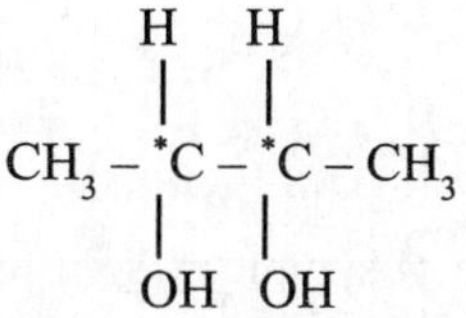 $CH_3-\overset{*}{C}H(OH)-\overset{*}{C}H(OH)-CH_3$

has two similar chiral carbon atoms. Therefore, it has three stereoisomers d, l and meso form. Out of these, only two (d-and l) are optically active.

106. Rotation of B through 180° within the plane of paper gives D which is an enantiomer of A.

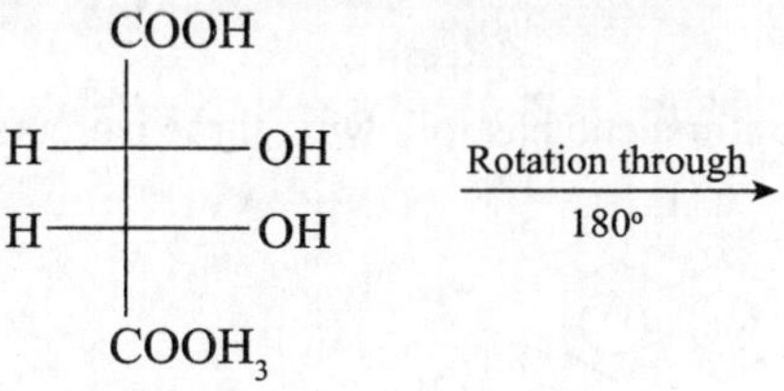

COOH; H—OH; H—OH; $COOH_3$ (B) —Rotation through 180°→

$COOH_3$; OH—H; OH—H; COOH (D) mirror $COOH_3$; H—OH; H—OH; COOH (A)

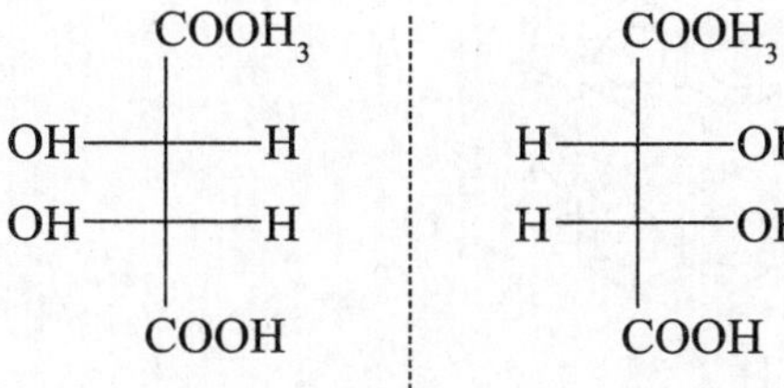

Thus, A and B are enantiomers.

107. 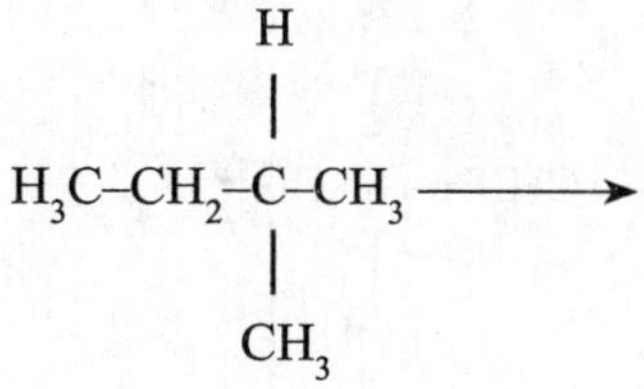 $H_3C-CH_2-CH(CH_3)-CH_3 \longrightarrow$

$CH_3-CH_2-\overset{*}{C}H(CH_3)-CH_2Cl + CH_3-CH_2-CCl(CH_3)-CH_3 +$
I II

$CH_3-\overset{*}{C}HCl-CH(CH_3)-CH_3 + CH_2Cl-CH_2-CH(CH_3)-CH_3$
III IV

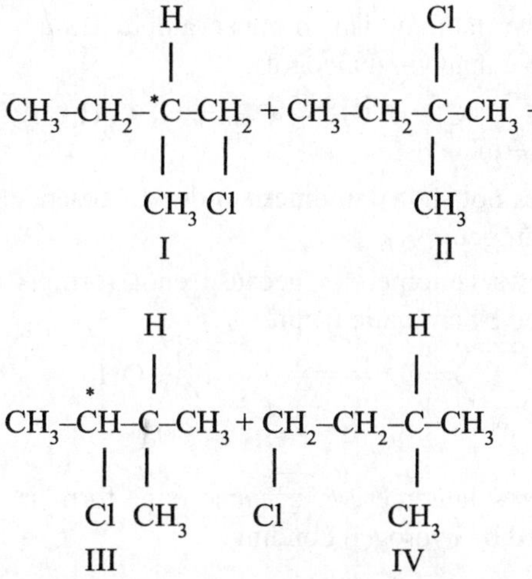

I and III exists as enantiomers.

115. 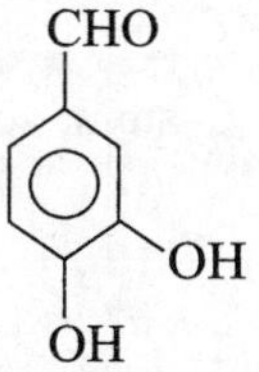 (benzene ring with CHO, OH at 3-position, OH at 4-position)

Protocatechualdehyde is 3,4-dihydroxybenzaldehyde not 2,3-dihydroxy benzaldehyde.

116. There are three structural isomers 1, 2-, 1, 3- and 1,4-dimethylcyclohexane and six geometrical isomers: Cis and trans 1, 2-, 1, 3- and 1, 4-dimethylcyclohexane.

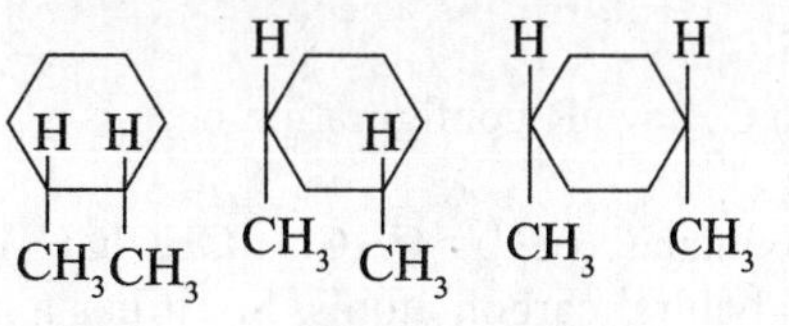

Cis – 1, 2-dimethyl cyclohexane; Cis – 1, 3-dimethyl cyclohexane; Cis – 1, 4-dimethyl cyclohexane

124. 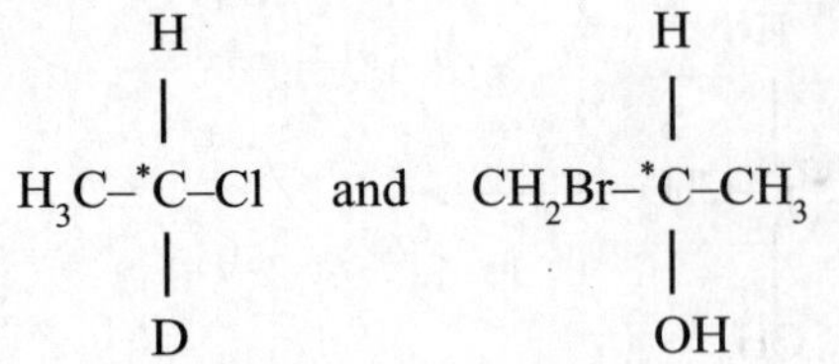 $H_3C-{}^*C(H)(D)-Cl$ and $CH_2Br-{}^*C(H)(OH)-CH_3$

125. (a) shows tautomerism because aldehydes are more stable than vinyl alcohols.

$C_6H_5 - CH = CH - OH \rightleftharpoons C_6H_5 - CH_2 - CH = O$

(b) does not show tautomerism due to absence of α–hydrogen.

(c) shows tautomerism because enol form is stabilized by aromatic nature.

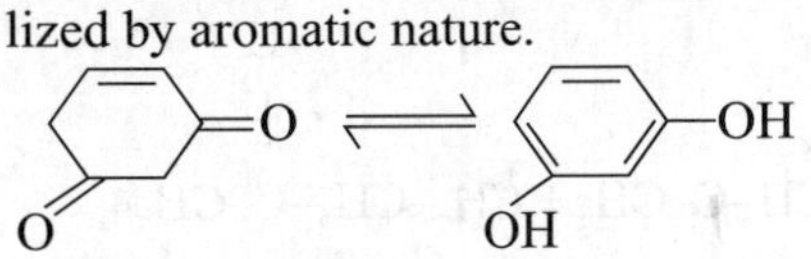

(cyclohexenedione) $\rightleftharpoons$ (OH, OH benzene)

(d) shows tautomerism because enol form is stabilized by hydrogen bonding.

130. The plane of polarized light is rotated by optically active compound, that is, it should be chiral.

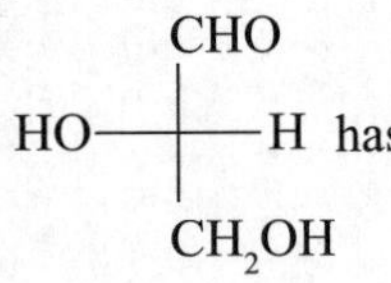

Therefore, HO–C(CHO)(CH$_2$OH)–H has chiral C-atom, so it is optically active.

131. Twisted boat is chiral as it does not have plane of symmetry.

132. 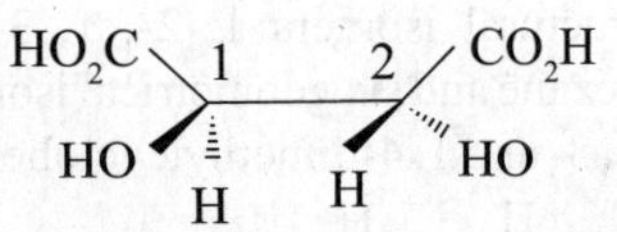 HO_2C–C1(HO)(H)–C2(HO)(H)–CO_2H

Both C_1 and C_2 have R-configuration here.

133. 2,3-dichlorobutane (CH_3 *CHCl *CHClCH_3) has two identical chiral carbon atoms. So, it has a meso-isomer.

136. Due to presence of two asymmetric carbon atoms in 2,3-dichlorobutane,

$CH_3 - {}^*CH(Cl) - {}^*CH(Cl) - CH_3$,

It exhibits optical isomerism.

137. 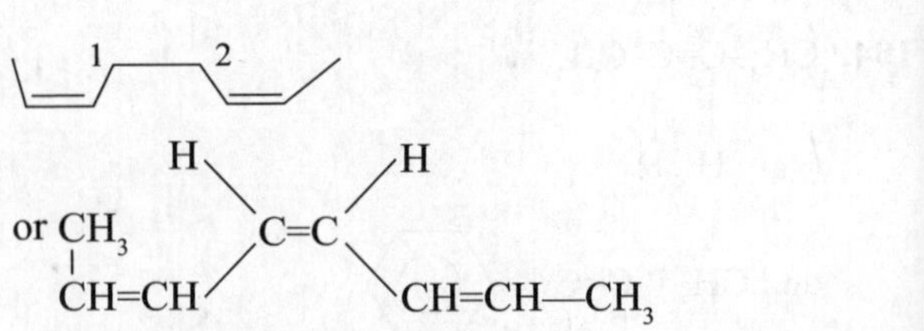 (1, 2 hexatriene skeleton)

or CH_3–CH=CH–C(H)=C(H)–CH=CH–CH_3

It is cis-bis (propenyl) ethane or 2 'Z' 4 'Z', 6 'Z' hexa 2, 4, 6 tri–ene.

138.

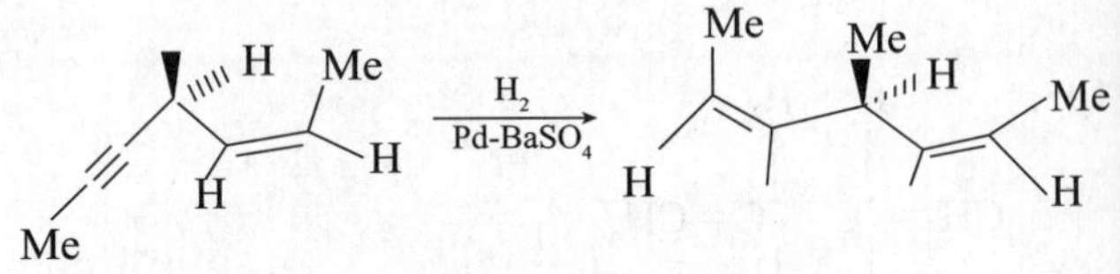

(Me–C≡C–CH(Me)(H)–CH=CH–Me) $\xrightarrow[\text{Pd-BaSO}_4]{H_2}$ (Me–CH=CH–CH(Me)(H)–CH=CH–Me)

As the product has a plane of symmetry so it is optically inactive.

139. This is gauche conformation as follows:

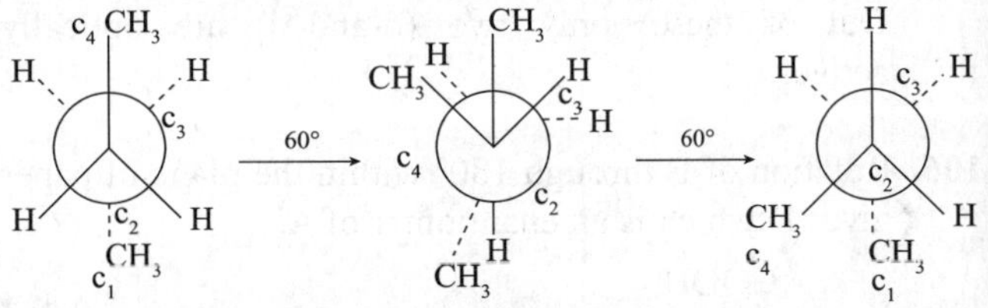

(Newman projections: C_4 CH_3 / H, H, H, H / C_2 CH_3 C_1 $\xrightarrow{60°}$ CH_3, H, H, CH_3, H, H $\xrightarrow{60°}$ H, H, H, CH_3 C_4, H, CH_3 C_1)

141. Methyl anthracene has following three isomers.

(I) 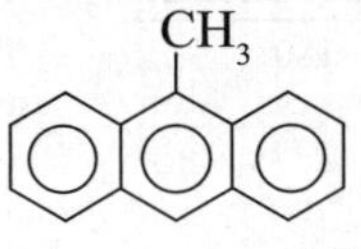 9-methylanthracene (CH_3)

(II) 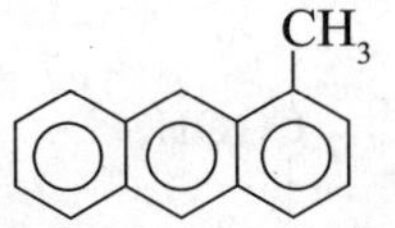 1-methylanthracene (CH_3)

(III) 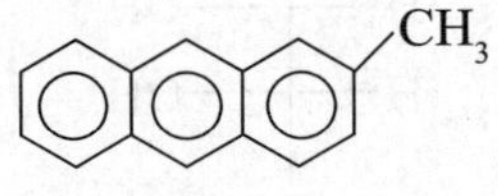 2-methylanthracene (CH_3)

142. 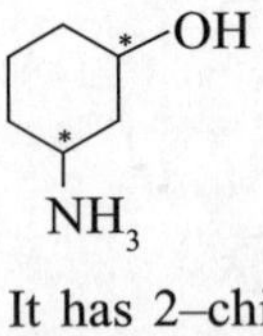 (cyclohexane with *OH and *NH_3)

It has 2–chiral C–atoms and have no plane of symmetry it has 4 stereo isomers

143. $CH_3-\overset{*}{C}H(NH_2)-CH = CH_2$

has one chiral C–atoms so it has 2 optical isomers but it can not show geometrical isomerism.

144. It has one chiral C–atom.

145. This compound has 3p° and one s°–C–atom.

PREVIOUS YEARS' QUESTIONS

1. But-2-ene exhibits cis-trans-isomerism due to
(a) Rotation around $C_3 - C_4$ sigma bond
(b) Rotation around $C_1 - C_2$ bond
(c) Restricted rotation around C = C bond
(d) Rotation around $C_2 - C_3$ double bond

[2000]

2. Dihedral angle in staggered form of ethane is
(a) 0° (b) 60°
(c) 120° (d) 180°

[2000]

3. The (R) – and (S) – enantiomers of an optically active compound differ in
(a) their optical rotation of plane polarized light
(b) their reactivity with a chiral reagents
(c) their solubility in achiral reagents
(d) their melting points

[2000]

4. A compound with molecular formula C_7H_{16} shows optical isomerism, the compound will be
(a) 2-methylhexane
(b) 2,2-dimethylpentane
(c) 2,3-dimethylpentane
(d) None of these

[2001]

5. $CH_3 - CH_2 - CH(Cl) - CH_3$

obtained by chlorination of n-butane, will be
(a) meso-form (b) racemic mixture
(c) d-form (d) l-form

[2001]

6. The incorrect IUPAC name is
(a) $CH_3 - CH(CH_3) - CH(CH_2CH_3) - CH_3$
2,3-dimethyl pentane
(b) $CH_3 - C(=O) - CH(CH_3) - CH_3$
2-methyl-3-butanone
(c) $CH_3 - CH(Cl) - CH(Br) - CH_3$
2-bromo-3-chloro butane
(d) $CH_3 - C \equiv CCH(CH_3)_2$
4-methyl-2-pentyne

[2001]

7. IUPAC name of the following is
$CH_2 = CH - CH_2 - CH_2 - C \equiv CH$
(a) 1-hexyne-5-ene (b) 1, 5-hexenyne
(c) 1, 5-hexynene (d) 1-hexene-5-yne

[2002]

8. Geometrical isomers are differ in
(a) Length of carbon chain
(b) Spatial arrangement of atoms
(c) Position of functional group
(d) Position of atoms

[2002]

9. Name of the compound given below is

CH_3, H_3C, CH_3, CH_3

(a) 3-methyl-4-ethyloctane
(b) 4-ethyl-3-methyloctane
(c) 2, 3-diethylheptane
(d) 5-ethyl-6-methyloctane

[2003]

10. Which of the following pairs of compounds are enantiomers?

(a)
```
     CH3               CH3
 H ──┼── OH      HO ──┼── H
HO ──┼── H       HO ──┼── H
     CH3               CH3
            and
```

(b)
```
     CH3               CH3
 H ──┼── OH      HO ──┼── H
HO ──┼── H        H ──┼── OH
     CH3               CH3
            and
```

(c)
```
     CH3               CH3
 H ──┼── OH       H ──┼── OH
HO ──┼── H        H ──┼── OH
     CH3               CH3
            and
```

(d)
```
     CH3               CH3
HO ──┼── H       HO ──┼── H
 H ──┼── OH      HO ──┼── H
     CH3      and      CH3
```

[2003]

11. The molecular formula of diphenyl methane

$C_6H_5-CH_2-C_6H_5$ is $C_{13}H_{12}$.

How many structural isomers are possible when one of the hydrogen is replaced by a chlorine atom?

(a) 4 (b) 6
(c) 7 (d) 8

[2004]

12. The chirality of the compound

```
       Br
       |
       C ▪▪▪H
      /  \
   H3C    Cl
```

(a) E (b) R
(c) Z (d) S

[2005]

13. IUPAC names of some compounds are given. Which one of the following is not correct?

(a) $CH_3-C\equiv C-CH(CH_3)_2$
4-methyl-2-pentyne

(b) $CH_3-CH_2-\underset{\underset{CH_2}{||}}{C}-\underset{\underset{CH_3}{|}}{CH}-CH_3$
2-ethyl-3-methyl-but-1-ene

(c) $CH_3-CH_2-CH_2-\underset{\underset{CH_2CH_3}{|}}{\overset{\overset{CH_3}{|}}{CH}}-CH-CH_2CH_3$
3-ethyl-4-methylheptane

(d) $CH_3-\underset{\underset{OH}{|}}{CH}-\underset{\underset{CH_3}{|}}{CH}-CH_3$
3-methyl-2-butanol

[2005]

14. Which one of the following pairs represents stereoisomerism?

(a) Chain isomerism and rotational isomerism
(b) Optical isomerism and geometrical isomerism
(c) Structural isomerism and geometrical isomerism
(d) Linkage isomerism and geometrical isomerism

[2005]

15. Which of the following is not chiral?

(a) 3-bromopentane (b) 2-hydroxypropanoic acid
(c) 2-butanol (d) 2,3-dibromopentane

[2006]

16. The IUPAC name of $CH_3CH_2CH(CH_3)CH(CH_3)COCl$ (skeletal structure with Cl and O)

(a) 2-ethyl-3methylbutanoyl chloride
(b) 2, 3-dimethylpentanoyl chloride
(c) 3, 4-dimethylpentanoyl chloride
(d) 1-chloro-1-oxo-2, 3-dimethylpentane

[2006]

17. $CH_3CHCl-CH_2-CH_3$ has a chiral centre. Which one of the following represents its R configuration?

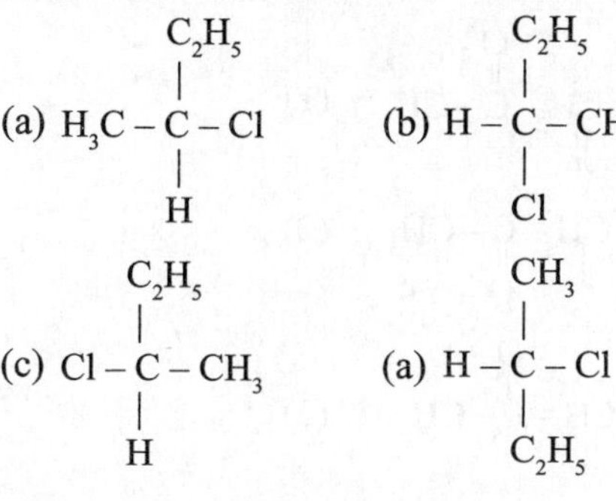

[2007]

18. If there is no rotation of plane polarized light by a compound in a specific solvent, thought to be chiral, it may mean that:

(a) the compound is certainly a chiral
(b) the compound is certainly meso
(c) there is no compound is the solvent
(d) the compound may be a racemic mixture

[2007]

19. Base strength of:

(a) $H_3C\overset{\oplus}{C}H_2$ (b) $H_2C=\overset{\oplus}{C}H_2$
(c) $H-C\equiv C^{\oplus}$

is in the order of:

(a) (a) > (c) > (b) (b) (a) > (b) > (c)
(c) (b) > (a) > (c) (d) (c) > (b) > (a)

[2008]

20. How many stereoisomers does this molecule have?

$CH_3CH=CHCH_2CHBrCH_3$

(a) 8 (b) 2
(c) 4 (d) 6

[2008]

21. The IUPAC name of the compound having the formula $CH\equiv C-CH=CH_2$ is

(a) 1 - butyn - 3 - ene
(b) but - 1 - yne - 3 - ene
(c) 1 - butene - 3 yne
(d) 3 - butene - 1 - yne

[2009]

22. Which of the following compound will exhibit cis-trans (geometrical) isomerism?

(a) 2 - Butene (b) Butanol
(c) 2- Butyne (d) 2 - Butenol

[2009]

23. In the following, the most stable conformation of n- butane is

(a)
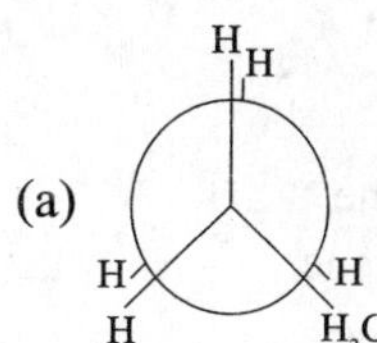

(b)
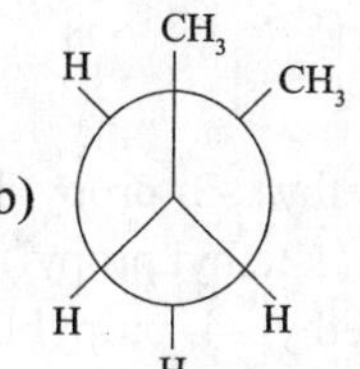

(c)
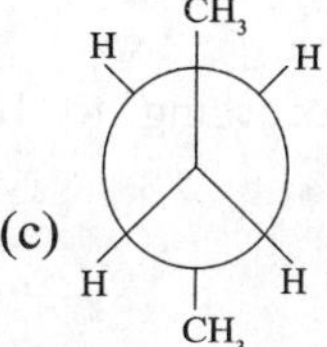

(d)
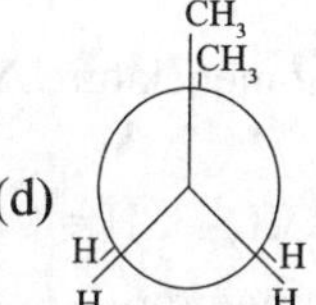

24. Which of the following conformers for ethylene glycol is most stable?

(a)
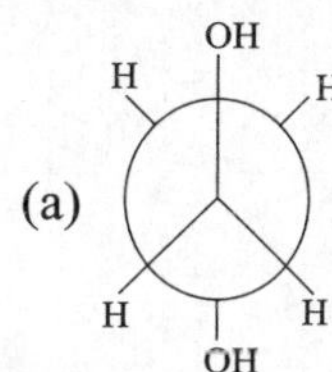

(b)
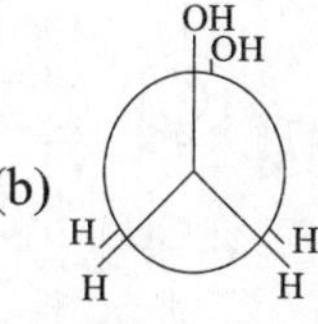

(a)
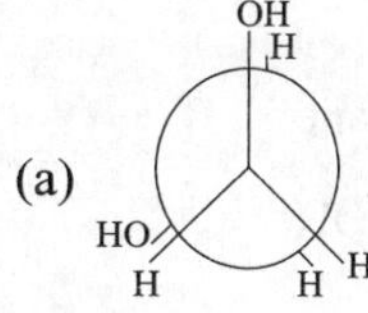

(b)
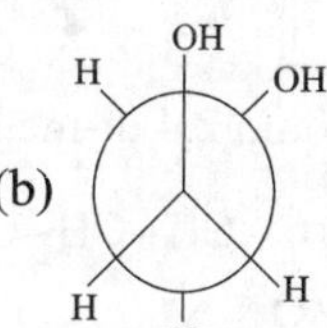

[2010]

25. The IUPAC name of the compound $CH_3CH=CHC\equiv CH$ is

(a) Pent -4- yn -2-ene
(b) Pent -3- en -1-yne
(c) Pent -2- en -4-ene
(d) Pent -1- yn -3-ene

[2010]

26. The correct order of increasing bond length of C –H, C – O, C – C and C = C is

(a) $C-H<C=C<C-O<C-C$
(b) $C-H<C-C<C-C<C=C$
(c) $C-O<C-H<C-C<C=C$
(d) $C-C<C=C<C-O<C-H$

[2011]

27. The correct IUPAC name of the compound [skeletal structure] is

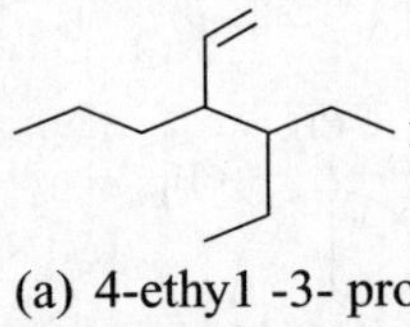

(a) 4-ethy1 -3- propyl hex-1- ene
(b) 3- (1- ethyl propyl) hex -1- ene
(c) 3-ethyl -4- propyl hex -5- ene
(d) 3-ethyl -4- ethenyl heptane

[2011]

28. Which nomenclature is not according to IUPAC system ?

(a) $Br - CH_2 - CH = CH_2$
1-Bromo prop-2-exe

(b) 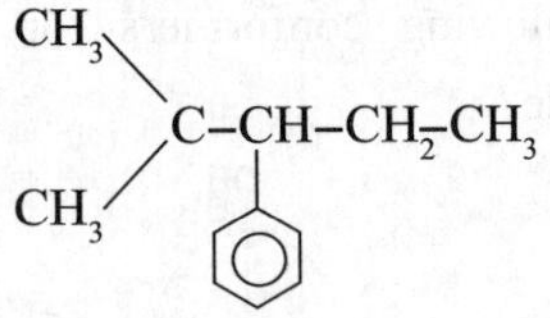 $(CH_3)_2C(C_6H_5)-CH-CH_2-CH_3$

(c) $CH_3-CH_2-C(CH_3)(Br)-CH_2-CH(CH_3)-CH_3$
4-Bromo 2,4 di-methye hexane

(d) $CH_3-C(=O)-CH_2-CH_2-CH_2-COOH$
5-Oxohexanoic acid

[2012]

29. Which of the following acid does not exhibit optical isomereism ?

(a) malic acid (b) lactic acid
(c) tartaric acid (d) x-amino acid

[2012]

30. The structure of isobutyl group an organic compound is ?

(a) $(H_3)(CH_3)CH-CH_2-$ (b) $CH_3-CH_2-CH_2-CH_2-$
(c) $CH_3-CH(-)-CH_2-CH_3$ (d) $H_3C-C(CH_3)_2-$

[2013]

31. The order of stability of the following tautometric compound is ?

$H_2C=C(OH)-CH_2-C(=O)-CH_3$ (1)
⇅
$CH_3-C(=O)-CH_2-C(=O)-CH_3$ (2)
⇅
$CH_3-C(OH)=CH-C(=O)-CH_3$ (3)

(a) $1 > 2 > 3$ (b) $2 > 1 > 3$
(c) $2 > 3 > 1$ (d) $3 > 2 > 1$

[2012]

32. Which of the given compounds can exhibit tautomerism?

(1) [4,4-dimethylcyclohexa-2,5-dienone] (2) [3,3-dimethylcyclohexanone] (3) [2,2-dimethylcyclohexanone]

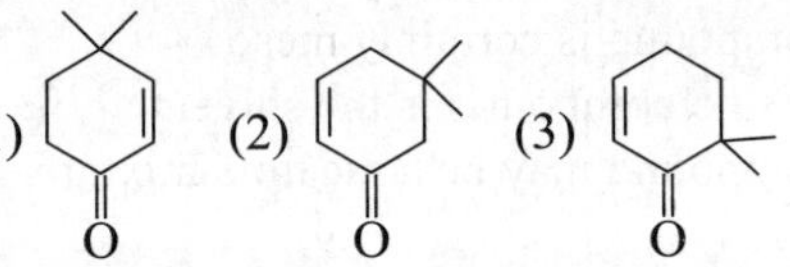

(a) 1 and 2 (b) 2 and 3
(c) 1 and 3 (d) 1, 2 and 3

[2015]

33. The enolicform of ethylacetoacetate as shown below has:

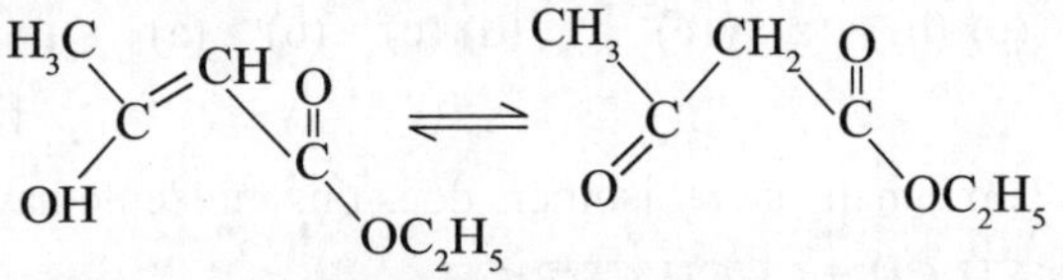

$H_3C-C(OH)=CH-C(=O)-OC_2H_5 \rightleftharpoons CH_3-C(=O)-CH_2-C(=O)-OC_2H_5$

(a) 16 sigma bonds and 1 Pi- Bond
(b) 18 sigma bonds and 2 Pi- Bond
(c) 9 sigma bonds and 2 Pi- Bond
(d) 9 sigma bonds and 1 Pi- Bond

[2015]

34. The number of structural isomers possible from the molecular formula C_3H_9N is:

(a) 2 (b) 3
(c) 4 (d) 5

[2015]

35. Which of the following biphenyls is optically active.

(a) 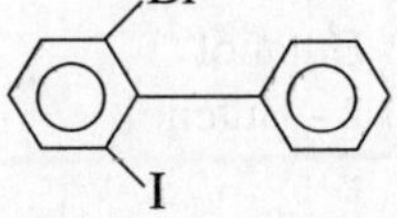 [biphenyl with Br and I on one ring] (b) 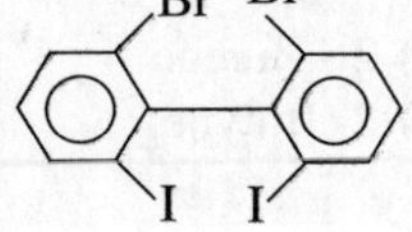 [biphenyl with Br, I on each ring]

(c) (d) NO_2 I

[2016]

36. The Correct statement regarding the comparison of staggered and edipsed conformation of ethane is ?

(a) The eclipsed conformation of ethane is more stable than staggared conformation because eclipsed conformation has no torsional strain.

(b) The eclipsed conformation of ethene is more stable than staggered conformation even though the eclipsed conformation has torsional strain.

(c) The staggered conformation of ethane is less stable than eslipsed conformation because, staggered conformation has torsional strain.

(d) The staggered conformation of ethane is more stable than eclipsed conformation because staggered conformation has no torsional strain.

37. In which of the following molecules all atoms are co-planar ?

(a) (b)

(c) $(CH_3)_2C=C(CN)_2$ (d)

[2016]

38. Which among the given molecules can exhibit Tautomerism ?

(1) (2) Ph Ph (3)

(a) 3 Only (b) Both 1 and 2

(c) Both 1 and 2 (d) Both 2 and 3

[2016]

ANSWER KEYS

1. (c)	2. (b)	3. (a)	4. (c)	5. (b)	6. (b)	7. (d)	8. (b)	9. (b)	10. (b)
11. (a)	12. (b)	13. (c)	14. (b)	15. (a)	16. (b)	17. (c)	18. (b)	19. (b)	20. (c)
21. (c)	22. (a)	23. (c)	24. (d)	25. (b)	26. (a)	27. (a)	28. (a)	29. (a)	30. (a)
31. (d)	32. (d)	33. (b)	34. (c)	35. (b)	36. (d)	37. (b)	38. (a)		

HINTS AND EXPLANATIONS

1. Due to presence of >C = C< in butene-2, it shows restricted rotation so it can give two types of arrangements around the space of >C=C< i.e., cis and trans-forms.

2. In the staggered form of ethane, dihedral angle is 60°.

3. (R)-and (S)- enantiomers differ only in their behaviour towards plane polarized light. (R) means rectus showing clock wise rotation while (S) means sinster showing anti-clock wise rotation.

4. $CH_3 - CH(CH_3) - \overset{*}{C}H(CH_3) - CHCH_3$

As 2,3-dimethylpentane has a chiral carbon atom so it is optically active.

5. Here the product will be racemic in nature. The reaction occurs in following steps:

$Cl_2 \rightarrow Cl^{\bullet} + Cl^{\bullet}$

$CH_3CH_2CH_2CH_3 + Cl^{\bullet} \rightarrow CH_3\,{}^{\bullet}CHCH_2CH_3 + HCl$

$CH_3\,{}^{\bullet}CHCH_2CH_3 + Cl_2 \rightarrow$

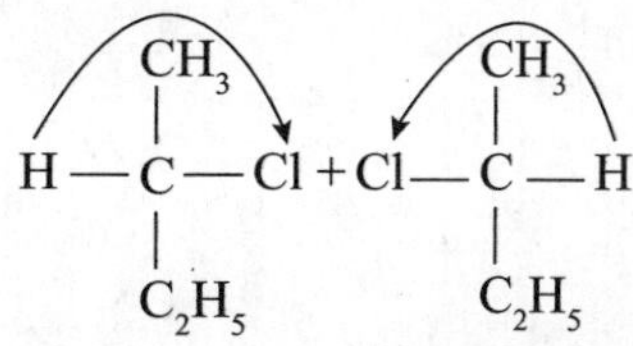

Racemic mixture

6. Incorrect IUPAC name is 2-methy1-3-butanone as the correct name is 3-methy1butan-2-one.

7. 1 2 3 4 5 6
$CH_2 = CH - CH_2 - CH_2 - C \equiv CH$
The IUPAC name of the compound is 1-hexene-5-yne (it is better to name it as hex-1-ene-5-yne).

8. Geometrical isomers differ in Spatial arrangement of atoms or groups i.e., cis and trans-type.

9. CH_3, H_3C, CH_3, CH_3

The name of this compound is 4-ethyl-3-methyl octane.

10. Enantiomers are mirror image to each other which are non-super imposable having same physical and chemical properties but different biochemical properties.

e.g.,

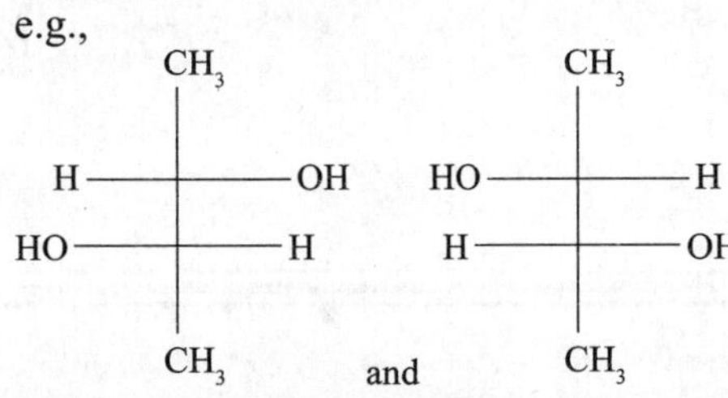

11. Diphencyl methane has four isomers in form of mono chloro derivatives as follows:

Cl, CH_2

Cl, CH_2

Cl, CH_2

CH_2, Cl

12. Br, C, H, H_3C, Cl

The chirality of this compound is 'R' as the rotation is clockwise here.

13. As the correct name should be 4-ethyl-3-methyl heptane.

14. Optical isomerism and geometrical isomerism represent stereoisomerism.

15. Br
$CH_3 - CH_2 - CH - CH_2 - CH_3$
3-bromopentane is not a chiral compound

16. Cl, O

The IUPAC name of this compound is 2,3-dimethyl pentanoyl chloride.

17. It is a case of R-configuration.

C_2H_5
$Cl - C^* - CH_3$
H

18. If there is no rotation of plane polarized light by a compound in a specific solvent, thought to be chiral, it may mean that the compound is certainly meso. Meso form is optically inactive due to mutual cancellation or internal compensation.

19. As here acidic nature decreases with the decrease of s- percentage as follows:

$$\underset{sp\ \ sp}{H - C \equiv C - H} > \underset{sp^2\ \ sp^2}{CH_2 = CH_2} > \underset{sp^3\ \ sp^3}{CH_3 - CH_3}$$

(Acidic character)

Since strength of conjugate base is inversely proportional to the strength of its acid hence

$$C^{\ominus} \equiv C - H < {}^{\ominus}CH = CH_2 < {}^{\ominus}CH_2 - CH_3$$

(Basic character)

20. $CH_3CH = CHCH_2CHBrCH_3$

This given compound can also be written as follows:

CH_3, H, H
$C = C$
H, $CH_2 - C^* - CH_3$
Br

As it fulfills the conditions to show both these isomerisms so geometrical isomerism (cis-trans form) and optical isomerism is possible in this compound

Number of optical $= 2^n = 2^1 = 2$
(Here n = number of asymmetric carbon)
Therefore, total number of stereoisomers
$= 2 + 2 = 4$

21. $\overset{1}{H}C \equiv \overset{3}{C} - \overset{2}{C}H = \overset{1}{C}H_2$ 1-Butene-3-yne

22. Out of these compounds only but -2- ene can show cis and trans geometrical isomerism.

$H-C-CH_3$ / $H-C-CH_3$ (cis)

$H-C-CH_3$ / H_3C-C-H (trans)

23. 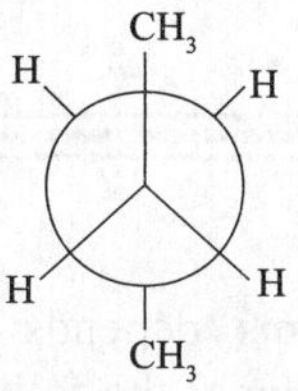 (Newman projection: CH_3, H, H front; H, H, CH_3 back — anti)

Is most stable due to less repulsion between two methyl group

24. The conformation is most stable because of intermolecular H- bonding

25. $\overset{5}{C}H_3\overset{4}{C}H = \overset{3}{C}H\overset{2}{C} \equiv \overset{1}{C}H$
Pent -3-en-1-yen
If a molecular contains both carbon–carbon double or triple bonds, the two are treated at par in seeking the lowest number combination. However, if the sum of numbers turns out to be the same starting from either of the carbon chain, then lowest number is given to the C = C double bond.

26. C–H C=C C–C C–O
1.09A°, 1.34A°, 1.54A°, 1.43A°
Hence the correct order is:
C–H<C=C<C–O<C–C

27. 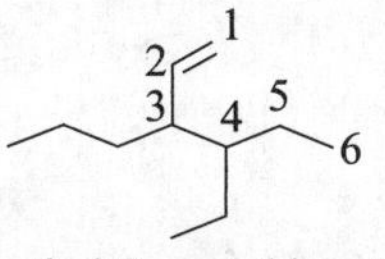 (skeletal structure numbered 1–6)

4-ethyl-3-propyl hex-1-ene

28. In case of This compound the corect name is given below:

$\overset{3}{C}H_2(Br)-\overset{2}{C}H=\overset{1}{C}H_2$

3-bromo propene

29. It can show geomatrical isomerism and not optical

H–C–COOH
‖
H–C–COOH

maleic acid

30. Here option ('a') is Iso-butyl
('b') is n-butyl (c) is s-butyl
And (d) is t-butyl Gp-respectively.

31. Here the stability is as follows:

III >	II >	I
Conjugate Make it stable	H-bonding	No conjugation Less stable

32. These may show tautomerism as follows:

(1) 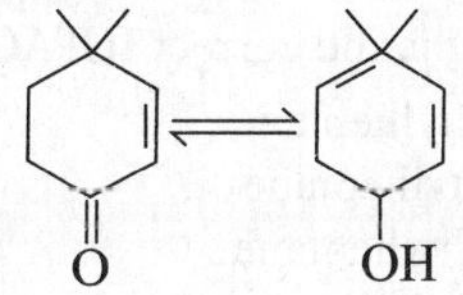 (keto ⇌ enol: O ⇌ OH)

(2) 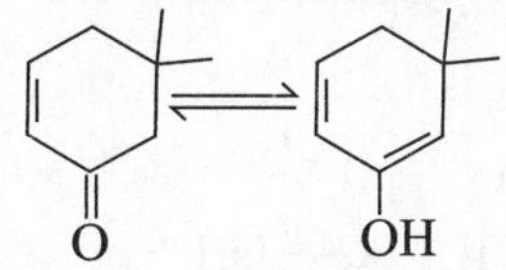 (keto ⇌ enol: O ⇌ OH)

(3) 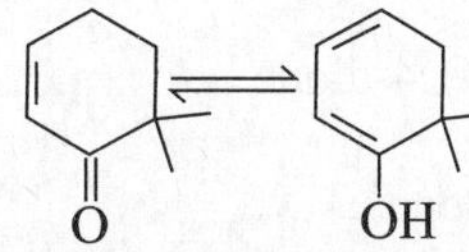 (keto ⇌ enol: O ⇌ OH)

Due to conjugation α–H is used.

33. As ethylacetoacetate exists in keto form so it must have 18σ and 2π bonds.

34. C_3H_9N has 4 structural isomers. See text part.

35. Ortho substituted biphenyl can show optical isomerism it miror image is non-superimposable. It is due to restricted rotation around C-C bond

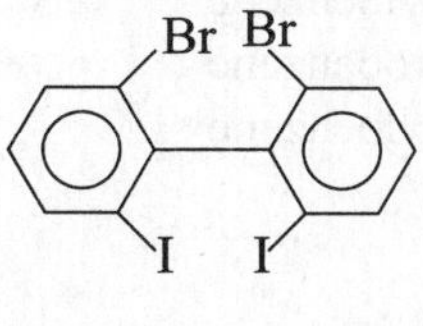

(biphenyl with Br, Br and I, I at ortho positions)

36. Staggered form is more stable as it does not have torsional strain.

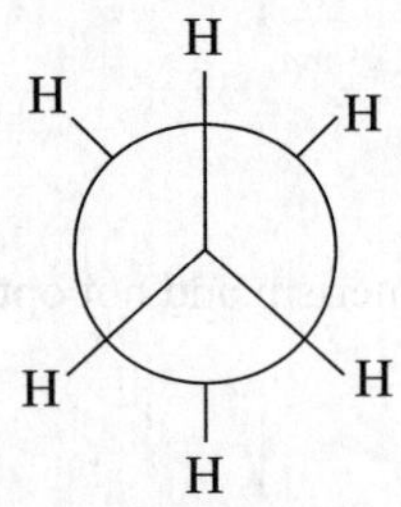

Staggered

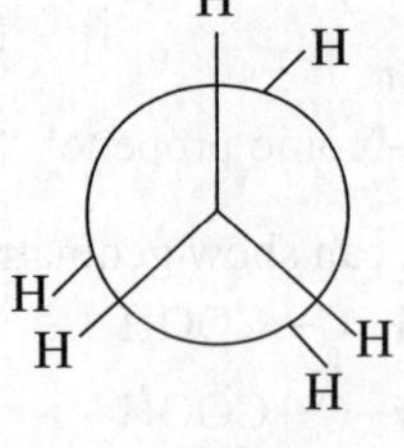

Eclipsed

37. For co-planarity in a molecule all Carbon atoms must be sp^2-hybriditred.

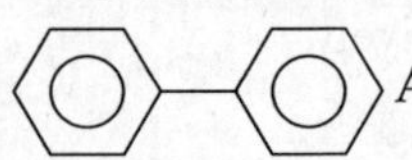
All C-atoms are sp^2 (coplanar)

38. Tautomerism is possible when α-H on saturated α-C-atom is present.

O, H → OH

O, Ph, Ph — No α-H

α–H are on bridge C-atoms so can't be migrated

NCERT Exemplar

1. Which of the following is the correct IUPAC name?

(a) 3-Ethyl-4, 4-dimethylheptane
(b) 4,4-Dimethyl-3-ethylheptane
(c) 5-Ethyl-4, 4-dimethylheptane
(d) 4,4-Bis(methyl)-3-ethylheptane

2. The IUPAC name for is

$$CH_3-\overset{\overset{\displaystyle O}{\|}}{C}-CH_2-CH_2-\overset{\overset{\displaystyle O}{\|}}{C}-OH$$

(a) 1-hydroxypentane-1,4-dione
(b) 1,4-dioxopentanol
(c) 1-carboxybutan-3-one
(d) 4-oxopentanoic acid

3. The IUPAC name for:

Cl, NO_2, CH_3

(a) 1-Chloro-2-nitro-4-methylbenzene
(b) 1-Chloro-4-methyl-2-nitrobenzene
(c) 2-Chloro-1-nitro-5-methylbenzene
(d) *m*-Nitro-*p*-chlorotoluene

4. Electronegativity of carbon atoms depends upon their state of hybridization. In which of the following compounds, the carbon marked with asterisk is most electronegative?

(a) $CH_3-CH_2-{}^*CH_2-CH_3$
(b) $CH_3-{}^*CH=CH-CH_3$
(c) $CH_3-CH_2-C\equiv{}^*CH$
(d) $CH_3-CH_2-CH={}^*CH_2$

5. In which of the following, functional group isomerism is **not** possible?

(a) Alcohols
(b) Aldehydes
(c) Alkyl halides
(d) Cyanides

6. Correct IUPAC name for $H_3C-\underset{\displaystyle C_3H_3}{\underset{|}{CH}}-\underset{\displaystyle C_3H_3}{\underset{|}{CH}}-CH_3$ is __________.

(a) 2- ethyl-3-methylpentane
(b) 3,4- dimethylhexane
(c) 2-*sec*-butylbutane
(d) 2, 3-dimethylbutane

ANSWER KEYS

1. (b) 2. (d) 3. (b) 4. (c) 5. (c) 6. (b)

HINTS AND EXPLANATIONS FOR SELECTIVE QUESTIONS

2. Functional group has high priority in naming the compound.

3. Nomenclature is done in alphabetical order.

AIIMS ESSENTIALS

ASSERTION AND REASON

In the following questions, two statements Assertion (A) and Reason (R) are given. Mark

(a) if A and R both are correct and R is the correct explanation of A;

(b) if A and R both are correct but R is not the correct explanation of A;

(c) A is true but R is false;

(d) A is false but R is true,

(e) Both A and R are false.

1. (A) : Compound possessing same molecular formulae but differing in their physical and chemical properties are called isomers.
   (R) : Rearrangement reactions produce isomers.
2. (A) : Tetrahedral complexes show geometrical isomerism.
   (R) : The relative positions of the ligands attached to the central metal atom are same with respect to each other.
3. (A) : Metamers can also be chain or position isomer.
   (R) : The term tautomerism was introduced to explain the reactivity of a substance according to two possible structures
4. (A) : Trans-1,2-dichloroethene has two-fold simple axis of relation.
   (R) : The rotation of the molecular model through 180° about the axis passing through the geometrical centre and perpendicular to the plane of the molecule produces an arrangement which is in distinguishable from the original.
5. (A) : cis-2-butene gives meso-2,3-butanediol with dilute alkene $KMnO_4$ solution.
   (R) : Dilute alkaline $KMnO_4$ solution gives trans addition with alkenes.
6. (A) : Meso-tartaric acid is optically active.
   (R) : Optically active molecule is a molecule that cannot be superimposed on its mirror image.
7. (A) : Cyclohexanone exhibits keto-enol tautomerism.
   (R) : In cyclohexanone, one form contains the keto-group (C = O) while other contains enolic group (–C = C – OH)
8. (A) : Enol form of cyclohexane-1,3,5 trione is more stable than its keto form.
   (R) : Enol form contains α-hydrogen atoms.
9. (A) : 6, 6'-dinitrodiphenic acid shows enantiomerism.
   (R) : Enantiomers rotate the plane of polarized light in the opposite direction but to the same extent.
10. (A) : Alkenes containing more than three carbons exhibit chain isomerism.
    (R) : All the carbon atoms in alkene are sp^3 - hybridised.
11. (A) : When even number of cummulated bonds are present, the compound is optically active. For eg.,

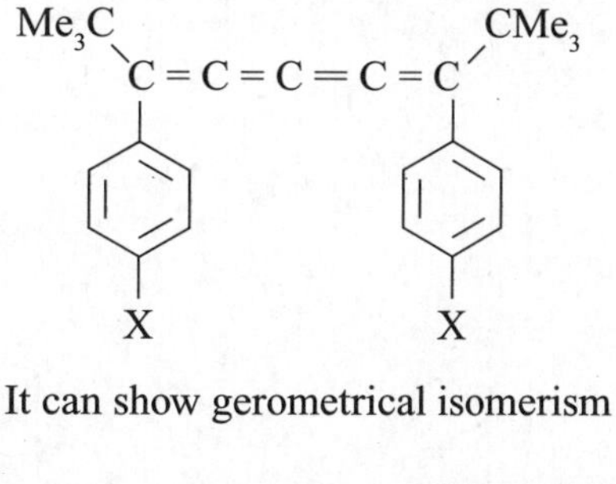

Me_3C and CMe_3 ; C = C = C = C = C, each terminal carbon bearing a p-X-phenyl group (X, X)

    (R) : It can show geometrical isomerism also.

12. (A) : Lactic acid shows geometrical isomerism.
    (R) : It has a chiral C-atom.

13. (A) : Cis-1,3-dimethylcyclo-hexane is achiral in its chair conformation.
    (R) : It has plane of symmetry passing through carbon-1 and carbon-4.

14. (A) : Conformation are not different compound and true isomers
    (R) : As in most cases they are interconvertable even at room temperature.

15. (A) : Cyclobutane is less stable than cyclopentane.
    (R) : Presence of 'bent bond' causes loss of orbital overlap.

16. (A) : Molecules that are not superimposable on their mirror images are chiral
    (R) : The term tautomerism was introduced to explain the reactivity of a substance

ANSWER KEYS

1. (b)	2. (d)	3. (b)	4. (a)	5. (c)	6. (d)	7. (a)	8. (b)	9. (b)	10. (c)
11. (b)	12. (d)	13. (c)	14. (a)	15. (a)	16. (c)				

CHAPTER **11B**

Organic Chemistry - II

Chapter Outline

Bond Fission

- Bond fission is the process of breaking of covalent bonds present in organic molecules.
- It occurs in the following two ways:

Homolytic Fission or Homolysis

- Bond fission occurs symmetrically in homolysis and as a result free radicals are formed.

  $A^{o} \nmid {}^{x}A \longrightarrow A^{o} + {}^{x}A$

- More energy is needed for homolysis than for heterolysis bond fission. For example,

  $Cl - Cl \longrightarrow Cl^{o} + {}^{x}Cl$

  $CH_3 - CH_3 \longrightarrow CH_3^{o} + {}^{x}CH_3$

- Photochlorination of methane involves homolytic fission.
- When homolysis occurs in ethane CH_3^{o} free radicals are formed, in which C-atom is sp^2 hybridized.

Heterolytic Fission or Heterolysis

- Bond fission occurs **asymmetrically** in heterolysis and as a result, ions are formed.

  More energy is needed for heterolysis than for homolysis. For example,

  $CH_3 \overset{X}{\underset{O}{-}} X \longrightarrow CH_3^{+} + X^{-}\overset{X}{\bullet}$

  Carbocation

  $CH_3 \overset{X}{\underset{O}{-}} H \longrightarrow CH_3^{-}\overset{X}{\bullet} + H^{+}$

  Carbanion

- The cleavage of bond becomes easier if electronegativity difference between the atoms is greater. For example, in case of C_2H_5OH, the bond most easily cleaved is –O–H bond.

Electrophile or Electrophillic Species

- Electrophile are electron deficient species which have a tendency to take up electrons, and therefore, they act as Lewis acids.
- These are of following types:

Positively Charged Electrophiles

H^+, X^+, R^+, N^+O_2, N^+O, S^+O_3H.

Exception M^+ (group IA), M^{2+} (group IIA), Al^{+3}, NH_4^+, NR_4^+ (as all have octet state)

Neutral electrophiles These are of the following types:

(a) Neutral electrophiles with an incomplete **octet state** have a central atom which is electron deficient, that is, it does not have an octet state.

For example, BeX_2, $ZnCl_2$, BH_3, BX_3, AlX_3, FeX_2.

$\dot{C}H_3$, $\ddot{C}H_2$, $\ddot{C}X_2$, $(CH_3)_3B$,

(b) Neutral electrophile with an extended **octet state** have a central atom with vacant dorbital, which can extend its octet state. For example, $SnCl_4$, PCl_5, IF_7.

(c) Neutral electrophiles with central metal atom compromising with two or more electronegative atoms. For example, CO_2, SO_2, CS_2.

Nucleophile or Nucleophilic Species

Nucleophile are species that have lone pairs of electrons, available for donation and therefore, they act as Lewis bases.

These are of the following types:

Negatively charged

H^-, OH^-, $R-O^-$, $\bar{C}H_3$, X^-, SH^-, R^--S^-.

Neutral

The central atom with lone pair of electron to donate. For example,

$\geq\ddot{N}$, (NH_3, RNH_2 etc.)

$-\ddot{O}-$ (H_2O, HOR, $R-O-R$)

$-\ddot{S}-$ ($R-SH$, $R-S-R$, H_2S)

$\geq\ddot{P}$

- Ambident nucleophiles have two nucleophilic centres, one of which is negatively charged and the other ane is neutral. For example,

  $\bar{C}\equiv\ddot{N}$, $\bar{O}-\ddot{N}=O$,

  Species behaving both like Electrophile and Nucleophile: Carbon atom with multiple bonds with O, N or S. For example, $R-CHO$, $R-C\equiv N$,

  $R-\overset{O}{\overset{\|}{C}}-R$, $R-\overset{O}{\overset{\|}{C}}-OH$, $R-\overset{O}{\overset{\|}{C}}-Cl$

Reaction Intermediates

Reaction intermediates are species which are formed during the reactions but cannot be isolated, as they are extremely reactive, unstable and therefore, short-lived. These are of further types:

Charged Reaction Intermediates

Carbocation or carbonium ion

- Carbo stands for 'carbon' and 'onium' stands for 'positive charge'.
- It has at least one carbon atom with a positive charge and 6 electrons.

$$-\overset{|}{\underset{|}{C}}{}^+$$

It is sp^2 hybridized and has a trigonal coplanar shape.

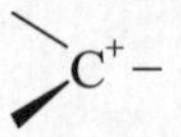

$$>C^+-$$

- It is diamagnetic in nature.

Preparation of carbocation

1. *By heterolytic fission*

$$-\overset{|}{\underset{|}{C}}\overset{x}{\underset{o}{-}}X \longrightarrow -\overset{|}{\underset{|}{C}}{}^+ + X^{-}{}^{x}_{o}$$

2. *By protonation of alcohols*

$$R-\ddot{O}-H + H^+ \longrightarrow R-\overset{+}{\ddot{O}}<^{H}_{H} \longrightarrow R-CH_2^+ + H_2O$$

3. *By protonation of alkene*

$$>C=C< + H^+ \longrightarrow >\underset{H}{\underset{|}{C}}-\overset{\oplus}{C}<$$

Types of carbocation

1. *Non-classical carbocation*

- Here, positive charged is delocalized. For example,

  $CH_2=CH-\overset{+}{C}H_2$

  Allyl carbocation

  $\overset{+}{C}H_2$ attached to a benzene ring

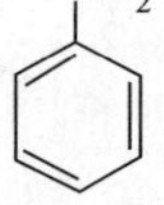

  Benzyl carbocation

2. *Classical carbocation*

- Here, positive charge is localized. For example,

$$H-\overset{H}{\underset{H}{C^+}},\quad \begin{matrix}CH_3\\CH_3\end{matrix}>\underset{H}{C^+},\quad \begin{matrix}CH_3\\CH_3\\CH_3\end{matrix}>C^+$$

Order of stability of carbocation

- The stability is decided by resonance, hyperconjugation, +I effect or the presence of electron releasing groups.

  Stability ∝ Resonance

  ∝ Hyperconjugation

  ∝ +I effect

  ∝ Number of alkyl groups

- $(Ph)_3\overset{\oplus}{C} > (Ph)_2\overset{\oplus}{C}H > (CH_3)_3\overset{\oplus}{C} > Ph\overset{\oplus}{C}H_2 > \overset{\oplus}{C}H_2-CH=CH_2$

$> R_2\overset{\oplus}{C}H > R\overset{\oplus}{C}H_2 > \overset{\oplus}{C}H_3 > CH_2=\overset{\oplus}{C}H- >$ (phenyl cation)

Reactions of carbocation Carbocations are reaction intermediates in following reactions:

- SN^1 (unimolecular nucleophillic substitution)
- E^1 reaction unimolecular elimination
- Electrophilic addition
- Molecular rearrangements like pinacole–pinacolone etc.

Carbanion or carbanium ion

It is anion having a negatively charged carbon atom with 8 electrons.

- It is formed as a result of heterolytic fission.

$CH_3 \overset{x}{\underset{o}{-}} H \longrightarrow \overset{-}{C}H_3{}^{x}_{o} + H^+$

- $R-CH_2 \overset{x}{\underset{o}{-}} MgBr \longrightarrow R-\overset{-}{C}H_2{}^{x}_{o} + \overset{+}{Mg}Br$
- It is sp^3 hybridized with pyramidal shape.
- It is diamagnetic in nature.

Order of stability of carbanion It is stabilized by resonance, hyperconjugation and electron attracting groups (–I effect) however it is de-stabilized by electron releasing groups like alkyl groups. Following gives the order of stability for various carbanions.

- $HC\equiv\overset{-}{C} > (Ph)_3\overset{-}{C} > (Ph)_2\overset{-}{C}H > Ph\overset{-}{C}H_2 >$

$\overset{-}{C}H_2-CH=CH_2 >$ (phenyl carbanion) $> \overset{\ominus}{C}H=CH_2 > \overset{\ominus}{C}H_3 > CH_3-\overset{\ominus}{C}H_2 >$

$(CH_3)_2\overset{-}{C}H > (CH_3)_3\overset{-}{C}$

- Stability of carbanion ∝ s % (percentage of s character)

$$\underset{sp}{R-C\equiv\ddot{\overset{-}{C}}} > \underset{sp^2}{R-CH=\ddot{\overset{-}{C}}H} > \underset{sp^3}{R-CH_2-\ddot{\overset{-}{C}}H_2}$$

Carbene

Carbene is a special form of carbon in which carbon atom is bivalent with two free or odd electrons.

$-\ddot{C}-$ $\quad -\dot{\underset{\cdot}{C}}-$

For example,

(1) Methylene

$H-\ddot{C}-H \quad H-\dot{\underset{\cdot}{C}}-H \quad CH_2\colon$

(2) Chloromethylene

$Cl-\ddot{C}-Cl \quad Cl-\dot{\underset{\cdot}{C}}-Cl \quad CCl_2\colon$

Types of carbene

Carbene are of the following types:

1. *Singlet carbine*

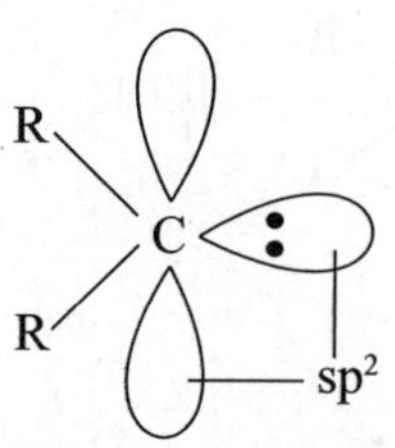

- It has two unshared electrons with opposite spins in one orbital.
- It is less stable, has more energy and is more reactive.
- It is sp^2 hybridized and has a bent shape.
- It shows similarity to carbocations.

2. *Triplet carbene*

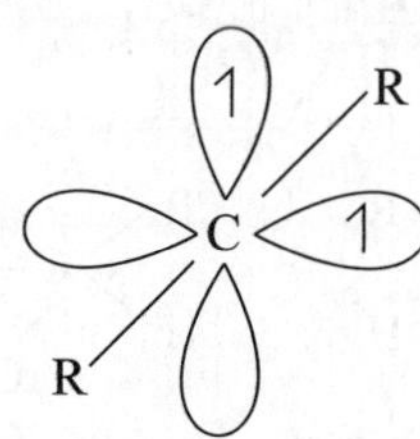

- It has two electrons with the same spin present in different orbitals, making it a free di-radical.
- It is sp hybridized with a linear shape.
- It is more stable and has less energy.

Formation of carbenes Carbenes are prepared as follows:

1. *From diazomethane*

$\underset{\text{Diazomethane}}{H_2C-\overset{+}{N}=\overset{-}{N}} \xrightarrow[-N_2]{hv} :CH_2$

2. *From ketene*

$CH_2=C=O \xrightarrow{hv} :CH_2 + CO\uparrow$

3. *From haloforms*

$CHX_3 \xrightarrow{\Delta \text{ alc. KOH}} \overset{-}{C}X_3 + H^+$

$\downarrow$

$\underset{\text{Halo carbene}}{\ddot{C}X_2} + \overset{-}{X}$

4. From carbonions: $CCl_3-C(=O)O^{-} \xrightarrow[-CO_2]{\Delta} \overset{\ominus}{C}Cl_3 \xrightarrow{-Cl^-} :CCl_2$

Reactions of carbenes

Carbenes show the following reactions:

1. *Addition with alkene*

$$\text{Cis-alkene (RHC=CHR)} + {\uparrow\downarrow}CH_2 \xrightarrow{\text{singlet carbene}} \text{Cis form (cyclopropane, } C\ H_2\text{)}$$

$$\text{RHC=CHR} + {\uparrow\downarrow}CH_2 \longrightarrow \text{Cis form} + \text{trans } (CH_2)$$

$$\text{trans alkene} + {\uparrow\downarrow}CH_2 \rightarrow \text{trans form } (CH_2)$$

2. *Addition of carbene at* 1, 2 *position in dienes*

$$H_2C=C=CH_2 + :CH_2 \longrightarrow H_2C—C=CH_2 \ (\text{ring with } CH_2)$$

3. *Insertion reaction* Here, carbene inserts itself in the chain which results in the increase in chain length.

$$-C-C-H + :CH_2 \longrightarrow -C-CH_2-C-$$

4. *Ring expansion reaction* (addition of halogen car-bene)

For example,

$$\text{Pyrrole (NH)} + :CCl_2 \xrightarrow{CHCl_3/C_2H_5ONa} \text{(dichlorocyclopropane adduct, } C(Cl)Cl\text{)} \xrightarrow{-HCl} \text{3-chloropyridine}$$

- *Carbenes as reaction intermediates:* Carbenes are reaction intermediate in Carbyl Amine reaction, Reimer–Tiemann reaction, Wittig reaction and Wolf rearrangement.

Stability of carbenes

- $\ddot{C}H_2 > \ddot{C}F_2 > \ddot{C}Cl_2 > \ddot{C}Br_2 > \ddot{C}I_2$

Stability decreases for singlet carbenes.

Nitrene: It is a neutral monovalent species of nitrogen which is just like carbene shown as given below:

Alkyl or aryl nitrene, $R—\ddot{N}:$ or $Ar—\ddot{N}:$

Acyl nitrene, $R—CO—\ddot{N}:$ or $Ar—CO—\ddot{N}:$

Formation of nitrenes

$$R—\ddot{N}=\overset{+}{N}=\overset{-}{\ddot{N}}: \xrightarrow{hv} R—\ddot{N}: + N\equiv N$$

Alkyl or aryl nitrene

$$R-N=C=O \xrightarrow{hr} R-\ddot{N}: + CO$$

$$R-\underset{\underset{O}{\|}}{C}-NH\,Br \xrightarrow[-BH]{B^{\ominus}} R-\overset{\overset{O}{\|}}{C}-\overset{\ominus}{\ddot{N}}-Br \xrightarrow{-Br^{\ominus}} R-\underset{\underset{O}{\|}}{C}-\ddot{N}:$$

Acyl Nitrene

Benzyne (1, 2 do-dehydro benzene)

Formation of benzyne

$$\text{Benzene (C}_6H_5X) \xrightarrow[\text{St. Base}]{NaNH_2 \text{ or } K.NH_2} \text{Benzyne} + X^-$$

- In benzyne, due to poor overlapping of porbitals triple bond is partial due to poor overlapping of porbitals as porbitals lie outside the plane of the ring in benzyne, so these are not properly aligned for overlapping.

sp^2

π-bond

sp^2

Free Radicals These are atoms or groups with a free or odd electron. They are neutral in nature, highly reactive and very short lived. Free radicals are formed as a result of homolytic fission.

$$A - A \longrightarrow A^o + {}^oA$$

- The presence of sunlight or heat is needed to prepare them. For example,

$$Cl - Cl \xrightarrow{\text{Sunlight}} Cl^o + {}^oCl$$

$$C_2H_5 - C_2H_5 \xrightarrow{\text{Sunlight}} C_2H_5{}^o + {}^oC_2H_5$$

$$CH_2 - CH_2 - CH_3 \xrightarrow{hv} CH_3 - {}^oCH - CH_3 + H^o$$

Features of free radicals

- It has one unpaired electron, so it is paramagnetic in nature.
- It may be sp^2 hybridized with a flat or trigonal coplanar structure.

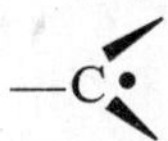

- Free radical shows disproportion reaction.

$$C_2H_5{}^o + {}^oC_2H_5 \longrightarrow C_2H_4 + C_2H_6$$

- Free radical does not have enantiomorphs due to continuous reversible exchange (that is, umbrella and it's reversal)

Order of stability of free radicals

Stability ∝ Resonance
∝ Ryperconjugation
∝ Presence of electron releasing groups

Following is the order of stability of free radicals:

$(Ph)_3\overset{o}{C} > (Ph)_2\overset{o}{C}H > Ph\overset{o}{C}H_2 >$ allyl > tertiary > secondary>

primary > vinyl free radical > $HC \equiv \overset{o}{C}$

- Vinyl free radical is the least stable, as it has no resonance or hyperconjugation.

ELECTRON DISPLACEMENT EFFECTS

Electron displacement effects are of the following types:

Inductive Effect

- Inductive effect is a permanent but weak effect involving displacement of σ electrons towards more electronegative species. It can be transmitted through a chain of C-atoms. (negligible at 4th on words C-atom.)

$$\overset{+\delta\delta\delta}{C} \rightarrow\!\!\!\rightarrow\!\!\!\rightarrow \overset{+\delta\delta}{C} \rightarrow\!\!\!\rightarrow \overset{+\delta}{C} \rightarrow \overset{-\delta}{C}$$

↓
Negligible onwards

Inductive effect can be of two types:

–I effect It is shown by electron attracting or withdrawing species. Here, sigma electrons are displaced away from the carbon atom towards this species.

$$\overset{+\delta}{C} \rightarrow \overset{-\delta}{X}$$

e^- attracting species

Following is the decreasing order of the strength of –I effect of various species:

–I gps in decreasing order of strength—

$-\overset{+}{N}R_3, -\overset{+}{S}R_2, -\overset{+}{N}H_3, \overset{\oplus}{O}R_2$ $-NO_2$, $-SO_2R$, –CN, $-SO_3H$, –COOH –F, –Cl, –Br, –I, –OAr, –COOR, –OR, –COR, –SH, –OH, $-NH_2$ –C≡CR, $C_6H_5^-$, –CH=CHR

- It increases acidic nature and stability of a carbanion.
- It decreases basic nature and stability of a carbocation.

+I Effect It is shown by electron releasing species. Here sigma electrons are displaced towards the carbon atom by the electron releasing species.

$$\overset{-\delta}{C} \leftarrow \overset{+\delta}{Y}$$

e^- releasing group

Following is the decreasing order of strength of + I effect: O^- > COO^- > tertiary alkyl group > secondary alkyl group > primary alkyl group > dentirium > hydrogen

- It increases basic nature and stability of a carbocation.
- It decreases acidic nature and stability of a carbanion.

- Inductive effect is responsible for high melting point, boiling point and dipole moment also, as it causes the compound to develop polarity.

Application of inductive effect

Stability of carbocations

Stability ∝ + I effect (electron releasing groups)

Stability $\propto \frac{1}{-\text{I effect}}$ (e^- withdrawing groups)

Order of stability:
tertiary > secondary > primary carbocation
For example,

$(CH_3)_3C^+ > (CH_3)_2 CH^+ > CH_3CH_2^+ > CH_3^+$

For example,

$CH_3CH_2^+ > CH_2^+ - CH_2 - C_6H_5 >$
$CH_2Cl - CH_2^+ > CH_2NO_2CH_2^+$

For example,

$CH_3CH^+OCH_3 > CH_3C^+HCH_3 > CH_3CH^+COCH_3 >$

Properties related to stability of carbocation

- Reactivity of –OH group in alcohol α stability of carbocation. For example,

(1) [structure: 1-butanol] (2) [structure: 2-propanol] (3) [structure: 2-methyl-1-propanol... OH] (4) [structure: tert-butanol]

The reactivity follows the order:
4 > 3 > 2 > 1

- Dehydration of alcohol ∝ Stability of carbocation. that is, t > s > p
- Reactivity of 'X'(halogen) in R – X ∝ Stability of carbocation.

Stability of carbanion

Stability $\propto \frac{1}{+\text{I effect}}$.

Stability ∝ –I effect

For example,

$\bar{C}H_3 > CH_3 - \bar{C}H_2 > (CH_3)_2 - \bar{C}H > (CH_3)_3\bar{C}$

- Here, +I effect increases the reactivity and hence decreases the stability of the carbanion.

For example,

(1) $CH_3 - \overset{\ominus}{C}H(NO_2)$ (2) $\overset{\ominus}{C}H_2 - CH_2(NO_2)$

(3) $CH_2(CH_3) - \bar{C}H_2$ (4) $CH_2(Cl) - \bar{C}H_2$

Stability follows the order:
1 > 2 > 4 > 3

Some important relations

Acidic nature of R – OH or Ester formation of R – OH or Reactivity of H– atom in R – OH ∝ solubility of alkoxide ion ∝ $\frac{1}{+\text{I effect}}$

For example,

$CH_3\text{–}OH > CH_3 - CH_2 - OH > (Me)_2CHOH > (Me)_3C - OH$

Basic nature of aliphatic amines

Basic nature of amines ∝ +I effect

$\propto \frac{1}{-\text{I effect}}$.

If R is methyl group, the basic nature of amines decreases as Secondary > primary >tertiary >NH_3 > $C_6H_5 - NH_2$ > ...

- The reason of lesser stability of tertiary amines is the steric hindrance caused by 3 bulky alkyl groups to donate electron by nitrogen atom. Moreover, poor solvation effect also decreases basic nature in this case.

⇒ *If R is other than methyl* the basic nature follows the order:

Secondary amine > tertiary amine > primary amine > NH_3

For example,
$(Pr)_2NH > (Pr)_3N > Pr.NH_2 > NH_3$
Here, Pr represents the propyl group (C_3H_7)

- Electron donating groups like (–R, –OR, –OH) increase the basic nature of aliphatic amines.
- Electron attracting groups like ($-NO_2$, –CN) decrease the basic nature of aliphatic amines.
For example, $CH_3 - CH_2 - NH_2 > NH_3 > Cl - NH_2$
- Basic nature ∝ Stability of conjugate acid formed

Acidic nature of carboxylic acids: The acidic nature of R-COOH depends upon case of H^+ and the stability of $R\text{–}COO^-$ formed.

$$R-C(=O)O^- \longleftrightarrow R-C(O^-)=O$$

$$= \left[R-C\overset{O}{\underset{O}{\lessgtr}}\right]^-$$

Resonance stabilized hybrid

- *Effect of electron withdrawing groups on acidic nature (– I effect)*

$$\left[X \leftarrow C\begin{matrix} \nearrow O \\ \searrow O \end{matrix}\right]^-$$

X (electron attracting groups) disperses the negative charge over the molecule and increases its stability and acidic nature.

- *Effect of electron releasing groups on acidic nature (+I effect)*

$$\left[Y \rightarrow C\begin{matrix} \nearrow O \\ \searrow O \end{matrix}\right]^-$$

Y (electron releasing groups) disperses less negative charge to decrease its stability and acidic nature.

Following are a few examples of order of acidic strength:

- $X - CH_2 - COOH > CH_3COOH > Y - CH_2 - COOH$
- $CH_2FCOOH > CH_2ClCOOH > CH_2BrCOOH > CH_2ICOOH$
- $CF_3COOH > CCl_3COOH > CBr_3COOH$
- $CCl_3COOH > CHCl_2COOH > CH_2ClCOOH$
- *In case of α, β, γ substituted acids*

$$-C-C-\underset{\alpha}{\overset{X}{C}}-COOH > -C-\underset{\beta}{\overset{X}{C}}-C-COOH > \ldots\ \gamma \ldots$$

- At high temperature, CH_3COOH is more acidic than CCl_3COOH. This is due to thermodynamic reason as Δ G is 1.5 kJ for CH_3COOH Δ G is 27 kJ for CCl_3COOH

Following are a few more examples of orders of acidic strength:

- $HCOOH > C_6H_5COOH > C_6H_5 - CH_2 - COOH > CH_3COOH$
- $COOH - COOH > COOH - CH_2 - COOH > HCOOH > CH_3COOH > C_2H_5COOH$

COOH–COOH	>	$CH_2(COOH)_2$	>	$(CH_2)_2(COOH)_2$	>
Oxalic Acid		Malonic Acid		Succinic Acid	

$(CH_2)_3(COOH)_2$	>	$(CH_2)_4(COOH)_2$	>	$(CH_2)_n(COOH)_2$
Glutaric Acid		Adipic Acid		Palmitic Acid

- Acidic nature ∝ s % (s character)

$$\underset{sp}{CH \equiv C} - COOH > \underset{sp^2}{CH_2 = CH} - COOH > \underset{sp^3}{CH_3\,CH_2}COOH$$

Electromeric Effect

$$X \overset{\curvearrowright}{=} Y \xrightarrow[\text{Reagent removed}]{\text{Polar reagent added}} X^+ - Y^-$$

- Electromeric effect is a temporary but more powerful effect than inductive effect.
- It involves the transfer of π e⁻ by the demand of attacking reagent towards more electronegative species.
- It is represented by the symbol (⌒).
- It always facilitates the reaction and does not inhibit it.
- It is shown by compounds having > C = C <, >C = O, –C ≡ N.
- It is .represented by the E (either +E or –E)

+E effect Here, the transfer of π electrons occur towards the attacking reagent.

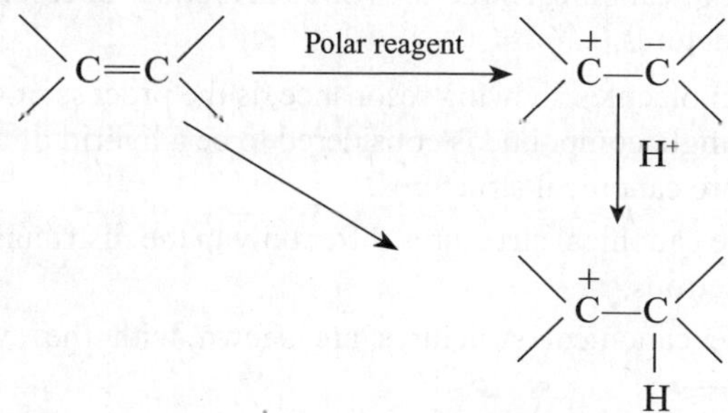

- If both the groups are same then the electromeric effect can be from any side.
- If both the groups are different then the electromeric effect is favoured towards the direction favouring the +I effect.

For example,

$$\begin{matrix} CH_3 & & H \\ & C=C & \\ H & & H \end{matrix} \longleftrightarrow \begin{matrix} CH_3 & & H \\ & \overset{+}{C}-\overset{-}{C} & \\ H & & H \end{matrix}$$

It is possible as (+I) favours it.

- **–E effect** Here the transfer to π electrons take place away from the attacking reagent.

$$>C=O \longleftrightarrow >\overset{+}{C}-\overset{-}{O} \xrightarrow{\overline{C}N^-} >\underset{CN}{C}-\overset{-}{O}$$

For example,

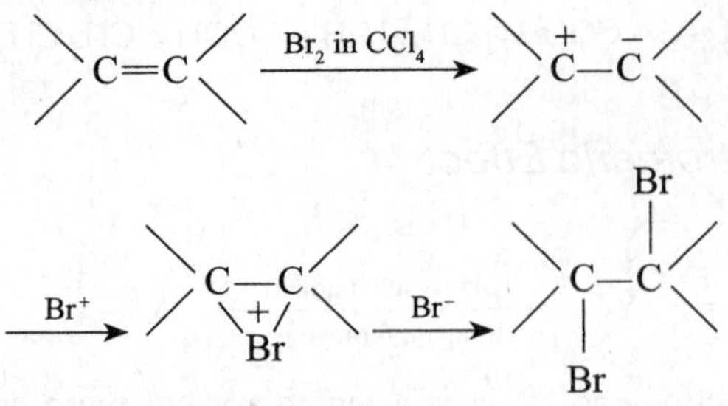

$$>C=C< \xrightarrow{Br_2 \text{ in } CCl_4} >\overset{+}{C}-\overset{-}{C}<$$

$$\xrightarrow{Br^+} >C-C< \text{ (bridged by } \overset{+}{Br}) \xrightarrow{Br^-} >C(Br)-C(Br)<$$

Cyclic bromonium ion

- As Br^+ is large in size and have lone pair of electrons so cyclic brominium ion is possible.
- Electromeric effect is helpful to explain the mechanism of electrophillic and nuelcophillic addition reactions.

Resonance

- When all the properties of a molecule cannot be explained by a single structure due to the delocalization of π electrons then such a molecule is represented by a number of canonical structures or contributing or resonating structures.
- In molecules showing resonance, is the process in which a single compound is considered to be a hybrid of two or more canonical structures.
- The canonical structures differ only in the distribution of electrons.
- The canonical structures are shown with the symbol (⟷).
- The canonical structures have the same arrangement of atoms.
- All canonical structures are planar or nearly planar.
- Total number of paired and unpaired electrons are same in all canonical structures of a particular compound.
- A few examples of position and arrangement of atoms are same in canonical structures are given below:

For example,

(1) $CH_2{=}CH-\ddot{\underset{..}{Cl}}: \longleftrightarrow \overset{-}{C}H_2-CH{=}\overset{+}{\underset{..}{Cl}}:$

(2) $CH_2 = CH - CH = CH_2$

$\updownarrow$

$CH_2^+ - CH = CH - CH_2^-$

$\updownarrow$

$CH_2^- - CH = CH - CH_2^+$

(3) 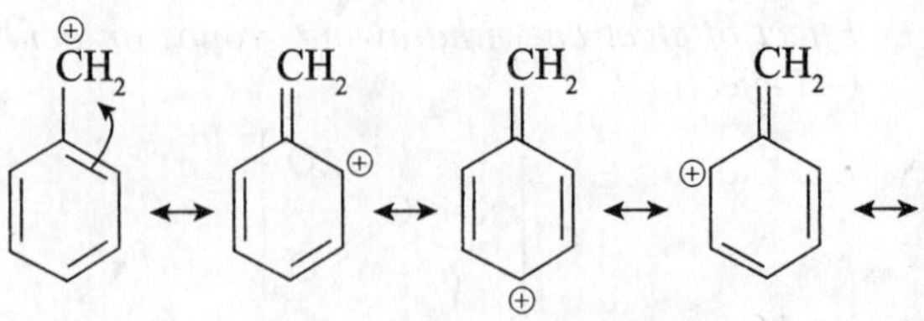 $\overset{\oplus}{C}H_2$ (benzyl cation) ⟷ CH_2 ⟷ CH_2 ⟷ CH_2 ⟷ (ring structures with ⊕ at ortho, para and ortho positions)

(4) 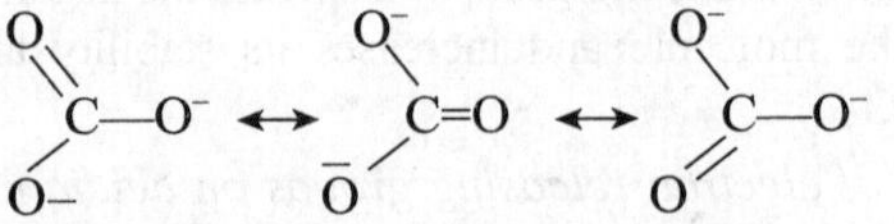 $O{=}C(-O^-)(-O^-)$ ⟷ $^-O-C(=O)-O^-$ ⟷ $^-O-C(-O^-)=O$

Effect of resonance in a molecule brings about changes in the bond length, bond order etc for example, in benzene carbon–carbon bond length is 1.39 Å. This value is intermediate of C–C (1.54 Å) and C = C (1.34 Å). Here bond order is 1.5.

Resonance energy

A Resonance hybrid in always more stable than any of its canonical structure due to the complete delocalisation of π e^-. The stability is measured in terms of Resonance energy which is the difference in internal energies of Resonance hybrid (actual molecule) and the most stable canonical structure.

Resonance Energy = Energy of most stable Canonical structure–Energy of resonance hybrid or actual molecule.

- Resonance energy ∝ Number of Canonical structure
  Resonance energy ∝ Stability
  Resonance energy $\propto \dfrac{1}{\text{Reactivity}}$
  Resonance energy = (Expected heat of hydrogenation) – (Actual heat of hydrogenation).
- Due to high resonance energy, benzene is quite stable and undergoes electrophilic substitution reactions but it does not undergo addition reaction despite having 3 double bonds (due to delocalization of π electrons or resonance).
- The resonance energy of benzene has 36 kcal/mole.
- The resonance energy of CO_2 is 154.9 kJ/mole.
- Resonance differs from tautomerism. Tautomers differ in the arrangement of atoms but in resonance, the arrangement of atoms remains unaltered.

Stability of different canonical structure

A non-polar structure is always more stable than a polar structure.

For example,

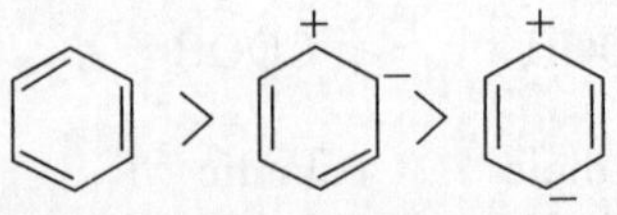

(benzene) > (polar structure with + and – adjacent) > (polar structure with + and – para)

As opposite charges are apart, so it is less stable.

- For polar structure, stability can be inferred applying the following rules:
  1. Greater the number of covalent bonds greater will be the stability
     For example,
     $CH_3-\overset{+}{C}\equiv O > CH_3-\overset{+}{C}=O$
     More number of covalent bonds
  2. The canonical structure in which positive charge in on positive atom and negative charge on negative atom is more stable.
     For example, 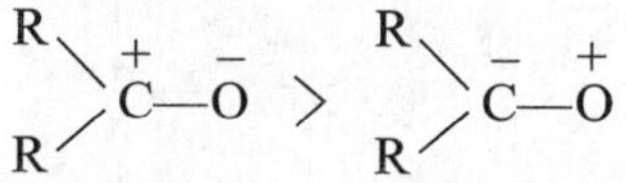
     $R_2\overset{+}{C}-\overset{-}{O} > R_2\overset{-}{C}-\overset{+}{O}$
  3. The canonical structure in which each atom has octet state is the more stable state.
     For example,
     $\underset{8}{CH_3-C\equiv\overset{+}{O}} > \underset{6}{CH_3-\overset{+}{C}=O}$
  4. If the same charges are closer, the particular structure will be unstable.
     For example,
     $:\overset{-}{N}=\overset{+}{N}=O \longrightarrow N\equiv\overset{+}{N}-\ddot{\overset{\cdot\cdot}{O}}:$
     Most stable
     $\longleftrightarrow :\overset{\cdot\cdot}{\overset{+}{N}}=\overset{+}{N}\equiv O$
     Unstable

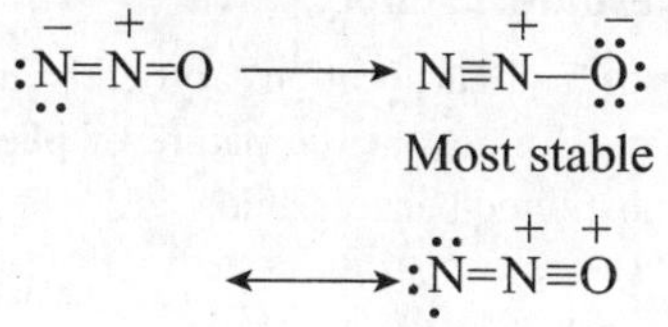

Types of resonance

Isovalent resonance When in resonating structures, the same number of bonds and same type of charges are present, it is called isovalent resonance.

For example, SO_2, NO_2, CO_3^{-2}

Heterovalent resonance When in resonating structures, different number of bonds and charges are present, it is called hetrovalent resonance.

For example, buta-1, 3 diene, vinyl cyanide.

Hyperconjugation

- This concept was introduced by Baker and Nathan. It is called **no-bond resonance** or **σ-π conjugation**. The magnitude of effect of hyperconjugation is smaller than resonance, so it is also called secondary resonance.
- "Alkyl groups with atleast one α H-atom attached with an unsaturated C-atom, is able to release electrons as in electromeric effect (delocalization)".

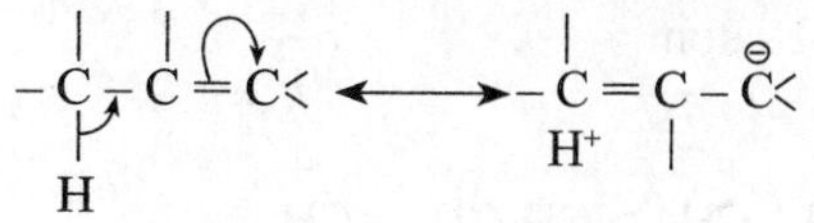

$$-\underset{H}{\overset{|}{C}}-\overset{|}{C}=C< \longleftrightarrow -\overset{|}{C}=C-\overset{\ominus}{C}< \quad H^+$$

e.g.,

$$H-\underset{H}{\overset{H}{C}}-CH=CH_2 \leftrightarrow H-\underset{H}{\overset{H^+}{C}}=CH-\overset{\ominus}{C}H_2 \leftrightarrow$$

$$\overset{+}{H}\ \underset{H}{\overset{H}{C}}=CH-\overset{\ominus}{C}H_2 \leftrightarrow H-\underset{H^+}{\overset{H}{C}}=CH-\overset{\ominus}{C}H_2$$

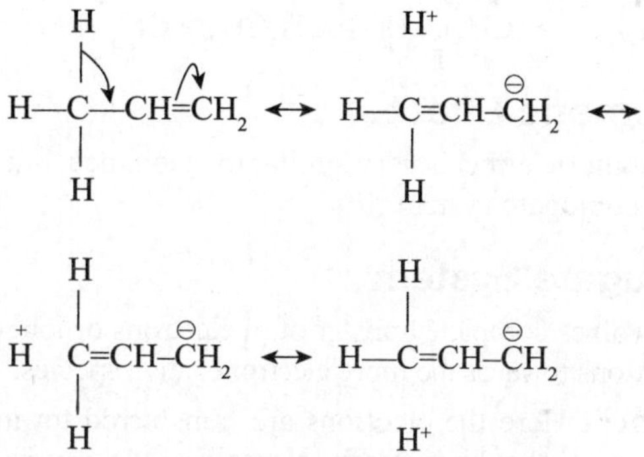

Hyperconjugation ∝ Number of α H-atoms

$$CH_3- > CH_3-CH_2- > (CH_3)_2CH- > (CH_3)_3C-$$

3α–H	2α–H	1α–H	Noα–H
Minimum hyperconjugation			Maximum hyperconjugation

- Hyperconjugation causes the release of electrons to the same magnitude as occurs in + inductive effect; however, the direction of the release of electron is opposite to that of + inductive effect.
- The magnitude of effect of hyperconjugation is smaller than resonance so it is also called secondary resonance.
- Hyperconjugation increases the stability in alkenes by decreasing heat of hydrogenation, for example, pent-2-ene is more stable than 1-pentene. Similarly, 2-butene or isobutene is more stable than butene-1

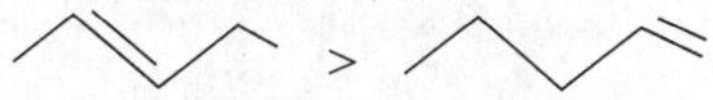

(pent-2-ene) > (pent-1-ene)
5 α-H-atoms 2 α-H-atoms

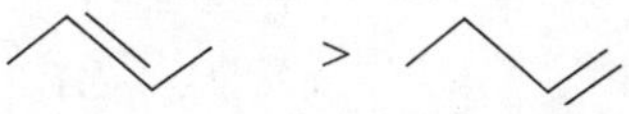

(but-2-ene) > (but-1-ene)
6 α-H-atoms 2 α-H-atoms

- Hyperconjugation results in a change in bond length, like resonance.

For example,

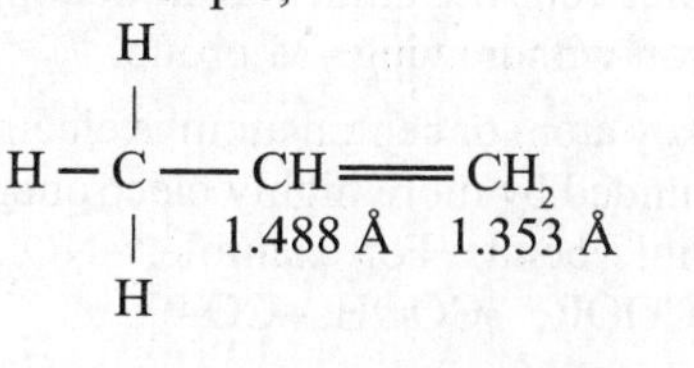

$$H-\underset{H}{\overset{H}{C}}-CH=CH_2$$

1.488 Å 1.353 Å

- Stability of carbocation and free radical
  $\propto$ Hyperconjugation
  $\propto$ Number of α-H-atom
  For example,

$$(CH_3)_3\overset{+}{C}- > (CH_3)_2\overset{+}{C}H- > CH_3\overset{+}{C}H_2 > \overset{+}{C}H_3$$

$$(CH_3)_3\overset{o}{C}- > (CH_3)_2\overset{o}{C}H- > CH_3\overset{o}{C}H_2 > \overset{o}{C}H_3$$

Mesomeric Effect

- Mesomeric effect is very similar to resonance, but occurs in a conjugate system.

Conjugate System

- It involves complete transfer of pi electrons or lone-pair of electrons towards the more electronegative species.

+M effect Here the electrons are transferred towards the conjugate system by electron releasing (+M) groups.

- In the +M group, the key atom or central atom has lone pair of electrons to donate, for example,
  $-NH_2$, $-NHR$, $-NR_2$, $-OH$, $-OR$, $-O-CO-R$, $X-$ (Cl, Br, I)
- When such groups are attached to benzene ring they increase the rate of electrophilic substitution, that is, they are ring activating and ortho- and para-directing groups as they increase electron density at ortho- and para-positions as shown below:

H—C—H ⟷ H—C H ⊕ ⊖ ⟷ H⊕ H—C—H ⊖ ⟷ ⊕ H C—H ⊖

—M effect Here, the electrons are attracted from conjugate system by the electron withdrawing –M group.

- In the –M group, the key atom or central atom is electron deficient and is surrounded by more highly electronegative atoms with multiple bonds. For example, $-NO_2$, $-SO_3H$, $-CN$, $-CX_3$, $-COOR$, $-COOH$, $-CO-$
- When such groups are attached to a benzene ring, they decrease the rate of electrophilic substitution since they are ring deactivating and meta-directing groups. They decrease the electron density at ortho- and para-position as shown below:

X—C—X ⟷ X—C X⊖ ⊕ ⟷ X⊖ X—C—X ⊕ ⟷ X⊖ C—X ⊕

- In these cases, electron density at meta-position remains unaffected, hence the electrophilic substitution can take place only at this position.

> **Mesomeric effect**
>
> Explains the stability of aryl carbocation, aryl carbanion, basic nature of aromatic amines, acidic nature of phenol and acidic nature of substituted benzoic acid.

Consequences of mesomeric effect

Stability of substituted benzyl carbocations:

The stability of a substituted benzyl carbocation depends on the nature of the group attached to the benzene ring as a consequence of which, +I, –I , +R or –R effect may occur. Here, R represents mesomeric effect.

- *Case-1 Groups causing – I and – R effects*
  m-derivative > p-derivative > o-derivative ⟶
  Decreasing order of stability
  For example,

$\overset{+}{C}H_2$, NO_2 (1) $\quad$ $\overset{+}{C}H_2$, NO_2 (2) $\quad$ $\overset{+}{C}H_2$, NO_2 (3)

Stability decreases as 2 > 3 > 1

- *Case 2 Groups causing + I and + R effetcs*

o-derivative > p-derivative > m-derivative

Decreasing order of stability

For example

(1) (2) (3)

1 > 3 > 2

- *Case 3 Groups causing + R and – I effect*

p-derivative > o-derivative > m-derivative

Decreasing order of stability

For example

(1) (2) (3)

3 > 1 > 2

- In case of halogen derivatives, stability depends only on the strength of –I effect caused by the group.

p-derivative > m-derivative > o-derivative

Decreasing order of stability

For example,

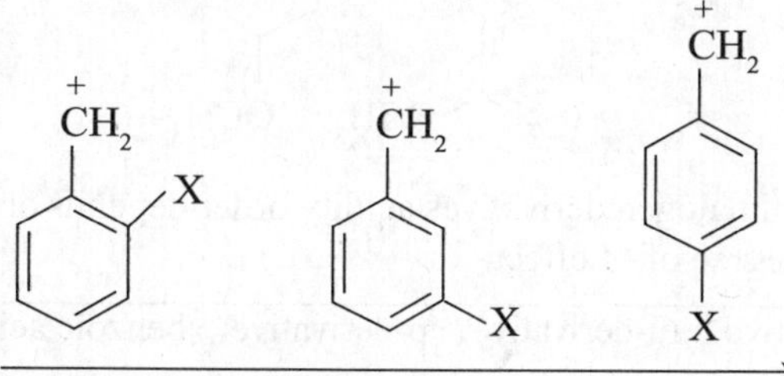

Stability in Increasing order

Stability of substituted benzyl carbanions

- *Case 1 Groups causing – R and – I effects*

o-derivative > p-derivative > m-derivative

Decreasing order of stability

For example,

- *Case 2 Groups causing + R and + I effects*

m-derivative > p-derivative > o-derivative

Decreasing order of stability

For example,

- *Case 3 Groups causing + R and – I effects*

m-derivative > o-derivative > p-derivative

Decreasing order of stability

For example,

Decreasing order of stability

Acidity of phenol Phenol is weakly acidic in nature due to the formation of phenoxide ion, which is resonance stabilized, that is, more the stability of phenoxide ion, more will be the acidity of phenol.

$$C_6H_5OH \longrightarrow C_6H_5O^{\ominus} + H^{\oplus}$$

Resonance Stabilisation of Phenoxide ion.

Acidity of substituted phenols It depends on the stability of the phenoxide ion, as acidity is the function of the stability of acid anion formed.

- *Case* 1 *Groups causing – R and – I effects*

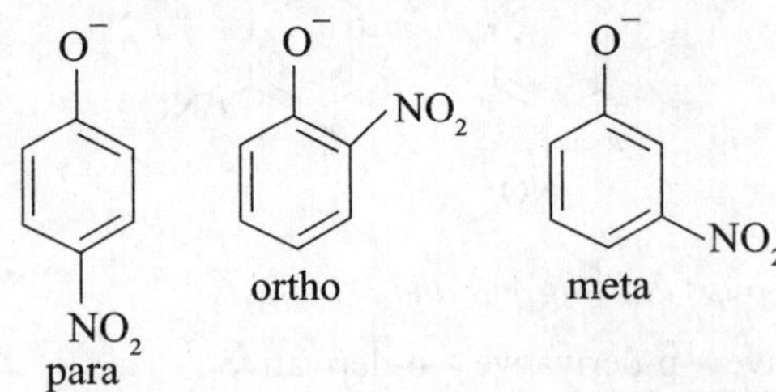

p-derivative > o-derivative > m-derivative > phenol

Acidity in decreasing order

- *Case* 2 *Groups causing + R and + I effects*

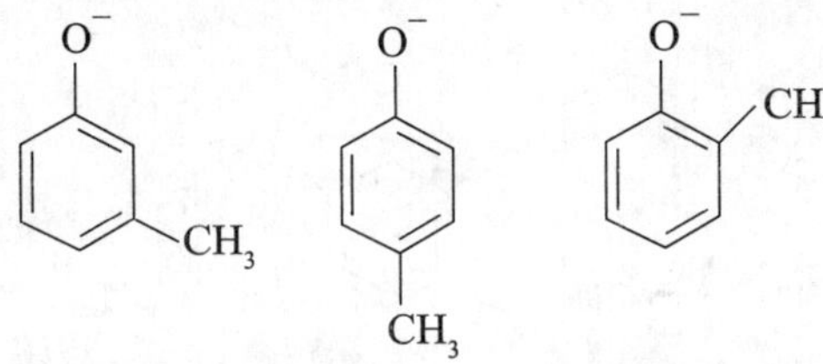

Phenol > m-derivative > p-derivative > o-derivative

Acidity in decreasing order

- *Case* 3 *Groups causing + R and – I effects.* Here the order m > o > p

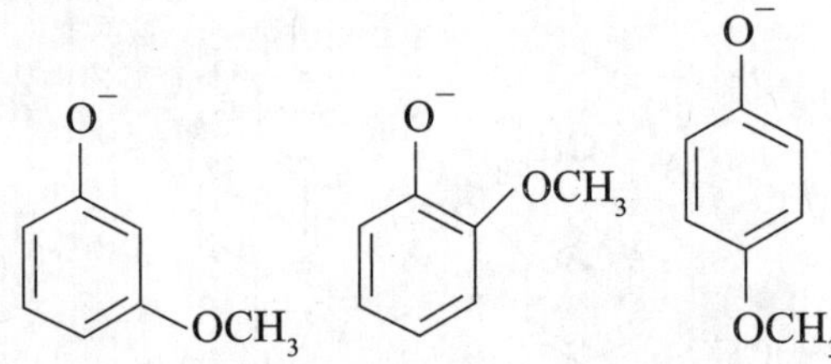

- In case of halogen substituted phenol (except fluoro), acidity depends on –I effect only and follows the given order:

  o-derivative > m-derivative > p-derivative > phenol

  Acidity in decreasing order

  In case of fluoro group, acidity order is as follows:

  m-derivative > p-derivative > o-derivative > phenol

  Acidity in decreasing order

- *Acidity of monosubstituted benzoic acid derivatives* It is increased by electron withdrawing groups, which stabilize $C_6H_5COO^-$ ion while electron releasing group decrease acidity by decreasing the stability of $C_6H_5COO^-$.
- An ortho substituted benzoic acid is stronger than m- or p-substituted benzoic acid. This is called **ortho effect.** Here bulky alkeyl groups coused steric imbibition of Resonance to increase acidic nature.
- *Case* 1 *Groups causing – R and – I effects*

  o-derivative > p-derivative > m-derivative > benzoic acid

  Acidity in decreasing order

  For example,

  o-nitrobenzoic acid (COOH, NO_2) > p-nitrobenzoic acid (COOH, NO_2) > m-nitrobenzoic acid (COOH, NO_2) > benzoic acid (COOH)

- *Case* 2 *Groups causing + R and + I effects*

  o-derivative > Benzoic acid > m-derivative > p-derivative

  Acidity in decreasing order

  For example,

  o-toluic acid (COOH, CH_3) > benzoic acid (COOH) > m-toluic acid (COOH, CH_3) > p-toluic acid (COOH, CH_3)

  Due to ortho effect

- *Case* 3 *Groups causing + R and – I group.*

  o-derivative > benzoic acid > m-derivative > p-derivative

  Acidity in decreasing order

  For example,

  o-methoxybenzoic acid (COOH, OCH_3) > m-methoxybenzoic acid (COOH, OCH_3) > p-methoxybenzoic acid (COOH, OCH_3)

- In case of halogen derivatives acidity order depends only on the degree of –I effect.

  o-derivative > m-derivative > p-derivative > benzoic acid

  Acidity in decreasing order

Basicity of compounds containing nitrogen: It is decided by the relative availability of the non-bonding electrons (lone pain) on the nitrogen atom to a proton donor or Lewis acid and by the stabilization of the positively charged nitrogen atom by solvation or, in some special cases, by resonance.

- **Basicity of aliphatic amines** It is measured by the position of equilibrium in reaction with water. The further the reaction goes to the right, the greater is the basic strength.

$$>\ddot{N} + HOH \rightleftharpoons >\overset{\oplus}{N}-H + \overset{\ominus}{O}H$$

Ammonium ion

Basicity $\propto$ Stability of ammonium ion

Stability of ammonium cation depends on the following two factors:

1. Lesser the magnitude of positive charge on the nitrogen atom, greater will be the stability
   ↓
   Basicity $\propto$ +I power of group
   Basicity $\propto \dfrac{1}{-\text{I power of group}}$
2. Greater the degree of salvation, higer is the stability of the cation
   Basicity $\propto$ Stability of cation
   $\propto$ Degree of solvation

The degree of solvation of ammonia, primary amines and secondary amines is almost the same but degree of solvation of tertiary amines is extremely low, to the extent of being almost zero.

- In aqueous solution, order of basicity for aliphatic amines—when alkyl group is methyl—is as follows:
  Secondary amine > primary amine > tertiary amine > NH_3
  Basicity of aliphatic amines in decreasing order →

Facts To Remember

In gaseous state, the basic strength follows the order:
Tertiary amine > secondary amine > primary amine
In case R is other than $-CH_3$ like Et Pr then the order is s° > t°> p°

- *Basicity of aromatic amines* It depends on electron density of nitrogen; higher the electron density of nitrogen, more will be the basicity. In aromatic amines, the lone pair of electrons present on nitrogen is delocalized, hence electron density decreases due to resonance so basicity decreases with resonance.

$$\text{Basicity} \propto \frac{1}{\text{Number of resonating structures}} \rightarrow$$

For example,

$C_6H_5NH_2 > (C_6H_5)_2NH > (C_6H_5)_3N$
4 7 10

Basicity of substituted anilines: A para-substituted aniline is more basic than ortho- substituted aniline and this effect is called **para effect.**

- *Case 1 groups causing – R and – I effects*
  aniline > m-derivative > p-derivative > o-derivative
  Basicity in decreasing order →
  For example, 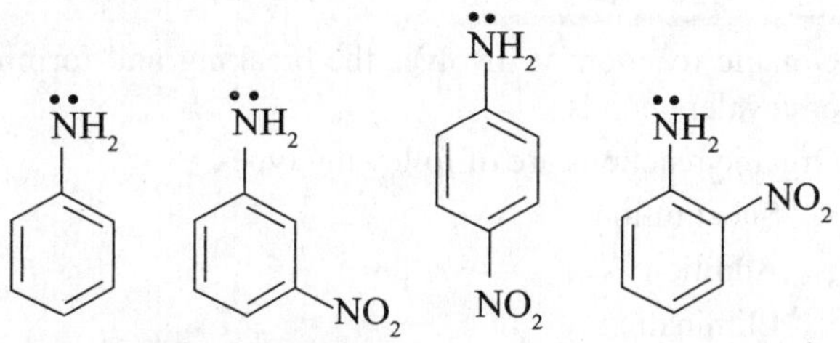
  aniline ($\ddot{N}H_2$), m-nitroaniline (NO_2), p-nitroaniline (NO_2), o-nitroaniline (NO_2)
  Basicity in decreasing order →
- *Case 2 Groups causing + R and + I effects*
  p-derivative > o-derivative > m-derivative > aniline
  Basicity in decreasing order →
  For example, 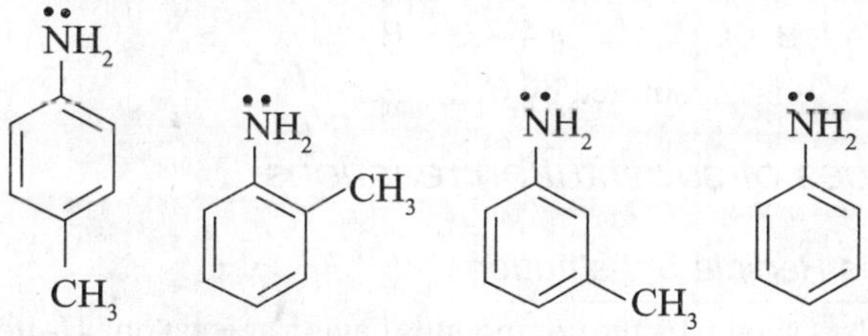
  p-toluidine ($\ddot{N}H_2$, CH_3), o-toluidine ($\ddot{N}H_2$, CH_3), m-toluidine ($\ddot{N}H_2$, CH_3), aniline ($\ddot{N}H_2$)
  Basicity in decreasing order →
- *Case 3 Groups causing + R and – I effects*
  p-derivative >aniline > o-derivative > m-derivative
  Basicity in decreasing order →
  For example, 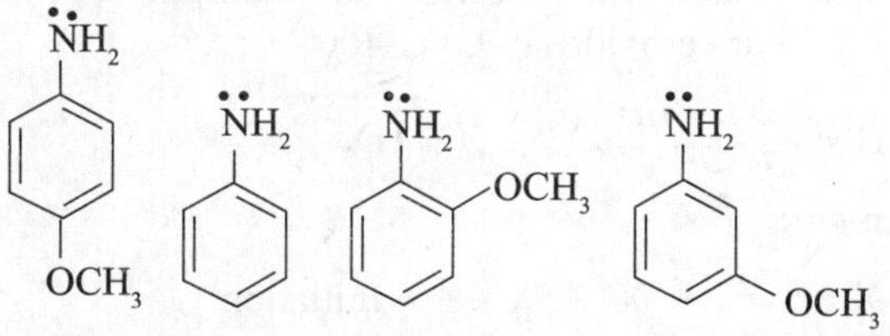
  p-anisidine ($\ddot{N}H_2$, OCH_3), aniline ($\ddot{N}H_2$), o-anisidine ($\ddot{N}H_2$, OCH_3), m-anisidine ($\ddot{N}H_2$, OCH_3)
  Basicity in decreasing order →

Facts To Remember

Basicity of some nitrogen compounds in decreasing order:
Guanidine > Aliphatic amines > NH_3 > Aromatic amines > Imine > Amide > Cyanide
Basicity in decreasing order →

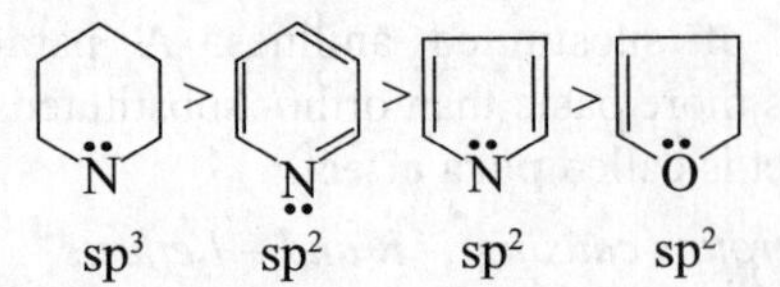

Reaction Mechanisms

- Organic reaction: It involves the breaking and forming of covalent bonds.

  Organic reactions are of following types:
  - Substitution
  - Addition
  - Elimination
  - Rearrangement
  - Pyrolysis and decomposition

Substitution or Displacement Reaction

In substitution reaction, one atom or group is substituted by another atom or group. For example,

$$A-B+C \longrightarrow A-C+B$$

Here, B is substituted by C.

Types of substitution reactions

Free Radicle Substitution

- It is shown by the compounds having at least one H-atom on sp^3 hybridized carbon-atom, that is, in the saturated compounds.
- Here substitution occurs due to a free radical, which means that a weak free radical substituent is substituted by a stronger one.
- It takes place in presence of sunlight or heat ($\geq 500°C$) or Peroxide (R–O–O–R)

$$R-H+X_2 \xrightarrow{\text{Sunlight}} R-X+HX$$

Mechanism:

$$X-X \longrightarrow X^o + {}^oX \quad \text{Initiation}$$

$$R-H \xrightarrow{X^o} R^o + H^o - X \quad \text{Propagation}$$

$$R^o + X_2 \longrightarrow R-X+X^o$$

$$R^o + R^o \longrightarrow R-R \quad X^o + X^o \rightarrow X-X \text{ Termination}$$

- Order of reactivity for halogenation towards free radical substitution: $F_2 > Cl_2 > Br_2 > I_2$
- Order of reactivity of alkanes towards free radical substitution: The order of substitution of H-atom in alkanes is decided by the stability of free radicals, that is,

  benzyl > allyl > $3^o > 2^o > 1^o > CH_3{}^o$

  For example,

$$(CH_3)_3CH + Cl-Cl \xrightarrow{hv} \underset{35\%}{(CH_3)_3C.Cl} + \underset{65\%}{CH_3-CH(CH_3)-CH_2Cl}$$

- In case of chlorination

$$\text{Reactivity} \propto \frac{1}{\text{Selectivity (that is, the percentage of product formed)}}$$

- The reaction rate of formation of alkyl radical by a Cl° is

  tertiary > secondary > primary
  5 3.8 1

$$\underset{\text{Toluene}}{C_6H_5CH_3} \xrightarrow[-HCl]{Cl_2,\ hv} \underset{\text{Benzyl chloride}}{C_6H_5CH_2Cl}$$

- Cl_2, SO_2Cl_2, $(CH)_3COCl$ are a few examples of **chlorinating reagents.**
- Bromine and n-bromosuccinamide (NBS) are examples of **brominating reagents.**
- Bromination is possible only in case of benzyl, Allyl group compounds.

$$\underset{\text{Isobutane}}{(CH_3)_3C-H} \xrightarrow[127°C]{Br_2/hv} \underset{\text{Tertiary butyl bromide}}{Me_3C-Br}$$

$$\underset{\text{N-bromosuccinamide}}{\text{N—Br}} \xrightarrow[CCl_4 \text{ or } CS_2]{hv\Delta} N^o + Br^o$$

$$\underset{\text{Propene}}{CH_2=CH-CH_3} \xrightarrow{Br^o} \underset{\text{Allyl free radical}}{CH_2=CH-\dot{C}H_2} + HBr$$

$$\text{N-Bromosuccinimide (N–Br)} + HBr \longrightarrow \text{Succinimide (N–H)} + Br_2$$

$$CH_2 = CH - \dot{C}H_2 + Br_2 \longrightarrow$$

$$\begin{array}{c} CH_2 = CH - CH_2 + Br^{o} \\ \quad\quad\quad\ \ \text{Br} \end{array}$$

$$\text{Cyclohexene} \xrightarrow[CCl_4]{NBS/hv} \text{3-bromocyclohexene}$$

Cyclohexene 3-bromocyclohexene

Electrophilic Substitution

- In electrophilic substitution, an electrophile substitutes another molecule or group from an organic compound. A weak electrophile present in the organic compound is substituted by a stronger electrophile. Positively charged species like NO_2^+ and X^+ act as electrophiles.
- It is a two step process proved by isotopic tracer technique.
- *Activating groups* These groups increase electron density at ortho- and para-positions, which means that these groups are ortho- and para-directing. These are highly ring activating groups so electrophilic substitution occurs at a very fast rate.

  Some activating groups in decreasing order are given below:

  $-NH_2$, –NHR, $-NR_2$, –OH, –OR, –NHCOR, φ, –R

  Deactivating groups: These groups are meta-directing and deactivate benzene ring, that is, they decrease electron density at ortho- and para-positions thereby making electrophilic substitution slower and occur only at the meta-position.

  Some deactivating groups in decreasing order are shown below:

  $\overset{\oplus}{N}R_3$, $-NO_2$, –CN, –COOH, –COOR, $-SO_3H$, –CHO, –COR

  Halogenation Here, substituent or electrophile is X^+. The hydrogen atom of benzene ring is substituted by X^+.
- The reaction occurs in cold and dark conditions or in presence of halogen carriers like Fe, $FeCl_3$, $SnCl_4$, ICl as shown below:

$$C_6H_6\ (\text{H}) + Cl—Cl \xrightarrow{FeCl_3} C_6H_5Cl + HCl$$

Mechanism of halogenation

$$\underset{\text{Halogen}}{X–X} + \underset{\text{Halogen carrier}}{FeX_3} \longrightarrow X^+ + \underset{\text{Electrophile}}{FeX_4^-}$$

$$\text{Benzene} + \underset{\text{Electrophile}}{X^+} \longrightarrow \underset{\pi\text{-complex}}{[C_6H_6 \to X^+]}$$

Rate determining step (r.d.s.)

$$\left[\text{H, X arenium ion (+ ortho)} \longrightarrow \text{(+ para)} \longrightarrow \text{(+ ortho')} \right]$$

$$\cong \ \text{H, X} \ (\text{+ delocalized})$$

σ-complex (Resonance hybrid)

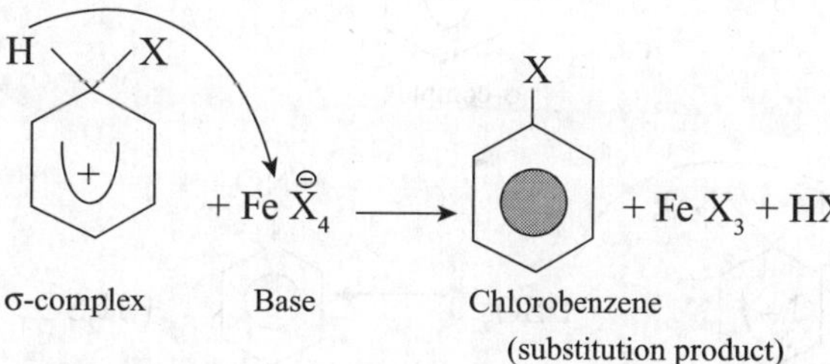

- Here, $AlCl_3$ and anhydrous $ZnCl_2$ can also be used as halogen carriers to polarize the attacking molecules (X_2).
- In molecules like phenol and aniline, there is no need for halogen carriers because both OH and NH_2 groups are highly ring activating.

Nitration

Here, the electrophile is NO_2^+ and the nitrating mixture is $HNO_3 + H_2SO_4$ or $HNO_3 + (CH_3CO)_2O$

$$C_6H_6\ (\text{H}) + HO—NO_2 \xrightarrow[H_2SO_4]{\Delta\ \text{conc.}} C_6H_5NO_2 + H_2O$$

Mechanism of nitration

$$H_2SO_4 + H–\ddot{O}–\overset{+}{N}\begin{matrix} ^{\nearrow O} \\ _{\searrow O^-} \end{matrix} \longrightarrow \underset{\text{Protonated nitric acid}}{H–\overset{+}{\underset{|}{O}}(H)–\overset{+}{N}\begin{matrix} ^{\nearrow O} \\ _{\searrow O^-} \end{matrix}} + \underset{\text{Base}}{HSO_4^-}$$

$$H-\overset{+}{O}(H)-\overset{+}{N}(=O)-O^- \rightleftharpoons H_2O + \overset{+}{N}O_2$$

Nitronium ion (Electrophile)

$$C_6H_6 + \overset{+}{N}O_2 \longrightarrow \text{(π-complex)} \; NO_2^+ \xrightarrow{\text{(r.d.s.)}} \left[\text{resonance structures of } C_6H_6(H)(NO_2)^+\right] \equiv \sigma\text{-complex}$$

σ-complex

$$\sigma\text{-complex} + HSO_4^- \longrightarrow C_6H_5NO_2 + H_2SO_4$$

σ-complex Nitrobenzene (Substitution product)

Sulphonation Here, electrophile is SO_3.

- It is catalysed in the presence of $HgSO_4$ or I_2.
- H_2SO_4 is the main sulphonating agent. However, $ClSO_3H$ dissolved in CCl_4 can also be used in some cases.

$$C_6H_6 + HO-SO_3H \xrightarrow[\text{Boil}]{\Delta} C_6H_5SO_3H + H_2O$$

Mechanism of Sulphonation

Concentrated sulphuric acid releases a molecule of SO_3.

$$H_2SO_4 + H_2SO_4 \rightleftharpoons H_3O^+ + HSO_4^- + SO_3$$

The electrophile attacks the benzene ring to form a σ-complex.

$$C_6H_6 + \overset{+}{S}O_3^- \longrightarrow \text{(π-complex)} + \overset{+}{S}O_3^- \longrightarrow \left[\text{resonance structures of } C_6H_6(H)(SO_3^-)^+\right] \equiv \sigma\text{-complex}$$

σ-complex (Resonance stabilized)

$$\sigma\text{-complex} + HSO_4^- \xrightarrow{\text{R.D.S}} C_6H_5SO_3^- + H_2SO_4$$

$$C_6H_5SO_3^- + H_3O^+ \longrightarrow C_6H_5SO_3H + H_2O$$

Benzene sulphonic acid (Substitution product)

Abstraction of H^+ is rate determining step here.

Friedel crafts reaction

In this reaction, the commonly used electrophiles are:

$$R^+,\ RCO^+,\ {}^+\overset{O}{\overset{\|}{C}}-H,\ {}^+\overset{O}{\overset{\|}{C}}-X,\ HC^+=NH$$

- This reaction is mainly catalysed by anhydrous $AlCl_3$. However, $FeCl_3$, $SnCl_4$, BF_3, $ZnCl_2$ can also be used.

  The decreasing order of these is as follows:

  $AlCl_3 > FeCl_3 > BF_3 > TiCl_3 > ZnCl_2 > SnCl_4$

Alkylation

Here, electrophile is R^+

$$C_6H_6 + R^+ \longrightarrow C_6H_6(H)(R)^{\oplus} \xrightarrow{-H^+} C_6H_5R$$

σ-complex

Mechanism of alkylation

$$\underset{\text{Alkyl halide}}{RCH_2Cl} + \underset{\text{Lewis acid}}{AlCl_3} \longrightarrow \underset{\text{Electrophile}}{RCH_2^+} + \underset{\text{Base}}{AlCl_4^-}$$

$$C_6H_6 + \overset{+}{C}H_2R \longrightarrow [C_6H_6 \rightarrow \overset{+}{C}H_2R] \xrightarrow{\text{(r.d.s)}} \sigma\text{-complex}$$

σ-complex
(Resonance stabilized)

$$\underset{\sigma\text{-complex}}{C_6H_6^+(CH_2R)} + \underset{\text{Base}}{AlCl_4^-} \longrightarrow \underset{\substack{\text{Alkyl benzene}\\ \text{(Substitution product)}}}{C_6H_5CH_2R} + HCl + AlCl_3$$

- To form a more stable carbocation, rearrangement must take place in this order:
  tertiary > secondary > primary
- R^+ is obtained from groups like R–X, RCH = CHR, R – OH and so on.

  Examples,

1. $C_6H_6 + C_2H_5Br \longrightarrow C_6H_5C_2H_5 + HBr$
2. $C_6H_6 + CH_3—CH_2—CH_2Cl \longrightarrow C_6H_5CH(CH_3)_2$
3. $C_6H_6 + (CH_3)_2CHCH_2Cl \longrightarrow C_6H_5C(CH_3)_3 + HCl$

- From alkene (in presence of H_2SO_4)

  Examples,

1. $C_6H_5CH{=}CH_2 + C_6H_6 \xrightarrow[H_2SO_4]{\Delta\ conc.} C_6H_5—CH(C_6H_5)—CH_3$
2. $C_6H_6 + (CH_3)_2C{=}CH_2 \xrightarrow[H_2SO_4]{\Delta\ conc.} C_6H_5C(CH_3)_3$
3. $C_6H_6 + (CH_3)_2CH = CH_2OH \xrightarrow[H_2SO_4]{\Delta\ conc.} C_6H_5C(CH_3)_3$

Some specific examples

1. $\Phi - H + CH_2Cl_2 \xrightarrow[AlCl_3]{Anhy.} (\Phi)_2CH_2 + 2HCl$
2. $\Phi - H + CHCl_3 \xrightarrow[AlCl_3]{Anhy.} (\Phi)_3CH + 3HCl$

- *Acylation* Acylation comprises two processes namely acetylation and benzylation. In acetylation, the electrophile is CH_3CO^+ and in benzylation the electrophile is $C_6H_5CO^+$.

  Examples,

1. $C_6H_6 + CH_3\,CO\,Cl \longrightarrow C_6H_5CO\,CH_3 + HCl$
2. $C_6H_6 + (CH_3CO)_2O \longrightarrow C_6H_5CO\,CH_3 + CH_3COOH$

Mechanism of acylation

$$R–CO–Cl + AlCl_3 \longrightarrow R–C^+O + AlCl_4^-$$

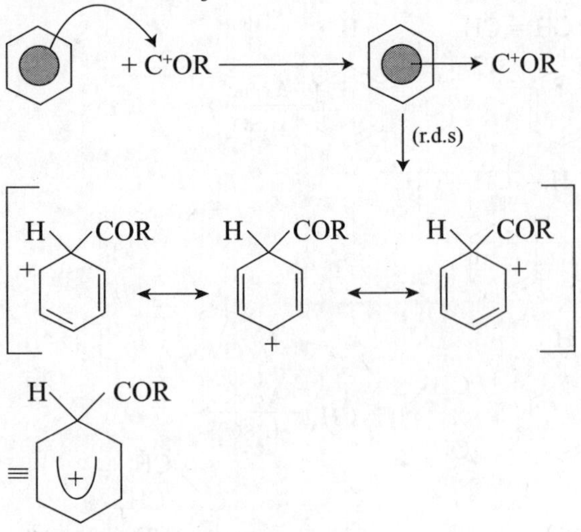

$$C_6H_6 + C^+OR \longrightarrow C_6H_6 \rightarrow C^+OR$$

(r.d.s)

H COR H COR H COR

+ ⟷ ⟷ +

\+

H COR

≡ (+)

σ-complex
(Resonance stabilized)

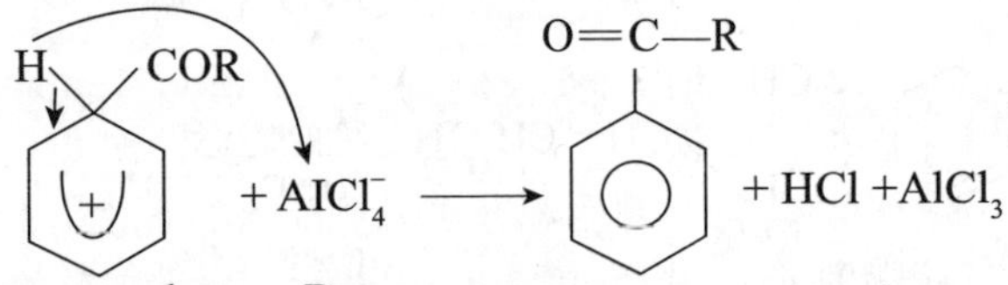

H COR O=C—R

(+) $+ AlCl_4^-$ ⟶ $+ HCl + AlCl_3$

σ-complex Base

Nucleophillic substitution:

- Here a nucleophile substitutes another molecule or group from an organic compound. A weak nucleophile is substituted by a stronger nucleophile.

  $R - X + OH^- \longrightarrow R - OH + X^-$

  It is of two types: SN^1 and SN^2

  SN^1 or unimolecular nucleophilic substitution reaction

  S : Substitution

  N : Nucleophillic

  1 : Unimolecular (1st Order)

- It is a first order reaction
- It takes place in two steps and the first step is the rate determining step.

  $R - X \xrightarrow[\text{Slow step}]{\text{Heterolytic}} R^+ + X^-$ [ionization]

  $R^+ + Y^- \xrightarrow{\text{fast step}} R - Y$
  Stronger nucleophile

  As rate is determined by the slow step, therefore,

  $\frac{-dx}{dt} \propto [R - X]^1$

  This implies a first order or unimolecular reaction

Important Facts About SN^1

- No inversion takes place here.
- A nucleophile can attack the carbon atom of the substrate from both rear and front sides.
- The order of reactivity of alkyl halides towards SN^1 reaction is

  $3° > 2° > 1° > CH_3X >$ vinyl
- It is favoured by using low concentration of mild nucleophiles.
- It is favoured by the solvent of high polarity like water.
- It is catalyzed by Lewis acids like Ag^+, $AlCl_3$, $ZnCl_2$
- In case of optically active halides, partial racemization is possible, also retention due to planar geometry of reaction intermediates.
- In SN^1 reaction mechanism the leaving group leaves the molecule before the incoming group has bonded itself to the molecule. This leads to the formation of carbocation.
- Carbocation is the reaction intermediate and it can go rearrangement to attain a state of maximum stability.
- A primary alkyl halide may also undergo SN^1 reaction in aqueous formic acid.
- The products of SN^1 reactions are almost racemic.
- The leaving group tendency order of nucleophiles in increasing order.

  $OH^-, NH^-, OR^-, F^-, Cl^-, Br^-, I^-, ToSo^-$

  For example,

  $(CH_2)_3C - Br + NaOH \longrightarrow (CH_3)_3COH + NaBr$

SN^2 or Bimolecular Nucleophilic Substitution

S : Substitution

N : Nucleophillic

2 : Bimolecular (2nd Order)

Important Facts About SN^2

- It is a bimolecular and a second order reaction.

  $\frac{-dx}{dt} \propto [R - X]^1 [Nu]^1$
- Here the reaction intermediate is in an unstable transition state.
- In this case, the stronger nucleophile attacks only the rear end of the molecule.
- Here reaction occurs in presence of polar aprotic solvent like acetone, DMF or DMSO (which favours formation of transition state)

- Here configuration changes i.e., optical inversion occurs.
- **Steric hindrance** is the rate determining factor here.
- The order of reactivity of alkyl halide is as follows:

  $CH_3–X > p^o > s^o > t^o$

  R–I > R–Br > RCl > R–P

  For example,

$$\text{Me}-\underset{\text{Et}}{\overset{\text{H}}{\underset{|}{\overset{|}{\text{C}}}}}-\text{Br} \xrightarrow[\text{SN}^2]{\text{NaOH}} \text{HO}-\underset{\text{Et}}{\overset{\text{H}}{\underset{|}{\overset{|}{\text{C}}}}}-\text{Me}$$

Addition Reactions

- Unsaturated compounds having π-bonds as shown below undergo this type of resonance

$$>C{=}C<,\quad -C\equiv C-,\quad -\overset{\text{O}}{\overset{\|}{\text{C}}}-,\quad -C\equiv N$$

- Here loss of one π-bond and gain of two σ- bonds take place.
- A less stable compound changes into a more stable product.

  Addition reactions are of following types:

Electrophilic addition This reaction is given by alkene and alkynes. ($>C = C<, -C \equiv C-$)

- It is a two step reaction in which the attacking reagent is an electrophile.
- If the reagent used is a weak electrophile, then the reaction is catalysed by strong acids like concentrated H_2SO_4

$$R-CH=CH_2 + H-Nu \longrightarrow R-\underset{\text{Nu}}{\underset{|}{\text{CH}}}-CH_3.$$

Mechanism

$$R-\underset{\text{H}}{\underset{|}{\text{C}}}=C\begin{smallmatrix}\text{H}\\ \text{H}\end{smallmatrix} \xrightarrow{H^+} R-CH-CH_2 \;(\text{bridged by } H^+)$$

π–complex

$$\longrightarrow R-\overset{+}{C}H-CH_3 \xrightarrow{Nu^-} R-\underset{\text{Nu}}{\underset{|}{\text{CH}}}-CH_3$$

- If the carbocation is not stable then a one time rearrangement of atoms is also possible.

Some Examples of electrophilic addition

- *Reaction with HX*

  Order of reactivity for HX:

  HF > HBr > HCl

Addition according to Markovnikov's rule

- In case of unsymmetrical alkenes or alkynes, addition takes place according to Markovnikov's rule where negative part of the attacking reagent is attached to the unsaturated carbon atom having the least number of hydrogen atoms binding to it.

$$(CH_3)_2C{=}CH{-}CH_3 + HCl \xrightarrow{\text{According to Markovnikov's rule}} (CH_3)_2\underset{\text{Cl}}{\text{C}}{-}CH_2{-}CH_3$$

$$(CH_3)_2C{=}CH_2 + HCl \xrightarrow{\text{Peroxide}} (CH_3)_3C{-}Cl$$

$$(CH_3)_2C{=}CH{-}CH_3 + HCl \longrightarrow (CH_3)_2\underset{\text{Cl}}{\text{C}}{-}CH_2{-}CH_3$$

$$CH_2{=}CH{-}CH_2{-}CH{=}CH_2 + HCl \longrightarrow CH_3{-}\underset{\text{Cl}}{\text{CH}}{-}CH_2{-}\underset{\text{Cl}}{\text{CH}}{-}CH_3$$

Peroxide effect or Khrasch effect or addition according to anti-Markovnikov's rule

- In case of unsymmetrical alkenes and alkynes, addition with HBr in presence of peroxide or peroxy acids takes place according to anti-Markovnikov's rule, that is, the negative part of the attacking reagent will be added to the unsaturated carbon atom having more hydrogen atoms.
- It involves free radical addition mechanism.
- Since HF and HCl have high bond energies peroxide effect is not possible in their respective cases. The energy present is not sufficient for this mechanism.
- HI, being very reactive, forms I_2 in place of I^o so no such effect is possible in this case.

$$RCH{=}CH_2 + HBr \xrightarrow{\text{Peroxide}} RCH_2CH_2Br$$

Mechanism

$$R{-}O{-}O{-}R \longrightarrow 2RO^o$$

$$R{-}O^o + H{-}Br \longrightarrow R{-}OH + Br^o$$

$R–CH = CH_2 + Br^o \longrightarrow R–{}^oCH–CH_2Br$ (s more stable) $\xrightarrow{H^o} R – CH_2 – CH_2Br$

For example,

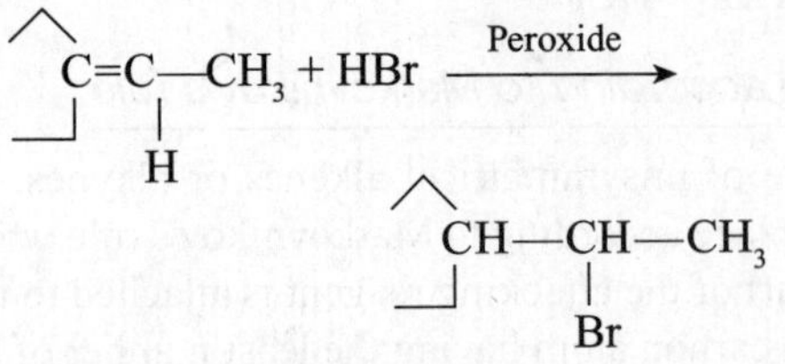

$$C{=}C{-}CH_3 + HBr \xrightarrow{\text{Peroxide}} CH{-}CH{-}CH_3$$

(H on the alkene carbon; Br on the middle carbon of the product)

Addition of X_2

$$>C{=}C< + Br – Br \xrightarrow{Br^+} >C{-}C< \text{(bridged } Br^+\text{)} \xrightarrow[\text{Trans-1, 2 addition}]{Br^-} >C(Br){-}C(Br)<$$

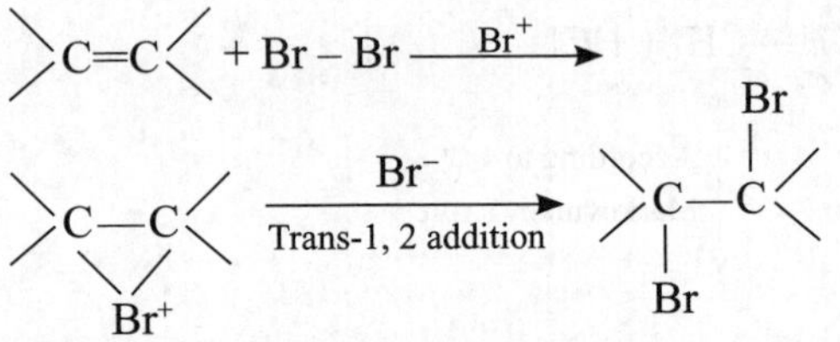

Some Important Reactions

- Cis alkene + syn addition ⟶ meso product
- Trans alkene + anti addition ⟶ meso product
- Cis alkene + anti addition ⟶ racemic prodcuct
- Trans alkene + syn addition ⟶ racemic product

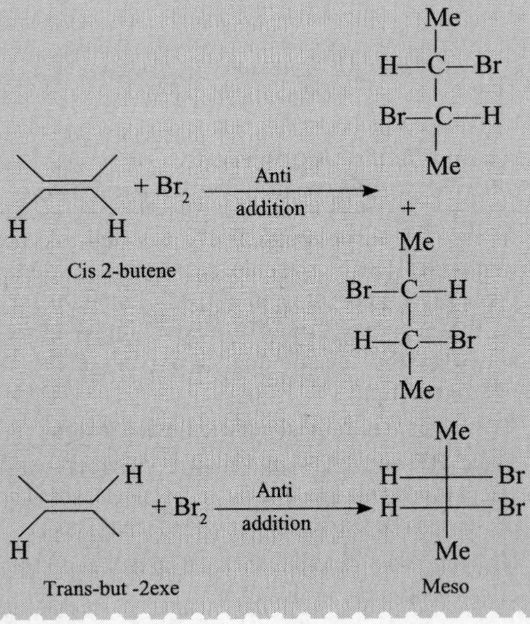

Cis 2-butene $+ Br_2 \xrightarrow{\text{Anti addition}}$ Me–C(H)(Br)–C(Br)(H)–Me + Me–C(Br)(H)–C(H)(Br)–Me

Trans-but -2exe $+ Br_2 \xrightarrow{\text{Anti addition}}$ Me–C(H)(Br)–C(H)(Br)–Me (Meso)

Nucleophilic Addition Reaction

- Mainly the carbonyl compounds with > C=O group undergo nucleophilic addition reaction.
- In this case, the attacking reagent is a nucleophile therefore, the nuclephile will be added first during addition.

Mechanism

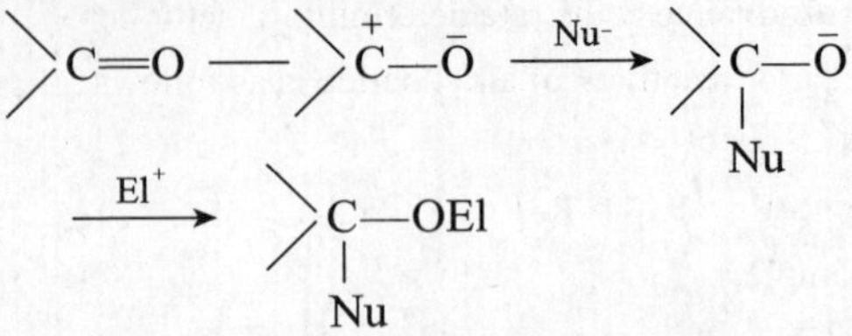

$$>C{=}O \longrightarrow >\overset{+}{C}{-}\overset{-}{O} \xrightarrow{Nu^-} >C(Nu){-}\overset{-}{O} \xrightarrow{El^+} >C(Nu){-}OEl$$

This mechanism is illustrated in the following examples:

- Reaction with HCN or cyanohydrin formation

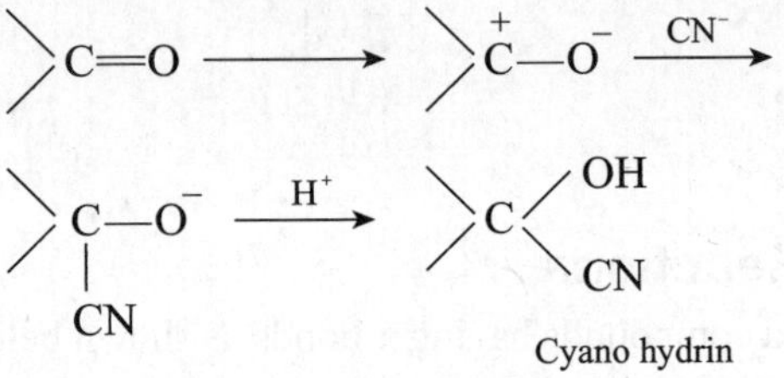

$$>C{=}O \longrightarrow >\overset{+}{C}{-}O^- \xrightarrow{CN^-} >C(CN){-}O^- \xrightarrow{H^+} >C(OH)(CN)$$

Cyano hydrin

- Reactivity of carbonyl compounds:
  Higher the positive charge on ($> C^+ – O^-$) stronger is the reactivity.
  $$\text{Reactivity} \propto \frac{1}{+I\ \text{effect}}$$
  Therefore, 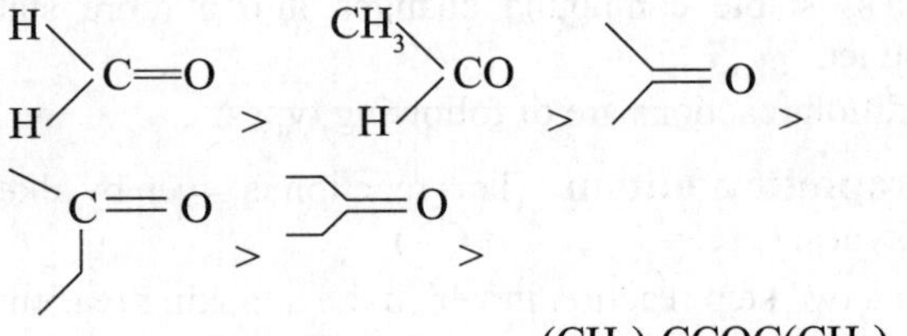
  HCHO > CH_3(H)CO > CH_3COCH_3 > $CH_3COC_2H_5$ > $C_2H_5COC_2H_5$ > ... $(CH_3)_3CCOC(CH_3)_3$ Unreactive due to steric hinderance

Specific example

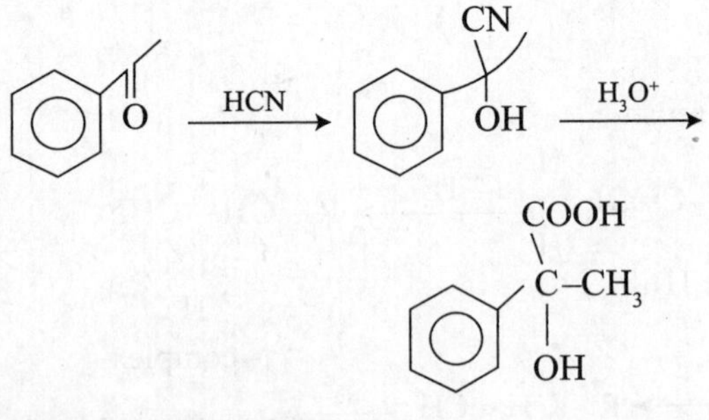

$$C_6H_5COCH_3 \xrightarrow{HCN} C_6H_5C(CN)(OH)CH_3 \xrightarrow{H_3O^+} C_6H_5C(COOH)(OH)CH_3$$

Elimination Reaction

- Elimination reaction is shown by the compounds bonded to leaving groups such as X, OH, OR, N^+R_3.
- Here, removal of atoms or groups in the form of small molecules like H_2O, NH_3, HX etc., leads to formation of

multiple bonds. In this way, it is the inverse of addition reaction.

- These reactions are mostly endothermic and materialize in the presence of heat.
- The product obtained is less stable.
- It involves the loss of 2 sigma (σ) bonds and formation of 1 pi (π) bond.

Following are the different types of elimination reactions:

α-elimination or 1,1 elimination reaction

- It is a rare reaction where both the species removed are from the same carbon atom.
- It is shown by geminal dihalides or trihalides in the presence of alcoholic bases like KOH to give carbenes.

For examples,

$$CHX_3 \xrightarrow[\text{Alc. KOH}]{\Delta} \ddot{C}X_2 + X^- + H^+$$

$$CH_2X_2 \xrightarrow[\text{Alc. KOH}]{\Delta} \ddot{C}HX + X^- + H^+$$

Carbenes

α–β or 1, 2 elimination reaction

Here, atoms or groups from α, β positions of the compound are eliminated as follows:

$$\underset{\beta}{CH_3 - CH_2} - \underset{\alpha}{CH_2} - L \longrightarrow CH_3 - CH = CH_2 + H^+ + L^-$$

It is of following types:

E_1 : Elimination, unimolecular and first order

E_{1cb} : Elimination, unimolecular with conjugation base

E_2 : Elimination, bimolecular or second order

E_1 Type

Here, the elimination reaction is acid catalysed leading to the formation of a carbocation which can undergo rearrangement once if it is less stable. In this way, it resembles SN^1.

Rate ∝ [subtract]^1

- The reaction occurs in the presence of polar protic solvents.
- *Dehydration of alcohol:* Dehydration of alcohol is decided by the stability of the carbocation (t > s > p). Some common **dehydrating agents** are anhydrous PCl_5, concentrated H_2SO_4, $KHSO_4$, anhydrous Al_2O_3, anhydrous $ZnCl_2$, P_2O_5 etc.

Mechanism of dehydration

$$H_2SO_4 \longrightarrow H^+ + HSO_4^-$$

$$-\underset{H}{\overset{|}{C}} - \underset{H}{\overset{OH}{C}} - \xrightarrow[\text{slow step}\ -H_2O]{H^+} -\underset{H}{\overset{|}{C}} - \overset{+}{C} -$$

$$\xrightarrow[\text{fast step}]{-H^+} \;>C=C<$$

As rate is determined by the slowest step, so,

$$\frac{-dx}{dt} \alpha \left(-\underset{H}{\overset{|}{C}} - \overset{OH}{C} - \right)^1$$

Saytzeff rule

- According to this rule, a β-hydrogen atom is removed from the β-carbon atom having lesser number of hydrogen atoms so that a more stable alkene is formed.
- Stability of alkene in decreasing order is given below:

$$R_2C=CR_2 > R_2C=CHR > R_2C=CH_2 > RCH=CHR > RCH=CH_2 > H_2C=CH_2$$

For example,

$$(CH_3)_3C-\underset{OH}{CH}-CH_3 \xrightarrow[-H_2O]{H+}$$

$$(CH_3)_3C-\overset{+}{CH}-CH_3 \xrightarrow[\text{Rearrangement}]{\bar{C}H_3}$$

$$H_3C\diagdown \overset{\oplus}{C}(t^o) - C(H)(CH_3)_2$$

$\xrightarrow{-H^+}$

$$(H_3C)_2C=C(CH_3)_2$$

- The following example exhibits a very specific case:

$$Me_2CH-CH(OH)-CH_2-CH=CH_2 \xrightarrow[350^oC]{\Delta\ Al_2O_3}$$

$$Me_2CH-CH=CH-CH=CH_2$$

As it is a conjugated diene, it is more stable and hence it is the major product.

or

$$Me_2C=CH-CH-CH=CH_2$$

This product is ignored as it is less stable.

E_2 type

- It is base catalysed reaction leading to the formation of an intermediary unstable state similar to that formed by SN^2 mechanism. Since it is a second order reaction, the rate of reaction is given as

$$\frac{-dx}{dt} \propto (R-X)^1\,(Base)^1$$

Rate of reaction ∝ Leaving group power

- The leaving group should be antiperiplanar (180° in same plane but anti-conformation)

For example,

$$CH_3-CH_2-CH_2-Br + C_2H_5\bar{O} \longrightarrow CH_3-CH=CH_2 + C_2H_5OH + Br^-$$

Dehydrohalogenation reaction

- Aliphatic halides (R–X) show this reaction.
- The trend followed by the aliphatic halides towards this reaction is in the following order:

tertiary > secondary > primary

- Tertiary and secondary aliphatic halides show E_1 elimination reaction.
- Primary aliphatic halides exhibit E_1 elimination when the β carbon atom is tertiary or quaternary and it exhibits E_2 elimination reaction when the β carbon atom is primary or secondary.

Following is the diagrammatic representation of this reaction: shown by R – X (Aliphatic halides)

t° > s° > p°

E_1 only (for t°, s°)

p°: when β-C-atom is 3°, 4° → E_1; when β-C-atom is 1°, 2° → E_2 (only)

Conditions favouring E_1 reactions

Alkyl group The preference towards alkyl group is as follows: 3° > 2° > 1°

Base Strength

Weak bases like $\frac{NaOH}{\Delta}$, $\frac{KOH}{\Delta}$, $\frac{C_2H_5O^-}{\Delta}$

Catalyst Lewis acids like $AlCl_3$

Conditions favouring E_2 reactions

Alkyl group The preference towards alkyl group is as follows: 1° > 2° > 3°

Base strength Strong bases like alcoholic KOH, $(CH_3)_3COOK$

Catalyst Phase transfer catalyst

Examples of E_1 reactions

1. $(CH_3)_3C-Br$ or $(CH_3)_2CH-CH_2-Br \xrightarrow{\text{Alc. KOH}} (CH_3)_2C=CH_2$

2. $CH_3CH_2CH(X)CH_3 \xleftrightarrow[\text{Alc. KOH}]{\ominus}$ $CH_3CH=CHCH_3$ + $CH_3CH_2CH=CH_2$

3. $\varphi-CH_2-CH(Br)-CH_3 \xrightarrow{\text{Alc. KOH}} \varphi-CH=CH-CH_3$

4. $\varphi-CH(Br)-CH_3$ or $\varphi-CH_2-CH_2(Br) \xrightarrow{\text{Alc. KOH}} \varphi-CH=CH_2$

Examples of E_2 reactions

$$Me-C(H)(Br)-C(Br)(H)-CH_3 \xrightarrow[\text{or NaI/acetone}]{\text{Zn dust}}$$

$CH_3 - CH = CH - CH_3 + ZnBr_2$
or
$2NaBr + I_2$

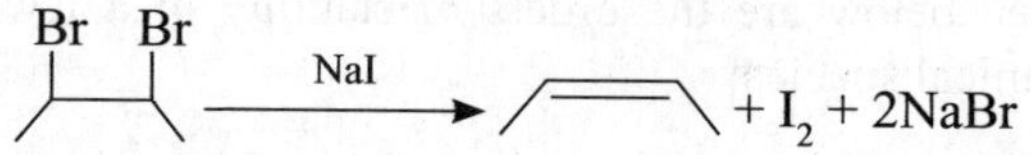

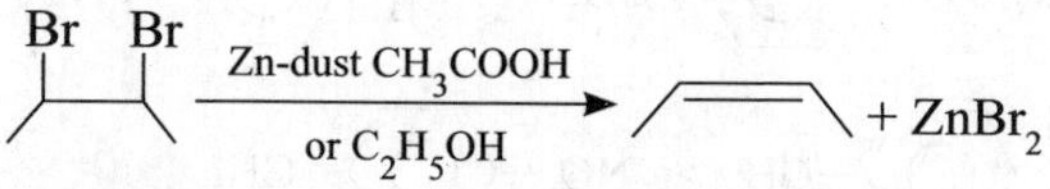

Br Br (vicinal dibromide) $\xrightarrow{NaI}$ alkene $+ I_2 + 2NaBr$

Br Br (vicinal dibromide) $\xrightarrow[\text{or } C_2H_5OH]{\text{Zn-dust } CH_3COOH}$ alkene $+ ZnBr_2$

E_{1cb} reaction

- This type of reaction occurs only when β^- carbon atom is bonded to electron attracting groups like >C=O, $-NO_2$, –CN, $-SO_3H$ attached to it. A strongly acidic β hydrogen also aids its easy removal.
- A weak leaving group helps in the easy formation of a carbanion.

For example,

$$CF_3 - CHCl_2 \xrightarrow{C_2H_5O^-} F_2C = CCl_2$$

$$\varphi - CH_2 - CH_2F \xrightarrow{C_2H_5O^-} \varphi - CH = CH_2$$

Hofmann's rule

- According to Hofmann's rule, the removal of a β-hydrogen atom is only from β- carbon atom which has more hydrogen atoms attached to it. This removal is possible in the presence of bulky leaving groups like N^+R_3, S^+R_2.

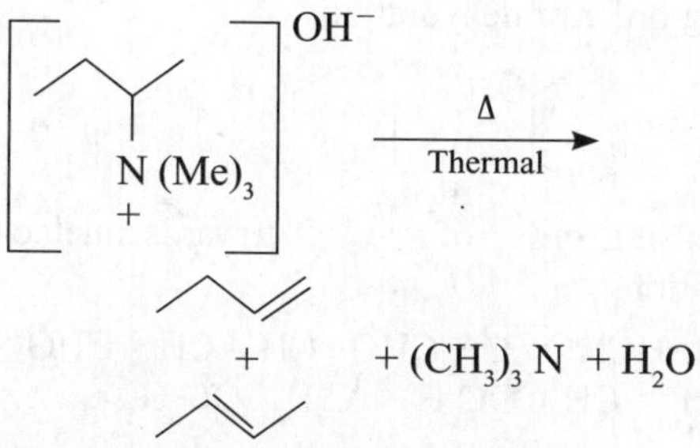

$[\text{2-butyl-}\overset{+}{N}(Me)_3]\ OH^- \xrightarrow[\text{Thermal}]{\Delta}$ but-1-ene + but-2-ene $+ (CH_3)_3N + H_2O$

Here, butene-1 is major product.

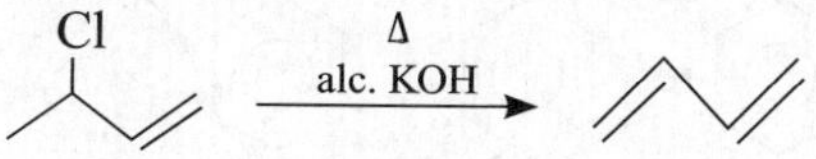

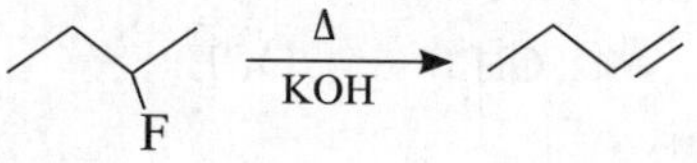

3-chlorobut-1-ene (Cl) $\xrightarrow[\text{alc. KOH}]{\Delta}$ buta-1,3-diene

2-fluorobutane (F) $\xrightarrow[\text{KOH}]{\Delta}$ but-1-ene

Thermal or pyrolytic elimination It is in accordance with Hoffmann rule.

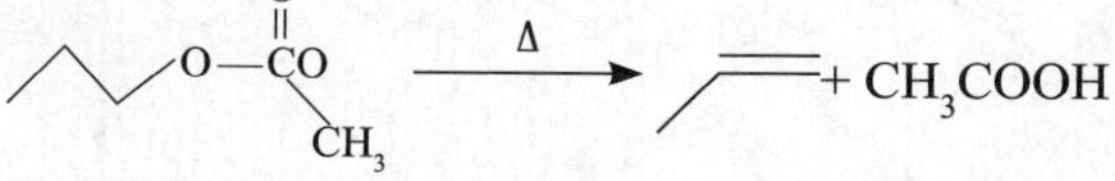

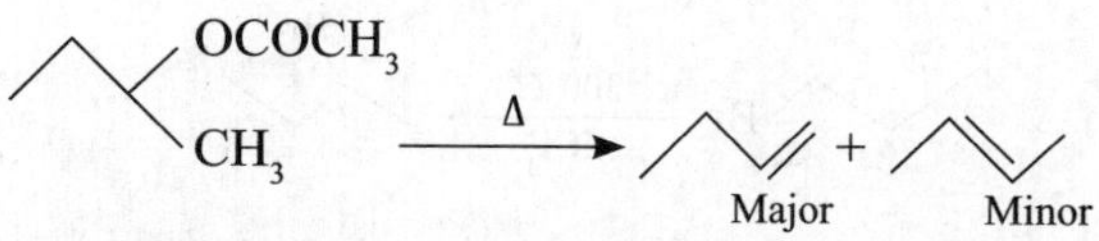

$$CH_3CH_2CH_2-O-\overset{O}{\overset{||}{C}}-CH_3 \xrightarrow{\Delta} CH_3CH=CH_2 + CH_3COOH$$

2-methylbutan-2-yl acetate ($OCOCH_3$, CH_3) $\xrightarrow{\Delta}$ alkene (Major) + alkene (Minor)

- **Cope reaction** Here, a tertiary amine oxide is converted into an alkene on being heating and following α–β-elimination reaction.

For example,

(1) $C_6H_{11}-CH_2-\ddot{N}R_2 \xrightarrow{[O], H_2O_2 \text{ or } H_2SO_5}$

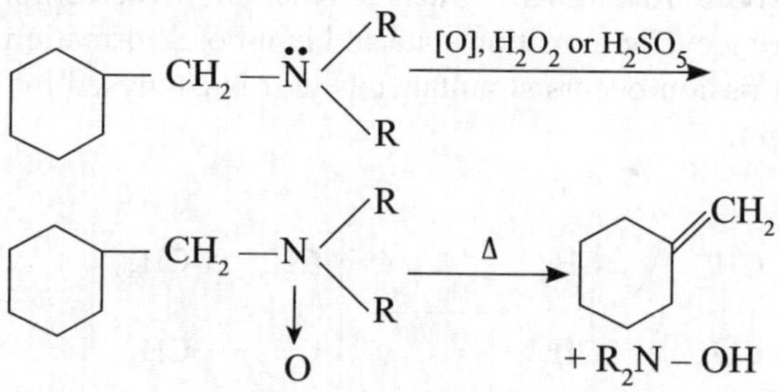

$C_6H_{11}-CH_2-\overset{\to O}{N}R_2 \xrightarrow{\Delta}$ methylenecyclohexane ($=CH_2$) $+ R_2N - OH$

(2) $\phi - \underset{CH_3}{\underset{|}{CH}} - \underset{CH_3}{\underset{|}{CH}} - \underset{CH_3}{\underset{|}{\overset{CH_3}{\overset{|}{N}}}} \xrightarrow{CF_3CO_3H}$

$\phi - \overset{\beta}{\underset{CH_3}{\underset{|}{CH}}} - \overset{\alpha}{\underset{CH_3}{\underset{|}{CH}}} - \underset{CH_3}{\underset{|}{\overset{CH_3}{\overset{|}{\overset{+}{N}}}}} - O^-$

$\downarrow \Delta$

$\phi - \underset{CH_3}{\underset{|}{C}} = \underset{CH_3}{\underset{|}{CH}} - (CH_3)_2N - OH$

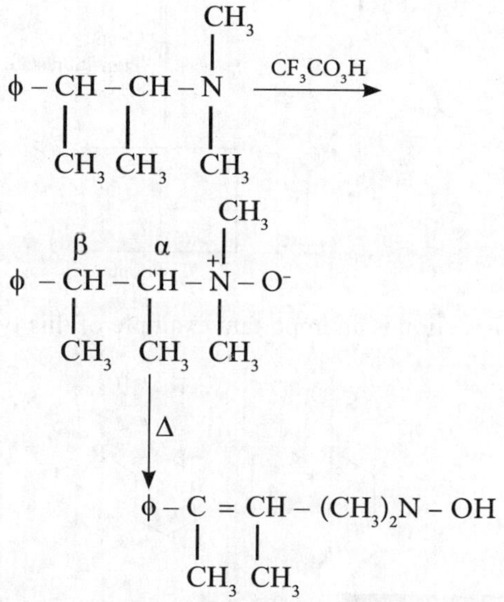

Rearrangement Reactions It involves migration of an atom or group from one atom to another atom.

e.g.,

$$R-\overset{O}{\underset{||}{C}}-NH_2 \xrightarrow[\Delta]{Br_2 + KOH} R-NH_2 + K_2CO_3 + 2H_2O$$

In this reaction (Hofmann bromamide reaction) R migrates from C–atom to N–atom.

Some other e.g., are Pinacole - pinacolone, curtius Rearrangement etc.

Isomerisation Reactions: Here one compound changes into another compound having some molecular formula by the action of heat etc.

$$\text{1-bromobutane} \xrightarrow[\text{AlCl}_3]{\Delta,\ 300°C} \text{2-bromobutane}$$

$$\text{1-butene} \xrightarrow[\text{AlCl}_3]{\Delta,\ 300°C} \text{2-butene}$$

Pericyclic Reactions: Such a reaction involves formation of a cyclic transition state and in it bond formation and bond fission occurs simultaneously. It is catalysed by heat or light.

e.g.,

$$CH_2{=}CH_2 + CH_2{=}CH_2 \xrightarrow{h\nu} [\text{Cyclic Transition state}] \rightarrow \text{Cyclobutane}$$

Diel-alder Reaction is an important example of this type of reaction

$$\text{Butadiene} + CH_2{=}CH{-}R \xrightarrow{\Delta} \text{4-R-cyclohexene}$$

POINTS TO REMEMBER

In order to make a carbocation more stable, rearrangement is made and the migrating aptitude of groups is as follows:

$$H^- > 3° > 2° > 1° > \overset{\ominus}{C}H_3$$

- In case of same nucleophilic centre the basicity increases. However, nucleophilicity decreases as the size of the nucleophile (Nu^-) approaching the substrate increases.

  $MeO^- < Me_2CHO^- < Me_3CO^-$
  (Basicity increases)

  $Me_3CO^- < Me_2CHO^- < MeCH_2O^-$
  (Nucleophilicity increases)

- $CH_3O^- > HO^- > CH_3COO^- > H_2O$
  (Nucleophilicity, basicity decreases)

- Given below are the orders of stability of a few chemical species:

(i) $CH_3{-}O{-}C_6H_4{-}\overset{+}{C}H_2 > CH_3{-}C_6H_4{-}\overset{+}{C}H_2 > C_6H_5{-}\overset{+}{C}H_2 > NO_2{-}C_6H_4{-}\overset{+}{C}H_2$

(ii) $CH_3{-}O{-}C_6H_4{-}\overset{-}{C}H_2 < CH_3{-}C_6H_4{-}\overset{-}{C}H_2 < C_6H_5{-}\overset{-}{C}H_2 < NO_2{-}C_6H_4{-}\overset{-}{C}H_2$

(iii) $CH_3{-}O{-}C_6H_4{-}\overset{\circ}{C}H_2 > CH_3{-}C_6H_4{-}\overset{\circ}{C}H_2 > C_6H_5{-}\overset{\circ}{C}H_2 > NO_2{-}C_6H_4{-}\overset{\circ}{C}H_2$

- In case of SN^2 reactions the nucleophilicity decreases as follows:

  $RS^- > ArS^- > CN^- > OH^- > N_3^- > Br^- > ArO^- > Cl^- > CH_3COO^- > H_2O$

- Decreasing order of dehydration

  cyclohexa-2,4-dien-1-ol > cyclohex-2-en-1-ol > cyclohexanol

- The decreasing order of > C=O towards nucleophilic addition.

  $CCl_3CHO > HCHO > CH_3CHO > CH_3{-}CH_2{-}CHO > CH_3COCH_3 > CH_3COC_6H_5 > C_6H_5COC_6H_5 > \ldots\ldots$

  $p\text{-}NO_2C_6H_4CHO > p\text{-}XC_6H_4CHO > p\text{-}CH_3C_6H_4CHO > p\text{-}OCH_3C_6H_4CHO >$

- **Free radical stability**

  cyclohexadienyl radical > cyclohex-2-enyl (allylic) radical > cyclohexyl radical > cyclohexenyl (vinylic) radical > phenyl radical

CHAPTER-END EXERCISES

PRACTICE QUESTIONS – I

1. In CH_3CH_2OH, the bond that undergoes heterolytic cleavage most readily is:
(a) O –H (b) C – H
(c) C – O (d) C – C

2. The four C – H bonds of methane are broken one at a time. The cleavage of which of the following bonds has the minimum bond dissociation energy?
(a) C – H (b) CH – H
(c) CH_2 – H (d) CH_3 – H

3. Which one of the following behaves both as a nucleophile and as an electrophile?
(a) $H_3C - NH_2$ (b) $H_3C - OH$
(c) $H_3C - C \equiv N$ (d) C_2H_5ONa

4. Pick the strongest nucleophile.
(a) $:\overset{\ominus}{C}H_3$ (b) $\overset{\ominus}{N}H_2$
(c) $H - C \equiv \overset{\ominus}{C}:$ (d) $H^{\ominus}$

5. Which of the following has the highest nucleophi-licity?
(a) F^- (b) OH^-
(c) CH_3^- (d) NH_2^-

6. The decreasing order of nucleophilicity of HS^-, $RCOO^-$, RCOOH and ROH is:
(a) $RCOO^- > HS^- > RCOOH > ROH$
(b) $HS^- > RCOO^- > RCOOH > ROH$
(c) $HS^- > RCOO^- > ROH > RCOOH$
(d) $RCOO^- > HS^- > ROH > RCOOH$

7. Which of the following is strongest nucleophile?
(a) Br^- (b) $:OH^-$
(c) $:\bar{C}N$ (d) $C_2H_5\bar{O}:$

8. Which of the following is the strongest nucleophile?
(a) NH_2^- (b) $CH_3 - CH_2^-$
(c) $HC \equiv C^-$ (d) $H_2C = CH^-$

9. Among the following, the strongest nucleophiles is:
(a) C_2H_5SH (b) CH_3COO^-
(c) CH_3NH_2 (d) $NCCH_2^-$

10. Phenol is more reactive than benzene towards eletrophillic substitution due to:
(a) Strong mesomeric effect
(b) Hyperconjugative effect
(c) Inductive effect only
(d) Hydrogen bonding

11. Free radicals are involved in mechanism of:
(a) Kharash effect
(b) Cracking of alkanes
(c) Halogenation of alkanes
(d) All of these

12. Abnormality high heat of formation and shortening of bond length are criteria of:
(a) Hybridisation
(b) Resonance
(c) Electron delocalization
(d) Ionization

13. Presence of a $-NO_2$ group around a carbonium ion:
(a) Eeduces its stability
(b) Increases its stability
(c) Makes no change in its stability
(d) None of these

14. The kind of delocalization involving sigma bond orbitals is called:
(a) Inductive effect
(b) Hyperconjugation effect
(c) Electromeric effect
(d) Mesomeric effect

15. Reaction not involving a carbanion is:
(a) Aldol condensation
(b) Claisen condensation
(c) Pinacol pinacolone rearrangement
(d) Michael reaction

16. The most stable carbonium ion is:
(a) $C_6H_5 - \overset{\oplus}{C}(C_6H_5) - C_6H_5$ (b) $CH_3 - \overset{\oplus}{C}(CH_3) - CH_3$
(c) $CH_3 - \overset{\oplus}{C}H - CH_3$ (d) $CH_3 - \overset{\oplus}{C}H_2 - CH_3$

17. 2–methyl but–2–ene is less reactive than but–2–ene. This is due to
(a) Resonance (b) Hyperconjugation
(c) Electromeric effect (d) Inductive effect

18. A solution of (+2)–2–chloro–2–phenylethane in toluene racemises slowly in the presence of small amounts of $SbCl_5$, due to the formation of:

(a) carbanion (b) carbene
(c) free radical (d) carbocation

19. The resonance structures
(a) Differ only in the arrangement of electrons
(b) Differ in number of paired and unpaired electrons
(c) Differ largely in their energy content
(d) Do not lie in the same plane

20. Resonance structures of a molecule does not have
(a) Identical arrangement of atoms
(b) Nearly the same energy content
(c) Same number of paired electrons
(d) Identical bonding

21. Which of the following statements is not correct?
(a) Cyclopentadiene has an acidic hydrogen
(b) Squaric acid is highly acidic
(c) Aniline is more basic than ammonia
(d) Aniline is less basic than ammonia

22. Which of the following is singlet carbene?
(a) $(CH_3)_3\overset{+}{C}$ (b) $C_2H_5\ddot{C}-H$
(c) $CH_3-\dot{C}HCH_3$ (d) $CH_2=CH-\overset{+}{C}H_2$

23. Which of the following species is carbene?
(a) $\bar{:}CH_3$ (b) $:CCl_2$
(c) $CH_2=C=O$ (d) $R-\underset{\underset{O}{\|}}{C}-R$

24. The electromeric effect involves
(a) A slight displacement of σ – electrons
(b) A slight displacement of π – electrons
(c) The complete transfer of σ – electrons
(d) The complete transfer of π – electrons

25. Chlorine gas is exposed to ultraviolet light in a closed container. The reactive species expected to be formed are
(a) Cl (b) Cl^+
(c) Cl^- (d) Cl

26. Which of the following species on photolysis does give a carbene?
(a) $CH_3-\overset{\overset{O}{\|}}{C}-CH_3$ (b) $CH_2=C=O$
(c) CCl_4 (d) $CHCl_3$

27. When a dextro rotatory alkyl halide is subjected to SN^1 reaction, the product obtained is
(a) Dextro rotatory
(b) Leavo rotatory
(c) A racemic mixture
(d) Meso.

28. The reaction $CHBr + OH^- \longrightarrow CH_3OH + Br^-$ follows:
(a) SN^1 mechanism
(b) SN^2 mechanism
(c) Either of the above two
(d) None of these

29. The compound that will react most readily with NaOH to form methanol is
(a) $(CH_3)_4N^+I^-$ (b) CH_3OCH_3
(c) $(CH_3)_3S^+I^-$ (d) $(CH_3)_3CCl$

30. The reaction is an example of

$$C_6H_6 \xrightarrow[\text{Anhy. Alcl}_3]{R-\overset{\|}{\underset{O}{C}}-Cl} C_6H_5COR + HCl$$

(a) Electrophilic substitution reactions
(b) Nucleophilic substitution reactions
(c) Elimination reactions
(d) Addition reactions

31. In the presence of peroxide, hydrogen chloride and hydrogen iodide do not undergo anti–Markownikoff's addition to alkenes because
(a) Both are highly ionic
(b) One is oxidizing and the other is reducing
(c) One of the steps is endothermic in both the cases
(d) All the steps are exothermic in both the cases

32. Which of the following is not polar?
(a) Tert–Butyl free radical
(b) Tert–Butyl carbocation
(c) Tert–Butyl carbanion
(d) Allyl cabanion

33. Which of the following molecules has all the effects: inductive, mesomeric and Baker Nathan effect?
(a) C_2H_5Cl
(b) $CH_3-CH=CH_2$
(c) $CH_2=CH-CH=CH_2$
(d) $CH_3-CH=CH-\underset{\underset{O}{\|}}{C}-CH_3$

34. During dehydration of tert–butanol, which of the following carbocation is more likely to be formed as an intermediate?
(a) $CH_3CH_2CH_2\overset{+}{C}H_2$ (b) $CH_3CH_2\overset{\oplus}{C}HCH_3$
(c) Both (a) and (b) (d) None of these

35. Which of the following is fast debrominated?

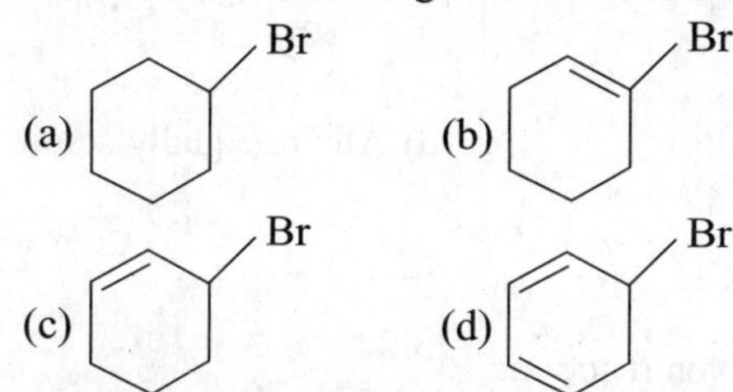

36. Tertiary alkyl halides are practically inert to substitution by SN^2 mechanism because of:
(a) Insolubility (b) Instability
(c) Inductive effect (d) Steric hindrance

37. Which of the following carbon atoms is most electronegative?

III II I
$CH - CH - C \equiv CH$

(a) III
(b) II
(c) I
(d) All are equally electronegative

38. Resonance energy per benzene ring is maximum for:
(a) Naphthalene (b) Anthracene
(c) Benzene (d) Phenanthrene

39. In buta–1, 3–diene, the $C_2 - C_3$ bond length is
(a) 1.20 Å (b) 1.35 Å
(c) 1.54 Å (d) 1.46 Å

40. Most stable carbanion is
(a) $(CH_3)_3C^-$ (b) $C_6H_5\underline{C}H_2^-$
(c) $(CH_3)_2CH^-$ (d) CH_3CH_2

41. The most stable carbonium ion is
(a) $(CH_3)_2CH^+$ (b) $CH_3CH_2^+$
(c) $(CH_3)_3C - CH_2^+$ (d) $(CH_3)_3C^+$

42. Which of the following compounds possesses the C – H bond with the lowest bond dissociation energy?
(a) N–pentane (b) Toluene
(c) Benzene (d) 2, 2–dimethylpropane

43. Which of the following intermediate has the complete octet around the carbon atom?
(a) Free radical (b) Carbene
(c) Carbanion (d) Carbonium ion

44. Amongst the following, the compound that can most readily get sulphonated is:
(a) Benzene (b) Toluene
(c) Nitrobenzene (d) Chlorobenzene

45. The major product obtained on treatment of $CH_3CH_2CH(F)CH_3$ with CH_3O^- / CH_3OH is
(a) $CH_3CH_2CH(OCH_3)CH_3$
(b) $CH_3CH = CHCH_3$
(c) $CH_3CH_2CH = CH_2$
(d) $CH_3CH_2CH_2CH_2OCH_3$

46. Among the following which one does not act as an intermediate in Hoffmann rearrangement?
(a) RNCO (b) $RCO\ddot{N}$
(c) $RCO\ddot{N}HBr$ (d) RNC

47. Pyridine is less basic than triethylamine because:
(a) Pyridine has aromatic character
(b) Nitrogen in pyridine is sp^2 hybridized
(c) Pyridine is a cyclic system
(d) In pyridine, lone pair of nitrogen is delocalized.

48. The reaction of HBr with:

$CH_3 - \underset{}{\overset{CH_3}{\overset{|}{C}}} = CH_2$ in the presence of peroxide will give

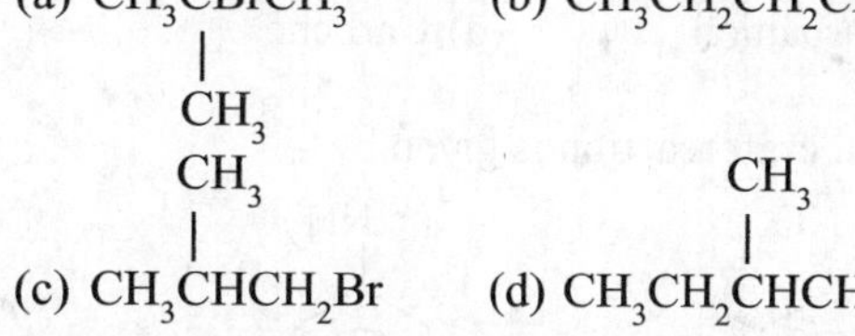

(a) $CH_3\underset{\underset{CH_3}{|}}{C}BrCH_3$ (b) $CH_3CH_2CH_2CH_2Br$

(c) $CH_3\overset{\overset{CH_3}{|}}{C}HCH_2Br$ (d) $CH_3CH_2\overset{\overset{CH_3}{|}}{C}HCH_3$

49. Which of the following undergoes nucleophillic substitution exclusively by SN^1 mechanism?
(a) Ethyl chloride (b) Isopropyl chloride
(c) Chlorobenzene (d) Benzyl chloride

50. Which one of the following compounds is most acidic?
(a) $Cl - CH_2 - CH_2 - OH$

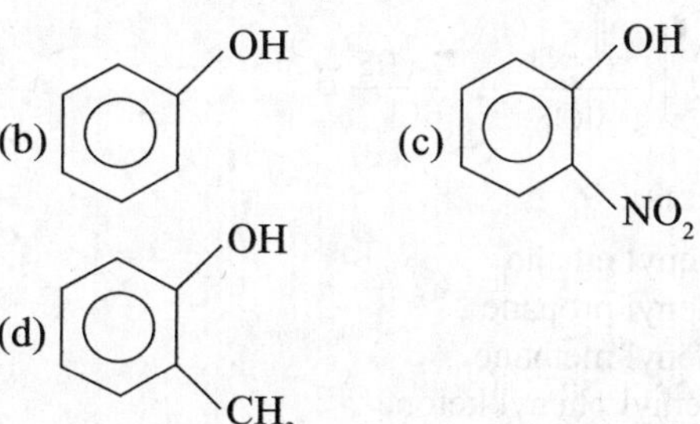

51. In which of the following carbocation rearrangement occurs most fast?

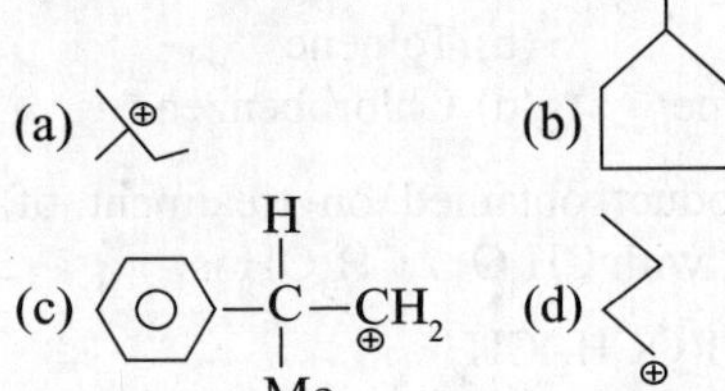

52. Among the following acids which has the lowest pKa value?
(a) CH_3COOH
(b) $HCOOH$
(c) $(CH_3)_2CH–COOH$
(d) CH_3CH_2COH

53. The elimination reaction

$$\underset{\overset{|}{\oplus N\,Me_3}}{CH_3 - CH_2.CH - CH_3} \xrightarrow{\Delta} \underset{\text{(major)}}{CH_3CH_2 - CH = CH_2}$$

(major)is governed by
(a) The Saytzev rule (b) The Hoffmann rule
(c) Both (a) and (b) (d) Can't be said

54. The reaction intermediate in E_{1Cb} is
(a) Carbonium ion (b) Carbon free radical
(c) Carbanion (d) Carbene

55. Friedal craft reaction is given by –

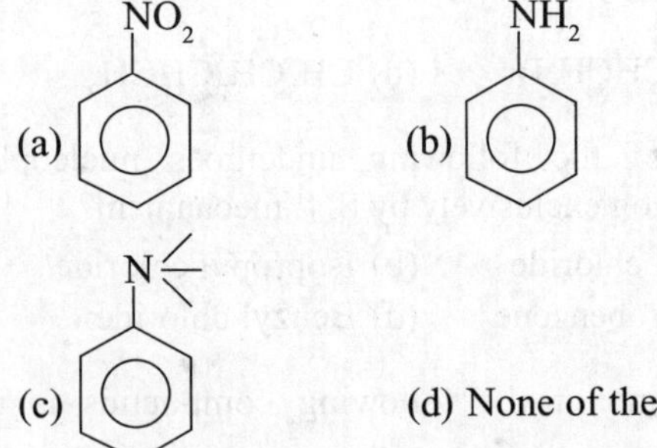

(d) None of these

56. 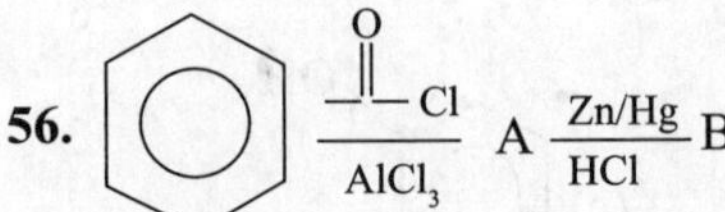

Here B is
(a) Phenyl ethane
(b) Phenyl propane
(c) Phenyl methane
(d) Methyl phenyl ketone

57. Which is most acidic here?
(a) O^- with OCH_3 at ortho position (benzene ring)
(b) O^- with OCH_3 at meta position (benzene ring)
(c) O^- with OCH_3 at para position (benzene ring)
(d) All are equally acidic

58. In the given reaction:

$$\underset{\underset{CH_3}{|}}{\overset{\overset{OH}{|}}{CH_3 - CH - CH - CH_3}} \xrightarrow{\text{conc. } H_2SO_4/\Delta}$$

Number of alkenes formed will be:
(a) One (b) Two
(c) Three (d) Four

59. Which one of the following pairs is correctly matched?
(a) Saytzeff rule, least substituted alkene
(b) Hoffmann rule, most substituted alkene
(c) E_1 cb reaction, Hoffmann elimination
(d) E_1 reaction, Hoffmann elimination

60. Anti–Markownikoff addition of HBr is not observed in
(a) Propene (b) Butene 1
(c) Pent–2–ene (d) But–2–ene

61. Which is most acidic

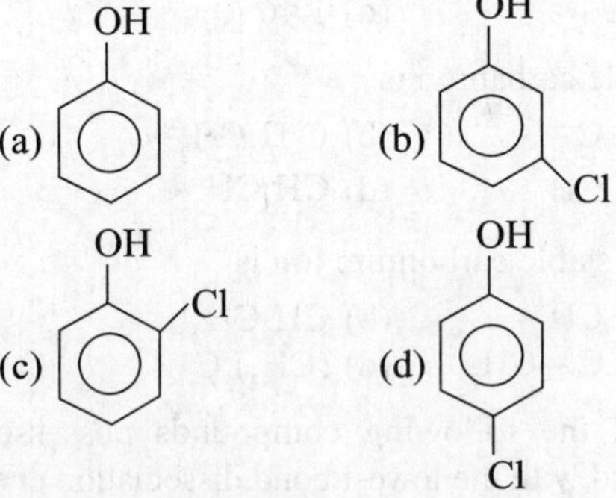

62. Which among the given acids has lowest pKa value?
(a) Chloroacetic acid (b) Bromoacetic acid
(c) Nitroacetic acid (d) Cyanoacetic acid

63. Which one of the following is least acidic?
(a) Phenol (b) O–fluorophenol
(c) M–fluorophenol (d) P–fluorophenol

64. Which is most acidic in nature?

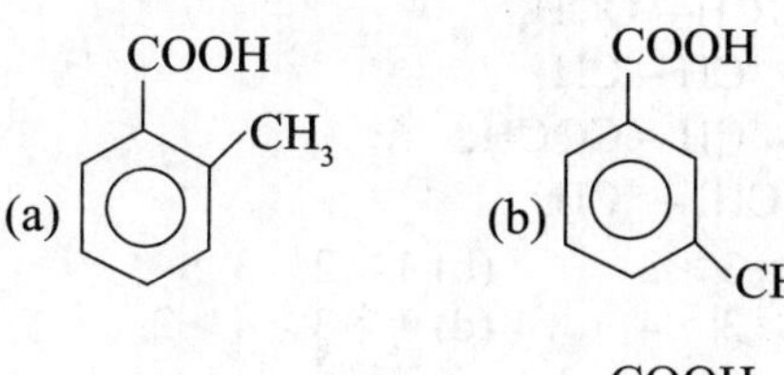

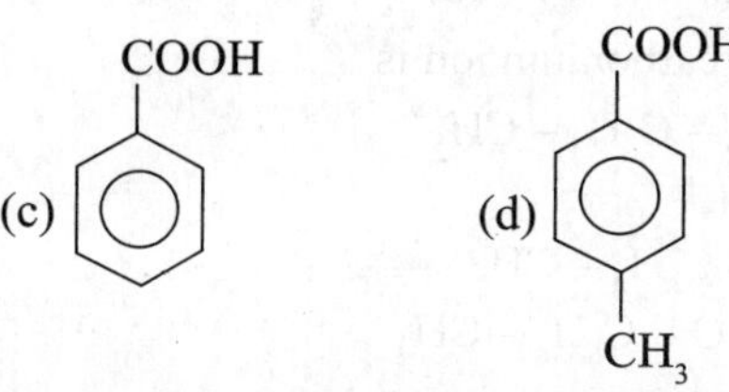

65. Out of these acids, the strongest acid is
(a) O–nitrobenzoic acid
(b) P – nitrobenzoic acid
(c) M – nitrobenzoic acid
(d) Benzoic acid

66. Which one of the following is most basic?

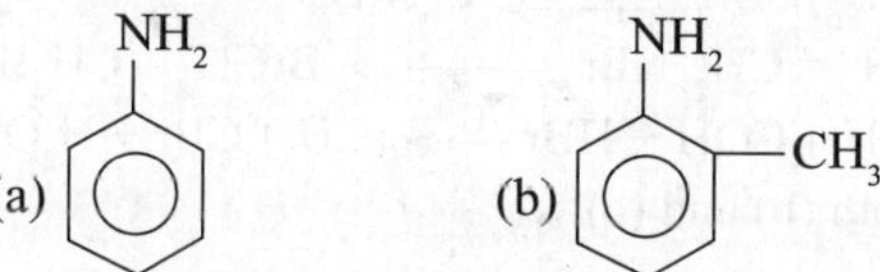

(c) CH_3 – [ring with NH_2] – CH_3 (d) [ring with NH_2, CH_3, CH_3]

67. Which one of the following is a sec–allylic carbocation?
(a) $\overset{\oplus}{C}H_2 - CH{=}CH_2$
(b) $CH_3 - CH{=}CH - \overset{\oplus}{C}H_2$
(c) $C_6H_5 - CH{=}\overset{\oplus}{C}H$
(d) $CH_3 - \overset{\oplus}{C}H - CH{=}CH_2$

68. Which of the following hydrolyzes fastest?
(a) CH_3Cl (b) CH_3CH_2Cl
(c) CH_3OCH_2Cl (d) $(CH_3)_2CHCl$

69. When pent–2–ene is treated with HI in presence of a peroxy acids the major product obtained is
(a) 2–Iodo–pentane (b) 3–Iodo–pentane
(c) Both of these (d) None of these

70. Which is correct about SN^1 reaction?
(a) Here reaction intermediate is carbocation
(b) It is favoured by polar solvents
(c) It is favoured by stability of carbocation
(d) All of these

PRACTICE QUESTIONS – II

71. Among the given compounds, the most susceptible to nucleophilic attack at the carbonyl group is
(a) MeCOCl
(b) MeCHO
(c) MeCOOMe
(d) Me CO – O – CO Me

72. The most stable free radical among the following is
(1) $C_6H_5\dot{C}H_2$
(2) $CH_3 - \dot{C}H - CH_3$
(3) $C_6H_5\dot{C}HCH_3$
(4) $CH_3\dot{C}H_2$
(a) 3 > 1 > 4 > 2
(b) 1 > 3 > 2 > 4
(c) 3 > 1 > 2 > 4
(d) 3 > 2 > 1 > 4

73. Arrange the following groups in order of decreasing inductive effects:
NO_2, $C(CH_3)_3$, CH_3, OCH_3, Br
(a) $NO_2 > Br > OCH_3 > C(CH_3)_3 > CH_3$
(b) $NO_2 > Br > OCH_3 > CH_3 > C(CH_3)_3$
(c) $NO_2 > OCH_3 > Br > C(CH_3)_3 > CH_3$
(d) $NO_2 > OCH_3 > C(CH_3)_3 > Br > CH_3$

74. +R power for the following groups decreases in the order
(1) $-NH_2$ (2) $-O^{\ominus}$
(3) – OH (4) $-NHCOCH_3$
(a) 2 > 1 > 3 > 4
(b) 2 > 3 > 1 > 4
(c) 1 > 2 > 3 > 4
(d) 3 > 2 > 1 > 4

75. In which case, the intermediate involved is incorrect?
(a) Pinacol – pinacolone rearrangement – carbocation
(b) Hofmann's bromamide reaction – Nitrene

(c) Aldol condensation – carbocation
(d) Anti Markonikoff's addition of HBr to propene – Free radical.

76. Arrange the following free radicals in order of stability:
(1) Benzyl (2) Allyl
(3) Methyl (4) Vinyl
(a) 4 > 3 > 2 > 1 (b) 1 > 2 > 3 > 4
(c) 1 > 3 > 4 > 2 (d) 4 > 3 > 1 > 2

77. The compound which gives the most stable carbonium ion on dehydration is

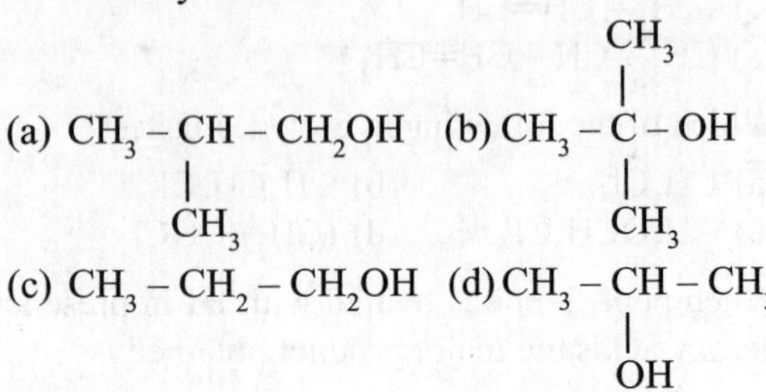

78. The order of effectiveness of o– and p–directing groups is
(a) OH > NR_2 > $NH.COCH_3$ > Cl > CH_3
(b) OH > Cl > CH_3 > NR_2 > $NH.COCH_3$
(c) OH > $NH.COCH_3$ > Cl > NR_2 > CH_3
(d) OH > CH_3 > $NH.COCH_3$ > Cl > NR_2

79. Resonance energy per benzene ring decreases in the order
(a) Naphthalene > Benzene > Anthracene > Phenanthrene
(b) Benzene > Naphthalene > Anthracene > Phenanthrene
(c) Benzene > Naphthalene > Phenanthrene > Anthracene
(d) All have equal resonance energy

80. Which of the following has the most acidic hydrogen?
(a) 3–hexanone
(b) 2, 4–hexanedione
(c) 2, 5– hexanedione
(d) 2, 3– hexanedione

81. Arrange the following free radicals in increasing order of stability:

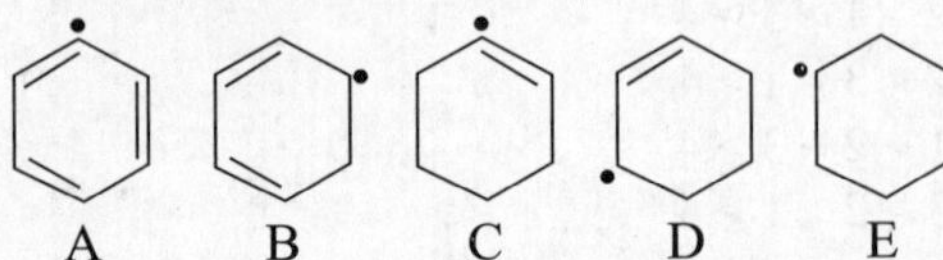

(a) A < B < C < D < E
(b) E < C < D < B < A
(c) C < A < B < E < D
(d) B < A < D < E < C

82. Which is the decreasing order of stability of the ions?
(1) $CH-{}^+CH-OCH_3$
(2) $CH_3-{}^+CH-CH_3$
(3) $CH_3-{}^+CH-COCH_3$
(4) $CN-CH_2-{}^+CH_2$
(a) 1 > 4 > 3 > 2 (b) 1 > 2 > 3 > 4
(c) 2 > 1 > 3 > 4 (d) 1 > 3 > 4 > 2.

83. Most stable carbonium ion is
(a) $p-NO_2-C_6H_4-CH_2^+$
(b) $C_6H_5CH_2^+$
(c) $p-Cl-C_6H_4-CH_2^+$
(d) $p-CH_3O-C_6H_4-CH_2^+$

84. Arrange $CCl_3(CH_2)_3NH_2$ (a), $CCl_3(CH_2)_2NH_2$ (b) and $CCl_3CH_2NH_2$ (c)in order of decreasing basic nature.
(a) C > B > A (b) C > A > B
(c) A > C > B (d) B > A > C

85. Which of the following reaction does not involve a carbocation as intermediate?
(a) $C_6H_6 + Br_2 \xrightarrow{AlBr_3} C_6H_5Br$
(b) $CH_2 = CH_2 + Br_2 \longrightarrow BrCH_2-CH_2Br$
(c) $(CH_3)_3COH + HBr \xrightarrow{H^+} (CH_3)_3CBr + H_2O$
(d) Both (b) and (c)

86. Most stable carbocation is
(a) $CH_2 = \overset{+}{C}H$ (b) $CH_2 = CH - \overset{+}{C}H_2$
(c) cyclohexenyl–$\overset{+}{C}H_2$ (d) cyclohexenyl(+)–CH_2

87. In the following compounds

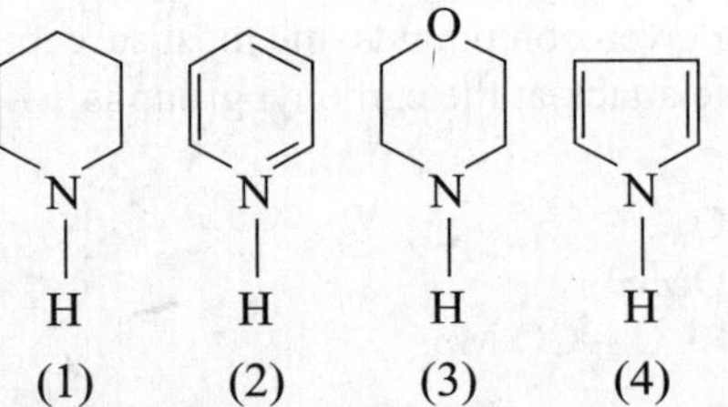

The order of basicity is
(a) 4 > 1 > 3 > 2 (b) 3 > 1 > 4 > 2
(c) 2 > 1 > 3 > 4 (d) 1 > 3 > 2 > 4

88. Which of the following will most readily be dehydrated in acidic condition?

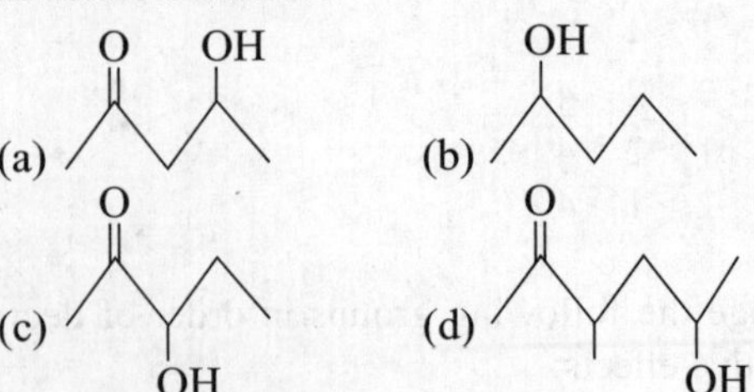

89. Which of the following will undergo dehydration most rapidly?

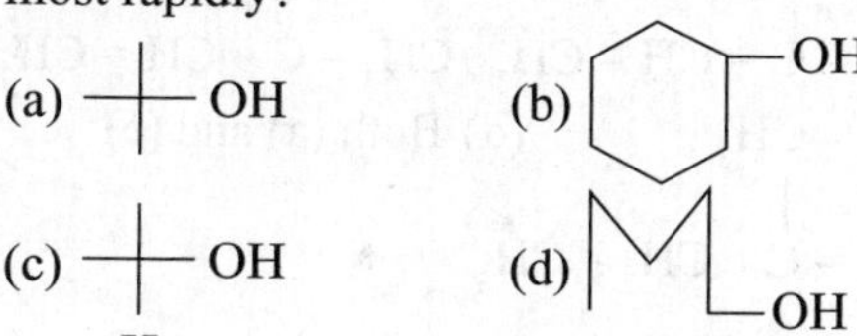

90. Consider the following nucleophiles:

H_2O, CH_3COO^-, $\bar{O}H$, CH_3O^-
1 2 3 4

The correct order of decreasing nucleophilicity is,

(a) 1 > 2 > 3 > 4 (b) 4 > 3 > 2 > 1
(c) 4 > 1 > 2 > 3 (d) 4 > 2 > 3 > 1

91. Stability of

(1) $CH_3-CH=CH-CH_3$ (2) $CH_3-C(CH_3)=C(CH_3)-CH_3$

(3) $CH_3-C(CH_3)=CH_2$ (4) $CH_3-C(CH_3)=CH-CH_3$

(a) 1 > 3 > 4 > 2 (b) 1 > 2 > 3 > 4
(c) 4 > 3 > 2 > 1 (d) 2 > 4 > 3 > 1.

92. Arrange methyl amine (a), pyridine (b) and guanidine (c) in decreasing order of basic strength

(a) C > A > B (b) A > B > C
(c) B > A > C (d) B > C > A

93. Arrange the following in decreasing order of basicity:

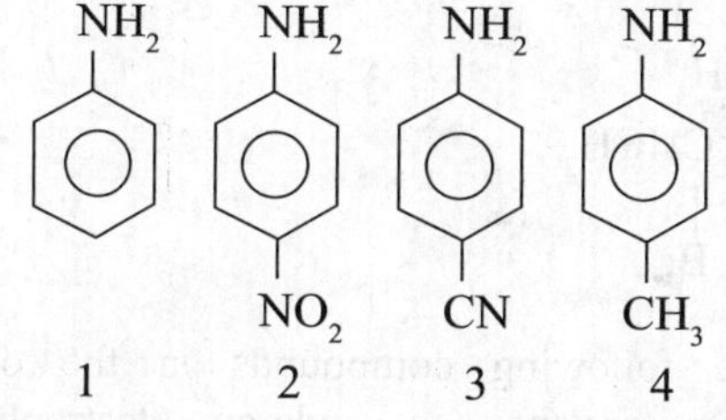

(a) 4 > 1 > 3 > 2
(b) 4 > 1 > 2 > 3
(c) 4 > 3 > 1 > 2
(d) 1 > 2 > 3 > 4

94. The basic nature of the following heterocyclic compounds decreases in the order:

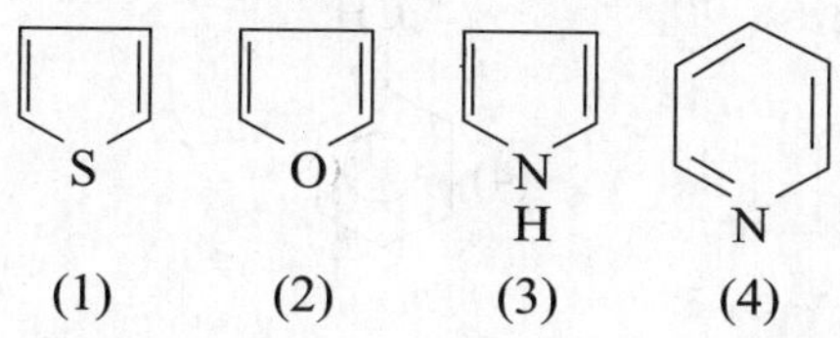

(a) 1 > 2 > 3 > 4 (b) 4 > 1 > 2 > 3
(c) 4 > 2 > 1 > 3 (d) 4 > 1 > 3 > 2

95. Which of the following alkyl halides would be the most reactive in an SN^2 reaction?

(a) $CH_3CH_2CH_2CH_2Br$
(b) $CH_3-CH(CH_3)-CH_2Br$
(c) $C_6H_5-C(CH_3)_2-Br$
(d) $C_6H_5CH_2CH(Br)CH_3$

96. The most stable resonating structure is?

(a) $CH_2=C_6H_4=C(H)-O^{\ominus}$ (b) $^{\ominus}CH_2-C_6H_4-C(=O)H$
(c) $^{\oplus}CH_2-C_6H_4-C(=O)H$ (d) $CH_2=C_6H_4-C(=O)H$

97. In which of the following compounds resonance does not lead to stability of the compound?

(a) cyclopropenyl anion (H, :−) (b) cyclopropenyl cation (H, +)
(c) benzene (d) $CH=CH-CH=CH_2$

98. Arrange in order of decreasing trend towards S_E reactions:

(1) Chlorobenzene
(2) Benzene
(3) Anilinium chloride
(4) Toluene

(a) 4 > 2 > 1 > 3 (b) 1 > 2 > 3 > 4
(c) 2 > 1 > 3 > 4 (d) 3 > 1 > 2 > 4

99. Hyper condugation may occur in?

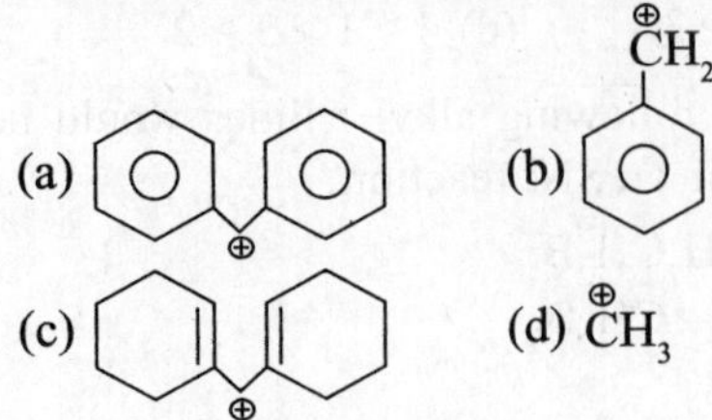

(d) $\overset{\oplus}{C}H_3$

100. 3–phenylpropene on reaction with HBr gives (as a major product)

(a) $C_6H_5CH_2CH(Br)CH_3$
(b) $C_6H_5CH(Br)CH_2CH_3$
(c) $C_6H_5CH_2CH_2CH_2Br$
(d) $C_6H_5CH(Br)CH = CH_2$

101. Which of the following statements are correct?

1. SN^1 mechanism is most commonly given by tertiary alkyl halides.
2. SN^1 mechanism proceeds through the formation of a carbocation.
3. SN^2 mechanism involves retention of configuration.
4. SN^2 mechanism proceeds through the formation of a transition state.

(a) 1, 2 and 4 (b) 1, 3 and 4
(c) 2, 3 and 4 (d) 1, 2, 3 and 4

102. The most unlikely representation of resonance structures of p–nitrophenoxide ion is

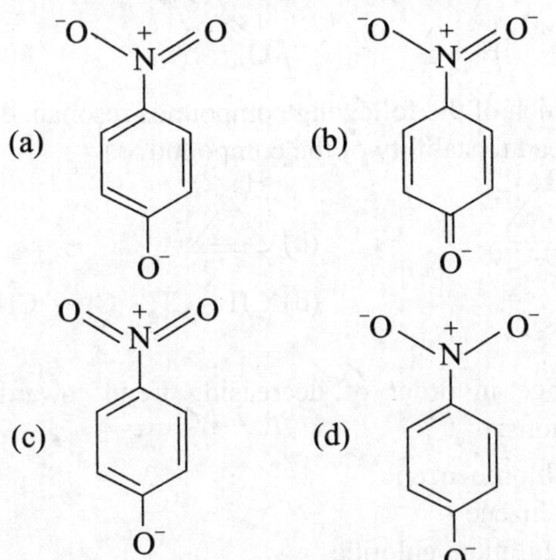

103. The major product obtained in the reaction

$$CH_3-\underset{H}{\overset{CH_3}{C}}-\underset{H}{\overset{Cl}{C}}-CH_3 \xrightarrow[-HCl]{OEt^-}$$

(a) $CH_3-\overset{CH_3}{C}=CH-CH_3$ (b) $CH_3-\overset{CH_3}{C}-CH=CH_2$
(c) $CH_2=\overset{CH_3}{C}-CH_2-CH_3$ (d) Both (a) and (c)

104. In which of the following reactions are free radical intermediates obtained?

(a) $CH_3CH=CH_2 + HBr \longrightarrow CH_3\underset{Br}{C}HCH_3$
(b) $CH_3\underset{OH}{C}H-CH_3 + H_2SO_4 \longrightarrow CH_3CH{=}CH_2 + H_2O$
(c) $CH_2{=}CHCH_3 + Hg(OAc)_2 \longrightarrow CH_3-\underset{OH}{C}H-CH_2OAc + CH_3COOH$
(d) $CH_3CH=CH_2 + HBr \xrightarrow{R-O-O-R} CH_3CH_2CH_2Br$

105. Which of the following alkyl halides would be the most reactive in an SN^1 reaction?

(a) $C_6H_5CH_2CH_2CH_2Br$
(b) $C_6H_5-\underset{CH_3}{C}H-CH_2Br$
(c) $C_6H_5-\underset{CH_3}{\overset{CH_3}{C}}-Br$
(d) $C_6H_5CH_2\underset{Br}{C}HCH_3$

106. Arrange the following compounds in the order of increasing tendency to undergo electrophillic substitution:

(1) nitrobenzene (NO_2) (2) benzene
(3) phenol (OH) (4) toluene (CH_3)

(5) 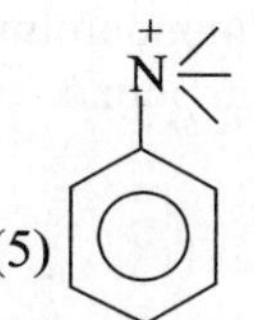

(a) 5 < 1 < 2 < 4 < 3
(b) 3 > 2 > 1 > 4 > 5
(c) 1 > 2 > 5 > 3 > 4
(d) 5 < 1 < 4 < 2< 3

107. Which of the following is correctly matched?

(a) $CH_3 - CH_2 - CH_2 - \overset{\oplus}{N}(CH_3)_3\overset{\ominus}{O}H$.$E_2$ reaction

(b) $CH_3 - \underset{CH_3}{\overset{Br}{\overset{|}{\underset{|}{C}}}} - CH_3$ E_2reaction

(c) $CH_3 - CH_2 - CH_2OH$, E_1 reaction

(d) $CH_3 - \underset{CH_3}{\overset{OH}{\overset{|}{\underset{|}{C}}}} - CH_3$ E_1 cb reaction

108. Which is dehydrated to a maximum extent using conc. H_2SO_4?

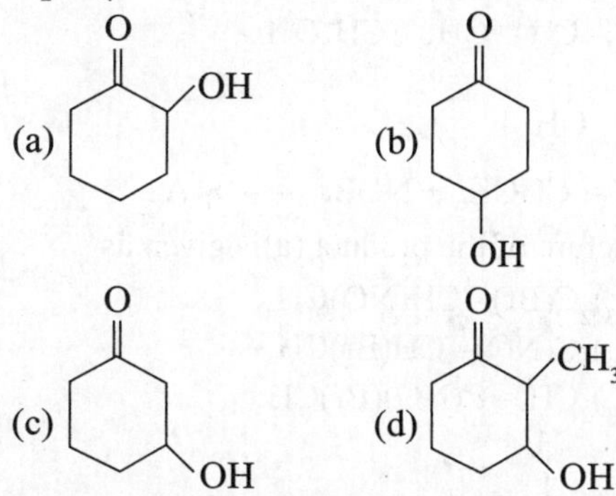

109. Dehydrobromination (–HBr) of the following in increasing order is

(1) Br (2) Br

(3) Br

(a) 1 < 2 < 3 (b) 3 < 2 < 1
(c) 2 = 3 < 1 (d) 3 < 1 < 2

110. Arrange these compounds in order of increasing SN^2 reaction rate

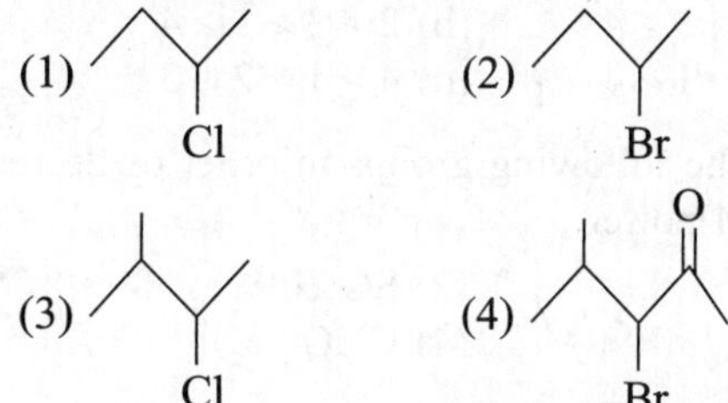

(a) 3 < 1 < 2 < 4 (b) 3 < 2 < 1 < 4
(c) 4 < 3 < 1 < 2 (d) 2 > 3 > 1> 4

111. Arrange the following compounds in the order of increasing tendency to undergo electrophilic substitution:

(1) Nitrobenzene (2) Benzene
(3) Phenol (4) Toluene
(5) Trimethyl phenyl ammonium ion

(a) 5 < 1 < 2 < 4 < 3 (b) 3 > 2 > 1 > 4 > 5
(c) 1 > 2 > 5 > 3 > 4 (d) 5< 2 < 4 < 1 < 3

112. Which of the following represent the correct order of nucleophillic addition for (1) HCHO, (2) CH_3COCH_3, (3) CH_3CHO, (4) $CH_3COC_2H_5$

(a) 1 > 2 > 3 > 4 (b) 1 > 3 > 2 > 4
(c) 4 > 2 > 3 > 1 (d) 1 > 4 > 3 > 2

113. Which nitrogen atom in LSD is most basic?

1 H–N; O ‖ 3 C–N(Et)$_2$; 2 N–CH_3

(a) 1 (b) 3
(c) 2 (d) all are equally basic

114. Consider the following carbanions:

(1) $CH_3 - \overset{\ominus}{C}H_2$ (2) $CH_2 = \overset{\ominus}{C}H$
(3) $CH \equiv \overset{\ominus}{C}$

Correct order of stability of these carbanions in decreasing order is

(a) 1 > 2 > 3 (b) 2 > 1 > 3
(c) 3 > 2 > 1 (d) 3 > 1 > 2

115. Consider the following compounds:

(1) $C_6H_5 - NH_2$
(2) $C_6H_5 - NH - C_6H_5$
(3) $(C_6H_5)_3N$
(4) $CH_3 - CH_2 - NH_2$

Arrange these compounds in decreasing order of their basicity.

(a) 1 > 2 > 3 > 4 (b) 2 > 3 > 1 > 4
(c) 3 > 2 > 1 > 4 (d) 4 > 1 > 2 > 3

116. Arrange the following groups in order of decreasing –R (or –M)power:

(1) NO_2 (2) SO_3H
(3) CF_3 (4) CHO

(a) 1 > 3 > 2 > 4
(b) 1 > 2 > 3 > 4
(c) 1 > 4 > 3 > 2
(d) 4> 3 > 2 > 1

117. 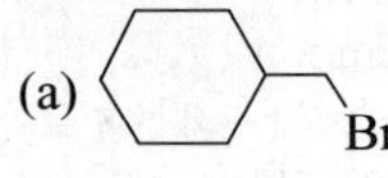 HBr —— X (predominant)

Here X is

(a) (cyclohexyl)–CH_2Br (b) Br–(cyclohexyl)–CH_3

(c) (cyclohexyl)–Br (d) 1-bromo-1-methylcyclohexane

118. Which hydrogen atom can be abstracted easily from given the compound?

$$-\overset{}{\underset{H}{C}}=\underset{H}{C}-\overset{|}{\underset{|}{C}}-H$$

(a) α (b) β
(c) γ (d) can't be said

119. In which cases, free radicals can be formed by hemolytic fission?

(a) $CH_3\overset{O}{\overset{\|}{C}}CH_3 \xrightarrow{hv}$

(b) $R-\ddot{N}=\ddot{N}-R \xrightarrow{hv}$

(c) $C_6H_5CH_2\overset{O}{\overset{\|}{C}}-\ddot{O}: \xrightarrow{hv}$

(d) In all cases

120. The major product obtained on the monobromination (with $Br_2/FeBr_3$) of the following compound A is

(3-methylanisole: OCH_3, CH_3)

(a) (OCH_3, Br, CH_3) (b) (OCH_3, CH_3, Br)

(c) (OCH_3, Br, CH_3) (d) (OCH_3, Br, CH_3)

121. The major product formed in the following reaction is $CH_3CH(Cl)CH_2-CH_2OH$ aq. KOH

(a) $CH_3CH=CH-CH_2OH$
(b) $CH_2=CH-CH_2-CH_2OH$
(c) $CH_3-CH-CH_2$ (ring: O — CH_2)
(d) $CH_3-\underset{OH}{\underset{|}{CH}}-CH_2-CH_2OH$

122. $(H_3C)_2C=CHCH_3 + NOBr \longrightarrow A$

The structure of the product (a)is given as

(a) $(CH_3)_2C(Br)-CH(NO)CH_3$
(b) $(CH_3)_2C(NO)-CH(Br)CH_3$
(c) $(CH_3)_2CH-C(NO)(Br)CH_3$
(d) $H_3C-\overset{H}{\overset{|}{\underset{NO}{\underset{|}{C}}}}-\underset{CH_3}{\underset{|}{CH}}-CH_3$

PRACTICE QUESTIONS – III

123. Match the following:

List I (Reagents)	**List II (Nature)**
1. (CN)	(1) Neutral nucleophile
2. (ONO)	(2) electrophile
3. NH_3	(3) Ambident nucleo phile
4. $AlCl_3$	(4) Both El and Nu

The correct matching is:

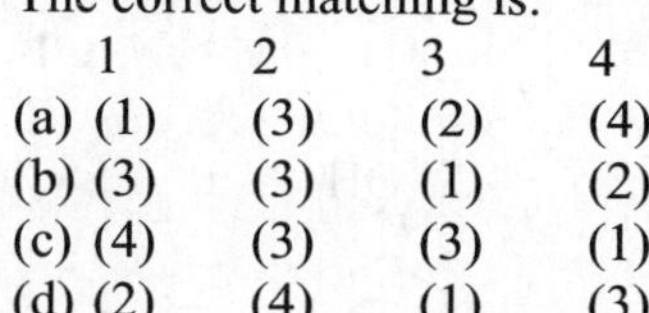

	1	2	3	4
(a)	(1)	(3)	(2)	(4)
(b)	(3)	(3)	(1)	(2)
(c)	(4)	(3)	(3)	(1)
(d)	(2)	(4)	(1)	(3)

124. The correct order of stability for these compounds

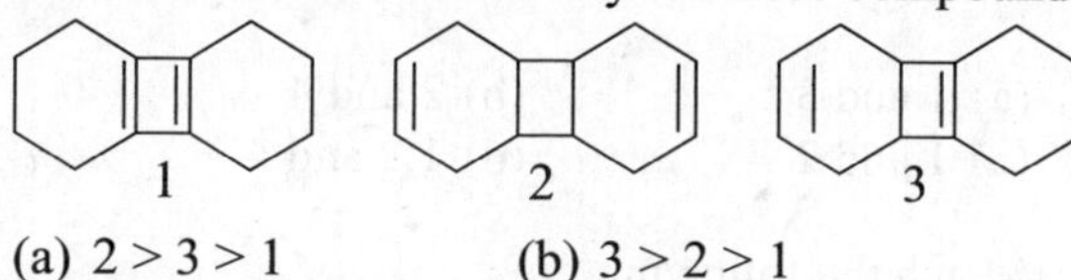

(a) $2 > 3 > 1$ (b) $3 > 2 > 1$
(c) $1 > 2 > 3$ (d) $3 > 1 > 2$

125. When a methyl radical is formed from CH_3Cl, select the correct statement:

(1) Bond angle of 109º28’ is retained
(2) Number of sigma bonds is three
(3) Carbon undergoes geometric change from tetrahedral to planar
(4) Hybridization changes sp^3 to sp^2

(a) 2, 3 and 4 (b) 1, 3 and 4
(c) 2 and 4 (d) 3 and 4

126. The correct increasing order of the reactivity of halides for SN_1 reaction is

(a) $CH_3-CH_2-X < (CH_3)_2CH-X < CH_2=CH-CH_2-X < PhCH_2-X$
(b) $(CH_3)_2CH-X < CH_3-CH_2-X < CH_2=CH-CH_2-X < PhCH_2-X$
(c) $PhCH_2-X < (CH_3)_2CH-X < CH_3-CH_2-X < CH_2=CH-CH_2-X$
(d) $CH_2=CH-CH_2-X < PhCH_2-X < (CH_3)_2CH-X < CH_3-CH_2-X$

127. Arrange the following carbanions in the decreasing order of stability.

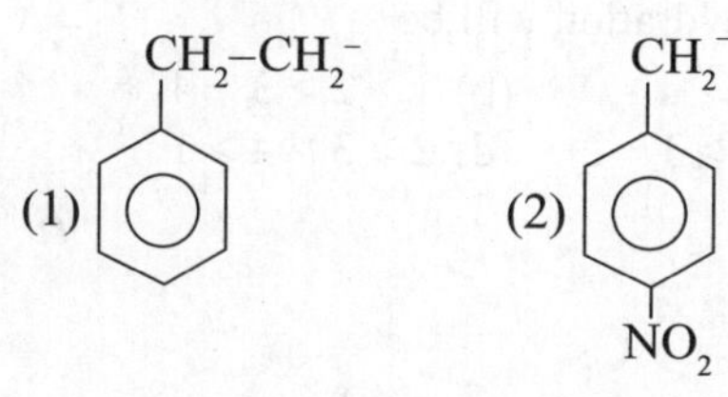

(3) CH_2^- NO_2 (4) CH_2^- OCH_3

(a) $4 > 2 > 3 > 1$ (b) $4 > 3 > 2 > 1$
(c) $3 > 2 > 1 > 4$ (d) $4 > 2 > 1 > 3$

128. The stability of the following carbocations decreases in the order.

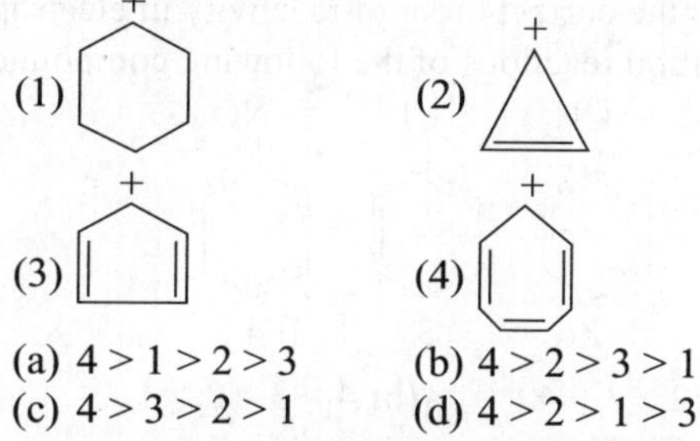

(a) $4 > 1 > 2 > 3$ (b) $4 > 2 > 3 > 1$
(c) $4 > 3 > 2 > 1$ (d) $4 > 2 > 1 > 3$

129. Which of the following statements are correct?

(1) NH_2^- is better nucleophile than NH_3 but latter (NH_3)is better nucleophile than NH_4^+
(2) $C_6H_5O^-$ is better nucleophile than

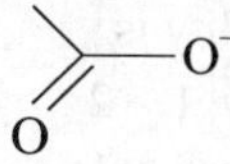

(3) OH^- is better nucleophile than SH^- and H_2O, but H_2O is better nucleophile than H_3O^+
(4) ClO^- is weaker nucleophile than ClO_4^-

(a) 1, 2 and 3 (b) 1, 3 and 4
(c) 2, 3 and 4 (d) 1,2 ,3 and 4

130. Consider the following structures

1. $CH_2=\overset{+}{C}H$ 2. $CH_3-\overset{+}{C}H_2$
3. $CH_2=CH-\overset{+}{C}H_2$ 4. $C_6H_5-\overset{+}{C}H-C_6H_5$

The correct sequence of these carbocations in the decreasing order of their stability is

(a) 1, 2, 3, 4 (b) 1, 3, 2, 4
(c) 4, 2, 3, 1 (d) 4, 3, 2, 1

131. The correct order of basicity of the following compound is

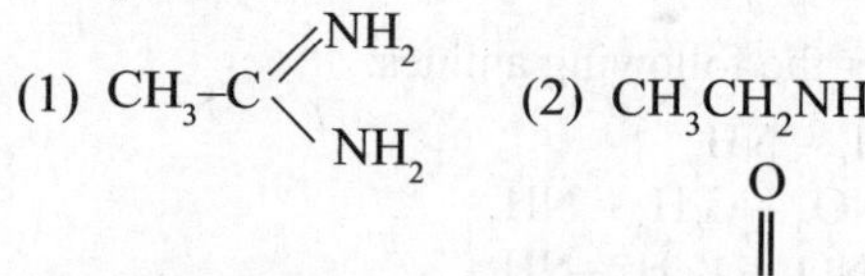

(3) $(CH_3)_2NH$ (4) $CH_3-\overset{O}{\overset{\|}{C}}-NH_2$
(5) CH_3CN

(a) $2 > 1 > 3 > 4 > 5$ (b) $1 > 3 > 2 > 4 > 5$
(c) $3 > 1 > 2 > 5 > 4$ (d) $1 > 3 > 2 > 5 > 4$.

132. Consider the following carbocations and arranged them in the increasing order of their stability:

(1) 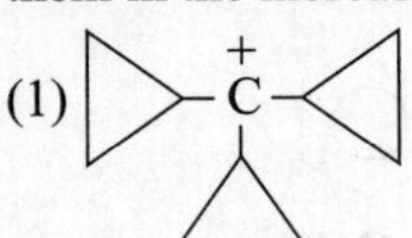
(2) $CH_3 - \overset{+}{C}H - CH_3$

(3) $C_6H_5 - \overset{+}{C}H_2$ (4) $C_6H_5 - \overset{+}{C}H - C_6H_5$
(5) $CH_3 - CO - C^{+}H_2$

(a) $5 < 2 < 3 < 4 < 1$ (b) $5 < 3 < 2 < 4 < 1$
(c) $5 < 1 < 2 < 4 < 3$ (d) $3 < 2 < 4 < 1 < 5$

133. Identify the correct order of reactivity in electrophilic substitution reactions of the following compounds:

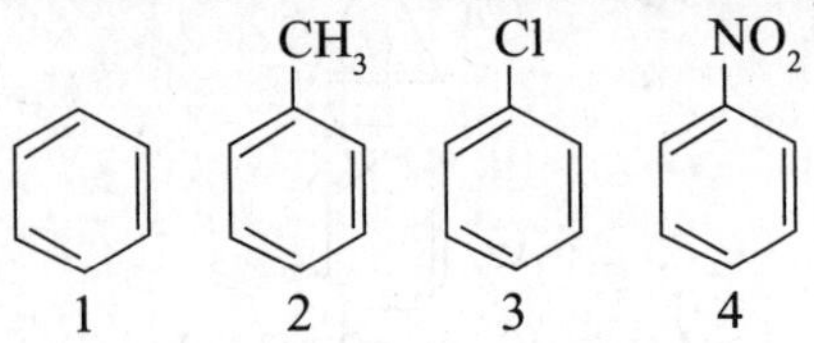

(a) $1 > 2 > 3 > 4$ (b) $4 > 3 > 2 > 1$
(c) $2 > 1 > 3 > 4$ (d) $2 > 3 > 1 > 4$

134. In the following groups:
–OAc (1), –OMe (2)
$-OSO_2Me$ (3), $-OSO_2CF_3$(4)
the order of the leaving group ability is
(a) $1 > 2 > 3 > 4$ (b) $4 > 3 > 1 > 2$
(c) $3 > 2 > 1 > 4$ (d) $2 > 3 > 4 > 1$.

135. In which of the following compounds, steric inhibition of Resonance occurs?

1. 2.

3. 4.

(a) 1, 2, and 3 (b) 1, 2
(c) 1 only (d) 2, 4

136. Consider the following amines:
(1) $C_6H_5 - NH_2$
(2) $o\text{–}NO_2 - C_6H_4 - NH_2$
(3) $m\text{–}NO_2 - C_6H_4 - NH_2$
(4) $p - NO_2 - C_6H_4 - NH_2$

Arrange these compounds in decreasing order of basicity.
(a) $1 > 2 > 3 > 4$ (b) $2 > 1 > 3 > 4$
(c) $1 > 3 > 4 > 2$ (d) $1 > 3 > 2 > 4$

137. Which will undergo Friedal–Craft alkylation reaction?

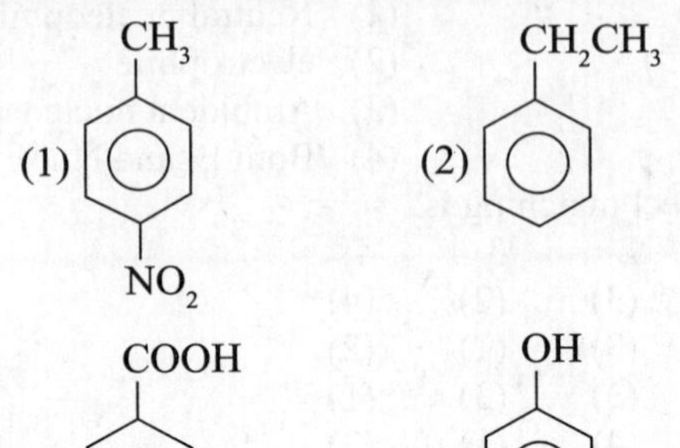

(3) COOH (4) OH

(a) 1 and 3 (b) 2 and 4
(c) 1 and 2 (d) 1,2 and 4

138. Match the following.

List I	List II
1. Carbocation	(1) delocalization of π e^-
2. Resonance	(2) coplanar
3. NO_2	(3) pyramidal
4. Carbanion	(4) –I, –R group

The correct matching is:

	1	2	3	4
(a)	(2)	(3)	(4)	(1)
(b)	(2)	(1)	(4)	(3)
(c)	(1)	(4)	(2)	(3)
(d)	(2)	(1)	(3)	(4)

139. Four structures (1) – (4) of different alcohols are given below:

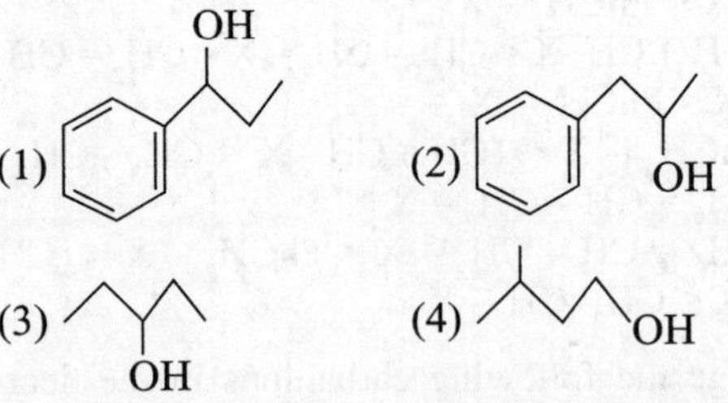

The order of facility, from fastest to slowest, of acid catalysed dehydration will be
(a) $2 > 1 > 3 > 4$ (b) $1 > 2 > 3 > 4$
(c) $4 > 3 > 2 > 1$ (d) $2 > 3 > 4 > 1$

140. Benzene + $Cl-CH_2CH_2-CH_3$ $\xrightarrow{AlCl_3}$ P $\xrightarrow[(ii)\ H_3O^+]{(i)\ O_2/\Delta}$ Q + Phenol

(a) n-propylbenzene and CH_3CH_2CHO

(b) n-propylbenzene and CH_3COCH_3

(c) isopropylbenzene and CH_3COCH_3

(d) isopropylbenzene and CH_3CH_2CHO

141. Match the following:

List I (Type of reaction)	List II (Phenomenon)
1. SN_2	(1) Walden inversion
2. SN^1	(2) Carbanion intermediate
3. E_2	(3) Anti-periplanar configuration
4. E_{1Cb}	(4) Carbocation intermediate

The correct matching is:

	1	2	3	4
(a)	(1)	(4)	(3)	(4)
(b)	(1)	(2)	(3)	(4)
(c)	(3)	(2)	(1)	(4)
(d)	(4)	(3)	(2)	(1)

142. The correct order of decreasing acidity of the acids given below is

1. $Cl_3CCH = CH - CH_2 - COOH$
2. $H_3CCH = CH - CH_2 - COOH$
3. $Cl_3CCH = CH - COOH$
4. $H_3CCH_2CH_2CH_2COOH$

(a) $1 > 3 > 2 > 4$ (b) $3 > 1 > 2 > 4$
(c) $3 > 4 > 1 > 2$ (d) $3 > 1 > 4 > 2$

143. In the following benzyl/allyl system

$R - CH = CH_2$ or C_6H_5-R

(R is alkyl group) decreasing order of inductive effect is

(a) $(CH_3)_3C- > (CH_3)_2CH - > CH_3CH_2-$
(b) $CH_3CH_2- > (CH_3)_2CH - > (CH_3)_3C-$
(c) $(CH_3)_2CH - > CH_3CH_2- > (CH_3)_3C-$
(d) $(CH_3)_3C- > CH_3CH_2- > (CH_3)_2CH -$

144. Maximum dehydration takes place that of

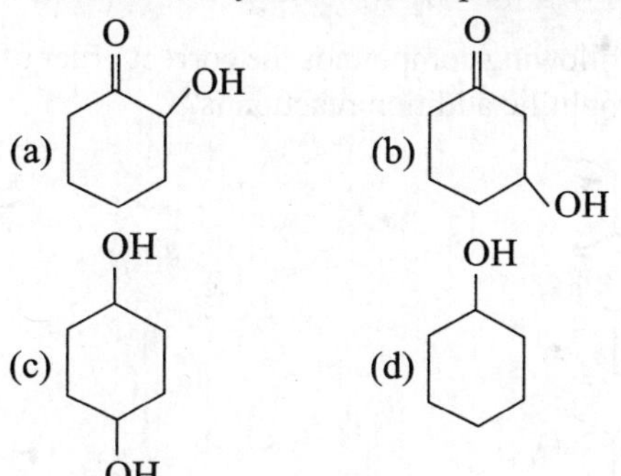

145. SN^1 reaction is feasible in

(a) $(CH_3)_3C-Cl$ + KOH
(b) $CH_3CH_2CH_2CH_2-Cl$ + KOH ⟶
(c) C_6H_5-Cl + KOH ⟶
(d) $C_6H_5-CH_2CH_2Cl$ + KOH ⟶

146. 1-methylcyclobutyl $-\overset{\oplus}{C}H_2$ (Me, $CH_2^{\oplus}$) The rearrangement of this carborium ion gives rise to mainly?

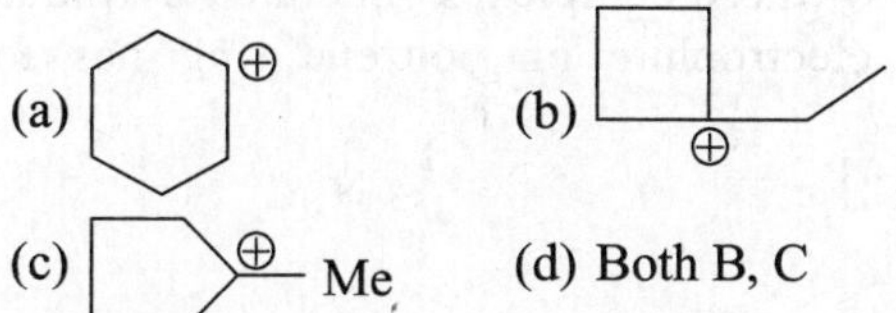

(d) Both B, C

147. Which of the following is the correct order of acidic strength for following?

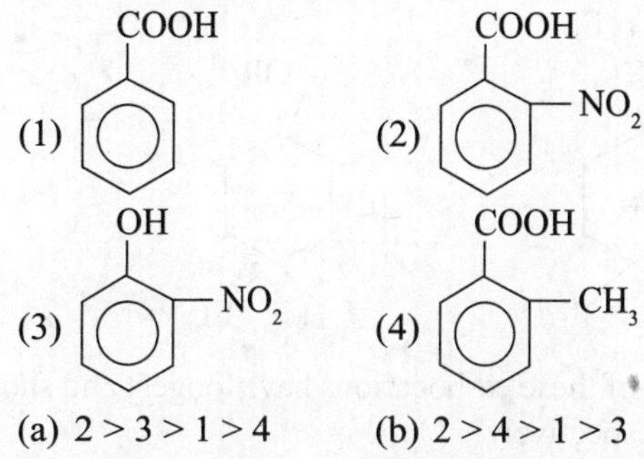

(a) $2 > 3 > 1 > 4$ (b) $2 > 4 > 1 > 3$
(c) $2 > 3 > 4 > 1$ (d) $4 > 3 > 1 > 2$

148. The correct order of increasing basic nature for the bases NH_3, CH_3NH_2, and $(CH_3)_2NH$ is

(a) $CH_3NH_2 < NH_3 < (CH_3)_2NH$
(b) $(CH_3)_2NH < NH_3 < CH_3NH_2$
(c) $NH_3 < CH_3NH_2 < (CH_3)_2NH$
(d) $CH_3NH_2 < (CH_3)_2NH < NH_3$

149. For the following compounds the correct order of case of electrophillic addition reaction is?

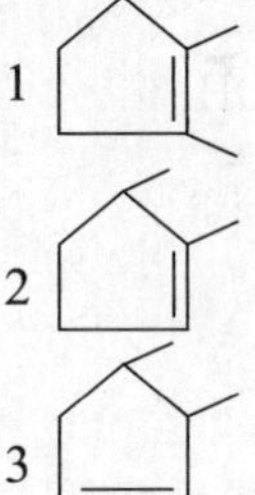

(a) 1 > 2 > 3 (b) 1 > 2 > 2
(c) 2 > 3 > 1 (d) 1 > 2 > 3

150. Consider the acidity of the carboxylic acids:

(1) PhCOOH
(2) o–$NO_2C_6H_4COOH$
(3) p–$NO_2C_6H_4COOH$
(4) m–$NO_2C_6H_4COOH$

Which of the following order is correct?

(a) 1 > 2 > 3 > 4
(b) 2 > 4 > 3 > 1
(c) 2 > 4 > 1 > 3
(d) 2> 3 > 4 > 1

151. Which of these complexes generated by the attack of an electrophite on benzene ring has lowest energy?

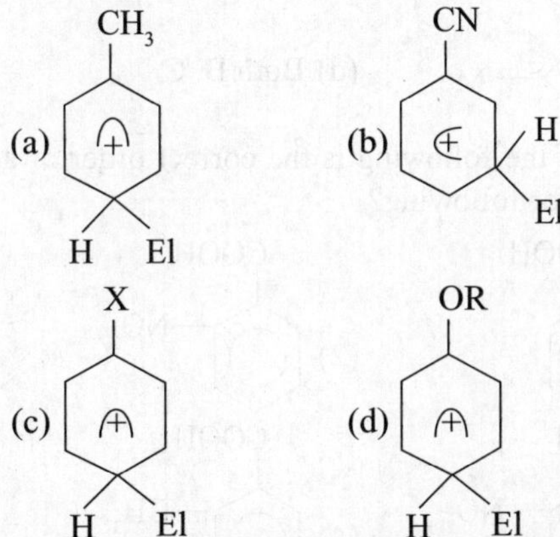

152. Which of these carbocations have longest and shortest lives respectively?

(1) $\Delta-\overset{\oplus}{C}H-\Delta$ (2) $CH_3-\overset{\oplus}{C}H-NO_2$

(3) $Ø_3\overset{+}{C}$ (4) $(CH_3)_2\overset{+}{C}H$

(5) $\overset{\oplus}{C}H_2$–C_6H_4–NO_2 (para)

(a) 1, 2 (b) 3, 2
(c) 1, 5 (d) 3, 5

153. The total number of lone pair of electron in this molecule is?

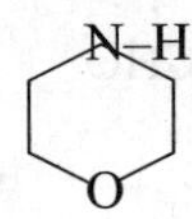

(a) 3 (b) 4
(c) 2 (d) 1

154. 2–methylbutane on reacting with bromine in the presence of sunlight gives mainly

(a) 1–bromo–2–methylbutane
(b) 2–bromo–2–methylbutane
(c) 2–bromo–3–methylbutane
(d) 1–bromo–3–methylbutane

155. For the following compounds the increasing order of acidic nature is

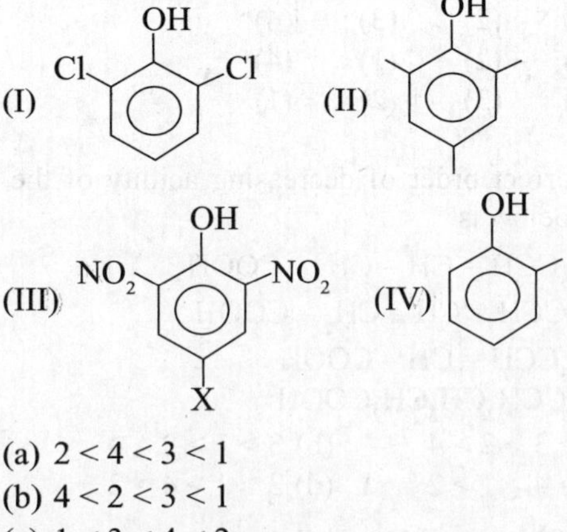

(a) 2 < 4 < 3 < 1
(b) 4 < 2 < 3 < 1
(c) 1 < 3 < 4 < 2
(d) 2 < 4 < 1 < 3

156. Which of the following carbocations can rearrange to form a more stable form?

(1) 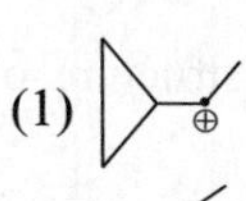 (2)

(3) (4)

(a) 2, 3 (b) 2, 3, 4
(c) 1, 2, 3 (d) 1, 2, 3, 4

157. The decreasing order of nucleophilicity among the nucleophiles

(1) $CH_3C(=O)-O^-$ (2) CH_3O^-

(3) CN^- (4) $H_3C-C_6H_4-SO_2-O^-$

(a) 1, 2, 3, 4 (b) 4, 3, 2, 1
(c) 2, 3, 1, 4 (d) 3, 2, 1, 4

158. The reaction is fastest when X is

$R-C(=O)X + Nu^{\ominus} \longrightarrow R-C(=O)Nu + X^{\ominus}$

(a) Cl (b) NH_2
(c) OC_2H_5 (d) OCOR

159. Elimination of bromine from 2–bromobutane results in the formation of

(a) equimolar mixture of 1 and 2–butene
(b) predominantly 2–butene
(c) predominantly 1–butene
(d) predominantly 2–butyne

160. The increasing order of the rate of HCN addition to compounds A – D is

(a) HCHO (b) CH_3COCH_3
(c) $PhCOCH_3$ (d) PhCOPh
(a) a < b < c < d (b) d < b < c < a
(c) d < c < b < a (d) c < d < b < a

161. $CH_3Br + Nu^- \longrightarrow CH_3 - Nu + Br^-$

The decreasing order of the rate of the above reaction with nucleophiles (Nu^-) a to d is

[Nu^- = (a) PhO^- (b) AcO^- (c) HO^- (d) CH_3O^-]

(a) d > c > a > b (b) d > c > b > a
(c) a > b > c > d (d) b > d > c > a

162. The correct order of increasing acid strength of the compound:

(1) CH_3COOH (2) $MeOCH_2CO_2H$
(3) CF_3CO_2H (4) $(Me)_2 CH CO_2H$

(a) 2 < 4 < 1 < 3
(b) 4 < 1 < 3 < 2
(c) 4 < 1 < 2 < 3
(d) 1 < 4 < 3 < 2

163. Reaction of trans–2–phenyl–1–bromocyclopentane on reaction with alcoholic KOH produces

(a) 4–phenylcyclopentene
(b) 2–phenylcyclopentene
(c) 1–phenylcyclopentene
(d) 3–phenylcyclopentene

164. The alkene formed as a major product in the given elimination reaction is

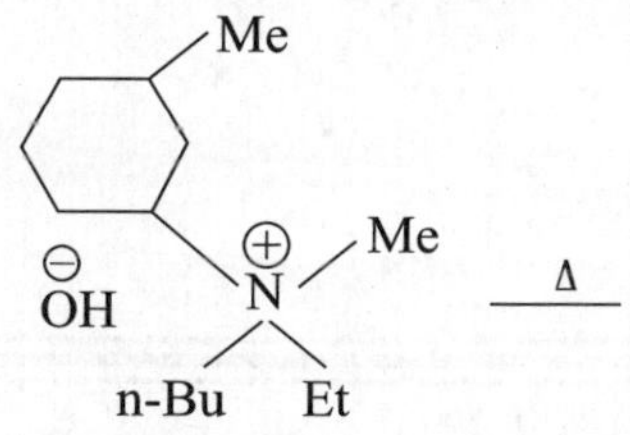

(a) Me–CH_2–CH=CH_2 (b) $CH_2 = CH_2$

(c) 1-methylcyclohexene (Me) (d) 4-methylcyclohexene (Me)

165. The electrophile involved in the given reaction is

Phenol (OH) + $CHCl_3$ + NaOH $\longrightarrow$ (O^-Na^+, CHO)

(a) dichloromethyl cation ($\overset{\oplus}{C}HCl_2$)
(b) dichlorocarbene ($:CCl_2$)
(c) trichloromethyl anion ($\overset{\ominus}{C}Cl_3$)
(d) formyl cation ($\overset{\oplus}{C}HO$)

166. The increasing order of stability of the following free radicals is

(a) $(CH_3)_2\dot{C}H < (CH_3)_3\dot{C} < (C_6H_5)_2\dot{C}H < (C_6H_5)_3\dot{C}$
(b) $(C_6H_5)_3\dot{C} < (C_6H_5)_2\dot{C}H < (CH_3)_3\dot{C} < (CH_3)_2\dot{C}H$
(c) $(C_6H_5)_2\dot{C}H < (C_6H_5)_3\dot{C} < (CH_3)_3\dot{C} < (CH_3)_2\dot{C}H$
(d) $(CH_3)_2\dot{C}H < (CH_3)_3\dot{C} < (C_6H_5)_3\dot{C} < (C_6H_5)_2\dot{C}H$

167. Which of the following is the correct order of decreasing SN[2] reactivity?

(a) $RCH_2X > R_2CHX > R_3CX$
(b) $R_3CX > R_2CHX > RCH_2X$
(c) $R_2CHX > R_3CX > RCH_2X$
(d) $RCH_2X > R_3CX > R_2CHX$

168. Which carbocation is most stable one here?

(a)

(b) 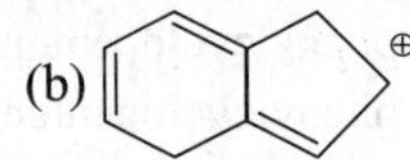

(c)

(d) 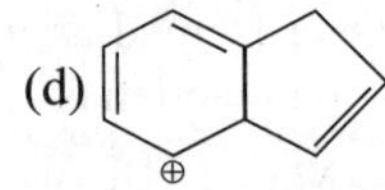

169. The most stable canonical structure of this molecule can be given as

(a)

(b) 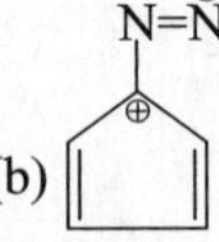

(c) 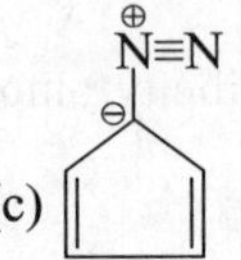

(d) 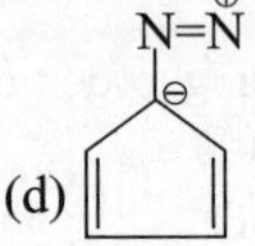

170. The order of SN[2] reactivity for these substrate under same conditions can be given as

(1)

(2) 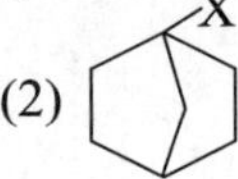

(3)

(a) 1 = 2 = 3
(b) 1 > 2 > 3
(c) 3 > 2 > 1
(d) 3 > 1 > 2

Answer Keys

1. (a)	2. (a)	3. (c)	4. (b)	5. (c)	6. (c)	7. (c)	8. (b)	9. (a)	10. (a)
11. (d)	12. (b)	13. (a)	14. (b)	15. (c)	16. (a)	17. (b)	18. (d)	19. (a)	20. (d)
21. (c)	22. (b)	23. (b)	24. (d)	25. (a)	26. (b)	27. (c)	28. (b)	29. (a)	30. (a)
31. (c)	32. (c)	33. (d)	34. (d)	35. (d)	36. (d)	37. (c)	38. (c)	39. (d)	40. (b)
41. (d)	42. (b)	43. (c)	44. (b)	45. (b)	46. (d)	47. (d)	48. (c)	49. (d)	50. (c)
51. (c)	52. (b)	53. (b)	54. (c)	55. (d)	56. (a)	57. (b)	58. (b)	59. (c)	60. (d)
61. (c)	62. (c)	63. (a)	64. (c)	65. (a)	66. (c)	67. (d)	68. (c)	69. (d)	70. (d)
71. (b)	72. (c)	73. (b)	74. (a)	75. (c)	76. (b)	77. (b)	78. (a)	79. (c)	80. (b)
81. (b)	82. (b)	83. (d)	84. (b)	85. (b)	86. (d)	87. (d)	88. (a)	89. (a)	90. (b)
91. (d)	92. (a)	93. (a)	94. (d)	95. (a)	96. (a)	97. (a)	98. (a)	99. (c)	100. (b)
101. (a)	102. (c)	103. (a)	104. (d)	105. (c)	106. (a)	107. (a)	108. (d)	109. (a)	110. (a)
111. (a)	112. (b)	113. (c)	114. (c)	115. (d)	116. (a)	117. (d)	118. (c)	119. (d)	120. (b)
121. (d)	122. (a)	123. (b)	124. (b)	125. (a)	126. (a)	127. (c)	128. (d)	129. (d)	130. (d)
131. (b)	132. (a)	133. (c)	134. (b)	135. (b)	136. (c)	137. (b)	138. (b)	139. (b)	140. (c)
141. (a)	142. (b)	143. (a)	144. (b)	145. (a)	146. (c)	147. (b)	148. (c)	149. (a)	150. (d)
151. (d)	152. (a)	153. (a)	154. (b)	155. (d)	156. (b)	157. (c)	158. (a)	159. (b)	160. (c)
161. (b)	162. (c)	163. (c)	164. (d)	165. (b)	166. (a)	167. (a)	168. (b)	169. (c)	170. (d)

HINTS AND EXPLANATIONS FOR SELECTIVE QUESTIONS

3. $CH_3-C\equiv N \longrightarrow CH_3-\overset{+}{C}=\overset{-}{N}$
4. NH_2^- is the strongest nucleophile among these.
5. As the electronegativity of the atom decreases (F > O > N > C), its tendency to donate electron pair increases i.e., nucleophilicity increases. Thus CH_3^- has the highest nucleophilicity.
7. The strength of nucleophile depends upon the nature of alkyl group R on which nucleophile has to attack and also on the nature of solvent. The order of strength of nucleophiles follows the order $CN^- > I^- > C_6H_5O^- > OH^- > Br^- > Cl^-$.
8. The least stable carbanion is the strongest nucleophile. $CH_3-CH_2^-$ is the strongest nucleophile.
10. Due to strong mesomeric effect.
11. Free radicals are involved in all these reactions.
12. Resonance causes shortening of bond length and high heat of formation by providing stability to the molecule.
13. Electron attacking groups such as $-NO_2$ make a carbonium ion less stable (–I effect).
14. It is Hyperconjugation effect.
15. Pinacol pinacolone rearrangement involves carboniumion while others involve carbanion.
17. 2–methyl but–2–ene and but–2–ene contain 9 alpha H atoms and 6 alpha H atoms respectively. Greater the number of alpha H atoms, greater is delocalisation and greater is the stability.
18. $SbCl_5$ will attract Cl^- ion to form $SbCl_6^-$ and $C_6H_5\underset{\oplus}{C}HCH_2$ (carbocation).
22. An intermediate neutral species having divalent carbon atom with 6 valence electrons out of which two are present in same orbital with opposite spins is called singlet carbene.
28. 1° alkyl halides undergo substitution by SN^2 mechanism.
29. $(CH_3)_3N^+ I^-$ is most reactive because CH_3 group attached to N^+ is more electron deficient and is more reactive towards nucleophiles. Moreover, the compound contains I^- which is a better leaving group.
32. Tert. Butyl carbanion (sp^3 hybridization) is pyramidal.
33. This compound has all the three effects.
34. As tertiary butyl carbocation is formed here.
37. Greater the s–character, more electro negative is the carbon i.e., 1.
39. Due to resonance, C_2-C_3 bond is little shorter than C – C single bond length of 1.54 Å in ethane. So the most appropriate value is 1.46 Å.
40. Benzyl carbanion due to resonance stabilization.
41. Being 3° carbocation, $(CH_3)_3C^+$ is the most stable.
42. C – H bond dissociation energy is the lowest where free radical produced is the most stable.
43. Carbanions have complete octet, i.e., 8 electrons.
48. $CH_3-\underset{\underset{CH_3}{|}}{C}=CH_2 \xrightarrow[\text{anti. Markownikoff addition}]{\text{HBr / peroxide}} CH_3-\underset{\underset{CH_3}{|}}{CH}CH_2-Br$
49. SN^1 reaction is favoured by heavy group on the carbon atom attached to halogens and nature of carbonium ion in substrate is

Benzyl > allyl > tertiary > secondary > primary > methylhalides

51. $C_6H_5-\overset{\overset{H}{|}}{\underset{\underset{CH_3}{|}}{C}}-\overset{\oplus}{C}H_2 \xrightarrow[\text{Shift}]{H^{\ominus}} C_6H_5-\overset{\oplus}{\underset{\underset{CH_3}{|}}{C}}-CH_3$
52. Strongest acid has lowest pKa value.
55. As all are m–directing groups.
71. In MeCHO, positive charge is maximum on carbonyl carbon atom, so this compound will be most susceptible to nucleophilic attack among the given compounds.
77. As a tertiary carbocation will be formed i.e., $(CH_3)_3C^+$ here.
80. Because of the presence of active methylene group, 2, 4–hexanedione is most acidic.
84. In trichloromethyl amines the $-CCl_3$ group is situated near to the $-NH_2$ group and withdraws electrons more efficiently and thus decreases its basicity.

85. As it involves free radical mechanism.

86. It is stabilized by inductive effect and hyper conjugation (with – CH_3 group) and by resonance with double bond. Moreover, it is a tertiary carbocation.

88. The product obtained on dehydration of (a) is conjugated and is more stable. Therefore, it is most readily dehydrated.

89. 'A' gives the most stable 3° carbocation, so it undergoes dehydration fastest.

93. Electron releasing group ($-CH_3$) increases basic nature while electron withdrawing ($-NO_2$, – CN) decreases the basic nature of amines. –I and –R effect of –CN is lesser than $-NO_2$, so III is more basic than II.

94. In pyridine, lone pair is not involved in conjugation with the ring and is fully available on nitrogen, so it is most basic. In all the other three, one lone pair is involved in resonance with the ring and other is present on the heteroatom. With increase of electronegativity (S < N < O), availability of lone pair on the hetero atom decreases so basic nature also decreases in the same order.

95. As it is a primary alkyl halide.

97. 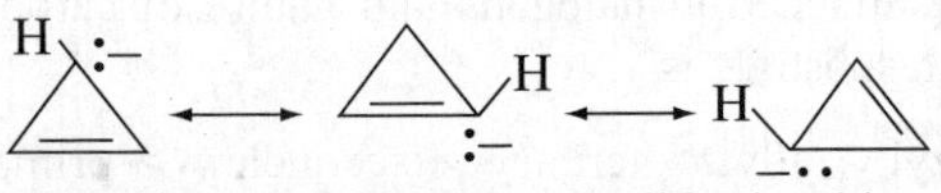 H

Although resonance is possible in this compound but as this compound is anti–aromatic because it has 4π electrons so here resonance does not contribute towards stability.

99. As it has α–H atoms so hypercondugation occurs in it.

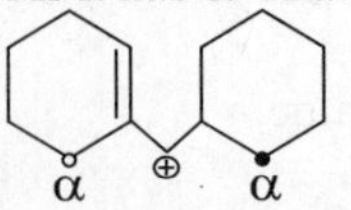

α ⊕ α

100. $C_6H_5CH = CH - CH_3 + HBr \longrightarrow$

Br

|

$C_6H_5CHCH_2CH_3$

101. SN^2 mechanism involves change of configuration and not retention of configuration so it is incorrect.

102. Among the given structures, the most unlikely structure is given in option (c) because in this structure N is shown to pentavalent.

105. As more stable carbocation is formed in this case.

126. Reactivity of halides towards SN^1 mechanism is Benzyl > allyl > 3° > 2° > 1°

128. Carbocation (3) is antiaromatic and hence is least stable. Carbocation (1), (2) and (4) are all secondary but (2) and (4) are aromatic. Therefore (2) and (4) are more stable than (1). Further, (4) is more stable than (2) because of extensive delocalisation due to resonance.

132. (1) is most stable of these and its stability is due to the conjugation between the bent orbitals of the cyclopropyl ring and vacant p–orbital of cationic carbon.

133. CH_3^- group is activating while Cl^- and NO_2^- groups are deactivating. Therefore 2 should be most reactive, followed by 1.

134. Weaker bases are better leaving groups. The basic strength of the given groups is in the order: –OMe (II) > –OAc (I) > $-OSO_2Me$ (III) > $-OSO_2CF_3$ (IV)

135. Steric in hibition of Resonance occurs when ortho positions are occupied by bulky groups, hare in compounds I, II it is possible.

137. $-NO_2$ and –COOH group being powerful electron withdrawing groups, reduce the electron density in the benzene ring and do not favour Friedal–Craft alkylation while CH_2CH_3 and OH groups being electron donating increase the electron density in the benzene ring and favour Friedal–Craft alkylation.

139. The facility by which the dehydration takes place depends on the stability of the carbocation formed on protonation and dehydration.

140. It is cumene hydroperoxide rearrangement reaction.

142. Here, the presence of a double bond followed by three chlorine atoms in the adjacent carbon makes (3) the strongest acid. In (1), the acid strength is little decreased due to the intervening $-CH_2-$ group.

143. This is a case of hyperconjugation when there is benzylic or allylic group.

$(CH_3)_3C- > (CH_3)_2CH > CH_3CH_2 > CH_3$

144. >C = O group is electron withdrawing hence $-CH_2$ group in between –OH and >C=O is acidic, so dehydration is maximum.

145. For 1°, 2°, 3° alkyl halides. The case of (i) SN_1 reaction increases as 1° < 2° < 3°.

152. As $\Delta-\overset{\oplus}{C}H(\Delta)$ is most stable and $CH_3-\overset{\oplus}{C}H(NO_2)$ is least stable so they will have longest and shortest life spans here.

153. It has 3 lone pair of electrons

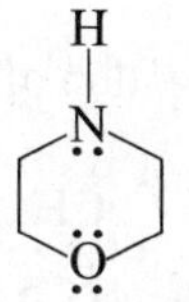

154. $(CH_3)_2CHCH_2CH_3 \xrightarrow{Br_2/hv\ 127°C} (CH_3)_2C(Br)CH_2CH_3$

157. Strong bases are generally good nucleophiles.

158. Cl^- is the best leaving group among the given options

159. $CH_3CH_2CH(Br)CH_3$ (2-bromobutane) $\xrightarrow{KOH(alc.)}$ $CH_3CH=CHCH_3$ 2-butene (80%) + $CH_3CH_2CH=CH_2$ 1-butene (20%)

In elimination reaction of alkyl halide, major product is obtained according to Saytzeff's rule, which states that when two alkenes may be formed, the alkene which is most substituted one predominates.

161. $C_6H_5O^-$ possesses less nucleophilicity due to stabilized nature of phenoxide ion. CH_3OH is weaker acid than CH_3COOH.

164. It will follow Hoffmann elimination rule.

165. The order of stability of free radicals is as follows: Tertiary > secondary > primary

167. More is the stearic hindrance at the carbon bearing the halogen, lesser is the SN^2 reactivity.

PREVIOUS YEARS' QUESTIONS

1. Among the following compounds, the decreasing order of reactivity towards electrophillic substitution is

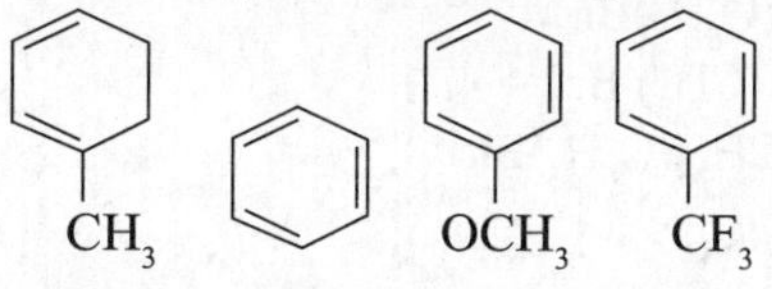

(1) $C_6H_5CH_3$ (2) C_6H_6 (3) $C_6H_5OCH_3$ (4) $C_6H_5CF_3$

(a) 3 > 1 > 2 > 4
(b) 4 > 1 > 2 > 3
(c) 1 > 2 > 3 > 4
(d) 2 > 1 > 3 > 4

[2000]

2. In HS^-, I^-, RNH_2 and NH_3 order of proton accepting tendency will be

(a) $I^- > NH_3 > RNH_2 > HS^-$
(b) $HS^- > RNH_2 > NH_3 > I^-$
(c) $RNH_2 > NH_3 > HS^- > I^-$
(d) $NH_3 > RNH_2 > HS^- > I^-$

[2001]

3. 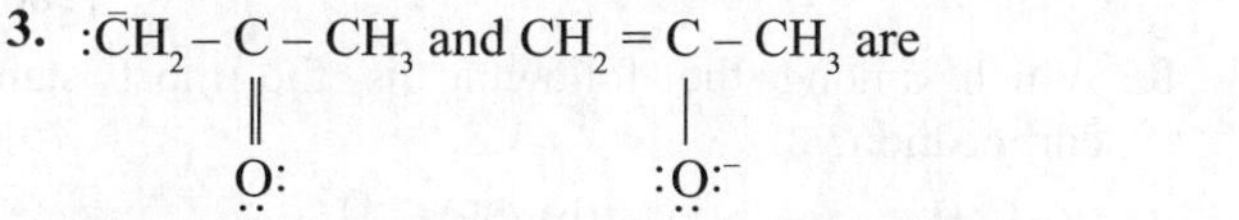 $:\bar{C}H_2-C(=\ddot{O}:)-CH_3$ and $CH_2=C(-\ddot{O}:^-)-CH_3$ are

(a) tautomers (b) optical isomers
(c) geometrical isomers (d) resonating structures

[2002]

4. Which one of the following is a free radical substitution reaction?

(a) $CH_3CHO + HCN \rightarrow CH_3CH(OH)CN$

(b) 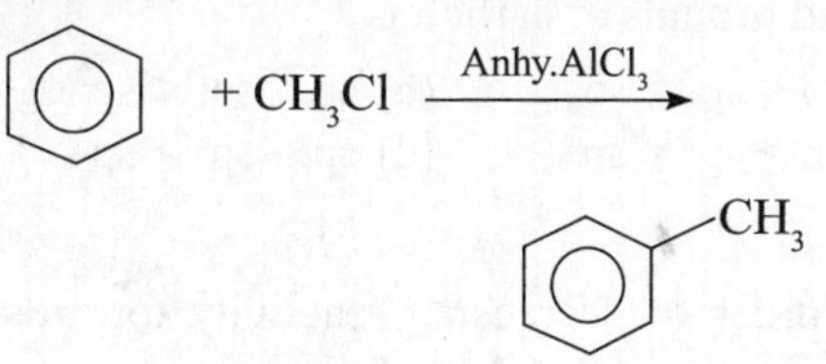 $C_6H_6 + CH_3Cl \xrightarrow{Anhy.AlCl_3} C_6H_5CH_3$

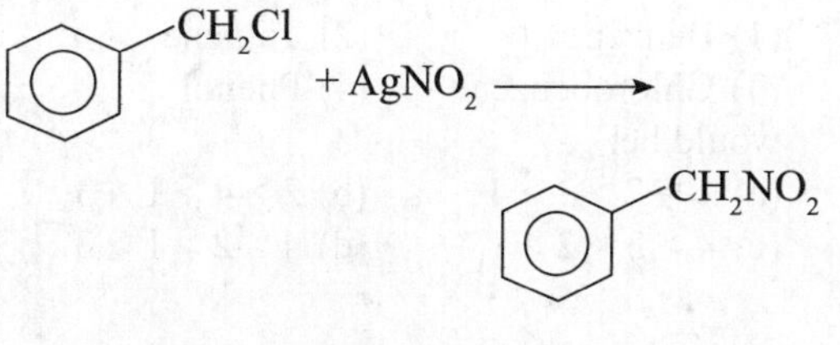

(c) $C_6H_5CH_2Cl + AgNO_2 \longrightarrow C_6H_5CH_2NO_2$

(d) $C_6H_5CH_3 + Cl_2 \xrightarrow{\text{Boiling}} C_6H_5CH_2Cl$

[2003]

5. The correct order of reactivity towards the electrophillic substitution of the compounds (1) aniline, (2) benzene, (3) nitrobenzene is:

(a) $1 > 2 > 3$ (b) $2 > 3 > 1$
(c) $2 < 3 > 1$ (d) $3 > 2 > 1$

[2003]

6. Which of the following is least reactive in a nucleophillic substitution reaction?

(a) $(CH_3)_3C - CI$ (b) $CH_2 = CHCI$
(c) CH_3CH_2CI (d) $CH_2 = CHCH_2CI$

[2004]

7. Which of the following undergoes nucleophillic substitution exclusively by SN^1 mechanism?

(a) Ethyl chloride
(b) Isopropyl chloride
(c) Chlorobenzene
(d) Benzyl chloride

[2005]

8. Which among the following is the most stable corbocation?

(a) C^+H_3 (b) $CH_3C^+H_2$
(c) $CH_3 - C^+(CH_3)$ (d) $CH_3C^+(CH_3)_2$

[2006]

9. The correct order regarding the electronegativity of hybrid orbitals of carbon is

(a) $sp < sp^2 > sp^3$ (b) $sp < sp^2 < sp^3$
(c) $sp > sp^2 < sp^3$ (d) $sp > sp^2 > sp^3$

[2006]

10. The order of decreasing reactivity towards an electrophillic reagent, for the following

(1) Benzene (2) Toluene
(3) Chlorobenzene (4) Phenol

would be:

(a) $1 > 2 > 3 > 4$ (b) $2 > 4 > 1 > 3$
(c) $4 > 3 > 2 > 1$ (d) $4 > 2 > 1 > 3$

[2007]

11. The stability of carbanions in the following:

(a) $RC \equiv \overset{\ominus}{C}$ (b) $C_6H_5^{\ominus}$ [phenyl anion]
(c) $R_2C = \overset{\ominus}{C}H$ (d) $R_3C = \overset{\ominus}{C}H_2$

is in the order of:

(a) (d) > (b) > (c) > (a)
(b) (a) > (c) > (b) > (d)
(c) (a) > (b) > (c) > (d)
(d) (b) > (c) > (d) > (a)

[2008]

12. The state of hybridization of C2, C3, C5 and C6 of the hydrocarbon,

$$\overset{7}{CH_3} - \overset{6}{C}(CH_3)_2 - \overset{5}{CH} = \overset{4}{CH} - \overset{3}{CH}(CH_3) - \overset{2}{C} \equiv \overset{1}{CH}$$

is in the sequence

(a) sp^3, sp^2, sp^2 and sp
(b) sp, sp^2, sp^2 and sp^3
(c) sp, sp^2, sp^3 and sp^2
(d) sp, sp^3, sp^2 and sp^3

[2009]

13. Which one is most reactive towards S_N1 reaction?

(a) $C_6H_5CH_2Br$
(b) $C_6H_5CH(C_6H_5)Br$
(c) $C_6H_5\ CH(CH_3)\ Br$
(d) $C_6H_5\ C(CH_3)(C_6H_5)Br$

[2010]

14. Which one is most reactive towards electrophilic reagent?

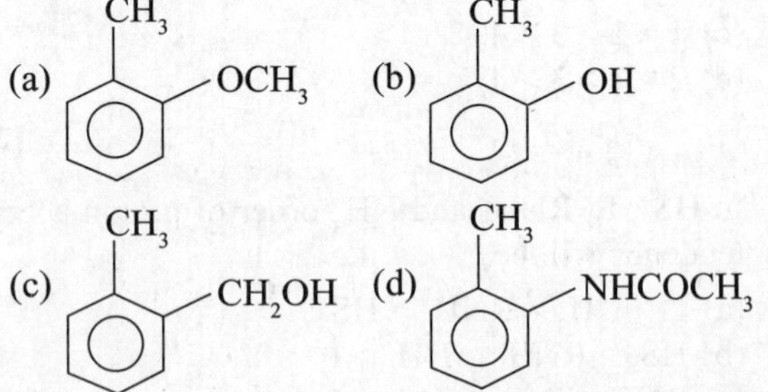

[2010]

15. Which one of the following compounds has the most acidic nature?

(a) 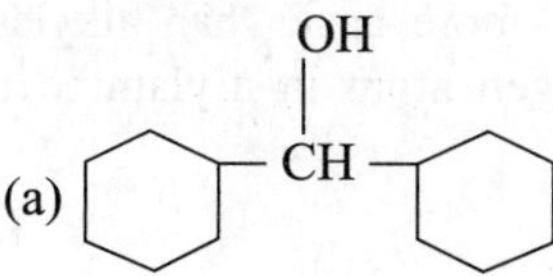

(b) $C_6H_5CH_2OH$

(c) C_6H_5OH (phenol)

(d) cyclohexanol (OH)

[2010]

16. Which of the following species is not electrophilic in nature?

(a) Cl^+ (b) BH_3

(c) H_3O^+ (d) N^+O_2

[2010]

17. Which one of the following is most reactive towards electrophilic reagent?

(a) CH_3 / CH_2OH (b) CH_3 / OCH_3

(c) CH_3 / OH (d) CH_3 / $NHCOCH_3$

[2011]

18. Among The following compounds one that is most reactive towards electrophillic nitration is ?

(a) benzene (b) toluene

(c) benzoic acid (d) nitro benzene

[2012]

19. The correct order of decreasing acidic strength of Trichloro aceticacid (A) Tri Fluoro aceticacid (B) aceticacid (C) and formic acid (D) is ?

(a) B > A > D > C (b) A > B > C > D

(c) B > D > C > A (d) A > C > B > D

[2012]

20. The radical $C_6H_5\overset{o}{C}H_2$ is aromatic because it has

(a) 6p orbitals and 6 un-paired electrons

(b) 7p orbitals and 6 un-paired electrons

(c) 7p orbitals and 7 un-paired electrons

(d) 6p orbitals and 7 un-paired electrons

[2013]

21. Which of the following organic compounds has same hybridisation as its combustion (CO_2) product ?

(a) ethane (b) ethene (c) ethyne (d) ethanol

[2014]

22. Which one is most reactive towards nucleophillic addition reaction ?

(a) C_6H_5CHO (b) $C_6H_5COCH_3$

(c) CHO / NO_2 (d) CHO / CH_3

[2014]

23. Consider the following compound:

(1) $H_3C—C(CH_3)_2—\overset{o}{C}H—C_6H_5$

(2) $(Ph)_3\overset{o}{C}$ (3)

Hyper conjugation occurs in

(a) 1 only (b) 2 only (c) 3 only (d) 1 and 3

[2014]

24. The enthalpy of hydrogenation of these compound will be in the order as

(1) CH_3, CH, CH_3 (2) H_3C, CH_2, CH_3

(3) H_2C, CH_2, CH_2

(a) 1 > 2 > 3 (b) 2 > 1 > 3

(c) 2 > 3 > 1 (d) 3 > 2 > 1

[2014]

25. Which of the following is the most correct electron displacement for a nucleophillic reaction to take place ?

(a) $H_3C \rightarrow CH{=}CH-CH_2-Cl$ (with H atoms on C=C; curved arrow from C–Cl bond to Cl)

(b) $CH_3 \rightarrow CH{=}CH-CH_2-Cl$ (curved arrows from C=C to C and from C–Cl bond to Cl)

(c) $CH_3 \rightarrow CH{=}CH-CH_2-Cl$ (curved arrows from C=C bond and from C–Cl bond to Cl)

(d) $H_3C \rightarrow CH{=}CH-CH_2-Cl$ (curved arrow from C=C bond to C–CH₂ bond)

[2014]

26. Which of the following statements is not correct for a nucleophile ?

(a) nucleophile is a lewis acid

(b) nucleophiles are not electron seeking

(c) ammonia is a nucleophile

(d) nucleophile attack low electron density site.

[2015]

27. The correct statement regarding the basicity of Arylamines is

(a) Arylamines are generally more basic than alkylamines because the nitrogen lone pair electrons are not delocalesed by interaction with the aromatic ring π electron system.

(b) Arylamines are generally less basic than alkylamines because the nitrogen lone pair electrons are delocalised by interaction with the aromatic ring π electron system.

(c) Arylamines are generally more basic than alkylamines because of aryl group.

(d) aryl are generally more basic than alkylamines because the nitrogen atom in arylamine is Sp-hybridised.

[2015]

28. For the following reactions:

(1) $CH_3CH_2CH_2Br + KOH \longrightarrow CH_3-CH{=}CH_2 + KBr + H_2O$

(2) $(CH_3)_2CH-Br + KOH \longrightarrow (H_3C)(CH_3)CH-OH + KB_2$

(3) cyclohexene $+ Br_2 \longrightarrow$ 1,2-dibromocyclohexane

Which of the following statement is correct:

(a) (1) is elimination reaction (2) is substitution and (3) is addition reaction

(b) (1) is elimination reaction (2) and (3) are substitution

(c) (1) is substitution, (2) and (3) are addition reaction

(d) (1) and (2) are elimination reaction and (3) is addition reaction

[2016]

29. The correct order of strength of the carboxylic acids

(1) cyclohexyl–COOH (2) tetrahydropyran-3-yl–COOH (3) tetrahydropyran-4-yl–COOH

(a) 1 > 2 > 3 (b) 2 > 3 > 1

(C) 2 > 1 > 3 (d) 3 > 2 > 1

[2016]

Answer Keys

1. (a)	2. (c)	3. (d)	4. (d)	5. (a)	6. (b)	7. (d)	8. (d)	9. (d)	10. (d)
11. (c)	12. (d)	13. (d)	14. (b)	15. (c)	16. (c)	17. (c)	18. (b)	19. (a)	20. (c)
21. (c)	22. (c)	23. (c)	24. (d)	25. (c)	26. (a)	27. (b)	28. (a)	29. (b)	

HINTS AND EXPLANATIONS

1. III is most reactive because $-OCH_3$ group is strongly activating due to release of electrons towards benzene ring through resonance (+R effect).

2. Since a strong base has higher tendency to accept the proton hence the increasing order of acid strength is:

   $RNH_3^+ < NH_4^+ < H_2S < I^-$

   As strength of base is inversely proportional to the strength of acid so the decreasing order of base strength is

   $RNH_2 > NH_3 > HS^- > I^-$

3. $^-{:}CH_2-\underset{\underset{O:}{\|}}{C}-CH_3$ and $CH_2=\underset{\underset{:\ddot{O}:^-}{|}}{C}-CH_3$ are

   Since these structures involve only movement of electrons but not of atoms so these are resonating structure which differ only in electronic distribution.

4. Toluene on boiling with chlorine at higher temperature forms benzyl chloride by involving free radical substitution.

   The reaction occurs in following steps:

   $Cl-Cl \rightarrow 2Cl^\bullet$

   $C_6H_5CH_3 + Cl^\bullet \longrightarrow C_6H_5\dot{C}H_2 + HCl$

   $C_6H_5\dot{C}H_2 + Cl_2 \longrightarrow C_6H_5CH_2Cl + Cl^\bullet$

   $Cl^\bullet + Cl^\bullet \longrightarrow Cl_2$

5. The correct order is: I > II > III as $-NH_2$ is ring activating while $-NO_2$ group is ring deactivating.

6. Chlorine atom of vinyl chloride ($CH_2=CHCl$) is least reactive towards nucleophillic substitution reaction as it shows following resonating structures due to +M effect of chlorine atom in which chlorine atom acquires double bond character so its substitution becomes quite difficult.

   $$\underset{(I)}{CH_2=CH-\ddot{\underset{..}{Cl}}:} \longleftrightarrow \underset{(II)}{\overset{\delta-}{C}H_2-CH=\overset{\delta+}{Cl}}$$

7. Benzyl chloride undergoes nucleophillic substitution reaction exclusively by SN[1] reaction as SN[1] reaction is favoured by heavy electron releasing groups on the carbon atom attached to halogens i.e., on the stability of carbocation.

8. Here $(CH_3)_3C+$ is the most stable carbocation due to maximum hyperconjugation and +I effect by three methyl groups.

   The decreasing order of stability of these carbocations is as follows:

   D > C > B > A.

9. As the s-character increases in hybridized orbitals, the electronegativity increases so

sp	>	sp^2	>	sp^3
50%		33%		25%

10. As the attack of the electrophile is favoured by electron releasing groups (+I) i.e., $OH > CH_3 > H > Cl$

    Hence the reactivity order is as follows:

    Phenol > Toluene > Benzene > Chloro benzene.

11. As higher the number of electron releasing group, lower is the stability of carbanion. Hence the correct order of stability is:

    $RC\equiv\overset{\ominus}{C} > C_6H_5^{\ominus} > R_2C=\overset{\ominus}{C}H > R_3C-\overset{\ominus}{C}H_2$

12. $$\underset{7}{CH_3}-\overset{6}{\underset{CH_3}{\overset{CH_3}{C}}}-\underset{sp^3}{\overset{5}{CH}}=\underset{sp^2}{\overset{4}{CH}}-\underset{sp^3}{\overset{3}{\overset{CH_3}{CH}}}-\overset{2}{C}\equiv\underset{sp}{\overset{1}{CH}}$$

13. $$\underset{\text{(more reactive)}}{C_6H_5-\underset{CH_3}{\overset{C_6H_5}{C}}-Br} \xrightarrow{-Br^-} \underset{\substack{\text{(more stable)}\\\text{(intermediate)}}}{C_6H_5-\underset{CH_3}{\overset{C_6H_5}{C}}{}^{\oplus}}$$

14. CH_3 / OH (o-cresol) has OH group which shows more +M effect.

15. (Phenol) $\xrightarrow{-OH^+}$ (Phenoxide $O^\ominus$)
(more acidic) (more stable)

16. H_3O^+ Can't be an electro philehere

17. +R effect of OH > OCH_3

18. The reactivity order for nitration is as follows:

CH_3 (toluene) > H (benzene) > COOH (benzoic acid) > NO_2 (nitrobenzene)

19. $\Longrightarrow$ Acidic nature a - I a $\frac{1}{+I}$

$F_3C-COOH > CCl_3COOH > HCOOH > CH_3-C(=O)-OH$
-I (Max) (B) > -I (Less) (A) > (D) > (C)

$\xrightarrow{+I}$
Acidic nature Dee as follows

20. It is aronatic in nature as it contains 7ρ-orbitals and 7 un paired e^-. Here 7un-paired e^- undergo delocalisation in ρ-orbitals

21. In CO_2 and ethyne HC≡CH C-atam is sp-hybridised.

22. $\Longrightarrow$ Reactivity for nueleophillic addition
Reactivity a - I (E.W. Gp)
a $\frac{1}{+I\ (E.R.Gp)}$

CH_2O (NO_2 substituted) > CHO > CHO (CH_3 substituted) > $C(=O)-CH_3$
(-I, -R) Strong (+I)

23. Hyperconjugation a Number of - a Hatan

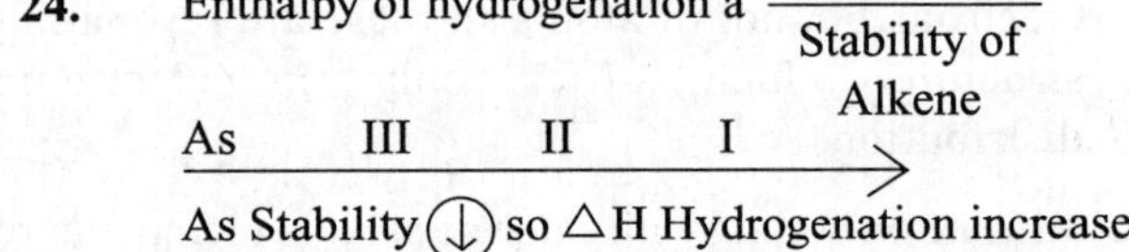

— NO Hyperconjugation

24. Enthalpy of hydrogenation a $\frac{1}{\text{Stability of Alkene}}$

As III II I $\longrightarrow$
As Stability (↓) so ΔH Hydrogenation increase

25. Option 'C' represents the correct representation

26. Since Nucleophile is a lewis base and not a lewis acid

27.
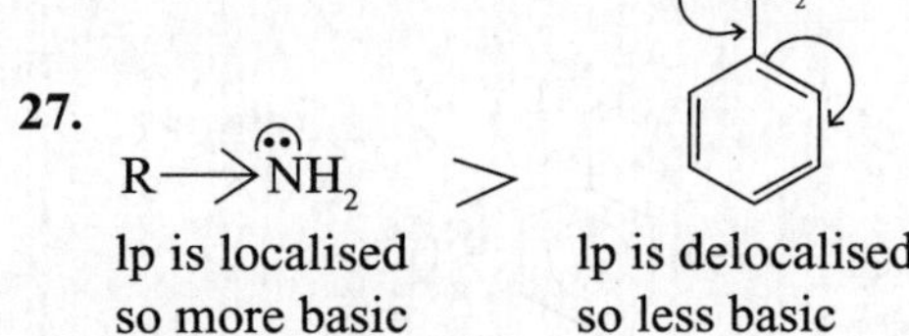

lp is localised so more basic; lp is delocalised so less basic

28. See text part

$I^{st} \longrightarrow$ Elimination
II $\longrightarrow$ Substitation
III $\longrightarrow$ Addition

29. Acidic nature α –I effect of O-atom

(II) COOH > (III) COOH > COOH
More - I Less - I
$\longrightarrow$
Acidic Nature Decreases

NCERT EXEMPLAR

1. What is the correct order of decreasing stability of the following cations:

$CH_3-\overset{\oplus}{C}H-CH_3$ (1.) $CH_3-\overset{\oplus}{C}H-OCH_3$ (2.)

$CH_3-\overset{\oplus}{C}H-CH_2-OCH_3$ (3.)

(a) 2 > 1 > 3 (b) 2 > 3 > 1
(c) 3 > 1 > 2 (d) 1 > 2 > 3

2. In which of the following compounds the carbon marked with asterisk is expected to have greatest positive charge?
(a) $^*CH_3-CH_2-Cl$
(b) $^*CH_3-CH_2-Mg^+Cl^-$
(c) $^*CH_3-CH_2-Br$
(d) $^*CH_3-CH_2-CH_3$

3. Ionic species are stabilized by the dispersal of charge. Which of the following carboxylate ion is the most stable?
(a) $CH_3-\overset{O}{\overset{\|}{C}}-O^-$
(b) $Cl-CH_2-\overset{O}{\overset{\|}{C}}-O^-$
(c) $F-CH_2-\overset{O}{\overset{\|}{C}}-O^-$
(d) $F_2CH-\overset{O}{\overset{\|}{C}}-O^-$

4. Electrophilic addition reactions proceed in two steps. The first step involves the addition of an electrophile. Name the type of intermediate formed in the first step of the following addition reaction. $H_3C-HC=CH_2 + H_+ \longrightarrow ?$
(a) 2° Carbanion
(b) 1° Carbocation
(c) 2° Carbocation
(d) 1° Carbanion

5. Covalent bond can undergo fission in two different ways. The correct representation involving a heterolytic fission of CH_3-Br is

$CH_3-Br \longrightarrow \overset{\oplus}{C}H_3 + Br^{\ominus}$

$CH_3-Br \longrightarrow \overset{\oplus}{C}H_3 + Br^{\ominus}$

$CH_3-Br \longrightarrow \overset{\ominus}{C}H_3 + Br^{\oplus}$

$CH_3-Br \longrightarrow \dot{C}H_3 + \dot{Br}$

6. The addition of HCl to an alkene proceeds in two steps. The first step is the attack of H^+ ion to portion which can be shown as:
(a) $H^+ \quad >C=C<$
(b) $H^+ \quad >C=C<$
(c) $H^+ \quad >C=C<$
(d) All of these are possible

ANSWER KEYS

1. (a) 2. (a) 3. (d) 4. (c) 5. (b) 6. (b)

HINTS AND EXPLANATIONS FOR SELECTIVE QUESTIONS

1. Apply the concepts of +I & -I effects and +R effect.
2. Apply concept of electronegativity.
6. $>C=C< + H^+ \longrightarrow >\overset{+}{C}-\underset{H}{\underset{|}{C}}<$

AIIMS ESSENTIALS

Assertion and Reason

In the following questions, two statements Assertion (A) and Reason (R) are given. Mark

(a) if A and R both are correct and R is the correct explanation of A;

(b) if A and R both are correct but R is not the correct explanation of A;

(c) A is true but R is false;

(d) A is false but R is true;

(e) Both A and R are false.

1. (A) : Tropylium cation is more stable than $(CH_3)_3$ $C^{\oplus}$ z
(R) : It is stabilized by both resonance effect and inductive effect.

2. (A) : Solvolysis of $CH_2 = CHCH_2Cl$ is much faster compared to $(CH_3)_2CHCl$
(R) : In $CH_2 = CHCH_2Cl$, solvolysis occurs through SN^2 mechanism.

3. (A) : In SN^1 mechanism, the product with inversion of configuration is obtained in higher amount compared to the product with the retention of configuration
(R) : Front side attack of nucleophile is hindered due to the presence of leaving group in the vicinity.

4. (A) : Benzene reacts with CO and HCl in presence of $AlCl_3$ to give benzaldehyde.
(R) : The electrophilic reagent formed is $H-\underset{\underset{O}{\|}}{C}-Cl$ and it is an electrophillic substitution reaction.

5. (A) : Neopentyl chloride undergoes SN^2 reaction easily.
(R) : It is a primary alkyl halide.

6. (A) : Pyrrole is weaker base than its hydrogenated product pyrrolidine.
(R) : Lone pairs of electrons on nitrogen are delocalized in pyrrole.

7. (A) : The hydrogen of the $-CH_2$ group of 1,3-cyclopentadiene are acidic and this hydrocarbon is nearly 10^{30} times more acidic than ordinary alkanes.
(R) : In cyclopentadienyl anion, all five carbons are equivalent as demonstrated by labeling experiments.

8. (A) : PH_3 is stronger nucleophile than NH_3
(R) : PH_3 is stronger base than NH_3

9. (A) : In the transition state of SN^2 reaction the central carbon atom with three non reacting groups. Nucleophile and the leaving group remain approximately in the same plane.
(R) : Back side attack of the nucleophile to substrate brings out this geometry.

10. (A) : Phenoxide ion ($C_6H_5^-O^{\ominus}$) is more reactive than phenol towards electrophiles
(R) : Phenoxide ion is more stable than phenol.

11. (A) : The Friedal craft alkylation of nitrobenzene gives meta alkylated nitrobenzene.
(R) : Nitro groups are meta directing.

12. (A) : $HC \equiv C^-$ is more stable than $CH_2 = CH^-$.
(R) : $HC \equiv C^-$ has more s-character than $CH_2 = CH^-$.

13. (A) : There is evidence for the existence of N-methyl acetamide in two structural forms I and II and shown

$$\underset{\text{(I)}}{CH_3 = \overset{\overset{O}{\|}}{C} - \underset{\underset{H}{|}}{\overset{\overset{CH_3}{|}}{NH}} - CH_3} \qquad \underset{\text{(II)}}{CH_2 = \overset{\overset{OH}{|}}{C} - \underset{\underset{CH_3}{|}}{NH}}$$

(R) : Rotation about the carbon-nitrogen bond is restricted.

14. (A) : Pyrrole is a relatively non basic amine
(R) : In pyrrole, nitrogens is sp^3 hybridised.

15. (A) : The carbocation $CF_3 - {}^{\oplus}CH_2$ is less stable ${}^{\oplus}CF_3$.
(R) : In case of $CF_3 - {}^{\oplus}CH_2$, CF_3 is strong electron withdrawing, therefore increases +ve charge whereas in ${}^{\oplus}CF_3$, long pair of 'F' overlap with vacant p-orbital of carbon reducing +ve charge by pπ-pπ bonding or back bonding.

16. (A) : In contrast to the six equivalent bonds in benzene, the C – C bonds naphthalene come in two lengths:

$C_1 - C_2$ is considerably shorter than $C_2 - C_3$ bond.

(R) : Out of the resonating structures of naphthalene $C_1 - C_2$ bond is double in two of the structure while $C_2 - C_3$ is double in one.

17. (a) : CH_3OH is a nuclophile.

(R) : CH_3OH forms sodium methoxide on reaction with NaH.

18. (A) : The acidic strength of the following is $HF > H_2O > NH_3 > CH \equiv CH$

(R) : More stable the conjugate base, of the corresponding acid, more the acidic strength.

19. (A) : $C_6H_5N^{\oplus}(CH_3)_3$ undergoes 100% meta nitration although amino group is ortho para directing.

(R) : Withdrawal of electrons from an aromatic ring always favours meta substitution.

20. (A) : Heterolytic fission of propane forms ethyl carbocation and $^{\ominus}CH_3$ carbanion.

(R) : Ethyl carbocation is stabilized by inductive effect and $^{\ominus}CH_3$ is most stable carbanion.

21. (A) : Tertiary carbonium are generally formed more easily than primary carbonium ions.

(R) : Hyperconjugative as well as inductive effect due to additional alkyl group stabilize tertiary carbonium ion.

22. (A) : Benzyl carbanion is more stable than propyl carbanion.

(R) : The carbon atom in carbanion is trivalent.

23. (A) : Same number of electron pairs are present in resonance structures.

(R) : Resonance structures differ in the location of electrons around the constituent atoms.

24. (a) : In benzyne, two out of six carbon atoms are sp hybridized.

(R) : Benzyne as one $C \equiv C$ bond.

25. (A) : When $CH_2 = CH - COOH$ is reacted with HBr, then $\underset{\displaystyle Br}{\underset{|}{CH_2}} - CH_2 - COOH$ is obtained

(R) : The carbocation formed has the stability order

$\overset{\oplus}{C}H_2 - CH_2COOH > CH_3 - \overset{\oplus}{C}H - COOH$

26. (A) : The acetate ion is a weaker base than the ethoxide ion.

(R) : In carboxylic acids, the carbonyl group is polarized and so the carbon of the carbonyl group bears a +ve charge.

ANSWER KEYS

1. (a)	2. (c)	3. (a)	4. (a)	5. (d)	6. (a)	7. (b)	8. (c)	9. (a)	10. (b)
11. (d)	12. (a)	13. (d)	14. (b)	15. (a)	16. (a)	17. (b)	18. (d)	19. (a)	20. (a)
21. (a)	22. (b)	23. (d)	24. (a)	25. (a)	26. (b)				

CHAPTER 11C

Organic Chemistry: Purification and Characterization of Carbon Compounds

CHAPTER OUTLINE

■ Purification ■ Detection of Elements or Qualitative Analysis ■ Estimation of Elements or Quantitative Analysis ■ Molecular Weight Determination ■ Empirical and Molecular Formulas ■ Modern Methods of Structure Elucidation

PURIFICATION

- The organic compounds obtained from natural sources or prepared in laboratory are generally impure, so their purification is essential.
- For the purification of solid organic compounds, methods like sublimation, crystallization, fractional crystallization, chromatography, steam distillation are used.
- In case of liquid organic compounds, methods like distillation, fractional distillation, steam distillation, vacuum distillation and chromatography are used for purification.

Sublimation

- In sublimation, a solid is converted directly into gaseous state on heating and vice versa in case of cooling.

  $$\text{Solid} \underset{\text{Cold}}{\overset{\Delta}{\rightleftharpoons}} \text{gaseous}$$

  For example, camphor, naphthalene, benzoic acid, iodine, ammonium chloride, anthracene, anthroquinone, phthalic anhydride are purified by sublimation.

- It is suitable for the solids which have non-volatile impurities and undergo sublimation on heating.

Crystallization

- Crystallization is based on the fact that during the crystal formation, impurities are left out as they are insoluble or feebly soluble in the used solvent. For crystallization, a suitable solvent like H_2O, R–OH, R–O–R, CCl_4, C_6H_6, acetone or petroleum are used. For example,
  1. A mixture of benzoic acid and naphthalene is separated by hot water in which only benzoic acid dissolves.
  2. The impurities of common salt in sugar can be removed by hot ethanol as only sugar dissolves in it.

Fractional Crystallization

- Fractional crystallization involves repeated crystallization.
- It is used when the substances differ slightly in the solubilities. For example, in a mixture of $KClO_3$ and KCl, KCl is more soluble than $KClO_3$ in cold water.

Distillation

It involves the conversion of a liquid into gaseous state by heating and vice versa.

Simple Distillation

The substances which do not decompose on heating at their boiling point and have non-volatile impurities are purified by this method. It is used when liquids differ in their boiling point by 30–50 K.

For example,

1. Ether and toluene
2. Hexene and toluene
3. Benzene and aniline

Fractional Distillation

- Fractional distillation is used when the substances have nearly same boiling points (5°C to 10°C difference only or 10–15 K). It involves repeated distillations and condensations.
- Here, a fractional column is used to increase cooling surface area and to provide hurdles for descending liquids and ascentiny vapour which would facilitate collection of fractions.
- The more volatile liquid (with a lower boiling point) distils off before the less volatile liquid (with higher boiling point).

  For example, acetone with a boiling point of 329 K distills before and methanol which has a boiling point of 338 K.

  Acetone (329 K) and Methyl alcohol (338 K)

  Benzene (353 K) and Toluene (384 K)

  Crude petroleum into gasoline, kerosene or diesel.
- It cannot be used for azeotropic mixture as the components distil at the same temperature. For example, ethyl alcohol (95.87% by weight) and water (4.13% by weight).

In case of such a mixture we use an another liquid with lower b.p then these two liquids for e.g., Benzene. In this type of azeotropic distillation, pure liquid is obtained after two or more repeated distillations. For e.g., 100% pure ethanol is obtained from a mixture of ethanol and water in 3rd distillation at nearly 351 K.

Steam Distillation

- The substances (solid or liquid) insoluble in water but volatile in steam, having a high vapour pressure and non-volatile impurities are purified by steam distillation method.
- By this method, only non-volatile impurities can be removed. For example, aniline, nitrobenzene, salycylic acid, salycyldehyde, o-hydroxy phenol, terpenes, essential oils, acetophenone etc.
- A mixture of 'ortho' and 'para' nitrophenols is separated by steam distillation. Ortho-nitrophenol is more volatile due to intramolecular hydrogen bonding.

In this kind of distillation, the mixture boils at the temperature at which the sum of VP of organic substance and VP of pure water is equal to that of atmospheric pressure. Here the mixture boils at less BP than that of BP of substance and water. This means, it also prevents any decomposition so it is also just like distillation under reduced pressure. For example, → Aniline can be purified by steam at 371.5 K i.e., quite below than its boiling point (453 K).

Vacuum Distillation or Distillation Under Reduced Pressure

- Vacuum distillation is used for those substances which decompose on heating at their boiling points or just below their boiling points.

  It is based on the assumption that a liquid boils at a temperature when its V.P becomes equal to that of the atmospheric pressure. It means if pressure is reduced distillation can occur below the BP of the substance.
- Here the pressure is reduced during distillation to avoid decomposition, so it is also called distillation at reduced pressure. This method is used during the manufacture of glycerol, sugar and nitric acid.

  Glycerol can be purified at 453 K by using a pressure of 12 mm Hg.

Chromatography

- Chromatography was introduced by Tswett in 1906.
- Here the separation and purification is carried out based on the difference in movement of components of a mixture through a stationary phase under the influence of a mobile phase.
- It is used for the separation of any substance which is present in a very small amount or which cannot be purified by other methods like distillation or crystallization.

Types of chromatography On the basis of nature of both the phases, chromatography is of following types:

Adsorption or column chromatography

- Here the mobile phase is a liquid while the stationary phase is a solid substance.
- It is used for large scale separation when components have different relative adsorption solubilities.
- The stationary phase is made of solid alumina (Al_2O_3), silica (SiO_2), starch, and charcoal while mobile phase is made of a pure solvent or a mixture of solvents of different polarity.

Thin layer chromatography

- Here the mobile phase is a liquid while the stationary phase is a solid substance.
- It is used in qualitative analysis for the identification and characterization of organic compounds.

Paper chromatography or partition chromatography

- Here both the phases are of liquids.
- It is used in the qualitative and quantitative analysis of sugars, polar organic and inorganic compounds, and α-amino acids.

Gas–liquid chromatography

- Here the mobile phase is a gas while the stationary phase is a liquid.
- It is used in the qualitative and quantitative analysis of inorganic compounds.

Some Separation Methods Based on Chemicals

- Phenol and acids can be separated using aqueous $NaHCO_3$ as only the acid will react with it and produce effervescences of CO_2.
- From pyroligneous acid [CH_3OH + CH_3COOH + CH_3COCH_3], CH_3COOH is separated in the form of CH_3COOCa, by using $Ca(OH)_2$.
- A mixture of primary, secondary and tertiary amines can be separated by using benzene sulphonyl chloride (Hinsberg Reagent method) and diethyloxalate (Hoffmann's method).
- Absolute alcohol can be separated from rectified spirit by using active quick lime.
- Benzene can be separated from Thiophene by using con. H_2SO_4 as thiophene undergoes sulphonation more readily than benzene to give Thiophene-2-sulphonic acid which dissolves in con. H_2SO_4 while benzene remains insoluble. From here pure benzene is separated out by washing it with H_2O to remove H_2SO_4, followed by drying over anhy. $CaCl_2$ and distillation.

Detection of Elements or Qualitative Analysis

Test for Carbon and Hydrogen

- The test for the detection of carbon and hydrogen is possible by heating the organic compound with dry CuO.
- The carbon is oxidized in to CO_2, which turns lime water milky.
- Hydrogen is oxidized into water which turns white anhydrous copper sulphate to blue colour.

Lassaigne's Test

Preparation of Lassaigne's extract or sodium extract

The given organic compound in small quantity is fused with a small piece of sodium in an ignition tube. The tube is first heated gently and then strongly, till it becomes red hot. Its contents are then poured in distilled water in a beaker. The solution is boiled and filtered. The filtrate is known as sodium extract.

- When an organic compound is fused with Na, a covalent organic compound is converted into an ionic compound. The nitrogen, sulphur and halogens in the organic compound are converted into NaCN, Na_2S and NaX respectively (where X = halogen).
- When N and S both are present in a compound, NaCNS is also formed.

$$Na + C + N \longrightarrow NaCN$$
$$2Na + S \longrightarrow Na_2S$$
$$Na + X \longrightarrow NaX$$
$$Na + C + N + S \longrightarrow NaCNS$$

- Lithium is not used in Lassaignes test as it reacts slowly. Moreover, its compounds are covalent.
- Potassium cannot be used in this test as its reaction is vigorous and violent.

Test for Nitrogen

- Nitrogen detection test is carried out in the following steps:

  Take sodium extract and add one drop NaOH and 1 mL $FeSO_4$ solution (fresh). Now boil it. After cooling, add aqueous $FeCl_3$ and acidifiy it. A deep blue colour is formed which is ferriferrocyanide.

$$FeSO_4 + 2NaOH \longrightarrow Fe(OH)_2 + Na_2SO_4$$
$$6NaCN + Fe(OH)_2 \longrightarrow Na_4Fe(CN)_6 + 2NaOH$$
$$3Na_4Fe(CN)_6 + 4FeCl_3 \longrightarrow \underset{\text{Iron (III)}}{Fe_4[Fe(CN)_6]_3} + 12NaCl$$

Ferriferrocyanide (hexa cyano ferrate II)
(prussian blue)

- NH_4Cl, $NaNO_3$, NH_2NH_2 which are inorganic compounds without having C-atom so they do not give a positive response to this test.
- This test can be used for detection of urea, aniline, acetamide, benzamide, azobenzene, glycine and other organic compounds.

Facts to Remember

Here only fresh $FeSO_4$ must be used as $FeSO_4$ on keeping oxidised into yellow coloured basic ferric sulphate which interferes light blue or green colour.

$$FeSO_4 + O_2 + 2H_2O \rightarrow 4Fe(OH)SO_4$$

Prussian blue can also be represented as

$$Na\overset{III}{Fe}[\overset{II}{Fe}(CN)_6]$$ and it can be obtained as follows.

$$2Na_4[\overset{II}{Fe}(CN)_6 + Fe_2(SO_4)_3 \rightarrow 2Na\overset{III}{Fe}[\overset{II}{Fe}(CN)_6] + 3Na_2SO_4$$

Test for Sulphur

(a) When lead acetate is added to sodium extract, a black precipitate of lead sulphide is obtained.

$$Na_2S + (CH_3COO)_2Pb \longrightarrow \underset{\text{Black}}{PbS} + 2CH_3COONa$$

(b) When sodium nitropruside solution is added to sodium extract a violet colour is formed.

$$Na_2S + Na_2[Fe(CN)_5NO] \longrightarrow \underset{\text{Sodium thionitro prusside (violet)}}{Na_4[Fe(CN)_5NOS]}$$

Middleton's test This is another test to detect the presence of sulphur. In this test, the organic compound when reacted with $Na_2CO_3 + ZnO$ forms ZnS in place of NaCNS.

Nitrogen and Sulphur Present Together

While carrying out the below mentioned test, if a red colour appears, it means both nitrogen and sulphur are present.

$$3NaCNS + FeCl_3 \longrightarrow \underset{\text{Ferric sulphocyanide (Red)}}{Fe(CNS)_3} + 3NaCl$$

Silver coin + Sodium extract ⟶ Coin turns black

Test for Phosphorus

- The test for the detection of phosphorus is carried out as mentioned below:

  Firstly, the organic compound is fused with sodium peroxide so that phosphorus is converted into sodium phosphate which is then extracted with water and boiled with concentrated HNO_3. Finally, ammonium molybdate is added to get a yellow precipitate of ammonium phosphomolybdate.

$$2P + 5Na_2O_2 \longrightarrow 2Na_3PO_4 + 2Na_2O$$

$$Na_3PO_4 + 3HNO_3 \xrightarrow{\Delta} H_3PO_4 + 3NaNO_3$$

$$H_3PO_4 + 12(NH_4)_2MoO_4 + 21HNO_3 \longrightarrow \underset{\text{Ammonium Phosphomolybdate (Yellow precipitate)}}{(NH_4)_3PO_4 \cdot 12MoO_3} + 21\,NH_4NO_3 + 12H_2O$$

Test for Halogens

- The test for determining the presence of halogens is carried out in the following steps:

Sodium extract is boiled with few drops of HNO_3 (to decompose NaCN to HCN gas, Na_2S to H_2S gas and to neutralized free alkali). The solution is then cooled and silver nitrate solution is added. A precipitate of AgX is formed as a result.

$$NaCN + AgNO_3 \longrightarrow AgCN\downarrow + NaNO_3$$

$$Na_2S + 2AgNO_3 \longrightarrow Ag_2S + 2NaNO_3$$

$$NaCN + HNO_3 \xrightarrow{\Delta} 2NaNO_3 + HCN\uparrow$$

$$Na_2S + 2HNO_3 \xrightarrow{\Delta} 2NaNO_3 + H_2S\uparrow$$

$$Na + AgNO_3 \longrightarrow AgX\downarrow + NaNO_3$$

- If a white precipitate, soluble in NH_4OH and insoluble in dilute HNO_3 is formed, chlorine is confirmed to be present.

$$NaCl + AgNO_3 \longrightarrow \underset{\text{Silver chloride White ppt.}}{AgCl} + NaNO_3$$

$$AgCl + 2NH_4OH \longrightarrow \underset{\text{Diammine silver chloride}}{Ag(NH_3)_2Cl} + 2H_2O$$

- If a light yellow precipitate, partially soluble in NH_4OH is formed, bromine is confirmed to be present.

$$NaBr + AgNO_3 \longrightarrow \underset{\text{Silver bromide Light yellow ppt.}}{AgBr} + NaNO_3$$

- If a dark yellow precipitate, insoluble in NH_4OH, is formed, iodine is confirmed to be present.

$$NaI + AgNO_3 \longrightarrow \underset{\text{Silver iodide Dark yellow ppt.}}{AgI} + NaNO_3$$

Beilstein's test: This test confirms the absence or presence of halogens by heating the given organic compound on copper wire.

- If Cu wire with the compound does not give green or bluish green flame, halogen is absent.
- If a bluish green flame (due to vapours of CuX_2) is observed, halogens may or may not be present, as pyridine and thiourea also impart green colour to the flame when subjected to this test due to formation of volatile $Cu(CN)_2$.
- Beilstein's test is not given by fluorine, as CuF_2 is non-volatile.

$CHCl_3$ or CCl_4 layer test (for Br and I): Take 2 or 3 mL of sodium extract, H_2SO_4 and 1mL $CHCl_3$ in a test tube. Add excess of chlorine water to it and shake throughly.

- If the colour of the CCl_4 layer becomes orange to brown, bromine is present.

- If the colour of the CCl_4 layer turns violet, iodine is present.

$2NaBr + Cl_2 \longrightarrow 2NaCl + Br_2$

$2NaI + Cl_2 \longrightarrow 2NaCl + I_2$

Estimation of Elements or Quantitative Analysis

Estimation of Carbon and Hydrogen

- The estimation of carbon and hydrogen is carried out by **Liebig's combustion method.**

$$\underset{}{\text{Org. compound}} + \underset{\text{Dry}}{2CuO} \xrightarrow{\Delta} \underset{\text{Sparingly}}{CO_2} + H_2O$$

$$C + 2CuO \xrightarrow{\Delta} CO_2 + 2Cu$$

$$CO_2 + Ca(OH)_2 \longrightarrow \underset{\text{Milky}}{CaCO_3\downarrow} + H_2O$$

$$\big\downarrow (H_2O + CO_2 \text{ excess})$$

$$\underset{\text{(Soluble)}}{Ca(HCO_3)_2}$$

$$H_2 + CuO \xrightarrow{\Delta} Cu + H_2O$$

$$CuSO_4 + 5H_2O \longrightarrow \underset{\text{Blue}}{CuSO_4.5H_2O}$$

$$C\% = \frac{12}{44} \times \frac{\text{Wt. of } CO_2}{\text{Wt. of org. compound}} \times 100$$

$$H\% = \frac{2}{18} \times \frac{\text{Wt. of } H_2O}{\text{Wt. of org. compound}} \times 100$$

Here, the evolved CO_2 is absorbed in a U-tube having KOH or NaOH+CaO (Ascarite) and the increase in the mass of this tube is equal to the amount of CO_2 produced. Here H_2O is also absorbed in another tube having anhy. $CaCl_2$. An increase in the mass of this tube is equal to mass of H_2O produced here.

Estimation of Nitrogen

Many methods are used to find out the percentage of nitrogen but the following two methods are commonly used.

Duma's method: This method can be used for the estimation of nitrogen in all types of organic compounds.

- By using this method, the amount of nitrogen present in the compound is estimated by calculating the amount of N_2 gas evolved.

$$N\% = \frac{28}{22400} \times \frac{\text{Volume of } N_2 \text{ at S.T.P}}{\text{Wt. of org. compound}} \times 100$$

OR

$$N\% = \frac{\text{Volume of } N_2 \text{ at STP}}{8 \times \text{Wt of org. compound}}$$

Here the evolved N_2 is collected over concn. KOH solution in a Nitrometer and it's value is find out as follows:

$$\frac{P_1V_1}{T_1} = \frac{P_2V_2}{T}$$ Here (1) → Given conditions

(2) → S.T.P conditions

Kjeldahl's method: This method is applicable only for those compounds where nitrogen is attached directly to either carbon or hydrogen.

- It is based on the fact that if an organic compound contains nitrogen then on heating it with concentrated H_2SO_4 in the presence of $CuSO_4$ and K_2SO_4 it gives ammonium sulphate. Ammonium sulphate on heating with KOH or NaOH liberates ammonia, which reacts with H_2SO_4.

  $(NH_4)_2SO_4 + 2NaOH \longrightarrow Na_2SO_4 + 2H_2O + 2NH_3$

  $2NH_3 + H_2SO_4 \longrightarrow (NH_4)_2SO_4$

- This method is genrally used to find out the percentage of nitrogen in organic compounds used as food or fertilizers.

  Percentage of nitrogen (N%) = $\frac{1.4 \times N \times V}{\text{Wt of org. compound}}$

  Here, N = Normality of the acid used to neutralize ammonia

  V = Volume of the acid used to neutralize ammonia

By this method, we cannot find N%. In case of heterocylic compounds having N-atom in the ring like Pyridine, Pyrole, Nitro and diazo compounds as N_2 cannot be converted into $(NH_4)_2\ SO_4$ in a complete manner.

Estimation of Halogens

Carius method: Carius method is suitable for those organic compounds which decompose easily.

- Here, the organic compound is heated with fuming HNO_3 and a few crystals of $AgNO_3$ in a sealed tube which results in the formation of precipitates of AgX.

  Organic compound + $HNO_3 \longrightarrow HX$

  $HX + AgNO_3 \longrightarrow HNO_3 + \underset{\text{Ppt.}}{AgX}$

$$X\% = \frac{\text{At. wt of X}}{\text{Mol. wt AgX}} \times \frac{\text{Wt of AgX}}{\text{Wt of org. compound}} \times 100$$

Atomic weight of chlorine = 35.5

Atomic weight of bromine = 80

Atomic weight of iodine = 127

Atomic weight of silver = 108

It is also estimated by Piria and Schiff's method.

Estimation of Phosphorus

- When a known mass of a given organic compound is heated with fuming HNO_3 in a Carius tube, phosphorus is oxidized into H_3PO_4. Phosphoric acid thus formed is precipitated as magnesium ammonium phosphate by adding magnesia mixture (a solution containing $MgCl_2$, NH_4Cl and NH_4OH).

$$MgCl_2 + NH_4Cl + H_3PO_4 \longrightarrow \underset{\text{Magnesium ammonium phosphate}}{MgNH_4PO_4} + 3HCl$$

- This precipitate is filtered, washed, dried and then ignited to produce magnesium pyrophosphate.

$$2MgNH_4PO_4 \xrightarrow{\Delta} \underset{\text{Magnesium pyrophosphate}}{Mg_2P_2O_7} + 2NH_3 + H_2O$$

- It is weighed and the percentage of phosphorus is found out by using the relation mentioned.

$$P\% = \frac{66}{222} \times \frac{\text{Mass of } Mg_2P_2O_7 \text{ formed}}{\text{Mass of substance taken}}$$

Estimation of Sulphur

Sulphur is estimated by **Carius method.**

- Organic compound with C, H, S + $HNO_3 \xrightarrow{\text{Heat}} CO_2 + H_2O + H_2SO_4$

$$H_2SO_4 + BaCl_2 \longrightarrow 2HCl + \underset{\text{White ppt.}}{BaSO_4}$$

$$S\% = \frac{32}{233}\frac{\text{Wt. of } BaSO_4}{\text{Wt. of org. compound}} \times 100$$

Molecular Weight Determination

It is possible by using these methods.

Silver salt method: It is based on the fact that most of the organic acids forms insoluble silver salts which on ignition give metallic 'Ag' as residue. Now by finding the mass of silver salt used (W) and mass of silver left (w g)as residue the molar mass of organic compound can be found as follows:

Molecular weight of acid = Equivalent weight × Basicity
= E × n

Here, n = Basicity

Platinichloride method: It is based on the fact that most of the organic bases give insoluble chloro platinates on reaction with chloroplatinic acid. This salton ignition give metallic 'Pt' as residue. Now molar mass can be found by using mass of chloroplatinate (W) and residue (w) as follows:

Molecular weight of base = Equivalent weight of base × Acidity

$$\text{Molar Mass} = \frac{1}{2}\left[\frac{W}{w} \times 195 - 410\right] \times n$$

Volumetric method It is based on the fact that acid and base react in equivalent proportions.

Molecular weight of acid = Equivalent weight × Basicity

Molecular weight of base = Equivalent weight × Acidity

Victor Mayer Method In this method, a known mass of a volatile compound is heated in a Victor Meyer's tube and the vapours of the compound displace an equal amount of air which is collected over water. The volume of this air is then reduced at S.T.P by using the relation

$$\frac{P_1V_1}{T_1} = \frac{P_2V_2}{T_2}$$

Here $P_2 = 760$ mm.
$T_2 = 273$ K
V_2 = Volume of air displaced at S.T.P

$$\text{molar mass} = \frac{\text{Mass of the compound used}}{\text{Vol. of Air displaced at S.T.P } (V_2)} \times 22400$$

Alternatively, we can also find molar mass by finding vapour density as follows:

$$\text{Vapour Density} = \frac{\text{Wt of certain vol. of vapour at NTP}}{\text{Wt of same vol of } H_2 \text{ at NTP}}$$

$$= \frac{w}{V_2 \times 0.00009}$$

$$V.D = \frac{w \times 22400}{V_2 \times 2}\left[\text{As 1 cm}^3 \text{ of } H_2 \text{ at N.T.P has mass } \frac{2}{22400}\text{gm}\right]$$

then Mol. wt = 2 × V.D

Empirical and Molecular Formulas

Empirical Formula

Empirical formula gives the simplest ratio of the number of atoms of different elements present in one molecule of a compound.

- It does not represent the actual number of atoms of different elements present in one molecule of the compound.

- Calculation of the empirical formula involves the following steps:
  1. First, determine the percentage composition by weight of each element present in the compound.

$$C\% = \frac{12}{44} \times \frac{\text{Wt of } CO_2}{\text{Wt of org. comp.}} \times 100$$

$$H\% = \frac{2}{18} \times \frac{\text{Wt of } H_2O}{\text{Wt of org. comp.}} \times 100$$

Duma's method

$$N\% = \frac{28}{22400} \times \frac{\text{Vol. of } N_2 \text{ at STP}}{\text{Wt of org. comp.}} \times 100$$

$$N\% = \frac{\text{Vol of } N_2 \text{ at STP}}{8 \times \text{Wt of org. comp.}}$$

Kjeldahl's method

$$\text{Percentage of nitrogen (N\%)} = \frac{1.4 \times N \times V}{\text{Wt of org. compound}}$$

Here, N = Normality of acid used to neutralize ammonia

V = Volume of acid used to neutralize ammonia

  2. Now the percentage of each element is divided by its atomic weight to get the relative number of atoms of each element.
  3. These relative numbers obtained, are divided by the smallest number amongst all values calculated for relative number of atoms to get the simplest ratio numbers.
  4. If the simplest ratio number is not a whole number, it should be multiplied by a suitable integer to get a whole number.
  5. The ratio of these simple whole numbers give the empirical formula of the compound.

Molecular Formula

It shows the actual number of atoms of different elements present in one molecule of the compound.

- $n = \frac{\text{Molecular weight}}{\text{Empirical formula weight}}$
- Molecular formula = Empirical formula × n.
- Molecular weight of a substance can be determined by methods like vapour density method, elevation in boiling point method, depression in freezing point method.
- The molecular weight of volatile compounds is determined by Victor Meyer's method where Molecular weight = 2 × Vapour density
- Molecular weight = Empirical formula weight × n.
- For some compounds, the molecular formula and the empirical formula may be the same.

Modern Methods of Structure Elucidation

- Due to the rapid development in the field of scientific instrumentation, it is now quite easy to find the complete structure of an organic compound with maximum accuracy.
- In these methods, only a small amount of the organic compound is needed.
- Molecular weight and molecular formula can be readily determined with the help of mass spectrometry.
- The nature and type of functional groups present in a particular molecule can be easily found from nuclear magnetic resonance (NMR), electronic (uv) or infra-red (IR) spectroscopies.
- At last, the complete three dimensional structure of a molecule, including its bond lengths and bond angles, can be determined by X-rays diffraction technique.
- This kind of formula which gives the three dimension arrangement of its atoms, including the functional groups present, bond length and bond angles is known as **structural formula**.

Mass Spectroscopy

- In a mass spectrometer, the molccules are bombarded with a stream of high energy electrons. As a result these molecules get ionized and cleaved into a number of fragments, some of which are positive ions.
- Each ion has a particular mass to charge ratio (that is, m/e value). For most ions, the charge is one. So, m/e ratio simply represents the mass of the ions. For example, in case of 2, 2-dimethylpropane (neo-petane), m/e values for different ions can be given as

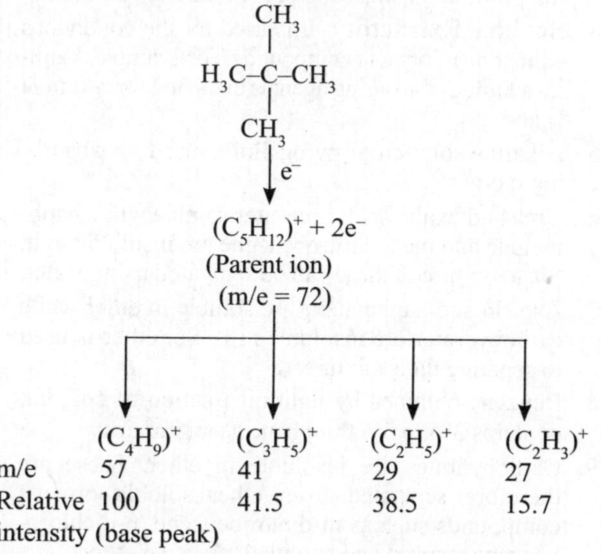

- A set of ions can be analysed by using a mass spectrometer in such a way that a specific signal is generated for each value of m/e. The intensity of each specific signal corresponds to the relative abundance of the ion producing that specific signal.
- The largest peak in this spectrum is known as **base peak.** Its intensity is assumed to be 100 and the intensities of other peaks is expressed relative to it.
- When a graph is plotted between different value of m/e and relative intensities of the signal, it is known as **mass spectrum.**
- This spectrum helps in finding the exact molecular weight of the compound.
- When one electron is removed from the parent molecule (A), then the parent ion (or molecular ion) A^+ is formed.

$$A + e^- \longrightarrow \underset{\substack{\text{Molecular ion} \\ \text{(Parent ion)}}}{A^+} + 2e^-$$

- m/e value gives the molecular weight of this compound.
- Sometimes, the A^+ peak is the base peak and it can be easily recognized.
- If it is not the base peak then it has to be identified or located. Once identified, it gives the most accurate mass of the compound.

Points To Remember

1. **Seeding** It is the addition of pure substances crystals in solution before crystallization.
2. **Mixed Melting Point Method** It is used to find whether the substances are pure and identical or not.
3. **Siwoloboff's Method** It is used to find the boiling point of a compound present in least amount.
4. **Soxhlet Extractor** It is used for the continuous extraction of organic compounds. For example, Vanillin a famous flavouring agent is extracted form vanilla beans.
5. Alkaline solution of pyrogallol is used for absorbing oxygen.
6. Urea and oxalic acids are water soluble while naphthalene and meta dinitrobenzene are insoluble in it. So, a mixture of these can be separated using water.
7. Aspirin and naphthalene are soluble in ether while carbohydrates are insoluble in it. So, ether is used to separate their mixture.
8. Benzene obtained by light oil fraction of coal tar, contains 3% to 5% thiophene as an impurity.
9. Carbohydrates are insoluble in ether. These are therefore separated from ether soluble organic compounds such as m-dinitrobenzene, p-dichlorobenzene, aspirin and naphthalene using ether.
10. **Nitrogen Rule:** All organic compounds containing odd number of nitrogen atoms (that is, 1, 3, 5, 7,......) have odd molecular masses while those compounds which contain an even number of nitrogen atoms (that is, 2, 4, 6, 8,....) have even molecular masses.
11. **Ter Meulen's Method** It is used for the estimation of nitrogen. **Stepanow method, Schiff and Piria method** are used for halogens, **Messenger's method** is used for the estimation of sulphur.
12. Oxygen cannot be detected by any chemical method.
13. The Various substances on the developed T.L.C can be identified by their Retention factor (R_f). R_f -Value is defined as the ratio of the distance travelled by the component to the distance travelled by the solvent.
14. If Na is in excess NaCNS changes into Na_2S, NaCN so blood red colour is not possible.
$$2Na + NaSCN \xrightarrow{\Delta} Na_2S + NaCN$$
15. **Differential extraction:** By it an organic compound is seperated from its aqueous solution by shaking it with a suitable solvent like Benzene, Chloroform, ether etc.
Here the solvent must be miscible with water and the compound must be highly soluble in it.
It is better here to use the solvent in several extractions to get the more amount of product to be extracted.
e.g., Benzoic acid by using benzene.
16. Molar mass ≠ 2 × Density

Tool Box and Summary of Important Relations

$$C\% = \frac{12}{44} \times \frac{\text{Wt of } CO_2}{\text{Wt of org. compound}} \times 100$$

$$H\% = \frac{2}{18} \times \frac{\text{Wt of } H_2O}{\text{Wt of org. compound}} \times 100$$

$$N\% = \frac{28}{22400} \times \frac{\text{Volume of } N_2 \text{ at STP}}{\text{Wt of org. compound}} \times 100$$

OR

$$N\% = \frac{\text{Vol of } N_2 \text{ at STP}}{8 \times \text{Wt of org. compound}}$$

$$\text{Percentage of nitrogen (N\%)} = \frac{1.4 \times N \times V}{\text{Wt of org. compound}}$$

$$X\% = \frac{\text{At. wt of X}}{\text{Mol. wt AgX}} \times \frac{\text{Wt of AgX}}{\text{wt of org. compound}}$$

$$S\% = \frac{32}{233} \times \frac{\text{Wt. BaSO}_4}{\text{Wt. of organic compound}} \times 100$$

$$P\% = \frac{62}{222} \times \frac{\text{Mass of Mg}_2\text{P}_2\text{O}_7\text{ formed}}{\text{Mass of organic compound}} \times 100$$

Silver Salt Method

Molecular weight of acid = Equivalent weight × Basicity

$= E \times n$

$= n\,[W/w \times 108 - 107]$

Here, n = Basicity

Platinichloride Method

Molecular weight of base = Equivalent weight of base × Acidity

$$= B \times n = n \times \frac{1}{2}\left[\frac{W}{w} \times 195 - 410\right]$$

$$= n\left[\frac{W}{w} \times \frac{195}{2} - 205\right]$$

Volumetric Method

Molecular weight of acid

= equivalent weight × basicity

$= n\,[\frac{m}{v} \times 1000]$

Victor Meyer's Method

Vapour Density

$$= \frac{\text{Wt of certain vol of vapour at N.T.P}}{\text{Wt of same vol. of H}_2\text{ at N.T.P}}$$

- $n = \frac{\text{Molecular weight}}{\text{Empirical formula weight}}$
- Molecular formula = Empirical formula × n.

  Molecular Weight = 2 × Vapour density
- Molecular weight = Empirical formula weight × n.

Solved Examples

1. 0.24 g of an organic compound on combustion produced 0.036 g water and 0.22 g of carbon dioxide. Calculate the percentage composition of the compound.

Solution:

Mass of substance = 0.24 g

Mass of water = 0.036 g

Mass of carbon dioxide = 0.22 g

$\% \text{ of H} = \frac{2}{188} \times 0.36 \times \frac{100}{0.24} = 1.66\%$

$\% \text{ of C} = \frac{12}{44} \times 0.22 \times \frac{100}{0.24} = 25\%$

$\% \text{ of O} = 100 - (25 + 1.66) = 73.34\%$

2. In a Duma's experiment, 0.1877 g of an organic substance yielded 31.7 cc of nitrogen gas at 14°C and 758 mm Hg pressure. Calculate the percentage of the nitrogen in the compound. (Aqueous tension at 14°C = 12 mm).

Solution:

Volume of N_2 at NTP $= \frac{V_1 + (P - p)\,273}{(t + 273) \times 760}$

Substituting the values V_1

$$= \frac{31.7 \times (758 - 12) \times 273}{(14 + 273) \times 760} = 29.6 \text{ mL}$$

Mass of 29.6 mL of nitrogen

$= \frac{28}{22400} = 29.6$ g

$\% \text{ of nitrogen} = \frac{28}{22400} \times 29.6 \times \frac{100}{0.1877} = 19.72$

3. 0.257 g of a nitrogeneous compound was digested with sulphuric acid and then distilled with excess of strong alkali. The gas evolved was absorbed in 50 mL N/10 H_2SO_4. At the end of the experiment, the acid required 20.2 mL of N/10 NaOH for neutralization. Determine the percentage of nitrogen in the substance.

Solution:

23.2 mL of N/10 NaOH ≅ 23.2 mL of N/10 H_2SO_4

Volume of N/10 H_2SO_4 neutralized by

NH_3 = 50 – 23.2 = 26.8 mL

26.8 mL of N/10 H_2SO_4 ≅ 26.8 mL of N/10

NH_3 ≅ 26.8 mL of N/10 nitrogen

26.8 mL of N/10 nitrogen will contain

$= \frac{14}{10} \times \frac{26.8}{1000} =$ g nitrogen

The mass of nitrogen was originally present in 0.257 g of the organic substance.

So, % of nitrogen $= \frac{14}{10} \times \frac{26.8}{1000} \times \frac{100}{0.257} = 14.6\%$

4. 0.50 g of a substance containing bromine gave 0.86 g of silver bromide. Find the percentage of bromine in the compound.

Solution:

% of bromine $= \frac{80}{188} \times \frac{0.86}{0.50} \times 100 = 73.19\%$

5. In an experiment, 0.36 g of an organic compound gave 0.35 g of $BaSO_4$. Calculate the percentage of sulphur in the compound.

Solution:

$= \frac{32}{233} \times \frac{0.35}{0.36} \times 100 = 13.35\%$

6. 0.30 g of an organic compound gave 0.550 g of magnesium pyrophosphate by usual analysis. Calculate the percentage of phosphorus in the compound.

Solution:

As 222 g of $Mg_2P_2O_7$ contain 62 g of phosphorus

So 0.55 g of $Mg_2P_2O_7$ contain $= \frac{62}{222} \times 0.55$ g

% of phosphorus $= \frac{62}{222} \times 0.55 \times \frac{100}{0.30} = 51.20\%$

7. A certain organic compound was found to contain 33.6% C, 5.6% H and 49.6% chlorine Find the empirical formula of the organic compound.

Solution:

Table 11.1

Element	%	Atomic ratio	Simplest ratio
C	33.8	33.8 / 12 = 2.8	2.8/0.7 = 4
H	5.6	5.6 / 1 = 5.6	5.6/0.7 = 8
Cl	49.6	49.6 /35.5 =1.4	1.4/0.7 = 2
O	11.0	11 / 16 = 0.7	0.7/0.7 = 1

Thus, $C_4H_8Cl_2O$ is the empirical formula of the compound.

8. A well known antibiotic penicillin, contains C 57.49 %, H 5.39%, N 8.39%, S 9.58%; the remainder being oxygen. Given that penicillin contains one atom of sulphur per molecule, calculate the molecular formula of the antibiotic.

Solution:

% C = 57.49

% H = 5.39

% N = 8.39

% S = 9.58

% O = 100 – (57.49 + 5.39 + 8.39 + 9.58)

=19.15

Table 11.2

Element	%	Relative number of atoms	Simple ratio
C	57.49	57.49/12 = 4.79	16
H	5.39	5.39/1 = 5.39	18
N	8.39	8.39/14 = 0.599	2
S	9.58	9.58/32 = 0.299	1
O	19.15	19.15/16 = 1.196	4

Empirical formula is $C_{16}H_{18}N_2SO_4$

As the given antibiotic contains only one atom of sulphur per molecule. The calculated empirical formula $C_{16}H_{18}N_2SO_4$, contains only one atom of sulphur. Hence, the same is the molecular formula of the antibiotic.

9. A compound has molar mass of 147 g. It contains 49% C and 2.72% H. On Carius estimation 2.561 mg of the compound gave 5 mg of AgCl. Find the empirical and molecular formula.

Solution:

Table 11.3

Element	%	Atomic ratio	Simple ratio
C	49	49/12 = 4.1	3
H	2.72	2.72/1 = 2.72	2
Cl	$\frac{35.5/143.5) \times 5 \times 100}{2.581}$ = 47.9	47.9/35.5 = 1.34	1

C_3H_2Cl is the empirical formula.

n = 147/73.5 = 2

The molecular formula is $C_6H_4Cl_2$.

10. Methyl orange, an acid–base indicator, is the sodium salt of an acid that contains C, H, N, S and O. Quantitative analysis gave 51.4% C, 4.3% H, 12.8% N, 9.8% S and 7.0% Na. What is the empirical formula of methyl orange?

Solution:

% C = 51.4

% H = 4.3

% N = 12.8

% S = 9.8

% Na = 7.0

% O = 14.7

TABLE 11.4

Element	%	Relative number of atoms	Simple ratio
C	51.4	51.4/12 = 4.28	14
H	4.3	4.3/1 = 4.3	14
N	12.8	12.8/14 = 0.914	3
S	9.8	9.8/32 = 0.306	1
Na	7.0	7/23 = 0.304	1
O	14.7	14.7/16 = 0.918	3

So, empirical formula is $C_{14}H_{14}N_3SO_3Na$.

CHAPTER-END EXERCISES

PRACTICE QUESTIONS – I

1. In Carius tube, the compound $ClCH_2COOH$ was heated with fuming HNO_3 and $AgNO_3$. After filtration and washing, a white precipitate was formed. The precipitate is:

(a) Ag_2SO_4 (b) $AgNO_3$
(c) AgCl (d) $ClCH_2COOAg$

2. The most satisfactory method to separate sugaris to use:

(a) Sublimation
(b) Fractional crystallization
(c) Benedict's reagent
(d) Chromatography

3. The compound that does not give a blue colour in Lassaigne's test is:

(a) Aniline (b) Glycine
(c) Hydrazine (d) Urea

4. Impure glycerine is purified by:

(a) Vacuum distillation
(b) Simple distillation
(c) Steam distillation
(d) None of the above

5. When pieces of human hair are heated strongly with soda-lime, smell of ammonia can be detected. Which one of the following conclusion can be drawn from this observation?

(a) Hair contains amino acids
(b) Ammonium salt is present in hair
(c) Ammonia is present in the human hair
(d) None of the above

6. Nitrogen in an organic compound can be estimated by:

(a) Duma's method only
(b) Kjeldahl's method only
(c) Both (a) and (b)
(d) None of these

7. In Lassaigne's test for nitrogen, the blue colour is due to the formation of:

(a) Ferriferrocyanide
(b) Sodium cyanide
(c) Potassium fericyanide
(d) Sodium ferrocyanide

8. Carbon and hydrogen are estimated by:

(a) Leibig method (b) Duma's method
(c) Carius method (d) Kjeldahl's method

9. Aniline is usually purified by:

(a) Chromatography technique
(b) By addition of oxalic acid
(c) Fractional crystallization
(d) Steam distillation

10. In Kjeldhal's method, nitrogen present is estimated as:

(a) NO (b) NH_3
(c) NO_2 (d) N_2

11. Molecular mass of a volatile substance may be obtained by:

(a) Liebig's method
(b) Kjeldhal's method
(c) Duma's method
(d) Victor Meyer's method

12. Which of the following has molecular weight of 92?

(a) Benzene (b) Propene
(c) Toluene (d) Methylene

13. Which of the following compounds does not show Lassaigne's test for nitrogen?

(a) Urea (b) Azobenzene
(c) Hydrazine (d) Phenylhydrazine

14. In a hydrocarbon, mass ratio of hydrogen and carbon is 1 : 3, the empirical formula of hydrocarbon is:

(a) CH_2 (b) CH_4
(c) CH_3 (d) CH

15. A mixture of benzene and chloroform is separated by:

(a) Sublimation
(b) Separation funnel
(c) Crystallization
(d) Distillation

16. Which method is used for the separation of two immiscible liquids?

(a) Chromatography
(b) Fractionating column
(c) Separating funnel
(d) Fractional distillation

17. The empirical formula of an acid is CH_2O_2, the probable molecular formula of the simplest acid may be:

(a) CH_2O (b) CH_2O_2
(c) $C_2H_4O_2$ (d) $C_3H_6O_4$

18. During formation of sodium extract, generally the substance that ignites is:

(a) Na (b) O_2
(c) CO_2 (d) H_2

19. The equivalent mass of an acid is equal to:

(a) Molecular mass × acidity
(b) Molecular mass × basicity
(c) Molecular mass/acidity
(d) Molecular mass/basicity

20. Fractional distillation is used to separate liquids which differ in their boiling points by:

(a) 5°C (b) 10°C
(c) 30°C–80°C (d) Less than 20°C

21. Thiophene can be removed from commercial benzene by:

(a) Steam distillation
(b) Shaking with conc. H_2SO_4
(c) Shaking with NaOH solution
(d) Shaking with ether

22. The Beilstein test for organic compounds is used to detect:

(a) Nitrogen (b) Sulphur
(c) Carbon (d) Halogens

23. Which of the following is the best scientific method to test presence of water in a liquid?

(a) Taste
(b) Smell
(c) Use of litmus paper
(d) Use of anhydrous copper sulphate.

24. Which of the following has high vapour pressure at temperature below its melting point?

(a) Benzoic acid (b) Citric acid
(c) Salicylic acid (d) All

25. When petroleum is heated gradually, first batch of vapours evolved will be rich in:

(a) Kerosene (b) Petroleum ether
(c) Diesel (d) Lubricating oil

26. Absolute alcohol is prepared by:

(a) Fractional distillation
(b) Kolbe's method
(c) Vacuum distillation
(d) Azeotropic distillation

27. In paper chromatography,

(a) Mobile phase is liquid and stationary phase is solid
(b) Mobile phase is solid and stationary phase in solid
(c) Mobile phase is liquid and stationary phase is liquid
(d) Mobile phase is solid and stationary phase is liquid.

28. Azeotropic mixture of two liquids can be separated by:

(a) Azeotropic distillation
(b) Fractional distillation
(c) Simple distillation
(d) Steam distillation

29. A mixture of benzoic acid and naphthalene can be separated by crystallization from:

(a) Cold water (b) Ether
(c) Hot water (d) Benzene

30. Aniline is insoluble in water and possesses a vapour pressure of 10–15 mm Hg at 373 K. It can be conveniently purified by:

(a) Simple distillation (b) Steam distillation
(c) Crystallization (d) Sublimation

31. Which of the following compound will not produce a white precipitate of AgCl on warming with ethanolic silver nitrate solution?
(a) $CH_3 - CH_2Cl$
(b) $CH_2 = CHCH_2Cl$
(c) CH_2Cl
(d) 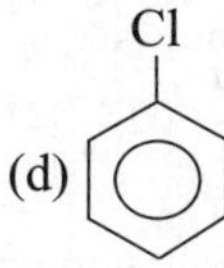

32. In Kjeldahl's method of estimation of nitrogen, K_2SO_4 acts as:
(a) Catalytic agent
(b) Boiling point elevator
(c) Hydrolysis agent
(d) Oxidizing agent

33. There is no direct test for the detection of which of the following when present in an organic compound?
(a) Cl
(b) O
(c) S
(d) N

34. Liquids which decompose below their normal boiling points can be distilled at lower temperature by:
(a) Decreasing the pressure
(b) Heating in sand bath
(c) Increasing the pressure
(d) Heating in water bath

35. Separation of two substances by fractional crystallization depends upon their difference in:
(a) Viscosity
(b) Density
(c) Solubility
(d) Crystalline shape

36. The compound formed in the positive test for nitrogen with the Lassaigne solution of an organic compound is:
(a) $Fe(CN)_3$
(b) $Fe_4[Fe(CN)_6]_3$
(c) $Na_4[Fe(CN)_5NOS]$
(d) $Na_3[Fe(CN)_6]$

37. Absolute alcohol cannot be obtained by simple fractionation of a solution of water and alcohol because:
(a) Ethanol molecules are solvated
(b) Boiling points are very near
(c) Ethanol remains dissolved in water
(d) They form a constant boiling mixture

38. In steam distillation, the vapour pressure of the volatile organic compound is:
(a) More than atmospheric pressure
(b) Equal to atmospheric pressure
(c) Less than atmospheric pressure
(d) None of these

39. The purity of organic compounds is determined by:
(a) Mixed melting point
(b) Molecular weight
(c) Melting point
(d) Density

40. In Kjeldahl's method of estimation of nitrogen, copper sulphate acts as:
(a) Catalytic agent
(b) Silver spiral
(c) Hydrolyzing agent
(d) Oxidizing agent

41. An organic dibasic acid is used in volumetric analysis. The percentage of elements are C = 26.7%, H = 2.2% and O = 71.1%. Find the empirical formula of the acid.
(a) CH_2O
(b) CHO_2
(c) $C_2H_2O_2$
(d) CH_4O

42. The molecular formula weight by empirical formula weight of a compound is 3. The empirical formula is CH_2. The vapour density of the compound is:
(a) 21
(b) 11.5
(c) 33
(d) 43

43. Chromatographic technique is used for the separation of:
(a) Plant pigments
(b) Dye stuffs
(c) Small samples of mixtures
(d) All of these

44. For a diacidic base, the molecular weight is 82. The weight of chloroplatinate is 0.984 g. The weight of platinum is:
(a) 0.39
(b) 0.68
(c) 7.8
(d) 1.8

45. The function of boiling the sodium extract with conc. HNO_3 before testing for halogens is:
(a) To make the solution acidic
(b) To make the solution clear
(c) To convert Fe^{2+} to Fe^{3+}
(d) To destroy CN^- and S^{2-} ions which will otherwise give precipitate

Practice Questions – II

46. A dibasic acid containing C, H and O was found to contain C = 26.7% and H = 2.2%. The vapour density of diethyl ester of the acid was found to be 73. What is the molecular formula of the acid?

(a) CH_2O_2 (b) $C_2H_2O_4$
(c) $C_3H_3O_4$ (d) $C_4H_4O_4$

47. Two elements X (atomic weight = 75) and Y (atomic weight = 16) combine to give a compound having 75.8% X. The formula of the compound is:

(a) X_2Y_2 (b) X_2Y_3
(c) X_2Y (d) XY

48. The percentage of sulphur in an organic compound whose amount of 0.32 g produces 0.233 g of $BaSO_4$ (Atomic weight of Ba = 137, S = 32) is:

(a) 1.0 (b) 10.0
(c) 25.3 (d) 32.1

49. 116 mg of a compound on vapourisation in a Victor Meyer's apparatus displaces 44.8 mL of air measured at STP. The molecular mass of the compound is:

(a) 58 g (b) 48 g
(c) 116 g (d) 44.8 g

50. Two organic compounds (X) and (Y) both containing only C and H yields on analysis, the same percentage composition by mass C = 92.3% and H = 7.7%. (X) decolourizes bromine water and (Y) does not. Identify (X) and (Y).

(a) $X = C_2H_2$, $Y = C_3H_8$
(b) $X = C_2H_4$, $Y = C_2H_6$
(c) $X = C_6H_6$, $Y = C_2H_2$
(d) $X = C_2H_2$, $Y = C_6H_6$

51. In the reaction, $CH_4 + 2O_2 \longrightarrow CO_2 + 2H_2O$, the amount of oxygen needed to completely burn 4.0 g of CH_4 is:

(a) 4 (b) 8 g
(c) 16 g (d) 32 g

52. 0.1914 g of an organic acid is dissolved in about 20 mL of water 25 mL of 0.12 N NaOH is required for the complete neutralization of the acid solution. The equivalent weight of the acid is:

(a) 63.8 (b) 64.0
(c) 65.8 (d) 66.2

53. An alkane has a C/H ratio of 5.1428 by mass. Its molecular formula is:

(a) C_5H_{12} (b) C_6H_{14}
(c) C_8H_{18} (d) C_7H_{10}

54. 0.759 g of a silver salt of a dibasic organic acid on ignition left 0.463 g of metallic silver. The equivalent weight of the acid is:

(a) 70 (b) 108
(c) 60 (d) 50

55. Which of the following compounds is violet in colour?

(a) $Na_4\,[Fe(CN)_5\,NSO_3]_3$
(b) $Na_3\,[Fe(CN)_6\,NSO]_2$
(c) $Na_2\,[Fe(CN)_5\,NSO]_3$
(d) $Na_4\,[Fe(CN)_5\,NOS]$

56. An organic compound on heating with CuO produces CO_2 but not water. It may be:

(a) CH_4 (b) CCl_4
(c) $CHCl_3$ (d) C_2H_5I

57. The empirical formula of a compound is CH_2. One mole of the compound has a mass of 42 g. Its molecular formula is:

(a) C_3H_6 (b) C_3H_8
(c) C_2H_2 (d) CH_2

58. In the estimation by Duma's method, 0.59 g. of an organic compound gave 112 mL nitrogen at NTP. The percentage of nitrogen in the compound is about:

(a) 23.7 (b) 16.8
(c) 33.07 (d) 27.5

59. 60 g of a compound on analysis produced 24 g carbon, 4 g hydrogen and 32 g oxygen. The empirical formula of the compound is:

(a) CH_4O (b) CH_2O_2
(c) $C_2H_4O_2$ (d) CH_2O

60. Mark the incorrect statement in Kjeldahl's method of estimation of nitrogen.

(a) Copper sulphate or mercury acts as a catalyst
(b) Nitrogen is quantitatively decomposed to give ammonium sulphate
(c) Nitrogen gas is collected over caustic potash solution
(d) Potassium sulphate is used as boiling point elevator of H_2SO_4

61. An organic compound is found to have the formula $C_5H_{10}ONCl$. The percentage of nitrogen present in it is:

(a) 23.36% (b) 10.3%
(c) 41.05% (d) 5.06%

62. 500 mL of a hydrocarbon gas burnt in excess of oxygen, gave 2500 mL of CO_2 and 3000 mL of water vapour (at constant T and P). The formula of the hydrocarbon is:

(a) C_3H_6 (b) C_2H_4
(c) C_5H_{12} (d) CH_4

63. The empirical formula of a compound is CH_2. One mole of the compound has a mass of 42 g, its molecular formula is:

(a) C_2H_2 (b) C_2H_4
(c) C_3H_8 (d) C_4H_8

64. Which of the following statements is wrong?

(a) In the estimation of carbon, an organic compound is heated with CaO in a combustion tube.
(b) Using Lassaigne's test nitrogen and sulphur present in an organic compound can be tested.
(c) Using Beilstein's test the presence of halogen in a compound can be tested
(d) Lassaigne's tests fails to identify nitrogen in diazo compound.

65. Sorbic acid consists of C = 64.3%, H = 7.2% and 28.5% oxygen. Find the empirical formula.

(a) $C_2H_4O_2$ (b) $C_3H_4O_2$
(c) C_3H_4O (d) CH_2O

66. A mixture contains four solid organic compounds (A), (B), (C) and (D). On heating, only (C) changes from solid to vapour state. The compound (C) can be separated from the mixture by:

(a) Sublimation
(b) Crystallization
(c) Distillation
(d) Fractional distillation

67. 0.24 g of a volatile liquid on vaporization gives 45 mL of vapours at STP. What will be the vapour density of the substance? (Density of $H_2 = 0.089\ gL^{-1}$)

(a) 99.39 (b) 39.95
(c) 95.53 (d) 59.93

68. If a compound on analysis was found to contain C = 18.5%, H = 1.55%, Cl = 55.04% and O = 24.81% then its empirical formula is:

(a) CHClO
(b) CH_2ClO
(c) C_2H_5OCl
(d) ClC_2H_5O

69. A compound with empirical formula CH_2O has a vapour density of 30. Its molecular formula is

(a) $C_3H_6O_3$ (b) $C_2H_4O_2$
(c) $C_2H_2O_2$ (d) $C_6H_{12}O_6$

70. The identity of two different solid organic compounds having the same melting point can be best ascertained by determining their:

(a) Dipole moment
(b) Solubility
(c) Melting point of their 1 : 1 mixture
(d) Refractive indices

71. 0.5 g of an organic compound in Kjeldahl experiment needs 29 ml of $\frac{N}{5}$ H_2SO_4 for complete neutralization of NH_3. The N% is:

(a) 16.2 (b) 8.1
(c) 24.3 (d) 32.4

72. When thiourea is heated with metallic sodium, the compound which cannot be formed is:

(a) Na_2SO_4 (b) Na_2S
(c) NaCN (d) NaCNS

73. In Lassaigne's test, which of the following organic compound will produce a blood red colour when its sodium extract is treated with $FeCl_3$ solution?

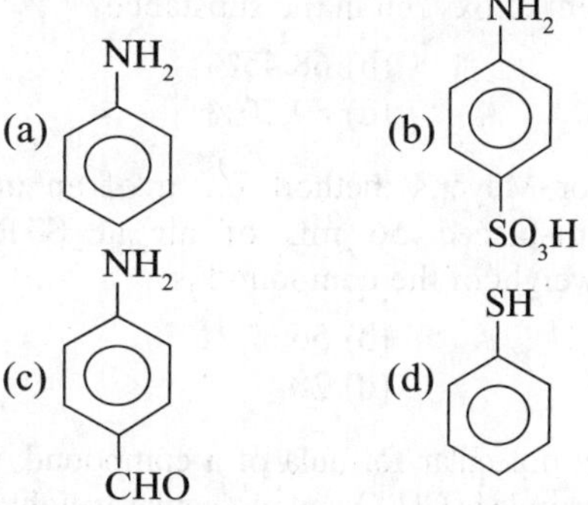

74. When 0.25 g of an organic compounds is heated with HNO_3 and $AgNO_3$ in a carius tube, it gives 0.35 g of silver chloride. The percentage of chlorine in the compound is:

(a) 36.6% (b) 45.3%
(c) 34.6% (d) 54.8%

75. If 0.1 g of an organic compound containing phosphorus gave 0.222 g of $Mg_2P_2O_7$, then the percentage of phosphorus in the compound is:

(a) 31 (b) 0.2
(c) 66 (d) 62

Practice Questions – III

76. The empirical formula of a compound is CH_2O. If 0.0832 mole of the compound contains 1.0 g of hydrogen, then the molecular formula of the compound is:

(a) $C_6H_{12}O_6$ (b) $C_{12}H_{22}O_{11}$
(c) $C_5H_{10}O_5$ (d) $C_3H_6O_3$

77. 0.765 g of an acid gives 0.535 g of CO_2 and 0.14 g of H_2O. Then the ratio of the percentage of carbon and hydrogen is:

(a) 1 : 9 (b) 20 : 16
(c) 18 : 11 (d) 19 : 2

78. 5.0 g of a certain metal, X (atomic mass = 27) is converted into 61.7% crystalline sulphate containing 48.6% by mass of water of crystallization. The simplest formula of the compound is:

(a) $X_2(SO_4)_3 18H_2O$
(b) $X_2(SO_4)_3 14H_2O$
(c) $X_2(SO_4)_3 16H_2O$
(d) $X_2(SO_4)_3 12H_2O$

79. If 0.2 gram of an organic compound containing carbon, hydrogen and oxygen, on combustion, yielded 0.147 gram carbon dioxide and 0.12 gram water. What will be the content of oxygen in the substance?

(a) 73.29% (b) 68.45%
(c) 85.23% (d) 89.50%

80. Using Victor Meyer's method, 0.2 g of an organic substance displaced 56 mL of air at STP. The molecular weight of the compound is:

(a) 112 (b) 56
(c) 80 (d) 28

81. What is the molecular formula of a compound, whose empirical formula is CH_2O and molecular weight is 90?

(a) $C_2H_6O_2$ (b) $C_6H_6O_3$
(c) $C_3H_6O_3$ (d) $C_4H_8O_4$

82. An organic compound has C and H percentage in the ratio 6 : 1 and C% and O% in the ratio 3 : 4. The compound is:

(a) HCHO (b) CH_3OH
(c) CH_3CH_2OH (d) $(COOH)_2$

83. A complex compound of cobalt with the composition Co = 22.58%, H = 5.79%, N = 32.2%, O = 12.20% and Cl = 27.17% on heating, looses ammonia to the extent of 32.63% of its mass. The number of molecules of ammonia present in one molecule of the cobalt complex is (At. mass of Co = 58.9)

(a) 7 (b) 5
(c) 8 (d) 12

84. In the enzyme peroxidase anhydrase, the percentage of selenium (Se) by mass is 0.5%. If atomic mass of Se is 78.4 U, then minimum molecular mass of peroxidase anhydrase enzyme is:

(a) 1.568×10^3 U (b) 1.568×10^4 U
(c) 1.568 U (d) 2.136×10^4 U

85. A gaseous hydrocarbon has 85% carbon and vapour density of 28. The possible formula of the hydrocarbon will be:

(a) C_4H_8 (b) C_2H_4
(c) C_2H_2 (d) C_3H_6

86. An organic compound contains 49.3% carbon, 6.84% hydrogen and its vapour density is 73. Molecular formula of the compound is:

(a) $C_3H_{10}O_2$ (b) $C_3H_5O_2$
(c) $C_6H_{10}O_4$ (d) $C_4H_{10}O_2$

87. An organic compound has C and H percentage in the ratio 6 : 1 and C and O percentage in the 3 : 4. The compound is:

(a) $(COOH)_2$ (b) CH_3OH
(c) HCHO (d) CH_3CH_2OH

88. An organic compound containing carbon, hydrogen and oxygen contains 52.2% carbon and 13.04% hydrogen. Vapour density of the compound is 23. Its molecular formula will be:

(a) C_4H_8O (b) C_3H_8O
(c) C_2H_6O (d) $C_5H_{10}O$

89. 9.9 g of an amide with molecular formula $C_4H_5N_xO_y$ on heating with alkali liberated 1.7 g of ammonia. If the percentage of oxygen is 32.33%, then the ratio of N and O atoms in the compound is:

(a) 2 : 1 (b) 1 : 2
(c) 2 : 5 (d) 2 : 3

90. On combustion, 0.2 g of an organic compound containing carbon, hydrogen and oxygen yields 0.147 g carbon dioxide and 0.12 g water. The percentage of carbon, hydrogen and oxygen respectively in the compound is:

(a) 22.5, 8.7 and 68.8
(b) 20, 6.7 and 73.3
(c) 18.2, 9.5 and 72.3
(d) 28.9, 12.8 and 58.8

91. In an organic compound of molar mass 108 g mol^{-1} C, H and N atoms are present in 9 : 1 : 3.5 by weight. The molecular formula of the compound can be:

(a) $C_3H_6N_2$ (b) $C_6H_8N_2$
(c) $C_5H_6N_3$ (d) $C_4H_{12}N_2$

92. Bottles containing C_6H_5I and $C_6H_5CH_2I$ lost their original labels. They were labelled A and B for testing. A and B were separately taken in test tubes and boiled with NaOH solution. The end solution in each tube was made acidic with dilute HNO_3 and some $AgNO_3$ solution added. Solution B gave a yellow precipitate. Which one of the following statements is true for the experiment?

(a) A was $C_6H_5CH_2I$
(b) A was C_6H_5I
(c) B was C_6H_5I
(d) Addition of HNO_3 was unnecessary

93. The ammonia evolved from the treatment of 0.30 g of an organic compound for the estimation of nitrogen was passed in 100 mL of 0.1 M sulphuric acid. The excess of acid required 20 mL of 0.5 M sodium hydroxide solution for complete neutralization. The organic compound is:

(a) Benzamide (b) Acetamide
(c) Thiourea (d) Urea

94. In Kjeldahl method, the gas evolved from 1.325g sample of a fertilizer is passed into 50 ml of 0.2030 N H_2SO_4. 25.32 ml of 0.198 N NaOH is needed for the titration of the un-used acid. Find the N% here?

(a) 10.66% (b) 5.3%
(c) 16.29% (d) 25.73%

95. An organic compound having molecular mass 60 is found to contain C = 20%, H = 6.67% and N = 46.67% while rest is oxygen. On heating, it gives NH_3 along with a solid residue. The solid residue give violet colour with alkaline copper sulphate solution. The compound is:

(a) CH_3CONH_2
(b) CH_3NCO
(c) $(NH_2)_2$ CO
(d) $CH_3CH_2CONH_2$

96. 0.21g of an organic substance having C, H N, O on combustion gave 0.462 g CO_2, 0.1215 g H_2O. When 0.104 g of it was distilled with NaOH the evolved NH_3 was neutralised by 15 ml of 0.05 N H_2SO_4. The empirical formula is?

(a) $C_6H_7NO_2$ (b) $C_7H_9NO_2$
(c) C_7H_9NO (d) $C_6H_5NO_2$

97. If in an organic compound the weights of C, H, O are taken in 6:1:8 ratio. The molecular formula of the compound is:

(a) $C_2H_4O_2$ (b) $C_2H_4O_4$
(c) $C_3H_6O_3$ (d) CH_2O_2

98. Which of the following is not correctly matched?

Compound	Method used for purification or separation
(a) Glycerol	Vaccum distillation
(b) Iodine	Sublimation
(c) Phenol	Ferric chloride
(d) o, p Nitro phenols	Fractional distillation

99. Which of the following compounds can show red colour in Lassaigne Test?

(1) NH_2CONH_2
(2) NaCNS
(3) NH_2–CS–NH_2
(4) 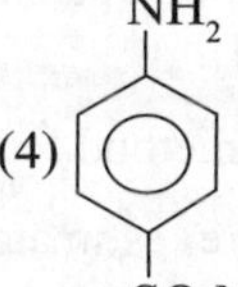

(a) 2, 3 (b) 1, 2, 3
(c) 2, 3, 4 (d) 3, 4

100. Detection of chlorine is possible without preparing sodium extract in:

(1) CCl_4
(2) CH_2Cl (benzene ring)
(3) Cl, NO_2, NO_2, NO_2 (benzene ring)
(4) $CHCl_3$

(a) 3 only (b) 2, 3
(c) 3, 4 (d) 2, 3, 4

Answer Keys

1. (c)	2. (d)	3. (c)	4. (a)	5. (a)	6. (c)	7. (a)	8. (a)	9. (d)	10. (b)
11. (d)	12. (c)	13. (c)	14. (b)	15. (d)	16. (c)	17. (b)	18. (d)	19. (d)	20. (b)
21. (b)	22. (d)	23. (d)	24. (d)	25. (b)	26. (d)	27. (c)	28. (a)	29. (c)	30. (b)
31. (d)	32. (b)	33. (b)	34. (a)	35. (c)	36. (b)	37. (d)	38. (c)	39. (a)	40. (a)
41. (b)	42. (a)	43. (d)	44. (b)	45. (d)	46. (b)	47. (b)	48. (b)	49. (a)	50. (d)
51. (c)	52. (a)	53. (b)	54. (a)	55. (d)	56. (b)	57. (a)	58. (a)	59. (d)	60. (c)
61. (b)	62. (c)	63. (d)	64. (a)	65. (c)	66. (a)	67. (d)	68. (a)	69. (b)	70. (c)
71. (a)	72. (a)	73. (b)	74. (c)	75. (d)	76. (d)	77. (a)	78. (a)	79. (a)	80. (c)
81. (c)	82. (a)	83. (b)	84. (b)	85. (a)	86. (c)	87. (c)	88. (c)	89. (b)	90. (b)
91. (b)	92. (b)	93. (d)	94. (b)	95. (c)	96. (b)	97. (c)	98. (d)	99. (c)	100. (b)

Hints and Explanations for Selective Questions

1. In Carius method, Cl is converted into AgCl.

2. Chromatography (actually paper chromatography)

3. As it does not have carbon atom.

5. Hair contains amino acids which upon fusion with soda-lime (NaOH + CaO) evolve NH_3.

6. Both Duma's method and Kjeldahl's method.

8. C and H in organic compound are estimated by Leibig's method.

9. Steam distillation, as aniline is steam volatile.

12. Molecular weight of toluene ($C_6H_5CH_3$)
$= 12 \times 7 + 8 \times 1 = 92$

13. Hydrazine (NH_2NH_2) does not contain C. On fusion with Na metal, it cannot form NaCN. So, hydrazine does not show Lassaigne's test.

14. The mass ratio of H and C is 1 : 12. Since in the given compound, mass ratio of H to C is 1 : 3, therefore, there are four atoms of H for each atom of C.

15. Distillation (particularly fractional distillation) because the boiling point of benzene (80°C) and chloroform (61.5°C) are close to each other.

16. Separating funnel, since the two liquids are immiscible.

17. An acid contain COOH group, so simplest acid containing this group is HCOOH, that is, CH_2O_2.

18. Hydrogen of the organic compound ignites to form H_2O.

19. Equivalent wt = Molecular wt/Basicity

20. When boiling point of two liquids differ atleast by 10°C–20°C

23. Anhydrous $CuSO_4$ turns blue in presence of water.

$$\underset{\text{White}}{CuSO_4} + 5\,H_2O \longrightarrow \underset{\text{Blue}}{CuSO_4 \cdot 5H_2O}$$

24. All exhibit sublimation.

25. The first distillation product of petroleum is petroleum ether. Note that natural gases comes out first during extraction of petroleum.

26. Azeotropic distillation, since alcohol and water form a constant boiling mixture (azeotrope).

28. Azeotropic mixture of two liquids can be separated indirectly by azeotropic distillation.

29. Naphthalene does not dissolve in hot water but benzoic acid does.

30. A liquid which is immiscible in water and has a vapour pressure of 10–15 mm Hg at 373 K can be conveniently purified by steam distillation.

32. Addition of K_2SO_4 increases the boiling point of H_2SO_4.

34. Boiling point gets lowered at low pressure.

36. If nitrogen is present in organic compound then sodium extract contains $Na_4[Fe(CN)_6]$

$$Na + C + N \xrightarrow{\text{Fuse}} NaCN$$

$FeSO_4 + 6NaCN \longrightarrow \underset{(A)}{Na_4[Fe(CN)_6]} + Na_2SO_4$

(A) changes to prussian blue $Fe_4[Fe(CN)_6]_3$ on reaction with $FeCl_3$.

$4FeCl_3 + 3Na_4[Fe(CN)_6] \rightarrow Fe_4[Fe(CN)_6]_3 + 12NaCl$

38. P (mixture) = P (compound) + P (steam)
= 1 atm (at boiling point)

40. To increase the rate of reaction so that $(NH_4)_2SO_4$ is formed easily.

47. $X : Y = \frac{75.8}{75} : \frac{24.2}{16} = 2.3$

So, empirical formula = X_2Y_3

48. Percentage of S = $\frac{32}{233} \times \frac{0.233}{0.32} \times 100 = 100.$

49. 44.8 mL of air displaced by = 116 mg = $\frac{116}{1000}$ g

So, 22400 mL of air displaced by

$= \frac{116}{1000} \times \frac{22400}{44.8} = 58g$

51. $CH_4 + 2O_2 \longrightarrow CO_2 + 2H_2O$

Now 16 g CH_4 requires 64 g O_2 for complete combustion.

Thus, 4.0 g would require = $\frac{64 \times 4}{16} = 16g\ O_2$

52. 25 mL of 0.12 N NaOH ≅ 0.1914 g of acid as 1000 mL of 1 N NaOH

$= \frac{0.1914 \times 1000}{25 \times 0.12} = 63.08g$

53. C/H ratio by mass in

$C_6H_{14} = \frac{6 \times 10}{14 \times 1} = \frac{72}{14} = 5.1428$

54. Equivalent weight of silver salt = Equivalent weight of Ag

$\frac{0.759}{E} = \frac{0.463}{108}$

Equivalent weight of Ag salt = 177

Equivalent weight of acid = 177 – 108 + 1 = 70

55. $Na_2S + Na_2[Fe(CN)_5NO] \longrightarrow Na_4[Fe(CN)_5NOS]$

57. Molar wt = mass of 1 mole of the compound = 42

Empirical formula wt of CH_2 = 12 + 2 = 14

Molar formula = $CH_2 \times \frac{42}{14} = C_3H_6$

58. % of $N_1 = \frac{28 \times 112 \times 100}{22400 \times 0.59} = 23.7$

59. $C : H : O = \frac{24}{12} : \frac{4}{1} : \frac{32}{16} = 1 : 2 : 1$

Empirical formula = CH_2O

61. Molecular mass of $C_5H_{10}ONCl$

= 60 + 10 + 16 + 14 + 35.5 = 135.5

$N\% = \frac{14 \times 100}{135.5} = 10.3\%$

62. $C_xH_y + [x + y/4]O_2 \longrightarrow xCO_2 + y/2\ H_2O$

500		0	0
		500x	$\frac{y}{2} \times 500$

given, 500x = 2500 or x = 5

$\frac{y}{2} \times 500 = 3000$ or y = 12.

Thus the hydrocarbon is C_5H_{12}

63. Mass of 1 mole of compound = molar wt. of compound = 56 mol

Mole formula

= Empirical formula × $\frac{\text{Mohr wt}}{\text{Empiricak formula wt}}$

$= CH_2 \times \frac{56}{14} = C_4H_8$

64. In estimation of C, organic compound is heated with CuO.

$2CuO + C \longrightarrow 2Cu + CO_2$

65.

Element	%	Atomic ratio	Simplest ratio
C	64.3	64.3 / 12 = 5.4	3
H	7.2	7.2 / 1 = 7.2	4
O	28.5	28.5 / 16 = 1.8	1

The empirical formula of compound is C_3H_4O.

67. V.D. = $\frac{\text{Wt of 4.5 mL of vapours at NTP}}{\text{Wt of 4.5 mL of } H_1 \text{ at NTP}}$

$= \frac{0.24}{45 \times 0000089}$

(as density of $H_2 = 0.089\ g\ L^{-1} = 0.000089$ g/mL)

= 59.93

68. $C : H : Cl : O = \frac{18.5}{12} : \frac{1.55}{1} : \frac{55.04}{25.5} : \frac{24.81}{16}$

$= 1 : 1 : 1 : 1$

So, empirical formula = CHClO

69. Molecular weight = 2 × V.D. = 2 × 30 = 60

Molecular formula = $\frac{\text{Molar wt} \times \text{Empirical formula}}{\text{Empirical formula wt}}$

$= \frac{60}{30} \times CH_2O = C_2H_4O_2$

71. $N\% = \frac{1.4 \times 1V_1V_1}{W_{O.C}} = \frac{1.4 \times 29}{0.5 \times 5} = 16.24\%$

72. Thiourea (NH_2CSNH_2) contains both N and S. Hence, depending upon the amount of Na used, all the three, that is, Na_2S, NaCN, and NaCNS are formed, but not Na_2SO_4.

74. % of chlorine = $\frac{35.5}{143.5} \times \frac{\text{wt of AgCl}}{\text{wt of O.C.}} \times 100$

$= \frac{35.5}{143.5} \times \frac{0.35}{0.25} \times 100 = 34.6\%$

75. $P\% = \frac{62}{222} \times \frac{\text{wt of } Mg_2P_2O_7}{\text{wt of OC}} \times 100$

$= \frac{62}{222} \times \frac{0.222}{0.1} \times 100 = 62$

76. 0.0833 mole of compound contains H = 1.0 g

1 mole of compound will contain H

$= \frac{1}{0.8832} = 62g$

No. of H atoms in one mole of compound

= 12/1 = 12

Since the empirical formula of compound (CH_2O) contains only 2H atoms, therefore, molecular formula must be six times empirical formula, that is,

$6 \times CH_2O = C_6H_{12}O_6$

77. % of C $= \frac{12}{44} \times \frac{0.535}{0.765} \times 100 = 19.07\%$

% of H $= \frac{2}{18} \times \frac{0.138}{0.765} \times 100 = -2.004$

C : H = 19 : 2

78. % of M in crystalline sulphate $= \frac{5 \times 100}{61.7} = 8.2\%$ of water = 48.6% (given)

% of SO_4^{2-} =100 – (48.6 + 8.2)

= 43.2%

X : SO_4^{2-} : H_2O $\frac{8.2}{27} : \frac{43.2}{96} : \frac{48.6}{18}$

= 0.3 : 0.45 : 2.7

= 1 : 1.5 : 9 = 2 : 3 : 18

Molecular formula = $X_2(SO_4)_3.18H_2O$

79. % of C $= \frac{12}{44} \times \frac{0.147}{0.2} \times 100 = 20.045$

% of H $= \frac{2}{18} \times \frac{0.12}{0.2} \times 100 = 6.666$

% of O = (100 – 20.045 – 0.6666)

= 73.289 = 73.29 (approx)

80. 56 mL of air at STP is displaced by compound = 0.2 g

22400 mL of air will by displaced by compound

$= \frac{0.2 \times 22400}{56} = 80$ g

Mol. wt of compound = 80.

81. Empirical formula = CH_2O

Molecular weight = 90

Empirical formula weight of CH_2O

= 12 + (1 × 2) + 16 = 30

$n = \frac{\text{molecular weight}}{\text{empirical formula weight}} = \frac{90}{30} = 3$

Molecular formula = $(CH_2O)_3 = C_3H_6O_3$

82. % ratio of C : H : : 6 : 1 and C : O : : 3 : 4

% ratio of C : H : O : : 6 : 1 : 8 %/at. wt

% of C $= \frac{6}{15} \times 100 = 40$ $\qquad \frac{40}{12} = 3.33$

% of H $= \frac{1}{15} \times 100 = 6.66$ $\qquad \frac{6.66}{1} = 6.66$

% of O $= \frac{8}{15} \times 100 = 53.3$ $\qquad \frac{53.3}{16} = 3.33$

Simplest ratio of C : H : O : : 1 : 2 : 1,

that is, CH_2O

83. Ratio of Co : H : N : O : Cl : NH_3

$= \frac{22.58}{58.9} : \frac{5.79}{1} : \frac{32.2}{14} : \frac{12.26}{16} : \frac{27.17}{35.5} : \frac{32.63}{17}$

= 0.383 : 5.74 : 2.3 : 0.766 : 0.766 : 1.94

= 1 : 15 : 6 : 2 : 2 : 5

Thus, the complex contains six N-atoms out of which 5 are present in form of NH_3.

84. % of Se $= \frac{\text{Atomic wt of Se}}{\text{Min. molecular}}$ wt × 100

Min. molecular wt $= \frac{\text{Atomic wt of Se}}{\text{\% of Se}} \times 100$

$= \frac{78.4 \times 100}{0.5} = 1.568 \times 10^1$

85. C : H $= \frac{85}{12} : \frac{15}{1} = 1 : 2$

Empirical formula = CH_2

So empirical formula weight

= 12 + 2 × 1 = 14

Molecular wt = 2 × V. D. = 2 × 28 = 56

as $n = \frac{56}{14} = 4$

So, molecular formula

= n × empirical formula

= 4 × CH_2 = C_4H_8

86. Percentage of O = 100 – 49.3 – 6.84

= 43.86

C : H : O $= \frac{49.3}{12} : \frac{6.84}{1} : \frac{43.86}{16} = 3 : 5 : 4$

So, empirical formula = $C_3H_5O_4$

and weight = 73

molecular weight = 2 × V.D. = 2 × 73 = 146

So, molecular formula $= \frac{146}{73}(C_3H_5O_4) = C_6H_{10}O_8$

87. C : H = 6 : 1

C : O = 3 : 4

So, H : C : P = 1 : 6 : 8

% of H $= \frac{1}{15} \times 100 = \frac{100}{15}$

% of C $= \frac{6}{15} \times 100 = \frac{600}{15}$

% of O $= \frac{8}{15} \times 100 = \frac{800}{15}$

Atomic masses of H = 1, C = 12, O = 16

Element	%	Atomic ratio	Simplest ratio
H	$\frac{100}{15}=\frac{100}{15}$	1	2
C	$\frac{600}{15\times 12}=\frac{50}{15}$	11	1
O	$\frac{800}{15\times 16}=\frac{50}{15}$	16	1

Empirical formula = H_2CO

So, the compound is HCHO.

88. % of O = 100 – (52.2 + 13.04) = 34.76

$C:H:O=\frac{52.2}{12}:\frac{13.04}{1}:\frac{34.76}{16}=2:6:1$

Empirical formula = C_2H_6O

Empirical formula wt

$= 2 \times 12 + 6 \times 1 + 1 \times 16 = 46$

Molecular wt = 2 × V. D. = 2 × 23 = 46

Molecular formula = empirical formula = C_2H_6O

89. 1 mole of NH_3 (17 g) will be obtained from 1 mole of amide $=\frac{9.9}{1.7}\times 17 = 99\text{g}$

Mol wt of amide = 99

$\%\text{ of C}=\frac{12\times 4}{99}\times 100 = 48.48$

$\%\text{ of H}=\frac{5\times 1}{99}\times 100 = 5.05$

$\%\text{ of N}=\frac{14x}{99}\times 100=\frac{1400x}{99}$

$\%\text{ of O}=\frac{16y}{99}\times 100=\frac{1600y}{99}$

But % of O = 32.33 (given)

$\frac{1600y}{99}=32.33 \text{ or } y=\frac{32.53\times 99}{1600}=2$

Further % of N = 100 – (48.48 + 5.05 + 32.33)

But $\%\text{ N}=\frac{1400x}{99}$ (as calculated above)

$\frac{1400x}{99}=14.14$

$x=\frac{14.14\times 99}{1400}=1$

x : y, that is, N : O = 1: 2.

90. $C\%=\frac{12}{44}\times\frac{\text{wt of CO}_1}{\text{wt of OC}}\times 100$

$=\frac{12}{44}\times\frac{0.147}{0.2}\times 100 = 20$

$H\%=\frac{2}{18}\times\frac{\text{wt of H}_1\text{O}}{\text{wt of OC}}\times 100$

$=\frac{2}{18}\times\frac{0.12}{0.2}\times 100 = 6.7$

O% = 100 – (20 + 6.7) = 73.3

91. Ratio of masses = C : H : N = 9 : 1 : 3.5

Ratio of atom $=\frac{91}{12}:\frac{1}{1}:\frac{3.5}{14}$

$=1:\frac{1}{1}\times\frac{12}{9}:\frac{3.5}{14}\times\frac{12}{9}$

= 1 : 1.33 : 0.33 = 3 : 4 : 1

Empirical formula of the compound C_3H_4N

Empirical mass = 12 × 3 + 4 × 1 + 14 × 1

= 36 + 4 + 14 = 54

But molecular mass = 108

$n=\frac{\text{Mol mass}}{\text{Empirical formula mass}}=\frac{108}{54}=2$

So, molecular formula = $(C_3H_4N)_2 = C_6H_8N_2$

92. As B gives yellow precipitate with $AgNO_3/HNO_3$, B must be $C_6H_5CH_2I$ and hence, A is C_6H_5I.

94. As 25.32 ml 0.198 N NaOH ≡ 25.32 ml 0.198 NH_2SO_4

≡ 24.696 ml 0.2030 NH_2SO_4

≈ 24.7 ml 0.2030 NH_2SO_4

$N\%=\frac{1.4\times 0.2030\times 24.7}{1.325}=5.3\%$

95.

Element	%	Relative number of atom	Simplest ratio
C	20	1.67	1
H	6.67	6.67	4
N	46.67	3.33	2
O	26.66	1.67	1

The molecular formula is CH_4N_2O. So the compound is H_2NCONH_2.

$$2NH_2CONH_2 \xrightarrow{\Delta} \underset{\text{Biuret}}{NH_2CONHCONH_2} + NH_3\uparrow$$

Biuret gives violet colour with alkaline copper sulphate solution.

97. Here wt. Ratio of C:H:O = 6:1:8

Atomic Ratio of C:H:O $=\frac{6}{12}:\frac{1}{1}:\frac{8}{16}$

$=\frac{1}{2}:1:\frac{1}{2}$

= 1 : 2 : 1

Hence Emp. formula = CH_2O

This means m.f can be $C_3H_6O_3$.

Previous Years' Questions

1. An organic compound containing C, H and N gave the following results on analysis C = 40%, H = 13.33% and N = 46.67%. Its empirical formula would be
(a) CH_5N (b) CH_4N
(c) C_2H_7N (d) $C_2H_7N_2$
[2002]

2. In steam distillation of toluene, the pressure of toluene in vapour is
(a) Equal to pressure of barometer
(b) Less than pressure of barometer
(c) More than vapour pressure of toluene in simple distillation
(d) Equal to vapour pressure of toluene in simple distillation
[2001]

3. Camphor is often used in molecular mass determination because:
(a) It is readily available
(b) It has a very high cryoscopic constant
(c) It is volatile
(d) It is solvent for organic substances
[2004]

4. The best method for the separation of naphthalene and benzoic acid from their mixture is
(a) Chromatography (b) Crystallisation
(c) Sublimation (d) Distillation
[2005]

5. An organic compound contains carbon, hydrogen and oxygen. Its elemental analysis gave C, 38.71 % and H, 9.67 %. The empirical formula of the compound would be:
(a) CHO (b) CH_4O
(c) CH_3O (d) CH_2O
[2008]

6. In Dumas' method of estimation of nitrogen 0.35 g of an organic compound gave 55 mL of nitrogen collected at 300 K temperature and 715 mm pressure. The percentage composition of nitrogen in the compound would be (aqueous tension at 300 K = 15 mm)
(a) 15.45 (b) 17.45
(c) 12.45 (d) 16.45
[2011]

7. The Lassaigne's extract is boiled with conc. HNO_3 testing for halogens. By doing so, it
(a) Decomposes Na_2S and NaCN, if formed
(b) Increase the concentration of NO_3^- ions
(c) Increases the solubility product of AgCl
(d) Helps in the precipitation of AgCl
[2011]

8. In Duma's method for estimation of nitrogen 0.25g of an organic compound gave fumes of nitrogen collected at 300k temperature and 725mm pressure, if the aqueous tension at 300k is 25mm the percentage of nitrogen in the compound is?
(a) 15.76 (b) 16.76
(c) 17.36 (d) 18.20
[2015]

Answer Keys

1. (b) 2. (b) 3. (b) 4. (b) 5. (c) 6. (d) 7. (a) 8. (b)

Hints and Explanations

1.

Element	%	At. Wt.	Relative no. of moles	Molar ratio
C	40.00	12	40/12 = 3.33	3.33/3.33 = 1
H	13.33	1	13.33/1 = 6.60	13.33/3.33 = 4
N	46.67	14	46.67/14 = 3.33	3.33/3.33 = 1

So empirical formula of the compound is CH_4N.

2. As in steam distillation of toluene, the pressure of toluene in vapour is less than pressure of barometer.

3. Camphor is often used in molecular mass determination because of its high cryoscopic constant (37.7). This method is known as Rast's camphor method.

4. As both benzoic acid and naphthalene are solid and have sublime nature so crystallization is the best method for their separation.

5.

Element	% composition	Mole ratio	Simple ratio
C	38.71	38.71/12 = 3.22	3.22/3.22 = 1
H	9.67	9.67/1 = 9.67	9.67/3.22 = 3
O	51.62	51.62/16 = 3.22	3.22/3.22 = 1

Thus empirical formula of the compound is CH_3O.

6. $\frac{P_1V_1}{T_1} = \frac{P_2V_2}{T_2}$, on calculation

$V_2 = 46.09$ ml

$\% \text{ of N} = \frac{28 \times 46.09 \times 100}{22400 \times 0.35} = 16.45$

7. It helps in decomposing NaCN into $HCN_{(g)}$ and Na_2S into $H_2S_{(g)}$.

8. $P_{\text{dry gas}} = P_{\text{moist gas}} - P_{\text{Aq. Tension}}$

$= 725 - 25 = 700$ mm

$W_{N_2} = \frac{PVM}{RT} = \frac{700 \times 40 \times 28}{760 \times 1000 \times 300 \times 0.0821}$

$= 0.42$g

$\% \text{ N} = \frac{0.042}{0.25} \times 100$

$= 16.573\%$

NCERT Exemplar

1. The fragrance of flowers is due to the presence of some steam volatile organic compounds called essential oils. These are generally insoluble in water at room temperature but are miscible with water vapour in vapour phase. A suitable method for the extraction of these oils from the flowers is:
   (a) Distillation
   (b) Crystallisation
   (c) Distillation under reduced pressure
   (d) Steam distillation

2. During hearing of a court case, the judge suspected that some changes in the documents had been carried out. He asked the forensic department to check the ink used at two different places. According to you which technique can give the best results?
   (a) Column chromatography
   (b) Solvent extraction
   (c) Distillation
   (d) Thin layer chromatography

3. The principle involved in paper chromatography is:
   (a) Adsorption (b) Partition
   (c) Solubility (d) Volatility

Answer Keys

1. (d) 2. (d) 3. (b)

Hints and Explanations for Selective Questions

1. Essential oils are insoluble in water, also have high vapour pressure so it can be separated by steam distillation.

3. In paper chromatography separation of mixture components depends upon portioning between water at stationary phase and mobile phase.

AIIMS ESSENTIALS

Assertion and Reason

In the following questions, two statements Assertion (A) and Reason (R) are given. Mark

(a) If A and R both are correct and R is the correct explanation of A;

(b) If A and R both are correct and R is not the correct explanation of A;

(c) A is true but R is false;

(d) A is false but R is true,

(e) Both A and R are false.

1. (A) : Lithium can be used is place of sodium in Lassaigne's test.

(R) : It generally forms covalent compounds.

2. (A) : 1°, 2°, 3° Amine's can be distinguished by diethyl oxalate.

(R) : 1° amines form N- alkyl oxamide solid product, 2° amine form oxamic ester which is liquid, 3° amine do not react.

3. (A) : Halogen free amides also give green colouration with Beilstein's reagent

(R) : They form volatile copper cynadies.

4. (A) : Silver salt method is used to determine molecular mass of organic esters.

(R) : Silver salt of the organic acid on ignition gives silver metal, and thus from the weights of silver salt taken and silver formed, molecular mass of the organic acid can be determined.

5. (A) : Lassaigne's test is not shown by diazonium salts.

(R) : Diazonium salts lose N_2 on heating much before they have a chance to react with fused sodium metal.

6. (A) : Mixed melting point can be used to test the purity of an organic compound.

(R) : Impurities raise the melting point of an organic compound.

7. (A) : p- $C_6H_5SO_3H$ gives blood red colouration while performing Lssaigne's test for nitrogen.

(R) : Sodium fusion extract containing NaCNS gives blood red colour on treatment with $FeCl_3$.

8. (A) : During test for nitrogen with Lassaigne extract on adding $FeCl_3$ solution, sometimes a red precipitate is formed

(R) : This indicated that sulphur is also present

9. (A) : Glycerol can be purified by distillation under reduced pressure.

(R) : Liquid organic compounds are purified by distillation.

10. (A) : A mixture of p- methylbenzoic acid and picric acid can be separated by $NaHCO_3$ solution.

(R) : p- Methylbenzoic acid is soluble in $NaHCO_3$ and it gives effervescence of CO_2.

11. (A) : Duma's method is more applicable to nitrogen containing organic compounds than Kjedahl's method.

(R) : Kjeldahl's method does not give satisfactory results for compounds in which nitrogen is linked to oxygen.

12. (A) : Thiophence present in commercial benzene as an impurity can be removed by shaking the mixture with cold conc.H_2SO_4

(R) : Thiophence is a heterocyclic aromatic compound.

13. (A) : During digestion with conc. H_2SO_4, nitrogen of the compound is converted to $(NH_4)_2SO_4$.

(R) : $(NH_4)_2SO_4$ on heating with alkali gives NH_3 gas.

14. (A) : Hydrazine contains nitrogen but does not gives Lassaigne's test for nitrogen.

(R) : Hydrazine reacts with fused sodium to give H_2 gas.

15. (A) : Sodium fusion extract of a compound gives black precipitate with lead acetate.

(R) : Sulphur containing compounds form Na_2S in sodium fusion extract.

Answer Keys

1. (d)	2. (a)	3. (a)	4. (d)	5. (a)	6. (c)	7. (a)	8. (a)	9. (b)	10. (d)
11. (b)	12. (b)	13. (b)	14. (b)	15. (b)					

CHAPTER 12

Hydrocarbons and Petroleum

Chapter Outline

■ Alkanes ■ Alkenes or Olefins ■ Alkynes ■ Benzene ■ Toluene Or Methyl Benzene ■ Petroleum

Hydrocarbons: These are compounds having only C, and H-atoms. Natural gas, Petroleum and coal are their major sources.

Alkanes

$$-\overset{|}{\underset{|}{C}}\overset{1.54\ Å}{—}\overset{|}{\underset{|}{C}}-$$
$$sp^3 \quad sp^3$$

- The general formula of alkanes is C_nH_{2n+2}.
- Their main sources are petroleum, natural gas and coal.
- Alkanes are also known as paraffins due to their reduced activity or almost inert nature.
- They have a C–C bond length is 1.54 Å and that of C–H bond is 1.11 Å.
- Alkanes exhibit a C–C bond energy of 80 kcal per mol and that of C–H bond is calculated to be 97 kcal per mol.
- They are non-polar in nature.
- Alkanes show chain, position and conformational isomerism.

  e.g., [skeletal structures]

- Liquefied Petroleum Gas (LPG) is composed of propane, butane and isobutane.

Methods of Preparation of Alkanes

By decarboxylation of sodium salts of faty acids

When R–COOH heated with Sodalime (4 : 1 Mixture of CaO + NaOH) alkane is formed. Here alkane has one C–atom less than R–COOH as one CO_2 molecule is climinated.

$$R\text{–}COOH \xrightarrow[-H_2O]{NaOH} R\text{–}COONa \xrightarrow[NaOH + CaO]{\Delta} R\text{–}H + Na_2CO_3$$

R–may be → CH_3, C_2H_5, C_6H_5 etc.

In case of HCOONa: →

$$HCOONa + NaOH \xrightarrow[CaO]{\Delta} H_2 + Na_2CO_3$$

Here CaO is used to make the reaction more mild and to keep NaOH dry. It also helps in easy fusion.

By Sabatier and Senderen reaction

- Here, the catalysts used are Raney Ni [An alloy of Ni–Al] and Pd–Pt (Adam's catalyst).
- The temperature range suited for the reaction is 200–300°C.
- In case of Raney, Ni reaction is posible at room temperature also.

$$-C\equiv C- \xrightarrow{H_2/Ni} \underset{H}{>}C=C\underset{H}{<} \xrightarrow[Ni]{H_2,\ \Delta} \overset{H}{\underset{H}{>}}C-C\overset{H}{\underset{H}{<}}$$

For example, $CH_3\text{–}C\equiv CH \xrightarrow[Ni]{H_2} CH_3 - CH{=}CH_2 \xrightarrow[Ni]{H_2} CH_3\text{–}CH_2\text{–}CH_3$

[skeletal alkene] $\xrightarrow{H_2/N_i}$ [skeletal alkane]

[skeletal alkene] $\xrightarrow{H_2/N_i}$ [skeletal alkane]

From the reduction of R–X

- Here, the reducing agent employed is usually Zn + NaOH or HCl or Zn–Cu/acid, Na/alcohol. $LiAlH_4$, $NaBH_4$, $(Ph)_3$ SnH etc

$$R–X + H–H \xrightarrow{Zn/HCl} R–H + HX$$

Mechanism

$$Zn \rightarrow Zn^{+2} + 2e^-$$

$$R – X + e^- \rightarrow R^\circ + X^-$$

$$R^\circ + e^- \rightarrow R:^-$$

$$R:^- + C_2H_5OH \rightarrow R – H + C_2H_5O^-$$

$$R–I + HI \xrightarrow{\Delta, 150°C} R–H + I_2 \uparrow$$

$$R–X \xrightarrow[H_2]{LiAlH_4 \text{ or } (Ph)_3SnH} R–H + HX$$

FACTS TO REMEMBER

Order of Reduction is RI > R – Br > R – Cl > R – F
$LiAlH_4$ is effective in case of P° and S° halides while $(Ph)_3$ SnH is effective far all types of halides.

$$(CH_3)_3C–Br \xrightarrow[Ether]{LiAlH_4} (CH_3)_2C=CH_2$$

2-Methyl propene

By Wurtz reaction

- In Wurtz reaction, an alkyl halide is treated with sodium metal in the presence of dry ether to give alkanes.

$$R–X + 2Na + X–R \xrightarrow{\text{Dry ether}} R–R + 2NaX$$

- If the 'R' group attached is same, then only one alkane is formed.

e.g., $2CH_3Cl \xrightarrow[Ether]{2Na} C_2H_6 + 2NaCl$

- If the 'R' group attached is different, then it leads to the formation of three alkanes.

$$R–X + 2Na + X–R' \xrightarrow[-2NaX]{\text{Dry ether}} R–R + R'–R' + R–R'$$

For example,

Cl + Cl → (Dry either) + +

The mechanism involved in the reaction is

$$R–X + Na \longrightarrow \dot{R} + NaX$$

$$\dot{R} + \dot{R} \longrightarrow R–R$$

Here disproportion is also possible

$$C_2H_5^\circ + C_2H_5^\circ \longrightarrow C_2H_4 + C_2H_6$$

In case of t°–halide Reaction is not possible as major product is according to dehydrohalogenation which means the reaction holds only far a P° or S° halide.

In place of Na here Activated Cu, Ag, Fe can also be used.

It is good far symmetrical alkanes (C_2H_6 C_4H_{10}) and CH_4 can not be obtained here.

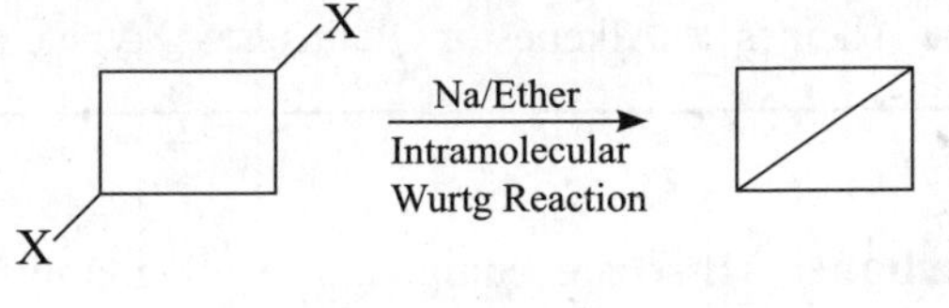

Cl Cl Cl Cl $\xrightarrow[Na/Ether]{\Delta}$

Spiro [2, 2] pentane

By Frankland reaction

- In Frankland reaction, an alkyl halide is treated with zinc dust in a closed tube to prepare alkanes.

$$R–X + Zn + X–R \xrightarrow{\Delta} R–R + ZnX_2$$

- Rest of the process is same as depicted above in Wurtz reaction.

Corey–house synthesis: This method is suitable for the preparation of alkanes possessing odd number of carbon atoms also (un-symetrical alkanes).

$$R–X \xrightarrow[Ether \; - LiX]{Li} R–Li \xrightarrow{CuI} R_2CuLi \xrightarrow{R'–X} R–R' + R–Cu + LiX$$

This process can be illustrated in the following example:

$\xrightarrow[CuI]{BrLi}$ $[\ ]_2$ CuLi

C_2H_5I CH_3I

2 methyl butane

Here R–X must be p° for better yield.

Kolbe synthesis

$$2R\text{–}COONa \xrightarrow[\text{At anode}]{\text{Electrolytic hydrolysis}} R\text{–}R + 2CO_2 + 2NaOH + H_2\uparrow$$

Mechanism of Kolbe's electrolysis

$$2R\text{–}COO\,Na \longrightarrow 2R\text{–}COO^- + 2Na^+$$

At Anode: $2R\text{–}COO^- - 2e^- \longrightarrow 2R\text{–}COO^\circ$

$$2R\text{–}COO^\circ \longrightarrow 2R^\circ + 2CO_2$$

$$R^\circ + R^\circ \longrightarrow R\text{–}R$$

At Cathode: $2H^+ + 2e^- \longrightarrow H_2\uparrow$

e.g. $\underset{\text{Sodiom acetate}}{2CH_3\text{–}COONa} \longrightarrow CH_3\text{–}CH_3 + 2CO_2 + 2NaOH + H_2$

- The pH increases during the reaction due to the formation of strong bases. (NaOH or KOH)
- If two different acid salts are used, a mixture of three alkanes will be formed.
- If we take a mixture of sodium acetate and sodium propanoate then ethane, propane, butane will be formed.

$$CH_3\text{–}COONa + C_2H_5COONa \xrightarrow{\text{Electro}} C_2H_6 + C_3H_8 + C_8H_{10} + 2CO_2 + 2NaOH + H_2$$

- This process involves both, ionic as well as free radical mechanism.
- Methane cannot be prepared by this method.

By decomposition of grignard reagent

- Any compound possessing an active hydrogen atom will show this reaction with Grignard reagent.

$R\text{–}MgX + H_2O \longrightarrow R\text{–}H + Mg.X.OH$

or H_2NR or

or HOR NH_2 etc

or $HC \equiv CR$

or H – NH – R

$$H\text{—}O\text{—}\overset{\overset{\displaystyle C}{\|}}{C}\text{—}R \text{ etc}$$

Birch reduction: In this process, terminal alkenes are reduced to alkanes.

$$R\text{–}CH{=}CH_2 + H_2 \xrightarrow[\text{MeOH}]{\text{Na/Liq.}NH_3} R\text{–}CH_2\text{–}CH_3$$

By the reduction of R–OH, R–CHO, R–CO–R, R –COOH

- Here, the reagent used is HI and red phosphorus at 150°C–200°C.

(1) $R-OH + 2HI \xrightarrow{\Delta} R-H + H_2O + I_2$

(2) $R-CH=O + 4HI \xrightarrow[-H_2O,-2I_2]{\Delta} R-CH_3$

(3) $\begin{matrix}R\\R\end{matrix}\!\!>\!C{=}O + 4HI \xrightarrow[-H_2O,\,2I_2]{\Delta} \begin{matrix}R\\R\end{matrix}\!\!>\!CH_2$

(4) $R\text{–}\overset{\overset{\displaystyle O}{\|}}{C}\text{–}OH + 6HI \xrightarrow[-2H_2O,-3I_2]{\Delta} R\text{–}CH_3$

By the hydrolysis of methanides

$$Be_2C + 4H\cdot OH \rightarrow CH_4 + 2Be(OH)_2$$

$$Al_4C_3 + 12H_2O \rightarrow 3CH_4 + 4Al(OH)_3$$

Wolf–Kishner reduction: Here, carbonyl compounds are reduced into alkanes in the presence of hydrazine and a base. The bases used are sodamide, C_2H_5ONa.

$$>C{=}O \xrightarrow[-H_2O]{NH_2.NH_2} >C{=}N.NH_2 \xrightarrow[-N_2]{\text{Base}} >CH_2$$

For example,

(1) $CH_3CHO \xrightarrow[-N_2]{\text{Base}} CH_3\text{—}CH=NNH_2 \xrightarrow[-N_2]{\text{Base}} CH_3\text{–}CH_3$

(2) 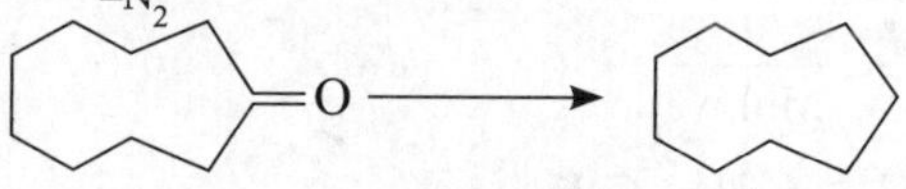 Cyclononanone ⟶ Cyclononane

(3) 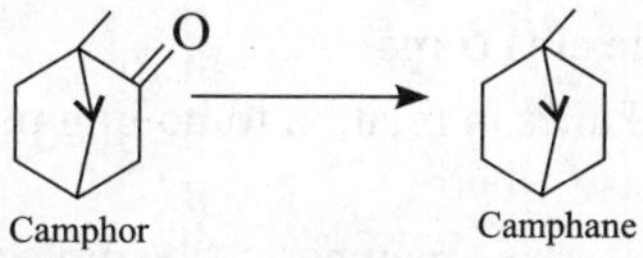 Camphor ⟶ Camphane

Clemmensen reduction: Here, carbonyl compounds are reduced into alkanes by Zn–Hg/HCl as shown below:

$$>C{=}O + 4\,[H] \xrightarrow{\text{Zn–Hg/HCl}} >CH_2$$

For example,

(1) $CH_3\text{–}CHO \xrightarrow[\text{Zn.Hg/HCl}]{4\,[H]} CH_3\text{–}CH_3$

(2) $\begin{matrix}CH_3\\CH_3\end{matrix}\!\!>\!C{=}O \xrightarrow[\text{Zn. Hg/HCl}]{4\,[H]} \begin{matrix}CH_3\\CH_3\end{matrix}\!\!>\!CH_2$

(3) 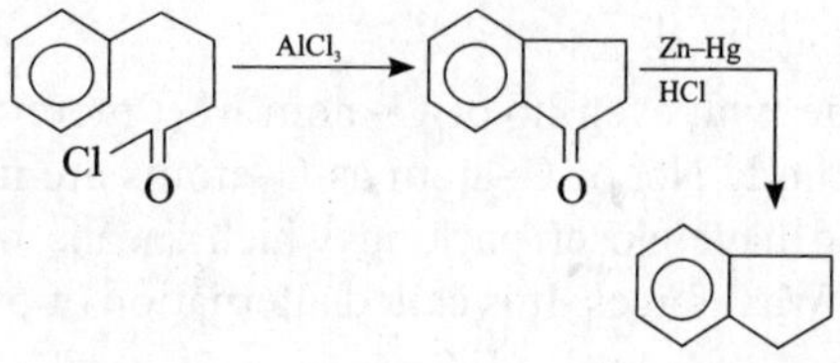 (acid chloride) $\xrightarrow{AlCl_3}$ (indanone) $\xrightarrow[\text{HCl}]{\text{Zn–Hg}}$ Indane

(4) [Reaction scheme: benzene-substituted 4-oxo acid $\xrightarrow{(1)\ Zn\text{–}Hg,\ (2)\ SOCl_2}$ acid chloride $\xrightarrow[-HCl]{AlCl_3}$ tetralone $\xrightarrow[-H_2O]{Zn\text{–}Hg}$ tetralin $\xrightarrow{\Delta,\ Se}$ Naphthalene]

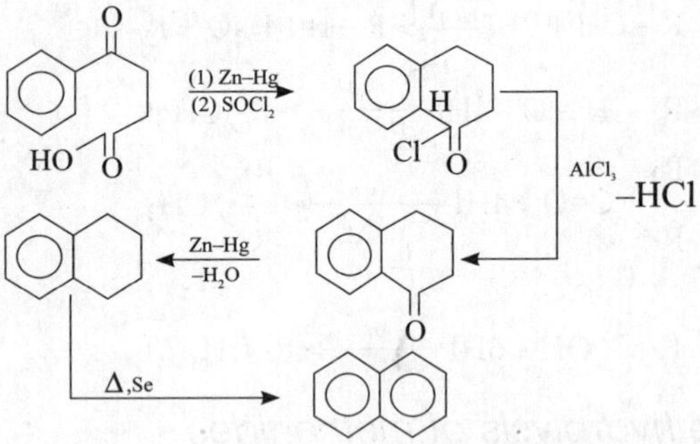

Naphthalene

Physical Properties of Alkanes

- Physical state
  $C_1 - C_4$——colourless gases
  $C_5 - C_{17}$——colourless liquids
  C_{18} and above are colourless waxy solids
- Density ∝ Molecular weight [$CH_4 < C_2H_6 <$]
- All alkanes are lighter than water and their limiting density is 0.8g /cm^3.
- Cyclo alkanes flot on water as they are lighter than water
- Alkanes are insoluble in polar solvents but soluble in non-polar solvents like CCl_4, ether etc.
- Solubility $\propto \dfrac{1}{\text{Mol. wt}}$

  $CH_4 > C_2H_6 > C_3H_8 > C_4H_{10} >$
- Boiling point ∝ Surface area α mol.wt.
  The trend shown by alkanes in relation to boiling points is as shown: normal > iso > neo
  when, for each successive member, the molecular formula differs by a (–CH_2–) group and subsequently the boiling point differs by 20°C – 30°C.

  $CH_4 < C_2H_6 < C_3H_8 <$ [isobutane]

  [n-butane] < [Neopentane] < [Isopentane] < [n-pentane]

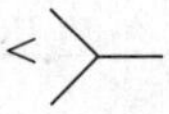

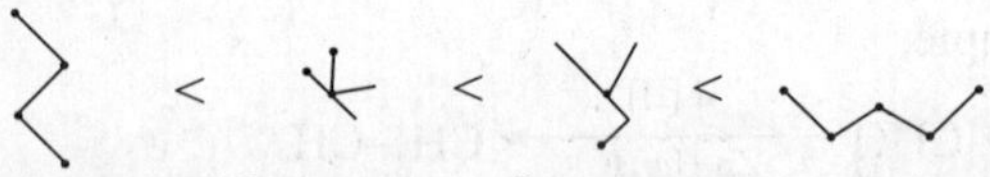

  n-butane Neopentane Isopentane n-pentane
- Melting point of even number alkane > Melting point of odd number alkane
  that is, $C_{15} < C_{16} > C_{17}$

In case of alkane with even no.of C–atom M.P more than next alkane with odd No. of C–atom as C–atoms are more symmetrical and have closer packing which means more stranger Vander Waal forces. It is called alternation or oscillation effect.

Chemical Properties of Alkanes

Alkanes are inert towards acids, bases, oxidizing agents etc. due to non-polar nature of C–C and C–H bonds and its inability to donate or accept electrons.

Halogenation: Here H–atoms of alkanes are substituted by X–atom ion.

- Reactivity of halogens is $F_2 > Cl_2 > Br_2 > I_2$
- Replacement of H-atom is easy and follows the trend: t > s > p > methyl
- It occurs in hυ, dark, and at high temperatures in the range of 1250°C–4000°C
- Presence of O_2 suppresses the reaction.
- It is initiated by dibenzoyl peroxide, $Pb(Et)_4$.
  For example,
  $$CH_4 + Cl_2 \xrightarrow{h\nu} \underset{\text{Final product}}{CCl_4}$$
  $$C_2H_6 + Br_2 \xrightarrow{h\nu} \underset{\text{Final product}}{C_2H_5Br} \xrightarrow{Br_2} C_2Br_6$$
  $$2R\text{–}X + HgF_2 \longrightarrow 2R\text{–}F + HgX_2$$
  (X = Br, I)

Nitration

$$R\text{–}H + HNO_3 \xrightarrow[\text{High temp. } 400°C]{\Delta} R.NO_2 + H_2O$$

- This reaction is used for hexane, heptane.
- In case of lower alkanes, only vapour phase nitration takes place as follows:

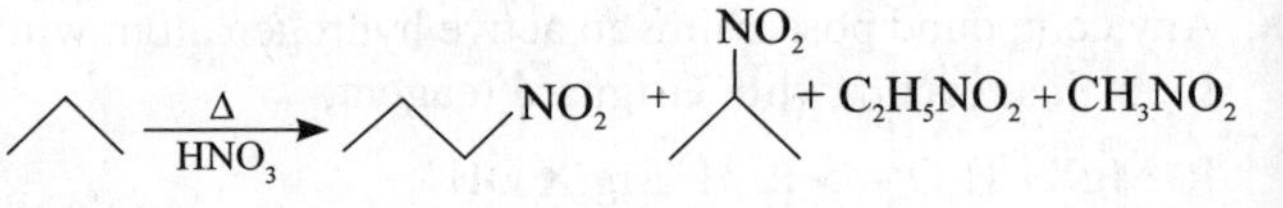

[propane] $\xrightarrow[HNO_3]{\Delta}$ [1-nitropropane]NO_2 + [2-nitropropane, NO_2] $+ C_2H_5NO_2 + CH_3NO_2$

Sulphonation

$$R\text{–}OH + H_2SO_4 \xrightarrow[\text{Prolonged heating}]{\Delta SO_3} R\text{–}SO_3H + H_2O$$

- This reaction is given by alkanes having minimum 6 carbon atoms like hexane, heptane.
- Lower members react with SO_3 to form sulphonic acids.
  $$R\text{–}H + SO_3 \xrightarrow[\text{Prolong}]{\Delta} R\text{–}SO_3H$$
  ↳CH_3, C_2H_5 etc.

Oxidation reactions

(a) Combustion or complete oxidation

$$(C_nH_{2n+2}) + \frac{(3n+1)}{2} O_2 \longrightarrow nCO_2 + (n+1) H_2O$$

e.g., $C_2H_6 + \frac{7}{2} O_2 \xrightarrow{\Delta} 2CO_2 + 3H_2O + 1538\ KJ$

(b) Incomplete combustion

$$CH_4 + O_2 \xrightarrow{\text{Burn}} \underset{\text{Carbon black}}{C} + 2H_2O$$

- Carbon black is used in printing ink.

$$2CH_4 + 3O_2 \xrightarrow{\text{Burn}} 2CO + 4H_2O$$

(c) Catalytic oxidation

$$CH_4 + [O] \xrightarrow[\text{100 atm 200°C}]{\text{Cu–tube}} CH_3OH$$

$$CH_3\text{–}(CH_2)_n\text{–}CH_3 \xrightarrow[\text{100°C –160°C}]{O_2} (CH_3)\text{–}(CH_2)_n\text{–}COOH$$

Mn-Stearate

$$2CH_4 + 3O_2 \xrightarrow{(CH_3COO)_2Mn} 2HCOOH + 2H_2O$$

$$2CH_3 - CH_3 + 3O_2 \xrightarrow{(CH_3COO)_2Mn} 2CH_3COOH + 2H_2O$$

(d) Chemically controlled oxidation is shown by alkanes having tertiary carbon atom.

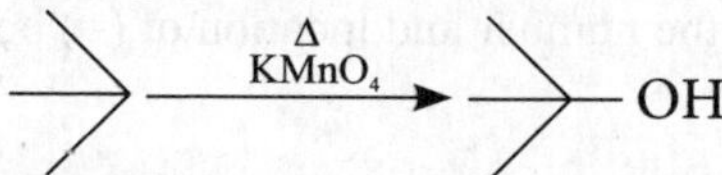

Isobutane Isobutyl alcohal

Pyrolysis or cracking: Here, higher alkane splits into lower alkane, alkene and hydrogen when heated strongly at high temperatures in the absence of air.

$$R\text{–}H \xrightarrow[\text{Absence of } O_2]{\Delta} \underset{\text{Lower alkane}}{R\text{–}H} + \underset{\text{Alkene}}{R{=}R} + H_2\uparrow$$

It occurs by free radical mechanism.
For example,

$$CH_4 \xrightarrow{1000°C} C + 2H_2$$

$$C_2H_6 \xrightarrow[Cr_2O_3 + Al_2O_3]{500°C} CH_4 + CH_2{=}CH_2 + H_2$$

$$CH_3\text{–}CH_2\text{–}CH_3 \xrightarrow[\text{770–1000 K}]{\Delta} CH_3\text{–}CH = CH_2 + CH_2 = CH_2 + C_2H_6 + CH_4 + H_2$$

- During pyrolysis C–C bonds break instead of C–H bonds, as bond energy of C–H > C–C.

Isomerization: In isomerization, higher alkanes are heated with aluminium chloride at high temperatures to convert them into their respective isomers.

$$\text{Normal chain isomer} \xrightarrow[\text{Anhy.AlCl}_3\text{+HBr, 200°C}]{\text{Anhy.AlCl}_3\text{+HCl}} \text{Iso isomer}$$

e.g.,

(butane) $\xrightarrow{\text{Anhy. AlCl}_3 + \text{HCl}}$ Iso-butane

e.g., n-heptane $\xrightarrow{\text{AlCl}_3 + \text{HCl}}$ 2, 2, 3 tri methyl butane

2-methyl pentane $\xrightarrow{\text{Only AlCl}_3}$ 2, 3-dimethyl butane

It occurs by 1, 2 hydride or methyl shift.

Aromatization

$$\text{Aliphatic alkanes} \xrightarrow[\substack{\text{10-20 atm 200°C} \\ \text{600°C} \\ \text{catalyst}}]{\Delta} \text{Aromatic alkanes}$$

For example,

$$\underset{\text{Normal hexane}}{CH_3\text{–}(CH_2)_4\text{–}CH_3} \xrightarrow[Cr_2O_3\text{–}Al_2O_3]{\Delta} C_6H_6 + 4H_2\uparrow$$

$$\underset{\text{Normal heptane}}{CH_3\text{–}(CH_2)_5\text{–}CH_3} \xrightarrow[Cr_2O_3\text{–}Al_2O_3]{\Delta} C_6H_5CH_3 + 4H_2$$

Alkylation: In this reaction, the H-atom of an alkyl group is substituted by an alkyl group.

e.g., $(CH_3)_3C\text{–}H + CH_2{=}CH\text{–}CH_3 \xrightarrow[\text{Con. } H_2SO_4]{\Delta} (CH_3)_3C\text{–}CH_2\text{–}CH_2\text{–}CH_3$

2, 2 Di methyl pentane

It involves Carbonation formation and it is useful in increasing the octane number of gasoline etc.

Specific reaction

(1) $R(H)CH_2 + 2[O] \xrightarrow[\text{270°C, 100 atm}]{\text{MoO}} R(H)C(OH)_2 \xrightarrow{-H_2O} RCHO$

(2) $6CH_4 + 2O_2 \xrightarrow{1500°C} 2CH{\equiv}CH + 2CO_2 + 10H_2\uparrow$

- $\underset{\text{Cellulose}}{(C_6H_{10}O_5)_n} + nH_2O \longrightarrow 3n - CH_4 + 3nCO_2$

- Chlorosulphonation (Reed Reaction)

$$2\,(\text{propane}) + 2SO_2 + 2Cl_2 \xrightarrow[\text{uv}]{50°C} (\text{propyl})SO_2Cl + (\text{isopropyl})SO_2Cl + 2HCl$$

FACTS TO REMEMBER

Cracking, Isomerination and Aromatization are widely used to increase the octane rating of gasoline.
Nitration, Sulphonation, Pyrolysis all occur with free Radical Mechanism.

ALKENES OR OLEFINS

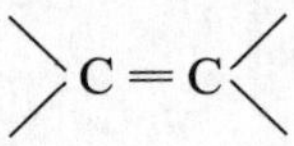

>C=C<

- Alkenes are unsaturated hydrocarbons having >C=C< with a general formula C_nH_{2n} .
- The word olefin means oil making. Lower members of this group give oily products on halogenation.
- Here, unsaturated carbon atoms are sp^2 hybridized with a trigonal planar geometry.
- The C=C bond length is 1.34 Å and the bond energy is 143.1 kcal per mol.
- Here C–H bond length is 1.1 Å and its bond energy is 98.7 kcal per mol.
- Alkenes exhibit chain, position, ring chain, functional, geometrical and optical isomerisms.

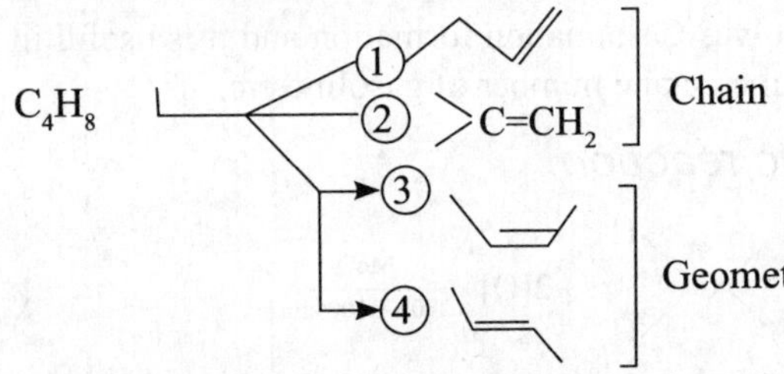

C_4H_8 — (1), (2) >C=CH_2 : Chain; (3), (4): Geometrical

Here 3,or 4 w.r.t 1 are position isomers and w.r.t (2) these are chain isomers.

All these four isomers are Ring Chain and functional isomers of [cyclobutane] and [methylcyclopropane, CH_3]

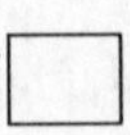

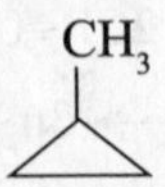

- Stability α $\frac{1}{\text{Heat of hydrogenation}}$

symmetrical alkene > unsymmetrical alkene

trans alkene > cis alkene

Stability in decreasing order:

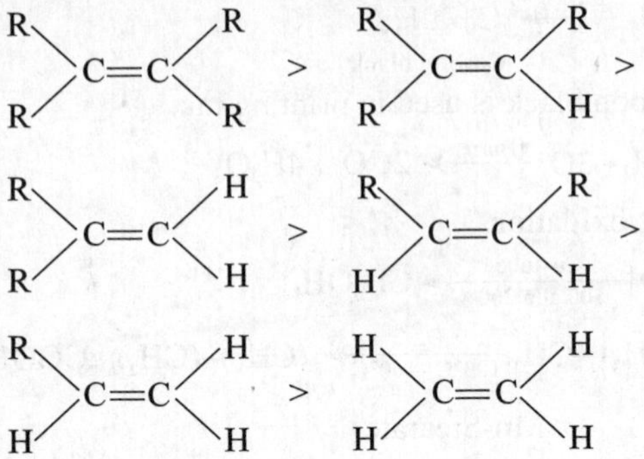

$R_2C=CR_2$ > $R_2C=CRH$ > $R_2C=CH_2$ > RHC=CHR (trans) > RHC=CH_2 > $H_2C=CH_2$

Test of Unsaturation

1. Alkenes decolourize Br_2 in CCl_4 or in water.
2. They also decolourize 1% alkaline $KMnO_4$ (Baeyer's Reagent).
3. Ozonolysis gives the number and location of (=) bonds in alkene.

Methods of Preparation of Alkenes

By α, β elimination reaction

By dehydrohalogenation of haloalkanes: Here, the reagents used are $C_2H_5O^-$, alc. KOH, $NaNH_2$, KNH_2, $(Me)_3COK$

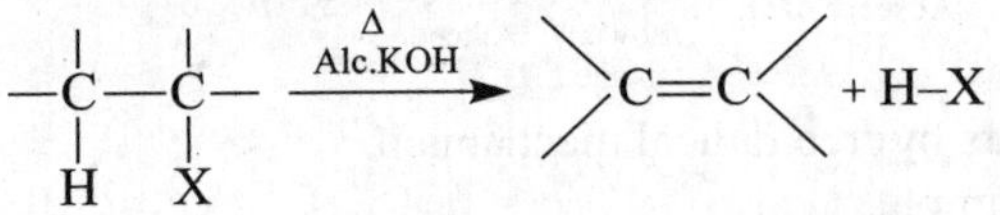

$$-\underset{H}{C}-\underset{X}{C}- \xrightarrow[\text{Alc.KOH}]{\Delta} >C=C< + H\text{–}X$$

For example, (1) 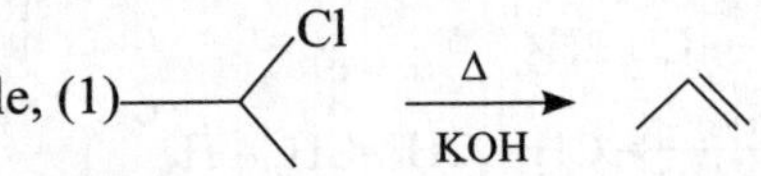 2-chloropropane (Cl) $\xrightarrow[\text{KOH}]{\Delta}$ propene

(2) chlorocyclohexane (Cl) $\xrightarrow[\text{KOH}]{\Delta}$ cyclohexene

(3) 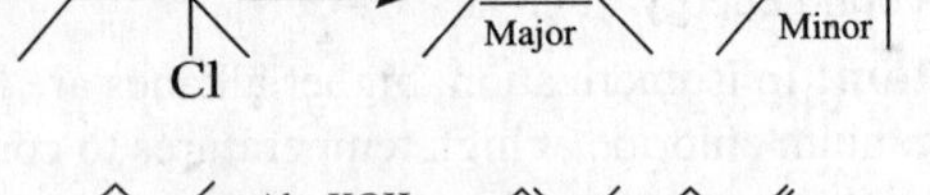 (Cl) ⟶ Major + Minor

(4) 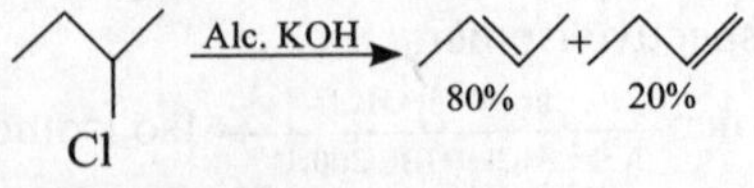 (Cl) $\xrightarrow{\text{Alc. KOH}}$ 80% + 20%

(5) 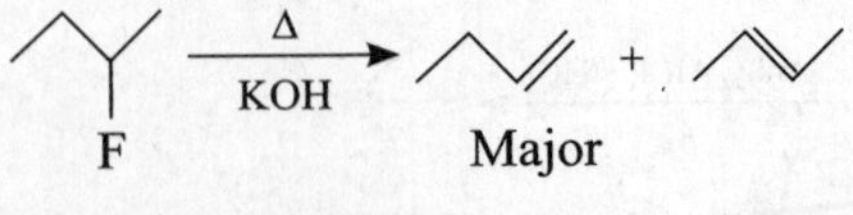 (F) $\xrightarrow[\text{KOH}]{\Delta}$ Major +

(6) 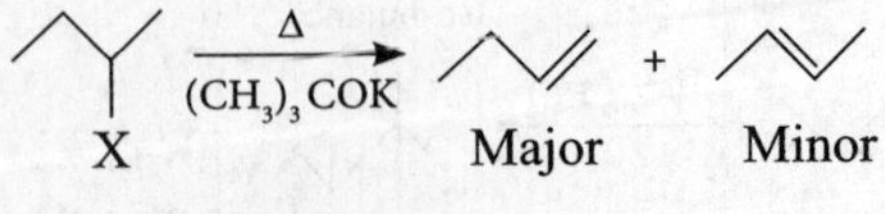 (X) $\xrightarrow[(CH_3)_3COK]{\Delta}$ Major + Minor

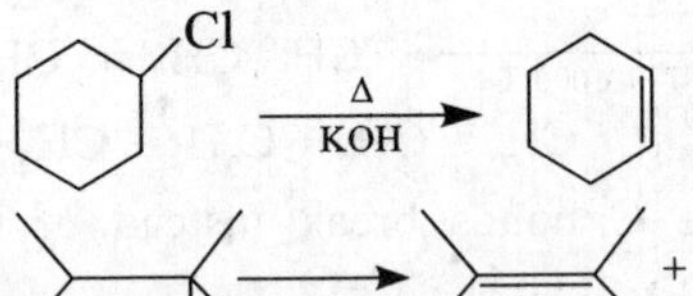

Major Product in eq. No. V, VI are according to Hoffmann Rule.

By dehydration of alcohol: Here dehydrating agents are H_2SO_4 at 170°C, H_3PO_4 at 200°C, P_2O_5, $ZnCl_2$ at 350°C, BF_3, dry HCl, $KHSO_4$.

$$\gt C(H)-C(OH)\lt \xrightarrow{\Delta} \gt C=C\lt + H_2O$$

For example,

(1) $CH_3CH_2OH \xrightarrow[170°C]{\Delta,\text{Conc. } H_2SO_4} CH_2=CH_2+H_2O$

(2) $CH_3-CH(H)-CH(OH)-CH_3 + H_2SO_4 \xrightarrow[\Delta]{170°C} CH_3-CH=CH-CH_3 + H_2SO_4 + H_2O$

(3) $CH_3-C(OH)(CH_3)-CH_3 + H_2SO_4 \xrightarrow[170°C]{\Delta} CH_3-C(CH_3)=CH_2 + H_2SO_4 + H_2O$

Isobutlene (major product)

(4) $CH_3-CH(CH_3)-CH_2OH \xrightarrow[H^+,\ -H_2O]{\Delta}$

$CH_3-CH(CH_3)-\overset{\oplus}{C}H_2\ (p^0) \xrightarrow[\text{shift}]{1,2\ \text{Hydride}} CH_3-\overset{\oplus}{C}(CH_3)-CH_3 \xrightarrow[-H^+]{\text{fast step}} (CH_3)_2C=CH_2$

(5) $CH_3-C(CH_3)_2-CH_2OH \xrightarrow[H^+,\ -H_2O]{\Delta}$

$CH_3-C(CH_3)_2-\overset{\oplus}{C}H_2 \xrightarrow{1,2\ \text{Methyl Shift}} CH_3-\overset{\oplus}{C}(CH_3)-CH(H)-CH_3 \xrightarrow[\text{fast}]{-H^+} CH_3-C(CH_3)=CH-CH_3$

2-Methyl but 2-ene

From dehalogenation of dihalogen derivative

From geminal dihalides When geminal dihalides are heated with zinc dust and alcohols alkenes are formed.

$$R-CHX_2 + 2Zn + X_2CH-R \xrightarrow[-2ZnX_2]{\Delta} R-CH=CH-R$$

For example,

$$CH_3-CHCl_2 + 2Zn + Cl_2CH-CH_3 \xrightarrow[-ZnCl_2]{\Delta} CH_3-CH=CH-CH_3$$

But 2-ene

- If we take two different types of geminal dihalides, we obtain three different types of alkenes in this reaction.

$$CH_2Cl_2 + CH_3-CHCl_2 \xrightarrow[2Zn]{\Delta} CH_2=CH_2 + CH_2=CH-CH_3 + CH_3-CH=CH-CH_3 + Zn\,Br_2$$

From vicinal dihalides When vicinal dihalides are heated with Zn dust, alkene having same number of carbon atoms is obtained.

$$R-CH(X)-CH(X) + \text{Zn dust} \xrightarrow{\Delta,\ 300°C} R-CH=CH-H + ZnX_2$$

For example,

(1) $H-CH(Cl)-CH(Cl)-H + \text{Zn dust} \xrightarrow{\Delta,\ 300°C} H-CH=CH-H + ZnCl_2$

(2) 1, 2-dibromopropane + Zn $\xrightarrow{\Delta}$ Propene

(3) 1,2-dibromocyclohexane $\xrightarrow[Zn]{\Delta}$ cyclohexene + $ZnBr_2$

By Kolbe's electrolytic reaction

$$\text{Potassium succinate} \xrightarrow{2H_2O} C_2H_4 + 2CO_2 + 2KOH + H_2$$

($C_2H_4 + 2CO_2$: Anode; H_2: Cathode)

By partial reduction of alkynes

$$R\text{–}C \equiv CH + H_2 \xrightarrow[\text{catalyst}]{\text{Lindlar}} R\text{–}CH\text{=}CH_2$$

- Lindlar catalyst is Pd-$BaSO_4$ or $CaCO_3$ poisoned with sulphur or quinoline. It is used to prevent excessive reduction. It is Syn addition and gives cis alkene while with Na or Li/NH_3 addition is Anti.

$$R\text{—}C\equiv C\text{—}R \xrightarrow[\text{Pd/BaSO}_4]{H_2} \text{(R)(H)C=C(R)(H)}$$
Cis-alkene

Birch Red^n — Na, liq. NH_3 ↓

(R)(H)C=C(H)(R)
trans-alkene

From Grignard reagent

$$R\text{–}MgX + CH_2\text{=}CH\text{–}X' \longrightarrow R\text{–}CH\text{=}CH_2 + Mg\begin{smallmatrix}X\\X'\end{smallmatrix}$$

From other organometallic compound

$$CH_2\text{=}CH\text{–}Cl + CuR_2 \xrightarrow{\text{Alc.KOH}} R\text{–}CH\text{=}CH_2 + R\text{–}Cu\text{–}Cl$$

Here, R may be CH_3, C_6H_6 etc.

By heating tetraalkyl ammonium halide or hydroxide

$$(C_2H_5)_4\,N.OH \xrightarrow{\Delta} C_2H_4 + H_2O + (C_2H_5)_3N$$

From esters by pyrolysis: When esters are heated in presence of liquid N_2 and glass wool, the alkyl part of ester converts into the respective alkene, while alkanoate part of the ester converts into respective acid.

$$CH_3\text{—}CO\text{—}O\text{—}CH_2\text{—}CH_2\text{—}H \xrightarrow[\text{Liquid } N_2]{\text{Glass wool, 450°C}} CH_3\text{–}COOH + CH_2\text{=}CH_2$$

Wittig reaction: Here, methylene triphenyl phosphorane or phosphorus ylide is treated with a carbonyl compound to prepare an alkene.

$$>C = O + (Ph)_3\,P = C< \longrightarrow >C = C< + (Ph)_3PO$$

For example,

(1) $CH_3\text{–}CHO + (Ph)_3\,P\text{=}CH_2 \longrightarrow CH_3\text{–}CH\text{=}CH_2 + (Ph)_3PO$
Acetaldehyde → Propene

(2) $CH_3\text{–}CO\text{–}CH_3 + (Ph)_3\,P\text{=}CH_2 \longrightarrow CH_3\text{–}C(CH_3)\text{=}CH_2 + (Ph)_3PO$
Acetone → Isobutene

Physical Properties of Alkenes

- Alkenes are colourless, odourless and lighter than water.
- Alkenes are insoluble in water but soluble in organic solvents.
- Physical state

C_1 to C_4	colourless gases
C_5 to C_{16}	colourless liquids
C_{17} and higher	colourless waxy solids

The B.P of an alkene is more than that of a corresponding alkenc due to more polarizability of π electrons.

- Cis alkenes have higher values of boiling point, dipole moment, reactivity, heat of hydrogenation and refractive index than the corresponding trans alkenes.
- Trans alkenes have higher melting point than their corresponding cis alkenes. This is due to compact packing of their crystal lattice.

Chemical Properties of Alkenes

Addition reactions of alkenes: As alkenes have π-bonds so they are quite reactive, as πe⁻ are easily available when every required.

It means C = C acts as a source of electrons on which electrophiles can easily attack.

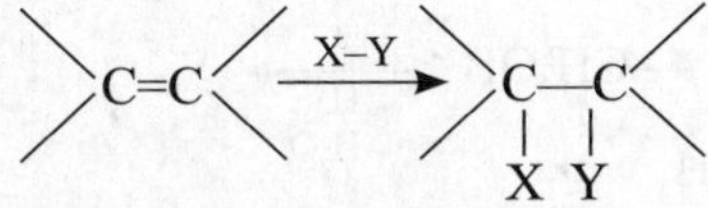

Alkenes show electrophilic addition reaction mainly with HX, H_2, X_2, H–OH, H–OSO_3H.

Reaction with H_2

- This reaction occurs at the surface of catalysts like Ni, Pd or Pt in a cissyn addition manner.

$$>C=C< \xrightarrow[150°C]{Ni/H_2} \text{—}\overset{|}{\underset{H}{C}}\text{—}\overset{|}{\underset{H}{C}}\text{—}$$

- Reducing agents like Wilkinson catalyst $[RhCl(Ph)_3P]$ and $NaBH_4$, $PtCl_2$ can also be used here.
- Birch reagent Na/liquid NH_3 with alcohol can also be used in case of terminal alkenes.

CH_3—C—H ‖ CH_3—C—H (Cis but 2-ene) $\xrightarrow[\text{Syn add}]{D_2/Ni}$ 'Meso' (CH_3, H—C—D, H—C—D, CH_3)

$\xrightarrow[\text{liq } NH_3,\ \text{Anti}]{D_2/Na}$ (CH_3, H—C—D, D—C—H, CH_3) + (CH_3, D—C—H, H—C—D, CH_3)

Racemic Mixture

In case of trans but 2–ene, nature of products is opposite to cis but 2–ene.

Reaction with X_2:

It involves trans addition (Stereospecific)

>C=C< + Br_2 $\xrightarrow[\text{CCl}_4 \text{ in dark}]{\text{Room temperature}}$ —C(Br)—C(Br)—

Here colour of Br_2/H_2O disappears.

C is but 2 – ene $\xrightarrow[\text{Anti}]{Br_2}$ (Me, H—C—Br, Br—C—H, Me) + (Me, Br—C—H, H—C—Br, Me)

Racemic Mixture of 2, 3 di bromo butane

trans-but 2-ene $\xrightarrow[\text{Anti}]{Br_2}$ (Me, H—C—Br, H—C—Br, Me)

Meso 2, 3 di-bromo butane

Here reactivity order for X_2 is $Cl_2 > Br_2 > I_2$.

Reaction with H_2O

$$>C=C< + H_2O \xrightarrow[\text{or } H_2SO_4]{\Delta,\ \text{acid } HNO_3} >C(H)—C(OH)<$$

For example

(1) $CH_3 - CH = CH_2 \xrightarrow{H_2O} CH_3 - CH(OH) - CH_3$

(2) Octahydronaphthalene $\xrightarrow[H^+]{H_2O}$ decalol (OH at ring junction)

2-Methylbut-2-ene $\xrightarrow[H^+]{H_2O}$ 2-Methyl butan 2-ol

In all these addition occur according to Markwonikoff's Rule (Regioselective by nature)

Reaction with R–OH

$$R–CH = CH_2 + R'–OH \xrightarrow{H_2SO_4} R–CH(OR')–CH_3$$

Propene $\xrightarrow{CH_3OH}$ $(CH_3)_2CH–OCH_3$

2-Methoxy propane

Methylcyclohexene $\xrightarrow{CH_3OH}$ 1-methoxy-1-methylcyclohexane ($-OCH_3$)

Addition of O_2 to alkenes

$C_2H_4 + 1/2\ O_2 \xrightarrow[400°C]{Ag}$ Oxirane $\xrightarrow{H_2O}$ $CH_2(OH)–CH_2(OH)$

$C_2H_4 \xrightarrow{C_6H_5COOOH}$ Epoxyethane or oxirane $\xrightarrow{H_2O}$ $CH_2(OH)–CH_2(OH)$

Propene $\xrightarrow{C_6H_5COOOH}$ 1, 2 Epoxy Propane

Reaction with H_2SO_4

$$>C=C< + H_2SO_4 \longrightarrow >\underset{H}{C}-\underset{OSO_3H}{C}< \xrightarrow[H_2O]{-H_2SO_4} -\underset{H}{C}-\underset{OH}{C}-$$

For example,

$$CH_3-CH=CH_2 \xrightarrow[\text{M. Rule}]{HO.SO_3H} CH_3-\underset{OSO_3H}{CH}-CH_3 \xrightarrow{H.OH} CH_3-\underset{OH}{CH}-CH_3$$

Addition With HX

$$R-CH=CH_2 \xrightarrow[\text{M. R}]{H-X} R-\underset{X}{CH}-CH_3$$

Reactivity order for HX is HI > HBr > HCl > HF.

$$CH_3-CH_2=CH_2 \xrightarrow{HCl} CH_3-\underset{Cl}{CH}-CH_3$$

$CH_3-CH_2=CH_2$ with HCl / Peroxide gives the same product, $CH_3-\underset{Cl}{CH}-CH_3$ (use only Markwonikoff's Rule)

$$CH_3-CH=CH_2 \xrightarrow[\text{M.R}]{HBr} CH_3-\underset{Br}{CH}-CH_3$$

$$CH_3-CH=CH_2 \xrightarrow[\text{Peroxide}]{HBr} CH_3-CH_2-CH_2Br$$

Anti M.R

No Anti M.R is followed for HF, HCl and HI.
[See in G.O.C II]

Addition with hypohalous acid (HXO or $\overset{+}{X}\cdot\overset{-}{OH}$)

In addition reaction with HXO ($X_2 + H_2O$), alkenes give holohydrins according to Markwonikoff's Rule and here attacking reagent is X^+ (Halonium ion).

$$R-CH=CH_2 \xrightarrow{HOCl} R-\underset{OH}{CH}-CH_2Cl$$

R may be $-CH_3$, $-C_2H_5$, $-C_6H_5$ etc.

⇒ With NOCl addition occurs as follows

$$R-CH=CH_2 \xrightarrow{\overset{+}{N}O\overset{-}{Cl}} R-\underset{Cl}{CH}-CH_2NO$$

Oxidation or hydroxylation

- Here, hydroxylation occurs by using 1% alkaline $KMnO_4$ (Baeyer's reagent), OsO_4, pyridine or $NaHSO_3/H_2O$
- On hydroxylation, all of these reagents give syn addition.

For example,

$$>C=C< \xrightarrow[\text{alk.}KMnO_4]{\text{1\% cold}} >\underset{OH}{C}-\underset{OH}{C}<$$

- With OsO_4, pyridine, $NaHSO_3/H_2O$

$$>C=C< + OsO_4 \xrightarrow{\text{Pyridine}} \text{cyclic osmate ester} \xrightarrow{H_2O \mid NaHSO_3} -\underset{OH}{C}-\underset{OH}{C}- + OsO_2$$

Cis 1, 2 di–O1s

- Anti-hydroxylation is not common and is as follows:

$$>C=C< \xrightarrow{\text{Peroxy acid}} >\underset{OH}{C}-\overset{OH}{C}<$$

trans-1, 2-diols

e.g., cyclohexene $\xrightarrow[KMnO_4]{\text{1\% alk}}$ cis-cyclohexane-1,2-diol

cyclohexene $\xrightarrow{RCO_3H}$ trans-cyclohexane-1,2-diol or (HO, OH)

- **Stereochemistry of hydroxylation**

cis alkene + cis mode reagent $\xrightarrow{\text{1\% alk. }KMnO_4 \text{ or } OsO_4 / NaHSO_3}$ meso product

trans alkene + cis mode reagent ⟶ racemic mixture

cis alkene + trans mode reagent $\xrightarrow{C_6H_5-\overset{O}{\overset{\|}{C}}-O-OH \text{ or } H-COOOH}$ racemic mixture

trans alkene + trans mode reagent ⟶ meso product

Combustion:

$$C_nH_{2n} + 3n/2\ O_2 \xrightarrow{\Delta} nCO_2 + nH_2O$$

$$C_2H_4 + 3O_2 \xrightarrow{\Delta} 2CO_2 + 2H_2O$$

Substitution reaction: Alkenes can undergo substitution only at allylic position and at a high temperature of 450°C–500°C.

$$\text{Propene} \xrightarrow[\text{Cl}_2, 500°C]{\Delta} CH_2—CH=CH_2 \text{ (with Cl on C-1)}$$

3-Chloropropene

Cyclohexene + CH_2-CO / CH_2-CO >N — Br (N-bromosuccinimide) ⟶ 3-Bromocyclohexene + Succinimide

N-bromosuccinimide

Succinimide

Isomerization

$$\text{But-1-ene} \xrightarrow[300°C]{\Delta,\ AlCl_3} \text{But-2-ene} + \text{Isobutene}$$

Polymerization: The most common catalyst used for polymerization of olefins is Zeigler natta catalyst [$TiCl_4 + (Et)_3Al$].

$$nH_2C=CH_2 \xrightarrow[\text{High temperature}]{\Delta} [-CH_2-CH_2-]n$$

Polythene

$$n\phi-CH=CH_2 \xrightarrow{\text{High pressure}} [\phi-CH-CH_2-]_n$$

Poly styrene (rubber)

where $\varphi = C_6H_5$ or (phenyl ring)

$$n\ CF_2=CF_2 \Longrightarrow [-CF_2-CF_2-]_n$$

Tetra fluoro ethylene — Teflon (PTFE)

Oxymercuration–demercuration: It involves synthesis of alcohols or ethers from alkenes in accordance with Markovnikov's rule as follows:

$$\underset{\text{Alkene}}{R-\overset{H}{C}=\overset{H}{C}-H} \xrightarrow[\text{(ii) } NaBH_4, OH^-]{\text{(i) } Hg(OAC)_2/THF\ H_2O} \underset{\text{Alcohol}}{R-\underset{OH}{CH}-CH_3}$$

For example,

(1)
$$\underset{\text{2 methylbut-1-ene}}{CH_3-CH_2-\overset{CH_3}{C}=CH_2} \xrightarrow[THF_1-H_2O]{Hg(OAc)_2} CH_3CH_2-\overset{CH_3}{\underset{OH}{C}}-\underset{HgOAc}{CH_2}$$

$$\xrightarrow[OH]{NaBH_4} CH_3-CH_2-\overset{CH_3}{\underset{OH}{C}}-CH_3 + Hg$$

2-methyl butan-2-01

(2)
$$\underset{\text{Alkene}}{R-\overset{H}{C}=\overset{H}{C}-H} \xrightarrow[\text{(2) } NaBH_4, OH^-]{\text{(1) } Hg(OAC)_2/THF,\ R\ OH} \underset{\text{Ether}}{R-\underset{OR}{CH}-CH_3}$$

(3)
$$\underset{\text{Propene}}{CH_3-\overset{H}{C}=\overset{H}{C}-H} \xrightarrow[\text{(2) } NaBH_4, OH^-]{\text{(1) } Hg(OAC)_2/THF, CH_3OH} \underset{\text{2-methoxy propane}}{CH_3-\underset{OCH_3}{CH}-CH_3}$$

FACTS TO REMEMBER

Here no rearrangement occurs and oxidation number of Hg becomes zero.

Ozonolysis

- It is an important reaction to find the number of double bonds present in alkenes and their location.

$$>C=C< + O_3 \xrightarrow[\text{Inert solvent, low temp.}]{CCl_4} \underset{\text{Ozonide}}{>C(-O-)(O-O)C<} \xrightarrow[-H_2O_2]{Zn/H_2O} 2\ \underset{\text{Carbonyl compound}}{>C=O}$$

$$\underset{\text{Ethene}}{H_2C=CH_2} + O_3 \xrightarrow{CCl_4} H_2C(-O-)(O-O)CH_2 \xrightarrow[-H_2O_2]{Zn/H_2O} 2\ \underset{\text{Formaldehyde}}{H_2C=O}$$

- If zinc is not used, carbonyl compounds get oxidized into acids.

For example,

$$\underset{\text{Ethene}}{H_2C=CH_2} + O_3 \xrightarrow{CCl_4} H_2C(-O-)(O-O)CH_2 \xrightarrow{-H_2O_2} \underset{\text{Formic acid}}{2HCOOH}$$

$$CH_3-CH=CH_2 \xrightarrow[\text{(2) } H_2O/Zn]{\text{(1) } O_3/CCl_4} CH_3CHO + CH_2O$$

$$CH_3-CH=CH_2 \xrightarrow[H_2O]{O_3/CCl_4} CH_3COOH + HCOOH$$

$$CH_2{=}CH{-}CH{=}CH_2 \xrightarrow[H_2O/Zn]{O_3/CCl_4} \underset{CHO}{\underset{|}{CHO}} + 2HCHO$$

$$\text{Cyclohexa-1,3-diene} \xrightarrow[H_2O/Zn]{O_3/CCl_4} 2\ CHO{-}CH_2{-}CHO$$

In case alkene is to be decided by the ozonolysis products do as follows

$$>C{=}\boxed{O + O}{=}C{-}C{=}\boxed{O + O}{=}C< \longrightarrow$$

$$>C{=}C{-}C{=}C<$$

e.g., An alkene on ozonolysis gives one molecule each of HCHO CH_3CHO and CHO–CHO the alkene is

$$CH_3{-}CH{=}\boxed{O + O}{=}CH{-}CH{=}\boxed{O + O}{=}CH_2$$

$$\downarrow$$

$$CH_3{-}CH{=}CH{-}CH{=}CH_2$$

Penta 1, 3 diene

FACTS TO REMEMBER

With Polyhalogen compounds like CHX_3, CCl_4 addition occurs according to Anti Markwonikoff's rule

$$R{-}CH{=}CH_2 \xrightarrow{CHCl_3} R{-}CH_2{-}CH_2.CCl_3$$

$$R{-}CH{=}CH_2 \xrightarrow{CCl_4} R{-}\underset{Cl}{\underset{|}{CH}}{-}CH_2.CCl_3$$

Hydroboration oxidation It is used to prepare alcohols in accordance with anti-Markovnikov's rule.

$$\underset{\text{Alkene}}{R{-}CH{=}CH_2} \xrightarrow[H_2O_2{-}NaOH]{B_2H_6} R{-}CH_2{-}CH_2{-}OH$$

Mechanism

$$R{-}C{\equiv}CH + H{-}\overset{H}{\overset{|}{B}}{-}H \longrightarrow R{-}\overset{H}{\overset{|}{C}}{=}CH{-}B\begin{matrix}H\\H\end{matrix}$$

$$\downarrow$$

$$R{-}CH{=}\overset{OH}{\overset{|}{CH}} \xleftarrow[H_2O_2]{NaOH} [R{-}CH{=}CH]_3B$$

$$\downarrow \text{Tautomerisation}$$

$$R{-}CH_2{-}CHO$$

This is shown in the following examples:

$$\underset{\text{Ethene}}{H_2C{=}CH_2} \xrightarrow[H_2O_2{-}NaOH]{B_2H_6} \underset{\text{Ethanol}}{H_3C{-}CH_2{-}OH}$$

For example,

$$\underset{\text{Propene}}{H_3C{-}CH{=}CH_2} \xrightarrow[H_2O_2{-}NaOH]{B_2H_6} \underset{\text{Propan-1-ol}}{H_3C{-}CH_2{-}CH_2{-}OH}$$

Oxidative cleavage by Hot alk. KM_nO_4

Alkenes on oxidative cleavage by hot alk. KM nO_4 give Acid, ketones etc.

$$\begin{matrix}H\\R\end{matrix}>C{=}C<\begin{matrix}H\\R'\end{matrix} \xrightarrow[\text{alk }KMnO_4]{\text{Hot}} R{-}COOH + R'{-}COOH$$

$$\begin{matrix}R\\R\end{matrix}>C{=}C<\begin{matrix}H\\R'\end{matrix} \xrightarrow[\text{alk }KMnO_4]{\text{Hot}} R{-}COOH + \begin{matrix}R\\R\end{matrix}>C{=}O$$

$$R{-}CH{=}CH_2 \xrightarrow[\text{alk }KMnO_4]{\text{Hot}} R{-}COOH + HCOOH$$

$$HCOOH \xrightarrow{(O)} H_2O + CO_2$$

HCOOH further decomposes or oxidises into CO_2 and water.

FACTS TO REMEMBER

Ozonide can also be reduced by using $(CH_3)_2$ S as follows:

$$\text{Ozonide} \xrightarrow{(CH_3)_2S} >C{=}O + >C{=}O + \begin{matrix}CH_3\\CH_3\end{matrix}>S{=}O$$

Dimethyl sulphoxide (D.M.S.O)

Alkylation of alkenes

$$\begin{matrix}CH_3\\CH_3\end{matrix}>C{=}CH_2 + H{-}\overset{CH_3}{\overset{|}{C}}{<}\begin{matrix}CH_3\\CH_3\end{matrix} \xrightarrow[H_2SO_4]{H{-}F \text{ or}}$$

$$\underset{\text{Isooctane}}{CH_3{-}\overset{CH_3}{\overset{|}{CH}}{-}CH_2{-}\overset{CH_3}{\overset{|}{\underset{CH_3}{\underset{|}{C}}}}{-}CH_3}$$

$$\begin{matrix}CH_2\\\|\\CH_2\end{matrix} + CH.(CH_3)_3 \xrightarrow{HF} \underset{\text{Major}}{CH_3{-}\overset{CH_3}{\overset{|}{CH}}{-}\underset{CH_3}{\underset{|}{CH}}{-}CH_3} + \underset{\text{Minor}}{CH_3{-}CH_2{-}C(CH_3)_3}$$

These reactions occur with C^+ formation and 1,2, shift of H – or $-CH_3$ group.

ALKYNES

sp sp
—C≡C—

- Alkynes have a general formula of $[C_nH_{2n-2}]$.
  For example,

CH≡CH	Ethyne
CH_3–C≡CH	Propyne
CH_3–C≡C–CH_3	Butyne-2.

Hexa 1, 4 di-yne

- Alkynes show chain, position, functional isomerism (with alkadienes).

  and (Position Isomers)

- All terminal alkynes are weakly acidic in nature and the acidic nature decreases as follows:
  R–C≡H > R–CH=CH_2 > R–CH_3
- CH≡CH > CH_3–C≡CH > CH_3–CH_2–C≡CH > .....
- C_2H_2 is a linear compound, which is poisonous and has garlic like smell due to the presence of impurities of AsH_3, PH_3, H_2S.
- C_2H_2 acts as a dibasic acid with very strong bases like NH_2^-, CH_3^-.
- Alkynes are less reactive to electrophilic addition as greater activation energy is needed due to formation of a less stable cyclic reaction intermediate species.

C=C (bridged by El^+)

Test of Terminal Alkynes

- Terminal alkynes give a white precipitate of R–C≡ CAg with ammonical a silver nitrate, $AgNO_3$.
- They give a red precipitate of R–C≡C.Cu with ammonical cuprous chloride.

Methods of Preparation of Alkynes

From carbides: Carbides on hydrolysis give alkynes as follows

$$CaO + 3C \xrightarrow[-CO]{1800\text{–}2100°C} CaC_2 \xrightarrow{2H_2O} C_2H_2 + Ca(OH)_2$$

$$Mg_2C_3 \xrightarrow{4H_2O} CH_3\text{–}C{\equiv}CH + 2Mg(OH)_2$$

Berthelot's synthesis

$$2C + H_2 \xrightarrow[\text{Arc } 1200°C]{\text{Electric}} CH{\equiv}CH$$

From CH_4

$$2C_2H_4 \xrightarrow[1200°C]{\text{Elec arc}} C_2H_2 + 3H_2$$

$$6CH_4 + O_2 \xrightarrow{1500°C} 2C_2H_2 + 2CO + 10H_2$$

By heating CHX_3 with silver powder

$$CHX_3 + 6Ag + X_3CH \xrightarrow{\Delta} HC{\equiv}CH + 6AgX$$

Kolbe's method: Sodium or potassium maleate or fumerate on electrolysis of its aqueous solution gives acetylene at the anode. Here the solution is basic and PH > 7.

$$\underset{\text{Sodium maleate or fumerate}}{CHCOONa \| CHCOONa} \xrightarrow{\text{Electric}} \underset{\text{Anode}}{C_2H_2 + 2CO_2} + \underset{\text{(Cathode)}}{2KOH + H_2}$$

By alkylation of acetylene or terminal alkynes

$$R\text{–}C{\equiv}CH \xrightarrow{NaNH_2} R\text{–}C{\equiv}CNa \xrightarrow{R'X} R\text{–}C{\equiv}C\text{–}R' + NaX$$

For example,

(1) C.Na≡CH + (1-bromopropane) $\xrightarrow{}$ Butyne

(2) Ethyne $\xrightarrow[\text{Br}]{NaNH_2}$ Butyne $\xrightarrow[\text{Br}]{NaNH_2}$ Hex-3-yne

$$CH_3\text{—}C{\equiv}CH \xrightarrow[-NH_3]{NaNH_2} CH_3\text{—}C{\equiv}\overset{-}{C}\cdot\overset{+}{Na} \xrightarrow{CH_3Cl} CH_3\text{—}C{\equiv}C\text{—}CH_3 \xrightarrow[-NaCl]{C_2H_5Cl} CH_3\text{—}C{\equiv}C\text{—}CH_2\text{—}CH_3$$

Pent 2-Yne

By elimination reaction: Here, dihalides undergo α, β-elimination reaction with $NaNH_2$ or alcoholic KOH to give alkynes.

$$R\text{–}CH_2\text{–}CX_2\text{–}R' \xrightarrow{2Na.NH_2} R\text{–}C{\equiv}C\text{–}R' + 2NaX + 2NH_3$$

$$R-\underset{H}{\overset{X}{C}}-\underset{H}{\overset{X}{C}}-R' \xrightarrow{2Na.NH_2} R-C\equiv C-R' + 2NaX + 2NH_3$$

For example,

$$\underset{CH_2Br}{\overset{CH_2Br}{|}} \xrightarrow[-2KBr,\ -2H_2O]{\Delta,\ 2\,KOH} CH\equiv CH$$

$$CH_3-CH_2-CHCl_2 \xrightarrow[2\,KOH]{\Delta} CH_3-C\equiv CH + 2KCl + 2H_2O$$

1-Butene $\xrightarrow{Br_2}$ 1,2-dibromobutane $\longrightarrow$ Butyne

From tetrahalogen compounds

$$R-\underset{H}{\overset{X}{C}}-\underset{H}{\overset{X}{C}}-H + 2Zn \xrightarrow{\Delta\ alc.} R-C\equiv CH + ZnX_2$$

For example,

$$CHBr_2-CHBr_2 + 2Zn \xrightarrow[CH_3OH]{\Delta} CH\equiv CH + 2Zn\,Br_2$$

$$CH_3-\overset{Cl}{\overset{|}{C}}Cl-CHCl_2 \xrightarrow[2\,Zn]{\Delta} CH_3-C\equiv CH + 2ZnCl_2$$

Physical Properties

- Alkynes are lighter than water.
- Alkynes having more than 3 carbon atoms are liquid while those with more than 14 carbon atoms are solids.
- Acetylene is a colourless gas with garlic like smell and with a boiling point of –84°C. It is poisonous and lighter than air.
- Alkynes have higher BP, MP than corresponding alkanes, alkenes due to more close packed arrangement (Linear structure) and more stronger interaction.

Chemical Properties

Hydrogenation

$$R-C\equiv C-R \xrightarrow[H_2]{Pt,\ Pd,\ Ni,\ etc.} R-CH=\overset{H}{\overset{|}{C}}-R \xrightarrow{H_2} R-CH_2-CH_2-R$$

$$CH\equiv CH \xrightarrow[Ni]{H_2} CH_2=CH_2 \xrightarrow[Ni]{H_2} CH_3-CH_3$$

$$CH_3-C\equiv CH \xrightarrow[Ni]{H_2} CH_3-CH=CH_2 \xrightarrow[Ni]{H_2} CH_3-CH_2-CH_3$$

- Lindlar catalyst ($PdCO_3/BaCO_3$ poisoned with quinoline or lead acetate) is used for selective hydrogenation to prepare alkene only.

1-ethynylcyclohexanol (C≡CH, OH) $\xrightarrow[\text{Lead acetate/quinoline}]{H_2Pd/CaCO_3}$ (CH = CH_2, OH) 1-vinyl cyclo hexanol

Stereochemistry of the reaction

$$R-C\equiv C-R + H_2 \xrightarrow[\text{syn addition}]{\text{Lindlar catalyst}} \text{(R)(H)C=C(R)(H) cis}$$

$$C_6H_5-C\equiv C-Me \xrightarrow{\text{Lindlar catalyst}} \text{(}C_6H_5\text{)(H)C=C(Me)(H)}$$

- Nickel boride (P-2) also gives syn addition.
- In case of Na/NH_3, hydrogenation is anti-addition. For example,

$$C_6H_5-C\equiv C-Me \xrightarrow{Na/NH_3} \text{(}C_6H_5\text{)(H)C=C(H)(Me)}$$

Reaction with halogen

$$R-C\equiv C-R \xrightarrow[\text{equiv. }CCl_4]{X_2\text{ one molar}} \text{(X)(R)C=C(R)(X)} \xrightarrow{X_2} R-\underset{X}{\overset{X}{C}}-\underset{X}{\overset{X}{C}}-R$$

- Here, addition is anti and reactivity order for X_2 is $Cl_2 > Br_2 > I_2$

For example,

$$HOOC-C\equiv C-COOH \xrightarrow{Br_2} \text{(HOOC)(Br)C=C(Br)(COOH)}$$

- $CH \equiv CH \xrightarrow{Cl_2} CHCl = CHCl$

$\xrightarrow{Cl_2} \underset{\text{Westrone}}{CHCl_2 - CHCl_2} \xrightarrow{BaCl_2} \underset{\text{Westrosol}}{CCl_2 = CHCl}$

- Both westrone and westrosol are good industrial solvents for oil, fat and waxes.

Addition of HX to alkynes: Addition occurs according to Markonikov's rule and anti-Markovnikov's rule.

$$-C{\equiv}C- \xrightarrow{HX} (H)C{=}C(X) \xrightarrow{HX} \underset{\text{Gemdi halide}}{(H)(H)C{-}C(X)(X)}$$

Reactivity order is HI > HBr > HCl

For example, $CH_3{-}C \equiv CH \xrightarrow{HCl} CH_3{-}CCl_2{-}CH_3$

$$C_4H_9 - C \equiv C - H \xrightarrow{HBr} C_4H_9 - C(Br) = CH_2 \xrightarrow{HBr} C_4H_9{-}C(Br)(Br){-}CH_3$$

- If we use CH_3COBr/alumina, the rate of reaction increases.

$$C_5H_{11}{-}C{\equiv}CH \xrightarrow{HBr} \underset{82\%}{C_6H_{11}{-}C(Br){=}CH_2}$$

$$CH{\equiv}CH \xrightarrow{HCl} \underset{\text{Vinyl chloride}}{CH_2{=}CH{-}Cl} \xrightarrow{\text{Polymerization}} \underset{\text{Poly vinyl chloride (PVC)}}{\left(-CH_2 - CH(Cl) - \right)_n}$$

Reaction with HCN

$$CH{\equiv}CH \xrightarrow[HCN]{Ba(CN)_2} \underset{\text{Vinyl cyanide}}{CH_2{=}CH{-}CN} \xrightarrow{\text{Polymerization}} \left(-CH_2 - CH(CN) - \right)_n$$

Orlon or acrylonitrile (synthetic fiber) or poly vinyl cyanide

Reaction with HOX:

Addition occurs according to Markonikov's rule and here X^+ is attacking reagent.

$$R - C \equiv CH \xrightarrow{\overset{-}{HO}\overset{+}{X}} R - C(OH) = CH{-}X \xrightarrow{\overset{-}{HO}\overset{+}{X}} R - C(OH)_2 - CHX_2 \xrightarrow{-H_2O} R - C(=O) - CHX_2$$

For example,

$$CH \equiv CH \xrightarrow{HOCl} CHOH = CHCl \xrightarrow{HOCl} CH(OH)_2 - CHCl_2 \xrightarrow{-H_2O} \underset{\text{Dichloroacetaldehyde}}{CHO - CHCl_2}$$

Hydration of alkynes: Alkynes give Carbonyl Compounds on hydration by 1% $HgSO_4$ and di/H_2SO_4 as follows.

$$R{-}C{\equiv}C{-}R \xrightarrow[\text{Slow}]{Hg^{2+}} R{-}CH{=}C(OH){-}R \xrightarrow{\text{Tautomerization}} R{-}CH_2{-}C(=O){-}R$$

It is called kucherov reaction.

This is shown in following examples:

(1) alkyne $\xrightarrow[H^+]{1\%\ Hg^{2+}}$ enol (OH) $\xrightarrow[\text{or } CH_3CHO]{\text{Tautomerisation}}$ carbonyl (O)

(2) propyne $\rightarrow$ enol (OH) $\rightarrow$ ketone (O)

(3) $C_6H_5{-}C \equiv C{-}CH_3 \xrightarrow[H^+]{Hg^{+2}} C_6H_5{-}C(=O){-}CH_2{-}CH_3$

Reaction with alcohol

$$CH \equiv CH \xrightarrow[HgSO_4]{C_2H_5OH} \underset{\text{Ethyl vinyl ether}}{CH_2 = CH - OC_2H_5} \xrightarrow{H_2O} CH_2 = CH_2 + C_2H_5OH$$

Reaction with acetic acid

$$CH \equiv CH \xrightarrow[HgSO_4]{CH_3COOH} \underset{\text{Vinylacetate}}{CH_2{=}CH - OOCCH_3}$$

$$\xrightarrow{\text{Polymerization}} \underset{\text{Poly vinyl acetate (adhesive)}}{\left(-CH_2 - CH(OOC\ CH_3) - \right)_n}$$

Reaction with $AsCl_3$

$$CH \equiv CH \xrightarrow{AsCl_3} \underset{\text{Lewisite or Lewisgas (wargas) (β-chlorovinyl dichloroarsine)}}{CHCl = CHAsCl_2}$$

- Lewisite (β-chloro vinyl dichloro arsine) is a poisonous gas which on inhalation causes instant death.Its antidote is British anti-Lewisite (BAL).

$CH_2{-}SH$ | $CH{-}SH$ | $CH_2{-}OH$ 3–Hydroxy propan 1, 2di–thiol

Hydroboration: It is used to prepare ketones from alkynes except ethyne which gives acetaldehyde.

$$3R - C \equiv C - CH_3 \xrightarrow[THF]{BH_3} \left[RCH = \overset{|}{C} - CH_3\right]_3 B \xrightarrow{OH^-/H_2O_2} RCH = \underset{OH}{\underset{|}{C}} - CH_3 \xrightarrow{\text{Rearrangement}} RCH_2 - CO - CH_3$$

$$3H - C \equiv C - H \xrightarrow[THF]{BH_3} \left[CH_2 = \overset{|}{C} - H\right]_3 B \xrightarrow{OH^-/H_2O_2} CH_2 = \underset{OH}{\underset{|}{C}} - H \xrightarrow{\text{Rearrangement}} CH_3 - CHO$$

Isomerization

Butyne-1 $\xrightarrow{\text{alc. KOH}}$ Butyne-2

But-2-yne $\xrightarrow{NaNH_2}$ (terminal alkyne) C.Na $\xrightarrow[-NaOH]{H.OH}$ But-1-yne

Substitution

$$HC \equiv CH \xrightarrow[-2NaOH]{2NaOCl} C.Cl \equiv C.Cl$$

$$R - C \equiv CH \xrightarrow[NaOH]{NaOCl} R - C \equiv C - Cl$$

Combustion

$$C_nH_{(2n-2)} + \frac{3(n-1)}{2} O_2 \xrightarrow{\Delta} nCO_2 + (n-1) H_2O$$

For example,

$$C_2H_2 + \frac{5}{2} O_2 \xrightarrow{\Delta} 2CO_2 + H_2O$$

$$C_3H_4 + 4O_2 \xrightarrow{\Delta} 3CO_2 + 2H_2O$$

Oxidation

(a) By $K_2Cr_2O_7 / H_2SO_4$

$$CH \equiv CH + [O] + H_2O \longrightarrow CH_3{-}COOH$$

(b) By alkaline $KMnO_4$

$$CH \equiv CH + 4[O] \xrightarrow{\text{Heat}} \underset{\text{Oxalic acid}}{COOH{-}COOH}$$

(c) With hot acidic or alkaline $KMnO_4$

Here, oxidative cleavage occurs to give acids.

$$R - C \equiv C - R' \rightarrow R - \overset{O}{\overset{\|}{C}} - OH + R' - \overset{O}{\overset{\|}{C}} - OH$$

- If triple bond is at corner, formic acid is formed, which oxidizes into CO_2 and water.

For example,

$$CH_3{-}C \equiv CH \xrightarrow{\text{Hot alk. } KMnO_4} CH_3COOH + HCOOH$$

$$HCOOH \xrightarrow{(O)} H_2O + CO_2$$

$$CH_3{-}C \equiv C{-}CH_3 \xrightarrow{\text{Hot alk. } KMnO_4} 2CH_3COOH$$

(d) With seleniun dioxide

$$R{-}C \equiv CH + 2[O] \xrightarrow{SeO_2} R - CO - CHO$$

For example,

$$CH \equiv CH \xrightarrow{SeO_2} \underset{\text{Glyoxal}}{CHO - CHO}$$

$$CH_3{-}C \equiv CH \xrightarrow{SeO_2} CH_3{-}\overset{O}{\overset{\|}{C}}{-}CHO$$

Ozonolysis

$$R - C \equiv C - R + O_3 \xrightarrow{CCl_4} \underset{\text{Ozonide}}{R - C(O)(O{-}O)C - R}$$

$$\xrightarrow[\substack{H_2O \\ -H_2O_2}]{Zn/CH_3COOH} R - \overset{O}{\overset{\|}{C}} - \overset{O}{\overset{\|}{C}} - R$$

- If the decomposition is carried out by H_2O_2 and Zn/CH_3COOH is not used then the dicarbonyl compounds formed undergo further oxidation to give acids.

For example,

$$CH\equiv CH \xrightarrow[Zn/H_2O]{O_3/CCl_4} \begin{matrix} CHO \\ | \\ CHO \end{matrix} \xrightarrow{[O]} \begin{matrix} COOH \\ | \\ COOH \end{matrix}$$

$$CH_3-C\equiv C-CH_3+O_3 \xrightarrow{CCl_4} CH_3-\underset{\text{Ozonide}}{C(O_3)}-CH_3$$

$$\xrightarrow[H_2O]{Zn/CH_3COOH} \underset{\text{But-2, 3-dione}}{CH_3-\overset{O}{\overset{\|}{C}}-\overset{O}{\overset{\|}{C}}-CH_3}$$

- $$\text{Ozonide} \xrightarrow{H_2O_2} CH_3-\overset{O}{\overset{\|}{C}}-\overset{O}{\overset{\|}{C}}-CH_3 \xrightarrow{[O]} 2CH_3-\overset{O}{\overset{\|}{C}}-OH$$

$$CH_3-C\equiv C-CH_2-CH_3 \xrightarrow[H_2O/Zn]{O_3/CCl_4} \underset{\text{Pentan 2, 3 di-one}}{CH_3-\overset{O}{\overset{\|}{C}}-\overset{O}{\overset{\|}{C}}-CH_2-CH_3}$$

$$CH_3-C\equiv C-CH_2-CH_3 \xrightarrow[H_2O/Zn]{O_3/CCl_4} CH_3COOH + CH_3-CH_2-COOH$$

$$HC\equiv C-CH_2-CH_3 \xrightarrow[\text{(II) } H_2O/Zn]{\text{(I) } O_3/CCl_4} \underset{\text{2-oxo butanal}}{CHO-\overset{O}{\overset{\|}{C}}-CH_2-CH_3}$$

$$HC\equiv C-CH_2-CH_3 \xrightarrow[H_2O]{O_3/CCl_4} HCOOH + CH_3-CH_2-COOH$$

Here HCOOH does not further oxidise into CO_2 and H_2O as in case of oxidative cleavage by hot. $KmnO_4$.

Reaction with HCHO

$$CH\equiv CH \xrightarrow[\text{Under Press}]{HCHO} \underset{\text{Propargyl alcohol}}{CH\equiv C.CH_2OH} \xrightarrow[\text{Reduction}]{\text{Partial}} \underset{\text{Alkyl alcohol (Prop 2-en-ol)}}{CH_2=CH-CH_2OH}$$

With H_2O_2

$$CH\equiv CH \xrightarrow{H_2O_2} \underset{\text{Glyoxal}}{\begin{matrix} CHO \\ | \\ CHO \end{matrix}} \xrightarrow{\text{Oxid}^n} \begin{matrix} COOH \\ | \\ COOH \end{matrix}$$

Polymerization

Cyclic polymerization: Here alkynes polymerize to give aromatic compounds as follows:

- $$\underset{\text{Ethyne}}{3C_2H_2} \xrightarrow{\text{Redhottube}} \underset{\text{Benzene}}{C_6H_6}$$
- $$4C_2H_2 \xrightarrow{Ni(CN)_2} \text{Cyclo octa-1, 3, 5, 7-tetra-ene}$$
- $$\underset{\text{Propyne}}{3CH_3-C\equiv CH} \xrightarrow{\text{Red hot tube}} \text{Mesitylene}$$
- $$\underset{\text{But-2-yne}}{3CH_3-C\equiv C-CH_3} \xrightarrow{\text{Red hot tube}} \text{Hexamethyl benzene}$$

Linear polymerization

$$2CH\equiv CH \xrightarrow{Cu_2Cl_2NH_4Cl} \underset{\text{Vinyl acetylene}}{CH_2=CH-C\equiv CH}$$

$$CH_2=CH-C\equiv CH + CH\equiv CH \rightarrow \underset{\text{Di Vinyl acetylene}}{CH_2=CH-C\equiv C-CH=CH_2}$$

$$CH_2=CH-C\equiv CH \xrightarrow{HCl} \underset{\text{Chloroprene or 2-chloro-buta-1,3-diene}}{CH_2=CH-CHCl=CH_2}$$

- Addition of HCl occurs at triple bonded carbon atom instead of double bonded carbon atom, due to the formation of a more stable conjugated diene product which is resonance stabilised.

Acidity of terminal alkynes: Terminal alkynes are slightly acidic in nature and acidic nature in them decreases with the increase of No. of C–atoms. It is conformed by the following reactions:

$$CH\equiv CH > CH_3-C\equiv CH > CH_3-CH_2-C\equiv CH$$

With ammonical silver nitrate: Here, terminal alkynes react with it as follows:

$$R{-}C\equiv CH + Ag(NH_3)_2^+ + OH^- \xrightarrow{\text{Solution in nitric acid}} R{-}C\equiv C.Ag + H_2O + 2NH_3$$

White ppt.

For example,

$$CH\equiv CH + 2AgNO_3 + 2NH_4OH \longrightarrow AgC\equiv CAg + 2NH_4NO_3 + 2H_2O$$

White ppt. of silver Acetilide

With ammonical cuprous chloride

$$R{-}C\equiv CH + Cu(NH_3)_2^+ + OH^- \longrightarrow R{-}C\equiv C.Cu\downarrow + H_2O + 2NH_3$$

Red ppt.

For example,

$$CH\equiv CH + Cu_2Cl_2 + 2NH_4OH \longrightarrow Cu.C\equiv C.Cu + 2NH_4Cl + 2H_2O$$

Cuprous acetilide Red ppt.

Replacement of terminal H-atom from terminal alkynes

$$R-C\equiv CH \xrightarrow[\text{Liq } NH_3]{NaNH_2} R-C\equiv C.Na \xrightarrow{R'-CH_2X} R-C\equiv C-CH_2R'$$

$$CH\equiv CH \xrightarrow[-2NH_3]{2NaNH_2} Na.C\equiv C.Na \xrightarrow[-2NaCl]{2CH_3Cl} CH_3{-}C\equiv C{-}CH_3$$

But 2–yne

$$CH_3{-}C\equiv CH \xrightarrow[-NH_3]{NaNH_2} CH_3{-}C\equiv C.Na \xrightarrow[-NaCl]{CH_3Cl} CH_3{-}C\equiv C{-}CH_3$$

But 2–yne

$$CH_3{-}C\equiv C.Na \xrightarrow[-NaCl]{C_2H_5Cl} CH_3{-}C\equiv C{-}CH_2{-}CH_3$$

Pent 2–yne

Formation of heterocyclic compounds

- When ethyne is passed over iron pyrite at 300ºC, it reacts with sulphur atom of iron pyrite to give thiophene.

$$CH\equiv CH + S + CH\equiv CH \longrightarrow \text{Thiophene}$$

Thio Phene

$$CH\equiv CH + H_2S + CH\equiv CH \xrightarrow[675\,K]{\Delta,\ Al_2O_3} \text{Thiophene}$$

$$CH\equiv CH + HCN + CH\equiv CH \xrightarrow{\text{Red Hot tube}} \text{Pyridine}$$

Pyridine

- Ethyne on reaction with ammonia gives pyrrole.
- $$CH\equiv CH + NH_3 + CH\equiv CH \longrightarrow \text{Pyrrole}$$

Pyrrole

Reaction with nitrogen

$$CH\equiv CH + N_2 \xrightarrow{\text{electric arc}} 2HCN$$

Aromatic Hydrocarbons

The word Aromatic has been derived from Greek word Aroma which means pleasant smell. Benzene and the compounds that resemble benzene in their chemical behaviour are called Aromatic compounds or Benzenoid compounds.

Aromatic hydrocarbons or Arenes have a general formula C_nH_{2n6y} (y represents No. of benzene Rings). and n > 6. Their main sources are coaltar coal, Petroleum etc.

Benzene Toluenes Xylene

Mesitylene Cumene (Isopropyl benzene)

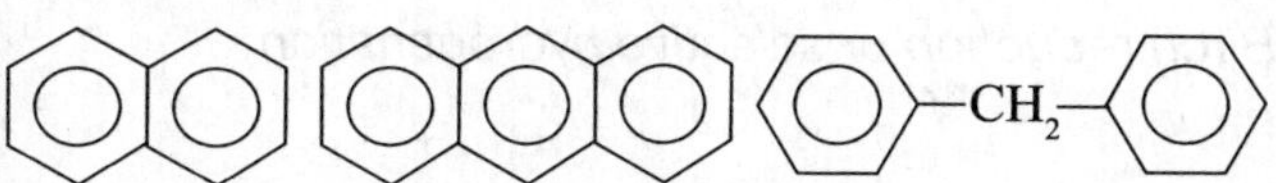

Naphthalene Anthracene Diphenyl Methane

Aromaticity and Hunckel (4n + 2)Rule

Aromaticity is defined as "An aromatic compound having a reasonably planar cyclic structure, with (4n + 2) πe^- clouds and having unusual stablity because of delocalisation of π–electrons.

According to Huckels Rule,

Any compound can be aromatic when:

(1) It has a cyclic planar structure (Necessary for complete delocalisation of πe^-)

e.g.,

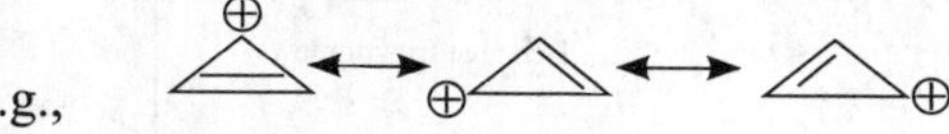

(2) It has a conjugate system.

(3) It has a high resonance energy.

(4) It has No. of πe^- according to (4n + 2) rule i.e., 2, 6, 10, 14, 18, 22 etc. (Here n = No.of cyclic planar ring).

In case any compound has πe^- according to 4n rule it is Anti aromatic.

e.g., (1) When n = 0, No. of πe^- = 2

e.g.,

e.g., (2) When n = 1, No. of πe^- = 6

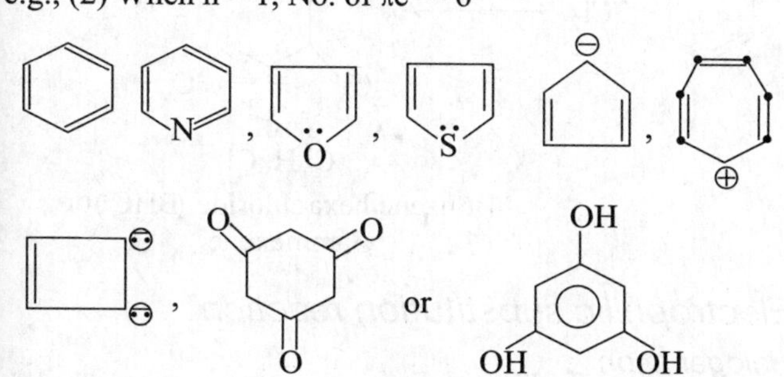

e.g., When n = 2, No. of πe^- = 10

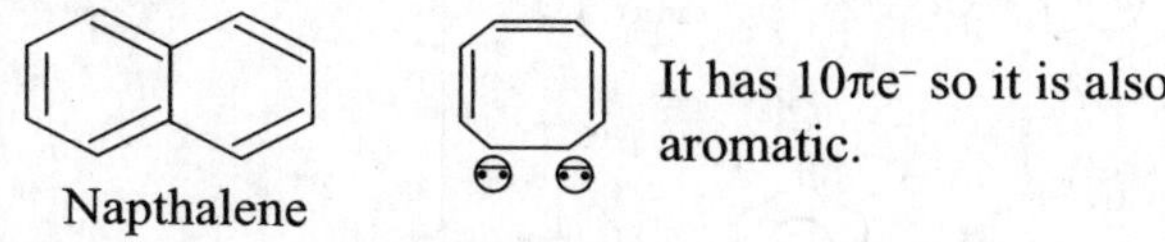

It has 10πe^- so it is also aromatic.

Napthalene

When n = 3, No. of πe^- = 14

Phenanthrene

Anthracene

Anti Aromatic compounds having 4, 8 πe^-.

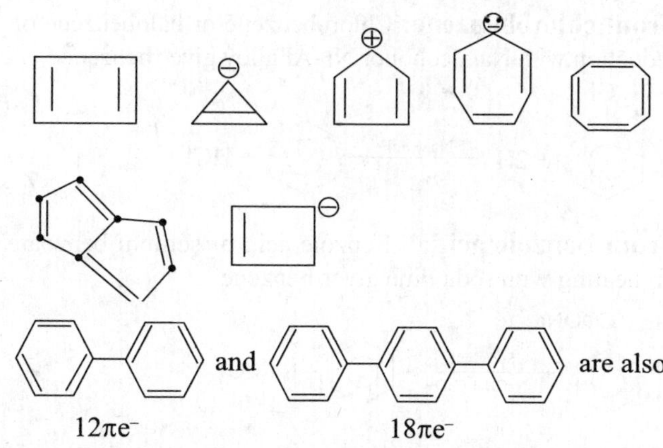

and are also

12πe^- 18πe^-

Aromatic in nature as each benzene ring has 6πe^- which are involved in resonance. Some Non-Planar molecules which are non aromatic are as follows

Cyclooctatetraene Cyclodecapentaene (10) Annulene

BENZENE

- Benzene is structurally depicted as shown below:

- Benzene was discovered by Faraday in 1825.
- The structure of benzene was given by Kekule in 1865.

Methods of Preparation of Benzene

From light oil fraction: The light oil fraction of coal tar contains benzene, toluene, xylene. This mixture on fractional distillation gives benzene as one of the product.

From petroleum

$$\text{h–Hexane} \xrightarrow[\text{High Pressure, } -4H_2]{CrO_3,\ Al_2O_3,\ 600°C} \text{Benzene}$$

From acetylene: Acetylene on passing through red hot tube polymerizes into benzene.

$$3C_2H_2 \xrightarrow[1500°C–2000°C]{\text{Red hot tube}} \text{Benzene}$$

From phenol: Phenol on distillation with zinc dust gives benzene.

$$\text{Phenol (OH)} + \text{Zn (dust)} \xrightarrow{\Delta} \text{Benzene} + \text{ZnO}$$

From chlorobenzene: Chlorobenzene or halobenzene on reduction with Na/alcohol or Ni–Al alloy gives benzene.

$$C_6H_5Cl + 2H \xrightarrow{\text{Na + acl.}} C_6H_6 + HCl$$

From benzoic acid: Benzoic acid or sodium benzoate on heating with soda lime gives benzene.

$$C_6H_5COONa \xrightarrow[\text{NaOH + CaO}]{\text{Soda lime}} C_6H_6 + Na_2CO_3$$

From benzene diazonium chloride: Benzene diazonium chloride on reaction with hypo phophorus acid gives benzene.

$$C_6H_5N=N-Cl \xrightarrow[H_3PO_2\ \Delta\ H_2O]{\text{Abs. } C_2H_5OH} C_6H_6 + HCl + N_2 + H_3PO_3$$

From benzene sulphonic acid: Benzene sulphonic acid on treatment with steam gives benzene.

$$C_6H_5SO_3H + HOH \xrightarrow{\Delta} C_6H_6 + H_2SO_4$$

Physical Properties

- Benzene is a colourless volatile, liquid.
- It is immiscible in water.
- It is highly inflammable (burns with a sooty flame).
- Its freezing point is 5.5ºC and boiling point is 80ºC.
- Benzene is lighter than water.
- It is a good solvent for fat, rubber and resin.

Chemical Properties

Benzene has three alternative double bonds, however, it fails to show the usual addition reactions or the test of unsaturation with bromine water or Baeyer's reagent. It shows electrophilic substitution reaction due to the phenomenon of resonance or complete delocalization of six π electrons.

Addition reaction

Hydrogenation

$$C_6H_6 + 3H_2 \xrightarrow{\text{Ni, 200°C}} \text{Cyclohexane}$$

Birch reduction or selective hydrogenation

$$C_6H_6 + H_2 \xrightarrow{\text{Ni, liq } NH_3} \text{1, 4 - dihydrobenzene}$$

Ozonolysis: Benzene on ozonolysis gives glyoxal as shown:

$$C_6H_6 + 3O_3 \longrightarrow \text{Benzene triozonide} \xrightarrow[\text{Zn}]{H_2O} 3\ \begin{matrix} CHO \\ | \\ CHO \end{matrix}\ \text{(Glyoxal)}$$

With chlorine: Benzene on chlorination in presence of sunlight gives gammexane or 666 or γ-Lindane, a famous insecticide.

$$C_6H_6 + 3Cl_2 \xrightarrow{\text{Sun light}} C_6H_6Cl_6$$

Benzene hexachloride (BHC) or Gammexene

Electrophilic substitution reaction

Halogenation

$$C_6H_6 + Cl_2 \xrightarrow[\text{carrier}]{\text{Cold and dark halogens}} C_6H_5Cl + HCl$$

- $C_6H_6 + Br_2 \xrightarrow{FeBr_3} C_6H_5Br + HCl$
- Iodination of benzene is a slow reversible process, so it is carried out in presence of oxidants like HIO_3 or HgO to check HI formation.

Nitration

$$C_6H_6 + \text{Conc. } HNO_3 \xrightarrow{\text{conc. } H_2SO_4,\ 50°C} C_6H_5NO_2 + H_2O$$

- $$C_6H_6 + \text{Furming } 3HNO_3 \xrightarrow{\text{Conc. } H_2SO_4,\ 50°C} C_6H_3(NO_2)_3$$

Tri nitro benzene (explosive)

Sulphonation

$$C_6H_6 + \text{Conc. } H_2SO_4 \xrightarrow{80°C} C_6H_5SO_3H + H_2O$$

- $$C_6H_6 + H_2S_2O_7 \text{ (oleum)} \xrightarrow{80°C} C_6H_4(SO_3H)_2 + H_2O$$

m-benzene disulphonic acid

Friedel–Crafts reaction: Here alkylation or acylation takes place in presence of catalysts like anhydrous $AlCl_3$, $FeCl_3$, BF_3, $ZnCl_2$ or $SnCl_4$.

- *Alkylation*: For alkylation, alkyl halide in presence of anhydrous aluminium chloride; alcohol alkene in presence of conc. H_2SO_4 can be used.

$$C_6H_6 + CH_3X \xrightarrow{AlCl_3} C_6H_5CH_3 + HX$$

Toluene

- Here, one time rearrangement of carbocation is possible as shown in the reaction given below:

$$C_6H_6 + CH_3CH_2CH_2Cl \xrightarrow{AlCl_3} C_6H_5CH(CH_3)_2$$

Cumine or isopropyl benzene

$$CH_3-CH_2-\overset{+}{C}H_2 \ (\text{'p'}) \xrightarrow{\text{Rearrangement by } H^{\ominus} \text{ shift}} CH_3-\overset{+}{C}H-CH_3 \ (\text{'s'})$$

More stable carbocation

- $$C_6H_6 + (CH_3)_2CH-CH_2OH \xrightarrow{\text{Conc. } H_2SO_4} C_6H_5C(CH_3)_3 + H_2O$$

Ter tiary butyl benzene

- $$C_6H_6 + C_6H_5-CH(OH)-CH_3 \xrightarrow{\text{Conc.} H_2SO_4} C_6H_5-CH(C_6H_5)-CH_3 + H_2O$$

- $$C_6H_6 + CH_3CH=CH_2 \xrightarrow{\text{conc. } H_2SO_4} C_6H_5CH(CH_3)_2$$

Acylation: For acylation, acid halides or acid anhydrides can be used with anhydrous $AlCl_3$.

$$C_6H_6 + ROCl \xrightarrow{AlCl_3} C_6H_5COR + HCl$$

- $$C_6H_6 + (RCO)_2O \xrightarrow{AlCl_3} C_6H_5COR + RCOOH$$

- Here, R may be CH_3COCl, C_6H_5COCl.

Gatterman aldehyde reaction

$$C_6H_5-H + HCN + HCl \xrightarrow{AlCl_3} C_6H_5-CH=NH \xrightarrow{HOH} C_6H_5-CHO + NH_3$$

Chloromethylation

$$C_6H_5-H + CH_2O + HCl \xrightarrow{AlCl_3} C_6H_5-CH_2Cl + H_2O$$

Chloromethyl benzene

- C_6H_5–H + Cl – CO – NH_2 (Chloroformamide) $\xrightarrow[-HCl]{AlCl_3}$ C_6H_5–$CONH_2$ (Benzamide)

- C_6H_5–H + Cl – CO – Cl (Chloroformyl) $\xrightarrow[-HCl]{AlCl_3}$ C_6H_5–COCl (Benzoyl chloride)

Gattermann Koch or formylation

C_6H_5–H + CO + HCl (Dry) $\xrightarrow[-HCl]{AlCl_3}$ C_6H_5–CHO (Benzaldehyde)

- In Friedel–Crafts reaction, the electrophiles are R^+, RC^+O, C^+ H=NH, C^+OCl, C^+ONH_2, HC^+O.

Mercuration

C_6H_5–H + $(CH_3COO)_2Hg$ $\xrightarrow{AlCl_3}$ C_6H_5–$HgOOC{-}CH_3$ + CH_3COOH

Phyenyl mercuric acetate

Reduction

$2\,C_6H_5$–H + 12HI $\xrightarrow{250°C}$ Cycolhexane + Methyl cyclopentane (Me) + $6I_2$

Oxidation

C_6H_5–H + 9 [O] $\xrightarrow[450-500°C]{V_2O_5}$ Maleic anhydride (CH–CO–O–CO–CH, CH=CH) + $2CO_2$ + $2H_2O$

Maleic anhydride

Combustion

$$2C_6H_6 + 15O_2 \xrightarrow{\Delta} 12CO_2 + 6H_2O$$

Various Structural Formulae of Benzene

Landenberg prism formula

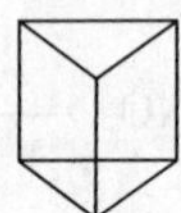

Clausing diagonal formula

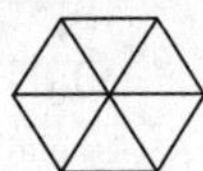

Dewars parallel's formula

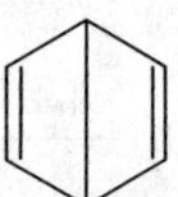

Baeyer and Armstrong centric formula

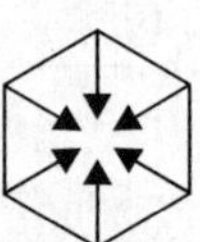

Resonance Hybrid of Kekule's structure of Benzene

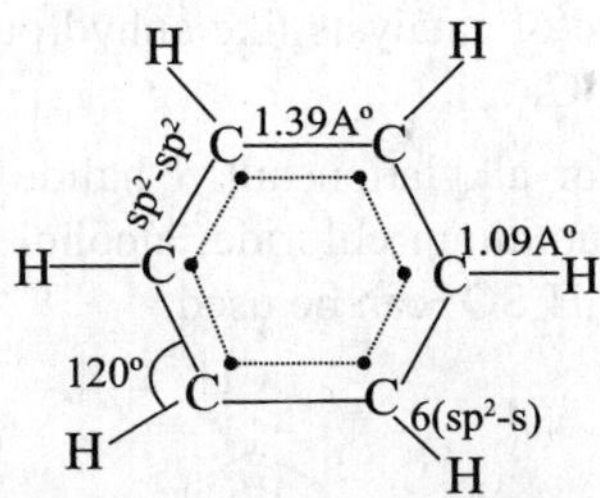

Toluene or Methyl Benzene

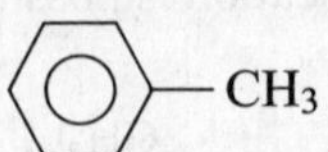

- Toluene was first obtained by the dry distillation of tolubalsam. It is also known as Toluol.

Methods of Preparation of Toluene

From light oil fraction of coal tar: The light oil fraction contains mainly benzene, toluene and xylene. All the three can be separated by fractional distillation where toluene is collected at 110°C.

From petroleum

n-Heptane $\xrightarrow[-H_2]{Cr_2O_3,\ Al_2O_3,\ 600°C}$ methylcyclohexane $\xrightarrow[-3H_2]{\Delta}$ toluene

Friedel–Crafts reaction

$$C_6H_6 + CH_3X \xrightarrow{\text{Anhy. } AlCl_3} C_6H_5CH_3 + HX$$

By Wurtz –Fittig reaction

$$C_6H_5Br + 2Na + BrCH_3 \xrightarrow{\text{Ether}} C_6H_5CH_3 + 2NaBr$$

By the reduction of benzaldehyde

$$C_6H_5CHO + 4[H] \xrightarrow{ZnHg + HCl} C_6H_5CH_3 + 2H_2O$$

From toluidine

$$CH_3C_6H_4NH_2 \xrightarrow[HCl]{\text{Dry } NaNO_2} CH_3C_6H_4N{=}N{-}Cl \xrightarrow{C_2H_5OH} C_6H_5CH_3 + N_2 + CH_3CHO + HCl$$

From cresol

$$CH_3C_6H_4OH + Zn\ (\text{dust}) \xrightarrow{\Delta} C_6H_5CH_3 + ZnO$$

From toluene sulphonic acid

$$CH_3C_6H_4SO_3H + HOH \longrightarrow C_6H_5CH_3 + H_2SO_4$$

From Grignard reagent

$$C_6H_5MgX + CH_3X \longrightarrow C_6H_5CH_3 + MgX_2$$

From decarboxylation of sodium toluate

$$C_6H_4\begin{matrix} CH_3 \\ COONa \end{matrix} + NaOH \xrightarrow[CaO]{\text{Soda lime}} C_6H_5CH_3 + Na_2CO_3$$

Physical Properties

- Toluene is a colourless liquid.
- Its boiling point is 111°C.
- It is immiscible in water.
- Toluene is used as a solvent and also in the synthesis of other compounds.

Chemical Properties

Reactions due to benzene ring: Here, methyl group activates the benzene ring, therefore, further substitution takes place at ortho (o) and para (p) positions and the rate of substitution is more fast than in benzene.

H—C—H, H, H^+, H^+, ⊕

e– density (↑) at o, p positions.

Halogenation

Toluene + Cl_2 —(Cold dark halogen carrier (FeX_3), –HCl)→ Ortho-Chlorotoluene + Para-Chlorotoluene

Nitration

Toluene —(Fumed HNO_3, Fumed H_2SO_4)→ o-nitrotoluene + p-nitrotoluene → 2,4-dinitrotoluene → 2, 4, 6-Trinitrotoluene (T.N.T)

NOTE: *Amatol* It is an explosive. It is a mixture of TNT and ammonium nitrate (NH_4NO_3).

Sulphonation

Toluene —(Conc. H_2SO_4, 50°C)→ Ortho-toluene Sulphonic acid + Para-toluene Sulphonic acid

Friedel–Crafts reaction

Toluene + CH_3X —(Anhy. $AlCl_3$)→ 'o' and 'p' Xylene

Toluene —($AlCl_3$, CH_3COOH)→ ortho-methyl acetophenone + para-methyl acetophenone

Reactions due to methyl group or side chain reactions

Halogenation

$C_6H_5-CH_3$ —(Cl_2, hν)→ $C_6H_5-CH_2Cl$ (Benzyl chloride) —(Cl_2, –HCl)→

$C_6H_5-CHCl_2$ (Benzal chloride) —(Cl_2, hν, –HCl)→ $C_6H_5-CCl_3$ (Benzo tri chloride)

$C_6H_5-CH_2Cl$ —(2H.OH)→ $C_6H_5-CH_2OH$ (Benzyl alcohol)

- $C_6H_5-CHCl_2$ —(2H.OH)→ $C_6H_5-CH(OH)_2$ —(–H_2O)→ C_6H_5-CHO (Benzaldehyde)

$C_6H_5-CCl_3$ —(HOH)→ $C_6H_5-C(OH)_3$ —(–H_2O)→ C_6H_5-COOH (Benzoic acid)

Oxidation

$$C_6H_5CH_3 \xrightarrow[\text{Etard Reaction}]{2[O],\ CrO_2Cl_2,\ C_2H_5OH} C_6H_5CHO + H_2O$$

Combustion

$$C_6H_5CH_3 + 9O_2 \xrightarrow{\Delta} 7CO_2 + 4H_2O$$

Reaction with hot alkaline $KMnO_4$ or $Na_2Cr_2O_7$ Here, alkyl group having α-hydrogen atom gets oxidized into COOH group.

$$C_6H_5-CH_3 + [O] \xrightarrow{\text{Hot alk.}KMnO_4} C_6H_5-COOH$$

- $$C_6H_5-CH_2CH_3 + [O] \xrightarrow{\text{Hot alk. }KMnO_4} C_6H_5-COOH$$
- $$C_6H_5-CH(CH_3)_2 + [O] \xrightarrow{\text{Hot alk.}KMnO_4} C_6H_5-COOH$$
- If alkyl group has no α-hydrogen atom or benzylic hydrogen atom the benzene ring gets oxidized into – COOH group (Conditionally) otherwise no reaction in normal way.

$$CH_3-C(CH_3)_2-C_6H_5 \xrightarrow{K_2Cr_2O_7 + H_2SO_4} CH_3-C(CH_3)_2-COOH$$

No benzylic H-atom

$$\underset{\text{O-Xylene}}{C_6H_4(CH_3)_2} \xrightarrow{\text{Hot alk. }KMnO_4} \underset{\text{Phthalic acid}}{C_6H_4(COOH)_2}$$

Hydrogenation

$$C_6H_5CH_3 + 3H_2 \longrightarrow \underset{\text{Methyl cyclohexane}}{C_6H_{11}CH_3}$$

TABLE 12.1 Fractional Distillation of Petroleum

S.No.	Fraction	Boiling range in K	Composition or number of C-atoms	Uses
1	Gaseous hydrocarbons	110–303	$C_1 - C_4$	Preparation of gasoline, carbon black
2	Crude naphthalene	303–424	$C_5 - C_{10}$	
	(a) Light petrol	303–362	$C_5 - C_7$	As a useful solvent for fats, oils
	(b) Petrol or gasoline	362–394	$C_7 - C_9$	Fuel for motors, production of petrol gas
3	Kerosene	424–514	$C_{10} - C_{13}$	As fuel for stoves, in production of oil gas
4	Fuel oils like diesel, gas oil	514–624	$C_{13} - C_{18}$	As a fuel
5	Lubricants for example motor oil, grease, medicinal oil	> 540	$C_{15} - C_{18}$	As transformer oil, paint oil
6.	Paraffin wax like petroleum wax, petroleum jelly	> 672	$C_{18} - C_{30}$	In the manufacture of candles, ointments
7	Heavy oil and bitumen	Forming residue	$> C_{30}$	Used mainly for road surfacing

PETROLEUM

Petroleum (Petra = rock, oleum = oil) It is a viscous, dark coloured fluorescent liquid which is a mixture of aliphatic hydrocarbons (mainly C_1–C_{40} alkanes), cycloalkanes or aromatic hydrocarbons. The various theories given on the origin of petroleum are:

1. Mendeleef's carbide theory
2. Engler theory
3. Modern theory

Synthetic petrol is manufactured by the following methods:

Bergius method

$$\text{Coal} + H_2 \xrightarrow{Fe_2O_3,\ 250\ \text{atm},\ 400°C\text{–}500°C} \text{Crude oil}$$

Crude oil → 150°C → Gasoline; 150 – 200°C → Middle oil; 200 – 300°C → Heavy oil

Fischer Tropsch method

$$C + H_2O \xrightarrow{1200°C} \underset{\text{water gas}}{CO + H_2} \xrightarrow{\Delta,\ H_2} \text{Mixture of hydrocarbons}$$

In half volume of water gas under volume

Refining of petroleum: Refining of petroleum is the separation of useful fractions of petroleum and the removal of various undesirable impurities. It is mainly carried out by fractional distillation of crude oil.

Liquefied petroleum gas (LPG): It is a mixture of propane and butane, the main component is butane. It smells like ethanthiol (C_2H_5SH), which is, ethyl mercapton. This smell helps in detecting leakage of gas.

Natural gas: It has 80% methane, 10% C_2H_6 and 10% long chain hydrocarbons.

$$\text{Natural gas} \xrightarrow[\text{Limited air}]{\Delta} \underset{\text{Carbon black or lamp black}}{C} + 2H_2O$$

Compressed natural gas (CNG): It has 95% CH_4, 5% C_2H_6, C_3H_8, N_2, He and CO_2. LNG is liquefied natural gas.

Knocking

Knocking is metallic sound produced due to pre-ignition of fuel in the cylinder of an internal combustion engine.
The following hydrocarbons are arranged in a decreasing order of the knocking produced:
Straight chain alkanes > branched chain alkane > olefins > cycloalkenes > atomatic hydrocarbons
Tetraethyl lead (TEL) is used as an anti-knocking agent. It has $(C_2H_5)_4Pb$ (63%), $C_2H_4Br_2$ (26%), $C_2H_4Cl_2$ (9%) and dyes (2%).

Octane number

- The quality of an automobile fuel is measured in terms of octane number. The hydrocarbon iso-octane (2, 2, 4-trimethyl pentane) which has good resistance to knocking even in highest compression motors is arbitrarily given an octane number of 100; whereas n-heptane which knocks the maximum is arbitrarily given the value of zero. Thus, higher the octane number, better will be the fuel.

$CH_3–CH_2–CH_2–CH_2–CH_2–CH_2–CH_3$

n-heptane

octane number = 0

$$CH_3-\overset{\overset{CH_3}{|}}{\underset{\underset{CH_3}{|}}{C}}-CH_2-\overset{\overset{CH_3}{|}}{CH}-CH_3$$

2, 2, 4-tri methyl pentane 91 octane number = 100

- The octane number of gasoline is defined as the percentage of iso-octane present in a mixture of iso-octane and n-heptane, when the mixture has the same knocking in the engine, as the gasoline under examination.
- Thus, a motor fuel is said to have an octane number of 80 when it is as good as a mixture of 20 % n-heptane and 80% of iso-octane.
- Octane number can be increased by:
  1. increasing the percentage of branched or cycloalkanes,
  2. by addition of BTX,
  3. by addition of TEL,
  4. by addition of power alcohol (C_2H_5OH).

Cetane number: It is the percentage of cetane (hexadecane) in a given diesel sample. For example, if a diesel sample contains 75% cetane and 25% α-methyl naphthalene, its cetane number is 75.

Flash point: Flash point or the ignition temperature may be defined as the minimum temperature at which an oil gives off enough vapours to form a momentary flash of light when a naked flame is brought near its surface.

- A fuel used in cold climates must have a low boiling point whereas the fuel used in tropical climate must have a high boiling point. Therefore, the flash point fixed for fuels in cold countries is low whereas in hot countries it is high. It is measured using **Abel's apparatus**.
- Gasoline, which has a pungent odour due to the pressence of sulphur compounds is called **sour gasoline**. The odour is removed by **doctor sweetening process**.

Cracking: In this process, the less valuable higher fraction of petroleum nun is converted into more valuable lower fractions by heating. The process involved for this is as follows:

Liquid phase cracking

$$\text{Heavy oil}(C_{18}-C_{43}) \xrightarrow[\text{catalyst 1000–1200 lbs psi}]{475\text{–}530°C} (C_4-C_{10})$$

Catalyst: Silica, titanium dioxide, zinc oxide, ferric oxide, or alumina is used as. High pressure keeps the oil in liquid state. Octane number of product: 65–70.

Vapour phase cracking

$$\text{Kerosene oil or gas oil} \xrightarrow[\text{50–150 lbs psi}]{600\text{–}800°C} \text{Lower fractions}$$

Cracking in presence of hydrogen

$$R-R+H_2 \xrightarrow{\text{Catalyst}} 2RH$$

$$ROH + H_2 \longrightarrow RH + H_2O$$

$$R_2S + 2H_2 \longrightarrow 2RH + H_2S$$

$$R_2NH + 2H_2 \longrightarrow 2RH + NH_3$$

Following are the reactions that take place during cracking:

1. Carbonization

$$CH_4 \xrightarrow{1000°C} C + 2H_2$$

2. Dehydrogenation

$$CH_3-CH_3 \xrightarrow{450°C} CH_2=CH_2+H_2$$

3. Polymerization

$$2(CH_3)_2C=CH_2 \longrightarrow (CH_3)_3C-CH_2-\underset{\displaystyle CH_3}{\underset{|}{C}}=CH_2 \xrightarrow{H_2} (CH_3)_3C-CH_2CH(CH_3)_2$$

4. Alkylation

$$(CH_3)_2CH-CH_3 + CH_2=CH_2 \longrightarrow (CH_3)_3C-CH_2-CH_3$$

5. Aromatization

$$CH_3-(CH_2)_4-CH_3 \longrightarrow C_6H_6 + 4H_2$$

6. Chain fission

$$C_{16}H_{34} \longrightarrow C_8H_{18} + C_8H_{16} \text{ or } C_6H_{14} + C_{10}H_{20}$$

FACTS TO REMEMBER

- Alkanes are inert at room temperature as (C–C), (C–H) bonds are non-polar.
- Liquid paraffins when burnt cannot be extinguished by water as it is lighter than water, so it will float over water.

The acid catalysed hydration of alkene is Regioselective and for various alkenes order of hydration is

$$R_2C=CH_2 > R-CH=CH_2 > \underset{\displaystyle CH_2}{\underset{\|}{CH_2}}$$

⇒ Viscosity of Alkanes α Mol.wt. α Branching

E.g.,: n-Heptane > 2, 3 dimethyl butane > Isohexane > n-Hexane > n-pentane

⇒ Presence of O_2 inhibits Halogenation as it forms a less Reactive Peroxy Radicle (R–O–O)

⇒ Thermal Chlorination in case of an alkane is known as McBee Reaction

⇒ C_2H_4 on reaction with S_2Cl_2 (Sulphur meno chlorides gives Mustard gas. (wargas)

$$2\begin{array}{c}CH_2\\ \|\\ CH_2\end{array} \xrightarrow{S_2Cl_2} \begin{array}{c}CH_2Cl \qquad CH_2Cl\\ | \qquad\qquad |\\ CH_2-S-CH_2\end{array} + \text{'S'}$$

Mustard gas

⇒ Anti-Markonikoff addition of HBr on propene can be prevented by using inhibitors like catehol or diphenylamine.

$$\Rightarrow CH_3-CH=CH_2 \xrightarrow[\substack{SO_2Cl_2\\ \text{Peroxide}}]{\Delta\ 475\ K} \underset{\displaystyle Cl}{\underset{|}{CH_2}}-CH=CH_2 + SO_2 + HCl$$

⇒ Decreasing order of acidic nautre and increasing order of conjugate base strength:

$$H_2O > R-OH > \underset{\displaystyle CH}{\underset{|||}{CH}} > NH_3 > \underset{\displaystyle CH_2}{\underset{\|}{CH_2}} > CH_4 > C_2H_6$$

$$\overset{\ominus}{O}H < R-\overset{\ominus}{O} < HC\equiv\overset{\ominus}{C} < \overset{\ominus}{N}H_2 < CH_2=\overset{\ominus}{C}H < \overset{\ominus}{C}H_3 < \overset{\ominus}{C}_2H_5$$

$CH_2=CH—CH=CH_2$ Buta 1, 3 di-ene

Cl_2, 1, 2 Addition At Low Δ → $CH_2—CH—CH=CH_2$ (Cl on C1 and C2)

Cl_2, 1, 4 addition At Higher Δ → $ClCH_2—CH=CH—CH_2Cl$

$CH-R$ || CH_2, Δ → (cyclohexene with R)

- Buna-N [General purpose rubber acrylonitrile] is a synthetic rubber which is a polymer of butadiene and acrylo nitrile.
- Buna-S [General purpose rubber styrene] is a synthetic rubber which is a polymer of butadiene and styrene.
- Natural rubber is a polymer of isoprene, that is, cis-2-methyl buta-1, 3,-diene.
- Guta pircha It is a polymer of trans-2-methyl buta-1, 3,-diene.
- Neoprene is a polymer of 2-chloroprene.
- On heating ethyne in the presence of spongy copper or Cu_2O, a cork like substance **cuprene** is formed which is used in the manufacture of linoleum.
- The light oil fraction of coal tar is composed of the hydrocarbons benzene (72%), toluene (13%) and xylene (4%), and is therefore known as BTX.
- When toluene is treated with Cl_2 in presence of a Lewis acid catalyst ($FeCl_3$), substitution takes place at ortho and para positions. These products thus formed do not give white precipitate with alcoholic $AgNO_3$.
- Benzene hexachloride (BHC) can exist in nine stereo-isomeric forms.
- Chloramine-T is the sodium salt of n-chloro-p-toluene sulphonamide.
- Actylene is used as an anaesthetic under the name Naraceylene.

CHAPTER-END EXERCISES

PRACTICE QUESTIONS – I

1. Natural gas is composed primarily of:
(a) Methane (b) n-butane
(c) n-octane (d) A mixture of alkanes

2. Alkanes mainly undergo reactions having:
(a) Free radical reaction
(b) Ionic elimination
(c) Electrophillic addition
(d) Heat/photochemical substitution

3. Carbon black, which is used in making printer's ink is obtained by the decomposition of:
(a) Acetylene (b) Benzene
(c) Carbon tetrahedral (d) Methane

4. Addition of HCN to ethyne in the presence of $Ba(CN)_2$ as a catalyst gives:
(a) Vinyl cyanide (b) Ethyl cyanide
(c) 1,1-dicyanoethane (d) Divinyl cyanide

5. The reagent used for the following conversion:
$CH_3CH_2COOH \longrightarrow CH_3CH_2CH_3$ is
(a) $LiAlH_4$
(b) Amalgamated zinc and concerntrated HCl
(c) Red phosphorus and concentrated HI
(d) Soda lime

6. Acetylene reacts with acetic acid in presence of Hg^+ ions at room temperature to give:
(a) Ethyl acetate (b) Acetaldehyde
(c) Vinyl acetate (d) Methyl acetate

7. Which of the following is used for the conversion of 2-hexyne into trans-2-hexane?
(a) $H_2/Pd/BaSO_4$ (b) H_2, PtO_2
(c) $NaBH_4$ (d) $Li–NH_3/C_2H_5OH$

8. The addition of HCl to 3, 3, 3-trichloropropene gives:
(a) $Cl_3CCH_2CH_2Cl$
(b) $Cl_2CHCH(Cl)CH_2Cl$
(c) $Cl_2CHCH_2CHCl_2$
(d) $Cl_3CCH_2CH_2Cl$

9. Br–(zigzag chain)–Br $\xrightarrow[R-O-R]{Mg}$ Product. Here the Product is
(a) Br–(zigzag chain)–Mg Br
(b) $CH_3–CH=CH_2$

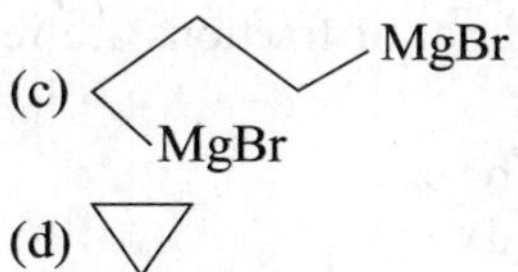

10. Isopropyl bromide on Wurtz reaction gives:
(a) Hexane
(b) Propane
(c) 2, 3-dimethyl butane
(d) Neo-hexane

11 The product formed by acid catalysed hydration of 2-phenyl propene is:
(a) 3-phenyl-2-propanol
(b) 1-phenyl-2-propanol
(c) 2-phenyl-2-propanol
(d) 2-phenyl-1-propanol

12. Reactivity of chlorine and bromine towards methane at 275°C differs by a factor nearly:
(a) 10^3 (b) 10^4
(c) 10^2 (d) $> 10^5$

13. Among the following compounds, the strongest acid is:
(a) $HC \equiv CH$ (b) C_6H_6
(c) C_2H_6 (d) CH_3OH

14. Antidote of Lewsite is:
(a) 3, 3-dimercaptopropan-1-ol
(b) 2-mercaptopropan-1, 3-diol
(c) 2, 3-dimercaptopropan-1-ol
(d) 2, 2-dimercaptopropan-1-ol

15. The number of possible enantiomeric pairs that can be produced during monochlorination of 2-methyl butane are:
(a) 2 (b) 3
(c) 4 (d) 1

16. Acetylene and HCHO react in the presence of copper acetylide catalyst to form:
(a) 2-butyne-1, 4-diol
(b) but-2-yne-1, 2-diol
(c) but-1-yne-1, 4-diol
(d) None of these

17. The intermediate product formed during the addition of HCl to propene in presence of peroxide is:
(a) $CH_3\dot{C}HCH_2Cl$ (b) $CH_3\overset{+}{C}HCH_3$
(c) $CH_3\dot{C}H_2CH_2$ (d) $CH_3CH_2\overset{+}{C}H_2$

18. Which will have smallest heat of hydrogenation per mole?
(a) Cis-2-butene (b) Trans-2-butene
(c) 1, 3-butadiene (d) 1-butene

19. When cyclohexane is poured on water, it floats because:
(a) Cyclohexane is in boat form
(b) Cyclohexane is in chair form
(c) Cyclohexane is in crown form
(d) Cyclohexane is less dense than water

20. Both methane and ethane can be prepared in one step by the reaction of:
(a) C_2H_4 (b) CH_3Br
(c) CH_3OH (d) CH_3CH_2OH

21. 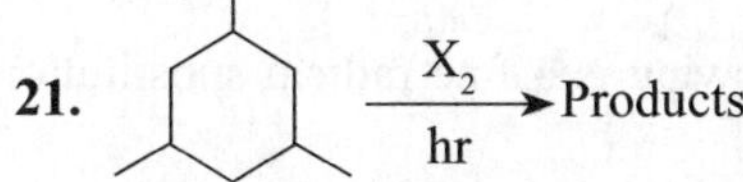

How many total products are formed here on mono halogen ation?
(a) 2 (b) 3
(c) 4 (d) 6

22. Which one of these is not compatible with arenes?
(a) Greater stability
(b) Resonance
(c) Delocalization of π-electrons
(d) Electrophillic addition

23. $(CH_3)_3CMgCl$ on reaction with D_2O produces:
(a) $(CH_3)_3CD$ (b) $(CH_3)_3COD$
(c) $(CD_3)_3CD$ (d) $(CD_3)_3COD$

24. The treatment of benzene with isobutene in the presence of sulphuric acid gives:
(a) Isobutyl benzene
(b) n-butyl benzene
(c) tert-butyl benzene
(d) Cumene

25. Addition of water to acetylene compounds is catalysed by .... and .... :
(a) Ba^{+2} salt and $HgSO_4$
(b) Hg^{+2} salt and conc. acid
(c) Hg^{+2} salt and dil. acid
(d) H_2O/H^+

26. Toluene on oxidation with chromyl chloride produces :
(a) Benzaldehyde (b) Benzoic acid
(c) Acetophenone (d) Benzyl alcohol

27. On the addition of HBr to propene in the absence of peroxides, the first step involves the addition of:
(a) H^+ (b) Br^-
(c) $\dot{H}$ (d) $\dot{B}r$

28. When ethyl bromide and n-propyl bromide is allowed to react with sodium, in ether, they form:
(a) Mixture of four alkanes
(b) Mixture of three alkanes
(c) Mixture of two alkanes
(d) Single alkane

29. Which of the following will have the least hindered rotation around carbon–carbon bond?
(a) Ethane (b) Ethylene
(c) Acetylene (d) Hexachloroethane

30. Which of the following is a free radical substitution reaction?
(a) Benzene $+ CH_3Cl \xrightarrow{\text{Anhy. } AlCl_3}$ Toluene (C_6H_5–CH_3)
(b) C_6H_5–$CH_2Cl + AgNO_2 \longrightarrow C_6H_5$–$CH_2NO_2$
(c) C_6H_5–$CH_3 + Cl_2 \xrightarrow{\text{Boiling}} C_6H_5$–$CH_2Cl$
(d) Both (b) and (c)

31. The reaction conditions leading to provide the best yield of C_2H_5Cl are:
(a) C_2H_6 (excess) $+ Cl_2 \xrightarrow{\text{uv light}}$
(b) $C_2H_6 + Cl_2 \xrightarrow{\text{dark, room temperature}}$
(c) $C_2H_6 + Cl_2$ (excess) $\xrightarrow{\text{uv light}}$
(d) $C_2H_6 + Cl_2 \xrightarrow{\text{uv light}}$

32. Octane number of gasoline can be increased by the addition of BTX. BTX stands for:
(a) Butane, TEL, xylene
(b) Butane, tetramethyl lead, xylene
(c) Benzene, toluene, xylene
(d) Benzene, TEL, xylene

33. $C_6H_5-CH(CH_3)-CH_2-CH_3 \xrightarrow[\text{hr}]{X_2}$ Products $\xrightarrow{\text{fractional Distillation}}$ fractions

The no. of products and no. of fractions are respectively?
(a) 6, 6 (b) 6, 5
(c) 6, 4 (d) 6, 3

34. Which of the following alkane can be prepared in large amounts by Wurtz reaction?
(a) $(CH_3)_2CHCH_2CH(CH_3)_2$
(b) $CH-CH(CH_3)-CH_2CH_2CH(CH_3)_2$
(c) $(CH_3)_3CCH_2CH_3$
(d) $CH_3CH(CH_3)CH_2CH_3$

35. The highest boiling point is expected for:
(a) iso-octane
(b) n-octane
(c) 2, 2, 3, 3-tetramethyl butane
(d) n-butane

36. In which reaction, the product from anti-Markovnikov's rule is observed?
(a) $CH_3CH{=}CH_2 \xrightarrow{HCl}$
(b) $ClCH{=}CH_2 \xrightarrow{HCl}$
(c) $CH_3OCH{=}CH_2 \xrightarrow{HCl}$
(d) None of these

37. Acidic hydrogen is present in:
(a) Ethyne (b) Ethene
(c) Benzene (d) Ethane

38. Which is not a product when (cyclohexanone ring bearing HO, OH and OH substituents) is treated with HIO_4?
(a) HCOOH (b) H_2O
(c) HCHO (d) CO_2

39. Anti-Markovnikov's addition of HBr is not observed in:
(a) Propene (b) but-1-ene
(c) but-2-ene (d) pent-3-ene

40. Which is an incorrect statement regarding oxymercuration–demercuration?
(a) Here rearrangement takes place
(b) In the second step demercuration occurs that is, $NaBH_4$ reduces–$HgOAc_2$ group to hydrogen

(c) The net reaction is addition of water according to Markovnikov's rule
(d) In the first step oxymercuration occurs, that is, water and $Hg(OAc)_2$ add to double bond

41. Baeyer's reagent is:
(a) Acidified permanganate solution
(b) Alkaline permanganate solution
(c) Neutral permanganate solution
(d) Aqueous bromine solution

42. $CH_2{=}CH{-}CH_3 \xrightarrow[HOBr]{NBS}$
(a) C – C – C with Br, Br, Br
(b) C – C – C with Br, OH, Br
(c) C – C – C with OH, Br, Br
(d) C – C – C with OH, OH, Br

43. The compound with the highest boiling point is:
(a) n-pentane
(b) 2-methyl butane
(c) 2, 2-dimethyl propane
(d) n-hexane

44. Which of the following will not give an alkene when treated with $Ph_3P{=}CH_2$?
(a) $R_2C{=}C{=}O$ (b) RNCO
(c) RNC (d) $R_2C{=}NR$

45. Which of the following will show geometrical isomerism?
(a) 2-methyl butene
(b) Propene
(c) Vinyl chloride
(d) but-2-ene

46. In the given reaction:
$CH_3{-}C{\equiv}C{-}CH_3 \xrightarrow{Na/NH_3(1)}$ [X] will be
(a) Butane (b) trans-2-butene
(c) cis-2-butene (d) 1-butene

47. Propyne $\xrightarrow[Hg_2SO_4]{HgSO_4}$ A. Compound A is:
(a) 2-propanal (b) Propane
(c) Propene (d) 2-propanone

48. In the given reaction (P) will be:

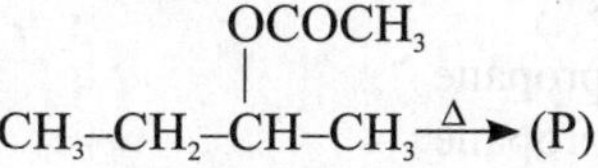

(a) 2-butanol (b) 2-butene
(c) 1-butene (d) 1-butanol

49. Which hydrocarbon has the highest octane number?
(a) Methane (b) Ethane
(c) Iso-octane (d) Triptane

50. In the reaction sequence (A) will be:

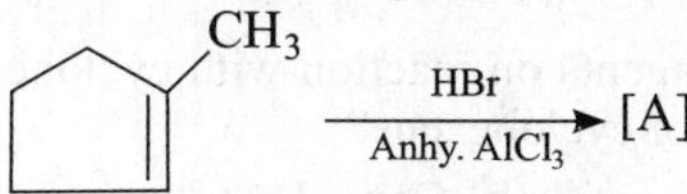

(a) 1-bromo-2- ethyl cyclopentane
(b) 1-bromo-1-methyl cyclopentane
(c) 1-bromo-5-methyl cyclopentane
(d) 5-bromo-1-methyl cyclopentane

51. Which of the following gives propyne on hydrolysis?
(a) Al_4C_3 (b) Mg_2C_3
(c) B_4C (d) La_4C_3

52. Which among the following reagents converts cyclohexene into trans diol?
(a) Baeyer reagent
(b) OsO_4
(c) I_2/CH_3COOAg (wet)
(d) I_2/CH_3COOAg (dry)

53. The product formed on reaction of HBr with propene is:
(a) 1-bromoproapne (b) 2-bromopropane
(c) 3-bromopropane (d) No reaction occurs

54. Molecular weight of an unknown compound [A] is 82. Compound [A] will be:
(a) $CH_3{-}CH_2{-}C{\equiv}C{-}CH_2{-}CH_3$
(b) $CH_2{=}CH_2{-}CH_2{-}CH_2{-}CH{=}CH_2$
(c) 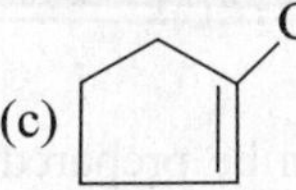
(d) All of these

55. There is free rotation about the carbon-carbon σ-bond in ethane. This means that the energy difference between the staggered and eclipsed conformers is:
(a) 3 kJ mol^{-1} (b) 23 kJ mol^{-1}
(c) 13 kJ mol^{-1} (d) 17 kJ mol^{-1}

56. Potassium salt of 2, 3-dimethyl maleic acid on electrolysis gives:
(a) Ethyene (b) Propyne
(c) 2-butyne (d) 1-butyne

57. Which of the following shows acidic character?
(a) $CH_3–CH_3$ (b) $CH_3–C≡CH$
(c) $CH_2=CH_2$ (d) $CH_3≡C–CH_3$

58. Which among the following alkyne will give an aldehyde on hydroboration oxidation reaction?
(a) CH≡CH (b) $CH_3–CH_2–C≡C–H$
(c) $CH_3–C≡C–H$ (d) all of these

59. Which of the reagents on reaction with cyclohexanol gives best yield of cyclohexene?
(a) Conc. H_3PO_4 (b) Conc. HCl
(c) Conc. HBr (d) Conc. HCl with $ZnCl_2$

60. Reductive ozonolysis of the alkene, $CH_3–CH=C=CH_2$ will give:
(a) Only CO_2
(b) Only CH_2O
(c) Only $CH_3–CHO$
(d) Mixture of CH_3CHO, HCHO and CO_2

61. In the preparation of alkene from alcohol using Al_2O_3, which is the most effective factor?
(a) Porosity of Al_2O_3
(b) Temperature
(c) Surface area of Al_2O_3
(d) Concentration

62. Which has the maximum boiling point as well as melting point out of the following:

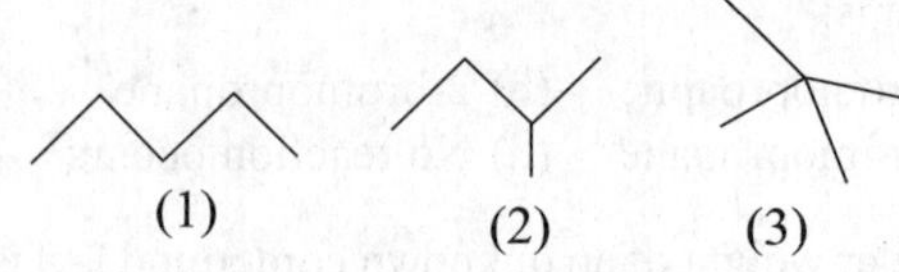
(1) (2) (3)

(a) 1 in both the cases (b) 1, 2
(c) 1, 3 (d) 2, 1

63. In the given reaction, [A] will be

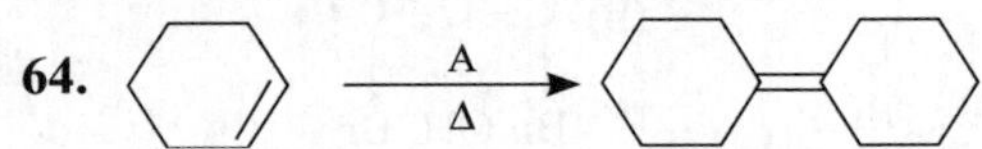

(a) Meso-2, 3-dibromobutane
(b) Racemic mixture of 2, 3-dibromobutane
(c) Meso as well as racemic mixture
(d) Reaction is not possible

64. [cyclohexene] $\xrightarrow[\Delta]{A}$ [cyclohexylidenecyclohexane]

In the above reaction, A will be
(a) t-BuOK (b) alcoholic KOH
(c) Et_3N (d) conc. H_2SO_4

65. $R–CH=CH_2 + CCl_4 \xrightarrow{\text{Peroxide}}$ A. Here, A is
(a) R CHCH$_2$ CCl$_3$ (with Cl on the CH)
(b) R CHCH$_2$ Cl (with CCl$_3$ on the CH)
(c) Both are correct
(d) Reaction does not take place

PRACTICE QUESTIONS – II

66. Benzyl chloride ($C_6H_5CH_2Cl$) can be prepared from toluene by chlorination with
(1) SO_2Cl_2 (2) $SOCl_2$
(3) Cl_2 (4) NaOCl
(a) 1 and 3 (b) 1 and 4
(c) 2 and 3 (d) 4

67. In the given reaction,

$$C_7H_8 \xrightarrow{3Cl\ \Delta} A \xrightarrow{Br_2/Fe} B \xrightarrow{Zn/HCl} C$$

The compound C is
(a) o-bromotoluene
(b) m-bromotoluene
(c) p-bromotoluene
(d) 3-bromo-2, 2, 6-trichlorotoluene

68. 0.037 g of an alcohol, R–OH was added to CH_3MgI and the gas evolved measured 11.2 cm^3 at STP. What is the molecular mass of R–OH?
(a) 46 (b) 60
(c) 74 (d) 88

69. $(H_3C)_2CH–I \xrightarrow[\text{(ii) } SO_2Cl_2,\ D]{\text{(i) Alc KOH}}$?
The product formed is
(a) 1, 2-dichloroethane
(b) 3-chloropropene
(c) 1-chloro-2-iodopropane
(d) 1-iodo-2-chloropropane

70. In which of the following compounds, the weight percentage of C and H are 80 and 20 respectively?

(a) C_2H_6 (b) CH_4
(c) C_6H_6 (d) C_2H_3

71. The relative rates of hydrogenation is in the order of:

(a) $R_2C{=}CHR > RCH{=}CHR > RCH{=}CH_2 > CH_2{=}CH_2$
(b) $CH_2{=}CH_2 > RCH{=}CH_2 > RCH{=}CHR > R_2C{=}CHR$
(c) $RCH{=}CHR > R_2C{=}CHR > RCH{=}CH_2 > CH_2{=}CH_2$
(d) $R_2C{=}CHR > CH_2{=}CH_2 > RCH{=}CHR > RCH{=}CH_2$

72. $\begin{matrix} CHCOONa \\ \| \\ CHCOONa \end{matrix} \xrightarrow{\text{Electrolysis}} [A] \xrightarrow[\text{liquid } NH_3]{I_2} [B]$

Here, [B] is:

(a) $CH{\equiv}CH$ (b) $NaC{\equiv}CNa$
(c) $I{-}C{\equiv}C{-}I$ (d) $IC{\equiv}CNa$

73. In the reaction of phenol with $CHCl_3$ and aqueous NaOH at 70°C (343 K), the electrophile attacking the ring is:

(a) $CHCl_3$ (b) $CHCl_2$
(c) CCl_2 (d) $COCl_2$

74. $C-C-C\equiv C \underset{B}{\overset{A}{\rightleftharpoons}} C-C\equiv C-C$

Here, [A] and [B] are:

(a) alc. KOH and $NaNH_2$
(b) $NaNH_2$ and Lindlar catalyst
(c) $NaNH_2$ and alcoholic KOH
(d) Lindlar catalyst and $NaNH_2$

75. Benzene reacts with CH_3COCl in the presence of anhydrous $AlCl_3$ to give:

(a) $C_6H_5CH_3$ (b) C_6H_5Cl
(c) $C_6H_5O_2Cl$ (d) $C_6H_5COCH_3$

76. In this reaction, A is:

$\xrightarrow{Cl_2,H_2O}$ A

(a) Cl, OH (b) OH, Cl
(c) OH, Cl (d) OH, Cl

77. $CH\equiv CH + 2HCHO \xrightarrow{CH_2O} X$ is:

(a) $HOH_2C{-}C{\equiv}C{-}CH_2OH$
(b) $HOH_2C{-}C{\equiv}C{-}CH_2OCH_3$
(c) Both are true
(d) None

78. What will be the product in the following reaction?

CH_2 $\xrightarrow{NBS}$

(a) Br, CH_3 (b) CH_3, Br
(c) CH_2Br (d) CH_3, Br

79. Which is most easily dehydrohalogenated?

—Cl 1 —Cl 2 —Cl 3

(a) 1 (b) 2
(c) 3 (d) Cannot be determined

80. Chlorobenzene can be prepared by reacting aniline with:

(a) Nitrous acid followed by heating with cuprous chloride
(b) Chlorine in presence of anhydrous aluminium chloride
(c) Cuprous chloride
(d) Hydrochloric acid

81. Which of the following compounds does not dissolve in concentrated H_2SO_4 even on warming?

(a) Aniline (b) Benzene
(c) Ethylene (d) Hexane

82. When propyne is treated with aqueous H_2SO_4 in presence of $HgSO_4$ the major product is

(a) Acetone
(b) Propanol
(c) Propanal
(d) Propyl hydrogen sulphate

83. $R-C\equiv C-R \xrightarrow[\text{Lindlar catalyst}]{H_2} ?$

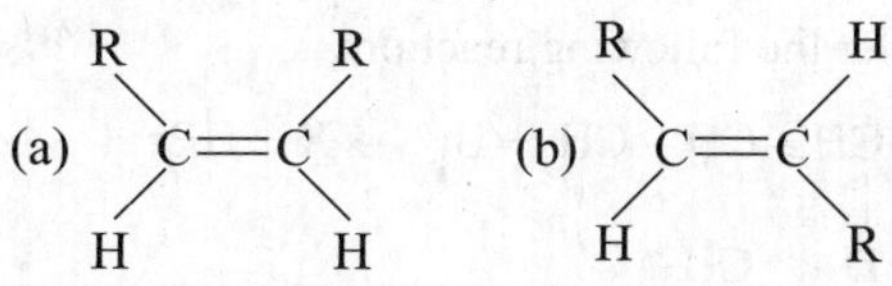

(c) Both (a) And (b) (d) $R{-}CH_2{-}CH_2{-}R$

84. Which is the decreasing order of strength of bases: OH^-, NH_2^-, $HC\equiv C^-$ and $CH_3CH_2^-$?

(a) $CH_3CH_2^- > NH_2^- > HC\equiv C^- > OH^-$
(b) $HC\equiv C^- > CH_3CH_2^- > NH_2^- > OH^-$
(c) $OH^- > NH_2^- > HC\equiv C^- > CH_3CH_2^-$
(d) $NH_2^- > HC\equiv C^- > OH^- > CH_3CH_2^-$

85. 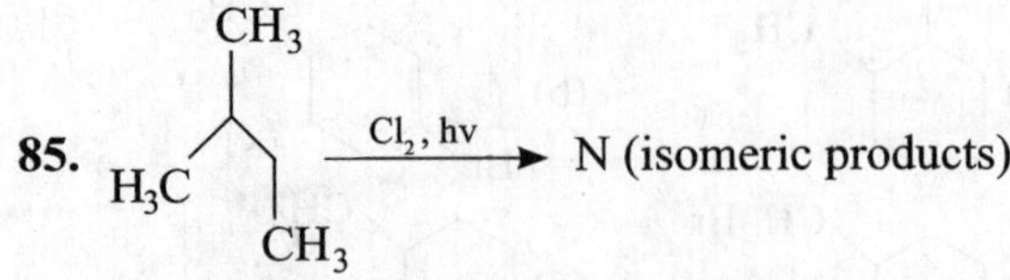

$C_5H_{11}Cl \xrightarrow[\text{distillation}]{\text{Fractional}} M$ (isomeric products)

(a) 6, 6 (b) 6, 4
(c) 4, 4 (d) 3, 3

86. $CH_3-CH=CH_2 + NOCl \longrightarrow P$. Identify P.

(a) $CH_3-CH(Cl)-CH_2(NO)$ (b) $CH_3-CH(NO)-CH_2(Cl)$
(c) $CH_3-CH_2-CH_2(Cl)$ (d) $CH_2(NO)-CH_2-CH_2(Cl)$

87. $Ph-C\equiv C-CH_3 \xrightarrow{Hg^{2+}/H^+} A$. Here, A is:

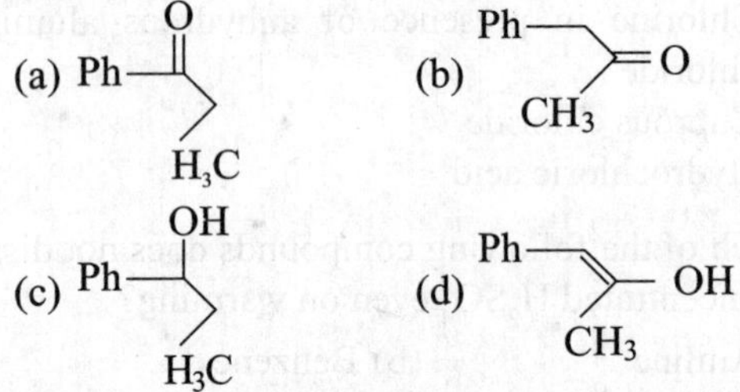

88. Which of the following are arromatic here:

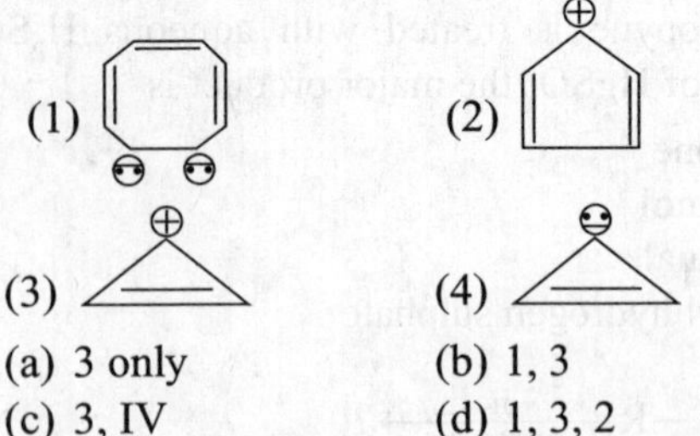

(a) 3 only (b) 1, 3
(c) 3, IV (d) 1, 3, 2

89. Consider the following reaction:

$CH_3-CH(D)-CH(CH_3)-CH_3 + Br^\circ \rightarrow X + HBr$

Identify the structure of the major product X:

(a) $H_3C-CH(D)-CH(CH_3)-\overset{\circ}{C}H_2$
(b) $H_3C-CH(D)-\overset{\circ}{C}(CH_3)-CH_3$
(c) $H_3C-\overset{\circ}{C}(D)-CH(CH_3)-CH_3$
(d) $H_3C-\overset{\circ}{C}H-CH(CH_3)-CH_3$

90. Wurtz reaction of methyl iodide yields an organic compound X. Which one of he following reactions also yields X?

(a) $CHCl_3 \xrightarrow{\text{Ag powder, }\Delta}$
(b) $C_2H_5Cl + C_2H_5ONa \longrightarrow$
(c) $C_2H_5Cl + LiAlH_4 \longrightarrow$
(d) $C_2H_5Cl + Mg \xrightarrow{\text{Dry ether}}$

91. Identify the reagent from the following list which can easily distinguish between 1-butyne and 2-butyne?

(a) Bromine, CCl_4
(b) H_2, Lindlar catalyst
(c) Dilute H_2SO_4, $HgSO_4$
(d) Ammonical cuprous chloride

92. In the presence of peroxide, and hydrogen chloride, hydrogen iodide does not undergo anti-Markovnikov's addition to alkenes because:

(a) Both are highly ionic
(b) One is oxidizing and other is reducing
(c) One of the steps is endothermic in both the cases
(d) All the steps are exothermic in both the cases

93. The reaction of propene with HOCl proceeds via the addition of:

(a) H^+ in the first step
(b) Cl^+ in the first step
(c) OH^- in the first step
(d) Cl^+ and OH^- in a single step

94. (Structure: Me, Me, H on central carbon; Me–C≡C– substituent; –CH=CH–Me with H, Me, H labels)

Hydrogenation of the above compound in the presence of poisoned palladium catalyst gives:

(a) Optically active compound
(b) An optically inactive compound
(c) A racemic mixture
(d) A diastereomeric mixture

95. The reaction of HBr with:

$CH_3-C(CH_3)=CH_2$ in the presence of peroxide will give

(a) $CH_3CBr(CH_3)CH_3$ (b) $CH_3CH_2CH_2CH_2Br$

(c) $CH_3CH(CH_3)CH_2Br$ (d) $CH_3CH_2CH(CH_3)CH_3$

96. Ozonolysis of mesitylene gives:

(a) $CH_3–CO–CHO$ (b) $CH_3–C(=O)–C(=O)–CH_3$

(c) CHO–CHO (d) Both a, b

97. Propyne and propene can be distinguished by:

(a) Conc. H_2SO_4 (b) Br_2 in CCl_4

(c) Diluted H_2SO_4 (d) $AgNO_3$ in ammonia

98. The product(s) obtained via oxymercuration ($HgSO_4$ + H_2SO_4) of but-1-yne would give:

(a) $CH_3CH_2–C(=O)–CH_3$

(b) $CH_3CH_2CH_2–CHO$

(c) $CH_3CH_2CHO + HCHO$

(d) $CH_3CH_2COOH + HCOOH$

99. The reaction of:

$CH_3-CH=CH-C_6H_4-OH$

with HBr gives:

(a) $CH_3-CHBrCH_2-C_6H_4-OH$

(b) $CH_3CH_2CHBr-C_6H_4-OH$

(c) $CH_3CHBrCH_2-C_6H_4-Br$

(d) $CH_3CH_2CHBr-C_6H_4-Br$

100. Ethene and ethyne can be distinguished by:

(a) Br_2 water

(b) $KMnO_4$ solution

(c) Cuprous chloride solution

(d) Any of the above

101. 2, 3-dimethyl hexane contains …. tertiary…. secondary and ….. primary carbon atoms, respectively.

(a) 2, 4, 3 (b) 3, 2, 4

(c) 4, 3, 2 (d) 2, 2, 4

102. One mole of an unsaturated hydrocarbon on ozonolysis gives one mole each of CH_3CHO, HCHO and OHC.CHO. The hydrocarbon is:

(a) $CH_3–CH_2C≡C–CH_3$

(b) $CH≡C–CH_2CH_2–CH_3$

(c) $CH_3–CH=CH–CH=CH_2$

(d) $CH_2=CH–CH_2–CH=CH_2$

103. Which among the following will give a precipitate with ammonical silver nitrate?

(a) 2-butene (b) 2-butyne

(c) Chlorobenzene (d) 3-methyl-1-butyne

104. Which of the following sequence of reagents can be used for the conversion of $C_6H_5CH_2CH_3$ into C_6H_5 $CH=CH_2$?

(a) $SOCl_2$; H_2O (b) SO_2Cl_2; alc. KOH

(c) Cl_2/hv; H_2O (d) $SOCl_2$; alc. KOH

105. An alkene on reductive ozonolysis gives two molecules of $CH_2(CHO)_2$. The alkene is:

(a) 2, 4-hexadiene

(b) 1, 3-cyclohexadiene

(c) 1, 4-cyclohexadiene

(d) 1-methyl-1, 3-cyclopentadiene

106. Bromide is converted into ethane by heating it in an ether medium with:

(a) Al (b) Zn

(c) Na (d) Cu

107. In the solvolysis of 3-methyl-3-bromohexane, which of the following statements is not correct?

(a) It involves carbocation intermediate

(b) The intermediate involves sp^2 carbon

(c) Polar solvents accelerates the reaction

(d) It involves inversion of configuration

108. Which one of the following compounds would have the highest heat of hydrogenation?

(a) $(CH_3)_2C=C(CH_3)_2$

(b) $H_3C–CH_2–CH=CH_2$

(c) $H_3C–CH=CH–CH_3$

(d) $H_2C=CH_2$

109. When 2-methyl butan-1-ol is dehydrated to give an alkene, the preferred product is:

(a) But-1-ene

(b) But-2-ene

(c) 2-methyl but-1-ene

(d) 2-methyl but-2-ene

110. A hydrocarbon of molecular formula, C_8H_{18} gives only one monochloro derivative. The hydrocarbon is:
(a) 2-methyl heptane
(b) Octane
(c) 2, 2, 3, 3-tetramethyl butane
(d) 2, 2, 4-trimethyl pentane

111. Chloroethane reacts with magnesium in dry ether to form X. When X is hydrolyzed, a carbon compound Y and Z are formed. Which of the following is Y?
(a) C_2H_4 (b) C_2H_2
(c) C_2H_6 (d) C_6H_6

112. Which one of the following compound yields only one product on monobromination?
(a) Neopentane
(b) Aniline
(c) Phenol
(d) Toluene

113. A hydrocarbon of molecular formula C_7H_{12} on catalytic hydrogenation over platinum gives C_7H_{16}. The parent hydrocarbon adds bromine and also reacts with $[Ag(NH_3)_2]OH$ to give a precipitate. The parent hydrocarbon is:
(a) $(CH_3)_3\ CCH_2C \equiv CH$
(b) $CH_3CH = CHCH_2CH = CH_2$
(c) $CH_3CH_2C \equiv CCH(CH_3)_2$
(d) $CH_3CH = CHCH = CHCH_3$

114. In order to obtain trans Me–CH=CH–Me from MeC≡CMe, the reagent(s) to be used would includes:
(a) $LiAlH_4$ in EtO_2
(b) H_2/Pd, $BaSO_4$
(c) H_2/Pd, $Pb(OAc)_2$
(d) Sodium in liquid ammonia or sodium in EtOH

115. Cold and dil. $KMnO_4$ combines with but-2-yne to form:
(a) Butane-1, 3-diol (b) Butane-2, 3-diol
(c) Butane-2, 3-dione (d) Butan-2-one

116. The products formed by the ozonolysis–hydrolysis of a compound of formula, C_5H_8 are $CH_3CH_2CH_2COOH$ and CO_2. The compound is:
(a) Pent-2-yne
(b) Pent-1-yne
(c) Penta-1, 3-diene
(d) Penta-1, 4-diene

117. A hydrocarbon C_6H_{12} decolourizes bromine solution and yields n-hexane on hydrogenation. On oxidation with $KMnO_4$ it forms two different monobasic acids of the type RCOOH. The compound is:
(a) Hex-2-ene (b) Hex-3-ene
(c) Cyclohexene (d) Hex-1-ene

118. The ozonolysis of 1, 2-dimethyl benzene (o-xylene) gives:
(a) Methyl glyoxal (b) Glyoxal
(c) Dimethyl glyoxal (d) All of these

119. The most reactive species among the following towards sulphonation is:
(a) Toluene
(b) Nitrobenzene
(c) 1, 3-dimethyl benzene
(d) Chlorobenzene

120. 18 C–H and 7C–C sigma bonds are present in:
(a) n-heptane
(b) Cyclohexane
(c) 3, 3-dimethyl pentane
(d) 2, 2, 3-trimethyl pentane

121. Match the following:

List 1 (Petroleum fractions)	List 2 (Approximate composition)
1. Gasoline	(1) C_{10} to C_{15}
2. Kerosene	(2) C_6 to C_8
3. Lubricating oil	(3) C_{30} to C_{40}
4. Pitch and asphalt	(4) C_{17} to C_{20}

The correct matching is:

	1	2	3	4
(a)	(2)	(1)	(3)	(4)
(b)	(1)	(3)	(4)	(2)
(c)	(1)	(2)	(4)	(3)
(d)	(2)	(1)	(4)	(3)

122. Which one of the following is aromatic in nature?

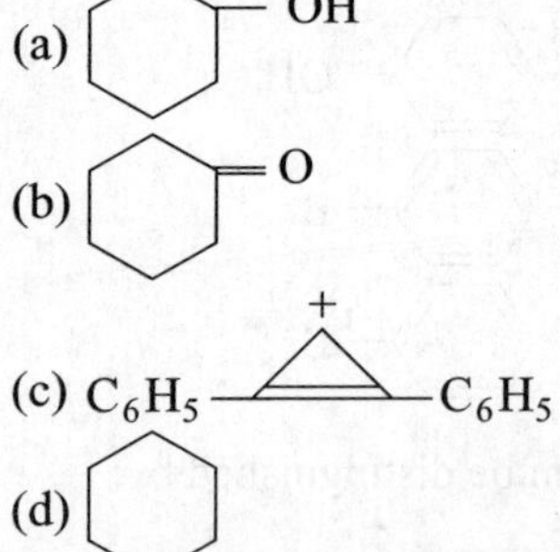

123. Cyclooctatetraene is not aromatic. The most important reason for this is that:
(a) It is a planar molecule
(b) It has eight π-electrons
(c) Its structure cannot be described by more than the canonical forms
(d) Its structure is not that of regular octagon

124. Which of the following reactions will yield 2, 2-dibromopropane?

(a) $CH_3CH = CHBr + HBr \longrightarrow$
(b) $HC \equiv CH + 2HBr \longrightarrow$
(c) $CH_3CH = CH_2 + HBr \longrightarrow$
(d) $CH_3C \equiv CH + 2HBr \longrightarrow$

125. Toluene reacts with methyl chloride in presence of anhydrous aluminium chloride to form mainly m-xylene. This is because:

(a) CH_3- group has +I effect
(b) CH_3- group is meta directing
(c) M-xylene is thermodynamically most stable of the other xylenes
(d) Hyperconjugation effect of CH_3- group

Practice Questions – III

126. Match the following:

List 1	List 2
1. Benzene	(1) Phosgene
2. Ethylene	(2) Silver mirror
3. Acetaldehyde	(3) Mustard gas
4. Chloroform	(4) $(4n + 2)$ π-electrons
	(5) Carbylamine

The correct matching is:

	1	2	3	4
(a)	(4)	(3)	(1)	(2)
(b)	(4)	(3)	(2)	(1)
(c)	(3)	(4)	(2)	(1)
(d)	(2)	(1)	(4)	(3)

127. Ozonolysis of 2, 3-dimethyl-1-butene followed by reduction with zinc and water gives:

(a) Methanoic acid and 3-methyl-2-butanone
(b) Methanal and 3-methyl-2-butanone
(c) Methanal and 2-methyl-3-butanone
(d) Methanoic acid and 2-methyl-3-butanone

128.

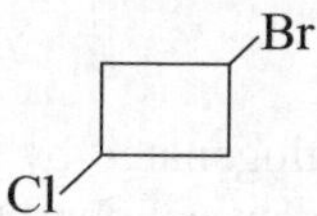

when treated with two equivalents Na in dry ether gives:

(a) 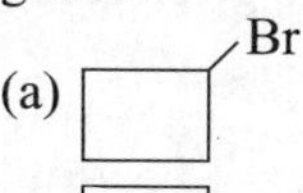

(b) 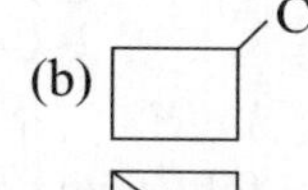

(c)

(d)

129. The number of structural and configurational isomers of a bromo compound, C_5H_9Br, formed by the addition of HBr to 2-pentyne respectively are:

(a) 1 and 2 (b) 2 and 4
(c) 4 and 2 (d) 2 and 1

130. Compare the C–H bond lengths of the following hydrocarbons:

1. $CH_3–CH_3$ 2. $CH_2=CH_2$
3. $CH \equiv CH$

The decreasing order of C–H bond lengths is:

(a) 3, 2, 1 (b) 1, 2, 3
(c) 2, 1, 3 (d) 3, 1, 2

131. A hydrocarbon C_6H_{10} does not react with ammonical solution of Cu^+ ions; it adsorbs 2 mol of hydrogen on catalytic hydrogenation to give 2-methylpentane. The hydrocarbon adds one molecule of water on treatment with Hg^{2+}/H_2SO_4. The structure of given compound is:

(a) $H_2C = CH - CH = C(CH_3)_2$
(b) $H_3C - C \equiv C - CH(CH_3)_2$
(c) $H_2C = C(CH_3) - CH_2 - CH = CH_2$ (with CH_3 on the second carbon)
(d) $HC \equiv C - CH_2 - CH(CH_3)_2$

132. Among the following compounds which has more than one type of hybridization for carbon atom?

1. $CH_3 - CH_2 - CH_2 - CH_3$
2. $CH_3 - CH = CH - CH_3$
3. $CH_2 = CH - C \equiv CH$
4. $CH_2 = CH_2$

(a) 2 only (b) 2 and 3
(c) 1 and 3 (d) 3 and 4

133. Match the following:

List 1	List 2
1. Ethane	(1) 2sp carbons
2. Ethylene	(2) $6sp^2$ carbons
3. Acetylene	(3) $2sp^3$ carbons
4. Benzene	(4) $2sp^2$ carbons
	(5) 1sp and 1sp2 carbons

The correct matching is:

	1	2	3	4
(a)	(2)	(3)	(4)	(5)
(b)	(3)	(1)	(2)	(5)

(c) (4) (5) (3) (2)
(d) (3) (4) (1) (2)

134. In the following sequence of reactions, what is D?

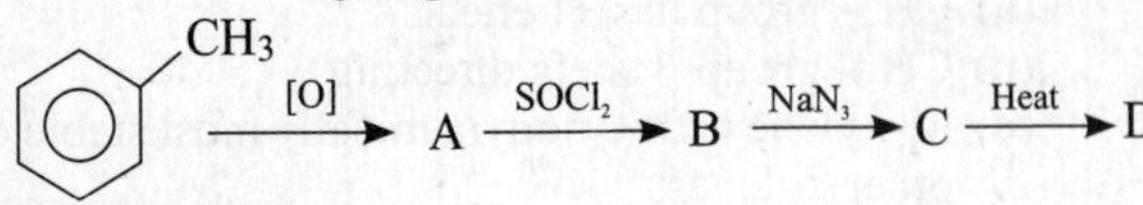

(a) Primary amine
(b) An amide
(c) Phenyl isocyanatc
(d) A chain lengthened hydrocarbon

135. Among the following, the anti-aromatic compound(s) is/are:

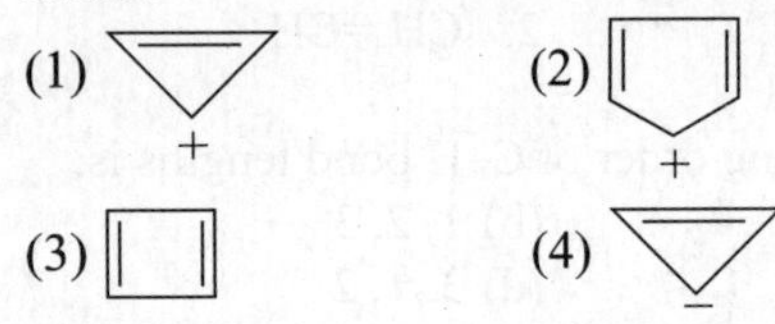

(a) 1, 2 and 3 (b) 2, 3 and 4
(c) 1 and 4 (d) 1, 3 and 4

136. In the following reaction:

$(CH_3)_2CHC{\equiv}CH \xrightarrow{\text{Step I}}$

$(CH_3)_2CH{-}\overset{\displaystyle Br}{\overset{|}{C}}{=}CH_2 \xrightarrow{\text{StepII}} (CH_3)_2CH{-}\overset{\displaystyle Br}{\overset{|}{CH}}{=}CH_2Br$

Which of the following sets of reagents can be used for step 1 and step 2?

Step 1	**Step 2**
(1) HBr	HBr and peroxide
(2) HBr and peroxide	HBr
(3) Br_2	HBr
(4) Br_2	HBr and peroxide

Select the correct answer using the codes given below:

(a) 1 alone (b) 1 and 2
(c) 1, 2 and 3 (d) 1, 2 and 4

137. In the following sequence of reactions the end product P is:

$C_6H_5C_2H_5 \xrightarrow{Cl_2/FeCl_3} \dots\dots \xrightarrow{Cl_2/\Delta} \dots\dots \xrightarrow{\Delta/\text{alc. KOH}} P$

(a) $CH=CH_2$ on a benzene ring bearing Cl (Cl at para position)

(b) $C_6H_5{-}\overset{\displaystyle Cl}{\overset{|}{C}}{=}CHCl$

(c) $C_6H_5{-}CH{=}CHCl$

(d) $CH{=}CH_2$ on a benzene ring bearing Cl (Cl at meta position)

138. The heat of hydrogenation of benzene is 51 kcal/ mol. The resonance energy of benzene is 36 kcal/ mol. The heat of hydrogenation of cyclohexene is approximately:

(a) 9 kcal/mol (b) 19 kcal/mol
(c) 29 kcal/mol (d) 27 kcal/mol

139. The heat of hydrogenation of benzene is 51 kcal/ mol and its resonance energy is 36 kcal/mol. Then the heats of hydrogenation of cyclohexadiene and cyclohexene are respectively:

(a) 58 kcal, 29 kcal
(b) 28 kcal, 59 kcal
(c) 58 kcal, 49 kcal
(d) 29 kcal, 48 kcal

140. Consider the following alkenes:

1. $H_2C = C\,(CH_2CH_3)\,CH(CH_3)_2$
2. $(CH_3)_2C = C(CH_3)\,CH_2\,CH_3$
3. $CH_3CH = C(CH_3)\,CH(CH_3)_2$

The correct sequence of increasing order of stability of these alkenes is

(a) 3, 1, 2 (b) 1, 3, 2
(c) 1, 2, 3 (d) 2, 1, 3

141. Consider the following compounds:

1. $CH_3 - CH_2 - CHCl - CH_3$
2. $CH_2 = CH - CH_2 - CH_2Cl$
3. $CH_3 - CH_2 - CH_2 - CH_2Cl$

These compounds are dehydrohalogenated by treatment with a strong base under identical conditions. The correct sequence of the increasing order of reactivity of these compounds in the given reaction is

(a) 3, 1, 2 (b) 3, 2, 1
(c) 1, 2, 3 (d) 2, 1, 3

142. Consider the following compounds:

1. $CH_3CH_2CH_2CH_2OH$
2. $CH_3CH_2CH(OH)CH_3$
3. $(CH_3)_3COH$

These compounds are dehydrated by treatment with sulphuric acid. The correct sequence of increasing order of the reactivity of these three compounds towards dehydration is

(a) 1, 3, 2 (b) 1, 2, 3
(c) 2, 1, 3 (d) 3, 1, 2

143. Match the following:

List 1 (Reagent)	**List 2 (Electrophiles)**
1. $Cl_2 + AlCl_3$	(1) $\overset{\oplus}{NO_2}$
2. $HNO_3 + H_2SO_4$	(2) $Cl \ldots\ldots Cl \ldots\ldots AlCl_3$ or Cl^+
3. $H_2S_2O_7$	(3) SO_3H (or H_2SO_4+SO_3)
4. Br_2 + Fe	(4) SO_3
	(5) $Br \ldots Br \ldots FeBr_3$ or Br^+

The correct matching is:

	1	2	3	4
(a)	(4)	(3)	(2)	(1)
(b)	(2)	(1)	(4)	(5)
(c)	(2)	(1)	(3)	(5)
(d)	(2)	(3)	(1)	(4)

144. Consider the following sequence of reactions and decide the end product Y.

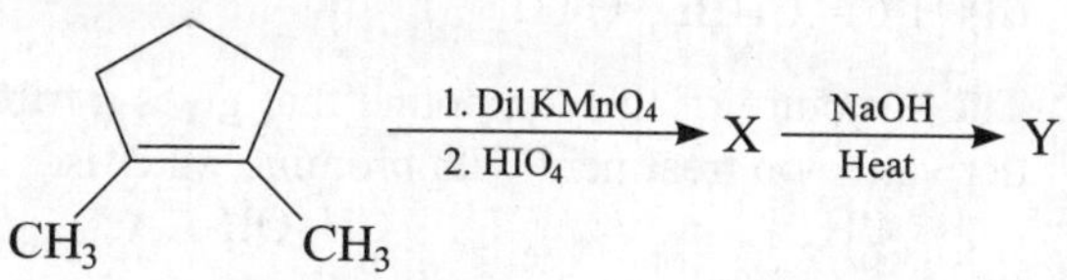

(1,2-dimethylcyclopentene) $\xrightarrow[\text{2. HIO}_4]{\text{1. Dil KMnO}_4}$ X $\xrightarrow[\text{Heat}]{\text{NaOH}}$ Y

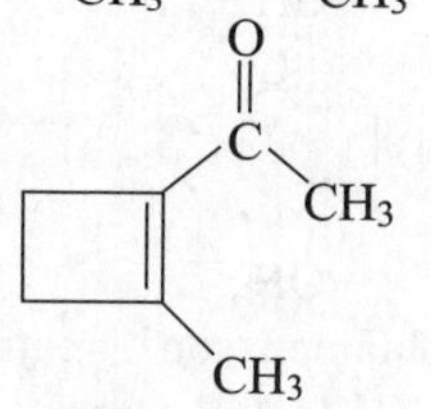

CH₃ CH₃

(a) 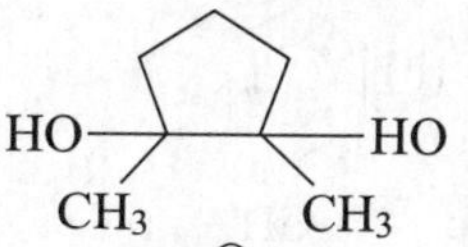 HO– –HO (cyclopentane ring with two C(CH₃)) CH₃ CH₃

(b) (cyclobutene bearing –C(=O)–CH₃ and CH₃)

(c) (cyclohexenone, O, with CH₃)

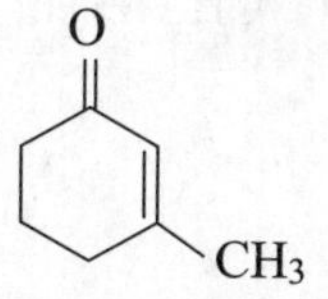

(d) $CH_3COCH_2CH_2CH_2COCH_3$

145. Place the following compounds in the decreasing order of their relative reactivity with electrophile (E^+).

1. p-nitrochlorobenzene
2. 2, 4-dinitrochlorobenzene
3. chlorobenzene

(a) $3 > 1 > 2$
(b) $1 > 2 > 3$
(c) $2 > 1 > 3$
(d) $3 > 2 > 1$

146. Consider the given statements about the molecule:
$(H_3C)_2CH–CH=CH–C≡C–CH=CH_2$

1. Three carbons are sp^3 hybridized
2. Three carbons are sp^2 hybridized
3. Two carbons are sp hybridized

Of these statements
(a) 1 and 2 are correct
(b) 1 and 3 are correct
(c) 2 and 3 are correct
(d) All are correct

147. Which one of the following would, on ozonolysis, yields CH_3COCH_3 and $CH_2(CHO)_2$ in the ratio 1:1?
(a) $(CH_3)_2$ CH–CH=CH–CH=CH–CH=C$(CH_3)_2$
(b) $(CH_3)_2$C=CH–CH_2–CH=CH–CH_2–CH=C$(CH_3)_2$
(c) $(CH_3)_2$C=CH–CH=CH–CH_2–CH_2–CH=C$(CH_3)_2$
(d) $(CH_3)_2$C=CH–CH_2–C(CH_3)=CH–CH_2–CH=CH–CH_3

148. Viscosity coefficients of some liquids are given below:

Liquid	η in millipoise at 30°C
$CH_3(CH_2)_3CH_3$	2.11
$CH_3(CH_2)_4CH_3$	2.89
$CH_3(CH_2)_5CH_3$	3.68

The order of viscosity coefficient of the liquids:
(1) $CH_3–CH_2–CH_2–CH_2–CH_2–CH_3$
(2) $CH_3–CH_2–CH(CH_3)–CH_2–CH_3$
(3) $CH_3–CH(CH_3)–CH(CH_3)–CH_3$

(a) $1 > 2 > 3$ (b) $1 < 2 < 3$
(c) $1 > 2 = 3$ (d) remains same

149. On vigorous oxidation by permanganate solution, $(CH_3)_2C = CHCH_2CHO$ gives:
(a) $(CH_3)_2CO$ and $OHCCH_2CHO$
(b) $(CH_3)_2C(OH) – CH(OH)CH_2CHO$
(c) $(CH_3)_2CO$ and $OHCCH_2COOH$
(d) $(CH_3)_2CO$ and $CH_2(COOH)_2$

150. 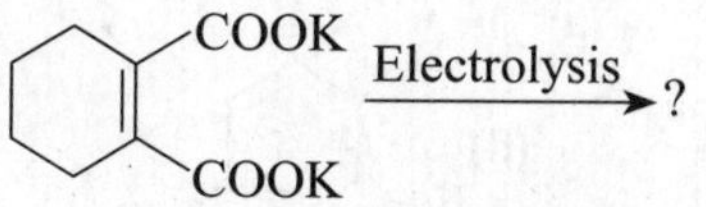 (cyclohexene-1,2-dicarboxylate: COOK, COOK) $\xrightarrow{\text{Electrolysis}}$?

Here the compound formed is?

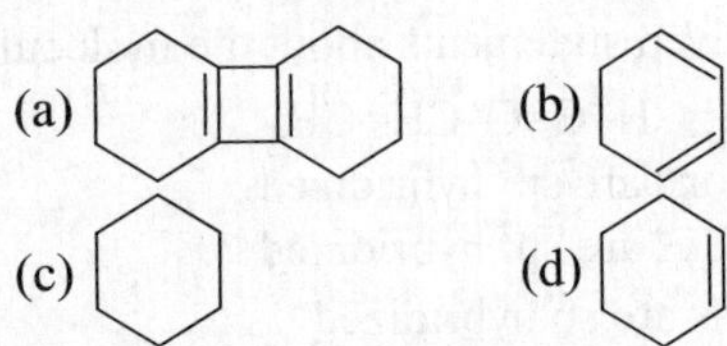

151. When 2- methylbutane is chlorinated, the percentage of $(CH_3)_2$ —CH—CH_2—CH_2Cl is nearly assuming reactivity ration of 3°H : 2°H : 1°H = 5:3 · 8:1

(a) 28% (b) 35%
(c) 23% (d) 14%

152. Which of these compounds can give 5–keto –2– methyl hexanal on ozonolysis?

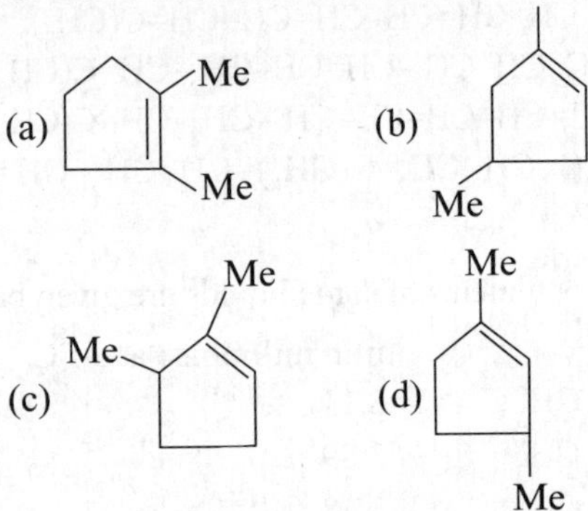

153. The reaction between HI and C_2H_4 in C_2H_5OH gives 'predominantly' C_2H_5I, whereas the reaction with HCl under the same conditions gives predominantly $(C_2H_5)_2O$. Identify the correct order of nucleophilicity of the nucleophiles involved in the above reactions.

(a) $I^- > EtO^- > Cl^-$ (b) $I^- > Cl^- > EtOH$
(c) $EtOH > Cl^- > I^-$ (d) $I^- > EtOH > Cl^-$

154. The major product obtained on the monobromination (with $Br_2/FeBr_3$) of the following compound A is:

OCH_3
CH_3
(A)

(a) OCH_3, Br, CH_3 (b) OCH_3, CH_3, Br
(c) OCH_3, Br, CH_3 (d) OCH_3, Br, CH_3

155. Identify (Y) in the following reaction:

$$CH \equiv CH \xrightarrow{O_3} (X) \xrightarrow{Zn/CH_3OH} (Y)$$

(a) $CH_3 - CH_3$ (b) $CH_2OH - CH_2OH$
(c) CH_3COOH (d) C_2H_5OH

156. Reaction of one molecule of HBr with one molecule of 1, 3-butadiene at 40°C gives predominantly:

(a) 3-bromobutene under kinetically controlled conditions
(b) 1-bromo-2-butene under thermodynamically controlled conditions
(c) 3-bromobutene under thermodynamically controlled conditions
(d) 1-bromo-2-butene under kinetically controlled conditions.

157. HBr reacts with $CH_2 = CH - OCH_3$ under anhydrous conditions at room temperature to give:

(a) CH_3CHO and CH_3Br
(b) $BrCH_2CHO$ and CH_3OH
(c) $BrCH_2 - CH_2 - OCH_3$
(d) $H_3C - CHBr - OCH_3$

158. The structure of the compound that gives a tribromo derivative on treatment with bromine water is:

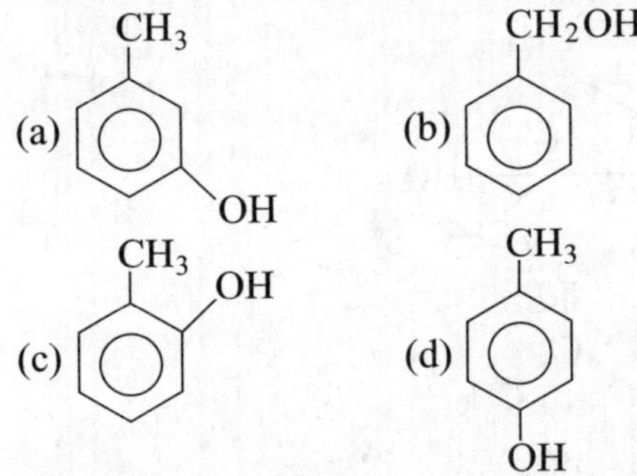

159. The major product obtained on treatment of $CH_3CH_2CH(F)CH_3$ with CH_3O^-/CH_3OH is:

(a) $CH_3CH_2CH(OCH_3)CH_3$
(b) $CH_3CH = CHCH_3$
(c) $CH_3CH_2CH = CH_2$
(d) $CH_3CH_2CH_2CH_2OCH_3$

160. Which of the following is the order of decreasing reactivity in a Diels - Alder reaction?

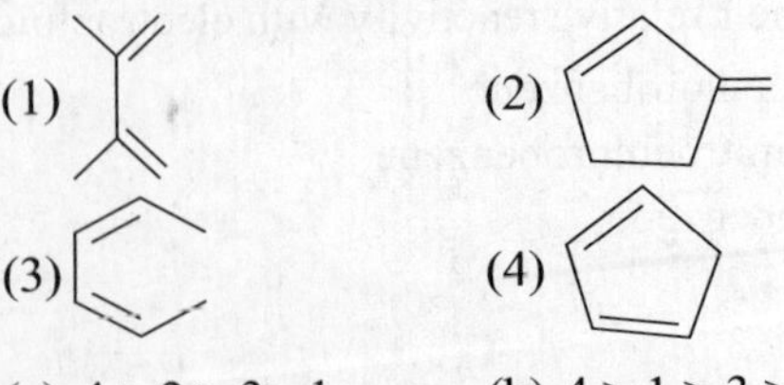

(a) 4 > 2 > 3> 1 (b) 4 > 1 > 3 > 2
(c) 4 > 1 > 2 > 3 (d) 1 > 4 > 3 > 2

161. The addition of Br_2 to Z-2 butene gives:

(a) (R, R)-2,3-dibromobutane only
(b) (S, S)-2,3-dibromobutane only
(c) (R, S)-2,3-dibromobutane only
(d) A mixture of (R, R) and (S, S)-2,3-dibromobutanes (50% : 50%)
(e) (R, S)-1,2-dibromobutane

162. The treatment of CH_3MgX with $CH_3C \equiv C - H$ produces:

(a) $CH_3 - CH = CH_2$
(b) $CH_3C \equiv C - CH_3$
(c) $CH_3 - \overset{H}{\overset{|}{C}} = \overset{H}{\overset{|}{C}} - CH_3$
(d) CH_4

163. Toluene in nitrated and the resulting product is reduced with tin and hydrochloric acid. The product so obtained is diazotized and then heated with cuprous bromide. The reaction mixture so formed contains:

(a) Mixture of o- and p-bromotoluenes
(b) Mixture of o- and p-dibromobenzenes
(c) Mixture of o- and p-bromoanilines
(d) Mixture of o- and m-bromotoluenes

164. In the following sequence of reactions, the alkene affords the compound 'B'

$$CH_3CH=CHCH_3 \xrightarrow{O_3} (A) \xrightarrow[Zn]{H_2O} (B)$$

The compound (B) is

(a) CH_3CH_2CHO (b) CH_3COCH_3
(c) $CH_3CH_2COCH_3$ (d) CH_3CHO

165. The hydrocarbon which can react with sodium in liquid ammonia is:

(a) $CH_3CH_2CH_2C \equiv CCH_2CH_2CH_3$
(b) $CH_3CH_2C \equiv CH$
(c) $CH_3CH = CHCH_3$
(d) $CH_3CH_2C \equiv CCH_2CH_3$

166. In electrophile, $E^{\oplus}$ attacks the benzene ring to generate the intermediate σ-complex. Of the following, which σ-complex is of lowest energy?

(a) 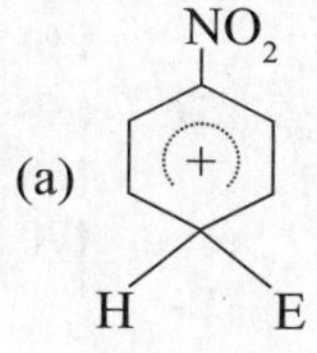

(b) 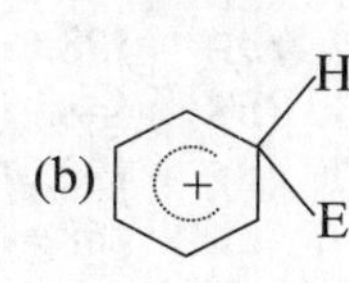

(c) [σ-complex: NO_2 on ring, H and E on adjacent carbon, + in ring]

(d) [σ-complex: NO_2 on ring, H and E on meta carbon, + in ring]

167. Which of the following most readily undergoes E2 elimination with a strong base?

(a) 2- bromo - 3 - methylbutane
(b) 2 - bromopentane
(c) 2- bromo-2- methylbutane
(d) 1- bromo-2, 2- dimethylpropane

168. The correct decreasing order towords reaction with HBr for the given all kones is given as:

1: (isobutylene skeletal structure)

2: $CH_3 - \underset{OMe}{\underset{|}{C}} = CH_2$

3: $CH_3 - \overset{NHMe}{\overset{|}{C}} = CH_2$

4: $CH_3 - \overset{CH_2-O-Me}{\overset{|}{CH}} = CH_2$

(a) 3 > 1 > 2 > 4
(b) 3 > 2 > 1 > 4
(c) 2 > 3 > 1 > 4
(d) 4 > 2 > 1 > 3

169. 2- methyl pent -2- ene on ozonolysis gives:

(a) Propanal only
(b) Propanal and methanal
(c) Propan -2 - one and propanal
(d) Propan -2- one and ethanal

170. Diel-alder reaction is fastest in:

(a) (3,4-dimethylenecyclobutene) + $CH_2=CH-NO_2$
(b) (pyrrole) N – H + $CH_2=CH-CN$
(c) (1,2-dimethylenecyclohexane) + $CH_2=CH-NO_2$
(d) (5,6-dimethylenecyclohexene) + $CH_2=CH-NO_2$

171. NBS → Ⓐ $\xrightarrow{HBr, H_2O_2}$ Ⓑ $\xrightarrow{\Delta, Mg}$ Ⓒ

(a) CH_3 (b) (c) (d) Mg Br, Br

172. Which of the species shown below is the most stable form of the intermediate in the electrophilic addition of Cl_2 in water to cyclohexene to form a halohydrin?

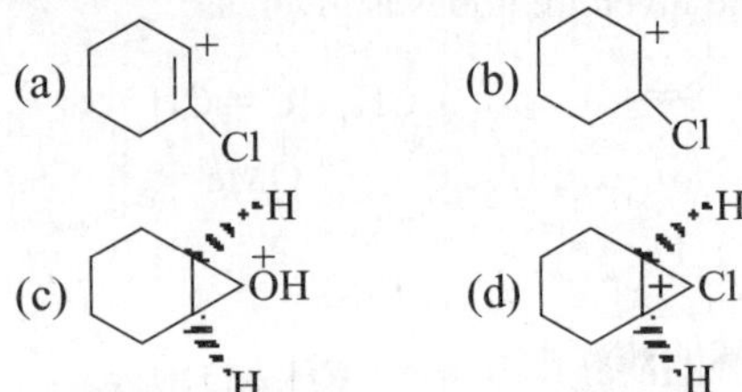

173. Which of these are aromatic here?

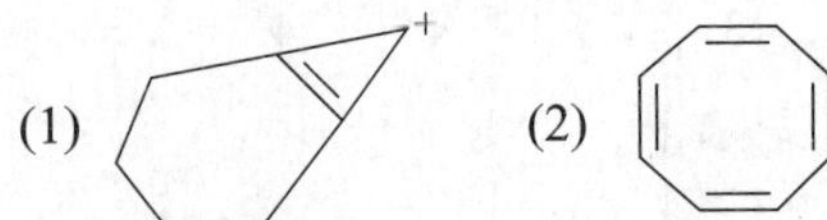

(3) (4)

(a) 1, 2 (b) 1, 3
(c) 2, 3 (d) 1, 4

174. Hexa 2, 4 di-one on ozonolysis by using O_3/CCl_4 following by H_2O/Zn gives:

(a) Acetaldehyde (b) Glyoxal
(c) Formaldehyde (d) Both A and B

175. $\xrightarrow{mCPBA}$?

Here the product are mCPBA stands for m- chloro perbenzoic acid

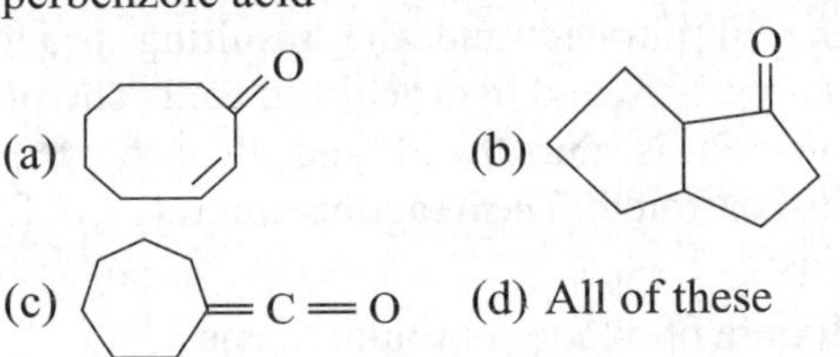

(d) All of these

Answer Keys

1. (a)	2. (d)	3. (d)	4. (a)	5. (c)	6. (c)	7. (d)	8. (a)	9. (d)	10. (c)
11. (d)	12. (d)	13. (d)	14. (c)	15. (a)	16. (a)	17. (b)	18. (b)	19. (d)	20. (b)
21. (b)	22. (d)	23. (a)	24. (c)	25. (c)	26. (a)	27. (a)	28. (b)	29. (a)	30. (c)
31. (a)	32. (c)	33. (b)	34. (a)	35. (b)	36. (b)	37. (a)	38. (c)	39. (c)	40. (a)
41. (b)	42. (b)	43. (d)	44. (c)	45. (d)	46. (b)	47. (d)	48. (c)	49. (d)	50. (b)
51. (b)	52. (d)	53. (b)	54. (d)	55. (c)	56. (c)	57. (b)	58. (d)	59. (a)	60. (d)
61. (b)	62. (b)	63. (a)	64. (d)	65. (a)	66. (a)	67. (b)	68. (c)	69. (b)	70. (a)
71. (b)	72. (d)	73. (c)	74. (c)	75. (d)	76. (a)	77. (a)	78. (a)	79. (a)	80. (a)
81. (d)	82. (a)	83. (a)	84. (a)	85. (b)	86. (b)	87. (a)	88. (b)	89. (b)	90. (c)
91. (d)	92. (c)	93. (b)	94. (b)	95. (c)	96. (a)	97. (d)	98. (a)	99. (b)	100. (c)
101. (d)	102. (c)	103. (d)	104. (b)	105. (c)	106. (c)	107. (d)	108. (d)	109. (d)	110. (c)
111.(c)	112. (a)	113. (a)	114. (d)	115. (c)	116. (b)	117. (a)	118. (d)	119. (c)	120. (d)
121.(d)	122. (c)	123. (b)	124. (d)	125. (c)	126. (b)	127. (b)	128. (d)	129. (b)	130. (b)
131.(b)	132. (b)	133. (d)	134. (c)	135. (b)	136. (a)	137. (a)	138. (c)	139. (a)	140. (b)
141. (a)	142. (b)	143. (b)	144. (c)	145. (a)	146. (b)	147. (b)	148. (d)	149. (d)	150. (a)
151. (d)	152. (d)	153. (d)	154. (b)	155. (b)	156. (b)	157. (d)	158. (a)	159. (b)	160. (b)
161. (d)	162. (d)	163. (a)	164. (d)	165. (b)	166. (b)	167. (c)	168. (b)	169. (c)	170. (c)
171. (a)	172. (d)	173. (b)	174. (d)	175. (d)					

HINTS AND EXPLANATIONS FOR SELECTIVE QUESTIONS

8. $Cl_3C-CH=CH_2 \xrightarrow[\text{anti-Markovnikov's rule}]{HCl} Cl_3C-CH=$

CH_2Cl

9. Br Br (1,3-dibromopropane) $\xrightarrow[\substack{Mg \\ R-O-R}]{\Delta}$? △ + $MgBr_2$

11. It will follow anti-Markovnikov's rule.

$$CH_3-\underset{\displaystyle |}{\overset{Ph}{C}}=CH_2 \xrightarrow{HOH} CH_3-\overset{Ph}{CH}-CH_2OH$$

12. Reactivity of Cl-atom is nearly 3,75,000 times more than Br-atom.

14. BAL is an antidote for Lewisite. It is 2, 3-dimercapto-propan-1-ol

CH_2OH
|
$CH.SH$
|
CH_2SH

15. As it will give four monochlorinated products on chlorination. But out of these only two are chiral, that is, two enantiomeric pairs.

16. $H_2C=O + HC\equiv CH + O=CH_2$

$\xrightarrow{\Delta \downarrow \text{Pressure, } CuC\equiv C.Cu}$

$HOH_2C-C\equiv C-CH_2OH$

But-2-yne-1, 4 diol

17. Peroxide effect is not applicable to the addition of HCl. The reactions takes place via more stable intermediate carbocation.

18. Here, trans-2-butene will have the lowest heat of hydrogenation.

27. In the absence of peroxides, electrophilic addition is observed. The first step is the addition of H^+ to alkene.

28. When two different alkyl halides are treated with sodium metal, three type of alkanes are formed. Here butane, pentane and hexane are formed.

29. In case of ethane and hexachloroethane, there is a less hindered rotation because of the presence of C–C single bond. Out of these, ethane has least hindered rotation because the chlorine atom is bigger than the hydrogen atom.

31. C_2H_6 must be in excess to get the best yield of C_2H_5Cl.

35. n-octane and 2, 2, 3, 3-tetramethyl butane both contain eight carbons but n-octane has a straight chain while 2, 2, 3, 3-tetramethyl butane has a branched chain.

37. Acidic hydrogen is the hydrogen attached to ≡C (triple bonded carbon).

39. Markovnikov's and anti-Markovnikov's addition, both are not observed in symmetrical alkenes. Out of these, only but-2-ene is a symmetrical alkene.

41. Baeyer's reagent is alkaline permanganate solution.

43. Higher the number of C-atoms, higher is the boiling point, that is, n-hexane.

47. $CH_3-C\equiv CH \xrightarrow[H_2SO_4]{HgSO_4} CH_3-\overset{O}{\overset{\|}{C}}-CH_3$

2-propanone

48. It follows Hoffmann rule. So, butene-1 is the major product.

50. It follows Markovnikov's rule.

51. $Mg_2C_3 + 4H_2O \longrightarrow CH_3-C\equiv CH + 2Mg(OH)_2$

53. $CH_3-CH=CH_2 + HBr \rightarrow CH_3-\underset{Br}{\underset{|}{CH}}-CH_3$

2-bromopropane

55. The staggered conformer is more stable than the eclipsed conformer by about 12.6 kJ mol^{-1}. As this energy difference is small, there is no effective energy barrier to rotation about a σ bond.

57. ≡C–H is acidic in nature, because higher the s-character in the orbital used by carbon bonded to the hydrogen, more acidic is the hydrogen.

58. As all of these are terminal alkynes so they will give aldehydes on hydroboration reaction.

59. As concentrated H_3PO_4 acts as a dehydrating agent so it will convert cyclohexanol into cyclohexene.

60. $CH_3-CH=C=CH_2 \xrightarrow[H_2O/Zn]{O_3/CCl_4} CH_3CHO + HCHO + CO_2$

(O O O O marked at the cleavage points)

67. Toluene ($C_6H_5CH_3$) $\xrightarrow[hr]{3Cl_2}$ $C_6H_5CCl_3$ $\xrightarrow{Br_2/Fe}$ m-Br-$C_6H_4CCl_3$ $\xrightarrow{Zn/HCl}$ m-Br-$C_6H_4CH_3$

m – fromoToluene

69. $(CH_3)_2CH-I \xrightarrow[KoH]{alc} CH_2=CH-CH_3 \xrightarrow{SO_2Cl_2} Cl-CH_2-CH=CH_2$

3 – Chlaropropene

70. For C_2H_6, total molecular weight = 30 g
out of 30 g ⟶ 24 g is the carbon weight, that is, 80%
same as out of 30 g ⟶ 6 g H_2 weight, that is, 20%.

73. In Reimer–Tiemann reaction,: CCl_2 is the attacking species.

74. Sodamide converts 1-yne into 2-yne while alcoholic KOH converts 2-yne into 1-yne.

75. $C_6H_6 + Cl–CO–CH_3 \xrightarrow[HgSO_4]{Anhy.\ AlCl_3} C_6H_5\ COCH_3 + HCl$

78. As NBS is a selective brominating reagent, it normally brominates the ethylenic compounds at the allylic position.

80. $C_6H_5NH_2 \xrightarrow{HONO,\ HCl} C_6H_5NH_2Cl \xrightarrow{CuCl} C_6H_5Cl$

81. $CH_2=CH_2 + H_2SO_4 \longrightarrow CH_3CH_2OSO_3H$
$C_6H_6 + H_2SO_4 \longrightarrow C_6H_5SO_3H + H_2O$
$C_6H_{14} + H_2SO_4 \longrightarrow$ No reaction
$C_6H_5NH_2 + H_2SO_4 \longrightarrow C_6H_5\ NH_3\ HSO_4^-$
Only hexane does not dissolve in concentrated H_2SO_4 even on warming.

82. $CH_3C{\equiv}CH + H_2O \xrightarrow[HgSO_4]{H_2SO_4} [CH_3CH(OH){=}CH_2] \longrightarrow CH_3COCH_3$
Acetone

86. NOCl ⟶ Markovnikov's addition

88. As both contains π e⁻ according to Huckel's Rule so they are arromatic

89. A tertiary carbon free radical (b) is more stable and is the major product.

91. Ammonical cuprous chloride will give red ppt. with 1-butyne(a terminal alkyne) and not with 2-butyne.

92. In both the cases, one of the steps is endothermic, that is, the reaction of HCl with carbon radical in case of HCl and addition of iodine radical to double bond in case of HI.

$\dot{I} + CH_2{=}CH_2 \rightarrow CH_2(I)-\dot{C}H_2$

$\Delta H = +46.0$ kJ/mole

93. HOCl has $Cl^+ + OH^-$.

$CH_3–CH{=}CH_2 + Cl^+ \xrightarrow{\text{Electrophilic addition}} CH_3–CH^+–CH_2–Cl \xrightarrow{OH^-} CH_3–CH(OH)–CH_2–Cl$

94. Due to cis-addition of H_2 to the triple bond, the reduced product has a plane of symmetry and hence, is optically inactive.

95. $CH_3–C(CH_3){=}CH_2 \xrightarrow[\text{anti, Markovinkoff addition}]{HBr/peroxide} CH_3–CH(CH_3)–CH_2–Br$

96. Mesitylene (1,3,5-trimethylbenzene) $\xrightarrow[H_2O/Zn]{O_3/CCl_4} 3CH_3–C(=O)–CHO$

97. Propyne reacts with $AgNO_3$ to give white precipitate.

$CH_3–C{\equiv}CH \xrightarrow{AgNO_3\ \text{in ammonia}} CH_3–C{\equiv}C–Ag$
(White ppt)

On the oth-r hand, propene does not react with $AgNO_3$ in ammonia.

98. $CH_3CH_2C{\equiv}CH + H_2O \xrightarrow[M.\ R.]{HgSO_4 + H_2SO_4} CH_3CH_2–C(OH){=}CH_2 \xrightarrow{Tautomerises} CH_3CH_2–C(=O)–CH_3$

Butan-2-one

99. $CH_3–CH{=}CH–C_6H_4–OH \xrightarrow{HBr} CH_3–CH_2–CH(Br)–C_6H_4–OH$

100. Terminal alkynes (having acidic H) react with ammonical cuprous chloride solution to form red precipitate.

$2H–C≡C–H + Cu_2Cl_2 + 2NH_4OH \longrightarrow$
Acetylene

$2CuC≡CCu + 2NH_4Cl + 2H_2O$
Copper acetylide
(Red ppt.)

101. p t t s s p

$CH_3–CH(CH_3)–CH(CH_3)–CH_2–CH_2–CH_3$ (the two CH_3 branches: p, p)

So, it has two tertiary, two secondary and four primary carbon atoms.

103. As 3-methyl-1-butyne is a terminal alkyne so, it can give a precipitate with ammonical silver nitrate.

104. $C_6H_5CH_2CH_3 \xrightarrow{SO_2Cl_2} C_6H_5–CH(Cl)–CH_3 \xrightarrow{\text{Alc. KOH}} C_6H_5–CH=CH_2$

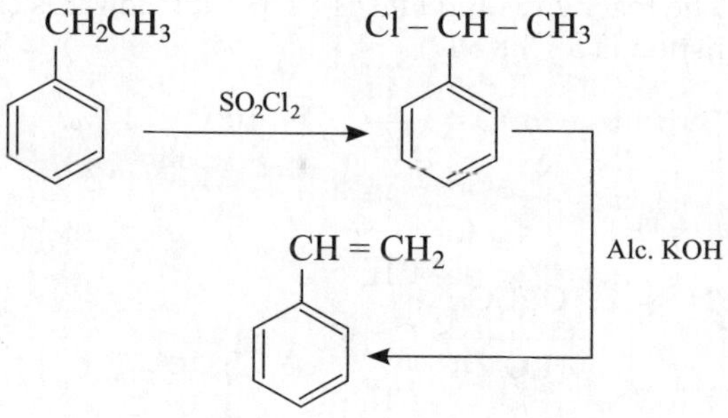

106. $CH_3Br + 2Na + BrCH_3 \xrightarrow{\text{Dryether}} CH_3–CH_3 + 2NaBr$

This reaction is called Wurtz reaction.

108. The greater is number of alkyl groups attached to the doubly bonded atoms, the more stable is the alkene.

111. $C_2H_5Cl + Mg \xrightarrow{\text{Dryether}} \underset{(X)}{C_2H_5MgCl} \xrightarrow{H_2O} \underset{(Y)}{C_2H_6} + \underset{(Z)}{Mg.(Cl).OH}$

112. $CH_3–C(CH_3)_2–CH_3$ has all twelve equivalent primary hydrogens

Hence, it forms only one product on monobromination.

115. $CH_3–C≡C–CH_3 \xrightarrow[\text{Cold}]{KMno_4} CH_3–\overset{O}{\overset{||}{C}}–\overset{O}{\overset{||}{C}}–CH_3 + MnO_2 + KOH$

117. This hydrocarbon cannot be a cycloalkene, because C_6H_{12} has only one double bond and gives n-hexane on hydrogenation. Hex-1-ene with $KMnO_4$ gives pentanoic acid and 1 mol of CO_2.

120. $CH_3–C(CH_3)_2–CH(CH_3)–CH_2–CH_3$

It has 7σ C–C bonds and 18σ C–H bonds.

[number of C–C sigma bonds = number of carbons–1] and

[number of C–H sigma bonds = number of hydrogens]

123. For a species to be aromatic in character, it must have a total of (4n + 2) π-electrons, when n=0, 1, 2, 3…… It has eight π-electrons, which is not a Huckel number of π-electrons for any value of 'n'.

124. $CH_3–C≡CH \xrightarrow[\text{Markovnikov's rule}]{\text{Dryether}} CH_3–C(Br)=CH_2$

$CH_3–C≡CH_2 \xrightarrow{IIBr} CH_3–CBr_2–CH_3$

2, 2-dibromopropane

128. 1-bromo-3-chlorocyclobutane + Na $\xrightarrow{\text{Dry ether}}$ radical (Cl) + Na $\xrightarrow{-NaCl}$ bicyclo [1, 1, 0] butane

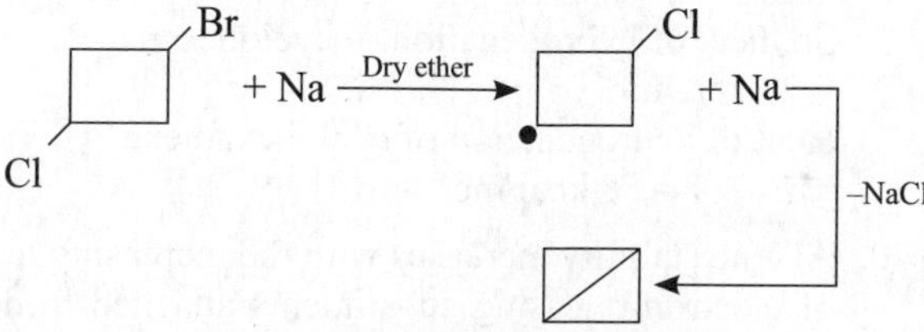

129. $CH_3–CH=C(Br)–CH_2–CH_3$

$CH_3–C(Br)=CH–CH_2CH_3$

pent-2-yne on addition of HBr gives two structural isomers (A and B). Each of these two isomers can exist as a pair of cis, trans isomers and, therefore, there are four configurational isomers.

136. In the step (1), addition of HBr occurs in accordance with Markovnikov's rule and in step (2), addition of HBr occurs in presence of peroxide in accordance with peroxide effect (or anti-Markovnikov's rule).

137. 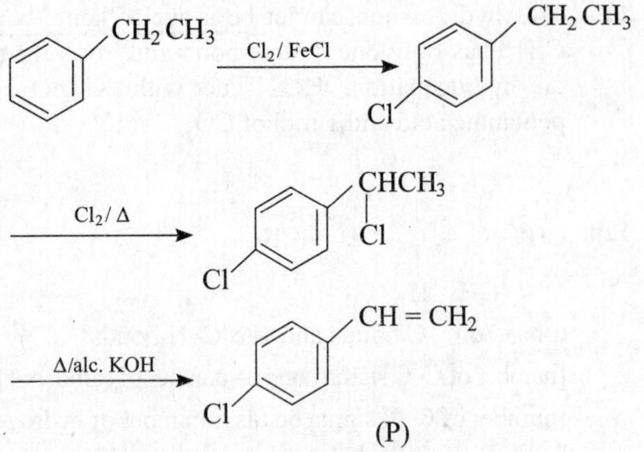 Ethylbenzene $\xrightarrow{Cl_2/FeCl}$ 4-chloroethylbenzene (CH_2CH_3, Cl) $\xrightarrow{Cl_2/\Delta}$ Cl–C_6H_4–CHClCH_3 $\xrightarrow{\Delta/alc.\ KOH}$ Cl–C_6H_4–CH = CH_2 (P)

138. Since, the resonance energy of benzene is 36 kcal/mol, therefore, benzene contains 36 kcal less energy than predicted. In other words, benzene is more stable by 36 kcal than cyclohexatriene.
So, expected heat of hydrogenation of cyclohexatriene = 51 + 36 kcal/mol
= 87 kcal/mol
Hence, heat of hydrogenation per double bond
= 87/3 = 29 kcal/mol
So, heat of hydrogenation of cyclohexene
= 29 kcal/mol.

139. Expected heat of hydrogenation of cyclohexatriene = 51 + 36 = 87 kcal
Hence, heat of hydrogenation per double bond = 87 kcal/3 = 29 kcal
So, heat of hydrogenation of cyclohexene
= 29 kcal/mol
Heat of hydrogenation of cyclohexadiene
= 2 × 29 = 58 kcal/mol

140. Alkene stability increases with the increasing number of electron-releasing substituents attached to doubly bonded carbons.
Alkene 1 is disubstituted
Alkene 2 is tetrasubstituted
Alkene 3 is trisubstituted.

142. The increasing order of stabilities of carbonium ions is primary < secondry < tertiary. Since, carbonium ion is formed as an intermediate in the dehydration of alcohol, so ease of dehydration in the increasing order is primary alcohol < secondry alcohol < tertiary alcohol.

147. $(CH_3)_2$=CH–CH_2–CH–CH=CH–CH_2–CH=C$(CH_3)_2$ + $3O_3$ $\xrightarrow{Zn/H_2O}$

$(CH_3)_2C$=O + OHC–CH_2–CHO
+
OHC–CH_2–CHO + O=C$(CH_3)_2$

149. $(H_3C)C_2$ = CH – CH_2 – CHO $\xrightarrow{(O)\ KMno_4}$ $\begin{matrix}CH_3\\CH_3\end{matrix}$>C = O + $CH_2(COOH)_2$ (COOH–CH_2–COOH)

151. Chlorination of 2- methylbutane gives four mono - chlorinated products.

CH_3–CH(CH_3)–CH_2CH_3 + Cl_2 ⟶

CH_2Cl–CH(CH_3) - CH_2 – CH_3 + CH_3–CCl(CH_3)–CH_2 – CH_3
+ CH_3–CH(CH_3)–CHCl–CH_3 + CH_3–CH(CH_3)–CH_2 – CH_2Cl

The reactivity ratio of 3°H : 2°H : 1°H towards chlorination is 5 : 3.8 : 1

$$\text{Percentage of D} = \frac{3}{31.6} \times 100 = 14\%$$

152. 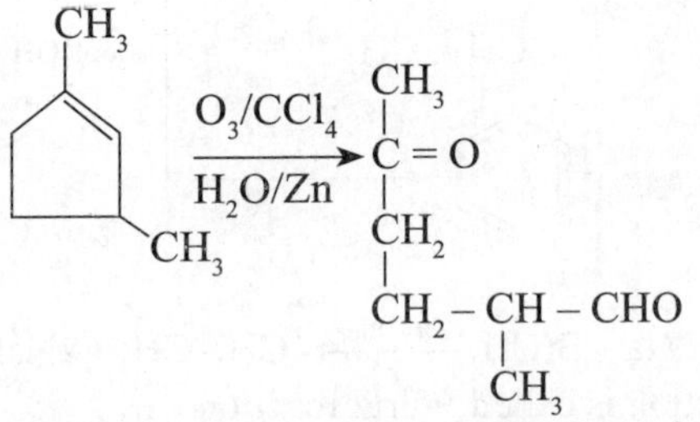 1,3-Dimethylcyclopentene (CH_3, CH_3) $\xrightarrow[H_2O/Zn]{O_3/CCl_4}$ CH_3–C(=O)–CH_2–CH_2 – CH(CH_3) – CHO

154. 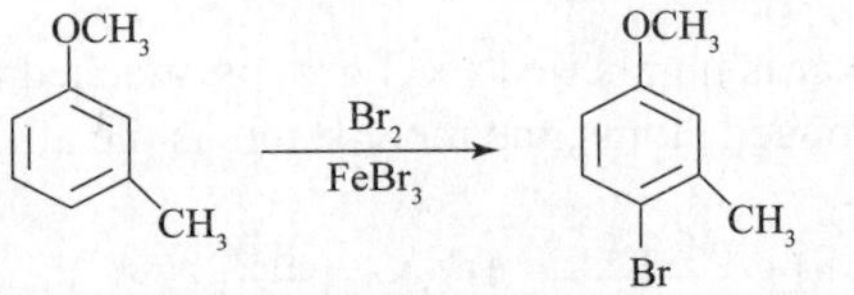 3-Methylanisole (OCH_3, CH_3) (A) $\xrightarrow[FeBr_3]{Br_2}$ 4-bromo-3-methylanisole (OCH_3, CH_3, Br)

As methyl group is ortho- or para-directing.

155. HC ≡ CH $\xrightarrow{O_3}$ OCH – CHO (X)
$\xrightarrow[\text{Reduction}]{Zn/CH_3OH}$ H_2OCH_2 – CH_2OH
Ethane-1,2-diol

156. 1, 2-addition product is a kinetically controlled product while 1, 4-addition product is a thermody-namically controlled product and formed at a comparatively higher temperature.

CH_2=CH–CH=CH_2 $\xrightarrow{HBr,\ 40°C}$ $CH_2(Br)CH$=$CHCH_3$ + $CH_3CH(Br)$–CH=CH_2
1, 4-addition 1, 2-addition

Therefore, 1-bromo-2-butene will be the main product under thermodynamically controlled conditions.

158. m-cresol (CH_3, OH) $\xrightarrow{Br_2}$ 2,4,6-tribromo-3-methylphenol (Br, Br, Br, CH_3, OH)

159. According to Saytzeff's rule,

$$CH_3CH_2CH(F)CH_3 \xrightarrow{CH_3O^-/CH_3OH} CH_3CH_2^a - C^+H - CH_3^b$$

$-H_a^-$ → $CH_3CH = CHCH_3$ (Major)

$-H_b^-$ → $CH_3CH_2CH = CH_2$ (Minor)

160. The most reactivity diene has the double bonds locked in the s -cis conformation, whereas the least reactive diene cannot achieve the required s- cis conformation because it is locked in the s -trans conformation.

> > >

162. Here methane is formed as follows:

$$CH_3 - MgX + CH_3 - C \equiv C - H \rightarrow CH_4$$

163. Toluene (CH_3) $\xrightarrow{\text{Nitration}}$ o-nitrotoluene (CH_3, NO_2) + p-nitrotoluene (CH_3, NO_2)

o-nitrotoluene $\xrightarrow{Sn/HCl}$ o-toluidine (CH_3, NH_2) $\xrightarrow{NaNO_2 / HCl}$ (CH_3, $N_2^+Cl^-$) $\xrightarrow{CuBr}$ o-bromotoluene (CH_3, Br)

p-nitrotoluene $\xrightarrow{Sn/HCl}$ p-toluidine (CH_3, NH_2) $\xrightarrow{NaNO_2 / HCl}$ (CH_3, $N_2^+Cl^-$) $\xrightarrow{CuBr}$ p-bromotoluene (CH_3, Br)

164. Here compound (B) is ethanal (CH_3CHO)

$$H_3C - CH = CH - CH_3 \xrightarrow{O_3}$$

ozonide (A): $H_3C - CH$ (–O–) $CH - CH_3$ with O—O bridge $\xrightarrow{H_2O/Zn}$

$$2CH_3 - CHO$$
(B)

165. $CH_3CH_2 - C \equiv H \xrightarrow[\Delta]{Na/liq.\ NH_3} CH_3CH_2C \equiv C^{\ominus}\ Na^{\oplus}$

As it is a terminal alkyne, having acidic hydrogen so it reacts with Na in liquid ammonia.

166. As $-NO_2$ is electron withdrawing so it will destabilize σ-complex.

167. 2– bromo -2- methylbutane is the only listed tertiary alkyl bromide. Tertiary alkyl bromides are usually most reactive in E2 mechanisms.

169. $CH_3 - C(CH_3) = CH - CH_2 - CH_3 + O_3 \longrightarrow$ ozonide ($(CH_3)_2C$ (–O–) $CH - CH_2 - CH_3$ with O—O bridge)

$$\xrightarrow{Zn / H_2O} (CH_3)_2C = O + CH_3CH_2CHO$$

Hence the product are propan -2- one and propanal

171. but-1-ene $\xrightarrow{NBs}$ 3-bromobut-1-ene (Br) $\xrightarrow[H_2O_2]{HBr}$ 1,3-dibromobutane (Br, Br) $\xrightarrow{Mg}$ $MgBr_2$ + methylcyclopropane (CH_3)

174. diozonide (O, O, O) $\xrightarrow[(2)\ H_2O/Zn]{(1)\ O_3/CCl_4}$

$$2CH_3CHO + \underset{CHO}{\overset{CHO}{|}}$$

PREVIOUS YEARS' QUESTIONS

1. In Friedal Craft's alkylation, besides $AlCl_3$ the other reactants are:
(a) $C_6H_6 + CH_3Cl$ (b) $C_6H_6 + CH_4$
(c) $C_6H_6 + NH_3$ (d) $C_6H_6 + CH_3COCl$
[2000]

2 Which of the following reagents convert propene to 1-propanol?
(a) B_2H_6, H_2O_2, OH^- (b) H_2O, H_2SO_4
(c) Aqueous KOH (d) $MgSO_4$, $NaBH_4 / H_2O$
[2000]

3. Among the following alkenes,
(1) 1-butene
(2) cis-2-butene
(3) trans-2-butene
The decreasing order of stability is
(a) $3 > 2 > 1$ (b) $1 > 2 > 3$
(c) $2 > 1 > 3$ (d) $3 > 1 > 2$
[2000]

4. Which alkene on ozonolysis gives:
CH_3CH_2CHO and $CH_3 - CO - CH_3$?
(a) $CH_3CH_2CH{=}C(CH_3)_2$
(b) $CH_3CH_2CH = CHCH_3$
(c) $CH_3CH_2CH = CHCH_2CH_3$
(d) $CH_3 - \underset{\displaystyle CH_3}{\underset{|}{C}} = CHCH_3$
[2001]

5. The compound:

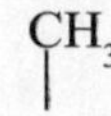

$CH_3 - C = CH - CH_3$ on reaction with $NaIO_4$ in the presence of $KMnO_4$ gives:
(a) $CH_3COCH_3 + CH_3COOH$
(b) CH_3COCH_3
(c) $CH_3COCH_3 + CH_3CHO$
(d) $CH_3CHO + CO_2$
[2003]

6. Using anhydrous $AlCl_3$ as catalyst, which one of the following reactions produce ethyl benzene (PhEt)?
(a) $H_3C - CH_2OH + C_6H_6$
(b) $CH_3 - CH = CH_2 + C_6H_6$
(c) $H_2C = CH_2 + C_6H_6$
(d) $H_3C - CH_3 + C_6H_6$
[2004]

7. Reaction of HBr with propene in the presence of peroxide gives:
(a) Isopropyl bromide (b) 3-bromo propane
(c) Allyl bromide (d) n-propyl bromide
[2004]

8. Products of the following reaction,
$CH_3C \equiv CCH_2CH_3 \xrightarrow[(2)\ \text{hydrolysis}]{(1)\ O_3} ?$
(a) $CH_3COOH + CO_2$
(b) $CH_3COOH + HOOC.CH_2CH_3$
(c) $CH_3CHO + CH_3CH_2CHO$
(d) $CH_3COOH + CH_3COCH_3$
[2005]

9. Which one of the following alkenes will react faster with H_2 under catalytic hydrogenation conditions?

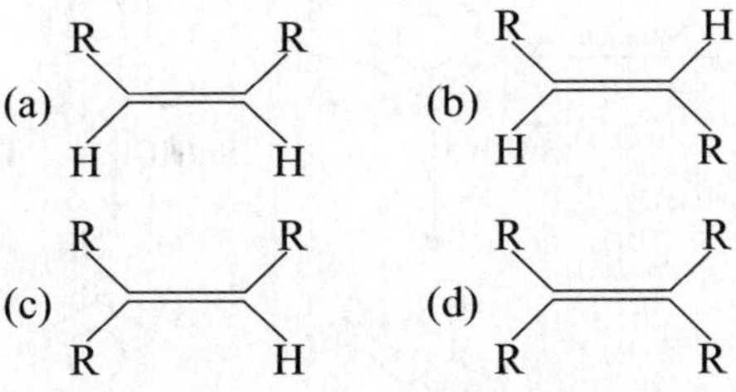

[2005]

10. Predict the product C obtained in the following reaction butyne-1.
$CH_3CH_2 - C \equiv CH + HCl \longrightarrow B \xrightarrow{HI} C$
(a) $CH_3 - \underset{\displaystyle Cl}{\underset{|}{CH}} - CH_2CH_2I$
(b) $CH_3 - CH_2 - CH_2 - \overset{\displaystyle I}{\overset{|}{\underset{\displaystyle Cl}{\underset{|}{C}}}} - H$

(c) $CH_3 - CH_2 - \underset{}{\overset{I}{\overset{|}{CH}}} - CH_2Cl$

(d) $CH_3CH_2 - \underset{\underset{Cl}{|}}{\overset{\overset{I}{|}}{C}} - CH_3$

[2007]

11. Which of the compounds with molecular formula C_5H_{10} yields acetone on ozonolysis?

(a) 2-Methyl-2-butene
(b) 3-Methyl-1-butene
(c) Cyclopentane
(d) 2-Methyl-1-butene

[2007]

12. $H_3C - \underset{\underset{CH_3}{|}}{CH} - CH = CH_2 + HBr \rightarrow (X)$

Here (X) (predominantly) is:

(a) $CH_3 - \underset{\underset{Br}{|}}{CH} - \underset{\underset{CH_3}{|}}{CH} - CH_3$

(b) $CH_3 - \underset{\underset{CH_3}{|}}{CH} - \underset{\underset{Br}{|}}{CH} - CH_3$

(c) $CH_3 - \underset{\underset{CH_3}{|}}{CH} - CH_2 - CH_2Br$

(d) $CH_3 - \underset{\underset{CH_3}{|}}{\overset{\overset{Br}{|}}{C}} - CH_2CH_3$

[2008]

13. Benzene reacts with CH_3Cl in presence of Anhydrous $AlCl_3$ to form?

(a) Chlorobenzene (b) Xylene
(c) Benzylchloride (d) Toluene

[2009]

14. Nitrobenzene can be prepared from benzene by using a mixture of conc. HNO_3 and con. H_2SO_4. In the nitrating mixture, HNO_3 acts as a:

(a) Base (b) Acid
(c) Reducing agent (d) Catalyst

[2009]

15. The reaction of Toluene with Cl_2 in Pressence of $fecl_3$ gives 'x' and reaction in presence of light gives 'y' thus 'x' and 'y' are:

(a) x = Benzyl chloride y = m-chlorotoluene
(b) x = y = O – chloro toluene
(c) x = m-chlorotoluene, y = p-chlorotoluene
(d) x = o– and p- chloro toluene, y = Trichloro methyl benzene

[2010]

16. Liquid Hydrocarbons can be converted to a mixture of gasesons hydro carbons by:

(a) Hydroysis
(b) Oxidation
(c) Cracking
(d) Distillation under reduced pressure

[2010]

17. Which one of the following is most reactive towards electrophilic attack?

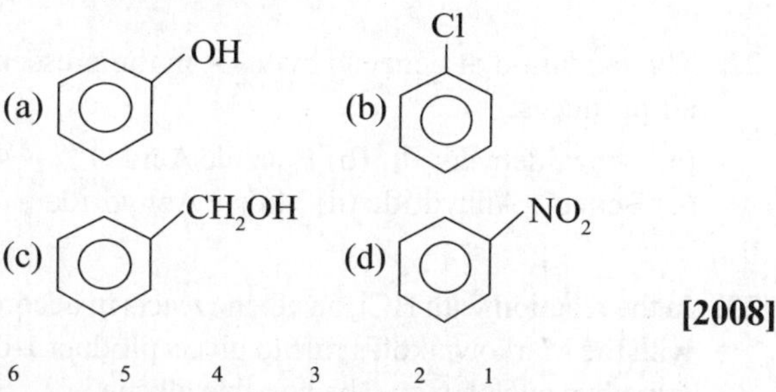

[2008]

18. $\overset{6}{C}H_3 - \overset{5}{C}H = \overset{4}{C}H - \overset{3}{C}H_2 - \overset{2}{C} \equiv \overset{1}{C}H$

The state of hybridisation of carbons 1, 3 and 5 are in the following sequence:

(a) sp, sp^2, sp^3 (b) sp^3, sp^2, sp
(c) sp^2, sp, sp^3 (d) sp, sp^3, sp^2

[2011]

19. Which of the following reagent will be able to distinguish between 1-butyne and 2-butyne ?

(a) $NaNH_2$ (b) HCl
(c) N_2 (d) O_2

[2012]

20. The reaction of $C_6H_5CH{=}CH\text{-}CH_3$ with HBr produces:

(a) $C_6H_5-CH(Br)-CH_2-CH_3$

(b) $C_6H_5-CH_2-CH(Br)-CH_3$

(c) $C_6H_5CH_2CH_2CH_2Br$

(d) p-Br-C_6H_4-$CH{=}CH-CH_3$

[2015]

21. 2, 3 dimethyl but 2-ene can be prepared by heating which of the following compounds with a strong acid?

(a) $(CH_3)_3C-CH{=}CH_2$

(b) $(CH_3)_2CH-CH(CH_3)-CH{=}CH_2$

(c) $(CH_3)_2C{=}CH\text{-}CH_2\text{-}CH_3$

(d) $(CH_3)_2CH\text{-}CH_2\text{-}CH{=}CH_2$

[2015]

22. The oxidation of benzene by V_2O_5 in the pressence of air produces:

(a) Benzaldehyde (b) Benzoic Acid

(c) Benzoic Anhydride (d) Maleic Anhydride

[2015]

23. In the reaction with HCl, an alkene reacts in accordance with the Markownikoff's rule to give a product 1-chloro 1-methye cyclohexone the possible alkene is ?

(a) methylenecyclohexane (CH_2) (b) 1-methylcyclohexene (CH_3)

(c) 3-methylcyclohexene (CH_3) (d) Both A and B

[2015]

24. In the reaction:

$$HC\equiv CH \xrightarrow[(2)\ CH_3CH_2Br]{(1)\ NaNH_2/NH_{3\,liq}} X$$

$$X \xrightarrow{\quad "\quad} Y$$

Here 'X' and 'Y' are ?

(a) X= 2-butyne, Y=3-hexyne

(b) X=2-butyne, Y=2-hexyne

(c) X=1-butyne, X=2-hexyne

(d) X=1-butyne, Y=3-hexyne

[2016]

25. Which of the following compounds will not produce propene by reaction with HBr followed by elimination or direct only by elimination reaction?

(a) $H_2C{=}C{=}O$ (b) $CH_3\text{-}CH_2\text{-}CH_2Br$

(c) $CH_3\text{-}CH_2\text{-}CH_2OH$ (d) cyclopropane (triangle)

26. The compound that will react most readily with gaseous bromine has the formula:

(a) C_2H_2 (b) C_2H_4 (c) C_3H_6 (d) C_4H_{10}

[2016]

27. The Given reaction

benzene + cyclohexene $\xrightarrow[0^oC]{HF}$ P

The product P is?

(a) 1-fluoro-1-phenylcyclohexane (F) (b) fluoro-phenylcyclohexane (F)

(c) phenylcyclohexane (d) diphenylcyclohexane

[2016]

28. Consider the nitration of benzene using mixed concentration H_2SO_4 and HNO_3. If a large amount of $KHSO_4$ is added to the mixture the rate of nitration will be ?

(a) slower (b) faster (c) doubled (d) unchanged

[2016]

Answer Keys

1. (a)	2. (a)	3. (a)	4. (a)	5. (a)	6. (a)	7. (d)	8. (b)	9. (a)	10. (d)
11. (a)	12. (d)	13. (d)	14. (a)	15. (d)	16. (c)	17. (a)	18. (d)	19. (a)	20. (a)
21. (a)	22. (d)	23. (d)	24. (d)	25. (a)	26. (d)	27. (c)	28. (a)		

Hints and Explanations

1. Benzene reacts with alkyl halides in the presence of $AlCl_3$ (Friedal craft's alkylation).

$$C_6H_6 + CH_3Cl \xrightarrow{\text{Anhyd. } AlCl_3} C_6H_5CH_3 + HCl$$

2. Propene can be converted into propanol as follows:

$$\underset{\text{Propene}}{CH_3CH = CH_2} \xrightarrow{HBr,\ (C_6H_5COO)_2O}$$

$$\underset{\text{n-propyl bromide}}{CH_3CH_2CH_2 - Br} \xrightarrow{\text{aq KOH}}$$

$$\underset{\text{1-propanol}}{CH_3CH_2CH_2 - OH}$$

3. As the stability of an alkene depends upon the heat of hydrogenation of that alkene.

Stability α 1/Heat of hydrogenation

Hence order of stability is given as:

trans-2-butene > cis-2-butene > 1-butene.

Heat of hydrogenation (kJ/mol) are 115.5, 119.6 and 126.8 respectively.

4. 2-methyl-pentene on ozonolysis gives propanol and propanol-2 as follows:

$$CH_3CH_2CH = C(CH_3)_2 \xrightarrow{\text{Ozonolysis}}$$

$$CH_3CH_2CHO + O = C(CH_3)_2$$

5. $(CH_3)_2C = CH - CH_3$ on reaction with $NaIO_4$ in the presence of $KMnO_4$ gives acetone and acetic acid.

$$(CH_3)_2C = CH - CH_3 + 3[O] \xrightarrow[\text{in presence of } KMnO_4]{NaIO_4}$$

$$(CH_3)_2C = O + CH_3COOH$$

6. CH_3CH_2OH reacts with $AlCl_3$ to form ethyl carbocation which then reacts with benzene to form ethyl benzene as follows:

$$CH_3CH_2OH + AlCl_3 \rightarrow CH_3CH_2 - OHAlCl_3 \rightarrow CH_3\ {}^{+}CH_2 + HOAlCl_3^{-}$$

$$C_6H_6 + CH_3CH_2^{+} \rightarrow C_6H_5CH_2CH_3 + H^{+}$$

7. The reaction of HBr with propene in the presence of peroxide gives n-propyl bromide.

It is an example of Anti Markowni Koff's addition reaction.

$$CH_3 - CH = CH_2 + HBr \xrightarrow{\text{Peroxide}} \underset{\text{n-propyl bromide}}{CH_3 - CH_2 - CH_2Br}$$

8. Pentyne-2 gives acetic acid and propanoic acid as follows:

$$CH_3C \equiv CCH_2CH_3 \xrightarrow[(2)\ H_2O]{(1)\ O_3/CCl_4}$$

$$CH_3COOH + CH_3CH_2COOH$$

9. As more is the heat of hydrogenation of alkene, easier is its hydrogenation or more is its reactivity.

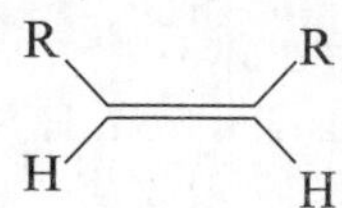

10. $CH_3 - CH_2 - C \equiv CH + HCl \longrightarrow$

$$CH_3 - CH_2 - \underset{\displaystyle Cl}{\underset{|}{C}} = CH_2 \xrightarrow{HI}$$

$$CH_3 - CH_2 - \overset{\displaystyle I}{\overset{|}{\underset{\displaystyle Cl}{\underset{|}{C}}}} - CH_3$$

11. 2-methyl-but-2-ene on ozonolysis gives acetone as follows:

$$\underset{\text{2-methyl-2-butene}}{H_3C - \underset{\displaystyle CH_3}{\underset{|}{C}} = CH - CH_3} \xrightarrow[(2)\ Zn/H_2O_2]{(1)\ O_3}$$

$$CH_3COCH_3 + CH_3CHO$$

12. The reaction occurs as follows:

$$CH_3-\underset{\displaystyle CH_3}{CH}-CH=CH_2 \xrightarrow{H^{\delta+}-Br^{\delta-}} (A)$$

3-methyl but-1-ene

Mechanism:

$$CH_3-\overset{CH_3}{\underset{H}{C}}-CH=CH_2+H^+ \rightarrow CH_3-\overset{CH_3}{\underset{H}{C}}-{}^+CH-CH_3$$

2° carbocation (less stable)

$$\xrightarrow{\text{1, 2 Hydride shift}} CH_3-\overset{CH_3}{C^+}-CH_2-CH_3$$

3° carbocation (more stable)

$$\xrightarrow{+Br^-} CH_3-\overset{CH_3}{\underset{Br}{C}}-CH_2-CH_3$$

2-Bromo-2-methylbutane
(Major product)

13. $C_6H_6 \xrightarrow[\text{AlCl}_3\ \text{Anhy}]{CH_3Cl} C_6H_5CH_3 + HCl$

14. As HNO_3 accepts H+ from H_2SO_4 so behaves like a base

15. $C_6H_5CH_3 \xrightarrow{Cl_2/FeCl_3}$ o-chlorotoluene + p-chlorotoluene (Cl)

$C_6H_5CH_3 \xrightarrow[-3HCl]{3Cl_2/light} C_6H_5CCl_3$

17. As –OH group is highly ring activating so phenol is most reactive towards electrophile attack here. As OH^- increases electron density in benzene ring which facilitates electrophillic attack.

In the hydrocarbon

18. In the following hydrocarbon

$$\overset{6}{CH_3}-\overset{5}{CH}=\overset{4}{CH}-\overset{3}{CH_2}-\overset{2}{C}\equiv\overset{1}{CH}$$

$sp^3 \quad sp^2 \quad sp^2 \quad sp^3 \quad sp \quad sp$

The state of hybridisation of carbons 1, 3 and 5 are sp, sp^3 and sp^2 respectively.

19. $\underset{\text{But-1-yne}}{CH_3CH_2-C\equiv CH} \xrightarrow[-H^+ \text{ or } -NH_3]{NaNH_2} CH_3-CH_2-C\equiv \overset{\ominus}{C}\,\overset{+}{Na}$

$\underset{\text{But-2-yne}}{CH_3-C\equiv C-CH_3} \xrightarrow{\text{"}} XXX$

20. $C_6H_5CH=CH-CH_3 \xrightarrow[\text{M.A.}]{\overset{\oplus}{H}-\overset{\ominus}{Br}} C_6H_5-\underset{Br}{CH}-CH_2-CH_3$ (steps 1st, 2nd, 3rd)

It is a case of markonikov's addition

21. $(CH_3)_2\underset{CH_3}{C}-CH=CH_2 \xrightarrow[\text{S.A}]{H^+} CH_3-\overset{CH_3}{\underset{CH_3}{C}}-\overset{\oplus}{CH}-CH_3$ (s°)

$\xrightarrow{CH_3\ \text{Shift}} CH_3-\underset{CH_3}{\overset{\oplus}{C}}-\overset{H}{\underset{CH_3}{C}}-CH_3 \xrightarrow{-H^+} (CH_3)_2C=C(CH_3)_2$

22. $C_6H_6 \xrightarrow[\frac{9}{2}O_2]{V_2O_5,\ \Delta}$ $\begin{matrix}CH-CO\\ \|\\ CH-CO\end{matrix}\!\!>O + 2CO_2 + 2H_2O$

Maleic Anhydride

23. Methylenecyclohexane ($=CH_2$) $\xrightarrow{H^+}$ 1-methylcyclohexyl cation ($\oplus$, CH_3) $\xrightarrow{Cl^-}$ 1-chloro-1-methylcyclohexane (Cl)

1-Methylcyclohexene (CH_3) $\xrightarrow{H^+}$ 1-methylcyclohexyl cation ($\oplus$) $\xrightarrow{Cl^-}$ 1-chloro-1-methylcyclohexane (Cl)

24. $HC\equiv CH \xrightarrow{NaNH_2/liq.\text{-}NH_3} HC\equiv \overset{\ominus}{C} \xrightarrow{C_2H_5Br} \underset{\text{1-Butyne}}{HC\equiv C-CH_2-CH_3}$

$\xrightarrow{NaNH_2} \overset{\ominus}{C}\equiv C-CH_2-CH_3 \xrightarrow{C_2H_5Br} \underset{\text{3-Hexyne}}{CH_3-CH_2-C\equiv C-CH_2-CH_3}$

25. As $\underset{\text{Ketene}}{H_2C=C=O} \longrightarrow :CH_2 + CO$

So it can't give propene.

26. In case of gaseous Br_2 Br^o is formed so the most suitable and fast reaction is possible with C_4H_{10} (Saturated alkane)

27. It is a case of Friedal craft reaction as follows:

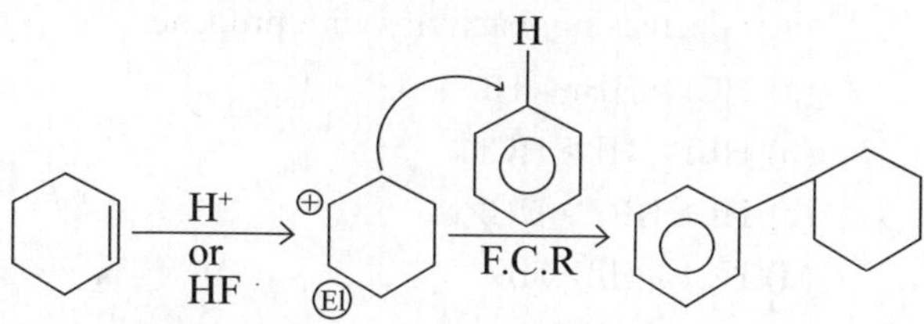

28. An adding large amount of $KHSO_4$ The concentration. of HSO_4^- increases so NO_2^+ decreases which means rate slows down.

NCERT EXEMPLAR

1. Arrange the following in decreasing order of their boiling points.

(A) *n*–butane (B) 2–methylbutane

(C) *n*-pentane (D) 2,2–dimethylpropane

(a) A > B > C > D

(b) B > C > D > A

(c) D > C > B > A

(d) C > B > D > A

2. Arrange the halogens F_2, Cl_2, Br_2, I_2, in order of their increasing reactivity with alkanes.

(a) $I_2 < Br_2 < Cl_2 < F_2$

(b) $Br_2 < Cl_2 < F_2 < I_2$

(c) $F_2 < Cl_2 < Br_2 < I_2$

(d) $Br_2 < I_2 < Cl_2 < F_2$

3. The increasing order of reduction of alkyl halides with zinc and dilute HCl is.

(a) R–Cl < R–I < R–Br

(b) R–Cl < R–Br < R–I

(c) R–I < R–Br < R–Cl

(d) R–Br < R–I < R–Cl

4. The correct IUPAC name of the following alkane is:

$H_3C-CH_2-\underset{\underset{CH_3\ \ CH_3}{CH}}{CH}-CH_2-CH_2-\underset{\underset{CH_3}{CH_2}}{CH}-CH_2-CH_3$

(a) 3,6 – Diethyl – 2 – methyloctane

(b) 5 – Isopropyl – 3 – ethyloctane

(c) 3 – Ethyl – 5 – isopropyloctane

(d) 3 – Isopropyl – 6 – ethyloctane

5. The addition of HBr to 1-butene gives a mixture of products A, B and C

(A): C bearing Br (up), H (wedge down), H_3C_2 (left, dashed), CH_3 (right)

(B): C bearing C_2H_3 (up), Br (down), H (left, dashed), CH_3 (right)

$CH_3-CH_2-CH_2-Br$ (C)

The mixture consists of

(a) A and B as major and C as minor products

(b) B as major, A and C as minor products

(c) B as minor, A and C as major products

(d) A and B as minor and C as major products

6. Which of the following will not show geometrical isomerism?

(a) $(F)(Cl)C=C(H)(D)$

(b) $(F)(Cl)C=C(F)(Cl)$

(c) $(H_3C)(H_5H_2)C=C(C_2H_5)(CH_3)$

(d) $(CH_3)(CH_3)C=C(CH_3)(C_2H_5)$

7. Arrange the following hydrogen halides in order of their decreasing reactivity with propene.

(a) HCl > HBr > HI

(b) HBr > HI > HCl

(c) HI > HBr > HCl

(d) HCl > HI > HBr

8. Arrange the following carbanions in order of their decreasing stability.

(A) $H_3C-C\equiv C^-$ (B) $H-C\equiv C^-$ (C) $H_3C-CH_2^-$

(a) A > B > C

(b) B > A > C

(c) C > B > A

(d) C > A > B

9. Arrange the following alkyl halides in decreasing order of the rate of β- elimination reaction with alcoholic KOH.

(A) $CH_3-CH(CH_3)-CH_2Br$ (B) CH_3-CH_2-Br

(C) $CH_3-CH_2-CH_2-Br$

(a) A > B > C

(b) C > B > A

(c) B > C > A

(d) A > C > B

10. Which of the following reactions of methane is incomplete combustion:

(a) $2CH_4 + O_2 \xrightarrow{Cu/523\ K/\ 100\ atm} 2CH_3OH$

(b) $CH_4 + O_2 \xrightarrow{Mo_2O_3} HCHO + H_2O$

(c) $CH_4 + O_2 \longrightarrow C(s) + 2H_2O\ (l)$

(d) $CH_4 + 2O_2 \longrightarrow CO_2\ (g) + 2H_2O\ (l)$

Answer Keys

1. (d)	2. (a)	3. (b)	4. (a)	5. (a)	6. (d)	7. (c)	8. (b)	9. (d)	10. (c)

Hints and Explanations for Selective Questions

1. Boiling point decreases with branching.

5. Apply Markovnikov's rule.

8. sp-hybridized carbon atom is more electronegative than sp^3-hybridized carbon atom.

10. Reactivity will be more with more number of β-substituents, hence more stable alkene will be formed on β-elimination..

AIIMS ESSENTIALS

Assertion and Reason

In the following questions, two statements Assertion (A) and Reason (R) are given. Mark

(a) If A and R both are correct and R is the correct explanation of A;

(b) if A and R both are correct but R is not the correct explanation of A;

(c) A is true but R is false;

(d) A is false but R is true;

(e) Both A and R are false.

1. (A) : Alkanes undergo free radical substitution reactions.

(R) : Alkanes undergo hemolytic fission.

2. (A) : Treatment of 1,3-dichloro propane on reaction with alc. KOH gives $CH_2 = C = CH_2$
(R) : It is nucleophilic elimination reaction.

3. (A) : $CH \equiv CH$ reacts with HCl in the presence of $HgCl_2$ while $CH_2 = CH_2$ does not.
(R) : There is more unsaturation in $CH \equiv CH$ than in $CH_2 = CH_2$.

4. (A) : The melting point of neopentane is higher than n-pentane but boiling point of neopentane is lower than n-pentane.
(R) : Melting point depends upon packing of molecules whereas boiling point depends upon surface area. Neopentane fits into crystal lattice readily but has minimum surface area.

5. (A) : Benzene does not decolourise alkaline $KMnO_4$.
(R): benzene is stabilized by resonance and π-electrons are delocalized.

6. (A) : Benzene reacts with Cl_2 in presence of sunlight to form benzene hexachloride (BHC).
(R) : BHC or Gammaxane or 666 is used as insecticide.

7. (A) : Parafix was is mixture of hydrocarbons.
(R) : All hydrocarbons are combustible.

8. (A) : Addition of HBr to 1-butene gives two optical isomers.
(R) : The product contains one chiral carbon atom.
$CH_2 = CH - CH_2 - CH_3 + HBr \rightarrow$

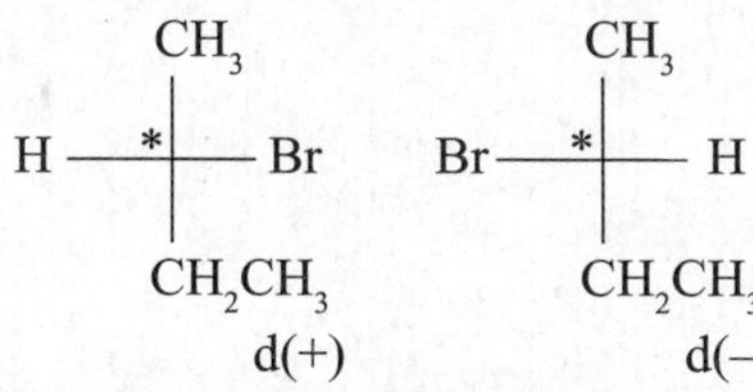

9. (A) : Benzene on heating with conc. H_2SO_4 gives benzene sulphonic acid (used in forming detergents) which when heated with super heated steam under pressure gives benzene.
(R) : Sulphonation is a reversible process.

10. (A) : In α, β unsaturated compounds with C = 1, and C = O conjugated, attack of nucleophile takes place on C = C.
(R) : The C = O bond is stronger than C = C.

11. (A) : Toluene undergoes nitration much more readily than benzene.
(R) : It is due to electron releasing nature of $-CH_3$ group which increases electron density on benzene and electrophillic substitution reaction like nitration becomes faster.

12. (A) : $CH_3 - \overset{\overset{CH_3}{|}}{C} = CH - CH_3$ is more stable than $CH_3 - \underset{\underset{CH_3}{|}}{CH} - CH = CH_2$
(R) : More alkyl substituted alkenes are more stable due to hyperconjugation.

13. (A) : Treatment of 1,3-dichloropropane on treatment with Zn dust gives cyclopropane.
(R) : The reaction of alkyl halide with Zn dust is dehydrogenation and called Frankland reaction.

14. (A) : $CH_3 - C \equiv C - CH_3$ is more reactive than $CH \equiv CH$ towards HCl.
(R) : The carbocation formed is more stable in the case of $CH_3 - C \equiv C - CH_3$ than $CH \equiv CH$.

15. (A) : Addition of HCl to acetylene in presence of $HgCl_2$ give vinyl chloride.
(R) : $HgCl_2$ acts as positive catalyst.

16. (A) : Friedal Crafts Acylation of benzene with acetic anhydride in presence of anhydrous $AlCl_3$ yields acetophenone and not poly substituted products.
(R) : It is due to stearic hindrance of bulky acyl group $CH_3 - \overset{||}{C} -$ and also, it is a deactivating group

17. (A) : When $CH_2 = \underset{\underset{Br}{|}}{C}H - COOH$ is reacted with HBr, then $CH_2 - CH_2 - COOH$ is obtained
(R) : The carbocation formed has the stability order
$\overset{\oplus}{C}H_2 - CH_2COOH > CH_3 - \overset{\oplus}{C}H - COOH$

18. (A) : n-butane on heating in presence of $AlCl_3$ gives isobutene.
(R) : n-butane and isobutane are isomers.

19. (A) : Alkyl iodides are more reactive than alkyl chlorides for elimination reactions.
(R) : I is better leaving group than Cl^-.

20. (A) : Addition of Br_2 to 1- butene gives two optical isomers

(R) : The product contains one asymmetric carbon

21. (A) : Dimethyl sulphide is commnly used for the reduction of an ozonide of an alkene to get the carbonyl compound

(R) : IT reduces the ozonide giving water soluble dimethyl sulphoxide and excess of it evaporates.

22. (A) : Addition of bromine to trans -2-butene yields meso-2, 3- dibromo butane.

(R) : Bromine addition to an alkene is an electrophillic addition

23. (A) : Benzene reacts with CO and HCl in presence of $AlCl_3$ to give benzaldehyde.

(R) : The electrophillic reagent formed is $H-\underset{\underset{O}{\|}}{C}-Cl$

ANSWER KEYS

1. (a)	**2. (a)**	**3. (b)**	**4. (a)**	**5. (a)**	**6. (b)**	**7. (b)**	**8. (a)**	**9. (a)**	**10. (c)**
11. (a)	**12. (a)**	**13. (b)**	**14. (a)**	**15. (c)**	**16. (a)**	**17. (a)**	**18. (b)**	**19. (a)**	**20. (a)**
21. (a)	**22. (c)**	**23. (a)**	**24. (b)**						

CHAPTER 13

Environmental Chemistry

Chapter Outline

■ Environment ■ Environmental Pollution and Pollutants ■ Control of Environmental Pollution and Green Chemistry

Environment

- The term 'environment' simply means surroundings conditions in which we live.
- Environment constitutes air, water, soil, the atmosphere and the plants around us.
- Environmental chemistry deals with the changes in chemical dynamics taking place around us and the damage it causes to the environment.
- Environment consists of four segments:
  1. Atmosphere
  2. Hydrosphere
  3. Lithosphere
  4. Biosphere

Atmosphere

It is the layer of air enveloping the earth at a height of around 1600 km and it consists of many gases.The major components of the atmosphere are nitrogen, oxygen and water vapour. The minor components are argon and carbon dioxide, the trace components include inert gases like methane.

The atmosphere maintains the heat balance on the earth by absorbing electromagnetic radiations coming from sun and transmitting ultra violet, visible and infra red radiations.

Hydrosphere

It contains all types of water resources like oceans, seas, rivers, reservoirs, lakes, polar ice caps and ground water.

Lithosphere

The landmass available on the earth's surface excluding the water bodies constitute this segment of the environment. It is the solid component of earth having soil, rocks and mountains.

The solid, thick, uppermost part of the earth is called **crust**. The inner layers of the lithosphere contain minerals and the deep inner layers contain natural gas and oil.

Biosphere

It includes all living organisms and their interactions with the environment, atmosphere, hydrosphere and lithosphere. For example, the levels of oxygen and carbon dioxide depend on plants.

Environmental Pollution and Pollutants

- **Environmental pollution** is caused by the addition of any undesirable substance to air, water or soil, naturally or by human activity to such an extent that its adverse effects are observed on human beings, animals and plants.
- **Pollutant** is that substance whose presence in undesirably higher concentrations causes pollution in turn, adversely affects the environment. Hg, Pb, CO, SO_2 are some examples of pollutants.
- **Contaminant** is a substance which is not present in nature but is introduced into the environment by human activity and has adverse effects on environment. For example, methyl isocyanate (MIC).
- **Receptor** is the medium effected by the pollutant. Human eyes are receptor for the smoke released by automobiles which cause irritation to eyes.
- **Sink** is the medium which interacts with the long lived pollutant and removes pollution. Sea water acts as a sink for carbon dioxide. The main causes of pollution are increase in population and depletion of natural resources, industrialization, urbanization and deforestation.

TABLE 13.1 Regions of Atmosphere

	Troposphere[a]	Stratosphere[b]	Mesosphere	Thermosphere
Altitude (km)	0–10	10–50	50–85	85–500
Temperature range	15°C–56°C	56°C–2°C	2°C–92°C	92°C–1200°C
Gases/species	N_2, O_2, CO_2, Water vapours	N_2, O_2, O_3, O^-	N_2, O_2, O_2^+, NO^+	O_2^+, O^+, NO^+, e^-

[a] *Troposhere constitutes the major portion of the atmosphere*

[b] *Also called ozonosphere as ozone layer is found here*

Types of Pollutants

Primary pollutants These pollutants after their formation, enter the environment and remain there unchanged. For example, NO, NO_2, CO, SO_2.

Secondary pollutants They are formed as a result of the chemical reactions between primary pollutants present in the atmosphere and those in the hydrosphere. For example, peroxy acylnitrate (PAN).

Biodegradable pollutants

This kind of pollutants can be easily decomposed with the help of microorganisms and are not harmful. For example, domestic sewage, cow dung.

Non-biodegradable pollutants

These pollutants cannot be decomposed and their presence is very harmful for animals and human beings. For example, DDT, mercury.

Types of Pollution

Air pollution It is caused by the addition of undesirable substances into the atmosphere either naturally or by human activity. It is of following two types:

Tropospheric pollution It is caused by gaseous air pollutants like SO_2, NO_2, CO_2, or H_2S.

Stratospheric pollution The damage done to the ozone layer by the detrimental action of compounds like nitric acid and chloroform carbons constitute this particular type of pollution.

Some common air pollutants are as follows:

- Oxides of carbon like CO and CO_2.
- Oxides of nitrogen like N_2O, NO.
- Oxides of sulphur like SO_2.
- Chlorofluoro carbons (freons).
- Hydrocarbons like methane and butane.
- Metals like lead and mercury.
- Organic pollutants like benzopyrene, biocides.
- Most of the air pollution is because of automobiles in which the carbon fuels undergo incomplete combustion and liberate carbon monoxide.

$$2C + O_2 \longrightarrow 2CO$$

- Due to degradation of organic substances, methane gas will be liberated which oxidizes into CO.

$$2CH_4 + 3O_2 \longrightarrow 2CO + 4H_2O$$

- Carbon monoxide combines with haemoglobin to form carboxy haemoglobin due to which the haemoglobin loses the oxygen carrying capacity.
- Oxides of nitrogen like N_2O, NO and NO_2 are liberated into air during the combustion of fossil fuels.
- Nitrogen oxide coming out of the supersonic jets directly enters the stratosphere and decomposes the ozone present there.

$$NO + O_3 \longrightarrow NO_2 + O_2$$

$$NO_2 + O_3 \longrightarrow NO + 2O_2$$

- SO_2 gas is released into the atmosphere directly by burning sulphur or by roasting sulphide ores or by burning fuels containing the sulphur and it causes respiratory tract diseases. SO_2 bleaches the chlorophyll and thus prevents photosynthesis.
- Carcinogenic benzopyrene is released into the air by diesel engines.
- Methane gas is released into air by the degradation of biomass.

- Peroxyacyl nitrate (PAN) and peroxy benzoyl nitrate (PBN) are the air pollutants forming smog (smoke + fog) known as **photochemical smog**. The London smog which caused the death of several people was formed by the above pollutants.
- When the oxides of nitrogen and sulphur combine with rain water, it forms acid rain.

Acid rain According to Robert Augus the pH of rain water is normally 5.6 due to dissolution of CO_2 in it.

$$CO_2 + H_2O \rightleftharpoons H_2CO_3 \rightleftharpoons H^+ + HCO_3^-$$

Acid rain is due to the large scale emission of acidic gaseous oxides (SO_2, NO_2 etc) into atmosphere by thermal power plants, automobiles and industries. The oxides of nitrogen and sulphur dissolve in rain water forming nitric acid and sulphuric acid resulting in acid rain.

$$4NO_2 + 2H_2O + O_2 \longrightarrow 4HNO_3$$

$$SO_2 + ½\ O_2 \longrightarrow SO_3$$

$$SO_3 + H_2O \longrightarrow H_2SO_4$$

- The pH of the acid rain is a about 4–5.

Effects of acid rain Acid rain changes the pH of the soil which in term affects the fertility of the soil.

- It is toxic to vegetation and plant life.
- Life of buildings is considerably reduced by acid rain.
- The Taj Mahal is affected by it due to the reaction between rain and marble ($CaCO_3$).

$$CaCO_3 + H_2SO_4 \longrightarrow CaSO_4 + CO_2 + H_2O$$

It corrodes water pipes due to which heavy metals (Fe, Cu, Pb etc) mix in water and cause toxis effects.

Greenhouse effect or Global warming

- The phenomenon of abnormal heating up of the earth's surface due to the presence of excess of greenhouse gases is called **greenhouse effect**.
- Carbon dioxide is the main greenhouse gas. Some other greenhouse gases are CFCs (17%), O_3 (18%), NO (4%) and water vapour (2%).
- Here due to selective energy absorption by these gases the heat from the sun can reach the earth but can not be radiated back into the space. It causes an increase in the temperature of the earth i.e., Global Warming.
- It results in melting of ice caps, decrease in ground water, unseasonal rains, effects on agriculture due to rapid depletion of surface water.

Depletion of ozone layer O_3 layer is present in stratosphere and it prevent U.V rays to reach earth by converting them into Infra Red Rays.

- Chlorofluoro carbons (CF_2Cl_2) called freons, absorb the ultra violet radiation and get photolysed to liberate chlorine atoms. The chlorine atoms catalyse the decomposition of ozone resulting in the depletion of ozone layer.

$$CF_2Cl_2 \xrightarrow{hv} CF_2Cl + \overset{o}{Cl}$$

$$CFCl_3 \xrightarrow[uv]{hv} CFCl_2 + \overset{o}{Cl}$$

$$\overset{o}{Cl} + O_3 \longrightarrow ClO^o + O_2$$

$$ClO^o + [O] \longrightarrow Cl^o + O_2$$

- The presence of oxides of nitrogen in the atmosphere increases the decomposition of ozone.

$$NO + O_3 \longrightarrow NO_2 + O_2$$

$$O_2 \xrightarrow{hv} O + O$$

$$NO_2 + O \longrightarrow NO + O_2$$

It has been noticed that one molecule of CFC can destroy more than one thousand O_3 molecules in the stratosphere due to this reaction a huge ozonehole was created in the ozone layer over Antarctica. However, in other parts of the stratosphere ozone hole is not observed due to the fact that both ClO and $Cl^°$ can be consumed as follows.

$$ClO^o + NO_2 \longrightarrow ClONO_2$$

$$Cl^o + CH_4 \longrightarrow \overset{o}{C}H_3 + HCl$$

- Due to the damage done to the a ozone layer, the ultraviolet light from the sun directly falls on the earth, causing skin cancer, irritation to the eyes and is harmful to vegetation also.

Controlling air pollution

- The microorganisms and enzymes which can naturally degrade pollutants must be developed. Such control measures are called **bioremedies**.
- Clean and green, environment friendly (eco-friendly) technology must be developed.
- By dissolving or absorbing harmful gases and chemicals.
- A large scale plantation of trees must be carried out.
- Industries must be situated far away from urban area.

- Use of vehicles should be limited.
- Petrol without load and Diesel with less 'S' must be used.
- Automobiles must be fitted with tune ups (for high air-fuel ratio) and catalytic convertors (to change CO into CO_2 and NO_2 into N_2).
- Begasse and Rice husk should not be used as fuel.
- Fly ash (coal-thermal plants) should be removed by wet method and should be used in building materials.
- Industrial wastes can be checked by using tall chimneys, wet scrubbers, cyclone collectors bag filters, electro-static precipitators etc.

Particulate pollutants These are the minute solid particles or liquid droplets present in air. These are present in vehicle emissions, smoke particles from fires, dust particles and ash from industries.

Particulates in the atmosphere may be of two types: viable and non-viable.

The viable particulates are bacteria, algae, fungi, moulds etc.

Types There are four main types of non-viable particulate matter in the atmosphere. These are as follows:

1. *Smoke*: It contains very small soot particles which may be liquid or solid. These are produced due to burning and combustion of organic matter e.g., Oil smoke, tobacco smoke and carbon smoke etc.
2. *Dust*: These are fine particles produced by natural disintegration of rocks and soil or by mechanical process of grinding, crushing etc.

   Particles from soils and minerals have calcium, aluminium and silicon compounds e.g., very small solid particles are carbon black, silver iodide combustion nuclei and sea salt nuclei and larger particles are cement dust, soil dust, foundry dust and pulverized coal.
3. *Smog*: The word smog is derived from smoke and fog. It is the most common example of air pollution that occurs in many cities throughout the world.

   There are following two types of smog:
   (i) Classical smog occurs in cool humid climate. It is a mixture of smoke, fog and sulphur dioxide. Chemically it is a reducing mixture and so it is also known as reducing smog.
   (ii) Photochemical smog occurs in warm, dry and sunny climate. The main components of the photochemical smog result from the action of sunlight on unsaturated hydrocarbons and nitrogen oxides produced by automobiles and factories.

Since this smog has high concentration of oxidizing agents so it is also known as oxidizing agent.

Formation of photochemical smog: When fossil fuels are burnt, a variety of pollutants are emitted into the earth's troposphere. Two of the pollutants that are emitted are hydrocarbons (unburnt fuels) and nitric oxide (NO).

When such pollutants build up to sufficiently high levels, a chain reaction takes place due to their interaction with sunlight in which NO is oxidised into NO_2. This NO_2 in turn absorbs energy from sunlight and breaks up into nitric oxide and free oxygen atom.

$$NO_2\,(g) \xrightarrow{hv} NO\,(g) + O\,(g)$$

Oxygen atoms are highly reactive so combine with the O_2 in air to produce ozone as follows:

$$O\,(g) + O_2\,(g) \rightleftharpoons O_3\,(g)$$

The ozone produced here reacts rapidly with the NO (g) formed to regenerate NO_2 which is a brown gas and at sufficiently high levels can contribute to haze.

$$NO\,(g) + O_3\,(g) \longrightarrow NO_2\,(g) + O_2\,(g)$$

As both NO_2 and O_3 are strong oxidizing agents so can react with the unburnt hydrocarbons in the polluted air to produce chemical like formaldehyde, acrolein and peroxyacetyl nitrate (PAN) as follows:

$$NO_2 \xrightarrow{hv} NO + O$$

$$O + O_2 \longrightarrow O_3$$

$$O_3 + NO \longrightarrow NO_2 + O_2$$

$$R\text{—}\underset{\underset{O}{\|}}{C}{}^{\circ} + O_2 \longrightarrow R\text{—}\underset{\underset{O}{\|}}{C}\text{—}O\text{—}O\text{—}(RCO)_3$$

$$R\text{–}CO_3^{\circ} + NO_2 \longrightarrow RCO_3NO_2$$

$$R\text{–}CO_3^{\circ} + \text{Hydrocarbon} \longrightarrow R\text{–}CHO,\ R_2CO$$

$$\underset{\text{PAN}}{CH_3\text{—}\underset{\underset{O}{\|}}{C}\text{—}O\text{—}O\text{—}NO_2},\quad \underset{\text{P FN}}{H\text{—}\underset{\underset{O}{\|}}{C}\text{—}O\text{—}O\text{—}NO_2}$$

(Peroxy farmyl nitrate)

$$Ø\text{—}\underset{\underset{O}{\|}}{C}\text{—}O\text{—}O\text{—}NO_2$$

PB_ZN

Peroxy benzoyl nitrate

Effects of photochemical smog: The main components of photochemical smog are ozone, nitric oxide, acrolein, formaldehyde and peroxyacetyl nitrate (PAN).

- Photochemical smog causes serious health problems. Both ozone and PAN act as powerful eye irritants.
- Photochemical smog leads to cracking of rubber and extensive damage to plant life. It also causes corrosion of metals, stones, building materials, rubber and painted surfaces.
- Ozone and nitric oxide irritate the nose and throat and their high concentration causes headache, chest pain, dryness of the throat, cough and difficulty in breathing.

Control of photochemical smog: Many techniques are used to control the formation of photochemical smog.

- If we control the primary precursors of photochemical smog like NO_2 and hydrocarbons, the secondary precursors like ozone and PAN, the photochemical smog will automatically be reduced.
- Usually catalytic converters can be used in the automobiles, which can prevent the liberation of nitrogen oxide and hydrocarbons to the atmosphere.
- Some plants like Pinus, Juniparus, Quercus, Pyrus and Vitis can metabolise nitrogen oxide so their plantation will help in controlling photochemical smog
   (i) *Fumes*: These are condensed vapours. Fumes of metals are the well known particulates of this type. These are normally liberated from chemical or metallurgical processes.
   (ii) *Mists*: These are those particles that are produced by spray liquids and formed by the condensation of vapours in air. Common examples of this type, portions of insecticides and herbicides that miss their targets and travel through the air.

Harmful effects of particulate pollutants

- The effect of particulate pollutants are largely dependent on the particle size. Air borne particles such as dust fumes, mist etc. are dangerous for human health.
- Particulate pollutants bigger than 5 microns are likely to lodge in the nasal passage, whereas particles of about 1.0 micron enter into lungs easily.

Water pollution: The phenomenon by which the quality of water in the hydrosphere deteriorates due to the contamination of water by foreign substances is called water pollution. Following are the changes that occur in polluted water:

- Change in the colour and increase in the salinity of water.
- The water develops a bad taste and an offensive odour.
- Uncontrolled growth of weeds in water.
- Decrease in the number of aquatic animals.

TABLE 13.2 Types of Water Pollution and Pollutants

Type	Pollutants
Ground water pollution	Pesticides, fertilizers
Surface water pollution	Industrial, agriculture wastes
Lake water pollution	Organic wastes, industrial wastes
River water pollution	Industrial waste, domestic sewage
Sea water pollution	Oily waste, radioactive waste

Effects of water pollution

- Polluted water is non-potable.
- Use of water polluted by human and animal waste leads to diseases like cholera, jaundice, typhoid and diarrhoea.
- Aquatic life is damaged.
- Tourists avoid dirty beaches.

Some water pollutants with their side effects are discussed in the following two categories:

Inorganic pollutants

- Cyanides, H_2S, NO_2, CO_2 and sulphites change the pH of water and make it toxic for aquatic animals. Heavy metals like Hg and Pb also make water toxic.
- Algal nutrients like CO_2, H_2, O_2, N_2, nitrates, sulphate, phosphates and micronutrients like B, Cl_2, Cu, Zn, V, Mn, Fe cause eutrophication.
- If the fluoride content in drinking water is more than 3 ppm, it is harmful and causes a disease called **fluorosis**. Here, the fluoride reacts with calcium present in teeth and bones to form calcium fluoride due to which the teeth become yellow and the bones become weak.

$$Ca + F_2 \longrightarrow CaF_2$$

- The presence of fluoride can be detected by decreasing the colour of Zirconium–alizarin-S dye by removing zirconium from the dye in the form of colourless zirconium fluoride.
- Salts, trace elements (Cu, Zn, As) and metals coming from chromium platting industry are also pollutants.
- Metals and complex compounds decrease the growth of algae and hamper photosynthesis in plants.

Organic pollutants

- Organic pollutants include plastics, fibres, detergents, paints, dyes, pharmaceuticals, pesticides, fungicides, weedicides and insecticides.
- The domestic waste and industrial waste degrades in the presence of micro organisms there by reducing the amount of dissolved oxygen in water.
- Wastes coming from industries and agriculture fields make the water toxic.
- Sewage from domestic, commercial, food processing and industrial effluents consume oxygen present in water.
- The pesticides and insecticides like DDT, malathion, chlorophenoxy group compounds, dithiocarbonate derivatives, phenyl mercury acetate are also some water pollutants.

Bioamplification: It is the process of increasing the concentration of pollutants in higher animals and human beings or lower animals.

The carriers through which bioamplification takes place constiture food chains. For example, plants, animals, fish and birds.

Eutrophication: It is the over nutrition provided to ponds and takes in the form of organic substances generated from agricultural and industrial wastes. This leads to rapid proliferation of unwanted algae and other plants which subsequently fill the pond and dry it up.

Removal of pollutants from water: Major water pollutants sources like sewage waste, industrial waste water are made pollution free as follows:

From domestic water: Treatment of domestic sewage water is done as follows:

- The water is first passed through sieves and then decanted.
- The decanted water is subjected to oxidation by aerobic bacteria which removes all soluble organic matter from it.
- The phosphates present in domestic sewage water can be removed by adding lime in the form of calcium phosphate.
- Finally, it is chlorinated and released into ponds.

Treatment of industrial waste water: It is treated as follows:

- Organic substances like endrin and DDT can be removed by adsorption using activated charcoal.
- Some dyes and chlorinated pesticides can be removed by passing the water through ion-exchange resins. However, fluoride ion cannot be removed by ion exchange owing to its presence in minute quantity.
- The fluoride ion in water can be reduced from 5–12 ppm to 1 ppm by passing the waste water through activated carbon so that fluoride ions are adsorbed.
- The deactivated carbon can be made active again by washing it first with 4% NaOH and then by 10% H_3PO_4.
- When the water containing fluorine is treated with bleaching powder, lime and alum in the same order, fluoride ions get precipitated as calcium aluminium fluoride.

Dissolved oxygen (DO): It is the amount of oxygen present in water. The optimum value of dissolved oxygen desired for good quality water is 4–6 mg/L (or 4 – 6 ppm).

- Lower DO value indicates water pollution and on basis of the DO value the extent of water pollution is estimated.
- DO is consumed by the oxidation of organic substance in presence of certain microorganisms or reducing agents as follows:

$$CH_2O + O_2 \xrightarrow{\text{Microorganisms}} CO_2 + H_2O$$

$$4Fe^{2+} + O_2 + 10H_2O \longrightarrow Fe(OH)_3 + 8H^+$$

Reducing agent

$$2SO_3^{2-} + O_2 \longrightarrow 2SO_4^{2-}$$

- DO is also consumed due to bio-oxidation of nitrogenous substance.

$$2NH_4^+ + 2O_2 \longrightarrow 2H^+ + NO_3^- + H_2O$$

Chemical oxygen demand (COD)

- It is the amount of oxygen required to oxidize the organic and inorganic substances present in water. It is the amount of O_2 in PPM consumed by the pollutants. COD for pure water is 4 PPM.
- COD is an important water quality parameter. COD is the index of the organic content of water.
- COD can be determined by oxidizing the organic matter with acidified (50% H_2SO_4) potassium dichromate solution.

$$K_2Cr_2O_7 + 4H_2SO_4 \longrightarrow K_2SO_4 + Cr_2(SO_4)_3 + 4H_2O + 3[O]$$

Biochemical oxygen demand (BOD): It is the amount of oxygen used by the microorganisms present in water at 20°C for five days. It is expressed in ml/L If BOD is less than 5 ppm, the water is pure, if it is greater than 5 ppm, the water is polluted.

$$BOD = \frac{\text{No. of miligrams of } O_2 \text{ needed}}{\text{No. oflitres of the sample}}$$

In case of polluted water BOD is > 17 PPM.

Soil or land pollution

- The major components of soil are mineral matter (gravel, sand, clay), organic matter (humus) and biological species (algae, bacteria).
- When the quality of soil is damaged due to excessive mining of minerals or addition of industrial wastes, it leads to soil pollution.

Types of sand or soil pollution

(1) Negative Soil Pollution: It is reduction in soil productivity because of erosion and over use.

(2) Positive Soil Pollution: It is the reduction in soil productivity due to addition of unwanted substances like pesticides, fertilizers etc.

(3) Landscape or Third Pollution: Here fertile land becomes barren one due to addition of dumping wastes like ash, garbage, broken cans, bottles etc.

Soil salination It is the process of increasing salt concentration of soil to make it barren. It is known as Halomorphic.

Control of Environmental Pollution and Green Chemistry

Environmental pollution can be controlled by implimenting these techniques:

(1) By plantation or vegetation.
(2) By management of wastes.
(3) By Recyling of waste materials into useful materials.
(4) By Burning and Incineration.
(5) By Sewage Treatment
(6) By Dumping
(7) By using environmental loving processes like Green Chemistry etc.

Green chemistry Although chemistry has a vital role in our lives by providing us medicines, food preservatives, fertilizers, textiles, cosmatics, polymer etc. Yet some chemical processes are creating quite hazardous wastes that damage our environment and cause various types of pollutions. Green chemistry is developing to design and implement such chemical processes that reduce wastes and minimize or eliminate the creation of hazardous substances. The Following twelve principles form the foundations of green chemistry:

(1) Prevent wastes
(2) Use less hazardous methods/processes
(3) Maximize atom economy
(4) Minimize derivatives
(5) Use safer solvents
(6) Use catalysis
(7) Use renewable feedstocks
(8) Design safer chemicals
(9) Design for energy efficiency
(10) Design for degration
(11) Monitor pollution in real time
(12) Avoid accidents

Examples

1. Recently, a new method of converting glycerol into a by-product of producing biodisel fuel from vegetable oils into propylene glycol (Anti-freezing agent for Automobiles) is developed.

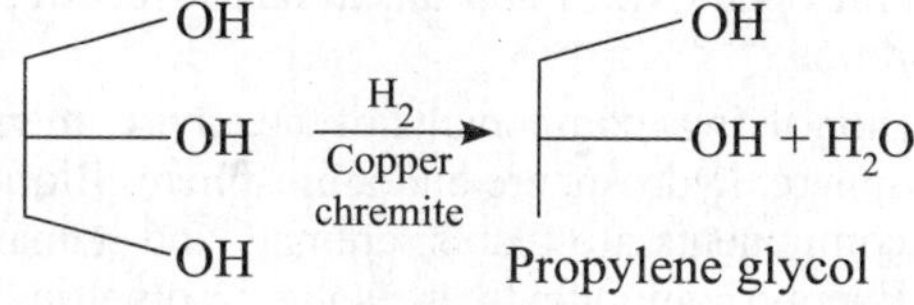

2. Synthesis of ibuprofen.

Major sources of soil pollution

- *Industrial Water*: It is poisonous and biodegradable.
- *Urban Wastes* These include polythene bags, plastic goods, paper.
- *Agriculture Pollutants*: These includes fertilizers, insecticides, herbicides, fungicides.
- *Radioactive Pollutants*: Nuclear wastes from nuclear power plants and nuclear tests also cause soil pollution.
- Soil conditioners having toxic metals like Pb, Hg, Cd also pollute the soil.

Points to Remember

1. Thermosphere is the region where the temperature increases with increase in altitude.
2. Sound waves cannot propagate in mesosphere.
3. Polluted atmosphere and hydrosphere affects biosphere.
4. Many people in Japan's Minamta Island died due to a disease called Minamita by eating fish contaminated by mercury.
5. Spilling of oil into seas causes water pollution.
6. **Speciation** is the identification of different pollutants as inorganic, organic or organo-metallic origin.

7. The toxic effects of pollutant depends on category to which it belongs. For example, mercury compounds are more toxic than lead compounds.

- Pollen and spores cause allergic reactions or hay fever in certain people.
- Mine dust destroys vegetation and causes many deformities in animals and human beings.
- Mercury poisoning produces a crippling and often fatal disease called **minamata disease**.
- Spraying of DDT on crops causes pollution of air, water and soil.
- **Halomorphic** is a process of increasing salt concentration of soil.
- Photochemical oxidants are secondary pollutants such as ozone, PAN, aldehydes and phenols. These are produced due to photochemical reactions between nitrogen oxides and unsaturated (reactive) hydrocarbons.
- Abiotic (non-living) components of the environment are lithosphere, hydrosphere and atmosphere. Biotic (living) components are plants, animals and human beings. Energy components are solar, geometrical, thermochemical, hydroelectric and nuclear energy.
- Polychlorinated biphenyls (PCBS) are the recently formed chemical compounds which are used as fluids in transformers and capacitors. Being resistant to oxidation, these are released in the atmosphere as vapours. They mix with rain water and thus contaminate water. These have been found to be carcinogenic and are thus a source of pollution.
- Threshold limit value (TLV) indicates the permissible level of the pollutant or toxic substance in atmosphere to which a healthy person can be exposed to, without causing any adverse effect.
- BOD_5 means the amount of oxygen consumed in 5 days at 20°C (293 K) and it is reported in ppm. Pure water has BOD_5 of less that 5 ppm whereas highly polluted water has BOD_5 value of more than 17 ppm. The untreated sewage has BOD_5 of 100–400 ppm.
- Particulates are small solid particles and liquid droplets suspended in air. These are of two types—viable and non-viable.
- Viable particulates are small sized microorganisms dispersed in the air which may cause diseases. For example, fungi, bacteria.
- Non-viable particulates are formed by the condenstion of small species or disintegration of large species. For example, mist, smoke, fume, dust.
- The particulate pollutants like lead, carbon and metals are released from the automobile exhausts causing respiratory disorders.
- Particulates can be removed from air by using gravity setting chamber, cyclone collector, wet scrubbers and electrostatic precipitators.

CHAPTER-END EXERCISES

Practice Questions – I

1. Which among these gives the measure of organic pollution in water?

(a) COD (b) DO
(c) pH (d) Salinity

2. The medium which interacts with the long lived pollutant is:

(a) Source (b) Receptor
(c) Sink (d) All

3. Which of the following is a contaminant?

(a) CO (b) CO_2
(c) SO_2 (d) Leakage of MIC

4. Which statement is correct about biodegradable materials?

(a) These are toxic in nature
(b) These can be broken down by bacteria
(c) These are used for converting waste to greenery
(d) These spoil the biological environment

5. Fluoride in water is removed using precipitation method by:

(a) Lime and alum
(b) Lime, alum and bleaching powder
(c) Activated carbon
(d) Defluoron

6. DDT is:
   (a) A non degradable pollutant
   (b) A fertilizer
   (c) An antibiotic
   (d) Dichloro difluoro titanium

7. Cyclon collector is used for minimizing:
   (a) Water pollution (b) Sound pollution
   (c) Soil pollution (d) Air pollution

8. In Antarctica, ozone depletion is due to the formation of compound.
   (a) SO_2 and SO_3 (b) Chlorine nitrate
   (c) Acrolein (d) Formaldehyde

9. Household fuel (LPG) mainly contains:
   (a) CH_4 (b) C_2H_2
   (c) C_2H_4 (d) C_4H_{10}

10. $CFCl_3$ is responsible for the decomposition of ozone into oxygen. Which of the following reacts with ozone to form oxygen?
    (a) Cl_2 (b) Cl^-
    (c) F^- (d) Cl^*

11. Biosphere includes:
    (a) Plants and animals
    (b) Rocks and minerals
    (c) Atmosphere and lithosphere
    (d) Water sources

12. The major components of the atmosphere are:
    (a) CO_2 and N_2 (b) O_3 and SO_2
    (c) CO and CO_2 (d) N_2 and O_2

13. The upper most region of the atmosphere is:
    (a) Troposphere (b) Exosphere
    (c) Stratosphere (d) Ionosphere

14. The point of temperature inversion between troposphere and ionosphere is called:
    (a) Mesopause (b) Stratopause
    (c) Ionopause (d) Tropopause

15. BOD is connected with:
    (a) Microbes (b) Organic matter
    (c) Both (a) and (b) (d) None of these

16. Biodegradable pollutant is:
    (a) DDT (b) Domestic waste
    (c) Aluminium foil (d) Mercury salt

17. Minamata disease is caused due to the presence of …… in water.
    (a) Hg (b) Pb
    (c) Cd (d) As

18. Which one of the following is an invaluable source of energy but does not cause pollution?
    (a) Fossil fuels (b) Nuclear energy
    (c) Sun (d) Petroleum

19. Water is often treated with chlorine to:
    (a) Kill germs
    (b) Remove hardness
    (c) Increase oxygen content
    (d) Remove suspended particles

20. Major sources of NOx pollutants are:
    (a) Natural gas
    (b) Gasoline
    (c) Combustion of coal and oil
    (d) All of these

21. Ozone hole refers to:
    (a) Hole in ozone layer
    (b) Reduction in the thickness of ozone layer in stratosphere
    (c) Reduction in the thickness of ozone layer in troposphere
    (d) Increased concentration of ozone

22. Which of the following is a primary pollutant?
    (a) PAN (b) Aldehydes
    (c) CO (d) H_2SO_4

23. Which of the following is not regarded as a pollutant?
    (a) CO_2 (b) O_3
    (c) NO_2 (d) hydrocarbons

24. Which of the following is not a greenhouse gas?
    (a) Water vapour (b) CO_2
    (c) O_2 (d) CH_4

25. The greenhouse effect is:
    (a) Rise in temperature of the earth
    (b) Rise in pressure on the earth
    (c) Decrease in oxygen content of the earth's atmosphere
    (d) Decrease in CO_2 content of earth's atmosphere

26. Which of the following is responsible for depletion of the ozone layer in the upper strata of the atmosphere?
    (a) Freons
    (b) Ferrocene

(c) Fullerenes
(d) Polyhalogens

27. Ozone in stratosphere is depleted by:

(a) C_7F_{16} (b) C_6F_6
(c) CF_2Cl_2 (d) $C_6H_6Cl_6$

28. Which of the following is not an air pollutant?

(a) N_2O (b) N_2
(c) CO (d) NO

29. Control of pollution is possible by:

(a) Developing bio remedies
(b) Planting trees
(c) Controlling the population growth
(d) All of these

30. By green chemistry, we mean:

(a) Performing only those reactions which are of biological origin
(b) The use of non toxic reagents and solvents to produce environmental friendly products
(c) Producing chemicals of our daily use from greenhouse gases
(d) Performing chemical processes which use green plants

31. Acid rain:

(a) Retards the growth of trees
(b) Effects big marble constructions
(c) Results in loss of flora and fauna
(d) All of these

32. The industrial usage of fluorocarbons is very high because:

(a) They are unstable
(b) They are gases
(c) They can be manufactured cheaply
(d) Their reactivity is high

33. Hydrosphere includes various forms of water as:

(a) Polar ice caps and ground water
(b) Oceans and lakes
(c) Sea and rivers
(d) All of these

34. The presence of fluoride ion in polluted water can be detected by using:

(a) KI + starch
(b) Zirconium alizarin-S dye
(c) $K_2Cr_2O_7$ + 50 % H_2SO_4
(d) Sodium nitroprusside

Practice Questions – II

35. The phosphates present in sewage water which cause eutrophication can be removed by adding:

(a) Alum (b) Chlorine
(c) Lime (d) Permanganate

36. Ozone is an important constituent of stratosphere because it:

(a) Removes poisonous gases of the atmosphere by reacting with them
(b) Destroys bacteria which are harmful to human life
(c) Prevents the formation of smog over large cities
(d) Absorbs ultraviolet radiation which is harmful to human life

37. SO_2 is a dangerous air pollutant and harms plant life. What changes in the plant indicate its toxic effects?

(a) Falling of leaves
(b) Bleaching of leaves
(c) Darkening of leaves
(d) Withering of leaves

38. The excessive use of phosphates as water softeners leads to aquatic pollution called:

(a) Eutrophication (b) Phosphorylation
(c) Deoxygenation (d) Denitrification

39. Spraying of DDT causes pollution of:

(a) Air (b) Air and water
(c) Air and soil (d) Air, water and soil

40. Which of the following causes water pollution?

(a) Fly ash (b) Pesticides
(c) Auto exhausts (d) Aeroplanes

41. Ozone layer is present in:

(a) Stratosphere (b) Troposphere
(c) Mesosphere (d) Exosphere

42. Human ear can tolerate noise upto:

(a) 100 decible (b) 80 decible
(c) 60 decible (d) 40 decible

43. Sound pollution is mainly due to:
(a) Industrial estates (b) Airport, railway stations
(c) Sound systems (d) All of these

44. Ozone hole is maximum over:
(a) Africa (b) Europe
(c) Antarctica (d) India

45. Ozone layer of stratosphere requires protection from indiscriminate use of:
(a) Aerosols and high flying jets
(b) Balloons
(c) Pesticides
(d) Atomic explosions

46. Sound pollution does not cause:
(a) Mental disorder (b) Hearing problems
(c) Headache (d) Skin diseases

47. Proper management for disposal of household and industrial wastes can be done by:
(a) Sewage treatment
(b) Recycling the waste material to give useful products again
(c) Burning and incineration of combustible waste
(d) All of these

48. Contaminant is:
(a) A pollutant released from the industries
(b) A pollutant
(c) A component originally not present in environment but released into environment by human activity
(d) A pollutant released into the environment in natural calamites

49. Untreated domestic sewage has a low:
(a) Persentage of dissolved oxygen
(b) Persentage of suspended organic matter
(c) BOD value
(d) Amount of dissolved salts

50. Presence of which of the following in the atmosphere will cause the depletion of ozone?
(a) Chlorofluoro carbons (b) Cl_2
(c) NO (d) Any one of these

51. There is a possibility of melting of polar ice caps and increase in the level of sea water due to:
(a) Greenhouse effect
(b) Acid rain
(c) Depletion of ozone layer
(d) Any one of these

52. Which of the following gases cause pollution when present in the exhaust fumes of vehicles?
(a) Water vapours (b) CO_2
(c) CO (d) C_2H_6

53. The poisonous gas present in the exhaust fumes of car is:
(a) CO_2 (b) CO
(c) CH_4 (d) C_2H_2

54. Incomplete combustion of petrol or diesel in automobile engines can be best detected by testing the fuel gases for the presence of:
(a) Sulphur dioxide
(b) Nitrogen dioxide
(c) Carbon monoxide
(d) Carbon monoxide and water vapour

55. Which one of the following regions of atmosphere contains ozone?
(a) Troposphere (b) Stratosphere
(c) Mesosphere (d) Thermosphere

56. The greenhouse effect is caused by:
(a) CO (b) NO
(c) NO_2 (d) CO_2

57. Surface water contains:
(a) Only salt
(b) Organic matter
(c) Salt and organic matter
(d) Suspended impurity

58. An object is located at a height of 5 km from the surface of the earth. The object is located in which part of atmosphere?
(a) Thermosphere (b) Mesosphere
(c) Stratosphere (d) Troposphere

59. Among the following, all except which causes pollution?
(a) Nuclear power plant
(b) Thermal power plant
(c) Hydro-electric plant
(d) Automobiles

Practice Questions – III

60. Which factor among the following is the most harmful for life on the globe?
(a) Soil erosion (b) Nuclear fall out
(c) Increasing desert (d) Deforestation

61. Which air pollutant is not released by automobiles?
(a) Fly ash (b) CO
(c) SO_2 (d) Hydrocarbons

62. Taj Mahal is threatened due to the effect of:
(a) Hydrogen (b) Oxygen
(c) Chlorine (d) Sulphur dioxide

63. Photochemical oxidants such as PAN and PBN are formed:
(a) By the action of hydrogen sulphide on hydrocabons in the presence of sunlight
(b) By the action of carbon dioxide on hydrocarbons in the presence of sunlight
(c) By the action of nitrogen oxides on hydrocarbons in the presence of sunlight
(d) None of these

64. Which of the following statements are true?
(1) London smog is oxidizing in nature.
(2) London smog is a mixture of smoke and fog.
(3) Photochemical smog causes irritation in eyes.
(4) Photochemical smog results in the formation of PAN.
(a) 1, 2 and 3 (b) 1, 2 and 4
(c) 2, 3 and 4 (d) 1, 3 and 4

65. Which of the following statements are correct?
(1) Ammonia acts as a sink for NOx.
(2) The average residence time of NO is one month.
(3) SOx can be removed from flue gases by passing through a solution of citrate ions.
(4) Limestone acts as a sink for SOx.
(a) 1, 3 and 4 (b) 1, 2 and 4
(c) 2, 3 and 4 (d) 1, 2 and 3

66. Fluorocarbons are effective scavengers for ozone due to:
(a) Photolytic production of oxides of nitrogen
(b) Photolytic decomposition of O_3 by Cl into O_2
(c) Photolytic decomposition of O_3 producing O_2
(d) Photolytic reaction of O_2 producing Cl radicals

67. Which of the following statements are true?
(1) Temperature inversion is caused by smog.
(2) A product of photochemical smog is PAN.
(3) O_3 is involved in photochemical smog.
(4) O_3 layer is destroyed by fluorocarbons.
(a) 1, 2 and 3 (b) 1, 2 and 4
(c) 2, 3 and 4 (d) 1, 3 and 4

68. Which of the following statements are incorrect about ozone layer?
1. It is beneficial because ozone cuts-off the ultraviolet radiation of the sun
2. It is harmful because ozone cuts out the important radiation of the sun which are vital for photosynthesis
3. It is harmful because ozone is dangerous to living organisms
4. It is beneficial because oxidation reaction can proceed faster in the presence of ozone
(a) 1, 2 and 3 (b) 1, 2 and 4
(c) 3 and 4 only (d) 2, 3 and 4

69. About 20 km above the earth, there is an ozone layer. Which of the following statements about ozone and ozone layer are false?
(1) The conversion of ozone to oxygen is an endothermic reaction.
(2) Ozone layer is beneficial to us because ozone cuts out the ultraviolet radiation of the sun.
(3) Ozone layer is harmful to us because it cuts off radiations useful for photosynthesis.
(4) Ozone has a triatomic linear molecule.
(a) 1, 2 and 3 (b) 1, 3 and 4
(c) 2, 3 and 4 (d) 3 and 4 only

70. Ozone is formed in the upper layer of the atmosphere from oxygen by the action of:
(a) Cosmic rays (b) Ultraviolet rays
(c) Free radicals (d) Nitrogen oxides

71. Pick up the incorrect statements here.
(1) Classical smog is good for health but not photo chemical smog.
(2) During formation of smog the level of ozone in the atmosphere goes down.
(3) Classical smog has an oxidizing character while the photochemical smog is reducing in character.
(4) Photochemical smog occurs in day time whereas the classical smog occurs in early morning hours.
(a) 1, 2 and 3 (b) 1, 3 and 4
(c) 2 and 3 only (d) 3 and 4 only

72. Which statements are correct here?

(1) H_2SO_4 particulates have size of 500–1000 nm.

(2) Soot particles have a diameter of about 5 nm.

(3) Fly ash particles have diameter of 5×10^5 nm.

(a) 1 and 2 (b) 2 and 3

(c) 1 and 3 (d) 1, 2 and 3

73. Use of chlorofluoro carbons is not encouraged because

(a) They are harmful to the eyes of people that use it

(b) They damage the refrigerators and air conditioers

(c) They eat away the ozone in the atmosphere

(d) They destroy the oxygen layer

74. Consider the following statements:

(1) PAN is secondary air pollutant.

(2) BOD and COD values are indicators of water po-lution level.

Which of the statements given above is/are correct?

(a) Both 1 and 2 (b) 1 only

(c) 2 only (d) Neither 1 nor 2

75. Match the List 1 and List 2 and pick the correct matching from the codes given below.

List 1	List 2
(A) Peroxy acetyl nitrate	1. Waste incineration
(B) Polycyclic aromatic hydrocarbons	2. Global warming
(C) Dioxins	3. Photochemical smog
(D) Indigo	4. Carcinogens
(E) IR active molecules	5. Vat dye

(a) A 3, B 4, C 1, D 5, E 2

(b) A 2, B 3, C 5, D 4, E 1

(c) A 5, B 3, C 1, D 2, E 4

(d) A 1, B 2, C 3, D 4, E 5

76. When rain is accompanied by a thunderstorm, the collected rain water will have a pH value:

(a) Which depends upon the amount of dust in the air

(b) Slightly lower than that of rain water without the thunderstorm

(c) Slightly higher than that when the thunderstorm is not here

(d) Uninfluenced by occurrence of thunderstorm

77. Smog is essentially caused by presence of:

(a) O_2 and N_2

(b) O_2 and O_3

(c) O_3 and N_2

(d) Oxides of sulphur and nitrogen

78. Select the correct statements in the following:

(1) Chlorofluorocarbons are responsible for ozone layer depletion.

(2) Green house effect is responsible for global warming.

(3) Ozone layer does not permit infrared radiation from the sun to react the earth.

(4) Acid rains is mostly because of oxides of nitrogen and sulphur.

(a) 1, 2

(b) 2, 3

(c) 1, 2, 4

(d) 1, 3, 4

Answer Keys

1. (a)	2. (c)	3. (d)	4. (b)	5. (b)	6. (a)	7. (d)	8. (b)	9. (d)	10. (d)
11. (a)	12. (d)	13. (b)	14. (d)	15. (c)	16. (b)	17. (a)	18. (c)	19. (a)	20. (d)
21. (b)	22. (c)	23. (a)	24. (c)	25. (a)	26. (a)	27. (c)	28. (b)	29. (d)	30. (b)
31. (d)	32. (c)	33. (d)	34. (b)	35. (c)	36. (d)	37. (b)	38. (a)	39. (d)	40. (b)
41. (a)	42. (c)	43. (d)	44. (c)	45. (a)	46. (d)	47. (d)	48. (c)	49. (a)	50. (d)
51. (a)	52. (c)	53. (b)	54. (c)	55. (b)	56. (d)	57. (c)	58. (d)	59. (c)	60. (b)
61. (a)	62. (d)	63. (c)	64. (c)	65. (a)	66. (b)	67. (c)	68. (d)	69. (b)	70. (b)
71. (a)	72. (d)	73. (c)	74. (a)	75. (a)	76. (b)	77. (d)	78. (c)		

Hints and Explanations for Selective Questions

13. The uppermost region of the atmosphere is exosphere.

14. Tropopause is the point of temperature inversion between troposphere and ionosphere.

15. BOD is connected with microbes and organic matter.

19. Water is treated with chlorine to kill germs.

22. CO is a primary pollutant.

23. CO_2 is generally not regarded as a pollutant.

24. O_2 is not a green house gas.

28. N_2 is not an air pollutant.

30. Green chemistry means the use of non toxic reagents and solvents to produce environmental friendly products.

39. Spraying of DDT produces air, water and soil pollution.

40. Pesticides cause water pollution.

41. Ozone layer is present in stratospherc (at an altitude of 25–0 km).

44. Ozone hole is maximum over Antarctica.

45. Aerosols use CFCs and high flying jets release NO which are responsible for depletion of ozone layer.

53. CO is poisonous and causes air pollution. It is present in the exhaust fumes of cars.

54. Incomplete combustion of petrol or diesel in automobile engines produce CO only.

55. Ozone is present in stratosphere (about 20 km above the surface of the earth).

56. CO_2 is responsible for the greenhouse effect.

57. Surface water contains salt and organic matter.

64. As London smog is reducing in nature.

65. As the average residence time of NO is four days.

68. Ozone layer is beneficial because ozone cuts-off the ultraviolet radiation of the sun.

70. Ultraviolet rays act on oxygen molecules to combine with oxygen atoms forming ozone in the upper atmosphere.

73. Chlorofluorocarbon is used in air conditioning and is domestic refrigerators for cooling purposes. Its main drawback in this, it is responsible for ozone depletion.

76. Normal rain water has a pH value of 5.6 thundertorms lower this value by repoducing nitric acid which makes rain water acidic. Therefore rain water accompanying thunderstorm has a pH lower than 5.6.

77. Oxides of sulphur and nitrogen along with dust and smoke combine with condensed water vapours to give dark for called smog.

78. As ozone layer does not permit ultraviolet radiation from sun to earth the earth.

NCERT Exemplar

1. Which of the following gases is not a green house gas?

(a) CO (b) O_3
(c) CH_4 (d) H_2O vapour

2. Photochemical smog occurs in warm, dry and sunny climate. One of the following is not amongst the components of photochemical smog, identify it.

(a) NO_2 (b) O_3
(c) SO_2 (d) Unsaturated hydrocarbon

3. Which of the following statements is not true about classical smog?

(a) Its main components are produced by the action of sunlight on emissions of automobiles and factories.
(b) Produced in cold and humid climate.
(c) It contains compounds of reducing nature.
(d) It contains smoke, fog and sulphur dioxide.

4. Biochemical Oxygen Demand, (BOD) is a measure of organic material present in water. BOD value less than 5 ppm indicates a water sample to be __________.

(a) Rich in dissolved oxygen.
(b) Poor in dissolved oxygen.
(c) Highly polluted.
(d) Not suitable for aquatic life.

5. Which of the following statements is wrong?

(a) Ozone is not responsible for green house effect.
(b) Ozone can oxidise sulphur dioxide present in the atmosphere to sulphur trioxide.

(c) Ozone hole is thinning of ozone layer present in stratosphere.
(d) Ozone is produced in upper stratosphere by the action of UV rays on oxygen.

6. Sewage containing organic waste should not be disposed in water bodies because it causes major water pollution. Fishes in such a polluted water die because of:
(a) Large number of mosquitoes.
(b) Increase in the amount of dissolved oxygen.
(c) Decrease in the amount of dissolved oxygen in water.
(d) Clogging of gills by mud.

7. Which of the following statements about photochemical smog is wrong?
(a) It has high concentration of oxidising agents.
(b) It has low concentration of oxidising agent.
(c) It can be controlled by controlling the release of NO_2, hydrocarbons, ozone etc.
(d) Plantation of some plants like pinus helps in controlling photochemical smog.

8. The gaseous envelope around the earth is known as atmosphere. The lowest layer of this is extended upto 10 km from sea level, this layer is ________.
(a) Stratosphere (b) Troposphere
(c) Mesosphere (d) Hydrosphere

9. Dinitrogen and dioxygen are main constituents of air but these do not react with each other to form oxides of nitrogen because ________.
(a) The reaction is endothermic and requires very high temperature.
(b) The reaction can be initiated only in presence of a catalyst.
(c) Oxides of nitrogen are unstable.
(d) N_2 and O_2 are unreactive.

10. The pollutants which come directly in the air from sources are called primary pollutants. Primary pollutants are sometimes converted into secondary pollutants. Which of the following belongs to secondary air pollutants?
(a) CO
(b) Hydrocarbon
(c) Peroxyacetyl nitrate
(d) NO

11. Which of the following statements is correct?
(a) Ozone hole is a hole formed in stratosphere from which ozone oozes out.
(b) Ozone hole is a hole formed in the troposphere from which ozone oozes out.
(c) Ozone hole is thinning of ozone layer of stratosphere at some places.
(d) Ozone hole means vanishing of ozone layer around the earth completely.

12. Which of the following practices will not come under green chemistry?
(a) If possible, making use of soap made of vegetable oils instead of using synthetic detergents.
(b) Using H_2O_2 for bleaching purpose instead of using chlorine based bleaching agents.
(c) Using bicycle for travelling small distances instead of using petrol/ diesel based vehicles.
(d) Using plastic cans for neatly storing substances.

Answer Keys

1. (a)	2. (c)	3. (a)	4. (a)	5. (a)	6. (c)	7. (b)	8. (b)	9. (a)	10. (c)
11. (c)	12. (d)								

Hints and Explanations for Selective Questions

1. Greenhouse gases absorb solar energy near the Earth's surface and radiate it back to the Earth.

4. BOD < 5 ppm assumes water is pure and is rich in dissolved oxygen.
BOD >17 ppm assumes water is highly polluted.

6. Decrease oxygen amount in water as organic waste consume large amount of oxygen.

12. Plastic is non-biodegradable polymer which doesnot come under green chemistry.

AIIMS ESSENTIALS

Assertion and Reason

In the following questions, two statements Assertion (A) and Reason (R) are given. Mark

(a) if A and R both are correct and R is the correct explanation of A;

(b) if A and R both are correct but R is not the correct explanation of A;

(c) A is true but R is false;

(d) A is false but R is true,

(e) Both A and R are false.

1. (A) : Inhabitants close to very busy airports are likely to experience health hazards.
(R) : Sound level of jet aeroplanes usually exceeds 160 dB.

2. (A) : There is an adverse effect of pollutants on Taj. Mahal
(R) : Discolouring of Taj. Mahal is due to nitrogen oxides.

3. (A) : CO_2 causes green house effect.
(R) : Other gases do not show such effect.

4. (A) : α, β and γ rays are emitted by disintegration of atomic nuclei of radioactive elements.
(R) : Nuclear fall out is shown by radioactive elements.

5. (A) : Methylmercury is a highly persistent kind of pollution that accumulates in food chains.
(R) : Mercury pollution of responsible for Minamata disease

6. (A) : Water pollutants are measured by BOD.
(R) : If BOD is more, the water is polluted.

7. (A) : Eutraphication shows increase in productivity in water.
(R) : With increasing eutrophication, the diversity of the phytoplankton increases.

8. (A) : Presently, the global atmosphere is warming up.
(R) : The depletion of stratospheric ozone layer has resulted in increase in ultraviolet radiations reaching the earth.

9. (A) : Suspended particulate matter (SPM) is an important pollutant released by diesel vehicles.
(R) : Catalytic converters greatly reduce pollution caused by automobiles.

10. (A) : Ozone in the upper atmosphere (stratosphere) is beneficial but it s harmful in the lower atmosphere.
(R) : Ozone absorbs UV light and is a strong oxidising agent.

11. (A) : Photochemical smog is formed by oxides of sulphur, smoke and dust particles
(R) : It is also known as Los Angles smog.

12. (A) : Acid rain has a pH less than 5.
(R) : Oxides of nitrogen and sulphur combine with rain water to produce acidic rain.

13. (A) : CO combines with haemoglobin.
(R) : It has affinity for haemoglobin.

Answer Keys

1. (a) **2.** (c) **3.** (c) **4.** (b) **5.** (b) **6.** (a) **7.** (b) **8.** (b) **9.** (b) **10.** (a)
11. (d) **12.** (a) **13.** (a)